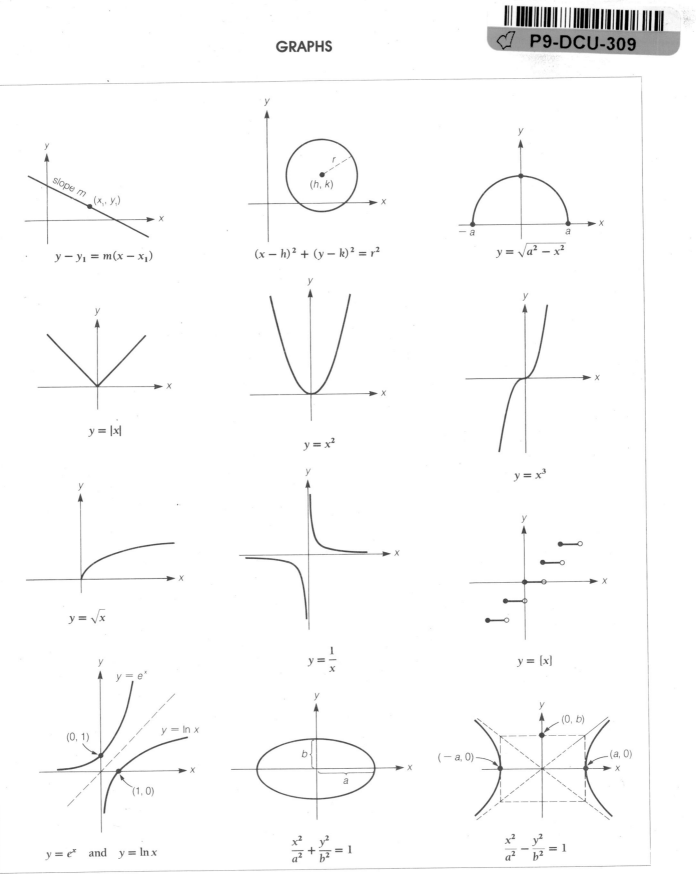

$y - y_1 = m(x - x_1)$

$(x - h)^2 + (y - k)^2 = r^2$

$y = \sqrt{a^2 - x^2}$

$y = |x|$

$y = x^2$

$y = x^3$

$y = \sqrt{x}$

$y = \dfrac{1}{x}$

$y = [x]$

$y = e^x$ and $y = \ln x$

$\dfrac{x^2}{a^2} + \dfrac{y^2}{b^2} = 1$

$\dfrac{x^2}{a^2} - \dfrac{y^2}{b^2} = 1$

PRECALCULUS

Fourth Edition

2^{14}
$\dfrac{26}{8}$

$\begin{array}{r} 3\,29 \\ -\ 29\,5 \\ \hline 34 \end{array}$

PRECALCULUS
A PROBLEMS-ORIENTED APPROACH

Fourth Edition

DAVID COHEN

Department of Mathematics
University of California
Los Angeles

WEST PUBLISHING COMPANY

Minneapolis / St. Paul   New York   Los Angeles   San Francisco

FOR MY SON ANDY

Copyediting: Susan Gerstein and Pamela McMurry
Composition: The Clarinda Company
Cover Image: From *Symmetry in Chaos* by Michael Field and Martin Golubitsky. Published by Oxford University Press. © M. Field and M. Golubitsky, 1992.
Technical Illustrations: TECH*arts,* Boulder, Colorado
Text Design: Janet Bollow
Production, Prepress, Printing, and Binding by West Publishing Company.

WEST'S COMMITMENT TO THE ENVIRONMENT

In 1906, West Publishing Company began recycling materials left over from the production of books. This began a tradition of efficient and responsible use of resources. Today, up to 95 percent of our legal books and 70 percent of our college and school texts are printed on recycled, acid-free stock. West also recycles nearly 22 million pounds of scrap paper annually—the equivalent of 181,717 trees. Since the 1960s, West has devised ways to capture and recycle waste inks, solvents, oils, and vapors created in the printing process. We also recycle plastics of all kinds, wood, glass, corrugated cardboard, and batteries, and have eliminated the use of styrofoam book packaging. We at West are proud of the longevity and the scope of our commitment to the environment.

LIBRARY OF CONGRESS CATALOGING-IN-PUBLICATION DATA

Cohen, David, 1942 Sept. 28-
    Precalculus / David Cohen. — 4th ed.
      p.   cm.
    Updated ed. of: Precalculus, a problems-oriented approach. c1990.
    Includes index
    ISBN 0-314-01255-9 (hard : alk. paper)
    1. Mathematics.   I. Cohen, David, 1942 Sept. 28- Precalculus, a
problem-oriented approach. II. Title.
QA39.2.C64   1993
512.1—dc20   92-41652
                          ∞ CIP

PRECALCULUS

Fourth Edition

$$2^{17}$$
$$\frac{26}{8}$$

$$\begin{array}{r} 3{:}29 \\ -\ 29\,5 \\ \hline 34 \end{array}$$

PRECALCULUS
A PROBLEMS-ORIENTED APPROACH

Fourth Edition

DAVID COHEN

Department of Mathematics
University of California
Los Angeles

WEST PUBLISHING COMPANY

Minneapolis / St. Paul   New York   Los Angeles   San Francisco

FOR MY SON ANDY

Copyediting: Susan Gerstein and Pamela McMurry
Composition: The Clarinda Company
Cover Image: From *Symmetry in Chaos* by Michael Field and Martin Golubitsky. Published by Oxford University Press. © M. Field and M. Golubitsky, 1992.
Technical Illustrations: TECH*arts,* Boulder, Colorado
Text Design: Janet Bollow
Production, Prepress, Printing, and Binding by West Publishing Company.

WEST'S COMMITMENT TO THE ENVIRONMENT

In 1906, West Publishing Company began recycling materials left over from the production of books. This began a tradition of efficient and responsible use of resources. Today, up to 95 percent of our legal books and 70 percent of our college and school texts are printed on recycled, acid-free stock. West also recycles nearly 22 million pounds of scrap paper annually—the equivalent of 181,717 trees. Since the 1960s, West has devised ways to capture and recycle waste inks, solvents, oils, and vapors created in the printing process. We also recycle plastics of all kinds, wood, glass, corrugated cardboard, and batteries, and have eliminated the use of styrofoam book packaging. We at West are proud of the longevity and the scope of our commitment to the environment.

LIBRARY OF CONGRESS CATALOGING-IN-PUBLICATION DATA

Cohen, David, 1942 Sept. 28-
    Precalculus / David Cohen. — 4th ed.
      p.   cm.
    Updated ed. of: Precalculus, a problems-oriented approach. c1990.
    Includes index
    ISBN 0-314-01255-9 (hard : alk. paper)
    1. Mathematics.   I. Cohen, David, 1942 Sept. 28- Precalculus, a problem-oriented approach.   II. Title.
QA39.2.C64   1993
512.1—dc20   92-41652
                                           ∞ CIP

CONTENTS

TABLES

ANSWERS TO SELECTED EXERCISES   A-51

INDEX   I-1

PREFACE

This text is for students who are preparing to take calculus or other courses requiring a background in precalculus mathematics. As in the earlier editions, my goal has been to create a book that is *accessible* to the student. The presentation is student-oriented in three specific ways. First, I've tried to talk to, rather than lecture at, the student. Second, examples are consistently used to introduce, to explain, and to motivate concepts. And third, all of the initial exercises for each section are carefully correlated with the worked examples in that section.

AUDIENCE   In writing *Precalculus,* I have assumed that the students have been exposed to intermediate algebra, but that they have not necessarily mastered that subject. Also, for many students, there may be a gap of several years between their last mathematics course and the present one. For these reasons, the review material in parts of the first two chapters and the appendix is unusually thorough.

FEATURES   **1.** *Word problems and applications.* Word problems and strategies for solving them are explained and developed throughout the book. Maximum–minimum problems relating to quadratic functions are discussed in detail in Section 4.4. The preceding section introduces some strategies for approaching these problems. To ensure that precalculus students gain appropriate practice and experience with these important strategies, the initial exercises in Section 4.3 make specific references to the corresponding worked examples in the text. In general, applications are integrated throughout the text. Two complete sections (5.5 and 5.6) are devoted to applications of the exponential function.
2. *Emphasis on graphing.* Graphs and techniques for graphing are developed throughout the text, and graphs are used to explain and reinforce algebraic concepts. (See, for example, Sections 3.3, 5.3, and 7.3.)
3. *Calculator Exercises.* There are two broad categories of calculator exercises in this text:
(i) OPTIONAL GRAPHING CALCULATOR EXERCISES
Over the past several years, all of us in the mathematics teaching community have become increasingly aware of the graphing calculator and its potential for making a positive impact in our teaching. While it is clear that vast reforms lie ahead, many of the specific details are still evolving in workshops and in classrooms across the country. And indeed, at present, even within a given school, some instructors are teaching the course with the graphing calculator, while others are not. Thus, for this 1993 edition of *Precalculus,* I have labeled the graphing calculator

exercises "optional," and I have placed them at the ends of the appropriate sections. Although these class-tested exercises can be adapted for use with any graphing calculator, they were written specifically for use with the Texas Instruments TI-81 Graphics Calculator. Since many of these exercises contain carefully detailed instructions for using this calculator, a minimum of class time is required for discussing its operation. Additionally, Section A.2 of the appendix (written by Professor Mickey Settle) contains a complete introduction to the basic features of the calculator.

(ii) (ORDINARY) CALCULATOR EXERCISES

As in the previous edition, there are many calculator exercises integrated throughout the regular exercise sets in the text. Some of these exercises reinforce or supplement the core material; some of them contain surprising results that motivate subsequent theoretical questions; and a few demonstrate that the use of a calculator cannot replace thinking or the need for mathematical proofs.

4. *Analytic Geometry.* The basic equations for lines and circles are introduced in Chapter 2 and used throughout the book. In Chapter 10, conic sections are discussed in greater detail than in most other precalculus books. The unifying focus–directrix property of the conics is developed in Section 10.6. In the past, analytic geometry traditionally served as the capstone in the student's preparation for calculus. It was here that the student had a real opportunity to sharpen his or her algebraic and analytic skills. Chapter 10 is intended to reaffirm, rather than dismiss, that tradition.

5. *Trigonometry.* The development of most topics in this text proceeds from the specific to the general, and the trigonometry follows this pattern, too. Chapter 6 introduces the right-triangle definitions (in Section 6.1) before the unit circle definitions (in Section 6.4).

The unit-circle definitions of the trigonometric functions involve simple geometric ideas that should be presented as such, and not cluttered with excessive notation or terminology (which will never appear in subsequent science or calculus courses). In presenting these definitions in Sections 6.4 and 7.2, the text sets a pace that is deliberately measured. Rather than rushing on, we take time to help the student understand the definitions. Experience shows that this pays dividends later on when the student studies the more analytical portions of trigonometry in Chapter 7.

6. *Graded Exercise Sets.* As a convenience for the instructor, most of the exercise sets (except for the chapter review sets) are divided into three categories: A, B, and C. The group-A exercises are based directly on the examples and definitions in that section of the text. Moreover, these problems treat topics in roughly the same order in which they appear in the text. The group-B exercises serve several key functions. Some of them require students to use several different techniques or topics in a single solution. Some group-B problems, while not conceptually more difficult than their group-A counterparts, require lengthier calculations or algebraic manipulations. Finally, some group-B exercises are included simply because they are interesting or because they allow the students to see the material from a different perspective. The group-C exercises contain the more challenging problems. (In many cases, though, detailed hints are provided.)

7. *End-of-Chapter Material.* Each chapter concludes with a detailed chapter summary, an extensive chapter-review exercise set, and a chapter test. Additionally, in Chapters 2 through 12, there are *Writing Mathematics* sections. In the *Writing Mathematics* questions, the student is asked to organize his or her thoughts and respond in complete sentences. Some of the questions are simply *true-or-false* questions that can be explained in just a sentence or two. In other cases, a more elaborate response is required. For example, in Chapter 2 (page 92), the *Writing Mathematics* section describes a geometric technique for solving quadratic equations and the student is asked to explain why the method is valid. In Chapter 8, the *Writing Mathematics* section describes a new proof for the law of cosines from the *College Mathematics Journal* and the student is asked to fill in the details. (See page 496.)

CHANGES IN THIS EDITION

Comments and suggestions from students, instructors, and reviewers have helped me to revise this text in a number of ways that I believe will make the book more useful to the instructor and more accessible to the student. As already mentioned, two new features of the book are the graphing calculator exercises and the *Writing Mathematics* sections. Other major changes occur in the following areas.

Chapter 1   *Algebra Background for Precalculus*   There are three structural changes that have shortened this chapter. The geometric material on rectangular coordinates has been moved into Chapter 2; the review of exponents and radicals is given in two appendices at the back of the book; and the introduction to the complex-number system now appears at the beginning of Chapter 11, *Roots of Polynomial Equations.* The section on factoring (Section 1.3) has been revised. There are more examples and exercises, and many of these emphasize patterns.

Chapter 2   *Coordinates and Graphs*   This is a self-contained review or reintroduction to elementary analytic geometry. Section 2.2 (*Graphs and Equations, a Second Look*) emphasizes the connection between graphing equations and solving equations. This section also contains a discussion of the six basic graphs:

$$y = |x| \qquad y = x^3 \qquad y = \sqrt{x}$$
$$y = x^2 \qquad y = 1/x \qquad y = \sqrt{1-x^2}$$

(In the previous edition, these graphs were discussed later, in the context of functions.) The material on symmetry (in Section 2.4) has been expanded to include symmetry about the line $y = x$. This simplifies the presentation of inverse functions in the next chapter.

Chapter 3   *Functions*   New material has been added on the average rate of change of a function. (See pages 113–114.) Section 3.3, *Techniques in Graphing*, has been streamlined and the presentation is now, I believe, easier to teach and easier to learn. (Some of the material on translation and coordinates is presented again in Chapter 10, but from a different viewpoint.) Following NCTM guidelines, exercises on iteration of functions are now included. (See, for example, Exercises 49–52 on page 138.)

Chapter 4   *Polynomial and Rational Functions. Applications to Optimization*   The material on graphing polynomials in factored form has been revised, now making use of the earlier work (in Chapter 1) on solving inequalities. Also, more details are given about how to analyze a graph in the vicinity of an x-intercept.

Chapter 5   *Exponential and Logarithmic Functions*   The chapter introduces exponential functions from a more practical or contemporary viewpoint. We indicate how these functions are used in modeling real-life situations. See pages 227 and 228 and Example 3 on page 277. In Section 5.2, the constant e is now introduced both in terms of the limit of $[1 + (1/n)]^n$ and in terms of slopes of tangent lines. In the previous edition, the applications were scattered throughout the sections. Now they are presented in the last two sections of the chapter: Section 5.5 discusses compound interest; Section 5.6 uses the function $\mathcal{N} = \mathcal{N}_0 e^{kt}$ to model population growth and radioactive decay.

Chapter 6   *Trigonometric Functions of Angles*   The section "Algebra and the Trigonometric Functions" from the third edition has been expanded and reorganized. Some of this material now appears much earlier (in Section 6.2). The treatment of the law of sines and the law of cosines has been moved into Chapter 8, along with several other applications.

Chapter 7   *Trigonometric Functions of Real Numbers*   The graphs of the trigonometric functions are explained in Sections 7.3 through 7.5. In Section 7.3, the initial discussion of periodic functions is now illustrated with examples from astronomy, medicine, and music. The discussion in Section 7.3 is limited to the graphs of $y = \sin x$ and $y = \cos x$. The approximation $\sin x \approx x$ is introduced both for its utility and for the insight it yields about the shape of $y = \sin x$.

Chapter 8   *Additional Topics in Trigonometry*   This chapter collects the applications that were spread over three different chapters in the previous edition: the laws of sines and cosines, vectors, and polar coordinates. The treatment of polar coordinates is expanded from one to two sections. In drawing polar curves, we make use of the Cartesian graphs for sine and cosine that were studied in the previous chapter. Also in this chapter, there is a new section on parametric equations.

Chapter 10   *Analytic Geometry*   There are two new sections here, as well as some reorganization. The new sections are *The Focus–Directrix Property of Conics* (10.6) and *The Conics in Polar Coordinates* (10.7). The material on the parabola has been split into two sections. The second section, labeled "optional," shows how to find the tangent to a parabola using the methods of algebra rather than calculus. The section on rotation of axes has been moved to the end of the chapter.

Chapter 11   *Roots of Polynomial Equations*   The chapter now begins with an introduction to the complex-number system. (In the previous edition, this introduction appeared at the end of Chapter 1.) The chapter ends with DeMoivre's theorem and the calculation of nth roots. (In the previous edition, this material was in the last chapter, *Additional Topics in Algebra*.)

SUPPLEMENTARY MATERIALS

1. The *Student's Solutions Manual,* by Ross Rueger, contains complete solutions for the odd-numbered exercises and for all test questions at the end of each chapter.
2. The *Instructor's Solutions Manual,* by Ross Rueger, contains answers or solutions for every even-numbered exercise in *Precalculus*.
3. The *Computer-Generated Testing Programs,* WestTest™ 3.0, are available to schools adopting *Precalculus*. (There are versions for both the Macintosh and the IBM PCs or compatibles running DOS or Windows 3.0.)
4. The *Test Bank,* by Charles Heuer, is a package that contains five versions of a

chapter test, as well as two multiple-choice versions, for each chapter in *Precalculus.*

5. *GraphToolz,* by Tom Saxton, is a software program for graphing and evaluating functions. This ingenious, easy-to-use software, along with *Drill and Enrichment Exercises,* by David Cohen, can make many of the topics in *Precalculus* really seem to come alive. *GraphToolz* is free to qualified adopters. (Available for the MacIntosh family of computers.)

6. *Transparency masters* for many of the key figures or tables appearing in the text are available to schools adopting *Precalculus.*

7. *College Algebra: In Simplest Terms* (a series of video tapes produced by Annenberg/CPB Collection) is available to qualified adopters. This series covers the major topics in college algebra, and it includes some excellent applications of precalculus mathematics.

ACKNOWLEDGMENTS Many students and teachers from both colleges and high schools have made useful constructive suggestions about the text and exercises, and I thank them for that. Special thanks go to Professor Charles Heuer for his careful work in checking the text and the exercise solutions for accuracy. I am grateful to Professor Mickey Settle for allowing us to reproduce his copyrighted work, *Using the TI-81 Graphics Calculator,* in Section A.2 of the appendix. In preparing and revising the manuscript, I received valuable suggestions and comments from the following reviewers.

Wayne Andrepont
University of Southwestern Louisiana

Edward Bradley
California Polytechnic State University,
San Luis Obispo

Curtis Chipman
Oakland University

Philip Farmer
Diablo Valley College

Ray Glenn
Tallahassee Community College

Harvey Greenwald
California Polytechnic State University, San Luis Obispo

John Gregory
Southern Illinois University

Giles Maloof
Boise State University

I would like to thank Professor Ross Rueger, who wrote the supplementary manuals and created the answer section for the text; and Susan Gerstein and Pam McMurry, who copyedited the manuscript. Finally, to Stacy Lenzen, Deanna Quinn, Peter Marshall, and Jane Bass at West Publishing Company, thank you for all your work and help in bringing this manuscript into print.

David Cohen
Lunada Bay, California, 1993

PRECALCULUS

Fourth Edition

CHAPTER ONE

ALGEBRA BACKGROUND FOR PRECALCULUS

There are topics [in algebra] *whose consideration prepares a student for a deeper understanding.*

Leonhard Euler (1707–1783) in *Introduction to Analysis of the Infinite*, translated by John D. Blanton (New York: Springer–Verlag, 1988)

Perhaps Pythagoras was a kind of magician to his followers because he taught them that nature is commanded by numbers. There is a harmony in nature, he said, a unity in her variety, and it has a language: numbers are the language of nature.

Jacob Bronowski in *The Ascent of Man* (Boston: Little, Brown, and Company, 1973)

INTRODUCTION

In Chapters 1 and 2 we review several key topics from algebra and coordinate geometry that form the foundation for our work in precalculus. Although you are probably already familiar with some of this material from previous courses, do not be lulled into a false sense of security. Now, in your second exposure to these topics, you really have the opportunity to master them. Take advantage of this opportunity; it will pay great dividends both in this course and in calculus.

1.1 SETS OF REAL NUMBERS

Natural numbers have been used since time immemorial; fractions were employed by the ancient Egyptians as early as 1700 B.C.; and the Pythagoreans, in ancient Greece, about 400 B.C., discovered numbers, like $\sqrt{2}$, which cannot be fractions.

Stefan Drobot in *Real Numbers* (Englewood Cliffs, NJ: Prentice-Hall, Inc., 1964)

What secrets lie hidden in decimals?

Stephen P. Richards in *A Number for Your Thoughts* (New Providence, NJ: S. P. Richards, 1982)

Here, as in your previous mathematics courses, most of the numbers we deal with are *real numbers*. These are the numbers used in everyday life, in the sciences, in industry, and in business. Perhaps the simplest way to define a real number is this: A **real number** is any number that can be expressed in decimal form. Some examples of real numbers are

$$7 \ (= 7.000\ldots) \qquad -\tfrac{2}{3} \ (= -0.\overline{6}) \qquad \sqrt{2} \ (= 1.4142\ldots)$$

(The bar above the 6 in the decimal $-0.\overline{6}$ indicates that the 6 repeats indefinitely.)

Certain sets of real numbers are referred to often enough to be given special names. These are summarized in the box that follows.

PROPERTY SUMMARY	SETS OF REAL NUMBERS	
NAME	DEFINITION AND COMMENTS	EXAMPLES
Natural numbers	These are the ordinary counting numbers, 1, 2, 3, and so on.	1, 4, 29, 1066
Integers	These are the natural numbers along with their negatives and zero.	$-26, 0, 1, 1993$
Rational numbers	As the name suggests, these are the real numbers that are *ratios* of two integers (with nonzero denominators, of course). It can be proved that a real number is rational if and only if its decimal expansion terminates (e.g., 3.15) or repeats (e.g., $2.\overline{43}$).	$4 \left(= \frac{4}{1}\right), -\frac{2}{3},$ $1.7 \left(= \frac{17}{10}\right),$ $4.\overline{3}, 4.1\overline{73}$
Irrational numbers	These are the real numbers that are not rational. Section A.4 of the Appendix contains a proof of the fact that the number $\sqrt{2}$ is irrational. The proof that π is irrational is more difficult. The first person to prove that π is irrational was the Swiss mathematician J. H. Lambert (1728–1777).	$\sqrt{2}, 3 + \sqrt{2}, 3\sqrt{2},$ $\pi, 4 + \pi, 4\pi$

As you've seen in previous courses, the real numbers can be represented as points on a *number line,* as shown in Figure 1. As indicated in Figure 1, the point associated with the number zero is referred to as the **origin.**

FIGURE 1

The fundamental fact here is that there is a **one-to-one correspondence** between the set of real numbers and the set of points on the line. This means that each real number is identified with exactly one point on the line; conversely, with each point on the line we identify exactly one real number. The real number associated with a given point is called the **coordinate** of the point. As a practical matter, we're usually more interested in relative locations than precise locations on a number line. For instance, since π is approximately 3.1, we show π slightly to the right of 3 in Figure 2. Similarly, since $\sqrt{2}$ is approximately 1.4, we show $\sqrt{2}$ slightly less than halfway from 1 to 2 in Figure 2.

It is often convenient to use number lines that show reference points other than the integers used in Figure 2. For instance, Figure 3(a) displays a number line with reference points that are multiples of π. In this case, of course, it is the integers that we then locate approximately. For example, in Figure 3(b) we show the approximate location of the number 1 on such a line.

FIGURE 2

FIGURE 3

Two of the most basic relations for real numbers are **less than** and **greater than**, symbolized by $<$ and $>$, respectively. For ease of reference, we review these and two related symbols in the box that follows.

▶ PROPERTY SUMMARY	NOTATION FOR LESS THAN AND GREATER THAN	
NOTATION	**DEFINITION**	**EXAMPLES**
$a < b$	a is less than b. On a number line, oriented as in Figure 1, 2, or 3, the point a lies to the left of b.	$2 < 3$; $-4 < 1$
$a \leq b$	a is less than or equal to b.	$2 \leq 3$; $3 \leq 3$
$b > a$	b is greater than a. On a number line oriented as in Figure 1, 2, or 3, the point b lies to the right of a. [$b > a$ is equivalent to $a < b$.]	$3 > 2$; $0 > -1$
$b \geq a$	b is greater than or equal to a.	$3 \geq 2$; $3 \geq 3$

In general, relationships involving real numbers and any of the four symbols $<, \leq, >,$ and $\geq$ are called **inequalities**. One of the simplest uses of inequalities is in defining certain sets of real numbers called *intervals*. Roughly speaking, any uninterrupted portion of the number line is referred to as an **interval**. In the definitions that follow, you'll see notation such as $a < x < b$. This means that *both* of the inequalities $a < x$ and $x < b$ hold; in other words, the number x is between a and b.

(a) The open interval (a, b) contains all real numbers from a to b, excluding a and b.

(b) The closed interval $[a, b]$ contains all real numbers from a to b, including a and b.

FIGURE 4

> **DEFINITION   Open Intervals and Closed Intervals**
>
> The **open interval** (a, b) consists of all real numbers x such that $a < x < b$. See Figure 4(a).
>
> The **closed interval** $[a, b]$ consists of all real numbers x such that $a \leq x \leq b$. See Figure 4(b).

Notice that the brackets in Figure 4(b) are used to indicate that the numbers a and b are included in the interval $[a, b]$, whereas the parentheses in Figure 4(a) indicate that a and b are excluded from the interval (a, b). At times you'll see notation such as $[a, b)$. This stands for the set of all real numbers x such that $a \leq x < b$. Similarly, $(a, b]$ denotes the set of all numbers x such that $a < x \leq b$.

▮▬▮ **EXAMPLE 1** Show each interval on a number line, and specify inequalities describing the numbers x in each interval.

$$[-1, 2] \qquad (-1, 2) \qquad (-1, 2] \qquad [-1, 2)$$

Solution   See Figure 5.

FIGURE 5

$[-1, 2]$	$(-1, 2)$	$(-1, 2]$	$[-1, 2)$
$-1 \leq x \leq 2$	$-1 < x < 2$	$-1 < x \leq 2$	$-1 \leq x < 2$

▮▮▮

FIGURE 6
The set of all real numbers x such that
$x > 2$.

In addition to the four types of intervals shown in Figure 5, we can also consider **unbounded intervals.** These are intervals that extend indefinitely in one direction or the other, as shown, for example, in Figure 6. We also have a convenient notation for unbounded intervals; for example, we indicate the unbounded interval in Figure 6 with the notation $(2, \infty)$.

COMMENT AND CAUTION   The symbol ∞ is read *infinity*. It is not a real number, and its use in the context $(2, \infty)$ is only to indicate that the interval has no right-hand boundary. In the box that follows we define the five types of unbounded intervals. Note that the last interval, $(-\infty, \infty)$, is actually the entire real-number line.

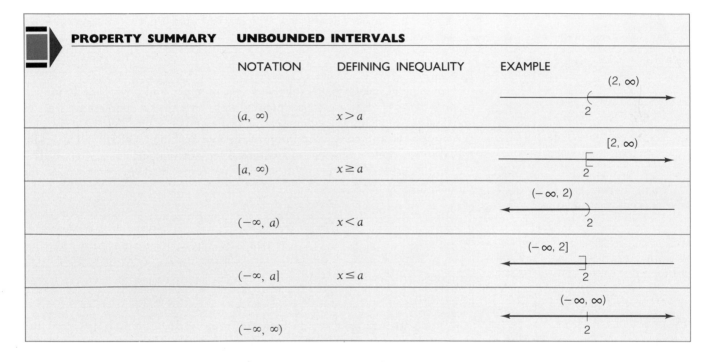

PROPERTY SUMMARY   UNBOUNDED INTERVALS

NOTATION	DEFINING INEQUALITY	EXAMPLE
(a, ∞)	$x > a$	$(2, \infty)$
$[a, \infty)$	$x \geq a$	$[2, \infty)$
$(-\infty, a)$	$x < a$	$(-\infty, 2)$
$(-\infty, a]$	$x \leq a$	$(-\infty, 2]$
$(-\infty, \infty)$		$(-\infty, \infty)$

FIGURE 7
$(-\infty, 4]$

■ ▮▮ ■ **EXAMPLE 2** Indicate each set of real numbers on a number line:
(a) $(-\infty, 4]$;   **(b)** $(-3, \infty)$.

Solution
(a) The interval $(-\infty, 4]$ consists of all real numbers that are less than or equal to 4. See Figure 7.
(b) The interval $(-3, \infty)$ consists of all real numbers that are greater than -3. See Figure 8.   ▮▮▮

FIGURE 8
$(-3, \infty)$

We conclude this section by mentioning that our treatment of the real-number system has been rather informal, and we have not derived any of the rules of arithmetic and algebra using the most basic properties of the real numbers. However, we do list those basic properties and derive some of their consequences in Secion A.3 of the Appendix. (For example, in the appendix we prove that the product of two negative numbers is positive.

EXERCISE SET 1.1

A

In Exercises 1–12, determine whether the number is a natural number, an integer, a rational number, or an irrational number. (Some numbers fit in more than one category.) The following facts will be helpful in some cases: Any number of the form $\sqrt{n}$, where n is a natural number that is not a perfect square, is irrational. Also, the sum, difference, product, and quotient of an irrational number and a nonzero rational are all irrational. (For example, the following four numbers are irrational: $\sqrt{6}$, $\sqrt{10} - 2$, $3\sqrt{15}$, and $-5\sqrt{3}/2$.)

1. (a) 7   (b) -7
2. (a) -203 (b) $\frac{204}{2}$
3. (a) $\frac{27}{4}$ (b) $\sqrt{\frac{27}{4}}$
4. (a) $\sqrt{7}$ (b) $4\sqrt{7}$
5. (a) 10^6 (b) $10^6/10^7$
6. (a) 8.7   (b) $8.\overline{7}$
7. (a) 8.74   (b) $8.\overline{74}$
8. (a) $\sqrt{99}$ (b) $\sqrt{99} + 1$
9. $3\sqrt{101} + 1$
10. $(3 - \sqrt{2}) + (3 + \sqrt{2})$
11. $\dfrac{\sqrt{5} + 1}{4}$
12. $\dfrac{0.1234}{0.5677}$

In Exercises 13–22, draw a number line similar to the one shown in Figure 1. Then indicate the approximate location of the given number. Where necessary, make use of the approximations $\sqrt{2} \approx 1.4$ and $\sqrt{3} \approx 1.7$. (The symbol $\approx$ means is approximately equal to.)

13. $\frac{11}{4}$
14. $-\frac{7}{8}$
15. $1 + \sqrt{2}$
16. $1 - \sqrt{2}$
17. $\sqrt{2} - 1$
18. $-\sqrt{2} - 1$
19. $\sqrt{2} + \sqrt{3}$
20. $\sqrt{2} - \sqrt{3}$
21. $(1 + \sqrt{2})/2$
22. $(2\sqrt{3} + 1)/2$

In Exercises 23–32, draw a number line similar to the one shown in Figure 3(a). Then indicate the approximate location of the given number.

23. $\pi/2$
24. $3\pi/2$
25. $\pi/6$
26. $7\pi/4$
27. -1
28. 3
29. $\pi/3$
30. $\frac{3}{2}$
31. $2\pi + 1$
32. $2\pi - 1$

33. Use a calculator to show the approximate location of the irrational number $(\sqrt{139} - 5)/3$ on a number line.
34. Use a calculator to show the approximate location of the irrational number $\sqrt{\dfrac{3 - \sqrt{5}}{2}}$ on a number line.

In Exercises 35–44, say whether the statement is true or false. (In Exercises 41–44, do not use a calculator or tables; use instead the approximations $\sqrt{2} \approx 1.4$ and $\pi \approx 3.1$.)

35. $-5 < -50$
36. $0 < -1$
37. $-2 \leq -2$
38. $\sqrt{7} - 2 \geq 0$
39. $\frac{13}{14} > \frac{15}{16}$
40. $0.\overline{7} > 0.7$
41. $2\pi < 6$
42. $2 \leq (\pi + 1)/2$
43. $2\sqrt{2} \geq 2$
44. $\pi^2 < 12$

In Exercises 45–58, show the given interval on a number line.

45. $(2, 5)$
46. $(-2, 2)$
47. $[1, 4]$
48. $\left[-\frac{3}{2}, \frac{1}{2}\right]$
49. $[0, 3)$
50. $(-4, 0]$
51. $(-3, \infty)$
52. $(\sqrt{2}, \infty)$
53. $[-1, \infty)$
54. $[0, \infty)$
55. $(-\infty, 1)$
56. $(-\infty, -2)$
57. $(-\infty, \pi]$
58. $(-\infty, \infty)$

B

From grade school on, we all acquire a good deal of experience in working with rational numbers. Certainly we can tell when two rational numbers are equal, although it may require some calculation, as, for example, in the case of $\frac{129}{31} = \frac{2193}{527}$. In Exercises 59 and 60, you'll see that in working with irrational numbers, it's sometimes hard to recognize when two numbers are equal. For each equation in Exercises 59 and 60, use a calculator to verify that the quantities on both sides agree, to six decimal places. (In each case it can be shown that the two quantities are indeed equal.)

59. (a) $\sqrt{6} + \sqrt{2} = 2\sqrt{2 + \sqrt{3}}$
    (b) $\sqrt{3 + \sqrt{5}} + \sqrt{3 - \sqrt{5}} = \sqrt{10}$
    (c) $\sqrt{\sqrt{6 + 4\sqrt{2}}} = \sqrt{2 + \sqrt{2}}$
60. (a) $\sqrt{7 - \sqrt{8 - 2\sqrt{7}}} = 1$
    (b) $\sqrt[3]{2 + \sqrt{5}} + \sqrt[3]{2 - \sqrt{5}} = 1$
    (c) $\sqrt[5]{176 + 80\sqrt{5}} = 1 + \sqrt{5}$

61. The value of the irrational number π, correct to ten decimal places (without rounding off), is 3.1415926535. By using a calculator, determine to how many decimal places each of the following quantities agrees with π.
    (a) $(4/3)^4$ This is the value used for π in the Rhind papyrus, an ancient Babylonian text written about 1650 B.C.
    (b) 22/7   Archimedes (287–212 B.C.) showed that $223/71 < \pi < 22/7$. The use of the approximation 22/7 for π was introduced to the western world through the writings of Boethius (ca 480–520), a Roman philosopher, mathematician, and statesman.

Among all fractions with numerators and denominators less than 100, the fraction 22/7 is the best approximation to π.

(c) 355/113   This approximation for π was obtained in fifth-century China by Zu Chong-Zhi (430–501) and his son. According to David Wells in *The Penguin Dictionary of Curious and Interesting Numbers* (New York: Viking Penguin Inc., 1986), "This is the best approximation of any fraction below 103993/33102."

(d) $\dfrac{63}{25}\left(\dfrac{17+15\sqrt{5}}{7+15\sqrt{5}}\right)$ This approximation for π was obtained by the Indian mathematician Scrinivasa Ramanujan (1887–1920).

Remark: A simple approximation that agrees with π through the first 14 decimal places is $\dfrac{355}{113}\left(1-\dfrac{0.0003}{3533}\right)$. This approximation was also discovered by Ramanujan.) For a fascinating account of the history of π, see the book by Petr Beckman, *A History of π,* third edition (New York: St. Martin's Press, 1974), and for a very contemporary look at π, see Richard Preston's article, "The Mountains of Pi" in *The New Yorker* (March 2, 1992; pp. 36–67).

C

*In Exercises 62–64, give an example of irrational numbers a and b such that the indicated expression is (**a**) rational and (**b**) irrational.*

62. $a+b$ **63.** ab **64.** a/b

65. (a) Give an example in which the result of raising a rational number to a rational power is an irrational number.

(b) Give an example in which the result of raising an irrational number to a rational power is a rational number.

66. Can an irrational number raised to an irrational power yield an answer that is rational? This problem shows that the answer here is "yes." (However, if you study the following solution very carefully, you'll see that even though we've answered the question in the affirmative, we've not pinpointed the specific case in which an irrational number raised to an irrational power is rational.)

(a) Let $A=\left(\sqrt{2}\right)^{\sqrt{2}}$. Now, either A is rational or A is irrational. If A is rational, we are done. Why?

(b) If A is irrational, we are done. Why?
   Hint: Consider $A^{\sqrt{2}}$.
   Remark: For more about this problem and related questions, see the article "Irrational Numbers," by J. P. Jones and S. Toporowski in *American Mathematical Monthly,* vol. 80 (1973), pp. 423–424.

1.2   ABSOLUTE VALUE

There has been a real need in analysis for a convenient symbolism for "absolute value" … and the two vertical bars introduced in 1841 by Weierstrass, as in $|z|$, have met with wide adoption; …

Florian Cajori in *A History of Mathematical Notations, vol. 1* (La Salle, Ill.: The Open Court Publishing Company, 1928)

As an aid in measuring distances on the number line, we review the concept of *absolute value.* We begin with a definition of absolute value that is geometric in nature. Then, after you have developed some familiarity with the concept, we explain a more algebraic approach that is often useful in analytical work.

DEFINITION   Absolute Value (geometric version)

The **absolute value** of a real number x, denoted by $|x|$, is the distance from x to the origin.

For instance, because the numbers 5 and -5 are both five units from the origin, we have $|5| = 5$ and $|-5| = 5$. Here are three more examples:

$$|17| = 17 \qquad \left|-\frac{2}{3}\right| = \frac{2}{3} \qquad |0| = 0$$

In dealing with an expression such as $|-5 + 3|$, the convention is to compute the quantity $-5 + 3$ first and then to take the absolute value. We therefore have, in this case,

$$|-5 + 3| = |-2| = 2$$

EXAMPLE 1 Evaluate each expression.

(a) $5 - |6 - 7|$ **(b)** $||-2| - |-3||$

Solution

(a) $5 - |6 - 7| = 5 - |-1|$ **(b)** $||-2| - |-3|| = |2 - 3|$
$$= 5 - 1 = 4 \qquad\qquad\qquad\qquad = |-1| = 1$$

As we said at the beginning of this section, there is an equivalent, more algebraic way to define absolute value. According to this equivalent definition, the value of $|x|$ is x itself when $x \geq 0$, and the value of $|x|$ is $-x$ when $x < 0$. We can write this symbolically as follows:

DEFINITION   Absolute Value (algebraic version)

$$|x| = \begin{cases} x & \text{when } x \geq 0 \\ -x & \text{when } x < 0 \end{cases}$$

EXAMPLE
$$|-7| = -(-7) = 7$$

By looking at examples with specific numbers, you should be able to convince yourself that both definitions yield the same results. We use the algebraic definition of absolute value in Examples 2 and 3.

EXAMPLE 2 Rewrite each expression in a form that does not contain absolute values.

(a) $|\pi - 4| + 1$
(b) $|x - 5|$ given that $x \geq 5$
(c) $|t - 5|$ given that $t < 5$

Solution

(a) The quantity $\pi - 4$ is negative (since $\pi \approx 3.14$), and therefore its absolute value is equal to $-(\pi - 4)$. In view of this, we have

$$|\pi - 4| + 1 = -(\pi - 4) + 1 = -\pi + 5$$

(b) Since $x \geq 5$, the quantity $x - 5$ is nonnegative, and therefore its absolute value is equal to $x - 5$ itself. Thus, we have

$$|x - 5| = x - 5 \qquad \text{when } x \geq 5$$

(c) Since $t < 5$, the quantity $t - 5$ is negative. Therefore its absolute value is equal to $-(t - 5)$, which in turn is equal to $5 - t$. In view of this, we have

$$|t - 5| = 5 - t \qquad \text{when } t < 5$$ ▮▮▮

▮▮▮ **EXAMPLE 3**   Simplify the expression $|x - 1| + |x - 2|$, given that x is in the open interval $(1, 2)$.

Solution   Since x is greater than 1, the quantity $x - 1$ is positive and, consequently,

$$|x - 1| = x - 1$$

On the other hand, we are also given that x is less than 2. Therefore the quantity $x - 2$ is negative, and we have

$$|x - 2| = -(x - 2) = -x + 2$$

Putting things together now, we can write

$$|x - 1| + |x - 2| = (x - 1) + (-x + 2)$$
$$= -1 + 2 = 1$$ ▮▮▮

The solutions in Examples 2 and 3 used the algebraic version of the definition of absolute value. In the error box that follows, the first two items indicate possible misuses of this definition. *Suggestion:* Before reading further, cover up the columns headed "Correction" and "Comment" in the box and try to decide for yourself where the error lies.

Errors to Avoid		
ERROR	CORRECTION	COMMENT
$\|3 - \sqrt{2}\| \neq -(3 - \sqrt{2}) = -3 + \sqrt{2}$	$\|3 - \sqrt{2}\| = 3 - \sqrt{2}$	Because the quantity $3 - \sqrt{2}$ is positive, the appropriate formula is $\|x\| = x$, not $\|x\| = -x$.
$\|x - y\| \neq -(x - y) = -x + y$	$\|x - y\| = \begin{cases} x - y & \text{if } x - y \geq 0 \\ -(x - y) & \text{if } x - y < 0 \end{cases}$	The error arises in assuming that the quantity $x - y$ is negative. Unless additional information is available, $\|x - y\|$ cannot be simplified.
$\|a + b\| \neq \|a\| + \|b\|$	$\|a + b\| = \begin{cases} a + b & \text{if } a + b \geq 0 \\ -(a + b) & \text{if } a + b < 0 \end{cases}$	The absolute value of a sum is, in general, not equal to the sum of the individual absolute values. For example, if $a = -1$ and $b = 2$, then $\|a + b\| = \|-1 + 2\| = \|1\| = 1$, but $\|a\| + \|b\| = \|-1\| + \|2\| = 3$.

In the box that follows, we list several basic properties of the absolute value. Each of these properties can be derived from the definitions. (With the exception of the *triangle inequality,* we shall omit the derivations. For a proof of the triangle inequality, see Exercise 71.)

PROPERTIES OF ABSOLUTE VALUE

1. For all real numbers x, we have
 (a) $|x| \geq 0$;
 (b) $x \leq |x|$ and $-x \leq |x|$.

2. For all real numbers a and b, we have
 (a) $|ab| = |a||b|$ and $|a/b| = |a|/|b|$ $(b \neq 0)$;
 (b) $|a + b| \leq |a| + |b|$ (the triangle inequality).

▌▐▌ EXAMPLE 4 Write the expression $|-2 - x^2|$ in an equivalent form that does not contain absolute values.

Solution We have

$$|-2 - x^2| = |-1(2 + x^2)|$$
$$= |-1||2 + x^2| \qquad \text{using Property 2(a)}$$
$$= 2 + x^2$$

The last equality follows from the fact that x^2 is nonnegative for any real number x and, consequently, the quantity $2 + x^2$ is positive. ▮▮▮

If we think of the real numbers as points on a number line, the distance between two numbers a and b is given by the absolute value of their difference. For instance, as indicated in Figure 1, the distance between 5 and 7, namely 2, is given by either $|5 - 7|$ or $|7 - 5|$. For reference, we summarize this simple but important fact as follows.

Distance $= |5 - 7| = |7 - 5| = 2$

3   4   5   6   7   8

FIGURE 1

DISTANCE ON A NUMBER LINE

The **distance** between a and b is $|a - b| = |b - a|$.

▌▐▌ EXAMPLE 5 Rewrite each of the following statements using absolute values.

(a) The distance between 12 and -5 is 17.
(b) The distance between x and 2 is 4.
(c) The distance between x and 2 is less than 4.
(d) The point t is more than five units from the origin.

Solution

(a) $|12 - (-5)| = 17$ or $|-5 - 12| = 17$
(b) $|x - 2| = 4$ or $|2 - x| = 4$
(c) $|x - 2| < 4$ or $|2 - x| < 4$
(d) $|t| > 5$ ▮▮▮

▌▐▌ EXAMPLE 6 In each case, the set of real numbers satisfying the given inequality is one or more intervals on the number line. Show the intervals on a number line.

(a) $|x| < 2$ **(b)** $|x| > 2$ **(c)** $|x - 3| < 1$ **(d)** $|x - 3| \geq 1$

Solution

(a) The given inequality tells us that the distance from x to the origin is less than two units. So, as indicated in Figure 2(a), the point x must lie in the open interval $(-2, 2)$.

(b) The condition $|x| > 2$ means that x is more than two units from the origin. Thus, as indicated in Figure 2(b), the point x lies either to the right of 2 or to the left of -2.

(c) The given inequality tells us that x must be less than one unit away from 3 on the number line. Looking one unit to either side of 3, then, we see that x must lie between 2 and 4 and x cannot equal 2 or 4. See Figure 2(c).

(d) The inequality $|x - 3| \geq 1$ says that x is at least one unit away from 3 on the number line. This means that either $x \geq 4$ or $x \leq 2$, as shown in Figure 2(d). [Here's an alternative way of thinking about this: The numbers satisfying the given inequality are precisely those numbers that *do not* satisfy the inequality in part (c). So for part (d), we need to shade that portion of the number line that was not shaded in part (c).]

FIGURE 2

(a) $|x| < 2$ (b) $|x| > 2$ (c) $|x - 3| < 1$ (d) $|x - 3| \geq 1$

EXERCISE SET 1.2

A

In Exercises 1–16, evaluate each expression.

1. $|3|$
2. $3 + |-3|$
3. $|-6|$
4. $-6 - |-6|$
5. $|-1 + 3|$
6. $|-6 + 3|$
7. $\left|-\frac{4}{5}\right| - \frac{4}{5}$
8. $\left|\frac{4}{5}\right| - \frac{4}{5}$
9. $|-6 + 2| - |4|$
10. $|-3 - 4| - |-4|$
11. $||-8| + |-9||$
12. $||-8| - |-9||$
13. $\left|\dfrac{27 - 5}{5 - 27}\right|$
14. $\dfrac{|27 - 5|}{|5 - 27|}$
15. $|7(-8)| - |7||-8|$
16. $|(-7)^2| + |-7|^2 - (-|-3|)^3$

In Exercises 17–24, evaluate each expression, given that $a = -2$, $b = 3$, and $c = -4$.

17. $|a - b|^2$
18. $a^2 - |bc|$
19. $|c| - |b| - |a|$
20. $|b + c| - |b| - |c|$
21. $|a + b|^2 - |b + c|^2$
22. $\dfrac{|a| + |b| + |c|}{|a + b + c|}$
23. $\dfrac{a + b + |a - b|}{2}$
24. $\dfrac{a + b - |a - b|}{2}$

In Exercises 25–38, rewrite each expression in a form that does not contain absolute values.

25. $\left|\sqrt{2} - 1\right| - 1$
26. $\left|1 - \sqrt{2}\right| + 1$
27. $|x - 3|$ given that $x \geq 3$
28. $|x - 3|$ given that $x < 3$
29. $|t^2 + 1|$
30. $|x^4 + 1|$
31. $\left|-\sqrt{3} - 4\right|$
32. $\left|-\sqrt{3} - \sqrt{5}\right|$
33. $|x - 3| + |x - 4|$ given that $x < 3$
34. $|x - 3| + |x - 4|$ given that $x > 4$
35. $|x - 3| + |x - 4|$ given that $3 < x < 4$
36. $|x - 3| + |x - 4|$ given that $x = 4$
37. $|x + 1| + 4|x + 3|$ given that $-\frac{5}{2} < x < -\frac{3}{2}$
38. $|x + 1| + 4|x + 3|$ given that $x < -3$

In Exercises 39–50, rewrite each statement using absolute values, as in Example 5.

39. The distance between x and 4 is 8.
40. The distance between x and 4 is at least 8.
41. The distance between x and 1 is $\frac{1}{2}$.
42. The distance between x and 1 is less than $\frac{1}{2}$.
43. The distance between x and 1 is at least $\frac{1}{2}$.

44. The distance between x and 1 exceeds $\frac{1}{2}$.
45. The distance between y and -4 is less than 1.
46. The distance between x^3 and -1 is at most 0.001.
47. The number y is less than three units from the origin.
48. The number y is less than one unit from the number t.
49. The distance between x^2 and a^2 is less than M.
50. The sum of the distances of a and b from the origin is greater than or equal to the distance of $a + b$ from the origin.

In Exercises 51–62, the set of real numbers satisfying the given inequality is one or more intervals on the number line. Show the intervals on a number line.

51. $|x| < 4$
52. $|x| < 2$
53. $|x| > 1$
54. $|x| > 0$
55. $|x - 5| < 3$
56. $|x - 4| < 4$
57. $|x - 3| \le 4$
58. $|x - 1| \le \frac{1}{2}$
59. $|x + \frac{1}{3}| < \frac{3}{2}$
60. $|x + \frac{\pi}{2}| > 1$
61. $|x - 5| \ge 2$
62. $|x + 5| \ge 2$

B

63. In parts (a) and (b), sketch the interval or intervals corresponding to the given inequality.
    (a) $|x - 2| < 1$
    (b) $0 < |x - 2| < 1$
    (c) In what way do your answers in (a) and (b) differ? (The distinction is important in the study of *limits* in calculus.)

64. Given two real numbers a and b, the notation **max(a, b)** denotes the larger of the two numbers. For instance, if $a = 1$ and $b = 2$, then $\max(1, 2) = 2$. In cases where $a = b$, then $\max(a, b)$ denotes the common value of a and b. It can be shown (see Exercise 69) that $\max(a, b)$ can be expressed in terms of absolute value as follows:

$$\max(a, b) = \frac{a + b + |a - b|}{2}$$

Verify this equation in each of the following cases.
    (a) $a = 1$ and $b = 2$
    (b) $a = -1$ and $b = -5$
    (c) $a = b = 10$

65. Given two real numbers a and b, the notation **min(a, b)** denotes the smaller of the two numbers. In cases where $a = b$, then $\min(a, b)$ denotes the common value of a and b. It can be shown (see Exercise 70)

that min (a, b) can be expressed in terms of absolute value as follows:

$$\min(a, b) = \frac{a + b - |a - b|}{2}$$

Verify this equation in each of the following cases.
    (a) $a = 6$ and $b = 1$
    (b) $a = 1$ and $b = -6$
    (c) $a = b = -6$

66. Show that for all real numbers a and b, we have

$$|a| - |b| \le |a - b|$$

Hint: Beginning with the identity $a = (a - b) + b$, take the absolute value of each side and then use the triangle inequality.

67. Show that

$$|a + b + c| \le |a| + |b| + |c|$$

for all real numbers a, b, and c. *Hint:* The left-hand side can be written $|a + (b + c)|$. Now use the triangle inequality.

68. Explain why there are no real numbers that satisfy the equation $|x^2 + 4x| = -12$.

C

69. (As background for this exercise, you'll need to have worked Exercise 64.) Prove that

$$\max(a, b) = \frac{a + b + |a - b|}{2}$$

Hint: Consider three separate cases: $a = b$; $a > b$; $b > a$.

70. (As background for this exercise, you'll need to have worked Exercise 65.) Prove that

$$\min(a, b) = \frac{a + b - |a - b|}{2}$$

71. Complete the following steps to prove the triangle inequality.
    (a) Let a and b be real numbers. Which property in the box on page 9 tells us that $a \le |a|$ and $b \le |b|$?
    (b) Add the two inequalities in part (a) to obtain $a + b \le |a| + |b|$.
    (c) In a similar fashion, add the two inequalities $-a \le |a|$ and $-b \le |b|$ and deduce that $-(a + b) \le |a| + |b|$.
    (d) Why do the results in parts (b) and (c) imply that $|a + b| \le |a| + |b|$?

1.3 POLYNOMIALS AND FACTORING

The Cartesian use of letters near the beginning of the alphabet for parameters and those near the end as unknown quantities, the adaptation of exponential notation to these, and the use of the Germanic symbols + and − all combined to make Descartes' algebraic notation look like ours, for, of course, we took ours from him.

Carl B. Boyer, *A History of Mathematics,* second edition, revised by Uta C. Merzbach (New York: John Wiley & Sons, Inc., 1991)

As background for our work on polynomials, we first review the terms *constant* and *variable.* By way of example, consider the familiar expression for the area of a circle of radius r, namely, πr^2. Here π is a constant; its value never changes throughout the discussion. On the other hand, r is a variable; we can substitute any positive number for r to obtain the area of a particular circle. More generally, by a **constant** we mean either a particular number (such as π, or -17, or $\sqrt{2}$) or a letter whose value remains fixed (although perhaps unspecified) throughout a given discussion. In contrast, a **variable** is a letter for which we can substitute any number selected from a given set of numbers. The given set of numbers is called the **domain** of the variable.

Some expressions will make sense only for certain values of the variable. For instance, $1/(x-3)$ will be undefined when x is 3 (for then the denominator is zero). So in this case we would agree that the domain of the variable x consists of all real numbers except $x=3$. Similarly, throughout this chapter we adopt the following convention:

The Domain Convention

The domain of a variable in a given expression is the set of all real-number values of the variable for which the expression is defined.

In algebra it's customary (but certainly not mandatory) to use letters near the end of the alphabet for variables; letters from the beginning of the alphabet are generally used for constants. So, for example, in the expression $ax + b$, the letter x is the variable and a and b are constants.

■■■■ EXAMPLE 1 Specify the variable, the constants, and the domain of the variable for each expression.

(a) $3x + 4$ (b) $\dfrac{1}{(t-1)(t+3)}$ (c) $ay^2 + by + c$ (d) $4x + 3x^{-1}$

Solution

	VARIABLE	CONSTANTS	DOMAIN
(a) $3x + 4$	x	$3, 4$	The set of all real numbers.
(b) $\dfrac{1}{(t-1)(t+3)}$	t	$1, -1, 3$	The set of all real numbers except $t = 1$ and $t = -3$.

(c) $ay^2 + by + c$ y a, b, c The set of all real numbers.

(d) $4x + 3x^{-1}$ x $4, 3$ The set of all real numbers except $x = 0$. (Recall, $x^{-1} = 1/x$.) ∎

The expressions in parts (a) and (c) of Example 1 are *polynomials*. By a **polynomial in x**, we mean an expression of the form

$$a_n x^n + a_{n-1} x^{n-1} + \cdots + a_1 x + a_0$$

where n is a nonnegative integer and $a_n \neq 0$. The individual expressions $a_k x^k$ making up the polynomial are called **terms**. In this chapter, the **coefficients** a_k will always be real numbers. For example, the terms of the polynomial $x^2 - 7x + 3$ are x^2, $-7x$, and 3; the coefficients are 1, -7, and 3. In writing a polynomial, it's customary (but not mandatory) to write the terms in order of decreasing powers of x. For instance, we would usually write $x^2 - 7x + 3$ rather than $x^2 + 3 - 7x$. The highest power of x in a polynomial is called the **degree** of the polynomial. For example, the degree of the polynomial $x^2 - 7x + 3$ is 2. In the case of a polynomial consisting only of a nonzero constant a_0, we say that the degree is zero (because $a_0 = a_0 x^0$). No degree is defined for the polynomial whose only term is zero.

Some additional terminology is useful in describing polynomials that have only a few terms. A polynomial with only one term (such as $3x^2$) is a **monomial**; a polynomial with only two terms (such as $2x + 3$) is a **binomial**; and a polynomial with only three terms (such as $5x^2 - 6x + 3$) is a **trinomial**. The table that follows provides examples of this terminology.

| | | If Polynomial: | | |
Expression	Polynomial?	Degree	Terms	Coefficients
$2x^3 - 3x^2 + 4x - 1$	yes	3	$2x^3, -3x^2, 4x, -1$	$2, -3, 4, -1$
$t^2 + 1$	yes	2	$t^2, 1$	$1, 1$
-12	yes	0	-12	-12
$2x^{-4} + 5$	no			
$\dfrac{1}{2x - 3}$	no			
$\sqrt{4x^2 + 1}$	no			
$\sqrt{2}\,x + 1$	yes	1	$\sqrt{2}\,x, 1$	$\sqrt{2}, 1$

In your courses in algebra, you learned to add, subtract, and multiply polynomials. In the multiplication of polynomials, several particular types of products occur so frequently that it is well worth your time to memorize the results. In the box that follows, we display these special products. (Exercises 9 and 10 will ask you to verify these formulas.) As you memorize these formulas, keep in mind that it's the *form*, or *pattern*, that is important—not the specific choice of letters.

PROPERTY SUMMARY   SPECIAL PRODUCTS

1. $(A - B)(A + B) = A^2 - B^2$

2. (a) $(A + B)^2 = A^2 + 2AB + B^2$
   (b) $(A - B)^2 = A^2 - 2AB + B^2$

3. (a) $(A + B)^3 = A^3 + 3A^2B + 3AB^2 + B^3$
   (b) $(A - B)^3 = A^3 - 3A^2B + 3AB^2 - B^3$

4. (a) $(A + B)(A^2 - AB + B^2) = A^3 + B^3$
   (b) $(A - B)(A^2 + AB + B^2) = A^3 - B^3$

▮▮▮ EXAMPLE 2   Use the special products to compute each product.

(a) $(5xy - 4)(5xy + 4)$ (b) $\left(3\sqrt{xy} - z\right)\left(3\sqrt{xy} + z\right)$
(c) $(4x^3 - 3)^2$ (d) $(2x + 5)^3$
(e) $(x - 2)(x^2 + 2x + 4)$

Solution

(a) $(5xy - 4)(5xy + 4) = (5xy)^2 - 4^2 = 25x^2y^2 - 16$ using special product 1

(b) $\left(3\sqrt{xy} - z\right)\left(3\sqrt{xy} + z\right) = \left(3\sqrt{xy}\right)^2 - z^2$ using special product 1
$$= 9xy - z^2$$

(c) $(4x^3 - 3)^2 = (4x^3)^2 - 2(4x^3)(3) + 3^2$ special product 2(b)
$$= 16x^6 - 24x^3 + 9$$

(d) $(2x + 5)^3 = (2x)^3 + 3(2x)^2(5) + 3(2x)(5)^2 + (5)^3$ special product 3(a)
$$= 8x^3 + 60x^2 + 150x + 125$$

(e) $(x - 2)(x^2 + 2x + 4) = x^3 - 2^3 = x^3 - 8$ special product 4(b)   ▮▮▮

There are many cases in algebra in which the process of *factoring* simplifies the work at hand. To **factor** a polynomial means to write it as a product of two or more nonconstant polynomials. For instance, a factorization of $x^2 - 9$ is given by

$$x^2 - 9 = (x - 3)(x + 3)$$

In this case, $x - 3$ and $x + 3$ are the **factors** of $x^2 - 9$.

There is one convention that we need to agree on at the outset. If the polynomial or expression that we wish to factor contains only integer coefficients, then the factors (if any) should involve only integer coefficients. For example, according to this convention, we will not consider the following type of factorization in this section:

$$x^2 - 2 = \left(x - \sqrt{2}\right)\left(x + \sqrt{2}\right)$$

because it involves coefficients that are irrational numbers. (We should point out, however, that factorizations such as this are useful at times, particularly in calculus.) As it happens, $x^2 - 2$ is an example of a polynomial that cannot be

factored using integer coefficients. We say in such a case that the polynomial is
irreducible over the integers.

If the coefficients of a polynomial are rational numbers, then we do allow factors with rational coefficients. For instance, the factorization of $y^2 - \frac{1}{4}$ over the rational numbers is given by

$$y^2 - \tfrac{1}{4} = \left(y - \tfrac{1}{2}\right)\left(y + \tfrac{1}{2}\right)$$

We'll consider five techniques for factoring in this section. (These techniques will be used throughout the text when we simplify expressions and when we solve equations.) In Table 1, we summarize these techniques. Notice that three of the formulas in the table are just restatements of special product formulas. Remember, it is the form or pattern in the formula that is important—not the specific choice of letters.

The idea in factoring is to use one or more of these techniques until each of the factors obtained is irreducible. The examples that follow show how this works in practice.

■■■ EXAMPLE 3 Factor: **(a)** $x^2 - 49$; **(b)** $(2a - 3b)^2 - 49$.

Solution

(a) $\begin{aligned} x^2 - 49 &= x^2 - 7^2 \\ &= (x - 7)(x + 7) \qquad \text{difference of squares} \end{aligned}$

(b) Notice that the pattern or form is the same as in part (a); it's a difference of squares. We have

$$\begin{aligned} (2a - 3b)^2 - 49 &= (2a - 3b)^2 - 7^2 \\ &= [(2a - 3b) - 7][(2a - 3b) + 7] \qquad \text{difference of squares} \\ &= (2a - 3b - 7)(2a - 3b + 7) \end{aligned}$$

■■■

TABLE 1

Basic Factoring Techniques

Technique	Example or Formula	Remark
Common factor	$3x^4 + 6x^3 - 12x^2 = 3x^2(x^2 + 2x - 4)$ $4(x^2 + 1) - x(x^2 + 1) = (x^2 + 1)(4 - x)$	In any factoring problem, the first step always is to look for the common factor of highest degree.
Difference of squares	$x^2 - a^2 = (x - a)(x + a)$	There is no corresponding formula for a sum of squares; $x^2 + a^2$ is irreducible over the integers.
Trial and error	$x^2 + 2x - 3 = (x + 3)(x - 1)$	In this example, the only possibilities, or trials, are: **(a)** $(x - 3)(x - 1)$ **(b)** $(x - 3)(x + 1)$ **(c)** $(x + 3)(x - 1)$ **(d)** $(x + 3)(x + 1)$ By inspection or by carrying out the indicated multiplications, we find that only case (c) checks.
Difference of cubes Sum of cubes	$x^3 - a^3 = (x - a)(x^2 + ax + a^2)$ $x^3 + a^3 = (x + a)(x^2 - ax + a^2)$	Verify these formulas for yourself by carrying out the multiplications. Then memorize the formulas.
Grouping	$\begin{aligned} x^3 - x^2 + x - 1 &= (x^3 - x^2) + (x - 1) \\ &= x^2(x - 1) + (x - 1) \cdot 1 \\ &= (x - 1)(x^2 + 1) \end{aligned}$	This is actually an application of the common factor technique.

❚▩❚ **EXAMPLE 4**   Factor: **(a)** $2x^3 - 50x$;   **(b)** $3x^5 - 3x$.

Solution

(a) $2x^3 - 50x = 2x(x^2 - 25)$ common factor
$ = 2x(x - 5)(x + 5)$ difference of squares

(b) $3x^5 - 3x = 3x(x^4 - 1)$ common factor
$ = 3x[(x^2)^2 - 1^2]$
$ = 3x(x^2 - 1)(x^2 + 1)$ difference of squares
$ = 3x(x - 1)(x + 1)(x^2 + 1)$ difference of squares, again   ▮▮▮

❚▩❚ **EXAMPLE 5**   Factor: **(a)** $x^2 - 4x - 5$;   **(b)** $(a + b)^2 - 4(a + b) - 5$.

Solution

(a) $x^2 - 4x - 5 = (x - 5)(x + 1)$ trial and error

(b) Note that the form of this expression is

$$(\quad)^2 - 4(\quad) - 5$$

This is the same *form* as the expression in part (a). So, we need only replace x with the quantity $a + b$ in the solution for part (a). This yields

$$(a + b)^2 - 4(a + b) - 5 = [(a + b) - 5][(a + b) + 1]$$
$$= (a + b - 5)(a + b + 1) \qquad ▮▮▮$$

❚▩❚ **EXAMPLE 6**   Factor: $2z^4 + 9z^2 + 4$.

Solution We use trial and error to look for a factorization of the form $(2z^2 + ?)(z^2 + ?)$. There are three possibilities:

(i) $(2z^2 + 2)(z^2 + 2)$ **(ii)** $(2z^2 + 1)(z^2 + 4)$ **(iii)** $(2z^2 + 4)(z^2 + 1)$

Each of these yields the appropriate first term and last term, but (after checking) only possibility (ii) yields $9z^2$ for the middle term. The required factorization is then $2z^4 + 9z^2 + 4 = (2z^2 + 1)(z^2 + 4)$.

QUESTION   Why didn't we consider any possibilities with subtraction signs in place of addition signs?   ▮▮▮

❚▩❚ **EXAMPLE 7**   Factor: **(a)** $x^2 + 9$;   **(b)** $x^2 + 2x + 3$.

Solution

(a) The expression $x^2 + 9$ is irreducible over the integers. (This can be discovered by trial and error.) If the given expression had instead been $x^2 - 9$, then it could have been factored as a difference of squares. Sums of squares, however, cannot in general be factored over the integers.

(b) The expression $x^2 + 2x + 3$ is irreducible over the integers. (Check this for yourself by trial and error.)   ▮▮▮

❚▩❚ **EXAMPLE 8**   Factor: $x^2 - y^2 + 10x + 25$.

Solution Familiarity with the special products suggests that we try grouping the terms this way:

$$(x^2 + 10x + 25) - y^2$$

Then we have

$$
\begin{aligned}
x^2 - y^2 + 10x + 25 &= (x^2 + 10x + 25) - y^2 \\
&= (x + 5)^2 - y^2 \\
&= [(x + 5) - y][(x + 5) + y] \qquad \text{difference of squares} \\
&= (x - y + 5)(x + y + 5)
\end{aligned}
$$

■■■ **EXAMPLE 9**   Factor: $ax + ay^2 + bx + by^2$.

Solution   We factor a from the first two terms and b from the second two terms to obtain

$$ax + ay^2 + bx + by^2 = a(x + y^2) + b(x + y^2)$$

We now recognize the quantity $(x + y^2)$ as a common expression that can be factored out. We have then

$$a(x + y^2) + b(x + y^2) = (x + y^2)(a + b)$$

The required factorization is therefore

$$ax + ay^2 + bx + by^2 = (x + y^2)(a + b)$$

As you may wish to check for yourself, this factorization can also be obtained by trial and error.   ■■■

■■■ **EXAMPLE 10**   Factor: **(a)** $t^3 - 125$;   **(b)** $8 + (a - 2)^3$.

Solution

(a)
$$
\begin{aligned}
t^3 - 125 &= t^3 - 5^3 \\
&= (t - 5)(t^2 + 5t + 25) \qquad \text{difference of cubes}
\end{aligned}
$$
(b)
$$
\begin{aligned}
8 + (a - 2)^3 &= 2^3 + (a - 2)^3 \\
&= [2 + (a - 2)][2^2 - 2(a - 2) + (a - 2)^2] \qquad \text{sum of cubes} \\
&= a(4 - 2a + 4 + a^2 - 4a + 4) \\
&= a(a^2 - 6a + 12)
\end{aligned}
$$
As you can check now, the expression $a^2 - 6a + 12$ is irreducible over the integers. Therefore the required factorization is $a(a^2 - 6a + 12)$.   ■■■

For some calculations (particularly in calculus), it's helpful to be able to factor an expression involving fractional exponents. For instance, suppose (as in the next example), we want to factor the expression

$$x(2x - 1)^{-1/2} + (2x - 1)^{3/2}$$

The common expression to factor out here is $(2x - 1)^{-1/2}$; the technique is to *choose the expression with the smaller exponent.* (*Note:*   For a review of fractional exponents, see Section A.6 in the Appendix.)

■■■ **EXAMPLE 11**   Factor: $x(2x - 1)^{-1/2} + (2x - 1)^{3/2}$.

Solution
$$
\begin{aligned}
x(2x - 1)^{-1/2} + (2x - 1)^{3/2} &= (2x - 1)^{-1/2}[x + (2x - 1)^{3/2 + 1/2}] \\
&= (2x - 1)^{-1/2}[x + (2x - 1)^2] \\
&= (2x - 1)^{-1/2}(4x^2 - 3x + 1)
\end{aligned}
$$
■■■

We conclude this section with examples of four common errors to avoid in factoring. The first two errors (in the box that follows) are easy to detect; simply multiplying out the supposed factorizations shows that they do not check. The third error may result from a lack of familiarity with the basic factoring techniques listed on page 15. The fourth error indicates a misunderstanding of what is required in factoring. In factoring, the final quantity must be expressed as a *product*, not a sum, of terms or expressions.

Errors to Avoid

ERROR	CORRECTION	COMMENT
$x^2 + 6x + 9 \neq (x-3)^2$	$x^2 + 6x + 9 = (x+3)^2$	Check the middle term.
$x^2 + 64 \neq (x+8)(x+8)$	$x^2 + 64$ is irreducible over the integers.	A sum of squares is, in general, irreducible over the integers. (A difference of squares, however, can always be factored.)
$x^3 + 64$ is irreducible over the integers.	$x^3 + 64 = (x+4)(x^2 - 4x + 16)$	Although a sum of squares is irreducible, a sum of cubes can be factored.
$x^2 - 2x + 3 \neq x(x-2) + 3$	$x^2 - 2x + 3$ is irreducible over the integers.	The polynomial $x^2 - 2x + 3$ is the *sum* of the expression $x(x-2)$ and the constant 3. But, by definition, the factored form of a polynomial must be a *product* (of two or more nonconstant polynomials).

EXERCISE SET 1.3

A

In Exercises 1–6, specify the domain of each variable.

1. (a) $2x^2 - 3x + 5$ (b) $2x^{1/2} - 3x + 5$

2. (a) $y - 1$ (b) $\dfrac{2y}{y - 1}$

3. (a) $ax + b$ (b) $ax^{1/3} + b$

4. (a) $x + x^{-1}$ (b) $t^{-1} + 2t^{-2}$

5. (a) $4\sqrt{t} + t^2$ (b) $\dfrac{4}{(t-1)\sqrt{t}}$

6. (a) $\dfrac{1}{x}$ (b) $\dfrac{1}{(x-1)(x+2)(x-3)}$

For Exercises 7 and 8, specify the degree and the (nonzero) coefficients of each polynomial.

7. (a) 4 (b) $4x^3$ (c) $x^6 + 4x^3 - x^2 + 2$

8. (a) 0 (b) $1 - x + 6x^5$ (c) $\sqrt{2}x^2 - 2\sqrt{2}$

In Exercises 9 and 10, you are asked to verify the special products in the box on page 14. As an example, here is a verification for special product 4(a).

EXAMPLE

$$(A + B)(A^2 - AB + B^2) = A(A^2 - AB + B^2)$$
$$+ B(A^2 - AB + B^2)$$
$$= A^3 - A^2B + AB^2 + A^2B$$
$$- AB^2 + B^3$$
$$= A^3 + (AB^2 - AB^2)$$
$$+ (A^2B - A^2B) + B^3$$
$$= A^3 + 0 + 0 + B^3 = A^3 + B^3$$

9. Verify special products 1, 2(a), and 2(b), given on page 14.

10. Verify special products 3(a), 3(b), and 4(b), given on page 14.

For Exercises 11–22, use the special product formulas to compute the products.

11. (a) $(x - y)(x + y)$ (b) $(x^2 - 5)(x^2 + 5)$

12. (a) $(3y - a^2)(3y + a^2)$
    (b) $\left(3\sqrt{y} - a^2\right)\left(3\sqrt{y} + a^2\right)$

13. (a) $(A - 4)(A + 4)$
    (b) $[(a + b) - 4][(a + b) + 4]$

14. (a) $\left(\sqrt{ab} + \sqrt{c}\right)\left(\sqrt{ab} - \sqrt{c}\right)$
    (b) $(a^{1/2} + b^{1/2})(a^{1/2} - b^{1/2})$

15. (a) $(x - 8)^2$ (b) $(2x^2 - 5)^2$

16. (a) $(2^m + 1)^2$ (b) $(a^m - a^{-m})^2$

17. (a) $\left(\sqrt{x} + \sqrt{y}\right)^2$ (b) $\left(\sqrt{x+y} - \sqrt{x}\right)^2$

18. (a) $(2x + y)^3$ (b) $(x - 2y)^3$

19. (a) $(a + 1)^3$ (b) $(3x^2 - 2a^2)^3$

20. (a) $(x - y)(x^2 + xy + y^2)$
    (b) $(x^2 - y^2)(x^4 + x^2y^2 + y^4)$

21. (a) $(x + 1)(x^2 - x + 1)$
    (b) $(x^2 + 1)(x^4 - x^2 + 1)$

22. (a) $(5 - y)(25 + 5y + y^2)$
    (b) $(3 + x)(9 - 3x + x^2)$

In Exercises 23–82, factor each polynomial or expression. If a polynomial is irreducible, state this. (In Exercises 23–28, the factoring techniques are specified.)

23. (Common factor and difference of squares)
    (a) $x^2 - 64$ (b) $7x^4 + 14x^2$
    (c) $121z - z^3$ (d) $a^2b^2 - c^2$

24. (Common factor and difference of squares)
    (a) $1 - t^4$ (b) $x^6 + x^5 + x^4$
    (c) $u^2v^2 - 225$ (d) $81x^4 - x^2$

25. (Trial and error)
    (a) $x^2 + 2x - 3$ (b) $x^2 - 2x - 3$
    (c) $x^2 - 2x + 3$ (d) $-x^2 + 2x + 3$

26. (Trial and error)
    (a) $2x^2 - 7x - 4$ (b) $2x^2 + 7x - 4$
    (c) $2x^2 + 7x + 4$ (d) $-2x^2 - 7x + 4$

27. (Sum or difference of cubes)
    (a) $x^3 + 1$ (b) $x^3 + 216$
    (c) $1000 - 8x^6$ (d) $64a^3x^3 - 125$

28. (Grouping)
    (a) $x^4 - 2x^3 + 3x - 6$
    (b) $a^2x + bx - a^2z - bz$

29. (a) $144 - x^2$
    (b) $144 + x^2$
    (c) $144 - (y - 3)^2$

30. (a) $4a^2b^2 + 9c^2$
    (b) $4a^2b^2 - 9c^2$
    (c) $4a^2b^2 - 9(ab + c)^2$

31. (a) $h^3 - h^5$
    (b) $100h^3 - h^5$
    (c) $100(h + 1)^3 - (h + 1)^5$

32. (a) $x^4 - x^2$
    (b) $3x^4 - 48x^2$
    (c) $3(x + h)^4 - 48(x + h)^2$

33. (a) $x^2 - 13x + 40$
    (b) $x^2 - 13x - 40$

34. (a) $x^2 + 10x + 16$
    (b) $x^2 - 10x + 16$

35. (a) $x^2 + 5x - 36$
    (b) $x^2 - 13x + 36$

36. (a) $x^2 - x + 6$
    (b) $x^2 + x - 6$

37. (a) $3x^2 - 22x - 16$
    (b) $3x^2 - x - 16$

38. (a) $x^2 - 4x + 1$
    (b) $x^2 - 4x - 3$

39. (a) $6x^2 + 13x - 5$
    (b) $6x^2 - x - 5$

40. (a) $16x^2 + 18x - 9$
    (b) $16x^2 - 143x - 9$

41. (a) $t^4 + 2t^2 + 1$
    (b) $t^4 - 2t^2 + 1$
    (c) $t^4 - 2t^2 - 1$

42. (a) $t^4 - 9t^2 + 20$
    (b) $t^4 - 19t^2 + 20$
    (c) $t^4 - 19t^2 - 20$

43. (a) $4x^3 - 20x^2 - 25x$
    (b) $4x^3 - 20x^2 + 25x$

44. (a) $x^3 + x^2 + x$
    (b) $x^3 + 2x^2 + x$

45. (a) $ab - bc + a^2 - ac$
    (b) $(u + v)x - xy + (u + v)^2 - (u + v)y$

46. (a) $3(x + 5)^3 + 2(x + 5)^2$
    (b) $a(x + 5)^3 + b(x + 5)^2$

47. $x^2z^2 + xzt + xyz + yt$

48. $a^2t^2 + b^2t^2 - cb^2 - ca^2$

49. $a^4 - 4a^2b^2c^2 + 4b^4c^4$

50. $A^2 - B^2 + 16A + 64$

51. $A^2 + B^2$

52. $x^2 + 64$

53. $x^3 + 64$

54. $27 - (a - b)^3$

55. $(x + y)^3 - y^3$

56. $(a + b)^3 - 8c^3$

57. $x^3 - y^3 + x - y$

58. $8a^3 + 27b^3 + 2a + 3b$

59. (a) $p^4 - 1$
    (b) $p^8 - 1$

60. (a) $p^4 + 4$
    (b) $p^4 - 4$

61. $x^3 + 3x^2 + 3x + 1$

62. $-1 + 6x - 12x^2 + 8x^3$

63. $x^2 + 16y^2$

64. $4u^2 + 25v^2$

65. $\frac{25}{16} - c^2$

66. $\frac{81}{4} - y^2$

67. $z^4 - \frac{81}{16}$

68. $\frac{(a + b)^2}{4} - \frac{a^2b^2}{9}$

69. $\frac{125}{m^3n^3} - 1$

70. $\frac{x^3}{8} - \frac{512}{x^3}$

71. $\frac{1}{4}x^2 + xy + y^2$

72. $x^2 + x + 1$

73. $64(x - a)^3 - x + a$

74. $64(x - a)^4 - x + a$

75. $ax^2 + (a + b)x + b$

76. $(5a^2 - 11a + 10)^2 - (4a^2 - 15a + 6)^2$

77. $(x + 1)^{1/2} - (x + 1)^{3/2}$

78. $(x^2 + 1)^{3/2} + (x^2 + 1)^{7/2}$

79. $(x + 1)^{-1/2} - (x + 1)^{-3/2}$

80. $(x^2 + 1)^{-2/3} + (x^2 + 1)^{-5/3}$

81. $(2x + 3)^{1/2} - \frac{1}{3}(2x + 3)^{3/2}$

82. $(ax + b)^{-1/2} - \sqrt{ax + b}/b$

B

In Exercises 83–87, factor each expression.

83. $A^3 + B^3 + 3AB(A + B)$

84. $p^3 - q^3 - p(p^2 - q^2) + q(p - q)^2$

85. $2x(a^2 + x^2)^{-1/2} - x^3(a^2 + x^2)^{-3/2}$

86. $\frac{1}{2}(x - a)^{-1/2}(x + a)^{-1/2} - \frac{1}{2}(x + a)^{1/2}(x - a)^{-3/2}$

87. $y^4 - (p + q)y^3 + (p^2q + pq^2)y - p^2q^2$

88. (a) Factor: $x^4 + 2x^2y^2 + y^4$.
   (b) Factor: $x^4 + x^2y^2 + y^4$.   *Hint:*   Add and subtract a term. [Keep part (a) in mind.]
   (c) Factor: $x^6 - y^6$ as a difference of squares.
   (d) Factor: $x^6 - y^6$ as a difference of cubes. [Use the result in part (b) to obtain the same answer as in part (c).]

89. The expression $\dfrac{x^2 - 16}{x - 4}$ is undefined when $x = 4$. In this exercise, we investigate the values of the expression when x is very close to 4.
   (a) Complete the tables.

x	$\dfrac{x^2 - 16}{x - 4}$		x	$\dfrac{x^2 - 16}{x - 4}$
3.9			4.1	
3.99			4.01	
3.999			4.001	
3.9999			4.0001	
3.99999			4.00001	

   (b) On the basis of the tables in part (a), what value does the expression $\dfrac{x^2 - 16}{x - 4}$ seem to be approaching as x gets closer and closer to 4? This "target value" is referred to as the *limit of* $\dfrac{x^2 - 16}{x - 4}$ as x approaches 4. (The notation of a limit is made more precise in calculus.)
   (c) How could factoring have been used to obtain this limit without the work in part (a)?

90. The expression $\dfrac{x^3 - 8}{x - 2}$ is undefined when $x = 2$. In this exercise, we investigate the values of the expression when x is very close to 2.
   (a) Complete the tables.

x	$\dfrac{x^3 - 8}{x - 2}$		x	$\dfrac{x^3 - 8}{x - 2}$
1.9			2.1	
1.99			2.01	
1.999			2.001	
1.9999			2.0001	
1.99999			2.00001	

   (b) On the basis of the tables in part (a), what "target value," or limit, does the expression $\dfrac{x^3 - 8}{x - 2}$ seem to be approaching as x gets closer and closer to 2?
   (c) How could factoring have been used to obtain this limit without the work in part (a)?

C

In Exercises 91–96, factor the expressions.

91. $x^4 + 64$ *Hint:*   Add and subtract the term $16x^2$.
92. $x^4 - 15x^2 + 9$ *Hint:*   Add and subtract a term.
93. $(x + y)^2 + (x + z)^2 - (z + t)^2 - (y + t)^2$
94. $(a - a^2)^3 + (a^2 - 1)^3 + (1 - a)^3$
95. $(b - c)^3 + (c - a)^3 + (a - b)^3$
96. $(a + b + c)^3 - a^3 - b^3 - c^3$

1.4   QUADRATIC EQUATIONS

The formula for the quadratic equation seems first to have been discovered by the Moslems around 900 A.D., although quadratic equations had been solved by the Babylonians 3000 years earlier.

Charles Robert Hadlock in *Field Theory and Its Classical Problems* (Washington, D.C.: Mathematical Association of America, 1978)

An equation that can be written in the form

$$ax^2 + bx + c = 0 \qquad \text{with } a, b, \text{ and } c \text{ real numbers and } a \neq 0$$

is called a **quadratic equation**. Examples of quadratic equations are

$$x^2 - 8x - 9 = 0 \qquad 2y^2 - 5 = 0 \qquad 3m^2 = 1 - m$$

Recall that a **solution**, or **root**, of an equation is a value for the unknown that makes the equation a true statement. For example, the value $x = -1$ is a solution of the quadratic equation $x^2 - 8x - 9 = 0$, because when x is replaced by -1 in this equation, we have

$$(-1)^2 - 8(-1) - 9 = 0$$
$$1 + 8 - 9 = 0$$
$$0 = 0 \qquad \text{True}$$

When applicable, one of the simplest techniques for solving quadratic equations involves factoring. This method relies on the following familiar and important property of the real-number system.

PROPERTY SUMMARY   ZERO-PRODUCT PROPERTY OF REAL NUMBERS

$pq = 0 \qquad$ if and only if $\qquad p = 0$ or $q = 0$ (or both)

For example, to solve the equation $x^2 - 2x - 3 = 0$ by factoring, we have

$$x^2 - 2x - 3 = 0$$
$$(x - 3)(x + 1) = 0$$
$$x - 3 = 0 \quad | \quad x + 1 = 0$$
$$x = 3 \quad | \qquad x = -1$$

As you can easily check, the values $x = 3$ and $x = -1$ both satisfy the given equation.

The zero-product property remains valid when we have a product of three or more factors equal to zero. We use this fact in the next example.

EXAMPLE 1 Solve each equation.

(a) $6x^3 - x^2 - 2x = 0$
(b) $y^3 + 3y^2 - y - 3 = 0$

Solution

(a) We can factor out the common term x from the left-hand side of the given equation. This yields

$$x(6x^2 - x - 2) = 0$$
$$x(2x + 1)(3x - 2) = 0$$
$$x = 0 \quad | \quad 2x + 1 = 0 \quad | \quad 3x - 2 = 0$$
$$\quad | \qquad x = -\tfrac{1}{2} \quad | \qquad x = \tfrac{2}{3}$$

The solutions of the equation are then $x = 0$, $x = -\tfrac{1}{2}$, and $x = \tfrac{2}{3}$.

(b)
$$y^3 + 3y^2 - y - 3 = 0$$
$$y^2(y + 3) - (y + 3) = 0$$
$$(y + 3)(y^2 - 1) = 0$$
$$(y + 3)(y - 1)(y + 1) = 0$$

| $y + 3 = 0$ | $y - 1 = 0$ | $y + 1 = 0$ |
| $y = -3$ | $y = 1$ | $y = -1$ |

As you can now easily check, each of the three values $y = -3$, $y = 1$, and $y = -1$ satisfies the given equation.   ■■■

In Example 1 we solved two equations by using the elementary factoring techniques from Section 1.3. Of course, not all equations are so readily solved by factoring. Consider, for example, the quadratic equation $x^2 - 2x - 4 = 0$. There are only three possible factorizations with integer coefficients, but none yields the appropriate middle term, $-2x$, when multiplied out:

$$(x - 4)(x + 1) \qquad (x + 4)(x - 1) \qquad (x - 2)(x + 2)$$

Moreover, it is not always obvious whether a particular equation, such as $x^2 + 156x + 5963 = 0$, can be solved by factoring. (We will come back to this equation in Exercise 39.)

The technique of *completing the square* provides a systematic approach for solving quadratic equations. We'll demonstrate this technique by solving the equation $x^2 - 2x - 4 = 0$. First, we rewrite the equation in the form

$$x^2 - 2x = 4 \tag{1}$$

with the x-terms isolated on the left-hand side of the equation. To complete the square, we follow these two steps:

STEP 1 Take half of the coefficient of x and square it.
STEP 2 Add the number obtained in Step 1 to both sides of the equation.

For equation (1), the coefficient of x is -2. Taking half of -2 and then squaring it gives us $(-1)^2$, or 1. Now, as directed in Step 2, we add 1 to both sides of equation (1). This yields

$$x^2 - 2x + 1 = 4 + 1$$
$$(x - 1)^2 = 5$$
$$x - 1 = \pm\sqrt{5}$$
$$x = 1 \pm \sqrt{5}$$

We have now obtained the two solutions, $1 + \sqrt{5}$ and $1 - \sqrt{5}$. (Exercise 6 will ask you to verify that these values indeed satisfy the original equation.) In the box that follows, we summarize the technique of completing the square, and we indicate how the process got its name.

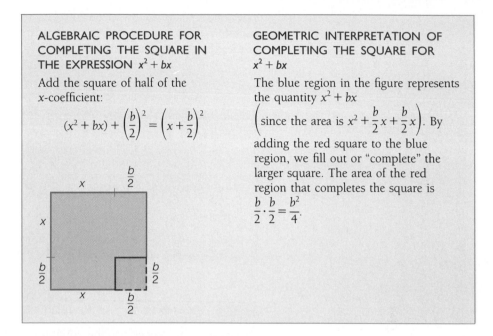

ALGEBRAIC PROCEDURE FOR COMPLETING THE SQUARE IN THE EXPRESSION $x^2 + bx$

Add the square of half of the x-coefficient:

$$(x^2 + bx) + \left(\frac{b}{2}\right)^2 = \left(x + \frac{b}{2}\right)^2$$

GEOMETRIC INTERPRETATION OF COMPLETING THE SQUARE FOR $x^2 + bx$

The blue region in the figure represents the quantity $x^2 + bx$ $\left(\text{since the area is } x^2 + \frac{b}{2}x + \frac{b}{2}x\right)$. By adding the red square to the blue region, we fill out or "complete" the larger square. The area of the red region that completes the square is $\frac{b}{2} \cdot \frac{b}{2} = \frac{b^2}{4}$.

Rather than repeat the process of completing the square for each new equation to be solved, we can instead use a general formula for solving any quadratic equation. The derivation of this **quadratic formula** runs as follows. We start with the general quadratic equation $ax^2 + bx + c = 0$ $(a \neq 0)$. In order to use the procedure for completing the square described in the preceding box, we first divide both sides of the given equation by a, so that the coefficient of x^2 is 1:

$$x^2 + \frac{b}{a}x + \frac{c}{a} = 0$$

Subtracting c/a from both sides yields

$$x^2 + \frac{b}{a}x = -\frac{c}{a}$$

Now, to complete the square, we add $\left[\frac{1}{2}(b/a)\right]^2$, or $b^2/4a^2$, to both sides. That gives us

$$x^2 + \frac{b}{a}x + \frac{b^2}{4a^2} = \frac{b^2}{4a^2} - \frac{c}{a}$$

$$\left(x + \frac{b}{2a}\right)^2 = \frac{b^2 - 4ac}{4a^2}$$

$$x + \frac{b}{2a} = \pm\sqrt{\frac{b^2 - 4ac}{4a^2}} = \pm\frac{\sqrt{b^2 - 4ac}}{2|a|}$$

$$= \pm\frac{\sqrt{b^2 - 4ac}}{2a}$$

This last equality follows from the fact that for any real number $a \neq 0$, the expressions $\pm 2|a|$ and $\pm 2a$ both represent the same two numbers. We now conclude that the solutions are

$$x = -\frac{b}{2a} + \frac{\sqrt{b^2 - 4ac}}{2a} \qquad \text{and} \qquad x = -\frac{b}{2a} - \frac{\sqrt{b^2 - 4ac}}{2a}$$

These solutions are usually written in the more compact form displayed in the following box.

The Quadratic Formula

The solutions of the equation $ax^2 + bx + c = 0$ $(a \neq 0)$ are given by

$$x = \frac{-b \pm \sqrt{b^2 - 4ac}}{2a}$$

EXAMPLE 2 Use the quadratic formula to solve $2x^2 = 4 - x$.

Solution We first rewrite the equation $2x^2 + x - 4 = 0$, so that it has the form $ax^2 + bx + c = 0$. By comparing these last two equations, we see that $a = 2$, $b = 1$, and $c = -4$. Therefore,

$$x = \frac{-b \pm \sqrt{b^2 - 4ac}}{2a} = \frac{-1 \pm \sqrt{1^2 - 4(2)(-4)}}{2(2)} = \frac{-1 \pm \sqrt{33}}{4}$$

Thus, the two solutions are $\dfrac{-1 + \sqrt{33}}{4}$ and $\dfrac{-1 - \sqrt{33}}{4}$.

EXAMPLE 3 Figure 1 shows a right circular cylinder with radius r and height h. The formula for the total surface area S of the cylinder is

$$S = 2\pi r^2 + 2\pi rh \tag{2}$$

Solve equation (2) for r in terms of h and S.

FIGURE 1

Solution By rewriting equation (2) in the following form, we can see that the equation is quadratic in r, and we can determine what values to use for a, b, and c in the quadratic formula:

$$(2\pi)r^2 + (2\pi h)r - S = 0$$

We have $a = 2\pi$, $b = 2\pi h$, and $c = -S$, and, therefore,

$$r = \frac{-2\pi h \pm \sqrt{4\pi^2 h^2 - 4(2\pi)(-S)}}{4\pi} = \frac{-2\pi h \pm 2\sqrt{\pi^2 h^2 + 2\pi S}}{4\pi}$$

$$= \frac{-\pi h \pm \sqrt{\pi^2 h^2 + 2\pi S}}{2\pi}$$

This last expression gives us two solutions to equation (2). In the present context, however, the radius r cannot be negative, so we discard the "minus" option within the plus-or-minus sign, since that would lead to a negative value for r. Thus we have

$$r = \frac{-\pi h + \sqrt{\pi^2 h^2 + 2\pi S}}{2\pi}$$

QUESTION How do we know that this last expression is indeed positive? ■■■

In Examples 2 and 3 we found that each quadratic equation had two real roots (although in Example 3 we were interested only in the positive real root). As the next example indicates, however, it is also possible for quadratic equations to have only one real root, or even no real roots. [In Example 4(b) we make use of the complex number system; if you need a review of this topic, see Section 11.1.]

■■■ **EXAMPLE 4** Use the quadratic formula to solve each equation:

(a) $4x^2 - 12x + 9 = 0$ **(b)** $x^2 + x + 2 = 0$.

Solution

(a) Here $a = 4$, $b = -12$, and $c = 9$, so

$$x = \frac{12 \pm \sqrt{(-12)^2 - 4(4)(9)}}{2(4)} = \frac{12 \pm \sqrt{144 - 144}}{8}$$

$$= \frac{12 \pm 0}{8} = \frac{12}{8} = \frac{3}{2}$$

In this case our only solution is $\frac{3}{2}$. We refer to $\frac{3}{2}$ as a **double root**. (*Note:* We've used the quadratic formula here only for an illustration; in this case we can solve the original equation more efficiently by factoring.)

(b) Since $a = 1$, $b = 1$, and $c = 2$, we have

$$x = \frac{-1 \pm \sqrt{(1)^2 - 4(1)(2)}}{2(1)} = \frac{-1 \pm \sqrt{-7}}{2}$$

Now, if we confine ourselves to the real-number system, the equation has no solutions, because the expression $\sqrt{-7}$ is undefined within the real-number system. However, if we take a broader point of view and work within the complex number system, then there are indeed two solutions. As you'll see in Section 11.1 (or as you perhaps already know from a previous course) these two nonreal complex solutions can be written

$$\frac{-1 + i\sqrt{7}}{2} \quad \text{and} \quad \frac{-1 - i\sqrt{7}}{2}$$ ■■■

The quantity $b^2 - 4ac$ that appears under the radical sign in the quadratic formula is called the **discriminant**. In Examples 2 and 3, the discriminant was positive and, consequently, each equation had two real solutions. In Example 4(a), in which the discriminant was zero, we obtained only one (real) solution. In Example 4(b), the discriminant was negative and, consequently, we obtained two solutions involving the imaginary unit i. These observations are generalized in the box that follows.

The Discriminant $b^2 - 4ac$

Consider the quadratic equation $ax^2 + bx + c = 0$, where a, b, and c are real numbers and $a \neq 0$. The expression $b^2 - 4ac$ is called the **discriminant**.

1. If $b^2 - 4ac > 0$, then the equation has two distinct real roots.

2. If $b^2 - 4ac = 0$, then the equation has exactly one real root.

3. If $b^2 - 4ac < 0$, then the equation has no real root. There are two (nonreal) complex-number roots. (These two roots are *complex conjugates,* that is, they have the form $A \pm Bi$, where A and B are real numbers and $B \neq 0$.)

■■■ **EXAMPLE 5**

(a) Compute the discriminant to determine how many real solutions there are for the equation $x^2 + x - 1 = 0$.

(b) Find a value for k such that the quadratic equation $x^2 + \sqrt{2}x + k = 0$ has exactly one real solution.

Solution

(a) Here $a = 1$, $b = 1$, and $c = -1$. Therefore,

$$b^2 - 4ac = 1^2 - 4(1)(-1) = 5$$

Since the discriminant is positive, the equation has two distinct real solutions.

(b) For the equation to have exactly one real solution, the discriminant must be zero, that is,

$$b^2 - 4ac = 0$$
$$\left(\sqrt{2}\right)^2 - 4(1)(k) = 0$$
$$2 - 4k = 0$$
$$k = \tfrac{1}{2}$$

The required value for k is $\tfrac{1}{2}$.   ■■■

The techniques presented in this section can often be used to solve equations that contain radicals. As a simple example, consider the equation

$$x - 2 = \sqrt{x}$$

By squaring both sides, we obtain

$$x^2 - 4x + 4 = x$$
$$x^2 - 5x + 4 = 0$$
$$(x - 4)(x - 1) = 0$$

Therefore,

$$x = 4 \qquad \text{or} \qquad x = 1$$

The value $x = 4$ checks in the original equation, but the value $x = 1$ does not check. (Verify this.) We say in this case that the value $x = 1$ is an **extraneous root** or **extraneous solution** of the original equation. [This extraneous root arises from the fact that if $a^2 = b^2$, it is not necessarily true that $a = b$. For instance, $(-3)^2 = 3^2$, but certainly $-3 \neq 3$.] In summary, whenever we square both sides of an equation (or raise both sides to an even integral power), there is a possibility of introducing extraneous roots. For this reason, it is essential to check all candidates for solutions obtained in this manner.

Our final example in this section involves an equation containing several square-root expressions. Before squaring both sides in such cases, it is usually a good idea to see if we can first isolate one of the radical expressions on one side of the equation.

EXAMPLE 6 Solve: $\sqrt{9+x} + \sqrt{1+x} - \sqrt{x+16} = 0$.

Solution
$$\sqrt{9+x} + \sqrt{1+x} = \sqrt{x+16} \qquad \text{adding } \sqrt{x+16} \text{ to both sides in preparation for squaring}$$

$$\left(\sqrt{9+x} + \sqrt{1+x}\right)^2 = \left(\sqrt{x+16}\right)^2$$
$$9 + x + 2\sqrt{9+x}\,\sqrt{1+x} + 1 + x = x + 16$$
$$2\sqrt{9+x}\,\sqrt{1+x} = 6 - x$$
$$\left(2\sqrt{9+x}\,\sqrt{1+x}\right)^2 = (6-x)^2$$
$$4(9 + 10x + x^2) = 36 - 12x + x^2$$
$$3x^2 + 52x = 0$$
$$x(3x + 52) = 0$$
$$x = 0 \qquad \bigg| \qquad 3x + 52 = 0$$
$$x = -\tfrac{52}{3}$$

We have found two possible solutions, 0 and $-\frac{52}{3}$. Since these were obtained by squaring both sides of an equation, we need to check to see if they satisfy the original equation. With $x = 0$, the original equation becomes $\sqrt{9} + \sqrt{1} - \sqrt{16} = 0$, or $3 + 1 - 4 = 0$, which is certainly true. On the other hand, with $x = -\frac{52}{3}$, the quantities beneath the radicals in the original equation are negative and, consequently, the (real) square roots are undefined. So $-\frac{52}{3}$ is an extraneous solution, and $x = 0$ is the only solution of the given equation.

EXERCISE SET 1.4

A

In Exercises 1–5, determine if the given value is a solution of the equation.

1. $2x^2 - 6x - 36 = 0$; $x = -3$
2. $(y - 4)(y - 5) = 0$; $y = -1$
3. $4x^2 - 1 = 0$; $x = -\frac{1}{4}$
4. $m^2 + m - \frac{5}{16} = 0$; $m = \frac{1}{4}$
5. $x^2 - 2x - 6 = 0$; $x = -1 - \sqrt{7}$
6. Verify that the numbers $1 + \sqrt{5}$ and $1 - \sqrt{5}$ both satisfy the equation $x^2 - 2x - 4 = 0$.

In Exercises 7–24, solve each equation by factoring.

7. $x^2 - 5x - 6 = 0$
8. $x^2 - 5x = -6$
9. $x^2 - 100 = 0$
10. $4y^2 + 4y + 1 = 0$
11. $25x^2 - 60x + 36 = 0$
12. $144 - t^2 = 0$
13. $10z^2 - 13z - 3 = 0$
14. $3t^2 - t - 4 = 0$
15. $(x + 1)^2 - 4 = 0$
16. $x^2 + 3x - 40 = 0$
17. $x(2x - 13) = -6$
18. $x(3x - 23) = 8$
19. $x^3 + x^2 - 156x = 0$
20. $8x^3 - 18x^2 - 5x = 0$
21. $8y^3 - 4y^2 - 6y + 3 = 0$
22. $y^4 + 2y^3 - 9y^2 - 18y = 0$

23. $-y^3 + y^2 + 2y = 2$

24. $y^6 - 9y^4 - 4y^2 = -36$

In Exercises 25–36, use the quadratic formula to solve the equation.

25. $x^2 + 10x + 9 = 0$

26. $x^2 - 10x + 25 = 0$

27. $x^2 - x - 5 = 0$

28. $x^2 - x + 5 = 0$

29. $2x^2 + 3x - 4 = 0$

30. $4x^2 - 3x - 9 = 0$

31. $12x^2 + 32x + 5 = 0$

32. $10x^2 - x - 1 = 0$

33. $2x^2 = x + 5$

34. $3 - 2x = -4x^2$

35. $-6x^2 + 12x = -1$

36. $-\sqrt{2}x^2 + x = -\sqrt{2}$

In Exercises 37 and 38, use the quadratic formula and a calculator to solve for x. Round each answer to two decimal places. (Note: After completing Exercises 37 and 38, compare the two equations and their solutions. These exercises show that a slight change in one of the coefficients can sometimes radically alter the nature of the solutions.)

37. $x^2 + 3x + 2.249 = 0$

38. $x^2 + 3x + 2.251 = 0$

In Exercises 39 and 40, use the quadratic formula and a calculator to solve for x. (These exercises provide examples of equations that could be solved by factoring, but for which the quadratic formula is clearly more efficient.)

39. $x^2 + 156x + 5963 = 0$

40. $52x^2 - 165x + 108 = 0$

In Exercises 41–52, solve the equations using any method you choose.

41. $(3x - 2)^2 = 3x - 2$

42. $3(x + 1)^2 - 4(x + 1) = 0$

43. $3x^2 + 4x - 3 = 0$

44. $13x^2 = 52$

45. $x^2 = 24$

46. $x^2 + 16x = 0$

47. $x(x - 1) = 1$

48. $x(x - 1) = -4$

49. $\frac{1}{2}x^2 - x - \frac{1}{3} = 0$

50. $x^2 + 34x + 288 = 0$

51. $2\sqrt{5}x^2 - x - 2\sqrt{5} = 0$

52. $\sqrt{2}x^2 + x = 10\sqrt{2}$

In Exercises 53–58, solve for x in terms of the other letters.

53. $2y^2x^2 - 3yx + 1 = 0$ $(y \neq 0)$

54. $(ax + b)^2 - (bx + a)^2 = 0$ $(a \neq \pm b)$

55. $(x - p)^2 + (x - q)^2 = p^2 + q^2$

56. $21x^2 - 2kx - 3k^2 = 0$ $(k > 0)$

57. $12x^2 = ax + 20a^2$ $(a > 0)$

58. $3Ax^2 - 2Ax - 3Bx + 2B = 0$ $(A \neq 0)$

In Exercises 59–64, solve for the indicated letter.

59. $2\pi r^2 + 2\pi rh = 20\pi$,   for r

60. $2\pi y^2 + \pi yx = 12$;   for y

61. $-16t^2 + v_0 t = 0$;   for t

62. $-\frac{1}{2}gt^2 + v_0 t + h_0 = 0$;   for t

63. $x^3 + bx^2 - 2b^2x = 0$;   for x

64. $\frac{1}{y^3} + \frac{b}{y^2} - \frac{2b^2}{y} = 0$;   for y

In Exercises 65–72, use the discriminant to determine how many real roots the equation has.

65. $x^2 - 12x + 16 = 0$

66. $2x^2 - 6x + 5 = 0$

67. $4x^2 - 5x - \frac{1}{2} = 0$

68. $4x^2 - 28x + 49 = 0$

69. $x^2 + \sqrt{3}x + \frac{3}{4} = 0$

70. $\sqrt{2}x^2 + \sqrt{3}x + 1 = 0$

71. $y^2 - \sqrt{5}y = -1$

72. $\frac{m^2}{4} - \frac{4m}{3} + \frac{16}{9} = 0$

In Exercises 73–76, find values for k such that the equation has exactly one real root.

73. $x^2 + 12x + k = 0$

74. $3x^2 + (\sqrt{2k})x + 6 = 0$

75. $x^2 + kx + 5 = 0$

76. $kx^2 + kx + 1 = 0$

In Exercises 77–88, determine all the real number solutions for the equation.

77. (a) $\sqrt{x - 8} = 4$ (b) $\sqrt{x - 8} = 2x + 1$

78. (a) $\sqrt{3x + 1} = 4$ (b) $\sqrt{3x + 1} = 2x - 1$

79. $\sqrt{1 - 3x} = 2$

80. $\sqrt{x^2 + 5x - 2} = 2$

81. $\sqrt{x^4 - 13x^2 + 37} = 1$ Hint: Let $x^2 = t$ and $x^4 = t^2$

82. $\sqrt{y + 2} = y - 4$

83. $\sqrt{1 - 2x} + \sqrt{x + 5} = 4$

84. $\sqrt{x - 5} - \sqrt{x + 4} + 1 = 0$

85. $\sqrt{3 + 2t} + \sqrt{-1 + 4t} = 1$

86. $\sqrt{2t + 5} - \sqrt{8t + 25} + \sqrt{2t + 8} = 0$

87. $\sqrt{2x + 1} + \sqrt{x + 4} = 1$

88. $\sqrt{2x + 6} + \sqrt{x + 4} - \sqrt{8x + 9} = 0$

B

In Exercises 89–98, find all the real solutions of the equation. You'll first want to eliminate the fractions by multiplying both sides by the least common denominator. [For example, in Exercise 89, first multiply both sides by $x(x + 5)$.] Be sure to check your answers for these exercises, because multiplying both sides of an equation by an expression involving the variable may introduce extraneous roots.

89. $\frac{3}{x + 5} + \frac{4}{x} = 2$

90. $\frac{5}{x + 2} - \frac{2x - 1}{5} = 0$

91. $1 - x - \frac{2}{6x + 1} = 0$

92. $\frac{x^2 - 3x}{x + 1} = \frac{4}{x + 1}$

93. $\dfrac{3x^2 - 6x - 3}{(x+1)(x-2)(x-3)} + \dfrac{5 - 2x}{x^2 - 5x + 6} = 0$

94. $\dfrac{6}{x^2 - 1} + \dfrac{x}{x+1} = \dfrac{3}{2}$ **95.** $\dfrac{2x}{x^2 - 1} - \dfrac{1}{x+3} = 0$

96. $\dfrac{3}{x^2 - x - 2} - \dfrac{4}{x^2 + x - 6} = \dfrac{1}{3x + 3}$

97. $\dfrac{x-1}{x+1} - \dfrac{x+1}{x+3} + \dfrac{4}{x^2 + 4x + 3} = 0$

98. $\dfrac{x}{x-2} + \dfrac{x}{x+2} = \dfrac{8}{x^2 - 4}$

In Exercises 99–104, find all the real solutions of the equation.

99. $\sqrt{\sqrt{x} + \sqrt{a}} + \sqrt{\sqrt{x} - \sqrt{a}} = \sqrt{2\sqrt{x} + 2\sqrt{b}}$

100. $x = \sqrt{3x + x^2} - 3\sqrt{3x + x^2}$

101. $\dfrac{\sqrt{x} - a}{\sqrt{x}} - \dfrac{\sqrt{x} + a}{\sqrt{x} - b} = 0$ $(a > 0, b > 0)$
Suggestion: Let $\sqrt{x} = t$.

102. $x - \sqrt{x^2 - x} = \sqrt{x}$ $(x > 0)$
Suggestion: If $x \neq 0$, then you can divide through by $\sqrt{x}$ to obtain a simpler equation.

103. $\sqrt{x^2 - x - 1} - \dfrac{2}{\sqrt{x^2 - x - 1}} = 1$
Hint: Let $t = x^2 - x - 1$.

104. $\sqrt{x^2 + 3x - 4} - \sqrt{x^2 - 5x + 4} = x - 1$ $(x > 4)$
Hint: Factor the expressions beneath the radicals. Then note that $\sqrt{x - 1}$ is a factor of both sides of the equation.

105. For a certain right circular cylinder, the height is $3\,\text{m}$ and the total surface area is $8\pi\,\text{m}^2$. Find the radius of the cylinder.

106. A ball is thrown straight upward. Suppose that the height of the ball at time t is given by $h = -16t^2 + 96t$, where h is in feet and t is in seconds, with $t = 0$ corresponding to the instant that the ball is first tossed.
(a) How long does it take before the ball lands?
(b) At what time is the height 80 ft? Why does this question have two answers?

107. During a flu epidemic in a small town, a public health official finds that the total number of people P who have caught the flu after t days is closely approximated by the formula

$$P = -t^2 + 26t + 106 \qquad (1 \leq t \leq 13)$$

(a) How many have caught the flu after 10 days?
(b) After approximately how many days will 250 people have caught the flu?

108. The radius of a circle is r units. By how many units should the radius be increased so that the area increases by b square units?

109. A piece of wire L in. long is cut into two pieces. Each piece is then bent to form a square. If the sum of the areas of the two squares is $5L^2/128$, how long is each piece of wire?

110. (a) Show that the sum of the roots of the equation $x^2 + px + q = 0$ is $-p$.
(b) Show that the product of the roots of the equation $x^2 + px + q = 0$ is q.

111. In this section we solved quadratic equations by factoring, by completing the square, and by using the quadratic formula. This exercise shows how to solve a quadratic equation by the *method of substitution*. As an example, we use the equation

$$x^2 + x - 1 = 0 \qquad\qquad (1)$$

(a) In equation (1), make the substitution $x = y + k$. Show that the resulting equation can be written

$$y^2 + (2k + 1)y = 1 - k - k^2 \qquad (2)$$

(b) Find a value for k so that the coefficient of y in equation (2) is 0. Then, using this value of k, show that equation (2) becomes $y^2 = \frac{5}{4}$.
(c) Solve the equation $y^2 = \frac{5}{4}$. Then use the equation $x = y + k$ to obtain the solutions of equation (1).

112. Use the substitution method (explained in Exercise 111) to solve the quadratic equation $2x^2 - 3x + 1 = 0$.

113. For which values of A and B will the roots of the equation $x^2 + Ax + B = 0$ be A and B?

114. For which values of k will the roots of the equation $x^2 = 2x(3k + 1) - 7(2k + 3)$ be equal?

115. Let α and β be the roots of the quadratic equation $ax^2 + bx + c = 0$. Verify each of the following statements.
(a) $\alpha + \beta = -b/a$ (b) $\alpha\beta = c/a$
(c) $\alpha^2 + \beta^2 = \dfrac{b^2 - 2ac}{a^2}$ *Suggestion:* Use (a) and (b) along with the fact that $\alpha^2 + \beta^2 = (\alpha + \beta)^2 - 2\alpha\beta$.
(d) $\dfrac{1}{\alpha^2} + \dfrac{1}{\beta^2} = \dfrac{b^2 - 2ac}{c^2}$ *Suggestion:* Add the fractions $1/\alpha^2$ and $1/\beta^2$.

116. Here is an outline for a slightly different derivation of the quadratic formula. The advantage to this method is that fractions are avoided until the very last step. Fill in the details.
(a) Beginning with $ax^2 + bx = -c$, multiply both sides by $4a$. Then add b^2 to both sides.
(b) Now factor the resulting left-hand side and take square roots.

C

117.   Consider the quadratic equation $x^2 + px + q = 0$, where both p and q are nonzero.
   (a) If one root is three times the other, show that $p^2/q = 16/3$.
   (b) If one root is n times the other, show that $p^2/q = (n + 1)^2/n$. (Assume $n \neq \pm 1$.)

In Exercises 118 and 119, solve each equation for x in terms of the other letters.

118. $(a + b + c)x^2 - 2(a + b)x + (a + b - c) = 0$
         $(c > 0,\ a + b + c \neq 0)$ *Hint:*   Let $d = a + b$.

119. $ab(a + b)x^2 - (a^2 + b^2)x - 2(a + b) = 0$
         (Assume that a and b are nonzero and $a \neq -b$.)

1.5   INEQUALITIES

The fundamental results of mathematics are often inequalities *rather than* equalities.

E. Beckenbach and R. Bellman in *An Introduction to Inequalities* (New York: Random House, 1961)

If we replace the equal sign in an equation with any one of the four symbols $<$, $\leq$, $>$, or $\geq$, we obtain an **inequality.** As with equations in one variable, a real number is a **solution** of an inequality if we obtain a true statement when the variable is replaced by the real number. For example, the value $x = 5$ is a solution of the inequality $2x - 3 < 8$, because when $x = 5$ we have

$$2(5) - 3 < 8$$
$$7 < 8 \qquad \text{which is true}$$

We also say in this case that the value $x = 5$ **satisfies** the inequality. To **solve** an inequality means to find all of the solutions. The set of all solutions of an inequality is called (naturally enough) the **solution set.**

   Recall that two equations are said to be equivalent if they have exactly the same solutions. Similarly, two inequalities are **equivalent** if they have the same solution set. Most of the procedures used for solving inequalities are similar to those for equalities. For example, adding or subtracting the same number on both sides of an inequality produces an equivalent inequality. We need to be careful, however, in multiplying or dividing both sides of an inequality by the same nonzero number. For instance, suppose that we start with the inequality $2 < 3$ and multiply both sides by 5. That yields $10 < 15$, which is certainly true. But if we multiply both sides of the inequality $2 < 3$ by -5, we obtain $-10 < -15$, which is false. Multiplying both sides of an inequality by the same *positive* number preserves the inequality, whereas multiplying by a *negative* number reverses the inequality. In the following box, we list some of the principal properties of inequalities. In general, whenever we use Property 1 or 2 in solving an inequality, we obtain an equivalent inequality. Also, note that each property can be rewritten to reflect the fact that $a < b$ is equivalent to $b > a$. For example, Property 3 can just as well be written this way: If $b > a$ and $c > b$, then $c > a$.

PROPERTY SUMMARY	PROPERTIES OF INEQUALITIES	
	PROPERTY	EXAMPLE
	1. If $a < b$, then $a + c < b + c$ and $a - c < b - c$.	If $x - 3 < 0$, then $(x - 3) + 3 < 0 + 3$ and, consequently, $x < 3$.
	2. (a) If $a < b$ and c is positive, then $ac < bc$ and $a/c < b/c$.	If $\frac{1}{2}x < 4$, then $2\left(\frac{1}{2}x\right) < 2(4)$ and, consequently, $x < 8$.
	(b) If $a < b$ and c is negative, then $ac > bc$ and $a/c > b/c$.	If $-\frac{x}{5} < 6$, then $(-5)\left(-\frac{x}{5}\right) > (-5)(6)$ and, consequently, $x > -30$.
	3. The transitive property: If $a < b$ and $b < c$, then $a < c$.	If $a < x$ and $x < 2$, then $a < 2$.

EXAMPLE 1 Solve: $4t + 8 \le 7(1 + t)$.

Solution   Our work follows the pattern that we would use to solve the equation $4t + 8 = 7(1 + t)$. We have

$$4t + 8 \le 7(1 + t)$$
$$4t + 8 \le 7 + 7t$$
$$-3t \le -1 \qquad \text{subtracting } 7t \text{ and } 8 \text{ from both sides}$$
$$t \ge \tfrac{1}{3} \qquad \text{dividing by } -3 \text{ (this reverses the inequality)}$$

FIGURE 1

$t \ge \tfrac{1}{3}$

The solution set is therefore $\left[\tfrac{1}{3}, \infty\right)$. See Figure 1.

In the next example, we solve the inequality

$$-\frac{1}{2} < \frac{3 - x}{-4} < \frac{1}{2}$$

By definition, this is equivalent to the pair of inequalities

$$-\frac{1}{2} < \frac{3 - x}{-4} \qquad \text{and} \qquad \frac{3 - x}{-4} < \frac{1}{2}$$

One way to proceed here would be first to determine the solution set for each inequality. Then the set of real numbers common to both solution sets would be the solution set for the original inequality. However, the method shown in Example 2 is more efficient.

EXAMPLE 2 Solve: $-\dfrac{1}{2} < \dfrac{3 - x}{-4} < \dfrac{1}{2}$.

Solution   We begin by multiplying through by -4. Remember, this will reverse the inequalities:

$$2 > 3 - x > -2$$

Next, with a view toward isolating x, we first subtract 3 to obtain

$$-1 > -x > -5$$

Finally, multiplying through by -1, we have

$$1 < x < 5$$

The solution set is therefore the interval $(1, 5)$.   ■■■

The next example refers to the Celsius and Fahrenheit scales for measuring temperature.* The formula relating the temperature readings on the two scales is

$$F = \frac{9}{5}C + 32$$

■■■■ EXAMPLE 3   Over the temperature range $32° \leq F \leq 39.2°$ on the Fahrenheit scale, water contracts (rather than expands) with increasing temperature. What is the corresponding temperature range on the Celsius scale?

Solution

$$32 \leq F \leq 39.2 \qquad\qquad \text{given}$$

$$32 \leq \frac{9}{5}C + 32 \leq 39.2 \qquad \text{substituting } \frac{9}{5}C + 32 \text{ for } F$$

$$0 \leq \frac{9}{5}C \leq 7.2 \qquad\qquad \text{subtracting 32}$$

$$0 \leq C \leq \frac{5}{9}(7.2) \qquad\quad \text{multiplying by } \frac{5}{9}$$

$$0 \leq C \leq 4$$

Thus, a range of $32°F - 39.2°F$ on the Fahrenheit scale corresponds to $0°C - 4°C$ on the Celsius scale.   ■■■

In the next three examples, we solve inequalities that involve absolute values. The following theorem is very useful in this context.

Theorem: Absolute Value and Inequalities
If $a > 0$, then
$\qquad |u| < a \qquad$ if and only if $\qquad -a < u < a$
and
$\qquad |u| > a \qquad$ if and only if $\qquad u < -a \quad$ or $\quad u > a$

You can see why this theorem is valid if you think in terms of distance and position on a number line. The condition $-a < u < a$ means that u lies between

---

*The Celsius scale was devised in 1742 by the Swedish astronomer Anders Celsius. The Fahrenheit scale was first used by the German physicist Gabriel Fahrenheit in 1724.

FIGURE 2

$-a$ and a, as indicated in Figure 2. But this is the same as saying that the distance from u to zero is less than a, which in turn can be written $|u| < a$. (The second part of the theorem can be justified in a similar manner.)

EXAMPLE 4 Solve: **(a)** $|x| < 1$; **(b)** $|x| \geq 1$.

Solution

(a) By the theorem we just discussed, the condition $|x| < 1$ is equivalent to

$$-1 < x < 1$$

The solution set is therefore the open interval $(-1, 1)$.

(b) In view of the second part of the theorem, the inequality $|x| \geq 1$ is satisfied when x satisfies either of the inequalities

$$x \leq -1 \qquad \text{or} \qquad x \geq 1$$

The solution sets for these last two inequalities are $(-\infty, -1]$ and $[1, \infty)$, respectively. Consequently, the solution set for the given inequality $|x| \geq 1$ consists of the two intervals $(-\infty, -1]$ and $[1, \infty)$.

In Example 4(b), we found that the solution set consisted of two intervals on the number line. We have a convenient notation for describing such sets. Given any two sets A and B, we define the set $A \cup B$ (read **A union B**) to be the set of all elements that are in A or in B (or in both). For example, if $A = \{1, 2, 3\}$ and $B = \{4, 5\}$, then $A \cup B = \{1, 2, 3, 4, 5\}$. As another example, the union of the two closed intervals $[3, 5]$ and $[4, 7]$ is given by

$$[3, 5] \cup [4, 7] = [3, 7]$$

because the numbers in the interval $[3, 7]$ are precisely those numbers that are in $[3, 5]$ or $[4, 7]$ (or in both). Using this notation, we can write the solution set for Example 4(b) as

$$(-\infty, -1] \cup [1, \infty)$$

EXAMPLE 5 Solve: $|x - 3| < 1$.

Solution We'll show two methods.

FIRST METHOD We use the theorem preceding Example 4. With $u = x - 3$ and $a = 1$, the theorem tells us that the given inequality is equivalent to

$$-1 < x - 3 < 1$$

or (by adding 3)

$$2 < x < 4$$

The solution set is therefore the open interval $(2, 4)$.

ALTERNATIVE METHOD The given inequality tells us that x must be less than one unit away from 3 on the number line. Looking one unit to either side of 3, then, we see that x must lie strictly between 2 and 4. The solution set is therefore $(2, 4)$, as obtained using the first method.

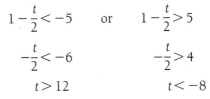

EXAMPLE 6 Solve: $\left|1-\dfrac{t}{2}\right|>5$.

Solution Referring again to the theorem, we use the fact that $|u|>a$ means that either $u<-a$ or $u>a$. So, in the present example, the given inequality means that either

$$1-\frac{t}{2}<-5 \qquad \text{or} \qquad 1-\frac{t}{2}>5$$

$$-\frac{t}{2}<-6 \qquad\qquad -\frac{t}{2}>4$$

$$t>12 \qquad\qquad t<-8$$

This tells us that the given inequality is satisfied when t is in either of the intervals $(12,\infty)$ or $(-\infty,-8)$. Thus the solution set is $(-\infty,-8)\cup(12,\infty)$. ∎

EXERCISE SET 1.5

A

In Exercises 1–34, solve the inequality and specify the answer using interval notation.

1. $x+5<4$
2. $2x-7<11$
3. $1-3x\le 0$
4. $6-4x\le 22$
5. $4x+6<3(x-1)-x$
6. $2(t-1)-3(t+1)\le -5$
7. $1-2(t+3)-t\le 1-2t$
8. $t-4[1-(t-1)]>7+10t$
9. $\dfrac{3x}{5}-\dfrac{x-1}{3}<1$
10. $\dfrac{2x+1}{2}+\dfrac{x-1}{3}<x+\dfrac{1}{2}$
11. $\dfrac{x-1}{4}-\dfrac{2x+3}{5}\le x$
12. $\dfrac{x}{2}-\dfrac{8x}{3}+\dfrac{x}{4}>\dfrac{23}{6}$
13. $-2\le x-6\le 0$
14. $-3\le 2x+1\le 5$
15. $-1\le\dfrac{1-4t}{3}\le 1$
16. $\dfrac{2}{3}\le\dfrac{5-3t}{-2}\le\dfrac{3}{4}$
17. $0.99<\dfrac{x}{2}-1<0.999$
18. $\dfrac{9}{10}<\dfrac{3x-1}{-2}<\dfrac{91}{100}$
19. (a) $|x|\le\frac{1}{2}$
    (b) $|x|>\frac{1}{2}$
20. (a) $|x|>2$
    (b) $|x|\le 2$
21. (a) $|x|>0$
    (b) $|x|<0$
22. (a) $|t|\ge 0$
    (b) $|t|\le 0$
23. (a) $x-2<1$
    (b) $|x-2|<1$
    (c) $|x-2|>1$
24. (a) $x-4\ge 4$
    (b) $|x-4|\ge 4$
    (c) $|x-4|\le 4$
25. (a) $1-x\le 5$
    (b) $|1-x|\le 5$
    (c) $|1-x|>5$
26. (a) $3x+5<17$
    (b) $|3x+5|<17$
    (c) $|3x+5|>17$
27. (a) $a-x<c$
    (b) $|a-x|<c$
    (c) $|a-x|\ge c$
28. (Assume $b<c$ throughout this exercise.)
    (a) $|x-a|+b<c$ (b) $|x+a|+b>c$
    (c) $|x+a|+b<c$
29. $\left|\dfrac{x-2}{3}\right|<4$
30. $\left|\dfrac{4-5x}{2}\right|>1$
31. $\left|\dfrac{x+1}{2}-\dfrac{x-1}{3}\right|<1$
32. $\left|\dfrac{3(x-2)}{4}+\dfrac{4(x-1)}{3}\right|\le 2$
33. (a) $|(x+h)^2-x^2|<3h^2$ $(h>0)$
    (b) $|(x+h)^2-x^2|<3h^2$ $(h<0)$
34. (a) $|3(x+2)^2-3x^2|<\frac{1}{10}$
    (b) $|3(x+2)^2-3x^2|<\epsilon$ $(\epsilon>0)$
35. The surface temperature of the moon varies over the interval $-280°\le F\le 260°$ on the Fahrenheit scale. What is the corresponding interval on the Celsius scale? (Round off the values that you obtain to the nearest 5°C.)
36. Data from the *Apollo 11* moon mission in July 1969 showed temperature readings on the lunar surface varying over the interval $-183°\le C\le 112°$ on the Celsius scale. What is the corresponding interval on the Fahrenheit scale? (Round off the numbers you obtain to the nearest integers.)
37. From the cloud tops of Venus to the planet's surface, temperature readings range over the interval

$-25° \leq C \leq 475°$ on the Celsius scale. What is the corresponding range on the Fahrenheit scale?

38. Measurements sent back to Earth from the *Viking I* spacecraft in June 1976 indicated temperature readings on the surface of Mars in the range $-139° \leq C \leq -28°$ on the Celsius scale. What is the corresponding range on the Fahrenheit scale? (Round off the numbers you obtain to the nearest integers.)

39. The temperature of the variable star Delta Cephei varies over the interval $5100° \leq C \leq 6500°$ on the Celsius scale. What is the corresponding range on the Fahrenheit scale? (Round off the numbers in your answer to the nearest 100°F.)

40. Data from the *Mariner 10* spacecraft (launched November 3, 1973) indicate that the surface temperature on the planet Mercury varies over the interval $-170° \leq C \leq 430°$ on the Celsius scale. What is the corresponding interval on the Fahrenheit scale? (Round off the values that you obtain to the nearest 10°F.)

B

41. Given two positive numbers a and b, we define the **geometric mean** and the **arithmetic mean** as follows:

$$\text{G.M.} = \sqrt{ab} \qquad \text{A.M.} = \frac{a+b}{2}$$

(a) Complete the table, using a calculator as necessary so that the entries in the third and fourth columns are in decimal form.

a	b	$\sqrt{ab}$ (G.M.)	$(a+b)/2$ (A.M.)	Which is larger, G.M. or A.M.?
1	2			
1	3			
1	4			
2	3			
3	4			
5	10			
9	10			
99	100			
999	1000			

(b) Prove that the following inequality is valid for all positive numbers a and b.

$$\sqrt{ab} \leq \frac{a+b}{2} \qquad \text{(The geometric mean is less than or equal to the arithmetic mean.)}$$

Hint: Use the following property of inequalities. If x and y are positive, then the inequality $x \leq y$ is equivalent to $x^2 \leq y^2$.

42. Given two positive numbers a and b, we define the **root mean square** as follows:

$$\text{R.M.S.} = \sqrt{\frac{a^2 + b^2}{2}}$$

(a) Complete the following table. (Use a calculator, as necessary, so that the entries in the third and fourth columns are in decimal form.)

a	b	$(a+b)/2$ (A.M.)	$\sqrt{(a^2+b^2)/2}$ (R.M.S.)	Which is larger, A.M. or R.M.S.?
1	2			
1	3			
1	4			
2	3			
3	4			
5	10			
9	10			
99	100			
999	1000			

(b) Prove that the following inequality is valid for all positive numbers a and b.

$$(a+b)/2 \leq \sqrt{(a^2+b^2)/2} \qquad \text{(The arithmetic mean is less than or equal to the root mean square.)}$$

[Use the hint in Exercise 41(b).]

43. (a) Complete the following table, using a calculator and rounding off to four decimal places.

x	y	$\dfrac{x}{y} + \dfrac{y}{x}$	True or False: $\dfrac{x}{y} + \dfrac{y}{x} \geq 2$
1	1		
2	3		
3	5		
4	7		
5	9		
9	10		
49	50		
99	100		

(b) Prove that for all positive numbers x and y, we have $\dfrac{x}{y} + \dfrac{y}{x} \geq 2$. *Hint:* Use the inequality in Exercise 41(b). Take $a = x/y$ and $b = y/x$.

44. Solve the inequality $|x - 1| + |x - 2| < 3$.
Hint: Begin by considering three cases: $x < 1$; $1 \leq x < 2$; $x \geq 2$.

45. Suppose that $|x - 2| < 0.1$ and $|y - 3| < 0.01$. Show that $|xy - 6| \leq 0.321$. *Hint:* First check that $xy - 6 = x(y - 3) + 3(x - 2)$. Then use the triangle inequality (page 9).

46. Let a and b be positive numbers. In Exercises 41 and 42, you were asked to show that

$$\sqrt{ab} \leq \frac{a + b}{2} \leq \sqrt{\frac{a^2 + b^2}{2}}$$

This exercise shows how to establish these important inequalities using geometric, rather than algebraic, methods. In the following figure, $\overline{CF}$ is the diameter of a semicircle with center E. Let $CD = a$ and $DF = b$.

(a) Explain why $CE = EH = (a + b)/2$.
(b) Show that $DG = \sqrt{ab}$ and $DH = \sqrt{(a^2 + b^2)/2}$.
(c) Explain why $DG \leq EH \leq DH$, and conclude from this that $\sqrt{ab} \leq (a + b)/2 \leq \sqrt{(a^2 + b^2)/2}$.

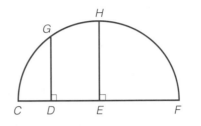

47. Let a, b, c, and d be positive real numbers. Use the inequality in Exercise 41(b) to prove each of the following inequalities.
(a) $\sqrt{abcd} \leq (ab + cd)/2$
(b) $\sqrt{ab} + \sqrt{cd} \leq \sqrt{(a + c)(b + d)}$

48. Let x denote the width of a rectangle with a perimeter of 30 ft.
(a) Show that the area A (in square feet) of the rectangle is given by $A = x(15 - x)$.
(b) Use the inequality in Exercise 41(b) to show that $A \leq 225/4$.
(c) For which value of x is $A = 225/4$? What are the dimensions of the rectangle in this case?

49. Let x denote the width of a rectangle with area 25 ft².
(a) Show that the perimeter P (in feet) of the rectangle is given by $P = 2x + (50/x)$.
(b) Use the inequality in Exercise 41(b) to show that $P \geq 20$.
(c) For which value of x is $P = 20$? What are the dimensions of the rectangle in this case?

1.6 MORE ON INEQUALITIES

The symbols of inequality $>$ and $<$ were introduced by [Thomas] Harriot [1560–1621]. The signs $\geq$ and $\leq$ were first used about a century later by the Parisian hydrographer Pierre Bouguer.

Florian Cajori in *A History of Mathematics*, 4th ed. (New York: Chelsea Publishing Company, 1985)

It is rather surprising to think that the man who surveyed and mapped Virginia was one of the founders of algebra as we know the subject today. Such however is the case, for Thomas

Harriot was sent by Sir Walter Raleigh to accompany Sir Richard Grenville (1585) to the New World, where he made the survey of that portion of American territory.

David Eugene Smith in *History of Mathematics*, vol. I (New York. Ginn and Co., 1923)

FIGURE 1

In this section we are going to solve inequalities involving polynomials and quotients of polynomials. Some observations about first-degree polynomials and inequalities will be helpful in introducing the main ideas. Consider, for example, the inequality $2x - 5 < 0$, along with its solution set $\left(-\infty, \frac{5}{2}\right)$. Let us call a solution of the corresponding equality $2x - 5 = 0$ a **key number** for the given inequality. So, in this case, the only key number is $x = \frac{5}{2}$. As indicated in Figure 1, the key number $\frac{5}{2}$ divides the number line into two intervals, namely, $\left(-\infty, \frac{5}{2}\right)$ and $\left(\frac{5}{2}, \infty\right)$. Notice that on each of the intervals determined by the key number, the algebraic sign of $2x - 5$ is constant. (That is, for $x < \frac{5}{2}$ the value of $2x - 5$ is always negative; for $x > \frac{5}{2}$ the value of $2x - 5$ is always positive.) More generally, it can be shown that this same type of behavior regarding *persistence of sign* occurs with all polynomials, and indeed with quotients of polynomials as well. This important fact, along with the definition of a key number, is presented in the box that follows.

Key Numbers and Persistence of Sign

Let P and Q be polynomials with no common factor (other than constants), and consider the following four inequalities:

$$\frac{P}{Q} < 0 \qquad \frac{P}{Q} \leq 0 \qquad \frac{P}{Q} > 0 \qquad \frac{P}{Q} \geq 0$$

The **key numbers** for each of these inequalities are the real numbers for which $P = 0$ or $Q = 0$. It can be proved that the algebraic sign of P/Q is constant on each of the intervals determined by these key numbers.

We'll show how this result is applied by solving the polynomial inequality

$$x^3 - 2x^2 - 3x > 0$$

First, using factoring techniques from Section 1.3, we rewrite the inequality in the equivalent form

$$x(x + 1)(x - 3) > 0$$

The key numbers, then, are the solutions of the equation $x(x + 1)(x - 3) = 0$. That is, the key numbers are $x = 0$, $x = -1$, and $x = 3$. Next, we locate these numbers on a coordinate line. As indicated in Figure 2, this divides the number line into four distinct intervals.

FIGURE 2

$(-\infty, -1) \quad (-1, 0) \qquad (0, 3) \qquad (3, \infty)$

$-1 \qquad 0 \qquad\qquad 3$

Now, according to the result stated in the box just prior to this example, no matter what x-value we choose in the interval $(-\infty, -1)$, the resulting sign of $x^3 - 2x^2 - 3x \, [= x(x + 1)(x - 3)]$ will always be the same. Thus, to see what that sign is, we first choose any convenient *test number* in the interval $(-\infty, -1)$, say, $x = -2$. Then, using $x = -2$, we determine the sign of $x(x + 1)(x - 3)$ simply by considering the sign of each factor, as indicated in Table 1.

TABLE I

On the interval $(-\infty, -1)$, the sign of $x^3 - 2x^2 - 3x$ [$= x(x + 1)(x - 3)$] is negative because it is the product of three negative factors.

Interval	Test Number	x	$x + 1$	$x - 3$	$x(x + 1)(x - 3)$
$(-\infty, -1)$	-2	neg.	neg.	neg.	neg.

Looking at Table 1, we conclude that the values of $x^3 - 2x^2 - 3x$ are negative *throughout* the interval $(-\infty, -1)$, and consequently, no number in this interval satisfies the given inequality. Next, we carry out similar analyses for the remaining three intervals, as shown in the Table 2. (You should verify for yourself that the entries in the table are correct.)

TABLE 2

On the interval $(-1, 0)$, the product $x(x + 1)(x - 3)$ is positive because it has two negative factors and one positive factor. On $(0, 3)$, the product is negative because it has two positive factors and one negative factor. And on $(3, \infty)$, the product is positive because all the factors are positive.

Interval	Test Number	x	$x + 1$	$x - 3$	$x(x + 1)(x - 3)$
$(-1, 0)$	$-\frac{1}{2}$	neg.	pos.	neg.	pos.
$(0, 3)$	1	pos.	pos.	neg.	neg.
$(3, \infty)$	4	pos.	pos.	pos.	pos.

Looking at the two tables now, we conclude that the sign of $x(x + 1)(x - 3)$ is positive throughout both of the intervals $(-1, 0)$ and $(3, \infty)$, and consequently, all the numbers in these intervals satisfy the given inequality. Moreover, our work also shows that the other two intervals that we considered are not part of the solution set. As you can readily check, the key numbers themselves do not satisfy the given inequality for this example. In summary, then, the solution set for the inequality $x^3 - 2x^2 - 3x > 0$ is $(-1, 0) \cup (3, \infty)$. Furthermore, it is important to notice that the work we just carried out also provides us with three additional pieces of information.

The solution set for $x^3 - 2x^2 - 3x \geq 0$ is $[-1, 0] \cup [3, \infty)$.

The solution set for $x^3 - 2x^2 - 3x < 0$ is $(-\infty, -1) \cup (0, 3)$.

The solution set for $x^3 - 2x^2 - 3x \leq 0$ is $(-\infty, -1] \cup [0, 3]$.

In the box that follows, we summarize the steps for solving polynomial inequalities.

> ### Steps for Solving Polynomial Inequalities
>
> **1.** If necessary, rewrite the inequality so that the polynomial is on the left-hand side and zero is on the right-hand side.
>
> **2.** Find the key numbers for the inequality and locate them on a number line.
>
> **3.** List the intervals determined by the key numbers.
>
> **4.** From each interval, choose a convenient test number. Then use the test number to determine the sign of the polynomial throughout the interval.
>
> **5.** Use the information obtained in the previous step to specify the required solution set. [Don't forget to take into account whether the original inequality is strict ($<$ or $>$) or nonstrict ($\leq$ or $\geq$).]

EXAMPLE 1 Solve: $x^4 \leq 14x^3 - 48x^2$.

Solution First, we rewrite the inequality so that zero is on the right-hand side. Then we factor the left-hand side as follows:

$$x^4 - 14x^3 + 48x^2 \leq 0$$
$$x^2(x^2 - 14x + 48) \leq 0$$
$$x^2(x - 6)(x - 8) \leq 0$$

$(-\infty, 0)$ $(0, 6)$ $(6, 8)$ $(8, \infty)$

FIGURE 3

From this last line, we see that the key numbers are $x = 0$, 6, and 8. As indicated in Figure 3, these numbers divide the number line into four distinct intervals. We need to choose a test number from each interval and see whether the polynomial is positive or negative on the interval. This work is carried out in the following table.

Interval	Test Number	x^2	$x - 6$	$x - 8$	$x^2(x - 6)(x - 8)$
$(-\infty, 0)$	-1	pos.	neg.	neg.	pos.
$(0, 6)$	1	pos.	neg.	neg.	pos.
$(6, 8)$	7	pos.	pos.	neg.	neg.
$(8, \infty)$	9	pos.	pos.	pos.	pos.

The table shows that the quantity $x^2(x - 6)(x - 8)$ is negative only for x-values in the interval $(6, 8)$. Also, as noted at the start, the quantity is equal to zero when $x = 0, 6,$ or 8. Thus, the solution set consists of the numbers in the closed interval $[6, 8]$, along with the number 0. We can write this set

$$[6, 8] \cup \{0\}$$

where $\{0\}$ denotes the set that has zero as its only member.

EXAMPLE 2 Solve: **(a)** $x^2 - x + 15 > 0$;   **(b)** $x^2 - x + 15 < 0$.

Solution

(a) The equation $x^2 - x + 15 = 0$ has no real solution (because the discriminant of the quadratic is -59, which is negative). So there is no key number, and, consequently, the polynomial $x^2 - x + 15$ never changes sign. To see what that sign is, choose the most convenient test number, namely, $x = 0$, and evaluate the polynomial: $0^2 - 0 + 15 > 0$. So the polynomial is positive for every value of x, and the solution set is $(-\infty, \infty)$, the set of all real numbers.

(b) Our work in part (a) shows that no real number satisfies the inequality $x^2 - x + 15 < 0$. ▪▪▪

The technique used in the previous examples can also be used to solve inequalities involving quotients of polynomials. For these cases recall that the definition of a key number also includes the x-values for which the denominator is zero. For example, the key numbers for the inequality $\dfrac{x+3}{x-4} \geq 0$ are -3 and 4.

$(-\infty, -3)$ $(-3, 4)$ $(4, \infty)$

-3 4

FIGURE 4

EXAMPLE 3 Solve: $\dfrac{x+3}{x-4} \geq 0$.

Solution The key numbers are -3 and 4. As indicated in Figure 4, these numbers divide the number line into three intervals. In the table that follows, we've chosen a test number from each interval and determined the sign of the quotient $(x + 3)/(x - 4)$ for each interval.

Interval	Test Number	$x + 3$	$x - 4$	$\dfrac{x+3}{x-4}$
$(-\infty, -3)$	-4	neg.	neg.	pos.
$(-3, 4)$	0	pos.	neg.	neg.
$(4, \infty)$	5	pos.	pos.	pos.

From these results, we conclude that the solution set for $\dfrac{x+3}{x-4} \geq 0$ contains the two intervals $(-\infty, -3)$ and $(4, \infty)$. However, we still need to consider the two endpoints -3 and 4. As you can easily check, the value $x = -3$ does satisfy the given inequality, but $x = 4$ does not. In summary, then, the solution set is $(-\infty, -3] \cup (4, \infty)$. ▪▪▪

EXAMPLE 4 Solve: $\dfrac{2x+1}{x-1} - \dfrac{2}{x-3} < 1$.

Solution Our first inclination here might be to multiply through by $(x - 1)(x - 3)$ to eliminate fractions. This strategy is faulty, however, since we don't know whether the quantity $(x - 1)(x - 3)$ is positive or negative. Thus, we begin by rewriting the inequality in an equivalent form, with zero on the right-hand side and a single fraction on the left-hand side.

$$\frac{2x+1}{x-1} - \frac{2}{x-3} - 1 < 0$$

$$\frac{(2x + 1)(x - 3) - 2(x - 1) - 1(x - 1)(x - 3)}{(x - 1)(x - 3)} < 0$$

$$\frac{x^2 - 3x - 4}{(x - 1)(x - 3)} < 0 \qquad \text{Check the algebra!}$$

$$\frac{(x + 1)(x - 4)}{(x - 1)(x - 3)} < 0 \qquad (1)$$

The key numbers are those x-values for which the denominator or the numerator is zero. By inspection then, we see that these numbers are -1, 4, 1, and 3. As Figure 5 indicates, these numbers divide the number line into five distinct intervals. Now, just as in the previous examples, we choose a test number from each interval and determine the sign of the quotient for that interval. (You should check each entry in the following table for yourself.)

FIGURE 5

Interval	Test Number	$(x + 1)(x - 4)$	$(x - 1)(x - 3)$	$\dfrac{(x + 1)(x - 4)}{(x - 1)(x - 3)}$
$(-\infty, -1)$	-2	pos.	pos.	pos.
$(-1, 1)$	0	neg.	pos.	neg.
$(1, 3)$	2	neg.	neg.	pos.
$(3, 4)$	$\frac{7}{2}$	neg.	pos.	neg.
$(4, \infty)$	5	pos.	pos.	pos.

From these results, we can see that the quotient on the left-hand side of inequality (1) is negative (as required) on the two intervals $(-1, 1)$ and $(3, 4)$. Now we need to check the endpoints of these intervals. When $x = -1$ or $x = 4$, the quotient is zero, and so, in view of the original inequality, we exclude these two x-values from the solution set. Furthermore, the quotient is undefined when $x = 1$ or $x = 3$, so we must also exclude those two values from the solution set. In summary, then, the solution set is $(-1, 1) \cup (3, 4)$. ▮▮▮

EXERCISE SET 1.6

A

Solve the inequalities in Exercises 1–62. *Suggestion: A calculator is useful for approximating the key numbers in Exercises 19, 20, 37, 46, and 48.*

1. $x^2 + x - 6 < 0$
2. $x^2 + 4x - 32 < 0$
3. $x^2 - 11x + 18 > 0$
4. $2x^2 + 7x + 5 > 0$
5. $9x - x^2 \leq 20$
6. $3x^2 + x \leq 4$
7. $x^2 - 16 \geq 0$
8. $24 - x^2 \geq 0$
9. $16x^2 + 24x < -9$
10. $x^4 - 16 < 0$
11. $x^3 + 13x^2 + 42x > 0$
12. $2x^3 - 9x^2 + 4x \geq 0$
13. $225x \leq x^3$
14. $81x^2 \leq x^6$
15. $2x^2 + 1 \geq 0$
16. $1 + x^2 < 0$
17. $12x^3 + 17x^2 + 6x < 0$
18. $8x^4 < x^2 - 2x^3$
19. $x^2 + x - 1 > 0$
20. $2x^2 + 9x - 1 > 0$
21. $x^2 - 8x + 2 \leq 0$
22. $3x^2 - x + 5 \leq 0$
23. $(x - 1)(x + 3)(x + 4) \geq 0$
24. $x^4(x - 2)(x - 16) \geq 0$
25. $(x + 4)(x + 5)(x + 6) < 0$

26. $\left(x - \frac{1}{2}\right)\left(x + \frac{1}{2}\right)\left(x + \frac{3}{2}\right) < 0$

27. $(x - 2)^2(3x + 1)^3(3x - 1) > 0$

28. $(2x - 1)^3(2x - 3)^5(2x - 5) > 0$

29. $(x - 3)^2(x + 1)^4(2x + 1)^4(3x + 2) \le 0$

30. $x^4 - 25x^2 + 144 \le 0$ 31. $20 \ge x^2(9 - x^2)$

32. $x^2(3x^2 + 11) \ge 4$

33. $9(x - 4) - x^2(x - 4) < 0$

34. $(x + 1)^2 - 5(x + 1) > 14$

35. $4(x^2 - 9) - (x^2 - 9)^2 > -5$

36. $x(1 - x^2)^4 + (x + 3)(1 - x^2)^4 \ge 0$

37. $(x - 4)(2x^2 - 6x - 1) < 0$

38. $(x + 2)^3(x^2 - 4x - 2) < 0$

39. $x^3 + 2x^2 - x - 2 > 0$

40. $2x^4 + x^3 - 16x - 8 > 0$

41. $\dfrac{x - 1}{x + 1} \le 0$ 42. $\dfrac{x + 4}{2x - 5} \le 0$

43. $\dfrac{2 - x}{3 - 2x} \ge 0$ 44. $\dfrac{x^2 - 1}{x^2 + 8x + 15} \ge 0$

45. $\dfrac{x^2 - 8x - 9}{x} < 0$ 46. $\dfrac{x^2 - 3x + 1}{1 - x} < 0$

47. $\dfrac{2x^3 + 5x^2 - 7x}{3x^2 + 7x + 4} > 0$ 48. $\dfrac{x^2 - x - 1}{x^2 + x - 1} > 0$

49. $\dfrac{x}{x + 1} > 1$ 50. $\dfrac{2x}{x - 2} < 3$

51. $\dfrac{1}{x} \le \dfrac{1}{x + 1}$ 52. $\dfrac{2}{x} < \dfrac{x}{2}$

53. $\dfrac{x + 2}{x + 5} \le 1$ 54. $\dfrac{2}{x - 4} - \dfrac{1}{x - 3} \ge \dfrac{2}{3}$

55. $\dfrac{1}{x - 2} - \dfrac{1}{x - 1} \ge \dfrac{1}{6}$ 56. $\dfrac{2x}{x + 5} + \dfrac{x - 1}{x - 5} < \dfrac{1}{5}$

57. $\dfrac{1 + x}{1 - x} - \dfrac{1 - x}{1 + x} < -1$ 58. $\dfrac{x + 1}{x + 2} > \dfrac{x - 3}{x + 4}$

59. $\dfrac{3 - 2x}{3 + 2x} > \dfrac{1}{x}$ 60. $\dfrac{x}{x - 2} - \dfrac{3}{x + 1} \ge 2$

61. $1 + \dfrac{1}{x} \ge \dfrac{1}{1 + x}$ 62. $x - \dfrac{10}{x - 1} \ge 4$

For Exercises 63 and 64, determine the domain of the variable in each expression. (Recall the domain convention from Section 1.3: The domain is the set of all real-number values for the variable for which the expression is defined.)

63. (a) $\sqrt{x^2 - 4x - 5}$ (b) $\sqrt{1/(x^2 - 4x - 5)}$

64. (a) $\sqrt{\dfrac{x + 2}{x - 4}}$ (b) $\sqrt[3]{\dfrac{x + 2}{x - 4}}$

B

65. For which values of b will the equation $x^2 + bx + 1 = 0$ have real solutions?

66. The sum of the first n natural numbers is given by

$$1 + 2 + 3 + \cdots + n = \frac{n(n + 1)}{2}$$

For which values of n will the sum be less than 1225?

67. For which values of a is $x = 1$ a solution of the inequality $\dfrac{2a + x}{x - 2a} < 1$?

68. Solve: $\dfrac{ax + b}{\sqrt{x}} > 2\sqrt{ab}$ $(a > 0, \ b > 0)$.

69. The two shorter sides in a right triangle have lengths x and $1 - x$ $(x > 0)$. For which values of x will the hypotenuse be less than $\sqrt{17}/5$?

70. A piece of wire 12 cm long is cut into two pieces. Denote the lengths of the two pieces by x and $12 - x$. Both pieces are then bent into squares. For which values of x will the combined areas of the squares exceed 5 cm²?

71. Let V and S denote the volume and total surface area, respectively, of a right circular cylinder of radius r and height 1 unit. For which r-values will the ratio V/S be less than $\frac{1}{3}$?

72. Let V and S denote the volume and total surface area, respectively, of a right circular cone of radius r and height 1 unit. For which r-values will the ratio V/S be less than $\frac{4}{27}$?

73. (a) Use a calculator to complete the table. (Round off the results to two decimal places.)

x	1	2	5	10	100	200	10,000
$\dfrac{x^2 + 1000}{2x^2 + x}$							

(b) For which positive real numbers x will we have $\dfrac{x^2 + 1000}{2x^2 + x} < 0.5$?

(c) What is the smallest *natural number* n for which $\dfrac{n^2 + 1000}{2n^2 + n} < 0.5$?

C

74. Find a nonzero value for c so that the solution set for the inequality

$$x^2 + 2cx - 6c < 0$$

is the open interval $(-3c, c)$.

75. Solve: $(x - a)^2 - (x - b)^2 > (a - b)^2/4$. Assume that $a > b$.

76. Solve: $x^3 + (1/x^3) \geq 3$. (Use a calculator to approximate the key numbers.)

▮▮ CHAPTER ONE SUMMARY OF PRINCIPAL TERMS AND NOTATION

TERMS OR NOTATIONS	PAGE REFERENCE	COMMENTS
1. Natural numbers, integers, rational numbers, and irrational numbers	2	The box on page 2 provides both definitions and examples. Also note the theorem in the box, which explains how to distinguish between rationals and irrationals in terms of their decimal representations.
2. $a < b$ $b > a$	3	a is less than b. b is greater than a.
3. (a, b) $[a, b]$	3	The open interval (a, b) consists of all real numbers between a and b, excluding a and b. The closed interval $[a, b]$ consists of all real numbers between a and b, including a and b.
4. (a, ∞)	4	The unbounded interval (a, ∞) consists of all real numbers x such that $x > a$. The infinity symbol, ∞, does not denote a real number. It is used in the context (a, ∞) to indicate that the interval has no right-hand boundary. For the definitions of the other types of unbounded intervals, see the box on page 4.
5. $|x|$	6, 7	The absolute value of a real number x is the distance of x from the origin. This is equivalent to the following algebraic definition: $$|x| = \begin{cases} x & \text{if } x \geq 0 \\ -x & \text{if } x < 0 \end{cases}$$
6. Constant, variable, and the domain convention	12	By a constant we mean either a particular number (such as -8 or π) or a letter whose value remains fixed (although perhaps unspecified) throughout a given discussion. In contrast, a variable is a letter for which we can substitute any number from a given set of numbers. The given set is called the domain of the variable. According to the domain convention, the domain of a variable in a given expression consists of all real numbers for which the expression is defined.
7. Polynomial	13	A polynomial is an expression of the form $a_n x^n + a_{n-1} x^{n-1} + \cdots + a_1 x + a_0$. The individual expressions $a_k x^k$ making up the polynomial are called *terms*. The numbers a_k are called *coefficients*.
8. Degree of a polynomial	13	For a polynomial that is not simply a constant, the degree is the highest power (that is, the exponent) of x that appears. The degree of a nonzero constant polynomial is zero. Degree is undefined for the zero polynomial.
9. Special products	14	The box on page 14 lists four basic types of products that you should memorize because they occur so frequently.
10. Factoring techniques	15	The techniques for factoring listed in the box on page 15 are fundamental for simplifying expressions and for solving equations.

TERMS OR NOTATIONS	PAGE REFERENCE	COMMENTS
11. Quadratic equation	20, 21	A quadratic equation is an equation that can be written in the form $ax^2 + bx + c = 0$ where $a \neq 0$.
12. Zero-product property	21	This property of real numbers (and of complex numbers in general) can be stated as follows. If $pq = 0$, then $p = 0$ or $q = 0$; conversely, if $p = 0$ or $q = 0$, then $pq = 0$. (We used this property in solving quadratic equations by factoring.)
13. Quadratic formula: $$x = \frac{-b \pm \sqrt{b^2 - 4ac}}{2a}$$	24	This is the quadratic formula; it provides the solutions of the quadratic equation $ax^2 + bx + c = 0$. The formula is derived on pages 23-24 by means of the useful technique of completing the square.
14. Discriminant	25	The discriminant of the quadratic equation $ax^2 + bx + c = 0$ is the number $b^2 - 4ac$. As indicated in the box on page 26, the discriminant provides information about the roots of the equation.
15. Extraneous solution (or extraneous root)	26	In solving equations, certain processes (such as squaring both sides) can lead to answers that do not check in the original equation. These numbers are called extraneous solutions (or extraneous roots).
16. Solution set of an inequality	30	This is the set of numbers that satisfy the inequality.
17. $A \cup B$	33	The set $A \cup B$ consists of all elements that are in at least one of the two sets A and B.
18. Key numbers of an inequality	37	Let P and Q denote polynomials and consider the following four inequalities: $$\frac{P}{Q} < 0 \qquad \frac{P}{Q} \leq 0 \qquad \frac{P}{Q} > 0 \qquad \frac{P}{Q} \geq 0$$ The key numbers for each of these inequalities are the real numbers for which $P = 0$ or $Q = 0$. It can be proved that the algebraic sign of P/Q is constant on each of the intervals determined by these key numbers.

CHAPTER ONE   REVIEW EXERCISES

Factor each expression in Exercises 1–20.

1. $a^2 - 16b^2$

2. $x^2 + ax + yx + ay$

3. $8 - (a + 1)^3$

4. $x^2 - 18x + 81$

5. $a^2x^3 + 2ax^2b + b^2x$

6. $8a^2x^2 + 16a^3$

7. $8x^2 + 6x + 1$

8. $12x^2 - 2x - 4$

9. $a^4x^4 - x^8a^8$

10. $2x^2 - 2bx + ax - ab$

11. $8 + 12a + 6a^2 + a^3$

12. $(x^2 + 2x - 8)^2 - (2x + 1)^2$

13. $4x^2y^2z^3 - 3xyz^3 - z^3$

14. $(x + y - 1)^2 - (x - y + 1)^2$

15. $1 - x^6$

16. $12x^3 + 44x^2 - 16x$

17. $a^2x^2 + 2abx + b^2 - 4a^2b^2x^2$

18. $a^2 + 2ab + b^2 + a + b$

19. $a^2 - b^2 + ac - bc + a^2b - b^2a$

20. $5(a + 1)^2 + 29(a + 1) - 144$

Use the special products formulas in Section 1.3 to compute the products in Exercises 21–34.

21. $(9x - y)(9x + y)$

22. $(3x + 1)^2$

23. $(2ab - 3)^2$

24. $(ax + b)^2(ax - b)^2$

25. $(2a + 3)^3$

26. $(x^2 - 2)^3$

27. $(1 - 3a)(1 + 3a + 9a^2)$

28. $(x^n + y^n)(x^{2n} - x^n y^n + y^{2n})$

29. $(x^{1/2} - y^{1/2})(x^{1/2} + y^{1/2})$

30. $(\sqrt{x} - \sqrt{3})(\sqrt{x} + \sqrt{3})$

31. $(x^{1/3} + y^{1/3})^3$

32. $(x^{1/3} + y^{1/3})(x^{2/3} - x^{1/3}y^{1/3} + y^{2/3})$

33. $(x^2 - 3x + 1)^2$ *Hint:*   Rewrite the expression as $[x^2 - (3x - 1)]^2$.

34. $(x^2 + x - 1)(x^2 + x + 1)$

For Exercises 35–39, rewrite the statements using absolute values and inequalities or equalities.

35. The distance between x and 6 is 2.

36. The distance between x and a is less than $\frac{1}{2}$.

37. The distance between a and b is 3.

38. The distance between x and -1 is 5.

39. The distance between x and 0 exceeds 10.

40. What can you say about x if $|x - 5| = 0$?

Rewrite each of the expressions in Exercises 41–46 in a form not containing absolute values.

41. $\left|\sqrt{6} - 2\right|$

42. $\left|2 - \sqrt{6}\right|$

43. $|x^4 + x^2 + 1|$

44. (a) $|x - 3|$ if $x < 3$
    (b) $|x - 3|$ if $x > 3$

45. $|x - 2| + |x - 3|$ if:
    (a) $x < 2$ (b) $2 < x < 3$ (c) $x > 3$

46. $|x + 2| + |x - 1|$ if:
    (a) $x < -2$ (b) $-2 < x < 1$ (c) $x > 1$

In Exercises 47–52, sketch each interval on a number line.

47. $(3, 5)$

48. $(3, 5]$

49. The set of all negative real numbers that are in the interval $[-5, 2]$.

50. $(-\infty, 4)$

51. $[-1, \infty)$

52. $\left(0, \sqrt{2}\right] \cup \left(\sqrt{2}, 2\right]$

In Exercises 53–57, sketch the intervals described by the given inequalities.

53. $|x - 6| < 3$

54. $\left|x - \frac{1}{2}\right| < 1$

55. $|x + 1| \geq 1$

56. $|x| \geq 5$

57. (a) $0 < |x - 4| < 5$ (b) $|x - 4| < 5$

58. Polynomials can be used to approximate more complicated expressions. For example, in calculus it is shown that when x is close to zero, the expression $\sqrt{1 + x}$ can be approximated by the polynomial $1 + \frac{1}{2}x$. Complete the following table to see evidence of this. (Use a calculator to compute $\sqrt{1 + x}$; report the values to six decimal places without rounding off.)

x	$1 + \dfrac{x}{2}$	$\sqrt{1 + x}$

59. Using calculus, it can be shown that when x is close to zero, the expression $1/\sqrt{1 - x^2}$ can be approximated by the polynomial $1 + \frac{1}{2}x^2 + \frac{3}{8}x^4$. Complete the table to see evidence of this. (Use a calculator and round off your answers to six significant digits.)

x	$1 + \dfrac{x^2}{2} + \dfrac{3x^4}{8}$	$1/\sqrt{1 - x^2}$

60. (a) Use a calculator to evaluate the expression
$$\frac{\sqrt{3 - \sqrt{5}}}{\sqrt{2} + \sqrt{7 - 3\sqrt{5}}}.$$

(b) Use a calculator to evaluate the expression $\sqrt{5}/5$. If you do the calculator work carefully, your results in parts (a) and (b) will suggest (but not prove!) that the two expressions are equal. Follow steps (c), (d), and (e) to prove that the two expressions are indeed equal.

(c) Multiply the expression in part (a) by
$$\frac{\sqrt{2} - \sqrt{7 - 3\sqrt{5}}}{\sqrt{2} - \sqrt{7 - 3\sqrt{5}}}, \text{ which equals 1. After}$$
carrying out the indicated operations, you should obtain
$$\frac{\sqrt{6 - 2\sqrt{5}} - 2\sqrt{9 - 4\sqrt{5}}}{-5 + 3\sqrt{5}}.$$

(d) Verify that $\sqrt{6 - 2\sqrt{5}} = \sqrt{5} - 1$. Also verify that $\sqrt{9 - 4\sqrt{5}} = \sqrt{5} - 2$.

(e) Use the results in part (d) to simplify the expression in part (c). The final result, after rationalizing the denominator, should be $\sqrt{5}/5$.

In Exercises 61–78, find all the real solutions of each equation.

61. $12x^2 + 2x - 2 = 0$

62. $4y^2 - 21y = 18$

63. $\frac{1}{2}x^2 + x - 12 = 0$

64. $x^2 + \frac{13}{2}x + 10 = 0$

65. $\frac{x}{5-x} = \frac{-2}{11-x}$

66. $\frac{x^2}{(x-1)(x+1)} = \frac{4}{x+1} + \frac{4}{(x-1)(x+1)}$

67. $\frac{1}{3x-7} - \frac{2}{5x-5} - \frac{3}{3x+1} = 0$

68. $4x^2 + x - 2 = 0$

69. $x^5 - 2x^3 - 2x = 0$

70. $x^3 - 6x^2 + 7x = 0$

71. $\sqrt{4 - 3x} = 5$

72. $\sqrt{x} = \sqrt{x + 27} - 1$

73. $\sqrt{4x + 3} = \sqrt{11 - 8x} - 1$

74. $2 - \sqrt{3\sqrt{2x - 1}} + x = 0$

75. $\frac{2}{\sqrt{x^2 - 36}} + \frac{1}{\sqrt{x + 6}} - \frac{1}{\sqrt{x - 6}} = 0$

76. $\sqrt{x} - \sqrt{1 - x} + \sqrt{x - 1} = 0$

77. $\sqrt{x + 7} - \sqrt{x + 2} = \sqrt{x - 1} - \sqrt{x - 2}$

78. $\frac{\sqrt{x} - 4}{\sqrt{x} - 18} = 3$

Solve the inequalities in Exercises 79–99.

79. $-1 < \frac{1 - 2(1 + x)}{3} < 1$

80. $3 < \frac{x - 1}{-2} < 4$

81. $|x| \le \frac{1}{2}$

82. $|t - 2| < 1$

83. $|x + 4| < \frac{1}{10}$

84. $|3 - 5x| < 2$

85. $|2x - 1| \ge 5$

86. $x^2 - 21x + 108 \le 0$

87. $x^2 + 3x - 40 < 0$

88. $x^2 \ge 15x$

89. $x^2 - 6x - 1 < 0$

90. $(x + 12)(x - 1)(x - 8) < 0$

91. $x^4 - 34x^2 + 225 < 0$

92. $\frac{x + 12}{x - 5} > 0$

93. $\frac{(x - 7)^2}{(x + 2)^3} \ge 0$

94. $\frac{(x - 6)^2(x - 8)(x + 3)}{(x - 3)^2} \le 0$

95. $\frac{x^2 - 10x + 9}{x^3 + 1} \le 0$

96. $\frac{3x + 1}{x - 4} < 1$

97. $\frac{1 - 2x}{1 + 2x} \le \frac{1}{2}$

98. $x^2 + \frac{1}{x^2} > 3$ *Suggestion:* Use a calculator to evaluate the key numbers.

99. $\sqrt{x} - \frac{5}{\sqrt{x}} \le 4$

For Exercises 100–103, find the values of k for which the roots of the equations are real numbers.

100. $kx^2 - 6x + 5 = 0$

101. $x^2 + x + k^2 = 0$

102. $kx(x + 2) = -1$

103. $x^2 + (k + 1)x + 2k = 0$

104. Determine p if the larger root of the equation $x^2 + px - 2 = 0$ is p.

105. A piece of wire x cm long is bent into a square. For which values of x will the area be (numerically) greater than the perimeter?

106. A piece of wire $6x$ cm long is bent into an equilateral triangle. For which values of x will the area be (numerically) less than the perimeter?

If an object is thrown vertically upward from a height of h_0 feet with an initial speed of v_0 ft/sec, then its height h (in feet) after t seconds is given by

$$h = -16t^2 + v_0 t + h_0$$

Make use of this formula in working Exercises 107–109.

107. A ball is thrown vertically upward from ground level with an initial speed of 64 ft/sec.
(a) At what time will the height of the ball be 15 ft? (Two answers.)
(b) For how long an interval of time will the height exceed 63 ft?

108. One ball is thrown vertically upward from a height of 50 ft with an initial speed of 40 ft/sec. At the same instant, another ball is thrown vertically upward from a height of 100 ft with an initial speed of 5 ft/sec. Which ball hits the ground first?

109. An object is projected vertically upward. Suppose that its height is H ft at t_1 sec and again at t_2 sec. Express the initial speed in terms of t_1 and t_2.
Answer: $16(t_1 + t_2)$

Exercise 110 (including the clever solution that is outlined) appears in the book The USSR Olympiad Problem Book *by D. O. Shklarsky et al. (San Francisco: W. H. Freeman and Co., 1962)*

110. Follow steps (a)–(d) to solve the equation
$$\sqrt{a - \sqrt{a + x}} = x \qquad (a \ge 1)$$
(a) By squaring as usual, obtain the equation
$$x^4 - 2ax^2 - x + a^2 - a = 0.$$

(b) Although the equation obtained in part (a) is a fourth-degree equation in x, it is only a quadratic in a. Solve for a in terms of x to obtain $a = x^2 + x + 1$ or $a = x^2 - x$.

(c) Solve each of the equations in part (b) for x in terms of a.

(d) Check your results in part (c) to eliminate any extraneous roots.

CHAPTER ONE TEST

In Exercises 1–4, factor each expression. (If an expression is irreducible over the integers, state this.)

1. (a) $64 - (x - 2)^3$ (b) $w^2y^2 + 16$
2. (a) $2x^2 + 3x - 8$ (b) $6x^2y^2z + xyz - z$
3. $6x(1 - x^2)^{-1/2} - 3x^2(1 - x^2)^{-3/2}$
4. $9ax + 6bx - 6ay - 4by$
5. Use one of the special product formulas to compute $(2a + 3b^2)^3$.
6. Simplify the expression $|x - 6| + |x - 7|$, given that $6 < x < 7$.
7. Specify the intervals that are described by the inequalities.
    (a) $|x - 4| < \frac{1}{10}$ (b) $x \geq -2$

In Exercises 8–11, find all of the real solutions of each equation.

8. (a) $x^2 + 4x = 5$ (b) $x^2 + 4x = 1$
9. $\sqrt{5 - 2x} - \sqrt{2 - x} - \sqrt{3 - x} = 0$
10. $2x^3 + x^2 - 3x = 0$
11. $y^3 + 3y^2 - 8y - 24 = 0$
12. (a) Solve for x: $|3x - 1| \leq 2$.
    (b) Solve for x and write the answer using interval notation: $|x - 8| \leq 1$.

In Problems 13–16, solve the inequalities.

13. $4(1 + x) - 3(2x - 1) \geq 1$
14. $\dfrac{3}{5} < \dfrac{3 - 2x}{-4} < \dfrac{4}{5}$
15. $(x - 4)^2(x + 8)^3 \geq 0$
16. $\dfrac{1}{x} + \dfrac{1}{x + 1} + \dfrac{1}{x + 2} \geq 0$

CHAPTER TWO

COORDINATES AND GRAPHS

... [Descartes and Fermat] *were moved by the same spirit: they wanted to show how the Renaissance algebra of Cardan and his successors could be applied to the geometry of the Greeks, ...*

From *A Source Book in Mathematics, 1200–1800*, D. J. Struik, ed. (Princeton: Princeton University Press, 1986)

The mathematics I was taught in my freshman year was algebra (old stuff), trigonometry (too much of it), and analytical geometry (a revelation).

Paul R. Halmos in *I Want To Be a Mathematician* (N.Y.: Springer-Verlag, 1985)

INTRODUCTION

In Chapter 1 we saw an important connection between algebra and geometry: The points on a line can be labeled using the real numbers. Just as each point on a line can be labeled with a unique real number, each point in a *plane* can be labeled with a unique *pair* of real numbers. The details of this are reviewed in Section 2.1, Rectangular Coordinates. These ideas are then used in the next three sections to graph equations. The basic theme in the work on graphing equations is this: We start with a given equation relating two variables, say, x and y. By choosing values for x, corresponding y-values are then determined. The resulting (x, y) number pairs (called *ordered pairs*) are then interpreted as points in the plane. This provides another basic connection between algebra and geometry: We can associate a geometric picture, a graph, with an equation in two variables. Sometimes it is easier to work directly from a graph, other times from the equation. But in any event, having the latitude to think either algebraically or geometrically (or both) will clarify many problems later on.

2.1 RECTANGULAR COORDINATES

The name coordinate *does not appear in the work of Descartes. This term is due to Leibniz, and so are* abscissa *and* ordinate *(1692).*

David M. Burton in *The History of Mathematics, an Introduction,* 2nd ed. (Dubuque, Iowa: Wm. C. Brown Publishers, 1991)

In previous courses you learned to work with a rectangular coordinate system such as that shown in Figure 1. In this section we review some of the most basic formulas and techniques that are useful here.

The point of intersection of the two perpendicular number lines, or **axes**, is called the **origin** and is denoted by the letter O. The horizontal and vertical axes are often labeled the **x-axis** and the **y-axis,** respectively, but any other variables

FIGURE 1

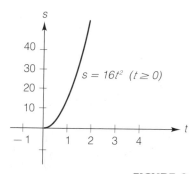

FIGURE 2

A graph of the formula $s = 16t^2$ in a t-s coordinate system. [The formula relates the distance s (in feet) and the time t (in seconds) for an object falling in a vacuum.]

will do just as well for labeling the axes. For instance, Figure 2 shows a t-s coordinate system. (We'll discuss curves, or graphs, like the one in Figure 2 in later sections.)

Notice that in Figures 1 and 2 the axes divide the plane into four regions, or **quadrants**, labeled I through IV, as shown in Figure 1. Unless indicated otherwise, we assume that the same unit of length is used on both axes. In Figure 1, the same scales are used on both axes; in Figure 2, the scale used on the s-axis is different than the scale on the t-axis.

Now look at the point P in Figure 3(a). Starting from the origin O, one way to reach P is to move three units in the positive x-direction and then two units in the positive y-direction. That is, the location of P relative to the origin and the axes is "right 3, up 2." We say that the **coordinates** of P are $(3, 2)$. The first number within the parentheses conveys the information "right 3," and the second number conveys the information "up 2." We say that the **x-coordinate** of P is 3 and the **y-coordinate** of P is 2. Likewise, the coordinates of point Q in Figure 3(a) are $(-2, 4)$. With this coordinate notation in mind, observe in Figure 3(b) that $(3, 2)$ and $(2, 3)$ represent different points; that is, the order in which the two numbers appear within the parentheses affects the location of the point. Figure 3(c) displays various points with given coordinates; you should check for yourself that the coordinates correspond correctly to the location of each point.

Some terminology and notation: The x-y coordinate system that we have described is often called a **Cartesian coordinate system**. The term *Cartesian* is used in honor of René Descartes, the seventeenth-century French philosopher and mathematician. The coordinates (x, y) of a point P are referred to as an **ordered pair**. Recall, for example, that $(3, 2)$ and $(2, 3)$ represent different points; that is, the order of the numbers matters. The x-coordinate of a point is sometimes referred to as the **abscissa** of the point; the y-coordinate is the **ordinate**. The notation $P(x, y)$ means that P is a point whose coordinates are (x, y). At times, we abbreviate the phrase *the point whose coordinates are (x, y)* to simply *the point (x, y)*.

The remainder of our work in this section depends on a key result from elementary geometry, the Pythagorean theorem. For reference, we state this theorem and its converse in the box that follows. (See Exercises 60 and 61 for two proofs of the theorem.)

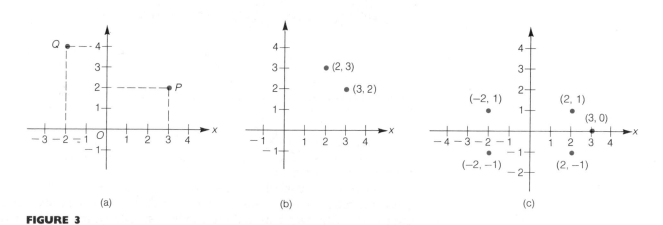

(a)   (b)   (c)

FIGURE 3

FIGURE 4

The Pythagorean Theorem and Its Converse

1. Pythagorean Theorem
(See Figure 4) In a right triangle, the lengths of the sides are related by the equation

$$a^2 + b^2 = c^2$$

where a and b are the lengths of the sides forming the right angle and c is the length of the hypotenuse (the side opposite the right angle).

2. Converse
If the lengths of the sides of a triangle are related by an equation of the form $a^2 + b^2 = c^2$, then the triangle is a right triangle, and c is the length of the hypotenuse.

■■■ **EXAMPLE 1**   Use the Pythagorean theorem to calculate the distance d between the points $(2, 1)$ and $(6, 3)$.

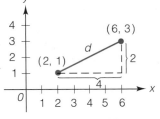

FIGURE 5

Solution   We plot the two given points and draw a line connecting them, as shown in Figure 5. Then we draw the broken lines as shown, parallel to the axes, and apply the Pythagorean theorem to the right triangle that is formed. The base of the triangle is four units long. You can see this by simply counting spaces or by using absolute value, as discussed in Section 1.2: $|6 - 2| = 4$. The height of the triangle is found to be two units, either by counting spaces or by computing the absolute value: $|3 - 1| = 2$. Thus, we have

$$d^2 = 4^2 + 2^2 = 20$$
$$d = \sqrt{20} = \sqrt{4}\sqrt{5} = 2\sqrt{5}$$ ■■■

The method used in Example 1 can be applied to derive a general formula for the distance d between any two points (x_1, y_1) and (x_2, y_2) (see Figure 6). Just as before, we draw in the right triangle and apply the Pythagorean theorem. We have

$$d^2 = |x_2 - x_1|^2 + |y_2 - y_1|^2$$
$$= (x_2 - x_1)^2 + (y_2 - y_1)^2 \qquad \text{(Why?)}$$

FIGURE 6

and therefore

$$d = \sqrt{(x_2 - x_1)^2 + (y_2 - y_1)^2}$$

This last equation is referred to as the **distance formula**. For reference, we restate it in the box that follows.

The Distance Formula

The distance d between the points (x_1, y_1) and (x_2, y_2) is given by

$$d = \sqrt{(x_2 - x_1)^2 + (y_2 - y_1)^2}$$

Examples 2 through 4 demonstrate some simple calculations involving the distance formula.   *Note:*   In computing the distance between two given points, it does not matter which one you treat as (x_1, y_1) and which as (x_2, y_2) because quantities such as $(x_2 - x_1)$ and $(x_1 - x_2)$ are negatives of each other, and so their squares are equal.

EXAMPLE 2 Calculate the distance between the points $(2, -6)$ and $(5, 3)$.

Solution Substituting $(2, -6)$ for (x_1, y_1) and $(5, 3)$ for (x_2, y_2) in the distance formula, we have

$$d = \sqrt{(5 - 2)^2 + [3 - (-6)]^2}$$
$$= \sqrt{3^2 + 9^2} = \sqrt{90}$$
$$= \sqrt{9}\sqrt{10} = 3\sqrt{10}$$

You should check for yourself that the same answer is obtained using $(2, -6)$ as (x_2, y_2) and $(5, 3)$ as (x_1, y_1).

EXAMPLE 3 Is the triangle with vertices $D(-2, -1)$, $E(4, 1)$, $F(3, 4)$ a right triangle?

Solution First, we sketch the triangle in question (see Figure 7).

From the sketch it appears that angle E could be a right angle, but certainly this is not a proof. We need to use the distance formula to calculate the lengths of the three sides and then check whether any relation of the form $a^2 + b^2 = c^2$ holds. The calculations are as follows:

$$DE = \sqrt{[4 - (-2)]^2 + [1 - (-1)]^2} = \sqrt{36 + 4} = \sqrt{40}$$
$$EF = \sqrt{(4 - 3)^2 + (1 - 4)^2} = \sqrt{1 + 9} = \sqrt{10}$$
$$DF = \sqrt{[3 - (-2)]^2 + [4 - (-1)]^2} = \sqrt{25 + 25} = \sqrt{50}$$

Because $\left(\sqrt{40}\right)^2 + \left(\sqrt{10}\right)^2 = \left(\sqrt{50}\right)^2$, we are indeed guaranteed that $\triangle DEF$ is a right triangle. (In Section 2.3, you'll see that this result can be obtained more efficiently using the concept of "slope.")

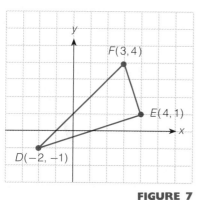

$F(3, 4)$

$E(4, 1)$

$D(-2, -1)$

FIGURE 7

We can use the distance formula to obtain the equation of a circle. Figure 8 shows a circle with center (h, k) and radius r. By definition, a point (x, y) is on this circle if and only if the distance from (x, y) to (h, k) is r. Thus we have

$$\sqrt{(x - h)^2 + (y - k)^2} = r$$

or, equivalently,

$$(x - h)^2 + (y - k)^2 = r^2 \tag{1}$$

(These last two equations are equivalent because two nonnegative quantities are equal if and only if their squares are equal.)

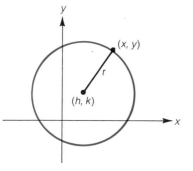

(x, y)

r

(h, k)

FIGURE 8

The work in the previous paragraph tells us two things. First, if a point (x, y) lies on the circle in Figure 8, then x and y together satisfy equation (1). Second, if a pair of numbers x and y satisfies equation (1), then the point (x, y) lies on the circle in Figure 8. (Does it sound to you as if the previous two sentences say the same thing? They don't! Think about it.) Equation (1) is called the **standard form for the equation of a circle**. For reference, we record the result in the box that follows.

> ### The Equation of a Circle in Standard Form
>
> The equation of a circle with center (h, k) and radius r is
>
> $$(x - h)^2 + (y - k)^2 = r^2$$

▌▬▌ EXAMPLE 4 Write the equation of the circle with center $(-2, 5)$ and radius 3.

Solution In the equation $(x - h)^2 + (y - k)^2 = r^2$, we substitute the given values $h = -2$, $k = 5$, and $r = 3$. This yields

$$[x - (-2)]^2 + (y - 5)^2 = 3^2$$
$$(x + 2)^2 + (y - 5)^2 = 9$$

This is the standard form for the equation of the given circle. An alternative form of this answer is found by carrying out the indicated algebra. We have

$$x^2 + 4x + 4 + y^2 - 10y + 25 = 9$$

and, therefore,

$$x^2 + 4x + y^2 - 10y + 20 = 0$$

The disadvantage to this alternative form is that information regarding the center and radius is no longer readily visible. **▬▬▬**

When the equation of a circle is not in standard form, the technique of completing the square (discussed in Section 1.3) can be used to convert the equation to standard form. Example 5 shows how this is done.

▌▬▌ EXAMPLE 5 Determine the center and radius of the circle given by

$$4x^2 - 24x + 4y^2 + 16y + 51 = 0$$

Solution First, divide through by 4 so that the coefficients of x^2 and y^2 are both 1:

$$x^2 - 6x + y^2 + 4y + \frac{51}{4} = 0$$

Now, in preparation for completing the squares, we write

$$(x^2 - 6x + \underline{\quad}) + (y^2 + 4y + \underline{\quad}) = -\frac{51}{4}$$

To complete the square in x, we need to add $\left(-\frac{6}{2}\right)^2$, or 9. To complete the square in y, we need to add $\left(\frac{4}{2}\right)^2$, or 4. Thus, we have

$$(x^2 - 6x + 9) + (y^2 + 4y + 4) = -\frac{51}{4} + 9 + 4$$

or

$$(x - 3)^2 + (y + 2)^2 = -\frac{51}{4} + \frac{52}{4} = \frac{1}{4}$$

$$(x - 3)^2 + (y + 2)^2 = \left(\frac{1}{2}\right)^2$$

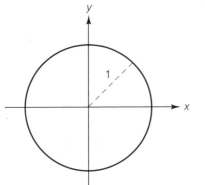

FIGURE 9
The unit circle $x^2 + y^2 = 1$.

This is the equation in standard form. The center of the circle is $(3, -2)$; the radius is $\frac{1}{2}$. ∎

For a large portion of our work in trigonometry later in this book, we will refer to the **unit circle.** This is the circle with center $(0, 0)$ and radius 1 (see Figure 9). To obtain an equation for this circle, we substitute the values $h = 0$, $k = 0$, and $r = 1$ in the standard form for the equation of a circle. This gives us

$$(x - 0)^2 + (y - 0)^2 = 1^2$$

or, more simply,

$$x^2 + y^2 = 1$$

This is the equation for the unit circle; we will use this equation in the next example.

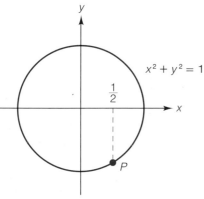

FIGURE 10

EXAMPLE 6 As indicated in Figure 10, the point P lies in the fourth quadrant, on the unit circle. If the x-coordinate of P is $\frac{1}{2}$, determine the y-coordinate of P.

Solution Since P lies on the unit circle, the coordinates of P satisfy the equation $x^2 + y^2 = 1$. We are given that the x-coordinate of P is $\frac{1}{2}$. Substituting the value $x = \frac{1}{2}$ in the equation $x^2 + y^2 = 1$ gives us

$$\left(\frac{1}{2}\right)^2 + y^2 = 1$$

$$y^2 = 1 - \frac{1}{4} = \frac{3}{4} \qquad \text{and therefore} \qquad y = \pm\frac{\sqrt{3}}{2}$$

We want to choose the negative root here because P lies in quadrant IV (in which all y-coordinates are negative). Thus, the y-coordinate of P is $-\sqrt{3}/2$. ∎

In Example 7 we use a simple result that you may recall from previous courses: the midpoint formula. This result is summarized in the box that follows. (For a proof of the formula, see Exercise 59.)

The Midpoint Formula	
	EXAMPLE
The midpoint of the line segment joining the points $P(x_1, y_1)$ and $Q(x_2, y_2)$ is	The midpoint of the line segment joining $(2, -15)$ and $(4, 5)$ is
$\left(\dfrac{x_1 + x_2}{2}, \dfrac{y_1 + y_2}{2}\right)$	$\left(\dfrac{2 + 4}{2}, \dfrac{-15 + 5}{2}\right) = (3, -5)$

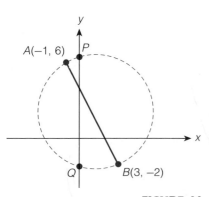

FIGURE 11
$\overline{AB}$ is a diameter. What are the y-intercepts of the circle?

EXAMPLE 7 Points $A(-1, 6)$ and $B(3, -2)$ are the endpoints of a diameter of a circle, as shown in Figure 11. Determine the coordinates of the points P and Q where the circle intersects the y-axis. (The y-coordinates of the points P and Q are called the y-intercepts of the circle. Both y-intercepts and x-intercepts are discussed more generally in the next section.)

Solution The x-coordinates of both P and Q are zero. So, our strategy here is as follows: If we knew the equation of the circle, then we could find the required y-coordinates by setting $x = 0$ in the equation of the circle. To write the equation of the circle, we need to know the center and the radius. We can find these by using the midpoint formula and the distance formula.

The center of the circle is the midpoint of diameter $\overline{AB}$. Therefore, the coordinates of the center are

$$x = \frac{-1 + 3}{2} = 1 \quad \text{and} \quad y = \frac{6 + (-2)}{2} = 2$$

So the center is the point $(1, 2)$. Since the radius r is the distance from the center to point $A(-1, 6)$, we have

$$r = \sqrt{[1 - (-1)]^2 + (2 - 6)^2} = \sqrt{4 + 16} = \sqrt{20}$$

Therefore,

$$r^2 = 20$$

It follows now that the equation of the circle is

$$(x - 1)^2 + (y - 2)^2 = 20$$

For the y-intercepts, we set $x = 0$ in this last equation to obtain

$$(y - 2)^2 = 19$$
$$y - 2 = \pm\sqrt{19}$$
$$y = 2 \pm \sqrt{19}$$

Thus, the y-intercepts are $2 + \sqrt{19}$, which is positive, and $2 - \sqrt{19}$, which is negative. Consequently, the coordinates of P and Q are $P\left(0, 2 + \sqrt{19}\right)$ and $Q\left(0, 2 - \sqrt{19}\right)$.

EXERCISE SET 2.1

A

1. Plot the points $(5, 2)$, $(-4, 5)$, $(-4, 0)$, $(-1, -1)$, and $(5, -2)$.

2. Draw the square $ABCD$ whose vertices (corners) are $A(1, 0)$, $B(0, 1)$, $C(-1, 0)$, and $D(0, -1)$.

3. (a) Draw the right triangle PQR with vertices $P(1, 0)$, $Q(5, 0)$, and $R(5, 3)$.
   (b) Use the formula for the area of a triangle, $A = \frac{1}{2}bh$, to find the area of triangle PQR in part (a).

4. (a) Draw the trapezoid $ABCD$ with vertices $A(0, 0)$, $B(7, 0)$, $C(6, 4)$, and $D(4, 4)$.
   (b) Compute the area of the trapezoid. (See the inside front cover of this book for the appropriate formula.)

In Exercises 5–10, calculate the distance between the given points.

5. (a) $(0, 0)$ and $(-3, 4)$ (b) $(2, 1)$ and $(7, 13)$

6. (a) $(-1, -3)$ and $(-5, 4)$ (b) $(6, -2)$ and $(-1, 1)$

7. (a) $(-5, 0)$ and $(5, 0)$ (b) $(0, -8)$ and $(0, 1)$

8. (a) $(-5, -3)$ and $(-9, -6)$
   (b) $\left(\frac{9}{2}, 3\right)$ and $\left(-2\frac{1}{2}, -1\right)$

9. $\left(1, \sqrt{3}\right)$ and $\left(-1, -\sqrt{3}\right)$

10. $(-3, 1)$ and $(374, -335)$ (Use a calculator.)

11. Which point is farther from the origin?
    (a) $(3, -2)$ or $\left(4, \frac{1}{2}\right)$ (b) $(-6, 7)$ or $(9, 0)$

12. Use the distance formula to show that, in each case, the triangle with given vertices is an isosceles triangle.
    (a) $(0, 2)$, $(7, 4)$, $(2, -5)$
    (b) $(-1, -8)$, $(0, -1)$, $(-4, -4)$
    (c) $(-7, 4)$, $(-3, 10)$, $(1, 3)$

13. In each case, determine if the triangle with the given vertices is a right triangle. *Hint:* Find the lengths of the sides and then use the *converse* of the Pythagorean theorem.
    (a) $(7, -1)$, $(-3, 5)$, $(-12, -10)$
    (b) $(4, 5)$, $(-3, 9)$, $(1, 3)$
    (c) $(-8, -2)$, $(1, -1)$, $(10, 19)$

14. (a) Two of the three triangles specified in Exercise 13 are right triangles. Find their areas.
    (b) Calculate the area of the remaining triangle in Exercise 13 by using the following formula for the area A of a triangle with vertices (x_1, y_1), (x_2, y_2), and (x_3, y_3):
    $$A = \tfrac{1}{2}\left| x_1 y_2 - x_2 y_1 + x_2 y_3 - x_3 y_2 + x_3 y_1 - x_1 y_3 \right|$$
    The derivation of this formula is given in Exercise 62.
    (c) Use the formula given in part (b) to check your answers in part (a).

15. Use the formula given in Exercise 14(b) to calculate the area of the triangle with vertices $(1, -4)$, $(5, 3)$, and $(13, 17)$. Conclusion?

16. The coordinates of points A, B, and C are $A(-4, 6)$, $B(-1, 2)$, and $C(2, -2)$.
    (a) Show that $AB = BC$ by using the distance formula.
    (b) Show that $AB + BC = AC$ by using the distance formula.
    (c) What can you conclude from parts (a) and (b)?

In Exercises 17–24, determine the center and the radius for the circle. Also, find the y-coordinates of the points (if any) where the circle intersects the y-axis.

17. $(x - 3)^2 + (y - 1)^2 = 25$
18. $x^2 + (y + 1)^2 = 20$
19. $x^2 + y^2 = \sqrt{2}$
20. $x^2 + y^2 - 10x + 2y + 17 = 0$
21. $x^2 + y^2 + 8x - 6y = -24$
22. $4x^2 - 4x + 4y^2 - 63 = 0$
23. $9x^2 + 54x + 9y^2 - 6y + 64 = 0$
24. $3x^2 + 3y^2 + 5x - 4y = 1$

In Exercises 25–32, you are given either the x- or the y-coordinate of a point P on the unit circle, and you are given the quadrant in which P is located. Find the remaining coordinate (as in Example 6). In Exercises 31 and 32, use a calculator and round off to three decimal places.

25. (a) $x = \tfrac{3}{5}$; I
    (b) $x = \tfrac{3}{5}$; IV
26. (a) $x = -\tfrac{5}{13}$; II
    (b) $x = -\tfrac{5}{13}$; III
27. (a) $y = -\sqrt{2}/2$; III
    (b) $y = -\sqrt{2}/2$; IV
28. (a) $y = \sqrt{3}/2$; II
    (b) $y = -\sqrt{3}/2$; III
29. (a) $x = -\tfrac{2}{3}$; II
    (b) $x = -\tfrac{2}{3}$; III
30. (a) $y = \tfrac{2}{3}$; I
    (b) $y = \tfrac{2}{3}$; II
31. (a) $x = 0.652$; I
    (b) $y = 0.652$; I
32. (a) $x = -0.293$; II
    (b) $y = -0.293$; III

In Exercises 33 and 34, determine the equation of the circle in standard form, given the coordinates of the diameter $\overline{PQ}$.

33. $P(-4, -2)$ and $Q(6, 4)$
34. $P(1, -3)$ and $Q(-5, -5)$

In Exercises 35 and 36, find the midpoint of the line segment joining points P and Q.

35. (a) $P(3, 2)$ and $Q(9, 8)$
    (b) $P(-4, 0)$ and $Q(5, -3)$
    (c) $P(3, -6)$ and $Q(-1, -2)$
36. (a) $P(12, 0)$ and $Q(12, 8)$
    (b) $P\left(\tfrac{3}{5}, -\tfrac{2}{3}\right)$ and $Q(0, 0)$
    (c) $P(1, \pi)$ and $Q(3, 3\pi)$

37. The coordinates of A and B are $A(-1, 2)$ and $B(5, -3)$. If B is the midpoint of line segment $\overline{AC}$, what are the coordinates of C?

38. The coordinates of the points S and T are $S(4, 6)$ and $T(10, 2)$. If M is the midpoint of $\overline{ST}$, find the midpoint of $\overline{SM}$.

39. (a) Sketch the parallelogram with vertices $A(-7, -1)$, $B(4, 3)$, $C(7, 8)$, and $D(-4, 4)$.
    (b) Compute the midpoints of the diagonals $\overline{AC}$ and $\overline{BD}$.
    (c) What conclusion can you draw from part (b)?

40. The vertices of $\triangle ABC$ are $A(1, 1)$, $B(9, 3)$, and $C(3, 5)$.
    (a) Find the perimeter of $\triangle ABC$.
    (b) Find the perimeter of the triangle that is formed by joining the midpoints of the three sides of $\triangle ABC$.
    (c) Compute the ratio of the perimeter in part (a) to the perimeter in part (b).
    (d) What theorem from geometry provides the answer for part (c) without using the results in (a) and (b)?

41. This exercise refers to $\triangle ABC$ whose vertices are specified in Exercise 40.
    (a) Compute the sum of the squares of the lengths of the sides of $\triangle ABC$.
    (b) Compute the sum of the squares of the lengths of the three medians. (A **median** is a line segment drawn from a vertex to the midpoint of the opposite side.)

(c) Check that the ratio of the answer in part (a) to that in part (b) is 4/3. *Remark:* As Exercise 56 asks you to show, this ratio is 4/3 for any triangle.

42. (a) Sketch the circle of radius 1 centered at the origin.
    (b) Write the equation for this circle.
    (c) Does the point $\left(\frac{3}{5}, \frac{4}{5}\right)$ lie on this circle?
    (d) Does the point $\left(-1/2, \sqrt{3}/2\right)$ lie on this circle?

43. The center of a circle is the point $(3, 2)$. If the point $(-2, -10)$ lies on this circle, find the equation of the circle.

44. Find the equation of the circle tangent to the x-axis and with center $(3, 5)$. *Hint:* First draw a sketch.

45. Find the equation of the circle tangent to the y-axis and with center $(3, 5)$.

46. Find the equation of the circle passing through the origin and with center $(3, 5)$.

47. Use the Pythagorean theorem to find the length a in the figure. Then find b, c, d, e, f, and g.

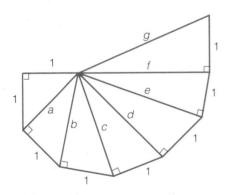

48. (A numerologist's delight)   Using the Pythagorean theorem and your calculator, compute the area of a right triangle in which the lengths of the hypotenuse and one leg are 2045 and 693, respectively.

B

In Exercises 49 and 50, determine the equation of the circle that satisfies the given conditions. Write the equation in standard form.

49. The circle passes through the origin and is concentric with the circle $x^2 - 6x + y^2 - 4y + 4 = 0$.

50. The circle passes through $(-4, 1)$, and its center is the midpoint of the line segment joining the centers of the two circles $x^2 + y^2 - 6x - 4y + 12 = 0$ and $x^2 + y^2 - 14x + 47 = 0$.

51. Find a value for t such that points $(0, 2)$ and $(12, t)$ are 13 units apart. *Hint:* By the distance formula, $13 = \sqrt{(12 - 0)^2 + (t - 2)^2}$. Now square both sides and solve for t.

52. (a) Find values for t such that the points $(-2, 3)$ and $(t, 1)$ are six units apart.
    (b) Can you find a real number t such that the points $(-2, 3)$ and $(t, 1)$ are one unit apart? Why or why not?

53. The diagonals of a parallelogram bisect each other. Steps (a), (b), and (c) outline a proof of this theorem.
    (a) In the parallelogram $OABC$ shown in the figure, check that the coordinates of B must be $(a + b, c)$.
    (b) Use the midpoint formula to calculate the midpoints of diagonals $\overline{OB}$ and $\overline{AC}$.
    (c) The two answers in part (b) are identical. This shows that the two diagonals do indeed bisect each other, as we wished to prove.

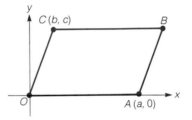

54. Prove that in a parallelogram, the sum of the squares of the lengths of the diagonals equals the sum of the squares of the lengths of the four sides. (Use the figure in Exercise 53.)

55. Use the figure in Exercise 53 to prove the following theorem: If the lengths of the diagonals of a parallelogram are equal, then the parallelogram is a rectangle. *Hint:* Equate the two expressions for lengths of the diagonals and conclude that $ab = 0$. The case in which $a = 0$ can be discarded.

56. The sum of the squares of the sides of any triangle equals 4/3 of the sum of the squares of the medians. Prove this, taking the vertices to be $(0, 0)$, $(2, 0)$, and $(2a, 2b)$. *Note:* The expression *squares of the sides* means *squares of the lengths of the sides.*

57. In $\triangle ABC$, let M be the midpoint of side $\overline{BC}$. Prove that $AB^2 + AC^2 = 2(BM^2 + AM^2)$. *Hint:* Let the coordinates be $A(0, 0)$, $B(2, 0)$, and $C(2a, 2b)$.

58. If the point (x, y) is equidistant from the points $(-3, -3)$ and $(5, 5)$, show that x and y satisfy the equation $x + y = 2$. *Hint:* Use the distance formula.

59. Suppose that the coordinates of points P, Q, and M are

    $$P(x_1, y_1) \qquad Q(x_2, y_2) \qquad M\left(\frac{x_1 + x_2}{2}, \frac{y_1 + y_2}{2}\right)$$

    Follow steps (a) and (b) to prove that M is the midpoint of the line segment from P to Q.

(a) By computing both of the distances PM and MQ, show that $PM = MQ$. (This shows that M lies on the perpendicular bisector of line segment $\overline{PQ}$, but it does not show that M actually lies *on* $\overline{PQ}$.)

(b) Show that $PM + MQ = PQ$. (This shows that M does lie on $\overline{PQ}$.)

60. This problem outlines one of the shortest proofs of the Pythagorean theorem. The proof was discovered by the Hindu mathematician Bhāskara (1114–ca. 1185). (For other proofs, see the next exercise and Exercise 88 on page 96.) In the figure we are given a right triangle ACB with the right angle at C, and we want to prove that $a^2 + b^2 = c^2$. In the figure, $\overline{CD}$ is drawn perpendicular to $\overline{AB}$.

(a) Check that $\angle CAD = \angle DCB$ and that $\triangle BCD$ and $\triangle BAC$ are similar.

(b) Use the result in part (a) to obtain the equation $a/y = c/a$, and conclude that $a^2 = cy$.

(c) Show that $\triangle ACD$ is similar to $\triangle ABC$, and use this to deduce that $b^2 = c^2 - cy$.

(d) Combine the two equations deduced in parts (b) and (c) to arrive at $a^2 + b^2 = c^2$.

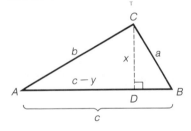

61. One of the oldest and simplest proofs of the Pythagorean theorem is found in the ancient Chinese

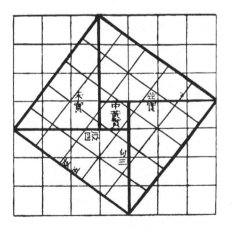

A diagram accompanying a proof of the "Pythagorean" theorem in the ancient Chinese text *Chou Pei Suan Ching* [from *Science and Civilisation in China*, vol. 3, by Joseph Needham (Cambridge: Cambridge University Press, 1959)].

text *Chou Pei Suan Ching*. This text was written during the Han period (206 B.C.–A.D. 222), but portions of it may date back to 600 B.C. The proof in *Chou Pei Suan Ching* is based on the diagram at the bottom of the left column. In this exercise we explain the details of the proof.

(a) Starting with the right triangle in Figure A, we make four replicas of this triangle and arrange them to form the pattern shown in Figure B. Explain why the outer quadrilateral in Figure B is a square.

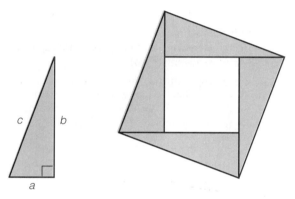

FIGURE A   FIGURE B

(b) The unshaded region in the center of Figure B is a square. What is the length of each side?

(c) The area of the outer square in Figure B is $(\text{side})^2 = c^2$. This area can also be computed by adding up the areas of the four right triangles and the inner square. Compute the area in this fashion. After simplifying, you should obtain $a^2 + b^2$. Now conclude that $a^2 + b^2 = c^2$, since both expressions represent the same area.

C

62. This problem indicates a method for calculating the area of a triangle when the coordinates of the three vertices are given.

(a) Calculate the area of $\triangle ABC$ in the figure.
Hint: First calculate the area of the rectangle enclosing $\triangle ABC$, and then subtract the areas of the three right triangles.

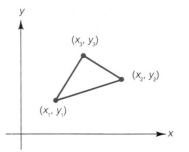

(b) Calculate the area of the triangle with vertices $(1, 3)$, $(4, 1)$, and $(10, 4)$.   *Hint:*   Work with an enclosing rectangle and three right triangles, as in part (a).

(c) Using the same technique that you used in parts (a) and (b), show that the area of the triangle in the following figure is given by

$$A = \tfrac{1}{2}(x_1y_2 - x_2y_1 + x_2y_3 - x_3y_2 + x_3y_1 - x_1y_3)$$

Remark: If we use absolute value signs instead of the parentheses, then the formula will hold regardless of the relative positions or quadrants of the three vertices. Thus, the area of a triangle with vertices (x_1, y_1), (x_2, y_2), (x_3, y_3) is given by

$$A = \tfrac{1}{2}|x_1y_2 - x_2y_1 + x_2y_3 - x_3y_2 + x_3y_1 - x_1y_3|$$

63. Refer to the following figure.

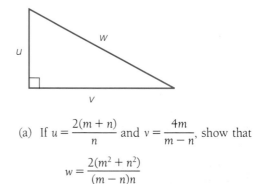

(a) If $u = \dfrac{2(m + n)}{n}$ and $v = \dfrac{4m}{m - n}$, show that

$$w = \frac{2(m^2 + n^2)}{(m - n)n}$$

(b) Show that $\tfrac{1}{2}uv = u + v + w$ (that is, the perimeter is numerically equal to the area).

(c) Give an example of a right triangle in which the length of each side is an integer and the perimeter is numerically equal to the area.

▶ OPTIONAL TI-81 GRAPHING CALCULATOR EXERCISES

EXERCISES FOR SECTION 2.1

---

1. This exercise shows how to graph circles. As an example, we will graph the circle $x^2 + y^2 = 4$. As preparation, press the Y= key and delete any formulas that appear. Next, press the ZOOM key and then the "6" key to obtain the *standard viewing rectangle*. (The standard viewing rectangle extends from -10 to $+10$ in both the x- and the y-directions.)

(a) Solve the equation $x^2 + y^2 = 4$ for y. This yields $y = \pm\sqrt{4 - x^2}$. Using the Y= key, enter the equations $Y_1 = \sqrt{4 - x^2}$ and $Y_2 = -\sqrt{4 - x^2}$. Now press the GRAPH key. As you can see, some adjustment needs to be made. Go on to part (b).

(b) Press the RANGE key and set $x_{min} = -2$, $x_{max} = 2$, $y_{min} = -2$, and $y_{max} = 2$. Now press the GRAPH key. The "circle" is larger, but it appears distorted.

This is because different scales have been used on the x- and the y-axes. Go on to part (c).

(c) Press the ZOOM key and then the "5" key. This ensures that the same scale is used on both axes. The resulting graph is our completed view of the circle $x^2 + y^2 = 4$. As the graph shows, the center of the circle is $(0, 0)$ and the radius is 2.

(d) Here is a shortcut that could have been used in part (a) for entering the second equation $Y_2 = -\sqrt{4 - x^2}$. Press the Y= key and, for the moment, delete the existing expression for Y_2. Now, rather than typing in the expression $-\sqrt{4 - x^2}$, we will enter the simpler (but equivalent) expression $-Y_1$. This is done as follows: With the cursor on the line corresponding to Y_2, press the "negative sign" key $(-)$ on the bottom row of the calculator. Next, press the "2nd" key and then the VARS key. In the resulting menu, note that the first item is Y_1. Also note that the cursor is on this item. Thus, to select Y_1, just press ENTER. This takes you back to the $Y=$ menu, and your screen display should read

$$Y_1 = \sqrt{(4 - x\hat{\ }2)}$$
$$Y_2 = -Y_1$$

Now press the GRAPH key to see the circle.

In Exercises 2–5, graph the circles using the technique explained in Exercise 1.

2. $x^2 + y^2 = 9$

3. $x^2 + y^2 = 81$

4. $(x - 2)^2 + (y - 1)^2 = 16$ *Hint:* Solve for y to obtain the two equations $y = 1 + \sqrt{16 - (x - 2)^2}$ and $y = 1 - \sqrt{16 - (x - 2)^2}$. For the RANGE settings, use $x_{min} = -2$, $x_{max} = 6$, $y_{min} = -3$, and $y_{max} = 5$.

5. $(x + 3)^2 + (y + 2)^2 = 1$

You know from the text that the graph of the equation $x^2 + y^2 = 9$ is a circle with center $(0, 0)$ and radius 3. In Exercises 6–9, use the calculator to investigate what happens when the equation is changed slightly.

6. (a) Graph the equation $x^2 + 2y^2 = 9$. The resulting curve is an *ellipse*. (We'll study this curve in a later chapter.)
   (b) On the same set of axes, graph the ellipse $x^2 + 2y^2 = 9$ and the circle $x^2 + y^2 = 9$. At which points do the curves appear to intersect?
   (c) Carry out the calculations to show that the points you have indicated in part (b) indeed lie on both graphs.

7. (a) Graph the equation $x^2 - y^2 = 9$. The resulting curve is a *hyperbola*. (We'll study this curve in a later chapter.)
   (b) On the same set of axes, graph the hyperbola $x^2 - y^2 = 9$ and the circle $x^2 + y^2 = 9$. At which points do the curves appear to intersect?
   (c) Carry out the calculations to show that the points you have indicated in part (b) indeed lie on both graphs.

8. (a) Graph the equation $x^2 - y^2 = 0$.
   (b) On the same set of axes, graph the equations $x^2 - y^2 = 9$ and $x^2 - y^2 = 0$.

9. Graph each equation: (a) $x^2 + y^3 = 9$;
                          (b) $x^3 + y^2 = 9$.

10. In this exercise, you will use the DRAW key to draw the triangle in Figure 7 on page 51. The vertices of this triangle are $D(-2, -1)$, $E(4, 1)$, and $F(3, 4)$.
    (a) Press the $Y=$ key and delete or deselect any functions that appear. Then press ZOOM 6.
    (b) We will start with a 10-by-10 viewing rectangle. To arrange this, press the RANGE key and enter the following values: $x_{min} = -5$, $x_{max} = 5$, $y_{min} = -5$, and $y_{max} = 5$. Now press the GRAPH key. The coordinate system shown on the screen will accommodate all of the given points, but it will not show the true proportions of $\triangle DEF$. (Why?) Press ZOOM 5. Finally, press the "2nd" key followed by the CLEAR key to return to the home screen.
    (c) To draw the segment connecting the points D and E, press the "2nd" key and then the DRAW key. The DRAW menu (with seven choices) will appear on the screen. Use the "down" arrow to move the cursor to choice 2, then press ENTER. The screen display will show

        Line (

        Now, enter the coordinates of the points D and E (separated by commas) so that the screen display looks as follows:

        Line $(-2, -1, 4, 1)$

        (To access the comma, press the ALPHA key and then the key used for the decimal point.) Next, press ENTER. The line segment $\overline{DE}$ will appear on the screen.
    (d) Now we will draw $\overline{EF}$. Press CLEAR, then "2nd," then DRAW. Move the cursor to choice 2 and press ENTER. The screen display should read as follows:

        Line $(-2, -1, 4, 1)$
        Line (

        In the second line of the display, enter the coordinates of the points E and F so that the screen display now reads

        Line $(-2, -1, 4, 1)$
        Line $(4, 1, 3, 4)$

        Press ENTER. The line segments $\overline{DE}$ and $\overline{EF}$ will both appear on the screen.
    (e) Use the procedure in part (d) to add the line segment $\overline{DF}$ to the picture. Check to see that your picture agrees with Figure 7 on page 51.

2.2 GRAPHS AND EQUATIONS, A SECOND LOOK

He [Pierre de Fermat (1601–1665)] invented analytic geometry in 1629 and described his ideas in a short work entitled Introduction to Plane and Solid Loci, *which circulated in manuscript form from early 1637 on but was not published during his lifetime. The credit for this achievement has usually been given to Descartes on the basis of his* Geometry, *which was published late in 1637.... However, nothing that we would recognize as analytic geometry can be found in Descartes's essay, except perhaps the idea of using algebra as a language for discussing geometric problems.*

George F. Simmons in *Calculus Gems, Brief Lives and Memorable Moments* (New York: McGraw-Hill, Inc.: 1992)

In this section we look at the connection between two basic skills that you've worked on in previous courses:

- graphing equations
- solving equations

We also present a list of six basic graphs to memorize. These graphs will provide handy examples when we discuss functions in the next chapter. We begin by recalling the definition of the *graph* of an equation.

> **DEFINITION** The Graph of an Equation
>
> The **graph** of an equation in two variables is the set of all points whose coordinates satisfy the equation.

Suppose, for example, that we want to graph the equation $y = 3x - 2$. We begin by noting that the domain of the variable x in the expression $3x - 2$ is the set of all real numbers. Now we choose (at will) various values for x and in each case compute the corresponding y-value from the equation $y = 3x - 2$. For example, if x is zero, then $y = 3(0) - 2 = -2$. Table 1 summarizes the results of some of these calculations. The first line in Table 1 tells us that the point with coordinates $(0, -2)$ is on the graph of $y = 3x - 2$. Reading down the table, we see that some other points on the graph are $(1, 1)$, $(2, 4)$, $(3, 7)$, $(-1, -5)$, and $(-2, -8)$.

We now plot (locate) the points we have determined and note, in this example, that they all appear to lie on a straight line. We draw the line indicated; this is the graph of $y = 3x - 2$ (see Figure 1). In Example 1, we ask a simple question that will test your understanding of the process we've just described and of graphing in general.

TABLE 1

$y = 3x - 2$

x	y
0	−2
1	1
2	4
3	7
−1	−5
−2	−8

EXAMPLE 1 Does the point $\left(1\frac{2}{3}, 2\frac{2}{3}\right)$ lie on the graph of $y = 3x - 2$?

Solution Looking at the graph in Figure 1, we certainly see that the point $\left(1\frac{2}{3}, 2\frac{2}{3}\right)$ is very close to the line, but from this visual inspection alone, we cannot be certain that this point actually lies on the line. To settle the question, then, we check to see if the values $x = 1\frac{2}{3} = \frac{5}{3}$ and $y = 2\frac{2}{3} = \frac{8}{3}$ together satisfy the equation $y = 3x - 2$. We have

FIGURE I
$y = 3x - 2$

FIGURE 2
$y = \dfrac{3x^2 - 5x + 2}{x - 1}$

FIGURE 3
$y = 2x^3 - 2x^2 - x$

$$\frac{8}{3} \overset{?}{=} 3\left(\frac{5}{3}\right) - 2$$

$$\frac{8}{3} \overset{?}{=} 3 \qquad \text{No!}$$

Since the coordinates $\left(1\frac{2}{3}, 2\frac{2}{3}\right)$ evidently do not satisfy the equation $y = 3x - 2$, we conclude that the point $\left(1\frac{2}{3}, 2\frac{2}{3}\right)$ is not on the graph. [The calculation further tells us that the point $\left(1\frac{2}{3}, 3\right)$ *is* on the graph.] ▮▮▮

The next example illustrates a type of graph that often appears in beginning calculus courses (in connection with the notion of a *limit*).

EXAMPLE 2 Graph the equation: $\quad y = \dfrac{3x^2 - 5x + 2}{x - 1}$.

Solution We need to do two things here before we set up a table of values. First, note that the domain of the expression on the right-hand side of the given equation consists of all real numbers except $x = 1$. Thus, whatever the appearance of the resulting graph, it cannot contain a point with x-coordinate 1. Next, can we simplify the given expression? (If so, that would save work when we make the calculations for our table.) In this case, we can simplify things as follows:

$$y = \frac{3x^2 - 5x + 2}{x - 1} = \frac{(3x - 2)(x - 1)}{x - 1} = 3x - 2$$

So, to graph the given equation we need only graph the simpler equation $y = 3x - 2$, remembering that the point with an x-coordinate of 1 must be excluded from the graph. We've previously done the work for $y = 3x - 2$ (see Figure 1). Thus, the required graph is obtained by deleting the point $(1, 1)$ from the graph in Figure 1. This is shown in Figure 2. [The open circle in Figure 2 is used to indicate that the point $(1, 1)$ does not belong to the graph.] ▮▮▮

When we graph an equation, it's helpful to know where the curve or line crosses the x- or y-axis. By a **y-intercept** of a graph, we mean the y-coordinate of a point where the graph crosses the y-axis. For instance, returning to Figure 1, the y-intercept is -2. Notice that this y-intercept can be obtained algebraically just by setting x equal to zero in the given equation $y = 3x - 2$. In a similar fashion, an **x-intercept** of a graph is the x-coordinate of a point where the graph crosses the x-axis. In Figure 1 we can see that the x-intercept is a number between 0 and 1. To determine this x-intercept, set y equal to zero in the equation $y = 3x - 2$. That yields $0 = 3x - 2$ and, consequently, $x = \frac{2}{3}$. So, the x-intercept of the graph in Figure 1 is $\frac{2}{3}$.

EXAMPLE 3 Figure 3 shows a graph of the equation $y = 2x^3 - 2x^2 - x$. Determine the x-intercepts of the graph.

Solution At the points where the graph crosses the x-axis, the y-coordinates are zero. Setting y equal to zero in the given equation yields

$$2x^3 - 2x^2 - x = 0$$
$$x(2x^2 - 2x - 1) = 0$$

Thus, $x = 0$ or $2x^2 - 2x - 1 = 0$. This last equation can be solved by using the quadratic formula. As you should check for yourself, the roots are $(1 \pm \sqrt{3})/2$. We've now found the three x-intercepts of the graph in Figure 3: they are 0, $(1 + \sqrt{3})/2$, and $(1 - \sqrt{3})/2$. *Suggestion:* Use your calculator to check that these last two values are consistent with Figure 3.   ▪▪▪

▌▬▌ **EXAMPLE 4**   Figure 4 shows a graph of the equation $x^2y - x^3 + 3 = 0$.

(a) From the portion of the graph shown in Figure 4, it appears that there is no y-intercept. Use the given equation to show that this is indeed the case.
(b) As indicated in Figure 4, there is an x-intercept between 1 and 2. Find this intercept. Give two forms for the answer: an exact expression and a calculator approximation rounded off to two decimal places.

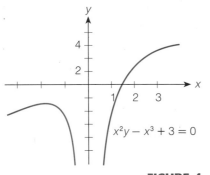

FIGURE 4
Does the graph ever cross the y-axis?

Solution

(a) If there were a y-intercept (perhaps somewhere below the portion of the graph displayed in Figure 4), we could find it by setting x equal to zero in the given equation. But with $x = 0$, the given equation reads $3 = 0$, which is clearly impossible. We thus conclude that there is no y-intercept; the graph never crosses the y-axis.
(b) The x-intercepts of the graph are determined by setting y equal to zero in the given equation. That gives us $x^2(0) - x^3 + 3 = 0$, and therefore $x^3 = 3$. Consequently, we have one x-intercept:

$$x = \sqrt[3]{3}$$
$$\approx 1.44 \qquad \text{using a calculator}$$

Notice that the value $x = 1.44$ is consistent with the graph shown in Figure 4.   ▪▪▪

In elementary graphing there is always the question of how many points must be plotted before the essential features of a graph are clear. As you'll see in later sections of this text, there are a number of techniques and concepts that make it unnecessary to plot a large number of points. For now, however, we point out that some equations arise frequently enough to make it worth memorizing the basic shapes and features of their graphs. Figure 5 displays six such equations and their graphs. Exercises 37 and 38 at the end of this section ask you to set up tables and verify for yourself that the graphs in Figure 5 are indeed correct. From now on (with the exception of Exercises 37 and 38), if you need to sketch a graph of one of these basic equations, you should do so from memory.

Each graph in Figure 5 is determined from a single equation. In some applications, however, a graph may be determined by several equations, as illustrated by the next two examples.

▌▬▌ **EXAMPLE 5**   The variable y is defined in terms of the variable x as follows:

$$y = \begin{cases} x^2 & \text{if} \quad x < 2 \\ 1/x & \text{if} \quad x \geq 2 \end{cases}$$

continued following Figure 5

FIGURE 5

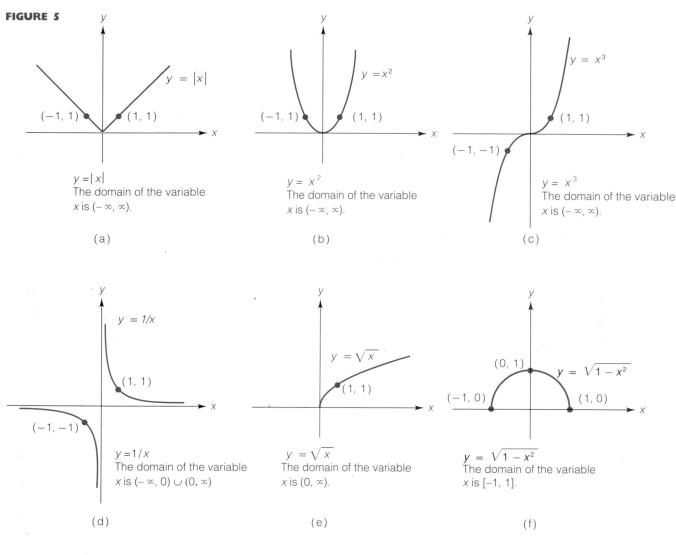

$y = |x|$
The domain of the variable
x is $(-\infty, \infty)$.

(a)

$y = x^2$
The domain of the variable
x is $(-\infty, \infty)$.

(b)

$y = x^3$
The domain of the variable
x is $(-\infty, \infty)$.

(c)

$y = 1/x$
The domain of the variable
x is $(-\infty, 0) \cup (0, \infty)$

(d)

$y = \sqrt{x}$
The domain of the variable
x is $(0, \infty)$.

(e)

$y = \sqrt{1 - x^2}$
The domain of the variable
x is $[-1, 1]$.

(f)

(a) Find y when $x = 3/2$, when $x = 2$, and when $x = 3$.

(b) Sketch the graph of $y = \begin{cases} x^2 & \text{if} \quad x < 2 \\ 1/x & \text{if} \quad x \geq 2 \end{cases}$

FIGURE 6

A graph of $y = \begin{cases} x^2, & x < 2 \\ \dfrac{1}{x}, & x \geq 2 \end{cases}$

Solution

(a) To find y when $x = 3/2$, do we use the equation $y = x^2$ or the equation $y = 1/x$? According to the given inequalities, we should use the equation $y = x^2$ whenever the x-values are less than 2. So, for $x = 3/2$ (which surely is less than 2), we have $y = (3/2)^2 = 9/4$. On the other hand, the inequalities tell us to use the equation $y = 1/x$ whenever the x-values are greater than or equal to 2. So when x is 2, we have $y = 1/2$, and when x is 3, we have $y = 1/3$.

(b) For the graph, we look back at Figures 5(b) and (d) and choose the appropriate portion of each. The result is displayed in Figure 6. The open circle in the figure is used to indicate that the point $(2, 4)$ does not belong to the graph. The filled-in circle, on the other hand, is used to indicate that the point $\left(2, \frac{1}{2}\right)$ does belong to the graph. ■■■

EXAMPLE 6 Sketch each graph: **(a)** $y = \begin{cases} |x| & \text{if } -1 \le x \le 1, \\ \sqrt{x} & \text{if } 1 \le x \le 4; \end{cases}$

(b) $y = \begin{cases} |x| & \text{if } -1 \le x < 1, \\ \sqrt{x} & \text{if } 1 < x \le 4. \end{cases}$

Solution

(a) We want to "paste together" portions of the graphs of $y = |x|$ and $y = \sqrt{x}$. The appropriate portions of each graph are indicated in color in Figures 7(a) and 7(b). Using Figures 7(a) and 7(b) as a guide, we sketch the required graph as shown in Figure 7(c).

FIGURE 7

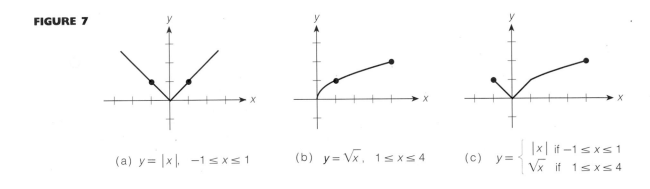

(a) $y = |x|, \ -1 \le x \le 1$ (b) $y = \sqrt{x}, \ 1 \le x \le 4$ (c) $y = \begin{cases} |x| & \text{if } -1 \le x \le 1 \\ \sqrt{x} & \text{if } 1 \le x \le 4 \end{cases}$

(b) Comparing the given equations and inequalities with those in part (a), we see that there is only one difference: in part (b), the number 1 is excluded from the domain of the variable x. To graph the equations in part (b), we need only take the graph obtained in Figure 7(c) and delete the point that corresponds to $x = 1$. That is, we delete the point $(1, 1)$ (see Figure 8). ■■■

FIGURE 8

$y = \begin{cases} |x| & \text{if } -1 \le x < 1 \\ \sqrt{x} & \text{if } 1 < x \le 4 \end{cases}$

We conclude this section by comparing the graphs in Figures 6, 7(c), and 8. Each graph was obtained by combining portions of the basic graphs in Figure 5. In Figures 6 and 8 the graphs have a break or gap in them. We say that the graph (or more precisely, the *function*) in Figure 6 is *discontinuous* when $x = 2$. (The two pieces do not meet at $x = 2$.) The function in Figure 8 is discontinuous when $x = 1$. (There is a gap or missing point when $x = 1$.) In Figure 7(c), however, the two portions of the graph form a graph with no break or gap. We say that the function in Figure 7(c) is *continuous* when $x = 1$. A rigorous definition of continuity is properly a subject for calculus. However, even at the intuitive level at which we've presented the idea here, you'll find that the concept is useful in helping you to organize your thoughts about graphs.

EXERCISE SET 2.2

A

In Exercises 1–6, determine whether the point lies on the graph of the equation, as in Example 1.   (Note: You are not asked to draw the graph.)

1. $(8, 6)$; $y = \frac{1}{2}x + 3$
2. $\left(\frac{3}{5}, -\frac{17}{5}\right)$; $y = -\frac{2}{3}x - 3$
3. $(4, 3)$; $3x^2 + y^2 = 52$
4. $(4, -2)$; $3x^2 + y^2 = 52$
5. $(a, 4a)$; $y = 4x$
6. $(a - 1, a + 1)$; $y = x + 2$

In Exercises 7–12, specify the domain of the variable x and then graph the equation, as we did in Example 2.

7. $y = \dfrac{x^2 + x - 20}{x - 4}$

8. $y = \dfrac{6x^2 + 11x + 4}{3x + 4}$

9. $y = \dfrac{x^2 - 9}{x - 3}$

10. $y = \dfrac{x - x^2}{x}$

11. $\dfrac{-x^3 + 4x^2 + x - 4}{x^2 - 1}$

12. $\dfrac{x^2 - 4}{(x + 2)(2 - x)}$

In Exercises 13–18, the graph of the equation is a straight line. Graph the equation after finding the x- and y-intercepts.

13. $3x + 4y = 12$
14. $3x - 4y = 12$
15. $y = 2x - 4$
16. $x = 2y - 4$
17. $x + y = 1$
18. $2x - 3y = 6$

In Exercises 19–30, determine any x- or y-intercepts for the graph of the equation.   Note: You're not asked to draw the graph.

19. (a) $y = x^2 + 3x + 2$ (b) $y = x^2 + 2x + 3$
20. (a) $y = x^2 - 4x - 12$ (b) $y = x^2 - 4x + 12$
21. (a) $y = x^2 + x - 1$ (b) $y = x^2 + x + 1$
22. (a) $y = 6x^3 + 9x^2 + x$ (b) $y = 9x^3 + 6x^2 + x$

23. $\dfrac{x}{2} + \dfrac{y}{3} = 1$

24. $y = |x - 1|$

25. $3x - 5y = 10$
26. $xy = 1$
27. $y = x^3 - 8$
28. $y = (x - 8)^3 + 1$
29. $y = \sqrt{x - 1} - x + 3$
30. $y = 7 + x - \sqrt{5 - x}$

In Exercises 31–36, each figure shows a graph of an equation. Find the x- and y-intercepts of the graph. If an intercept involves a radical, give both the radical form of the answer and a calculator approximation rounded to two decimal places. (Check to see that your answer is consistent with the given figure.)

31.

32.

33.

34.

35.

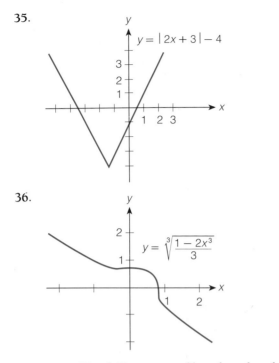

$y = |2x + 3| - 4$

36.

$y = \sqrt[3]{\dfrac{1 - 2x^3}{3}}$

In Exercises 37 and 38, set up a table and graph each equation. *Suggestion:* In Exercises 38(b) and (c), a calculator is helpful.

37. (a) $y = |x|$ (b) $y = x^2$ (c) $y = x^3$
38. (a) $y = 1/x$ (b) $y = \sqrt{x}$ (c) $y = \sqrt{1 - x^2}$

In Exercises 39–44, sketch the graphs, as in Examples 5 and 6.

39. $y = \begin{cases} |x| & \text{if } x \le 0 \\ x^2 & \text{if } x > 0 \end{cases}$

40. (a) $y = \begin{cases} x^3 & \text{if } x \le 0 \\ |x| & \text{if } x > 0 \end{cases}$

    (b) $y = \begin{cases} |x| & \text{if } x \le 0 \\ x^3 & \text{if } x > 0 \end{cases}$

41. (a) $y = \begin{cases} \sqrt{x} & \text{if } 0 \le x \le 1 \\ 1/x & \text{if } 1 < x < 2 \end{cases}$

    (b) $y = \begin{cases} \sqrt{x} & \text{if } 0 \le x < 1 \\ 1/x & \text{if } 1 < x < 2 \end{cases}$

42. (a) $y = \begin{cases} \sqrt{1 - x^2} & \text{if } -1 \le x \le 0 \\ x^2 & \text{if } 0 < x \le 2 \end{cases}$

    (b) $y = \begin{cases} \sqrt{1 - x^2} & \text{if } -1 \le x < 0 \\ x^2 & \text{if } 0 < x \le 2 \end{cases}$

43. $y = \begin{cases} x^3 & \text{if } -1 \le x < 0 \\ \sqrt{x} & \text{if } 0 \le x < 1 \\ 1/x & \text{if } 1 \le x \le 3 \end{cases}$

44. $y = \begin{cases} |x| & \text{if } -2 \le x \le -1 \\ \sqrt{1 - x^2} & \text{if } -1 < x < 1 \\ |x| & \text{if } 1 \le x \le 2 \end{cases}$

B

45. The figure shows the graph of the equation $F = \frac{9}{5}C + 32$, which relates temperature on the Fahrenheit scale to temperature on the Celsius scale.
    (a) Use the graph to estimate the Celsius temperature corresponding to 0°F.
    (b) Use the equation to find the exact Celsius temperature corresponding to 0°F. Check that your answer is consistent with your estimate in part (a).

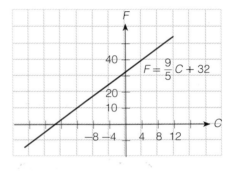

46. Draw the following graphs on the same set of axes and in each case include only the portion of the graph between $x = 0$ and $x = 1$. Draw the graphs carefully enough to ensure that you can make accurate comparisons among them.
    (a) $y = x$ (b) $y = x^2$ (c) $y = x^3$ (d) $y = x^4$
    What is the pattern here? What would you say that $y = x^{100}$ must look like in this interval?

47. The accompanying figure shows the graph of $y = \sqrt{x}$. Use this graph to estimate the following quantities (to one decimal place).
    (a) $\sqrt{2}$ (b) $\sqrt{3}$
    (c) $\sqrt{6}$ *Hint:* $\sqrt{ab} = \sqrt{a}\sqrt{b}$

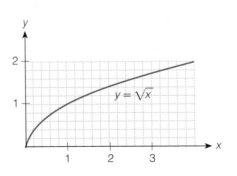

48. The following figure shows the graph of $y = \sqrt[3]{x}$. Use the graph to estimate the following quantities. In each case, compare your estimate with a value obtained from a calculator.
    (a) $\sqrt[3]{2}$ (b) $\sqrt[3]{5}$ (c) $\sqrt[3]{6.5}$

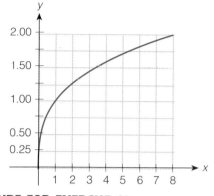

FIGURE FOR EXERCISE 48

49. In a certain biology experiment, the number N of bacteria increases with time t as indicated in the following figure.
    (a) How many bacteria are initially present when $t = 0$?
    (b) Approximately how long does it take for the original colony to double in size?
    (c) For which value of t is the population approximately 2500?
    (d) During which time interval does the population increase more rapidly, between $t = 0$ and $t = 1$ or between $t = 3$ and $t = 4$?

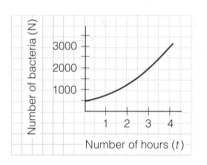

50. Graph the equation $(y - x^2)(y - |x|) = 0$ for $-1 \le x \le 1$.

51. Graph the equation $y^2 - x^2 y - x^3 y + x^5 = 0$ for $-1 \le x \le 1$.   *Hint:*   Factor the left-hand side of the equation.

52. Graph the equation $y^2 - x^2 y - \sqrt{x}\,xy + x^2\sqrt{x} = 0$ for $0 \le x \le 1$. (Use the hint in Exercise 51.)

In Exercises 53 and 54, determine the x-coordinates of the points A and B. (Each dashed line is parallel to one of the coordinate axes.)

53.

54.

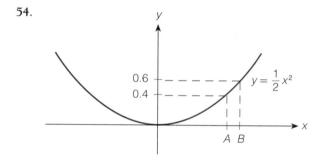

C

55. Suppose that the circle $x^2 + 2Ax + y^2 + 2By = C$ has two x-intercepts, x_1 and x_2, and two y-intercepts, y_1 and y_2. Prove each statement.
    (a) $\dfrac{x_1 + x_2}{y_1 + y_2} = \dfrac{A}{B}$ (b) $x_1 x_2 - y_1 y_2 = 0$
    (c) $x_1 x_2 + y_1 y_2 = -2C$

56. In the accompanying figure, the x-coordinates of the points A and B are a and $1/a$, respectively. (Assume that $0 < a < 1$.) If P denotes an arbitrary point on the semicircle, show that

    $$PA/PB = a$$

    What is remarkable about this result?

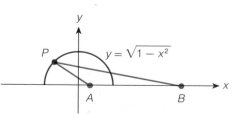

OPTIONAL TI-81 GRAPHING CALCULATOR EXERCISES

EXERCISES FOR SECTION 2.2

Note: As preparation for this set of exercises, read the instructions in Appendix A-2 (at the back of this book) for using the ZOOM BOX feature of the calculator.

1.   In this exercise we graph the following three equations.

$$Y_1 = |x| \qquad Y_2 = x^2 \qquad Y_3 = x^3$$

(These are three of the basic six equations that are graphed in the text on page 63.) Press the Y= key and type in these three equations.

(a)   In the Y= menu, select (only) the function $Y_1 = |x|$. Then press ZOOM 6 to see the graph in the standard viewing rectangle. Check that the graph you obtain is similar to the one shown in Figure 5 in the text on page 63.

(b)   In the Y= menu, select (only) the function $Y_2 = x^2$. Then press ZOOM 6 to see the graph in the standard viewing rectangle. Notice how flat the graph appears to be near the origin. For a closer look, use the ZOOM BOX. For the upper left corner of the box, use a point as close as possible to $(-2, 4)$; for the lower right corner, use a point as close as possible to $(2, -1)$.

(c)   In the Y= menu, select (only) the function $Y_3 = x^3$. Then press ZOOM 6 to see the graph in the standard viewing rectangle. Notice how flat the graph appears to be near the origin. For a closer look, use the ZOOM BOX. For the lower left corner of the box, use a point as close as possible to $(-2, -8)$; for the upper right corner, use a point as close as possible to $(2, 8)$.

(d)   Compare the graphs of $Y_2 = x^2$ and $Y_3 = x^3$ on the interval $[0, 1]$. In the Y= menu, select both of these functions and then press ZOOM 6. Next, press the RANGE key and enter the settings $x_{min} = 0$, $x_{max} = 1$, $y_{min} = 0$, and $y_{max} = 1$. Now press the GRAPH key. Where do the graphs intersect? For x-values in the interval $(0, 1)$, which quantity is larger, x^2 or x^3?

(e)   Carry out an analysis similar to the one in part (d) to compare $Y_2 = x^2$ and $Y_3 = x^3$ when $x \geq 1$.

2.   In this exercise we graph the following three equations.

$$Y_1 = 1/x \qquad Y_2 = \sqrt{x} \qquad Y_3 = \sqrt{1 - x^2}$$

(These are three of the basic six equations that are graphed in the text on page 63.) Press the Y= key and type in these three equations.

(a)   In the Y= menu, select (only) the function

$Y_1 = 1/x$ Then press ZOOM 6 to see the graph in

the standard viewing rectangle. Next, for a better view, use the ZOOM BOX. For the upper left corner of the box, use a point as close as possible to $(-4, 4)$; for the lower right corner use a point as close as possible to $(4, -4)$. Check that the graph you obtain is similar to the one shown in Figure 5 in the text on page 63.

(b)   In the Y= menu, select (only) the function $Y_2 = \sqrt{x}$. Then press ZOOM 6 to see the graph in the standard viewing rectangle. Next, for a better view, use the ZOOM BOX. For the lower left corner of the box, use a point as close as possible to $(-1, -1)$; for the upper right corner, use a point as close as possible to $(9, 3)$. Check that the graph you obtain is similar to the one shown in Figure 5 in the text on page 63.

(c)   Press ZOOM 6 [with the function from part (b) still selected] to return to the view of the square root function in the standard viewing rectangle. Now press the TRACE key. You will see a blinking cursor located on the curve near the origin. The numbers on the bottom of the screen are the coordinates of the cursor. By repeatedly pressing the right-arrow key, move the cursor along the curve until the x-coordinate is 2. As you can see on the screen, the corresponding y-coordinate of the point on the curve is approximately 1.4142136. What is the *exact* value of the y-coordinate?

(d)   In the Y= menu, select (only) the function $Y_3 = \sqrt{1 - x^2}$. Then press ZOOM 6 to see the graph in the standard viewing rectangle. Next, for a better view, use the ZOOM BOX. For the lower left corner of the box, use a point as close as possible to $(-1, 0)$; for the upper right corner, use a point as close as possible to $(1, 1)$. This yields a larger picture, but the semicircle is still distorted because different scales are used on the axes. To correct for this, press ZOOM 5. (What defects still remain in the view shown on your calculator screen?)

3.   This exercise shows how to graph functions that are defined piecewise. As an example, we will use

$$y = \begin{cases} x^2 - 1 & \text{if } x \leq 2 \\ 5 - x & \text{if } x > 2 \end{cases}$$

As preparation, clear all functions that appear in the Y= menu, and press ZOOM 6 for the standard viewing rectangle. Now press the Y= key.

(a) In the $Y=$ menu, we are going to define Y_1 to be the function $x^2 - 1$ with the restriction $x \leq 2$. Begin by typing in the following sequence of keystrokes.

⟨ (⟩ ⟨ X|T ⟩ ⟨ ∧ ⟩ ⟨ 2 ⟩ ⟨ – ⟩ ⟨ 1 ⟩ ⟨) ⟩

Do not press ENTER now. Rather, continue with the following keystrokes:

⟨ (⟩ ⟨ X|T ⟩ ⟨ 2nd ⟩ ⟨ MATH ⟩ ⟨ 5 ⟩ ⟨ 2 ⟩ ⟨) ⟩

Your screen display in the $Y=$ menu should now read: $Y_1 = (x \hat{\ } 2 - 1)(x \leq 2)$. Press the GRAPH key to see the portion of $x^2 - 1$ corresponding to $x \leq 2$. In order to remove the extraneous points that appear in the graph, press the MODE key and select "Dot" rather than "Connected"; then press the ENTER key, then the GRAPH key.

(b) Next, in the $Y=$ menu, we are going to define Y_2 to be the function $5 - x$ with the restriction $x > 2$. The sequence of keystrokes is similar to that in part (a). In the Y_2 line, begin by typing

⟨ (⟩ ⟨ 5 ⟩ ⟨ – ⟩ ⟨ X|T ⟩ ⟨) ⟩. Do not press ENTER. Rather, continue with the following sequence of keystrokes.

⟨ (⟩ ⟨ X|T ⟩ ⟨ 2nd ⟩ ⟨ MATH ⟩ ⟨ 3 ⟩ ⟨ 2 ⟩ ⟨) ⟩

Your screen display for Y_2 should read: $Y_2 = (5 - x)(x > 2)$. By pressing the GRAPH key now, you will obtain the required graph.

4. Graph the function defined by

$$y = \begin{cases} -x^2 & \text{if } -2 \leq x \leq 0 \\ \frac{1}{2}x & \text{if } 0 < x \leq 4 \end{cases}$$

Hint: You need to create the following screen in the $Y=$ menu (as indicated in the previous exercise, the

symbols for inequality are accessed by pressing the "2nd" key and then the MATH key):

$$:Y_1 = (-x \hat{\ } 2)(x \geq -2)(x \leq 0)$$
$$:Y_2 = (x/2)(x > 0)(x \leq 4)$$

As in Exercise 3, you will want to view the graph in the "Dot" mode rather than the "Connected" mode.

In Exercises 5–12, graph the functions using the techniques described in Exercises 3 and 4.

5. $y = \begin{cases} x^2 & \text{if } 0 \leq x < 1 \\ 1/x & \text{if } 1 \leq x \leq 4 \end{cases}$

6. $y = \begin{cases} |x| & \text{if } -4 \leq x < 0 \\ x^2 & \text{if } 0 \leq x < 1 \\ 1/x & \text{if } 1 \leq x \leq 4 \end{cases}$

7. $y = \begin{cases} \sqrt{1 - x^2} & \text{if } -1 \leq x < 0 \\ \sqrt{x} & \text{if } 0 \leq x \leq 4 \end{cases}$

8. $y = \begin{cases} x^3 & \text{if } -1 \leq x < 0 \\ \sqrt{x} & \text{if } 0 \leq x \leq 4 \end{cases}$

9. $y = \begin{cases} -1/x & \text{if } x < -1 \\ x^2 & \text{if } -1 \leq x < 1 \\ 1/x & \text{if } x \geq 1 \end{cases}$

10. $y = \begin{cases} |x| & \text{if } x < -1 \\ x^2 & \text{if } -1 \leq x < 1 \\ |x| & \text{if } x \geq 1 \end{cases}$

11. $y = \begin{cases} x^2 & \text{if } -1 \leq x \leq 1 \\ (x - 2)^2 & \text{if } 1 < x \leq 3 \\ (x - 4)^2 & \text{if } 3 < x \leq 5 \end{cases}$

12. $y = \begin{cases} -x + 2 & \text{if } 0 \leq x \leq 2 \\ -x + 4 & \text{if } 2 < x \leq 4 \\ -x + 6 & \text{if } 4 < x \leq 6 \end{cases}$

2.3 EQUATIONS OF LINES

He [Pierre de Fermat (1601–1665)] introduced perpendicular axes and found the general equations of straight lines and circles and the simplest equations of parabolas, ellipses, and hyperbolas; and he further showed in a fairly complete and systematic way that every first- or second-degree equation can be reduced to one of these types.

George F. Simmons in *Calculus Gems, Brief Lives and Memorable Moments* (New York: McGraw-Hill Book Company, 1992)

In this section we take a systematic look at equations of lines and their graphs. We begin by recalling the concept of *slope*, which you've seen in previous courses. The slope of a nonvertical line is a number that measures the slant or direction of the line; it is defined as follows.

DEFINITION Slope

The **slope** of a nonvertical line passing through the two points (x_1, y_1) and (x_2, y_2) is the number m defined by

$$m = \frac{y_2 - y_1}{x_2 - x_1}$$

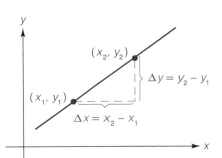

FIGURE 1

Note that the quantity $x_2 - x_1$ appearing in the definition of slope is the amount by which x changes as we move from (x_1, y_1) to (x_2, y_2) along the line. We denote this change in x by the symbol Δx (read *delta x*). Thus $\Delta x = x_2 - x_1$ (see Figure 1). Similarly, the symbol Δy is defined to mean the change in y: $\Delta y = y_2 - y_1$. Using these ideas, we can rewrite our definition of slope as $m = \Delta y/\Delta x$.

The slope of a line does not depend on which two particular points on the line are used in the calculation. To see why this is so, consider Figure 2. The two right triangles are similar (because the corresponding angles are equal). This implies that the corresponding sides of the two triangles are proportional, and so we have

$$\frac{a}{d} = \frac{b}{c}$$

Now notice that the left-hand side of this equation represents the slope $\Delta y/\Delta x$ calculated using the points A and D, and the right-hand side represents the slope calculated using the points B and C. Thus the values we obtain for the slope are indeed equal.

FIGURE 2

■■ EXAMPLE 1 Compute and compare the slopes of the three lines shown in Figure 3.

FIGURE 3

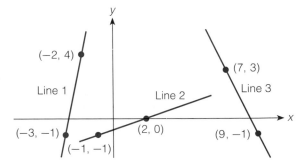

Solution First, we'll calculate the slope of line 1, using the formula $m = (y_2 - y_1)/(x_2 - x_1)$. Which point should we use as (x_1, y_1) and which as (x_2, y_2)? In fact, it does not matter how we label our points. Using $(-3, -1)$ for (x_1, y_1) and $(-2, 4)$ for (x_2, y_2), we have

$$m = \frac{y_2 - y_1}{x_2 - x_1} = \frac{4 - (-1)}{-2 - (-3)} = 5$$

So the slope of line 1 is 5. If, instead, we had used $(-2, 4)$ for (x_1, y_1) and $(-3, -1)$ for (x_2, y_2), then we'd have $m = (-1 - 4)/(-3 + 2) = 5$, the same result. This is not accidental because, in general,

$$\frac{y_2 - y_1}{x_2 - x_1} = \frac{y_1 - y_2}{x_1 - x_2}$$

(Exercise 63 asks you to verify this identity.) Next, we calculate the slopes of lines 2 and 3.

Slope of line 2: $\dfrac{0 - (-1)}{2 - (-1)} = \dfrac{1}{3}$ Slope of line 3: $\dfrac{3 - (-1)}{7 - 9} = -2$

Lines 1 and 2 both have positive slopes and slant upward to the right. Note that line 1 is steeper than line 2 and correspondingly has the larger slope, 5. Line 3 has a negative slope, -2, and slants downward to the right. ▮▮▮

The observations made in Example 1 are true in general. Lines with a positive slope slant upward to the right, the steeper line having the larger slope. Likewise, lines with a negative slope slant downward to the right. See Figure 4.

FIGURE 4   **FIGURE 5**

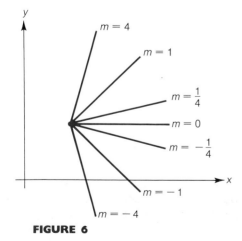

FIGURE 6

We have yet to mention slopes for horizontal or vertical lines. In Figure 5, line 1 is horizontal; it passes through the points (a, b) and (c, b). Note that the two y-coordinates must be the same in order for the line to be horizontal. Line 2 in Figure 5 is vertical; it passes through (d, e) and (d, f). Note that the two x-coordinates must be the same in order for the line to be vertical. For the slope of line 1 we have $m = \dfrac{b - b}{c - a} = \dfrac{0}{c - a} = 0$ (provided $a \neq c$). Thus the slope of line 1, a horizontal line, is zero. For the vertical line in Figure 5, the calculation of slope begins with writing $\dfrac{e - f}{d - d}$, but the denominator is zero, and since division by zero is undefined, we conclude that slope is undefined for vertical lines. We summarize these results in the box that follows. Figure 6 shows for comparison some values of m for various lines.

> **Horizontal and Vertical Lines**
>
> **1.** The slope of a horizontal line is zero.
>
> **2.** Slope is not defined for vertical lines.

We can use the concept of slope to find the equation of a line. Suppose we have a line with slope m and passing through the point (x_1, y_1), as shown in Figure 7(a). Let (x, y) be any other point on the line, as in Figure 7(b). Then the slope of the line is given by $m = (y - y_1)/(x - x_1)$, and therefore, $y - y_1 = m(x - x_1)$. Note that the given point (x_1, y_1) also satisfies this last equation, because in that case we have $y_1 - y_1 = m(x_1 - x_1)$, or $0 = 0$. The equation $y - y_1 = m(x - x_1)$ is called the **point–slope formula.** We have shown that any point on the line satisfies this equation. (Conversely, it can be shown that if a point satisfies this equation, then the point does lie on the given line.)

FIGURE 7

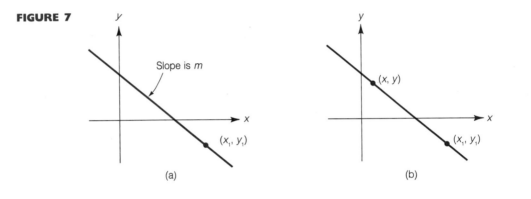

(a)

(b)

> **The Point–Slope Formula**
>
> The equation of a line with slope m passing through the point (x_1, y_1) is
>
> $$y - y_1 = m(x - x_1)$$

EXAMPLE 2 Write the equation of the line passing through $(-3, 1)$ with a slope of -2. Sketch a graph of the line.

Solution Since the slope and a point are given, we use the point–slope formula:

$$y - y_1 = m(x - x_1)$$
$$y - 1 = -2[x - (-3)]$$
$$y - 1 = -2x - 6$$
$$y = -2x - 5$$

This is the required equation. One way to graph this line is to pick values for x and then compute the corresponding y-values, as done in Section 2.2. The table and graph are displayed in Figure 8. Notice that if we know ahead of time that the graph is a line, a table as brief as that in Figure 8 is sufficient. ∎

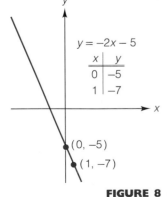

$y = -2x - 5$

x	y
0	-5
1	-7

$(0, -5)$

$(1, -7)$

FIGURE 8
$y = -2x - 5$

There is another way to go about graphing the line $y = -2x - 5$ in Example 2. Since slope is (change in y)/(change in x), a slope of -2 (or $-2/1$) can be interpreted as telling us that if we start at $(-3, 1)$ and let x increase by one unit, then y must decrease by two units to bring us back to the line. Following this path in Figure 9 takes us from $(-3, 1)$ to $(-2, -1)$. We now draw the line through these two points, as shown in Figure 9.

FIGURE 9

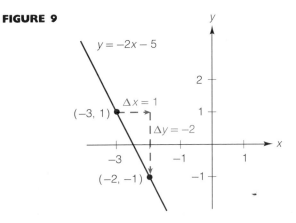

EXAMPLE 3 Find the equation of the line passing through the points $(-2, -3)$ and $(2, 5)$.

Solution The slope of the line is

$$m = \frac{y_2 - y_1}{x_2 - x_1} = \frac{5 - (-3)}{2 - (-2)} = \frac{8}{4} = 2$$

Knowing the slope, we can apply the point–slope formula, making use of either $(-2, -3)$ or $(2, 5)$. Using the point $(2, 5)$ as (x_1, y_1), we have

$$y - y_1 = m(x - x_1)$$
$$y - 5 = 2(x - 2)$$
$$y - 5 = 2x - 4$$
$$y = 2x + 1$$

Thus the required equation is $y = 2x + 1$. You should check for yourself that the same answer is obtained using the point $(-2, -3)$ instead of $(2, 5)$ in the last set of calculations. ■■■

EXAMPLE 4 Find the equation of the horizontal line passing through the point $(4, -2)$. See Figure 10.

Solution Since the slope of a horizontal line is zero, we have

$$y - y_1 = m(x - x_1)$$
$$y - (-2) = 0(x - 4)$$
$$y = -2$$

Thus the equation of the horizontal line passing through $(4, -2)$ is $y = -2$. ■■■

FIGURE 10

FIGURE 11

By using the point–slope formula exactly as we did in Example 4, we can show more generally that the equation of the horizontal line in Figure 11 is $y = b$. What about the equation of the vertical line in Figure 12 passing through the point (a, b)? Because slope is not defined for vertical lines, the point–slope formula is not applicable this time. However, note that as we move along the vertical line, the x-coordinate is always a; it is only the y-coordinate that varies. The equation $x = a$ expresses exactly these two facts; it says that x must always be a and it places no restrictions on y.

In the box that follows, we summarize our results concerning horizontal and vertical lines.

FIGURE 12

> **Equations of Horizontal and Vertical Lines**
>
> **1.** The equation of a horizontal line through the point (a, b) is $y = b$. (See Figure 11.)
>
> **2.** The equation of a vertical line through the point (a, b) is $x = a$. (See Figure 12.)

FIGURE 13

Another basic form for the equation of a line is the *slope–intercept form*. We are given the slope m and y-intercept b, as shown in Figure 13, and we want to find the equation of the line. To say that the line has a y-intercept of b is the same as saying that the line passes through $(0, b)$. The point–slope formula is applicable now, using the slope m and the point $(0, b)$. We have

$$y - y_1 = m(x - x_1)$$
$$y - b = m(x - 0)$$
$$y = mx + b$$

This last equation is called the **slope–intercept formula**.

> **The Slope–Intercept Formula**
>
> The equation of a line with slope m and y-intercept b is
>
> $$y = mx + b$$

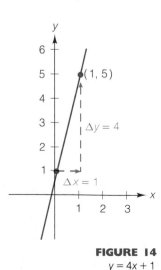

FIGURE 14
$y = 4x + 1$

■ ▬ ■   **EXAMPLE 5**   Write the equation of a line with slope 4 and y-intercept 1. Graph the line.

Solution   Substituting $m = 4$ and $b = 1$ in the equation $y = mx + b$ yields

$$y = 4x + 1$$

This is the required equation. We could draw the graph by first setting up a simple table, but for purposes of emphasis and review we proceed as we did just after Example 2. Starting from the point $(0, 1)$, we interpret the slope of 4 as saying that if x increases by 1, then y increases by 4. This takes us from $(0, 1)$ to the point $(1, 5)$ and the line can now be sketched as in Figure 14.   ▐▐▐

███ **EXAMPLE 6** Find the slope and y-intercept of the line $3x - 5y = 15$.

Solution   First we solve for y so that the equation is in the form $y = mx + b$:

$$3x - 5y = 15$$

$$-5y = -3x + 15 \qquad \text{and therefore} \qquad y = \frac{3}{5}x - 3$$

The slope m and the y-intercept b can now be read directly from the equation: $m = \frac{3}{5}$ and $b = -3$.   ▮▮▮

As a consequence of our work up to this point, we can say that the graph of any **linear equation** $Ax + By + C = 0$, where A and B are not both zero, is a line, since an equation of this type can always be rewritten in one of the following three forms: $y = mx + b$, or $x = a$, or $y = b$. In the box that follows, we summarize our basic results on equations of lines.

▶ **PROPERTY SUMMARY**   **EQUATIONS OF LINES**

EQUATION	COMMENT
$y - y_1 = m(x - x_1)$ (the point–slope formula)	This is the equation of a line with slope m, passing through the point (x_1, y_1).
$y = mx + b$ (the slope–intercept formula)	This is the equation of a line with slope m and y-intercept b.
$Ax + By + C = 0$	The graph of any equation of this form (where A and B are not both zero) is a line. Special cases: If $A = 0$, then the equation can be written $y = -C/B$, which is the equation of a horizontal line with y-intercept $-C/B$. If $B = 0$, then the equation can be written $x = -C/A$, which represents a vertical line with x-intercept $-C/A$.

We conclude this section by discussing two useful relationships regarding the slopes of parallel lines and the slopes of perpendicular lines. First, nonvertical parallel lines have the same slope. This should seem reasonable if you recall that slope is a number indicating the direction or slant of a line. The relationship concerning slopes of perpendicular lines is not so obvious. The slopes of two nonvertical perpendicular lines are negative reciprocals of each other. That is, if m_1 and m_2 denote the slopes of the two perpendicular lines, then $m_1 = -1/m_2$ or, equivalently, $m_1 m_2 = -1$. For example, if a line has a slope of $\frac{2}{3}$, then any line perpendicular to it must have a slope of $-\frac{3}{2}$. For reference we summarize these facts in the box that follows. (Proofs of these facts are outlined in detail in Exercises 61 and 62.)

Parallel and Perpendicular Lines

Let m_1 and m_2 denote the slopes of two nonvertical lines. Then:

1. The lines are parallel if and only if $m_1 = m_2$;

2. The lines are perpendicular if and only if $m_1 = -1/m_2$.

EXAMPLE 7 Are the two lines $3x - 6y - 8 = 0$ and $2y = x + 1$ parallel?

Solution By solving each equation for y, we can see what the slopes are:

$$3x - 6y - 8 = 0 \qquad\qquad 2y = x + 1$$
$$-6y = -3x + 8 \qquad\qquad y = \frac{1}{2}x + \frac{1}{2}$$
$$y = \frac{-3}{-6}x + \frac{8}{-6}$$
$$y = \frac{1}{2}x - \frac{4}{3}$$

From this we see that both lines have the same slope, namely, $m = \frac{1}{2}$. It follows therefore that the lines are parallel. ∎

EXAMPLE 8 Find the equation of the line parallel to $5x + 6y = 30$ and passing through the origin.

Solution First, find the slope of the line $5x + 6y = 30$:

$$5x + 6y = 30$$

$$6y = -5x + 30 \qquad \text{and therefore} \qquad y = -\frac{5}{6}x + 5$$

The slope is the x-coefficient in the last equation: $m = -\frac{5}{6}$. A parallel line will have the same slope. Since the required line is to pass through $(0,0)$, we have

$$y - y_1 = m(x - x_1)$$
$$y - 0 = -\frac{5}{6}(x - 0)$$
$$y = -\frac{5}{6}x$$

This is the equation of a line parallel to $5x + 6y = 30$ and passing through the origin, as required. See Figure 15. ∎

FIGURE 15

(graph: $5x + 6y = 30$ and $y = -\frac{5}{6}x$)

EXAMPLE 9 Find the equation of the line tangent to the circle $x^2 + y^2 = 25$ at the point $(3, -4)$ (see Figure 16). Write the answer in the form $y = mx + b$.

FIGURE 16

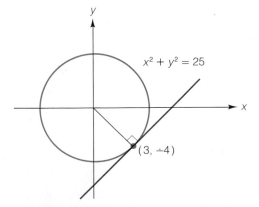

$x^2 + y^2 = 25$

$(3, -4)$

Solution We have one point on the tangent line, namely, $(3, -4)$. If we knew the slope, then we could apply the point–slope formula. From elementary geometry, we know that the tangent line is perpendicular to the radius. Since the radius passes through $(0, 0)$ and $(3, -4)$, its slope is

$$\frac{-4 - 0}{3 - 0} = -\frac{4}{3}$$

and consequently, the slope of the tangent is $\frac{3}{4}$. The point–slope formula is now applicable:

$$y - y_1 = m(x - x_1)$$
$$y - (-4) = \frac{3}{4}(x - 3)$$

Check now for yourself that this can be simplified to $y = \frac{3}{4}x - \frac{25}{4}$, which is the required equation of the tangent line. ∎∎∎

EXERCISE SET 2.3

A

In Exercises 1–3, compute the slope of the line passing through the two given points. In Exercise 3, include a sketch with your answers.

1. (a) $(-3, 2), (1, -6)$
   (b) $(2, -5), (4, 1)$
   (c) $(-2, 7), (1, 0)$
   (d) $(4, 5), (5, 8)$

2. (a) $(-3, 0), (4, 9)$
   (b) $(-1, 2), (3, 0)$
   (c) $\left(\frac{1}{2}, -\frac{3}{5}\right), \left(\frac{3}{2}, \frac{3}{4}\right)$
   (d) $\left(\frac{17}{3}, -\frac{1}{2}\right), \left(-\frac{1}{2}, \frac{17}{3}\right)$

3. (a) $(1, 1), (-1, -1)$
   (b) $(0, 5), (-8, 5)$
   (c) $(-1, 1), (1, -1)$
   (d) $(a, b), (b, a)$
   (assume $a \neq b$)

4. Compute the slope of the line in the following figure using each pair of points indicated.
   (a) *A* and *B*   (b) *B* and *C*   (c) *A* and *C*
   The principle involved here is that no matter which pair of points you choose, the slope is the same.

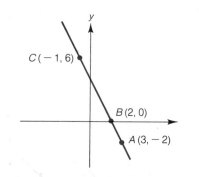

5. The slopes of four lines are indicated in the figure. List the slopes m_1, m_2, m_3, m_4 in order of increasing value.

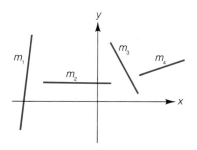

6. Refer to the accompanying figure.

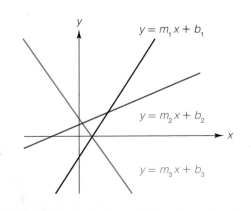

(a) List the slopes m_1, m_2, and m_3 in order of increasing size.

(b) List the numbers b_1, b_2, and b_3 in order of increasing size.

In Exercises 7–9, three points A, B, and C are specified. Determine if A, B, and C are collinear (lie on the same line) by checking to see whether the slope of $\overline{AB}$ equals the slope of $\overline{BC}$.

7. $A(-8, -2)$, $B\left(2, \frac{1}{2}\right)$, $C(11, -1)$

8. $A(4, -3)$, $B(-1, 0)$, $C(-4, 2)$

9. $A(0, -5)$, $B(3, 4)$, $C(-1, -8)$

10. If the area of the "triangle" formed by three points is zero, then the points must in fact be collinear. Use this observation, along with the formula in Exercise 14(b) of Exercise Set 2.1, to rework Exercise 9.

In Exercises 11 and 12, find the equation for the line having the given slope and passing through the given point. Write your answers in the form $y = mx + b$.

11. (a) $m = -5$; through $(-2, 1)$
    (b) $m = 4$; through $(4, -4)$
    (c) $m = \frac{1}{3}$; through $\left(-6, -\frac{2}{3}\right)$
    (d) $m = -1$; through $(0, 1)$

12. (a) $m = 22$; through $(0, 0)$
    (b) $m = -222$; through $(0, 0)$
    (c) $m = \sqrt{2}$; through $(0, 0)$

In Exercises 13 and 14, find the equation for the line passing through the two given points. Write your answer in the form $y = mx + b$.

13. (a) $(4, 8)$ and $(-3, -6)$
    (b) $(-2, 0)$ and $(3, -10)$
    (c) $(-3, -2)$ and $(4, -1)$

14. (a) $(7, 9)$ and $(-11, 9)$
    (b) $\left(\frac{5}{4}, 2\right)$ and $\left(\frac{3}{4}, 3\right)$
    (c) $(12, 13)$ and $(13, 12)$

In Exercises 15 and 16, write the equation of a vertical line passing through the given point. In Exercises 17 and 18, write the equation of a horizontal line passing through the given point.

15. $(-3, 4)$ 16. $(8, 5)$

17. $(-3, 4)$ 18. $(8, 5)$

19. Is the graph of the line $x = 0$ the x-axis or the y-axis?

20. Is the graph of the line $y = 0$ the x-axis or the y-axis?

In Exercises 21 and 22, find the equation of the line with the given slope and y-intercept.

21. (a) slope -4; y-intercept 7
    (b) slope 2; y-intercept $\frac{3}{2}$
    (c) slope $-\frac{4}{3}$; y-intercept 14

22. (a) slope 0; y-intercept 14
    (b) slope 14; y-intercept 0

In Exercises 23–28, find an equation for the line that is described, and sketch the graph. For Exercises 23–25, write the final answer in the form $y = mx + b$; for Exercises 26–28, write the answer in the form $Ax + By + C = 0$.

23. (a) Passing through $(-3, -1)$, with slope 4
    (b) Passing through $\left(\frac{5}{2}, 0\right)$, with slope $\frac{1}{2}$
    (c) With x-intercept 6, y-intercept 5
    (d) With x-intercept -2, slope $\frac{3}{4}$
    (e) Passing through $(1, 2)$ and $(2, 6)$

24. (a) Passing through $(-7, -2)$ and $(0, 0)$
    (b) Passing through $(6, -3)$, with y-intercept 8
    (c) Passing through $(0, -1)$ and with the same slope as the line $3x + 4y = 12$
    (d) Passing through $(6, 2)$ and with the same x-intercept as the line $-2x + y = 1$
    (e) With x-intercept -6, y-intercept $\sqrt{2}$

25. With the same x- and y-intercepts as the circle $x^2 + y^2 + 4x - 4y + 4 = 0$

26. Passing through $(3, -5)$ and through the center of the circle $4x^2 + 8x + 4y^2 - 24y + 15 = 0$

27. Passing through $(-3, 4)$ and parallel to the x-axis

28. Passing through $(-3, 4)$ and parallel to the y-axis

In Exercises 29 and 30, find the x- and y-intercepts of the line, and find the area and perimeter of the triangle formed by the line and the axes.

29. (a) $3x + 5y = 15$ (b) $3x - 5y = 15$

30. (a) $5x + 4y = 40$ (b) $2x + 4y = \sqrt{2}$

31. Determine whether each pair of lines is parallel, perpendicular, or neither.
    (a) $3x - 4y = 12$; $4x - 3y = 12$
    (b) $y = 5x - 16$; $y = 5x + 2$
    (c) $5x - 6y = 25$; $6x + 5y = 0$
    (d) $y = -\frac{2}{3}x - 1$; $y = \frac{3}{2}x - 1$
    (e) $-2x - 5y = 1$; $y - \frac{2}{5}x - 4 = 0$
    (f) $x = 8y + 3$; $4y - \frac{1}{2}x = 32$

32. Are the lines $y = x + 1$ and $y = 1 - x$ parallel, perpendicular, or neither?

In Exercises 33–38, find an equation for the line that is described. Write the answer in the two forms $y = mx + b$ and $Ax + By + C = 0$.

33. Parallel to $2x - 5y = 10$ and passing through $(-1, 2)$

34. Parallel to $4x + 5y = 20$ and passing through $(0, 0)$

35. Perpendicular to $4y - 3x = 1$ and passing through $(4, 0)$

36. Perpendicular to $x - y + 2 = 0$ and passing through $(3, 1)$

37. Parallel to $3x - 5y = 25$ and with the same y-intercept as $6x - y + 11 = 0$

38. Perpendicular to $9y - 2x = 3$ and with the same x-intercept as the circle $x^2 + y^2 - 12x - 2y = -36$

39. (a) Sketch the circle $x^2 + y^2 = 25$.
(b) Find the equation of the line tangent to this circle at the point $(-4, -3)$. Include a sketch.

40. (a) Tangents are drawn to the circle $x^2 + y^2 = 169$ at the points $(5, 12)$ and $(5, -12)$. Find the equations of the tangents.
(b) Find the coordinates of the point where these two tangents intersect.

B

41. Show that the slope of the line passing through the two points $(3, 9)$ and $(3 + h, (3 + h)^2)$ is $6 + h$.

42. Show that the slope of the line passing through the two points (x, x^2) and $(x + h, (x + h)^2)$ is $2x + h$.

43. Show that the slope of the line passing through the two points (x, x^3) and $(x + h, (x + h)^3)$ is $3x^2 + 3xh + h^2$.

44. Show that the slope of the line passing through the points $(x, \sqrt{x})$ and $(x + h, \sqrt{x + h})$ can be written
$$1/(\sqrt{x + h} + \sqrt{x})$$

45. Show that the slope of the line passing through the two points $\left(x, \dfrac{1}{x}\right)$ and $\left(x + h, \dfrac{1}{x + h}\right)$ is $\dfrac{-1}{x(x + h)}$.

46. Why is the letter m used to denote slope? Look up the verb *monter* in a French-to-English dictionary.

47. Sketch the curve $y = x^2$ and indicate the point $T(3, 9)$ on the graph. In each case that follows, you're given the x-coordinate of a point P on the curve. Find the corresponding y-coordinate, and then calculate the slope of the line through P and T. Display your results in a table.
(a) $x = 2.5$ (b) $x = 2.9$ (c) $x = 2.99$
(d) $x = 2.999$ (e) $x = 2.9999$
As P gets closer to T, what numerical value does the slope of $\overline{PT}$ seem to approach? What would you estimate to be the slope of the line that is tangent to the curve $y = x^2$ at point T?

48. Follow the instructions for Exercise 47 using the curve $y = \sqrt{x}$, the point $T(1, 1)$, and the following sequence of x-values.
(a) $x = 1.1$ (b) $x = 1.01$ (c) $x = 1.001$
(d) $x = 1.0001$ (e) $x = 1.00001$
(Round your answers to six decimal places.)

49. Find a point P on the curve $y = x^3$ such that the slope of the line passing through P and $(1, 1)$ is $\frac{3}{4}$.

50. Find a point P on the curve $y = \sqrt{x}$ such that the slope of the line through P and $(1, 1)$ is $\frac{1}{4}$.

51. Find a point P on the curve $y = 1/x$ such that the slope of the line through P and $\left(2, \frac{1}{2}\right)$ is $-\frac{1}{16}$.

52. Find a point P on the circle $x^2 + y^2 = 20$ such that the slope of the radius drawn from $(0, 0)$ to P is 2. (You will get two answers.)

53. A line with a slope of -5 passes through the point $(3, 6)$. Find the area of the triangle in the first quadrant formed by this line and the coordinate axes.

54. The y-intercept of the line in the figure is 6. Find the slope of the line if the area of the shaded triangle is 72 square units.

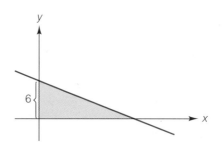

55. (a) Sketch the line $y = \frac{1}{2}x - 5$ and the point $P(1, 3)$. Follow parts (b)–(d) to calculate the perpendicular distance from point $P(1, 3)$ to the line.
(b) Find the equation of the line that passes through $P(1, 3)$ and is perpendicular to the line $y = \frac{1}{2}x - 5$.
(c) Find the coordinates of the point where these two lines intersect.
(d) Use the distance formula to find the perpendicular distance from $P(1, 3)$ to the line $y = \frac{1}{2}x - 5$.

56. (a) Show that the equation of the line tangent to the circle $x^2 + y^2 = a^2$ at the point (x_1, y_1) on the circle is
$$x_1 x + y_1 y = a^2$$
(b) Use the result in part (a) to rework Exercises 39(b) and 40(a).

57. Follow the instructions in parts (a)–(e) to determine the slope m of the line in the figure.

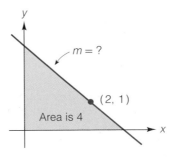

(a) Use the point–slope formula to show that the equation of the line is $y - 1 = m(x - 2)$.
(b) Show that the y-intercept for the line is $-2m + 1$. Show that the x-intercept is $(2m - 1)/m$.

(c) Show that the area of the triangle, being one-half base times height, is $\frac{1}{2}\left(\frac{2m-1}{m}\right)(-2m+1)$.

(d) Set the expression for area in part (c) equal to 4, as given in the figure, and simplify the resulting equation to obtain $4m^2 + 4m + 1 = 0$.

(e) Solve the equation for m. You should obtain $m = -\frac{1}{2}$.

58. A line with a negative slope passes through the point $(3, 1)$. The area of the triangle bounded by this line and the axes is 6 square units. Find the possible slopes of the line. *Hint:* See Exercise 57.

59. By analyzing sales figures, the accountant for College Stereo Company knows that 280 units of a compact disc player can be sold each month when the price is $P = \$195$ per unit. The figures also show that for each \$15 hike in the price, 10 fewer units are sold monthly.

(a) Let x denote the number of units sold per month and P the price per unit. Find an equation that expresses x in terms of P, assuming that this relationship is linear.
Hint: $\Delta x/\Delta P = -10/15 = -2/3$.

(b) Use the equation that you found in part (a) to determine how many units can be sold in a month when the price is \$270 per unit.

(c) What should the price be to sell 205 units per month?

60. Imagine that you own a grove of orange trees, and suppose that from past experience you know that when 100 trees are planted, each tree will yield approximately 240 oranges per year. Furthermore, you've noticed that when additional trees are planted in the grove, the yield per tree decreases. Specifically, you have noted that the yield per tree decreases by about 20 oranges for each additional tree planted.

(a) Let y denote the yield per tree when x trees are planted. Find a linear equation relating x and y.

(b) Use the equation in part (a) to determine how many trees should be planted to obtain a yield of 400 oranges per tree.

(c) If the grove contains 95 trees, what yield can you expect from each tree?

61. This exercise outlines a proof of the fact that two nonvertical lines are parallel if and only if their slopes are equal. The proof relies on the following observation for the given figure: The lines $y = m_1x + b_1$ and $y = m_2x + b_2$ will be parallel if and only if the two vertical distances AB and CD are equal. (In the figure, the points C and D both have x-coordinate 1.)

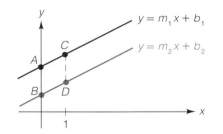

(a) Verify that the coordinates of A, B, C, and D are

$$A(0, b_1) \qquad B(0, b_2) \qquad C(1, m_1 + b_1) \qquad D(1, m_2 + b_2)$$

(b) Using the coordinates in part (a), check that $AB = b_1 - b_2$ and $CD = (m_1 + b_1) - (m_2 + b_2)$.

(c) Use part (b) to show that the equation $AB = CD$ is equivalent to $m_1 = m_2$.

62. This exercise outlines a proof of the fact that two nonvertical lines with slopes m_1 and m_2 are perpendicular if and only if $m_1m_2 = -1$. In the figure below, we've assumed that our two nonvertical lines $y = m_1x$ and $y = m_2x$ intersect at the origin. [If they did not intersect there, we could just as well work with lines parallel to these that do intersect at $(0, 0)$, recalling that parallel lines have the same slope.] The proof relies on the following geometric fact:

$$\overline{OA} \perp \overline{OB} \qquad \text{if and only if} \qquad OA^2 + OB^2 = AB^2$$

(a) Verify that the coordinates of A and B are $A(1, m_1)$ and $B(1, m_2)$.

(b) Show that

$$OA^2 = 1 + m_1^2$$
$$OB^2 = 1 + m_2^2$$
$$AB^2 = m_1^2 - 2m_1m_2 + m_2^2$$

(c) Use part (b) to show that the equation $OA^2 + OB^2 = AB^2$ is equivalent to $m_1m_2 = -1$.

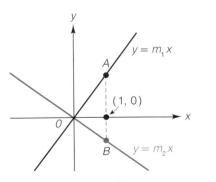

63. Verify the identity $(y_2 - y_1)/(x_2 - x_1) = (y_1 - y_2)/(x_1 - x_2)$. What does this identity tell you about calculating slope?

64. The following figure shows two tangent lines drawn from the point (a, b) to the circle $x^2 + y^2 = R^2$. Follow steps (a)–(d) to show that the equation of the line passing through the two points of tangency is

$$ax + by = R^2$$

(This problem, along with the following clever solution, appears in the classic text by Isaac Todhunter, *A Treatise on Plane Co-ordinate Geometry*, first published in 1855.)

(a) Let (x_1, y_1) be one of the points of tangency. Show that the equation of the tangent line through this point is $x_1 x + y_1 y = R^2$.

(b) Using the result in part (a), explain why $x_1 a + y_1 b = R^2$.

(c) In a similar fashion, explain why $x_2 a + y_2 b = R^2$, where (x_2, y_2) is the other point of tangency.

(d) The equation $ax + by = R^2$ represents a line. Explain why this line must pass through (x_1, y_1) and (x_2, y_2).

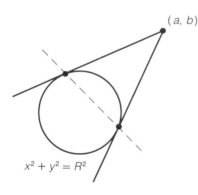

$x^2 + y^2 = R^2$

(a, b)

C

65. Given two lines L_1 and L_2 passing through the origin as shown, a third line L is said to bisect the area between L_1 and L_2 provided that for each point P on L the area of region A equals the area of region B. (In the following figure, each dashed line is parallel to one of the coordinate axes.)

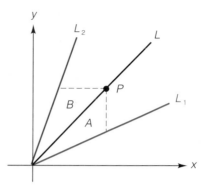

(a) Suppose that the equation of L_1 is $y = \frac{1}{2}x$ and the equation of L_2 is $y = 8x$. Find the equation of L.

(b) Suppose that the equations of L_1 and L_2 are $y = nx$ and $y = mx$, respectively. Find the equation of L.

66. The vertices of a triangle are $A(-4, 0)$, $B(2, 0)$, and $C(0, 6)$. Let M_1, M_2, and M_3 be the midpoints of $\overline{AB}$, $\overline{BC}$, and $\overline{AC}$, respectively. Let H_1, H_2, and H_3 be the feet of the altitudes on sides $\overline{AB}$, $\overline{BC}$, and $\overline{AC}$, respectively.

(a) Find the equation of the circle passing through M_1, M_2, and M_3.　　*Answer:* $3x^2 + 3y^2 + 3x - 11y = 0$

(b) Find the equation of the circle passing through H_1, H_2, and H_3.　　*Answer:* $3x^2 + 3y^2 + 3x - 11y = 0$

(c) Find the point P at which the three altitudes intersect.　　*Answer:* $\left(0, \frac{4}{3}\right)$

(d) Let N_1, N_2, and N_3 be the midpoints of $\overline{AP}$, $\overline{BP}$, and $\overline{CP}$, respectively. Show that the circle obtained in parts (a) and (b) passes through N_1, N_2, and N_3. *Note:* This circle is called the *nine-point circle* of $\triangle ABC$.

(e) The *circumcircle* of $\triangle ABC$ is the circle passing through the three points A, B, and C. Show that the coordinates of the center Q of the circumcircle are $Q\left(-1, \frac{7}{3}\right)$. *Hint:* Use the result from geometry that the perpendicular bisectors of the sides of the triangle intersect at the center of the circumcircle.

(f) For $\triangle ABC$, show that the radius of the nine-point circle is one-half the radius of the circumcircle.

(g) Show that the midpoint of line segment $\overline{QP}$ is the center of the nine-point circle.

(h) Find the point G where the three medians of $\triangle ABC$ intersect.

(i) Show that G lies on line segment $\overline{QP}$ and that $PG = 2(GQ)$.

OPTIONAL TI-81 GRAPHING CALCULATOR EXERCISES

EXERCISES FOR SECTION 2.3

1. (a) What are the equations of the four lines passing through the origin with slopes 1, 2, 3, and 10? Enter these equations in the $Y=$ menu and graph the four lines. As the slopes increase, what happens to the graphs?
   (b) What are the equations of the four lines passing through the origin with slopes -1, -2, -3, and -10? Graph these four lines. As the slopes decrease from -1 to -10, what happens to the graphs?

2. (a) The following four lines are parallel: $y = 2x - 1$, $y = 2x$, $y = 2x + 1$, $y = 2x + 2$. Without drawing the graphs, how do you know this to be true?
   (b) On the same set of axes, graph the four lines in part (a).

3. Graph the following set of parallel lines on the same set of axes: $y - 1 = -0.5(x - 2)$; $y - 1 = -0.5(x + 2)$; $y - 1 = -0.5(x - 4)$.

4. In each case, graph the pair of perpendicular lines on the same set of axes: (a) $y = 2x$ and $y = -\frac{1}{2}x$; (b) $y = \frac{8}{3}x$ and $y = -\frac{3}{8}x$; (c) $y = 3x - 4$ and $y = -\frac{1}{3}x - 4$.

5. In parts (a) through (d), solve each equation for y and then graph the line. In each case, make a note of the x- and y-intercepts.

   (a) $\dfrac{x}{2} + \dfrac{y}{3} = 1$ (b) $\dfrac{x}{-2} + \dfrac{y}{-3} = 1$

   (c) $\dfrac{x}{6} + \dfrac{y}{5} = 1$ (d) $\dfrac{x}{-6} + \dfrac{y}{-5} = 1$

   (e) Based on your results in parts (a) through (d), describe, in general, the graph of $\dfrac{x}{a} + \dfrac{y}{b} = 1$.

6. In this exercise, you'll draw Figure 16 on page 76. (As background, you should first reread Example 9 on page 76.) Press the $Y=$ key and delete any functions that are present. Then enter the following three functions: $Y_1 = \sqrt{25 - x^2}$; $Y_2 = -\sqrt{25 - x^2}$; $Y_3 = \frac{3}{4}x - \frac{25}{4}$. Now press ZOOM 6, ZOOM 5, and check that your result agrees with Figure 16.

For Exercises 7 and 8, you'll need to recall the following definitions and results from elementary geometry. In a triangle, a line segment drawn from a vertex to the midpoint of the opposite side is called a **median**. *The three medians in any triangle are* **concurrent;** *that is, they intersect in a single point. This point of intersection is called the* **centroid** *of the triangle. A line segment drawn from a vertex perpendicular to the opposite side is called an* **altitude**. *The three altitudes of a triangle are concurrent; the point where the altitudes intersect is called the* **orthocenter** *of the triangle.*

7. This exercise provides an example of the fact that the medians of a triangle are concurrent. As preparation,

press the $Y=$ key and clear any functions that are present. Press the RANGE key and set $x_{min} = -4$, $x_{max} = 2$, $y_{min} = 0$, and $y_{max} = 6$. Press ENTER.
   (a) Using the DRAW technique explained in Exercise 10 [parts (c) and (d)] on page 59, draw $\triangle ABC$ with vertices $A(-4, 0)$, $B(2, 0)$, and $C(0, 6)$. (*Note:* Since side $\overline{AB}$ coincides with the x-axis, you won't need to draw a line segment for this side.)
   (b) What are the coordinates of the midpoints of each side of the triangle? Use the DRAW technique to add the three medians to your picture. Note that the medians do appear to intersect in a single point. Use your graph to estimate (to the nearest integers) the coordinates of the centroid.
   (c) Using the techniques of Section 2.3 (and not your calculator), find the equations of the three medians.
   Answers: median from A to side $\overline{BC}$: $y = \frac{3}{5}x + \frac{12}{5}$; median from B to side $\overline{AC}$: $y = -\frac{3}{4}x + \frac{3}{2}$; median from C to side $\overline{AB}$: $y = 6x + 6$.
   Now (using simultaneous equations, from intermediate algebra) find the exact coordinates of the centroid. How do these results compare with your estimates in part (b)?

8. This exercise illustrates the fact that the altitudes of a triangle are concurrent. Retain the RANGE settings from the previous exercise. Again we'll be using $\triangle ABC$ with vertices $A(-4, 0)$, $B(2, 0)$, and $C(0, 6)$. (Note that one altitude of this triangle is just the portion of the y-axis extending from $y = 0$ to $y = 6$. We won't have to draw or graph this altitude; it will already be in the picture.)
   (a) Using the techniques of Section 2.3 (and not your calculator), find the equations of the three altitudes.
   Answers: altitude from A to side $\overline{BC}$: $y = \frac{1}{3}x + \frac{4}{3}$ altitude from B to side $\overline{AC}$: $y = y = \frac{-2}{3}x + \frac{4}{3}$ altitude from C to side $\overline{AB}$: $x = 0$.
   (b) Press the $Y=$ key and enter the altitude equations $Y_1 = (x + 4)/3$ and $Y_2 = (-2x + 4)/3$. Press GRAPH. Your figure will show the altitudes, but not $\triangle ABC$ itself. (When one uses the $Y=$ key to graph functions, the calculator screen deletes lines or curves drawn previously using the DRAW command.) Go on to part (c).
   (c) Press "2nd," DRAW, 1, ENTER, and CLEAR. Now we are ready to draw in the sides of $\triangle ABC$. Press "2nd," DRAW, 2, and then just as you did in the previous exercise, obtain the line segments $\overline{AC}$ and $\overline{BC}$. Note that the altitudes do appear to intersect in a single point. Use your graph to estimate the coordinates of the orthocenter.
   (d) Now (using simultaneous equations, from intermediate algebra) find the exact coordinates of the orthocenter. How do these results compare with your estimates in part (b)?

2.4 SYMMETRY AND GRAPHS

Symmetry is a working concept. If all the object is symmetrical, then the parts must be halves (or some other rational fraction) and the amount of information necessary to describe the object is halved (etc.).

Alan L. Macay, Department of Crystallography, University of London

As you will see throughout this text, there are a number of techniques that help us to understand the essential features of a graph. In the box that follows, we introduce three types of *symmetry* that are useful in graphing equations.

Three Types of Symmetry

TYPE OF SYMMETRY

1. Symmetry about the *x*-axis

EXAMPLE

FIGURE 1
Symmetry about the *x*-axis

DEFINITION

For each point (x, y) on the graph, the point $(x, -y)$ is also on the graph. The points (x, y) and $(x, -y)$ are *reflections of one another about* (or *in*) *the x-axis*. In Figure 1, the portions of the graph above and below the *x*-axis also are said to be reflections of one another about the *x*-axis.

2. Symmetry about the *y*-axis

FIGURE 2
Symmetry about the *y*-axis

For each point (x, y) on the graph, the point $(-x, y)$ is also on the graph. The points (x, y) and $(-x, y)$ are *reflections of one another about* (or *in*) *the y-axis*. In Figure 2, the portions of the graph to the right and left of the *y*-axis also are said to be reflections of one another about the *y*-axis.

3. Symmetry about the origin

FIGURE 3
Symmetry about the origin

For each point (x, y) on the graph, the point $(-x, -y)$ is also on the graph. The points (x, y) and $(-x, -y)$ are *reflections of one another about the origin*. In Figure 3, the first- and third-quadrant portions of the curve also are said to be reflections of one another about the origin. In terms of the two previous symmetries, the point $(-x, -y)$ can be obtained from (x, y) as follows: first reflect (x, y) about the *y*-axis to obtain $(-x, y)$, then reflect $(-x, y)$ about the *x*-axis to obtain $(-x, -y)$.

▮▮▮ **EXAMPLE I** A line segment $\mathscr{L}$ has endpoints $(1, 2)$ and $(5, 3)$. Sketch the reflection of $\mathscr{L}$ about: **(a)** the x-axis;   **(b)** the y-axis;   **(c)** the origin.

Solution

(a) First reflect the endpoints $(1, 2)$ and $(5, 3)$ about the x-axis to obtain the new endpoints $(1, -2)$ and $(5, -3)$, respectively. Then join these two points as indicated in Figure 4(a), in quadrant IV.

(b) Reflect the given endpoints about the y-axis to obtain the new endpoints $(-1, 2)$ and $(-5, 3)$; then join these two points as shown in Figure 4(a), in quadrant II.

(c) As described in the box preceding this example, reflection about the origin can be carried out in two steps: first reflect about the y-axis, then reflect about the x-axis. In part (b) we obtained the reflection of $\mathscr{L}$ in the y-axis. So now we need only reflect the line segment obtained in part (b) about the x-axis. See Figure 4(b). ▮▮▮

FIGURE 4

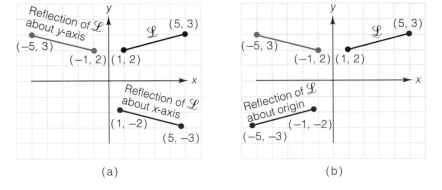

(a)   (b)

In the box that follows we list three rules for testing whether the graph of an equation possesses any of the types of symmetry we've been discussing. (The validity of each rule follows directly from the definitions of symmetry.)

Three Tests for Symmetry

1. The graph of an equation is symmetric about the y-axis if replacing x with $-x$ yields an equivalent equation.

2. The graph of an equation is symmetric about the x-axis if replacing y with $-y$ yields an equivalent equation.

3. The graph of an equation is symmetric about the origin if replacing x and y with $-x$ and $-y$, respectively, yields an equivalent equation.

▮▮▮ **EXAMPLE 2**

(a) Test for symmetry about the y-axis: $y = x^4 - 3x^2 + 1$.

(b) Test for symmetry about the x-axis: $x = \dfrac{y^2}{y - 1}$.

Solution

(a) Replacing x with $-x$ in the original equation gives us

$$y = (-x)^4 - 3(-x)^2 + 1 \quad \text{or} \quad y = x^4 - 3x^2 + 1$$

Since this last equation is the same as the original equation, we conclude that the graph is symmetric about the y-axis.

(b) Replacing y with $-y$ in the original equation gives us

$$x = \frac{(-y)^2}{(-y) - 1} \quad \text{or} \quad x = \frac{y^2}{-y - 1}$$

Since this last equation is not equivalent to the original equation, we conclude that the graph is not symmetric about the x-axis. ∎

EXAMPLE 3 Is the graph of $y = 2x^3 - x$ symmetric about the origin?

Solution Replacing x and y with $-x$ and $-y$, respectively, gives us

$$-y = 2(-x)^3 - (-x) = -2x^3 + x$$
$$y = 2x^3 - x \qquad \text{multiplying through by } -1$$

This last equation is identical to the given equation. Therefore the graph is symmetric about the origin. ∎

In the next two examples we use the notions of symmetry and intercepts as guides for drawing the required graphs.

EXAMPLE 4 Graph the equation: $y = -x^2 + 5$.

Solution The domain of the variable x in the expression $-x^2 + 5$ is the set of all real numbers. However, since the graph must be symmetric about the y-axis (why?), we need only to be able to sketch the graph to the right of the y-axis; the portion to the left will then be the mirror image of this (with the y-axis as the mirror). The x- and y-intercepts (if any) are computed as follows:

x-INTERCEPTS	y-INTERCEPTS
$-x^2 + 5 = 0$	$y = -(0)^2 + 5$
$x^2 = 5$	$y = 5$
$x = \pm\sqrt{5} \ (\approx \pm 2.2)$	

In Figure 5(a) (on the next page) we've set up a short table of values and sketched the graph for $x \geq 0$. The complete graph is then obtained by reflection about the y-axis, as shown in Figure 5(b). [The curve in Figure 5(b) is a *parabola*. This type of curve will be considered in detail in later chapters.] ∎

EXAMPLE 5 Graph: $y = -4/x$.

Solution Before doing any calculations, note that the domain of the variable x consists of all real numbers other than zero. So, however the resulting graph may look, it cannot contain a point with an x-coordinate of zero. In other words, the graph cannot cross the y-axis. (Check for yourself that the graph cannot cross the x-axis either.) Now, since the graph is symmetric about the origin (why?), we need only to be able to sketch the graph for $x > 0$; the portion cor-

FIGURE 5

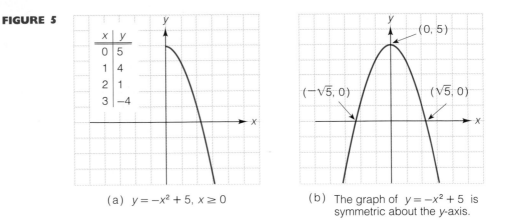

(a) $y = -x^2 + 5,\ x \geq 0$

(b) The graph of $y = -x^2 + 5$ is symmetric about the y-axis.

responding to $x < 0$ can then be obtained by reflection about the origin. In Figure 6(a) we've set up a table of values and sketched the graph for $x > 0$. (Note the fractional x-values at the end of the table. Since y is undefined when x is zero, it's informative to pick x-values near zero and look at the corresponding y-values.) In Figure 6(b) we have reflected the first-quadrant portion of the graph first about the y-axis and then about the x-axis to obtain the required reflection about the origin. Figure 7 shows our final graph of $y = -4/x$.   ▮▮▮

We can use the graph in Figure 6(a) to introduce the idea of an *asymptote*. A line is an **asymptote** for a curve if the distance between the line and the curve approaches zero as we move out farther and farther along the line. So, for the curve in Figure 6(a), both the x-axis and the y-axis are asymptotes. (We'll return to this idea several times later in the text. For other pictures of asymptotes, see Figure 2 in Section 4.6.)

Symmetry about the x-axis and symmetry about the y-axis are both examples of *symmetry about a line*. In these two cases, the axis of symmetry is one of the coordinate axes. Now we'll look at another example of this type of symme-

FIGURE 6

x	1	2	3	4	5	6	7	8	$\frac{1}{2}$	$\frac{1}{3}$
y	-4	-2	$-\frac{4}{3}$	-1	$-\frac{4}{5}$	$-\frac{2}{3}$	$-\frac{4}{7}$	$-\frac{1}{2}$	-8	-12

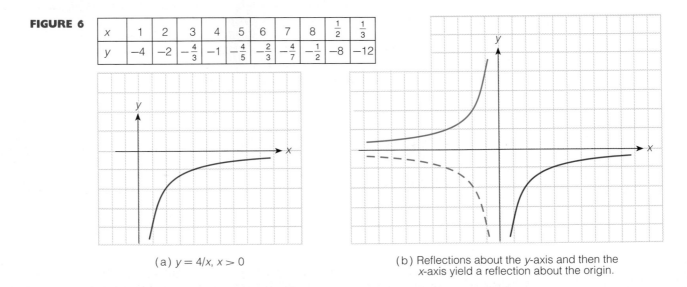

(a) $y = 4/x,\ x > 0$

(b) Reflections about the y-axis and then the x-axis yield a reflection about the origin.

FIGURE 7

The graph of $y = -4/x$ is symmetric about the origin.

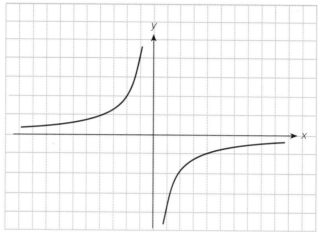

try: symmetry about the line $y = x$. This kind of symmetry will arise when we analyze inverse functions in the next chapter.

DEFINITION   Symmetry About the Line y = x

Refer to Figure 8. Two points P and Q are symmetric about the line $y = x$ provided that

1. $\overline{PQ}$ is perpendicular to the line $y = x$; and

2. the points P and Q are equidistant from the line $y = x$.

FIGURE 8

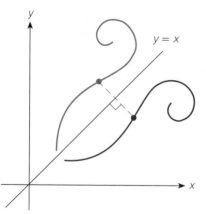

FIGURE 9

The two curves are symmetric about the line $y = x$.

This definition says that P and Q are symmetric about the line $y = x$ if $y = x$ is the perpendicular bisector of line segment $\overline{PQ}$. In Figure 8, we say that two points P and Q are **reflections** of each other about the line $y = x$ and that $y = x$ is the **axis of symmetry**. In addition, we say that two curves are symmetric about $y = x$ if each point on one curve is the reflection of a corresponding point on the other curve, and vice versa. See Figure 9.

EXAMPLE 6   Verify that the points $P(4, 1)$ and $Q(1, 4)$ are symmetric about the line $y = x$. See Figure 10.

Solution   We have to show that the line $y = x$ is the perpendicular bisector of the line segment $\overline{PQ}$. Now, the slope of $\overline{PQ}$ is

$$m = \frac{4 - 1}{1 - 4} = -1$$

FIGURE 10

We know that the slope of the line $y = x$ is 1. Since these two slopes are negative reciprocals, we conclude that the line $y = x$ is perpendicular to $\overline{PQ}$. Next, we must show that the line $y = x$ passes through the midpoint of $\overline{PQ}$. But (as you should check for yourself), the midpoint of $\overline{PQ}$ is $\left(\frac{5}{2}, \frac{5}{2}\right)$, which does lie on the line $y = x$. In summary, then, we've shown that the line $y = x$ passes through the midpoint of $\overline{PQ}$ and is perpendicular to $\overline{PQ}$. Thus P and Q are symmetric about the line $y = x$, as we wished to show. ▮▮▮

Just as we've shown that $(4, 1)$ and $(1, 4)$ are symmetric about the line $y = x$, we can also show in general that (a, b) and (b, a) are always symmetric about the line $y = x$. Figure 11 displays two examples of this, and Exercise 31 at the end of this section asks you to supply a proof for the general situation.

FIGURE 11
Two examples of the fact that (a, b) and (b, a) are symmetric about $y = x$.

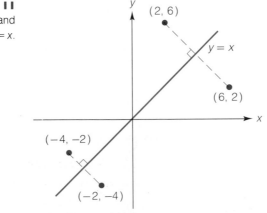

▮▮▮ **EXAMPLE 7** A line segment $\mathscr{L}$ has endpoints $(-4, 1)$ and $(-1, 2)$. Sketch the line segment that results from **(a)** a reflection in the line $y = x$; **(b)** a reflection in the line $y = x$ followed by a reflection in the y-axis; **(c)** a reflection in the y-axis followed by a reflection in the line $y = x$. Are the results in parts (b) and (c) the same?

Solution

(a) First reflect the endpoints $(-4, 1)$ and $(-1, 2)$ in the line $y = x$ to obtain the new endpoints $(1, -4)$ and $(2, -1)$, respectively. Then join these new endpoints as shown in Figure 12(a).

FIGURE 12

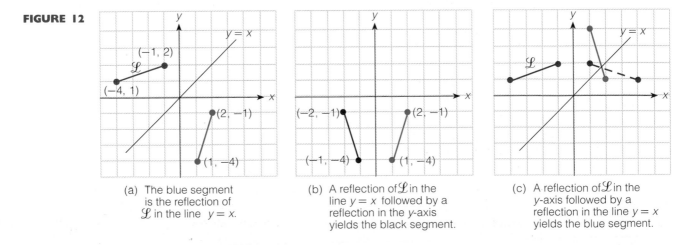

(a) The blue segment is the reflection of $\mathscr{L}$ in the line $y = x$.

(b) A reflection of $\mathscr{L}$ in the line $y = x$ followed by a reflection in the y-axis yields the black segment.

(c) A reflection of $\mathscr{L}$ in the y-axis followed by a reflection in the line $y = x$ yields the blue segment.

(b) In Figure 12(a) we found the reflection of $\mathcal{L}$ in the line $y = x$. Reflecting this line segment in the y-axis yields the result shown in Figure 12(b).

(c) In Figure 12(c), we've reflected $\mathcal{L}$ first in the y-axis and then in the line $y = x$. Note that the results here and in part (b) are not the same. The order in which we carry out these operations makes a difference. ▪▪▪

EXERCISE SET 2.4

A

In Exercises 1–6, the endpoints of a line segment $\overline{AB}$ are given. Sketch the reflection of $\overline{AB}$ about (a) *the x-axis;* (b) *the y-axis;* (c) *the origin;* (d) *the line $y = x$.*

1. $A(1, 4)$ and $B(3, 1)$
2. $A(-1, -2)$ and $B(-5, -2)$
3. $A(-2, -3)$ and $B(2, -1)$
4. $A(-3, -3)$ and $B(-3, -1)$
5. $A(0, 1)$ and $B(3, 1)$
6. $A(-2, -2)$ and $B(0, 0)$

In Exercises 7 and 8, fill in the entries in the tables with "yes" or "no" to indicate whether the graph of the equation possesses the given symmetry.

7.

	Symmetric about the x-axis	Symmetric about the y-axis	Symmetric about the origin
$y = x^2$			
$y = x^3$			
$y = \sqrt{x}$			

8.

	Symmetric about the x-axis	Symmetric about the y-axis	Symmetric about the origin
$y = |x|$			
$y = 1/x$			
$y = \sqrt{1 - x^2}$			

In Exercises 9–26, graph the equation after determining the x- and y-intercepts and whether the graph possesses any of the three types of symmetry discussed in this section.

9. $y = 4 - x^2$
10. $y = -x^3$
11. $y = -1/x$
12. $x = y^2 - 1$
13. $y = -x^2$
14. $y = 1/x^2$
15. $y = -1/x^3$
16. $y = |x| - 2$
17. $y = \sqrt{x^2}$
18. $y = x + 1$
19. $y = x^2 - 2x + 1$
20. $x = y^3 - 1$
21. $y^2 = 2x - 4$
22. $|y| = 2x - 4$
23. $y = 2x^2 + x - 4$
24. $y = 2x^2$
25. (a) $x + y = 2$
    (b) $|x| + y = 2$
26. (a) $x + |y| = 2$
    (b) $|x + y| = 2$

B

27. The accompanying figure shows the graphs of $y = \frac{3}{4}x - 2$ and $y = \left|\frac{3}{4}x - 2\right|$.

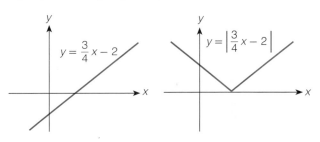

(a) Determine the x- and y-intercepts for each graph.
(b) Which portions of the two graphs are identical? (Give your answer in terms of an interval along the x-axis.)
(c) Explain how the graph of $y = \left|\frac{3}{4}x - 2\right|$ can be obtained from that of $y = \frac{3}{4}x - 2$ by means of reflection.

28. The accompanying figure shows the graphs of the two equations $y = x^2 - 6x + 8$ and $y = |x^2 - 6x + 8|$.

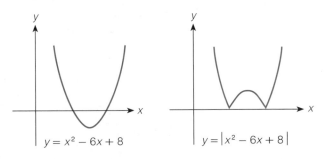

(a) Determine the x- and y-intercepts for each graph.

(b) Which portions of the two graphs are identical? (Give your answer in terms of intervals along the x-axis.)

(c) Explain how the graph of $y = |x^2 - 6x + 8|$ can be obtained from that of $y = x^2 - 6x + 8$ by means of reflection.

29. Reread the definition of symmetry about the line $y = x$ given on page 87. Then use the method shown in Example 6 to show that the points $(7, -1)$ and $(-1, 7)$ are symmetric about the line $y = x$.

30. Use the method shown in Example 6 to show that $(5, 2)$ and $(2, 5)$ are symmetric about $y = x$.

31. Use the method of Example 6 to show that the points (a, b) and (b, a) are symmetric about the line $y = x$.

32. Points P and Q are reflected in the line $y = x$ to obtain points P' and Q', respectively. Does the distance from P to Q equal the distance from P' to Q'?

33. Pick any three points on the line $y = \frac{1}{2}x + 3$. Reflect each point in the line $y = x$. Show that the three reflected points all lie on one line. What is the equation of that line?

34. Determine the intercepts and symmetry for the equation $y = x^{2/3}$, and then graph the equation. (Use a calculator.)

35. Find the intercepts and symmetry for $y = x^3 - 27x$. Then use a calculator (as necessary) to set up a table of x- and y-values, with x running from 0 to 6 in increments of $\frac{1}{2}$. Finally, graph the equation.

36. Consider the equation $y = 4x^2/(1 + x^2)$.

(a) Use a calculator to set up a table with x-values running from 0 to 5 in increments of $\frac{1}{2}$.

(b) Graph the equation using your table of values and symmetry.

C

In Exercises 37–40, use the following definition of symmetry about a line: Two points P and Q are symmetric about a line $\mathscr{L}$ provided that

1. $\overline{PQ}$ is perpendicular to $\mathscr{L}$; and
2. points P and Q are equidistant from $\mathscr{L}$.

In other words, P and Q are symmetric about the line $\mathscr{L}$ if $\mathscr{L}$ is the perpendicular bisector of line segment $\overline{PQ}$. We say that $\mathscr{L}$ is the axis of symmetry and that P and Q are reflections of each other through $\mathscr{L}$.

37. Determine values for m and b so that the points $(8, 2)$ and $(4, 8)$ are symmetric about the line $y = mx + b$.

38. Reflect the point $(2, 1)$ in the line $y = 3x$. What are the coordinates of the point you obtain?

39. (a) If the point (a, b) is reflected in the line $y = 3x$, show that the coordinates of the reflected point are
$$\left(\frac{3b - 4a}{5}, \frac{3a + 4b}{5} \right).$$

(b) If the point (a, b) is reflected in the line $y = mx$, what are the coordinates of the reflected point?

40. (a) Find the reflection of the point (a, b) in the line $y = -x$.   *Answer:* $(-b, -a)$

(b) Begin with the point (a, b). Reflect this point about the x-axis to obtain a point Q. Then reflect Q about the y-axis to obtain a point R. Then reflect R about the line $y = x$ to obtain a point S. Show that S is the reflection of (a, b) in the line $y = -x$.

OPTIONAL TI-81 GRAPHING CALCULATOR EXERCISES
EXERCISES FOR SECTION 2.4

In Exercises 1–10, graph the equation. Does the graph appear to possess any of the three types of symmetry defined on page 83? In cases where your answer is "yes," use the symmetry tests on page 84 to verify your answer. Note: Exercises 1 through 4, 9, and 10 contain detailed suggestions for the settings in the RANGE menu. In Exercises 5–8, you are on your own; try to use settings that show the essential features of the graph.

1. $y = x^2 - 3x$ *Suggestions:* After entering the given function in the $Y=$ menu, press the ZOOM key and then the "6" key to see the graph in the *standard viewing rectangle*. (Recall that the standard viewing rectangle extends from -10 to $+10$ in both the x- and

the y-directions.) Next, for a better view of the graph, press the RANGE key and make the following adjustments:
$x_{\min} = -2$, $x_{\max} = 5$, and $y_{\min} = -4$. (Retain the initial setting $y_{\max} = 10$.) Now, press the GRAPH key.

2. $y = x^3 - 3x$ *Suggestions:* Press the $Y=$ key. Rather than typing in the entire expression $x^3 - 3x$, it's easier just to edit the expression that you entered in Exercise 1. Using the arrow keys, move the cursor to the location of the exponent "2" that you entered in Exercise 1, and then type "3". Now press the ZOOM key and then the "6" key to see the graph in the

standard viewing rectangle. Finally, for a better view, press the RANGE key and set $x_{min} = -3$ and $x_{max} = 3$. Then press the GRAPH key. (Additional question: As the graph shows, one of the x-intercepts is located between $x = 1$ and $x = 2$. What is the *exact* value for that intercept?)

3. $y = 2^x$ *Suggestions:*   After entering the given function in the Y= menu, press the ZOOM key and then the "6" key to see the graph in the standard viewing rectangle. Next, for a better view, press the RANGE key and make the following adjustments: $x_{min} = -3$, $x_{max} = 3$, $y_{min} = 0$, $y_{max} = 8$. Now press the GRAPH key.

4. (a) $y = 2^x + 2^{-x}$ *Suggestions:*   First look at the graph in the standard viewing rectangle. Then make the following adjustments in the RANGE menu: $x_{min} = -3$, $x_{max} = 3$, $y_{min} = 0$, $y_{max} = 9$.
   (b) $y = 2^x - 2^{-x}$ *Suggestion:*   Use the standard viewing rectangle.

5. $y = \sqrt{|x|}$

6. $y = \sqrt{|x|^3}$

7. $y = 2x - x^3 - x^5 + x^7$

8. $y = |2x - x^3 - x^5 + x^7|$

9. $y = 1/(x^2 - x)$ *Hint:*   In the standard viewing rectangle (obtained by pressing the ZOOM key and then the "6" key, the graph is difficult to analyze because of the scale that is involved. For a better view, press the RANGE key and set $x_{min} = -2$ and $x_{max} = 2$. (Then press the GRAPH key.)

10. $y = 1/(x^3 - x)$ *Hint:*   Use the RANGE settings given in Exercise 9.

11. (a) Using the Y= key, enter the function $Y_1 = x^2 - 2x^3$. Next, press the ZOOM key and then the "6" key to obtain the graph of this equation in the standard viewing rectangle. Which type of symmetry does the graph appear to possess?
    (b) The graph in part (a) is deceiving because of the scale that is involved. For a different view, press the RANGE key and set $x_{min} = -2$ and $x_{max} = 2$. (Then press the GRAPH key.) Note that the graph no longer appears to be symmetric about the origin.
    (c) Use a symmetry test from page 84 to confirm that the graph is not symmetric about the origin.

CHAPTER TWO   SUMMARY OF PRINCIPAL TERMS AND FORMULAS

TERMS OR FORMULAS	PAGE REFERENCE	COMMENTS
1. Pythagorean theorem $a^2 + b^2 = c^2$	50	In a right triangle, the lengths of the sides are related by this equation, where c is the length of the hypotenuse. Conversely, if the lengths of the sides of a triangle are related by an equation of the form $a^2 + b^2 = c^2$, then the triangle is a right triangle, and c is the length of the hypotenuse.
2. $d = \sqrt{(x_2 - x_1)^2 + (y_2 - y_1)^2}$	50	d is the distance between (x_1, y_1) and (x_2, y_2).
3. $(x - h)^2 + (y - k)^2 = r^2$	52	This is the equation of a circle with center at (h, k) and radius r.
4. The unit circle	53	This refers to the circle with center $(0, 0)$ and radius 1. The equation of the unit circle is $x^2 + y^2 = 1$.
5. $\left(\dfrac{x_1 + x_2}{2}, \dfrac{y_1 + y_2}{2}\right)$	53	This is the midpoint of the line segment joining (x_1, y_1) and (x_2, y_2).
6. Graph of an equation	60	The graph of an equation in the variables x and y is the set of all points (x, y) with coordinates that satisfy the equation.
7. x-intercept and y-intercept	61	An x-intercept of a graph is the x-coordinate of a point where the graph intersects the x-axis. Similarly, a y-intercept is the y-coordinate of a point where the graph intersects the y-axis.
8. $m = \dfrac{y_2 - y_1}{x_2 - x_1}$	70	m is the slope of a nonvertical line passing through the points (x_1, y_1) and (x_2, y_2).

TERMS OR FORMULAS	PAGE REFERENCE	COMMENTS
9. $y - y_1 = m(x - x_1)$	72	This is the point–slope form for the equation of a line passing through the point (x_1, y_1) with slope m.
10. $y = mx + b$	74	This is the slope–intercept form for the equation of a line with slope m and y-intercept b.
11. $m_1 = m_2$	75	Nonvertical parallel lines have the same slope.
12. $m_1 = -1/m_2$	75	Two nonvertical lines are perpendicular if and only if the slopes are negative reciprocals of one another.
13. Symmetry about the x-axis	83	A graph is symmetric about the x-axis if, for each point (x, y) on the graph, the point $(x, -y)$ is also on the graph. The points (x, y) and $(x, -y)$ are *reflections* of each other about (or in) the x-axis.
14. Symmetry about the y-axis	83	A graph is symmetric about the y-axis if, for each point (x, y) on the graph, the point $(-x, y)$ is also on the graph. The points (x, y) and $(-x, y)$ are *reflections* of each other about (or in) the y-axis.
15. Symmetry about the origin	83	A graph is symmetric about the origin if, for each point (x, y) on the graph, the point $(-x, -y)$ is also on the graph. The points (x, y) and $(-x, -y)$ are *reflections* of each other about the origin.
16. Symmetry tests	84	(i) The graph of an equation is symmetric about the y-axis if replacing x with $-x$ yields an equivalent equation. (ii) The graph of an equation is symmetric about the x-axis if replacing y with $-y$ yields an equivalent equation. (iii) The graph of an equation is symmetric about the origin if replacing x and y with $-x$ and $-y$, respectively, yields an equivalent equation.
17. Symmetry about the line $y = x$	87	Two points P and Q are symmetric about the line $y = x$ provided that $\overline{PQ}$ is perpendicular to $y = x$ and the points P and Q are equidistant from $y = x$. Two points of the form (a, b) and (b, a) are always symmetric about the line $y = x$, and we say that these two points are *reflections* of each other in the line $y = x$.

WRITING MATHEMATICS

1. The following geometric method for solving quadratic equations of the form
    $x^2 - ax + b = 0$ is due to the British writer Thomas Carlyle (1795–1881):

    > To solve the equation $x^2 - ax + b = 0$, plot the points $A(0, 1)$ and $B(a, b)$.
    > Then draw the circle with $\overline{AB}$ as diameter. The x-intercepts of this circle
    > are the roots of the equation.

    Use the techniques of this chapter to verify for yourself that this method indeed
    yields the roots for the equation $x^2 - 6x + 5 = 0$. (You need to find the equation
    of the circle, determine the x-intercepts, and then check that the numbers you
    obtain are the roots of the given equation.) After you have done this, carefully
    write out your verification in detail (using complete sentences), as if you were
    explaining the method and why it works to a classmate. Be sure to make explicit
    reference to any formulas or equations that you use from the chapter.

2. In Section 2.3 we developed the point–slope formula, $y - y_1 = m(x - x_1)$, and the slope–intercept formula, $y = mx + b$. Although these two formulas look quite different from one another, both are merely restatements of the definition of slope. Write a paragraph (or two, at the most) to explain this last sentence. Make use of the following two figures.

 CHAPTER TWO   REVIEW EXERCISES

In Exercises 1–18, find an equation for the line satisfying the given conditions. For Exercises 1–9, write the answer in the form $y = mx + b$; in Exercises 10–18, write the answer in the form $Ax + By + C = 0$.

1. Passing through $(-4, 2)$ and $(-6, 6)$

2. $m = -2$; y-intercept 5

3. $m = \frac{1}{4}$; passing through $(-2, -3)$

4. $m = \frac{1}{3}$; x-intercept -1

5. x-intercept -4; y-intercept 8

6. $m = -10$; y-intercept 0

7. y-intercept -2; parallel to the x-axis

8. Passing through $(0, 0)$ and parallel to $6x - 3y = 5$

9. Passing through $(1, 2)$ and perpendicular to the line $x + y + 1 = 0$

10. Passing through $(1, 1)$ and through the center of the circle $x^2 - 4x + y^2 - 8y + 16 = 0$

11. Passing through the centers of the circles $x^2 + 4x + y^2 + 2y = 0$ and $x^2 - 4x + y^2 - 16y = 0$

12. $m = 3$; the same x-intercept as the line $3x - 8y = 12$

13. Passing through the origin and the midpoint of the line segment joining the points $(-2, -3)$ and $(6, -5)$

14. Tangent to the circle $x^2 + y^2 = 20$ at the point $(-2, 4)$.

15. Tangent to the circle $x^2 - 6x + y^2 + 8y = 0$ at the point $(0, 0)$

16. Passing through $(2, 4)$; the y-intercept is twice the x-intercept

17. Passing through $(2, -1)$; the sum of the x- and y-intercepts is 2 (There are two answers.)

18. Passing through $(4, 5)$; no x-intercept

19. Find the distance between the points $(-1, 2)$ and $(4, -10)$.

20. (a) Find the perimeter of the triangle with vertices $A(3, 1)$, $B(7, 4)$, and $C(-2, 13)$.
    (b) Find the perimeter of the triangle formed by joining the midpoints of the sides of the triangle in part (a).

21. Which point is farther from the origin, $(15, 6)$ or $(16, 2)$?

22. Find the midpoint of the line segment joining
    (a) $(4, 9)$ and $(10, -2)$;   (b) $\left(\frac{2}{3}, -1\right)$ and $\left(-\frac{3}{4}, -\frac{1}{3}\right)$.

In Exercises 23–32, test each equation for symmetry about the x-axis, the y-axis, and the origin.

23. $y = x^4 - 2x^2$

24. $y = x^4 - 2x^2 + 1$

25. $y = x^3 + 5x$

26. $y = x^3 + 5x + 1$

27. $y^2 = (x + y)^4$

28. $y = 2/(1 - x^2)$

29. $y = 3x - \dfrac{1}{x}$

30. $y = 2^x + 2^{-x}$

31. $y = 2^x - 2^{-x}$

32. $x^2 + y^2 + 2x = 2\sqrt{x^2 + y^2}$

For Exercises 33–40, tell whether each graph is symmetric about the x-axis, the y-axis, the origin, or the line $y = x$.

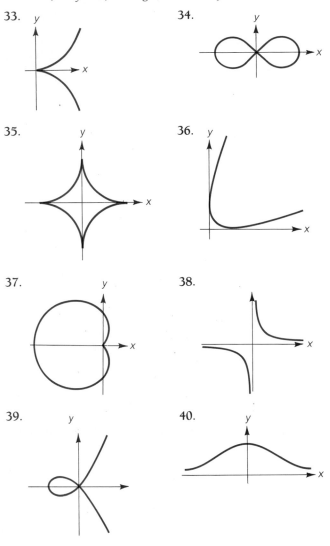

33.

34.

35.

36.

37.

38.

39.

40.

In Exercises 41–44, determine the coordinates of the point (or points) P satisfying the given conditions.

41. The point P lies on the line $y = -\frac{2}{3}x + 1$ and the distance from P to the origin is $\sqrt{10}$ units.

42. The point P lies on the line $y = 17 - x$ and the distance from P to the origin is 13 units.

43. The point P lies on the curve $y = x^2$ and the sum of the x- and y-coordinates of P is 2.

44. The point P lies on the curve $y = \sqrt{x}$ and the product of the x- and y-coordinates is $\frac{27}{64}$.

Graph the equations in Exercises 45–62.

45. $x = 9 - y^2$

46. $y = -2x + 6$

47. $x^2 + y^2 = 1$

48. $(x + 2)^2 + y^2 = 1$

49. $y = |x| + 1$

50. $y = 1 - |x|$

51. $y = x^2 - 4x + 4$

52. $y = -x$

53. $y = |x - 2| + 2$

54. $y = |x - 2| - 2$

55. $3x + y = 0$

56. $|x| + |y| = 1$

57. $x^2 - 4x + y^2 + 6y = 0$

58. $x^2 + 2x + y^2 + 8y = 8$

59. $y = \dfrac{x^2 - 16}{x - 4}$

60. $y = 1 + \dfrac{1}{x}$

61. $(4x - y + 4)(4x + y - 4) = 0$

62. $(y - x)(y - x + 2)(y - x - 2) = 0$

63. Find a value for t such that the slope of the line passing through $(2, 1)$ and $(5, t)$ is 6.

64. Express in terms of x the slope of the line passing through (x, x^2) and $(-3, 9)$.

65. Express in terms of x the slope of the line passing through $(-2, -8)$ and (x, x^3).

66. The vertices of a parallelogram are $(0, 0)$, $(5, 2)$, $(8, 7)$, and $(3, 5)$. Find the midpoint of each diagonal. Observation?

67. The vertices of a right triangle are $A(0, 0)$, $B(0, 2b)$, and $C(2c, 0)$. Let M be the midpoint of the hypotenuse. Compute the three distances MA, MB, and MC. What do you observe?

68. The vertices of parallelogram $ABCD$ are $A(-4, -1)$, $B(2, 1)$, $C(3, 3)$, and $D(-3, 1)$.
    (a) Compute the sum of the squares of the two diagonals.
    (b) Compute the sum of the squares of the four sides. What do you observe?

In Exercises 69–72, two points are given. In each case compute (a) the distance between the two points, (b) the slope of the line segment joining the two points, and (c) the midpoint of the line segment joining the two points.

69. $(2, 5)$ and $(5, -6)$

70. $\left(\frac{3}{2}, 1\right)$ and $\left(\frac{7}{2}, 9\right)$

71. $\left(\sqrt{3}/2, 1/2\right)$ and $\left(-\sqrt{3}/2, -1/2\right)$

72. $\left(\sqrt{3}, \sqrt{2}\right)$ and $\left(\sqrt{2}, \sqrt{3}\right)$

73. A line passes through the points $(1, 2)$ and $(4, 1)$. Find the area of the triangle bounded by this line and the coordinate axes.

74. Let $P_1(x_1, y_1)$ and $P_2(x_2, y_2)$ be two given points. Let Q be the point $\left(\frac{1}{3}x_1 + \frac{2}{3}x_2, \frac{1}{3}y_1 + \frac{2}{3}y_2\right)$
    (a) Show that the points P_1, Q, and P_2 are collinear. *Hint:* Compute the slope of $\overline{P_1 Q}$ and the slope of $\overline{QP_2}$.

(b) Show that $P_1Q = \frac{2}{3}P_1P_2$. (In other words, Q is on the line segment $\overline{P_1P_2}$ and two-thirds of the way from P_1 to P_2.)

75. (a) Let the vertices of $\triangle ABC$ be $A(-5,3)$, $B(7,7)$, and $C(3,1)$. Find the point on each median that is two-thirds of the way from the vertex to the midpoint of the opposite side. (Recall that a median of a triangle is a line segment drawn from a vertex to the midpoint of the opposite side.) What do you observe? *Hint:* Use the result in Exercise 74.

    (b) Follow part (a) but take the vertices to be $A(0,0)$, $B(2a,0)$, and $C(2b,2c)$. What do you observe? What does this prove?

76. Suppose that the circle $(x-h)^2 + (y-k)^2 = r^2$ has two x-intercepts, x_1 and x_2, and two y-intercepts, y_1 and y_2. Compute the quantity $x_1x_2 - y_1y_2$.

77. The point $(1, -2)$ is the midpoint of a chord of the circle $x^2 - 4x + y^2 + 2y = 15$. Find the length of the chord.

78. In the accompanying figure, points P and Q trisect the hypotenuse in $\triangle ABC$. Prove that the square of the hypotenuse is equal to $\frac{9}{5}$ the sum of the squares of the distances from the trisection points to the vertex A of the right angle. *Hint:* Let the coordinates of B and C be $(0, 3b)$ and $(3c, 0)$, respectively. Then the coordinates of P and Q are $(c, 2b)$ and $(2c, b)$, respectively.

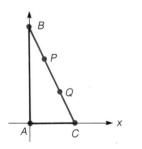

79. Figure A shows a triangle with sides of lengths s, t, and u and a median of length m. Prove that

$$m^2 = \frac{1}{2}(s^2 + t^2) - \frac{1}{4}u^2$$

Hint: Set up a coordinate system as indicated in Figure B. Then each of the quantities m^2, s^2, t^2, and u^2 can be computed in terms of a, b, and c.

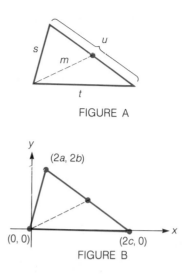

FIGURE A

FIGURE B

In Exercises 80 and 81, the endpoints of a line segment AB are given. Sketch the reflection of $\overline{AB}$ about (a) the line $y = x$; (b) the x-axis; (c) the y-axis; (d) the origin.

80. $A(-2, -3)$ and $B(0, 2)$

81. $A(3, 1)$ and $B(3, -2)$

82. In this exercise you'll derive a useful formula for the (perpendicular) distance d from the point (x_0, y_0) to the line $y = mx + b$. The formula is

$$d = \frac{|y_0 - mx_0 - b|}{\sqrt{1 + m^2}}$$

(a) Refer to the figure. Use similar triangles to show that

$$\frac{d}{AB} = \frac{1}{\sqrt{1 + m^2}}$$

Therefore, $d = AB/\sqrt{1 + m^2}$.

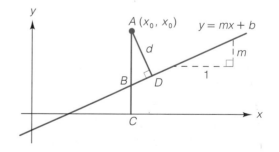

(b) Check that $AB = AC - BC = y_0 - mx_0 - b$.

(c) Conclude from parts (a) and (b) that

$$d = \frac{y_0 - mx_0 - b}{\sqrt{1 + m^2}}$$

For the general case (in which the point and line may not be situated as in our figure), we need to use the absolute value of the quantity in the numerator to assure that AB and d are nonnegative.

83. Use the formula given in Exercise 82 to find the distance from the point $(1, 2)$ to the line $y = \frac{1}{2}x - 5$.

84. Use the formula given in Exercise 82 to demonstrate that the distance from the point (x_0, y_0) to the line $Ax + By + C = 0$ is

$$d = \frac{|Ax_0 + By_0 + C|}{\sqrt{A^2 + B^2}}$$

85. Use the formula given in Exercise 84 to find the distance from the point $(-1, -3)$ to the line $2x + 3y - 6 = 0$.

86. Find the equation of the circle with center $(2, 3)$ that is tangent to the line $x + y - 1 = 0$. *Hint:* Use the formula given in Exercise 84 to find the radius.

87. In the figure, the circle is tangent to the x-axis, to the y-axis, and to the line $3x + 4y = 12$.
   (a) Find the equation of the circle. *Suggestion:* First decide what the relationships must be between h, k, and r in the equation $(x - h)^2 + (y - k)^2 = r^2$. Then find a way to apply the formula given in Exercise 82.
   (b) Let S, T, and U denote the points where the circle touches the x-axis, the line $3x + 4y = 12$, and the y-axis, respectively. Find the equation of the line through A and T; through B and U; through S and C.
   (c) Where do the line segments $\overline{AT}$ and $\overline{CS}$ intersect? Where do the line segments $\overline{AT}$ and $\overline{BU}$ intersect? What do you observe? *Remark:* The point determined here is called the *Gergonne point* of triangle ABC, so named in honor of its discoverer, French mathematician Joseph-Diez Gergonne (1771–1859).

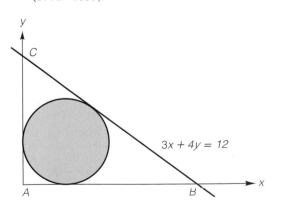

88. This exercise outlines a proof of the Pythagorean theorem that was discovered by James A. Garfield, twentieth president of the United States. Garfield published the proof in 1876, when he was Republican leader in the House of Representatives. We start with a right triangle with legs of length a and b and hypotenuse of length c. We want to prove that $a^2 + b^2 = c^2$.
   (a) Take two copies of the given triangle and arrange them as shown in Figure A. Explain why the angle marked θ is a right angle. (θ is the Greek letter "theta.")

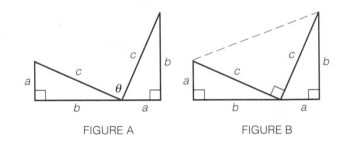

FIGURE A   FIGURE B

   (b) Draw the line segment indicated in Figure B. Notice that the outer quadrilateral in Figure B is a trapezoid. (Two sides are parallel.) The area of this trapezoid can be computed in two distinct ways: using the formula for the area of a trapezoid (given on the inside front cover of this book) or adding the areas of the three right triangles in Figure B. Compute the area of the trapezoid in each of these two ways. When you equate the two answers and simplify, you should obtain $a^2 + b^2 = c^2$, as required.

CHAPTER TWO TEST

In Problems 1–4, find an equation for the line satisfying the given conditions. Write the answer in the form $y = mx + b$.

1. Passing through $(1, -2)$ and $(3, 8)$

2. Passing through $(2, -1)$ and perpendicular to the line $5x + 6y = 30$

3. Passing through the origin and the midpoint of the line segment joining the points $(-5, -2)$ and $(3, 8)$

4. Tangent to the circle $x^2 - 8x + y^2 + 10y = 0$ at the origin

5. Which point is farther from the origin, $(3, 9)$ or $(5, 8)$?

6. Test each equation for symmetry about the x-axis, the y-axis, and the origin:
    (a) $y = x^3 + 5x$; (b) $y = 3^x + 3^{-x}$; (c) $y^2 = 5x^2 + x$.

In Problems 7–9, graph the equations and specify all x- and y-intercepts.

7. $3x - 5y = 15$

8. $2x^2 - 10x + 2y^2 + 6y = 17$

9. $y = |x + 1| + 2$

10. Find a value for t such that the slope of the line passing through $(-4, 3)$ and $(t, 13)$ is 2.

11. Find the point (or points) on the graph of the equation $y = x^2$ where the sum of the x- and y-coordinates is 12.

12. Does the point $\left(-\frac{1}{2}, 5\right)$ lie on the graph of the equation $y = 4x^2 - 8x$?

13. Sketch each graph.

    (a) $y = \dfrac{4x^2 - 1}{2x + 1}$ (b) $y = \begin{cases} \sqrt{1 - x^2} & \text{if } -1 \le x \le 0 \\ |x| & \text{if } 0 < x \le 2 \end{cases}$

    (c) $y = \begin{cases} \sqrt{x} & \text{if } 0 \le x \le 1 \\ 1/x & \text{if } 1 < x \le 4 \end{cases}$

14. The endpoints of line segment $\overline{AB}$ are $A(3, -1)$ and $B(-1, 2)$. Sketch the reflection of $\overline{AB}$ about (a) the x-axis; (b) the y-axis; (c) the origin; (d) the line $y = x$.

15. A line passes through the points $(6, 2)$ and $(1, 4)$. Find the area of the triangle bounded by this line and the x- and y-axes.

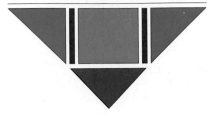

CHAPTER THREE

FUNCTIONS

From the beginning of modern mathematics in the 17th century the concept of function has been at the very center of mathematical thought.

Richard Courant and Fritz John in *Introduction to Calculus and Analysis* (New York: Wiley-Interscience, 1965)

The word "function" was introduced into mathematics by Leibniz, who used the term primarily to refer to certain kinds of mathematical formulas. It was later realized that Leibniz's idea of function was much too limited in scope, and the meaning of the word has since undergone many steps of generalization.

Tom M. Apostol in *Calculus*, 2nd ed. (New York: John Wiley & Sons, 1967)

INTRODUCTION

We begin the study of functions in this chapter. The first section deals with matters of definition and notation. In Sections 3.2 and 3.3 we discuss graphs of functions; several techniques are developed that allow us to graph functions with a minimum of calculation. In Sections 3.4 and 3.5 we find that two functions can be combined in various ways to produce new functions. One of these ways of combining two functions is known as *composition of functions*. As you'll see, this is the unifying theme between Sections 3.4 and 3.5.

3.1 THE DEFINITION OF A FUNCTION

There are numerous instances in mathematics and its applications in which one quantity corresponds to or depends on another according to some definite rule. Consider, for example, the equation $y = 3x - 2$. Each time that we select an x-value, a corresponding y-value is determined, in this case according to the rule *multiply by 3, then subtract 2*. In this sense, the equation $y = 3x - 2$ is an example of a *function*. It is a *rule* specifying a y-value corresponding to each x-value. It is useful to think of the x-values as inputs and the corresponding y-values as outputs. The function, or rule, then tells us what output results from a given input. This is indicated schematically in Figure 1. As another example, the area A of a circle depends on the radius r according to the rule or function $A = \pi r^2$. For each value of r there is a corresponding value for A obtained by the rule *square r and multiply the result by π*. In this case the inputs are the values of r and the outputs are the corresponding values of A.

For most of the functions studied in this text (and in beginning calculus), the inputs and outputs are real numbers and the function or rule is specified by means of an equation. This was the case with the two examples we have given; however, this need not always be the case. Consider, for example, the correspondences in Table 1, which indicate the number of moons each planet in the

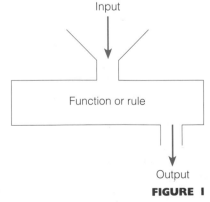

Input

Function or rule

Output

FIGURE 1

TABLE 1

Mercury	0
Venus	0
Earth	1
Mars	2
Jupiter	16
Saturn	18
Uranus	15
Neptune	8
Pluto	1

solar system was known to have as of 1992. If we think of the inputs as the planets listed in the left-hand column of Table 1 and the outputs as the numbers in the right-hand column, then these correspondences constitute a function. The rule for this function can be stated, *Assign to each planet in the solar system the number of moons it was known to have in 1992.*

The following is one definition of the term *function*. As you will see, the definition is broad enough to encompass all the examples we have just looked at.

DEFINITION   Function

Let A and B be two nonempty sets. A **function** from A to B is a rule of correspondence that assigns to each element in A exactly one element in B.

The set A in the definition just given is called the **domain** of the function. Think of the domain as the set of all possible inputs. The set of all outputs, on the other hand, is called the **range** of the function. When a function is defined by means of an equation, the letter representing elements from the domain (that is, the inputs) is called the **independent variable**. For example, in the equation $y = 3x - 2$, the independent variable is x. The letter representing elements from the range (that is, the outputs) is called the **dependent variable**. In the equation $y = 3x - 2$, the dependent variable is y; its value *depends* on x. This is also expressed by saying that y is a function of x.

For functions defined by equations, we'll agree on the following convention regarding the domain: Unless otherwise indicated, the domain is assumed to be the set of all real numbers that lead to unique real-number outputs. (This is essentially the domain convention described in Section 1.3.) Thus, the domain of the function defined by $y = 3x - 2$ is the set of all real numbers, whereas the domain of the function defined by $y = 1/(x - 5)$ is the set of all real numbers except 5. [The expression $1/(x - 5)$ is undefined when $x = 5$ because the denominator is then zero.]

■■■■ **EXAMPLE 1**   Find the domain of the function defined by each equation.

(a) $y = \sqrt{2x + 6}$ **(b)** $s = \dfrac{1}{t^2 - 6t - 7}$

Solution

(a) The quantity under the radical sign must be nonnegative, so we have

$$2x + 6 \geq 0$$
$$2x \geq -6$$
$$x \geq -3$$

The domain is therefore the interval $[-3, \infty)$.

(b) Since division by zero is undefined, the domain of this function consists of all real numbers t except those for which the denominator is zero. Thus to find out which values of t to exclude, we solve the equation $t^2 - 6t - 7 = 0$. We have

$$t^2 - 6t - 7 = 0$$
$$(t - 7)(t + 1) = 0$$
$$t - 7 = 0 \quad | \quad t + 1 = 0$$
$$t = 7 \quad | \quad t = -1$$

It follows now that the domain of the function defined by $s = 1/(t^2 - 6t - 7)$ is the set of all real numbers except $t = 7$ and $t = -1$. ∎

EXAMPLE 2 Find the range of the function defined by $y = \dfrac{x + 2}{x - 3}$.

Solution The range of this function is the set of all outputs y. One way to see what restrictions the given equation imposes on y is to solve the equation for x as follows:

$$y(x - 3) = x + 2 \qquad \text{multiplying by } (x - 3)$$
$$xy - 3y = x + 2$$
$$xy - x = 3y + 2$$
$$x(y - 1) = 3y + 2$$
$$x = \frac{3y + 2}{y - 1}$$

From this last equation we see that the value of y cannot be 1. (The denominator is zero when $y = 1$.) The range therefore consists of all real numbers except $y = 1$. ∎

We often use single letters to name functions. If f is a function and x is an input for the function, then the resulting output is denoted by $f(x)$. This is read *f of x* or *the value of f at x*. As an example of this notation, suppose that f is the function defined by

$$f(x) = x^2 - 3x + 1 \tag{1}$$

Then $f(-2)$ denotes the output that results when the input is -2. To calculate this output, just replace x with -2 throughout equation (1). This yields

$$f(-2) = (-2)^2 - 3(-2) + 1$$
$$= 4 + 6 + 1 = 11$$

That is, $f(-2) = 11$. Figure 2 summarizes this result and the notation.

$f(-2) = 11$

Output

Input

Name of function

FIGURE 2

EXAMPLE 3 Let $f(x) = \dfrac{1}{x - 1}$. Compute: **(a)** $f(0)$;   **(b)** $f(t)$;   **(c)** $f(x - 1)$.

Solution

(a) $f(0) = \dfrac{1}{0 - 1} = -1$

(b) $f(t) = \dfrac{1}{t-1}$

(c) Replace x with the quantity $x - 1$ throughout the given equation. This yields

$$f(x - 1) = \frac{1}{(x - 1) - 1} = \frac{1}{x - 2}$$

▊▊ **EXAMPLE 4** Let $g(x) = 1 - x^2$. Compute $g(x - 1)$.

Solution In the equation $g(x) = 1 - x^2$, we substitute the quantity $x - 1$ in place of each occurrence of x. This gives us

$$g(x - 1) = 1 - (x - 1)^2$$
$$= 1 - (x^2 - 2x + 1) = -x^2 + 2x$$

Thus, $g(x - 1) = -x^2 + 2x$. ▊▊

Here is a slightly different perspective on function notation that we can apply in Example 4 and in similar problems. Instead of writing $g(x) = 1 - x^2$, we can write $g(\ \ ) = 1 - (\ \ )^2$, with the understanding that whatever quantity goes in the parentheses on the left-hand side of the equation must also be placed in the parentheses on the right-hand side. In particular, if we want $g(x - 1)$, we simply write $x - 1$ inside each set of parentheses:

$$g(\ \ ) = 1 - (\ \ )^2 \qquad \text{and therefore} \qquad g(x - 1) = 1 - (x - 1)^2$$

From here on, the algebra is the same as in Example 4. We again obtain $g(x - 1) = -x^2 + 2x$.

In the box that follows, we indicate several errors to avoid when using function notation. The first error indicated in the box involves a confusion between a function and a number. For any real numbers r, a, and b, it is always true that $r(a + b) = ra + rb$; this is the familiar *distributive law* for real numbers. If f is a function, however, it is not in general true that $f(a + b) = f(a) + f(b)$.

Errors to Avoid	
ERROR	EXAMPLE USING $f(x) = x + 1$, $a = 2$, $b = 3$
$f(a + b) \not\equiv f(a) + f(b)$	$f(2 + 3) = f(5) = 6$; $f(2) + f(3) = 3 + 4 = 7$; therefore $f(2 + 3) \neq f(2) + f(3)$.
$f(ab) \not\equiv f(a) \cdot f(b)$	$f(2 \cdot 3) = f(6) = 7$; $f(2) \cdot f(3) = 3 \cdot 4 = 12$; therefore $f(2 \cdot 3) \neq f(2) \cdot f(3)$.
$f\left(\dfrac{1}{a}\right) \not\equiv \dfrac{1}{f(a)}$	$f\left(\dfrac{1}{2}\right) = \dfrac{1}{2} + 1 = \dfrac{3}{2}$; $\dfrac{1}{f(2)} = \dfrac{1}{2 + 1} = \dfrac{1}{3}$; therefore $f\left(\dfrac{1}{2}\right) \neq \dfrac{1}{f(2)}$.
$\dfrac{f(a)}{f(b)} \not\equiv \dfrac{a}{b}$	$\dfrac{f(2)}{f(3)} = \dfrac{2 + 1}{3 + 1} = \dfrac{3}{4} \neq \dfrac{2}{3}$

The next two examples using function notation involve calculations with a **difference quotient**. This is an expression of the form

$$\frac{f(x + h) - f(x)}{h} \qquad \text{or} \qquad \frac{f(x) - f(a)}{x - a}$$

For now we'll concentrate on the algebra used in calculating these quantities. (In the next section you'll see some applications; difference quotients are used to study the *average rate of change* of a function.

EXAMPLE 5 Let $f(x) = x^2 + 3x$. Compute $\dfrac{f(x) - f(2)}{x - 2}$.

Solution

$$\frac{f(x) - f(2)}{x - 2} = \frac{(x^2 + 3x) - [2^2 + 3(2)]}{x - 2}$$

$$= \frac{x^2 + 3x - 10}{x - 2}$$

$$= \frac{(x - 2)(x + 5)}{x - 2} = x + 5$$

The difference quotient is $x + 5$.

EXAMPLE 6 Let $G(x) = 2/x$. Find $\dfrac{G(x + h) - G(x)}{h}$.

Solution

$$\frac{G(x + h) - G(x)}{h} = \frac{[2/(x + h)] - [2/x]}{h}$$

An easy way to simplify this last expression is to multiply it by $\dfrac{(x + h)x}{(x + h)x}$, which equals 1. This yields

$$\frac{G(x + h) - G(x)}{h} = \frac{(x + h)x}{(x + h)x} \cdot \frac{\dfrac{2}{x + h} - \dfrac{2}{x}}{h}$$

$$= \frac{2x - 2(x + h)}{h(x + h)x}$$

$$= \frac{2x - 2x - 2h}{h(x + h)x} = \frac{-2h}{h(x + h)x}$$

$$= \frac{-2}{(x + h)x}$$

In Examples 1 through 6 we've considered functions that are defined by means of equations. It is important to understand, however, that not all equations and rules define functions. For example, consider the equation $y^2 = x$ and the input $x = 4$. Then we have $y^2 = 4$ and, consequently, $y = \pm 2$. So we have *two* outputs in this case, whereas the definition of a function requires that there be *exactly one* output. Example 7 provides some additional perspective on this situation.

EXAMPLE 7 Let $A = \{b, g\}$ and $B = \{s, t, u, z\}$. Which of the four correspondences in Figure 3 represent functions from A to B? For those correspondences that do represent functions, specify the range in each case.

FIGURE 3

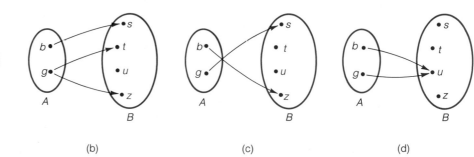

       (a)   (b)   (c)   (d)

Solution (a) This is not a function. The definition requires that *each* element in A be assigned an element in B. The element g in this case has no assignment.

(b) This is not a function. The definition requires *exactly one* output for a given input. In this case there are two outputs for the input g.

Example 7 continues on page 104.

■▶ **PROPERTY SUMMARY**	**TERMINOLOGY FOR FUNCTIONS**	
TERM	**DEFINITION AND COMMENTS**	**EXAMPLE**
Function	Given two nonempty sets A and B, a function f from A to B is a rule that assigns to each element x in A exactly one element $f(x)$ in B. We think of each element x in A as an input and the corresponding element $f(x)$ in B as an output.	The equation $f(x) = x^2$ defines a function. The rule in this case is: *For each real number x, compute its square.* Given the input $x = 3$, for example, the output is $f(3) = 9$.
Domain and range	The domain of a function is the set of all inputs; this is the set A in the definition. The range is the set of all outputs; see Figure 4.	The domain and the range of the function defined by $f(x) = x^2$ are the sets $(-\infty, \infty)$ and $[0, \infty)$, respectively.
Independent and dependent variables	The letter used to represent the elements in the domain of a function (the inputs) is the independent variable. The letter used for elements in the range (the outputs) is the dependent variable.	For the function $y = x^2$, the independent variable is x and the dependent variable is y. The value of y *depends* on x.

FIGURE 4

A function f from A to B

(c) and (d)   Both of these rules qualify as functions from A to B. For each input there is exactly one output. (Regarding the function in (d) in particular, notice that nothing in the definition of a function prohibits two different inputs from producing the same output.) For the function in (c), the outputs are s and z, and so the range is the set $\{s, z\}$. For the function in (d), the only output is u, consequently the range is the set $\{u\}$. ■■■

Because the function concept is so central to the rest of the material in this text, for reference we summarize the terminology introduced in this section. (See the box on page 103.)

EXERCISE SET 3.1

A

In Exercises 1–8, find the domain of each function.

1. (a) $y = -5x + 1$ (b) $y = \sqrt{-5x + 1}$
    (c) $y = \dfrac{x}{\sqrt{|-5x + 1|}}$

2. (a) $y = 3x + 12$ (b) $y = \sqrt{3x + 12}$
    (c) $y = 1/(3x + 12)$

3. (a) $f(x) = x^2 - 10x + 24$
    (b) $g(x) = \sqrt{x^2 - 10x + 24}$ *Hint:*   Section 1.6 explains a method for solving inequalities such as $x^2 - 10x + 24 \geq 0$.
    (c) $h(x) = \sqrt{x/(x^2 - 10x + 24)}$

4. (a) $y = x^2 - 3x + 4$ (b) $y = \sqrt{x^2 - 3x + 4}$
    (c) $y = \sqrt{x^2 - 3x - 4}$

5. (a) $y = \dfrac{x + 4}{x - 4}$ (b) $y = \dfrac{x^2 + 4}{x^2 - 4}$ (c) $y = \sqrt{\dfrac{x^2 + 4}{x^2 - 4}}$

6. (a) $y = 1/(x^3 - 16x)$ (b) $y = 1/(x^3 - 16x^2)$
    (c) $y = \sqrt[3]{x^3 - 16x}$

7. (a) $y = \sqrt{x^2 - 4x - 5}$ (b) $y = \dfrac{2x}{\sqrt{x^2 - 4x - 5}}$
    (c) $y = \sqrt[3]{x^2 - 4x - 5}$

8. (a) $y = x^3 - 14x^2 - 6x - 1$
    (b) $y = 1/(x^3 + x^2 - 2x - 2)$ (c) $y = 1/(x^3 + x^2 - 2x)$

In Exercises 9–12, find the range of each function.

9. (a) $y = \dfrac{x + 3}{x - 5}$ (b) $y = \dfrac{x - 5}{x + 3}$

10. (a) $y = \dfrac{3x - 2}{x + 4}$ (b) $y = \dfrac{1}{x + 4}$

11. (a) $y = x^2 + 4$ (b) $y = x^3 + 4$

12. (a) $y = \dfrac{ax + b}{x}$ (a and b are real numbers and $b \neq 0$)
    (b) $y = cx^3 + d$ ($c \neq 0$)

13. Let $A = \{x, y, z\}$ and $B = \{1, 2, 3\}$. Which of the rules displayed in the figure represent functions from A to B?

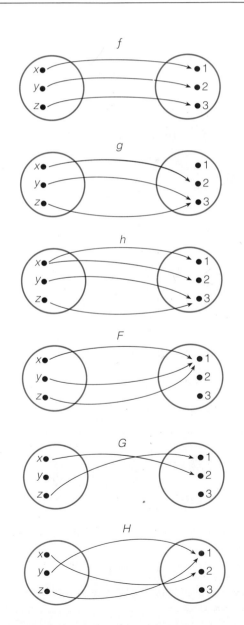

14. Let $D = \{a, b\}$ and $C = \{i, j, k\}$. Which of the rules displayed in the following figure represent functions from D to C?

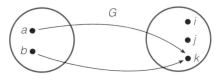

15. (a) Specify the range for each rule that represents a function in Exercise 13.
(b) Specify the range for each rule that represents a function in Exercise 14.

16. (a) Suppose that in Exercise 13 all the arrows were reversed. Which, if any, of the new rules would be functions from B to A?
(b) Suppose that in Exercise 14 all the arrows were reversed. Which, if any, of the new rules would be functions from C to D?

17. Each of the following rules defines a function with domain the set of all real numbers. Express each rule in the form of an equation.

EXAMPLE

The rule *For each real number, compute its square* can be written $y = x^2$.

(a) For each real number, subtract 3 and then square the result.
(b) For each real number, compute its square and then subtract 3 from the result.
(c) For each real number, multiply it by 3 and then square the result.

(d) For each real number, compute its square and then multiply the result by 3.

18. Each of the following rules defines a function with domain equal to the set of all real numbers. Express each rule in words.
(a) $y = 2x^3 + 1$ (b) $y = 2(x + 1)^3$
(c) $y = (2x + 1)^3$ (d) $y = (2x)^3 + 1$

19. Let $f(x) = x^2 - 3x + 1$. Compute the following.
(a) $f(1)$ (b) $f(0)$ (c) $f(-1)$
(d) $f\left(\frac{3}{2}\right)$ (e) $f(z)$ (f) $f(x + 1)$
(g) $f(a + 1)$ (h) $f(-x)$ (i) $|f(1)|$
(j) $f(\sqrt{3})$ (k) $f(1 + \sqrt{2})$ (l) $|1 - f(2)|$

20. Let $H(x) = 1 - x + x^2 - x^3$.
(a) Which number is larger, $H(0)$ or $H(1)$?
(b) Find $H\left(\frac{1}{2}\right)$. Does $H\left(\frac{1}{2}\right) + H\left(\frac{1}{2}\right) = H(1)$?

21. Let $f(x) = 3x^2$. Find the following.
(a) $f(2x)$ (b) $2f(x)$ (c) $f(x^2)$
(d) $[f(x)]^2$ (e) $f(x/2)$ (f) $f(x)/2$
For checking: No two answers are the same.

22. Let $f(x) = 4 - 3x$. Find the following.
(a) $f(2)$ (b) $f(-3)$ (c) $f(2) + f(-3)$
(d) $f(2 + 3)$ (e) $f(2x)$ (f) $2f(x)$
(g) $f(x^2)$ (h) $f(1/x)$ (i) $f[f(x)]$
(j) $x^2 f(x)$ (k) $1/f(x)$ (l) $f(-x)$
(m) $-f(x)$ (n) $-f(-x)$

23. Let $H(x) = 1 - 2x^2$. Find the following.
(a) $H(0)$ (b) $H(2)$
(c) $H(\sqrt{2})$ (d) $H\left(\frac{5}{6}\right)$
(e) $H(x + 1)$ (f) $H(x + h)$
(g) $H(x + h) - H(x)$ (h) $\dfrac{H(x + h) - H(x)}{h}$

24. (a) If $f(x) = 2x + 1$, does $f(3 + 1) = f(3) + f(1)$?
(b) If $f(x) = 2x$, does $f(3 + 1) = f(3) + f(1)$?
(c) If $f(x) = \sqrt{x}$, does $f(3 + 1) = f(3) + f(1)$?

25. Let $R(x) = (2x - 1)/(x - 2)$. Find the following.
(a) the domain and range of R (b) $R(0)$
(c) $R\left(\frac{1}{2}\right)$ (d) $R(-1)$
(e) $R(x^2)$ (f) $R(1/x)$
(g) $R(a)$ (h) $R(x - 1)$

26. Let $g(x) = 2$, for all x. Find each output.
(a) $g(0)$ (b) $g(5)$ (c) $g(x + h)$

27. Let $d(t) = -16t^2 + 96t$.
(a) Compute $d(1)$, $d\left(\frac{3}{2}\right)$, $d(2)$, $d(t_0)$.
(b) For which values of t is $d(t) = 0$?
(c) For which values of t is $d(t) = 1$?

28. Let $A(x) = |x^2 - 1|$. Compute $A(2)$, $A(1)$, and $A(0)$.

29. Let $g(t) = |t - 4|$. Find $g(3)$. Find $g(x + 4)$.

30. Let $f(x) = x^2/|x|$.
(a) What is the domain of f?
(b) Find $f(2)$, $f(-2)$, $f(20)$, and $f(-20)$.

(c) What is the range of f? *Hint:* Look over your results in part (b).

31. Compute $\dfrac{f(x+h) - f(x)}{h}$ for the function f specified in each case.
    (a) $f(x) = x^2$ (b) $f(x) = 2x^2 - 3x + 1$
    (c) $f(x) = x^3$

32. Compute $\dfrac{f(x) - f(a)}{x - a}$ for the three functions given in Exercise 31.

33. Let $f(x) = x/(x - 1)$. Compute each difference quotient.
    (a) $\dfrac{f(x) - f(a)}{x - a}$ (b) $\dfrac{f(x) - f(3)}{x - 3}$
    (c) $\dfrac{f(x+h) - f(x)}{h}$ (d) $\dfrac{f(3+h) - f(3)}{h}$

34. Let $g(t) = 1/(3t - 5)$. Compute each difference quotient.
    (a) $\dfrac{g(t) - g(a)}{t - a}$ (b) $\dfrac{g(t) - g(2)}{t - 2}$
    (c) $\dfrac{g(t+h) - g(t)}{h}$ (d) $\dfrac{g(2+h) - g(2)}{h}$

35. In each case a pair of functions f and g are given. Find all real numbers x_0 for which $f(x_0) = g(x_0)$.
    (a) $f(x) = 4x - 3$; $g(x) = 8 - x$
    (b) $f(x) = x^2 - 4$; $g(x) = 4 - x^2$
    (c) $f(x) = x^2$; $g(x) = x^3$
    (d) $f(x) = 2x^2 - x$; $g(x) = 3$

In Exercises 36–40, use a calculator for the computations.

36. Let $h(x) = \sqrt{x}$.
    (a) Compute $\dfrac{h(5) - h(1)}{5 - 1}$ and $\dfrac{1}{h(5) + 1}$. Round both answers to three decimal places.
    (b) Why are the two answers in part (a) the same?

37. When $1000 is deposited in a savings account at an annual rate of 12%, compounded quarterly, the amount in the account after t years is given by
    $$A(t) = 1000\left(1 + \frac{0.12}{4}\right)^{4t}$$
    (a) Compute $A(1) - A(0)$. This is the amount by which the account will grow in the first year.
    (b) Compute $A(10) - A(9)$. This is the amount by which the account will grow in the tenth year.

38. Let $f(n) = [1 + (1/n)]^n$.
    (a) Complete the table. (Round the results to three decimal places.)

n	1	2	5	10	15	20
$f(n)$						

(b) By trial and error, find the smallest natural number n such that $f(n) > 2.7$.
(c) Using your calculator, can you find a number n such that $f(n) \geq 2.8$?

39. Let $g(n) = n^{1/n}$.
    (a) Complete the table. (Round the results to four decimal places.)

n	2	3	4	5	6	7	8
$g(n)$							

(b) By trial and error, find the smallest natural number n such that $g(n) < 1.2$.

40. Consider the function f defined by
    $$f(x) = x^2 + \frac{2}{x^2} \quad x > 0$$
    (a) Complete the table. (Round the results to four decimal places.)

x	1	1.05	1.10	1.15	1.20	1.25
$f(x)$						

(b) Which x-value in the table yields the smallest value for $f(x)$?
(c) It can be shown using calculus that the input x yielding the smallest possible output $f(x)$ for this function is $x = 2^{1/4}$. Which x-value in the table is closest to this?
(d) Compute $f(2^{1/4})$. Which value of $f(x)$ in the table is closest to $f(2^{1/4})$?

In Exercises 41–44, refer to the following table. The left-hand column of the table lists four errors to avoid in working with function notation. In each case, use the function $f(x) = x^2 - 1$ and give a numerical example showing that the expressions on each side of the equation are not equal.

Error to Avoid	Numerical example showing that the equation is not, in general, valid
41. $f(a + b) \neq f(a) + f(b)$	
42. $f(ab) \neq f(a) \cdot f(b)$	
43. $f\left(\dfrac{1}{a}\right) \neq \dfrac{1}{f(a)}$	
44. $\dfrac{f(a)}{f(b)} \neq \dfrac{a}{b}$	

B

45. Let $f(x) = (x - a)/(x + a)$.
(a) Find $f(a)$, $f(2a)$, and $f(3a)$. Is it true that
$f(3a) = f(a) + f(2a)$?
(b) Show that $f(5a) = 2f(2a)$.

46. Let $M(x) = (x - a)/(x + a)$. Compute $M(1/x)$.

47. Let $\phi(y) = 2y - 3$. Show that $\phi(y^2) \neq [\phi(y)]^2$.

48. Let $k(x) = 5x^3 + \dfrac{5}{x^3} - x - \dfrac{1}{x}$. Show that $k(x) = k(1/x)$.

49. Let $f(x) = 2x + 3$. Find values for a and b such that the equation $f(ax + b) = x$ is true for all values of x.

50. If $p(x) = 2^x$, verify each identity: (a) $2p(x) = p(x + 1)$;
(b) $p(a + b) = p(a) \cdot p(b)$; (c) $p(x) \cdot p(-x) - 1 = 0$.

51. Let $f(t) = (t - x)/(t + y)$. Show that
$$f(x + y) + f(x - y) = \frac{-2y^2}{x^2 + 2xy}$$

52. Let $f(z) = \dfrac{3z - 4}{5z - 3}$. Find $f\left(\dfrac{3z - 4}{5z - 3}\right)$.

53. Let $F(x) = \dfrac{ax + b}{cx - a}$. Show that $F\left(\dfrac{ax + b}{cx - a}\right) = x$.
(Assume that $a^2 + bc \neq 0$.)

54. If $f(x) = -2x^2 + 6x + k$ and $f(0) = -1$, find k.

55. If $g(x) = x^2 - 3xk - 4$ and $g(1) = -2$, find k.

56. If $f(x) = 1/x^2$, show that
$$\frac{f(x + h) - f(x)}{h} = \frac{-2x - h}{(x + h)^2 x^2}$$

57. A function doesn't always have to be given by an algebraic formula. For example, let the function L be defined by the following rule: $L(x)$ *is the exponent to which 2 must be raised to yield x.* (For the moment, we won't concern ourselves with the domain and range.) Then $L(8) = 3$, for example, since the exponent to which 2 must be raised to yield 8 is 3 $(8 = 2^3)$. Find the following outputs.
(a) $L(1)$ (b) $L(2)$ (c) $L(4)$ (d) $L(64)$
(e) $L(\frac{1}{2})$ (f) $L(\frac{1}{4})$ (g) $L(\frac{1}{64})$ (h) $L(\sqrt{2})$
The function L is called a *logarithm function*. The usual notation for $L(x)$ in this example is $\log_2 x$. Logarithm functions will be studied in Chapter 5.

58. Let $f(x) = ax^2 + bx + c$. Show that
$$\frac{f(x + h) - f(x)}{h} = 2ax + ah + b$$

59. Let $q(x) = ax^2 + bx + c$. Evaluate
$$q\left(\frac{-b + \sqrt{b^2 - 4ac}}{2a}\right)$$

60. By definition, a **fixed point** for the function f is a number x_0 such that $f(x_0) = x_0$. For instance, to find any fixed points for the function $f(x) = 3x - 2$, we write $3x_0 - 2 = x_0$. On solving this last equation, we find that $x_0 = 1$. Thus, 1 is a fixed point for f. Calculate the fixed points (if any) for each function.
(a) $f(x) = 6x + 10$ (b) $g(x) = x^2 - 2x - 4$
(c) $S(t) = t^2$ (d) $R(z) = (z + 1)/(z - 1)$

61. Let $f(x) = \dfrac{3x - 4}{x - 3}$.
(a) Find $f[f(x)]$.
(b) Find $f[f(\frac{22}{7})]$. Try not to do it the hard way!

62. Consider the following two rules, F and G, where F is the rule that assigns to each person his or her mother, and G is the rule that assigns to each person his or her aunt. Explain why F is a function but G is not.

*In Exercises 63–65, use this definition: A **prime number** is a positive whole number with no factors other than itself and 1. For example, 2, 13, and 37 are primes, but 24 and 39 are not. By convention, 1 is not considered prime, so the list of the first few primes is as follows:*

2, 3, 5, 7, 11, 13, 17, 19, 23, 29, ...

63. Let G be the rule that assigns to each positive integer the nearest prime. For example, $G(8) = 7$, since 7 is the prime nearest 8. Explain why G is not a function. How could you alter the definition of G to make it a function?

64. Let f be the function that assigns to each natural number x the number of primes that are less than or equal to x. For example, $f(12) = 5$, because, as you can easily check, five primes are less than or equal to 12. Similarly, $f(3) = 2$, because two primes are less than or equal to 3. Find $f(8)$, $f(10)$, and $f(50)$.

65. If $P(x) = x^2 - x + 17$, find $P(1)$, $P(2)$, $P(3)$, and $P(4)$. Can you find a natural number x for which $P(x)$ is not prime?

66. If $P(x) = x^2 - x + 41$, find $P(1)$, $P(2)$, $P(3)$, and $P(4)$. Can you find a natural number x for which $P(x)$ is not prime?

C

67. Let $S(x) = \dfrac{3^x - 3^{-x}}{2}$ and $C(x) = \dfrac{3^x + 3^{-x}}{2}$. Show that the functions S and C possess the given properties.
(a) $S(0) = 0$; $C(0) = 1$; $S(1) = \frac{4}{3}$; $C(1) = \frac{5}{3}$
(b) $[C(x)]^2 - [S(x)]^2 = 1$
(c) $S(-x) = -S(x)$ and $C(-x) = C(x)$
(d) $S(x + y) = S(x)C(y) + C(x)S(y)$
(e) $C(x + y) = C(x)C(y) + S(x)S(y)$
(f) $S(2x) = 2S(x)C(x)$ and $C(2x) = [C(x)]^2 + [S(x)]^2$

(g) $S(3x) = 3S(x) + 4[S(x)]^3$
(h) $[S(x) + C(x)]^2 = S(2x) + C(2x)$

In Exercises 68–70, use the following definitions. Two positive integers are said to be **relatively prime** if they have no common factor other than 1. For example, 4 and 7 are relatively prime, but 4 and 18 are not relatively prime, nor are 5 and 10. The **Euler ϕ-function** is then defined as follows:
  (i) The domain of ϕ is the set of positive integers;
 (ii) $\phi(1) = 1$;
(iii) for $m > 1$, $\phi(m)$ is the number of positive integers that are less than m and relatively prime to m.

For example, to find $\phi(14)$, first make a list of the positive integers that are less than 14 and are relatively prime to 14: 1, 3, 5, 9, 11, 13. Now count how many numbers there are in this list. There are six numbers in the list, and therefore $\phi(14) = 6$.

68. (a) For the following tables, first verify for yourself that the completed entries are correct. Then complete the remaining entries.

m	1	2	3	4	5	6	7	8	9	10	11
$\phi(m)$	1	1	2	2	4						

m	12	13	14	15	16	17	18	19	20
$\phi(m)$			6						8

  (b) Use the results in part (a) to decide whether each of the following statements is true or false.
    (i) $\phi(2) \cdot \phi(5) = \phi(2 \cdot 5)$ (The dots denote multiplication.)
   (ii) $\phi(2) \cdot \phi(6) = \phi(2 \cdot 6)$
  (iii) $\phi(2) \cdot \phi(7) = \phi(2 \cdot 7)$
   (iv) $\phi(3) \cdot \phi(4) = \phi(3 \cdot 4)$
    (v) $\phi(3) \cdot \phi(5) = \phi(3 \cdot 5)$
   (vi) $\phi(3) \cdot \phi(6) = \phi(3 \cdot 6)$

69. (a) The factors of the number 6 are 1, 2, 3, and 6. Use a table in Exercise 68(a) to verify that
$\phi(1) + \phi(2) + \phi(3) + \phi(6) = 6$.
  (b) The factors of the number 15 are 1, 3, 5, and 15. Use the tables in Exercise 68(a) to verify that
$\phi(1) + \phi(3) + \phi(5) + \phi(15) = 15$.

70. Suppose that p is a prime number. Explain (in complete sentences) why $\phi(p) = p - 1$.

3.2 THE GRAPH OF A FUNCTION

In my own case, I got along fine without knowing the name of the distributive law until my sophomore year in college; meanwhile I had drawn lots of graphs.

Professor Donald E. Knuth (judged by many to be the world's preeminent computer scientist) in *Mathematical People* (Boston: Birkhäuser, 1985)

When the domain and range of a function are sets of real numbers, the function can be graphed in the same way in which equations were graphed in Chapter 2. In graphing functions, the usual practice is to reserve the horizontal axis for the independent variable and the vertical axis for the dependent variable. The function or rule then tells you how you must pick your y-coordinate, once you have selected an x-coordinate.

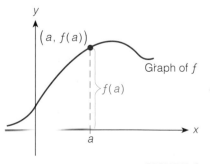

FIGURE 1

> **DEFINITION Graph of a Function**
>
> The **graph** of a function f in the x-y plane consists of those points (x, y) such that x is in the domain of f and $y = f(x)$. See Figure 1.

EXAMPLE 1 Graph the functions f and g defined as follows:
(a) $f(x) = x^2 - 1$;   (b) $g(x) = |x^2 - 1|$.

Solution

(a) The graph of this function is by definition the graph of the equation $y = x^2 - 1$. (Whether we label the vertical axis y or $f(x)$ is immaterial.) As preparation for drawing the graph, we first determine the domain of f, the x- and y-intercepts of the graph, and any symmetries the graph may have. As you can readily check, the results are as follows.

domain: $(-\infty, \infty)$	x-intercepts: $1, -1$
symmetry: about the y-axis	y-intercept: -1

The required graph is then sketched in Figure 2(a).

(b) The graph of g is shown in Figure 2(b). It is obtained from the graph of f in Figure 2(a) as follows. First, for $x \geq 1$ or $x \leq -1$, the graph of g is identical to that of f because in this case, we have

$$g(x) = |x^2 - 1| = x^2 - 1 = f(x)$$

Second, on the interval $(-1, 1)$, the graph of g is obtained by reflecting the graph of f in the x-axis because the quantity $x^2 - 1$ is negative on this interval and therefore

$$g(x) = |x^2 - 1| = -(x^2 - 1) = -f(x)$$ ■■■

FIGURE 2

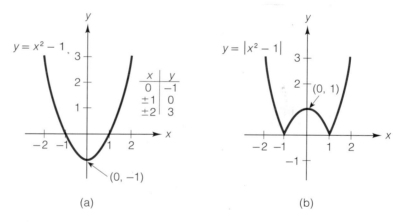

(a)   (b)

■■■■ **EXAMPLE 2** Specify the domain and the range of the function g graphed in Figure 3(a).

FIGURE 3

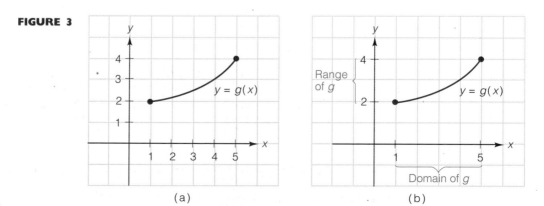

(a)   (b)

Solution   The domain of *g* is just that portion of the *x*-axis (the inputs) utilized in graphing *g*. As Figure 3(b) indicates, this amounts to all real numbers *x* from 1 to 5, inclusive: $1 \le x \le 5$. Recall from Section 1.1 that this set of numbers is denoted by [1, 5]. To find the range of *g*, we need to check which part of the *y*-axis is utilized in graphing *g*. As Figure 3(b) indicates, this is the set of all real numbers *y* between 2 and 4, inclusive: $2 \le y \le 4$. Our shorthand notation for this interval of numbers is [2, 4].   ▪▪▪

▪▪▪ EXAMPLE 3   The graph of a function *h* is shown in Figure 4. The open circle in the figure is used to indicate that the point (3, 3) does not belong to the graph of *h*.

(a) Specify the domain and the range of the function *h*.
(b) Determine each value: **(i)** $h(-2)$;   **(ii)** $h(3)$;   **(iii)** $|h(-4)|$

FIGURE 4

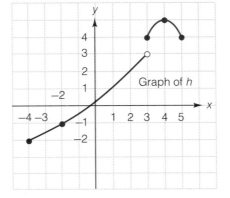

Graph of *h*

Solution

(a) The domain of *h* is the interval $[-4, 5]$. The range is the set $[-2, 3) \cup [4, 5]$.

(b) (i)   The function notation $h(-2)$ stands for the *y*-coordinate of that point on the graph of *h* whose *x*-coordinate is -2. Since the point $(-2, -1)$ is on the graph of *h*, we conclude $h(-2) = -1$.

(ii)   We conclude that $h(3) = 4$ because the point (3, 4) lies on the graph of *h*. Note that *h* would not be considered a function if the point (3, 3) were also part of the graph. (Why?)

(iii)   Since the point $(-4, -2)$ lies on the graph of *h*, we write $h(-4) = -2$. Thus $|h(-4)| = |-2| = 2$.   ▪▪▪

Most of the graphs that we looked at in Chapter 2 are graphs of functions. However, it's important to understand that not every graph represents a function. Consider, for example, the graph in Figure 5.

Figure 5 shows a vertical line intersecting the graph in two distinct points. The specific coordinates of the two points are unimportant. What the vertical line helps us to see is that two different *y*-values have been assigned to the same *x*-value, and therefore the graph cannot be the graph of a function $y = f(x)$. These remarks are generalized in the box following Figure 5.

FIGURE 5

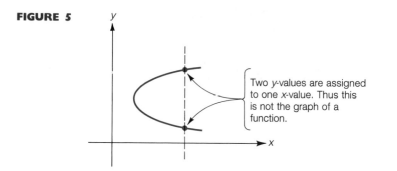

Two y-values are assigned to one x-value. Thus this is not the graph of a function.

Vertical Line Test

A graph in the x-y plane represents a function $y = f(x)$ provided that any vertical line intersects the graph in at most one point.

EXAMPLE 4 The vertical line test implies that the graph in Figure 6(a) represents a function and that the graph in Figure 6(b) does not. ∎

FIGURE 6

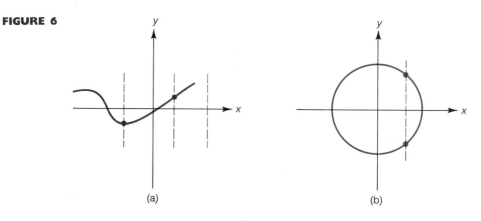

(a)   (b)

EXAMPLE 5 Figure 7 shows three closely related graphs. (Notice that the graphs are constructed from two of the six basic equations described in Section 2.2.) Which of the graphs represent functions?

FIGURE 7

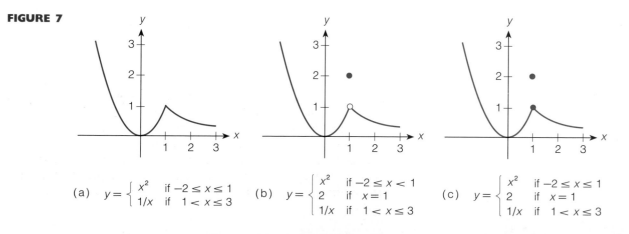

(a) $y = \begin{cases} x^2 & \text{if } -2 \le x \le 1 \\ 1/x & \text{if } 1 < x \le 3 \end{cases}$

(b) $y = \begin{cases} x^2 & \text{if } -2 \le x < 1 \\ 2 & \text{if } x = 1 \\ 1/x & \text{if } 1 < x \le 3 \end{cases}$

(c) $y = \begin{cases} x^2 & \text{if } -2 \le x \le 1 \\ 2 & \text{if } x = 1 \\ 1/x & \text{if } 1 < x \le 3 \end{cases}$

Solution The vertical line test tells us that the graphs in Figures 7(a) and 7(b) represent functions, but that the graph in Figure 7(c) does not. ▮▮▮

One of the most basic questions regarding a function is this: What are the highest and the lowest points on the graph? We will use Figures 8 and 9 to introduce some terminology that is useful here.

The points P and Q in Figures 8 and 9 are called *turning points*. At a **turning point,** the graph changes from rising to falling, or vice versa. In Figure 8, the highest point on the graph of the function G is $P(2, 102)$. We say that the *maximum value* of the function G is 102; and it occurs at $t = 2$. More generally, we say that $f(x_0)$ is the **maximum value** of a function f if the inequality $f(x_0) \geq f(x)$ holds for every x in the domain of f. Minimum values are defined similarly: $f(x_0)$ is the **minimum value** for a function f if the inequality $f(x_0) \leq f(x)$ holds for every x in the domain of f. Assuming that the domain of the function in Figure 8 is $[0, 6]$, the minimum value of the function G is 99, and it occurs when $t = 6$. In Figure 9, the minimum value of the function is -1, and it occurs when $x = 3$. Not every function has a maximum or minimum value. For example, the function $y = x^3$ possesses neither a maximum nor a minimum value. (Picture the graph.)

The function G in Figure 8 is said to be *increasing* on the open interval $(0, 2)$ and *decreasing* on the interval $(2, 6)$. In terms of the temperature interpretation for Figure 8, the patient's temperature is rising between noon and two o'clock and falling between two and six o'clock. In Figure 9, the function is decreasing on the interval $(-\infty, 3)$ and increasing on the interval $(3, \infty)$. For theoretical work, the terms increasing and decreasing can be defined in terms of inequalities. A function f is **increasing** on an interval if the following condition holds: If x_1 and x_2 are in the interval and $x_1 < x_2$, then $f(x_1) < f(x_2)$. Similarly, a function f is said to be **decreasing** on an interval if the following condition holds: If x_1 and x_2 are in the interval and $x_1 < x_2$, then $f(x_1) > f(x_2)$.

In the previous section we discussed difference quotients, and in this section we've worked with graphs of functions. The following definition ties together these two concepts.

DEFINITION   The Average Rate of Change of a Function

Refer to Figure 10. The **average rate of change** of a function f on the interval $[a, b]$ is the slope of the line segment joining the two points $(a, f(a))$ and $(b, f(b))$. If $y = f(x)$, the average rate of change is denoted by $\Delta y/\Delta x$ or $\Delta f/\Delta x$, and we have

$$\frac{\Delta y}{\Delta x} = \frac{f(b) - f(a)}{b - a}$$

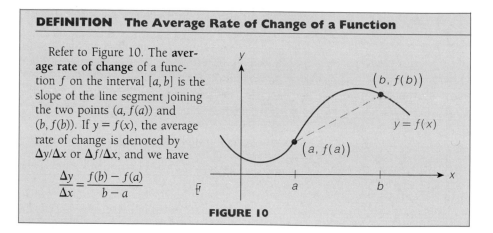

FIGURE 10

EXAMPLE 6   Compute and compare the average rates of change for the function $f(x) = x^2$ on the following intervals: (a) $[-2, 0]$;   (b) $[0, 1]$; (c) $[1, 2]$.

Solution

(a) On $[-2, 0]$: $\dfrac{\Delta f}{\Delta x} = \dfrac{f(0) - f(-2)}{0 - (-2)} = \dfrac{0^2 - (-2)^2}{2} = -2$

(b) On $[0, 1]$: $\dfrac{\Delta f}{\Delta x} = \dfrac{f(1) - f(0)}{1 - 0} = \dfrac{1^2 - 0^2}{1} = 1$

(c) On $[1, 2]$: $\dfrac{\Delta f}{\Delta x} = \dfrac{f(2) - f(1)}{2 - 1} = \dfrac{2^2 - 1^2}{1} = 3$

(Refer to Figure 11.) On the interval $[-2, 0]$, the function is decreasing and the average rate of change is negative. On both of the intervals $[0, 1]$ and $[1, 2]$, the function is increasing and the average rates of change are positive. The graph rises more steeply on the interval $[1, 2]$ than on $[0, 1]$ and, correspondingly, the average rate of change is greater for the interval $[1, 2]$ than for $[0, 1]$.   ∎

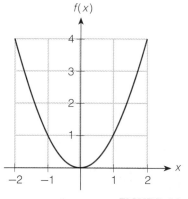

FIGURE 11
$f(x) = x^2$

A WORD OF CAUTION   Like other types of averages occurring in arithmetic and statistics, the average rate of change can mask or hide some details. For instance, just because the average rate of change is positive on an interval, that does not guarantee that the function is increasing throughout that interval. For example, look back at Figure 10; for the function shown there, the average rate of change on $[a, b]$ is positive (because the slope of the dashed line is positive). The function, however, is increasing only on a portion of $[a, b]$, not throughout the entire interval.

In applications, it's informative to keep track of the units associated with the average rate of change. Consider, for instance, the patient's fever graph in Figure 8 (on page 112). Let's compute the average rate of change from $t = 2$ hr to $t = 6$ hr (that is, from 2 P.M. to 6 P.M.). We have

$$\frac{\Delta G}{\Delta t} = \frac{G(6) - G(2)}{6 - 2} = \frac{99° - 102°}{4 \text{ hr}} = \frac{-3°}{4 \text{ hr}} = -\frac{3}{4} \text{ degree/hr}$$

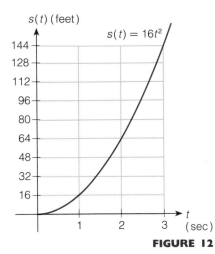

FIGURE 12

Interpretation: Between 2 P.M. and 6 P.M., The patient's temperature is dropping at an average rate of $\frac{3}{4}$ degree per hour.

As another example, consider the function in Figure 12, $s(t) = 16t^2 (t \geq 0)$. This function relates the distance $s(t)$ and the time t for an object falling in a vacuum. Here, t is measured in seconds, $s(t)$ is in feet, and $t = 0$ corresponds to the instant that the object begins to fall. For instance, after 1 second, the object falls a distance of $s(1) = 16(1^2) = 16$ ft. And after 2 seconds, the total distance will be $s(2) = 16(2^2) = 64$ ft. To see how the units work here, let's calculate the average rate of change $\Delta s/\Delta t$ from $t = 1$ to $t = 3$. We have

$$\frac{\Delta s}{\Delta t} = \frac{s(3) - s(1)}{3 \text{ sec} - 1 \text{ sec}} = \frac{16(3^2) \text{ ft} - 16(1^2) \text{ ft}}{2 \text{ sec}} = \frac{128 \text{ ft}}{2 \text{ sec}} = 64 \text{ ft/sec}$$

Notice that the units here have the form *distance per unit time*. So in this case, $\Delta s/\Delta t$ gives us the *average velocity* of the object. More generally, whenever we have a function expressing distance $s(t)$ in terms of time t, the **average velocity** over an interval is defined to be the average rate of change $\Delta s/\Delta t$ over that interval.

■■■ **EXAMPLE 7**   Refer to the distance function in Figure 12, $s(t) = 16t^2$.

(a) Find a general expression for the average velocity $\Delta s/\Delta t$ over the interval $[t_1, t_2]$. (Assume that $t_1 \neq t_2$.)

(b) Use the result in part (a) to compute the average velocity over the following intervals: (i) $t = 1$ to $t = 2$ seconds; (ii) $t = 1$ to $t = 1.5$ seconds; (iii) $t = 1$ to $t = 1.1$ seconds.

Solution

(a) By definition, the average velocity or average rate of change is

$$\frac{\Delta s}{\Delta t} = \frac{s(t_2) - s(t_1)}{t_2 - t_1}$$

$$= \frac{16t_2^2 - 16t_1^2}{t_2 - t_1} = \frac{16(t_2 - t_1)(t_2 + t_1)}{t_2 - t_1} \qquad \text{factoring}$$

$$= 16(t_2 + t_1) \qquad\qquad\qquad \text{simplifying}$$

So, we have $\Delta s/\Delta t = 16(t_2 + t_1)$. As explained previously, the units here are feet per second. Question: Where did we use the given condition $t_1 \neq t_2$?

(b) From $t = 1$ to $t = 2$: $\Delta s/\Delta t = 16(t_2 + t_1) = 16(2 + 1) = 48$ ft/sec

From $t = 1$ to $t = 1.5$: $\Delta s/\Delta t = 16(t_2 + t_1) = 16(1.5 + 1) = 40$ ft/sec

From $t = 1$ to $t = 1.1$: $\Delta s/\Delta t = 16(t_2 + t_1) = 16(1.1 + 1) = 33.6$ ft/sec

[*Note*   Exercise 51 asks you to carry out these calculations several steps further, using shorter and shorter time intervals. In calculus, this procedure leads to the concept of *instantaneous* (as opposed to *average*) *velocity*.]   ■■■

EXERCISE SET 3.2

A

In Exercises 1–8, the graph of a function is given. In each case, specify the domain and the range of the function. (The axes are marked off in one-unit intervals.)

1.

2.

3.

4.

5.

6.

7.

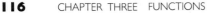

8.

In Exercises 9 and 10, refer to the graph of the function F in the figure. (Assume that the axes are marked off in one-unit intervals.)

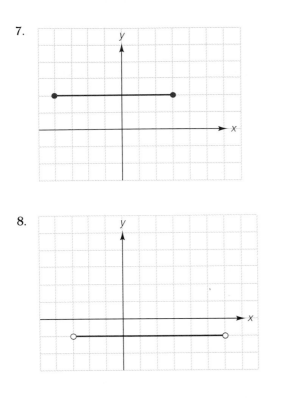

9. (a) Find $F(-5)$. (b) Find $F(2)$.
(c) Is $F(1)$ positive?
(d) For which value of x is $F(x) = -3$?
(e) Find $F(2) - F(-2)$.

10. (a) Find $F(4)$. (b) Find $F(-1)$.
(c) Is $F(-4)$ positive?
(d) For which value of x is $F(x) = 5$?
(e) Find $F(5) - F(-3)$.

11. The following figure displays the graph of a function f.

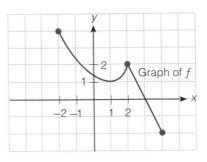

(a) Is $f(0)$ positive or negative?
(b) Find $f(-2)$, $f(1)$, $f(2)$, and $f(3)$.
(c) Which is larger, $f(2)$ or $f(4)$?
(d) Find $f(4) - f(1)$. (e) Find $|f(4) - f(1)|$.
(f) Write the domain and range of f using the interval notation $[a, b]$.

12. The following figure shows the graph of a function h.

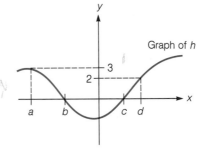

(a) Find $h(a)$, $h(b)$, $h(c)$, and $h(d)$.
(b) Is $h(0)$ positive or negative?
(c) For which values of x does $h(x) = 0$?
(d) Which is larger, $h(b)$ or $h(0)$?
(e) As x increases from c to d, do the corresponding values of $h(x)$ increase or decrease?
(f) As x increases from a to b, do the corresponding values of $h(x)$ increase or decrease?

In Exercises 13–15, refer to the graphs of the functions f and g in the figure. Assume that the domain of each function is $[-3, 3]$ and that the axes are marked off in one-unit intervals.

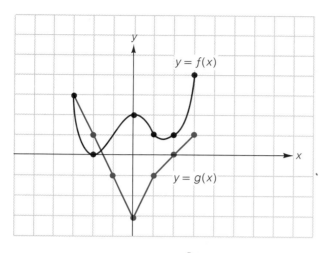

In Exercises 17–20, you are given functions with domain $[0, 4]$. Specify (a) the range of each function; (b) the maximum value of the function; (c) the minimum value; (d) interval(s) where the function is increasing; and (e) interval(s) where the function is decreasing.

13. (a) Which is larger, $f(-2)$ or $g(-2)$?
  (b) Compute $f(0) - g(0)$.
  (c) Which among the following three quantities is the smallest?

  $$f(1) - g(1) \qquad f(2) - g(2) \qquad f(3) - g(3)$$

  (d) For which value(s) of x does $g(x) = f(1)$?
  (e) Is the number 4 in the range of f or in the range of g?

14. (a) For the interval $[0, 3]$, is the quantity $g(x) - f(x)$ positive or negative?
  (b) For the interval $(-3, -2)$, is the quantity $g(x) - f(x)$ positive or negative?
  (c) Compute $\dfrac{f(x) - f(2)}{x - 2}$ when $x = 3$.
  (d) Compute $\dfrac{g(x) - g(-2)}{x + 2}$ when $x = -3$.

15. Complete the following table.

Function	$|x|$	x^2	x^3
Domain			
Range			
Turning point			
Maximum value			
Minimum value			
Interval(s) where increasing			
Interval(s) where decreasing			

16. Set up and complete a table like the one in Exercise 15 for the three functions $1/x$, $\sqrt{x}$, and $\sqrt{1 - x^2}$.

17.

18.

19.

20.

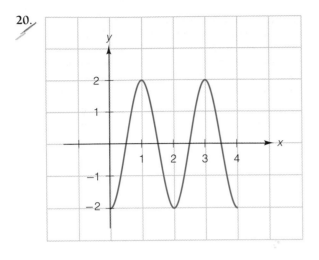

In Exercises 21 and 22, use the vertical line test to determine whether each graph represents a function $y = f(x)$.

21.

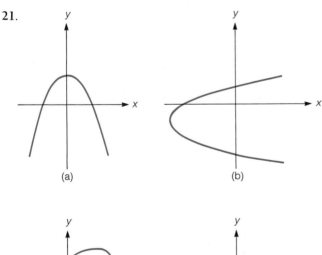

(a) (b)

(c) (d)

22.

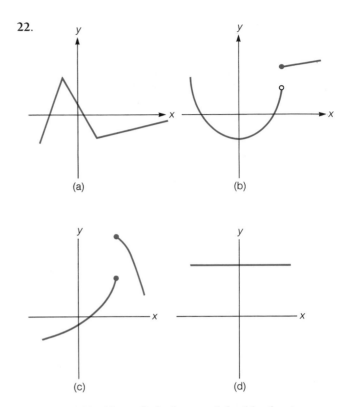

(a) (b)

(c) (d)

In Exercises 23–32, graph the function defined by the given rules.

23. $A(x) = \begin{cases} x^3 & \text{if } -2 \le x \le -1 \\ x^2 & \text{if } x > -1 \end{cases}$

24. $B(x) = \begin{cases} \sqrt{1-x^2} & \text{if } -1 \le x < 1 \\ 1/x & \text{if } x \ge 1 \end{cases}$

25. $C(x) = \begin{cases} x^3 & \text{if } x < 1 \\ \sqrt{x} & \text{if } x > 1 \end{cases}$

26. (a) $f(x) = \begin{cases} x^2/|x| & \text{if } x \ne 0 \\ 0 & \text{if } x = 0 \end{cases}$

(b) $F(x) = \begin{cases} x^2/|x| & \text{if } x \ne 0 \\ 1 & \text{if } x = 0 \end{cases}$

27. (a) $g(x) = \begin{cases} x/|x| & \text{if } x \ne 0 \\ 0 & \text{if } x = 0 \end{cases}$

(b) $G(x) = \begin{cases} x/|x| & \text{if } x \ne 0 \\ 1 & \text{if } x = 0 \end{cases}$

28. $U(x) = \begin{cases} 1 & \text{if } x \le -2 \\ -1 & \text{if } x > -2 \end{cases}$

29. $f(x) = \begin{cases} 1/x & \text{if } x < -1 \\ x & \text{if } -1 \le x \le 1 \\ 1/x & \text{if } x > 1 \end{cases}$

30. $g(x) = \begin{cases} 1/x & \text{if } x < -\frac{1}{2} \\ 1 & \text{if } -\frac{1}{2} \le x \le 1 \\ x^3 & \text{if } x > 1 \end{cases}$

31. $y = \begin{cases} x & \text{if } 0 \le x < 1 \\ x - 1 & \text{if } 1 \le x < 2 \\ x - 2 & \text{if } 2 \le x < 3 \end{cases}$

32. $T(x) = \begin{cases} 2x + 2 & \text{if } -1 \le x < 0 \\ -2x + 2 & \text{if } 0 \le x < 1 \\ 2x - 2 & \text{if } 1 \le x < 2 \\ -2x + 6 & \text{if } 2 \le x \le 3 \end{cases}$

33. Specify the domain and the range of the function whose graph is the horizontal line $y = 3$.

34. Specify the domain and the range of the function whose graph is the horizontal line $y = -\sqrt{2}$.

35. Let $f(x) = x^2$. Find the slope of the line passing through the points $(3, f(3))$ and $(4, f(4))$. What does this have to do with the average rate of change of f on the interval $[3, 4]$?

36. Let $g(x) = 1/x$. Find the slope of the line passing through the points $(2, g(2))$ and $(6, g(6))$. What does this have to do with the average rate of change of g on the interval $[2, 6]$?

In Exercises 37–46, compute the average rate of change of the function on the given interval. In Exercises 43–46, simplify your answer as much as possible.

37. $f(x) = x^2 + 2x$ on $[3, 5]$ **38.** $f(x) = \sqrt{x}$ on $[4, 9]$

39. $g(x) = 2x^2 - 4x$ on $[-1, 3]$ **40.** $g(x) = x^3 - x$ on $[1, 2]$

41. $h(t) = 2t - 6$ on $[5, 12]$ **42.** $h(t) = 16 - 7t$ on $\left[-\sqrt{5}, \sqrt{2}\right]$

43. $f(x) = 3/x$ on $[a, b]$ **44.** $f(x) = 3x^2$ on $[a, b]$

45. $F(x) = -2x^3$ on $[a, b]$ **46.** $F(x) = 4/(x - 1)$ on $[a, b]$

47. The following graph shows the temperature $G(t)$ of a solution during the first 8 minutes of a chemistry experiment.

Compute the average rate of change of temperature $\Delta G/\Delta t$ over the following intervals. (Be sure to specify the units as part of each answer.)
(a) $t = 0$ to $t = 3$ min
(b) $t = 3$ to $t = 6$ min
(c) $t = 6$ to $t = 8$ min

48. Iodine-131 is a radioactive substance. The accompanying graph shows how an initial one-gram sample decays over a 32-day period.

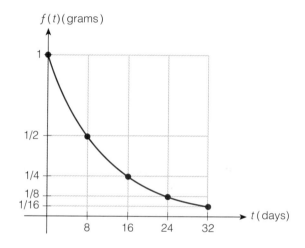

(a) Compute $\Delta f/\Delta t$ over each of the following intervals: $t = 0$ to $t = 8$; $t = 8$ to $t = 16$; $t = 16$ to $t = 24$; $t = 24$ to $t = 32$. (Be sure to specify the units with each answer.)
(b) Find the (arithmetic) average of the four answers in part (a).
(c) Compute $\Delta f/\Delta t$ over the interval from $t = 0$ to $t = 32$. Is the answer the same as that obtained in part (b)?

B

49. Let $f(x) = 1/x$. Find a number b such that the average rate of change of the function f on the interval $[1, b]$ is $-\frac{1}{5}$.

50. Let $f(x) = \sqrt{x}$. Find a number b such that $\Delta f/\Delta x$ on the interval $[1, b]$ is $\frac{1}{7}$.

51. As background for this exercise, reread Example 7.
(a) Let $s(t) = 16t^2$ (as in Example 7). Complete the following table.

Interval	$[1, 1.1]$	$[1, 1.01]$	$[1, 1.001]$	$[1, 1.0001]$	$[1, 1.00001]$
$\Delta s/\Delta t$					

(b) In part (a), as the right-hand endpoint gets closer and closer to 1, what value does the average velocity $\Delta s/\Delta t$ seem to be approaching? This "target value" is called the **instantaneous velocity** (or just **velocity**) at $t = 1$.

52. Near the surface of the moon, the distance traveled by a falling object is given by the formula $s(t) = \frac{8}{3}t^2$, where t is measured in seconds, $s(t)$ is in feet, and $t = 0$ corresponds to the instant that the object begins to fall.
    (a) Show that the average velocity over the interval $[t_1, t_2]$ is given by
    $$\frac{\Delta s}{\Delta t} = \frac{8}{3}(t_2 + t_1)$$
    (b) Use the result in part (a) to complete the following table. (Round off to four decimal places.)

Interval	[2, 2.1]	[2, 2.01]	[2, 2.001]	[2, 2.0001]	[2, 2.00001]
$\Delta s/\Delta t$					

    (c) In part (b), as the right-hand endpoint gets closer and closer to 2, what value does the average velocity seem to be approaching? (As mentioned in the previous exercise, this target value is called the *instantaneous velocity* (or *velocity*) when $t = 2$.)

53. Suppose that during the first 4 hours of a laboratory experiment, the temperature of a certain substance is given by the formula $f(t) = t^3 - 6t^2 + 9t$, where t is measured in hours with $t = 0$ corresponding to the time the experiment begins and $f(t)$ is the temperature (degrees Fahrenheit) of the substance at time t. Calculate the average rate of change of temperature between the following times.
    (a) $t = 0$ and $t = 1$ (b) $t = 1$ and $t = 2$
    (c) $t = 0$ and $t = 3$ (d) $t = 0$ and $t = 4$

54. (a) Let f be the temperature function given in Exercise 53. Compute $\Delta f/\Delta t$ on the interval $[0, t_1]$.
    (b) Use the result in part (a) to complete the following table. (Round off to four decimal places.)

Interval	[0, 0.1]	[0, 0.01]	[0, 0.001]	[0, 0.0001]
$\Delta f/\Delta t$				

    (c) In part (b), as the right-hand endpoint gets closer and closer to 0, what value does the average rate of change of temperature seem to be approaching? (This target value tells us the rate at which the temperature is changing at the instant that the experiment begins.)

55. A bug travels counterclockwise around the square in the figure, beginning from the point $(1, 0)$. As indicated in the figure, t denotes the distance the bug has traveled, and $P(t)$ denotes the bug's location when it has traveled the distance t. For example, when $t = 0$,

$P(t)$ is the point $(1, 0)$, and when $t = 1$, $P(t)$ is the point $(1, 1)$.
(a) A function S is defined as follows: $S(t)$ is the y-coordinate of the point $P(t)$. For example, if $t = \frac{1}{2}$, then $P(t)$ is the point $(1, \frac{1}{2})$ and, therefore, $S(t) = \frac{1}{2}$. Set up a table showing the values of $S(t)$ with t running from 0 to 3 in increments of $\frac{1}{4}$. Then graph the function S on the interval $0 \le t \le 3$.

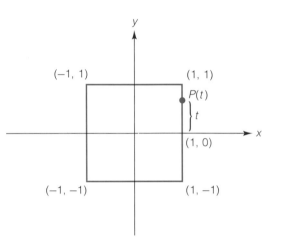

(b) Set up a table showing the values of $S(t)$ with t running from 3 to 5 in increments of $\frac{1}{4}$. Then graph the function S on the interval $3 \le t \le 5$.
(c) Graph the function S for $0 \le t \le 8$. This corresponds to one complete trip around the square by the bug.
(d) Graph the function S for $8 \le t \le 16$. How does your result relate to the graph in part (c)?

56. Let $P(t)$ be defined as in Exercise 55. Define a function C as follows: $C(t)$ is the x-coordinate of the point $P(t)$. Graph the function C for $0 \le t \le 8$.

57. The notation $[x]$ denotes the greatest integer that is less than or equal to x. For instance, $[3] = 3, [4\frac{1}{2}] = 4$, and $[-\frac{4}{3}] = -2$.
    (a) Complete the following table.

x	0	0.1	0.5	0.9	1.0
$[x]$					

    (b) Graph the **greatest integer function** $y = [x]$ on the interval $-2 \le x < 3$.
    (c) The domain of the greatest integer function is $(-\infty, \infty)$. What is the range?

OPTIONAL TI-81 GRAPHING CALCULATOR EXERCISES

EXERCISES FOR SECTION 3.2

In general, the techniques of calculus are required to obtain the precise coordinates of the turning points of a function. In Exercises 1–4, you will use the calculator to obtain the approximate coordinates of the turning points.

1. (a) Let $f(x) = x^3 - x + 1$. Enter this function in the Y= menu and press ZOOM 6 to see the graph in the standard viewing rectangle. How many turning points do there appear to be?

   (b) For a closer look at the turning points, press the RANGE key and make the following adjustments: $x_{min} = -2$, $x_{max} = 2$, $y_{min} = -2$, $y_{max} = 2$. Then press GRAPH.

   (c) Press the TRACE key. You will see a blinking cursor on the curve (on the y-axis). Press the right-arrow key (▶) repeatedly to move the cursor to the turning point in the first quadrant. (If you overshoot the mark, use the left-arrow key to back up.) The read-out at the bottom of the screen gives you the (approximate) coordinates of the turning point. Make a note of the coordinates (round off to three decimal places). Next, use the left-arrow key to move the cursor to the turning point in the second quadrant. Again, make a note of the coordinates.

   (d) Using calculus, it can be shown that the precise x-coordinates of the turning points are $\pm 1/\sqrt{3}$. Use your calculator to evaluate these numbers. To how many decimal places do your results in part (c) agree with these exact x-coordinates?

   (e) Use the graph in part (a) and the information given in part (d) to specify the intervals on which f is increasing and the intervals on which f is decreasing.

2. Follow the steps outlined in Exercise 1 to investigate the turning point(s) for the function $f(x) = x^3(x - 2)^2$. In part (b), use the following settings: $x_{min} = -1$, $x_{max} = 2.5$, $y_{min} = -0.5$, $y_{max} = 1.5$. For part (d) (for comparison), use the fact that the precise x-coordinates of the turning points are $x = 2$ and $x = \frac{6}{5}$.

3. Follow the steps outlined in Exercise 1 to investigate the turning point(s) for the function $f(x) = x/(2x^2 + 1)$. In part (b), use the following settings: $x_{min} = -2$, $x_{max} = 2$, $y_{min} = -0.5$, $y_{max} = 0.5$. For part (d) (for comparison), use the fact that the precise x-coordinates of the turning points are $x = \pm\sqrt{2}/2$.

4. Follow the steps outlined in Exercise 1 to investigate the turning point(s) for the function $f(x) = \sqrt[3]{x} - x^2$. In part (b), use the following settings: $x_{min} = -0.5$, $x_{max} = 1.5$, $y_{min} = -1$, $y_{max} = 1.5$. For

part (d) (for comparison), use the fact that the precise x-coordinate of the turning point is $1/\sqrt[5]{216}$.

5. The TI-81 calculator has a key for computing the average rate of change of a function on a given interval. As background for understanding how to use this key, refer to the following figure. In the figure, the number m is halfway between a and b, and h denotes the common distance from m to a or to b.

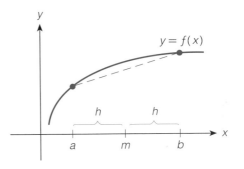

   (a) Use the figure to explain why the average rate of change of f on $[a, b]$ is given by

   $$\frac{\Delta f}{\Delta x} = \frac{f(m + h) - f(m - h)}{2h}$$

   or

   $$\frac{\Delta f}{\Delta x} = \frac{f(m + h) - f(m - h)}{b - a} \qquad (1)$$

   (b) In Example 6(c) in the text, we computed the average rate of change for the function $f(x) = x^2$ on the interval $[1, 2]$. Here is how to use the calculator and equation (1) to make that calculation. First of all, since $a = 1$ and $b = 2$, we have $m = 1.5$ and $h = 0.5$. On the calculator, use the following keystrokes to store the value of m:

   Next, press the MATH key and then 8. The screen display should then look like this:

   NDeriv(

   Now use the following sequence of keystrokes to tell the calculator that the function is x^2 and that the value of h is 0.5:

   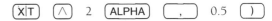

   (The key for the comma is in the bottom row on the calculator.) At this stage the screen display should look like this: NDeriv(X^2, 0.5). Finally, press the ENTER key. The calculator will display the number 3, which is the average rate of change of $f(x) = x^2$ on the interval $[1, 2]$.

In Exercises 6–10, use the calculator (as in Exercise 5) to compute the average rate of change for the function on the given interval.

6. $f(x) = 16x^2$ on $[1, 1.1]$ [Check that your answer is consistent with the result in Example 7(b) in the text.]

7. $f(x) = 3x^2 - 2x + 4$ on   (a) $[2, 3]$;   (b) $[0, \frac{2}{3}]$

8. $g(t) = 1/(1 + t^2)$ on   (a) $[0, 0.5]$;   (b) $[0, 0.1]$;   (c) $[0, 0.001]$

9. $y = (2.718)^x$ on   (a) $[-1, 1]$;   (b) $[-0.1, 0.1]$;   (c) $[-0.01, 0.01]$

10. $y = x^x$ on   (a) $[0.1, 0.3]$;   (b) $[0.1, 1]$;   (c) $[0.1, 5]$

3.3   TECHNIQUES IN GRAPHING

... geometrical figures are graphic formulas.

David Hilbert (1862–1943)

The simple geometric concepts of reflection and translation can be used to great advantage in graphing. We discussed the idea of reflection in the x-axis and in the y-axis in Section 2.4. By a **translation** of a graph, we mean a shift in its location such that every point of the graph is moved the same distance in the same direction. The size and the shape of a graph are unchanged by a translation.

EXAMPLE 1 The graph of a function G is the quarter-circle shown in Figure 1. In each case, sketch the resulting graph after the following operations are carried out on the graph of G:

(a) a translation of four units to the right followed by a translation of one unit vertically downward;

(b) a translation of three units to the right followed by a reflection in the y-axis;

(c) a reflection in the y-axis followed by a translation of three units to the right. [Note that these are the same two operations used in part (b), but here the order is reversed.]

$y = G(x)$

FIGURE 1

Solution

(a) Figure 2 shows the result of translating the graph of G four units to the right and then one unit down. (As you can check by drawing a sketch, the same end result is obtained if we first translate one unit down and then four units to the right. *Fact:* If several translations are to be carried out in succession, the end result will be the same, no matter in what order the individual translations are carried out.)

FIGURE 2
The red curve is the end result when the graph of G is translated four units to the right and then one unit down.

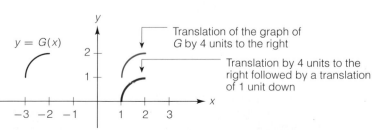

Translation of the graph of G by 4 units to the right

Translation by 4 units to the right followed by a translation of 1 unit down

(b) Figure 3(a) shows the graph of G after a translation of three units to the right. When this is followed by a reflection in the y-axis, we obtain the result shown in Figure 3(b).

FIGURE 3

Figure (a) shows the result of translating the graph of G three units to the right. When this is followed by a reflection in the y-axis, we obtain the red curve in Figure (b).

(a)   (b)

(c) Figure 4 shows what happens to the graph of G if we use the same two operations that were used in part (b) but reverse the order in which they are carried out. Note that the end result is quite different from that obtained in part (b). ▮▮▮

FIGURE 4

The reflection of the graph of G in the y-axis followed by a translation of three units to the right.

Given the graph of a function f, it is useful to know how to sketch efficiently the graphs of the following closely related functions. (In the following equations, c denotes a positive constant.)

$$y = f(x) + c \qquad y = f(x + c) \qquad y = -f(x)$$
$$y = f(x) - c \qquad y = f(x - c) \qquad y = f(-x)$$

Each of these can be graphed by translating or reflecting the graph of $y = f(x)$. In the box that follows, we summarize the techniques that are involved. (Exercises 45–48 will help you to see why these techniques are valid, and Section 10.2 will explain these techniques from a more general point of view.)

PROPERTY SUMMARY   **TRANSLATIONS AND REFLECTIONS**

(In items 1–4, the letter c denotes a positive constant.)

EQUATION	HOW TO OBTAIN THE GRAPH FROM THAT OF $y = f(x)$
1. $y = f(x) + c$	Translate c units vertically upward
2. $y = f(x) - c$	Translate c units vertically downward
3. $y = f(x + c)$	Translate c units to the left
4. $y = f(x - c)$	Translate c units to the right
5. $y = -f(x)$	Reflect in the x-axis
6. $y = f(-x)$	Reflect in the y-axis

In Figure 5 we show examples of the first four techniques in the box, the techniques involving translation. Figure 6 displays examples of the remaining two techniques, the ones involving reflection. You should look over these figures carefully before going on to Examples 2 through 5, in which the various graphing techniques are combined.

FIGURE 5

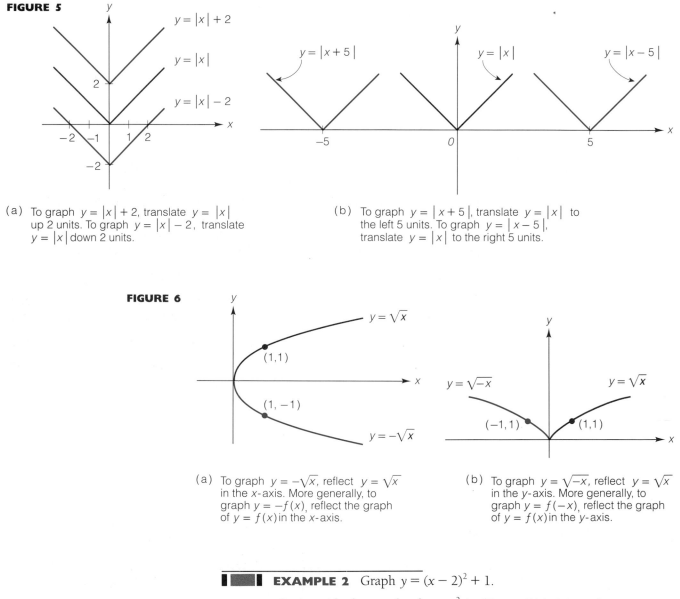

(a) To graph $y = |x| + 2$, translate $y = |x|$ up 2 units. To graph $y = |x| - 2$, translate $y = |x|$ down 2 units.

(b) To graph $y = |x + 5|$, translate $y = |x|$ to the left 5 units. To graph $y = |x - 5|$, translate $y = |x|$ to the right 5 units.

FIGURE 6

(a) To graph $y = -\sqrt{x}$, reflect $y = \sqrt{x}$ in the x-axis. More generally, to graph $y = -f(x)$, reflect the graph of $y = f(x)$ in the x-axis.

(b) To graph $y = \sqrt{-x}$, reflect $y = \sqrt{x}$ in the y-axis. More generally, to graph $y = f(-x)$, reflect the graph of $y = f(x)$ in the y-axis.

■■■ EXAMPLE 2 Graph $y = (x - 2)^2 + 1$.

Solution Begin with the graph of $y = x^2$ in Figure 7(a). Move the curve two units to the right to obtain the graph of $y = (x - 2)^2$; see Figure 7(b). Then move the curve in Figure 7(b) up one unit to obtain the graph of $y = (x - 2)^2 + 1$; see Figure 7(c). ■■■

FIGURE 7

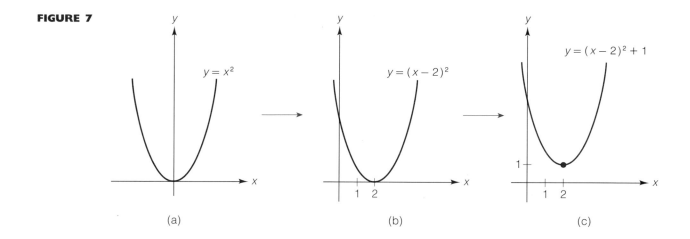

(a) (b) (c)

EXAMPLE 3 Graph the functions $y = \dfrac{1}{x-1}$ and $y = \dfrac{1}{x-1} + 1$.

Solution Begin with the graph of $y = 1/x$ in Figure 8(a). The x- and y-axes are asymptotes for this graph. Moving this graph to the right one unit yields the graph of $y = \dfrac{1}{x-1}$, shown in Figure 8(b). Note that the vertical asymptote moves one unit to the right also, but the horizontal asymptote is unchanged. Next, we move the graph in Figure 8(b) up one unit (why?) to obtain the graph of $y = \dfrac{1}{x-1} + 1$, as shown in Figure 8(c). (You should verify for yourself that the y-intercepts in Figures 8(b) and 8(c) are -1 and 0, respectively.) ∎

FIGURE 8

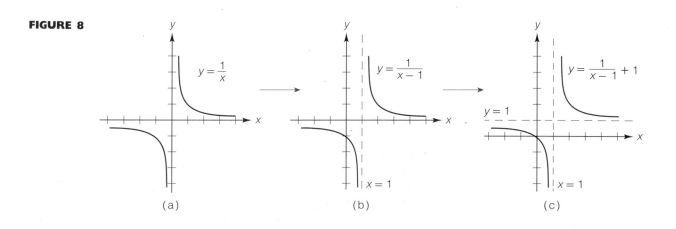

(a) (b) (c)

█ ▓ █ EXAMPLE 4 Graph each equation: **(a)** $y = -|x|$; **(b)** $y = -|x - 2| + 3$.

Solution See Figure 9.

FIGURE 9

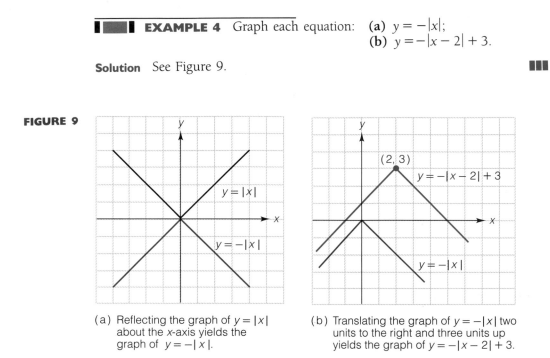

(a) Reflecting the graph of $y = |x|$ about the x-axis yields the graph of $y = -|x|$.

(b) Translating the graph of $y = -|x|$ two units to the right and three units up yields the graph of $y = -|x - 2| + 3$.

█ ▓ █ EXAMPLE 5 The graph of a function f is a line segment joining the points $(-3, 1)$ and $(2, 4)$. Graph each of the following functions.

(a) $y = f(-x)$ **(b)** $y = -f(x)$ **(c)** $y = -f(-x)$
(d) $y = f(x + 1)$ **(e)** $y = -f(x + 1)$ **(f)** $y = f(1 - x)$

Solution See Figure 10.

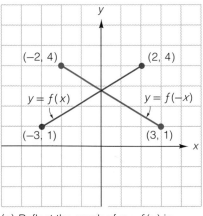

(a) Reflect the graph of $y = f(x)$ in the y-axis to obtain the graph of $y = f(-x)$.

(b) Reflect the graph of $y = f(x)$ in the x-axis to obtain the graph of $y = -f(x)$.

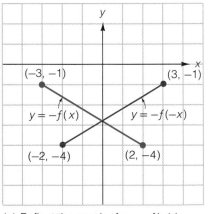

(c) Reflect the graph of $y = -f(x)$ in the y-axis to obtain the graph of $y = -f(-x)$. [Or, reflect the graph of $y = f(-x)$ in the x-axis.]

FIGURE 10

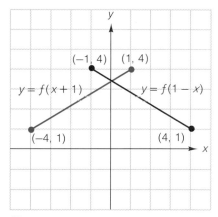

(d) Translate the graph of $y = f(x)$ one unit to the left to obtain the graph of $y = f(x + 1)$.

(e) Translate the graph of $y = -f(x)$ one unit to the left to obtain the graph of $y = -f(x + 1)$. [Or, reflect the graph of $y = f(x + 1)$ in the x-axis.]

(f) Reflect the graph of $y = f(x + 1)$ in Figure 10(d) in the y-axis to obtain the graph of $y = f(1 - x)$. [Reason: Replacing x with $-x$ in $y = f(x + 1)$ yields $y = f(1 - x)$.]

FIGURE 10 (continued)

EXERCISE SET 3.3

A

In Exercises 1 and 2, the right-hand column contains instructions for translating and/or reflecting the graph of $y = f(x)$. Match each equation in the left-hand column with an appropriate set of instructions in the right-hand column.

1.
(a) $y = f(x - 1)$ (A) Translate left 1 unit.
(b) $y = f(x) - 1$ (B) Reflect in the x-axis, then translate left 1 unit.
(c) $y = f(x) + 1$ (C) Translate right 1 unit.
(d) $y = f(x + 1)$ (D) Reflect in the x-axis, then translate up 1 unit.
(e) $y = f(-x) + 1$ (E) Reflect in the x-axis, then translate down 1 unit.
(f) $y = f(-x) - 1$ (F) Translate down 1 unit.
(g) $y = -f(x) + 1$ (G) Reflect in the x-axis, reflect in the y-axis, then translate up 1 unit.
(h) $y = -f(x + 1)$ (H) Translate left 1 unit, then reflect in the y-axis, then translate up 1 unit.
(i) $y = -f(x) - 1$ (I) Translate up 1 unit.
(j) $y = f(1 - x) + 1$ (J) Reflect in the y-axis, then translate up 1 unit.
    Hint: See Example 5, part (f)
(k) $y = -f(-x) + 1$ (K) Reflect in the y-axis, then translate down 1 unit.

2.
(a) $y = f(x + 2) + 3$ (A) Translate left 2 units, then translate down 3 units.
(b) $y = f(x + 3) + 2$ (B) Translate left 3 units, then translate up 2 units.
(c) $y = f(x - 2) + 3$ (C) Translate right 3 units, then translate up 2 units.
(d) $y = f(x - 2) - 3$ (D) Translate left 3 units, then translate down 2 units.
(e) $y = f(x + 2) - 3$ (E) Translate right 3 units and down 2 units.
(f) $y = f(x - 3) + 2$ (F) Reflect in the y-axis, then translate up 2 units.
(g) $y = f(x - 3) - 2$ (G) Reflect in the x-axis, then translate right 2 units.
(h) $y = f(x + 3) - 2$ (H) Reflect in the x-axis, then translate left 2 units.
(i) $y = -f(x + 2)$ (I) Translate left 2 units, then reflect in the y-axis.
(j) $y = -f(x - 2)$ (J) Translate right 2 units, then translate up 3 units.
(k) $y = f(2 - x)$ (K) Translate left 2 units, then translate up 3 units.
(l) $y = f(-x) + 2$ (L) Translate right 2 units and down 3 units.

In Exercises 3–24, sketch the graph of the function.

3. $y = x^3 - 3$ **4.** $y = x^2 + 3$
5. $y = (x + 4)^2$ **6.** $y = (x + 4)^2 - 3$
7. $y = (x - 4)^2$ **8.** $y = (x - 4)^2 + 1$

9. $y = -x^2$

10. $y = -x^2 - 3$

11. $y = -(x-3)^2$

12. $y = -(x-3)^2 - 3$

13. $y = \sqrt{x-3}$

14. $y = \sqrt{x-3} + 1$

15. $y = -\sqrt{x+1}$

16. $y = -\sqrt{x+1} + 1$

17. $y = \dfrac{1}{x+2} + 2$

18. $y = \dfrac{1}{x-3} - 1$

19. $y = (x-2)^3$

20. $y = (x-2)^3 + 1$

21. $y = -x^3 + 4$

22. $y = -(x-1)^3 + 4$

23. (a) $y = |x+4|$ (b) $y = |4-x|$
    (c) $y = -|4-x| + 1$

24. (a) $y = \sqrt{x+2}$ (b) $y = \sqrt{2-x}$
    (c) $y = -\sqrt{2-x}$

In Exercises 25–40, sketch the graph of the function, given that f, F, and g are defined as follows:

$$f(x) = |x| \qquad F(x) = 1/x \qquad g(x) = \sqrt{1-x^2}$$

25. $y = f(x-5)$

26. $y = -f(x-5)$

27. $y = f(5-x)$

28. $y = -f(5-x)$

29. $y = 1 - f(x-5)$

30. $y = f(-x)$

31. $y = F(x+3)$

32. $y = F(x) + 3$

33. $y = -F(x+3)$

34. $y = F(-x) + 3$

35. $y = g(x-2)$

36. $y = -g(x-2)$

37. $y = 1 - g(x-2)$

38. $y = g(-x)$

39. $y = g(2-x)$

40. $y = -g(2-x)$

In Exercises 41 and 42, refer to the graph of the function f shown in the figure to graph each function.

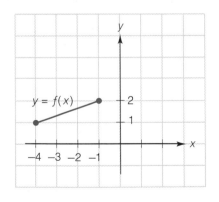

41. (a) $y = -f(x)$ (b) $y = f(-x)$
    (c) $y = -f(-x)$

42. (a) $y = f(x) - 2$ (b) $y = f(x-2)$
    (c) $y = f(x-2) - 2$

In Exercises 43 and 44, refer to the graph of the function g shown in the figure in order to graph each function.

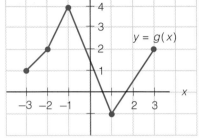

43. (a) $y = g(-x)$ (b) $y = -g(x)$
    (c) $y = -g(-x)$

44. $y = -g(x-3) - 1$

45. (a) Complete the following table.

x	x^2	$x^2 - 1$	$x^2 + 1$
0			
± 1			
± 2			
± 3			

    (b) Using the results in the table, graph the functions $y = x^2$, $y = x^2 - 1$, and $y = x^2 + 1$ on the same set of axes. How are the graphs related?

46. (a) Complete the given table.

x	x^2	$(x-1)^2$	$(x+1)^2$
0			
1			
2			
3			
-1			
-2			
-3			

    (b) Using the results in the table, graph the functions $y = x^2$, $y = (x-1)^2$, and $y = (x+1)^2$ on the same set of axes. How are the graphs related?

47. (a) Complete the following table. (Use a calculator where necessary.)

x	$\sqrt{x}$	$-\sqrt{x}$
0		
1		
2		
3		
4		
5		

(b) Using the results in the table, graph the functions $y = \sqrt{x}$ and $y = -\sqrt{x}$ on the same set of axes. How are the two graphs related?

48. (a) Complete the given tables. (Use a calculator where necessary.)

x	0	1	2	3	4	5
$\sqrt{x}$						

x	0	-1	-2	-3	-4	-5
$\sqrt{-x}$						

(b) Using the tables, graph the functions $y = \sqrt{x}$ and $y = \sqrt{-x}$ on the same set of axes. How are the graphs related?

B

For Exercises 49 and 50, assume that (a, b) is a point on the graph of $y = f(x)$, and specify the corresponding point on the graph of each equation. [For example, the point that corresponds to (a, b) on the graph of $y = f(x - 1)$ is $(a + 1, b)$.]

49. (a) $y = f(x - 3)$ (b) $y = f(x) - 3$
    (c) $y = f(x - 3) - 3$ (d) $y = -f(x)$
    (e) $y = f(-x)$ (f) $y = -f(-x)$
    (g) $y = f(3 - x)$ (h) $y = -f(3 - x) + 1$

50. (a) $y = f(-x) + 2$ (b) $y = -f(-x) + 2$
    (c) $y = -f(x - 3)$ (d) $y = 1 - f(x + 1)$
    (e) $y = f(1 - x)$ (f) $y = -f(1 - x) + 1$

51. Show that $\dfrac{x}{x - 1} = \dfrac{1}{x - 1} + 1$, provided that $x \neq 1$. Then use this fact to graph the function g defined by $g(x) = x/(x - 1)$.

52. Verify that $-x^2 + 2x + 2 = -(x - 1)^2 + 3$. Then use this fact to graph the function h defined by $h(x) = -x^2 + 2x + 2$.

53. What is the range of the function f defined by

$$f(x) = \frac{x^2}{1 - \sqrt{1 - x^2}}?$$ *Hint:* The function is easy to graph after rationalizing the denominator.

54. (a) A function f is said to be **even** if the equation $f(-x) = f(x)$ is satisfied by all values of x in the domain of f. Explain why the graph of an even function must be symmetric about the y-axis.
    (b) Show that each function is even by computing $f(-x)$ and then noting that $f(x)$ and $f(-x)$ are equal.
        (i) $f(x) = x^2$ (ii) $f(x) = 2x^4 - 6$
        (iii) $f(x) = 3x^6 - \dfrac{4}{x^2} + 1$

55. (a) A function f is said to be **odd** if the equation $f(-x) = -f(x)$ is satisfied by all values of x in the domain of f. Show that if (x, y) is a point on the graph of an odd function f, then the point $(-x, -y)$ is also on the graph. (This implies that the graph of an odd function must be symmetric about the origin.)
    (b) Show that each function is odd by computing $f(-x)$ as well as $-f(x)$ and then noting that the two expressions obtained are equal.
        (i) $f(x) = x^3$ (ii) $f(x) = -2x^5 + 4x^3 - x$
        (iii) $f(x) = \dfrac{|x|}{x + x^7}$

56. Is each function odd, even, or neither? (See Exercises 54 and 55 for definitions.)
    (a) $f(x) = \dfrac{1 - x^2}{2 + x^2}$ (b) $g(x) = \dfrac{x - x^3}{2x + x^3}$
    (c) $h(x) = x^2 + x$ (d) $F(x) = (x^2 + x)^2$
    (e) $G(x) = \begin{cases} 1 & \text{if } x > 0 \\ 0 & \text{if } x = 0 \\ -1 & \text{if } x < 0 \end{cases}$
    Suggestion for part (e): Look at the graph.

57. Suppose that the function f is increasing on the interval $(2, 4)$ and decreasing on the intervals $(-\infty, 2)$ and $(4, \infty)$. On what interval(s) is each function increasing?
    (a) $y = f(-x)$ (b) $y = -f(x)$

58. Suppose that the function g is decreasing on $(-\infty, 1)$ and increasing on $(1, \infty)$. On what interval(s) is the function $y = -f(-x)$ increasing?

OPTIONAL TI-81 GRAPHING CALCULATOR EXERCISES
EXERCISES FOR SECTION 3.3

In Exercises 1–3, you will use the calculator to provide examples of the six graphing techniques listed in the box on page 123.

1. (a) Press the Y= key and enter the two functions $Y_1 = x^{2/3}$ and $Y_2 = x^{2/3} + 3$. In the RANGE menu, use the settings $x_{min} = -10$, $x_{max} = 10$, $y_{min} = 0$, $y_{max} = 10$. Now press GRAPH and observe that Y_2 is obtained by translating Y_1 up three units. (This illustrates property 1 in the box on page 123.)

   Calculator note: If you enter the function Y_1 using the ∧ key, you'll find that the calculator graphs only the portion of the curve corresponding to $x \geq 0$. To avoid this problem, enter Y_1 using the following keystrokes:

   (MATH) 4 ((X|T) (x²))

   Another calculator note: After you've entered the function Y_1, you can use the VARS key to enter Y_2 simply as $Y_1 + 3$. To do this, begin in the Y= menu with the blinking cursor just to the right of the equal sign in the line for Y_2. Then use the following sequence of keystrokes:

   (2nd) (VARS) 1 (+) 3

   Now press the GRAPH key to obtain the required graphs.

   (b) Press the Y= key, and in the line for Y_2, replace the plus sign with a minus sign (so that Y_2 is now $Y_1 - 3$). In the RANGE menu, change the settings (from part (a)) so that $y_{min} = -5$ and $y_{max} = 5$. Now press the GRAPH key and observe that Y_2 is obtained by translating Y_1 down three units. (This illustrates property 2 in the box on page 123.)

2. Press the Y= key and enter the following three functions.

   $$Y_1 = \sqrt{1 - x^3} \quad Y_2 = \sqrt{1 - (x+2)^3} \quad Y_3 = \sqrt{1 - (x-2)^3}$$

   Press the RANGE key and use the following settings: $x_{min} = -7$; $x_{max} = 3$; $y_{min} = 0$; $y_{max} = 10$. Now press GRAPH and observe that Y_2 is obtained by translating Y_1 two units to the *left* and that Y_3 is obtained by

translating Y_1 two units to the *right*. (This illustrates properties 3 and 4 in the box on page 123.)

3. (a) Press the Y= key and enter the functions $Y_1 = 2^x$ and $Y_2 = -2^x$. Graph these two functions using the following RANGE settings: $x_{min} = -4$; $x_{max} = 4$; $y_{min} = -4$; $y_{max} = 4$. Observe that Y_2 is obtained by reflecting Y_1 in the x-axis. (This illustrates property 5 in the box on page 123.)

   (b) Retain the function Y_1 from part (a), and replace Y_2 with $Y_2 = 2^{-x}$. Press the GRAPH key and note that Y_2 is obtained by reflecting Y_1 in the y-axis. (This illustrates property 6 in the box on page 123.)

4. (a) Graph the function $f(x) = x/(x - 1)$ using the following RANGE settings: $x_{min} = -5$; $x_{max} = 5$; $y_{min} = -5$; $y_{max} = 5$.

   (b) From the screen display, it appears that the graph may be a translation of the basic graph $y = 1/x$. Prove that this is indeed the case by using algebra (not your calculator!) to show that

   $$\frac{x}{x-1} = \frac{1}{x-1} + 1$$

   What translations are involved here?

For Exercises 5–10, tell how the graph in part (a) should be translated and/or reflected to obtain the remaining graphs. Then use the calculator to draw the graphs and check your statements.

5. (a) $y = x^2$ (b) $y = (x+3)^2$
   (c) $y = -(x+3)^2$ (d) $y = (-x+3)^2$

6. (a) $y = x^3$ (b) $y = (x+2)^3$
   (c) $y = -(x+2)^3$ (d) $y = (-x+2)^3$

7. (a) $y = |x|$ (b) $y = |x+4|$
   (c) $y = |-x+4|$ (d) $y = |-x+4| - 2$

8. (a) $F(x) = \frac{1}{2}x^3 - \frac{3}{2}x^2$ (b) $y = F(-x)$
   (c) $y = -F(x)$ (d) $y = -F(-x)$

9. (a) $G(x) = 2/(1 + x^2)$ (b) $y = G(x+3)$
   (c) $y = G(x-3)$ (d) $y = -G(x) + 2$

10. (a) $y = \sqrt{x+2}$ (b) $y = \sqrt{2-x}$
    (c) $y = 3 - \sqrt{2-x}$

3.4 METHODS OF COMBINING FUNCTIONS

The notation ϕx to indicate a function of x was introduced by ... [John Bernoulli] in 1718, ... but the general adoption of symbols like f, F, ϕ, and ψ, ... to represent functions, seems to be mainly due to Euler and Lagrange.

W. W. Rouse Ball in *A Short Account of the History of Mathematics* (New York: Dover Publications, 1960)

Two given numbers a and b can be combined in various ways to produce a third number. For instance, we can form the sum $a + b$ or the difference $a - b$ or the product ab. Also, if $b \neq 0$, we can form the quotient a/b. Similarly, two functions can be combined in various ways to produce a third function. Suppose, for example, that we start with the two functions $y = x^2$ and $y = x^3$. It seems natural to define their sum, difference, product, and quotient as follows:

$$\text{sum:} \quad y = x^2 + x^3$$
$$\text{difference:} \quad y = x^2 - x^3$$
$$\text{product:} \quad y = x^2 x^3 = x^5$$
$$\text{quotient:} \quad y = \frac{x^2}{x^3} = \frac{1}{x}$$

Indeed, this is the idea behind the formal definitions we now give.

DEFINITIONS   Arithmetical Operations with Functions

Let f and g be two functions. Then the **sum** $f + g$, the **difference** $f - g$, the **product** fg, and the **quotient** f/g are functions defined by the following equations:

$$(f + g)(x) = f(x) + g(x) \tag{1}$$
$$(f - g)(x) = f(x) - g(x) \tag{2}$$
$$(fg)(x) = f(x) \cdot g(x) \tag{3}$$
$$(f/g)(x) = f(x)/g(x) \qquad \text{provided } g(x) \neq 0 \tag{4}$$

For the functions defined by equations (1), (2), and (3), the domain is the set of all inputs x belonging to both the domain of f and the domain of g. For the quotient function in equation (4), we impose the additional restriction that the domain exclude all inputs x for which $g(x) = 0$.

EXAMPLE 1   Let $f(x) = 3x + 1$ and $g(x) = x - 1$. Compute each of the following:

(a) $(f + g)(x)$;   **(b)** $(f - g)(x)$;   **(c)** $(fg)(x)$;   **(d)** $(f/g)(x)$.

Solution

(a) $(f + g)(x) = f(x) + g(x) = (3x + 1) + (x - 1) = 4x$

(b) $(f - g)(x) = f(x) - g(x) = (3x + 1) - (x - 1) = 2x + 2$

(c) $(fg)(x) = [f(x)][g(x)] = (3x + 1)(x - 1) = 3x^2 - 2x - 1$

(d) $(f/g)(x) = \dfrac{f(x)}{g(x)} = \dfrac{3x + 1}{x - 1} \qquad (x \neq 1)$

For the remainder of this section, we are going to discuss a method for combining functions known as **composition of functions**. As you will see, this method is based on the familiar algebraic process of substitution. Suppose, for example, that f and g are two functions defined by

$$f(x) = x^2 \qquad g(x) = 3x + 1$$

Choose any number in the domain of g, say $x = -2$. We can compute $g(-2)$:

$$g(-2) = 3(-2) + 1 = -5$$

Now let's use the output -5 that g has produced as an *input* for f. We obtain

$$f(-5) = (-5)^2 = 25$$

and, consequently,

$$f[g(-2)] = 25$$

So, beginning with the input -2, we've successively applied g and then f to obtain the output 25. Similarly, we could carry out this same procedure for any other number in the domain of g. Here is a summary of the procedure:

1. Start with an input x and calculate $g(x)$.
2. Use $g(x)$ as an input for f; that is, calculate $f[g(x)]$.

We use the notation $f \circ g$ to denote the function, or rule, that tells us to assign the output $f[g(x)]$ to the initial input x. In other words, $f \circ g$ denotes the rule consisting of two steps: *First apply g; then apply f.* We read the notation $f \circ g$ as f *circle g* or f *composed with g.* In Figure 1, we summarize these ideas.

FIGURE 1
Diagram for the function $f \circ g$

When we write $g(x)$, we assume that x is in the domain of the function g. Likewise, for the notation $f[g(x)]$ to make sense, the outputs $g(x)$ must themselves be acceptable inputs for the function f. Our formal definition, then, for the composite function $f \circ g$ is as follows.

DEFINITION   Composition of Functions: $f \circ g$

Given two functions f and g, the function $f \circ g$ is defined by

$$(f \circ g)(x) = f[g(x)]$$

The domain of $f \circ g$ consists of those inputs x (in the domain of g) for which $g(x)$ is in the domain of f.

▌�▮▌ **EXAMPLE 2**   Let $f(x) = x^2$ and $g(x) = 3x + 1$. Compute $(f \circ g)(x)$ and $(g \circ f)(x)$.

Solution

$$(f \circ g)(x) = f[g(x)] \qquad \text{definition of } f \circ g$$
$$= f(3x + 1) \qquad \text{definition of } g$$
$$= (3x + 1)^2 \qquad \text{definition of } f$$
$$= 9x^2 + 6x + 1$$

$$(g \circ f)(x) = g[f(x)] \qquad \text{definition of } g \circ f$$
$$= g(x^2) \qquad \text{definition of } f$$
$$= 3(x^2) + 1 \qquad \text{definition of } g$$
$$= 3x^2 + 1$$

Notice that the two results obtained in Example 2 are not the same. This shows that, in general, $f \circ g$ and $g \circ f$ represent different functions.

EXAMPLE 3 Let f and g be defined as in Example 2: $f(x) = x^2$ and $g(x) = 3x + 1$. Compute $(f \circ g)(-2)$.

Solution   We will show two methods.

FIRST METHOD   Using the formula for $(f \circ g)(x)$ developed in Example 2, we have

$$(f \circ g)(x) = 9x^2 + 6x + 1$$

and therefore,

$$(f \circ g)(-2) = 9(-2)^2 + 6(-2) + 1$$
$$= 36 - 12 + 1$$
$$= 25$$

ALTERNATIVE METHOD   Working directly from the definition of $f \circ g$, we have

$$(f \circ g)(-2) = f[g(-2)]$$
$$= f[3(-2) + 1]$$
$$= f(-5)$$
$$= (-5)^2 = 25$$

In Examples 2 and 3, the domain of both f and g is the set of all real numbers. And as you can easily check, the domain of both $f \circ g$ and $g \circ f$ is also the set of all real numbers. In Example 4, however, some care needs to be taken in describing the domain of the composite function.

EXAMPLE 4 Let f and g be defined as follows:

$$f(x) = x^2 + 1 \qquad g(x) = \sqrt{x}$$

Compute $(f \circ g)(x)$. Find the domain of $f \circ g$ and sketch its graph.

Solution

$$(f \circ g)(x) = f[g(x)]$$
$$= f(\sqrt{x}) = (\sqrt{x})^2 + 1 = x + 1$$

So we have $(f \circ g)(x) = x + 1$. Now what about the domain of $f \circ g$? Our first inclination may be to say (incorrectly!) that the domain is the set of all real numbers, since any real number can be used as an input in the expression $x + 1$. However, the definition of $f \circ g$ on page 132 tells us that the inputs for $f \circ g$ must first of all be acceptable inputs for g. Given the definition of g, then,

FIGURE 2

Graph of the function $f \circ g$ in Example 4

we must require that x be nonnegative. On the other hand, for any nonnegative input x, the number $g(x)$ will be an acceptable input for f. (Why?) In summary, then, the domain of $f \circ g$ is the interval $[0, \infty)$. The graph of $f \circ g$ is shown in Figure 2. ▪▪▪

One reason for studying the composition of functions is that it lets us express a given function in terms of simpler functions. This is often useful in calculus. Suppose, for example, that we wish to express the function C defined by

$$C(x) = (2x^3 - 5)^2$$

as a composition of simpler functions. That is, we want to come up with two functions f and g so that the equation

$$C(x) = (f \circ g)(x)$$

holds for every x in the domain of C.

We begin by thinking what we would do to compute $(2x^3 - 5)^2$ for a given value of x. First, we would compute the quantity $2x^3 - 5$, then we would square the result. Therefore, recalling that the rule $f \circ g$ tells us to do g first, we let $g(x) = 2x^3 - 5$. And then, since the next step is squaring, we let $f(x) = x^2$. Now let us see if these choices for f and g are correct; that is, let us calculate $(f \circ g)(x)$ and see if it really is the same as $C(x)$.

Using $f(x) = x^2$ and $g(x) = 2x^3 - 5$, we have

$$(f \circ g)(x) = f[g(x)] = f(2x^3 - 5) = (2x^3 - 5)^2 = C(x)$$

This shows that our choices for f and g were indeed correct, and we have expressed C as a composition of two simpler functions.*

Note that in expressing C as $f \circ g$, we chose g to be the "inner" function, that is, the quantity inside parentheses: $g(x) = 2x^3 - 5$. This observation is used in Example 5.

▪▪▪▪ **EXAMPLE 5** Let $s(x) = \sqrt{1 + x^4}$. Express the function s as a composition of two simpler functions f and g.

Solution Let g be the "inner" function; that is, let $g(x)$ be the quantity inside the radical:

$$g(x) = 1 + x^4$$

And let us take f to be the square root function:

$$f(x) = \sqrt{x}$$

Now we need to verify that these are appropriate choices for f and g; that is, we need to check that the equation $(f \circ g)(x) = s(x)$ is true for every x in the domain of s. We have

$$(f \circ g)(x) = f[g(x)] = f(1 + x^4) = \sqrt{1 + x^4} = s(x)$$

Thus, $(f \circ g)(x) = s(x)$, as required. ▪▪▪

---

*Other answers are possible, too. For instance, if $F(x) = (x - 5)^2$ and $G(x) = 2x^3$, then (as you should verify for yourself) $C(x) = F[G(x)]$.

EXERCISE SET 3.4

A

In Exercises 1–10, compute each expression, given that the functions f, g, h, k, and m are defined as follows:

$$f(x) = 2x - 1 \qquad k(x) = 2, \quad \text{for all } x$$
$$g(x) = x^2 - 3x - 6 \qquad m(x) = x^2 - 9$$
$$h(x) = x^3$$

1. (a) $(f + g)(x)$ (b) $(f - g)(x)$ (c) $(f - g)(0)$
2. (a) $(fh)(x)$ (b) $(h/f)(x)$ (c) $(f/h)(1)$
3. (a) $(m - f)(x)$ (b) $(f - m)(x)$
4. (a) $(fg)(x)$ (b) $(fg)\left(\frac{1}{2}\right)$
5. (a) $(fk)(x)$ (b) $(kf)(x)$ (c) $(fk)(1) - (kf)(2)$
6. (a) $(g + m)(x)$ (b) $(g + m)(x) - (g - m)(x)$
7. (a) $(f/m)(x) - (m/f)(x)$ (b) $(f/m)(0) - (m/f)(0)$
8. (a) $[h \cdot (f + m)](x)$ *Note:* h and $(f + m)$ are two functions; the notation $h \cdot (f + m)$ denotes the product function.
    (b) $(hf)(x) + (hm)(x)$
9. (a) $[m \cdot (k - h)](x)$ (b) $(mk)(x) - (mh)(x)$
    (c) $(mk)(-1) - (mh)(-1)$
10. (a) $(g + g)(x)$ (b) $(g - g)(x)$
    (c) $(kg)(x)$ (d) $(g + g)(-3) - (kg)(-3)$
11. Let $f(x) = 3x + 1$ and $g(x) = -2x - 5$. Compute the following.
    (a) $(f \circ g)(x)$ (b) $(f \circ g)(10)$
    (c) $(g \circ f)(x)$ (d) $(g \circ f)(10)$
12. Let $f(x) = 1 - 2x^2$ and $g(x) = x + 1$. Compute the following.
    (a) $(f \circ g)(x)$ (b) $(f \circ g)(-1)$
    (c) $(g \circ f)(x)$ (d) $(g \circ f)(-1)$
    (e) $(f \circ f)(x)$ (f) $(g \circ g)(-1)$
13. Compute $(f \circ g)(x)$, $(f \circ g)(-2)$, $(g \circ f)(x)$, and $(g \circ f)(-2)$ for each pair of functions.
    (a) $f(x) = 1 - x$; $g(x) = 1 + x$
    (b) $f(x) = x^2 - 3x - 4$; $g(x) = 2 - 3x$
    (c) $f(x) = x/3$; $g(x) = 1 - x^4$
    (d) $f(x) = 2^x$; $g(x) = x^2 + 1$
    (e) $f(x) = x$; $g(x) = 3x^5 - 4x^2$
    (f) $f(x) = 3x - 4$; $g(x) = \frac{1}{3}(x + 4)$
14. Let $h(x) = 4x^2 - 5x + 1$, $k(x) = x$, and $m(x) = 7$ for all x. Compute the following.
    (a) $h[k(x)]$ (b) $k[h(x)]$ (c) $h[m(x)]$
    (d) $m[h(x)]$ (e) $k[m(x)]$ (f) $m[k(x)]$
15. Let $F(x) = \dfrac{3x - 4}{3x + 3}$ and $G(x) = \dfrac{x + 1}{x - 1}$. Compute the following.

    (a) $(F \circ G)(x)$ (b) $F[G(t)]$ (c) $(F \circ G)(2)$
    (d) $(G \circ F)(x)$ (e) $G[F(y)]$ (f) $(G \circ F)(2)$
16. Let $f(x) = \dfrac{1}{x^2} + 1$ and $g(x) = \dfrac{1}{x - 1}$.
    (a) Compute $(f \circ g)(x)$.
    (b) What is the domain of $f \circ g$?
    (c) Graph the function $f \circ g$.
17. Let $M(x) = \dfrac{2x - 1}{x - 2}$.
    (a) Compute $M(7)$ and then $M[M(7)]$.
    (b) Compute $(M \circ M)(x)$.
    (c) Compute $(M \circ M)(7)$, using the formula you obtained in part (b). Check that your answer agrees with that obtained in part (a).
18. Let $F(x) = (x + 1)^5$, $f(x) = x^5$, and $g(x) = x + 1$. Which of the following is true for all x?

    $$(f \circ g)(x) = F(x) \qquad \text{or} \qquad (g \circ f)(x) = F(x)$$

19. Refer to the graphs of the functions f, g, and h to compute the required quantities. Assume that all the axes are marked off in one-unit intervals.

(a)

(b)

(c)

(a) $f[g(3)]$ (b) $g[f(3)]$ (c) $f[h(3)]$

(d) $(h \circ g)(2)$ (e) $h\{f[g(3)]\}$

(f) $(g \circ f \circ h \circ f)(2)$ *Note:* This means first do f, then h, then f, then g.

20. Let $f(x) = 2x - 1$. Find $f[f(x)]$. If $z = f[f(x)]$, find $f(z)$.

21. (a) Let $T(x) = 4x^3 - 3x^2 + 6x - 1$ and $I(x) = x$. Find $(T \circ I)(x)$ and $(I \circ T)(x)$.

    (b) Let $G(x) = ax^2 + bx + c$ and $I(x) = x$. Find $(G \circ I)(x)$ and $(I \circ G)(x)$.

    (c) What general conclusion do you arrive at from the results of parts (a) and (b)?

22. The figure shows the graphs of two functions, F and G. Use the graphs to compute each quantity.

    (a) $(G \circ F)(1)$ (b) $(F \circ G)(1)$

    (c) $(F \circ F)(1)$ (d) $(G \circ G)(-3)$

    (e) $(G \circ F)(5)$ (f) $(F \circ F)(5)$

    (g) $(F \circ G)(-1)$ (h) $(G \circ F)(2)$

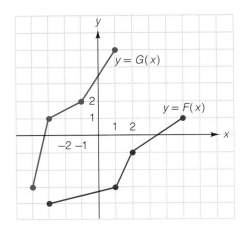

23. The domain of a function f consists of the numbers $-1, 0, 1, 2,$ and 3. The following table shows the output f assigns to each input.

x	-1	0	1	2	3
$f(x)$	2	2	0	3	1

The domain of a function g consists of the numbers $0, 1, 2, 3,$ and 4. The following table shows the output g assigns to each input.

x	0	1	2	3	4
$g(x)$	3	2	0	4	-1

Use this information to complete the tables for $f \circ g$ and $g \circ f$. *Note:* Two of the entries will be undefined.

x	0	1	2	3	4
$(f \circ g)(x)$					

x	-1	0	1	2	3	4
$(g \circ f)(x)$						

24. The following two tables show certain pairs of inputs and outputs for functions f and g.

x	0	$\pi/6$	$\pi/4$	$\pi/3$	$\pi/2$
$f(x)$	0	1/2	$\sqrt{2}/2$	$\sqrt{3}/2$	1

y	0	1/4	$\sqrt{2}/4$	1/2	$\sqrt{2}/2$	3/4	$\sqrt{3}/2$	1
$g(y)$	$\pi/2$	π	0	$\pi/3$	$\pi/4$	0	$\pi/6$	0

Use this information to complete the following table of values for $(g \circ f)(x)$.

x	0	$\pi/6$	$\pi/4$	$\pi/3$	$\pi/2$
$(g \circ f)(x)$					

25. Let $f(x) = 2x + 1$ and $g(x) = 3x - 4$.

    (a) Find $(f \circ g)(x)$ and graph the function $f \circ g$.

    (b) Find $(g \circ f)(x)$ and graph the function $g \circ f$.

26. Let $F(x) = x^2$ and $G(x) = x - 4$.

    (a) Find $F[G(x)]$ and graph the function $F \circ G$.

    (b) Find $G[F(x)]$ and graph the function $G \circ F$.

27. Let $g(x) = \sqrt{x} - 3$ and $f(x) = x - 1$.

    (a) Sketch a graph of g. Specify the domain and range.

    (b) Sketch a graph of f. Specify the domain and range.

    (c) Compute $(f \circ g)(x)$. Graph the function $f \circ g$ and specify its domain and range.

    (d) Find a formula for $g[f(x)]$. Which values of x are acceptable inputs here? That is, what is the domain of $g \circ f$?

    (e) Use the results of part (d) to sketch a graph of the function $g \circ f$.

28. Let $F(x) = -x^2$ and $G(x) = \sqrt{x}$. Determine the domains of $F \circ G$ and $G \circ F$.

29. Let $C(x) = (3x - 1)^4$. Express C as a composition of two simpler functions.

30. Let $C(x) = (1 + x^2)^3$. Find functions f and g so that $C(x) = (f \circ g)(x)$ is true for all values of x.

31. Express each function as a composition of two functions.

(a) $F(x) = \sqrt[3]{3x + 4}$ (b) $G(x) = |2x - 3|$
(c) $H(x) = (ax + b)^5$ (d) $T(x) = 1/\sqrt{x}$

32. Let $a(x) = x^2$, $b(x) = |x|$, and $c(x) = 3x - 1$. Express each of the following functions as a composition of two of the given functions.
(a) $f(x) = (3x - 1)^2$ (b) $g(x) = |3x - 1|$
(c) $h(x) = 3x^2 - 1$

33. Let $a(x) = 1/x$, $b(x) = \sqrt[3]{x}$, $c(x) = 2x + 1$, and $d(x) = x^2$. Express each of the following functions as a composition of two of the given functions.
(a) $f(x) = \sqrt[3]{2x + 1}$ (b) $g(x) = 1/x^2$
(c) $h(x) = 2x^2 + 1$ (d) $K(x) = 2\sqrt[3]{x} + 1$
(e) $l(x) = \dfrac{2}{x} + 1$ (f) $m(x) = \dfrac{1}{2x + 1}$
(g) $n(x) = x^{2/3}$

34. Express $G(x) = 1/(1 + x^4)$ as a composition of two simpler functions, one of which is $f(x) = 1/x$.

B

35. The circumference of a circle with radius r is given by $C(r) = 2\pi r$. Suppose that the circle is shrinking in size and that the radius at time t is given by

$$r = f(t) = \frac{1}{t^2 + 1}$$

Assume that t is measured in minutes and that r, or $f(t)$, is in feet. Find $(C \circ f)(t)$ and use this to find the circumference when $t = 3$ min.

36. A spherical weather balloon is being inflated in such a way that the radius is given by

$$r = g(t) = \tfrac{1}{2}t + 2$$

Assume that r is in meters and t is in seconds, with $t = 0$ corresponding to the time that inflation begins. If the volume of a sphere of radius r is given by

$$V(r) = \tfrac{4}{3}\pi r^3$$

compute $V[g(t)]$ and use this to find the time at which the volume of the balloon is 36π m^3.

37. Suppose that a manufacturer knows that the daily production cost to build x bicycles is given by the function C, where

$$C(x) = 100 + 90x - x^2 \qquad (0 \le x \le 40)$$

That is, $C(x)$ represents the cost in dollars of building x bicycles. Furthermore, suppose that the number of bicycles that can be built in t hr is given by the function f, where

$$x = f(t) = 5t \qquad (0 \le t \le 8)$$

(a) Compute $(C \circ f)(t)$.

(b) Compute the production cost on a day that the factory operates for $t = 3$ hr.
(c) If the factory runs for 6 hr instead of 3 hr, is the cost twice as much?

38. Suppose that in a certain biology lab experiment, the number of bacteria is related to the temperature of the environment by the function

$$N(T) = -2T^2 + 240T - 5400 \qquad (40 \le T \le 90)$$

Here, $N(T)$ represents the number of bacteria present when the temperature is T degrees Fahrenheit. Also, suppose that t hr after the experiment begins, the temperature is given by

$$T(t) = 10t + 40 \qquad (0 \le t \le 5)$$

(a) Compute $N[T(t)]$.
(b) How many bacteria are present when $t = 0$? When $t = 2$ hr? When $t = 5$ hr?

39. Let $g(x) = 4x - 1$. Find $f(x)$, given that the equation $(g \circ f)(x) = x + 5$ is true for all values of x.

40. Let $g(x) = 2x + 1$. Find $f(x)$, given that $(g \circ f)(x) = 10x - 7$.

41. Let $f(x) = -2x + 1$ and $g(x) = ax + b$. Find a and b so the equation $f[g(x)] = x$ holds for all values of x.

42. Let $f(x) = \dfrac{3x - 4}{x - 3}$.
(a) Compute $(f \circ f)(x)$.
(b) Find $f\left[f\left(\frac{113}{355}\right)\right]$. (Try not to do it the hard way.)

43. Let $f(x) = x^2$ and $g(x) = 2x - 1$.
(a) Compute $\dfrac{f[g(x)] - f[g(a)]}{g(x) - g(a)}$.
(b) Compute $\dfrac{f[g(x)] - f[g(a)]}{x - a}$.

44. Suppose that $y = f(x)$ and $g(y) = x$. Show that $(g \circ f)(x) - (f \circ g)(y) = x - y$.

45. Given three functions f, g, and h, we define the function $f \circ g \circ h$ by the formula

$$(f \circ g \circ h)(x) = f\{g[h(x)]\}$$

For example, if $f(x) = x^2$, $g(x) = x + 1$, and $h(x) = x/2$, we evaluate $(f \circ g \circ h)(x)$ as

$$(f \circ g \circ h)(x) = f\{g[h(x)]\}$$

$$= f\left[g\left(\frac{x}{2}\right)\right] = f\left(\frac{x}{2} + 1\right)$$

$$= \left(\frac{x}{2} + 1\right)^2 = \frac{x^2}{4} + x + 1$$

Using $f(x) = x^2$, $g(x) = x + 1$, and $h(x) = x/2$, compute each composition.

(a) $(g \circ h \circ f)(x)$ (b) $(h \circ f \circ g)(x)$
(c) $(g \circ f \circ h)(x)$ (d) $(f \circ h \circ g)(x)$
(e) $(h \circ g \circ f)(x)$

Hint for checking: No two answers are the same.

46. Let $F(x) = 2x - 1$, $G(x) = 4x$, and $H(x) = 1 + x^2$. Compute each composition.

(a) $(F \circ G \circ H)(x)$ (b) $(G \circ F \circ H)(x)$
(c) $(F \circ H \circ G)(x)$ (d) $(H \circ F \circ G)(x)$
(e) $(H \circ G \circ F)(x)$ (f) $(G \circ H \circ F)(x)$

47. Let $f(x) = x^2$, $g(x) = 1 - x$, and $h(x) = 3x$. Express each function as a composition of f, g, and h.

(a) $p(x) = 1 - 9x^2$ (b) $q(x) = 3 - 3x^2$
(c) $r(x) = 1 - 6x + 9x^2$ (d) $s(x) = 3 - 6x + 3x^2$

48. Suppose that three functions f', g, and g' are defined as follows (f' is read f prime):

$$f'(x) = 3x^2 \qquad g(x) = 2x^2 + 5 \qquad g'(x) = 4x$$

(a) Find $f'[g(x)] \cdot g'(x)$. (The dot denotes multiplication, not composition.)
(b) Evaluate $f'[g(-1)] \cdot g'(-1)$.
(c) Find $f'[g(t)] \cdot g'(t)$.

49. In the accompanying figure, each dashed line is parallel to one of the coordinate axes. What are the y-coordinates of the points P, Q, and R? (Express the answers in terms of the function f and the number a.)

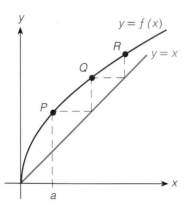

50. In the following figure, each dashed line is parallel to one of the coordinate axes. Determine the y-coordinates of the points P, Q, R, and S.

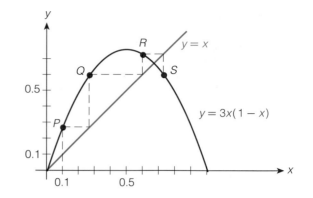

51. Suppose that we have a function f and an input x_0. The **iterates** of x_0 are the numbers $f(x_0)$, $f(f(x_0))$, $f(f(f(x_0)))$, and so on. The number $f(x_0)$ is called the **first iterate**, the number $f(f(x_0))$ is called the **second iterate**, and so on. For example, with $f(x) = x^2$ and $x_0 = 2$, the first three iterates are as follows.

first iterate: $f(2) = 2^2 = 4$
second iterate: $f(4) = 4^2 = 16$
third iterate: $f(16) = 16^2 = 256$

(a) Let $f(x) = \sqrt{x}$ and $x_0 = 16$. Compute the first six iterates. From the third iterate on, you'll need to use a calculator. (In reporting your final answers, round off to two decimal places. However, during the calculations, work with all of the decimal places that your calculator affords.)
(b) Extend your work in part (a) by computing the seventh through twelfth iterates. What "target number" do the iterates seem to be approaching? (Round off the answers to three decimal places.)

52. *The $3x + 1$ conjecture*   (As background for this exercise, you'll need to have worked Exercise 51.) Define a function f, with domain the positive integers, as follows:

$$f(x) = \begin{cases} 3x + 1 & \text{if } x \text{ is odd} \\ x/2 & \text{if } x \text{ is even} \end{cases}$$

(a) Compute $f(1)$, $f(2)$, $f(3)$, $f(4)$, $f(5)$, and $f(6)$.
(b) Compute the first three iterates of $x_0 = 1$.
(c) Compute the iterates of $x_0 = 3$ until you obtain the value 1. (After this, the iterates will recycle through the simple pattern obtained in part (b).)
(d) The $3x + 1$ *conjecture* asserts that for any positive integer x_0, the iterates eventually return to the value 1. Verify that this conjecture is valid for each of the following values of x_0:   2, 4, 5, 6, and 7.

Remark: At present, the $3x + 1$ conjecture is indeed a conjecture, not a theorem. No one yet has found a proof that the assertion is valid for *every* positive integer. Computer checks, however, have verified the conjecture on a case-by-case basis for values of x_0 ranging into the trillions. For a history of this problem and references, see the article, "The $3x + 1$ Problem and Its Generalizations" by Jeffrey C. Lagarias in *The American Mathematical Monthly* (vol. 92, Jan. 1985, pp. 3–23).

C

53. Let $i(x) = x$, $a(x) = -x$, $b(x) = 1/x$, and $c(x) = -1/x$. Assume that the domain of each function is $(0, \infty)$.

(a) Complete the following table, where the operation is composition of functions. For example, c is the proper entry in the second row, third column, since if you compute $(a \circ b)(x)$, you'll get $-1/x$, which is $c(x)$.

$\circ$	i	a	b	c
i				
a			c	
b				
c				

(b) According to your table, does $a \circ b = b \circ a$?
(c) Use your table to find a^2 (meaning $a \circ a$), b^2, c^2, and c^3.
(d) Use your table to find out if $(a \circ b) \circ c = a \circ (b \circ c)$.
(e) Let G denote the set consisting of the four functions you've worked with in this exercise. Now define a function f, with domain G, as follows: $f(t) = t^2$ for each element t in the set G. [For example, $f(a) = a^2 = a \circ a = i$.] What is the range of f?
(f) Verify that $f(a \circ b) = [f(a)] \circ [f(b)]$.
(g) Find out if $f(a \circ b \circ c) = [f(a)] \circ [f(b)] \circ [f(c)]$.

54. Let p and q be two numbers whose sum is 1. Define the function F by

$$F(x) = p - \frac{1}{x + q}$$

Show that $F\{F[F(x)]\} = x$.

OPTIONAL TI-81 GRAPHING CALCULATOR EXERCISES
EXERCISES FOR SECTION 3.4

1. The TI-81 can be used to evaluate expressions of the form $f(g(x_0))$. As an example, let's take the two functions $f(x) = 2x^3$ and $g(x) = 5 - 4x$. Suppose that we want to complete the following table.

x	0.1	0.2	0.3
$f(g(x))$			

Press the $Y=$ key and enter the functions $Y_1 = 5 - 4x$ and $Y_2 = 2x^3$. (We chose Y_1 to be $g(x)$ rather than $f(x)$ because in evaluating $f(g(x))$, we want to first evaluate $g(x)$.) Now press the "2nd" key and then the CLEAR key to return to the home screen. Next, use the keystrokes

to tell the calculator that the value of x is 0.1. The following sequence of keystrokes can now be used to compute $f(g(0.1))$:

(2nd) (VARS) 1 (ENTER)   This computes $g(0.1)$.

(STO▶) (X|T) (ENTER)   This tells the calculator that the new input is $g(0.1)$.

(2nd) (VARS) 2 (ENTER)   This computes $f(g(0.1))$

After completing this sequence, you should obtain the result 194.672. This is the required value for $f(g(0.1))$. Now repeat the procedure to compute the remaining two items in the table above.

In Exercises 2–4, complete the tables using the technique explained in Exercise 1.

2. $f(x) = 4x - 1$, $g(x) = x^2 - 2x$

(a)

x	1.4	1.5	1.6
$f(g(x))$			

(b)

x	1.4	1.5	1.6
$g(f(x))$			

3. $f(x) = 3x^2 - x + 4$, $g(x) = -x^2 + 5x + 1$

(a)

x	7	8	9
$f(g(x))$			

(b)

x	7	8	9
$g(f(x))$			

4. $f(x) = \sqrt{x} - 4$, $g(x) = 2x^2 - x$, $h(x) = (x-1)/(x+1)$

(a)

x	10	11	12
$h(g(f(x)))$			

(b)

x	10	11	12
$g(h(f(x)))$			

5.   As mentioned previously (Exercise 51 on page 138), the numbers

$$f(x_0),\ f(f(x_0)),\ f(f(f(x_0))),\ \ldots$$

are the *iterates* of x_0 for the function f. The number $f(x_0)$ is called the *first iterate,* the number $f(f(x_0))$ is called the *second iterate,* and so on. This exercise shows how to compute successive iterates literally at "the touch of a button." (The technique explained in Exercise 1 could also be used here, but the following method is much more efficient.) As an example, we will compute the iterates of $x_0 = 2$ for the function $f(x) = \frac{1}{2}(x + 5/x)$. As preparation, press (CLEAR).

(a)   First type "2" (the value of x_0) and then (ENTER). Next, use the following sequence of keystrokes:

(Note the screen display at this point; you've actually typed in the formula for the function f, but instead of using "x" for the input, you've used "ANS.") Now press (ENTER), and the value for $f(2)$ will appear on your screen. Furthermore, from here on, each time that you press (ENTER), the calculator will display the value of the next iterate. Compute the successive iterates in this way until the answers begin to repeat themselves. How many iterates are required before the answers begin to repeat? Make a note of the "target value" that you obtain. (Record all of the decimal places displayed on the calculator.)

(b)   Follow part (a) using $x_0 = 3$.

(c)   Follow part (a) using $x_0 = 25$.

(d)   Use your calculator to evaluate $\sqrt{5}$, and observe the connection with the previous results.   *Remark:* Using calculus, it can be shown that for every positive number x_0, the iterates approach $\sqrt{5}$.

6.   Let $f(x) = 4x(1 - x)$.

(a)   Use the technique described in Exercise 5 to compute the first ten iterates of $x_0 = 0.4$.

(b)   Compute the first ten iterates of $x_0 = 0.2$.

(c)   Do the iterates in parts (a) and (b) display the same type of behavior that you observed in Exercise 5?

7.   Let $f(x) = \dfrac{2x^3 + 30}{3x^2}$.

(a)   Compute the successive iterates of $x_0 = 5$ until the answers begin to repeat themselves. How many iterates are required before the answers begin to repeat? Make a note of the target value that you obtain.

(b)   Follow part (a) using $x_0 = 25$.

(c)   Use your calculator to evaluate $\sqrt[3]{30}$, and observe the connection with the previous results.   *Remark:* Using calculus, it can be shown that for every positive number x_0, the iterates approach $\sqrt[3]{30}$.

3.5   INVERSE FUNCTIONS

If anybody ever told me why the graph of $y = x^{1/2}$ is the reflection of the graph of $y = x^2$ in a 45° line, it didn't sink in. To this day, there are textbooks that expect students to think that it is so obvious as to need no explanation. This is a pity, if only because [in calculus] it is such a common practice to define the natural logarithm first and then define the exponential function as its inverse.

Professor Ralph P. Boas, recalling his student days in the article "Inverse Functions" in *The College Mathematics Journal* 16(1985): 42.

We shall introduce the idea of inverse functions through an easy example. Consider the function f defined by $f(x) = 2x$. A short table of values for f and the graph of f are displayed in Figure 1.

PROBLEM
SOLUTION

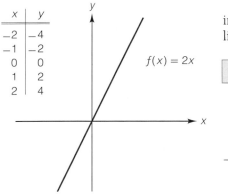

x	y
−2	−4
−1	−2
0	0
1	2
2	4

$f(x) = 2x$

FIGURE 1

Now we ask this question: What happens if we take the table for f and interchange the entries in the x and y columns to obtain a new table that looks like the following?

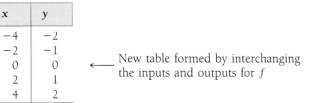

x	y
−4	−2
−2	−1
0	0
2	1
4	2

← New table formed by interchanging the inputs and outputs for f

Several observations are in order. Our new table does itself represent a function (in this case); for each input x there is exactly one output y. Let us call this new function g. Note that each output is half the corresponding input. So the function g can be described by the formula $g(x) = x/2$. In summary, we began with the "doubling function," $f(x) = 2x$; then, by interchanging the inputs with the outputs, we obtained the "halving function," $g(x) = x/2$.

The doubling function f and the halving function g are in a certain sense opposites; each reverses, or undoes, the effect of the other. That is, if you begin with x and then calculate $g[f(x)]$, you get x again; similarly, $f[g(x)]$ is also equal to x. (Verify these last two statements for yourself.)

The two functions f and g that we've been discussing are an example of a pair of *inverse functions*. The general definition for inverse functions is given in the box that follows.

DEFINITION Inverse Functions

Two functions f and g are **inverses** of one another provided that

$$f[g(x)] = x \qquad \text{for each } x \text{ in the domain of } g$$

and

$$g[f(x)] = x \qquad \text{for each } x \text{ in the domain of } f$$

■■■■ **EXAMPLE 1** Verify that the functions f and g are inverses, where

$$f(x) = \frac{1}{3}x + 2 \qquad \text{and} \qquad g(x) = 3x - 6$$

Solution In view of the definition, we must check that $f[g(x)] = x$ and $g[f(x)] = x$. We have

$$f[g(x)] = f(3x - 6)$$
$$= \tfrac{1}{3}(3x - 6) + 2 = x \qquad \text{(Check the algebra.)}$$

Thus, $f[g(x)] = x$. Now we still need to check that $g[f(x)] = x$. We have

$$g[f(x)] = g\left(\tfrac{1}{3}x + 2\right)$$
$$= 3\left(\tfrac{1}{3}x + 2\right) - 6 = x \qquad \text{(Again, check the algebra.)}$$

Having shown that $f[g(x)] = x$ and $g[f(x)] = x$, we conclude that f and g are indeed inverse functions. ■■■

■■■ **EXAMPLE 2**   Suppose that f and g are a pair of inverse functions. If $f(2) = 3$, what is $g(3)$?

Solution   We can use either of two methods.

FIRST METHOD   The quantities 3 and $f(2)$ are declared to be equal. Thus, whether we use 3 or $f(2)$ as an input for g, the result must be the same. We then have

$$3 = f(2)$$
$$g(3) = g[f(2)]$$
$$g(3) = 2 \qquad \text{using the fact that } g[f(x)] = x \text{ for any } x \text{ in the domain of } f$$

Thus $g(3) = 2$.

ALTERNATIVE METHOD   Think of a table of values for the function f. Since $f(2) = 3$, one entry in the table must look like this:

x		2	
y		3	

Now, g reverses the roles of x and y, so in the table for g, one entry must look like this:

x		3	
y		2	

Thus, $g(3) = 2$, as obtained previously. In Figure 2 we show a graphic interpretation of this method.   ■■■

FIGURE 2

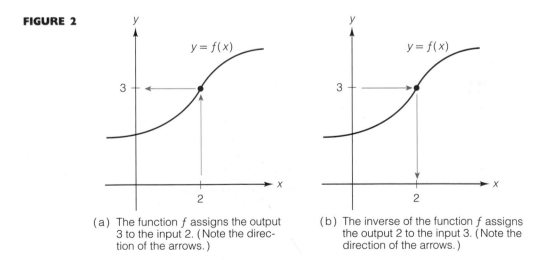

(a) The function f assigns the output 3 to the input 2. (Note the direction of the arrows.)

(b) The inverse of the function f assigns the output 2 to the input 3. (Note the direction of the arrows.)

It is customary to use the notation f^{-1} (read f *inverse*) for the function that is the inverse of f. So in Example 1, for instance, we have

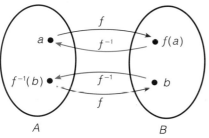

FIGURE 3

The action of inverse functions: The set
A in the figure is both the domain of f
and the range of f^{-1}. The set B is both
the domain of f^{-1} and the range of f.

$$f(x) = \frac{1}{3}x + 2 \qquad \text{and} \qquad f^{-1}(x) = 3x - 6$$

Note In the context of functions, f^{-1} does *not* in general mean $1/f$. (For an example of this, see Exercise 4 at the end of this section.) In the following box, we rewrite the defining equations for inverse functions, using the f^{-1} notation rather than g. Figure 3 provides a graphic summary of these equations.

$$f[f^{-1}(x)] = x \qquad \text{for each } x \text{ in the domain of } f^{-1}$$

and

$$f^{-1}[f(x)] = x \qquad \text{for each } x \text{ in the domain of } f$$

■■■ EXAMPLE 3 Suppose $f(x) = 4x^3 + 7$. Find $f[f^{-1}(5)]$. (Assume that f^{-1} exists and that 5 is in its domain.)

Solution By definition, $f[f^{-1}(x)] = x$. Substituting $x = 5$ on both sides of this identity directly gives us

$$f[f^{-1}(5)] = 5 \qquad \text{as required}$$

Note that we didn't need to make use of the formula $f(x) = 4x^3 + 7$, nor did we need to find a formula for $f^{-1}(x)$. ■■■

■■■ EXAMPLE 4 Solve the following equation for x, given that the domain of both f and f^{-1} is $(-\infty, \infty)$ and that $f(1) = -2$.

$$3 + f^{-1}(x - 1) = 4$$

Solution

$$
\begin{aligned}
3 + f^{-1}(x - 1) &= 4 \\
f^{-1}(x - 1) &= 1 \\
f[f^{-1}(x - 1)] &= f(1) \qquad \text{applying } f \text{ to both sides} \\
x - 1 &= f(1) \qquad \text{definition of inverse function} \\
x &= 1 + f(1) = 1 + (-2) = -1
\end{aligned}
$$

■■■

We'll postpone until the end of this section a discussion of which functions have inverses and which do not. For functions that do have inverses, however, there's a simple method that can often be used to determine those inverses. This method is illustrated in Examples 5 and 6.

■■■ EXAMPLE 5 Let $f(x) = 3x - 4$. Find $f^{-1}(x)$.

Solution We begin by rewriting the given equation as

$$y = 3x - 4$$

We know that f^{-1} interchanges the inputs and outputs of f. So to determine f^{-1}, we only need to switch the x's and y's in the equation $y = 3x - 4$. That gives us $x = 3y - 4$. Now we solve for y as follows:

$$x = 3y - 4$$
$$x + 4 = 3y$$
$$\frac{x + 4}{3} = y$$

Thus, the inverse function is $y = (x + 4)/3$. We can also write this as

$$f^{-1}(x) = \frac{x + 4}{3}$$

Actually, we should call this result just a *candidate* for f^{-1}, since certain technical matters remain to be discussed. However, if you compute $f[f^{-1}(x)]$ and $f^{-1}[f(x)]$ and find that they both do equal x, then the matter is settled, by definition. Exercise 13 at the end of this section asks you to carry out those calculations. ∎

Here is a summary of our procedure for calculating $f^{-1}(x)$.

To Find $f^{-1}(x)$ for the Function $y = f(x)$

1. Interchange x and y in the equation $y = f(x)$.

2. Solve the resulting equation for y.

EXAMPLE 6 Let $f(x) = \dfrac{2x - 1}{3x + 5}$. Find $f^{-1}(x)$.

Solution

STEP 1 Write the given function as $y = \dfrac{2x - 1}{3x + 5}$. Interchange x and y to get

$$x = \frac{2y - 1}{3y + 5}$$

STEP 2 Solve the resulting equation for y.

$$x = \frac{2y - 1}{3y + 5}$$
$$3xy + 5x = 2y - 1 \qquad \text{multiplying by } 3y + 5$$
$$3xy - 2y = -5x - 1 \qquad \text{rearranging}$$
$$y(3x - 2) = -5x - 1$$
$$y = \frac{-5x - 1}{3x - 2}$$

Thus, the inverse function is given by

$$f^{-1}(x) = \frac{-5x - 1}{3x - 2}$$

A certain symmetry always occurs when we graph a function and its inverse. As an example, consider the functions f and f^{-1} from Example 5:

$$f(x) = 3x - 4 \qquad f^{-1}(x) = \tfrac{1}{3}x + \tfrac{4}{3}$$

Figure 4 shows the graphs of f, f^{-1}, and the line $y = x$. Note that the graphs of f and f^{-1} are symmetric about the line $y = x$. This kind of symmetry, in fact, always occurs for the graphs of any pair of inverse functions.

FIGURE 4

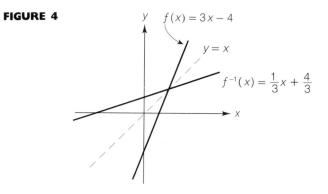

Why are the graphs of f and f^{-1} always mirror images of each other about the line $y = x$? First, recall that the function f^{-1} switches the inputs and outputs of f. Thus, (a, b) is on the graph of f if and only if (b, a) is on the graph of f^{-1}. But (as we know from Section 2.4) the points (a, b) and (b, a) are mirror images in the line $y = x$. It follows, therefore, that the graphs of f and f^{-1} are reflections of each other about the line $y = x$. For reference, we record this useful fact in the box that follows.

PROPERTY SUMMARY   THE GRAPHS OF f AND f^{-1}

The graphs of f and f^{-1} are symmetric about the line $y = x$.

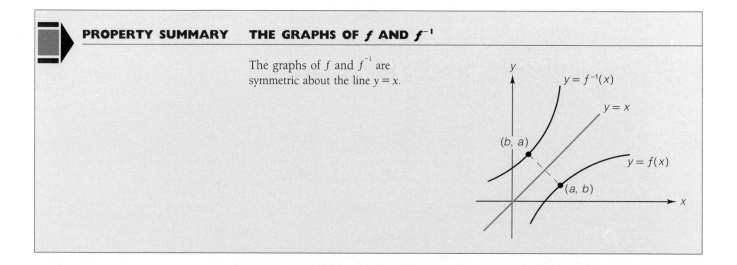

▌▐▌ **EXAMPLE 7** The graph of a function f consists of the line segment joining the points $(-2, -3)$ and $(-1, 4)$, as shown in Figure 5. Sketch a graph of f^{-1}.

Solution The graph of f^{-1} is obtained by reflecting the graph of f about the line $y = x$. The reflection of $(-2, -3)$ is $(-3, -2)$, and the reflection of $(-1, 4)$ is $(4, -1)$. To graph f^{-1}, then, we plot the reflected points $(-3, -2)$ and $(4, -1)$ and connect them with a line segment, as shown in Figure 6. For reference, Figure 6 also shows the graphs of f and $y = x$. ▐▐▐

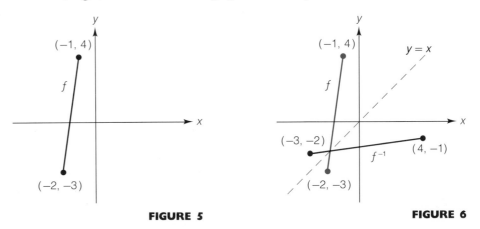

FIGURE 5   **FIGURE 6**

▌▐▌ **EXAMPLE 8** Let $g(x) = x^3$. Find $g^{-1}(x)$, and then, on the same set of axes, sketch the graphs of g, g^{-1}, and $y = x$.

Solution We begin by writing $y = x^3$, then switching x and y and solving for y. We first have

$$x = y^3$$

To solve this equation for y, we take the cube root of both sides to obtain

$$\sqrt[3]{x} = y$$

So the inverse function is $g^{-1}(x) = \sqrt[3]{x}$. We could graph the function g^{-1} by plotting points, but the easier way is to reflect the graph of $g(x) = x^3$ about the line $y = x$. See Figure 7. ▐▐▐

FIGURE 7

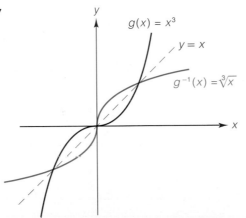

TABLE I

(a) $F(x) = x^2$ (b) Reversing the inputs and outputs

x	y
1	1
2	4
3	9
−1	1
−2	4
−3	9

x	y
1	1
4	2
9	3
1	−1
4	−2
9	−3

Earlier in this section, we mentioned that not every function has an inverse function. For instance, Tables 1(a) and (b) indicate what happens if we begin with the function defined by $F(x) = x^2$ and interchange the inputs and outputs. The resulting table, Table 1(b), does not define a function. For instance, the input 4 in Table 1(b) produces two distinct outputs, 2 and −2, but the definition of a function requires exactly one output for a given input.

Why didn't this difficulty arise when we looked for the inverse of $f(x) = 2x$ at the beginning of this section? The answer is that there is an essential difference between the functions defined by $f(x) = 2x$ and $F(x) = x^2$. With $f(x) = 2x$, distinct inputs never yield the same output. That is, if $x_1 \neq x_2$, then $f(x_1) \neq f(x_2)$. This condition guarantees that interchanging the inputs and outputs of f will yield a function. With $F(x) = x^2$, however, distinct inputs can yield the same output. For instance

$$2 \neq -2$$

but

$$F(2) = F(-2) \qquad \text{because } 2^2 = 4 = (-2)^2$$

It's useful to have a name for the type of function in which distinct inputs always yield distinct outputs. We call such a function a **one-to-one function**. So in view of the discussion in the previous paragraph, the function defined by $f(x) = 2x$ is one-to-one, but $F(x) = x^2$ is not. Using graphs, there is an easy way to tell which functions are one-to-one:

Horizontal Line Test

A function f is one-to-one if and only if each horizontal line intersects the graph of $y = f(x)$ in at most one point.

EXAMPLE 9 Determine if each function is one-to-one:

(a) $F(x) = 2x$; **(b)** $H(x) = x^2$; **(c)** $h(x) = \left(\sqrt{x}\right)^4$.

Solution We sketch the graphs and apply the horizontal line test, as in Figure 8. The horizontal line test tells us that F and h are one-to-one, but H is not.

FIGURE 8

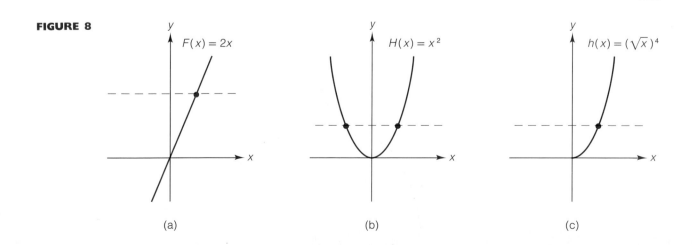

(a)   (b)   (c)

The following theorem tells us that the functions with inverses are precisely the one-to-one functions. (The proof of the theorem isn't difficult, but we'll omit giving it here.)

Theorem

A function f has an inverse function if and only if f is one-to-one.

■■■ **EXAMPLE 10** Which of the three functions in Example 9 have inverse functions?

Solution According to our theorem, the functions with inverses are those that are one-to-one. In Example 9, we found that F and h were one-to-one, but H was not. So F and h each have inverses, but H does not. ▪▪▪

EXERCISE SET 3.5

A

1. Verify that the given pairs of functions are inverse functions.
   (a) $f(x) = 3x$; $g(x) = x/3$
   (b) $f(x) = 4x - 1$; $g(x) = \frac{1}{4}x + \frac{1}{4}$
   (c) $g(x) = \sqrt{x}$; $h(x) = x^2$ [Assume that the domain of both g and h is $[0, \infty)$.]

2. Which pairs of functions are inverses?
   (a) $f(x) = -3x + 2$; $g(x) = \frac{2}{3} - \frac{1}{3}x$
   (b) $F(x) = 2x + 1$; $G(x) = \frac{1}{2}x - 1$
   (c) $G(x) = x^3$; $H(x) = 1 - x^3$
   (d) $f(t) = t^3$; $g(t) = \sqrt[3]{t}$

3. Let $f(x) = x^3 + 2x + 1$ and assume that the domain of f^{-1} is $(-\infty, \infty)$. Simplify each expression.
   (a) $f[f^{-1}(4)]$
   (b) $f^{-1}[f(-1)]$
   (c) $(f \circ f^{-1})(\sqrt{2})$
   (d) $f[f^{-1}(t + 1)]$
   (e) $f(0)$; use your answer to find $f^{-1}(1)$.
   (f) $f(-1)$; use your answer to find $f^{-1}(-2)$.

4. Let $f(x) = 2x + 1$.
   (a) Find $f^{-1}(x)$.
   (b) Calculate $f^{-1}(5)$ and $1/f(5)$. Are your answers the same?

5. Let $f(x) = 3x - 1$.
   (a) Compute $f^{-1}(x)$.
   (b) Verify that $f[f^{-1}(x)] = x$ and that $f^{-1}[f(x)] = x$.
   (c) On the same set of axes, sketch the graphs of f, f^{-1}, and the line $y = x$. Note that the graphs of f and f^{-1} are symmetric about the line $y = x$.

6. Follow Exercise 5, but use $f(x) = \frac{1}{3}x - 2$.

7. Follow Exercise 5, but use $f(x) = \sqrt{x - 1}$. [The domain of f^{-1} will be $[0, \infty)$.]

8. Follow Exercise 5, but use $f(x) = 1/x$.

9. Let $f(x) = \dfrac{x + 2}{x - 3}$.
   (a) Find the domain and range of the function f.
   (b) Find $f^{-1}(x)$.
   (c) Find the domain and range of the function f^{-1}. What do you observe?

10. Let $f(x) = \dfrac{2x - 3}{x + 4}$. Find $f^{-1}(x)$. Find the domain and range for f and f^{-1}. What do you observe?

11. Let $f(x) = 2x^3 + 1$. Find $f^{-1}(x)$.

12. In our discussion at the beginning of this section, we considered the functions $f(x) = 2x$ and $g(x) = x/2$. Verify that $f[g(x)] = x$ and that $g[f(x)] = x$. On the same set of axes, sketch the graphs of f, g, and $y = x$.

13. This exercise refers to the comments made at the end of Example 5. Compute $f[f^{-1}(x)]$ and $f^{-1}[f(x)]$ using the functions $f(x) = 3x - 4$ and $f^{-1}(x) = \frac{1}{3}x + \frac{4}{3}$. By actually carrying out the calculations, you can see in each case that the answer is indeed x. Then, on the same set of axes, sketch the graphs of f, f^{-1}, and the line $y = x$.

14. Let $f(x) = (x - 3)^2$ and take the domain to be $[3, \infty)$.
   (a) Find $f^{-1}(x)$. Give the domain of f^{-1}.
   (b) On the same set of axes, graph both f and f^{-1}.

15. Let $f(x) = (x - 3)^3 - 1$.
   (a) Compute $f^{-1}(x)$.
   (b) On the same set of axes, sketch the graphs of f, f^{-1}, and the line $y = x$. *Hint:* Sketch f using the ideas of Section 3.3. Then reflect f to get f^{-1}.

16. The following figure shows the graph of a function f. (The axes are marked off in one-unit intervals.) Graph each of the following functions.
   (a) $y = f^{-1}(x)$
   (b) $y = f(x - 2)$
   (c) $y = f(x) - 2$
   (d) $y = f(-x)$
   (e) $y = -f(x)$
   (f) $y = f^{-1}(x - 2)$
   (g) $y = f^{-1}(x) - 2$
   (h) $y = f^{-1}(x - 2) - 2$
   (i) $y = f^{-1}(-x)$
   (j) $y = -f^{-1}(x)$

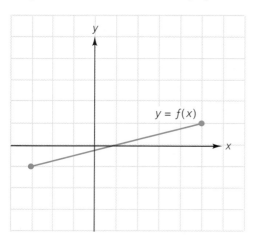

17. The following figure shows the graph of a function g. (The axes are marked off in one-unit intervals.) Graph each of the following functions.
   (a) $y = g^{-1}(x)$
   (b) $y = g^{-1}(x) - 1$
   (c) $y = g^{-1}(x - 1)$
   (d) $y = g^{-1}(-x)$
   (e) $y = -g^{-1}(x)$
   (f) $y = -g^{-1}(-x)$

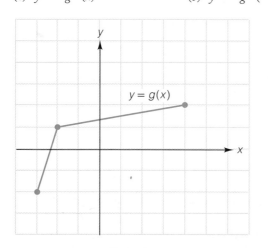

18. The following figure shows the graph of a function $y = h(x)$.
   (a) Sketch a graph that is the reflection of h about the y-axis. What is the equation of the graph you obtain?
   (b) On the same set of axes, sketch graphs of h, h^{-1}, and the line $y = x$.

(c) Let k be the function whose graph is the reflection of h in the x-axis. Graph the function k^{-1}.

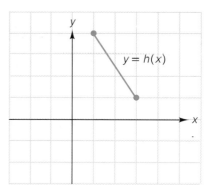

In Exercises 19–30, use the horizontal line test to determine whether the function is one-to-one (and therefore has an inverse).

19. $y = x^2 + 1$
20. $y = \sqrt{x}$
21. $f(x) = 1/x$
22. $g(x) = |x|$
23. $y = x^3$
24. $y = 1 - x^3$
25. $y = \sqrt{1 - x^2}$
26. $y = 3x$
27. $g(x) = 5$ (for all x)
28. $y = mx + b$ $(m \neq 0)$

29. $f(x) = \begin{cases} x^2 & \text{if } -1 \leq x \leq 0 \\ x^2 + 1 & \text{if } x > 0 \end{cases}$

30. $g(x) = \begin{cases} x^2 & \text{if } -1 \leq x < 0 \\ x^2 + 1 & \text{if } x \geq 0 \end{cases}$

31. Let $f(x) = \dfrac{3x - 2}{5x - 3}$. Show that $f[f(x)] = x$. This says that f is its own inverse.

32. Let $f(x) = \dfrac{ax + b}{cx - a}$. Show that f is its own inverse. (Assume that $a^2 + bc \neq 0$.)

In Exercises 33–36, assume that the domain of f and f^{-1} is $(-\infty, \infty)$. Solve the equation for x or for t (whichever is appropriate) using the given information.

33. (a) $7 + f^{-1}(x - 1) = 9$; $f(2) = 6$
    (b) $4 + f(x + 3) = -3$; $f^{-1}(-7) = 0$

34. (a) $f^{-1}(2x + 3) = 5$; $f(5) = 13$
    (b) $f(1 - 2x) = -4$; $f^{-1}(-4) = -5$

35. $f^{-1}\left(\dfrac{t + 1}{t - 2}\right) = 12$; $f(12) = 13$

36. $f\left(\dfrac{1 - 2t}{1 + 2t}\right) = 7$; $f^{-1}(7) = -3$

B

37. Let $f(x) = \sqrt{x}$.
    (a) Find $f^{-1}(x)$. What is the domain of f^{-1}? [The domain is not $(-\infty, \infty)$.]

(b) In each case, determine whether the given point lies on the graph of f or f^{-1}.
  (i) $(4, 2)$ (ii) $(2, 4)$ (iii) $(5, \sqrt{5})$
  (iv) $(\sqrt{5}, 5)$ (v) $(a, f(a))$ (Assume $a \geq 0$.)
  (vi) $(f(a), a)$ (vii) $(b, f^{-1}(b))$ (Assume $b \geq 0$.)
  (viii) $(f^{-1}(b), b)$

38. Let $F(x) = x^3 + 7x - 5$. Find $F^{-1}(-5)$. [Assume that the domain of F^{-1} is $(-\infty, \infty)$.]

39. In the following figure, determine the coordinates of the points A, B, C, and D. Express your answers in terms of the function f and the number a. (Each dashed line through A is parallel to an axis.)

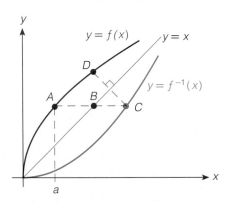

40. In the following figure, determine the coordinates of the points A, B, C, D, E, and F. Specify the answers in terms of the functions f and f^{-1} and the number b.

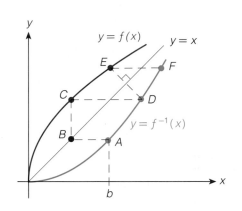

OPTIONAL TI-81 GRAPHING CALCULATOR EXERCISES

EXERCISES FOR SECTION 3.5

In Exercises 1–6, the two given functions are inverse functions. In each case, graph the pair of inverse functions on the same set of axes along with the line $y = x$. Note the symmetry. Make use of the ZOOM 5 setting. (Why?)

1. $y = 4x + 8$ and
    $y = \frac{1}{4}x - 2$

2. $y = 3x - 5$ and
    $y = (x + 5)/3$

3. $y = \sqrt{x - 2}, \quad x \geq 2$
    $y = x^2 + 2, \quad x \geq 0$
    Note: For instructions on graphing equations with restrictions on x, see Exercise 3 on page 68.

4. $y = \sqrt{2x + 3}, \quad x \geq -1.5$
    $y = (x^2 - 3)/2, \quad x \geq 0$

5. $y = \dfrac{1}{\sqrt{x}}$ and
    $y = \dfrac{1}{x^2}, \quad x > 0$

6. $y = \dfrac{1}{2x - 3}$ and
    $y = \dfrac{3x + 1}{2x}$

In Exercises 7–12, graph each function and then use the horizontal line test to determine whether the function is one-to-one.

7. $f(x) = x^2 + 2x$ 8. $f(x) = x^3 + 2x$

9. $f(x) = 2x^3 + x^2$ *Hint:*   First looks may be deceiving. After looking at the graph in the standard viewing rectangle (ZOOM 6), make use of the ZOOM BOX to look more closely at the graph.

10. (a) $g(x) = 0.01x^4 - 1$
     (b) $g(x) = 2x^5 + x - 1$

11. (a) $y = x^3 + x^2 + x$
     (b) $y = x^3 - x^2 + x$
     (c) $y = x^3 - x^2 - x$

12. (a) $y = x^x, \quad x > 1$
     (b) $y = x^x, \quad x > 0$

13. Let $F(x) = \sqrt{1 - x^2}$, with domain $-1 \leq x \leq 0$. Find $F^{-1}(x)$ and then graph both functions on the same set of axes. Check your results: Are the two graphs

symmetric about the line $y = x$? (Why will the ZOOM 5 setting be necessary here?)

14. Let $f(x) = 2x + 1$ and $g(x) = \frac{1}{4}x - 3$.
    (a) Compute each of the following: (i) $f(g(x))$;
        (ii) $g(f(x))$; (iii) $f^{-1}(x)$; (iv) $g^{-1}(x)$;
        (v) $f^{-1}(g^{-1}(x))$; (vi) $g^{-1}(f^{-1}(x))$.
    (b) On the same set of axes, graph the two answers that you obtained in (i) and (v) of part (a). Note that the graphs are *not* symmetric about $y = x$. The conclusion here is that the inverse function for $f(g(x))$ is not $f^{-1}(g^{-1}(x))$.
    (c) On the same set of axes, graph the two answers that you obtained in (i) and (vi) of part (a); also put the line $y = x$ into the picture. Note that the two graphs *are* symmetric about the line $y = x$. (If this symmetry does not appear in your graph, press ZOOM 5 so that the same scales are used on both axes.) The conclusion here is that the inverse function for $f(g(x))$ is $g^{-1}(f^{-1}(x))$. For reference, we summarize this fact about the inverse of a composite function as follows:

$$(f \circ g)^{-1} = g^{-1} \circ f^{-1}$$

15. Let $f(x) = \dfrac{3x}{1 + |x|}$.
    (a) Graph this function. Note that, according to the horizontal line test, the function is one-to-one.
    (b) Since the function is one-to-one, it has an inverse. Compute this inverse. *Hint:* Consider two cases. First, if $x \geq 0$, then the given equation is $y = 3x/(1 + x)$. Second, if $x < 0$, the given equation is equivalent to $y = 3x/(1 - x)$. (Why?) When you think you have found f^{-1}, graph it and f on the same set of axes and check that the graphs are symmetric about $y = x$.
        Answer: For the inverse, you should obtain
        $$y = \begin{cases} x/(3 - x) & \text{if } 0 \leq x < 3 \\ x/(3 + x) & \text{if } -3 < x < 0 \end{cases}$$

16. (a) Graph the following two functions on the same set of axes. (For Y_1 use the positive square root; for Y_2 use the negative square root. Use ZOOM 6 for the standard viewing rectangle.)

$$y = \pm\sqrt{\frac{6x^2 - x^3}{2 + x}}$$

(The resulting curve is called the *Trisectrix of Maclaurin*. The curve, studied by Colin Maclaurin in 1742, can be used to trisect an angle.) If you've entered the equations correctly, your graph should be similar to the following one.

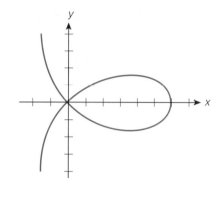

(b) For the rest of this problem, we focus on the first-quadrant portion of the trisectrix. To get a better look at it, press the Y= key and deselect Y_2. Now move the cursor to the end of the expression for Y_1 and add the restriction $(x \geq 0)$. (Recall that inequalities are accessed by pressing the "2nd" key followed by the MATH key.) Now press the GRAPH key. Finally, use the ZOOM BOX. For the lower left corner of the box, choose a point as close as possible to $(-1, 1)$; for the upper right corner, choose a point as close as possible to $(7, 3)$.

(c) Let's let $y = g(x)$ denote the function whose graph is the arch in the first quadrant, which you just drew in part (b). By looking carefully at the graph, estimate the largest value of c for which $y = g(x)$ is one-to-one on the closed interval $[0, c]$. (This is also the input for which g is maximum.) For example, from the picture, it's clear that the function is one-to-one on the interval $[0, 2]$, and it's also clear that the function is not one-to-one on the interval $[0, 5]$. So c is somewhere between 2 and 5. Now use the TRACE key and cursor arrows to obtain a more accurate value for c. Make a note of this estimate for c. Next, use the ZOOM BOX to take a closer look at the graph near its turning point. Then use the TRACE key to obtain a better estimate for c. Repeat this process several times, then go on to part (d).

(d) Using calculus, it can be shown that the exact value of c is $2\sqrt{3}$. Use your calculator to evaluate this expression. Did you come close to this value in part (c)?

▐■▌ CHAPTER THREE SUMMARY OF PRINCIPAL TERMS AND FORMULAS

TERMS OR FORMULAS	PAGE REFERENCE	COMMENTS
1. Function	99	Given two nonempty sets A and B, a function from A to B is a rule of correspondence that assigns to each element of A exactly one element in B.
2. Domain	99	The domain of a function is the set of all inputs for that function. If f is a function from A to B, then the domain is the set A.
3. Range	99	The range of a function from A to B is the set of all elements in B that are actually used as outputs.
4. $f(x)$	100	Given a function f, the notation $f(x)$ denotes the output that results from the input x.
5. Graph of a function	108	The graph of a function f consists of those points (x, y) such that x is in the domain of f and $y = f(x)$.
6. Vertical line test	111	A graph in the x-y plane represents a function $y = f(x)$ provided that any vertical line intersects the graph in at most one point.
7. Turning point	112	A turning point on a graph is a point where the graph changes from rising to falling, or vice versa. See, for example, Figures 8 and 9 on page 112.
8. Maximum value and minimum value	112	An output $f(a)$ is a maximum value of the function f if $f(a) \geq f(x)$ for every x in the domain of f. An output $f(b)$ is a minimum value if $f(b) \leq f(x)$ for every x in the domain of f.
9. Increasing function	112	A function f is increasing on an interval if the following condition holds: If a and b are in the interval and $a < b$, then $f(a) < f(b)$. Geometrically, this means that the graph is rising as we move in the positive x-direction.
10. Decreasing function	112	A function f is decreasing on an interval if the following condition holds: If a and b are in the interval and $a < b$, then $f(a) > f(b)$. Geometrically, this means that the graph is falling as we move in the positive x-direction.
11. Average rate of change	113	The average rate of change, $\Delta f/\Delta x$, of a function f on an interval $[a, b]$ is the slope of the line segment through the two points $(a, f(a))$ and $(b, f(b))$: $$\frac{\Delta f}{\Delta x} = \frac{f(b) - f(a)}{b - a}$$
12. Techniques for graphing functions	123	The box on page 123 lists six techniques involving translation and reflection. For a comprehensive example, see Example 5 on pages 126–127.
13. $f \circ g$	132	The composition of two functions f and g: $(f \circ g)(x) = f[g(x)]$
14. Inverse functions	141	Two functions f and g are inverses of one another provided that the following two conditions are met. First, $f[g(x)] = x$ for each x in the domain of g. Second, $g[f(x)] = x$ for each x in the domain of f.
15. f^{-1}	142	f^{-1} denotes the inverse function for f. *Note:* In this context, f^{-1} does not mean $1/f$. For an example, see Exercise 4 on page 148.
16. One-to-one function	147	A function f is said to be one-to-one provided that distinct inputs always yield distinct outputs. That is, if x_1 and x_2 are in the domain of f and $x_1 \neq x_2$, then $f(x_1) \neq f(x_2)$. The relationship between inverse functions and the notion of one-to-one is given by the following theorem: A function f has an inverse if and only if f is one-to-one.

TERMS OR FORMULAS	PAGE REFERENCE	COMMENTS
17. Horizontal line test	147	A function f is one-to-one if and only if each horizontal line intersects the graph of $y = f(x)$ in at most one point.

WRITING MATHEMATICS

1. Section 3.5 on inverse functions begins with the following quotation from Professor R. P. Boas:

   > If anybody ever told me why the graph of $y = x^{1/2}$ is the reflection of the graph of $y = x^2$ $[x \geq 0]$ in a 45° line, it didn't sink in. To this day, there are textbooks that expect students to think that it is so obvious as to need no explanation.

   Show your instructor that you are not in the dark about inverse functions. Write out an explanation of why the graphs of $y = x^{1/2}$ and $y = x^2$ $[x \geq 0]$ are indeed reflections of one another in a 45° line. (As preparation for your writing, you'll probably need to look over Sections 2.4 and 3.5 and make a few notes.)

2. Decide which of the following rules are functions. Write out your reasons in complete sentences. (In each case, assume that the domain is the set of students in your school.)
   (a) F is the rule that assigns to each person his or her brother.
   (b) G is the rule that assigns to each person his or her aunt.
   (c) H is the rule that assigns to each person his or her mother.
   (d) K is the rule that assigns to each person his or her mother or father.

3. Given the graph of two functions f and g, points on the graph of the composite function $f \circ g$ can be constructed geometrically. Use the following figure to explain (in complete sentences) how this works.   *Hint:*   Determine in succession the coordinates of A, B, C, and D.

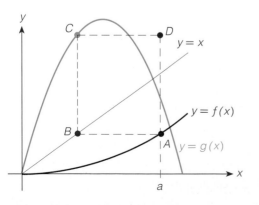

CHAPTER THREE REVIEW EXERCISES

1. (a) Find the domain of the function defined by
   $f(x) = \sqrt{15 - 5x}$.
   (b) Find the range of the function defined by
   $g(x) = (3 + x)/(2x - 5)$.

2. Let $f(x) = 3x^2 - 4x$ and $g(x) = 2x + 1$. Compute each of the following: (a) $(f - g)(x)$; (b) $(f \circ g)(x)$; (c) $f[g(-1)]$.

3. A **linear function** is a function defined by an equation of the form $F(x) = ax + b$, where a and b are real numbers. Suppose f and g are linear functions. Is the function $f \circ g$ a linear function?

4. A **quadratic function** is a function defined by an equation of the form $F(x) = ax^2 + bx + c$, where a, b, and c are real numbers and $a \neq 0$.
   (a) Give an example of quadratic functions f and g for which $f + g$ is also a quadratic function.
   (b) Give an example of quadratic functions f and g for which $f + g$ is not a quadratic function.
   (c) Suppose f and g are quadratic functions. Is the function $f \circ g$ a quadratic function?

5. (a) Compute $\dfrac{F(x) - F(a)}{x - a}$ given that $F(x) = 1/x$.

   (b) Compute $\dfrac{g(x + h) - g(x)}{h}$ for the function defined by $g(x) = x - 2x^2$.

6. The y-intercept for the graph of $y = f(x)$ is 4. What is the y-intercept for the graph of $y = -f(x) + 1$?

7. (a) Find $g^{-1}(x)$ given that $g(x) = \dfrac{1 - 5x}{3x}$.

   (b) The figure displays the graph of a function f. Sketch the graph of f^{-1}.

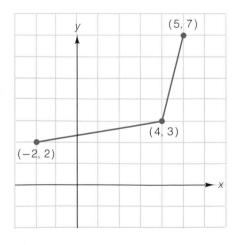

8. Refer to the function f in Exercise 7(b).
   (a) Compute $\Delta f/\Delta x$ on $[-2, 5]$.
   (b) Find the average rate of change of f^{-1} on $[2, 7]$.

9. Graph each function and specify the intercepts:
   (a) $y = |x + 2| - 3$; (b) $y = \dfrac{1}{x + 2} - 1$.

10. Refer to the following graph of $y = g(x)$. The domain of g is $[-5, 2]$.
    (a) What are the coordinates of the turning point(s)?
    (b) What is the maximum value of g?
    (c) Which input yields a minimum value for g?
    (d) On which interval(s) is g increasing?
    (e) Compute $\Delta g/\Delta x$ for each of the following intervals: $[-5, -3]$; $[-3, -2]$; $[-5, 2]$.

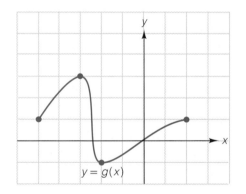

11. Let $f(x) = 3x^2 - 2x$. Find: (a) $f(-1)$; (b) $f(1 - \sqrt{2})$.

12. Graph the function G defined by
    $$G(x) = \begin{cases} \sqrt{1 - x^2} & \text{if } -1 \le x < 0 \\ \sqrt{x} & \text{if } x \ge 0 \end{cases}$$

13. Express the slope of a line passing through the points $(5, 25)$ and $(5 + h, (5 + h)^2)$ in terms of h.

14. Graph the function $y = |9 - x^2|$. Specify symmetry and intercepts.

15. The figure shows the graph of a function $y = f(x)$. Sketch the graph of $y = f(-x)$.

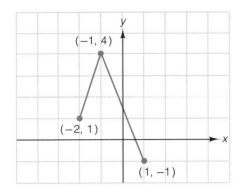

(−1, 4)

(−2, 1)

(1, −1)

16. Given that the domains of f and f^{-1} are both $(-\infty, \infty)$ and that $f^{-1}(1) = -4$, solve the following equation for t:

$$2 + f(3t + 5) = 3$$

17. Suppose that (a, b) is a point on the graph of $y = f(x)$. Match the functions defined in the left-hand column with the points in the right-hand column. For example, the appropriate match for (a) in the left-hand column is determined as follows. The graph of $y = f(x) + 1$ is obtained by translating the graph of $y = f(x)$ up one unit. Thus, the point (a, b) moves up to $(a, b + 1)$ and, consequently, (E) is the appropriate match for (a).

(a) $y = f(x) + 1$ (A) $(-a, b)$
(b) $y = f(x + 1)$ (B) (b, a)
(c) $y = f(x - 1) + 1$ (C) $(a - 1, b)$
(d) $y = f(-x)$ (D) $(-b, a + 1)$
(e) $y = -f(x)$ (E) $(a, b + 1)$
(f) $y = -f(-x)$ (F) $(1 - a, b)$
(g) $y = f^{-1}(x)$ (G) $(-a, -b)$
(h) $y = f^{-1}(x) + 1$ (H) $(-b, 1 - a)$
(i) $y = f^{-1}(x - 1)$ (I) $(b, -a)$
(j) $y = f^{-1}(-x) + 1$ (J) $(a, -b)$
(k) $y = -f^{-1}(x)$ (K) $(b + 1, a)$
(l) $y = -f^{-1}(-x) + 1$ (L) $(a + 1, b + 1)$
(m) $y = 1 - f^{-1}(x)$ (M) $(b, a + 1)$
(n) $y = f(1 - x)$ (N) $(b, 1 - a)$

In Exercises 18–40, sketch the graph and specify any x- or y-intercepts.

18. $y = 4 - x^2$

19. $y = -(x - 1)^2 + 2$

20. $y = \dfrac{1}{x} + 1$

21. $f(x) = \dfrac{1}{x + 1}$

22. $y = \dfrac{1}{x + 1} + 1$

23. $y = |x + 3|$

24. $g(x) = -\sqrt{x - 4}$

25. $h(x) = \sqrt{1 - x^2}$

26. $f(x) = \sqrt{1 - x^2} + 1$

27. $y = 1 - (x + 1)^3$

28. $y = 4 - \sqrt{-x}$

29. $y = (\sqrt{x})^2$

30. $f \circ g$, where $g(x) = x + 3$ and $f(x) = x^2$

31. $f \circ g$, where $g(x) = \sqrt{x - 1}$ and $f(x) = -x^2$

32. $f(x) = \begin{cases} \sqrt{1 - x^2} & \text{if } -1 \le x \le 0 \\ \sqrt{x + 1} & \text{if } x > 0 \end{cases}$

33. $F(x) = \begin{cases} -\sqrt{1 - x^2} & \text{if } -1 \le x < 0 \\ \sqrt{x} & \text{if } x \ge 0 \end{cases}$

34. $y = \begin{cases} |x - 1| & \text{if } 0 \le x \le 2 \\ |x - 3| & \text{if } 2 < x \le 4 \end{cases}$

35. $y = \begin{cases} 1/x & \text{if } 0 < x \le 1 \\ 1/(x - 1) & \text{if } 1 < x \le 2 \end{cases}$

36. $(y + |x| - 1)(y - |x| + 1) = 0$

37. f^{-1}, where $f(x) = \frac{1}{2}(x + 1)$

38. g^{-1}, where $g(x) = \sqrt[3]{x + 2}$

39. $f \circ f^{-1}$, where $f(x) = \sqrt{x - 2}$

40. $f^{-1} \circ f$, where $f(x) = \sqrt{x - 2}$

In Exercises 41–50, find the domain of the function.

41. $y = 1/(x^2 - 9)$

42. $y = x^3 - x^2$

43. $y = \sqrt{8 - 2x}$

44. $y = \dfrac{x}{6x^2 + 7x - 3}$

45. $y = \sqrt{|2 - 5x|}$

46. $y = (25 - x^2)/\sqrt{x^2 + 1}$

47. $y = \sqrt{x^2 - 2x - 3}$

48. $y = \sqrt{5 - x^2}$

49. $y = x + \dfrac{1}{x}$

50. $z = \dfrac{1}{t - \sqrt{t + 2}}$

In Exercises 51–56, determine the range of the function.

51. $y = \dfrac{x + 4}{3x - 1}$

52. $y = \dfrac{2x - 3}{x - 2}$

53. $f \circ g$, where $f(x) = 1/x$ and $g(x) = 3x + 4$

54. $g \circ f$, where $f(x) = \dfrac{x + 2}{x - 1}$ and $g(x) = \dfrac{x + 1}{x + 4}$

55. f^{-1}, where $f(x) = \dfrac{x}{3x - 6}$

56. f^{-1}, where $f(x) = \dfrac{5 - x}{1 + x}$

In Exercises 57–64, express the function as a composition of two or more of the following functions:

$$f(x) = 1/x \qquad g(x) = x - 1 \qquad F(x) = |x| \qquad G(x) = \sqrt{x}$$

57. $a(x) = \dfrac{1}{x - 1}$

58. $b(x) = \dfrac{1}{x} - 1$

59. $c(x) = \sqrt{x - 1}$

60. $d(x) = \sqrt{x} - 1$

61. $A(x) = 1/\sqrt{x} - 1$

62. $B(x) = |x - 2|$

63. $C(x) = \sqrt[4]{x} - 1$

64. $D(x) = 1/\sqrt{x - 3}$

For Exercises 65–98, compute the indicated quantity using the functions f, g, and F defined as follows:

$$f(x) = x^2 - x \qquad g(x) = 1 - 2x \qquad F(x) = \frac{x-3}{x+4}$$

65. $f(-3)$

66. $f(1 + \sqrt{2})$

67. $F\left(\frac{3}{4}\right)$

68. $f(t)$

69. $f(-t)$

70. $g(2x)$

71. $f(x - 2)$

72. $g(x + h)$

73. $g(2) - g(0)$

74. $f(x) - g(x)$

75. $|f(1) - f(3)|$

76. $f(x + h) - f(x)$

77. $f(x^2)$

78. $f(x)/x \quad (x \neq 0)$

79. $[f(x)][g(x)]$

80. $f[f(x)]$

81. $f[g(x)]$

82. $g[f(3)]$

83. $(g \circ f)(x)$

84. $(g \circ f)(x) - (f \circ g)(x)$

85. $(F \circ g)(x)$

86. $\dfrac{g(x + h) - g(x)}{h}$

87. $\dfrac{f(x + h) - f(x)}{h}$

88. $\dfrac{F(x) - F(a)}{x - a}$

89. $F^{-1}(x)$

90. $F[F^{-1}(x)]$

91. $F^{-1}[F(x)]$

92. $F^{-1}(0) - \dfrac{1}{F(0)}$

93. $(g \circ g^{-1})(x)$

94. $g^{-1}(x)$

95. $g^{-1}(-x)$

96. $\dfrac{g^{-1}(x + h) - g^{-1}(x)}{h}$

97. $F^{-1}\left[F\left(\frac{22}{7}\right)\right]$

98. $T^{-1}(x)$, where $T(x) = f(x)/x \quad (x \neq 0)$

In Exercises 99–112, refer to the graph of the function f in the figure. (The axes are marked off in one-unit intevals.)

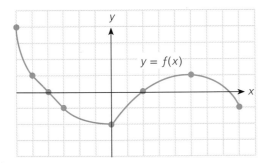

99. Is $f(0)$ positive or negative?

100. Specify the domain and range of f.

101. Find $f(-3)$.

102. Which is larger, $f\left(-\frac{5}{2}\right)$ or $f\left(-\frac{1}{2}\right)$?

103. Compute $f(0) - f(8)$.

104. Compute $|f(0) - f(8)|$.

105. Specify the coordinates of the turning points.

106. What are the minimum and the maximum values of f?

107. On which interval(s) is f decreasing?

108. For which x-values is it true that $1 \leq f(x) \leq 4$?

109. What is the largest value of $f(x)$ when $|x| \leq 2$?

110. Is f a one-to-one function?

111. Does f possess an inverse function?

112. Compute $f[f(-4)]$.

For Exercises 113–128, refer to the graphs of the functions f and g in the figure. Assume that the domain of each function is [0, 10].

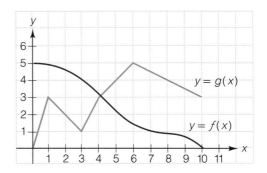

113. For which x-value is $f(x) = g(x)$?

114. For which x-values is it true that $g(x) \leq f(x)$?

115. (a) For which x-value is $f(x) = 0$?
(b) For which x-value is $g(x) = 0$?

116. Compute $f(0) + g(0)$.

117. Compute each of the following:
(a) $(f + g)(8)$;
(b) $(f - g)(8)$;
(c) $(fg)(8)$;
(d) $(f/g)(8)$.

118. Compute each of the following:
(a) $g[f(5)]$;
(b) $f[g(5)]$;
(c) $(g \circ f)(5)$;
(d) $(f \circ g)(5)$.

119. Which is larger, $(f \circ f)(10)$ or $(g \circ g)(10)$?

120. Compute $g[f(10)] - f[g(10)]$.

121. For which x-values is it true that $f(x) \geq 3$?

122. For which x-values is it true that $|f(x) - 3| \leq 1$?

123. What is the largest number in the range of g?

124. Specify the coordinates of the highest point on the graph of each of the following equations:
(a) $y = g(-x)$;
(b) $y = -g(x)$;
(c) $y = g(x - 1)$;
(d) $y = f(-x)$;
(e) $y = -f(x)$;
(f) $y = -f(-x)$.

125. On which intervals is the function g decreasing?

126. What are the coordinates of the turning points of g?

127. For which values of x in the interval $(4, 7)$ is the quantity $\dfrac{f(x) - f(5)}{x - 5}$ negative?

128. For which values of x in the interval $(0, 5)$ is the quantity $\dfrac{f(x) - f(2)}{x - 2}$ positive?

CHAPTER THREE TEST

1. Find the domain of the function defined by $f(x) = \sqrt{x^2 - 5x - 6}$.

2. Find the range of the function defined by $g(x) = \dfrac{2x - 8}{3x + 5}$.

3. Let $f(x) = 2x^2 - 3x$ and $g(x) = 2 - x$. Compute each of the following:
   (a) $(f - g)(x)$;   (b) $(f \circ g)(x)$;   (c) $f[g(-4)]$.

4. Let $f(t) = 2/t$. Compute $\dfrac{f(t) - f(a)}{t - a}$.

5. Let $g(x) = 2x^2 - 5x$. Compute $\dfrac{g(x + h) - g(x)}{h}$.

6. Find $g^{-1}(x)$ given that $g(x) = -4x/(6x + 1)$.

7. The graph of a function f is a line segment joining the two points $(-3, 1)$ and $(5, 6)$. Determine the slope of the line segment that results from graphing the equation $y = -f^{-1}(x)$.

8. Graph each function and specify the intercepts:

   (a) $y = -|x - 3| + 1$;   (b) $y = \dfrac{1}{x + 3} - 2$.

9. The figure shows the graph of a function g with domain $[-4, 2]$.

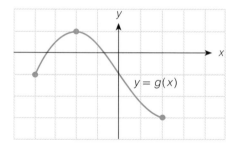

   (a) Specify the range of g.
   (b) What are the coordinates of the turning point?
   (c) What is the minimum value of g?
   (d) Which input yields a maximum value for g? What is that maximum value?
   (e) On which interval is g decreasing?
   (f) Compute $\Delta g/\Delta x$ on the interval $[-4, 2]$.

10. Let $f(x) = x^2 - 3x - 1$. Compute each of the following:
    (a) $f\left(-\tfrac{3}{2}\right)$;   (b) $f\left(\sqrt{3} - 2\right)$.

11. Graph the function F defined by

$$F(x) = \begin{cases} |x + 1| & \text{if } x < -1 \\ -x^2 + 1 & \text{if } x > -1 \end{cases}$$

What is the domain of this function?

12. The graph of f is a line segment joining the points $(1, 3)$ and $(5, -2)$. Sketch the graph of $y = f(-x)$.

13. Given that the domains of f and f^{-1} are both $(-\infty, \infty)$ and that $f^{-1}(-3) = 1$, solve the equation $5 + f(4t - 3) = 2$ for t.

14. Let $f(x) = 1/x$. Find a value for b so that on the interval $[1, b]$, we have $\Delta f/\Delta x = -\frac{1}{10}$.

15. During the first five hours of a lab experiment, the temperature of a solution is given by

$$F(t) = 0.16t^2 - 1.6t + 35 \quad (0 \le t \le 5)$$

where t is measured in hours, $t = 0$ corresponds to the time that the experiment begins, and $F(t)$ is the temperature (in degrees Celsius) at time t.
    (a) Find (and simplify) a formula for $\Delta F/\Delta t$ over the interval $[t_1, 5]$.
    (b) Use the result in part (a) to compute the average rate of change of temperature during the period from $t = 4$ hr to $t = 5$ hr. (Be sure to include the units as part of the answer.)

CHAPTER FOUR

POLYNOMIAL AND RATIONAL FUNCTIONS. APPLICATIONS TO OPTIMIZATION

INTRODUCTION

In Chapter 3, we studied some rather general rules for operating with functions and their graphs. Now, in Chapters 4 through 7, we will focus on several particular types of functions. In Section 4.1 we will discuss linear functions and their applications. As you'll see, these applications, although very elementary, are quite diverse, ranging from business and economics to physics and to statistics. Section 4.2 contains a discussion of quadratic functions. Section 4.3 could aptly be titled "Translating English into Algebra." In that section we will develop the background skills for solving the maximum and minimum problems in Section 4.4. (If a friend of yours is taking calculus, you will find that you can solve some of your friend's maximum and minimum problems without using calculus.) The last two sections of this chapter, 4.5 and 4.6, present a number of techniques for graphing polynomial and rational functions.

4.1 LINEAR FUNCTIONS

All decent functions are practically linear.

Professor Andrew Gleason

By a **linear function**, we mean a function defined by an equation of the form

$$f(x) = Ax + B$$

where A and B are constants. In this chapter, the constants A and B will always be real numbers. From our work in Chapter 2, we know that the graph of $y = Ax + B$ is a straight line.

EXAMPLE 1 Let f be a linear function. If $f(1) = 0$ and $f(2) = 3$, find an equation defining f.

Solution From the statement of the problem, we know that the graph of f is a straight line passing through the points $(1, 0)$ and $(2, 3)$. Thus the slope of the line is

$$m = \frac{y_2 - y_1}{x_2 - x_1} = \frac{3 - 0}{2 - 1} = 3$$

Now we can use the point–slope formula to find the required equation. We have

$$y - y_1 = m(x - x_1)$$
$$y - 0 = 3(x - 1)$$
$$y = 3x - 3$$

This is the equation defining f. If we wish, we can rewrite it using function notation: $f(x) = 3x - 3$. ▪▪▪

One basic application of linear functions that occurs in business and economics is *linear* or *straight-line depreciation*. In this situation we assume that the value V of an asset (such as a machine or an apartment building) decreases linearly over time t; that is, $V = mt + b$, where the slope m is negative.

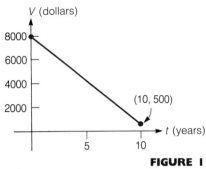

V (dollars)

(10, 500)

t (years)

FIGURE I

� ▬ ▌ **EXAMPLE 2** A factory owner buys a new machine for $8000. After ten years, the machine has a salvage value of $500.

(a) Assuming linear depreciation (as indicated in Figure 1), find a formula for the value V of the machine after t years, where $0 \le t \le 10$.

(b) Use the formula derived in part (a) to find the value of the machine after five years.

Solution

(a) We need to determine m and b in the equation $V = mt + b$. From Figure 1 we see that the V-intercept of the line segment is $b = 8000$. Furthermore, since the line segment passes through the two points $(10, 500)$ and $(0, 8000)$, the slope m is $\dfrac{8000 - 500}{0 - 10}$ or -750. So we have $m = -750$ and $b = 8000$, and the required equation is

$$V = -750t + 8000 \qquad (0 \le t \le 10)$$

(b) Substituting $t = 5$ in the equation $V = -750t + 8000$ yields

$$V = -750(5) + 8000 = -3750 + 8000 = 4250$$

Thus the value of the machine after five years is $4250. ▪▪▪

It is instructive to keep track of the units associated with the slope in Example 2. Repeating the slope calculation and keeping track of the units, we have

$$m = \frac{\$8000 - \$500}{0 \text{ years} - 10 \text{ years}} = -\$750/\text{year} \qquad (1)$$

Thus, slope is a *rate of change*. In this example, the slope represents the rate of change of the value of the machine. Notice that the calculations in (1) are exactly those that we would have used in the previous chapter to compute the *average* rate of change $\Delta V/\Delta t$. In the context of linear functions, we often suppress the word "average" and refer simply to the *rate of change*, because no matter what two points we choose on the graph of a linear function, the slope will always be the same.

As a second example of slope as a rate of change, let us suppose that a small manufacturer of handmade running shoes knows that her total cost in

(a)

(b)

FIGURE 2

dollars, $C(x)$, for producing x pairs of shoes each business day is given by the linear function

$$C(x) = 10x + 50$$

Figure 2(a) displays a table and graph for this function. Actually, since x represents the daily number of pairs of shoes, x can assume only whole-number values. So, technically, the graph that should be given is the one in Figure 2(b). However, the graph in Figure 2(a) turns out to be useful in practice, and we shall follow this convention.

The slope of the line $C(x) = 10x + 50$ in Figure 2(a) is 10, the coefficient of x. But to understand the units involved, let's calculate the slope using two of the points in Figure 2(a). Using the points $(0, 50)$ and $(1, 60)$ and keeping track of the units, we have

$$m = \frac{\$60 - \$50}{1 \text{ pair} - 0 \text{ pair}} = \frac{\$10}{1 \text{ pair}} = \$10/\text{pair}$$

Again the slope is a rate. In this case, the slope represents the rate of increase of cost; each additional pair of shoes produced costs the manufacturer \$10.

In the study of economics, an equation that gives the cost $C(x)$ for producing x units of a commodity is called a **cost equation** or **cost function**. When the graph of the cost equation is a line, we define the **marginal cost** as the additional cost to produce one more unit. Thus, in the preceding example, the marginal cost is \$10 per pair, and we see that the slope of the line in Figure 2(a) represents this marginal cost.

▌▐▌ **EXAMPLE 3** Suppose that the cost $C(x)$ in dollars of producing x bicycles is given by

$$C(x) = 625 + 45x$$

(a) Find the cost of producing 10 bicycles.
(b) What is the marginal cost?
(c) Use the answers in parts (a) and (b) to find the cost of producing 11 bicycles. Then check the answer by evaluating $C(11)$.

Solution

(a) Using $x = 10$ in the cost equation, we have

$$C(10) = 625 + 45(10) = 1075$$

Thus, the cost of producing ten bicycles is \$1075.
(b) Since C is a linear function, the marginal cost is the slope and we have

marginal cost = \$45 per bicycle

(c) According to the result in part (b), each additional bicycle costs \$45. Therefore, we can compute the cost of 11 bicycles by adding \$45 to the cost for 10 bicycles:

cost of 11 bicycles = cost of 10 bicycles + marginal cost
$$= \$1075 + \$45 = \$1120$$

So, the cost of producing 11 bicycles is $1120. We can check this result by using the cost function $C(x) = 625 + 45x$ to compute $C(11)$ directly:

$$C(11) = 625 + 45(11) = 625 + 495$$
$$= \$1120 \qquad \text{as obtained previously} \qquad ▮▮▮$$

Interpreting slope as a rate of change is not restricted to applications in business or economics. Suppose, for example, that you are driving a car at a steady rate of 50 mph. Using the distance formula from elementary mathematics,

$$\text{distance} = \text{rate} \times \text{time}$$

or

$$d = rt$$

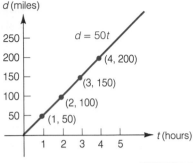

d (miles)

FIGURE 3

The slope of the line is the rate of change of distance with respect to time.

we have, in this case, $d = 50t$, where d represents the distance traveled (in miles) in t hours. In Figure 3, we show a graph of the linear function $d = 50t$. The slope of this line is 50, the coefficient of t. But 50 is also the given rate of speed, in miles per hour. So again, slope is a rate of change. In this case, slope is the **velocity**, or *rate of change of distance with respect to time*.

We conclude this section with an example that shows a statistical application of linear functions. Table 1 (on the next page) shows the evolution of the record for the one-mile run during the years 1911–1980. (In case you're wondering why the table begins with 1911, John Paul Jones was the first twentieth-century runner to break the previous century's record of 4:15.6, set in 1895.) In Figure 4(a) we've plotted the (x, y) pairs given in the table. The resulting plot is called a **scatter diagram**.

A striking feature of the scatter diagram in Figure 4(a) is that the records do not appear to be leveling off; rather, they seem to be decreasing in an approximately linear fashion. Using the *least-squares technique* from the field of statistics, it can be shown that the linear function that best fits this particular set of data is

$$f(x) = -0.41x + 1040.94 \qquad (2)$$

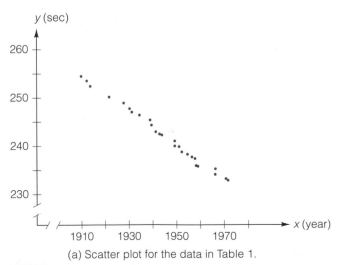

(a) Scatter plot for the data in Table 1.

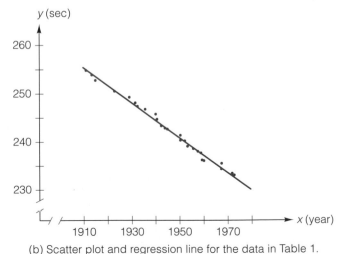

(b) Scatter plot and regression line for the data in Table 1.

FIGURE 4

The graph of this line is shown in Figure 4(b). The line itself is referred to as the **regression line** or the **least-squares line.** (The formulas for determining this line are given in Exercise 41.)

As an intuitive check on the least-squares line, let's use it to estimate what the record for the mile run might have been in 1985 (then we'll check our prediction against the actual record). With $x = 1985$, we have

$$f(1985) = -0.41 \ (1985) + 1040.94 = 227.09 \text{ seconds}$$

Thus (after converting to minutes and rounding off), our least-squares estimate for the 1985 record is 3:47.1. This agrees favorably with the actual record of 3:46.31 set by Steve Cram of Great Britain.

■■■ EXAMPLE 4 The June 1976 issue of *Scientific American* contained an article entitled "Future Performance in Footracing" by H. W. Ryder, H. J. Carr, and P. Herget. According to the authors of the article, "It appears likely that within 50 years the record [for the mile] will be down to 3:30." Use the regression line defined in equation (2) to make a projection for the mile time in the year 2026 (which will be 50 years after the article appeared). Is your projection close to the one given in the *Scientific American* article?

TABLE I

Evolution of the Record for the Mile Run, 1911–1980

x (year)	y (time)	
1911	4:15.4	(John Paul Jones, U.S.)
1913	4:14.6	(John Paul Jones, U.S.)
1915	4:12.6	(Norman Taber, U.S.)
1923	4:10.4	(Paavo Nurmi, Finland)
1931	4:09.2	(Jules Ladoumegue, France)
1933	4:07.6	(Jack Lovelock, New Zealand)
1934	4:06.8	(Glen Cunningham, U.S.)
1937	4:06.4	(Sidney Wooderson, Great Britain)
1942	4:06.2	(Gunder Haegg, Sweden)
1942	4:06.2	(Arne Andersson, Sweden)
1942	4:04.6	(Gunder Haegg, Sweden)
1943	4:02.6	(Arne Andersson, Sweden)
1944	4:01.6	(Arne Andersson, Sweden)
1945	4:01.4	(Gunder Haegg, Sweden)
1954	3:59.4	(Roger Bannister, Great Britain)
1954	3:58.0	(John Landy, Australia)
1957	3:57.2	(Derek Ibbotson, Great Britain)
1958	3:54.5	(Herb Elliott, Australia)
1962	3:54.4	(Peter Snell, New Zealand)
1964	3:54.1	(Peter Snell, New Zealand)
1965	3:53.6	(Michel Jazy, France)
1966	3:51.3	(Jim Ryun, U.S.)
1967	3:51.1	(Jim Ryun, U.S.)
1975	3:51.0	(Filbert Bayi, Tanzania)
1975	3:49.4	(John Walker, New Zealand)
1979	3:49.0	(Sebastian Coe, Great Britain)
1980	3:48.8	(Steve Ovett, Great Britain)

Solution   The given equation for the regression line is

$$f(x) = -0.41x + 1040.94$$

Setting $x = 2026$ and using a calculator, we have

$$f(2026) = -0.41(2026) + 1040.94 = 210.28 \text{ seconds}$$

Converting this result to minutes and then rounding off, our projection for the year 2026 is 3:30.3, which is essentially the same as the projection in *Scientific American*.   ∎

EXERCISE SET 4.1

A

In Exercises 1–10, find the linear functions satisfying the given conditions.

1. $f(-1) = 0$ and $f(5) = 4$

2. $f(3) = 2$ and $f(-3) = -4$

3. $g(0) = 0$ and $g(1) = \sqrt{2}$

4. The graph passes through the points $(2, 4)$ and $(3, 9)$.

5. $f(\frac{1}{2}) = -3$ and the graph of f is a line parallel to the line $x - y = 1$.

6. $g(2) = 1$ and the graph of g is perpendicular to the line $6x - 3y = 2$.

7. The graph of f is a horizontal line that passes through the larger of the two y-intercepts of the circle $x^2 - 2x + y^2 - 3 = 0$.

8. The x- and y-intercepts of the graph of g are 1 and 4, respectively.

9. The graph of the inverse function passes through the points $(-1, 2)$ and $(0, 4)$.

10. The x- and y-intercepts of the inverse function are 5 and -1, respectively.

11. Let $f(x) = 3x - 4$ and $g(x) = 1 - 2x$. Determine whether the function $f \circ g$ is linear.

12. Explain why there is no linear function with a graph that passes through all three of the points $(-3, 2)$, $(1, 1)$, and $(5, 2)$.

13. A factory owner buys a new machine for $20,000. After eight years, the machine has a salvage value of $1000. Find a formula for the value of the machine after t years, where $0 \le t \le 8$.

14. A manufacturer buys a new machine costing $120,000. It is estimated that the machine has a useful lifetime of ten years and a salvage value of $4000 at that time.
    (a) Find a formula for the value of the machine after t years, where $0 \le t \le 10$.
    (b) Find the value of the machine after eight years.

15. A factory owner installs a new machine costing $60,000. Its expected lifetime is five years, and at the end of that time the machine has no salvage value.
    (a) Find a formula for the value of the machine after t years, where $0 \le t \le 5$.
    (b) Complete the following depreciation schedule.

End of Year	Yearly Depreciation	Accumulated Depreciation	Value V
0	0	0	60,000
1			
2			
3			
4			
5		60,000	0

16. Let x denote a temperature on the Celsius scale, and let y denote the corresponding temperature on the Fahrenheit scale.
    (a) Find a linear function relating x and y; use the facts that 32°F corresponds to 0°C, and 212°F corresponds to 100°C. Write the function in the form $y = Ax + B$.
    (b) What Celsius temperature corresponds to 98.6°F?
    (c) Find a number z for which $z°F = z°C$.

17. Suppose that the cost $C(x)$, in dollars, of producing x electric fans is given by $C(x) = 450 + 8x$.
    (a) Find the cost to produce 10 fans.
    (b) Find the cost to produce 11 fans.
    (c) Use your answers in parts (a) and (b) to find the marginal cost. (Then check that your answer is the slope of the line.)

18. Suppose that the cost to a manufacturer of producing x units of a certain motorcycle is given by $C(x) = 220x + 4000$, where $C(x)$ is in dollars.

(a) Find the marginal cost.
(b) Find the cost of producing 500 motorcycles.
(c) Use your answers in parts (a) and (b) to find the cost of producing 501 motorcycles.

19. Suppose that the cost $C(x)$, in dollars, of producing x units of a certain cassette tape player is given by $C(x) = 400 + 50x$.
    (a) Compute $C(n + 1) - C(n)$.
    (b) What is the marginal cost?
    (c) What is the relationship between the answers in parts (a) and (b)?

20. Suppose that the cost $C(x)$, in dollars, of producing x cassette tapes is given by $C(x) = 0.5x + 500$.
    (a) Graph the given equation.
    (b) Compute $C(150)$.
    (c) If you add the marginal cost to the answer in part (b), you obtain a certain dollar amount. What does this amount represent?

21. The following graphs each relate distance and time for a moving object. Determine the velocity in each case.

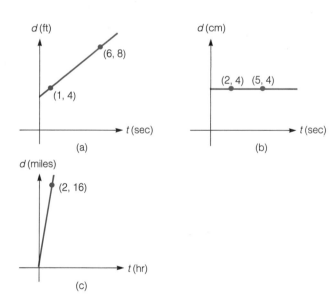

22. The distance d, in feet, covered by a particular object in t sec is given by the equation $d = 5t$.
    (a) Find the velocity of the object.
    (b) Find the distance covered in 15 sec.
    (c) Use your answers in parts (a) and (b) to find the distance covered in 16 sec. Check your answer by substituting $t = 16$ in the given distance formula.

23. Two points A and B move along the x-axis. After t sec, their positions are given by the equations

    A: $x = 3t + 100$
    B: $x = 20t - 36$

(a) Which point is traveling faster, A or B?
(b) Which point is farther to the right when $t = 0$?
(c) At what time t do A and B have the same x-coordinate?

24. A point moves along the x-axis, and its x-coordinate after t seconds is $x = 4t + 10$. (Assume that x is in centimeters.)
    (a) What is the velocity?
    (b) What is the x-coordinate when $t = 2$ sec?
    (c) Use your answers in parts (a) and (b) to find the x-coordinate when $t = 3$ sec. Check your answer by letting $t = 3$ in the given equation.

25. (a) On graph paper, plot the following points: $(1, 0)$, $(2, 3)$, $(3, 6)$, $(4, 7)$.
    (b) In your scatter diagram from part (a), sketch a line that, you feel, seems to best fit the data. Estimate the slope and y-intercept of the line.
    (c) The actual regression line in this case is $y = 2.4x - 2$. Add the graph of this line to your sketch from parts (a) and (b).

26. (a) On graph paper, plot the following points: $(1, 2)$, $(3, 2)$, $(5, 4)$, $(8, 5)$, $(9, 6)$.
    (b) In your scatter diagram from part (a), sketch a line that, you feel, seems to best fit the data.
    (c) Using your sketch from part (b), estimate the y-intercept and the slope of the regression line.
    (d) The actual regression line is $y = 0.518x + 1.107$. Check to see if your estimates in part (c) are consistent with the actual y-intercept and slope. Graph this line along with the points given in part (a). (In sketching the line, use the approximation $y = 0.5x + 1.1$.)

27. The following table shows the population of the city of Los Angeles over the years 1930–1988. The accompanying graph shows the corresponding scatter plot and regression line.

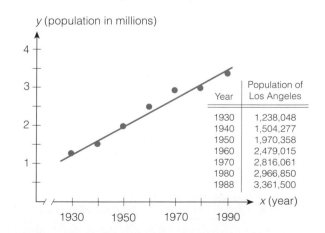

Year	Population of Los Angeles
1930	1,238,048
1940	1,504,277
1950	1,970,358
1960	2,479,015
1970	2,816,061
1980	2,966,850
1988	3,361,500

The equation of the regression line (after some rounding off) is

$$y = 37,025x - 70,226,566$$

(a) Predict whether the population of Los Angeles will reach 4 million by the year 2000.
(b) Make a projection for the population of Los Angeles in the year 2010. Round off your answer to the nearest 5,000.

28. The following table shows expenditures on pollution control and abatement in the United States during the period 1975–1985. (The figures include both government and private-sector expenditures.) The accompanying graph displays the scatter plot and regression line for the data.

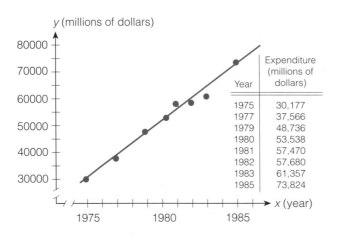

Year	Expenditure (millions of dollars)
1975	30,177
1977	37,566
1979	48,736
1980	53,538
1981	57,470
1982	57,680
1983	61,357
1985	73,824

The equation of the regression line (after rounding off) is

$$f(x) = (4198.8)x - 8262216$$

(a) Use the regression line (and a calculator) to make a projection for the expenditures in the year 1995.
(b) Find $f^{-1}(x)$.
(c) Use your answer in part (b) to find the year when the expenditures might reach 125 billion dollars.

29. Many companies in the United States manufacture or assemble their products abroad to save on labor costs. For example, the toy industry in the United States imports approximately 30% of its products from China (GI Joe and Teenage Mutant Ninja Turtles are assembled in China.) The following table shows the dollar value of these imports during the period 1986–1990, while the graph displays the scatter plot and regression line for the data.

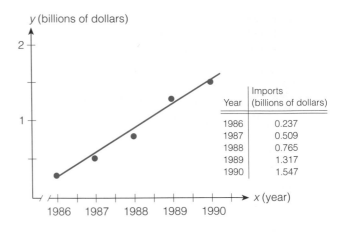

Year	Imports (billions of dollars)
1986	0.237
1987	0.509
1988	0.765
1989	1.317
1990	1.547

The equation of the regression line is

$$f(x) = (0.3428)x - 680.6114$$

(a) Use the regression line to make a projection for the value of the toy imports from China in 1996.
(b) Predict whether the value of the toy imports will exceed 5 billion dollars by the year 2000.

30. This exercise illustrates one of the pitfalls that can arise in using a regression line: large discrepancies can occur when the regression line is used to make long-term projections. The following table shows the population of California during the years 1860–1900.

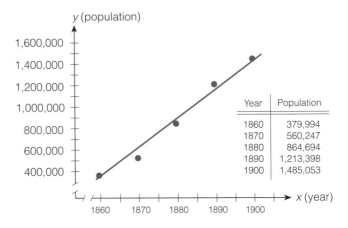

Year	Population
1860	379,994
1870	560,247
1880	864,694
1890	1,213,398
1900	1,485,053

The accompanying graph displays the scatter plot of the data and the regression line

$$f(x) = (28632.69)x - 52928780$$

As you can see from the scatter plot, the population growth over this period is very nearly linear.
(a) Use the regression line to estimate the population of California in 1990.
(b) According to the U.S. Bureau of the Census, the population of California in 1990 was 29,839,250. Is your estimate in part (a) close to this figure?

B

As background for Exercises 31 and 32, reread the definition of the iterates of x_0 for a function f in Exercise 51 on page 138.

31. Let $f(x) = \frac{1}{2}x + 2$.
     (a) Compute the first three iterates of $x_0 = 1$.
     (b) Refer to the following figure. Explain why the y-coordinates of the points A, B, and C are the iterates that you computed in part (a).

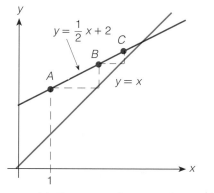

     (c) Determine the y-coordinate of the point where the lines $y = x$ and $y = \frac{1}{2}x + 2$ intersect.
     (d) If we continue the pattern of dashed lines in the figure, it appears that the iterates of $x_0 = 1$ will approach 4. Verify this empirically by using your calculator to compute the fourth through tenth iterates. (Round off the final answers to three decimal places.)

32. (a) Let $f(x) = \frac{1}{2}x + 2$ (as in Exercise 31). Compute the first ten iterates of $x_0 = 5$. (After the third iterate, round off the final answers to three decimal places.) What number do the answers seem to be approaching?
     (b) Let $g(x) = 2x + 1$. Compute the first ten iterates of $x_0 = 1$. Do you observe a pattern similar to the one in part (a)?

In Exercises 33–36, let f and g be the linear functions defined by

$$f(x) = Ax + B \quad (A \neq 0) \qquad \text{and} \qquad g(x) = Cx + D \quad (C \neq 0)$$

In each case, compute the average rate of change of the given function on an interval $[a, b]$.

33. (a) f (b) $f \circ f$
     (c) $g \circ f$ (d) $f \circ g$
34. (a) f^{-1} (b) g^{-1}
35. (a) $(f \circ g)^{-1}$ (b) $(g \circ f)^{-1}$
36. (a) $f^{-1} \circ g^{-1}$ (b) $g^{-1} \circ f^{-1}$
37. Show that the linear function $f(x) = mx$ satisfies the following identities:
     (a) $f(a + b) = f(a) + f(b)$;
     (b) $f(ax) = af(x)$.

38. Let f be a linear function such that
     $$f(a + b) = f(a) + f(b)$$
     for all real numbers a and b. Show that the graph of f passes through the origin.   *Hint:*   Let $f(x) = Ax + B$ and show that $B = 0$.

39. Find a linear function $f(x) = mx + b$ such that m is positive and $(f \circ f)(x) = 9x + 4$.

40. (a) Let f be a linear function. Show that
         $$f\left(\frac{x_1 + x_2}{2}\right) = \frac{f(x_1) + f(x_2)}{2}$$
         (In words: The output of the average is the average of the outputs.)
     (b) Show that the equation in part (a) does not hold for the function $f(x) = x^2$.

41. This exercise shows how to compute the slope and the y-intercept of the regression line. As an example, we'll work with the simple data set given in Exercise 25:

x	1	2	3	4
y	0	3	6	7

     (a) Let Σx denote the sum of the x-coordinates in the data set, and let Σy denote the sum of the y-coordinates. Check that $\Sigma x = 10$ and $\Sigma y = 16$.
     (b) Let Σx^2 denote the sum of the squares of the x-coordinates, and let Σxy denote the sum of the products of the corresponding x- and y-coordinates. Check that $\Sigma x^2 = 30$ and $\Sigma xy = 52$.
     (c) The slope m and the y-intercept b of the regression line satisfy the following pair of simultaneous equations [in the first equation, n denotes the number of points (x, y) in the data set]:
         $$\begin{cases} nb + (\Sigma x)m = \Sigma y \\ (\Sigma x)b + (\Sigma x^2)m = \Sigma xy \end{cases}$$
         So, in the present example, these equations become
         $$\begin{cases} 4b + 10m = 16 \\ 10b + 30m = 52 \end{cases}$$
         Solve this pair of equations for m and b, and check that your answers agree with the values in Exercise 25(c).

In Exercises 42–45, use the method described in Exercise 41 to find the equation of the regression line for the given data sets.

42.

x	2	4	8	10
y	-7	-5	-2	-1

43.

x	1	2	3	4	5
y	2	3	9	9	11

44.

x	1	2	3	4	5
y	16	13.1	10.5	7.5	2

45.

x	520	740	560	610	650
y	81	98	83	88	95

OPTIONAL TI-81 GRAPHING CALCULATOR EXERCISES
EXERCISES FOR SECTION 4.1

1. Reread Example 2. Then graph the function $f(x) = -750x + 8000$. Use the following settings in the RANGE menu so that your picture is similar to Figure 1 on page 160.

$$x_{min} = 0 \qquad y_{min} = 0$$
$$x_{max} = 10 \qquad y_{max} = 8000$$
$$x_{scl} = 5 \qquad y_{scl} = 2000$$

Now press the TRACE key and use the arrow keys to move the cursor to a point where the x-coordinate is as close to 5 as possible. What is the corresponding y-coordinate? How close is your result to the answer $4250 obtained in the text?

2. This exercise shows how to use the calculator to determine a regression line. We will use the data set in the following table, which shows the world grain consumption for the years 1970–1980.

Year	World Grain Consumption (million metric tons)
1970	1130
1972	1192
1974	1213
1976	1303
1978	1418
1980	1475

Source: Lester R. Brown, project director, *State of the World, 1989* (New York: W. W. Norton and Company, 1989).

(a) Clear any existing statistical data by first pressing the "2nd" key, then the STAT key, the left-hand arrow key, and the "2" key. After that, press ENTER.
(b) To enter the data in the table, begin by pressing the "2nd" key, then the STAT key, and then the left-hand arrow key. But do not press ENTER yet. In-

stead press the number "1" to select the edit option. As you can see, the screen display shows

DATA

$x_1 =$

$y_1 = 1$

For the x_1 value, type 1970, and then press ENTER to move the cursor down to the line for y_1. For the y_1 value, type 1130 and then press ENTER. The remaining values are then entered in the same fashion.

(c) After you have entered all of the data, press the "2nd" key and then the STAT key. Now use the arrow to move the cursor to item 2 (which is the choice for linear regression) and press ENTER. Then press ENTER again. You should see the following values displayed on your screen:

$$a = -69049.71429$$
$$b = 35.61428571$$

(There is also a value for r, the *coefficient of correlation*, but we will not need that here.) These are the values for the constants a and b in the regression line $y = a + bx$.

(d) Use the regression line to estimate the world grain consumption for the year 1985. (The actual value for 1985 was 1594 million metric tons. Is your estimate too high or too low?)

3. Linear functions can be used to approximate more complicated functions. This is one of the meanings or implications of the quotation on page 159. This exercise illustrates that idea. As preparation, clear any functions that are entered in the Y= menu. Press ZOOM 6 and then make the following adjustments in the RANGE window. $x_{min} = -2$; $x_{max} = 3$; $y_{min} = -3$; $y_{max} = 4$.

(a) Using calculus, it can be shown that the equation of the line that is tangent to the curve $y = x^2$ at the point $(1, 1)$ is $y = 2x - 1$. Verify this visually by entering the two functions $Y_1 = x^2$ and $Y_2 = 2x - 1$ and then pressing the GRAPH key. Note that the line is virtually indistinguishable from the curve in the immediate vicinity of $(1, 1)$.

(b) For numerical rather than visual evidence of how well the linear function $y = 2x - 1$ approximates the function $y = x^2$, use your calculator to complete the following tables.

x	0.9	0.99	0.999
$2x - 1$			
x^2			

x	1.1	1.01	1.001
$2x - 1$			
x^2			

4.2 QUADRATIC FUNCTIONS

After the linear functions, the next simplest functions are the **quadratic functions**, which are defined by equations of the form

$$f(x) = ax^2 + bx + c \qquad (a \neq 0)$$

where a, b, and c are constants and a is not zero. In this chapter the constants a, b, and c will always be real numbers. We will see that the graph of any quadratic function is a curve called a **parabola**, which is similar or identical in shape to the graph of $y = x^2$. Figure 1 displays the graphs of two typical parabolas.

FIGURE 1

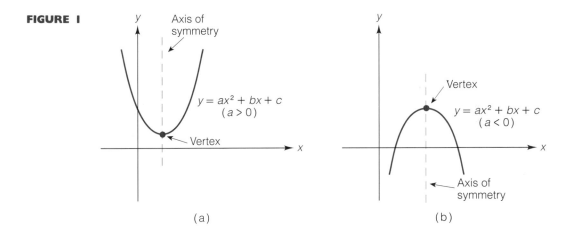

(a)   (b)

Subsequent examples will demonstrate that the parabola $y = ax^2 + bx + c$ opens upward when $a > 0$ and downward when $a < 0$. As Figure 1 indicates, the turning point on the parabola is called the **vertex**. The **axis of symmetry** of the parabola $y = ax^2 + bx + c$ is the vertical line passing through the vertex.

The methods that we will use for dealing with quadratic functions have already been developed in Chapters 1 and 3. In particular, the following two topics are prerequisites for understanding the examples in this section:

1. completing the square; for a review, see Section 1.4;
2. translations and reflections; for a review, see Section 3.3.

EXAMPLE 1   Graph the function $y = x^2 - 2x + 3$.

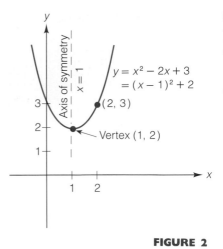

FIGURE 2

Solution   The idea here is to use the technique of completing the square; this will enable us to obtain the required graph simply by shifting the basic $y = x^2$ graph. We begin by writing the given equation

$$y = x^2 - 2x \qquad + 3$$

To complete the square for the x-terms we want to add 1. (Check this.) Of course, to keep the equation in balance we have to account for this by writing

$$y = (x^2 - 2x + 1) + 3 - 1 \qquad \text{adding zero to the right side}$$

or

$$y = (x - 1)^2 + 2$$

Now, as we know from Section 3.3, the graph of this last equation is obtained by moving the parabola $y = x^2$ one unit in the positive x-direction and two units in the positive y-direction. This shifts the vertex from the origin to the point $(1, 2)$. See Figure 2.

Note   As a guide to sketching the graph, you'll want to know the y-intercept. To find the y-intercept, substitute $x = 0$ in the given equation to obtain $y = 3$. Then, given the vertex $(1, 2)$ and the point $(0, 3)$, a reasonably accurate graph can be quickly sketched. [Actually, once you find that $(0, 3)$ is on the graph, you also know that the reflection of this point about the axis of symmetry is on the graph. This is why the point $(2, 3)$ is shown in Figure 2.] ▪▪▪

Now we want to compare the graphs of $y = x^2$, $y = 2x^2$, and $y = \frac{1}{2}x^2$. The last two equations were not specifically discussed in Chapters 2 or 3, so for the moment you can think about graphing them by first setting up tables. Figure 3 displays the graphs. All three graphs are parabolas that open upward, but the shapes are not identical. The parabola $y = 2x^2$ is narrower than $y = x^2$, while $y = \frac{1}{2}x^2$ is wider than $y = x^2$. These observations about shape would also apply to $y = -x^2$, $y = -2x^2$, and $y = -\frac{1}{2}x^2$; in these cases, though, the parabolas open downward rather than upward. The observations that we have just made are generalized in the box that follows.

FIGURE 3

▶ **PROPERTY SUMMARY**   **THE GRAPH OF $y = ax^2$**

1. The graph of $y = ax^2$ is a parabola with vertex at the origin. It is similar in shape to $y = x^2$.

2. The parabola $y = ax^2$ opens upward if $a > 0$, downward if $a < 0$.

3. The parabola $y = ax^2$ is narrower than $y = x^2$ if $|a| > 1$, wider than $y = x^2$ if $|a| < 1$.

▮▮▮ **EXAMPLE 2**   Sketch the graph of $y = 3(x - 1)^2$.

Solution   Because of the $x - 1$, we shift the basic parabola $y = x^2$ one unit to the right. The factor of 3 in the given equation tells us that we want to draw a parabola that is narrower than $y = x^2$ but that has the same vertex, $(1, 0)$. To see exactly how narrow to draw $y = 3(x - 1)^2$, we would like to know another point

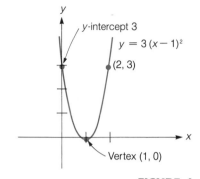

y-intercept 3

$y = 3(x - 1)^2$

(2, 3)

Vertex (1, 0)

FIGURE 4

on the graph other than the vertex $(1, 0)$. An easy point to obtain is the y-intercept. Setting $x = 0$ in the equation yields

$$y = 3(0 - 1)^2 = 3$$

Now that we know the vertex, $(1, 0)$, and the y-intercept, 3, a reasonably accurate graph can be sketched. See Figure 4. ∎∎∎

EXAMPLE 3 Sketch the graph of $y = -3(x - 1)^2$.

Solution In Example 2 we sketched the graph of $y = 3(x - 1)^2$. By reflecting that graph in the x-axis, we obtain the graph of $y = -3(x - 1)^2$. See Figure 5. ∎∎∎

Vertex (1, 0)

$y = -3(x - 1)^2$

y-intercept −3

(2, −3)

FIGURE 5

EXAMPLE 4 Graph the function $f(x) = -2x^2 + 12x - 16$ and specify the vertex, axis of symmetry, maximum or minimum value of f, and x- and y-intercepts.

Solution The idea is to complete the square, as in Example 1. We have

$$y = -2x^2 + 12x \qquad - 16$$
$$= -2(x^2 - 6x \qquad) - 16$$
$$= -2(x^2 - 6x + 9) - 16 + 18 \qquad \text{adding } (2)(9) = 18 \text{ to keep the equation in balance}$$
$$= -2(x - 3)^2 + 2 \tag{1}$$

From equation (1), we see that the required graph is obtained simply by shifting the graph of $y = -2x^2$ "right 3, up 2," so that the vertex is $(3, 2)$. As a guide to sketching the graph, we want to compute the intercepts. The y-intercept is -16. (Why?) For the x-intercepts, we replace y with 0 in equation (1) to obtain

$$-2(x - 3)^2 = -2$$
$$(x - 3)^2 = 1$$
$$x - 3 = \pm 1$$
$$x = 4 \quad \text{or} \quad 2$$

Thus the x-intercepts are $x = 4$ and $x = 2$. Knowing these intercepts and the vertex, we can sketch the graph as in Figure 6 (on the next page). You should check for yourself that the information accompanying Figure 6 is correct. ∎∎∎

FIGURE 6

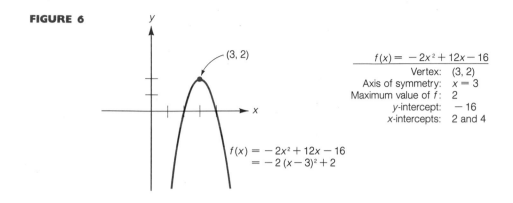

$$f(x) = -2x^2 + 12x - 16$$

Vertex:	$(3, 2)$
Axis of symmetry:	$x = 3$
Maximum value of f:	2
y-intercept:	-16
x-intercepts:	2 and 4

$$f(x) = -2x^2 + 12x - 16$$
$$= -2(x-3)^2 + 2$$

We conclude this section with several examples involving maximum and minimum values of functions. As background for this, we first summarize our basic technique for graphing parabolas.

PROPERTY SUMMARY   THE GRAPH OF THE PARABOLA $y = ax^2 + bx + c$

By completing the square, the equation of the parabola $y = ax^2 + bx + c$ can always be rewritten in the form

$$y = a(x - h)^2 + k$$

In this form, the vertex of the parabola is (h, k) and the axis of symmetry is the line $x = h$. The parabola opens upward if $a > 0$ and downward if $a < 0$.

■■■ **EXAMPLE 5**   Among all possible inputs for the function

$$g(x) = 3x^2 - 2x - 6$$

which yields the smallest output? What is this minimum output?

Solution   Think in terms of a graph. Since the graph of this function is a parabola opening upward, the input we want is the x-coordinate of the vertex. By completing the square (as shown in Example 4), we find that the given equation can be rewritten

$$g(x) = 3\left(x - \frac{1}{3}\right)^2 - \frac{19}{3}$$

(You should verify this for yourself, using Example 4 as a model.) Thus, the vertex of the parabola is $\left(\frac{1}{3}, -\frac{19}{3}\right)$. From this, and from the fact that the parabola opens upward, we conclude that the input $x = \frac{1}{3}$ produces the smallest output, and this minimum output is $g\left(\frac{1}{3}\right) = -\frac{19}{3}$.   ▬▬▬

For the previous two examples, keep in mind that we were able to find the maximum or minimum easily only because the functions were quadratics. In contrast, you cannot expect to find the minimum of $y = x^4 - 8x$ using the method of Examples 4 and 5 because it is not a quadratic function. In general, the techniques of calculus are required to find maxima and minima for functions other than quadratics. There are some cases, however, in which our present method can be adapted to functions that are closely related to quadratics, as shown in the next two examples.

EXAMPLE 6 Let $D = \sqrt[3]{x^2 - x + 1}$. For which value of x is D a minimum?

Solution The same x-value that makes D as small as it can be will also cause the quantity $D^3 = x^2 - x + 1$ to be its smallest. That is, we only need to find the x-value that minimizes the quantity $x^2 - x + 1$. Call this quantity y, and complete the square as follows:

$$y = x^2 - x \qquad + 1$$
$$= x^2 - x + \frac{1}{4} + 1 - \frac{1}{4}$$
$$= \left(x - \frac{1}{2}\right)^2 + \frac{3}{4}$$

This shows that the vertex of the parabola $y = x^2 - x + 1$ is $\left(\frac{1}{2}, \frac{3}{4}\right)$. Since this parabola opens upward, we conclude that the input $x = \frac{1}{2}$ will produce the minimum value for the quantity $x^2 - x + 1$. According to our initial remark, this same value, $x = \frac{1}{2}$, also minimizes the quantity $D = \sqrt[3]{x^2 - x + 1}$. (Our work also shows that the minimum value of D is $\sqrt[3]{\frac{3}{4}}$.) ■■■

EXAMPLE 7 Find the maximum value of the function $y = 16t^2 - 4t^4$.

Solution Although this isn't a quadratic function, we can complete the square as follows:

$$y = -4(t^4 - 4t^2 \qquad)$$
$$= -4(t^4 - 4t^2 + 4) + 16$$
$$= -4(t^2 - 2)^2 + 16$$
$$= 16 - 4(t^2 - 2)^2$$

From this last equation we see that y never exceeds 16 (because the quantity being subtracted from 16 is nonnegative). Furthermore, y does attain the value 16 when $t^2 = 2$ or $t = \pm\sqrt{2}$. It now follows that the maximum value of the given function is 16. ■■■

EXERCISE SET 4.2

A

In Exercises 1–18, graph the quadratic function. Specify the vertex, axis of symmetry, maximum or minimum value, and intercepts.

1. $y = (x + 2)^2$
2. $y = -(x + 2)^2$
3. $y = 2(x + 2)^2$
4. $y = 2(x + 2)^2 + 4$
5. $y = -2(x + 2)^2 + 4$
6. $y = x^2 + 6x - 1$
7. $f(x) = x^2 - 4x$
8. $F(x) = x^2 - 3x + 4$
9. $g(x) = 1 - x^2$
10. $y = 2x^2 + \sqrt{2}x$
11. $y = x^2 - 2x - 3$
12. $y = 2x^2 + 3x - 2$
13. $y = -x^2 + 6x + 2$
14. $y = -3x^2 + 12x$
15. $s = 16t^2$
16. $s = -\frac{1}{2}t^2 - t$
17. $s = 2 + 3t - 9t^2$
18. $s = -\frac{1}{4}t^2 + t - 1$

For Exercises 19–24, determine the input that produces the largest or smallest output (whichever is appropriate). State whether the output is largest or smallest.

19. $y = 2x^2 - 4x + 11$
20. $f(x) = 8x^2 + x - 5$
21. $g(x) = -6x^2 + 18x$
22. $s = -16t^2 + 196t + 80$
23. $f(x) = x^2 - 10$
24. $h(x) = x^2 - 10x$

In Exercises 25–30, find the maximum or minimum value for each function (whichever is appropriate). State whether the value is a maximum or minimum.

25. $y = x^2 - 8x + 3$
26. $y = \frac{1}{2}x^2 + x + 1$
27. $y = -2x^2 - 3x + 2$
28. $y = -\frac{1}{3}x^2 - 2x$
29. $f(t) = -12t^2 + 1000$
30. $g(t) = 400t^2$

31. How far from the origin is the vertex of the parabola $y = x^2 - 6x + 13$?

32. Find the distance between the vertices of the parabolas $y = -\frac{1}{2}x^2 + 4x$ and $y = 2x^2 - 8x - 1$.

For Exercises 33–38, the functions f, g, and h are defined as follows:

$$f(x) = 2x - 3 \qquad g(x) = x^2 + 4x + 1 \qquad h(x) = 1 - 2x^2$$

In each exercise, classify the function as linear, quadratic, or neither.

33. $f \circ g$
34. $g \circ f$
35. $g \circ h$
36. $h \circ g$
37. $f \circ f$
38. $h \circ h$

In Exercises 39 and 40, determine the inputs that yield the minimum values for each function. Compute the minimum value in each case.

39. (a) $f(x) = \sqrt{x^2 - 6x + 73}$
    (b) $g(x) = \sqrt[3]{x^2 - 6x + 73}$
    (c) $h(x) = x^4 - 6x^2 + 73$

40. (a) $F(x) = (4x^2 - 4x + 109)^{1/2}$
    (b) $G(x) = (4x^2 - 4x + 109)^{1/3}$
    (c) $H(x) = 4x^4 - 4x^2 + 109$

In Exercises 41 and 42, determine the maximum value for each function.

41. (a) $f(x) = \sqrt{-x^2 + 4x + 12}$
    (b) $g(x) = \sqrt[3]{-x^2 + 4x + 12}$
    (c) $h(x) = -x^4 + 4x^2 + 12$

42. (a) $F(x) = (27x - x^2)^{1/2}$
    (b) $G(x) = (27x - x^2)^{1/3}$
    (c) $H(x) = 27x^2 - x^4$

B

43. Let $f(x) = ax^2 + bx + c$. Compute the average rate of change $\Delta f / \Delta x$ for the interval $[x_1, x_2]$.

44. Let $f(x) = ax^2 + bx + c$. Show that
$$\frac{f(x + h) - f(x)}{h} = 2ax + ah + b$$

45. By completing the square, show that the coordinates of the vertex of the parabola $y = ax^2 + bx + c$ are $(-b/2a, -D/4a)$, where $D = b^2 - 4ac$.

46. Compute the average of the two x-intercepts of the graph of $y = ax^2 + bx + c$. (Assume $b^2 - 4ac > 0$.) How does your answer relate to the result in Exercise 45?

47. Let $f(x) = ax^2 + bx + c$.
    (a) Show that $f\left(\dfrac{-b}{2a} + h\right) = ah^2 - D/4a$, where D denotes the discriminant $b^2 - 4ac$.
    (b) Show that $f\left(\dfrac{-b}{2a} - h\right) = ah^2 - D/4a$.
    Remark: The results in parts (a) and (b) prove that the parabola is symmetric about the line $x = -b/2a$.

48. Find the x-coordinate of the vertex of the parabola $y = (x - a)(x - b)$, where a and b are constants.

As background for Exercises 49 and 50, read the definition of the iterates of x_0 in Exercise 51 on page 138.

49. The figure shows portions of the parabola $f(x) = x^2 - 0.6$ and the line $y = x$.

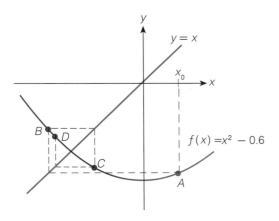

(a) Explain why the y-coordinates of A, B, C, and D are the first four iterates of x_0 for the function $f(x) = x^2 - 0.6$.
(b) Suppose that $x_0 = 0.2$. Compute the first four iterates. (For the third and fourth iterates, round off the final answers to three decimal places.)

50. The following figure shows portions of the parabola $f(x) = x^2 - 0.3$ and the line $y = x$.

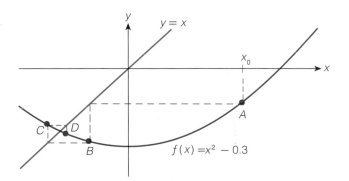

$y = x$

x_0

A

C D

B

$f(x) = x^2 - 0.3$

(a) Explain why the y-coordinates of A, B, C, and D are the first four iterates of x_0 for the function $f(x) = x^2 - 0.3$.

(b) Suppose that $x_0 = 0.4$. Compute the first four iterates. For the third and fourth iterates, round off the final answers to four decimal places.

(c) Use your calculator to compute the fifth through tenth iterates of $x_0 = 0.4$. Round off the final answers to four decimal places.

(d) Use the quadratic formula and your calculator to find the y-coordinate of the point in the third quadrant where the line $y = x$ intersects the parabola $f(x) = x^2 - 0.3$. Round off the answer to four decimal places. Note how close your answer is to the tenth iterate in part (c). (As the picture suggests, if the iteration process is continued indefinitely, the iterates approach this y-coordinate.)

In Exercises 51–54, find quadratic functions satisfying the given conditions.

51. The graph passes through the origin, and the vertex is $(2, 2)$.

52. The graph is obtained by translating $y = x^2$ four units in the negative x-direction and three units in the positive y-direction.

53. The vertex is $(3, -1)$, and one x-intercept is 1.

54. The axis of symmetry is the line $x = 1$. The y-intercept is 1. There is only one x-intercept.

55. Find the equation of the circle that passes through the origin and is centered at the vertex of the parabola $y = 2x^2 + 12x + 14$.

56. For which value of c will the minimum value of the function $f(x) = x^2 + 2x + c$ be $\sqrt{2}$?

C

57. Consider the quadratic function $y = px^2 + px + r$, where $p \neq 0$.
    (a) Show that if the vertex lies on the x-axis, then $p = 4r$.
    (b) Show that if $p = 4r$, then the vertex lies on the x-axis.

58. Let $g(x) = x^2 + 2(a + b)x + 2(a^2 + b^2)$, where a and b are constants.
    (a) Show that the coordinates of the vertex are
    $$\left(-(a + b), (a - b)^2\right)$$
    (b) Use the result in part (a) to explain why the graph of g has no x-intercepts unless $a = b$, in which case there is only one x-intercept.

59. As you have seen in this section, if the quadratic equation $f(x) = x^2 + Bx + C = 0$ has two real roots, there is a geometric interpretation: these roots are the x-intercepts for the graph of f. Less well known is the fact that there is a geometric interpretation when the quadratic equation $f(x) = x^2 + Bx + C = 0$ has nonreal complex roots. If these roots are $a + bi$ and $a - bi$, where a and b are real numbers, show that the coordinates of the vertex for the graph of f are (a, b^2). *Hint:* If r_1 and r_2 are the roots of the equation $x^2 + Bx + C = 0$, then $x^2 + Bx + C$ can be written $(x - r_1)(x - r_2)$.

OPTIONAL TI-81 GRAPHING CALCULATOR EXERCISES
EXERCISES FOR SECTION 4.2

Exercises 1 and 2 show how the shape of the parabola $y = ax^2$ changes as the quantity a changes. As preparation for Exercise 1, clear any functions that are entered in the Y= menu. Press ZOOM 6 and then in the RANGE window, set $x_{min} = -3$ and $x_{max} = 3$.

1. (a) Press the Y= key and define the following four functions.

   $$Y_1 = x^2 \qquad Y_3 = 3x^2$$
   $$Y_2 = 2x^2 \qquad Y_4 = 8x^2$$

Graph all four functions on the same set of axes. As the coefficient of x^2 increases from 1 to 8, how does the resulting graph change?

(b) Where do you think the graph of $y = \frac{1}{2}x^2$ would fit in with the graphs in part (a)? Press the $Y=$ key and redefine Y_4 to be $\frac{1}{2}x^2$. Now press the GRAPH key and see if you are right.

2. (a) Press the $Y=$ key and define the following four functions.

$$Y_1 = x^2 \qquad Y_3 = \frac{1}{3}x^2$$
$$Y_2 = \frac{1}{2}x^2 \qquad Y_4 = \frac{1}{8}x^2$$

Press ZOOM 6 to graph all four functions in the standard viewing rectangle. As the coefficient of x^2 decreases from 1 to $\frac{1}{8}$, how does the resulting graph change?

(b) Where do you think the graph of $y = (0.01)x^2$ would fit in with the graphs in part (a)? Press the $Y=$ key and redefine Y_4 to be $(0.01)x^2$. Now press the GRAPH key. Where is the graph of Y_4?

3. This exercise provides some additional perspective on Example 6.

(a) Press the $Y=$ key and enter the functions $Y_1 = x^2 - x + 1$ and $Y_2 = (x^2 - x + 1)^{1/3}$. Now press ZOOM 6 to see the graphs. According to Example 6, the same input (namely, $x = \frac{1}{2}$) minimizes both functions. To see this more clearly, press the RANGE key and set $x_{\min} = -2$; $x_{\max} = 2$; $x_{\text{scl}} = 0.5$; $y_{\min} = 0$; $y_{\max} = 2$.

(b) Press the ZOOM key, then the "2" key, and then move the cursor so that it appears to be on the lowest point of the graph of $Y_2 = (x^2 - x + 1)^{1/3}$. Now press ENTER. Next press the TRACE key and use the arrow keys to move the cursor to a point on the graph of Y_2 where the x-coordinate is as close to 0.5 as possible. The corresponding y-coordinate on the screen readout is a calculator approximation for the minimum value of the function. Make a note of this number, and then go on to part (c).

(c) According to the text, the minimum value for $Y_2 = (x^2 - x + 1)^{1/3}$ is $\sqrt[3]{3/4}$. Use your calculator to evaluate this expression. Is your approximation in part (b) close to this?

$\underline{4.3}$ **APPLIED FUNCTIONS: SETTING UP EQUATIONS**

I hope that I shall shock a few people in asserting that the most important single task of mathematical instruction in the secondary schools is to teach the setting up of equations to solve word problems.

George Polya (1887–1985)

Each problem that I solved became a rule which afterwards served to solve other problems.

René Descartes (1596–1650)

One of the first steps in problem solving often involves defining a function. The function then serves to describe or summarize the given situation in a way that is both concise and (we hope) revealing. In this section, we want to practice setting up equations that define such functions. Then, in Section 4.4, we will use this skill in solving a certain class of applied problems involving quadratic functions. For many of the examples we look at, we'll rely on the following four-step procedure to set up the required equation. You may eventually want to modify this procedure to fit your own style. The point, however, is that it is possible to approach these problems in a systematic manner. A word of advice: You're accustomed to working mathematics problems in which the answers are numbers. In this section, the answers are functions (or, more precisely, equations defining functions); you'll need to get used to this.

> **Steps for Setting Up Equations**
>
> STEP 1 After reading the problem carefully, draw a picture that conveys the given information.
>
> STEP 2 State in your own words, as specifically as you can, what the problem is asking for. (This usually requires rereading the problem.) Now, assuming that the problem asks you to find a particular quantity (or to find a formula for a particular quantity), assign a variable to denote that key quantity.
>
> STEP 3 Label any other quantities in your figure that appear relevant. Are there equations relating these quantities?
>
> STEP 4 Find an equation involving the key variable that you identified in step 2. (Some people prefer to do this right after step 2.) Now, as necessary, substitute in this equation using the auxiliary equations from step 3 to obtain an equation involving only the required variables.

EXAMPLE 1 The perimeter of a rectangle is 100 cm. Express the area of the rectangle in terms of the width x.

Solution Let's follow our four-step procedure.

STEP 1 See Figure 1.

STEP 2 We want to express the area of the rectangle in terms of x, the width. Let A stand for the area of the rectangle.

STEP 3 Call the length of the rectangle L, as indicated in Figure 2. Then, since the perimeter is given as 100 cm, we have

$$2x + 2L = 100$$
$$x + L = 50$$
$$L = 50 - x \tag{1}$$

STEP 4 The area of a rectangle equals width times length:

$$A = x \cdot L$$
$$= x(50 - x) \qquad \text{substituting for } L \text{ using equation (1)}$$
$$= 50x - x^2 \tag{2}$$

This is the required equation expressing the area of the rectangle in terms of the width x. To emphasize this dependence of A on x, we can use function notation to rewrite equation (2):

$$A(x) = 50x - x^2$$

The domain of this area function is the open interval $(0, 50)$. To see why this is so, first note that $x > 0$ because x denotes a width. Furthermore, in view of equation (1), we must have $x < 50$ (otherwise L, the length, would be zero or negative).

EXAMPLE 2 A rectangle is inscribed in a circle of diameter 8 cm. Express the perimeter of the rectangle as a function of its width x.

Perimeter = 100 cm

x

FIGURE 1

Perimeter = 100 cm

x

L

FIGURE 2

FIGURE 3

Solution   Let's follow our four-step procedure.

STEP 1 See Figure 3. Note that the diagonal of the rectangle is a diameter of the circle.

STEP 2 The problem asks us to come up with a formula or function that gives the perimeter of the rectangle in terms of x, the width. Let P denote the perimeter.

STEP 3 Let L denote the length of the rectangle, as shown in Figure 3. Then, by the Pythagorean theorem, we have

$$L^2 + x^2 = 8^2 = 64$$

This equation relates the length L and the width x. Rather than leaving the equation in this form, however, we'll solve for L in terms of x (because the instructions for the problem mention x, not L):

$$L^2 = 64 - x^2$$
$$L = \sqrt{64 - x^2} \tag{3}$$

STEP 4 The perimeter of the rectangle is the sum of the lengths of the four sides. Thus,

$$P = x + x + L + L = 2x + 2L \tag{4}$$

This equation expresses P in terms of x and L. However, the problem asks for P in terms of just x. Using equation (3) to substitute for L in equation (4), we have

$$P = 2x + 2\sqrt{64 - x^2} \tag{5}$$

This is the required equation. It expresses the perimeter P as a function of the width x, so if you know x, you can calculate P. To emphasize this dependence of P on x, we can employ function notation to rewrite equation (5):

$$P(x) = 2x + 2\sqrt{64 - x^2} \tag{6}$$

Before leaving this example, we need to specify the domain of the perimeter function in equation (6). An easy way to do this is to look back at Figure 3. Since x represents a width, we certainly want $x > 0$. Furthermore, Figure 3 tells us that $x < 8$, because in any right triangle, a leg is always shorter than the hypotenuse. Putting these observations together, we conclude that the domain of the perimeter function in equation (6) is the open interval $(0, 8)$.

Note Although it would make sense *algebraically* to use an input such as $x = -1$ in equation (6), it does not make sense in our *geometric* context, where x denotes the width. ∎

▌▆▌ **EXAMPLE 3**   Let $P(x, y)$ be a point on the curve $y = \sqrt{x}$. Express the distance from P to the point $(1, 0)$ as a function of x.

FIGURE 4

Solution

STEP 1 See Figure 4.

STEP 2 We want to express the length of the broken line in Figure 4 in terms of *x*. Call this length *D*.

STEP 3 There are no other quantities in Figure 4 that need labeling. But don't forget that we are given

$$y = \sqrt{x} \tag{7}$$

STEP 4 By the distance formula, we have

$$D = \sqrt{(x-1)^2 + (y-0)^2}$$
$$= \sqrt{x^2 - 2x + 1 + y^2} \tag{8}$$

Now we can use equation (7) to eliminate *y* in equation (8):

$$D(x) = \sqrt{x^2 - 2x + 1 + \left(\sqrt{x}\right)^2}$$
$$= \sqrt{x^2 - 2x + 1 + x}$$
$$= \sqrt{x^2 - x + 1} \tag{9}$$

Equation (9) expresses the distance as a function of *x*, as required. What about the domain of this distance function? Since the *x*-coordinate of a point on the curve $y = \sqrt{x}$ can be any nonnegative number, the domain of the distance function is $[0, \infty)$. ∎

EXAMPLE 4 A point $P(x, y)$ lies in the first quadrant on the parabola $y = 16 - x^2$, as indicated in Figure 5. Express the area of the triangular region in Figure 5 as a function of *x*.

Solution

STEP 1 See Figure 5.

STEP 2 We want to express the area of the shaded triangle in terms of *x*. Let *A* denote the area of this triangle.

STEP 3 Since the coordinates of *P* are (x, y), the base of our triangle is *x* and the height is *y*. Also, *x* and *y* are related by the given equation

$$y = 16 - x^2 \tag{10}$$

STEP 4 The area of a triangle equals $\frac{1}{2}$(base)(height):

$$A = \tfrac{1}{2}(x)(y)$$

so

$$A(x) = \tfrac{1}{2}(x)(16 - x^2) \qquad \text{substituting for } y \text{ using equation (10)}$$
$$= 8x - \tfrac{1}{2}x^3$$

This last equation expresses the area of the triangle as a function of *x*, as required. (Exercise 46 at the end of this section will ask you to specify the domain of this area function.) ∎

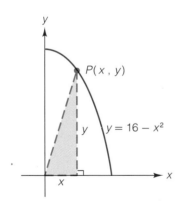

FIGURE 5

EXAMPLE 5 A piece of wire *x* inches long is bent into the shape of a circle. Express the area of the circle in terms of *x*.

Solution

STEP 1 See Figure 6.

STEP 2 We are supposed to express the area of the circle in terms of x, the circumference. Let A denote the area.

STEP 3 The general formula for circumference in terms of radius is $C = 2\pi r$. Since in our case the circumference is given as x, our equation becomes

$$x = 2\pi r \qquad \text{or} \qquad r = \frac{x}{2\pi} \qquad (11)$$

STEP 4 The general formula for the area of a circle in terms of the radius is

$$A = \pi r^2 \qquad (12)$$

This expresses A in terms of r, but we want A in terms of x. So we replace r in equation (12) by the quantity given in equation (11). This yields

$$A(x) = \pi\left(\frac{x}{2\pi}\right)^2 = \frac{\pi x^2}{4\pi^2} = \frac{x^2}{4\pi}$$

Thus, we have

$$A(x) = \frac{x^2}{4\pi}$$

This is the required equation. It expresses the area of the circle in terms of the circumference x. In other words, if we know the length of the piece of wire that is to be bent into a circle, we can use that length to calculate what the area of the circle will be. ■■■

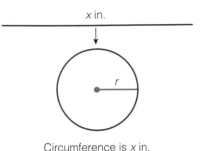

x in.

r

Circumference is *x* in.

FIGURE 6

■■■ **EXAMPLE 6** Figure 7 displays a right circular cylinder, along with the formulas for its volume V and total surface area S. Given that the volume is 10 cm³, express the surface area S as a function of r, the radius of the base.

Solution

STEP 1 See Figure 7.

STEP 2 We are given a formula that expresses the surface area S in terms of both r and h. We want to express S in terms of just r.

STEP 3 We are given that $V = 10$ and also that $V = \pi r^2 h$. Thus,

$$\pi r^2 h = 10$$

and consequently, expressing h in terms of r,

$$h = \frac{10}{\pi r^2} \qquad (13)$$

STEP 4 We take the given formula for S, namely,

$$S = 2\pi r^2 + 2\pi rh$$

Volume = 10 cm³

h

r

$V = \pi r^2 h$
$S = 2\pi r^2 + 2\pi rh$

FIGURE 7

and replace h with the quantity given in equation (13). We get

$$S(r) = 2\pi r^2 + 2\pi r \left(\frac{10}{\pi r^2} \right)$$

$$= 2\pi r^2 + \frac{20}{r}$$

This is the required equation. It expresses the total surface area in terms of the radius r. Since the only restriction on r is that it be positive, the domain of the area function is $(0, \infty)$.   ∎

We have followed the same four-step procedure in Examples 1 through 6. Of course, no single method can cover all possible cases. As usual, common sense and experience are often necessary. Also, you should not feel compelled to follow this procedure at any cost. Keep this in mind as you study the last two examples in this section.

▌▐▐▌ **EXAMPLE 7** Two numbers add up to 8. Express the product P of these two numbers in terms of a single variable.

Solution   If we call the two numbers x and $8 - x$, then their product P is given by

$$P(x) = x(8 - x) = 8x - x^2$$

That's it. This last equation expresses the product as a function of the variable x. Since there are no restrictions on x (other than it's being a real number), the domain of this function is $(-\infty, \infty)$.   ∎

▌▐▐▌ **EXAMPLE 8** In economics, the revenue R generated by selling x units at a price of p dollars per unit is given by

$$R = x \cdot p$$

In Figure 8, we are given a hypothetical function relating the selling price of a certain item to the number of units sold. Such a function is called a **demand function**. Express the revenue as a function of x.

Solution   More than anything else, this problem is an exercise in reading. After reading the problem several times, we find that it comes down to this:

given: $R = x \cdot p$ and $p = -\dfrac{1}{3}x + 4$ $(0 \le x \le 11)$

find: an equation expressing R in terms of x

In view of this, we write

$$R = x \cdot p$$

$$R(x) = x\left(-\frac{1}{3}x + 4 \right)$$

FIGURE 8

Thus, we have

$$R(x) = -\frac{1}{3}x^2 + 4x \qquad (0 \le x \le 11)$$

This is the required function. It allows us to calculate the revenue when we know the number of units sold. Note that this revenue function is a quadratic function.

Question for review and also preview   How would you find the maximum revenue in this case?   ▮▮▮

EXERCISE SET 4.3

A

1. (a) The perimeter of a rectangle is 16 cm. Express the area of the rectangle in terms of the width x. *Suggestion:*   First reread Example 1.
   (b) The area of a rectangle is 85 cm². Express the perimeter as a function of the width x.

2. A rectangle is inscribed in a circle of diameter 12 in.
   (a) Express the perimeter of the rectangle as a function of its width x. *Suggestion:*   First reread Example 2.
   (b) Express the area of the rectangle as a function of its width x.

3. A point $P(x, y)$ is on the curve $y = x^2 + 1$.
   (a) Express the distance from P to the origin as a function of x. *Suggestion:*   First reread Example 3.
   (b) In part (a), you expressed the length of a certain line segment as a function of x. Now express the slope of that line segment in terms of x.

4. A point $P(x, y)$ lies on the curve $y = \sqrt{x}$, as shown in the figure.

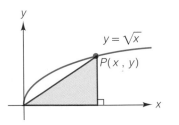

   (a) Express the area of the shaded triangle in terms of x. *Suggestion:*   First reread Example 4.
   (b) Express the perimeter of the shaded triangle in terms of x.

5. A piece of wire πy inches long is bent into a circle.
   (a) Express the area of the circle as a function of y. *Suggestion:*   First reread Example 5.

   (b) If the original piece of wire were bent into a square instead of a circle, how would you express the area in terms of y?

6. The volume of a right circular cylinder is 20 in³. Express the total surface area of the cylinder as a function of r, the radius of the base. *Suggestion:*   First reread Example 6.

7. Two numbers add to 16.
   (a) Express the product of the two numbers in terms of a single variable. *Suggestion:*   First reread Example 7.
   (b) Express the sum of the squares of the two numbers in terms of a single variable.
   (c) Express the difference of the cubes of the two numbers in terms of a single variable. (There are two answers.)
   (d) What happens when you try to express the average of the two numbers in terms of one variable?

8. The product of two numbers is 16. Express the sum of the squares of the two numbers as a function of a single variable.

9. Given a demand function $p = -\frac{1}{4}x + 8$, express the revenue as a function of x. (See Example 8 for terminology and definitions.)

10. In Example 1, we considered a rectangle with perimeter 100 cm. We found that the area of such a rectangle is given by $A(x) = 50x - x^2$, where x is the width of the rectangle. Compute the numbers $A(1)$, $A(10)$, $A(20)$, $A(25)$, and $A(35)$. Which width x seems to yield the largest area $A(x)$?

11. In Example 2, we considered a rectangle inscribed in a circle of diameter 8 cm. We found that the perimeter of such a rectangle is given by

$$P(x) = 2x + 2\sqrt{64 - x^2}$$

where x is the width of the rectangle.

(a) Use a calculator to complete the following table. Round off each answer to two decimal places.

x	1	2	3	4	5	6	7
$P(x)$							

(b) In your table, what is the largest value for $P(x)$? What is the width x in this case?

(c) Using calculus, it can be shown that among all possible widths x, the width $x = 4\sqrt{2}$ cm yields the largest possible perimeter. Use a calculator to compute the perimeter in this case, and check to see that the value you obtain is indeed larger than all of the values obtained in part (a).

12. In economics, the demand function for a given commodity tells us how the unit price p is related to the number of units x that are sold. Suppose we are given a demand function

$$p = 5 - \frac{x}{4} \qquad (p \text{ in dollars})$$

(a) Graph this demand function.

(b) How many units can be sold when the unit price is \$3? Locate the point on the graph of the demand function that conveys this information.

(c) To sell 12 items, how should the unit price be set? Locate the point on the graph of the demand function that conveys this information.

(d) Find the revenue function corresponding to the given demand function. (Use the formula $R = x \cdot p$.) Graph the revenue function.

(e) Find the revenue when $x = 2$, when $x = 8$, and when $x = 14$.

(f) According to your graph in part (d), which x-value yields the greatest revenue? What is that revenue? What is the corresponding unit price?

13. Let $2s$ denote the length of the side of an equilateral triangle.

(a) Express the height of the triangle as a function of s.

(b) Express the area of the triangle as a function of s.

(c) Use the function you found in part (a) to determine the height of an equilateral triangle, each side of which is 8 cm long.

(d) Use the function you found in part (b) to determine the area of an equilateral triangle, each side of which is 5 in. long.

14. If x denotes the length of a side of an equilateral triangle, express the area of the triangle as a function of x.

15. The height of a right circular cylinder is twice the radius. Express the volume as a function of the radius.

16. Using the information given in Exercise 15, express the radius as a function of the volume.

17. The volume of a right circular cylinder is 12π in.[3]
(a) Express the height as a function of the radius.
(b) Express the total surface area as a function of the radius.

18. The total surface area of a right circular cylinder is 14 in.[2] Express the volume as a function of the radius.

19. The volume V and the surface area S of a sphere of radius r are given by the formulas $V = \frac{4}{3}\pi r^3$ and $S = 4\pi r^2$. Express V as a function of S.

20. The base of a rectangle lies on the x-axis, while the upper two vertices lie on the parabola $y = 10 - x^2$. Suppose that the coordinates of the upper right vertex of the rectangle are (x, y). Express the area of the rectangle as a function of x.

21. The hypotenuse of a right triangle is 20 cm. Express the area of the triangle as a function of the length x of one of the legs.

22. (a) Express the area of the shaded triangle in the accompanying figure as a function of x.
(b) Express the perimeter of the shaded triangle as a function of x.

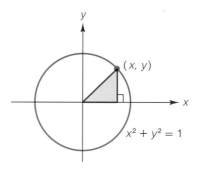

23. For the following figure, express the length of $\overline{AB}$ as a function of x. *Hint:* Note the similar triangles.

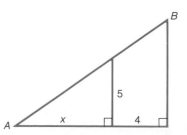

24. Five hundred feet of fencing is available to enclose a rectangular pasture alongside a river, which serves as one side of the rectangle (so only three sides require fencing). Express the area of the rectangular pasture as a function of x. (See the figure.)

25. After rereading Example 1, complete the following table. Which x-value in the table yields the largest area A? What is the corresponding value of the length L in that case?

x	5	10	20	24	24.8	24.9	25	25.1	25.2	45
$A(x)$										

26. After rereading Example 3, complete the following three tables. For each table, specify the x-value that yields the smallest distance D. (For Table 2, round off the values of D to two significant digits. For Table 3, round off the values of D to seven significant digits.)

TABLE 1

x	1	2	3	4	5
D					

TABLE 2

x	0.25	0.50	0.75
D			

TABLE 3

x	0.498	0.499	0.500	0.501	0.502
D					

27. (a) After rereading Example 4, complete the following three tables. For each table, specify the x-value that yields the largest area A. (For Tables 2 and 3, round off the values of A to six significant digits. In part (b), you'll see why we are asking for that many digits.)

TABLE 1

x	1	2	3	4
A				

TABLE 2

x	1.75	2.00	2.25	2.50	2.75
A					

TABLE 3

x	2.15	2.20	2.25	2.30	2.35
A					

(b) Using calculus, it can be shown that the x-value yielding the largest area A is $x = 4\sqrt{3}/3$. Which x-value in the tables is closest to this x-value? To six significant digits, what is the area A when $x = 4\sqrt{3}/3$?

B

28. A piece of wire 4 m long is cut into two pieces, and then each piece is bent into a square. Express the combined area of the two squares in terms of one variable.

29. A piece of wire 3 m long is cut into two pieces. Let x denote the length of the first piece and $3 - x$ the length of the second. The first piece is bent into a square and the second into a rectangle in which the width is half the length. Express the combined area of the square and the rectangle as a function of x. Is the resulting function a quadratic function?

In Exercises 30–33, refer to the following figure, which displays a right circular cone along with the formulas for the volume V and the lateral surface area S.

$$V = \frac{1}{3}\pi r^2 h$$
$$S = \pi r \sqrt{r^2 + h^2}$$

30. The volume of a right circular cone is $12\pi \text{ cm}^3$.
    (a) Express the height as a function of the radius.
    (b) Express the radius as a function of the height.

31. Suppose that the height and radius of a right circular cone are related by the equation $h = \sqrt{3}r$.
    (a) Express the volume as a function of r.
    (b) Express the lateral surface area as a function of r.

32. The volume of a right circular cone is 2 ft^3. Show that the lateral surface area as a function of r is given by
$$S = \frac{\sqrt{\pi^2 r^6 + 36}}{r}$$

33. In a certain right circular cone, the volume is numerically equal to the lateral surface area.
   (a) Express the radius as a function of the height.
   (b) Express the height as a function of the radius.

34. A line is drawn from the origin O to a point $P(x, y)$ in the first quadrant on the graph of $y - 1/x$. From point P, a line is drawn perpendicular to the x-axis, meeting the x-axis at B.
   (a) Draw a figure of the situation described.
   (b) Express the perimeter of $\triangle OPB$ as a function of x.
   (c) Try to express the area of $\triangle OPB$ as a function of x. What happens?

35. A piece of wire 14 in. long is cut into two pieces. The first piece is bent into a circle, the second into a square. Express the combined total area of the circle and the square as a function of x, where x denotes the length of the wire that is used for the circle.

36. A wire of length L is cut into two pieces. The first piece is bent into a square, the second into an equilateral triangle. Express the combined total area of the square and the triangle as a function of x, where x denotes the length of wire used for the triangle. (L is to be considered a constant here, not another variable.)

37. An athletic field with a perimeter of $\frac{1}{4}$ mile consists of a rectangle with a semicircle at each end, as shown in the figure. Express the area of the field as a function of r, the radius of the semicircle.

38. A square of side x is inscribed in a circle. Express the area of the circle as a function of x.

39. An equilateral triangle of side x is inscribed in a circle. Express the area of the circle as a function of x.

40. Refer to the accompanying figure. An offshore oil rig is located at point A, which is 10 miles out to sea. An oil pipeline is to be constructed from A to a point C on the shore and then to an oil refinery at point D, farther up the coast. If it costs \$8000 per mile to lay the pipeline in the sea and \$2000 per mile on land, express the cost of laying the pipeline in terms of x, where x is the distance from B to C.

41. An open-top box is constructed from a 6-by-8-inch rectangular sheet of tin by cutting out equal squares at each corner and then folding up the flaps, as shown in the figure. Express the volume of the box as a function of x, the length of the side of each cutout square.

42. Follow Exercise 41, but assume that the original piece of tin is a square, 12 in. on each side.

43. A Norman window is in the shape of a rectangle surmounted by a semicircle, as shown in the figure. Assume that the perimeter of the window is 32 ft.

   (a) Express the area of the window as a function of r, the radius of the semicircle.
   (b) The function you were asked to find in part (a) is a quadratic function, so its graph is a parabola. Does the parabola open upward or downward? Does it pass through the origin? Show that the vertex of the parabola is
$$\left(\frac{32}{\pi + 4}, \frac{512}{\pi + 4} \right)$$

44. Refer to the following figure. Express the lengths CB, CD, BD, and AB in terms of x. *Hint:* Recall the theorem from geometry stating that in a 30°–60°–90° right triangle, the side opposite the 30° angle is half the hypotenuse.

45. Refer to the figure; let s denote the ratio of y to z.
  (a) Express y as a function of s.
  (b) Express s as a function of y.
  (c) Express z as a function of s.
  (d) Express s as a function of z.

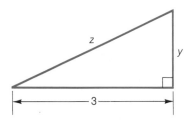

46. Refer to Example 4.
  (a) What is the x-intercept of the curve $y = 16 - x^2$ in Figure 5?
  (b) What is the domain of the area function in Example 4, assuming that the point P does not lie on the x- or y-axis?

47. The following figure shows the parabola $y = x^2$ and a line segment $\overline{AP}$ drawn from the point $A(0, -1)$ to the point $P(a, a^2)$ on the parabola.
  (a) Express the slope of $\overline{AP}$ in terms of a.
  (b) Show that the area of the shaded triangle in the figure is given by
  $$\text{area} = \frac{a^5}{2(a^2 + 1)}$$

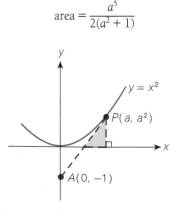

48. A rancher who wishes to fence off a rectangular area finds that the fencing in the east-west direction will require extra reinforcement owing to strong prevailing winds. Fencing in the east-west direction will therefore cost $12 per (linear) yard, as opposed to a cost of $8 per yard for fencing in the north-south direction. Given that the rancher wants to spend $4800 on fencing, express the area of the rectangle as a function of x, its width. The required function is in fact a quadratic, so its graph is a parabola. Does the parabola open upward or downward? By considering this graph, find which width x yields the rectangle of largest area. What is this maximum area?

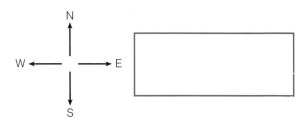

C
49. The following figure shows two concentric squares. Express the area of the shaded triangle as a function of x.

50. The following figure shows two concentric circles of radii r and R. Let A denote the area within the larger circle but outside the smaller one. Express A as a function of x (where x is defined in the figure).

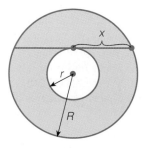

51. A straight line with slope m ($m < 0$) passes through the point $(1, 2)$ and intersects the line $y = 4x$ at a point in the first quadrant. Let A denote the area of the triangle bounded by $y = 4x$, the x-axis, and the given line of slope m. Express A as a function of m.

52. A line with slope m ($m < 0$) passes through the point (a, b) in the first quadrant and intersects the line $y = Mx$ ($M > 0$) at another point in the first quadrant. Let A denote the area of the triangle bounded by $y = Mx$, the x-axis, and the given line with slope m. Express A in terms of m, M, a, and b.

$$\text{Answer:} \quad A = \frac{M(am - b)^2}{2m(m - M)}$$

53. A line with slope m ($m < 0$) passes through the point (a, b) in the first quadrant. Express the area of the triangle bounded by this line and the axes in terms of m.

54. One corner of a page of width a is folded over and just reaches the opposite side, as indicated in the following figure. Express L, the length of the crease, in terms of x and a.

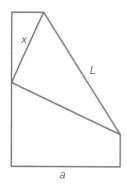

4.4 MAXIMUM AND MINIMUM PROBLEMS

... problems on maxima and minima, although new features in an English textbook, stand so little in need of apology with the scientific public that I offer none.

G. Chrystal in his preface to *Textbook of Algebra* (1886)

In daily life it is constantly necessary to choose the best possible (optimal) solution. A tremendous number of such problems arise in economics and in technology. In such cases it is frequently useful to resort to mathematics.

V. M. Tikhomirov in *Stories about Maxima and Minima* (translated from the Russian by Abe Shenitzer) (Providence, Rhode Island: The American Mathematical Society, 1990)

You have already seen several examples of maximum and minimum problems in the second section of this chapter. Before reading further in the present section, you should first review Examples 5 through 7 on pages 172–173.

Actually, we begin this section's discussion of maximum and minimum problems at a more intuitive level. Consider, for example, the following question: If two numbers add to 9, what is the largest possible value of their product? To gain some insight here, we carry out a few preliminary calculations (see Table 1).

We have circled the 20 in the right-hand column of Table 1 because it *appears* to be the largest product. We say "appears" because our table is incomplete. For instance, what if we allowed x- and y-values that are not whole numbers? Might we get a product exceeding 20? Table 2 shows the results of some additional calculations along these lines.

As you can see from Table 2, there is a product exceeding 20, namely, 20.25. Now the question is, if we further expand our tables, can we find yet an even larger product, one exceeding 20.25? And here we have come about as far as we want to go using this approach involving arithmetic and tables. For no matter what candidate we come up with for the largest product, there will always be the question of whether we might do still better using a larger table.

Nevertheless, this approach was useful, for it showed us what is really at the heart of a typical maximum or minimum problem. Essentially, we are trying

TABLE 1

x	y	$x + y$	xy
-2	11	9	-22
-1	10	9	-10
0	9	9	0
1	8	9	8
2	7	9	14
3	6	9	18
4	5	9	(20)

TABLE 2

x	y	$x + y$	xy
1	8	9	8
1.5	7.5	9	11.25
2	7	9	14
2.5	6.5	9	16.25
3	6	9	18
3.5	5.5	9	19.25
4	5	9	20
4.5	4.5	9	20.25

to sort through an infinite number of possible cases and pick out the required extreme case. In the example at hand, there are infinitely many pairs of numbers x and y adding to 9. We are asked to look at the products of all of these pairs and see which (if any!) is the largest. Example 1 shows how to apply our knowledge of quadratic functions to solve this problem in a definitive manner.

▌�▐ **EXAMPLE I**   Two numbers add to 9. What is the largest possible value for their product?

Solution   Call the two numbers x and $9 - x$. Then their product P is given by

$$P = x(9 - x) = 9x - x^2$$

The graph of this quadratic function is the parabola in Figure 1. Note the accompanying calculation for the vertex. As Figure 1 and the accompanying calculations show, the largest value of the product P is 20.25. This is the required solution. (Note from the graph that there is no smallest value of P.)

FIGURE I

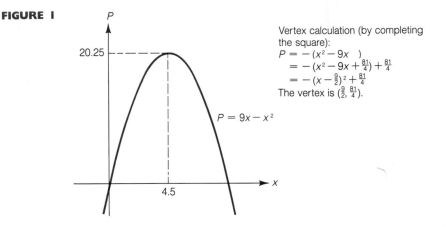

Vertex calculation (by completing the square):
$$P = -(x^2 - 9x \quad)$$
$$= -(x^2 - 9x + \tfrac{81}{4}) + \tfrac{81}{4}$$
$$= -(x - \tfrac{9}{2})^2 + \tfrac{81}{4}$$
The vertex is $(\tfrac{9}{2}, \tfrac{81}{4})$.

$P = 9x - x^2$

Example 1 illustrates the general strategy we will follow for solving the maximum and minimum problems in this section.

> **Strategy for Solving the Maximum and Minimum Problems in This Section**
>
> **1.** Express the quantity to be maximized or minimized in terms of a single variable. For instance, in Example 1, we found $P = 9x - x^2$. In setting up such functions, you'll want to keep in mind the four-step procedure used in Section 4.3.
>
> **2.** Assuming that the function you have determined is a quadratic, note whether its graph, a parabola, opens upward or downward. Check whether this is consistent with the requirements of the problem. For instance, the parabola in Figure 1 opens downward; so it makes sense to look for a *largest*, not a smallest, value of P. Now complete the square to locate the vertex. (If the function is not a quadratic but is closely related to a quadratic, these ideas may still apply. See, for instance, Examples 6 and 7 in Section 4.2.)
>
> **3.** After you have determined the vertex, you must relate that information to the original question. In Example 1, for instance, we were asked for the product P, not for x.

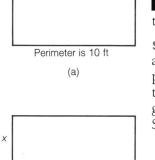

Perimeter is 10 ft

(a)

x

y

Perimeter is 10 ft

(b)

FIGURE 2

▮▮▮ EXAMPLE 2 Among all rectangles having a perimeter of 10 ft, find the dimensions (length and width) of the one with the greatest area.

Solution First we want to set up a function that expresses the area of the rectangle in terms of a single variable. In doing this, we'll be guided by the four-step procedure that we used in Section 4.3. Figure 2(a) displays the given information. Our problem is to determine the dimensions of the rectangle that has the greatest area. As Figure 2(b) indicates, we can label the dimensions x and y. Since the perimeter is given as 10 ft, we have

$$2x + 2y = 10$$
$$x + y = 5$$
$$y = 5 - x \tag{1}$$

Letting A denote the area of the rectangle, we can write

$$A = xy$$
$$A(x) = x(5 - x) \qquad \text{substituting for } y \text{ using equation (1)}$$
$$= 5x - x^2$$

This expresses the area of the rectangle in terms of the width x. Since the graph of this quadratic function is a parabola that opens downward, it does make sense to talk about the maximum. We can find the vertex of the parabola by completing the square:

$$A(x) = -(x^2 - 5x \qquad)$$
$$= -\left(x^2 - 5x + \frac{25}{4}\right) + \frac{25}{4} \qquad \text{adding 0}$$
$$= -\left(x - \frac{5}{2}\right)^2 + \frac{25}{4}$$

So the vertex is $\left(\frac{5}{2}, \frac{25}{4}\right)$ and, in particular, the width $x = \frac{5}{2}$ ft yields the maximum area. The problem asks us for the length and width of this rectangle, not its area. To compute the length y, we again use equation (1):

$$y = 5 - x = 5 - \frac{5}{2} = \frac{5}{2}$$

Thus, among all rectangles having a perimeter of 10 ft, the one with the greatest area is actually the square with dimensions of $2\frac{1}{2}$ ft by $2\frac{1}{2}$ ft. ▮▮▮

▮▮▮ EXAMPLE 3 Suppose that a baseball is tossed straight up and that its height as a function of time is given by the formula

$$h = -16t^2 + 64t + 6$$

In this formula, h is measured in feet and t in seconds, with $t = 0$ corresponding to the instant that the ball is released. What is the maximum height of the ball? When does the ball reach this height?

Solution The given function tells us how the height h of the ball depends on the time t. We want to know the largest possible value for h. Since the graph of the given function is a parabola opening downward, we can determine the largest value of h just by finding the vertex of the parabola. After completing the square, we find that the original equation can be rewritten

$$h = -16(t-2)^2 + 70 \qquad\qquad (2)$$

(You should verify this for yourself. If you get stuck, follow the model for completing the square in Example 4 of Section 4.2.) From equation (2) we see that the vertex of the parabola $h = -16t^2 + 64t + 6$ is $(2, 70)$. Therefore the maximum height of the ball is 70 ft, and the ball reaches this height at $t = 2\,\text{sec}$.

■■■

■■■ **EXAMPLE 4** Which point on the curve $y = \sqrt{x}$ is closest to the point $(1, 0)$?

Solution In Figure 3, we let D denote the distance from a point (x, y) on the curve to the point $(1, 0)$. We are asked to find out exactly which point (x, y) will make the distance D as small as possible. Using the distance formula, we have

$$D = \sqrt{(x-1)^2 + (y-0)^2}$$
$$= \sqrt{x^2 - 2x + 1 + y^2} \qquad\qquad (3)$$

This expresses D in terms of both x and y. To express D in terms of x alone, we use the given equation $y = \sqrt{x}$ to substitute for y in equation (3). This yields

$$D(x) = \sqrt{x^2 - 2x + 1 + \left(\sqrt{x}\right)^2}$$
$$= \sqrt{x^2 - x + 1}$$

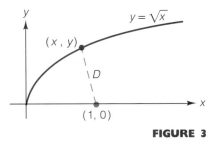

$y = \sqrt{x}$

(x, y)

D

$(1, 0)$

FIGURE 3

Since the same x-value that makes $D(x)$ its smallest will also cause the quantity $[D(x)]^2 = x^2 - x + 1$ to be its smallest, we need only find the x-value that minimizes the quantity $x^2 - x + 1$. Denoting this quantity by $f(x)$ for the moment (so that we have an equation whose graph is a parabola), we have

$$f(x) = x^2 - x + 1$$
$$= \left(x - \frac{1}{2}\right)^2 + \frac{3}{4} \qquad \text{completing the square}$$

From this last equation we conclude that the quantity $x^2 - x + 1$ will be smallest when $x = \frac{1}{2}$.

Now that we have x, we want to calculate y. (The problem did not ask for D; we needn't calculate it.) We know that $y = \sqrt{x}$ and $x = \frac{1}{2}$. Therefore,

$$y = \sqrt{\frac{1}{2}} = \frac{1}{\sqrt{2}} \cdot \frac{\sqrt{2}}{\sqrt{2}} = \frac{\sqrt{2}}{2}$$

Thus, among all points (x, y) on the curve $y = \sqrt{x}$, the point closest to $(1, 0)$ is $\left(\frac{1}{2}, \sqrt{2}/2\right)$. This is the required solution.

Question Why would it have made no sense to ask instead for the point farthest from $(1, 0)$?

■■■

■■■ **EXAMPLE 5** Suppose that you have 600 m of fencing with which to build two adjacent rectangular corrals. The two corrals are to share a common fence on one side, as shown in Figure 4. Find the dimensions x and y so that the total enclosed area is as large as possible.

Solution You have 600 m of fencing to be set up as shown in Figure 4. The question is how you should choose x and y so that the total area is a maximum. Since the total length of fencing is 600 m, we can relate x and y by writing

$$x + x + x + y + y = 600$$
$$3x + 2y = 600$$
$$2y = 600 - 3x$$
$$y = 300 - \frac{3}{2}x \qquad (4)$$

FIGURE 4

Letting A denote the total area, we have

$$A = xy$$
$$A(x) = x\left(300 - \frac{3}{2}x\right) \qquad \text{using equation (4) to substitute for } y$$
$$= 300x - \frac{3}{2}x^2$$

This last equation expresses the area as a function of x. Note that the graph of this function is a parabola opening downward, so it does make sense to talk about a maximum. We complete the square, as usual, to locate the vertex. As you can check, the result is

$$A(x) = -\frac{3}{2}(x - 100)^2 + 15000$$

Thus the x-coordinate of the vertex is 100; this is the x-value that maximizes the area. The corresponding y-value can now be calculated using equation (4):

$$y = 300 - \frac{3}{2}x = 300 - \frac{3}{2}(100) = 150$$

Thus, by choosing x to be 100 m and y to be 150 m, the total area in Figure 4 will be as large as possible. Incidentally, note that the exact location of the fence dividing the two corrals does not influence our work or final answer. ∎

EXAMPLE 6 Suppose that the following function relates the selling price, p, of an item to the quantity sold, x:

$$p = -\frac{1}{3}x + 40 \qquad (p \text{ in dollars})$$

For which value of x will the corresponding revenue be a maximum? What are the maximum revenue and the unit price in this case?

Solution First, let us recall the formula for revenue given on page 181:

$$R = \text{number of units} \times \text{price per unit}$$

Using this, we have

$$R = x \cdot p$$
$$R(x) = x\left(-\frac{1}{3}x + 40\right) = -\frac{1}{3}x^2 + 40x \qquad (5)$$

We want to know which value of x yields the largest revenue R. Since the graph of the revenue function in equation (5) is a parabola opening downward, the required x-value is the x-coordinate of the vertex of the parabola. We complete the square to locate the vertex:

$$R(x) = -\frac{1}{3}(x^2 - 120x \qquad)$$

$$= -\frac{1}{3}(x^2 - 120x + 3600) + 1200 \qquad \text{adding } 0 \ (= -1200 + 1200)$$

$$= -\frac{1}{3}(x - 60)^2 + 1200$$

The vertex is therefore $(60, 1200)$. Consequently, the revenue is a maximum when $x = 60$ items, and this maximum revenue is \$1200. For the corresponding unit price, we have

$$p(x) = -\frac{1}{3}x + 40$$

$$p(60) = -\frac{1}{3}(60) + 40 = 20$$

Thus the unit price corresponding to the maximum revenue is \$20. ▮▮▮

▮▮▮ **EXAMPLE 7** Figure 5 shows a graph of the function

$$f(x) = x^4 - 3x^2$$

From the graph, it is clear that the minimum of the function is slightly less than -2. Find the exact value of this minimum and the corresponding x-values at which it occurs.

Solution Although this is not a quadratic function, we can nevertheless use the technique of completing the square in this particular case. (You have already seen one instance of this in Example 7 of Section 4.2.) We have

$$f(x) = x^4 - 3x^2 + \frac{9}{4} - \frac{9}{4}$$

$$= \left(x^2 - \frac{3}{2}\right)^2 - \frac{9}{4}$$

$$= -\frac{9}{4} + \left(x^2 - \frac{3}{2}\right)^2$$

From this last equation we see that $f(x)$ is never less than $-\frac{9}{4}$ (because the quantity being added to $-\frac{9}{4}$ is nonnegative). Furthermore, $f(x)$ does attain the value $-\frac{9}{4}$ when $x^2 = \frac{3}{2}$; that is, when $x = \pm\sqrt{\frac{3}{2}}$. Thus the minimum value of f is $-\frac{9}{4}$ and the corresponding x-values are

$$x = \pm\frac{\sqrt{3}}{\sqrt{2}} = \pm\frac{\sqrt{6}}{2} \qquad \text{rationalizing the denominator}$$

Note Use your calculator to approximate these x-values, and then check that they are consistent with the graph in Figure 5. ▮▮▮

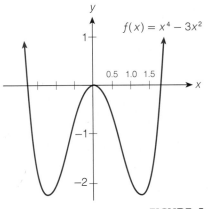

$f(x) = x^4 - 3x^2$

FIGURE 5

EXERCISE SET 4.4

A

1. Two numbers add to 5. What is the largest possible value of their product?

2. Find two numbers adding to 20 such that the sum of their squares is as small as possible.

3. The difference of two numbers is 1. What is the smallest possible value for the sum of their squares?

4. For each quadratic function, state whether it would make sense to look for a highest or a lowest point on the graph. Then determine the coordinates of that point.
   (a) $y = 2x^2 - 8x + 1$ (b) $y = -3x^2 - 4x - 9$
   (c) $h = -16t^2 + 256t$ (d) $f(x) = 1 - (x + 1)^2$
   (e) $g(t) = t^2 + 1$ (f) $f(x) = 1000x^2 - x + 100$

5. Among all rectangles having a perimeter of 25 m, find the dimensions of the one with the largest area.

6. What is the largest possible area for a rectangle whose perimeter is 80 cm?

7. What is the largest possible area for a right triangle in which the sum of the lengths of the two shorter sides is 100 in.?

8. The perimeter of a rectangle is 12 m. Find the dimensions for which the diagonal is as short as possible.

9. Two numbers add to 6.
   (a) Let T denote the sum of the squares of the two numbers. What is the smallest possible value for T?
   (b) Let S denote the sum of the first number and the square of the second. What is the smallest possible value for S?
   (c) Let U denote the sum of the first number and twice the square of the second number. What is the smallest possible value for U?
   (d) Let V denote the sum of the first number and the square of twice the second number. What is the smallest possible value for V?

10. Suppose that the height of an object shot straight up is given by $h = 512t - 16t^2$ (here h is in feet and t is in seconds). Find the maximum height and the time at which the object hits the ground.

11. A baseball is thrown straight up, and its height as a function of time is given by the formula $h = -16t^2 + 32t$ (where h is in feet and t is in seconds).
   (a) Find the height of the ball when $t = 1$ and when $t = \frac{3}{2}$.
   (b) Find the maximum height of the ball and the time at which that height is attained.
   (c) At what times is the height 7 ft?

12. Find the point on the curve $y = \sqrt{x}$ that is nearest to the point $(3, 0)$.

13. Which point on the curve $y = \sqrt{x - 2} + 1$ is closest to the point $(4, 1)$? What is this minimum distance?

14. Find the coordinates of the point on the line $y = 3x + 1$ closest to $(4, 0)$.

15. (a) What number exceeds its square by the greatest amount?
    (b) What number exceeds twice its square by the greatest amount?

16. Suppose that you have 1800 m of fencing with which to build three adjacent rectangular corrals, as shown in the figure. Find the dimensions so that the total enclosed area is as large as possible.

17. Five hundred feet of fencing is available for a rectangular pasture alongside a river, the river serving as one side of the rectangle (so only three sides require fencing). Find the dimensions yielding the greatest area.

18. Let $A = 3x^2 + 4x - 5$ and $B = x^2 - 4x - 1$. Find the minimum value of $A - B$.

19. Let $R = 0.4x^2 + 10x + 5$ and $C = 0.5x^2 + 2x + 101$. For which value of x is $R - C$ a maximum?

20. Suppose that the revenue generated by selling x units of a certain commodity is given by $R = -\frac{1}{5}x^2 + 200x$. Assume that R is in dollars. What is the maximum revenue possible in this situation?

21. Suppose that the function $p = -\frac{1}{4}x + 30$ relates the selling price p of an item to the number of units x that are sold. Assume that p is in dollars. For which value of x will the corresponding revenue be a maximum? What is this maximum revenue and what is the unit price?

22. The action of sunlight on automobile exhaust produces air pollutants known as *photochemical oxidants*. In a study of cross-country runners in Los Angeles, it was shown that running performances can be adversely affected when the oxidant level reaches 0.03 part per million. Suppose that on a given day, the oxidant level L is approximated by the formula

$$L = 0.059t^2 - 0.354t + 0.557 \qquad (0 \le t \le 7)$$

where t is measured in hours, with $t = 0$ corresponding to 12 noon, and L is in parts per million. At what time is the oxidant level L a minimum? At this time, is the oxidant level high enough to affect a runner's performance?

23. (a) Find the smallest possible value of the quantity $x^2 + y^2$ under the restriction that $2x + 3y = 6$.
    (b) Find the radius of the circle whose center is at the origin and that is tangent to the line $2x + 3y = 6$. How does this answer relate to your answer in part (a)?

B

24. Through a type of chemical reaction known as *autocatalysis,* the human body produces the enzyme trypsin from the enzyme trypsinogen. (Trypsin then breaks down proteins into amino acids, which the body needs for growth.) Let r denote the rate of the chemical reaction in which trypsin is formed from trypsinogen. It has been shown experimentally that $r = kx(a - x)$, where k is a positive constant, a is the initial amount of trypsinogen, and x is the amount of trypsin produced (so x increases as the reaction proceeds). Show that the reaction rate r is a maximum when $x = a/2$. In other words, the speed of the reaction is greatest when the amount of trypsin formed is half the original amount of trypsinogen.

25. (a) Let $x + y = 15$. Find the minimum value of the quantity $x^2 + y^2$.
    (b) Let C be a constant and $x + y = C$. Show that the minimum value of $x^2 + y^2$ is $C^2/2$. Then use this result to check your answer in part (a).

26. Suppose that A, B, and C are positive constants and that $x + y = C$. Show that the minimum value of $Ax^2 + By^2$ occurs when $x = BC/(A + B)$ and $y = AC/(A + B)$.

27. The following figure shows a square inscribed within a unit square. For which value of x is the area of the inner square a minimum? What is the minimum area?

Hint: Denote the lengths of the two segments that make up the base of the unit square by t and $1 - t$. Now use the Pythagorean theorem and congruent triangles to express x in terms of t.

28. (a) Find the coordinates of the point on the line $y = mx + b$ that is closest to the origin.

    Answer: $\left(\dfrac{-mb}{1 + m^2}, \dfrac{b}{1 + m^2} \right)$

    (b) Show that the perpendicular distance from the origin to the line $y = mx + b$ is $|b|/\sqrt{1 + m^2}$. *Suggestion:* Use the result in part (a).
    (c) Use part (b) to show that the perpendicular distance from the origin to the line $Ax + By + C = 0$ is $|C|/\sqrt{A^2 + B^2}$.

29. The point P lies in the first quadrant on the graph of the line $y = 7 - 3x$. From the point P, perpendiculars are drawn to both the x-axis and the y-axis. What is the largest possible area for the rectangle thus formed?

30. Show that the largest possible area for the shaded rectangle shown in the figure is $-b^2/4m$. Then use this to check your answer to Exercise 29.

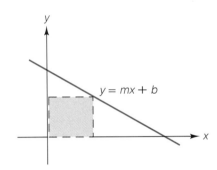

31. Show that the maximum possible area for a rectangle inscribed in a circle of radius R is $2R^2$. *Hint:* Maximize the square of the area.

32. An athletic field with a perimeter of $\frac{1}{4}$ mile consists of a rectangle with a semicircle at each end, as shown in the figure. Find the dimensions x and r that yield the greatest possible area for the rectangular region.

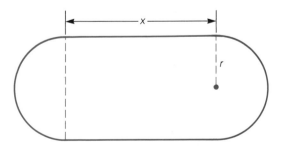

33. A rancher who wishes to fence off a rectangular area finds that the fencing in the east-west direction will require extra reinforcement owing to strong prevailing winds. Because of this, the cost of fencing in the east-west direction will be $12 per (linear) yard, as opposed to a cost of $8 per yard for fencing in the

north-south direction. Find the dimensions of the largest possible rectangular area that can be fenced for $4800.

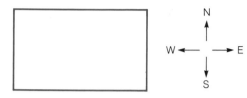

34. Let $f(x) = (x - a)^2 + (x - b)^2 + (x - c)^2$, where a, b, and c are constants. Show that $f(x)$ will be a minimum when x is the average of a, b, and c.

35. Let $y = a_1(x - x_1)^2 + a_2(x - x_2)^2$, where a_1, a_2, x_1, and x_2 are all constants. In addition, suppose that a_1 and a_2 are both positive. Show that the minimum of this function occurs when

$$x = \frac{a_1x_1 + a_2x_2}{a_1 + a_2}$$

36. Among all rectangles with a given perimeter P, find the dimensions of the one with the shortest diagonal.

37. By analyzing sales figures, the economist for a stereo manufacturer knows that 150 units of a tape player can be sold each month when the price is set at $p = \$200$ per unit. The figures also show that for each $10 hike in price, 5 fewer units are sold each month.
    (a) Let x denote the number of units sold per month and let p denote the price per unit. Find a linear function relating p and x.
        Hint: $\Delta p/\Delta x = 10/(-5) = -2$.
    (b) The revenue R is given by $R = xp$. What is the maximum revenue? At what level should the price be set to achieve this maximum revenue?

38. Let $f(x) = x^2 + px + q$, and suppose that the minimum value of this function is 0. Show that $q = p^2/4$.

39. For which numbers t will the value of the function $f(t) = t^2 - t^4$ be as large as possible?

40. Find the minimum value of the function $f(t) = t^4 - 8t^2$.

41. Among all possible inputs for the function $f(t) = -t^4 + 6t^2 - 6$, which ones yield the largest output?

42. Let $f(x) = x - 3$ and $g(x) = x^2 - 4x + 1$.
    (a) Find the minimum value of $g \circ f$.
    (b) Find the minimum value of $f \circ g$.
    (c) Are the results in parts (a) and (b) the same?

In Exercises 43–50, use the following result to lessen the amount of computation required: The x-coordinate of the vertex of the parabola $y = ax^2 + bx + c$ is given by the formula $x = -b/(2a)$. (As indicated in Exercise 45 in Section 4.2, this result can be derived by completing the square.)

43. A piece of wire 16 in. long is to be cut into two pieces. Let x denote the length of the first piece and $16 - x$ the length of the second. The first piece is to be bent into a circle and the second piece into a square.
    (a) Express the total combined area A of the circle and the square as a function of x.
    (b) For which value of x is the area A a minimum?
    (c) Using the x-value that you found in part (b), find the ratio of the lengths of the shorter to the longer piece of wire.   *Answer:* $\pi/4$

44. A 30-in. piece of string is to be cut into two pieces. The first piece will be formed into the shape of an equilateral triangle and the second piece into a square. Find the length of the first piece if the combined area of the triangle and the square is to be as small as possible.

C

45. Repeat Exercise 43, but assume that the length of the wire is L in.

46. Let $f(x) = x^2 + bx + 1$. Find a positive value for b such that the distance from the origin to the vertex of the parabola is as small as possible.

47. The next figure shows a rectangle inscribed in a given triangle of base b and height h. Find the ratio of the area of the triangle to the area of the rectangle when the area of the rectangle is a maximum.

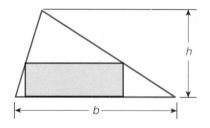

48. A Norman window is in the shape of a rectangle surmounted by a semicircle, as shown in the figure. Assume that the perimeter of the window is P, a constant. Show that the area of the window is a maximum when both x and r are equal to $P/(\pi + 4)$. Show that this maximum area is $\frac{1}{2}P^2/(\pi + 4)$.

49. A triangle is inscribed in a semicircle of diameter $2R$, as shown in the figure. Show that the smallest possible value for the area of the shaded region is $\frac{1}{2}(\pi - 2)R^2$.

Hint: The area of the shaded region is a minimum when the area of the triangle is a maximum. Find the value of x that maximizes the *square* of the area of the triangle. This will be the same x that maximizes the area of the triangle.

50. (a) Complete the following table. Which x-y pair in the table yields the smallest sum $x + y$?

x	0.5	1	1.5	2	2.5	3	3.5
y							
xy	12	12	12	12	12	12	12
$x + y$							

(b) Find two positive numbers with a product of 12 and as small a sum as possible. *Hint:* The quantity that you need to minimize is $x + (12/x)$, where $x > 0$. But

$$x + \frac{12}{x} = \left(\sqrt{x} - \sqrt{\frac{12}{x}}\right)^2 + 2\sqrt{12}$$

This last expression is minimized when the quantity within parentheses is zero. Why?

(c) Use a calculator to verify that the two numbers obtained in part (b) produce a sum that is smaller than any of the sums obtained in part (a).

51. What is the smallest possible value for the sum of a positive number and its reciprocal? *Hint:* After setting up the appropriate function, adapt the hint given in Exercise 50(b).

52. Let $f(x) = \dfrac{2x^2 - 4x + 1}{2x}$, and assume that $x > 0$.

(a) Complete the following two tables. In both cases, specify the x-value that yields the smallest output. (Round off the outputs in the tables to six decimal places.)

x	0.4	0.5	0.6	0.7	0.8	0.9
$f(x)$						

x	0.68	0.69	0.70	0.71	0.72	0.73
$f(x)$						

(b) Among all positive inputs for the given function f, find the one for which $f(x)$ is a minimum.

Hint: Write $f(x)$ as $\left(x + \dfrac{1}{2x}\right) - 2$, and then adapt the hint given in Exercise 50(b).

(c) Use a calculator to verify that the x-value obtained in part (b) produces an output that is smaller than any of the outputs obtained in part (a).

53. Let $G(x) = (a + x)(b + x)/x$, where a and b are positive constants and $x > 0$. Show that the minimum value of G is $\left(\sqrt{a} + \sqrt{b}\right)^2$ and that this minimum value occurs when $x = \sqrt{ab}$. *Hint:* Show that $G(x) = (a + b) + \left(\dfrac{ab}{x} + x\right)$. Then adapt the hint in Exercise 50(b).

54. A line with a positive slope m passes through the point $(-2, 1)$.

(a) Show that the area of the triangle bounded by this line and the axes is $\left(2m + \dfrac{1}{2m}\right) + 2$.

(b) For which value of m will the area be a minimum? *Hint:* Adapt the hint given in Exercise 50(b).

55. Let $f(x) = 1/(2x^2 - x + 1)$.

(a) Experiment with your calculator; what is the largest output you can find for the function f?

(b) Prove that for every value of x, we have $f(x) \le 8/7$.

56. Find the largest value of the function

$$f(x) = \frac{1}{x^4 + 2x^2 + 1}.$$

OPTIONAL TI-81 GRAPHING CALCULATOR EXERCISES
EXERCISES FOR SECTION 4.4

1. (a) In Example 1 in this section, we found that the maximum value of the function $P = 9x - x^2$ is 20.25. How was this value obtained?
   (b) Graph the function $P = 9x - x^2$ using the following settings in the RANGE menu: $x_{min} = 0$; $x_{max} = 10$; $x_{scl} = 1$; $y_{min} = 0$; $y_{max} = 24$; $y_{scl} = 2$.
   (c) Use the TRACE key and then the arrow keys to estimate the maximum value of the function. (Round off to three decimal places.) How close is your estimate to the value 20.25 determined in Example 1?

2. (a) Graph the parabola $y = -16x^2 + 64x + 6$ using the following settings in the RANGE menu: $x_{min} = 0$; $x_{max} = 4$; $x_{scl} = 1$; $y_{min} = 0$; $y_{max} = 80$; $y_{scl} = 10$.
   (b) Use the TRACE key and then the arrow keys to estimate the maximum value of the function. (Round off to three decimal places.) How close is your estimate to the value 70 determined in Example 3 in this section?

3. (a) According to a statement made in Example 4, the same x-value that minimizes the quadratic function $y = x^2 - x + 1$ will also minimize the function $y = \sqrt{x^2 - x + 1}$. Confirm this graphically as follows. Press the Y= key and enter the two functions $Y_1 = x^2 - x + 1$ and $Y_2 = \sqrt{x^2 - x + 1}$. Now press ZOOM 6 to see the graphs in the standard viewing rectangle. Next, for a closer look, make the following adjustments in the RANGE menu (and then press GRAPH): $x_{min} = 0$; $x_{max} = 1$; $y_{min} = 0.5$; $y_{max} = 1$.

   (b) As you can see from the graphs in part (a), it does appear that both curves "bottom out" at the same x-value. Use the TRACE key and the arrow keys to estimate this x-value. Round off the answer to three decimal places. How close is your estimate to the exact value $x = \frac{1}{2}$ obtained in Example 4?

4. (a) Graph the function $Y_1 = x^4 - 3x^2$ in the standard viewing rectangle (ZOOM 6). Note that the graph is symmetric about the y-axis. How could you have predicted this before drawing the graph?
   (b) To focus on the fourth-quadrant portion of the curve, press the RANGE key and make the following adjustments: $x_{min} = 0$; $x_{max} = 2$; $x_{scl} = 1$; $y_{min} = -3$; $y_{max} = 0$; $y_{scl} = 1$. Press GRAPH. Now use the TRACE key and the arrow keys to estimate the x-coordinate of the turning point. Round off the result to three decimal places.
   (c) According to Example 7 in this section, the exact value for the x-coordinate of this turning point is $x = \sqrt{6}/2$. Use your calculator to evaluate this expression. Round off to three decimal places. How close to this is your estimate in part (b)?

5. (a) Let $f(x) = 2x^2 - 6x + 5$ and $g(x) = 1/(2x^2 - 6x + 5)$. Use the calculator to find out whether the following statement is valid: The x-value that maximizes g is the same as the x-value that minimizes f.
   (b) Use the TRACE and arrow keys to estimate the maximum value of g.
   (c) What is the exact value for the maximum of g? Is your estimate in part (b) close to this?

4.5 POLYNOMIAL FUNCTIONS

We can rephrase the definitions of linear and quadratic functions using the terminology for polynomials given in Section 1.3.

A function f is **linear** if $f(x)$ is a **polynomial of degree 1**:

$$f(x) = a_1 x + a_0 \qquad (1)$$

A function f is **quadratic** if $f(x)$ is a **polynomial of degree 2**:

$$f(x) = a_2 x^2 + a_1 x + a_0 \qquad (2)$$

In view of equation (1), linear functions are sometimes called **polynomial functions of degree 1**. Similarly, in view of equation (2), quadratic functions are

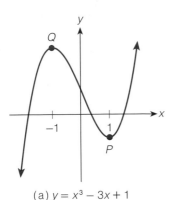

(a) $y = x^3 - 3x + 1$

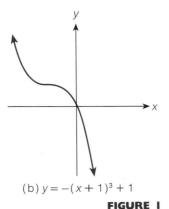

(b) $y = -(x + 1)^3 + 1$

FIGURE 1

polynomial functions of degree 2. More generally, by a **polynomial function of degree n**, we mean a function defined by an equation of the form

$$f(x) = a_n x^n + a_{n-1} x^{n-1} + \cdots + a_1 x + a_0 \qquad (3)$$

where n is a nonnegative integer and $a_n \neq 0$. Throughout the remainder of this chapter, the coefficients a_k will always be real numbers.

In principle, we can obtain the graph of any polynomial function by setting up a table and plotting a sufficient number of points. Indeed, this is just the way a graphing calculator or computer operates. However, to *understand* why the graphs look as they do, we want to discuss some additional methods for graphing polynomial functions.

There are three facts that we shall need. By way of example, look at the graphs of the polynomial functions in Figure 1. First, notice that both graphs are unbroken, smooth curves with no "corners." As is shown in calculus, this is true for the graph of every polynomial function. In contrast, the graphs in Figures 2 and 3 cannot represent polynomial functions. The graph in Figure 2 has a break in it, and the graph in Figure 3 has a **cusp.**

Now look back at the graph in Figure 1(a). Recall (from Section 3.2) that points such as P and Q are called **turning points.** These are points where the graph changes from rising to falling, or vice versa. It is a fact (proved in calculus) that the graph of a polynomial function of degreee n has *at most $n - 1$* turning points. For instance, in Figure 1(a), there are two turning points, while the degree of the polynomial is 3. However, as Figure 1(b) indicates, we needn't have any turning points at all.

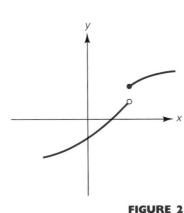

FIGURE 2
Since the graph has a break, it cannot
represent a polynomial function.

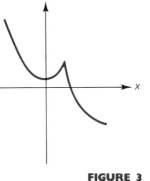

FIGURE 3
Since the graph has a cusp, it cannot
represent a polynomial function.

A third property of polynomial functions concerns their behavior when $|x|$ is very large. We'll illustrate this property using the function $y = x^3 - 3x + 1$, graphed in Figure 1(a). Now, in Figure 1(a), the x-values are relatively small; for instance, the x-coordinates of P and Q are 1 and -1, respectively. In Figure 4, however, we show the graph of this same function using units of 100 on the

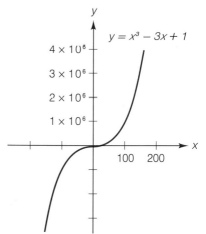

$y = x^3 - 3x + 1$

FIGURE 4

When $|x|$ is very large, the graph of $y = x^3 - 3x + 1$ appears indistinguishable from that of $y = x^3$.

x-axis. On this scale, the graph appears indistinguishable from that of $y = x^3$. In particular, note that as $|x|$ gets very large, $|y|$ grows very large.

It's easy to see why the function $y = x^3 - 3x + 1$ resembles $y = x^3$ when $|x|$ is very large. First, let's rewrite the equation $y = x^3 - 3x + 1$ as

$$y = x^3 \left(1 - \frac{3}{x^2} + \frac{1}{x^3} \right)$$

Now, when $|x|$ is very large, both $3/x^2$ and $1/x^3$ are close to zero. So we have

$$y \approx x^3 \, (1 - 0 + 0)$$
$$\approx x^3 \qquad \text{when } |x| \text{ is very large}$$

The same technique that we've just used in analyzing $y = x^3 - 3x + 1$ can be applied to any (nonconstant) polynomial function. The result is summarized in item 3 in the following box.

PROPERTY SUMMARY	**GRAPHS OF POLYNOMIAL FUNCTIONS**

1. The graph of a polynomial function of degree 2 or greater is an unbroken smooth curve. (For degrees 1 and 0, the graph is a line.)

2. The graph of a polynomial function of degree n has at most $n - 1$ turning points.

3. For the graph of any polynomial function (other than a constant function), as $|x|$ gets very large, $|y|$ grows very large. If

$$f(x) = a_n x^n + a_{n-1} x^{n-1} + \cdots + a_1 x + a_0 \quad (a_n \neq 0)$$

then

$$f(x) \approx a_n x^n \qquad \text{when } |x| \text{ is very large.}$$

EXAMPLE 1   A function f is defined by

$$f(x) = -x^3 + x^2 + 9x + 9$$

Which of the graphs in Figure 5 (on page 200) might represent this function?

Solution   When $|x|$ is very large, $f(x) \approx -x^3$. This rules out the graph in Figure 5(a). The graph in Figure 5(c) can also be ruled out, but for a different reason. That graph has four turning points, whereas the graph of the cubic function f can have at most two turning points. The graph in Figure 5(b), on the other hand, does have two turning points; furthermore, that graph does behave like $y = -x^3$ when $|x|$ is very large. The graph in Figure 5(b) might be (in fact, it is) the graph of the given function f.

FIGURE 5

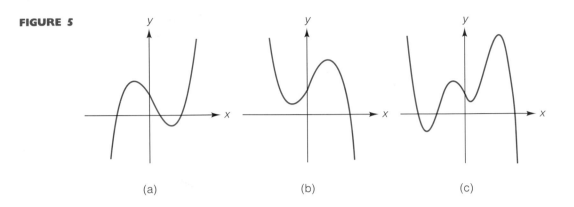

(a)   (b)   (c)

The simplest polynomial functions to graph are those of the form $y = x^n$. From our work in Chapter 2 we are already familiar with the graphs of $y = x$, $y = x^2$, and $y = x^3$. For reference, these are shown in Figure 6.

FIGURE 6

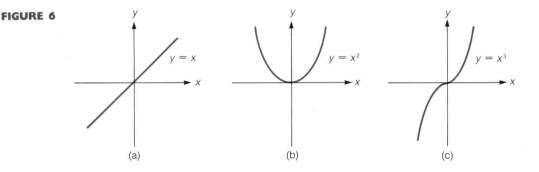

(a)   (b)   (c)

When n is greater than 3, the graph of $y = x^n$ resembles the graph of $y = x^2$ or $y = x^3$, depending on whether n is even or odd. Consider, for instance, the graph of $y = x^4$, shown in Figure 7 along with the graph of $y = x^2$. Just as with $y = x^2$, the graph of $y = x^4$ is a symmetric, U-shaped curve passing through the three points $(-1, 1)$, $(1, 1)$, and $(0, 0)$. However, in the interval $(-1, 1)$, the graph of $y = x^4$ is flatter than that of $y = x^2$. Similarly, the graph of $y = x^6$ in this interval would be flatter still. The data in Table 1 show why this is true. Figure 8 displays the graphs of $y = x^2$, $y = x^4$, and $y = x^6$ for the interval $0 \le x \le 1$.

FIGURE 7

TABLE 1

x	0.2	0.4	0.6	0.8	1.0
x^2	0.04	0.16	0.36	0.64	1.0
x^4	0.0016	0.0256	0.1296	0.4096	1.0
x^6	0.000064	0.004096	0.046656	0.262144	1.0

Incidentally, Figure 8 indicates one of the practical difficulties you may encounter in trying to draw an accurate graph of $y = x^n$. Suppose, for instance, that

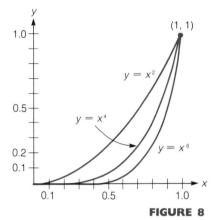

FIGURE 8

you want to graph $y = x^6$, and the lines you draw are 0.01 cm thick. Also suppose that you use the same scale on both axes, taking the common unit to be 1 cm. Then, in the first quadrant, your graph of $y = x^6$ will be indistinguishable from the x-axis when $x^6 < 0.01$, or $x < \sqrt[6]{0.01} \approx 0.46$ cm (using a calculator). This explains why sections of the graphs in Figure 8 appear horizontal.

For $|x| > 1$, the graph of $y = x^4$ rises more rapidly than that of $y = x^2$. Similarly, the graph of $y = x^6$ rises still more rapidly. This is shown in Figures 9(a) and (b). (Note the different scales used on the y-axes in the two figures.)

As mentioned earlier, when n is odd, the graph of $y = x^n$ resembles that of $y = x^3$. In Figure 10 we compare the graphs of $y = x^3$ and $y = x^5$. Notice that both curves pass through $(0, 0)$, $(1, 1)$, and $(-1, -1)$. For reasons similar to those explained for even n, the graph of $y = x^5$ is flatter than that of $y = x^3$ in the interval $-1 < x < 1$, and the graph of $y = x^7$ would be flatter still. For $|x| > 1$,

FIGURE 9

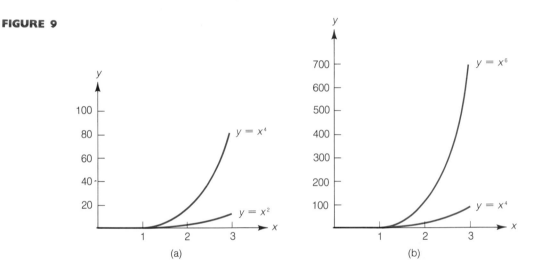

(a)

(b)

the graph of $y = x^5$ is steeper than that of $y = x^3$, and $y = x^7$ would be steeper still.

In Section 4.2, we observed the effect of the constant a on the graph of $y = ax^2$. Those same comments apply to the graph of $y = ax^n$. For instance, the graph of $y = \frac{1}{2}x^4$ is wider than that of $y = x^4$, while the graph of $y = -\frac{1}{2}x^4$ is obtained by reflecting $y = \frac{1}{2}x^4$ in the x-axis.

EXAMPLE 2 Sketch the graph of $y = (x + 2)^5$ and specify the y-intercept.

Solution The graph of $y = (x + 2)^5$ is obtained by moving the graph of $y = x^5$ two units to the left. As a guide to drawing the curve, we recall that $y = x^5$ passes through the points $(1, 1)$ and $(-1, -1)$. Thus, $y = (x + 2)^5$ must pass through $(-1, 1)$ and $(-3, -1)$, as shown in Figure 11 (on the next page). Although the curve rises and falls very sharply, it is important to realize that it is never really vertical. For instance, the curve eventually crosses the y-axis. To find the y-intercept, we set $x = 0$ to obtain $y = 2^5 = 32$. Thus, the y-intercept is 32.

FIGURE 10

FIGURE 11

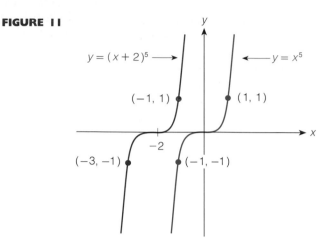

EXAMPLE 3   Graph the function $y = -2(x - 3)^4$.

Solution   We begin with the graph of $y = -2x^4$ in Figure 12(a). The points $(1, -2)$ and $(-1, -2)$ are obtained by substituting $x = 1$ and $x = -1$, respectively, in the equation $y = -2x^4$. Now if we replace x with $x - 3$ in the equation $y = -2x^4$, we have $y = -2(x - 3)^4$, whose graph is obtained by translating the graph in Figure 12(a) three units to the right. See Figure 12(b).   ■■■

FIGURE 12

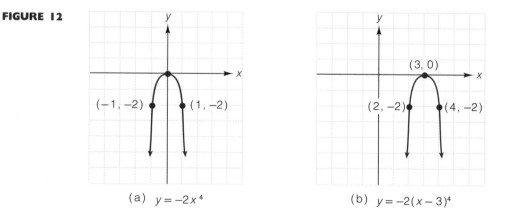

(a) $y = -2x^4$

(b) $y = -2(x - 3)^4$

We can use our work on solving inequalities (in Section 1.6) to graph polynomial functions that are in factored form. Consider, for example, the function

$$f(x) = x(x + 1)(x - 3)$$

First of all, by inspection, we see that $f(x) = 0$ when $x = 0$, $x = -1$, or $x = 3$. These are the x-intercepts for the graph. Also, note that the y-intercept is 0. (Why?) Next, we want to know what the graph looks like in the intervals between the x-intercepts. To do this, we solve the two inequalities $f(x) > 0$ and $f(x) < 0$ using the technique in Section 1.6. Table 2 shows the results. (You should check these results for yourself; if you need a review, the details of this very example are worked out on pages 37–38.)

Now we interpret the results in Table 2 graphically. When x is in either of the intervals $(-\infty, -1)$ or $(0, 3)$, the graph lies below the x-axis (because

TABLE 2

Interval	$x(x + 1)(x - 3)$
$(-\infty, -1)$	negative
$(-1, 0)$	positive
$(0, 3)$	negative
$(3, \infty)$	positive

$f(x) < 0$); and when x is in either of the intervals $(-1, 0)$ or $(3, \infty)$, the graph lies above the x-axis (because $f(x) > 0$). This information is summarized in Figure 13(a). The three dots in the figure indicate the x-intercepts of the graph. The shaded regions are the **excluded regions** through which the graph cannot pass. The graph must pass only through the unshaded regions (and through the three x-intercepts). In Figure 13(b) we have drawn a rough sketch of a curve satisfying these conditions.

FIGURE 13

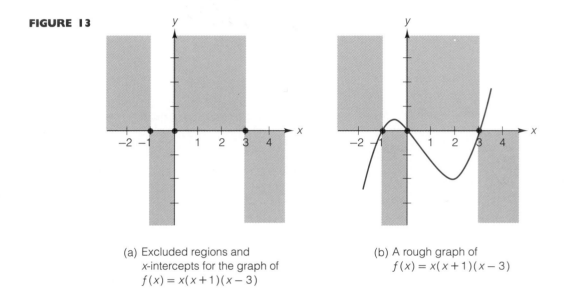

(a) Excluded regions and
    x-intercepts for the graph of
    $f(x) = x(x + 1)(x - 3)$

(b) A rough graph of
    $f(x) = x(x + 1)(x - 3)$

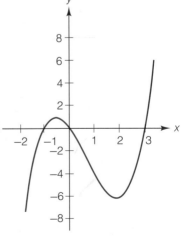

FIGURE 14

$f(x) = x(x + 1)(x - 3)$

Notice that to draw a smooth curve satisfying the conditions of Figure 13(a), we need at least two turning points: one between $x = -1$ and $x = 0$ and another between $x = 0$ and $x = 3$. On the other hand, since the degree of $f(x)$ is 3, there can be no more than two turning points. Thus Figure 13(b) has exactly two turning points. As another check on our rough sketch, we note that for large values of $|x|$, the graph indeed resembles that of $y = x^3$. (We are using the third property in the summary box on page 199.)

While the precise location of the turning points is a matter for calculus, we can nevertheless improve upon the sketch in Figure 13(b) by computing $f(x)$ for some specific values of x. Some reasonable choices in this case are the inputs -2, $-\frac{1}{2}$, 1, 2, and 4. As you can check, the resulting points on the graph are $(-2, -10)$, $\left(-\frac{1}{2}, \frac{7}{8}\right)$, $(1, -4)$, $(2, -6)$, and $(4, 20)$. We can now sketch the graph as shown in Figure 14.

In the example just concluded, the polynomial $f(x) = x(x + 1)(x - 3)$ has no *repeated factors*. That is, none of the factors is squared or cubed or raised to a higher power. Now let us look at two examples in which there are repeated factors. The observations we make will help us in determining the general shape of a graph without the need for plotting a large number of individual points. In Figure 15(a), we show the graph of $g(x) = x(x + 1)(x - 3)^2$. Notice that in the immediate vicinity of the intercept at $x = 3$, the graph has the same general shape as that of $y = A(x - 3)^2$. Similarly, in Figure 15(b), we show the graph of $h(x) = x(x + 1)(x - 3)^3$; notice that in the immediate vicinity of $x = 3$, the graph

FIGURE 15

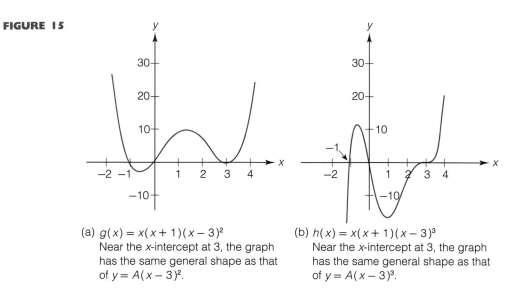

(a) $g(x) = x(x + 1)(x - 3)^2$
Near the x-intercept at 3, the graph has the same general shape as that of $y = A(x - 3)^2$.

(b) $h(x) = x(x + 1)(x - 3)^3$
Near the x-intercept at 3, the graph has the same general shape as that of $y = A(x - 3)^3$.

has the same general shape as that of $y = A(x - 3)^3$. In the box that follows, we state the general principle underlying these observations. (The principle can be justified using calculus.)

The Behavior of a Polynomial Function near an x-Intercept

Let $f(x)$ be a polynomial and suppose that $(x - a)^n$ is a factor of $f(x)$. [Furthermore, assume that none of the other factors of $f(x)$ contains $(x - a)$.] Then, in the immediate vicinity of the x-intercept at a, the graph of $y = f(x)$ closely resembles that of $y = A(x - a)^n$.

The principle that we have just stated is easy to apply because we already know how to graph functions of the form $y = A(x - a)^n$. The next example shows how this works.

EXAMPLE 4 Describe the behavior of each function in the immediate vicinity of the indicated x-intercept.

(a) $f(x) = \frac{1}{2}(x - 3)(x - 1)^3$; intercept: $x = 1$
(b) $g(x) = (x + 1)(x + 4)(x + 3)^2$; intercept: $x = -3$

Solution

(a) We make the following observation:

$$f(x) = \frac{1}{2}\underbrace{(x - 3)}(x - 1)^3$$

When x is close to 1, this factor is close to $1 - 3$, or -2.

So if x is very close to 1, we have the approximation

$$f(x) \approx \frac{1}{2}(1 - 3)(x - 1)^3 = \frac{1}{2}(-2)(x - 1)^3 = -(x - 1)^3$$

FIGURE 16
When x is close to 1, the graph of $f(x) = \frac{1}{2}(x-3)(x-1)^3$ closely resembles $y = -(x-1)^3$.

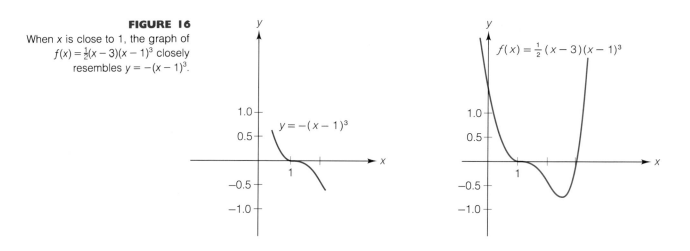

Thus, in the immediate vicinity of $x = 1$, the graph of f closely resembles $y = -(x-1)^3$. Notice the technique used to obtain this result. We retained the factor corresponding to the intercept $x = 1$, and we approximated the remaining factor using the value $x = 1$. See Figure 16.

(b) We use the approximation technique shown in part (a).

$$g(x) = \underbrace{(x+1)}\underbrace{(x+4)}(x+3)^2$$

When x is close to -3, this factor is close to $-3 + 1$, or -2.

When x is close to -3, this factor is close to $-3 + 4$, or 1.

Thus, when x is close to -3, we have the approximation

$$g(x) \approx (-3+1)(-3+4)(x+3)^2 = -2(x+3)^2$$

This tells us that in the immediate vicinity of $x = -3$, the graph of g resembles $y = -2(x+3)^2$. See Figure 17. ▮▮▮

FIGURE 17
When x is close to -3, the graph of $g(x) = (x+1)(x+4)(x+3)^2$ closely resembles $y = -2(x+3)^2$.

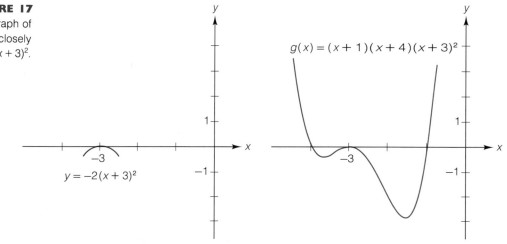

EXERCISE SET 4.5

A

In Exercises 1–4, give a reason (as in Example 1) why each graph cannot represent a polynomial function of degree 3.

1.

4.

2.

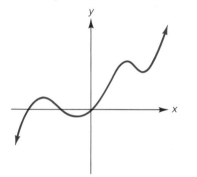

In Exercises 5–8, give a reason why each graph cannot represent a polynomial function that has highest-degree term $2x^5$.

5.

3.

6.

7.

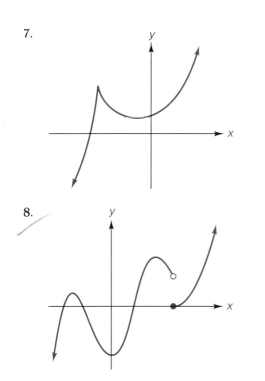

8.

In Exercises 9–20, sketch the graph of each function and specify all x- and y-intercepts.

9. $y = (x - 2)^2 + 1$	**10.** $y = -3x^4$
11. $y = -(x - 1)^4$	**12.** $y = -(x + 2)^3$
13. $y = (x - 4)^3 - 2$	**14.** $y = -(x - 4)^3 - 2$
15. $y = -2(x + 5)^4$	**16.** $y = -2x^4 + 5$
17. $y = \frac{1}{2}(x + 1)^5$	**18.** $y = \frac{1}{2}x^5 + 1$
19. $y = -(x - 1)^3 - 1$	**20.** $y = x^8$

In Exercises 21–28, (a) determine the x- and the y-intercepts and the excluded regions for the graph of the given function. Specify your results using a sketch similar to Figure 13(a). In Exercises 25–28, you will first need to factor the polynomial. (b) Graph each function.

21. $y = (x - 2)(x - 1)(x + 1)$

22. $y = (x - 3)(x + 2)(x + 1)$

23. $y = 2x(x - 2)(x - 1)$

24. $y = (x - 3)(x - 2)(x + 2)$

25. $y = x^3 - 4x^2 - 5x$ **26.** $y = x^3 - 9x$

27. $y = x^3 + 3x^2 - 4x - 12$ **28.** $y = x^3 - 5x^2 - x + 5$

In Exercises 29–38, (a) determine the x- and the y-intercepts and the excluded regions for the graph of the given function. Specify your results using a sketch similar to Figure 13(a). (b) Describe the behavior of the function at each x-intercept that corresponds to a repeated factor. Specify your results using a

sketch similar to the left-hand portion of Figure 17. (c) Graph each function.

29. $y = x^3(x + 2)$ **30.** $y = (x - 1)(x - 4)^2$

31. $y = 2(x - 1)(x - 4)^3$ **32.** $y = (x - 1)^2(x - 4)^2$

33. $y = (x + 1)^7(x - 1)(x - 3)$

34. $y = x^2(x - 4)(x + 2)$

35. $y = -x^3(x - 4)(x + 2)$ **36.** $y = 4(x - 2)^2(x + 2)^3$

37. $y = -4x(x - 2)^2(x + 2)^3$ **38.** $y = -3x^3(x + 1)^4$

B

In Exercises 39–44, six functions are defined as follows:

$$f(x) = x \qquad g(x) = x^2 \qquad h(x) = x^3$$
$$F(x) = x^4 \qquad G(x) = x^5 \qquad H(x) = x^6$$

Refer also to the following figure.

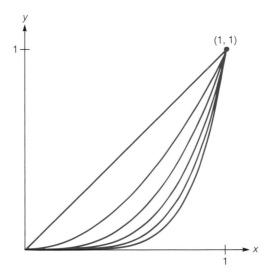

39. The six graphs in the figure are the graphs of the six given functions for the interval $[0, 1]$, but the graphs are not labeled. Which is which?

40. For which x-values in $[0, 1]$ will the graph of g lie strictly below the horizontal line $y = 0.1$? Use a calculator to evaluate your answer. Round off the result to two significant figures.

41. Follow Exercise 40, using the function H instead of g.

42. Find a number t in $[0, 1]$ such that the vertical distance between $f(t)$ and $g(t)$ is $\frac{1}{4}$.

43. Is there a number t in $[0, 1]$ such that the vertical distance between $g(t)$ and $F(t)$ is 0.26?

44. Find all numbers t in $[0, 1]$ such that $F(t) = G(t) + H(t)$.

45. Do the graphs of $y = x$ and $y = \frac{1}{100}x^2$ intersect anywhere other than at the origin? (See the following figure.)

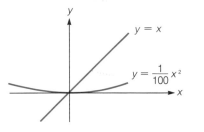

46. (a) Graph the function $y = 4x^2 - x^4$.
 (b) Find the coordinates of the turning points.
  Hint: See Example 7 in Section 4.2.

47. (a) Graph the function $D(x) = x^2 - x^4$.
 (b) Find the turning points of the graph. *Hint:* See Example 7 in Section 4.2.
 (c) On the same set of axes, sketch the graphs of $y = x^2$ and $y = x^4$ for $0 \le x \le 1$. What is the maximum vertical distance between the graphs?

48. (a) An open-top box is to be constructed from a 6-in.-by-8-in. rectangular sheet of tin by cutting out equal squares at each corner and then folding up the resulting flaps. Let x denote the length of the side of each cutout square. Show that the volume $V(x)$ is

$$V(x) = x(6 - 2x)(8 - 2x)$$

 (b) What is the domain of the volume function in part (a)? [The answer is *not* $(-\infty, \infty)$.]
 (c) Graph the volume function.

(d) Use your graph to estimate the maximum possible volume for the box.

49. A cylinder is inscribed in a sphere of radius 6 cm. Let r and h denote the radius and height of the cylinder, as shown in the figure. We are going to estimate the maximum possible volume for the cylinder.
 (a) Show that $h = 2\sqrt{36 - r^2}$.
 (b) Let V denote the volume of the cylinder. Check that $V = 2\pi r^2\sqrt{36 - r^2}$.

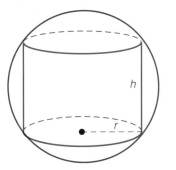

 (c) What is the domain of the volume function in part (b)?
 (d) The expression for V in part (b) is not a polynomial but the square of the expression is, so we'll work with the function $f(r) = 4\pi^2 r^4(36 - r^2)$. At the end, we can convert back to V by taking the square root of $f(r)$. Graph the function $f(r) = 4\pi^2 r^4(36 - r^2)$. (To fine-tune your graph, use a calculator to compute $f(r)$ for r running from 0 to 6 in increments of 0.5.) From the graph, estimate the maximum value of $f(r)$. Now estimate the maximum possible volume of the cylinder.

▶ OPTIONAL TI-81 GRAPHING CALCULATOR EXERCISES
EXERCISES FOR SECTION 4.5

1. According to property 3 on page 199, the graph of the polynomial function

$$f(x) = a_n x^n + a_{n-1}x^{n-1} + \cdots + a_0$$

resembles the graph of $y = ax^n$ when $|x|$ is large. In this exercise we look at some examples of this property.
 (a) Look at the graph of the function $Y_1 = x^2 + 6x + 2$ in the standard viewing rectangle (ZOOM 6). According to property 3, this graph resembles that of $y = x^2$ when $|x|$ is large. To see a vivid demonstration of this, make the following adjustments in the RANGE menu: $x_{min} = -1000$;

$x_{max} = 1000$; $x_{scl} = 100$; $y_{min} = -100,000$; $y_{max} = 100,000$; $y_{scl} = 10,000$. Now press GRAPH. As you can see, the resulting curve indeed resembles the familiar parabola $y = x^2$. Finally, press the Y= key, enter $Y_2 = x^2$, and press GRAPH. What do you observe?
 (b) Follow the procedure in part (a) for the function $Y_1 = -0.25x^3 + 3x + 4$. (After looking at Y_1 in the standard viewing rectangle, use the following RANGE adjustments for the long-range view: $x_{min} = -100$; $x_{max} = 100$; $x_{scl} = 10$; $y_{min} = -1000$; $y_{max} = 1000$; $y_{scl} = 100$.)

(c) Follow the general procedure in part (a) for the function

$$Y_1 = 2x^5 - 16x^4 - 26x^3 + 352x^2 - 72x - 1140$$

You'll find that the standard viewing rectangle does not show enough of the graph. So begin instead with the following RANGE settings: $x_{min} = -6$; $x_{max} = 8$; $x_{scl} = 1$; $y_{min} = -2000$; $y_{max} = 2000$; $y_{scl} = 200$. Then, for the long-range view, use these settings: $x_{min} = -250$; $x_{max} = 250$; $x_{scl} = 25$; $y_{min} = -10^{11}$; $y_{max} = 10^{11}$; $y_{scl} = 10^{10}$. *Note:* To enter a number such as 10^{11}, press the (EE) key and then type "11".

2. This exercise demonstrates the behavior of a polynomial function near its x-intercepts. Let $g(x) = x(x + 1)(x - 3)^2$.
   (a) First, without a calculator, use the estimation technique explained in Example 4 to verify the following statements.

   > Near $x = 0$, the graph of g resembles the line $y = 9x$.
   >
   > Near $x = -1$, the graph of g resembles the line $y = -16(x + 1)$.
   >
   > Near $x = 3$, the graph of g resembles the parabola $y = 12(x - 3)^2$.

   (b) Press the Y= key and enter the four functions

   $$Y_1 = 9x \qquad Y_2 = -16(x + 1)$$
   $$Y_3 = 12(x - 3)^2 \qquad Y_4 = x(x + 1)(x - 3)^2$$

   Now, since we do not want to activate these functions quite yet, move the cursor to each equal sign and press ENTER in each case. (This will deactivate each function. When you need to activate a function, just move the cursor back to the equal sign and press ENTER.) Next, press ZOOM 6. (No graph yet.) Then press the RANGE key and change the settings on x_{min} and x_{max} to read $x_{min} = -2$ and $x_{max} = 4$.

(c) In the Y= menu, activate Y_1 and Y_4. Press GRAPH. What do you observe?
(d) In the Y= menu, deactivate Y_1 and activate Y_2. Press GRAPH. What do you observe?
(e) In the Y= menu, deactivate Y_2 and activate Y_3. Press GRAPH. What do you observe?

Problems 3–5 are maximum–minimum problems, just as you practiced in Section 4.4, but now the function that you set up will turn out to be a polynomial of degree greater than two. To find the required maximum or minimum and/or input, first set up the appropriate function. (Your calculator won't help you with this part of the work.) Then use the calculator to graph the function. The ZOOM key and the TRACE key will then allow you to approximate the required values.

3. Which point on the curve $y = x^3$ is closest to the point $(4, 5)$? *Remark:* If you work with the *square* of the distance, then a polynomial function is directly involved. But in fact, with the graphing calculator, it is just as easy to work with the distance function itself.

4. The figure shows a rectangle with its base on the x-axis and its upper two vertices on the curve $y = 1 - x^4$. What is the maximum possible area for such a rectangle?

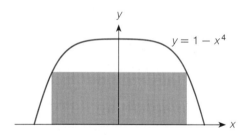

5. An open-top box is constructed from a 9-in.-by-12-in. rectangular sheet of tin by cutting out equal squares at each corner and then folding up the flaps. Find the largest possible volume for the box.

4.6 RATIONAL FUNCTIONS

We shall not attempt to explain the numerous situations in life sciences where the study of such graphs is important nor the chemical reactions which give rise to the rational functions $r(x)$.... Let it suffice to mention that, to the experimental biochemist, theoretical results concerning the shapes of rational functions are of considerable interest.

W. G. Bardsley and R. M. W. Wood in "Critical Points and Sigmoidicity of Positive Rational Functions," *The American Mathematical Monthly,* **92** (1985) 37–42.

After the polynomial functions, the next simplest functions are the **rational functions**. These are functions defined by equations of the form

$$y = \frac{f(x)}{g(x)}$$

where $f(x)$ and $g(x)$ are polynomials. In general, throughout this section, when we write a function such as $y = f(x)/g(x)$, we assume that $f(x)$ and $g(x)$ contain no common factors (other than constants). (Exercises 36 and 37 ask you to consider several cases in which $f(x)$ and $g(x)$ do contain common factors.) Also, for each of the examples that we discuss, the degree of $f(x)$ is less than or equal to the degree of $g(x)$. (A case in which the degree of $f(x)$ exceeds the degree of $g(x)$ is developed in the exercises.)

■■■ **EXAMPLE 1**   Specify the domain of the rational function defined by

$$y = \frac{3x - 2}{x^2 - 1}$$

Also, find the x- and y-intercepts (if any) for the graph of this function.

Solution   By factoring the denominator, we can rewrite the given equation

$$y = \frac{3x - 2}{(x - 1)(x + 1)}$$

Since the denominator is zero when $x = 1$ and when $x = -1$, it follows that the domain of the given function consists of all real numbers except 1 and -1. To determine the x-intercepts of the graph, we set $y = 0$ (as usual) in the given equation to obtain

$$\frac{3x - 2}{x^2 - 1} = 0 \qquad (x \neq 1, \ x \neq -1)$$

$$3x - 2 = 0 \qquad \begin{array}{l}\text{multiplying both sides of the equation by the}\\ \text{nonzero quantity } x^2 - 1, \text{ since } x \neq \pm 1\end{array}$$

$$x = \tfrac{2}{3}$$

Thus, the only x-intercept is $x = \tfrac{2}{3}$. For the y-intercept, we set $x = 0$ (as usual) in the given equation. As you can readily check, this yields $y = 2$. In summary then, the x- and y-intercepts are $\tfrac{2}{3}$ and 2, respectively.   ■■■

In the box that follows, we summarize the ideas used in Example 1 for determining the domain and x-intercepts of a rational function.

Let R be a rational function defined by

$$R(x) = \frac{f(x)}{g(x)}$$

where $f(x)$ and $g(x)$ are polynomials with no common factors (other than constants). Then the domain of R consists of all real numbers for which $g(x) \neq 0$; the x-intercepts (if any) are the real solutions of the equation $f(x) = 0$.

You are already familiar with the graph of one rational function, because in Chapter 2 we graphed $y = 1/x$. For convenience, this graph is shown again in

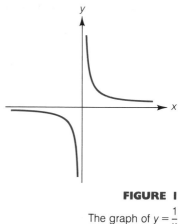

FIGURE 1

The graph of $y = \dfrac{1}{x}$

FIGURE 2

Curves with asymptotes

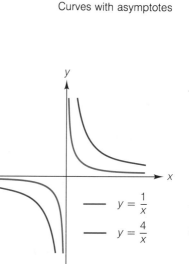

—— $y = \dfrac{1}{x}$

—— $y = \dfrac{4}{x}$

FIGURE 3

Figure 1. The graph in Figure 1 differs from the graph of every polynomial function in two important aspects. First, the graph has a break in it; it is composed of two distinct pieces or **branches**. (Recall from Section 4.5 that the graph of a polynomial function never has a break in it.) In general, the graph of a rational function has one more branch than the number of distinct real values for which the denominator is zero. For instance, referring to the function in Example 1, we can expect the graph of $y = \dfrac{3x - 2}{x^2 - 1}$ to have three branches.

The second way that the graph in Figure 1 differs from that of a polynomial function is related to *asymptotes*. A line is an **asymptote** for a curve if the distance between the line and the curve approaches zero as we move out farther and farther along the line. Thus, the *x*-axis is a horizontal asymptote for the graph in Figure 1, while the *y*-axis is a vertical asymptote. It can be shown that the graph of a polynomial function never has an asymptote. Figure 2 displays additional examples of curves with asymptotes.

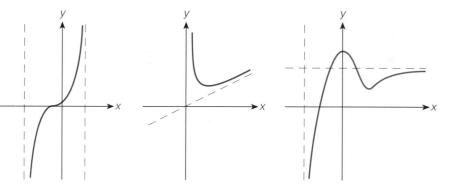

If k is a positive constant, the graph of $y = k/x$ resembles that of $y = 1/x$. Figure 3 shows the graphs of $y = 1/x$ and $y = 4/x$. Once we know about the graph of $y = k/x$, we can graph any rational function of the form

$$y = \frac{ax + b}{cx + d}$$

The next three examples show how this is done using the translation and reflection techniques developed in Chapter 3.

EXAMPLE 2 Graph $y = \dfrac{1}{x - 1}$.

Solution By translating the graph of $y = 1/x$ one unit to the right, we obtain the graph of $y = 1/(x - 1)$ shown in Figure 4. Notice that the horizontal asymptote is unchanged by the translation. However, the vertical asymptote moves one unit to the right. (The *y*-intercept is -1; why?)

EXAMPLE 3 Graph $y = \dfrac{2}{x - 1}$ and $y = \dfrac{-2}{x - 1}$.

Solution The graph of $y = 2/(x - 1)$ has the same basic shape and location as the graph of $y = 1/(x - 1)$ shown in Figure 4. As a further guide to sketching

$y = \dfrac{1}{x - 1}$

Asymptote $x = 1$

FIGURE 4

The graph of $y = \dfrac{1}{x - 1}$

$y = 2/(x - 1)$, we can pick several convenient x-values on either side of the asymptote $x = 1$ and then compute the corresponding y-values. After doing this, we obtain the graph shown in Figure 5(a). By reflecting this graph about the x-axis, we obtain the graph of $y = -2/(x - 1)$, which is shown in Figure 5(b).

FIGURE 5

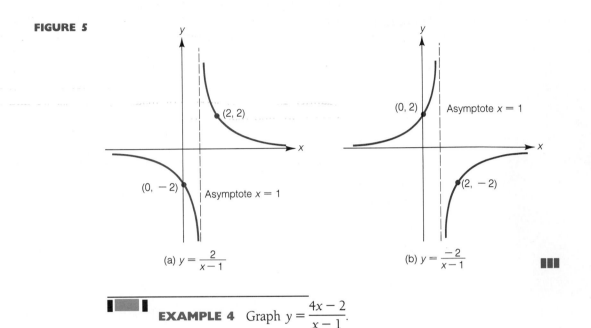

(a) $y = \dfrac{2}{x - 1}$ (b) $y = \dfrac{-2}{x - 1}$

EXAMPLE 4   Graph $y = \dfrac{4x - 2}{x - 1}$.

Solution   First, as you can readily check, the x- and y-intercepts are $\frac{1}{2}$ and 2, respectively. Next, using long division, we find that

$$\frac{4x - 2}{x - 1} = 4 + \frac{2}{x - 1}$$

We conclude that the required graph can be obtained by moving up the graph of $y = 2/(x - 1)$ four units in the y-direction (see Figure 6). Notice that the ver-

FIGURE 6

The graph of $y = \dfrac{4x - 2}{x - 1}$

Horizontal asymptote $y = 4$

Vertical asymptote $x = 1$

tical asymptote is still $x = 1$, but the horizontal asymptote is now $y = 4$ instead of the x-axis. ▮▮▮

Now let's look at rational functions of the form $y = 1/x^n$. First we'll consider $y = 1/x^2$. As with $y = 1/x$, the domain consists of all real numbers except $x = 0$. For $x \neq 0$, the quantity $1/x^2$ is always positive. This means that the graph will always lie above the x-axis. Furthermore, the graph will be symmetric about the y-axis. (Why?) As $|x|$ becomes very large, the quantity $1/x^2$ approaches zero; this is true whether x itself is positive or negative. So when $|x|$ is very large, we expect the graph of $y = 1/x^2$ to look as shown in Figure 7(a). On the other hand, when x is a very small fraction, close to zero, either negative or positive, the quantity $1/x^2$ is very large. For instance, if $x = \frac{1}{10}$, we find that $y = 100$, and if $x = \frac{1}{100}$, we find that $y = 10,000$. Thus, as x approaches zero, from the right or from the left, the graph of $y = 1/x^2$ must look as shown in Figure 7(b). Now, by plotting several points and taking Figures 7(a) and (b) into account, we obtain the graph of $y = 1/x^2$, shown in Figure 8. Also, by following a similar line of reasoning, we find that the graph of $y = 1/x^3$ looks as shown in Figure 9. (Note that $y = 1/x^3$ is symmetric about the origin.)

(a)

(b)

FIGURE 7

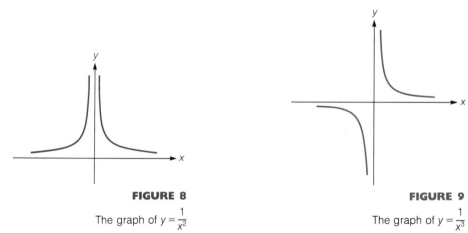

FIGURE 8

The graph of $y = \dfrac{1}{x^2}$

FIGURE 9

The graph of $y = \dfrac{1}{x^3}$

In general, when n is an even integer greater than 2, the graph of $y = 1/x^n$ resembles that of $y = 1/x^2$. When n is an odd integer greater than 3, the graph of $y = 1/x^n$ resembles that of $y = 1/x^3$.

▮▮▮ **EXAMPLE 5**   Graph $y = \dfrac{-1}{(x + 3)^2}$.

Solution   Refer to Figure 10 (on the next page). Begin with the graph of $y = 1/x^2$. By moving the graph three units to the left, we obtain the graph of $y = 1/(x + 3)^2$. Then by reflecting the graph of $y = 1/(x + 3)^2$ about the x-axis, we get the graph of $y = -1/(x + 3)^2$.

Question   What are the y-intercepts in Figures 10(b) and (c)? ▮▮▮

FIGURE 10

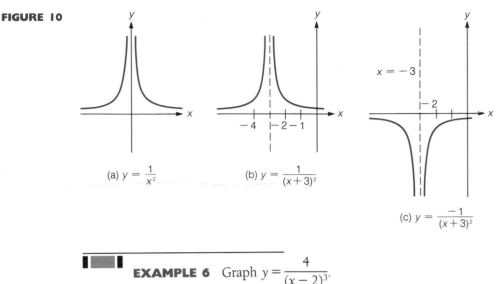

(a) $y = \dfrac{1}{x^2}$

(b) $y = \dfrac{1}{(x+3)^2}$

(c) $y = \dfrac{-1}{(x+3)^2}$

FIGURE 11

The graph of $y = \dfrac{4}{(x-2)^3}$

■ ▒ ■ EXAMPLE 6   Graph $y = \dfrac{4}{(x-2)^3}$.

Solution   Moving the graph of $y = 1/x^3$ two units to the right gives us the graph of the function $y = 1/(x-2)^3$. The graph of $y = 4/(x-2)^3$ will have the same basic shape and location. As a further guide to sketching the required graph, we can pick several convenient x-values near the asymptote $x = 2$ and compute the corresponding y-values. Using $x = 0$, $x = 1$, and $x = 3$, we find that the points $\left(0, -\tfrac{1}{2}\right)$, $(1, -4)$, and $(3, 4)$ are on the graph. With this information, the graph can be sketched as in Figure 11.   **■■■**

As we have seen in Examples 2 through 6, the vertical asymptotes of $y = f(x)/g(x)$ are found by solving the equation $g(x) = 0$. In Example 7 we demonstrate a method for determining the horizontal asymptote. The method involves dividing both the numerator and the denominator by the highest power of x that appears.

■ ▒ ■ EXAMPLE 7   Determine the horizontal asymptote for the graph of $y = \dfrac{3x^2}{x^2 - 4x + 1}$.

Solution   First we look for the highest power of x that appears in the given fractional expression; this is x^2. Now we divide both the numerator and the denominator by x^2. (You'll see the reason for doing this in a moment.) We have

$$y = \frac{3x^2}{x^2 - 4x + 1} = \frac{\dfrac{3x^2}{x^2}}{\dfrac{x^2 - 4x + 1}{x^2}} = \frac{3}{1 - \dfrac{4}{x} + \dfrac{1}{x^2}}$$

When $|x|$ grows very large, the two fractions $4/x$ and $1/x^2$ both approach zero. We therefore have

$$y \approx \frac{3}{1 - 0 + 0} = 3 \qquad \text{when } |x| \text{ is very large}$$

This tells us that the line $y = 3$ is a horizontal asymptote for the graph of the given function. (You will see another example of this type of calculation in Example 8.) ∎

In the previous section you learned a technique for determining the general shape of a polynomial function near an x-intercept. That same technique is also very useful in graphing rational functions in which both the numerator and denominator can be expressed as products of linear factors. Example 8 shows how this works.

EXAMPLE 8 Graph the function $y = \dfrac{(x-3)(x+2)}{(x+1)(x-2)}$.

Solution By inspection, we see that the x-intercepts are $x = 3$ and $x = -2$, the y-intercept is 3, and the vertical asymptotes are the lines $x = -1$ and $x = 2$. To find the horizontal asymptote, we write

$$y = \frac{x^2 - x - 6}{x^2 - x - 2}$$

$$= \frac{1 - \dfrac{1}{x} - \dfrac{6}{x^2}}{1 - \dfrac{1}{x} - \dfrac{2}{x^2}} \approx \frac{1 - 0 - 0}{1 - 0 - 0} \quad \text{when } |x| \text{ is large}$$

The horizontal asymptote is therefore the line $y = 1$.

Now we want to see how the graph looks in the immediate vicinity of the x-intercepts and the vertical asymptotes. To do this, we use the approximation technique explained in the previous section for polynomial functions. Let's start with the x-intercept at $x = 3$. As in Section 4.5, we'll retain the factor $x - 3$ and approximate the remaining factors using $x = 3$. So, for x near 3, we have

$$y = \frac{(x-3)(x+2)}{(x+1)(x-2)} \approx \frac{(x-3)(3+2)}{(3+1)(3-2)} = \frac{5}{4}(x-3)$$

Thus, in the immediate vicinity of the x-intercept $x = 3$, the required graph will closely resemble the line $y = \frac{5}{4}(x-3)$. The remaining calculations for approximating the graph near the other x-intercept and near the two vertical asymptotes are carried out in exactly the same manner. As Exercise 35 will ask you to verify, the results are:

x near 3: $\quad y \approx \frac{5}{4}(x-3)$	x near -1: $\quad y \approx \dfrac{4/3}{x+1}$
x near -2: $\quad y \approx -\frac{5}{4}(x+2)$	x near 2: $\quad y \approx \dfrac{-4/3}{x-2}$

We summarize these results as follows. As the graph passes through the points $(3, 0)$ and $(-2, 0)$, it resembles the lines $y = \frac{5}{4}(x-3)$ and $y = -\frac{5}{4}(x+2)$, respectively. Near the vertical asymptote $x = -1$, the graph has the same basic shape as $y = 1/(x+1)$. And near the vertical asymptote $x = 2$, the graph resembles $y = -1/(x-2)$. See Figure 12.

FIGURE 12

The graph of $y = \dfrac{(x-3)(x+2)}{(x+1)(x-2)}$ near its

x-intercepts and vertical asymptotes.

TABLE 1

When $|x|$ is large, y is less than 1.

x	y
10	0.95
100	0.99
-10	0.96
-100	0.9996

Finally, we want to find out how the graph approaches the horizontal asymptote $y = 1$. Perhaps the simplest way to decide this is to do some calculations. In Table 1 we have computed the outputs for some relatively large values of $|x|$. From the table, we see that when $|x|$ is large, y is less than 1. Graphically, this means that the curve approaches the asymptote $y = 1$ from below. In Figure 13 we summarize what we have discovered up to this point. Then, using Figure 13 as a guide, we can sketch the required graph, as shown in Figure 14.

Note A general method for finding the coordinates of the lowest point on the middle branch of the graph is studied in calculus. Those coordinates can also be found, however, by applying algebraic techniques. (See Exercise 38.) ▌▌▌

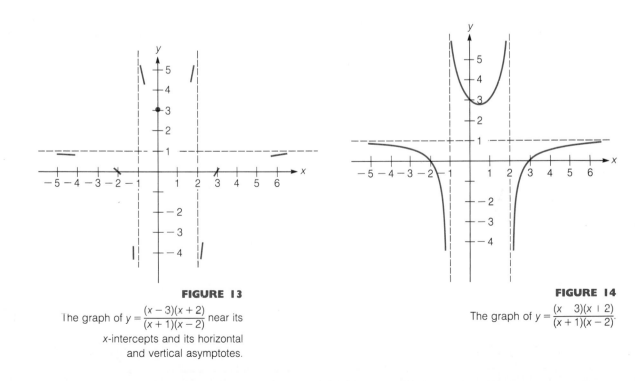

FIGURE 13

The graph of $y = \dfrac{(x-3)(x+2)}{(x+1)(x-2)}$ near its

x-intercepts and its horizontal
and vertical asymptotes.

FIGURE 14

The graph of $y = \dfrac{(x-3)(x+2)}{(x+1)(x-2)}$.

EXERCISE SET 4.6

A

In Exercises 1–6, find the domain and the x- and y-intercepts for each rational function.

1. $y = \dfrac{3x + 15}{4x - 12}$

2. $y = \dfrac{(x + 6)(x + 4)}{(x - 1)^2}$

3. $y = \dfrac{x^2 - 8x - 9}{x^2 - x - 6}$

4. $y = \dfrac{2x^2 + x - 5}{x^2 + 1}$

5. $y = \dfrac{(x^2 - 4)(x^3 - 1)}{x^6}$

6. $y = \dfrac{x^5 - 2x^4 - 9x + 18}{8x^3 + 2x^2 - 3x}$

In Exercises 7–28, sketch the graph of each rational function. Specify the intercepts and the asymptotes.

7. $y = \dfrac{1}{x + 4}$

8. $y = \dfrac{-1}{x + 4}$

9. $y = \dfrac{3}{x + 2}$

10. $y = \dfrac{-3}{x + 2}$

11. $y = \dfrac{-2}{x - 3}$

12. $y = \dfrac{x - 1}{x + 1}$

13. $y = \dfrac{x - 3}{x - 1}$

14. $y = \dfrac{2x}{x + 3}$

15. $y = \dfrac{4x - 2}{2x + 1}$

16. $y = \dfrac{3x + 2}{x - 3}$

17. $y = \dfrac{1}{(x - 2)^2}$

18. $y = \dfrac{-1}{(x - 2)^2}$

19. $y = \dfrac{3}{(x + 1)^2}$

20. $y = \dfrac{-3}{(x + 1)^2}$

21. $y = \dfrac{1}{(x + 2)^3}$

22. $y = \dfrac{-1}{(x + 2)^3}$

23. $y = \dfrac{-4}{(x + 5)^3}$

24. $y = \dfrac{x}{(x + 1)(x - 3)}$

25. $y = \dfrac{-x}{(x + 2)(x - 2)}$

26. $y = \dfrac{2x}{(x + 1)^2}$

27. (a) $\dfrac{3x}{(x - 1)(x + 3)}$

    (b) $\dfrac{3x^2}{(x - 1)(x + 3)}$

28. (a) $y = \dfrac{4x^2 + 1}{x^2 - 1}$

    (b) $y = \dfrac{4x^2 + 1}{x^2 + 1}$

29. (a) $f(x) = \dfrac{(x - 2)(x - 4)}{x(x - 1)}$

    (b) $g(x) = \dfrac{(x - 2)(x - 4)}{x(x - 3)}$

[Compare the graphs you obtain in parts (a) and (b). Notice how a change in only one constant can radically alter the nature of the graph.]

30. (a) $f(x) = \dfrac{(x - 1)(x + 2.75)}{(x + 1)(x + 3)}$

    (b) $g(x) = \dfrac{(x - 1)(x + 3.25)}{(x + 1)(x + 3)}$

[Compare the graphs you obtain in parts (a) and (b). Notice how a relatively small change in one of the constants can radically alter the graph.]

B

In Exercises 31–34, graph the functions.   Note: Each graph intersects its horizontal asymptote once. Find the intersection point before sketching your final graph.

31. $y = \dfrac{(x - 4)(x + 2)}{(x - 1)(x - 3)}$

32. $y = \dfrac{(x - 1)(x - 3)}{(x + 1)^2}$

33. $y = \dfrac{(x + 1)^2}{(x - 1)(x - 3)}$

34. $y = \dfrac{2x^2 - 3x - 2}{x^2 - 3x - 4}$

35. (This exercise refers to Example 8.) Let

$$y = \dfrac{(x - 3)(x + 2)}{(x + 1)(x - 2)}$$

Verify each of the following approximations.

(a) When x is close to -2, then $y \approx -\frac{5}{4}(x + 2)$.

(b) When x is close to -1, then $y \approx \dfrac{\frac{4}{3}}{x + 1}$.

(c) When x is close to 2, then $y \approx \dfrac{-\frac{4}{3}}{x - 2}$.

In Exercises 36 and 37, graph the functions. Notice in each case that the numerator and denominator contain at least one common factor. Thus, you can simplify each quotient; but don't lose track of the domain of the function as it was initially defined.

36. (a) $y = \dfrac{x + 2}{x + 2}$

    (b) $y = \dfrac{x^2 - 4}{x - 2}$

    (c) $y = \dfrac{x - 1}{(x - 1)(x - 2)}$

37. (a) $y = \dfrac{x^2 - 9}{x + 3}$

    (b) $y = \dfrac{x^2 - 5x + 6}{x^2 - 2x - 3}$

    (c) $y = \dfrac{(x - 1)(x - 2)(x - 3)}{(x - 1)(x - 2)(x - 3)(x - 4)}$

38. This exercise shows you how to determine the coordinates of the lowest point on the middle branch of the curve in Figure 14. The basic idea is as follows. Suppose that the required coordinates are (h, k). Then the horizontal line $y = k$ is the unique horizontal line intersecting the curve in one and only one point. (Any other horizontal line intersects the curve either in two points or not at all.) In steps (a) through (c) that follow, we use these observations to determine the point (h, k).

(a) Given any horizontal line $y = k$, its intersection with the curve in Figure 14 is determined by solving the following pair of simultaneous equations:

$$\begin{cases} y = k \\ y = \dfrac{x^2 - x - 6}{x^2 - x - 2} \end{cases}$$

In the second equation of the system, replace y with k and show that the resulting equation can be written

$$(k - 1)x^2 - (k - 1)x + (6 - 2k) = 0 \qquad (1)$$

(b) If k is indeed the required y-coordinate, then equation (1) must have exactly one real solution. Set the discriminant of the quadratic equation equal to zero to obtain

$$(k - 1)^2 - 4(k - 1)(6 - 2k) = 0$$

and deduce from this that $k = 1$ or $k = \frac{25}{9}$. The solution $k = 1$ can be discarded. (To see why, look at Figure 14.)

(c) Using the value $y = \frac{25}{9}$, show that the corresponding x-coordinate is $\frac{1}{2}$. Thus, the required point is $\left(\frac{1}{2}, \frac{25}{9}\right)$.

39. Graph the function $y = x/(x - 3)^2$. Use the technique explained in Exercise 38 to find the coordinates of any turning points on the graph.

40. Graph the function $y = 2/(x - x^2)$. Use the technique explained in Exercise 38 to find the coordinates of any turning points on the graph.

An asymptote that is neither horizontal nor vertical is called a **slant**, or **oblique, asymptote**. For example, as indicated in the following figure, the line $y = x$ is a slant asymptote for the graph of $y = (x^2 + 1)/x$. To understand why the line $y = x$ is an asymptote, we carry out the indicated division and write the function in the form

$$y = x + \frac{1}{x} \qquad (2)$$

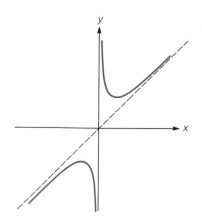

From equation (2), we see that if $|x|$ is very large, then $y \approx x + 0$; that is, $y \approx x$, as we wished to show. Equation (2) actually tells us more than this. When $|x|$ is very close to zero, equation (2) yields $y \approx 0 + (1/x)$. In other words, as we approach the y-axis, the curve looks more and more like the graph of $y = 1/x$. In general, if we have a rational function $f(x)/g(x)$ in which the degree of $f(x)$ is 1 greater than the degree of $g(x)$, then the graph has a slant asymptote that is obtained as follows. Divide $f(x)$ by $g(x)$ to obtain an equation of the form

$$\frac{f(x)}{g(x)} = (mx + b) + \frac{h(x)}{g(x)}$$

where the degree of $h(x)$ is less than the degree of $g(x)$. Then the equation of the slant asymptote is $y = mx + b$. (For instance, using the previous example, we have $mx + b = x$ and $h(x)/g(x) = 1/x$.) In Exercises 41–43, you are asked to graph functions that have slant asymptotes.

41. Let $y = F(x) = \dfrac{x^2 + x - 6}{x - 3}$.

(a) Use long division to show that

$$\frac{x^2 + x - 6}{x - 3} = (x + 4) + \frac{6}{x - 3}$$

(b) The result in part (a) shows that the line $y = x + 4$ is a slant asymptote for the graph of the function F. Verify this fact empirically by completing the following two tables. (Use a calculator.)

x	$x + 4$	$\dfrac{x^2 + x - 6}{x - 3}$
10		
100		
1000		

x	$x + 4$	$\dfrac{x^2 + x - 6}{x - 3}$
-10		
-100		
-1000		

(c) Determine the vertical asymptote and the x- and y-intercepts for the graph of F.

(d) Graph the function F. (Use the techniques in this section along with the fact that $y = x + 4$ is a slant asymptote.)

(e) Use the technique explained in Exercise 38 to find the coordinates of the two turning points on the graph of F.

42. (a) Show that the line $y = x - 2$ is a slant asymptote for the graph of $F(x) = x^2/(x + 2)$.

(b) Sketch the graph of F.

43. Show that the line $y = -x$ is a slant asymptote for the graph of $y = (1 - x^2)/x$. Then sketch the graph of this function.

OPTIONAL TI-81 GRAPHING CALCULATOR EXERCISES
EXERCISES FOR SECTION 4.6

Problems 1–7 are maximum–minimum problems, just as you practiced in Section 4.4, but now the function that you set up will turn out to be a rational function rather than a quadratic. To find the required maximum or minimum and/or input, first set up the appropriate function. (Your calculator won't help with this part of the work.) Then use the calculator to graph the function. The ZOOM key and the TRACE key will then allow you to approximate the required values.

1. The accompanying figure shows a line with slope m passing through the point $(2, 1)$. Find the slope of this line so that the area of the shaded triangle is a minimum. *Hint:* You should obtain the following function for the area A as a function of the slope.

$$A = \frac{-4m^2 + 4m - 1}{2m}$$

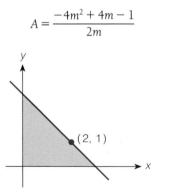

2. The product of two positive numbers is 6.
   (a) Express the sum S of the two numbers as a function of one variable. *Answer:* $S = (x^2 + 6)/x$
   (b) For which value of x will the sum be as small as possible? What is the other number in this case, and what is the corresponding minimum sum? *Hint for checking:* Using calculus, it can be shown that the exact value for x is $\sqrt{6}$. Evaluate this using a calculator. How close were you to this value?

3. Find the point P on the first quadrant portion of the parabola $y = x^2$, with the property that the slope of the line through $(0, -1)$ and P is a minimum. (After you've finished this problem, use the calculator to draw the parabola and the line through P and $(0, -1)$. What do you observe?)

4. What is the smallest possible value for the sum of a positive number and its reciprocal?

5. The area of a rectangle is $6 \, \text{m}^2$.
   (a) Express the perimeter of the rectangle as a function of the width x.
   (b) For which value of x is the perimeter a minimum? What is the length in this case?

6. A printer designing a small book decides to make the area of each page $60 \, \text{in}^2$. The margins on the top and bottom of the page are to be 2 in. and 1 in., respectively. The margins on the sides are each to be 1.5 in.
   (a) Express the area of the printed portion of the page as a function of x, the width of the whole page.
   (b) For which width x will the printed area be a maximum? What is the corresponding length in this case?
   (c) *Hint for checking:* Using calculus, it can be shown that the exact value for x is $2\sqrt{15}$. Evaluate this using a calculator. How close were you to this value in part (b)?

7. The volume of a right circular cylinder is to be $54\pi \, \text{in}^3$.
   (a) Express the height h of the cylinder in terms of r, the radius of the base.

(b) Express the total (outer) surface area of the cylinder in terms of r.

(c) Graph the function you found in part (b) and use the graph to estimate the value of r for which the surface area is a minimum.

(d) Using your formula from part (a) and the value of r that you obtained in part (c), find the corresponding value of h. How are the values of h and r related? *Hint for checking:* If you've worked out the correct formula in part (b), then the first-quadrant portion (the relevant portion) of the graph in (c) should be similar to the figure on the right.

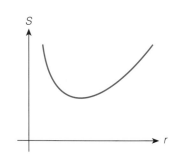

CHAPTER FOUR SUMMARY OF PRINCIPAL TERMS

TERMS	PAGE REFERENCE	COMMENTS
1. Linear function	159	A linear function is a function of the form $f(x) = Ax + B$. The graph of a linear function is a straight line. An important idea that arose in several of the examples is that the slope of a line can be interpreted as a rate of change. Two instances of this are marginal cost and velocity.
2. Quadratic function	169	A quadratic function is a function of the form $f(x) = ax^2 + bx + c$. The graph of a quadratic function is a parabola. See the boxes on pages 170 and 172.
3. Polynomial function	197	A polynomial function of degree n is a function of the form $$f(x) = a_n x^n + a_{n-1} x^{n-1} + \cdots + a_1 x + a_0$$ where n is a nonnegative integer and $a_n \neq 0$. Three basic properties of polynomial functions are summarized in the box on page 199.
4. Rational function	210	A rational function is a function of the form $y = f(x)/g(x)$, where $f(x)$ and $g(x)$ are polynomials.
5. Asymptote	211	A line is said to be an asymptote for a curve if the distance between the line and the curve approaches zero as we move out farther and farther along the line. See, for example, Figure 2 on page 211, in which the dashed lines are asymptotes.

WRITING MATHEMATICS

Consider the following two problems.

(A) The perimeter of a rectangle is $2\,\text{m}$.
  (i) Express the area of the rectangle as a function of the width w.
  (ii) Find the maximum possible area for such a rectangle.
(B) The area of a rectangle is $2\,\text{m}^2$.
  (i) Express the perimeter of the rectangle as a function of the width w.
  (ii) Find the minimum possible perimeter for such a rectangle.

1. After working out problem A for yourself, write out the solution in complete sentences, as if you were explaining it to a classmate or to your instructor. Be sure

to let the reader know where you are headed and why each of the main steps is necessary.

2. After working out part (i) of problem B for yourself, write out the solution in complete sentences.

3. Explain why the methods of Section 4.4 are not applicable for solving the second part of problem B.

4. The following result is known as the *arithmetic-geometric mean inequality:*

> For all nonnegative real numbers a and b, we have $a + b \geq 2\sqrt{ab}$, with equality holding when and only when $a = b$.

By working with a classmate or your instructor, find out how to apply this result to solve the second part of problem B. Then write out your solution in complete sentences. How does your final answer here compare with that in problem A?

CHAPTER FOUR   REVIEW EXERCISES

1. Find $G(0)$ if G is a linear function such that $G(1) = -2$ and $G(-2) = -11$.

2. (a) Let $f(x) = 3x^2 + 6x - 10$. For which input x is the value of the function a minimum? What is that minimum value?
   (b) Let $g(t) = 6t^2 - t^4$. For which input t is the value of the function a maximum?

3. Suppose the function $p = -\frac{1}{8}x + 100 \quad (0 \leq x \leq 12)$ relates the selling price p of an item to the quantity x that is sold. Assume that p is in dollars. What is the maximum revenue possible in this situation?

4. Graph the function $y = -1/(x + 1)^3$.

5. Graph the function $y = (x - 4)(x - 1)(x + 1)$.

6. Graph the function $f(x) = x^2 + 4x - 5$. Specify the vertex, the x- and y-intercepts, and the axis of symmetry.

7. A factory owner buys a new machine for $1000. After five years, the machine has a salvage value of $100. Assuming linear depreciation, find a formula for the value V of the machine after t years, where $0 \leq t \leq 5$.

8. Graph the function $y = 2(x - 3)^4$. Does the graph cross the y-axis? If so, where?

9. Graph the function $y = \dfrac{3x + 5}{x + 2}$. Specify all intercepts and asymptotes.

10. What is the largest area possible for a right triangle in which the sum of the lengths of the two shorter sides is 12 cm?

11. Graph the function $x/[(x + 2)(x - 4)]$.

12. Let $P(x, y)$ be a point [other than $(-1, -1)$] on the graph of $f(x) = x^3$. Express the slope of the line passing through the points P and $(-1, -1)$ as a function of x. Simplify your answer as much as possible.

13. A rectangle is inscribed in a circle. The circumference of the circle is 12 cm. Express the perimeter of the rectangle as a function of its width w.

14. Give a reason why each of the following two graphs cannot represent a a polynomial function with highest-degree term $-\frac{1}{3}x^3$.

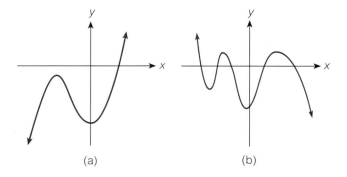

(a)   (b)

In Exercises 15–20, find equations for the linear functions satisfying the given conditions. Write each answer in the form $f(x) = mx + b$.

15. $f(3) = 5$ and $f(-2) = 0$.

16. $f(1) = 5$ and the graph of f passes through the origin.

17. $f(4) = -1$ and the graph of f is parallel to the line $3x - 8y = 16$.

18. The graph passes through $(6, 1)$ and the x-intercept is twice the y-intercept.

19. $f(-3) = 5$, and the graph of the inverse function passes through $(2, 1)$.

20. The graph of f passes through the vertices of the two parabolas $y = x^2 + 4x + 1$ and $y = \frac{1}{2}x^2 + 9x + \frac{81}{2}$.

In Exercises 21–26, graph the quadratic functions. In each case, specify the vertex and the x- and y-intercepts.

21. $y = x^2 + 2x - 3$

22. $f(x) = x^2 - 2x - 15$

23. $y = -x^2 + 2\sqrt{3}x + 3$

24. $f(x) = 2x^2 - 2x + 1$

25. $y = -3x^2 + 12x$

26. $f(x) = -4x^2 + 16x$

27. Find the distance between the vertices of the two parabolas $y = x^2 - 4x + 6$ and $y = -x^2 - 4x - 5$.

28. Find the value of a, given that the maximum value of the function $f(x) = ax^2 + 3x - 4$ is 5.

29. The sum of two numbers is $\sqrt{3}$. Find the largest possible value for their product.

30. The sum of two numbers is $\frac{2}{3}$. What is the smallest possible value for the sum of their squares?

31. Suppose that an object is thrown vertically upward (from ground level) with an initial velocity of v_0 ft/sec. It can be shown that the height h (in feet) after t sec is given by the formula $h = v_0 t - 16t^2$.
    (a) At what time does the object reach its maximum height? What is that maximum height?
    (b) At what time does the object strike the ground?

32. Let $f(x) = 4x^2 - x + 1$ and $g(x) = (x - 3)/2$.
    (a) For which input will the value of the function $f \circ g$ be a minimum?
    (b) For which input will the value of $g \circ f$ be a minimum?

33. (a) Let P be a point on the parabola $y = x^2$. Express the distance from P to the point $(0, 2)$ in terms of x.
    (b) Which point in the second quadrant on the parabola $y = x^2$ is closest to the point $(0, 2)$?

34. Find the coordinates of the point on the line $y = 2x - 1$ closest to $(-5, 0)$.

35. Find all values of b such that the minimum distance from the point $(2, 0)$ to the line $y = \frac{4}{3}x + b$ is 5.

36. What number exceeds one-half its square by the greatest amount?

37. Suppose that $x + y = \sqrt{2}$. Find the minimum value of the quantity $x^2 + y^2$.

38. For which numbers t will the value of $9t^2 - t^4$ be as large as possible?

39. Find the maximum area possible for a right triangle with a hypotenuse of 15 cm.   *Hint:* Let x denote the length of one leg. Show that the area is $A = \frac{1}{2}x\sqrt{225 - x^2}$. Now work with A^2.

40. For which point (x, y) on the curve $y = 1 - x^2$ is the sum $x + y$ a maximum?

41. Let $f(x) = x^2 - (a^2 + 2a)x + 2a^3$, where $0 < a < 2$. For which value of a will the distance between the x-intercepts of f be a maximum?

42. Suppose that the revenue R (in dollars) generated by selling x units of a certain product is given by $R = 300x - \frac{1}{4}x^2$ $(0 \le x \le 1200)$. How many units should be sold to maximize the revenue? What is that maximum revenue?

43. Suppose that the function $p = 160 - \frac{1}{5}x$ relates the selling price p of an item to the quantity x that is sold. Assume that p is in dollars. For which value of x will the revenue R $(= xp)$ be a maximum? What is the selling price p in this case?

44. A piece of wire 16 cm long is cut into two pieces. Let x denote the length of the first piece and $16 - x$ the length of the second. The first piece is formed into a rectangle in which the length is twice the width. The second piece of wire is also formed into a rectangle, but with the length three times the width. For which value of x is the total area of the two rectangles a minimum?

In Exercises 45–60, graph each function and specify the x- and y-intercepts and asymptotes, if any.

45. $y = (x + 4)(x - 2)$

46. $y = (x + 4)(x - 2)^2$

47. $y = -(x + 5)^3$

48. $y = -x(x + 1)$

49. $y = -x^2(x + 1)$

50. $y = -x^3(x + 1)$

51. $y = x(x - 2)(x + 2)$

52. $y = (x - 3)(x + 1)(x + 5)$

53. $y = \dfrac{3x + 1}{x}$

54. $y = \dfrac{1 - 2x}{x}$

55. $y = \dfrac{-1}{(x - 1)^2}$

56. $y = \dfrac{x + 1}{x + 2}$

57. $y = \dfrac{x - 2}{x - 3}$

58. $y = \dfrac{x}{(x - 2)(x + 4)}$

59. $y = \dfrac{x^2 - 2x + 1}{x^2 - 4x + 4}$

60. $y = \dfrac{x(x - 2)}{(x - 4)(x + 4)}$

61. Let $f(x) = x^2 + 2bx + 1$.
    (a) If $b = 1$, find the distance from the vertex of the parabola to the origin.
    (b) If $b = 2$, find the distance from the vertex of the parabola to the origin.
    (c) For which real numbers b will the distance from the vertex to the origin be as small as possible?

62. Find the range of the function $f(x) = -2x^2 + 12x - 5$. *Hint:* Look at the graph.

63. The range of the function $y = x^2 - 2x + k$ is the interval $[5, \infty)$. Find the value of k.

64. Find a value for b such that the range of the function $f(x) = x^2 + bx + b$ is the interval $[-15, \infty)$.

65. Find the range of the function $y = \dfrac{(x-1)(x-3)}{x-4}$.

    Hint: Solve the equation for x in terms of y using the quadratic formula. If you're careful with the algebra, you will find that the expression under the resulting radical sign is $y^2 - 8y + 4$. The range of the given function can then be found by solving the inequality $y^2 - 8y + 4 \geq 0$.

66. In the following figure, triangle OAB is equilateral and $\overline{AB}$ is parallel to the x-axis. Find the length of a side and the area of the triangle OAB.

    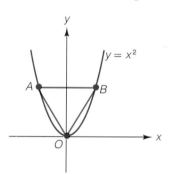

67. Let A denote the area of the right triangle in the first quadrant that is formed by the y-axis and the lines $y = mx$ and $y = m$. (Assume $m > 0$.) Express the area of the triangle as a function of m.

68. (a) Express the distance from the origin to the vertex of the parabola $y = x^2 - 2bx$ $(b > 0)$ as a function of b.
    (b) Let V denote the vertex of the parabola in part (a), and let A and B denote the points where the curve meets the x-axis. Express the area of $\triangle VAB$ as a function of b.
    (c) A circle is drawn with center V and passing through A and B. Then, through the smaller y-intercept of the circle, a tangent line is drawn. Express the x-intercept of this tangent line as a function of b.

69. In the accompanying figure, the radius of the circle is $OC = 1$. Express the area of $\triangle ABC$ as a function of x.

    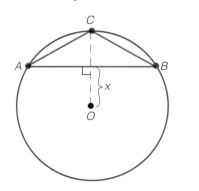

70. (a) Factor the expression $x^3 - 3x^2 + 4$. *Hint:* Subtract and add 1, then factor by grouping.
    (b) Use the factorization from part (a) to graph the function $y = x^3 - 3x^2 + 4$.

71. The following table shows the world energy production over the years 1982–1989. The scatter diagram and regression line are shown in the accompanying graph. The equation of the regression line (after some rounding off) is $y = 9.243x - 18044.316$.
    (a) Use this equation to estimate the world energy production in 1990.
    (b) According to the United States Energy Information Administration, the world energy production for 1990 was 345.06 quadrillion Btu. Is your estimate in part (a) close to this? Is your estimate too high or too low?

Year	Energy Production (quadrillion Btu)
1982	278.87
1983	280.80
1984	295.57
1985	302.86
1986	312.69
1987	320.52
1988	332.91
1989	340.47

World Energy Production, 1982–1989 (Energy from coal, natural gas, crude oil, nuclear power, hydroelectric power). *Source: Annual Energy Review 1991* (United States Energy Information Administration)

72. (a) Use the equation of the regression line in Exercise 71 to estimate what the world energy production might have been in 1975.
    (b) According to the United States Energy Information Administration, the world energy production for 1975 was 246.18 quadrillion Btu. Is your estimate in part (a) close to this? Is your estimate too high or too low?

73. (a) Draw a scatter diagram (i.e., plot the points) for the data in the following table.

x (year)	y (number of refugees)
1983	12083
1984	10285
1985	9350
1986	8713
1987	8606
1988	7818

Refugees admitted into the United States from Eastern Europe during the (fiscal) years 1983–1988.   *Source:*   U.S. State Department

   (b) In your scatter diagram, draw the line that you feel seems to fit the data best. Estimate the slope and the y-intercept of the line.

(c) Using the formulas in Exercise 41 of Section 4.1, it can be shown that the equation of the regression line is

$$f(x) = (-771.4)x + 1{,}541{,}090$$

Use this equation (and your calculator) to estimate (to the nearest 100) the number of refugees from Eastern Europe that will arrive in the United States in 1995, assuming that the current trend continues.
   (d) Find $f^{-1}(x)$.
   (e) Use your answer in part (d) to make a projection for the year when the number of refugees will dip below 1000.

74. (a) Let $f(x) = ax^2 + bx + c$ $(a \neq 0)$. Find a constant x_0 such that the following equation is an identity:

$$f(x_0 + x) = f(x_0 - x)$$

   (b) What is the geometric significance of your answer in part (a)?

75. In Section 4.2, we saw how the sign of the constant a influences the graph of the parabola $f(x) = ax^2 + bx + c$. In this exercise you'll see how the sign of b affects the graph.
    (a) On the same set of axes, graph the parabolas $y = x^2 + 4x + 1$ and $y = x^2 - 4x + 1$. What type of symmetry do you observe?
    (b) Let $f(x) = ax^2 + bx + c$. Compute $f(-x)$.
    (c) Use the result in part (b) to describe what happens to the graph of $f(x) = ax^2 + bx + c$ when the sign of b is reversed.

CHAPTER FOUR TEST

1. A linear function L satisfies the conditions $L(-2) = -4$ and $L(5) = 1$. Find $L(0)$.

2. (a) Let $F(x) = 4x - 2x^2$. On which interval is this function increasing? What is the maximum value for this function?
   (b) Let $G(t) = 9t^4 + 6t^2 + 2$. For which value of t is this function a minimum?

3. Let $f(x) = (x - 3)(x + 4)^2$.
   (a) Determine the intercepts and the excluded regions for the graph of f.
   (b) Determine the behavior of f when x is very close to -4.
   (c) Sketch the graph of f.

4. Graph each function: (a) $-1/(x - 3)^3$;   (b) $-1/(x - 3)^2$.

5. Graph the function $y = -x^2 + 7x + 6$. Specify the coordinates of the turning point, the intercepts, and the axis of symmetry.

6. Suppose that the function $p = -\frac{1}{6}x + 80$ $(0 \leq x \leq 400)$ relates the selling price of an item to the quantity x that is sold. Assume that p is in dollars. What is the maximum revenue possible in this situation? Which price p generates this maximum revenue?

7. The price of a new machine is $14,000. After ten years, the machine has a salvage value of $750. Assuming linear depreciation, find a formula for the value of the machine after t years, where $0 \leq t \leq 10$.

8. Graph the function $y = -\frac{1}{2}(3 - x)^2$. Specify the intercepts. *Hint:* First graph $y = -\frac{1}{2}(x + 3)^2$.

9. Graph the function $y = \dfrac{2x - 3}{x + 1}$. Specify the intercepts and asymptotes.

10. (a) Suppose that $P(x, y)$ is a point on the line $y = 3x - 1$ and Q is the point $(-1, 3)$. Express the length PQ as a function of x.
    (b) For which value of x is the length PQ a minimum?

11. Let $f(x) = \dfrac{x(x - 2)}{x^2 - 9}$

    (a) Find the vertical and the horizontal asymptotes.
    (b) Use a sketch to show the behavior of f near the x-intercepts.
    (c) Use a sketch to show the behavior of f near the asymptotes.
    (d) Sketch the graph of f.

12. A rectangle is inscribed in a semicircle of diameter 8 cm. (See the accompanying figure.) Express the area of the rectangle as a function of the width w of the rectangle.

13. Explain why each of the following graphs cannot represent a polynomial function with highest-degree term $-x^3$.

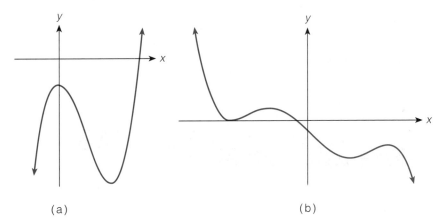

(a)   (b)

14. The following table shows the number of automobiles in use (worldwide) during the period from 1965 to 1980.

x (year)	y (automobiles in use) (millions)
1965	140
1970	195
1975	260
1980	321

Source: Lester R. Brown, project director, *State of the World, 1989* (New York: W. W. Norton and Company, 1989)

(a) Draw a scatter diagram (i.e., plot the points) for the data in the table. Then, draw the line that you feel seems to fit the data best.

(b) Using formulas from statistics, it can be shown that the equation for the regression line is $y = -23757 + 12.16x$. Use this equation to estimate the number of automobiles in use worldwide in the year 1985. (The actual estimate from the publication indicated above was 375 million. Is your estimate too high or too low?)

CHAPTER FIVE

EXPONENTIAL AND LOGARITHMIC FUNCTIONS

The exponential growth of population and its attendant assault on the environment is so recent that it is difficult for people to appreciate how much damage is being done.

Nathan Keyfitz in "The Growing Human Population," *Scientific American* **261** (September 1989), 119–126

... in 1958 ... C. D. Keeling of The Scripps Institution of Oceanography ... began a series of painstaking measurements of CO_2 concentration on a remote site at what is now the Mauna Loa Observatory in Hawaii. His observations, continued to the present, show an exponential growth of atmospheric carbon dioxide.... Keeling's observations have been duplicated at other stations in various parts of the world over shorter periods of time. In all sets of observations, the exponential increase is clear....

Gordon MacDonald in "Scientific Basis for the Greenhouse Effect," from *The Challenge of Global Warming,* Dean Edwin Abrahamson, ed. (Washington, D.C.: Island Press, 1989)

INTRODUCTION

In the previous chapter we saw that linear functions can be used to describe or *model* certain data sets. For instance, when we looked at the records for the one-mile run, we found that the times decreased in a linear fashion (see Figure 4 in Section 4.1). In this chapter we consider two other important types of functions that are often useful as models: the *exponential functions* and their inverses, the *logarithmic functions*. Figure 1 (on the next page) provides an example. Figure 1(a) shows the population of the United States over the years 1800–1900 along with a scatter plot of the data. The scatter plot makes it clear that the population did *not* increase in a linear fashion. Figure 1(b) shows the graph of a function that does fit this data very closely; this function is an exponential function. (The actual equation for this particular function is $y = ab^t$, where $a \approx 4.33 \times 10^{-15}$ and $b \approx 1.027$.)

As you will see in Sections 5.5 and 5.6, the applications of exponential functions are quite diverse and not restricted to population growth. For instance, in studying the greenhouse effect, scientists obtain graphs like the one in Figure 2. This graph shows the concentration of carbon dioxide in the atmosphere over the years 1800–1988. Again, the growth can be described or modeled very closely by an exponential function.

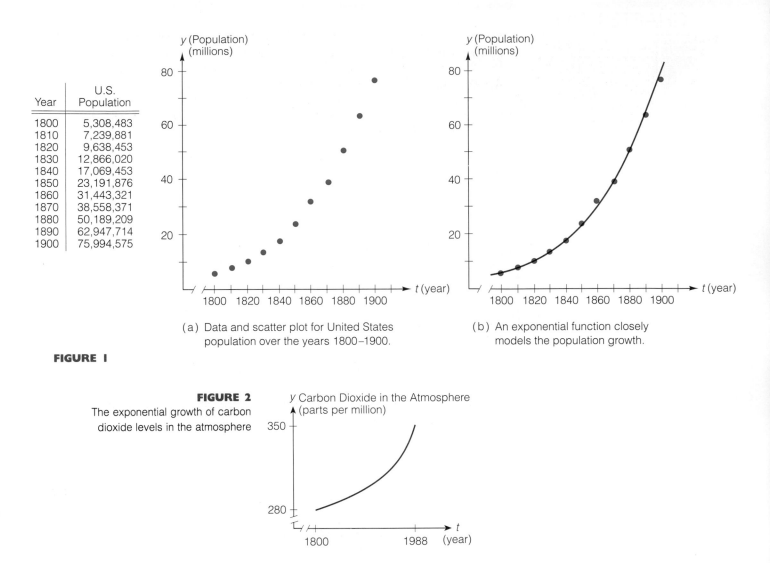

Year	U.S. Population
1800	5,308,483
1810	7,239,881
1820	9,638,453
1830	12,866,020
1840	17,069,453
1850	23,191,876
1860	31,443,321
1870	38,558,371
1880	50,189,209
1890	62,947,714
1900	75,994,575

(a) Data and scatter plot for United States population over the years 1800–1900.

(b) An exponential function closely models the population growth.

FIGURE 1

FIGURE 2

The exponential growth of carbon dioxide levels in the atmosphere

5.1   EXPONENTIAL FUNCTIONS

We begin with an example. Suppose that your mathematics instructor, in an effort to improve classroom attendance, offers to pay you each day for attending class! Suppose you are to receive 2¢ on the first day you attend class, 4¢ the second day, 8¢ the third day, and so on, as shown in Table 1. How much money will you receive for attending class on the 30th day?

As you can see by looking at Table 1, the amount y earned on day x is given by the rule, or *exponential function,*

$$y = 2^x$$

Thus on the 30th day (when $x = 30$), you will receive

$$y = 2^{30} \text{ cents}$$

TABLE 1

x (day number)	y (amount earned that day)
1	2¢ $(= 2^1)$
2	4¢ $(= 2^2)$
3	8¢ $(= 2^3)$
4	16¢ $(= 2^4)$
5	32¢ $(= 2^5)$
⋮	⋮
x	2^x

If you use a calculator, you will find this amount to be well over 10 million dollars! The point here is simply this: Although we begin with a small amount, $y = 2$¢, repeated doubling quickly leads to a very large amount. In other words, the exponential function grows very rapidly.

Before leaving this example, we mention a simple method for quickly estimating numbers such as 2^{30} (or any power of two) in terms of the more familiar powers of ten. Begin by observing that

$$2^{10} \approx 10^3 \qquad \text{(a useful coincidence, worth remembering)}$$

Now just cube both sides to obtain

$$(2^{10})^3 \approx (10^3)^3 \qquad \text{or} \qquad 2^{30} \approx 10^9$$

Thus 2^{30} is about one billion. To convert this number of cents to dollars, we divide by 100 or 10^2 to obtain

$$\frac{10^9}{10^2} = 10^7 \text{ dollars}$$

which is 10 million dollars, as mentioned before.

EXAMPLE 1 Estimate 2^{40} in terms of a power of 10.

Solution Take the basic approximation $2^{10} \approx 10^3$ and raise both sides to the fourth power. This yields

$$(2^{10})^4 \approx (10^3)^4$$

or

$$2^{40} \approx 10^{12}, \qquad \text{as required}$$

EXAMPLE 2 Estimate the power to which 10 must be raised to yield 2.

Solution We begin with our approximation

$$10^3 \approx 2^{10}$$

Raising both sides to the power $\frac{1}{10}$ yields

$$(10^3)^{1/10} \approx (2^{10})^{1/10}$$

and, consequently,

$$10^{3/10} \approx 2$$

Thus the power to which 10 must be raised to yield 2 is approximately $\frac{3}{10}$.

In previous algebra courses, you worked with the basic definitions and properties of exponents. As preparation for our discussion of exponential functions, we list those definitions and properties in the two boxes that follow. (If you need a more thorough review, see Appendices A−5 and A−6 at the back of this book.)

PROPERTY SUMMARY   DEFINITIONS FOR b^n

DEFINITIONS	EXAMPLES
1. If n is a natural number, then $b^n = \underbrace{b \cdot b \cdot b \cdot \;\cdots\; \cdot b}_{n \text{ factors}}$.	$4^3 = 4 \cdot 4 \cdot 4 = 64$
2. If $n = 0$ and $b \neq 0$, then $b^0 = 1$. (0^0 is undefined.)	$7^0 = 1$; $(x^2 + 3)^0 = 1$
3. If n is a natural number and $b \neq 0$, then $b^{-n} = \dfrac{1}{b^n}$.	$4^{-3} = \dfrac{1}{4^3} = \dfrac{1}{64}$; $x^{-2} = \dfrac{1}{x^2}$
4. Suppose n is a rational number, say, $n = p/q$ (reduced to lowest terms). Assume that q is positive and that $\sqrt[q]{b}$ is a real number. Then $b^{p/q} = \left(\sqrt[q]{b}\right)^p$ or, equivalently, $b^{p/q} = \sqrt[q]{b^p}$.	$8^{2/3} = \left(\sqrt[3]{8}\right)^2 = 2^2 = 4$ or, equivalently, $8^{2/3} = \sqrt[3]{8^2} = \sqrt[3]{64} = 4$

PROPERTY SUMMARY   PROPERTIES OF EXPONENTS

PROPERTY	EXAMPLES
1. $b^n b^m = b^{n+m}$	$b^5 b^6 = b^{11}$
2. $(b^m)^n = b^{mn}$	$(2^3)^4 = 2^{12}$; $\left((x+3)^3\right)^4 = (x+3)^{12}$
3. $\dfrac{b^m}{b^n} = b^{m-n}$	$\dfrac{b^6}{b^2} = b^4$; $\dfrac{x^2}{x^5} = x^{-3} = \dfrac{1}{x^3}$
4. $(ab)^n = a^n b^n$; $\left(\dfrac{a}{b}\right)^n = \dfrac{a^n}{b^n}$	$(2x^2)^3 = 2^3(x^2)^3 = 8x^6$; $\left(\dfrac{x^2}{y^3}\right)^4 = \dfrac{x^8}{y^{12}}$

The statements in the first box define b^n when n is a natural number, when n is an integer, and when n is a rational number. But what if the exponent n is an irrational number? Although a rigorous definition for irrational exponents requires concepts from calculus, we can nevertheless convey the basic idea by means of an example. (We need to do this before we give the general definition for exponential functions.)

How shall we assign a meaning to $2^{\sqrt{2}}$, for example? The basic idea is to evaluate the expression 2^x successively by using rational numbers x that are closer and closer to $\sqrt{2}$. Table 2 displays the results of some calculations along these lines.

TABLE 2

Values of 2^x for Rational Numbers x Approaching $\sqrt{2}\,(= 1.41421356\ldots)$

x	1.4	1.41	1.414	1.4142	1.41421	1.414213
2^x	$2.6\ldots$	$2.65\ldots$	$2.664\ldots$	$2.6651\ldots$	$2.66514\ldots$	$2.665143\ldots$

The data in the table suggest that as x approaches $\sqrt{2}$, the corresponding values of 2^x approach a unique real number, call it t, whose decimal expansion begins as 2.665. Furthermore, by continuing this process we could obtain (in theory, at least) as many places in the decimal expansion of t as we wished. The value of the expression $2^{\sqrt{2}}$ is then defined to be this number t. The following results (stated here without proof) summarize this discussion and also pave the way for the definition we will give for exponential functions.

PROPERTY SUMMARY   **REAL NUMBER EXPONENTS**

Let b denote an arbitrary positive real number. Then:

1. For each real number x, the quantity b^x is a unique real number.

2. When x is irrational, we can approximate b^x as closely as we wish by evaluating b^r, where r is a rational number sufficiently close to the number x.

3. The properties of rational exponents continue to hold for irrational exponents.

4. If $b^x = b^y$ and $b \neq 1$, then $x = y$.

▮▭▮ **EXAMPLE 3** Use the properties of exponents to simplify each expression: **(a)** $\left(3^{\sqrt{2}}\right)^{\sqrt{2}}$; **(b)** $\left(3^{\sqrt{2}}\right)^2$.

Solution

(a) $\left(3^{\sqrt{2}}\right)^{\sqrt{2}} = 3^{\sqrt{2} \times \sqrt{2}} = 3^2 = 9$

(b) $\left(3^{\sqrt{2}}\right)^2 = \left(3^2\right)^{\sqrt{2}}$ (Why?)

$\qquad\qquad\quad = 9^{\sqrt{2}}$ ▮▮▮

▮▭▮ **EXAMPLE 4** Solve the equation $4^x = 8$.

Solution First, let's estimate x, just to get a feeling for what kind of answer to expect. Since $4^1 = 4$, which is less than 8, and $4^2 = 16$, which is more than 8, we know that our final answer should be a number between 1 and 2. To obtain this answer, we take advantage of the fact that both 4 and 8 are powers of 2. Using this fact, we can write the given equation as

$$(2^2)^x = 2^3$$
$$2^{2x} = 2^3$$
$$2x = 3 \qquad \text{using Property 4 in the box above}$$
$$x = \frac{3}{2}$$

Note that the answer $x = \frac{3}{2}$ is indeed between 1 and 2, as we estimated.   ▮▮▮

For the remainder of this section, b denotes an arbitrary positive constant other than 1. In the box that follows, we define the exponential function with base b.

DEFINITION **The Exponential Function with Base *b***

EXAMPLES

Let *b* denote an arbitrary positive constant other than 1. The **exponential function with base *b*** is defined by the equation

$$y = b^x$$

Note In many scientific applications, functions of the form $y = ab^{kx}$ (where *a, b,* and *k* are constants) are also referred to as exponential functions.

1. The equations $y = 2^x$ and $y = 3^x$ define the exponential functions with bases 2 and 3, respectively.

2. The equation $y = (1/2)^x$ defines the exponential function with base 1/2.

3. The equations $y = x^2$ and $y = x^3$ do not define exponential functions.

To help with our analysis of exponential functions, let's set up a table and use it to graph the exponential function $y = 2^x$. This is done in Figure 1. In drawing a smooth and unbroken curve, we are actually relying on the results in the Property Summary on page 231. The key features of the exponential function $y = 2^x$ and its graph are:

1. The domain of $y = 2^x$ is the set of all real numbers. The range is the set of all *positive* real numbers.
2. The *y*-intercept of the graph is 1. The graph has no *x*-intercept.
3. For $x > 0$, the function increases or grows very rapidly. For $x < 0$, the graph rapidly approaches the *x*-axis; the *x*-axis is a horizontal asymptote for the graph. (Recall from Section 4.6 that a line is an **asymptote** for a curve if the separation distance between the curve and the line approaches zero as we move out farther and farther along the line.)

You should memorize the basic shape and features of the graph of $y = 2^x$ so that you can sketch it as needed without first setting up a table. The next example shows why this is useful.

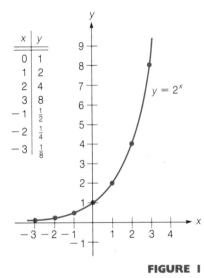

x	y
0	1
1	2
2	4
3	8
−1	$\frac{1}{2}$
−2	$\frac{1}{4}$
−3	$\frac{1}{8}$

FIGURE 1

EXAMPLE 5 Graph each of the following functions. In each case specify the domain, range, intercept(s), and asymptote.

(a) $y = -2^x$ **(b)** $y = 2^{-x}$ **(c)** $y = \left(\frac{1}{2}\right)^x$ **(d)** $y = 2^{-x} - 2$

Solution

(a) Recall that -2^x means $-(2^x)$, not $(-2)^x$. The graph of $y = -2^x$ is obtained by reflecting the graph of $y = 2^x$ in the *x*-axis. See Figure 2(a).

(b) Similarly, the graph of $y = 2^{-x}$ is obtained from the graph of $y = 2^x$ by reflection in the *y*-axis. See Figure 2(b).

(c) Next, regarding $y = \left(\frac{1}{2}\right)^x$, observe that

$$\left(\frac{1}{2}\right)^x = \frac{1^x}{2^x} = \frac{1}{2^x} = 2^{-x}$$

In other words, $y = \left(\frac{1}{2}\right)^x$ is really the same function as $y = 2^{-x}$, which we already graphed in Figure 2(b).

FIGURE 2

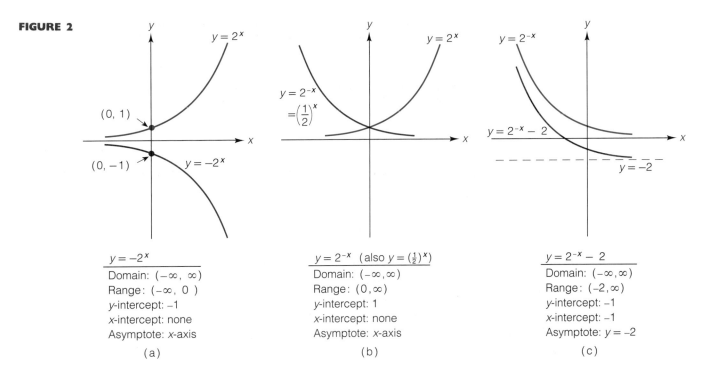

$y = -2^x$
Domain: $(-\infty, \infty)$
Range: $(-\infty, 0)$
y-intercept: -1
x-intercept: none
Asymptote: x-axis
(a)

$y = 2^{-x}$ (also $y = (\frac{1}{2})^x$)
Domain: $(-\infty, \infty)$
Range: $(0, \infty)$
y-intercept: 1
x-intercept: none
Asymptote: x-axis
(b)

$y = 2^{-x} - 2$
Domain: $(-\infty, \infty)$
Range: $(-2, \infty)$
y-intercept: -1
x-intercept: -1
Asymptote: $y = -2$
(c)

(d) Finally, to graph $y = 2^{-x} - 2$, take the graph of $y = 2^{-x}$ in Figure 2(b) and move it two units in the negative y-direction, as shown in Figure 2(c). Note that the asymptote and y-intercept will also move down two units. To find the x-intercept, we set $y = 0$ in the given equation to obtain

$$2^{-x} - 2 = 0$$
$$2^{-x} = 2^1$$
$$-x = 1 \qquad \text{using Property 4 on page 231}$$
$$x = -1$$

Thus, the x-intercept is -1.   ∎

In the next example, we apply our knowledge about the graph of $y = 2^x$ to solve an equation. In particular, we use the fact that the graph of $y = 2^x$ always lies above the x-axis; for no value of x is 2^x ever zero.

EXAMPLE 6 Solve the equation $x^2 2^x - 2^x = 0$.

Solution First, we factor the left-hand side of the equation; the common term is 2^x. This gives us

$$2^x(x^2 - 1) = 0$$

Since 2^x is never zero, we may now divide both sides of this last equation by 2^x (without losing a solution) to obtain

$$x^2 - 1 = 0$$
$$(x - 1)(x + 1) = 0$$

and, consequently,

$$x = 1 \quad \text{or} \quad x = -1$$

Thus, the required solutions are $x = 1$, $x = -1$. (You should check for yourself that each of these values satisfies the original equation.) ▪▪▪

Now what about exponential functions with bases other than 2? As Figure 3 indicates, the graphs are similar to $y = 2^x$. The graph of $y = 4^x$ rises more rapidly than $y = 2^x$ when x is positive. For negative x-values, the graph of $y = 4^x$ is below that of $y = 2^x$. You can see why this happens by taking $x = -1$, for example, and comparing the values of 4^x and 2^x. If $x = -1$, then

$$2^x = 2^{-1} = \frac{1}{2} \quad \text{but} \quad 4^x = 4^{-1} = \frac{1}{4}$$

Therefore $4^x < 2^x$ when $x = -1$. Notice also in Figure 3 that all three graphs have the same y-intercept of 1. This follows from the fact that $b^0 = 1$ for any positive number b.

FIGURE 3

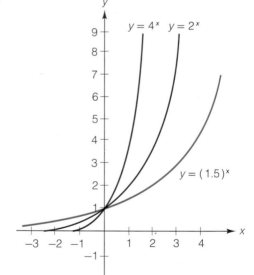

The exponential functions in Figure 3 all have bases larger than 1. To see examples in which the bases are in the interval $0 < b < 1$, we need only reflect the graphs in Figure 3 in the y-axis. For instance, the reflection of $y = 4^x$ in the y-axis gives us the graph of $y = \left(\frac{1}{4}\right)^x$. (We discussed the idea behind this in Example 5; see Figure 2(b), for instance.)

In the box that follows, we summarize what we've learned up to this point regarding the exponential function $y = b^x$.

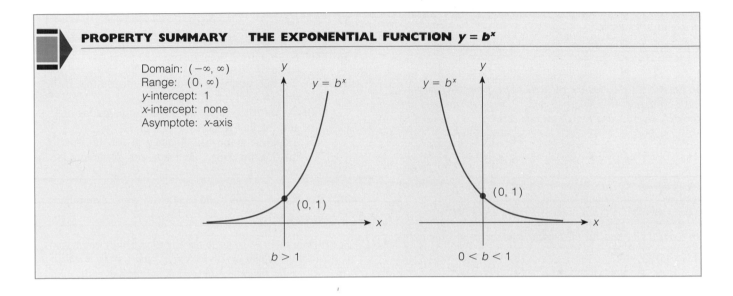

PROPERTY SUMMARY   THE EXPONENTIAL FUNCTION $y = b^x$

Domain: $(-\infty, \infty)$
Range: $(0, \infty)$
y-intercept: 1
x-intercept: none
Asymptote: x-axis

$y = b^x$ $(0, 1)$ $b > 1$

$y = b^x$ $(0, 1)$ $0 < b < 1$

EXAMPLE 7   Graph the function $y = -3^{-x} + 1$.

Solution   The required graph is obtained by reflecting and translating the graph of $y = 3^x$, as shown in Figure 4.

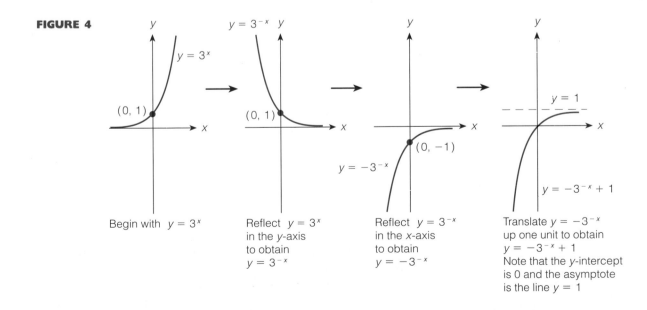

FIGURE 4

$y = 3^x$ $(0, 1)$

Begin with $y = 3^x$

$y = 3^{-x}$ $(0, 1)$

Reflect $y = 3^x$
in the y-axis
to obtain
$y = 3^{-x}$

$(0, -1)$ $y = -3^{-x}$

Reflect $y = 3^{-x}$
in the x-axis
to obtain
$y = -3^{-x}$

$y = 1$ $y = -3^{-x} + 1$

Translate $y = -3^{-x}$
up one unit to obtain
$y = -3^{-x} + 1$
Note that the y-intercept
is 0 and the asymptote
is the line $y = 1$

EXERCISE SET 5.1

A

In Exercises 1 and 2, estimate each quantity in terms of powers of 10, as in Example 1.

1. (a) 2^{30} (b) 2^{50}
2. (a) 2^{90} (b) 4^{20}

In Exercises 3–10, use the properties of exponents to simplify each expression. In Exercises 9 and 10, write the answers in the form b^n, where b and n are real numbers.

3. $\left(5^{\sqrt{3}}\right)^{\sqrt{3}}$

4. $\left(\sqrt{2}^{\sqrt{2}}\right)^{\sqrt{2}}$

5. $\left(4^{1+\sqrt{2}}\right)\left(4^{1-\sqrt{2}}\right)$

6. $\left(3^{2+\sqrt{5}}\right)\left(3^{2-\sqrt{5}}\right)$

7. $\dfrac{2^{4+\pi}}{2^{1+\pi}}$

8. $\dfrac{10^{\pi+2}}{10^{\pi-2}}$

9. $\left(\sqrt{5}^{\sqrt{2}}\right)^{2}$

10. $\left(\sqrt{3}^{\pi}\right)^{4}$

In Exercises 11 and 12, solve each equation, as in Example 4.

11. (a) $3^x = 27$ (b) $9^t = 27$
    (c) $3^{1-2y} = \sqrt{3}$ (d) $3^z = 9\sqrt{3}$

12. (a) $2^x = 32$ (b) $2^t = \frac{1}{4}$
    (c) $2^{3y+1} = \sqrt{2}$ (d) $8^{z+1} = 32\sqrt{2}$

In Exercises 13–16, specify the domain of the function.

13. $y = 2^x$

14. $y = \dfrac{1}{2^x}$

15. $y = \dfrac{1}{2^x - 1}$

16. $y = \dfrac{1}{2^x - 1}$

In Exercises 17–24, graph the pair of functions on the same set of axes.

17. $y = 2^x,\ y = 2^{-x}$ 18. $y = 3^x,\ y = 3^{-x}$
19. $y = 3^x,\ y = -3^x$ 20. $y = 4^x,\ y = -4^x$
21. $y = 2^x,\ y = 3^x$ 22. $y = \left(\frac{1}{3}\right)^x,\ y = 3^x$
23. $y = \left(\frac{1}{2}\right)^x,\ y = \left(\frac{1}{3}\right)^x$ 24. $y = \left(\frac{1}{2}\right)^{-x},\ y = \left(\frac{1}{3}\right)^{-x}$

For Exercises 25–32, graph the function and specify the domain, range, intercept(s), and asymptote.

25. $y = -2^x + 1$ 26. $y = -3^x + 3$
27. $y = 3^{-x} + 1$ 28. $y = 3^{-x} - 3$
29. $y = 2^{x-1}$ 30. $y = 2^{x-1} - 1$
31. $y = 3^{x+1} + 1$ 32. $y = 1 - 3^{x-1}$

For Exercises 33–36, solve the equation.

33. $3x(10^x) + 10^x = 0$ 34. $4x^2(2^x) - 9(2^x) = 0$
35. $3(3^x) - 5x(3^x) + 2x^2(3^x) = 0$
36. $\dfrac{(x+4)10^x}{x-3} = 2x(10^x)$

B

37. Let $f(x) = 2^x$. Show that $\dfrac{f(x+h) - f(x)}{h} = 2^x\left(\dfrac{2^h - 1}{h}\right)$.

38. Let $\phi(t) = 1 + a^t$. Show that $\dfrac{1}{\phi(t)} + \dfrac{1}{\phi(-t)} = 1$.

39. Let $f(x) = 2^x$ and let g denote the function that is the inverse of f.
    (a) On the same set of axes, sketch the graphs of f, g, and the line $y = x$.
    (b) Using the graph you obtained in part (a), specify the domain, range, intercept, and asymptote for the function g.

40. Let $S(x) = \dfrac{2^x - 2^{-x}}{2}$ and $C(x) = \dfrac{2^x + 2^{-x}}{2}$. Compute $[C(x)]^2 - [S(x)]^2$.

For Exercises 41–48, refer to the following figure, which shows portions of the graphs of $y = 2^x$, $y = 3^x$, and $y = 5^x$. In each case, (a) use the figure to estimate the indicated quantity, and (b) use a calculator to compute the indicated quantity, rounding off the result to two decimal places.

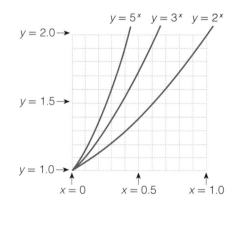

41. $\sqrt{2}$ Hint: $\sqrt{2} = 2^{\frac{1}{2}}$ 42. $\sqrt[5]{2}$
43. $2^{\frac{3}{5}}$ 44. $\sqrt[5]{8}$ Hint: $8^{\frac{1}{5}} = 2^{?}$
45. $\sqrt{3}$ 46. $\sqrt[3]{3}$
47. $5^{\frac{3}{10}}$ 48. $\sqrt[4]{5}$

In Exercises 49–52, refer to the following graph of the exponential function $y = 10^x$. Use the graph to estimate (to the nearest tenth) the solution of each equation. (After we've studied logarithmic functions later in the chapter, we will be able to obtain more precise solutions.)

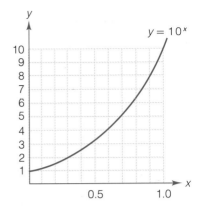

y = 10^x

49. $10^x = 2$

50. $10^x = 4$

51. $10^x = 5$

52. $10^x = 8$

C

53. This exercise serves as a preview for the work on logarithms in Section 5.3. Follow steps (a)–(f) to complete the table. (Notice that one entry in the table is already filled in. Reread Example 2 in the text to see how that entry was obtained.)
   (a) Fill in the entries in the right-hand column corresponding to $x = 1$ and $x = 10$.
   (b) Note that 4 and 8 are powers of 2. Use this information along with the approximation $10^{0.3} \approx 2$ to find the entries in the table corresponding to $x = 4$ and $x = 8$.
   (c) Find the entry corresponding to $x = 5$.

   Hint: $5 = \dfrac{10}{2} \approx \dfrac{10}{10^{0.3}}$.

(d) Find the entry corresponding to $x = 7$.
   Hint: $7^2 \approx 50 = 5 \times 10$; now make use of your answer in part (c).
(e) Find the entry corresponding to $x = 3$.
   Hint: $3^4 \approx 80 = 8 \times 10$.
(f) Find the entries corresponding to $x = 6$ and $x = 9$.
   Hint: $6 = 3 \times 2$ and $9 = 3^2$.

Remark This table is called a *table of logarithms to the base 10*. We say, for example, that the logarithm of 2 to the base 10 is (about) 0.3. We write this symbolically as $\log_{10} 2 \approx 0.3$.

x	Exponent to Which 10 Must Be Raised to Yield x
1	
2	≈ 0.3
3	
4	
5	
6	
7	
8	
9	
10	

5.2 THE EXPONENTIAL FUNCTION $y = e^x$

This is undoubtedly the most important function in mathematics.

Walter Rudin in *Real and Complex Analysis* (New York: McGraw-Hill, 1966)

From the standpoint of calculus and scientific applications, one particular base for exponential functions is by far the most useful. This base is a certain irrational number that lies between 2 and 3 and is denoted by the letter e. For purposes of approximation, you'll need to know that

$$e \approx 2.7$$

[The Swiss mathematician Leonhard Euler (1707–1783) introduced the letter e to denote this number. To six decimal places, the value of e is 2.718281....]

At the precalculus level, it's hard to escape the feeling that $y = 2^x$ or $y = 10^x$ is by far more simple and more natural than $y = e^x$. So as we work with

TABLE I

Values of $\left(1 + \dfrac{1}{x}\right)^{x}$

x	10	100	1000	10^4	10^5	10^6
$(1 + 1/x)^x$	2.5937424...	2.7048138...	2.7169239...	2.7181459...	2.7182682	2.7182804...

e and e^x in this chapter, you will need to take it on faith that, in the long run, the number e and the function $y = e^x$ make life simpler, not more complex.

There are several different ways that the number e can be defined. One way involves investigating the values of the expression

$$\left(1 + \frac{1}{x}\right)^{x}$$

as x becomes larger and larger. See Table 1.

The data in Table 1 suggest that the quantity $\left(1 + \dfrac{1}{x}\right)^{x}$ gets closer and closer to e as x becomes larger and larger. Indeed, in many calculus books, the number e is defined as the *limiting value* or *limit* of the quantity $\left(1 + \dfrac{1}{x}\right)^{x}$ as x grows ever larger. Now, admittedly, we have not defined here the meaning of *limiting value* or *limit*—that is a topic for calculus. Nevertheless, Table 1 should give you a reasonable, if intuitive, appreciation of the idea. In Figure 1, we give a graphical interpretation.

FIGURE I

The line $y = e$ is an asymptote for the graph of $y = \left(1 + \dfrac{1}{x}\right)^{x}$, $x > 0$.

As x increases without bound, the values of $\left(1 + \dfrac{1}{x}\right)^{x}$ approach the number e.

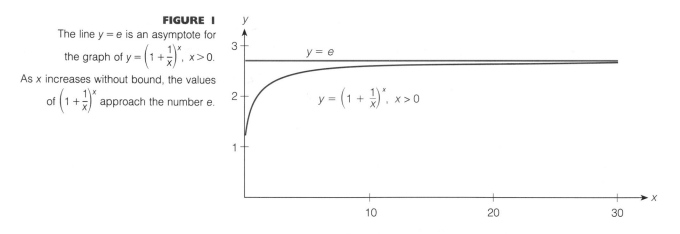

One more comment about Table 1 and the number e: In Section 5.5, you'll see that the data in Table 1 can be interpreted in terms of banking and compound interest. For this reason, the constant e is sometimes called the "banker's constant."

In calculus the number e is sometimes introduced in a way that involves slopes of lines. You know that the graph of each exponential function $y = b^x$ passes through the point $(0, 1)$. Figure 2(a) shows the exponential function $y = 2^x$ along with a line that is tangent to the curve at the point $(0, 1)$. By carefully measuring rise and run, it can be shown that the slope of this tangent line

FIGURE 2

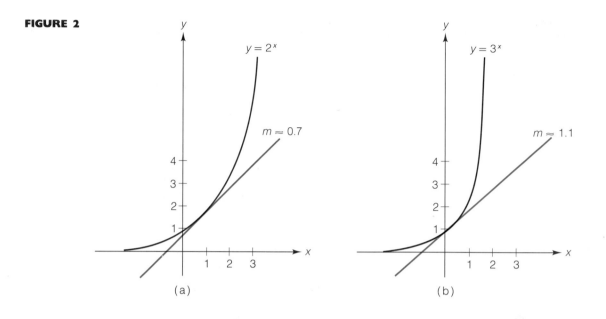

(a)   (b)

is about 0.7. Figure 2(b) shows a similar situation with the curve $y = 3^x$. Here the slope of the tangent line through $(0, 1)$ is approximately 1.1. Now, since the slope of the tangent to $y = 2^x$ is a bit less than 1, while that for $y = 3^x$ is a bit more than 1, it seems reasonable to suppose that there is a number between 2 and 3, call it e, with the property that the slope of the tangent through $(0, 1)$ is exactly 1. See Figure 3 and the summary box that follows.

Remark   It's certainly not obvious that this definition for e is equivalent to the definition involving the expression $\left(1 + \dfrac{1}{x}\right)^x$. The methods of calculus are required to demonstrate that the definitions are indeed equivalent. (For an informal justification, however, see Exercise 42.)

FIGURE 3

The slope of the tangent to the curve
$y = e^x$ at the point $(0, 1)$ is $m = 1$.

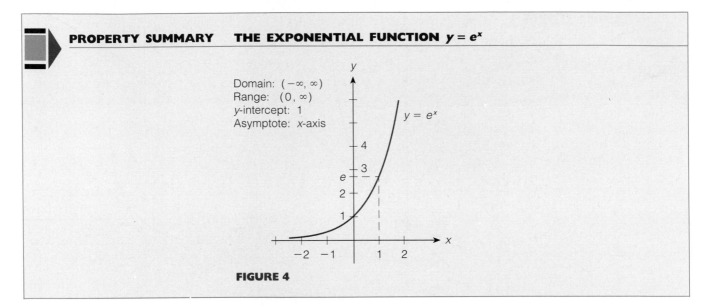

PROPERTY SUMMARY   THE EXPONENTIAL FUNCTION $y = e^x$

Domain: $(-\infty, \infty)$
Range: $(0, \infty)$
y-intercept: 1
Asymptote: x-axis

$y = e^x$

FIGURE 4

▮▮▮ **EXAMPLE** ▮ Graph each of the following functions, specifying the domain, range, intercept, and asymptote: **(a)** $y = e^{x-1}$;   **(b)** $y = -e^{x-1}$;   **(c)** $y = -e^{x-1} + 1$.

Solution

(a) We begin with the graph of $y = e^x$ in Figure 4. Moving the graph to the right one unit yields the graph of $y = e^{x-1}$, shown in Figure 5(a). The x-axis is still an asymptote for this translated graph, but the y-intercept will no longer be 1. To find the y-intercept, we replace x with 0 in the given equation to obtain

$$y = e^{-1} = \frac{1}{e} \approx 0.37 \qquad \text{using a calculator}$$

FIGURE 5

Domain: $(-\infty, \infty)$
Range: $(0, \infty)$
y-intercept: 1/e
Asymptote: x-axis

Domain: $(-\infty, \infty)$
Range: $(-\infty, 0)$
y-intercept: $-1/e$
Asymptote: x-axis

Domain: $(-\infty, \infty)$
Range: $(-\infty, 1)$
x-intercept: 1
y-intercept: $-1/e + 1$
Asymptote: $y = 1$

$y = e^{x-1}$

$y = -e^{x-1}$

$y = 1$

$y = -e^{x-1} + 1$

(a)

(b)

(c)

(b) Reflecting the graph from part (a) in the x-axis yields the graph of $y = -e^{x-1}$ [see Figure 5(b)]. Note that under this reflection, the y-intercept moves from $1/e$ to $-1/e$.

(c) Translating the graph in Figure 5(b) up one unit produces the graph of $y = -e^{x-1} + 1$, shown in Figure 5(c). Under this translation, the asymptote moves from $y = 0$ (the x-axis) to $y = 1$. Also, the y-intercept moves from $-1/e$ to $-1/e + 1$ (≈ 0.63). The x-intercept in Figure 5(c) is obtained by setting $y = 0$ in the equation $y = -e^{x-1} + 1$. This yields

$$0 = -e^{x-1} + 1$$
$$e^{x-1} = 1 = e^0$$
$$x - 1 = 0 \qquad \text{(Why?)}$$
$$x = 1$$

One of the characteristic features of exponential functions is their rapid rate of growth. You will see instances of this in the next example, in the exercises, and in the applications sections later in this chapter.

EXAMPLE 2 Let $f(x) = x^2$, $g(x) = 2^x$, and $h(x) = e^x$. Compute and compare the average rates of change for these functions on the following intervals: **(a)** $[0, 1]$; **(b)** $[9, 10]$.

Solution

(a) On the interval $[0, 1]$:

$$\frac{\Delta f}{\Delta x} = \frac{f(1) - f(0)}{1 - 0} = \frac{1^2 - 0^2}{1 - 0} = 1$$

$$\frac{\Delta g}{\Delta x} = \frac{g(1) - g(0)}{1 - 0} = \frac{2^1 - 2^0}{1 - 0} = 1$$

$$\frac{\Delta h}{\Delta x} = \frac{h(1) - h(0)}{1 - 0} = \frac{e^1 - e^0}{1 - 0} \approx 1.7$$

On the interval $[0, 1]$, the average rate of change for the functions $f(x) = x^2$ and $g(x) = 2^x$ is 1. The average rate of change for $h(x) = e^x$ on this interval is slightly larger; it is approximately 1.7.

(b) On the interval $[9, 10]$:

$$\frac{\Delta f}{\Delta x} = \frac{f(10) - f(9)}{10 - 9} = \frac{100 - 81}{1} = 19$$

$$\frac{\Delta g}{\Delta x} = \frac{g(10) - g(9)}{10 - 9} = \frac{2^{10} - 2^9}{1} = 512$$

$$\frac{\Delta h}{\Delta x} = \frac{h(10) - h(9)}{10 - 9} = \frac{e^{10} - e^9}{1} \approx 13{,}923 \qquad \text{using a calculator}$$

On the interval $[9, 10]$, the average rate of change for $g(x) = 2^x$ is more than 25 times as great as that for $f(x) = x^2$. The average rate of change for $h(x) = e^x$ is a whopping 733 times greater than that for $f(x) = x^2$. (Check these last two statements with your calculator.)

EXERCISE SET 5.2

A

In Exercises 1–12, graph the function and specify the domain, range, intercept(s), and asymptote.

1. $y = e^x$
2. $y = e^{-x}$
3. $y = -e^x$
4. $y = -e^{-x}$
5. $y = e^x + 1$
6. $y = e^{x+1}$
7. $y = e^{x+1} + 1$
8. $y = e^{x-1} - 1$
9. $y = -e^{x-2}$
10. $y = -e^{x-2} - 2$
11. $y = e - e^x$
12. $y = e^{-x} - e$

13. On the same set of axes, graph the functions $y = 2^x$, $y = e^x$, and $y = 3^x$.

14. On the same set of axes, graph the functions $y = 2^{-x}$, $y = e^{-x}$, and $y = 3^{-x}$.

In Exercises 15–22, answer true or false. You do not need a calculator for these exercises. Rather, use the fact that e is approximately 2.7.

15. $e < \frac{1}{2}$
16. $e < \frac{5}{2}$
17. $\sqrt{e} < 1$
18. $e^2 < 4$
19. $e^2 < 9$
20. $e^3 < 27$
21. $e^{-1} < 0$
22. $e^0 = 1$

In Exercises 23–30, refer to the following graph of $y = e^x$ $(-2 \le x \le 1)$. In each case, use the graph to estimate (to the nearest tenth) the indicated quantity. For Exercises 23–28, use a calculator to obtain a second estimate. Round off the calculator value to two decimal places.

23. $e^{0.2}$
24. $e^{0.6}$
25. $e^{4/5}$
26. e^{-1}
27. $\sqrt{e}$
28. $e^{7/10}$
29. The value of x for which $e^x = 0.8$
30. The value of x for which $e^x = 0.2$
31. Let $f(x) = x^4$, $g(x) = 2^x$, and $h(x) = e^x$. Compute and compare (as in Example 2) the average rates of change for these functions on the following intervals:
    (a) [2, 3];   (b) [8, 9];   (c) [14, 15].
32. Let $f(x) = x^6$, $g(x) = 2^x$, and $h(x) = e^x$. Compute and compare (as in Example 2) the average rates of change for these functions on the following intervals:
    (a) [1, 2];   (b) [16, 17];   (c) [27, 28].
33. The following figure shows portions of the graphs of the exponential functions $y = e^x$ and $y = \pi^x$.
    (a) Use the figure (not your calculator) to determine which number is larger, e^π or π^e.
    (b) Use your calculator to check your answer in part (a).

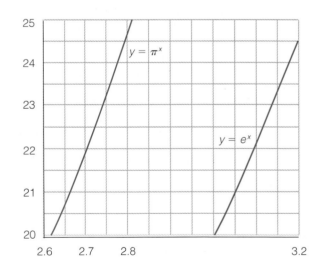

34. (a) Use your calculator to approximate the numbers $e\pi$ and $e + \pi$.
    Remark   It is not known whether these numbers are rational or irrational.
    (b) Use your calculator to approximate $\left(e^{\pi\sqrt{163}} - 744\right)^{1/3}$
    Remark   Contrary to the empirical evidence on your calculator, it is known that this number is irrational.

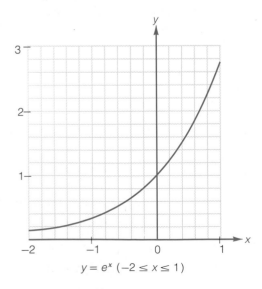

$y = e^x \ (-2 \le x \le 1)$

B

35. If $E(x) = e^x$, show that

$$\frac{E(x+h) - E(x)}{h} = e^x\left(\frac{e^h - 1}{h}\right)$$

36. The **hyperbolic sine** function S is defined by

$$S(x) = \frac{e^x - e^{-x}}{2}$$

(a) What is the domain of S?
(b) Find the y-intercept for the graph of S.
(c) Show that $S(-x) = -S(x)$. What does this say about the graph of S?
(d) Sketch the graph of S for $x \geq 0$. (Use your calculator to set up a table of values.) Then use the result in part (c) to complete the graph.

37. The **hyperbolic cosine** function C is defined by

$$C(x) = \frac{e^x + e^{-x}}{2}$$

(a) What is the domain of C?
(b) Find the y-intercept for the graph of C.
(c) Show that $C(-x) = C(x)$. What does this say about the graph of C?
(d) Sketch the graph of C for $x \geq 0$. (Use your calculator to set up a table of values.) Then use the result in part (c) to complete the graph.

38. Refer to the graph of $y = e^x$ in Figure 3.
(a) Show that the equation of the tangent line in Figure 3 is $y = x + 1$.
(b) When x is close to zero, e^x can be approximated by the quantity $x + 1$. (This is because the tangent line and the curve are virtually indistinguishable in the immediate vicinity of $x = 0$.) Complete the following tables to see just how good this approximation is.

x	0.3	0.2	0.1	0.01	0.001	0.0001
$x + 1$						
e^x						

x	-0.3	-0.2	-0.1	-0.01	-0.001	-0.0001
$x + 1$						
e^x						

39. (a) Use a calculator to complete the following table. Round off the answers to six decimal places.

x	1000	10^4	10^5	10^6	10^7
$(1 + 2/x)^x$					

(b) Use a calculator to compute e^2. What do you observe?

40. (a) Use a calculator to complete the following table. Round off the answers to six decimal places.

x	1000	10^4	10^5	10^6	10^7
$(1 + 3/x)^x$					

(b) Use a calculator to compute e^3. What do you observe?

41. Let $f(x) = e^x$. Let L denote the function that is the inverse of f.
(a) On the same set of axes, sketch the graphs of f and L.   *Hint:* You do not need the equation for $L(x)$.
(b) Specify the domain, range, intercept, and asymptote for the function L and its graph.
(c) Graph each of the following functions. Specify the intercept and asymptote in each case.
(i) $y = -L(x)$
(ii) $y = L(-x)$
(iii) $y = L(x - 1)$

42. The text showed two distinct ways to define the number e. One involved the expression $(1 + 1/x)^x$ and the other involved tangent lines. In this exercise you'll see that these two approaches are, in fact, related. For convenience, we'll write the expression $(1 + 1/x)^x$ using the letter n rather than x (so we can use x for something else in a moment). Then we have

$$\left(1 + \frac{1}{n}\right)^n \approx e \qquad \begin{array}{l}\text{as } n \text{ becomes larger and}\\ \text{larger, without bound}\end{array} \qquad (1)$$

(a) Define x by the equation $x = 1/n$. As n becomes larger and larger without bound, what happens to x? Complete the following table.

n	100	1000	10^6	10^9
x				

(b) Substitute $x = 1/n$ in approximation (1) to obtain

$$(1 + x)^{1/x} \approx e \qquad \text{as } x \text{ approaches zero}$$

Next, raise both sides to the power x to obtain

$$(1 + x) \approx e^x \qquad \text{as } x \text{ approaches zero} \qquad (2)$$

Complete the following table to see just how well the quantity $1 + x$ approximates e^x as x approaches zero.

x	$1 + x$	e^x	$e^x - (1 + x)$
0.1			
0.01			
0.001			

This suggests (but does not *prove*) that the line $y = 1 + x$ is tangent to the curve $y = e^x$ at $x = 0$.

C

43. Let S and C denote the functions defined in Exercises 36 and 37, respectively. Prove each of the following identities. [The identities in parts (b) and (c) have already appeared in Exercises 36 and 37, but they are included here for the sake of completeness.]
    (a) $[C(x)]^2 - [S(x)]^2 = 1$
    (b) $S(-x) = -S(x)$
    (c) $C(-x) = C(x)$
    (d) $S(x + y) = S(x)C(y) + C(x)S(y)$
    (e) $C(x + y) = C(x)C(y) + S(x)S(y)$
    (f) $S(2x) = 2S(x)C(x)$
    (g) $C(2x) = [C(x)]^2 + [S(x)]^2$

OPTIONAL TI-81 GRAPHING CALCULATOR EXERCISES

EXERCISES FOR SECTIONS 5.1 AND 5.2

1. As preparation, press ZOOM 6 and then make the following adjustments in the RANGE menu: $x_{min} = -2$, $x_{max} = 3$, $y_{min} = -1$, $y_{max} = 20$, and $y_{scl} = 2$.
   (a) Graph the functions $Y_1 = 2^x$, $Y_2 = 3^x$, and $Y_3 = 4^x$ on the same set of axes.
   (b) Where do you think the graph of $Y_4 = 75^x$ would fit into the picture that you obtained in part (a)? Add the graph of this function to the picture and see if you are right.

Instructions for Exercises 2–7:
(a) Without using the calculator, determine the intercepts and asymptotes for the graph of each function.
(b) Use the calculator to graph the function. (Choose appropriate RANGE settings or ZOOM settings so that the essential features of the graph are clear.) Then check to see if your answers in part (a) are consistent with the graph.

2. $y = -3^{x-2}$
3. $y = -3^{x-2} + 1$
4. $y = 4^{-x} - 4$
5. $y = 4 + \left(\frac{1}{4}\right)^x$
6. $y = 10^{x-1}$
7. $y = -10^{x-1} + 0.5$

8. If you add two quadratic functions, the result is again a quadratic (assuming that the x^2-terms don't add to zero). For example, if $f(x) = x^2 + 2x$ and $g(x) = 2x^2 - 1$, then the sum is $(f + g)(x) = 3x^2 + 2x - 1$, another quadratic. But in general, this is not the case with exponential functions. To explore a visual example, consider the function given by $y = 2^x + 2^{-x}$. (This is the sum of the two exponential functions 2^x and $\left(\frac{1}{2}\right)^x$.)
   (a) Graph the function $Y_1 = 2^x + 2^{-x}$. (Use ZOOM 6.) As you can see, the resulting graph is U-shaped.
   (b) Now add the graph of the function $Y_2 = x^2$ to your picture. Note (at this scale) how similar the graphs appear.
   (c) Actually, the graph of $Y_1 = 2^x + 2^{-x}$ rises much more steeply than $Y_2 = x^2$. To see this, make the following adjustments in the RANGE window and then press GRAPH: $y_{min} = 0$, $y_{max} = 100$, and $y_{scl} = 10$.

9. Follow parts (a) and (b) in Problem 8, but use the two functions $Y_1 = 2^x - 2^{-x}$ and $Y_2 = x^3$.

10. For this problem, first reread Example 2 in Section 5.1. (The example shows you how to find the approximate value of x for which $10^x = 2$.)
    (a) On the same set of axes, graph the two functions $Y_1 = 10^x$ and $y = 2$. (Use the following settings in the RANGE window: $x_{min} = 0$, $x_{max} = 1$, $x_{scl} = 0.1$, $y_{min} = 0$, $y_{max} = 10$, and $y_{scl} = 1$.
    (b) Explain how your graph in part (a) supports the answer $\frac{3}{10}$ that was obtained in Section 5.1.

11. For comparison, draw the graphs of 2^x, e^x, and 3^x on the same set of axes. Begin with the standard viewing rectangle. Then, to see the graphs more clearly when x is negative, use the following settings in the RANGE menu: $x_{min} = -1$, $x_{max} = 0$, $y_{min} = 0$, $y_{max} = 1$. To compare the graphs for positive values of x, use $x_{min} = 0$, $x_{max} = 2$, $y_{min} = 0$, $y_{max} = 10$. Note that the graph of e^x is bounded between the graphs of 2^x and 3^x, just as the number e is between 2 and 3. Also note in particular that we have

$$2^x < e^x < 3^x \qquad \text{when } x > 0$$

but

$$2^x > e^x > 3^x \qquad \text{when } x < 0$$

12. According to the text (p. 239), the line $y = x + 1$ is tangent to the graph of $y = e^x$ at the point $(0, 1)$. Confirm this visually by graphing the curve and the line on the same set of axes.

In Problems 13–18, first tell what translations or reflections are required to obtain the graph of the given function from that of $y = e^x$. Then graph the given function along with $y = e^x$, and check that the picture is consistent with your ideas.

13. $y = e^{-x}$

14. $y = -e^x$

15. $y = -e^{-x}$

16. (a) $y = e^x - 1$
(b) $y = e^{-x} - 1$

17. (a) $y = e^{x-1}$
(b) $y = e^{-x-1}$

18. (a) $y = e^{x-1} - 1$
(b) $y = e^{-x-1} - 1$

In Section 5.2, we saw that the number e can be defined as the "limit" of the expression $(1 + 1/x)^x$ as x becomes indefinitely large. We write this fact symbolically as follows:

$$\lim_{x \to \infty}\left(1 + \frac{1}{x}\right)^x = e$$

Actually, this is a particular instance of the following more general result: For each real number a, the limiting value (as x becomes indefinitely large) of $\left(1 + \dfrac{a}{x}\right)^x$ is e^a. In symbols,

$$\lim_{x \to \infty}\left(1 + \frac{a}{x}\right)^x = e^a.$$ *In Exercises 19–21, you'll use the calculator to draw graphs that illustrate this result.*

19. (a) Enter the functions $Y_1 = e^2$ and $Y_2 = \left(1 + \dfrac{2}{x}\right)^x$

($x > 0$); for Y_1, use the keystrokes (2nd) (e^x) 2. Now use ZOOM 6 to see the graphs in the standard viewing rectangle. Note that the line appears to be an asymptote for the curve. Go on to part (b).

(b) For more convincing evidence that the line is an asymptote, press the RANGE key and make the following changes: $x_{max} = 100$, $x_{scl} = 10$, $y_{min} = 0$, $y_{max} = 8$. Now press the GRAPH key to see evidence of the fact that

$$\lim_{x \to \infty}\left(1 + \frac{2}{x}\right)^x = e^2.$$

20. (a) Enter the functions $Y_1 = e^3$ and $Y_2 = \left(1 + \dfrac{3}{x}\right)^x$

($x > 0$). Press the RANGE key and use the following settings: $x_{min} = 0$, $x_{max} = 10$, $x_{scl} = 1$, $y_{min} = 0$, $y_{max} = 30$, $y_{scl} = 5$. Press the GRAPH key and note that the line appears to be an asymptote for the curve.

(b) For more convincing evidence that the line is an asymptote, press the RANGE key and make the following changes: $x_{max} = 200$, $x_{scl} = 20$. Now

press GRAPH to see evidence of the fact that

$$\lim_{x \to \infty}\left(1 + \frac{3}{x}\right)^x = e^3.$$

21. Graph the line $y = 1/e$ and the curve $y = \left(1 - \dfrac{1}{x}\right)^x$ to

demonstrate graphically that $\lim\limits_{x \to \infty}\left(1 - \dfrac{1}{x}\right)^x = 1/e$. Use the following RANGE settings: $x_{min} = 1$; $x_{max} = 50$; $x_{scl} = 5$; $y_{min} = 0$; $y_{max} = 0.5$; $y_{scl} = 0.1$.

22. This exercise shows how to use the calculator to fit an exponential curve $y = ab^x$ to a data set. (In Chapter 4 we used linear functions, rather than exponential functions, to model a data set. But as explained in the introduction to Chapter 5, there are cases where the exponential model is more appropriate.) We will use the following population data for the United States. (This is the data used in the chapter introduction.)

Year	U.S. Population
1800	5,308,483
1810	7,239,881
1820	9,638,453
1830	12,866,020
1840	17,069,453
1850	23,191,876
1860	31,443,321
1870	38,558,371
1880	50,189,209
1890	62,947,714
1900	75,994,575

(a) Clear any existing statistical data by first pressing the "2nd" key, then the STAT key, the left-hand arrow key, and the "2" key. After that, press ENTER.

(b) To enter the data in the table, begin by pressing the "2nd" key, then the STAT key, and then the left-hand arrow key, but do not press ENTER yet. Instead, press the number "1" to select the edit option. As you can see, the screen display shows

DATA

$x_1 =$

$y_1 = 1$

For the x_1-value, type 1800 and then press ENTER to move the cursor down to the line for y_1. For the y_1-value, type 5308483 and then press ENTER. The remaining values are then entered in the same fashion.

(c) After you have entered all of the data, press the "2nd" key and then the STAT key. Now use the arrow to move the cursor to item 4 (which is the choice for exponential regression) and press

ENTER. Then press ENTER again. You should see the following values displayed on your screen:

$$a = 4.3342717 \times 10^{-15}$$
$$b = 1.027380654$$

These are the values for the constants a and b in the regression curve $y = ab^x$. (There is also a value for r, a measure of how well the curve fits the data, but we do not need that here.)

(d) Press the Y= key and enter the function $Y_1 = ab^x$ using the values for a and b determined in part (c). Press the RANGE key and make the following adjustments to the standard viewing rectangle: $x_{min} = 1800$, $x_{max} = 1900$, $x_{scl} = 10$, $y_{min} = 0$, $y_{max} = 80,000,000$, $y_{scl} = 20,000,000$. Now, press the GRAPH key and check that your graph is consistent with Figure 1(b) in the chapter introduction.

5.3 LOGARITHMIC FUNCTIONS

[John Napier] hath set my head and hands a work with his new and admirable logarithms. I hope to see him this summer, if it please God, for I never saw book that pleased me better, or made me more wonder.

Henry Briggs, March 10, 1615

It has been thought that the earliest reference to the logarithmic curve was made by the Italian Evangelista Torricelli in a letter of the year 1644, but Paul Tannery made it practically certain that Descartes knew the curve in 1639.

Florian Cajori in *A History of Mathematics*, 4th ed. (New York: Chelsea Publishing Co., 1985)

For the number whose logarithm is unity, let e be written, ...

Leonhard Euler in a letter written in 1727 or 1728

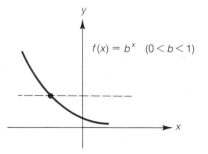

In the previous two sections we studied exponential functions. Now we consider functions that are inverses of exponential functions. These inverse functions are called *logarithmic functions*.

Having said this, let's back up for a moment to review briefly some of the basic ideas behind inverse functions (as discussed in Section 3.5). We start with a given function F, say, $F(x) = 3x$, for example, that is one-to-one. (That is, for each output there is but one input.) Then, by interchanging the inputs and outputs, we obtain a new function, the so-called inverse function. In the case of $F(x) = 3x$, it's easy to find an equation defining the inverse function. We just interchange x and y in the equation $y = 3x$ to obtain $x = 3y$. Solving for y in this last equation then gives us $y = \frac{1}{3}x$, which defines the inverse function. Using function notation, we can summarize the situation by writing $F(x) = 3x$ and $F^{-1}(x) = \frac{1}{3}x$.

In the preceding paragraph we saw that a particular linear function had an inverse, and we found a formula for that inverse. Now let's repeat that same reasoning beginning with an exponential function. First, we must make sure the exponential function f defined by $f(x) = b^x$ is one-to-one. We can see this by applying the horizontal line test, as indicated in Figure 1. Next, since $f(x) = b^x$ is one-to-one, it has an inverse function. Let's study this inverse.

We begin by writing the exponential function $f(x) = b^x$ in the form

$$y = b^x \tag{1}$$

FIGURE 1
The exponential function $y = b^x$ is one-to-one.

Then, in order to obtain an equation for f^{-1}, we interchange x and y in equation (1). This gives us

$$x = b^y \tag{2}$$

The crucial step now is to express equation (2) in words:

y is the exponent to which b must be raised to yield x. (3)

Statement (3) defines the function that is the inverse of $y = b^x$. Now we introduce a notation that will allow us to write this statement in a more compact form.

DEFINITION $\log_b x$

	EXAMPLES
We define the expression $\log_b x$ to mean "the exponent to which b must be raised to yield x." ($\log_b x$ is read *log base b of x* or *the logarithm of x to the base b.*)	**(a)** $\log_2 8 = 3$, since 3 is the exponent to which 2 must be raised to yield 8.
	(b) $\log_{10} \frac{1}{10} = -1$, since -1 is the exponent to which 10 must be raised to yield $\frac{1}{10}$.
	(c) $\log_5 1 = 0$, since 0 is the exponent to which 5 must be raised to yield 1.

TABLE I

Exponential Form of Equation	Logarithmic Form of Equation
$8 = 2^3$	$\log_2 8 = 3$
$\frac{1}{9} = 3^{-2}$	$\log_3 \frac{1}{9} = -2$
$1 = e^0$	$\log_e 1 = 0$
$a = b^c$	$\log_b a = c$

Using this notation, statement (3) becomes

$$y = \log_b x$$

Since (2) and (3) are equivalent, we have the following important relationship.

$$y = \log_b x \qquad \text{is equivalent to} \qquad x = b^y.$$

We say that the equation $y = \log_b x$ is in **logarithmic form** and that the equivalent equation $x = b^y$ is in **exponential form**. Table 1 displays some examples.

Let us now summarize our discussion up to this point.

1. According to the horizontal line test, the function $f(x) = b^x$ is one-to-one and therefore possesses an inverse. This inverse function is written

$$f^{-1}(x) = \log_b x$$

2. $\log_b a = c$ means that $a = b^c$.

To graph the function $y = \log_b x$, we recall from Chapter 3 that the graph of a function and its inverse are reflections of one another about the line $y = x$. Thus, to graph $y = \log_b x$, we need only reflect the curve $y = b^x$ about the line $y = x$. This is shown in Figure 2.

Note For the rest of this chapter we assume that the base b is greater than 1 when we use the expression $\log_b x$.

With the aid of Figure 2, we can make the following observations about the function $y = \log_b x$.

FIGURE 2

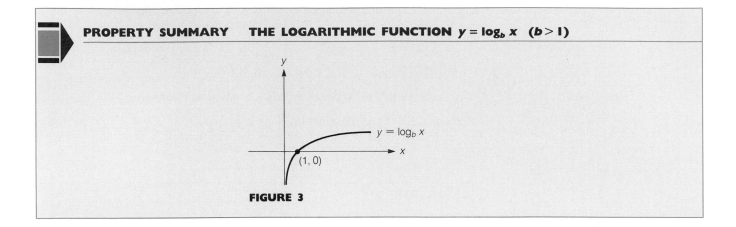

PROPERTY SUMMARY   THE LOGARITHMIC FUNCTION $y = \log_b x$ ($b > 1$)

FIGURE 3

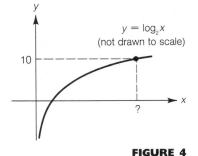

FIGURE 4

One aspect of the function $y = \log_b x$ may not be immediately apparent to you from Figure 2. The function grows or increases *very* slowly. Consider $y = \log_2 x$, for example. Let us ask how large x must be before the curve reaches the height $y = 10$ (see Figure 4).

To answer this question, we substitute $y = 10$ in the equation $y = \log_2 x$:

$$10 = \log_2 x$$

Writing this equation in exponential form yields

$$x = 2^{10} = 1024$$

In other words, we must go out beyond 1000 on the x-axis before the curve $y = \log_2 x$ reaches a height of 10 units. Exercise 49 at the end of this section asks you to show that the graph of $y = \log_2 x$ doesn't reach a height of 100 until x is greater than 10^{30}. (Numbers as large as 10^{30} rarely occur in any of the sciences. For instance, the distance in inches to the Andromeda galaxy is less than 10^{24}.) The point we are emphasizing here is this: The graph of $y = \log_2 x$ (or $\log_b x$) is always rising, but very slowly.

Before going on to consider some numerical examples, let us pause for a moment to think about the notation we have been using. In the expression

$$\log_b x$$

$\log_b$ is the name of a function and x is an input. To emphasize this, we might be better off writing $\log_b x$ as $\log_b(x)$, so that the similarity to the familiar $f(x)$ notation is clear. However, for historical reasons,* the convention is to suppress the parentheses, and we follow that convention here. In the box that follows, we indicate some errors that can occur upon forgetting that $\log_b$ is the name of a function, not a number.

---

*The notation *Log* was introduced in 1624 by the astronomer Johannes Kepler (1571–1630). In Leonhard Euler's text *Introduction to Analysis of the Infinite*, first published in 1748, appears the statement, "It has been customary to designate the logarithm of y by the symbol log y."

Errors to Avoid		
ERROR	CORRECTION	COMMENT
$\dfrac{\log_2 8}{\log_2 4} \neq \dfrac{8}{4}$	$\dfrac{\log_2 8}{\log_2 4} = \dfrac{3}{2}$	$\log_2$ is the name of a function. It is not a factor that that can be "cancelled" from the numerator and denominator.
$\dfrac{\log_2 16}{16} \neq \log_2$	$\dfrac{\log_2 16}{16} = \dfrac{4}{16} = \dfrac{1}{4}$	This "equation" is nonsense. On the left-hand side there is a quotient of two real numbers, which is a real number. On the right-hand side is the name of a function.

We conclude this section with a set of examples involving logarithms and logarithmic functions. In one way or another, every example makes use of the key fact that the equation $\log_b a = c$ is equivalent to $b^c = a$.

EXAMPLE 1   Which quantity is larger: $\log_3 10$ or $\log_7 40$?

Solution   First we estimate $\log_3 10$. This quantity represents the exponent to which 3 must be raised to yield 10. Since $3^2 = 9$ (less than 10), but $3^3 = 27$ (more than 10), we conclude that the quantity $\log_3 10$ lies between 2 and 3. In a similar way we can estimate $\log_7 40$; this quantity represents the exponent to which 7 must be raised to yield 40. Since $7^1 = 7$ (less than 40), while $7^2 = 49$ (more than 40), we conclude that the quantity $\log_7 40$ lies between 1 and 2. It now follows from these two estimates that $\log_3 10$ is larger than $\log_7 40$.   ▮▮▮

EXAMPLE 2   Evaluate $\log_4 32$.

Solution   Let $y = \log_4 32$. The exponential form of this equation is $4^y = 32$. Now, since both 4 and 32 are powers of 2, we can rewrite the equation $4^y = 32$ using the same base on both sides:

$$(2^2)^y = 2^5$$
$$2^{2y} = 2^5$$
$$2y = 5 \qquad \text{using Property 4 on page 231}$$
$$y = \frac{5}{2} \qquad \text{as required}$$ ▮▮▮

EXAMPLE 3   Graph the equations: **(a)** $y = \log_{10} x$; **(b)** $y = -\log_{10} x$.

Solution

(a) The function $y = \log_{10} x$ is the inverse function for the exponential function $y = 10^x$. Thus we obtain the graph of $y = \log_{10} x$ by reflecting the graph of $y = 10^x$ in the line $y = x$. See Figure 5(a) on the next page.

(b) To graph $y = -\log_{10} x$, we reflect $y = \log_{10} x$ in the x-axis. See Figure 5(b).   ▮▮▮

FIGURE 5

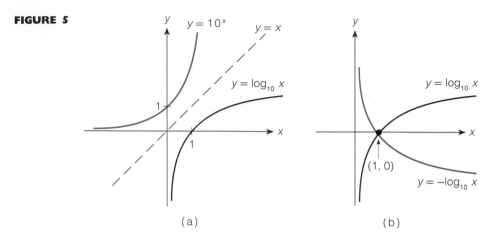

(a)   (b)

■▨■ EXAMPLE 4 Find the domain of the function $f(x) = \log_2(12 - 4x)$.

Solution As you can see by looking back at Figure 3 on page 248, the inputs for the logarithmic function must be positive. So, in the case at hand, we require that the quantity $12 - 4x$ be positive. Consequently, we have

$$12 - 4x > 0$$
$$-4x > -12$$
$$x < 3$$

Therefore the domain of the function $f(x) = \log_2(12 - 4x)$ is the interval $(-\infty, 3)$. ▪▪▪

The next example concerns the exponential function $y = e^x$ and its inverse function, $y = \log_e x$. Many books, as well as calculators, abbreviate the expression $\log_e x$ by $\ln x$, read *natural log of x*.* For reference and emphasis, we repeat this fact in the following box. (Incidentally, on most calculators, "log" is an abbreviation for $\log_{10}$.)

DEFINITION

$$\ln x \quad \text{means} \quad \log_e x.$$

■▨■ EXAMPLE 5 Graph the equations: **(a)** $y = \ln x$;
(b) $y = \ln(x - 1)$.

Solution

(a) The function $y = \ln x$ ($= \log_e x$) is the inverse of $y = e^x$. Thus its graph is obtained by reflecting $y = e^x$ in the line $y = x$, as in Figure 6(a).
(b) To graph $y = \ln(x - 1)$, we take the graph of $y = \ln x$ and move it one unit in the positive x-direction; see Figure 6(b). ▪▪▪

*According to the historian Florian Cajori, the notation $\ln x$ was used by (and perhaps first introduced by) Irving Stringham in his text *Uniplanar Algebra* (San Francisco: University Press, 1893).

FIGURE 6

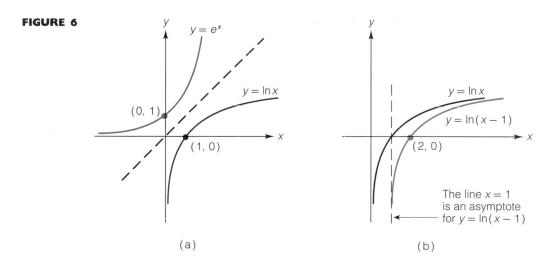

(a)   (b)

EXAMPLE 6 Simplify each expression: **(a)** $\ln e$; **(b)** $\ln 1$.

Solution

(a) The expression $\ln e$ denotes the exponent to which e must be raised to yield e. Clearly this exponent is 1. Thus $\ln e = 1$.

(b) Similarly, $\ln 1$ denotes the exponent to which e must be raised to yield 1. Since $e^0 = 1$, zero is the required exponent. Thus $\ln 1 = 0$. ■■■

EXAMPLE 7 Solve the equations: **(a)** $10^{2x} = 200$; **(b)** $e^{3t-1} = 2$.

Solution

(a) We write the equation $10^{2x} = 200$ in logarithmic form to obtain

$$2x = \log_{10} 200$$

$$x = \frac{\log_{10} 200}{2}$$

Without a calculator or tables, we leave the answer in this form. Alternatively, using a calculator we obtain $x \approx 1.15$.

(b) To solve $e^{3t-1} = 2$, we write the equation in logarithmic form:

$$3t - 1 = \log_e 2$$

that is,

$$3t - 1 = \ln 2$$

$$3t = 1 + \ln 2$$

$$t = \frac{1 + \ln 2}{3}$$

Without a calculator or tables, this is the final answer. On the other hand, using a calculator we obtain $t \approx 0.56$. ■■■

EXERCISE SET 5.3

A

Exercises 1–6 are review exercises dealing with inverse functions.

1. Which of the following functions is one-to-one and therefore has an inverse?
   (a) $y = x^2 + 1$ (b) $y = 3x$ (c) $y = (x + 1)^3$

2. Does each of the following functions have an inverse?
   (a) $f(x) = \begin{cases} x^2 & \text{if } -1 \le x \le 0 \\ x^2 + 1 & \text{if } x > 0 \end{cases}$

   (b) $g(x) = \begin{cases} x^2 & \text{if } -1 \le x < 0 \\ x^2 + 1 & \text{if } x \ge 0 \end{cases}$

3. Let $f(x) = \dfrac{2x - 1}{3x + 4}$. Find each quantity.
   (a) $f^{-1}(x)$ (b) $1/f(x)$
   (c) $f^{-1}(0)$ (d) $1/f(0)$

4. Let $f(x) = x^3 + 2x + 1$. Evaluate $f[f^{-1}(5)]$. [Assume that the domain of f^{-1} is $(-\infty, \infty)$.]

5. The graph of $y = f(x)$ is a line segment joining the two points $(3, -2)$ and $(-1, 5)$. What are the corresponding endpoints for the graph of $y = f^{-1}(x - 1)$?

6. Which (if either) of the following two conditions tells us that a function is one-to-one?
   (a) For each input there is exactly one output.
   (b) For each output there is exactly one input.

In Exercises 7 and 8, write each equation in logarithmic form.

7. (a) $9 = 3^2$ (b) $1000 = 10^3$
   (c) $7^3 = 343$ (d) $\sqrt{2} = 2^{1/2}$

8. (a) $\frac{1}{125} = 5^{-3}$ (b) $e^0 = 1$
   (c) $5^x = 6$ (d) $e^{3t} = 8$

9. Write each equation in exponential form.
   (a) $\log_2 32 = 5$ (b) $\log_{10} 1 = 0$
   (c) $\log_e \sqrt{e} = \frac{1}{2}$ (d) $\log_3 \frac{1}{81} = -4$
   (e) $\log_t u = v$

10. Complete the tables.
    (a)

x	1	10	10^2	10^3	10^4	10^{-1}	10^{-2}	10^{-3}
$\log_{10} x$								

   (b)

x	1	e	e^2	e^3	e^4	e^{-1}	e^{-2}	e^{-3}
$\ln x$								

11. Which quantity is larger, $\log_5 30$ or $\log_8 60$?

12. Which quantity is larger?
    (a) $\log_{10} 90$ or $\log_e e^5$ (b) $\log_2 3$ or $\log_3 2$

In Exercises 13 and 14, evaluate each expression.

13. (a) $\log_9 27$ (b) $\log_4(1/32)$ (c) $\log_5 5\sqrt{5}$

14. (a) $\log_{25}(1/625)$ (b) $\log_{16}(1/64)$
    (c) $\log_{10} 10$ (d) $\log_2 8\sqrt{2}$

In Exercises 15 and 16, solve each equation for x by converting to exponential form. In Exercises 15(b) and 16, give two forms for each answer: one involving e and the other a calculator approximation rounded off to two decimal places.

15. (a) $\log_4 x = -2$ (b) $\ln x = -2$

16. (a) $\log_5 x = e$ (b) $\ln x = -e$

In Exercises 17 and 18, find the domain of each function.

17. (a) $y = \log_4 5x$ (b) $y = \log_{10}(3 - 4x)$
    (c) $y = \ln(x^2)$ (d) $y = (\ln x)^2$
    (e) $y = \ln(x^2 - 25)$

18. (a) $y = \ln(2 - x - x^2)$ (b) $y = \log_{10} \dfrac{2x + 3}{x - 5}$

19. In the accompanying figure, what are the coordinates of the four points A, B, C, and D?

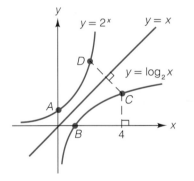

20. In the accompanying figure, what are the coordinates of the points A, B, C, and D?

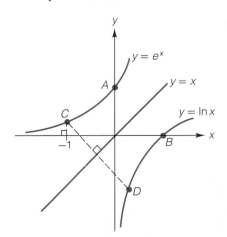

In Exercises 21–26, graph each function and specify the domain, range, intercept(s), and asymptote.

21. (a) $y = \log_2 x$ (b) $y = -\log_2 x$
(c) $y = \log_2(-x)$ (d) $y = -\log_2(-x)$

22. (a) $y = \ln x$ (b) $y = -\ln x$
(c) $y = \ln(-x)$ (d) $y = -\ln(-x)$

23. $y = -\log_3(x - 2) + 1$ **24.** $y = -\log_{10}(x + 1)$

25. $y = \ln(x + e)$ **26.** $y = \ln(-x) + e$

In Exercises 27 and 28, simplify each expression.

27. (a) $\ln e^4$ (b) $\ln(1/e)$ (c) $\ln\sqrt{e}$

28. (a) $\ln e$ (b) $\ln e^{-2}$ (c) $(\ln e)^{-2}$

In Exercises 29–36, find all the real-number solutions for each equation. In each case, give both the exact value of the answer and a calculator approximation rounded off to two decimal places.

29. $10^x = 25$ **30.** $10^x = 145$

31. $10^{x^2} = 40$ **32.** $(10^x)^2 = 40$

33. $e^{2t+3} = 10$ **34.** $e^{t-1} = 16$

35. $e^{1-4t} = 12.405$ **36.** $e^{3x^2} = 112$

In Exercises 37–40, use a calculator to match the equations with the graphs.

37. $f(x) = \sqrt{x}$, $g(x) = \ln x$

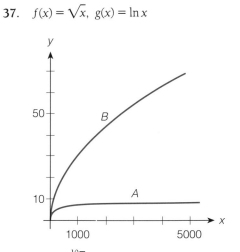

38. $f(x) = \sqrt[10]{x}$, $g(x) = \ln x$

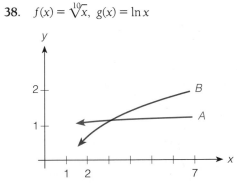

39. $f(x) = \sqrt[10]{x}$, $g(x) = \ln x$

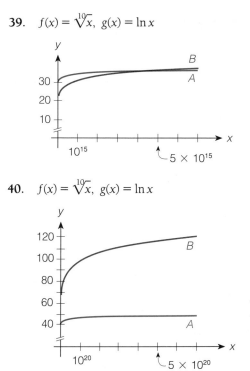

40. $f(x) = \sqrt[10]{x}$, $g(x) = \ln x$

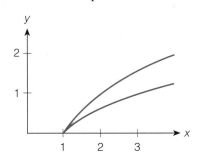

B

41. The accompanying figure shows portions of the graphs of $y = \log_2 x$ and $y = \log_3 x$. Which is which? Give your reasons in complete sentences.

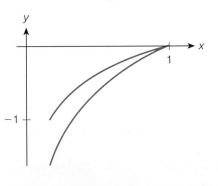

42. The accompanying figure shows portions of the graphs of $y = \log_3 x$ and $y = \log_6 x$. Which is which? Explain your reasons in complete sentences.

43. Let $f(x) = e^{x+1}$. Find $f^{-1}(x)$ and sketch its graph. Specify any intercept or asymptote.

44. Let $g(t) = \ln(t-1)$. Find $g^{-1}(t)$ and draw its graph. Specify any intercept or asymptote.

45. Sketch the region bounded by $y = e^x$, $y = e^{-x}$, the x-axis, and the vertical lines $x = \pm 1$. Why must the area of this region be less than two square units?

46. Solve $2^{2x} - 2^{x+1} - 15 = 0$ for x.   *Hint:*   First show the equation is equivalent to $(2^x)^2 - 2(2^x) - 15 = 0$.

47. Solve $e^{2x} - 5e^x - 6 = 0$ for x.

48. Solve $e^x - 3e^{-x} = 2$ for x.   *Hint:*   Multiply both sides by e^x.

49. Estimate a value for x such that $\log_2 x = 100$. Use the approximation $10^3 \approx 2^{10}$ to express your answer as a power of 10.   *Answer:* 10^{30}

50. (a) How large must x be before the graph of $y = \ln x$ reaches a height of $y = 100$?
    (b) How large must x be before the graph of $y = e^x$ reaches a height of (i) $y = 100$? (ii) $y = 10^6$?

In Exercises 51 and 52 use the following information on pH. Chemists define pH by the formula $pH = -\log_{10}[H^+]$, where $[H^+]$ is the hydrogen ion concentration measured in moles per liter. For example, if $[H^+] = 10^{-5}$, then $pH = 5$. Solutions with a pH of 7 are said to be neutral; *a pH below 7 indicates an* acid; *and a pH above 7 indicates a* base.

51. (a) For some fruit juices, $[H^+] = 3 \times 10^{-4}$. Determine the pH and classify these juices as acid or base.
    (b) For sulfuric acid, $[H^+] = 1$. Find the pH.

52. An unknown substance has a hydrogen ion concentration of 3.5×10^{-9}. Classify the substance as acid or base.

53. This exercise indicates one of the ways the natural logarithm function is used in the study of prime numbers. Recall that a prime number is a natural number greater than 1 with no factors other than itself and 1. For example, the first ten prime numbers are 2, 3, 5, 7, 11, 13, 17, 19, 23, and 29.
    (a) Let $P(x)$ denote the number of prime numbers that do not exceed x. For instance, $P(6) = 3$, since there are three prime numbers (2, 3, and 5) that do not exceed 6. Compute $P(10)$, $P(18)$, and $P(19)$.
    (b) According to the **prime number theorem**, $P(x)$ can be approximated by $x/\ln x$ when x is large, and in fact, the ratio $\dfrac{P(x)}{x/\ln x}$ approaches 1 as x grows larger and larger. Verify this empirically by completing

the following table. Round your results to three decimal places. (These facts were discovered by Carl Friedrich Gauss in 1792, when he was 15 years old. It was more than 100 years later, however, in 1896, before the prime number theorem was formally proved by the French mathematician J. Hadamard and also by the Belgian mathematician C. J. de la Vallée-Poussin).

x	$P(x)$	$\dfrac{x}{\ln x}$	$\dfrac{P(x)}{x/\ln x}$
10^2	25		
10^4	1229		
10^6	78498		
10^8	5761455		
10^9	50847534		
10^{10}	455052512		

(c) In 1808, A. M. Legendre found that he could improve upon Gauss's approximation for $P(x)$ by using the expression $\dfrac{x}{\ln x - 1.08366}$ rather than $\dfrac{x}{\ln x}$. Complete the following table to see how well Legendre's expression approximates $P(x)$. Round off your results to four decimal places.

x	$P(x)$	$\dfrac{x}{\ln x - 1.08366}$	$\dfrac{P(x)}{x/(\ln x - 1.08366)}$
10^2	25		
10^4	1229		
10^6	78498		
10^8	5761455		
10^9	50847534		
10^{10}	455052512		

54. The figure (on the next page) shows a portion of the graph of $y = e^x$ for $1.5 \le x \le 1.7$. Using the figure, estimate the quantity $\ln 5$ to two decimal places. Then use a calculator to determine the percent error in your approximation.
    Hint: Write the equation $x = \ln 5$ in exponential form.

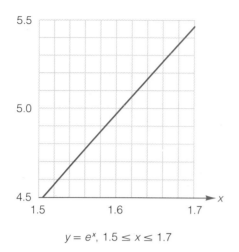

$y = e^x$, $1.5 \leq x \leq 1.7$

In Exercises 55 and 56, use logarithms (and a calculator) to find a natural number n satisfying the given conditions.

55. $e^n < 10^{12} < e^{n+1}$

56. $10^n < e^{100} < 10^{n+1}$

57. Using, as usual, x for the inputs and y for the outputs, graph the function defined by the equation $\ln y = x$.

C

58. (a) Find the domain of the function f defined by $f(x) = \ln(\ln x)$.
(b) Find $f^{-1}(x)$ for the function f in part (a).
(c) Find the domain of the function g defined by $g(x) = \ln[\ln(\ln x)]$.

OPTIONAL TI-81 GRAPHING CALCULATOR EXERCISES

EXERCISES FOR SECTION 5.3

1. (a) According to Section 5.3, what kind of symmetry would you see if you graphed the functions $Y_1 = e^x$ and $Y_2 = \ln x$ on the same set of axes?
(b) Graph those two functions on the same set of axes. Also, add the line $Y_3 = x$ to the picture. [Choose an appropriate viewing rectangle and then press ZOOM 5. (Why is the ZOOM 5 setting necessary?)]

2. Follow the steps in Exercise 1, but use the functions $Y_1 = 10^x$ and $Y_2 = \log_{10} x$, respectively.

3. Let $F(x) = e^x$, $G(x) = \ln x$, $H(x) = x - 2$, and $K(x) = x^2$. Specify the domain of each of the following functions and then check your answers by drawing the graphs. *Hint:* The answers for (c) and (d) are not the same.
(a) F (b) G (c) $G \circ F$ (d) $F \circ G$
(e) $G \circ H$ (f) $H \circ G$ (g) $G \circ K$ (h) $K \circ G$

4. Draw graphs to help answer the following questions. (You will need to experiment with the RANGE settings.)
(a) For which x-values do we have $\ln x < \log_{10} x$?
(b) For which x-values do we have $\ln x > \log_{10} x$?
(c) For which x-values do we have $e^x < 10^x$?
(d) For which x-values do we have $e^x > 10^x$?

5. This exercise uses the RANGE key and the TRACE key to demonstrate the very slow growth of the natural logarithm function $y = \ln x$. Begin by clearing all functions in the $Y=$ menu and pressing ZOOM 6.
(a) Press the $Y=$ key and enter the function $Y_1 = \ln x$. Press GRAPH and note how slowly the graph rises. We ask the question, How large must x be before the graph reaches a height of 10? Press the TRACE key and use the right-hand arrow key to move the cursor to the point on the graph where $x = 10$. As indicated on the screen display, the y-value is only about 2.3. Go on to part (b).
(b) Since we are trying to see when the graph of $Y_1 = \ln x$ reaches a height of 10, press the $Y=$ key and add the graph of (the horizontal line) $Y_2 = 10$ to the picture obtained in part (a). Next, press the RANGE key and make the following changes: $x_{min} = 0$, $x_{max} = 100$, and $x_{scl} = 10$. Press the GRAPH key. Now press the TRACE key and move the cursor to the point on the graph where $x = 100$. Note that the corresponding y-coordinate isn't even 5. Go on to part (c).
(c) Press the RANGE key and make the following changes: $x_{max} = 1000$ and $x_{scl} = 100$. Press the GRAPH key. Now press the TRACE key and, as before, determine the coordinates of the highest point shown on the graph. You'll find that at $x = 1000$, the value of y is almost 7. We are getting closer to 10.
(d) Repeat the process in part (c) using $x_{max} = 10000$ and $x_{scl} = 1000$. As you can see from the results, the curve has almost reached the height $y = 10$.
(e) The last step: in the RANGE menu, set $x_{max} = 100000$, $x_{scl} = 10000$, and $y_{max} = 15$. Press GRAPH. Now use the TRACE key and the arrow keys to estimate the value of x for which $\ln x = 10$. (*Question:* How would you find the *exact* value of x for which $\ln x = 10$?)

6. In this exercise we use logarithms and the graphing calculator to investigate the relationship between a planet's average distance from the Sun and its period (the time required for the planet to make one complete revolution around the Sun). As preparation, press the Y= key and clear any functions that are present. Press the MODE key and set the number of decimal places to 2 (in the FLOAT row). Then press ENTER and CLEAR.

(a) Complete the following table. The abbreviation AU stands for astronomical units, a unit of distance such that the Earth is 1 AU from the Sun.

Planet	x Avg. Distance from Sun (AU)	y Period (years)	ln x	ln y
Mercury	0.387	0.241		
Venus	0.723	0.615		
Earth	1.000	1.000		
Mars	1.523	1.881		
Jupiter	5.202	11.820		

(b) Obtain the regression line for the pairs of numbers (ln x, ln y). (Follow the procedure explained in Exercise 2 on page 168. You should obtain $a = 0$ and $b = 1.5$. Thus, the quantities ln x and ln y are related as follows:

$$\ln y = 1.5 \ln x$$

(c) Show that the equation obtained in part (b) can be simplified to

$$y = x^{1.5} \qquad (1)$$

(d) Equation (1) was obtained using data for the first five planets from the Sun. Complete the following table to see how well equation (1) fits the data for the remaining planets.

Planet	x (AU)	y (years) (Calculated from Eqn. (1))	y (years) (From astronomical observations)
Saturn	9.555		29.46
Uranus	19.22		84.01
Neptune	30.11		164.79
Pluto	39.44		248.50

Remark Equation (1) is *Kepler's third law of planetary motion.* According to George F. Simmons in *Calculus Gems* (New York: McGraw-Hill, Inc., 1992), "Kepler had struggled for more than twenty years to find this connection between a planet's *distance* from the sun and the *time* required to complete its orbit. He published his discovery in 1619 in a work entitled *Harmonices Mundi* (The Harmonies of the World)."

5.4 PROPERTIES OF LOGARITHMS

... he [John Napier] invented the word "logarithms," using two Greek words, ... arithmos, "number" and logos, "ratio." It is impossible to say exactly what he had in mind when making up this word.

Alfred Hooper in *Makers of Mathematics* (New York: Random House, Inc., 1948)

The teaching of logarithms as a computing device is vanishing from the schools....

The logarithmic function, however, will never die, for the simple reason that logarithmic and exponential variations are a vital part of nature and of analysis.

Howard Eves in *An Introduction to the History of Mathematics*, 6th ed. (Philadelphia: Saunders College Publishing, 1990)

A few basic properties of logarithms are used repeatedly. Our procedure in this section will be to state these properties, discuss their proofs, and then look at examples.

> **Properties of Logarithms**
>
> **1.** (a) $\log_b b = 1$ (b) $\log_b 1 = 0$
>
> **2.** $\log_b PQ = \log_b P + \log_b Q$
> *The log of a product is the sum of the logs of the factors.*
>
> **3.** $\log_b(P/Q) = \log_b P - \log_b Q$
> *The log of a quotient is the log of the numerator minus the log of the denominator.*
> As a useful particular case, we have $\log_b(1/Q) = -\log_b Q$.
>
> **4.** $\log_b P^n = n \log_b P$
>
> **5.** $b^{\log_b P} = P$
>
> **Note** P and Q are assumed to be positive in Properties 2–5.

In essence, each of these properties follows from the equivalence of the two equations $y = \log_b x$ and $b^y = x$. For instance, the equivalent exponential forms for properties 1(a) and 1(b) are $b^1 = b$ and $b^0 = 1$, respectively, both of which are certainly valid.

To prove Property 2, we begin by letting $x = \log_b P$. The equivalent exponential form of this equation is

$$P = b^x \tag{1}$$

Similarly, we let $y = \log_b Q$. The exponential form of this equation is

$$Q = b^y \tag{2}$$

If we multiply equation (1) by equation (2), we get

$$PQ = b^x b^y$$

and, therefore,

$$PQ = b^{x+y} \tag{3}$$

Next we write equation (3) in its equivalent logarithmic form. This yields

$$\log_b PQ = x + y$$

But, using the definitions of x and y, this last equation is equivalent to

$$\log_b PQ = \log_b P + \log_b Q$$

That completes the proof of Property 2.

The proof of Property 3 is quite similar to the proof given for Property 2. Exercise 72(a) asks you to carry out the proof of Property 3.

We turn now to the proof of Property 4. We begin by letting $x = \log_b P$. In exponential form, this last equation becomes

$$b^x = P \tag{4}$$

Now we raise both sides of equation (4) to the power n. This yields

$$(b^x)^n = b^{nx} = P^n$$

The logarithmic form of this last equation is

$$\log_b P^n = nx$$

or (from the definition of x)

$$\log_b P^n = n \log_b P$$

The proof of Property 4 is now complete.

Property 5 is again just a restatement of the meaning of $\log_b P$. To derive this property, let $x = \log_b P$. Therefore,

$$b^x = P$$

Now, in this last equation, we simply replace x with $\log_b P$. The result is

$$b^{\log_b P} = P \qquad \text{as required}$$

Now let's see how these properties are used. To begin with, we display some simple numerical examples in the box that follows.

PROPERTY SUMMARY   **PROPERTIES OF LOGARITHMS**

PROPERTY	EXAMPLE
$\log_b P + \log_b Q = \log_b PQ$	Simplify: $\log_{10} 50 + \log_{10} 2$ Solution: $\log_{10} 50 + \log_{10} 2 = \log_{10}(50 \cdot 2)$ $= \log_{10} 100$ $= 2$
$\log_b P - \log_b Q = \log_b \dfrac{P}{Q}$	Simplify: $\log_8 56 - \log_8 7$ Solution: $\log_8 56 - \log_8 7 = \log_8 \dfrac{56}{7}$ $= \log_8 8 = 1$
$\log_b P^n = n \log_b P$	Simplify: $\log_2 \sqrt[5]{16}$ Solution: $\log_2 \sqrt[5]{16} = \log_2(16^{1/5})$ $= \dfrac{1}{5} \log_2 16$ $= \dfrac{1}{5} \cdot 4 = \dfrac{4}{5}$
$b^{\log_b P} = P$	Simplify: $3^{\log_3 7}$ Solution: $3^{\log_3 7} = 7$

The preceding examples showed how we can simplify or shorten certain expressions involving logarithms. The next example is also of this type.

▮▮▮ EXAMPLE I   Express as a single logarithm with a coefficient of 1:

$$\tfrac{1}{2} \log_b x - \log_b(1 + x^2)$$

Solution $\frac{1}{2}\log_b x - \log_b(1 + x^2) = \log_b x^{1/2} - \log_b(1 + x^2)$ using Property 4 on page 257

$$= \log_b \frac{x^{1/2}}{1 + x^2}$$ using Property 3

This last expression is the required answer.   ▮▮▮

Property 2 says that the logarithm of a product of two factors is equal to the sum of the logarithms of the two factors. This can be generalized to any number of factors. For instance, with three factors, we have

$$\log_b(ABC) = \log_b[A(BC)]$$
$$= \log_b A + \log_b BC \qquad \text{using Property 2}$$
$$= \log_b A + \log_b B + \log_b C \qquad \text{using Property 2 again}$$

The next example makes use of this idea.

▮▮▮ **EXAMPLE 2** Express as a single logarithm with a coefficient of 1:

$$\ln(x^2 - 9) + 2\ln\frac{1}{x + 3} + 4\ln x \qquad (x > 3)$$

Solution $\ln(x^2 - 9) + 2\ln\dfrac{1}{x + 3} + 4\ln x = \ln(x^2 - 9) + \ln\left[\left(\dfrac{1}{x + 3}\right)^2\right] + \ln x^4$

$$= \ln(x^2 - 9) + \ln\frac{1}{(x + 3)^2} + \ln x^4$$

$$= \ln\left[(x^2 - 9) \cdot \frac{1}{(x + 3)^2} \cdot x^4\right]$$

$$= \ln\frac{(x^2 - 9)x^4}{(x + 3)^2}$$

This last expression can be simplified still further by writing $x^2 - 9$ as $(x - 3)(x + 3)$. Then a factor of $x + 3$ can be divided out of the numerator and denominator of the fraction. The result is

$$\ln(x^2 - 9) + 2\ln\frac{1}{x + 3} + 4\ln x = \ln\frac{(x - 3)x^4}{x + 3}$$

$$= \ln\frac{x^5 - 3x^4}{x + 3}$$ ▮▮▮

In the examples considered so far, we've used the properties of logarithms to shorten given expressions. We can also use these properties to expand an expression. (This is useful in calculus.)

▮▮▮ **EXAMPLE 3** Write each of the following quantities as sums and differences of simpler logarithmic expressions. Express each answer in such a way that no logarithm of products, quotients, or powers appears.

(a) $\log_{10}\sqrt{3x}$ **(b)** $\log_{10}\sqrt[3]{\dfrac{2x}{3x^2 + 1}}$ **(c)** $\ln\dfrac{x^2\sqrt{2x - 1}}{(2x + 1)^{3/2}}$

Solution

(a) $\log_{10}\sqrt{3x} = \log_{10}(3x)^{1/2} = \frac{1}{2}\log_{10}(3x)$
$$= \frac{1}{2}(\log_{10}3 + \log_{10}x)$$

(b) $\log_{10}\sqrt[3]{\dfrac{2x}{3x^2+1}} = \log_{10}\left[\left(\dfrac{2x}{3x^2+1}\right)^{1/3}\right]$

$$= \frac{1}{3}\log_{10}\frac{2x}{3x^2+1}$$
$$= \frac{1}{3}[\log_{10}2x - \log_{10}(3x^2+1)]$$
$$= \frac{1}{3}[\log_{10}2 + \log_{10}x - \log_{10}(3x^2+1)]$$

(c) $\ln\dfrac{x^2\sqrt{2x-1}}{(2x+1)^{3/2}} = \ln\dfrac{x^2(2x-1)^{1/2}}{(2x+1)^{3/2}}$

$$= \ln x^2 + \ln(2x-1)^{1/2} - \ln(2x+1)^{3/2}$$
$$= 2\ln x + \frac{1}{2}\ln(2x-1) - \frac{3}{2}\ln(2x+1)$$

EXAMPLE 4 Given that $\log_{10}A = a$, $\log_{10}B = b$, and $\log_{10}C = c$, express $\log_{10}\left(\dfrac{A^3}{B^4\sqrt{C}}\right)$ in terms of a, b, and c.

Solution $\log_{10}\left(\dfrac{A^3}{B^4\sqrt{C}}\right) = \log_{10}\left(\dfrac{A^3}{B^4C^{1/2}}\right)$

$$= \log_{10}A^3 - \log_{10}B^4C^{1/2}$$
$$= \log_{10}A^3 - (\log_{10}B^4 + \log_{10}C^{1/2})$$
$$= 3\log_{10}A - (4\log_{10}B + \frac{1}{2}\log_{10}C)$$
$$= 3\log_{10}A - 4\log_{10}B - \frac{1}{2}\log_{10}C$$
$$= 3a - 4b - \frac{1}{2}c$$

The properties of logarithms are also useful in solving equations, as the next examples demonstrate.

EXAMPLE 5 Figure 1 shows the graph of $y = 2^x - 3$. (The axes are marked off in one-unit intervals.) As indicated in the figure, the x-intercept of the curve is between 1 and 2. Determine both an exact expression for this x-intercept and a calculator approximation rounded off to two decimal places.

Solution Setting y equal to zero in the given equation yields $2^x - 3 = 0$, or $2^x = 3$. Notice that the unknown, x, appears in the exponent. To solve for x, we'll take the logarithm of both sides of the equation. (Base e logarithms are used in the following computations; Exercise 71 asks you to check that the use of base 10 logarithms produces an equivalent answer.) We have

$$\ln 2^x = \ln 3$$
$$x \ln 2 = \ln 3 \qquad \text{using Property 4 on page 257}$$
$$x = \frac{\ln 3}{\ln 2} \approx 1.58 \qquad \textit{Caution:} \quad \frac{\ln 3}{\ln 2} \neq \ln 3 - \ln 2$$

$y = 2^x - 3$

FIGURE 1

■|■■| EXAMPLE 6 Solve for x: $\log_3 x + \log_3(x + 2) = 1$

Solution Using Property 2, we can write the given equation as

$$\log_3[x(x + 2)] = 1 \qquad \text{or} \qquad \log_3(x^2 + 2x) = 1$$

Writing this last equation in exponential form yields

$$x^2 + 2x = 3^1$$
$$x^2 + 2x - 3 = 0$$
$$(x + 3)(x - 1) = 0$$

Thus, we have

$$x + 3 = 0 \qquad \text{or} \qquad x - 1 = 0$$

and, consequently,

$$x = -3 \qquad \text{or} \qquad x = 1$$

Now let us check these values in the original equation to see if they are indeed solutions.

If $x = -3$, the equation becomes

$$\underbrace{\log_3(-3)}_{\nwarrow} + \underbrace{\log_3(-1)}_{\nearrow} \overset{?}{=} 1$$

Neither expression is defined, since the domain of the logarithm function does not contain negative numbers.

If $x = 1$, the equation becomes

$$\log_3 1 + \log_3 3 \overset{?}{=} 1$$
$$0 + 1 \overset{?}{=} 1 \qquad \text{true}$$

Thus, the value $x = 1$ is a solution of the original equation, but $x = -3$ is not.

■■■

In the example we just concluded, an extraneous solution ($x = -3$) was generated along with the correct solution, $x = 1$. How did this happen? It happened because we used the property $\log_b PQ = \log_b P + \log_b Q$ in the second line of the solution. But this property is valid only when both P and Q are positive. In the box that follows, we generalize this remark about extraneous solutions.

PROPERTY SUMMARY   EXTRANEOUS SOLUTIONS FOR LOGARITHMIC EQUATIONS

Using the properties of logarithms (on page 257) to solve logarithmic equations may introduce extraneous solutions that do not check in the original equation. (This occurs because the logarithm function requires positive inputs, but in solving an equation, we may not know ahead of time the sign of an input involving a variable.) Therefore, it is always necessary to check any candidates for solutions that are obtained in this manner.

It is sometimes necessary to convert logarithms in one base to logarithms in another base. After the next example, we will state a formula for this. However, as the next example indicates, it is easy to work this type of problem from the basics, without relying on a formula.

■|■■| EXAMPLE 7 Express $\log_2 5$ in terms of base 10 logarithms.

Solution Let $z = \log_2 5$. The exponential form of this equation is

$$2^z = 5$$

We now take the base 10 logarithm of each side of this equation to obtain

$$\log_{10} 2^z = \log_{10} 5$$

$$z \log_{10} 2 = \log_{10} 5 \qquad \text{using Property 4}$$

$$z = \frac{\log_{10} 5}{\log_{10} 2}$$

Given our definition of z, this last equation can be written

$$\log_2 5 = \frac{\log_{10} 5}{\log_{10} 2}$$

This is the required answer. ▮▮▮

The method shown in Example 7 can be used to convert between any two bases. Exercise 72(b) at the end of this section asks you to follow this method, using letters rather than numbers, to arrive at the following general formula.

PROPERTY SUMMARY   **CHANGE OF BASE**

FORMULA

$$\log_a x = \frac{\log_b x}{\log_b a}$$

EXAMPLES

$$\log_2 3 = \frac{\log_{10} 3}{\log_{10} 2} = \frac{\ln 3}{\ln 2}$$

$$\log_{10} e = \frac{\ln e}{\ln 10} = \frac{1}{\ln 10}$$

Examples 1–7 have dealt with applications of the five properties of logarithms. However, you also need to understand what the properties *don't* say. For instance, Property 3 does not apply to an expression such as $\dfrac{\log_{10} 5}{\log_{10} 2}$. [Property 3 *would* apply if the expression were $\log_{10}(5/2)$.] In the box that follows, we list some errors to avoid in working with logarithms.

Errors to Avoid

ERROR	CORRECTION	COMMENT
$\log_b(x+y) \neq \log_b x + \log_b y$	$\log_b(xy) = \log_b x + \log_b y$	In general, there is no simple identity involving $\log_b(x+y)$. This is similar to the situation for $\sqrt{x+y}$ (which is not equal to $\sqrt{x} + \sqrt{y}$).
$\dfrac{\log_b x}{\log_b y} \neq \log_b x - \log_b y$	$\log_b\left(\dfrac{x}{y}\right) = \log_b x - \log_b y$	In general, there is no simple identity involving the quotient $(\log_b x)/(\log_b y)$.
$(\ln x)^3 \neq 3\ln x$	$(\ln x)^3 = (\ln x)(\ln x)(\ln x)$	$(\ln x)^3$ is not the same as $\ln(x^3)$. Regarding the latter, we do have $\ln(x^3) = 3\ln x$, for $x > 0$.
$\ln\dfrac{x}{2} \neq \dfrac{\ln x}{2}$	$\ln\dfrac{x}{2} = \ln x - \ln 2$, for $x > 0$ and $\dfrac{\ln x}{2} = \dfrac{1}{2}\ln x = \ln\sqrt{x}$, for $x > 0$	The confusion between $\ln\dfrac{x}{2}$ and $\dfrac{\ln x}{2}$ is sometimes due simply to careless handwriting.

EXERCISE SET 5.4

A

In Exercises 1–10, simplify the expression by using the definition and properties for logarithms.

1. $\log_{10} 70 - \log_{10} 7$
2. $\log_{10} 40 + \log_{10}(5/2)$
3. $\log_7 \sqrt{7}$
4. $\log_9 25 - \log_9 75$
5. $\log_3 108 + \log_3(3/4)$
6. $\ln e^3 - \ln e$
7. $-\frac{1}{2} + \ln \sqrt{e}$
8. $e^{\ln 3} + e^{\ln 2} - e^{\ln e}$
9. $2^{\log_2 5} - 3 \log_5 \sqrt[3]{5}$
10. $\log_b b^b$

In Exercises 11–19, write the expression as a single logarithm with a coefficient of 1.

11. $\log_{10} 30 + \log_{10} 2$
12. $2 \log_{10} x - 3 \log_{10} y$
13. $\log_5 6 + \log_5(1/3) + \log_5 10$
14. $p \log_b A - q \log_b B + r \log_b C$
15. (a) $\ln 3 - 2 \ln 4 + \ln 32$
    (b) $\ln 3 - 2(\ln 4 + \ln 32)$
16. (a) $\log_{10}(x^2 - 16) - 3 \log_{10}(x + 4) + 2 \log_{10} x$
    (b) $\log_{10}(x^2 - 16) - 3[\log_{10}(x + 4) + 2 \log_{10} x]$
17. $\log_b 4 + 3[\log_b(1 + x) - \frac{1}{2} \log_b(1 - x)]$
18. $\ln(x^3 - 1) - \ln(x^2 + x + 1)$
19. $4 \log_{10} 3 - 6 \log_{10}(x^2 + 1) + \frac{1}{2}[\log_{10}(x + 1) - 2 \log_{10} 3]$

In Exercises 20–26, write the quantity using sums and differences of simpler logarithmic expressions. Express the answer so that logarithms of products, quotients, and powers do not appear.

20. (a) $\log_{10} \sqrt{(x + 1)(x + 2)}$

    (b) $\ln \sqrt{\dfrac{(x + 1)(x + 2)}{(x - 1)(x - 2)}}$

21. (a) $\log_{10} \dfrac{x^2}{1 + x^2}$
    (b) $\ln \dfrac{x^2}{\sqrt{1 + x^2}}$

22. (a) $\log_b \dfrac{\sqrt{1 - x^2}}{x}$
    (b) $\ln \dfrac{x^3 \sqrt{4x + 1}}{\sqrt{2x - 1}}$

23. (a) $\log_{10} \sqrt{9 - x^2}$
    (b) $\ln \dfrac{\sqrt{4 - x^2}}{(x - 1)(x + 1)^{3/2}}$

24. (a) $\log_b \sqrt[3]{\dfrac{x + 3}{x}}$
    (b) $\ln \dfrac{1}{\sqrt{x^2 + x + 1}}$

25. (a) $\log_b \sqrt{x/b}$
    (b) $2 \ln \sqrt{(1 + x^2)(1 + x^4)(1 + x^6)}$

26. (a) $\log_b \sqrt[3]{\dfrac{(x - 1)^2(x - 2)}{(x + 2)^2(x + 1)}}$
    (b) $\ln \left(\dfrac{e - 1}{e + 1}\right)^{3/2}$

In Exercises 27 and 28, suppose that $\log_{10} A = a$, $\log_{10} B = b$, and $\log_{10} C = c$. Express the following logarithms in terms of a, b, and c.

27. (a) $\log_{10} AB^2C^3$
    (b) $\log_{10} 10\sqrt{A}$
    (c) $\log_{10} \sqrt{10ABC}$
    (d) $\log_{10}(10A/\sqrt{BC})$
28. (a) $\log_{10} A + 2 \log_{10}(1/A)$
    (b) $\log_{10}(A/10)$
    (c) $\log_{10} \dfrac{100A^2}{B^4\sqrt[3]{C}}$
    (d) $\log_{10} \dfrac{(AB)^5}{C}$

In Exercises 29 and 30, suppose that $\ln x = t$ and $\ln y = u$. Write each expression in terms of t and u.

29. (a) $\ln(ex)$
    (b) $\ln xy - \ln(x^2)$
    (c) $\ln \sqrt{xy} + \ln(x/e)$
    (d) $\ln(e^2x\sqrt{y})$
30. (a) $\ln(e^{\ln x})$
    (b) $e^{\ln(\ln xy)}$
    (c) $\ln\left(\dfrac{ex}{y}\right) - \ln\left(\dfrac{y}{ex}\right)$
    (d) $\dfrac{(\ln x)^3 - \ln(x^4)}{\left(\ln \dfrac{x}{e^2}\right)\ln(xe^2)}$

In Exercises 31 and 32, graph the equations and determine the x-intercepts (as in Example 5).

31. (a) $y = 2^x - 5$
    (b) $y = 2^{x/2} - 5$
32. (a) $y = 3^{x-1} - 2$
    (b) $y = 3^{1-x} - 4$

In Exercises 33–37, solve the equations. Express the answers in terms of natural logarithms.

33. $5 = 2e^{2x-1}$ Suggestion: First divide by 2.
34. $3e^{1+t} = 2$
35. $2^x = 13$
36. $5^{3x-1} = 27$
37. $10^x = e$

In Exercises 38–47, solve the equation for x; check the answers to remove any extraneous roots.

38. $\log_6 x + \log_6(x + 1) = 1$
39. $\log_9(x + 1) = \frac{1}{2} + \log_9 x$
40. $\log_2(x + 4) = 2 - \log_2(x + 1)$
41. $\log_{10}(2x + 4) + \log_{10}(x - 2) = 1$
42. $\ln x + \ln(x + 1) = \ln 12$
43. $\log_{10}(x + 3) - \log_{10}(x - 2) = 2$
44. $\ln(x + 1) = 2 + \ln(x - 1)$
45. $\log_b(x + 1) = 2 \log_b(x - 1)$
46. $\log_2(2x^2 + 4) = 5$
47. $\log_{10}(x - 6) + \log_{10}(x + 3) = 1$
48. Solve for x in terms of a:
$$\log_2(x + a) - \log_2(x - a) = 1$$
49. Solve for x in terms of y.
    (a) $\log_{10} x - y = \log_{10}(3x - 1)$
    (b) $\log_{10}(x - y) = \log_{10}(3x - 1)$
50. Solve for x in terms of b: $\log_b(1 - 3x) = 3 + \log_b x$.

In Exercises 51–56, express the quantity in terms of base 10 logarithms.

51. $\log_2 5$ **52.** $\log_5 10$ **53.** $\ln 3$

54. $\ln 10$ **55.** $\log_b 2$ **56.** $\log_2 b$

In Exercises 57–61, express the quantity in terms of natural logarithms.

57. $\log_{10} 6$ **58.** $\log_2 10$ **59.** $\log_{10} e$

60. $\log_b 2$, where $b = e^2$ **61.** $\log_{10}(\log_{10} x)$

62. Give specific examples showing that each statement is false.
(a) $\log(x + y) = \log x + \log y$
(b) $(\log x)/(\log y) = \log x - \log y$
(c) $(\log x)(\log y) = \log x + \log y$
(d) $(\log x)^k = k \log x$

63. True or false?
(a) $\log_{10} A + \log_{10} B - \frac{1}{2} \log_{10} C = \log_{10}(AB/\sqrt{C})$
(b) $\log_e \sqrt{e} = \frac{1}{2}$ (c) $\ln \sqrt{e} = \frac{1}{2}$
(d) $\ln x^3 = \ln 3x$ (e) $\ln x^3 = 3 \ln x$
(f) $\ln 2x^3 = 3 \ln 2x$ (g) $\log_a c = b$ means $a^b = c$.
(h) $\log_5 24$ is between 5^1 and 5^2.
(i) $\log_5 24$ is between 1 and 2.
(j) $\log_5 24$ is closer to 1 than to 2.
(k) The domain of $g(x) = \ln x$ is the set of all real numbers.
(l) The range of $g(x) = \ln x$ is the set of all real numbers.
(m) The function $g(x) = \ln x$ is one-to-one.

Use a calculator for Exercises 64 and 65.

64. (a) Check Property 2 using the values $b = 10$, $P = \pi$, and $Q = \sqrt{2}$.
(b) Let $P = 3$ and $Q = 4$. Show that $\ln(P + Q) \neq \ln P + \ln Q$.
(c) Check Property 3 using the values $b = 10$, $P = 2$, and $Q = 3$.
(d) If $P = 10$ and $Q = 20$, show that $\ln(PQ) \neq (\ln P)(\ln Q)$.
(e) Check Property 3 using natural logarithms and the values $P = 17$ and $Q = 76$.
(f) Show that $(\log_{10} 17)/(\log_{10} 76) \neq \log_{10} 17 - \log_{10} 76$.

65. (a) Check Property 4 using the values $b = 10$, $P = \pi$, and $n = 7$.
(b) Using the values given for b, P, and n in part (a), show that $\log_b P^n \neq (\log_b P)^n$.
(c) Verify Property 5 using the values $b = 10$ and $P = 1776$.
(d) Verify that $\ln 2 + \ln 3 + \ln 4 = \ln 24$.
(e) Verify that $\log_{10} A + \log_{10} B + \log_{10} C = \log_{10}(ABC)$, using the values $A = 11$, $B = 12$, and $C = 13$.
(f) Let $f(x) = e^x$ and $g(x) = \ln x$. Compute $f[g(2345.6)]$.

(g) Let $f(x) = 10^x$ and $g(x) = \log_{10} x$. Compute $g[f(0.123456)]$.

66. Use your calculator to evaluate $\left[\frac{1}{\pi} \ln(640320^3 + 744)\right]^2$.
Remark Contrary to the empirical evidence from your calculator, it is known that this number is irrational.

B

67. *The approximation* $\ln(1 + x) \approx x$: As indicated in the accompanying graph, the values of $\ln(1 + x)$ and x are very close to one another for small positive values of x. Using your calculator, complete the table to obtain numerical evidence of this. For your answers, report the first six decimal places of the calculator display (don't round off).

x		0.1	0.05	0.005	0.0005
$\ln(1 + x)$					

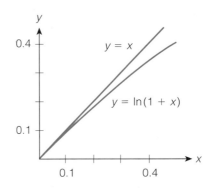

68. As indicated in the figure, the graph of $y = ae^{bx}$ passes through the two points $(2, 10)$ and $(8, 80)$. Follow steps (a) through (c) to determine the values of the constants a and b.

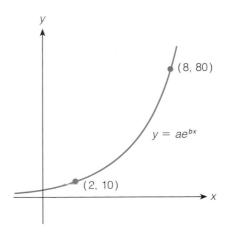

(a) Since the point $(8, 80)$ lies on the graph, the pair of values $x = 8$ and $y = 80$ satisfies the equation $y = ae^{bx}$, that is,

$$80 = ae^{8b} \qquad (1)$$

Similarly, since the point $(2, 10)$ lies on the graph, we have

$$10 = ae^{2b} \qquad (2)$$

Now use equations (1) and (2) to show that $b = (\ln 8)/6$.

(b) In equation (2), substitute for b using the expression obtained in part (a). Show that the resulting equation can be written $10 = a(e^{\ln 8})^{1/3}$.

(c) Use one of the properties of logarithms to simplify the right-hand side of the equation in part (b). Then solve for a. (You should obtain $a = 5$.)

In Exercises 69 and 70, you are given the coordinates of two points on the graph of the curve $y = ae^{bx}$. In each case, determine the values of a and b. Hint: Use the method explained in Exercise 68.

69. $(-2, 324)$ and $\left(\frac{1}{2}, \frac{4}{3}\right)$ **70.** $(1, 2)$ and $(4, 8)$

71. (a) Use base 10 logarithms to solve the equation $2^x - 3 = 0$.
    (b) Use your calculator to evaluate the answer in part (a). Also use the calculator to evaluate $(\ln 3)/(\ln 2)$.
    (c) Prove that $(\log_{10} 3)/(\log_{10} 2) = (\ln 3)/(\ln 2)$.

72. (a) Prove that $\log_b(P/Q) = \log_b P - \log_b Q$.
    Hint: Study the proof of Property 2 in the text.
    (b) Prove the **change of base formula**:

$$\log_a x = \frac{\log_b x}{\log_b a}$$

    Hint: Use the method of Example 7 in the text.

73. Show that $\log_b \dfrac{\sqrt{3} + \sqrt{2}}{\sqrt{3} - \sqrt{2}} = 2 \log_b(\sqrt{3} + \sqrt{2})$.

74. (a) Show that $\log_b(P/Q) + \log_b(Q/P) = 0$.
    (b) Simplify: $\log_a x + \log_{1/a} x$.

75. Simplify: $b^{3 \log_b x}$.

76. Let $\log_{10} 2 = a$ and $\log_{10} 3 = b$. Express each logarithm in terms of a and/or b.
    (a) $\log_{10} 4$ (b) $\log_{10} 8$ (c) $\log_{10} 5$
    (d) $\log_{10} 6$ (e) $\log_{10}(5/64)$ (f) $\log_{10} 108$
    (g) $\log_{10} \sqrt[3]{12}$ (h) $\log_{10} 0.0027$

77. Simplify: $\log_2 \sqrt[5]{4\sqrt{2}}$.

In Exercises 78 and 79, solve for x in terms of α and β.

78. $\alpha \ln x + \ln \beta = 0$ **79.** $3 \ln x = \alpha + 3 \ln \beta$

80. Is there a constant k such that the equation $e^x = 2^{kx}$ holds for all values of x?

81. Prove that $\log_b a = 1/(\log_a b)$.

82. Simplify: $(\log_2 3)(\log_3 4)(\log_4 5)$.

C

83. Prove that $(\log_a x)/(\log_{ab} x) = 1 + \log_a b$.

84. Simplify a^x when $x = [\log_b(\log_b a)]/\log_b a$.

85. If $a^2 + b^2 = 7ab$, and a and b are positive, show that

$$\log[\tfrac{1}{3}(a + b)] = \tfrac{1}{2}(\log a + \log b)$$

no matter which base is used for the logarithms (but it is understood that the same base is used throughout).

86. Solve for x (assuming $a > b > 0$):

$$(a^4 - 2a^2b^2 + b^4)^{x-1} = (a - b)^{2x}(a + b)^{-2}$$

Answer: $x = \dfrac{\ln(a - b)}{\ln(a + b)}$

87. Let $f(x) = \ln(x + \sqrt{x^2 + 1})$. Find $f^{-1}(x)$.

88. Suppose that $\log_{10} 2 = a$ and $\log_{10} 3 = b$. Solve for x in terms of a and b:

$$6^x = \frac{10}{3} - 6^{-x}$$

89. Let $y = \ln[\ln(\ln x)]$. First, complete the table using a calculator. After doing that, disregard the evidence in your table and prove (without a calculator, of course) that the range of the given function is actually the set of all real numbers.

x	100	1000	10^6	10^{20}	10^{50}	10^{99}
y						

5.5 COMPOUND INTEREST

S = Pe^{rt}. This result is remarkable both because of its simplicity and the occurrence of e. (Who would expect that number to pop up in finance theory?)

Philip Gillett in *Calculus and Analytic Geometry,* 3rd ed. (Lexington, Mass.: D.C. Heath & Company, 1988)

We begin this section by considering how money accumulates in a savings account. Eventually, this will lead us back to the number e and to functions of the form $y = ae^{bx}$.

The following idea from arithmetic is a prerequisite for our discussion. To increase a given quantity by, say, 15%, we multiply the quantity by 1.15. For instance, suppose that we want to increase $100 by 15%. The calculations can be written

$$100 + 0.15(100) = 100(1 + 0.15) = 100(1.15)$$

Similarly, to increase a quantity by 30%, we would multiply by 1.30, and so on. The next example displays some calculations involving percentage increase. The results may surprise you unless you're already familiar with this topic.

EXAMPLE 1 An amount of $100 is increased by 15% and then the new amount is increased by 15%. Is this the same as an overall increase of 30%?

Solution To increase $100 by 15%, we multiply by 1.15 to obtain $100(1.15). Now to increase this new amount by 15%, we multiply it by 1.15 to obtain

$$[(\$100)(1.15)](1.15) = \$100(1.15)^2 = \$132.25$$

Alternatively, if we increase the original $100 by 30%, we obtain

$$\$100(1.30) = \$130$$

Comparing, we see that the result of two successive 15% increases is greater than the result of a single 30% increase.

Now let's look at another example and use it to introduce some terminology. Suppose that you place $1000 in a savings account at 10% interest *compounded annually*. This means that at the end of each year, the bank contributes to your account 10% of the amount that is in the account at that time. Interest compounded in this manner is called **compound interest**. The original deposit of $1000 is called the **principal**, denoted P. The interest rate, expressed as a decimal, is denoted by r. Thus, $r = 0.10$ in this example. The variable A is used to denote the **amount** in the account at any given time. The calculations displayed in Table 1 (on the next page) show how the account grows.

We can learn several things from Table 1. First, consider how much interest is paid each year.

Interest paid for first year:	$1100 − $1000 = $100
Interest paid for second year:	$1210 − $1100 = $110
Interest paid for third year:	$1331 − $1210 = $121

Thus the interest earned each year is not a constant; it increases each year.

TABLE 1

Time Period	Algebra	Arithmetic
After 1 year	$A = P(1 + r)$	$A = 1000(1.10)$ $= \$1100$
After 2 years	$A = [P(1 + r)](1 + r)$ $= P(1 + r)^2$	$A = 1100(1.10)$ $= \$1210$
After 3 years	$A = [P(1 + r)^2](1 + r)$ $= P(1 + r)^3$	$A = 1210(1.10)$ $= \$1331$

If you look at the algebra in Table 1, you can see what the general formula should be for the amount after t years.

Compound Interest Formula (Interest compounded annually)

Suppose that a principal of P dollars is invested at an annual rate r that is compounded annually. Then the amount A after t years is given by

$$A = P(1 + r)^t$$

EXAMPLE 2 Suppose that $2000 is invested at $7\frac{1}{2}$% interest compounded annually. How many years will it take for the money to double?

Solution In the formula $A = P(1 + r)^t$, we use the given values $P = \$2000$ and $r = 0.075$. We want to find how long it will take for the money to double; that is, we want to find t when $A = \$4000$. Making these substitutions in the formula, we obtain

$$4000 = 2000(1 + 0.075)^t$$

and, therefore,

$$2 = 1.075^t \qquad \text{dividing by 2000}$$

We can solve this exponential equation by taking the logarithm of both sides. We use base e logarithms. (Base 10 would also be convenient here.) This yields

$$\ln 2 = \ln 1.075^t$$

and, consequently,

$$\ln 2 = t \ln 1.075 \qquad \text{(Why?)}$$

To isolate t, we divide both sides of this last equation by $\ln 1.075$. This yields

$$t = \frac{\ln 2}{\ln 1.075} \approx 9.6 \text{ years}$$

Now, assuming that the bank computes the compound interest only at the end of the year, we must round the preliminary answer of 9.6 years and say that when $t = 10$ years, the initial $2000 will have *more than* doubled. Table 2 adds some perspective to this. The table shows that after 9 years, something less than $4000 is in the account; whereas after 10 years, the amount exceeds $4000.

TABLE 2

$A = 2000(1.075)^t$

t (years)	A (dollars)
9	3834.48
10	4122.06

In Example 2, the interest was compounded annually. In practice, though, interest is usually computed more often. For instance, a bank may advertise a rate of 10% per year compounded semiannually. This means that after half a year, the interest is compounded at 5%, and then after another half year, the interest is again compounded at 5%. If you review Example 1, you'll see that one compounding at a rate of r is not the same as two compoundings, each at a rate of $r/2$. The formula $A = P(1 + r)^t$ can be generalized to cover such cases where interest is compounded more than once each year.

Compound Interest Formula
(Interest compounded n times per year)

Suppose that a principal of P dollars is invested at an annual rate r that is compounded n times per year. Then the amount A after t years is given by

$$A = P\left(1 + \frac{r}{n}\right)^{nt}$$

▌�î▌ **EXAMPLE 3**   Suppose that $1000 is placed in a savings account at 10% per annum. How much is in the account at the end of one year if the interest is **(a)** compounded once each year ($n = 1$)? **(b)** compounded quarterly ($n = 4$)?

Solution   We use the formula $A = P[1 + (r/n)]^{nt}$.

(a) For $n = 1$, we obtain

$$A = 1000\left(1 + \frac{0.10}{1}\right)^{(1)(1)}$$

$$= 1000(1.1)$$

$$= \$1100$$

(b) For $n = 4$, we obtain

$$A = 1000\left(1 + \frac{0.10}{4}\right)^{4(1)}$$

$$= 1000(1.025)^4$$

$$= \$1103.81 \qquad \text{using a calculator}$$

Notice that compounding the interest quarterly rather than annually yields the greater amount. This is in agreement with our observations in Example 1.   ▌▌▌

The results in Example 3 will serve to illustrate some additional terminology used by financial institutions. In that example, the interest for the year under quarterly compounding was

$$\$1103.81 - \$1000 = \$103.81$$

Now, $103.81 is 10.381% of $1000. We say in this case that the **effective rate** of interest is 10.381%. The given rate of 10% per annum compounded once a year is called the **nominal rate.*** The next example further illustrates these ideas.

▌▐▌ **EXAMPLE 4**   A bank offers a nominal interest rate of 12% per annum for certain accounts. Compute the effective rate if interest is compounded monthly.

---

*The nominal rate and the effective rate are also referred to as the *annual rate* and the *annual yield,* respectively.

Solution Let P denote the principal earning 12% ($r = 0.12$) compounded monthly ($n = 12$). Then with $t = 1$, our formula yields

$$A = P\left(1 + \frac{0.12}{12}\right)^{12(1)} = P(1.01)^{12}$$

Using a calculator to approximate the quantity $(1.01)^{12}$, we obtain

$$A \approx P(1.12683)$$

This shows that the effective interest rate is about 12.68%. ▪▪▪

There are two rather natural questions to ask when one first encounters compound interest calculations:

QUESTION 1 For a fixed period of time (say one year), does more and more frequent compounding of interest continue to yield greater and greater amounts?

QUESTION 2 Is there a limit on how much money can accumulate in a year when interest is compounded more and more frequently?

The answer to both of these questions is yes. If you look back over Example 1, you'll see evidence for the affirmative answer to question 1. For additional evidence, and for the answer to question 2, let's do some calculations. To keep things as simple as possible, suppose a principal of \$1 is invested for 1 year at the nominal rate of 100% per annum. (More realistic figures could be used here, but the algebra becomes more cluttered.) With these data, our formula becomes

$$A = 1\left(1 + \frac{1}{n}\right)^{n(1)} \qquad \text{or} \qquad A = \left(1 + \frac{1}{n}\right)^{n}$$

Now, as we know from Section 5.2, the expression $\left(1 + \frac{1}{n}\right)^{n}$ approaches the value e as n grows larger and larger. See Table 3 (on the next page).

Table 3 shows that the amount does increase with the number of compoundings. But assuming that the bank rounds off to the nearest penny, Table 3 also shows that there is no difference between compounding hourly, compounding each minute, and compounding each second. In each case, the amount, when rounded off, is \$2.72.

Some banks advertise interest compounded not monthly, daily, or even hourly, but *continuously*—that is, at each instant. The formula for the amount earned under continuous compounding of interest is as follows.

Compound Interest Formula
(Interest compounded continuously)

Suppose that a principal of P dollars is invested at an annual rate r that is compounded continuously. Then the amount A after t years is given by

$$A = Pe^{rt}$$

TABLE 3

Results of compounding interest more and more frequently

Number of Compoundings, n	Amount, $\left(1 + \dfrac{1}{n}\right)^n$
$n = 1$ (annually)	$\left(1 + \dfrac{1}{1}\right)^1 = 2$
$n = 2$ (semiannually)	$\left(1 + \dfrac{1}{2}\right)^2 = 2.25$
$n = 4$ (quarterly)	$\left(1 + \dfrac{1}{4}\right)^4 \approx 2.44$
$n = 12$ (monthly)	$\left(1 + \dfrac{1}{12}\right)^{12} \approx 2.61$
$n = 365$ (daily)	$\left(1 + \dfrac{1}{365}\right)^{365} \approx 2.7146$
$n = 8760$ (hourly)	$\left(1 + \dfrac{1}{8760}\right)^{8760} \approx 2.7181$
$n = 525,600$ (each minute)	$\left(1 + \dfrac{1}{525,600}\right)^{525,600} \approx 2.71827$
$n = 31,536,000$ (each second)	$\left(1 + \dfrac{1}{31,536,000}\right)^{31,536,000} \approx 2.71828$

▮▮▮ **EXAMPLE 5** A sum of $100 is placed in a savings account at 5% per annum compounded continuously. Assuming no subsequent withdrawals or deposits, when will the balance reach $150?

Solution Substitute the values $A = 150$, $P = 100$, and $r = 0.05$ in the formula $A = Pe^{rt}$ to obtain

$$150 = 100e^{0.05t}$$

and, therefore,

$$1.5 = e^{0.05t}$$

To solve this last equation for t, we rewrite it in its equivalent logarithmic form:

$$0.05t = \ln 1.5$$
$$t = \frac{\ln 1.5}{0.05} \text{ years}$$

Using a calculator, we find that $t \approx 8.1$ years. In other words, it will take slightly more than 8 years 1 month for the balance to reach $150. ▮▮▮

In the next example, we compare the nominal rate with the effective rate under continuous compounding of interest.

▌▌▌ EXAMPLE 6

(a) Given a nominal rate of 8% per annum compounded continuously, compute the effective interest rate.

(b) Given an effective rate of 8% per annum, compute the nominal rate.

Solution

(a) With the values $r = 0.08$ and $t = 1$, the formula $A = Pe^{rt}$ yields

$$A = Pe^{0.08(1)}$$

$$A \approx P(1.08329) \qquad \text{using a calculator}$$

This shows that the effective interest rate is approximately 8.33% per year.

(b) We now wish to compute the nominal rate r, given an effective rate of 8% per year. An effective rate of 8% means that the initial principal P grows to $P(1.08)$ by the end of the year. Thus, in the formula $A = Pe^{rt}$, we make the substitutions $A = P(1.08)$ and $t = 1$. This yields

$$P(1.08) = Pe^{r(1)}$$

Dividing both sides of this last equation by P, we have

$$1.08 = e^{r}$$

To solve this equation for r, we rewrite it in its equivalent logarithmic form:

$$r = \ln(1.08)$$

$$r \approx 0.07696 \qquad \text{using a calculator}$$

Thus, a nominal rate of about 7.70% per annum yields an effective rate of 8%. Table 4 summarizes these results. ▌▌▌

We now have two basic formulas for compound interest:

$$A = P\left(1 + \frac{r}{n}\right)^{nt} \qquad \text{for interest compounded } n \text{ times per year}$$

and

$$A = Pe^{rt} \qquad \text{for interest compounded continuously}$$

Comparison of Nominal and Effective Rates in Example 6

Nominal Rate (% per annum)	Effective Rate (% per annum)
8	8.33
7.70	8

At first glance, these formulas appear to be very different from one another: the second formula involves e, the first does not. However, appearances can be deceiving. As the next example shows, the equation $A = P\left(1 + \frac{r}{n}\right)^{nt}$ can be rewritten in a form that is entirely similar to $A = Pe^{rt}$.

▌▌▌ EXAMPLE 7 Find a constant k such that

$$P\left(1 + \frac{r}{n}\right)^{nt} = Pe^{kt} \tag{1}$$

Solution We want to solve for k in equation (1). After dividing both sides of the equation by P, we have

$$\left(1 + \frac{r}{n}\right)^{nt} = e^{kt}$$

Writing this last equation in logarithmic form yields

$$kt = \ln\left(1 + \frac{r}{n}\right)^{nt} = nt \ln\left(1 + \frac{r}{n}\right) \qquad \text{(Why?)}$$

$$k = n \ln\left(1 + \frac{r}{n}\right) \qquad \text{dividing by } t \qquad\qquad (2)$$

This shows that the formula $A = P\left(1 + \frac{r}{n}\right)^{nt}$ can be rewritten $A = Pe^{kt}$, where k is given by equation (2). ▮▮▮

Now we come to one of the remarkable and characteristic features of compound interest and of growth governed by the formula $A = Pe^{rt}$. By the **doubling time** we mean, as the name implies, the amount of time required for a given principal to double. The surprising fact here is that the doubling time does not depend on the principal P. To see why this is so, we begin with the formula

$$A = Pe^{rt}$$

We are interested in the time t at which $A = 2P$. Replacing A by $2P$ in the formula yields

$$2P = Pe^{rt}$$
$$2 = e^{rt}$$
$$rt = \ln 2$$
$$t = \frac{\ln 2}{r}$$

Denoting the doubling time by T_2, we have the following formula.

$$\boxed{\text{Doubling time} = T_2 = \frac{\ln 2}{r}}$$

As you can see, the formula for the doubling time T_2 does not involve P, but only r. Thus, at a given rate under continuous compounding, \$2 and \$2000 would both take the same amount of time to double. (This idea takes some getting used to.)

▮▮▮ **EXAMPLE 8** Compute the doubling time T_2 when a sum is invested at an interest rate of 4% per annum compounded continuously.

Solution $T_2 = \dfrac{\ln 2}{r} = \dfrac{\ln 2}{0.04} \approx 17.3$ years   using a calculator   ▮▮▮

There is a convenient approximation that allows us easily to estimate doubling times. Using a calculator, we see that

$$\ln 2 \approx 0.7$$

Using this approximation, we have the following rule of thumb for estimating doubling time.

$$T_2 \approx \frac{0.7}{r}$$

TABLE 5

t (years)	A (dollars)
0	1000
7	2000
14	4000
21	8000
28	16000

FIGURE 1
Exponential growth function

Let's use this rule to rework Example 8. With $r = 0.04$, we obtain

$$T_2 \approx \frac{0.7}{0.04} = \frac{70}{4} = 17.5 \text{ years}$$

Notice that this estimation is quite close to the actual doubling time obtained in Example 8.

The idea of doubling time is useful in graphing the function $A = Pe^{rt}$. In this discussion, we assume that P and r are constants, so that the amount A is a function of the time t. Suppose, for example, that a principal of $1000 is invested at 10% per annum compounded continuously. Then the function we wish to graph is

$$A = 1000e^{0.1t}$$

The doubling time in this situation is

$$T_2 \approx \frac{0.7}{r} = \frac{0.7}{0.1} = 7 \text{ years}$$

Table 5 shows the results of doubling a principal of $1000 every 7 years.

We'll use the data in Table 5 to graph the function $A = 1000e^{0.1t}$. We'll mark off units on the t-axis in multiples of 7; on the A-axis, we'll use multiples of 2000. Figure 1 shows the result of plotting the points from Table 5 and then joining them with a smooth curve. Notice that the domain in this context is $[0, \infty)$.

EXERCISE SET 5.5

1. You invest $800 at 6% interest compounded annually. How much is in the account after 4 years, assuming that you make no subsequent withdrawal or deposit?

2. A sum of $1000 is invested at an interest rate of $5\frac{1}{2}$% compounded annually. How many years will it take until the sum exceeds $2500? (First find out when the amount equals $2500; then round off as in Example 2.)

3. At what interest rate (compounded annually) will a sum of $4000 grow to $6000 in 5 years?

4. A bank pays 7% interest compounded annually. What principal will grow to $10,000 in 10 years?

5. You place $500 in a savings account at 5% compounded annually. After 4 years you withdraw all your money and take it to a different bank, which advertises a rate of 6% compounded annually. What is the balance in this new account after 4 more years? (As usual, assume that no subsequent withdrawal or deposit is made.)

6. A sum of $3000 is placed in a savings account at 6% per annum. How much is in the account after 1 year if the interest is compounded (a) annually? (b) semiannually? (c) daily?

7. A sum of $1000 is placed in a savings account at 7% per annum. How much is in the account after 20 years if the interest is compounded (a) annually? (b) quarterly?

8. Your friend invests $2000 at $5\frac{1}{4}$% per annum compounded semiannually. You invest an equal amount at the same yearly rate, but compounded daily. How much larger is your account than your friend's after 8 years?

9. You invest $100 at 6% per annum compounded quarterly. How long will it take for your balance to exceed $120? (Round off your answer to the next quarter.)

10. A bank offers an interest rate of 7% per annum compounded daily. What is the effective rate?

11. What principal should you deposit at $5\frac{1}{2}$% per annum compounded semiannually so as to have $6000 after 10 years?

12. You place a sum of $800 in a savings account at 6% per annum compounded continuously. Assuming that you make no subsequent withdrawal or deposit, how much is in the account after 1 year? When will the balance reach $1000?

13. A bank offers an interest rate of $6\frac{1}{2}$% per annum compounded continuously. What principal will grow to $5000 in 10 years under these conditions?

14. Given a nominal rate of 6% per annum, compute the effective rate under continuous compounding of interest.

15. Suppose that under continuous compounding of interest, the effective rate is 6% per annum. Compute the nominal rate.

16. You have two savings accounts, each with an initial principal of $1000. The nominal rate on both accounts is $5\frac{1}{4}$% per annum. In the first account, interest is compounded semiannually. In the second account, interest is compounded continuously. How much more is in the second account after 12 years?

17. You want to invest $10,000 for 5 years, and you have a choice between two accounts. The first pays 6% per annum compounded annually. The second pays 5% per annum compounded continuously. Which is the better investment?

18. Suppose that a certain principal is invested at 6% per annum compounded continuously.
    (a) Use the rule $T_2 \approx 0.7/r$ to estimate the doubling time.
    (b) Compute the doubling time using the formula $T_2 = (\ln 2)/r$.
    (c) Do your answers in (a) and (b) differ by more than 2 months?

19. A sum of $1500 is invested at 5% per annum compounded continuously.
    (a) Estimate the doubling time.
    (b) Compute the actual doubling time.
    (c) Let d_1 and d_2 denote the actual and estimated doubling times, respectively. Define d by

    $$d = |d_1 - d_2|$$

    What percent is d of the actual doubling time?

20. A sum of $5000 is invested at 10% per annum compounded continuously.
    (a) Estimate the doubling time.
    (b) Estimate the time required for the $5000 to grow to $40,000.

21. After carrying out the calculations in this problem, you'll see one of the reasons why the government imposes inheritance taxes and why laws are passed to prohibit savings accounts from being passed from generation to generation without restriction. Suppose that a family invests $1000 at 8% per annum

compounded continuously. If this account were to remain intact, being passed from generation to generation, for 300 years, how much would be in the account at the end of those 300 years?

22. A principal of $500 is invested at 7% per annum compounded continuously.
    (a) Estimate the doubling time.
    (b) Sketch a graph similar to the one in Figure 1, showing how the amount increases with time.

23. A principal of $7000 is invested at 5% per annum compounded continuously.
    (a) Estimate the doubling time.
    (b) Sketch a graph showing how the amount increases with time.

24. In one savings account, a principal of $1000 is deposited at 5% per annum. In a second account, a principal of $500 is deposited at 10% annum. Both accounts compound interest continuously.
    (a) Estimate the doubling time for each account.
    (b) On the same set of axes, sketch graphs showing the amount of money in each account over time. Give the (approximate) coordinates of the point where the two curves meet. In financial terms, what is the significance of this point? (In working this problem, assume that the initial deposits in each account were made at the same time.)
    (c) During what period of time does the first account have the larger balance?

25. A principal of $2000 is deposited in a savings account (Account 1) at 4% per annum, compounded quarterly. At the same time, a principal of $2000 is deposited in another savings account (Account 2) at 4% per annum compounded continuously.
    (a) Complete the following table to compare the results. (Round off the values to the nearest dollar.)

t (years)	1	2	3	4	10
A (Account 1)					
A (Account 2)					

    (b) Compute the doubling time for each account.
    (c) Compute the average rate of change $\Delta A/\Delta t$ for Account 1 over the period $t = 2$ years to $t = 3$ years.
    (d) Compute the average rate of change $\Delta A/\Delta t$ for Account 2 over the period $t = 2$ years to $t = 3$ years.

5.6 EXPONENTIAL GROWTH AND DECAY

Drawing by Professor Ann Jones, University of Colorado, Boulder. From the cover of *The Physics Teacher*, vol. 14, no. 7 (October 1976).

In newspapers and in everyday speech, the term *exponential growth* is used rather loosely to describe any situation involving rapid growth. In the sciences, however, **exponential growth** refers specifically to growth governed by functions of the form $y = ae^{bx}$, where a and b are positive constants. For example, since the function $A = Pe^{rt}$ (discussed in the previous section) has this general form, we say that money grows exponentially under continuous compounding of interest. Similarly, in the sciences, **exponential decay** refers specifically to decrease or decay governed by functions of the form $y = ae^{bx}$, where a is positive and b is negative. As examples in this section, we shall consider population growth, global warming, and radioactive decay.

Under ideal conditions involving unlimited food and space, a colony of bacteria increases according to the *growth law*

$$\mathcal{N} = \mathcal{N}_0 e^{kt}$$

In this formula, $\mathcal{N}$ is the population at time t and k is a positive constant related to the growth rate of the population. The number $\mathcal{N}_0$ is also a constant; it represents the size of the population at time $t = 0$. Sometimes, to emphasize the fact that $\mathcal{N}$ depends on (is a function of) t, we use function notation to rewrite the growth law as $\mathcal{N}(t) = \mathcal{N}_0 e^{kt}$.

EXAMPLE I Suppose that at the start of an experiment in a biology laboratory, 1200 bacteria are present in a colony. Four hours later, the population is found to be 2200.

(a) Determine the *growth constant k*.

(b) How many bacteria were there three hours after the experiment began? (Round off to the nearest 50.)

(c) When will the population reach 3000?

Solution

(a) The initial population is 1200, so we have

$$\mathcal{N} = 1200e^{kt} \tag{1}$$

We are also given that $\mathcal{N} = 2200$ when $t = 4$. Substituting these values in equation (1) yields $2200 = 1200e^{4k}$, and therefore $e^{4k} = 11/6$. Rewriting this last equation in logarithmic form, we have

$$4k = \ln(11/6)$$

and therefore

$$k = \frac{\ln(11/6)}{4} \tag{2}$$

(b) Substituting $t = 3$ in equation (1) yields

$$\mathcal{N} = 1200e^{3k} \qquad \text{where } k \text{ is given by equation (2)}$$
$$\approx 1900 \text{ bacteria} \qquad \text{using a calculator and rounding off}$$

(c) We want to find t when $\mathcal{N} = 3000$. Substituting $\mathcal{N} = 3000$ in equation (1) yields $3000 = 1200e^{kt}$, and therefore $e^{kt} = 5/2$. Rewriting this equation in logarithmic form, we have

$$kt = \ln(5/2)$$
$$t = \frac{\ln(5/2)}{k} \qquad \text{where } k \text{ is given by equation (2)}$$
$$t \approx 6 \text{ hours} \qquad \text{using a calculator and rounding off}$$

The population reaches 3000 approximately six hours after the start of the experiment. ▮▮▮

In the next example we predict the population of the world in the year 2000, assuming that the population grows exponentially. In this context, it can be shown that the constant k in the formula $\mathcal{N} = \mathcal{N}_0 e^{kt}$ represents the **relative growth rate** of the population $\mathcal{N}$.* Algebraically, we will work with k just as we did in the calculations for compound interest in the previous section, where k represented an interest rate.

▮▮▮ **EXAMPLE 2** Statistical projections from the United Nations indicate that the relative growth rate for the world's population during the years 1992–2000 will be approximately 1.6% per year. (This is down from the all-time high of 2% per year in 1970.) Use the formula $\mathcal{N} = \mathcal{N}_0 e^{kt}$ to predict the world population in the year 2000, given that the population in 1992 was 5.420 billion.

---

*The *relative growth rate* is $(\Delta \mathcal{N}/\mathcal{N})/(\Delta t)$; it is the fractional change in $\mathcal{N}$ per unit time. This is distinct from (but sometimes confused with) the *average growth rate,* $\Delta \mathcal{N}/\Delta t$.

Solution   Let $t = 0$ correspond to the year 1992. Then the year 2000 corresponds to the value $t = 8$. Our given data therefore are $k = 0.016$, $t = 8$, and $N_0 = 5.420$ (in units of one billion). Using these values in the formula $N = N_0 e^{kt}$, we obtain

$$N = 5.420 e^{(0.016)8}$$
$$\approx 6.160 \qquad \text{using a calculator and rounding off}$$

Thus, our prediction for the world's population in the year 2000 is 6.160 billion.   ∎∎∎

∎▪∎ **EXAMPLE 3**   The following statements about the increasing levels of carbon dioxide in the atmosphere appear in the book *The Challenge of Global Warming,* edited by Dean Edwin Abrahamson (Washington, D.C.: Island Press, 1989).

> Carbon dioxide, the single most important greenhouse gas, accounts for about half of the warming that has been experienced as a result of past emissions† and also for half of the projected future warming. The present [1988] concentration is now about 350 parts per million (ppm) and is increasing about 0.4% ... per year....

Assuming that the concentration of carbon dioxide in the atmosphere continues to increase exponentially at 0.4% per year, estimate when the concentration might reach 600 ppm. (This would be roughly twice the level estimated to exist prior to the Industrial Revolution.)

Solution   Let $t = 0$ correspond to the year 1988. In the formula $N = N_0 e^{kt}$, we want to determine the time t when $N = 600$. Using the values $N_0 = 350$, $N = 600$, and $k = 0.004$, we have

$$600 = 350 e^{0.004t} \qquad \text{or} \qquad \frac{600}{350} = e^{0.004t}$$

and, consequently,

$$\ln(600/350) = 0.004t$$
$$t = \frac{\ln(600/350)}{0.004} \approx 135 \qquad \text{using a calculator and rounding off}$$

Now, 135 years beyond our base year of 1988 is 2123. Rounding off once more, we summarize our result this way: If the carbon dioxide levels continue to increase exponentially at 0.4% per year, then the concentration will reach 600 ppm by approximately 2125.   ∎∎∎

In Examples 1 and 2, we used the function $N = N_0 e^{kt}$ to describe population growth. In Example 3 we used this function to model the increasing levels of carbon dioxide in the atmosphere. It is a remarkable fact that the same basic formula, but now with $k < 0$, describes radioactive decay. How do scientists know that this is the appropriate formula here? We mention two reasons—one practical, one theoretical. As early as 1902, E. Rutherford and F. Soddy carried out experiments in England to measure radioactive decay. They found that an equation of the form $N = N_0 e^{kt}$, with $k < 0$, indeed aptly described their data.

---

†The two principal sources of these emissions are the burning of fossil fuels and the burning of vegetation in the tropics.

Alternatively, on theoretical grounds, it can be argued that the decay rate at each instant must be proportional to the amount $\mathcal{N}$ present at that instant. Under this condition, calculus can be used to show that the law of exponential decay applies. (See Exercise 49 for some of the details.)

In discussing radioactive decay, it is convenient to introduce the term *half-life*. (As you'll see, this is analogous to the concept of *doubling time*, which we discussed in the previous section.)

DEFINITION Half-life

The **half-life** of a radioactive substance is the time required for half of a given sample to disintegrate. The half-life is an intrinsic property of the substance; it does not depend on the given sample size.

EXAMPLE

Iodine-131 is a radioactive substance with a half-life of 8 days. Suppose that 2 g are present initially. Then:

at $t = 0$, 2 g are present;

at $t = 8$ days, 1 g is left;

at $t = 16$ days, $\frac{1}{2}$ g is left;

at $t = 24$ days, $\frac{1}{4}$ g is left;

at $t = 32$ days, $\frac{1}{8}$ g is left.

TABLE I

t (days)	$\mathcal{N}$ (amount)
0	$\mathcal{N}_0$
8	$\mathcal{N}_0/2$
16	$\mathcal{N}_0/4$
24	$\mathcal{N}_0/8$
32	$\mathcal{N}_0/16$

Just as we used the idea of doubling time to graph an exponential growth function in the previous section, we can use the half-life to graph an exponential decay function. Consider, for example, the radioactive substance iodine-131, which has a half-life of eight days. Table 1 shows what fraction of an initial amount remains at eight-day intervals. Using the data in this table, we can draw the graph of the decay function $\mathcal{N} = \mathcal{N}_0 e^{kt}$ for iodine-131 (see Figure 1). Notice that we are able to construct this graph without specifically evaluating the decay constant k. (The next example shows how to determine k.)

FIGURE I
The exponential decay function for iodine-131

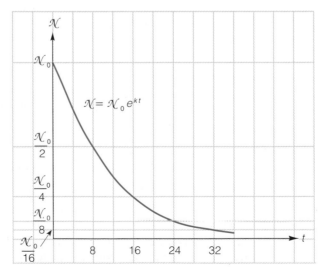

▌▬▌ **EXAMPLE 4** Hospitals utilize the radioactive substance iodine-131 in the diagnosis of conditions of the thyroid gland. The half-life of iodine-131 is eight days.

(a) Determine the *decay constant k* for iodine-131.
(b) If a hospital acquires 2 g of iodine-131, how much of this sample will remain after 20 days?
(c) How long will it be until only 0.01 g remains?

Solution

(a) The half-life is eight days. This means that when t is 8, the value of $\mathcal{N}$ is $\frac{1}{2}\mathcal{N}_0$. Using these values for t and $\mathcal{N}$ in the formula $\mathcal{N} = \mathcal{N}_0 e^{kt}$, we obtain

$$\tfrac{1}{2}\mathcal{N}_0 = \mathcal{N}_0 e^{k8} \qquad \text{and therefore} \qquad \tfrac{1}{2} = e^{8k}$$

We solve for k by writing this last equation in logarithmic form:

$$8k = \ln \tfrac{1}{2} \qquad \text{and therefore} \qquad k = \frac{\ln \frac{1}{2}}{8}$$

If a numerical value for k is required, we can use a calculator (or tables) to obtain $k \approx -0.0866$.

(b) We are given that $\mathcal{N}_0 = 2$ g, and we want to find the value of $\mathcal{N}$ when $t = 20$ days. First, before using algebra and a calculator, let's estimate the answer to get a feeling for the situation. The half-life is eight days. So after eight days, 1 g remains; after 16 days, 0.5 g remains; and after 24 days, 0.25 g remains. Since 20 is between 16 and 24, it follows that after 20 days, the amount remaining will be something between 0.5 g and 0.25 g. Now for the actual calculations: Substitute the values $\mathcal{N}_0 = 2$ and $t = 20$ in the decay law $\mathcal{N} = \mathcal{N}_0 e^{kt}$. This yields

$$\mathcal{N} = 2e^{20k} \qquad \text{where } k \text{ has the value determined in part (a)}$$

$$\approx 0.354 \text{ g} \qquad \text{using a calculator}$$

Thus, after 20 days, approximately 0.35 g of the iodine-131 remains. Note that this figure is indeed between 0.5 g and 0.25 g, as we first estimated.

(c) We are given that $\mathcal{N}_0 = 2$ g and we want to find the time t at which $\mathcal{N}$ is 0.01 g. Substituting the values $\mathcal{N}_0 = 2$ and $\mathcal{N} = 0.01$ into the decay law $\mathcal{N} = \mathcal{N}_0 e^{kt}$ yields

$$0.01 = 2e^{kt} \qquad \text{where } k \text{ has the value determined in part (a)}$$

To solve this equation for t, first divide both sides by 2 to obtain $0.005 = e^{kt}$. The logarithmic form of this last equation is $kt = \ln(0.005)$. Therefore

$$t = \frac{\ln(0.005)}{k}$$

$$\approx 61.2 \text{ days} \qquad \text{using a calculator and the value of } k \text{ from part (a)} \quad ▌▌▌$$

In Example 4(a), we found that the decay constant for iodine-131 is given by $k = \left(\ln \tfrac{1}{2}\right)/8$. Notice that the denominator in this expression is the half-life for iodine-131. By following the same reasoning used in Example 4(a), we find that the decay constant for any radioactive substance is always $\ln \tfrac{1}{2}$ divided by the half-life. For reference, this useful fact is restated in the box that follows.

PROPERTY SUMMARY   A FORMULA FOR THE DECAY CONSTANT k

FORMULA	EXAMPLE
$k = \dfrac{\ln \frac{1}{2}}{\text{half-life}}$	The half-life of strontium-90 is 28 years, and therefore $k = (\ln \frac{1}{2})/28$.

EXAMPLE 5 An article on nuclear energy appeared in the January 1976 issue of *Scientific American*. The author of the article is Hans Bethe (1906–1992), a Nobel prize winner in physics. At one point in the article, Professor Bethe discussed the disposal (through burial) of radioactive waste material from a nuclear reactor. The particular waste product under discussion is plutonium-239.

> … Plutonium-239 has a half-life of nearly 25,000 years, and 10 half-lives are required to cut the radioactivity by a factor of 1000. Thus, the buried wastes must be kept out of the biosphere for 250,000 years.

(a) Supply the detailed calculations to support the statement that 10 half-lives are required before the radioactivity is reduced by a factor of 1000.

(b) Show how the figure of 10 half-lives can be obtained by estimation, as opposed to detailed calculation.

Solution

(a) Let $\mathcal{N}_0$ denote the initial amount of plutonium-239 at time $t = 0$. Then the amount $\mathcal{N}$ present at time t is given by $\mathcal{N} = \mathcal{N}_0 e^{kt}$. We wish to determine t when $\mathcal{N} = \frac{1}{1000}\mathcal{N}_0$. First, since the half-life is 25,000 years, we have $k = (\ln \frac{1}{2})/25{,}000$. Now, substituting $\frac{1}{1000}\mathcal{N}_0$ for $\mathcal{N}$ in the decay law gives us

$$\frac{1}{1000}\mathcal{N}_0 = \mathcal{N}_0 e^{kt} \qquad \text{where } k \text{ is } (\ln \tfrac{1}{2})/25{,}000$$

$$\frac{1}{1000} = e^{kt}$$

$$kt = \ln \frac{1}{1000}$$

$$t = \frac{\ln \frac{1}{1000}}{k} = \frac{\ln \frac{1}{1000}}{(\ln \frac{1}{2})/25{,}000} = \frac{(\ln \frac{1}{1000})(25{,}000)}{\ln \frac{1}{2}}$$

Using a calculator, we obtain the value 249,144.6 years; however, given the time scale involved, it would be ludicrous to announce the answer in this form. Instead, we round off the answer to the nearest thousand years and say that after 249,000 years, the radioactivity will have decreased by a factor of 1000. Notice that this result confirms Professor Bethe's ballpark estimate of 10 half-lives, or 250,000 years.

(b) After 1 half-life: $\mathcal{N} = \dfrac{\mathcal{N}_0}{2}$

After 2 half-lives: $\mathcal{N} = \dfrac{1}{2}\left(\dfrac{\mathcal{N}_0}{2}\right) = \dfrac{\mathcal{N}_0}{2^2}$

After 3 half-lives: $\mathcal{N} = \dfrac{1}{2}\left(\dfrac{\mathcal{N}_0}{2^2}\right) = \dfrac{\mathcal{N}_0}{2^3}$

Following this pattern, we see that after 10 half-lives, we should have

$$\mathcal{N} = \frac{\mathcal{N}_0}{2^{10}}$$

But as we noted in the first section of this chapter, 2^{10} is approximately 1000. Therefore, we have

$$\mathcal{N} \approx \frac{\mathcal{N}_0}{1000} \qquad \text{after 10 half-lives}$$

This is in agreement with Professor Bethe's statement.   ▮▮▮

EXERCISE SET 5.6

In Exercises 1 and 2, assume that the populations grow exponentially, that is, according to the law $\mathcal{N} = \mathcal{N}_0 e^{kt}$.

1. At the start of an experiment, 2000 bacteria are present in a colony. Two hours later, the population is 3800.
    (a) Determine the growth constant k.
    (b) Determine the population five hours after the start of the experiment.
    (c) When will the population reach 10,000?

2. At the start of an experiment, 2×10^4 bacteria are present in a colony. Eight hours later, the population is 3×10^4.
    (a) Determine the growth constant k.
    (b) What was the population two hours after the start of the experiment?
    (c) How long will it take for the population to triple?

3. The figure shows the graph of an exponential growth function $\mathcal{N} = \mathcal{N}_0 e^{kt}$. Determine the values of $\mathcal{N}_0$ and k.

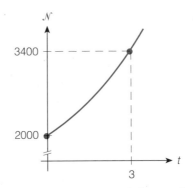

4. Suppose that you are helping a friend with his homework on the growth law, $\mathcal{N} = \mathcal{N}_0 e^{kt}$, and on his paper you see the equation $e^k = -0.75$. How do you know at this point that your friend must have made an error?

The statistics in Exercises 5–11 are taken from the 1992 World Population Data Sheet (Washington, D.C.: Population Reference Bureau, Inc., 1992). In Exercises 5–7(a), complete the tables, assuming that the populations grow exponentially and that the indicated relative growth rates are valid through the year 2000.

5.

Region	1992 Population (billions)	Percent of Population in 1992	Relative Growth Rate (percent per year)	Year 2000 Population (billions)	Percent of World Population in 2000
World	5.420	100	1.7	?	?
More developed regions	1.224	?	0.5	?	?
Less developed regions	4.196	?	2.0	?	?

6.

Country	1992 Population (millions)	Relative Growth Rate (percent per year)	Year 2000 Population (millions)	Percent Increase in Population
United States	255.6	0.8	?	?
People's Republic of China	1165.8	1.3	?	?
Mexico	87.7	2.3	?	?

7. (a)

Country	1992 Population (millions)	Relative Growth Rate (percent per year)	Year 2000 Population (millions)	Percent Increase in Population
Iraq	18.2	3.7	?	?
United Kingdom	57.8	0.3	?	?

(b) Iraq has one of the highest population growth rates in the world, and United Kingdom has one of the lowest. Use the data given in part (a) to compute an estimate for the year in which the two populations could be equal. (Round off your final answer to the nearest five years.) *Hint:* You need to solve the equation $(18.2)e^{0.037t} = (57.8)e^{0.003t}$.

8. In 1992, the countries Chad and Denmark each had populations of about 5.2 million. However, the relative growth rate for Chad was 2.5% per year, while that for Denmark was 0.1% per year.
(a) Assuming exponential growth and the given growth rates, make projections for the population of each nation in the year 2010.
(b) When might the population of Chad reach 10 million? What would be the population of Denmark at this time?

9. In 1992, Cyprus and Gaza each had populations of about 0.7 million. However, the relative growth rate for Cyprus was 1.1% per year, while that for Gaza was 4.6% per year.
(a) Assuming exponential growth and the given growth rates, make projections for the populations in the year 2000.
(b) When might the population of Cyprus reach 1 million? What would be the population of Gaza at this time?

10. The countries of Georgia and Tajikistan (former members of the Soviet Union) each had 1992 populations of 5.5 million. However, the relative growth rate for Georgia was 0.9% per year, while that for Tajikistan was 3.2% per year.
(a) Assuming exponential growth and the given growth rates, make projections for the population of each nation in the year 2010.
(b) In what year might the population of Georgia reach 10 million? What would be the population of Tajikistan at this time?

11. In 1992 the population of Kuwait was 1.4 million.
(a) Assuming exponential growth, what value for k would lead to a population of 2 million in 25 years (that is, by the year 2017)?
Remark The answer you obtain is, in fact, the current growth rate for Kuwait.
(b) What value for k would lead to a population of 2 million in the year 2100?

12. According to the United States Bureau of the Census, the population of the United States grew most rapidly during the period 1800–1810 and least rapidly during the period 1930–1940. Use the following data to compute the relative growth rate (percent per year) for each of these two periods. In 1800, the population was 5,308,483; in 1810, it was 7,239,881. In 1930, the population was 123,202,624; in 1940, it was 132,164,569.

13. According to the U.S. Bureau of the Census, in the year 1850 the population of the United States was 23,191,876; in 1900, the population was 62,947,714.
(a) Assuming that the population grew exponentially during this period, compute the growth constant k.
(b) Assuming continued growth at the same rate, predict the 1950 population.
(c) The actual population for 1950, according to the Bureau of the Census, was 150,697,361. How does this compare to your prediction in part (b)? Was the actual growth over the period 1900–1950 faster or slower than exponential growth with the growth constant k determined in part (a)?

In Exercises 14–16, use the population statistics given in the following table, provided by the U.S. Bureau of the Census.

State	1930	1940	1950
California	5,677,251	6,907,387	10,586,223
New York	12,588,066	13,479,142	14,830,192
North Dakota	680,845	641,935	619,636

14. (a) Assume that the population of California grew exponentially over the period 1930–1940. Compute the growth constant k and express it as a percent (per year).
    (b) Assume that the population of California grew exponentially over the period 1940–1950. Compute the growth constant k and express it as a percent.
    (c) What would the 1990 California population have been if the relative growth rate obtained in part (b) had remained in effect throughout the period 1950–1990?   *Hint:*   Let $t = 0$ correspond to 1950. Then find $\mathcal{N}$ when $t = 40$.
    (d) How does your prediction in part (c) compare to the actual 1990 population of 29,279,000?

15. Repeat Exercise 14 for New York: *Note:* The 1990 population of New York was 17,627,000.

16. Repeat Exercise 14 for North Dakota. *Note:* The 1990 population of North Dakota was 634,000.

17. Lester R. Brown, in his book *State of the World 1985* (New York: W. W. Norton and Co., 1985), makes the following statement:

    > The projected [*population*] growth for North America, all of Europe, and the Soviet Union is less than the additions expected in either Bangladesh or Nigeria.

    In this exercise, you are asked to carry out the type of calculations that could be used to support Lester Brown's projection for the period 1990–2025. The source of the data in the following table is the Population Division of the United Nations.

Region	1990 Population (millions)	Projected Relative Growth Rate (percent per year)	2025 Projected Population (millions)
North America	275.2	0.7	?
Soviet Union	291.3	0.7	?
Europe	499.5	0.2	?
Nigeria	113.3	3.1	?

   (a) Complete the table, assuming that the populations grow exponentially and that the indicated growth rates are valid over the period 1990–2025.
   (b) According to the projections in part (a), what will be the net increase in Nigeria's population over the period 1990–2025?
   (c) According to the projections in part (a), what will be the net increase in the combined populations of North America, the (former) Soviet Union, and Europe over the period 1990–2025?
   (d) Compare your answers in parts (b) and (c). Do your results support or contradict Lester Brown's projection?

18. As of 1992, nine countries in the world had populations exceeding 100 million. These countries are listed (in order of population, from largest to smallest) in the table. According to *World Population Profile: 1987*, published by the United States Bureau of the Census, "The latest projections suggest that India's population may surpass China's in less than 60 years, or before today's youngsters in both countries reach old age." Using the data in the table, estimate the populations of the nine countries in the year 2050. (Assume that the indicated growth rates are valid over the period 1992–2050.) Tabulate your final answers, listing the countries in order of population, from largest to smallest. Do your results for China and India support the projection by the Census Bureau?

Country	1992 Population (millions)	Relative Growth Rate (percent per year)
1. China	1165.8	1.2
2. India	882.6	2.0
3. United States	255.6	0.8
4. Indonesia	184.5	1.7
5. Brazil	150.8	1.9
6. Russia	149.3	0.2
7. Japan	124.4	0.3
8. Pakistan	121.7	3.1
9. Bangladesh	111.4	2.4

Source: *1992 World Population Data Sheet*, published by Population Reference Bureau, Inc., Washington, D.C.

In Exercises 19 and 20, use the half-life information to complete each table. (The formula $\mathcal{N} = \mathcal{N}_0 e^{kt}$ is not required.)

19. (a) Uranium-228: half-life = 550 seconds

t (seconds)	0	550	1100	1650	2200
$\mathcal{N}$ (grams)	8				

   (b) Uranium-238: half-life = 4.9×10^9 years

t (years)	0				
$\mathcal{N}$ (grams)	10	5	2.5	1.25	0.625

20. (a) Polonium-210: half-life = 138.4 days

t (days)	0	138.4	276.8	415.2	
N (grams)	0.4				0.025

(b) Polonium-214: half-life = 1.63×10^{-4} second

t (seconds)	0				6.52×10^{-4}
N (grams)	0.1	0.05	0.025	0.0125	

21. The half-life of iodine-131 is eight days. How much of a 1-g sample will remain after seven days?

22. The half-life of strontium-90 is 28 years. How much of a 10-g sample will remain after (a) 1 year? (b) 10 years?

23. The radioactive isotope sodium-24 is used as a tracer to measure the rate of flow in an artery or vein. The half-life of sodium-24 is 14.9 hours. Suppose that a hospital buys a 40-g sample of sodium-24.
(a) How much of the sample will remain after 48 hours?
(b) How long will it be until only 1 g remains?

24. The radioactive isotope carbon-14 is used as a tracer in medical and biological research. Compute the half-life of carbon-14 given that the decay constant k is -1.2097×10^{-4}. (The units for k here are such that your half-life answer will be in years.)

25. (a) The half-life of radium-226 is 1620 years. Draw a graph of the decay function for radium-226, similar to that shown in Figure 1.
(b) The half-life of radium-A is 3 min. Draw a graph of the decay function for radium-A.

26. (a) The half-life of thorium-232 is 1.4×10^{10} years. Draw a graph of the decay function.
(b) The half-life of thorium-A is 0.16 sec. Draw a graph of the decay function.

27. The half-life of plutonium-241 is 13 years.
(a) How much of an initial 2-g sample remains after 5 years?
(b) Find the time required for 90% of the 2-g sample to decay. *Hint:* If 90% has decayed, then 10% remains.

28. The half-life of radium-226 is 1620 years.
(a) How much of a 2-g sample remains after 100 years?
(b) Find the time required for 80% of the 2-g sample to decay.

29. The half-life of thorium-229 is 7340 years.
(a) Compute the time required for a given sample to be reduced by a factor of 1000. Show detailed calculations, as in Example 5(a).

(b) Express your answer in part (a) in terms of half-lives.
(c) As in Example 5(b), estimate the time required for a given sample of thorium-229 to be reduced by a factor of 1000. Compare your answer with that obtained in part (b).

30. The Chernobyl nuclear explosion (in the former Soviet Union, on April 25, 1987) released large amounts of radioactive substances into the atmosphere. These substances included cesium-137, iodine-131, and strontium-90. Although the radioactive material covered many countries, the actual amount and intensity of the fallout varied greatly from country to country, due to vagaries of the weather and the winds. One area that was particularly hard hit was Lapland, where heavy rainfall occurred just when the Chernobyl cloud was overhead.
(a) Many of the pastures in Lapland were contaminated with cesium-137, a radioactive substance with a half-life of 33 years. If the amount of cesium-137 was found to be ten times the normal level, how long would it take until the level returns to normal? (That is, compute the time required for the amount of cesium-137 to be reduced by a factor of 10.)
(b) Follow part (a), but assume that the amount of cesium-137 was 100 times the normal level. Remark Several days after the explosion, it was reported that the level of cesium-137 in the air over Sweden was 10,000 times the normal level. Fortunately there was little or no rainfall.

31. Strontium-90, with a half-life of 28 years, is a waste product from nuclear fission reactors. One of the reasons great care is taken in the storage and disposal of this substance stems from the fact that strontium-90 is, in some chemical respects, similar to ordinary calcium. Thus, strontium-90 in the biosphere, entering the food chain via plants or animals, would eventually be absorbed into our bones. (In fact, everyone already has a measurable amount of strontium-90 in their bones as a result of fallout from atmospheric nuclear tests.)
(a) Compute the decay constant k for strontium-90.
(b) Compute the time required if a given quantity of strontium-90 is to be stored until the radioactivity is reduced by a factor of 1000.
(c) Using half-lives, estimate the time required for a given sample to be reduced by a factor of 1000. Compare your answer with that obtained in (b).

32. (a) Suppose that a certain country violates the ban against above-ground nuclear testing and, as a result, an island is contaminated with debris containing the radioactive substance iodine-131. A team of scientists from the United Nations wants to visit the island to look for clues in determining which country was involved. However, the level of radioactivity from the iodine-131 is estimated to be 30,000 times the safe level. Approximately how long must the team wait before it is safe to visit the island? The half-life of iodine-131 is 8 days.

(b) Rework part (a), assuming instead that the radioactive substance is strontium-90 rather than iodine-131. The half-life of strontium-90 is 28 years. Assume, as before, that the initial level of radioactivity is 30,000 times the safe level. (This exercise underscores the difference between a half-life of 8 days and one of 28 years.)

33. The *mean life* μ of a radioactive particle or nucleus is defined by $\mu = -1/k$, where k is the decay constant. (It can be shown that the mean life is the average time that a particle survives before it decays.)

(a) Show that $\mu = \dfrac{\text{half-life}}{\ln 2}$.

(b) Krypton-91 has a half-life of 10 sec. Compute the mean life of krypton-91.

(c) How much of a 1-g sample of krypton-91 remains after a period of three mean lives?

34. (a) Show that after a period of three mean lives, approximately 5% of an initial sample $\mathcal{N}_0$ will remain.

(b) Show that after a period of four mean lives, less than 2% of an initial sample $\mathcal{N}_0$ will remain.

The statistics presented in Exercises 35(b)–38 are taken from the Global 2000 Report to the President. *The following statement by then President Jimmy Carter provides some background on the report.*

> I am directing the Council on Environmental Quality and the Department of State, working in cooperation with ... other appropriate agencies, to make a one-year study of the probable changes in the world's population, natural resources, and environment through the end of the century.
>
> President Jimmy Carter, May 23, 1977

35. In 1969 the United States National Academy of Sciences issued a report entitled *Resources and Man.* One conclusion in the report is that a world population of 10 billion "is close to (if not above) the maximum that an intensively managed world might hope to support with some degree of comfort and individual choice." (The figure "10 billion" is sometimes referred to as the *carrying capacity* of the Earth.)

(a) When the report was issued in 1969, the world population was about 3.6 billion, with a relative growth rate of 2% per year. Assuming continued exponential growth at this rate, estimate the year in

which the Earth's carrying capacity of 10 billion might be reached.

(b) The data in the following table, from the *Global 2000 Report,* show the world population estimates for 1975. Use the "low" figures to estimate the year in which the Earth's carrying capacity of 10 billion might be reached.

(c) Use the "high" figures to estimate the year in which the Earth's carrying capacity of 10 billion might be reached.

Projection	Population (billions)	Relative Growth Rate (percent per year)
Low	4.043	1.5
High	4.134	2.0

36. *Depletion of Nonrenewable Energy Resources:* Suppose that the world population grows exponentially. Then, as a first approximation, it is reasonable to assume that the use of nonrenewable energy resources, such as petroleum and coal, also grows exponentially. Under these conditions, the following formula can be derived (using calculus):

$$A = \frac{A_0}{k}(e^{kT} - 1)$$

where A is the amount of the resource consumed from time $t = 0$ to $t = T$, the quantity A_0 is the amount of the resource consumed during the year $t = 0$, and k is the relative growth rate of annual consumption.

(a) Show that solving the formula for T yields

$$T = \frac{\ln[(Ak/A_0) + 1]}{k}$$

This formula gives the "life expectancy" T for a given resource. In the formula, A_0 and k are as previously defined, and A represents the total amount of the resource available.

(b) In the *Global 2000 Report to the President,* it was estimated that the 1976 worldwide consumption of oil was 21.7 billion barrels and that the total remaining oil resources ultimately available were 1661 billion barrels. Compute the "life expectancy" for oil if the relative growth rate k is (i) 1%/year; (ii) 2%/year; (iii) 3%/year. (This last value is, in fact, the predicted rate for the next decade.)

(c) In the *Global 2000 Report,* it was estimated that the 1976 worldwide consumption of natural gas was approximately 50 tcf (trillions of cubic feet), while the total remaining gas ultimately available was 8493 tcf. Compute the "life expectancy" for natural gas if $k = 2\%$/year.

37. In this exercise, the term *nonrenewable energy resources* refers collectively to petroleum, natural gas, coal, shale oil, and uranium. In the *Global 2000 Report,* it was estimated that in 1976, the total worldwide consumption of these nonrenewable resources amounted to 250 quadrillion Btu. It was also estimated that in 1976, the total nonrenewable resources ultimately available were 161,241 quadrillion Btu. Use the formula for T in Exercise 36 to compute the life expectancy for these nonrenewable energy resources if (a) $k = 1\%$ per year; (b) $k = 2\%$ per year.

38. The data in the following table, from the *Global 2000 Report,* show the world reserves for selected minerals. The term *world reserves* refers to known resources available with current technology. Use the formula for T derived in Exercise 36 to estimate the "depletion date" for each mineral under these conditions (depletion date = $T + 1976$).

Mineral	1976 World Reserves	1976 Consumption	Relative Growth Rate (percent per year)
Fluorine (million short tons)	37	2.1	4.58
Silver (million troy ounces)	6,100	305	2.33
Tin (thousand metric tons)	10,000	241	2.05
Copper (million short tons)	503	8.0	2.94
Phosphate rock (million metric tons)	25,732	107	5.17
Aluminum in bauxite (million short tons)	5,610	18	4.29

39. In 1956, scientists B. Gutenberg and C. F. Richter developed a formula to estimate the amount of energy E released in an earthquake. The formula is

$$\log_{10} E = 11.4 + 1.5M$$

where E is the energy in units of ergs and M is the Richter magnitude.
(a) Solve the formula for E.
(b) The table shows the Richter magnitudes for two large earthquakes. If E_1 denotes the energy of the Alaskan quake and E_2 denotes the energy of the quake in Cairo, compute the ratio E_1/E_2 to compare the energies of the two earthquakes. Round off your answer to the nearest integer.

Date	Location	M (Magnitude)
March 28, 1964	Alaska	8.6
October 18, 1992	Cairo	5.9

40. The age of some rocks can be estimated by measuring the ratio of the amounts of certain chemical elements within the rock. The method known as the *rubidium–strontium method* will be discussed here. This method has been used in dating the moon rocks brought back on the Apollo missions.

Rubidium-87 is a radioactive substance with a half-life of 4.7×10^{10} years. Rubidium-87 decays into the substance strontium-87, which is stable (nonradioactive). We are going to derive the following formula for the age of a rock:

$$T = \frac{\ln[(\mathcal{N}_s/\mathcal{N}_r) + 1]}{-k}$$

where T is the age of the rock, k is the decay constant for rubidium-87, $\mathcal{N}_s$ is the number of atoms of strontium-87 now present in the rock, and $\mathcal{N}_r$ is the number of atoms of rubidium-87 now present in the rock.
(a) Assume that initially, when the rock was formed, there were $\mathcal{N}_0$ atoms of rubidium-87 and none of strontium-87. Then, as time goes by, some of the rubidium atoms decay into strontium atoms, but the total number of atoms must still be $\mathcal{N}_0$. Thus, after T years, we have

$$\mathcal{N}_0 = \mathcal{N}_r + \mathcal{N}_s$$

or, equivalently,

$$\mathcal{N}_s = \mathcal{N}_0 - \mathcal{N}_r \qquad (1)$$

However, according to the law of exponential decay for the rubidium-87, we must have $\mathcal{N}_r = \mathcal{N}_0 e^{kT}$. Solve this equation for $\mathcal{N}_0$ and then use the result to eliminate $\mathcal{N}_0$ from equation (1). Show that the result can be written

$$\mathcal{N}_s = \mathcal{N}_r e^{-kT} - \mathcal{N}_r \qquad (2)$$

(b) Solve equation (2) for T to obtain the formula given at the beginning of this exercise.

41. (Continuation of Exercise 40) The half-life of rubidium-87 is 4.7×10^{10} years. Compute the decay constant k.

42. (Continuation of Exercise 40) Analysis of lunar rock samples taken on the Apollo 11 mission showed the strontium-rubidium ratio to be

$$\frac{\mathcal{N}_s}{\mathcal{N}_r} = 0.0588$$

Estimate the age of these lunar rocks.

43. (Continuation of Exercise 40) Analysis of the so-called genesis rock sample taken on the Apollo 15 mission revealed a strontium-rubidium ratio of 0.0636. Estimate the age of this rock.

44. *Radiocarbon Dating:* Because rubidium-87 decays so slowly, the technique of rubidium-strontium dating is generally considered effective only for objects older than 10 million years. In contrast, archeologists and geologists rely on the *radiocarbon dating* method in assigning ages ranging from 500 to 50,000 years.

    Two types of carbon occur naturally in our environment—carbon-12, which is nonradioactive, and carbon-14, which has a half-life of 5730 years. All living plant and animal tissue contains both types of carbon, always in the same ratio. (The ratio is one part carbon-14 to 10^{12} parts carbon-12.) As long as the plant or animal is living, this ratio is maintained. When the organism dies, however, no new carbon-14 is absorbed, and the amount of carbon-14 begins to decrease exponentially. Since the amount of carbon-14 decreases exponentially, it follows that the level of radioactivity also must decrease exponentially. The formula describing this situation is

    $$\mathcal{N} = \mathcal{N}_0 e^{kT}$$

    where T is the age of the sample, $\mathcal{N}$ is the present level of radioactivity (in units of disintegrations per hour per gram of carbon), and $\mathcal{N}_0$ is the level of radioactivity T years ago, when the organism was alive. Given that the half-life of carbon-14 is 5730 years and that $\mathcal{N}_0 = 920$ disintegrations per hour per gram, show that the age T of a sample is given by

    $$T = \frac{5730 \ln(\mathcal{N}/920)}{\ln \frac{1}{2}}$$

In Exercises 45–48, use the formula derived in Exercise 44 to estimate the age of each sample. Note: Some technical complications arise in interpreting such results. Studies have shown that the ratio of carbon-12 to carbon-14 in the air (and therefore in living matter) has not in fact been constant over time. For instance, air pollution from factory smokestacks tends to increase the level of carbon-12. In the other direction, nuclear bomb testing increases the level of carbon-14.

45. Prehistoric cave paintings were discovered in the Lascaux cave in France. Charcoal from the site was analyzed and the level of radioactivity was found to be $\mathcal{N} = 141$ disintegrations per hour per gram. Estimate the age of the paintings.

46. Before radiocarbon dating was used, historians estimated that the tomb of Vizier Hemaka, in Egypt, was constructed about 4900 years ago. After radiocarbon dating became available, wood samples from the tomb were analyzed, and it was determined that the radioactivity level was 510 disintegrations per hour per gram. Estimate the age of the tomb on the basis of this reading and compare your answer to the figure already mentioned.

47. Analyses of the oldest campsites in the Western Hemisphere reveal a carbon-14 radioactivity level of $\mathcal{N} = 226$ disintegrations per hour per gram. Show that this implies an age of 11,500 years, to the nearest 500 years. (This would correspond to the last Ice Age. At that time, the sea level was significantly lower than today. It is believed that at the time of the last Ice Age, land stretched across what is now the Bering Strait, and humans first entered the Western Hemisphere by this route.)

48. The Dead Sea Scrolls are a collection of ancient manuscripts discovered in caves along the west bank of the Dead Sea. (The discovery occurred by accident when an Arab herdsman of the Taamireh tribe was searching for a stray goat.) When the linen wrappings on the scrolls were analyzed, the carbon-14 radioactivity level was found to be 723 disintegrations per hour per gram. Estimate the age of the scrolls using this information. Historical evidence suggests that some of the scrolls date back somewhere between 150 B.C. and A.D. 40. How do these dates compare with the estimate derived using radiocarbon dating?

49. Let $\mathcal{N} = \mathcal{N}_0 e^{kt}$. In this exercise we show that if Δt is very small, then $\Delta \mathcal{N}/\Delta t \approx k\mathcal{N}$. In other words, over very small intervals of time, the average rate of change of $\mathcal{N}$ is proportional to $\mathcal{N}$ itself.

    (a) Show that the average rate of change of the function $\mathcal{N} = \mathcal{N}_0 e^{kt}$ on the interval $[t, t + \Delta t]$ is given by

    $$\frac{\Delta \mathcal{N}}{\Delta t} = \frac{\mathcal{N}_0 e^{kt}(e^{k\Delta t} - 1)}{\Delta t} = \frac{\mathcal{N}(e^{k\Delta t} - 1)}{\Delta t}$$

    (b) In Section 5.2 we saw that $e^x \approx 1 + x$ when x is close to 0. Thus, if Δt is sufficiently small, we have $e^{k\Delta t} \approx 1 + k\Delta t$. Use this approximation and the result in part (a) to show that

    $$\frac{\Delta \mathcal{N}}{\Delta t} \approx k\mathcal{N} \qquad \text{when } \Delta t \text{ is sufficiently close to zero}$$

CHAPTER FIVE SUMMARY OF PRINCIPAL TERMS AND FORMULAS

TERMS OR FORMULAS	PAGE REFERENCE	COMMENTS
1. Exponential function with base b	232	The exponential function with base b is defined by the equation $y = b^x$. It is understood here that the base b is a positive number other than 1.
2. Asymptote	232	A line is an asymptote for a curve if the separation distance between the curve and the line approaches zero as we move out farther and farther along the line.
3. The number e	237	The irrational number e is one of the basic constants in mathematics, as is the irrational number π. To five decimal places, the value of e is 2.71828. In calculus, e is the base most commonly used for exponential functions. The graph of the exponential function $y = e^x$ is shown in the box on page 240.
4. $\log_b x$	247	The expression $\log_b x$ denotes the exponent to which b must be raised to yield x. The equation $\log_b x = y$ is equivalent to $b^y = x$.
5. $\ln x$	250	The expression $\ln x$ means $\log_e x$. Logarithms to the base e are known as *natural logarithms*. For the graph of $y = \ln x$, see Figure 6(a) in Section 5.3.
6. $\log_a x = \dfrac{\log_b x}{\log_b a}$	262	This is the change of base formula for converting logarithms from one base to another.
7. $A = P\left(1 + \dfrac{r}{n}\right)^{nt}$	268	This formula gives the amount A that accumulates after t years when a principal of P dollars is invested at an annual rate r that is compounded n times per year.
8. $A = Pe^{rt}$	269	This formula gives the amount A that accumulates after t years when a principal of P dollars is invested at an annual rate r that is compounded continuously.
9. $N = N_0 e^{kt}$ $(k > 0)$	275	This is the formula for exponential growth. (It can be shown that if a quantity N grows at a rate that is proportional to N itself, then N must follow this law of exponential growth.) In the formula, N is a function of the time t and N_0 denotes the value of N at time $t = 0$. The positive number k is the *growth constant*; it is related to (but not equal to) the rate at which N increases. *Optional note:* It can be shown that k is the fractional change in N per unit time: $k = \dfrac{\Delta N / N}{\Delta t}$.
10. $N = N_0 e^{kt}$ $(k < 0)$	275	This is the decay law for radioactive substances. In the formula, N denotes the amount of the substance present at time t, and N_0 is the amount present at time $t = 0$. The negative number k is the *decay constant*; it is related to (but not equal to) the rate at which N decreases. (See the optional note in item 9.)
11. Half-life	278	The half-life of a radioactive substance is the time required for half of a given sample to disintegrate. Notice that the half-life is an amount of time, not an amount of the substance.

WRITING MATHEMATICS

1. In Section 5.3, we defined logarithmic functions in terms of inverse functions.
   (a) Explain in general what is meant by a pair of inverse functions.
   (b) Assuming a knowledge of the function $f(x) = 2^x$, explain how to define and then graph the function $g(x) = \log_2 x$.

2. A student who was trying to simplify the expression $\dfrac{\ln(e^3)}{\ln(e^2)}$ wrote

   $$\frac{\ln(e^3)}{\ln(e^2)} = \ln(e^3) - \ln(e^2) = 3 - 2 = 1$$

   (a) What is the error here? What do you think is the most probable reason for making this error?
   (b) Give the correct solution.

3. A student who wanted to simplify the expression $\dfrac{100}{\log_{10} 100}$ wrote

   $$\frac{100}{\log_{10} 100} = \frac{\overset{1}{\cancel{100}}}{\log_{10} \underset{1}{\cancel{100}}} = \frac{1}{\log_{10}}$$

   Explain why this is nonsense, and then indicate the correct solution.

CHAPTER FIVE   REVIEW EXERCISES

1. Which is larger, $\log_5 126$ or $\log_{10} 999$?

2. Graph the function $y = 3^{-x} - 3$. Specify the domain, range, intercept(s), and asymptote.

3. Suppose that the population of a colony of bacteria increases exponentially. If the population at the start of an experiment is 8000, and 4 hours later it is 10,000, how long (from the start of the experiment) will it take for the population to reach 12,000? (Express the answer in terms of base e logarithms.)

4. Express $\log_{10} 2$ in terms of base e logarithms.

5. Let f be the function defined by

   $$f(x) = \begin{cases} 2^{-x} & \text{if } x < 0 \\ x^2 & \text{if } x \ge 0 \end{cases}$$

   Sketch the graph of f and then use the horizontal line test to determine whether f is one-to-one.

6. Estimate 2^{60} in terms of an integral power of 10.

7. Solve for x: $\ln(x+1) - 1 = \ln(x-1)$.

8. Suppose that $5000 is invested at 8% interest compounded annually. How many years will it take for the money to double? Make use of the approximations $\ln 2 \approx 0.7$ and $\ln 1.08 \approx 0.08$ to obtain a numerical answer.

9. On the same set of axes, sketch the graphs of $y = e^x$ and $y = \ln x$. Specify the domain and range for each function.

10. Solve for x: $xe^x - 2e^x = 0$.

11. Simplify: $\log_9\left(\frac{1}{27}\right)$.

12. Given that $\ln A = a$, $\ln B = b$, and $\ln C = c$, express $\ln\left[(A^2\sqrt{B})/C^3\right]$ in terms of a, b, and c.

13. The half-life of plutonium-241 is 13 years. What is the decay constant? Use the approximation $\ln\frac{1}{2} \approx -0.7$ to obtain a numerical answer.

14. Express as a single logarithm with a coefficient of 1: $3\log_{10} x - \log_{10}(1-x)$.

15. Solve for x, leaving your answer in terms of base e logarithms: $5e^{2-x} = 12$.

16. Let $f(x) = e^{x+1}$. Find a formula for $f^{-1}(x)$ and specify the domain of f^{-1}.

17. Suppose that in 1990 the population of a certain country was 2 million and increasing with a relative growth rate of 2% per year. Estimate the year in which the population will reach 3 million. (Use the approximation $\ln 1.5 \approx 0.40$ to obtain a numerical answer.)

18. Simplify: $\ln e + \ln \sqrt{e} + \ln 1 + \ln e^{\ln 10}$.

19. A principal of $1000 is deposited at 10% per annum compounded continuously. Estimate the doubling time and then sketch a graph showing how the amount increases with time.

20. Simplify: $\ln(\log_8 56 - \log_8 7)$.

In Exercises 21–32, graph the function and specify the asymptote(s) and intercept(s).

21. $y = e^x$
22. $y = -e^{-x}$
23. $y = \ln x$
24. $y = \ln(x + 2)$
25. $y = 2^{x+1} + 1$
26. $y = \log_{10}(-x)$
27. $y = (1/e)^x$
28. $y = \left(\frac{1}{2}\right)^{-x}$
29. $y = e^{x+1} + 1$
30. $y = -\log_2(x + 1)$
31. $y = \ln(e^x)$
32. $y = e^{\ln x}$

In Exercises 33–49, solve the equation for x. (When logarithms appear in your answer, leave the answer in that form, rather than using a calculator.)

33. $\log_4 x + \log_4(x - 3) = 1$
34. $\log_3 x + \log_3(2x + 5) = 1$
35. $\ln x + \ln(x + 2) = \ln 15$
36. $\log_6 \dfrac{x + 4}{x - 1} = 1$
37. $\log_2 x + \log_2(3x + 10) - 3 = 0$
38. $2 \ln x - 1 = 0$
39. $3 \log_9 x = \frac{1}{2}$
40. $e^{2x} = 6$
41. $e^{1 - 5x} = 3\sqrt{e}$
42. $2^x = 100$
43. $\log_{10} x - 2 = \log_{10}(x - 2)$
44. $\log_{10}(x^2 - x - 10) = 1$
45. $\ln(x + 2) = \ln x + \ln 2$
46. $\ln(2x) = \ln 2 + \ln x$
47. $\ln x^4 = 4 \ln x$
48. $(\ln x)^4 = 4 \ln x$
49. $\log_{10} x = \ln x$

50. Solve for x: $(\ln x)/(\ln 3) = \ln x - \ln 3$. Use a calculator to evaluate your result and round off to the nearest integer.   *Answer:*   206765

In Exercises 51–66, simplify the expression without using a calculator.

51. $\log_{10} \sqrt{10}$
52. $\log_7 1$
53. $\ln \sqrt[5]{e}$
54. $\log_3 54 - \log_3 2$
55. $\log_{10} \pi - \log_{10} 10\pi$
56. $\log_2 2$
57. 10^t, where $t = \log_{10} 16$
58. $e^{\ln 5}$
59. $\ln(e^4)$
60. $\log_{10}(10^{\sqrt{2}})$
61. $\log_{12} 2 + \log_{12} 18 + \log_{12} 4$
62. $(\log_{10} 8)/(\log_{10} 2)$
63. $(\ln 100)/(\ln 10)$
64. $\log_5 2 + \log_{1/5} 2$
65. $\log_2 \sqrt[7]{16\sqrt[3]{2\sqrt{2}}}$
66. $\log_{1/8}\left(\frac{1}{16}\right) + \log_5 0.02$

In Exercises 67–70, express the quantity in terms of a, b, and c, where $a = \log_{10} A$, $b = \log_{10} B$, and $c = \log_{10} C$.

67. $\log_{10} A^2 B^3 \sqrt{C}$
68. $\log_{10} \sqrt[3]{AC/B}$
69. $16 \log_{10} \sqrt{A} \sqrt[4]{B}$
70. $6 \log_{10}\left[B^{1/3}/(A\sqrt{C})\right]$

In Exercises 71–76, find consecutive integers n and n + 1 such that the given expression lies between n and n + 1. Do not use a calculator.

71. $\log_{10} 209$
72. $\ln 2$
73. $\log_6 100$
74. $\log_{10}\left(\frac{1}{12}\right)$
75. $\log_{10} 0.003$
76. $\log_3 244$

77. (a) On the same set of axes, graph the curves $y = \ln(x + 2)$ and $y = \ln(-x) - 1$. According to your graph, in which quadrant do the two curves intersect?
    (b) Find the x-coordinate of the intersection point.

78. The curve $y = ae^{bt}$ passes through the point $(3, 4)$ and has a y-intercept of 2. Find a and b.

79. A certain radioactive substance has a half-life of T years. Find the decay constant (in terms of T).

80. Find the half-life of a certain radioactive substance if it takes T years for one-third of a given sample to disintegrate. (Your answer will be in terms of T.)

81. A radioactive substance has a half-life of M minutes. What percentage of a given sample will remain after $4M$ minutes?

82. At the start of an experiment, the initial population in a colony is a bacteria. After b hours, there are c bacteria present. Determine the population d hours after the start of the experiment.

83. The half-life of a radioactive substance is d days. If you begin with a sample weighing b grams, how long will it be until c grams remain?

84. Find the half-life of a radioactive substance if it takes D days for P percent of a given sample to disintegrate.

In Exercises 85–90, write the expression as a single logarithm with a coefficient of 1.

85. $\log_{10} 8 + \log_{10} 3 - \log_{10} 12$
86. $4 \log_{10} x - 2 \log_{10} y$
87. $\ln 5 - 3 \ln 2 + \ln 16$
88. $\ln(x^4 - 1) - \ln(x^2 - 1)$
89. $a \ln x + b \ln y$
90. $\ln(x^3 + 8) - \ln(x + 2) - 2 \ln(x^2 - 2x + 4)$

In Exercises 91–98, write the quantity using sums and differences of simpler logarithmic expressions. Express the answer so that logarithms of products, quotients, and powers do not appear.

91. $\ln \sqrt{(x-3)(x+4)}$

92. $\log_{10} \dfrac{x^2 - 4}{x + 3}$

93. $\log_{10} \dfrac{3}{\sqrt{1 + x}}$

94. $\ln \dfrac{x^2\sqrt{2x + 1}}{2x - 1}$

95. $\log_{10} \sqrt[3]{x/100}$

96. $\log_{10} \sqrt{\dfrac{x^2 + 8}{(x^2 + 9)(x + 1)}}$

97. $\ln\left(\dfrac{1 + 2e}{1 - 2e}\right)^3$

98. $\log_{10} \sqrt[3]{x\sqrt{1 + y^2}}$

99. Suppose that A dollars are invested at $R\%$ compounded annually. How many years will it take for the money to double?

100. A sum of \$2800 is placed in a savings account at 9% per annum. How much is in the account after two years if the interest is compounded quarterly?

101. A bank offers an interest rate of 9.5% per annum compounded monthly. Compute the effective interest rate.

102. A sum of D dollars is placed in an account at $R\%$ per annum compounded continuously. When will the balance reach E dollars?

103. Compute the doubling time for a sum of D dollars invested at $R\%$ per annum compounded continuously.

104. Your friend invests D dollars at $R\%$ per annum compounded semiannually. You invest an equal amount at the same yearly rate, but compounded daily. How much larger is your account than your friend's after T years?

105. (a) You invest \$660 at 5.5% per annum compounded quarterly. How long will it take for your balance to reach \$1000? Round off your answer to the next quarter of a year.
   (b) You invest D dollars at $R\%$ per annum compounded quarterly. How long will it take for your balance to reach nD dollars? (Assume that $n > 1$.)

In Exercises 106–110, find the domain of each function.

106. (a) $y = \ln x$
   (b) $y = e^x$

107. (a) $y = \log_{10} \sqrt{x}$
   (b) $y = \sqrt{\log_{10} x}$

108. (a) $f \circ g$ where $f(x) = \ln x$ and $g(x) = e^x$
   (b) $g \circ f$ where f and g are as in part (a)

109. (a) $y = \log_{10}|x^2 - 2x - 15|$
   (b) $y = \log_{10}(x^2 - 2x - 15)$

110. $y = \dfrac{2 + \ln x}{2 - \ln x}$

111. Find the range of the function defined by $y = \dfrac{e^x + 1}{e^x - 1}$.

In Exercises 112–122, use the graph and the properties of logarithms to estimate the quantity to the nearest tenth.

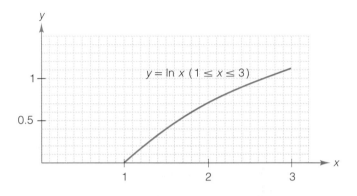

112. $\ln 2$

113. $\ln 0.5$

114. $\ln 3$

115. $\ln \frac{1}{9}$

116. $\ln 6$

117. $\ln 72$

118. $\ln 40.5$ *Hint:* $40.5 = \frac{81}{2}$

119. e *Hint:* If $x = e$, then $\ln x = 1$.

120. $e^{1.1}$ *Hint:* If $x = e^{1.1}$, then $\ln x = 1.1$.

121. $\log_2 3$

122. $\log_3 6$

123. The figure shows a portion of the graph of $y = e^x$, for $1.0970 \le x \le 1.1000$.
   (a) Use the graph to estimate $\ln 3$ to four decimal places.
   (b) Using your calculator, compute the percent error in the approximation in part (a).
   (c) Using the approximation in part (a), estimate the quantities $\ln \sqrt{3}$, $\ln 9$, and $\ln \frac{1}{3}$.

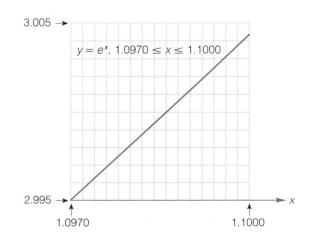

124. Given two positive numbers a and b, we define the geometric mean (G.M.) and the arithmetic mean (A.M.) as follows:

$$\text{G.M.} = \sqrt{ab} \qquad \text{A.M.} = (a+b)/2$$

If $a > b > 0$, we define the logarithmic mean (L.M.) by

$$\text{L.M.} = \frac{a-b}{\ln a - \ln b}$$

In Exercise 41 in Section 1.5, we saw that the geometric mean is always less than or equal to the arithmetic mean. In this exercise you will see that the logarithmic mean always lies between the geometric mean and the arithmetic mean. In symbols:
G.M. < L.M. < A.M.

(a) Complete the following table. Use a calculator and round off to three decimal places in all rows except for the first and the last. In the first row round off to four decimal places, and in the last round off to five decimal places.

a	b	$\sqrt{ab}$ (G.M.)	$\dfrac{a-b}{\ln a - \ln b}$ (L.M.)	$\dfrac{a+b}{2}$ (A.M.)
1.1	1.01			
2	1			
3	1			
10	5			
10	9			
10	9.9			

(b) In each of the rows that you completed in part (a), notice that the geometric mean is less than the logarithmic mean, which is in turn less than the arithmetic mean. To demonstrate that this relationship holds in general, we'll rely on the following inequalities, which hold for all values of x greater than 1:

$$\sqrt{x} < \frac{x-1}{\ln x} < \frac{x+1}{2} \qquad (1)$$

Although a rigorous proof of these inequalities requires calculus, the following figure suggests that the inequalities are correct.

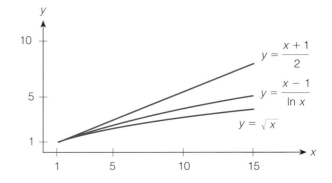

In the inequalities in (1), make the substitution $x = a/b$ and show that the resulting inequalities can be written in the form

$$\sqrt{ab} < \frac{a-b}{\ln a - \ln b} < \frac{a+b}{2} \qquad \text{as required}$$

Hint: After making the indicated substitution, multiply through by b.

CHAPTER FIVE TEST

1. Graph the function $y = 2^{-x} - 3$. Specify the domain, range, intercepts, and asymptote.

2. Suppose that the population of a colony of bacteria increases exponentially. At the start of an experiment there are 6000 bacteria, and one hour later the population is 6200. How long (from the start of the experiment) will it take for the population to reach 10,000? Give two forms for the answer: one in terms of base e logarithms and the other a calculator approximation rounded off to the nearest hour.

3. Which is larger, $\log_2 17$ or $\log_3 80$? Explain the reasons for your answer.

4. Express $\log_2 15$ in terms of base e logarithms.

5. Estimate 2^{40} in terms of an integral power of 10.

6. Suppose that 9500 is invested at 6% interest compounded annually. How long will it take for the amount in the account to reach $12,000?

7. (a) For which values of x is the identity $e^{\ln x} = x$ valid?
   (b) On the same set of axes, graph $y = e^x$ and $y = \ln x$.

8. Given that $\log_{10} A = a$ and $\log_{10} B = b$, express $\log_{10}(A^3/\sqrt{B})$ in terms of a and b.

9. Simplify each expression.
   (a) $\log_5(1/\sqrt{5})$
   (b) $\ln e^2 + \ln 1 - e^{\ln 3}$

10. Solve for x: $x^3 e^x - 4xe^x = 0$.

11. The half-life of a radioactive substance is four days.
    (a) Find the decay constant.
    (b) How much of an initial 2-g sample will remain after 10 days?

12. Express as a single logarithm with a coefficient of 1: $2\ln x - \ln\sqrt[3]{x^2 + 1}$.

13. Solve for x, leaving your answer in terms of base e logarithms: $-2e^{3x-1} = 9$.

14. Let $g(x) = \log_{10}(x - 1)$. Find a formula for $g^{-1}(x)$ and specify the range of g^{-1}.

15. A principal of \$12,000 is deposited at 6% per annum compounded continuously.
    (a) Estimate the doubling time.
    (b) Sketch a graph showing how the amount increases with time.

16. (a) For which values of x is it true that $\ln(x^2) = 2\ln x$?
    (b) For which values of x is it true that $(\ln x)^2 = 2\ln x$?

17. Let $f(x) = \begin{cases} e^x & \text{if } x < 0, \\ \sqrt{1 - x^2} & \text{if } 0 \le x < 1. \end{cases}$

    (a) Sketch the graph of f.
    (b) Is the function f one-to-one?

18. Solve for x: $\frac{1}{2} - \log_{16}(x - 3) = \log_{16} x$.

CHAPTER SIX

TRIGONOMETRIC FUNCTIONS OF ANGLES

Trigonometry, as we know the subject today, is a branch of mathematics that is linked with algebra. As such it dates back only to the eighteenth century.

   When treated purely as a development of geometry, however, it goes back to the time of the great Greek mathematician–astronomers. . . .

Alfred Hooper in *Makers of Modern Mathematics* (New York: Random House, Inc., 1948)

It is certain that the division of the ecliptic into 360 degrees . . . [was] adopted by the Greeks from Babylon. . . . [But] It was Hipparchus [ca. 180–ca. 125 B.C.] who first divided the circle in general into 360 parts or degrees, and the introduction of this division coincides with his invention of trigonometry.

Sir Thomas Heath in *A History of Greek Mathematics,* vol. II (London: Oxford University Press, 1921)

The origins of trigonometry are obscure. There are some problems in the Rhind papyrus [ca. 1650 B.C.] that involve the cotangent of the dihedral angles at the base of a pyramid, and . . . the Babylonian cuneiform tablet Plimpton 322 essentially contains a remarkable table of secants. It may be that modern investigations into the mathematics of ancient Mesopotamia will reveal an appreciable development of practical trigonometry.

Howard Eves in *An Introduction to the History of Mathematics,* 6th ed. (Philadelphia: Saunders College Publishing, 1990)

INTRODUCTION

In general, there are two approaches to trigonometry at the precalculus level. In a sense, these two approaches correspond to the two historical roots of the subject mentioned in the first of our opening quotations. One approach centers around the study of triangles. Indeed, the word "trigonometry" is derived from two Greek words, *trigonon,* meaning triangle, and *metria,* meaning measurement. It is with this triangle approach that we begin the chapter. The second approach, which is in a sense a generalization of the first, we will take up in Chapter 7. Before getting down to specifics, we offer the following bird's-eye view of Chapters 6 and 7 in terms of functions. As you'll see in subsequent chapters, a real familiarity with both approaches to trigonometry is necessary.

	INPUTS FOR THE FUNCTIONS	OUTPUTS	EMPHASIS
Chapter 6	angles	real numbers	The trigonometric functions are used to study triangles.
Chapter 7	real numbers	real numbers	The objects of study are the trigonometric functions themselves.

6.1 TRIGONOMETRIC FUNCTIONS OF ACUTE ANGLES

. . . [the English clergyman and mathematician] *William Oughtred* [1574–1660] *introduced a vast array of characters into mathematics. . . . In the* Trigonometria *(1657), he employed* ∠ *for angle.* . . .

Florian Cajori in *A History of Mathematical Notations*, vol. I (La Salle, Illinois: The Open Court Publishing Company, 1928)

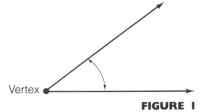

Vertex

FIGURE 1

In elementary geometry, an **angle** is a figure formed by two rays with a common endpoint. As indicated in Figure 1, the common endpoint is called the **vertex** of the angle. Recall that there are several conventions used in naming angles; Figure 2 indicates some of these. In Figure 2, the symbol θ is the lowercase Greek letter *theta*. Greek letters are often used to name angles. For reference, the Greek alphabet is given in the endpapers at the back of this book. The ∠ symbol that you see in Figure 2 stands for the word "angle." When three letters are used in naming an angle, as with ∠*ABC* in Figure 2, the middle letter always indicates the vertex of the angle.

There are several ways to indicate the size of (i.e., the amount of rotation in) an angle. In this chapter we use the familiar units of *degrees*. Recall that 360 degrees is the measure of an angle obtained by rotating a ray through one complete circle. The symbol for degrees is °. For comparison, Figure 3 (on the next page) displays angles of various degrees. In this section we will not make any distinction whether the rotation generating an angle is clockwise or counterclockwise; in a subsequent section this distinction will be important. The angles in Figures 3(a) and 3(b) are **acute angles**; these are angles with degree measure θ satisfying $0° < \theta < 90°$. The angle in Figure 3(d) is an **obtuse angle**, that is, an angle with degree measure θ satisfying $90° < \theta < 180°$.

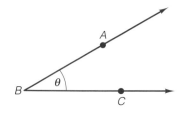

Notations for the angle at *B*:
∠*B*, θ, ∠*ABC*, ∠*CBA*

FIGURE 2

The degree is subdivided into two smaller units: the **minute**, which is $\frac{1}{60}$ of a degree; and the **second**, which is $\frac{1}{60}$ of a minute. The abbreviations for minutes and seconds are ′ and ″, respectively. Thus, we have

$1° = 60'$ There are 60 minutes in one degree.

and

$1' = 60''$ There are 60 seconds in one minute.

Until modern handheld calculators became commonplace (in the 1970s and 1980s), these subdivisions of the degree were widely used. Nowadays, however,

FIGURE 3

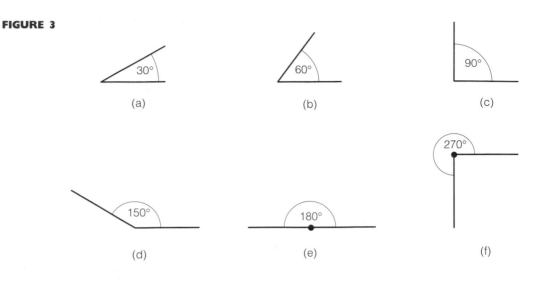

the convention in most scientific applications is to use ordinary decimal nota-tion for the fractional parts of a degree, and this is the convention we follow in this text. (We mention in passing that minutes and seconds are still widely used in some portions of astronomy and in navigation.)

One more convention: Suppose, for example, that the measure of an angle, say angle B, is 70°. We can write this as

measure $\angle B = 70°$ or, more concisely, $m \angle B = 70°$

For ease of notation and speech, however, we will often simply write $\angle B = 70°$.

We are going to define six functions, the six *trigonometric functions*. As in-dicated by the title of this section, the inputs for these functions (in this section) will be acute angles. The outputs will be real numbers. In a later section we'll make a transition so that angles of any size can serve as inputs, and in Chapter 7 we'll make another transition, this time to real-number inputs. But for now, as we have said, the inputs will be acute angles.

For reasons that are more historical* than mathematical, the trigonometric functions have names that are words rather than single letters, such as f. The use of words rather than single letters for naming functions is by no means peculiar to trigonometry. For instance, in almost every computer-programming language, the absolute value function is ABS. So, for example, we have

ABS(5) = 5   and   ABS(−5) = 5

The names of the six trigonometric functions, along with their abbreviations, are as follows.

---

*For a discussion of the names of the trigonometric functions, see Howard Eves, *An Introduction to the History of Mathematics,* 6th ed., pp. 236–237 (Philadelphia: Saunders College Publishing, 1990).

NAME OF FUNCTION	ABBREVIATION
cosine	cos
sine	sin
tangent	tan
secant	sec
cosecant	csc
cotangent	cot

FIGURE 4

Now, let θ be an acute angle in a right triangle, as shown in Figure 4. We define the six trigonometric functions of θ in the box that follows.

(Refer to Figure 4.)

$$\cos\theta = \frac{\text{length of side adjacent angle }\theta}{\text{length of hypotenuse}} \qquad \sec\theta = \frac{\text{length of hypotenuse}}{\text{length of side adjacent angle }\theta}$$

$$\sin\theta = \frac{\text{length of side opposite angle }\theta}{\text{length of hypotenuse}} \qquad \csc\theta = \frac{\text{length of hypotenuse}}{\text{length of side opposite angle }\theta}$$

$$\tan\theta = \frac{\text{length of side opposite angle }\theta}{\text{length of side adjacent angle }\theta} \qquad \cot\theta = \frac{\text{length of side adjacent angle }\theta}{\text{length of side opposite angle }\theta}$$

Our work for the next several sections will be devoted to exploring these definitions and their consequences. Two preliminary observations that will help you to understand and memorize the definitions are these.

1. An expression such as $\cos\theta$ really means $\cos(\theta)$, where cos or cosine is the name of the function and θ is an input. It is for historical rather than mathematical reasons that the parentheses are suppressed.

2. There are three pairs of reciprocals in the definitions: $\cos\theta$ and $\sec\theta$; $\sin\theta$ and $\csc\theta$; and $\tan\theta$ and $\cot\theta$.

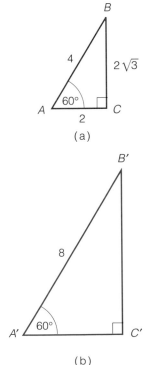

(a)

(b)

FIGURE 5

The definitions that we have just given for the trigonometric functions involve ratios of sides in a right triangle. The fact that *ratios* are involved is important. We'll use two examples to explain and emphasize this. Figure 5 shows two right triangles; both contain an angle of 60 degrees. In Figure 5(a) all three sides of the triangle are specified, while in Figure 5(b), only one side is specified. Using the data in Figure 5(a), we have

$$\tan 60° = \frac{\text{opposite}}{\text{adjacent}} = \frac{BC}{AC} = \frac{2\sqrt{3}}{2} = \sqrt{3}$$

So, from Figure 5(a), we have $\tan 60° = \sqrt{3}$. Now, the question is, will Figure 5(b) give us the same value for $\tan 60°$? [It had better, otherwise tan will not qualify as a function. (Why?)] Indeed, Figure 5(b) does give us the same value for $\tan 60°$, and it is not even necessary to use the fact that we are given a specific length in Figure 5(b). Elementary geometry* tells us that $\triangle ABC$ is similar

*If two angles of one triangle are congruent to two corresponding angles of another triangle, then the triangles are similar.

to $\triangle A'B'C'$. Then, because corresponding sides of similar triangles are proportional, we have

$$\frac{BC}{AC} = \frac{B'C'}{A'C'} \tag{1}$$

Now look at the ratio on the right-hand side of this last equation; it represents tan 60° as we would calculate it in Figure 5(b). The left-hand side of the equation, as we've already seen from Figure 5(a), represents tan 60° = $\sqrt{3}$. In summary, similar triangles and equation (1) tell us that we obtain the same value for tan 60° whether we use Figure 5(a) or 5(b).

The point of the example we've just concluded is this: For a given acute angle θ, the values of the trigonometric functions depend on the ratios of the lengths of the sides but not on the size of the particular right triangle in which θ resides. For emphasis, the next example makes this same point.

EXAMPLE 1   Figure 6 shows two right triangles. The first right triangle has sides 5, 12, and 13. The second right triangle is similar to the first (the angles are the same), but each side is 10 times longer than the corresponding side in the first triangle. Calculate and compare the values of sin θ, cos θ, and tan θ for both triangles.

FIGURE 6

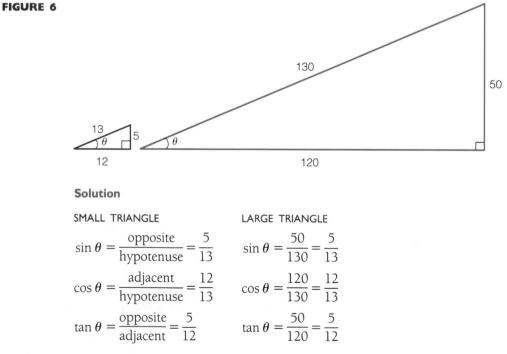

Solution

SMALL TRIANGLE	LARGE TRIANGLE
$\sin\theta = \dfrac{\text{opposite}}{\text{hypotenuse}} = \dfrac{5}{13}$	$\sin\theta = \dfrac{50}{130} = \dfrac{5}{13}$
$\cos\theta = \dfrac{\text{adjacent}}{\text{hypotenuse}} = \dfrac{12}{13}$	$\cos\theta = \dfrac{120}{130} = \dfrac{12}{13}$
$\tan\theta = \dfrac{\text{opposite}}{\text{adjacent}} = \dfrac{5}{12}$	$\tan\theta = \dfrac{50}{120} = \dfrac{5}{12}$

Observation   The corresponding values for sin θ, cos θ, and tan θ are the same for both triangles.

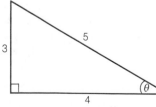

FIGURE 7

EXAMPLE 2   Let θ be the acute angle indicated in Figure 7. Determine the six quantities cos θ, sin θ, tan θ, sec θ, csc θ, and cot θ.

Solution   We use the definitions:

$$\cos\theta = \frac{\text{adjacent}}{\text{hypotenuse}} = \frac{4}{5} \qquad \sec\theta = \frac{\text{hypotenuse}}{\text{adjacent}} = \frac{5}{4}$$

$$\sin\theta = \frac{\text{opposite}}{\text{hypotenuse}} = \frac{3}{5} \qquad \csc\theta = \frac{\text{hypotenuse}}{\text{opposite}} = \frac{5}{3}$$

$$\tan\theta = \frac{\text{opposite}}{\text{adjacent}} = \frac{3}{4} \qquad \cot\theta = \frac{\text{adjacent}}{\text{opposite}} = \frac{4}{3}$$

Note the pairs of answers that are reciprocals. (As mentioned before, this helps in memorizing the definitions.)   ▌▌▌

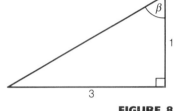

FIGURE 8

▌▐▌ **EXAMPLE 3**   Let β be the acute angle indicated in Figure 8. Find $\sin\beta$ and $\cos\beta$.

Solution   In view of the definitions, we need to know the length of the hypotenuse in Figure 8. If we call this length h, then by the Pythagorean theorem we have

$$h^2 = 3^2 + 1^2 = 10$$
$$h = \sqrt{10}$$

Therefore

$$\sin\beta = \frac{\text{opposite}}{\text{hypotenuse}} = \frac{3}{\sqrt{10}} = \frac{3\sqrt{10}}{10}$$

and

$$\cos\beta = \frac{\text{adjacent}}{\text{hypotenuse}} = \frac{1}{\sqrt{10}} = \frac{\sqrt{10}}{10} \qquad \text{▌▌▌}$$

In the next example we use a calculator to evaluate the trigonometric functions for a particular acute-angle input. All scientific calculators allow for two common units of angle measurement: the degree and the *radian*. Radian measure is discussed in Chapter 7; for the present chapter, though, we are using degree measure exclusively. So, before working through the next example on your own, you need to check that your calculator is set in the "degree mode" rather than the "radian mode." If necessary, consult the owner's manual for your calculator on this point.

One more calculator-related detail: For most calculators, when you want to evaluate a function, first you enter the input, then you press the appropriate function key. For example (assuming that your calculator is set in the degree mode), to evaluate sin 40°, first enter the input 40, then press the (SIN) key. Try this now. If the resulting display on your calculator begins 0.6427 . . . , then this was the appropriate sequence for your calculator. If not (but assuming that your calculator is indeed set in the degree mode), then you may have a calculator that requires you to enter the function *before* the input. (The Casio and Texas Instruments graphing calculators work that way, for example.) Again, consult your owner's manual on this point, as necessary.

▌▬▌ **EXAMPLE 4**   Use a calculator to evaluate cos 40°, sin 40°, and tan 40°. Round off the results to three decimal places.

Solution   With the calculator set in the degree mode, the sequence of keystrokes and the results (rounded off to three decimal places) are as follows. You should verify each of these results for yourself.

EXPRESSION	KEYSTROKES	OUTPUT
cos 40°	40 (COS)	0.766
sin 40°	40 (SIN)	0.643
tan 40°	40 (TAN)	0.839

In the next example, we use a calculator to evaluate the three functions secant, cosecant, and cotangent for a particular input. These three functions are referred to as the **reciprocal functions**. This is because, from the definitions on page 297, it follows that

$$\sec \theta = \frac{1}{\cos \theta} \qquad \csc \theta = \frac{1}{\sin \theta} \qquad \cot \theta = \frac{1}{\tan \theta}$$

So, for example, although most calculators do not have a key labeled "sec," we can nevertheless evaluate the secant function by using the cosine key (COS) and then the reciprocal key (1/x).

▌▬▌ **EXAMPLE 5**   Use a calculator to evaluate sec 10°, csc 10°, and cot 10°. Round off the results to three decimal places.

Solution   As in the previous example, we first check that the calculator is in the degree mode. Then, the sequence of keystrokes and the results are as follows (again, you should verify each of these results for yourself):

EXPRESSION	KEYSTROKES	OUTPUT
sec 10°	10 (COS) (1/x)	1.015
csc 10°	10 (SIN) (1/x)	5.759
cot 10°	10 (TAN) (1/x)	5.671

There are certain angles for which we can evaluate the trigonometric functions without using a calculator. Three such angles are 30°, 45°, and 60°. We focus on these particular angles for now, not because they are somehow more fundamental than others, but because a ready knowledge of their trigonometric values will provide a useful source of examples in our subsequent work.

In order to obtain the trigonometric values for angles of 30° and 60°, we rely on the following theorem from geometry. (For a proof of this theorem, see Exercise 41.)

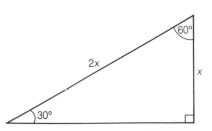

FIGURE 9

THEOREM   The 30°–60° Right Triangle

In a 30°–60° right triangle, the length of the side opposite the 30° angle is half the length of the hypotenuse. (See Figure 9.)

If we let y denote the length of the unmarked side in Figure 9, then (according to the Pythagorean theorem) we have $y^2 + x^2 = (2x)^2$ and, consequently,

$$y^2 = 4x^2 - x^2 = 3x^2$$
$$y = \sqrt{3}x$$

Our 30°–60° right triangle can now be labeled as in Figure 10. From this figure, the required trigonometric values can be written down by inspection. We will list only the values of the cosine, sine, and tangent, since the remaining three values are just the reciprocals of these.

FIGURE 10

$$\cos 30° = \frac{x\sqrt{3}}{2x} = \frac{\sqrt{3}}{2} \qquad\qquad \cos 60° = \frac{x}{2x} = \frac{1}{2}$$

$$\sin 30° = \frac{x}{2x} = \frac{1}{2} \qquad\qquad \sin 60° = \frac{x\sqrt{3}}{2x} = \frac{\sqrt{3}}{2}$$

$$\tan 30° = \frac{x}{x\sqrt{3}} = \frac{1}{\sqrt{3}} = \frac{\sqrt{3}}{3} \qquad \tan 60° = \frac{x\sqrt{3}}{x} = \sqrt{3}$$

There are two observations to be made here. First, the final answers do not involve x. Once again this shows that for a given angle, it is the *ratios* of the sides that determine the trigonometric values, not the size of the triangle. Second, notice that

$$\sin 30° = \cos 60° \qquad \text{and} \qquad \sin 60° = \cos 30°$$

It is easy to see that this is no coincidence. The terms "opposite" and "adjacent" are relative ones. As a glance at Figure 10 shows, the side opposite to one acute angle is automatically adjacent to the other acute angle. We'll return to this point again in the next section.

FIGURE 11

Another angle for which we can compute the trigonometric values without the aid of a calculator is 45°. Figure 11 shows a 45°–45° right triangle. Since the base angles of this triangle are equal, it follows from a theorem of geometry that the triangle is isosceles. That is, the triangle has two equal sides, as indicated by the use of the letter x twice in Figure 11. If we now use y to denote the length of the unmarked side in Figure 11, we have

$$y^2 = x^2 + x^2 = 2x^2 \qquad \text{and, consequently,} \qquad y = \sqrt{2}x$$

FIGURE 12

Our 45°–45° right triangle can now be labeled as in Figure 12, and the trigonometric values can then be obtained by inspection. As before, we list only the values for cosine, sine, and tangent, since the remaining three are just the reciprocals of these.

$$\cos 45° = \frac{x}{x\sqrt{2}} = \frac{1}{\sqrt{2}} = \frac{\sqrt{2}}{2}$$

$$\sin 45° = \frac{x}{x\sqrt{2}} = \frac{1}{\sqrt{2}} = \frac{\sqrt{2}}{2}$$

$$\tan 45° = \frac{x}{x} = 1$$

Table 1 summarizes the results we've now obtained for angles of 30, 45, and 60 degrees. These results should be memorized.

TABLE I

θ	$\sin \theta$	$\cos \theta$	$\tan \theta$
30°	$\frac{1}{2}$	$\frac{\sqrt{3}}{2}$	$\frac{\sqrt{3}}{3}$
45°	$\frac{\sqrt{2}}{2}$	$\frac{\sqrt{2}}{2}$	1
60°	$\frac{\sqrt{3}}{2}$	$\frac{1}{2}$	$\sqrt{3}$

EXERCISE 6.1

A

In Exercises 1–4, use the definitions (as in Example 2) to evaluate the six trigonometric functions of (a) θ; (b) β. In cases where a radical occurs in a denominator, rationalize the denominator.

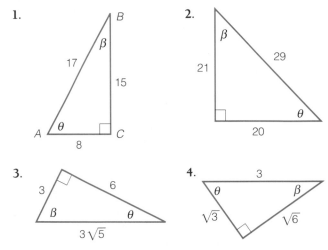

1.

2.

3.

4.

In Exercises 5–12, suppose that $\triangle ABC$ is a right triangle with $\angle C = 90°$.

5. If $AC = 3$ and $BC = 2$, find the following quantities:
(a) $\cos A$, $\sin A$, $\tan A$;   (b) $\sec B$, $\csc B$, $\cot B$.

6. If $AC = 6$ and $BC = 2$, find the following quantities:
(a) $\cos A$, $\sin A$, $\tan A$;   (b) $\sec B$, $\csc B$, $\cot B$.

7. If $AB = 13$ and $BC = 5$, compute the values of the six trigonometric functions of angle B.

8. If $AB = 3$ and $AC = 1$, compute the values of the six trigonometric functions of angle A.

9. If $AC = 1$ and $BC = \frac{3}{4}$, compute each quantity:
(a) $\sin B$, $\cos A$;   (b) $\sin A$, $\cos B$;
(c) $(\tan A)(\tan B)$.

10. If $AC = BC = 4$, compute the following:
(a) $\sec A$, $\csc A$, $\cot A$;   (b) $\sec B$, $\csc B$, $\cot B$;
(c) $(\cot A)(\cot B)$.

11. If $AB = 25$ and $AC = 24$, compute each of the required quantities: (a) $\cos A$, $\sin A$, $\tan A$;
(b) $\cos B$, $\sin B$, $\tan B$;   (c) $(\tan A)(\tan B)$.

12. If $AB = 1$ and $BC = \sqrt{3}/2$, compute the following:
(a) $\cos A$, $\sin B$;   (b) $\tan A$, $\cot B$;
(c) $\sec A$, $\csc B$.

In Exercises 13–18, use a calculator to compute $\cos \theta$, $\sin \theta$, and $\tan \theta$ for the given value of θ. Round off each result to three decimal places.

13. $\theta = 65°$ **14.** $\theta = 21°$ **15.** $\theta = 38.5°$
16. $\theta = 12.4°$ **17.** $\theta = 80.06°$ **18.** $\theta = 0.99°$

In Exercises 19–24, use a calculator to compute $\sec \theta$, $\csc \theta$, and $\cot \theta$ (as in Example 5) for the given value of θ. Round off each result to three decimal places.

19. $\theta = 20°$ **20.** $\theta = 40°$ **21.** $\theta = 17.5°$
22. $\theta = 18.5°$ **23.** $\theta = 1°$ **24.** $\theta = 89.9°$

In Exercises 25–36, verify that each of the following equations is correct by evaluating each side. Do not use a calculator. The purpose of Exercises 25–36 is twofold. First, doing the problems will help you to review the values in Table 1 of this section. Second, the exercises serve as an algebra review.   Note: In the exercises, notation such as $\sin^2 \theta$ stands for $(\sin \theta)^2$. This common notational convention will be discussed in the next section.

25. $\cos 60° = \cos^2 30° - \sin^2 30°$

26. $\cos 60° = 1 - 2 \sin^2 30°$

27. $\sin^2 30° + \sin^2 45° + \sin^2 60° = \frac{3}{2}$

28. $\sin 30° \cos 60° + \cos 30° \sin 60° = 1$

29. $2 \sin 30° \cos 30° = \sin 60°$

30. $2 \sin 45° \cos 45° = 1$

31. $\sin 30° = \sqrt{\frac{1}{2}(1 - \cos 60°)}$

32. $\cos 30° = \sqrt{\frac{1}{2}(1 + \cos 60°)}$

33. $\tan 30° = \dfrac{\sin 60°}{1 + \cos 60°}$ **34.** $\tan 30° = \dfrac{1 - \cos 60°}{\sin 60°}$

35. $1 + \tan^2 45° = \sec^2 45°$

36. $1 + \cot^2 60° = \csc^2 60°$

37. (a) Use a calculator to evaluate $\cos 30°$ and $\cos 45°$. Give your answers to as many decimal places as your calculator allows.
(b) In Table 1, there are expressions for $\cos 30°$ and $\cos 45°$. Use your calculator to evaluate these expressions, and check to see that the results agree with those in part (a).

38. (a) Use a calculator to evaluate $\tan 30°$ and $\tan 60°$. Give your answers to as many decimal places as your calculator allows.
(b) In Table 1, there are expressions for $\tan 30°$ and $\tan 60°$. Use your calculator to evaluate these expressions, and check to see that the results agree with those in part (a).

B

For Exercises 39 and 40, refer to the following figure. In each case, say which of the two given quantities is larger. If the quantities are equal, say so.

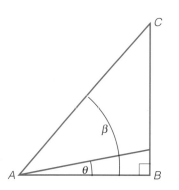

39. (a) $\cos\theta$, $\cos\beta$
(b) $\sec\theta$, $\sec\beta$

40. (a) $\tan\theta$, $\tan\beta$
(b) $\cot\theta$, $\cot\beta$

41. The following figure shows a 30°–60° right triangle, $\triangle ABC$. Prove that $AC = 2AB$. *Suggestion:* Construct $\triangle DBC$ as shown, congruent to $\triangle ABC$. Then note that $\triangle ADC$ is equilateral.

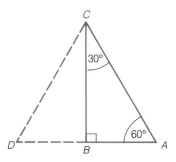

42. This exercise shows how to obtain radical expressions for $\sin 15°$ and $\cos 15°$. In the figure, assume that $AB = BD = 2$.

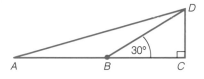

(a) In the right triangle BCD, note that $DC = 1$ because $\overline{DC}$ is opposite the 30° angle and $BD = 2$. Use the Pythagorean theorem to show that $BC = \sqrt{3}$.

(b) Use the Pythagorean theorem to show that $AD = 2\sqrt{2 + \sqrt{3}}$.

(c) Show that the expression for AD in part (b) is equal to $\sqrt{6} + \sqrt{2}$. *Hint:* Two nonnegative quantities are equal if and only if their squares are equal.

(d) Explain why $\angle BAD = \angle BDA$.

(e) According to a theorem from geometry, an exterior angle of a triangle is equal to the sum of the two nonadjacent interior angles. Apply this to $\triangle ABD$ with exterior angle $DBC = 30°$, and show that $\angle BAD = 15°$.

(f) Using the figure and the values that you have obtained for the lengths, conclude that

$$\sin 15° = \frac{1}{\sqrt{6}+\sqrt{2}} \qquad \cos 15° = \frac{2+\sqrt{3}}{\sqrt{6}+\sqrt{2}}$$

(g) Rationalize the denominators in part (f) to obtain

$$\sin 15° = \frac{\sqrt{6}-\sqrt{2}}{4} \qquad \cos 15° = \frac{\sqrt{6}+\sqrt{2}}{4}$$

(h) Use your calculator to check the results in part (g).

43. If an angle θ is an integral multiple of 3°, then the real number $\sin\theta$ is either rational or expressible in terms of radicals. You've already seen examples of this with the angles 30°, 45°, and 60°. The accompanying table gives the values of $\sin\theta$ for some multiples of 3°. Use your calculator to check the entries in the table. *Remark:* If θ is an integral number of degrees, but not a multiple of 3°, then the real number $\sin\theta$ cannot be expressed in terms of radicals within the real-number system.

θ	$\sin\theta$
3°	$\frac{1}{16}\left[(\sqrt{6}+\sqrt{2})(\sqrt{5}-1) - 2(\sqrt{3}-1)\sqrt{5+\sqrt{5}}\right]$
6°	$\frac{1}{8}(\sqrt{30-6\sqrt{5}} - \sqrt{5} - 1)$
9°	$\frac{1}{8}(\sqrt{10}+\sqrt{2} - 2\sqrt{5-\sqrt{5}})$
12°	$\frac{1}{8}(\sqrt{10+2\sqrt{5}} - \sqrt{15} + \sqrt{3})$
15°	$\frac{1}{4}(\sqrt{6}-\sqrt{2})$
18°	$\frac{1}{4}(\sqrt{5}-1)$

6.2 ALGEBRA AND THE TRIGONOMETRIC FUNCTIONS

Perhaps the first use of abbreviations for the trigonometric [functions] . . . goes back to the [Danish] physician and mathematician, Thomas Finck [1561–1646]. . . .

Florian Cajori in *A History of Mathematical Notations*, vol. II (Chicago: The Open Court Publishing Company, 1929)

Albert Girard (1595–1632), who seems to have lived chiefly in Holland, . . . interested himself in spherical trigonometry and trigonometry. In 1626, he published a treatise on trigonometry that contains the earliest use of our abbreviations sin, tan, *and* sec *for* sine, tangent, *and* secant.

Howard Eves in *An Introduction to the History of Mathematics*, 6th ed. (Philadelphia: Saunders College Publishing, 1990)

In this section, we are going to concentrate on the algebra involved in working with the trigonometric functions. This will help to pave the way both for the applications in the next section and for some of the more analytical parts of trigonometry in the next chapter. We begin by listing some common notational conventions.

Notational Conventions

1. The quantity $(\sin \theta)^n$ is usually written $\sin^n \theta$. For example, $(\sin \theta)^2$ is written $\sin^2 \theta$. The same convention also applies to the other five trigonometric functions.*

2. Parentheses are often omitted in multiplication. For example, the product $(\sin \theta)(\cos \theta)$ is usually written $\sin \theta \cos \theta$. Similarly, $2(\sin \theta)$ is written $2 \sin \theta$.

3. An expression such as $\sin \theta$ really means $\sin(\theta)$, where sin or sine is the name of the function and θ is an input. It is for historical rather than mathematical reasons that the parentheses are suppressed. An exception to this, however, occurs in expressions such as $\sin(A + B)$; here the parentheses are necessary.

█▬█ EXAMPLE 1

(a) Combine like terms: $3 \sin^2 \theta \cos \theta - 5 \sin^2 \theta \cos \theta$.
(b) Carry out the multiplication: $(2 \sin \theta - 3 \cos \theta)^2$.

Solution

(a) We do this in the same way we would simplify the algebraic expression $3S^2C - 5S^2C$. Since

$$3S^2C - 5S^2C = -2S^2C$$

we have

$$3 \sin^2 \theta \cos \theta - 5 \sin^2 \theta \cos \theta = -2 \sin^2 \theta \cos \theta$$

Note If the original expression had been $3 \sin^2 \theta \cos \theta - 5 \sin \theta \cos^2 \theta$, then there would be no like terms to combine.

(b) We do this in the same way that we would expand $(2S - 3C)^2$. Since

$$(2S - 3C)^2 = 4S^2 - 12SC + 9C^2$$

we have

$$(2 \sin \theta - 3 \cos \theta)^2 = 4 \sin^2 \theta - 12 \sin \theta \cos \theta + 9 \cos^2 \theta \qquad ▮▮▮$$

█▬█ EXAMPLE 2 Factor: $\tan^2 A + 5 \tan A + 6$.

---

*A single exception to this convention occurs when $n = -1$. The meaning of expressions such as $\sin^{-1} x$ will be explained in Sections 8 and 9 in Chapter 7.

Solution

PRELIMINARY SOLUTION   To help us focus on the algebra that is actually involved, let's replace each occurrence of the quantity tan A by the letter T. Then

$$T^2 + 5T + 6 = (T + 3)(T + 2)$$

ACTUAL SOLUTION

$$\tan^2 A + 5 \tan A + 6 = (\tan A + 3)(\tan A + 2)$$

Note   After you are accustomed to working with trigonometric expressions, you should be able to eliminate the preliminary step.   ▪▪▪

▮▬▮ **EXAMPLE 3**   Add $\sin\theta + \dfrac{1}{\cos\theta}$.

Solution

PRELIMINARY SOLUTION

$$S + \frac{1}{C} = \frac{S}{1} + \frac{1}{C} = \frac{S}{1} \cdot \frac{C}{C} + \frac{1}{C}$$

$$= \frac{SC}{C} + \frac{1}{C} = \frac{SC + 1}{C}$$

ACTUAL SOLUTION

$$\sin\theta + \frac{1}{\cos\theta} = \frac{\sin\theta}{1} + \frac{1}{\cos\theta} = \frac{\sin\theta}{1} \cdot \frac{\cos\theta}{\cos\theta} + \frac{1}{\cos\theta}$$

$$= \frac{\sin\theta\cos\theta}{\cos\theta} + \frac{1}{\cos\theta} = \frac{\sin\theta\cos\theta + 1}{\cos\theta}$$ ▪▪▪

There are numerous identities involving the trigonometric functions. Recall that an **identity** is an equation that is satisfied by all relevant values of the variables concerned. Two examples of identities are $(x - y)(x + y) = x^2 - y^2$ and $x^3 = x^4/x$. The first of these is true no matter what real numbers are used for x and y; the second is true for all real numbers except $x = 0$. For now, in the context of right-triangle trigonometry, we'll consider only a few of the most basic identities.

▮■ **PROPERTY SUMMARY**   **BASIC RIGHT-TRIANGLE IDENTITIES**

IDENTITY

1. $\sin^2\theta + \cos^2\theta = 1$

2. $\dfrac{\sin\theta}{\cos\theta} = \tan\theta$

3. $\sin(90° - \theta) = \cos\theta$; $\cos(90° - \theta) = \sin\theta$

4. $\sec\theta = \dfrac{1}{\cos\theta}$; $\csc\theta = \dfrac{1}{\sin\theta}$; $\cot\theta = \dfrac{1}{\tan\theta}$

EXAMPLES

1. $\sin^2 30° + \cos^2 30° = 1$
$\sin^2 19° + \cos^2 19° = 1$

2. $\dfrac{\sin 60°}{\cos 60°} = \tan 60°$

3. $\sin 70° = \cos 20°$; $\sin 20° = \cos 70°$

4. $\cot 40° = \dfrac{1}{\tan 40°}$

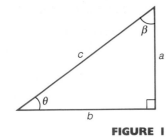

FIGURE 1

The fourth set of identities in the box follows directly from the definitions on page 297. For the proofs of the other identities, we refer to Figure 1.

PROOF THAT $\sin^2\theta + \cos^2\theta = 1$ Looking at Figure 1, we have $\sin\theta = a/c$ and $\cos\theta = b/c$. Thus

$$\sin^2\theta + \cos^2\theta = \left(\frac{a}{c}\right)^2 + \left(\frac{b}{c}\right)^2 = \frac{a^2}{c^2} + \frac{b^2}{c^2} = \frac{a^2 + b^2}{c^2}$$

$$= \frac{c^2}{c^2} \qquad \text{using } a^2 + b^2 = c^2, \text{ the Pythagorean theorem}$$

$$= 1 \qquad \text{as required}$$

PROOF THAT $(\sin\theta)/(\cos\theta) = \tan\theta$ Again with reference to Figure 1, we have, by definition, $\sin\theta = a/c$, $\cos\theta = b/c$ and $\tan\theta = a/b$. Therefore

$$\frac{\sin\theta}{\cos\theta} = \frac{a/c}{b/c} = \frac{a}{c} \times \frac{c}{b}$$

$$= \frac{a}{b} = \tan\theta \qquad \text{as required}$$

PROOF THAT $\sin(90° - \theta) = \cos\theta$ First of all, since the sum of the angles in any triangle is 180°, we have

$$\theta + \beta + 90° = 180°$$

$$\beta = 90° - \theta$$

Then $\sin(90° - \theta) = \sin\beta = b/c$. But also (by definition) $\cos\theta = b/c$. Thus

$$\sin(90° - \theta) = \cos\theta$$

since both expressions equal b/c. This is what we wanted to prove.

The proof that $\cos(90° - \theta) = \sin\theta$ is entirely similar and we shall omit it. We can conveniently summarize these last two results by recalling the notion of complementary angles. Two acute angles are said to be **complementary** provided their sum is 90°. Thus, the two angles θ and $90° - \theta$ are complementary. In view of this, we can restate the last two results as follows:

> If two angles are complementary, then the sine of (either) one equals the cosine of the other.

Incidentally, this result gives us an insight into the origin of the term *cosine*: it is a shortened form of the phrase *complement's sine*.

■■■ **EXAMPLE 4**   Suppose that B is an acute angle and $\cos B = \frac{2}{5}$. Find $\sin B$ and $\tan B$.

Solution   We'll show two methods. The first uses the identities $\sin^2 B + \cos^2 B = 1$ and $(\sin B)/(\cos B) = \tan B$. The second makes direct use of the Pythagorean theorem.

FIRST METHOD   Replace $\cos B$ with $\frac{2}{5}$ in the identity $\sin^2 B + \cos^2 B = 1$. This yields

$$\sin^2 B + \left(\frac{2}{5}\right)^2 = 1$$

$$\sin^2 B = 1 - \frac{4}{25} - \frac{21}{25}$$

$$\sin B = \sqrt{\frac{21}{25}} = \frac{\sqrt{21}}{5}$$

Notice that we've chosen the positive square root here. This is because the values of the trigonometric functions of an acute angle are by definition positive. *Caution:* In a later section, we'll extend the definitions of the trigonometric functions to include angles that are not acute. In those cases, some of the trigonometric values will not be positive; we will then need to pay more attention to the matter of signs.

For $\tan B$, we have

$$\tan B = \frac{\sin B}{\cos B} = \frac{\sqrt{21}/5}{2/5} = \frac{\sqrt{21}}{2}$$

ALTERNATE METHOD   Since $\cos B = \frac{2}{5} = $ adjacent/hypotenuse, we can work with a right triangle labeled as in Figure 2. Using the Pythagorean theorem, we have $2^2 + x^2 = 5^2$, from which it follows that $x = \sqrt{21}$. Consequently,

$$\sin B = \frac{\text{opposite}}{\text{hypotenuse}} = \frac{x}{5} = \frac{\sqrt{21}}{5}$$

and

$$\tan B = \frac{\text{opposite}}{\text{adjacent}} = \frac{x}{2} = \frac{\sqrt{21}}{2} \qquad \text{as obtained previously}$$

FIGURE 2

One of the many uses for the identities in this section is in simplifying trigonometric expressions. In the example that follows, we use the identity $\sin^2 A + \cos^2 A = 1$.

EXAMPLE 5   Combine and simplify: $\dfrac{\sin A}{\cos A} + \dfrac{\cos A}{\sin A}$.

Solution   The common denominator is $\cos A \sin A$. Therefore we have

$$\frac{\sin A}{\cos A} + \frac{\cos A}{\sin A} = \frac{\sin A}{\cos A} \cdot \frac{\sin A}{\sin A} + \frac{\cos A}{\sin A} \cdot \frac{\cos A}{\cos A}$$

$$= \frac{\sin^2 A}{\cos A \sin A} + \frac{\cos^2 A}{\sin A \cos A}$$

$$= \frac{\sin^2 A + \cos^2 A}{\cos A \sin A}$$

$$= \frac{1}{\cos A \sin A}$$

This is the required result. If we wish, we can write this answer in an alternative form that doesn't involve fractions:

$$\frac{1}{\cos A \sin A} = \frac{1}{\cos A} \cdot \frac{1}{\sin A} = \sec A \csc A$$

■■■

One of the most useful techniques for simplifying a trigonometric expression is first to rewrite it in terms of sines and cosines and then to carry out the usual algebraic simplifications. This is demonstrated in Examples 6 through 8.

▌■■▌ EXAMPLE 6 Simplify the expression $\csc A \tan A \cos^2 A$.

Solution We use the basic identities to express everything in terms of sines and cosines.

$$\csc A \tan A \cos^2 A = \frac{1}{\cancel{\sin A}} \cdot \frac{\cancel{\sin A}^{1}}{\cos A} \cdot \cos^2 A$$

$$= \frac{\cos^2 A}{\cos A} = \cos A$$

■■■

▌■■▌ EXAMPLE 7 Simplify the expression $\dfrac{1 - \tan^2 B}{1 + \tan^2 B}$.

Solution

$$\frac{1 - \tan^2 B}{1 + \tan^2 B} = \frac{1 - \dfrac{\sin^2 B}{\cos^2 B}}{1 + \dfrac{\sin^2 B}{\cos^2 B}}$$

Since $\tan B = \dfrac{\sin B}{\cos B}$,

it follows that $\tan^2 B = \dfrac{\sin^2 B}{\cos^2 B}$.

$$= \frac{\cos^2 B \left(1 - \dfrac{\sin^2 B}{\cos^2 B}\right)}{\cos^2 B \left(1 + \dfrac{\sin^2 B}{\cos^2 B}\right)}$$

multiplying both the numerator and the denominator by $\cos^2 B$

$$= \frac{\cos^2 B - \sin^2 B}{\cos^2 B + \sin^2 B}$$

$$= \frac{\cos^2 B - \sin^2 B}{1} = \cos^2 B - \sin^2 B$$

■■■

▌■■▌ EXAMPLE 8 Express in terms of sines and cosines and simplify: $\dfrac{\tan(90° - A)}{\cos A \sin A}$.

Solution

$$\frac{\tan(90° - A)}{\cos A \sin A} = \frac{\dfrac{\sin(90° - A)}{\cos(90° - A)}}{\cos A \sin A} \qquad \text{applying the identity } \tan \theta = \frac{\sin \theta}{\cos \theta}$$

$$= \frac{\dfrac{\cos A}{\sin A}}{\cos A \sin A} \qquad \text{(Why?)}$$

$$= \frac{\cos A}{\sin A} \cdot \frac{1}{\cos A \sin A} = \frac{1}{\sin^2 A}$$

The given expression is therefore equal to $1/\sin^2 A$. We can also write this as $\csc^2 A$. ∎

EXERCISE SET 6.2

A

In Exercises 1–12, carry out the indicated operations.

1. (a) $-SC + 12SC$
   (b) $-\sin \theta \cos \theta + 12 \sin \theta \cos \theta$

2. (a) $10SC + 4SC - 16SC$
   (b) $10 \sin \theta \cos \theta + 4 \sin \theta \cos \theta - 16 \sin \theta \cos \theta$

3. (a) $4C^3S - 12C^3S$
   (b) $4 \cos^3 \theta \sin \theta - 12 \cos^3 \theta \sin \theta$

4. (a) $-C^2S^2 + (2SC)^2$
   (b) $-\cos^2 \theta \sin^2 \theta + (2 \sin \theta \cos \theta)^2$

5. (a) $(1 + T)^2$
   (b) $(1 + \tan \theta)^2$

6. (a) $(3 - 2T)^2$
   (b) $(3 - 2 \tan \theta)^2$

7. (a) $(T + 3)(T - 2)$
   (b) $(\tan \theta + 3)(\tan \theta - 2)$

8. (a) $(S^2 - 3)(S^2 + 3)$
   (b) $(\sec^2 \theta - 3)(\sec^2 \theta + 3)$

9. (a) $\dfrac{S - C}{C - S}$
   (b) $\dfrac{\sin \theta - \cos \theta}{\cos \theta - \sin \theta}$

10. (a) $\dfrac{5 - 2T}{2T - 5}$
    (b) $\dfrac{5 - 2 \tan \theta}{2 \tan \theta - 5}$

11. (a) $C + \dfrac{2}{S}$
    (b) $\cos A + \dfrac{2}{\sin A}$

12. (a) $\dfrac{1}{S} - \dfrac{3}{C}$
    (b) $\dfrac{1}{\sin A} - \dfrac{3}{\cos A}$

In Exercises 13–18, factor each expression.

13. (a) $T^2 + 8T - 9$
    (b) $\tan^2 \beta + 8 \tan \beta - 9$

14. (a) $3S^2 + 2S - 8$
    (b) $3 \sec^2 \beta + 2 \sec \beta - 8$

15. (a) $4C^2 - 1$
    (b) $4 \cos^2 B - 1$

16. (a) $16S^3 - 9S^2$
    (b) $16 \sin^3 B - 9 \sin^2 B$

17. (a) $9S^2T^3 + 6ST^2$
    (b) $9 \sec^2 B \tan^3 B + 6 \sec B \tan^2 B$

18. (a) $5C^2c^2 - 15Cc$
    (b) $5 \csc^2 B \cot^2 B - 15 \csc B \cot B$

In Exercises 19–30, use the given information to determine the values of the remaining five trigonometric functions. (Assume that all of the angles are acute.) When radicals appear in a denominator, rationalize the denominator. In Exercises 19–24, use either of the methods shown in Example 4. In Exercises 25–30, use the second method shown in Example 4.

19. $\sin \theta = \frac{3}{4}$

20. $\sin \theta = \frac{2}{5}$

21. $\cos \beta = \sqrt{3}/5$

22. $\cos \beta = \sqrt{7}/3$

23. $\sin A = \frac{5}{13}$

24. $\cos A = \frac{8}{17}$

25. $\tan B = \frac{4}{3}$

26. $\tan B = 5$

27. $\sec C = \frac{3}{2}$

28. $\csc C = \sqrt{5}/2$

29. $\cot \alpha = \sqrt{3}/3$

30. $\cot \alpha = \sqrt{3}/2$

In Exercises 31–34, use a calculator to determine the values of the remaining five trigonometric functions. (In each case, θ denotes an acute angle.) Round off the final answers to three decimal places.

31. $\cos \theta = 0.4626$

32. $\sin \theta = 0.5917$

33. $\tan \theta = 1.1998$

34. $\sec \theta = 2.3283$

In Exercises 35–56, simplify each expression.

35. $\dfrac{\sin^2 A - \cos^2 A}{\sin A - \cos A}$ *Hint:*   Factor the numerator.

36. $\dfrac{\sin^4 A - \cos^4 A}{\cos A - \sin A}$

37. $\sin^2 \theta \cos \theta \csc^3 \theta \sec \theta$

38. $\sin \theta \csc \theta \tan \theta$

39. $\cot B \sin^2 B \cot B$

40. $\dfrac{3 \sin \theta + 6}{\sin^2 \theta - 4}$

41. $\dfrac{\cos^2 A + \cos A - 12}{\cos A - 3}$

42. $\dfrac{\cos A - 2 \sin A \cos A}{\cos^2 A - \sin^2 A + \sin A - 1}$

43. $\dfrac{\tan \theta}{\sec \theta - 1} + \dfrac{\tan \theta}{\sec \theta + 1}$

44. $\cot \theta + \dfrac{1 - 2 \cos^2 \theta}{\sin \theta \cos \theta}$

45. $\sec A \csc A - \tan A - \cot A$

46. $(\sec A + \tan A)(\sec A - \tan A)$

47. $\dfrac{\cot^2 \theta}{\csc^2 \theta} + \dfrac{\tan^2 \theta}{\sec^2 \theta}$

48. $\dfrac{\tan \theta + \tan \theta \sin \theta - \cos \theta \sin \theta}{\sin \theta \tan \theta}$

49. $\dfrac{\cos(90° - \theta)}{\cos \theta}$

50. $\sin^2(90° - \beta) + \cos^2(90° - \beta)$

51. $\dfrac{\cos^2(90° - A)}{\sin^2(90° - A)} - \dfrac{1}{\cos^2 A}$

52. $\dfrac{\cos(90° - A)}{\csc A} + \dfrac{\sin(90° - A)}{\sec A}$

53. $1 - \sin(90° - \theta)\cos \theta$

54. $\dfrac{\cos A \tan A}{\tan(90° - A)} - \dfrac{1}{\sin(90° - A)}$

55. $\dfrac{\dfrac{\cos \theta + 1}{\cos \theta} + 1}{\dfrac{\cos \theta - 1}{\cos \theta} - 1}$

56. $\dfrac{\dfrac{\cot \beta + 1}{\cot \beta} + 1}{\dfrac{\cot \beta - 1}{\cot \beta} - 1}$

B

57. Suppose that

$A \sin \theta + \cos \theta = 1$ and $B \sin \theta - \cos \theta = 1$

Show that $AB = 1$.   *Hint:*   Solve the first equation for A, the second for B, and then compute AB.

58. If $\sin \alpha + \cos \alpha = a$ and $\sin \alpha - \cos \alpha = b$, show that

$$\tan \alpha = \dfrac{a + b}{a - b}$$

59. If $a \sin^2 \theta + b \cos^2 \theta = 1$, show that

$$\sin^2 \theta = \dfrac{1 - b}{a - b} \quad \text{and} \quad \tan^2 \theta = \dfrac{b - 1}{1 - a}$$

60. This exercise shows how to obtain radical expressions for sin 18° and cos 18°, using the following figure.

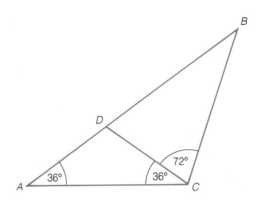

(a) Find $\angle B$, $\angle BDC$, and $\angle ADC$.

(b) Why does $AC = BC = BD$?

For the rest of this problem, assume $AD = 1$.

(c) Why does $CD = 1$?

(d) Let x denote the common lengths AC, BC, and BD. Use similar triangles to deduce that $x/(1 + x) = 1/x$. Then show that $x = (1 + \sqrt{5})/2$.

(e) In $\triangle BDC$, draw an altitude from B to $\overline{DC}$, meeting $\overline{DC}$ at F. Use right triangle BFC to conclude that $\sin 18° = 1/(1 + \sqrt{5})$.

(f) Rationalize the denominator in part (e) to obtain $\sin 18° = \tfrac{1}{4}(\sqrt{5} - 1)$.

(g) Use the identity $\sin^2 \theta + \cos^2 \theta = 1$, along with part (f), to show that

$$\cos 18° = \tfrac{1}{4}\sqrt{10 + 2\sqrt{5}}$$

(h) Use your calculator to check the results in parts (f) and (g).

61. Suppose that β is an acute angle and

$$\sin \beta = \dfrac{m^2 - n^2}{m^2 + n^2} \quad (m > n > 0)$$

Show that

$$\cos \beta = \dfrac{2mn}{m^2 + n^2} \quad \text{and} \quad \tan \beta = \dfrac{m^2 - n^2}{2mn}$$

6.3 RIGHT-TRIANGLE APPLICATIONS

We began this chapter in Section 6.1 by defining the six trigonometric functions of an acute angle θ. Then, in Section 6.2, we looked at some of the algebra that is involved in working with the trigonometric functions. Now we are ready to apply these ideas in solving some basic problems involving right triangles.

▮▮▮ **EXAMPLE 1** Use one of the trigonometric functions to find x in Figure 1.

Solution Relative to the given 30° angle, x is the adjacent side. The length of the hypotenuse is 100 cm. Since the adjacent side and the hypotenuse are involved, we use the cosine function here:

$$\cos 30° = \frac{\text{adjacent}}{\text{hypotenuse}} = \frac{x}{100}$$

Consequently,

$$x = 100 \cos 30° = (100) \cdot \left(\frac{\sqrt{3}}{2}\right) = 50\sqrt{3}\,\text{cm}$$

This is the result we are looking for.   ▮▮▮

FIGURE 1

We used the cosine function in Example 1 because the adjacent side and the hypotenuse were involved. We could instead use the secant. In that case, again with reference to Figure 1, the calculations would look like this:

$$\sec 30° = \frac{\text{hypotenuse}}{\text{adjacent}} = \frac{100}{x}$$

$$\frac{2}{\sqrt{3}} = \frac{100}{x}$$

$$2x = 100\sqrt{3}$$

$$x = \frac{100\sqrt{3}}{2} = 50\sqrt{3}\,\text{cm} \qquad \text{as obtained previously}$$

▮▮▮ **EXAMPLE 2** Find y in Figure 1.

Solution As you can see from Figure 1, the side of length y is opposite the 30° angle. Furthermore, we are given the length of the hypotenuse. Since the opposite side and the hypotenuse are involved, we use the sine function. This yields

$$\sin 30° = \frac{\text{opposite}}{\text{hypotenuse}} = \frac{y}{100}$$

and therefore

$$y = 100 \sin 30° = (100) \cdot \left(\tfrac{1}{2}\right) = 50\,\text{cm}$$

This is the required answer. Actually, we could have obtained this particular result much faster by recalling that in the 30°–60° right triangle, the side opposite

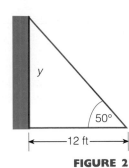

FIGURE 2

the 30° angle, namely y, is half the hypotenuse. That is, $y = \frac{100}{2} = 50$ cm, as obtained previously. ▮▮▮

▮▬▮ **EXAMPLE 3** A ladder, which is leaning against the side of a building, forms an angle of 50° with the ground. If the foot of the ladder is 12 ft from the base of the building, how far up the side of the building does the ladder reach? See Figure 2.

Solution In Figure 2 we have used y to denote the required distance. Notice that y is opposite the 50° angle, while the given side is adjacent to that angle. Since the opposite and adjacent sides are involved, we'll use the tangent function. (The cotangent function could also be used.) We have

$$\tan 50° = \frac{y}{12} \quad \text{and therefore} \quad y = 12\tan 50°$$

Without the use of a calculator or tables, this is our final answer. On the other hand, using a calculator we find that

$$y = 14 \text{ ft} \quad \text{to the nearest foot} \quad ▮▮▮$$

In the next example we derive a useful formula for the area of a triangle. The formula can be used when we are given two sides and the included angle, but not the height.

▮▬▮ **EXAMPLE 4** Show that the area of the triangle in Figure 3 is given by $A = \frac{1}{2}ab \sin θ$.

Solution In Figure 3, if h denotes the length of the altitude, then we have

$$\sin θ = \frac{h}{a}$$

and therefore

$$h = a \sin θ$$

FIGURE 3

This value for h can now be used in the usual formula for the area of a triangle:

$$A = \frac{1}{2}bh = \frac{1}{2}b(a \sin θ) = \frac{1}{2}ab \sin θ \quad ▮▮▮$$

The formula that we just derived in Example 4 is worth remembering. In words, the formula states that *the area of a triangle is equal to half the product of the lengths of two of the sides times the sine of the included acute angle.* In Section 6.4, we will see that the formula remains valid even when the included angle is not an acute angle.

▮▬▮ **EXAMPLE 5** Find the area of the triangle in Figure 4.

Solution In the formula $A = \frac{1}{2}ab \sin θ$, we let $a = 3$ cm, $b = 12$ cm, and $θ = 60°$. Then

$$A = \frac{1}{2}(3)(12) \sin 60° = (18)\frac{\sqrt{3}}{2}$$

$$= 9\sqrt{3} \text{ cm}^2$$

FIGURE 4 This is the required area. ▮▮▮

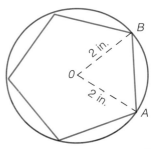

FIGURE 5

█ █ EXAMPLE 6 Figure 5 shows a regular pentagon inscribed in a circle of radius 2 in. (*Regular* means that all of the sides are equal and all of the angles are equal.) Find the area of the pentagon.

Solution The idea here is first to find the area of triangle *BOA* using the area formula from Example 4. Then, since the pentagon is composed of five such identical triangles, the area of the pentagon will be five times the area of triangle *BOA*. We will make use of the result from geometry that, in a regular *n*-sided polygon, the central angle is 360°/*n*. In our case, we therefore have

$$\angle BOA = \frac{360°}{5} = 72°$$

We can now find the area of triangle *BOA*:

$$\text{area} = \tfrac{1}{2}ab \sin \theta$$
$$= \tfrac{1}{2}(2)(2)\sin 72° = 2 \sin 72° \text{ in}^2$$

The area of the pentagon is five times this, or 10 sin 72° in² (Using a calculator, this is about 9.51 in².) **███**

Now we introduce some terminology that will be used in the next two examples. Suppose that a surveyor sights an object at a point above the horizontal, as indicated in Figure 6(a). Then the angle between the line of sight and the horizontal is called the **angle of elevation**. The **angle of depression** is similarly defined for an object below the horizontal, as shown in Figure 6(b).

FIGURE 6

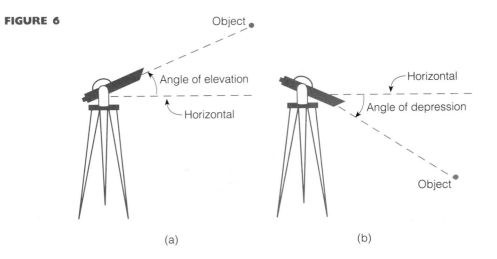

(a) (b)

█ █ EXAMPLE 7 A helicopter hovers 800 ft directly above a small island that is off the California coast. From the helicopter, the pilot takes a sighting to a point *P* directly ashore on the mainland, at the water's edge. If the angle of depression is 35°, how far off the coast is the island? See Figure 7.

FIGURE 7

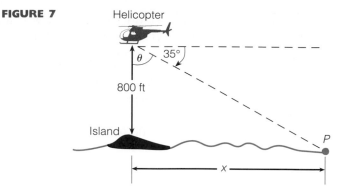

Solution   Let x denote the distance from the island to the mainland. Then, as you can see from Figure 7, we have $\theta + 35° = 90°$, from which it follows that $\theta = 55°$. Now we can write

$$\tan 55° = \frac{x}{800}$$

or

$$x = 800 \tan 55° \approx 1150 \text{ ft}$$ using a calculator and rounding to the nearest 50 feet   ∎

▌▌▌ EXAMPLE 8   Two satellite tracking stations, located at points A and B in California's Mojave Desert, are 200 miles apart. At a prearranged time, both stations measure the angle of elevation of a satellite as it crosses the vertical plane containing A and B. (See Figure 8.) If the angles of elevation from A and from B are α and β, respectively, express the altitude h of the satellite in terms of α and β.

Solution   We want to express the length $h = SC$ in Figure 8 in terms of the angles α and β. Note that $\overline{SC}$ is a side of both of the right triangles SCA and SCB. Working first in the right triangle SCB, we have

$$\cot \beta = \frac{CA + 200}{h}$$

or

$$h \cot \beta = CA + 200 \qquad\qquad (1)$$

We can eliminate CA from equation (1) as follows. Looking at right triangle SCA, we have

$$\cot \alpha = \frac{CA}{h} \qquad \text{and thus} \qquad CA = h \cot \alpha$$

Using this last equation to substitute for CA in equation (1), we obtain

$$h \cot \beta = h \cot \alpha + 200$$
$$h \cot \beta - h \cot \alpha = 200$$
$$h(\cot \beta - \cot \alpha) = 200$$

$$h = \frac{200}{\cot \beta - \cot \alpha} \text{ miles} \qquad \text{as required} \qquad ▌▌▌$$

FIGURE 8

▌▐▌ **EXAMPLE 9** The arc shown in Figure 9 is a portion of the unit circle, $x^2 + y^2 = 1$. Express the following quantities in terms of θ.

(a) OA **(b)** AB **(c)** OC **(d)** Area of $\triangle OAC$

Solution

(a) In right triangle OAB:

$$\cos \theta = \frac{OA}{OB} = \frac{OA}{1}$$

$$OA = \cos \theta$$

(b) In right triangle OAB:

$$\sin \theta = \frac{AB}{OB} = \frac{AB}{1}$$

$$AB = \sin \theta$$

(c) In right triangle OAC, we have $\cos \theta = OC/OA$, and therefore

$$OC = OA \cos \theta = (\cos \theta)(\cos \theta) = \cos^2 \theta$$

(d) Area $\triangle OAC = \frac{1}{2}(OA)(OC)(\sin \theta)$ using the formula area $= \frac{1}{2}ab \sin \theta$

$\quad = \frac{1}{2}(\cos \theta)(\cos^2 \theta)(\sin \theta)$ using the results from parts (a) and (c)

$\quad = \frac{1}{2}\cos^3 \theta \sin \theta$ ▌▌▌

FIGURE 9

EXERCISE SET 6.3

A

For Exercises 1–6, refer to the following figure. (However, each problem is independent of the others.)

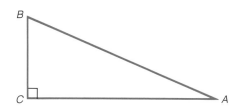

1. If $\angle A = 30°$ and $AB = 60$ cm, find AC and BC.
2. If $\angle A = 60°$ and $AB = 12$ cm, find AC and BC.
3. If $\angle B = 60°$ and $AC = 16$ cm, find BC and AB.
4. If $\angle B = 45°$ and $AC = 9$ cm, find BC and AB.
5. If $\angle B = 50°$ and $AB = 15$ cm, find BC and AC. (Round off your answers to one decimal place.)
6. If $\angle A = 25°$ and $AC = 100$ cm, find BC and AB. (Round off your answers to one decimal place.)
7. A ladder 18 ft long leans against a building. The ladder forms an angle of 60° with the ground.

(a) How high up the side of the building does the ladder reach? [Here and in part (b), give two forms for your answers: one with radicals and one (using a calculator) with decimals, rounded off to two places.]

(b) Find the horizontal distance from the foot of the ladder to the base of the building.

8. From a point level with and 1000 ft away from the base of the Washington Monument, the angle of elevation to the top of the monument is 29.05°. Determine the height of the monument to the nearest half foot.

9. Refer to the following figure (on the next page). At certain times, the planets Earth and Mercury line up in such a way that $\angle EMS$ is a right angle, as shown in the figure. At such times, $\angle SEM$ is found to be 21.16°. Use this information to estimate the distance MS of Mercury from the Sun. Assume that the distance from the Earth to the Sun is 93 million miles. (Round off your answer to the nearest million miles. Because Mercury's orbit is not exactly circular, the actual distance of Mercury from the Sun varies from about 28 million miles to 43 million miles.)

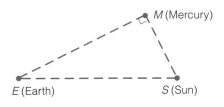

M (Mercury)

E (Earth)   S (Sun)

10. Determine the distance AB across the lake shown in the figure, using the following data: $AC = 400$ m, $\angle C = 90°$, and $\angle CAB = 40°$. Round off the answer to the nearest meter.

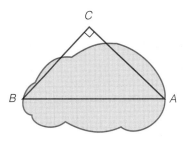

C

B   A

11. A building contractor wants to put a fence around the perimeter of a flat lot with the shape of a right triangle. One angle of the triangle is 41.4°, and the length of the hypotenuse is 58.5 m. Find the length of fencing required. Round off the answer to one decimal place.

12. Suppose that the contractor in Exercise 11 reviews his notes and finds that it is not the hypotenuse that is 58.5 m but rather the side opposite the 41.4° angle. Find the length of fencing required in this case. Again, round off the answer to one decimal place.

For Exercises 13 and 14, refer to following diagram for the roof of a house. In the figure, x is the length of a rafter measured from the top of a wall to the top of the roof; θ is the acute angle between a rafter and the horizontal; and h is the vertical distance from the top of the wall to the top of the roof.

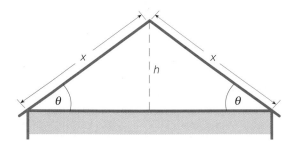

x   x
h
θ   θ

13. Suppose that $\theta = 39.4°$ and $x = 43.0$ ft.
  (a) Determine h. Round off the answer to one decimal place.
  (b) The *gable* is the triangular region bounded by the rafters and the attic ceiling. Find the area of the

gable. Round off the final answer to one decimal place.

14. Suppose that $\theta = 34°$ and $h = 36.5$ ft.
  (a) Determine x. Round off the answer to one decimal place.
  (b) Find the area of the gable. Round off the final answer to one decimal place. [See Exercise 13(b) for the definition of gable.]

In Exercises 15 and 16, find the area of the triangle. In Exercise 16, use a calculator and round off the final answer to two decimal places.

15.   **16.**

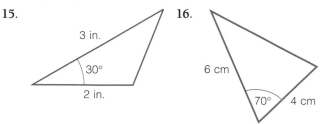

3 in.
30°
2 in.

6 cm
70°
4 cm

In Exercises 17–22, determine the area of the shaded region, given that the radius of the circle is 1 unit and the inscribed polygon is a regular polygon. Give two forms for each answer: an expression involving radicals or the trigonometric functions; a calculator approximation rounded off to three decimal places.

17.   **18.**

19.   **20.**

21.   **22.**

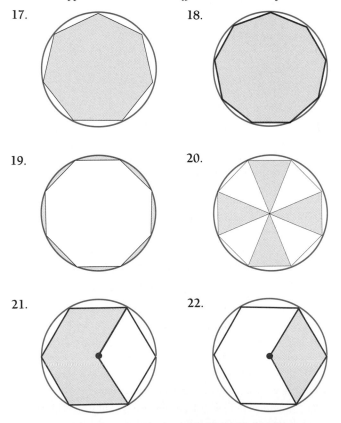

23. Show that the perimeter of the pentagon in Example 6 is 20 sin 36°. *Hint:* In Figure 5, draw a perpendicular from O to $\overline{AB}$.

24. In triangle OAB, lengths $OA = OB = 6$ in. and $\angle AOB = 72°$. Find AB. *Hint:* Draw a perpendicular from O to $\overline{AB}$. Round off the answer to one decimal place.

25. The accompanying figure shows two ships at points P and Q, which are in the same vertical plane with an airplane at point R. When the height of the airplane is 3500 ft, the angle of depression to P is 48° and that to Q is 25°. Find the distance between the two ships. Round off the answer to the nearest 10 feet.

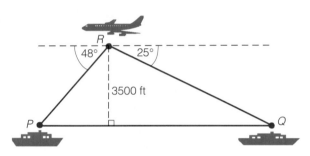

26. An observer in a lighthouse is 66 ft above the surface of the water. The observer sees a ship and finds the angle of depression to be 0.7°. Estimate the distance of the ship from the base of the lighthouse. Round off the answer to the nearest 5 feet.

27. From a point on ground level, you measure the angle of elevation to the top of a mountain to be 38°. Then you walk 200 meters farther away from the mountain and find that the angle of elevation is now 20°. Find the height of the mountain. Round off the answer to the nearest meter.

28. A surveyor stands 30 yards from the base of a building. On top of the building is a vertical radio antenna. Let α denote the angle of elevation when the surveyor sights to the top of the building. Let β denote the angle of elevation when the surveyor sights to the top of the antenna. Express the length of the antenna in terms of the angles α and β.

29. In $\triangle ACD$, you are given $\angle C = 90°$, $\angle A = 60°$, and $AC = 18$ cm. If B is a point on $\overline{CD}$ and $\angle BAC = 45°$, find BD. Express the answer in terms of a radical (rather than using a calculator).

30. The radius of the circle in the following figure is 1 unit. Express the lengths OA, AB, and DC in terms of α.

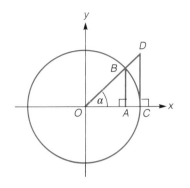

31. The arc in the next figure is a portion of the unit circle, $x^2 + y^2 = 1$.
    (a) Express the following angles in terms of θ: $\angle BOA$, $\angle OAB$, $\angle BAP$, $\angle BPA$. (Assume θ is in degrees.)
    (b) Express the following lengths in terms of $\sin \theta$ and $\cos \theta$: AO, AP, OB, BP.

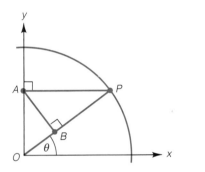

32. In the accompanying figure, suppose that $AD = 1$. Find the length of each of the other line segments in the figure. When radicals appear in an answer, leave the answer in that form, rather than using a calculator.

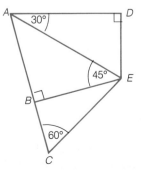

B

33. Refer to the figure. Express each of the following lengths as a function of θ.
(a) BC (b) AB (c) AC

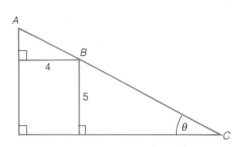

34. If $AB = 8$ in., express x as a function of θ.
Hint: First work Exercise 33.

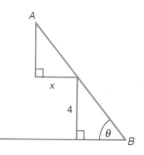

35. In the figure, line segment $\overline{BA}$ is tangent to the unit circle at A. Also, $\overline{CF}$ is tangent to the circle at F. Express the following lengths in terms of θ.
(a) DE (b) OE (c) CF
(d) OC (e) AB (f) OB

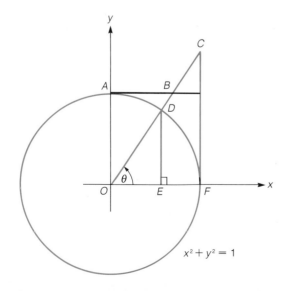

$$x^2 + y^2 = 1$$

36. At point P on the earth's surface, the moon is observed to be directly overhead, while at the same time at point T, the moon is just visible. See the figure.
(a) Show that $MP = \dfrac{OT}{\cos\theta} - OP$.
(b) Use a calculator and the following data to estimate the distance MP from the earth to the moon: $\theta = 89.05°$ and $OT = OP = 4000$ miles. Round off your answer to the nearest thousand miles. (Because the moon's orbit is not really circular, the actual distance varies from about 216,400 miles to 247,000 miles.)

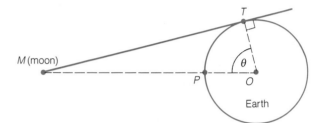

37. Refer to the accompanying figure. Let r denote the radius of the moon.
(a) Show that $r = \left(\dfrac{\sin\theta}{1 - \sin\theta}\right)PS$.
(b) Use a calculator and the following data to estimate the radius r of the moon: $PS = 238,857$ miles and $\theta = 0.257°$. Round off your answer to the nearest 10 miles.

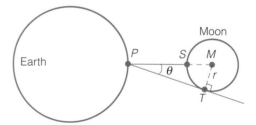

38. Figure A shows a regular hexagon inscribed in a circle of radius 1. Figure B shows a regular heptagon (seven-sided polygon) inscribed in a circle of radius 1. In Figure A, a line segment drawn from the center of the circle perpendicular to one of the sides is called an **apothem** of the polygon.

Figure A Figure B

(a) Show that the length of the apothem in Figure A is $\sqrt{3}/2$.

(b) Show that the length of one side of the heptagon in Figure B is $2\sin(180°/7)$.

(c) Use a calculator to evaluate the expressions in parts (a) and (b). Round off each answer to four decimal places, and note how close the two values are. Approximately two thousand years ago, Heron of Alexandria made use of this coincidence when he used the length of the apothem of the hexagon to approximate the length of the side of the heptagon. (The apothem of the hexagon can be constructed with ruler and compass; the side of the regular heptagon cannot.)

39. (a) Show that the area of a regular n-gon inscribed in a circle of radius 1 unit is given by

$$A_n = \frac{n}{2}\sin\frac{360°}{n}$$

(b) Use a calculator and the formula in part (a) to complete the following table.

n	5	10	50	100	1000	5000	10,000
A_n							

(c) Explain why the successive values of A_n in your table get closer and closer to π.

40. The figure shows a regular pentagon and a regular hexagon, with a common side of length 2 cm. Compute the area within the hexagon but outside of the pentagon. Round off the answer to two decimal places.

2 cm

C

41. In the accompanying figure, the smaller circle is tangent to the larger circle. Ray PQ is a common tangent and ray PR passes through the centers of both circles. If the radius of the smaller circle is a and the radius of the larger circle is b, show that $\sin\theta = (b-a)/(a+b)$. Then, using the identity $\sin^2\theta + \cos^2\theta = 1$, show that $\cos\theta = 2\sqrt{ab}/(a+b)$.

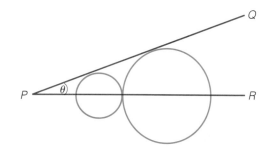

42. A vertical tower of height h stands on level ground. From a point P at ground level and due south of the tower, the angle of elevation to the top of the tower is θ. From a point Q at ground level and due west of the tower, the angle of elevation to the top of the tower is β. If d is the distance between P and Q, show that

$$h = \frac{d}{\sqrt{\cot^2\theta + \cot^2\beta}}$$

43. The following problem is taken from *An Elementary Treatise on Plane Trigonometry*, 8th ed., by R. D. Beasley (London: Macmillan and Co., 1884): The [angle of] elevation of a tower standing on a horizontal plane is observed; a feet nearer it is found to be 45°; b feet nearer still it is the complement of what it was at the first station; shew that the height of the tower is $ab/(a-b)$ feet.

$\underline{6.4}$ **TRIGONOMETRIC FUNCTIONS OF ANGLES**

Be clear about the signs of the functions. Your calculator will show them to you correctly, but you still need to be able to figure them out for yourself, quadrant by quadrant.

David Halliday and Robert Resnick in *Fundamentals of Physics*, 3rd ed. (New York: John Wiley and Sons, 1988)

Recall that the definitions of the trigonometric functions given in Section 6.1 apply only to acute angles. That is, as defined on page 297, the domains of trigonometric functions consist of all angles θ such that $0° < \theta < 90°$. In this section, we want to expand our definitions of the trigonometric functions to include angles of any size. To do this, we will need to be more explicit than we have been about angles and their size or measure.

For analytical purposes, we think of the two rays that form an angle as having been originally coincident; then while one ray is held fixed, the other is rotated to create the given angle. As Figure 1 indicates, the fixed ray is called the **initial side** of the angle, and the rotated ray is called the **terminal side**. By convention, we take the measure of an angle to be **positive** if the rotation is counterclockwise and **negative** if the rotation is clockwise. For example, the measure of the angle in Figure 2(a) is *positive* thirty degrees because the rotation is counterclockwise, whereas the measure of the angle in Figure 2(b) is *negative* thirty degrees because the rotation is clockwise. Figure 2(c) shows an angle of 390°.

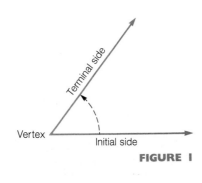

FIGURE 1

FIGURE 2
Angles generated by a counterclockwise rotation have positive measure. Angles generated by a clockwise rotation have negative measure.

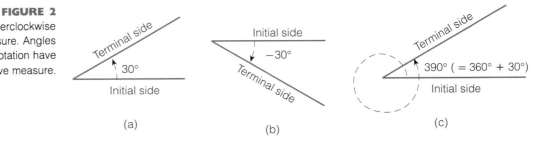

In our development of trigonometry, it will be convenient to have a *standard position* for angles. In a rectangular coordinate system, an angle is in **standard position** if the vertex is located at $(0, 0)$ and the initial side of the angle lies along the positive horizontal axis. Figure 3 shows examples of angles in standard position.

FIGURE 3
Examples of angles in standard position

Now we are ready to extend our definitions of the trigonometric functions to accommodate angles of any size. We begin by placing the angle θ in standard position and drawing in the **unit circle** $x^2 + y^2 = 1$, as shown in Figure 4. (Recall from Chapter 2 that the equation $x^2 + y^2 = 1$ represents a circle of radius 1, with center at the origin.) Notice the notation $P(x, y)$ in Figure 4; this stands for the point P, with coordinates (x, y), where the terminal side of angle θ intersects the unit circle. With this notation, we define the six trigonometric functions of θ as follows.

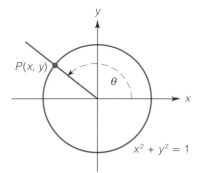

FIGURE 4

$P(x, y)$ denotes the point where the terminal side of angle θ intersects the unit circle.

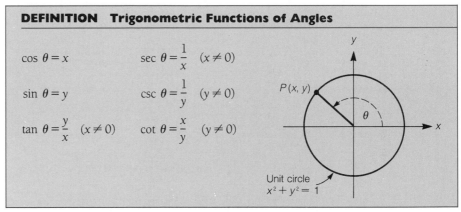

DEFINITION Trigonometric Functions of Angles

$$\cos \theta = x \qquad\qquad \sec \theta = \frac{1}{x} \quad (x \neq 0)$$

$$\sin \theta = y \qquad\qquad \csc \theta = \frac{1}{y} \quad (y \neq 0)$$

$$\tan \theta = \frac{y}{x} \quad (x \neq 0) \qquad\qquad \cot \theta = \frac{x}{y} \quad (y \neq 0)$$

Even before turning to some examples, we make three initial observations concerning our new definitions. The first two observations will help in memorizing the definitions. The third observation will help you to see why these new definitions are consistent with our previous work on right-triangle trigonometry.

1. $\cos \theta$ is the first coordinate of the point where the terminal side of angle θ intersects the unit circle; $\sin \theta$ is the second coordinate. (You can remember this by noting that, alphabetically, cosine comes before sine.)

2. Just as with the right-triangle definitions, we have the same three pairs of reciprocals: $\cos \theta$ and $\sec \theta$; $\sin \theta$ and $\csc \theta$; $\tan \theta$ and $\cot \theta$.

3. For the cases in which the angle θ is acute, these definitions are really equivalent to the original right-triangle definitions. Consider, for instance, the acute angle θ in Figure 5. According to our "new" definition of sine, we have

$$\sin \theta = y$$

On the other hand, applying the original right-triangle definition to Figure 5 yields

$$\sin \theta = \frac{\text{opposite}}{\text{hypotenuse}} = \frac{y}{1} = y$$

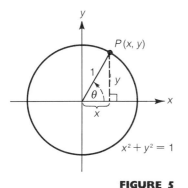

FIGURE 5

Thus, in both cases we obtain the same result. Using Figure 5, you should check for yourself that the same agreement also occurs for the other trigonometric functions.

▮▮▮ **EXAMPLE 1** Compute $\cos 90°$, $\sin 90°$, $\tan 90°$, $\sec 90°$, $\csc 90°$, and $\cot 90°$.

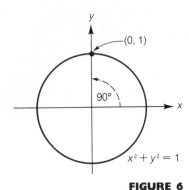

FIGURE 6

Solution We place the angle $\theta = 90°$ in standard position. Then, as Figure 6 indicates, the terminal side of the angle meets the unit circle at the point $(0, 1)$. Now we apply the definitions.

$$(0, 1)$$

By definition, cos 90° ⎯⎯↑ ↑⎯⎯ By definition, sin 90°
is this number.   is this number.

Thus,

$$\cos 90° = 0 \qquad \text{and} \qquad \sin 90° = 1$$

For the remaining trigonometric functions of 90°, we have

$$\tan 90° = \frac{y}{x} = \frac{1}{0} \qquad \text{undefined}$$

$$\sec 90° = \frac{1}{x} = \frac{1}{0} \qquad \text{undefined}$$

$$\csc 90° = \frac{1}{y} = \frac{1}{1} = 1$$

$$\cot 90° = \frac{x}{y} = \frac{0}{1} = 0$$

These are the required results.

▌▐ ▌ **EXAMPLE 2** Evaluate the trigonometric functions of $-180°$.

Solution The results can be read off from Figure 7.

FIGURE 7

$$\cos(-180°) = x = -1 \qquad \sec(-180°) = \frac{1}{x} = \frac{1}{-1} = -1$$

$$\sin(-180°) = y = 0 \qquad \csc(-180°) = \frac{1}{y} = \frac{1}{0} \qquad \text{undefined}$$

$$\tan(-180°) = \frac{y}{x} = \frac{0}{-1} = 0 \qquad \cot(-180°) = \frac{x}{y} = \frac{-1}{0} \qquad \text{undefined}$$

In the two examples just concluded, we evaluated the trigonometric functions for $\theta = 90°$ and $\theta = -180°$. In the same manner, the trigonometric functions can be evaluated just as easily for any angle that is an integral multiple of 90°. Table 1 shows the results of such calculations. Exercise 14 at the end of this section asks you to make these calculations for yourself.

TABLE I

θ	$\cos \theta$	$\sin \theta$	$\tan \theta$	$\sec \theta$	$\csc \theta$	$\cot \theta$
0°	1	0	0	1	undefined	undefined
90°	0	1	undefined	undefined	1	0
180°	-1	0	0	-1	undefined	undefined
270°	0	-1	undefined	undefined	-1	0
360°	1	0	0	1	undefined	undefined

▌▊▌ **EXAMPLE 3** Use Figure 8 to approximate the following trigonometric values to within successive tenths. Then use a calculator to check your answers. Round off the calculator values to two decimal places.

(a) cos 160° and sin 160°   **(b)** cos(−40°) and sin(−40°)

Solution

(a) Figure 8 shows that (the terminal side of) an angle of 160° in standard position meets the unit circle at a point in the second quadrant. Letting (x, y) denote the coordinates of that point, we have (from Figure 8)

$$-1.0 < x < -0.9 \qquad \text{and} \qquad 0.3 < y < 0.4$$

But, by definition, cos 160° = x and sin 160° = y and, consequently,

$$-1.0 < \cos 160° < -0.9 \qquad \text{and} \qquad 0.3 < \sin 160° < 0.4$$

The corresponding calculator results here are $\cos 160° \approx -0.94$ and $\sin 160° \approx 0.34$. Note that the estimations we obtained from Figure 8 are consistent with these calculator values. (When you check these calculator values for yourself, be sure that your calculator is set in the degree mode.)

(b) As you can verify using Figure 8, the terminal side of an angle of −40° intersects the unit circle at the same point as does the terminal side of an

FIGURE 8
Degree measure on the unit circle for angles in standard position

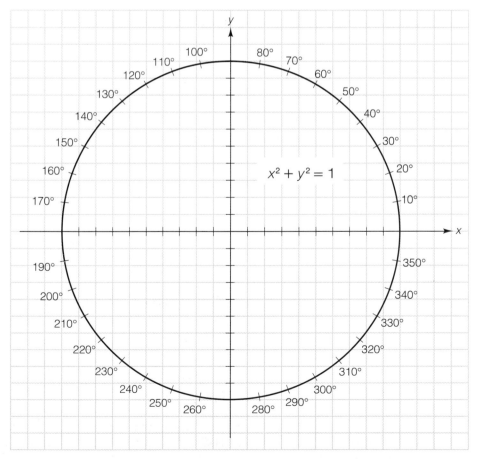

angle of 320°. Letting (x, y) denote the coordinates of that point, we have (from Figure 8)

$$0.7 < x < 0.8 \qquad \text{and} \qquad -0.7 < y < -0.6$$

and, consequently,

$$0.7 < \cos(-40°) < 0.8 \qquad \text{and} \qquad -0.7 < \sin(-40°) < -0.6$$

The corresponding calculator values here are $\cos(-40°) \approx 0.77$ and $\sin(-40°) \approx -0.64$. Again, note that the estimations we obtain from Figure 8 are consistent with these calculator values.   ▮▮▮

In evaluating the trigonometric functions, we'd like to take advantage, as much as possible, of the symmetry in Figure 8. To do this, we introduce the concept of a *reference angle*. (Even when we are using a calculator with the trigonometric functions, there are times, nevertheless, when we will need the reference angle concept. For instance, see Example 9 in Section 7.8.)

DEFINITION   The Reference Angle

Let θ be an angle in standard position, and suppose that θ is not a multiple of 90°. The **reference angle** associated with θ is the acute angle (with positive measure) formed by the x-axis and the terminal side of the angle θ.

In Figure 9 we show four examples of angles and their respective reference angles. The first part of Figure 9 shows how to find the reference angle for $\theta = 135°$. First we place the angle $\theta = 135°$ in standard position. Then we find the acute angle between the x-axis and the terminal side of θ. As you can see in this case, this acute angle is 45°; that is the reference angle associated with $\theta = 135°$. In the same way, you should work through the remaining three parts of Figure 9.

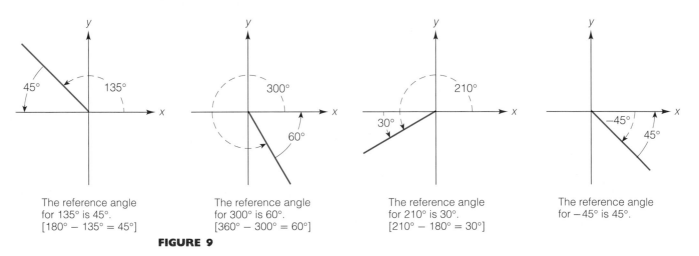

The reference angle for 135° is 45°.
[180° − 135° = 45°]

The reference angle for 300° is 60°.
[360° − 300° = 60°]

The reference angle for 210° is 30°.
[210° − 180° = 30°]

The reference angle for −45° is 45°.

FIGURE 9

Now let's look at an example to see how reference angles are used in evaluating the trigonometric functions. Suppose that we want to evaluate $\cos 150°$. In Figure 10(a) we've placed the angle $\theta = 150°$ in standard position. As you can see, the reference angle for 150° is 30°. By definition, the value of $\cos 150°$

is the *x*-coordinate of the point *P* in Figure 10(a). To find this *x*-coordinate, we reflect the line segment $\overline{OP}$ in the *y*-axis; the reflected line segment is the segment $\overline{OP'}$ in Figure 10(b). Since $\angle P'OA = 30°$, the *x*-coordinate of the point *P'* is by definition cos 30°, or $\sqrt{3}/2$. The *x*-coordinate of *P* is then the negative of this. That is, the *x*-coordinate of *P* is $-\sqrt{3}/2$. It follows now, again by definition, that the value of cos 150° is $-\sqrt{3}/2$.

FIGURE 10

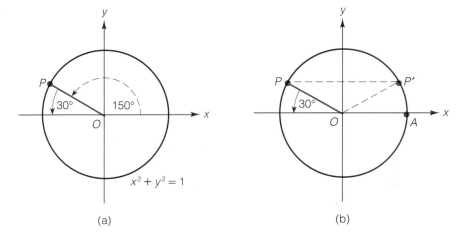

(a)   (b)

The same method that we have just used to evaluate cos 150° can be used to evaluate any of the trigonometric functions when the angles are not multiples of 90°. The following three steps summarize this method.

STEP 1 Determine the reference angle associated with the given angle.
STEP 2 Evaluate the given trigonometric function using the reference angle for the input.
STEP 3 Affix the appropriate sign to the number found in Step 2.

The next two examples illustrate this procedure.

EXAMPLE 4 Evaluate the following quantities.

(a) sin 135°   **(b)** cos 135°   **(c)** tan 135°

Solution As Figure 11 indicates, the reference angle associated with 135° is 45°.

(a) STEP 1 The reference angle is 45°.
STEP 2 $\sin 45° = \sqrt{2}/2$
STEP 3 sin θ is the *y*-coordinate. In the second quadrant, *y*-coordinates are positive. Thus, sin 135° is positive, since the terminal side of $\theta = 135°$ lies in the second quadrant. We therefore have

$$\sin 135° = \frac{\sqrt{2}}{2}$$

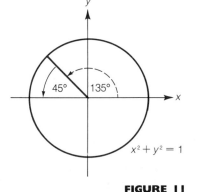

FIGURE 11

(b) STEP 1 The reference angle is 45°.
STEP 2 $\cos 45° = \sqrt{2}/2$
STEP 3 cos θ is the *x*-coordinate. In the second quadrant, *x*-coordinates are

negative. Thus, cos 135° is negative, since the terminal side of $\theta = 135°$ lies in the second quadrant. We therefore have

$$\cos 135° = -\frac{\sqrt{2}}{2}$$

(c)   STEP 1 The reference angle is 45°.
      STEP 2 $\tan 45° = 1$
      STEP 3 By definition, $\tan \theta = y/x$. Now, the terminal side of $\theta = 135°$ lies in the second quadrant, in which y is positive and x is negative. Thus, $\tan 135°$ is negative. We therefore have

$$\tan 135° = -1$$ ■■■

───────────────────────────

■■■ **EXAMPLE 5**   Evaluate each of the following quantities.

(a) $\cos(-120°)$ **(b)** $\cot(-120°)$ **(c)** $\sec(-120°)$

Solution   As Figure 12 shows, the reference angle for $-120°$ is 60°.

(a)   STEP 1 The reference angle for $-120°$ is 60°.
          STEP 2 $\cos 60° = \frac{1}{2}$
          STEP 3 $\cos \theta$ is the x-coordinate. In the third quadrant, the x-coordinates are negative. Thus, $\cos(-120°)$ is negative, since the terminal side of $\theta = -120°$ lies in the third quadrant. It now follows that

$$\cos(-120°) = -\frac{1}{2}$$

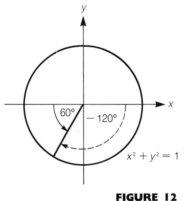

FIGURE 12

(b)   STEP 1 The reference angle for $-120°$ is 60°.
          STEP 2 $\cot 60° = \sqrt{3}/3$
          STEP 3 By definition, $\cot \theta = x/y$. Now, the terminal side of $\theta = -120°$ lies in the third quadrant, in which both x and y are negative. So x/y is positive and we have

$$\cot(-120°) = \frac{\sqrt{3}}{3}$$

(c)   We could follow our three-step procedure here, but in this case there is a faster method. In part (a) of this example, we found that $\cos(-120°) = -\frac{1}{2}$. Therefore $\sec(-120°) = -2$, because (according to the unit-circle definitions) $\sec \theta$ is the reciprocal of $\cos \theta$.   ■■■

In this section we have used the unit circle to generalize our definitions of the trigonometric functions to accommodate angles of any size. We conclude this section now with an application that relies on both the unit-circle definitions and the right-triangle definitions. Recall that in the previous section we found that the area of a triangle is given by the formula $A = \frac{1}{2}ab \sin \theta$, where a and b are the lengths of two sides and θ is the angle included between those two sides. When we derived this result in the previous section, we were assuming that θ was an acute angle. In fact, however, the formula is still valid when θ is an obtuse angle (an angle between 90° and 180°). To establish this result, we'll need to rely on the following identity.

$$\sin(180° - \theta) = \sin \theta$$

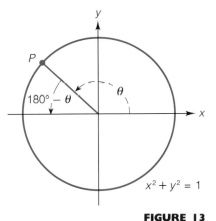

We will use Figure 13 to establish this identity for the case in which θ is an obtuse angle. (Exercise 53 shows you how to prove this identity when θ is an acute angle, and Section 7.6 discusses identities such as this from a more general point of view.) Figure 13 shows an obtuse angle θ in standard position. Let us apply our three-step procedure to determine $\sin \theta$. (As you'll see, the end result will be the required identity.)

STEP 1 As indicated in Figure 13, the reference angle for θ is $180° - \theta$.

STEP 2 The sine of the reference angle is $\sin(180° - \theta)$.

STEP 3 $\sin \theta$ is positive because the terminal side of θ lies in the second quadrant. Therefore, we have

$$\sin \theta = \sin(180° - \theta) \qquad \text{as required}$$

FIGURE 13

We are now prepared to show that the area A of the triangle in Figure 14(a) is given by $A = \frac{1}{2}ab \sin \theta$.

The area of any triangle is given by $A = \frac{1}{2}(\text{base})(\text{height})$. Thus, referring to Figure 14(b), we have

$$A = \frac{1}{2}bh \tag{1}$$

Also from Figure 14(b), we have

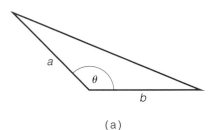

(a)

$$\sin(180° - \theta) = \frac{\text{opposite}}{\text{hypotenuse}} = \frac{h}{a}$$

$$h = a \sin(180° - \theta)$$

$$h = a \sin \theta \qquad \text{using the identity } \sin(180° - \theta) = \sin \theta \tag{2}$$

Now we can use equation (2) to substitute for h in equation (1). This yields

$$A = \frac{1}{2}b(a \sin \theta) = \frac{1}{2}ab \sin \theta$$

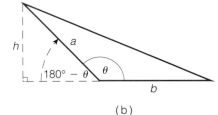

(b)

FIGURE 14

The formula we've just derived is a useful one. We will use it in later work to prove the *law of sines*. For reference, we summarize the result in the box that follows.

Formula for the Area of a Triangle

If a and b are the lengths of two sides of a triangle and θ is the angle included between those two sides, then the area of the triangle is given by

$$\text{area} = \frac{1}{2}ab \sin \theta$$

In Words The area of a triangle equals half the product of the lengths of two sides times the sine of the included angle.

EXERCISE SET 6.4

A

In Exercises 1–8, sketch each angle in standard position and specify the reference angle.

1. (a) 110°
   (b) −110°
2. (a) 160°
   (b) −160°
3. (a) 200°
   (b) −200°
4. (a) 225°
   (b) −225°
5. (a) 300°
   (b) −300°
6. (a) 325°
   (b) −325°
7. (a) 60°
   (b) −60°
8. (a) 460°
   (b) −460°

In Exercises 9–13, use the definitions (not a calculator) to evaluate the six trigonometric functions of the angle. If a value is undefined, state this.

9. 270°
10. 450°
11. −270°
12. −630°
13. 810°

14. Use the definitions of the trigonometric functions to complete the table. (When you are finished, check your results against the values shown in Table 1.)

θ	$\cos\theta$	$\sin\theta$	$\tan\theta$	$\sec\theta$	$\csc\theta$	$\cot\theta$
0°						
90°						
180°						
270°						
360°						

In Exercises 15–28, use Figure 8 to approximate within successive tenths the given trigonometric values. Then use a calculator to approximate the values to the nearest hundredth.

15. sin 10° and sin(−10°)
16. cos 10° and cos(−10°)
17. cos 80° and cos(−80°)
18. sin 80° and sin(−80°)
19. sin 120° and sin(−120°)
20. cos 120° and cos(−120°)
21. sin 150° and sin(−150°)
22. cos 150° and cos(−150°)
23. cos 220° and cos(−220°)
24. sin 220° and sin(−220°)
25. cos 310° and cos(−310°)
26. sin 310° and sin(−310°)
27. sin(40° + 360°)
28. cos(40° + 360°)

In Exercises 29–38, evaluate each expression using the method shown in Examples 4 and 5.

29. (a) cos 315°
    (b) cos(−315°)
    (c) sin 315°
    (d) sin(−315°)
30. (a) cos 240°
    (b) cos(−240°)
    (c) sin 240°
    (d) sin(−240°)
31. (a) cos 300°
    (b) cos(−300°)
    (c) sin 300°
    (d) sin(−300°)
32. (a) cos 150°
    (b) cos(−150°)
    (c) sin 150°
    (d) sin(−150°)
33. (a) cos 210°
    (b) cos(−210°)
    (c) sin 210°
    (d) sin(−210°)
34. (a) cos 585°
    (b) cos(−585°)
    (c) sin 585°
    (d) sin(−585°)
35. (a) cos 390°
    (b) cos(−390°)
    (c) sin 390°
    (d) sin(−390°)
36. (a) cos 405°
    (b) cos(−405°)
    (c) sin 405°
    (d) sin(−405°)
37. (a) sec 600°
    (b) csc(−600°)
    (c) tan 600°
    (d) cot(−600°)
38. (a) sec 330°
    (b) csc(−330°)
    (c) tan 330°
    (d) cot(−330°)

In Exercises 39 and 40, complete the tables.

39.

θ	$\sin\theta$	$\cos\theta$	$\tan\theta$
0°			
30°			
45°			
60°			
90°			
120°			
135°			
150°			
180°			

40.

θ	sin θ	cos θ	tan θ
180°			
210°			
225°			
240°			
270°			
300°			
315°			
330°			
360°			

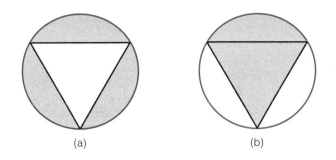

(a)   (b)

In Exercises 41–44, use the given information to determine the area of each triangle.

41. Two of the sides are 5 cm and 7 cm, and the angle between those sides is 120°.

42. Two of the sides are 3 m and 6 m, and the included angle is 150°.

43. Two of the sides are 21.4 cm and 28.6 cm, and the included angle is 98.5°. (Round off the final answer to one decimal place.)

44. Two of the sides are 5.98 cm and 8.05 cm, and the included angle is 107.1°. (Round off the answer to one decimal place.)

45. The figure shows an equilateral triangle inscribed in a circle of radius 12 cm. Use the method of Example 6 in Section 6.3 to compute the area of the triangle. Give two forms for your answer: one with a square root and the other a calculator approximation rounded off to two decimal places.

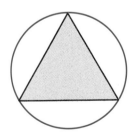

46. An equilateral triangle is inscribed in a circle of radius 8 cm. Compute the area of the shaded region in each of the following figures. Round off your final answers to three decimal places.

B

47. (a) Complete the following table, using the words "positive" or "negative" as appropriate.

	Terminal Side of Angle θ Lies in			
	Quadrant I	Quadrant II	Quadrant III	Quadrant IV
cos θ and sec θ	positive	negative		
sin θ and csc θ				
tan θ and cot θ				

(b) The mnemonic (memory device) ASTC (*all students take calculus*) is sometimes used to recall the signs of the trigonometric values in each quadrant:

A . . . . . . All are positive in the first quadrant.

S . . . . . . Sine is positive in the second quadrant.

T . . . . . . Tangent is positive in third quadrant.

C . . . . . . Cosine is positive in the fourth quadrant.

Check the validity of this mnemonic against your chart in part (a).

48. The value of tan θ is undefined when $\theta = 90°$. Use a calculator to complete the following tables to investigate the behavior of tan θ as θ approaches 90°. Round off each value to the nearest integer.

(a)

θ	89°	89.9°	89.99°	89.999°
tan θ				

(b)

θ	91°	90.1°	90.01°	90.001°
tan θ				

In Exercises 49–52, determine which of the two trigonometric values is the larger. Do not use a calculator.   Hint: Sketch the angles in standard position or refer to Figure 8.

49. (a) sin 10° or sin 70°
    (b) cos 10° or cos 70°

50. (a) tan 140° or tan 40°
    (b) tan 130° or tan 230°

51. (a) cos 140° or cos 100°
    (b) sin 140° or sin 100°

52. (a) sec 10° or sec 80°
    (b) csc 190° or csc 250°

53. In the text we derived the identity $\sin(180° - \theta) = \sin \theta$ in the case when θ is an obtuse angle. Now we'll derive the identity when θ is an acute angle.   Note: If θ is an acute angle, then $180° - \theta$ is an obtuse angle.
    (a) Sketch a figure showing the obtuse angle $180° - \theta$ in standard position. Note that the terminal side of the angle $180° - \theta$ lies in the second quadrant.
    (b) What is the reference angle for the angle $180° - \theta$?
    (c) Use steps (a) and (b) to conclude that $\sin(180° - \theta) = \sin \theta$.

54. Check that the identity $\sin(180° - \theta) = \sin \theta$ is valid in the three cases $\theta = 0°$, $\theta = 90°$, and $\theta = 180°$.

55. *Formula for* $\sin(\alpha + \beta)$ In the following figure, $\overline{AD} \perp \overline{BC}$ and $AD = 1$.

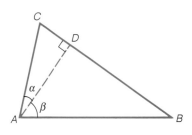

    (a) Show that $AC = \sec \alpha$ and $AB = \sec \beta$.
    (b) Show that

$$\text{area } \triangle ADC = \tfrac{1}{2} \sec \alpha \sin \alpha$$
$$\text{area } \triangle ADB = \tfrac{1}{2} \sec \beta \sin \beta$$
$$\text{area } \triangle ABC = \tfrac{1}{2} \sec \alpha \sec \beta \sin(\alpha + \beta)$$

(c) The sum of the areas of the two smaller triangles in part (b) equals the area of $\triangle ABC$. Use this fact and the expressions given in part (b) to show that

$$\sin(\alpha + \beta) = \sin \alpha \cos \beta + \cos \alpha \sin \beta$$

(d) Use the formula in part (c) to compute sin 75°.
    Hint: 75° = 30° + 45°
(e) Show that $\sin 75° \neq \sin 30° + \sin 45°$.
(f) Compute sin 105° and then check that $\sin 105° \neq \sin 45° + \sin 60°$.

For Exercises 56–59, let f, g, and h be the following three functions:

$$f(\theta) = \sin \theta \qquad g(\theta) = \cos \theta \qquad h(\theta) = \tan \theta$$

56. Evaluate each of the following (without using a calculator):   (a) $f(30°)$;   (b) $g(135°)$;   (c) $h(-60°)$.

57. If $\theta = 60°$, evaluate each of the following (without using a calculator):   (a) $h(\theta)$;   (b) $h(2\theta)$;   (c) $h(\theta/2)$.

58. If $\theta = 15°$, evaluate each of the following (without using a calculator):   (a) $f(-2\theta)$;   (b) $f(-4\theta)$;   (c) $[f(\theta)]^2 + [g(\theta)]^2$.

59. Use a calculator to evaluate each of the following. Round off your answers to three decimal places.
    (a) $f(27°)/g(27°)$ (b) $h(27°)$
    (c) $[f(27°)]^2 + [g(27°)]^2$

C

60. Use the figure to prove the following theorem: The area of a quadrilateral is equal to half the product of the diagonals times the sine of the included angle.
    Hint: Find the areas of each of the four triangles that make up the quadrilateral. Show that the sum of those areas is $\tfrac{1}{2}(\sin \theta)[qs + rq + rp + ps]$. Then factor the quantity within the brackets.

6.5 TRIGONOMETRIC IDENTITIES

A special name for the function which we call the sine is first found in the works of Āryabhata (c. 510) . . . .

It is further probable, from the efforts made to develop simple tables, that the Hindus were acquainted with the principles which we represent by the . . . [formula]
$\sin^2 \phi + \cos^2 \phi = 1. \ . \ . \ .$

David Eugene Smith in *History of Mathematics*, vol. II (New York: Ginn and Company, 1925)

As we mentioned in Section 6.2, there are many identities involving the trigonometric functions. In fact, that is one reason why these functions are so useful. In the box that follows, we list some of the same basic identities that we studied in Section 6.2. Keep in mind, however, that now the identities will be true for all angles for which the expressions are defined, rather than for acute angles only. This means that we will need to pay attention to the signs of the trigonometric values. (You'll see how this works in Examples 1 and 2.)

PROPERTY SUMMARY	BASIC TRIGONOMETRIC IDENTITIES
	1. $\sin^2 \theta + \cos^2 \theta = 1$
	2. $\dfrac{\sin \theta}{\cos \theta} = \tan \theta$
	3. $\sec \theta = \dfrac{1}{\cos \theta}; \quad \csc \theta = \dfrac{1}{\sin \theta}; \quad \cot \theta = \dfrac{1}{\tan \theta}$

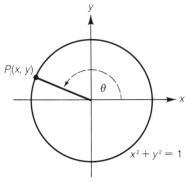

FIGURE 1

The second and the third identities in the box are immediate consequences of the unit-circle definitions of the trigonometric functions. To see why the first identity in the box is valid, consider Figure 1. Since the point $P(x, y)$ lies on the unit circle, we have

$$x^2 + y^2 = 1$$

But, by definition, $x = \cos \theta$ and $y = \sin \theta$. Therefore, we have

$$\cos^2 \theta + \sin^2 \theta = 1$$

which is essentially what we wished to show. Incidentally, you should also become familiar with the equivalent forms of this identity:

$$\cos^2 \theta = 1 - \sin^2 \theta \qquad \text{and} \qquad \sin^2 \theta = 1 - \cos^2 \theta$$

EXAMPLE 1 Given $\sin \theta = \frac{2}{3}$ and $90° < \theta < 180°$, find $\cos \theta$ and $\tan \theta$.

Solution Substituting $\sin \theta = \frac{2}{3}$ into the identity $\cos^2 \theta = 1 - \sin^2 \theta$ yields

$$\cos^2 \theta = 1 - \left(\frac{2}{3}\right)^2 = \frac{5}{9}$$

$$\cos \theta = \pm \sqrt{\frac{5}{9}} = \frac{\pm\sqrt{5}}{3}$$

To decide whether to choose the positive or negative value, note that the given inequality $90° < \theta < 180°$ tells us that the terminal side of θ lies in the second quadrant. Since in the second quadrant x-coordinates are negative, we choose the negative value here. Thus

$$\cos \theta = \frac{-\sqrt{5}}{3}$$

For $\tan \theta$, we have

$$\tan \theta = \frac{y}{x} = \frac{\sin \theta}{\cos \theta} = \frac{2/3}{-\sqrt{5}/3}$$

$$= -\frac{2}{\sqrt{5}} = -\frac{2\sqrt{5}}{5}$$

■■■ EXAMPLE 2 Suppose that

$$\cos \theta = t/2$$

where $270° < \theta < 360°$ and $t > 0$. Express the other five trigonometric values as functions of t.

Solution Replacing $\cos \theta$ with the quantity $t/2$ in the identity $\sin^2 \theta = 1 - \cos^2 \theta$ yields

$$\sin^2 \theta = 1 - \left(\frac{t}{2}\right)^2 = 1 - \frac{t^2}{4} = \frac{4 - t^2}{4}$$

and, consequently,

$$\sin \theta = \pm \frac{\sqrt{4 - t^2}}{2}$$

To decide whether to choose the positive or the negative value here, note that the given inequality $270° < \theta < 360°$ tells us that the terminal side of θ lies in the fourth quadrant. Since in the fourth quadrant y-coordinates are negative, we choose the negative value here. Thus

$$\sin \theta = -\frac{\sqrt{4 - t^2}}{2}$$

To obtain $\tan \theta$, we use the identity $\tan \theta = (\sin \theta)/(\cos \theta)$. This yields

$$\tan \theta = \frac{-\sqrt{4 - t^2}/2}{t/2} = -\frac{\sqrt{4 - t^2}}{t}$$

The remaining three values can now be found simply by taking reciprocals:

$$\sec \theta = \frac{1}{\cos \theta} = \frac{2}{t}$$

$$\csc \theta = \frac{1}{\sin \theta} = -\frac{2}{\sqrt{4 - t^2}} = -\frac{2\sqrt{4 - t^2}}{4 - t^2} \qquad \text{rationalizing the denominator}$$

$$\cot \theta = \frac{1}{\tan \theta} = -\frac{t}{\sqrt{4 - t^2}} = -\frac{t\sqrt{4 - t^2}}{4 - t^2} \qquad \text{rationalizing the denominator}$$

In the next five examples, we are asked to show that certain trigonometric equations are, in fact, identities. The procedures here will be very much like those in our work on simplifying trigonometric expressions in Section 6.2, but now we have an advantage. In each case we are given an answer toward which to work. The identities in the next five examples should not be memorized; they are too specialized. Instead, concentrate on the proofs themselves, noting where the fundamental identities (such as $\sin^2 \theta + \cos^2 \theta = 1$) come into play.

EXAMPLE 3 Prove that the equation $\csc A \tan A \cos A = 1$ is an identity.

Solution We begin with the left-hand side and express each factor in terms of sines or cosines:

$$\csc A \tan A \cos A = \frac{1}{\cancel{\sin A}} \cdot \frac{\cancel{\sin A}^{\,1}}{\cancel{\cos A}} \cdot \cancel{\cos A}^{\,1}$$

$$= 1 \qquad \text{as required}$$

EXAMPLE 4 Prove that $1 - \dfrac{\cot^2 \theta}{\csc^2 \theta} = \sin^2 \theta$.

Solution We work with the left-hand side, expressing everything in terms of sines and cosines and then simplifying (as in Section 6.2).

$$1 - \frac{\cot^2 \theta}{\csc^2 \theta} = 1 - \frac{(\cos^2 \theta)/(\sin^2 \theta)}{1/\sin^2 \theta}$$

Since $\tan \theta = (\sin \theta)/(\cos \theta)$, it follows that $\tan^2 \theta = (\sin^2 \theta)/(\cos^2 \theta)$ and consequently $\cot^2 \theta = (\cos^2 \theta)/(\sin^2 \theta)$.

$$= 1 - \frac{\cos^2 \theta}{1}$$

Multiplying both numerator and denominator of the fraction by $\sin^2 \theta$

$$= \sin^2 \theta$$

EXAMPLE 5 Prove that $\cos^2 B - \sin^2 B = \dfrac{1 - \tan^2 B}{1 + \tan^2 B}$.

Solution We begin with the right-hand side this time; it is the more complicated expression. As in the previous examples, we express everything in terms of sines and cosines.

$$\frac{1 - \tan^2 B}{1 + \tan^2 B} = \frac{1 - \dfrac{\sin^2 B}{\cos^2 B}}{1 + \dfrac{\sin^2 B}{\cos^2 B}} = \frac{\cos^2 B \left(1 - \dfrac{\sin^2 B}{\cos^2 B}\right)}{\cos^2 B \left(1 + \dfrac{\sin^2 B}{\cos^2 B}\right)}$$

$$= \frac{\cos^2 B - \sin^2 B}{\cos^2 B + \sin^2 B} = \frac{\cos^2 B - \sin^2 B}{1} = \cos^2 B - \sin^2 B$$

EXAMPLE 6 Prove that $\dfrac{\cos \theta}{1 - \sin \theta} = \dfrac{1 + \sin \theta}{\cos \theta}$.

Solution The suggestions given in the previous examples are not applicable here. Everything is already in terms of sines and cosines. Furthermore, neither side appears more complicated than the other. A technique that does work here is to begin with the left-hand side and multiply numerator *and* denominator by the same quantity, namely, $1 + \sin \theta$. Doing so gives us

$$\frac{\cos \theta}{1 - \sin \theta} = \frac{\cos \theta}{1 - \sin \theta} \cdot \frac{1 + \sin \theta}{1 + \sin \theta}$$

$$= \frac{\cos \theta(1 + \sin \theta)}{1 - \sin^2 \theta}$$

$$= \frac{\cos \theta(1 + \sin \theta)}{\cos^2 \theta} = \frac{1 + \sin \theta}{\cos \theta}$$

The general strategy for each of the proofs in Examples 3 through 6 was the same. In each case we worked with one side of the given equation, and we transformed it (into equivalent expressions) until it was identical to the other side of the equation. This is not the only strategy that can be used. For instance, an alternate way to establish the identity in Example 6 is as follows. The given equation is equivalent to

$$\frac{\cos \theta}{1 - \sin \theta} - \frac{1 + \sin \theta}{\cos \theta} = 0 \tag{1}$$

Now we show that equation (1) is an identity. To do this, we combine the two fractions on the left-hand side, using the common denominator $\cos \theta(1 - \sin \theta)$:

$$\frac{\cos \theta}{1 - \sin \theta} - \frac{1 + \sin \theta}{\cos \theta} = \frac{\cos^2 \theta - (1 + \sin \theta)(1 - \sin \theta)}{\cos \theta(1 - \sin \theta)}$$

$$= \frac{\cos^2 \theta - (1 - \sin^2 \theta)}{\cos \theta(1 - \sin \theta)}$$

$$= \frac{\cos^2 \theta + \sin^2 \theta - 1}{\cos \theta(1 - \sin \theta)} = \frac{1 - 1}{\cos \theta(1 - \sin \theta)} = 0$$

This shows that equation (1) is an identity, and thus the equation given in Example 6 is an identity.

Another strategy that can be used in establishing trigonometric identities is to work independently with each side of the given equation until a common expression is obtained. This is the strategy used in Example 7.

EXAMPLE 7   Prove that $\dfrac{1}{1 - \cos \beta} + \dfrac{1}{1 + \cos \beta} = 2 + 2 \cot^2 \beta.$

Solution

$$\begin{array}{ll}
\text{Left-hand} \\ \text{side}
\end{array}
= \frac{1(1 + \cos \beta) + 1(1 - \cos \beta)}{(1 - \cos \beta)(1 + \cos \beta)}
\qquad
\begin{array}{ll}
\text{Right-hand} \\ \text{side}
\end{array}
= 2 + \frac{2 \cos^2 \beta}{\sin^2 \beta}$$

$$= \frac{2}{1 - \cos^2 \beta}
\qquad\qquad\qquad\qquad\qquad\qquad
= \frac{2 \sin^2 \beta + 2 \cos^2 \beta}{\sin^2 \beta}$$

$$= \frac{2(\sin^2 \beta + \cos^2 \beta)}{1 - \cos^2 \beta}$$

$$= \frac{2}{1 - \cos^2 \beta}$$

We've now established the required identity by showing that both sides are equal to the same quantity. ∎

EXERCISE SET 6.5

A

In Exercises 1–8, use the given information to determine the remaining five trigonometric values. Rationalize any denominators that contain radicals.

1. (a) $\sin \theta = \frac{1}{5},\ 90° < \theta < 180°$
   (b) $\sin \theta = -\frac{1}{5},\ 180° < \theta < 270°$

2. (a) $\cos \theta = -\frac{3}{4},\ 90° < \theta < 180°$
   (b) $\cos \theta = \frac{3}{4},\ 270° < \theta < 360°$

3. (a) $\cos \theta = \frac{5}{13},\ 0° < \theta < 90°$
   (b) $\cos \theta = -\frac{5}{13},\ 180° < \theta < 270°$

4. (a) $\sin \theta = \sqrt{3}/6,\ 0° < \theta < 90°$
   (b) $\sin \theta = -\sqrt{3}/6,\ 270° < \theta < 360°$

5. $\csc A = -3,\ 270° < A < 360°$ *Hint:*   First find $\sin A$.

6. $\csc A = \frac{6}{5},\ 90° < A < 180°$

7. $\sec B = -\frac{3}{2},\ 180° < B < 270°$

8. $\sec B = \frac{25}{24},\ 270° < B < 360°$

In Exercises 9–14, use the given information to express the remaining five trigonometric values as functions of t or u. Assume that t and u are positive. Rationalize any denominators that contain radicals.

9. $\cos \theta = t/3,\ 270° < \theta < 360°$

10. $\cos \theta = -2t/5,\ 90° < \theta < 180°$

11. $\sin \theta = -3u,\ 180° < \theta < 270°$

12. $\sin \theta = -u^3,\ 270° < \theta < 360°$

13. $\cos \theta = u/\sqrt{3},\ 0° < \theta < 90°$

14. $\sin \theta = 2u/\sqrt{7},\ 0° < \theta < 90°$

In Exercises 15–36, prove that the equations are identities.

15. $\sin \theta \cos \theta \sec \theta \csc \theta = 1$

16. $\tan^2 A + 1 = \sec^2 A$

17. $(\sin \theta \sec \theta)/(\tan \theta) = 1$

18. $\tan \beta \sin \beta = \sec \beta - \cos \beta$

19. $(1 - 5\sin x)/\cos x = \sec x - 5\tan x$

20. $\dfrac{1}{\sin \theta} - \sin \theta = \cot \theta \cos \theta$

21. $\cos A(\sec A - \cos A) = \sin^2 A$

22. $\dfrac{\sin \theta}{\csc \theta} + \dfrac{\cos \theta}{\sec \theta} = 1$

23. $(1 - \sin \theta)(\sec \theta + \tan \theta) = \cos \theta$

24. $(\cos \theta - \sin \theta)^2 + 2\sin \theta \cos \theta = 1$

25. $(\sec \alpha - \tan \alpha)^2 = \dfrac{1 - \sin \alpha}{1 + \sin \alpha}$

26. $\dfrac{\sin B}{1 + \cos B} + \dfrac{1 + \cos B}{\sin B} = 2\csc B$

27. $\sin A + \cos A = \dfrac{\sin A}{1 - \cot A} - \dfrac{\cos A}{\tan A - 1}$

28. $(1 - \cos C)(1 + \sec C) = \tan C \sin C$

29. $\csc^2 \theta + \sec^2 \theta = \csc^2 \theta \sec^2 \theta$

30. $\cos^2 \theta - \sin^2 \theta = 1 - 2\sin^2 \theta$

31. $\sin A \tan A = \dfrac{1 - \cos^2 A}{\cos A}$

32. $\dfrac{\cot A - 1}{\cot A + 1} = \dfrac{1 - \tan A}{1 + \tan A}$

33. $\cot^2 A + \csc^2 A = -\cot^4 A + \csc^4 A$

34. $\dfrac{\cot^2 A - \tan^2 A}{(\cot A + \tan A)^2} = 2\cos^2 A - 1$

35. $\dfrac{\sin A - \cos A}{\sin A} + \dfrac{\cos A - \sin A}{\cos A} = 2 - \sec A \csc A$

36. $\tan A \tan B = \dfrac{\tan A + \tan B}{\cot A + \cot B}$

B

37. Only one of the following two equations is an identity. Decide which equation this is, and give a proof to show that it is, indeed, an identity. For the other equation, give an example showing that it is not an identity. (For example, to show that the equation $\sin \theta + \cos \theta = 1$ is not an identity, let $\theta = 30°$. Then the equation becomes $\frac{1}{2} + \frac{\sqrt{3}}{2} = 1$, which is false.)
   (a) $\dfrac{\csc^2 \alpha - 1}{\csc^2 \alpha} = \cos \alpha$
   (b) $(\sec^2 \alpha - 1)(\csc^2 \alpha - 1) = 1$

38. Follow the directions given in Exercise 37.
   (a) $(\csc \beta - \cot \beta)^2 = \dfrac{1 + \cos \beta}{1 - \cos \beta}$
   (b) $\cot \beta + \dfrac{\sin \beta}{1 + \cos \beta} = \csc \beta$

39. Prove the identity $\dfrac{\sin\theta}{1-\cos\theta}=\dfrac{1+\cos\theta}{\sin\theta}$ in two ways.

(a) Adapt the method of Example 6.

(b) Begin with the left-hand side and multiply numerator and denominator by $\sin\theta$.

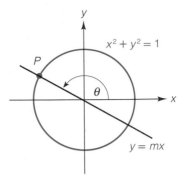

In Exercises 40–44, prove that the equations are identities.

40. $\dfrac{2\sin^3\beta}{1-\cos\beta}=2\sin\beta+2\sin\beta\cos\beta$

Hint: Write $\sin^3\beta$ as $(\sin\beta)(\sin^2\beta)$.

41. $\dfrac{\sec\theta-\csc\theta}{\sec\theta+\csc\theta}=\dfrac{\tan\theta-1}{\tan\theta+1}$

42. $1-\dfrac{\sin^2\theta}{1+\cot\theta}-\dfrac{\cos^2\theta}{1+\tan\theta}=\sin\theta\cos\theta$

43. $(\sin^2\theta)(1+n\cot^2\theta)=(\cos^2\theta)(n+\tan^2\theta)$

44. $(r\sin\theta\cos\phi)^2+(r\sin\theta\sin\phi)^2+(r\cos\theta)^2=r^2$

45. (a) Factor the expression $\cos^3\theta-\sin^3\theta$.

(b) Prove the identity

$$\frac{\cos\phi\cot\phi-\sin\phi\tan\phi}{\csc\phi-\sec\phi}=1+\sin\phi\cos\phi$$

46. Prove the following identities. (These two identities, along with $\sin^2\theta+\cos^2\theta=1$, are known as the *Pythagorean identities*. They will be discussed in the next chapter.)

(a) $\tan^2\theta=\sec^2\theta-1$ (b) $\cot^2\theta=\csc^2\theta-1$

47. If $\tan\alpha\tan\beta=1$ and α and β are acute angles, show that $\sec\alpha=\csc\beta$. *Hint:* Make use of the identities in Exercise 46.

48. Refer to the figure. Express the slope m of the line as a function of θ. *Hint:* What are the coordinates of the point P?

49. Suppose that $\sin\theta=\dfrac{p-q}{p+q}$, where p and q are positive and $90°<\theta<180°$. Show that $\tan\theta=\dfrac{q-p}{2\sqrt{qp}}$.

50. In this exercise, you will use the unit-circle definitions of sine and cosine, along with the identity $\sin^2\theta+\cos^2\theta=1$ to prove a surprising geometric result. In the figure, we show an equilateral triangle inscribed in the unit circle $x^2+y^2=1$. The vertices of the equilateral triangle are $A\left(-\frac{1}{2},-\frac{\sqrt{3}}{2}\right)$, $B(1,0)$, and $C\left(-\frac{1}{2},\frac{\sqrt{3}}{2}\right)$. Prove that for any point P on the unit circle, the sum of the squares of the distances from P to the three vertices is 6. *Hint:* Let the coordinates of P be $(\cos\theta,\sin\theta)$.

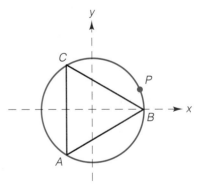

CHAPTER SIX SUMMARY OF PRINCIPAL TERMS AND FORMULAS

FORMULAS OR NOTATIONS	PAGE REFERENCE	COMMENT
1. $\cos\theta$ $\sin\theta$ $\tan\theta$ $\sec\theta$	297	When θ is an acute angle in a right triangle, the six trigonometric functions of θ are defined as follows. (In these definitions, the word "adjacent" means "the length of the side adjacent to angle θ." The same convention applies to the word "opposite.")

FORMULAS OR NOTATIONS	PAGE REFERENCE	COMMENT
csc θ cot θ		$\cos \theta = \dfrac{\text{adjacent}}{\text{hypotenuse}} \qquad \sec \theta = \dfrac{\text{hypotenuse}}{\text{adjacent}}$ $\sin \theta = \dfrac{\text{opposite}}{\text{hypotenuse}} \qquad \csc \theta = \dfrac{\text{hypotenuse}}{\text{opposite}}$ $\tan \theta = \dfrac{\text{opposite}}{\text{adjacent}} \qquad \cot \theta = \dfrac{\text{adjacent}}{\text{opposite}}$
2. $\sin^n \theta$	304	$\sin^n \theta$ means $(\sin \theta)^n$. The same convention also applies to the other five trigonometric functions.
3. $\sin^2 \theta + \cos^2 \theta = 1$ $(\sin\theta)/(\cos \theta) = \tan \theta$ $\sin(90° - \theta) = \cos \theta$ $\cos(90° - \theta) = \sin \theta$ $\sec \theta = 1/\cos \theta$ $\csc \theta = 1/\sin \theta$ $\cot \theta = 1/\tan \theta$	305	These are some of the most fundamental trigonometric identities. (We'll see many more in the following chapter.) With the exception of the third and fourth identities in this list, we discussed these identities in the context of both right-triangle and unit-circle trigonometry. Although the third and fourth identities were discussed only in the context of right triangles, you'll see in Chapter 7 that these two identities indeed hold for all angles.
4. $A = \frac{1}{2}ab \sin \theta$	312, 327	The area of a triangle equals half the product of the lengths of two sides times the sine of the included angle.
5. Initial side of an angle Terminal side of an angle	320	We think of the two rays that form an angle as having been originally coincident; then, while one ray is held fixed, the other is rotated to create the given angle. The fixed ray is the *initial side* of the angle, and the rotated ray is the *terminal side*. (See Figure 1 on page 320). The measure of an angle is *positive* if the rotation is counterclockwise and *negative* if the rotation is clockwise. (See Figure 2 on page 320.)
6. Standard position	320	In a rectangular coordinate system, an angle is in *standard position* if the vertex is located at $(0, 0)$ and the initial side of the angle lies along the positive horizontal axis. For examples, see Figure 3 on page 320.
7. $\cos \theta$ $\sin \theta$ $\tan \theta$ $\sec \theta$ $\csc \theta$ $\cot \theta$	321	If θ is an angle in standard position and $P(x, y)$ is the point where the terminal side of the angle meets the unit circle, then the six trigonometric functions of θ are defined as follows: $\cos \theta = x \qquad\qquad \sec \theta = \dfrac{1}{x} \quad (x \neq 0)$ $\sin \theta = y \qquad\qquad \csc \theta = \dfrac{1}{y} \quad (y \neq 0)$ $\tan \theta = \dfrac{y}{x} \ (x \neq 0) \quad \cot \theta = \dfrac{x}{y} \quad (y \neq 0)$
8. Reference angle	324	Let θ be an angle in standard position in an x-y coordinate system, and suppose that θ is not a multiple of 90°. The reference angle associated with θ is the acute angle (with positive measure) formed by the x-axis and the terminal side of the angle θ.

FORMULAS OR NOTATIONS	PAGE REFERENCE	COMMENT
9. The three-step procedure for evaluating the trigonometric functions.	325	The following three-step procedure can be used to evaluate the trigonometric functions for angles that are not multiples of 90°. STEP I Determine the reference angle associated with the given angle. STEP 2 Evaluate the given trigonometric function using the reference angle for the input. STEP 3 Affix the appropriate sign to the number found in Step 2. (See Examples 4 and 5 on pages 325–326.)

WRITING MATHEMATICS

1. A student who wanted to simplify the expression $\dfrac{30°}{\sin 30°}$ wrote

   $$\frac{30°}{\sin 30°} = \frac{\cancel{30}°}{\sin \cancel{30}°} = \frac{1}{\sin}$$

   Explain why this is nonsense and then indicate the correct solution.

2. Determine if each statement is *true* or *false*. In each case, write out your reason or reasons in complete sentences. If you draw a diagram to accompany your writing, be sure that you clearly label any parts of the diagram to which you refer in the writing.
   (a) If θ is an angle in standard position and $180° < \theta < 270°$, then $\tan \theta < \sin \theta$.
   (b) If θ is an acute angle in a right triangle, then $\sin \theta < 1$.
   (c) If θ is an acute angle in a right triangle, then $\tan \theta < 1$.
   (d) For all angles θ, we have $\sin \theta = \sqrt{1 - \cos^2 \theta}$.
   (e) For all acute angles θ, we have $\sin \theta = \sqrt{1 - \cos^2 \theta}$.
   (f) The formula for the area of a triangle, $A = \frac{1}{2}ab \sin \theta$, is not valid when θ is a right angle.
   (g) Every angle θ satisfying $0° < \theta < 180°$ is an allowable input for the sine function.
   (h) Every angle θ satisfying $0° < \theta < 180°$ is an allowable input for the tangent function.

3. A student who was asked to simplify the expression $\sin^2 33° + \sin^2 57°$ wrote $\sin^2 33° + \sin^2 57° = \sin^2(33° + 57°) = \sin^2 90° = 1^2 = 1$.
   (a) Where is the error?
   (b) What is the correct answer?

▌█ CHAPTER SIX REVIEW EXERCISES

For Exercises 1–16, evaluate each expression. (Don't use a calculator.)

1. $\sin 135°$
2. $\cos(-60°)$
3. $\tan(-240°)$
4. $\sin 450°$
5. $\csc 210°$
6. $\sec 225°$
7. $\sin 270°$
8. $\cot(-330°)$
9. $\cos(-315°)$
10. $\cos 180°$
11. $\cos 1800°$
12. $\sec 120°$
13. $\csc 240°$
14. $\csc(-45°)$
15. $\sec 780°$
16. $\sin^2 33° + \sin^2 57°$
17. Simplify: $\ln(\sin^2 17° + \cos^2 17°) - \cos(\ln e^\pi)$
18. If $\theta = 45°$, find each of the following.

(a) $\cos\theta$ (b) $\cos^3\theta$ (c) $\cos 2\theta$
(d) $\cos 3\theta$ (e) $\cos(2\theta/3)$ (f) $(\cos 3\theta)/3$
(g) $\cos(-\theta)$ (h) $\cos^3 5\theta$

In Exercises 19–31, the lengths of the three sides of a triangle are denoted a, b, and c; the angles opposite these sides are A, B, and C, respectively. In each exercise, use the given information to find the required quantities.

	GIVEN	FIND
19.	$B = 90°$, $A = 30°$ $b = 1$	a and c
20.	$B = 90°$, $A = 60°$, $a = 1$	c and b
21.	$B = 90°$, $\sin A = \frac{2}{5}$, $a = 7$	b
22.	$B = 90°$, $\sec C = 4$, $c = \sqrt{2}$	b
23.	$B = 90°$, $\cos A = \frac{3}{8}$	$\sin A$ and $\cot A$
24.	$B = 90°$, $b = 1$, $\tan C = \sqrt{5}$	a
25.	$b = 4$, $c = 5$, $A = 150°$	area of $\triangle ABC$
26.	$A = 120°$, $b = 8$, area of $\triangle ABC = 12\sqrt{3}$	c
27.	$c = 4$, $a = 2$, $B = 90°$	$\sin^2 A + \cos^2 B$
28.	$B = 90°$, $2a = b$	A
29.	$A = 30°$, $B = 120°$, $b = 16$	a
30.	$a = 7$, $b = 8$, $\sin C = \frac{1}{4}$	area of $\triangle ABC$
31.	$a = b = 5$, $\sin\left(\frac{1}{2}C\right) = \frac{9}{10}$	c and area of $\triangle ABC$

32. For the following figure, show that

$$y = x[\tan(\alpha + \beta) - \tan\beta]$$

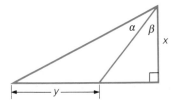

33. For the following figure, show that

$$y = \frac{x}{\cot\alpha - \cot\beta}$$

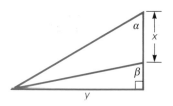

34. For the following figure, show that

$$\cot\theta = \frac{a}{b} + \cot\alpha$$

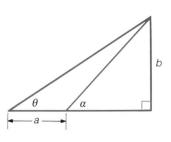

35. Suppose that θ is an acute angle in a right triangle and $\sin\theta = \dfrac{2p^2q^2}{p^4 + q^4}$. Find $\cos\theta$ and $\tan\theta$.

36. This problem is adapted from the text, *An Elementary Treatise on Plane Trigonometry*, by R. D. Beasley, first published in 1884.
   (a) Prove the following identity, which will be used in part (b): $1 + \tan^2\alpha = \sec^2\alpha$.
   (b) Suppose that α and θ are acute angles and
$$\tan\theta = \frac{1 + \tan\alpha}{1 - \tan\alpha}.$$
   Express $\sin\theta$ and $\cos\theta$ in terms of α.   *Hint:*   Draw a right triangle, label one of the angles θ, and let the lengths of the sides opposite and adjacent to θ be $1 + \tan\alpha$ and $1 - \tan\alpha$, respectively.   *Answer:*

$$\sin\theta = \frac{1}{\sqrt{2}}(\cos\alpha + \sin\alpha)$$

$$\cos\theta = \frac{1}{\sqrt{2}}(\cos\alpha - \sin\alpha)$$

In Exercises 37–46, convert each expression into one involving only sines and cosines and then simplify. (Leave your answers in terms of sines and/or cosines.)

37. $\dfrac{\sin A + \cos A}{\sec A + \csc A}$

38. $\dfrac{\csc A \sec A}{\sec^2 A + \csc^2 A}$

39. $\dfrac{\sin A \sec A}{\tan A + \cot A}$

40. $\cos A + \tan A \sin A$

41. $\dfrac{\cos A}{1 - \tan A} + \dfrac{\sin A}{1 - \cot A}$

42. $\dfrac{1}{\sec A - 1} \div \dfrac{1}{\sec A + 1}$

43. $(\sec A + \csc A)^{-1}[(\sec A)^{-1} + (\csc A)^{-1}]$

44. $\dfrac{\dfrac{\tan^2 A - 1}{\tan^3 A + \tan A}}{\dfrac{\tan A + 1}{\tan^2 A + 1}}$

45. $\dfrac{\dfrac{\sin A + \cos A}{\sin A - \cos A} - \dfrac{\sin A - \cos A}{\sin A + \cos A}}{\dfrac{\sin A + \cos A}{\sin A - \cos A} + \dfrac{\sin A - \cos A}{\sin A + \cos A}}$

46. $\dfrac{\sin A \tan A - \cos A \cot A}{\sec A - \csc A}$

In Exercises 47–50, refer to the following figure. The figure shows a highly magnified view of the point P, where the terminal side of an angle of 10° (in standard position) meets the unit circle.

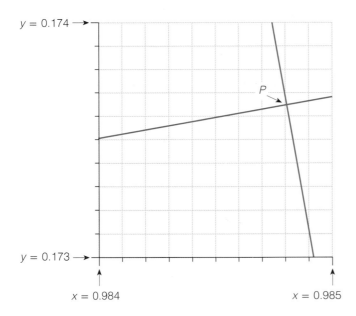

$y = 0.174$

$y = 0.173$

$x = 0.984$ $x = 0.985$

47. Using the figure, estimate the value of cos 10° to four decimal places. (Then use a calculator to check your estimate.)

48. Use your estimate in Exercise 47 to evaluate each of the following.
(a) cos 170°   (b) cos 190°   (c) cos 350°

49. Using the figure, estimate the value of sin 10° to four decimal places. (Then use a calculator to check your estimate.)

50. Use your estimate in Exercise 49 to evaluate each of the following.
(a) sin(−10°)   (b) sin(−190°)   (c) sin(−370°)

In Exercises 51–60, use the given information to find the required quantities.

	GIVEN	FIND
51.	$\cos\theta = \frac{3}{5}$, $0° < \theta < 90°$	$\sin\theta$ and $\tan\theta$
52.	$\sin\theta = -\frac{12}{13}$, $180° < \theta < 270°$	$\cos\theta$
53.	$\sec\theta = \frac{25}{7}$, $270° < \theta < 360°$	$\tan\theta$
54.	$\cot\theta = \frac{4}{3}$, $0° < \theta < 90°$	$\sin\theta$
55.	$\csc\theta = \frac{13}{12}$, $0° < \theta < 90°$	$\cot\theta$
56.	$\sin\theta = \frac{1}{5}$, $0° < \theta < 90°$	$\cos(90° - \theta)$
57.	$\cos\theta = 5t$, $0° < \theta < 90°$	$\tan(90° - \theta)$
58.	$\sin\theta = -\sqrt{5}\,u$, $90° < \theta < 270°$	$\sec\theta$
59.	$\tan\theta = \dfrac{\sqrt{2} - 1}{\sqrt{2} + 1}$, $0° < \theta < 90°$	$\sin\theta$
60.	$\cos\theta = \dfrac{2u}{u^2 + 1}$, $0° < \theta < 90°$	$\tan\theta$

61. Suppose θ is an acute angle and $\tan\theta + \cot\theta = 2$. Show that $\sin\theta + \cos\theta = \sqrt{2}$.

62. A 100-ft vertical antenna is on the roof of a building. From a point on the ground, the angles of elevation to the top and the bottom of the antenna are 51° and 37°, respectively. Find the height of the building.

63. In an isosceles triangle, the two base angles are each 35°, and the length of the base is 120 cm. Find the area of the triangle.

64. Find the perimeter and the area of a regular pentagon inscribed in a circle of radius 9 cm.

65. In triangle ABC, let h denote the length of the altitude from A to $\overline{BC}$. Show that $h = a/(\cot B + \cot C)$.

66. The length of each side of an equilateral triangle is $2a$. Show that the radius of the inscribed circle is $a/\sqrt{3}$ and the radius of the circumscribed circle is $2a/\sqrt{3}$.

67. From a helicopter h ft above the sea, the angle of depression to the pilot's horizon is θ. Show that $\cot\theta = R/\sqrt{2Rh + h^2}$, where R is the radius of the Earth.

68. In triangle ABC, angle C is a right angle. If a, b, and c denote the lengths of the sides opposite angles A, B, and C, respectively, show that

$$\frac{\sin A - \sin B}{\sin A + \sin B} = \frac{a - b}{a + b}$$

In Exercises 69–78, show that each equation is an identity.

69. $\sin^2(90° - \theta) \csc \theta - \tan^2(90° - \theta) \sin \theta = 0$

70. $\dfrac{1 - \sin \theta \cos \theta}{(\cos \theta)(\sec \theta - \csc \theta)} \times \dfrac{\sin^2 \theta - \cos^2 \theta}{\sin^3 \theta + \cos^3 \theta} = \sin \theta$

71. $\cos A \cot A = \csc A - \sin A$

72. $\sec A - 1 = (\sec A)(1 - \cos A)$

73. $\dfrac{\cot A - 1}{\cot A + 1} = \dfrac{\cos A - \sin A}{\cos A + \sin A}$

74. $\dfrac{\sin A}{\csc A - \cot A} = 1 + \cos A$

75. $\dfrac{1}{1 - \cos A} + \dfrac{1}{1 + \cos A} = 2 + 2 \cot^2 A$

76. $\dfrac{1}{\csc A - \cot A} - \dfrac{1}{\csc A + \cot A} = 2 \cot A$

77. $\dfrac{1}{1 + \sin^2 A} + \dfrac{1}{1 + \cos^2 A} + \dfrac{1}{1 + \sec^2 A}$
$$+ \dfrac{1}{1 + \csc^2 A} = 2$$

78. $\dfrac{\sec A - \tan A}{\sec A + \tan A} = \left(\dfrac{\cos A}{1 + \sin A}\right)^2$

79. The radius of the circle in the given figure is one unit. The line segment $\overline{PT}$ is tangent to the circle at P, and $\overline{PN}$ is perpendicular to $\overline{OT}$. Express each of the following lengths in terms of θ.
(a) PN (b) ON (c) PT
(d) OT (e) NA (f) NT

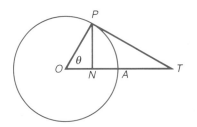

80. Refer to the figure accompanying Exercise 79.
(a) Show that the ratio of the area of triangle ONP to the area of triangle NPT is $\cot^2 \theta$.
(b) Show that the ratio of the area of triangle NPT to the area of triangle OPT is $\sin^2 \theta$.

81. Refer to the following figure. From a point R, two tangent lines are drawn to a circle of radius 1. These

tangents meet the circle at the points P and Q. At the midpoint of the circular arc $\overset{\frown}{PQ}$, a third tangent line is drawn, meeting the other two tangents at T and U, as shown. Express the area of triangle RTU as a function of θ (where θ is defined in the figure).

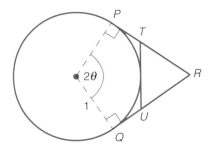

82. Suppose that in $\triangle ABC$, angle C is a right angle. Let a, b, and c denote the lengths of the sides opposite angles A, B, and C, respectively. Use the right-triangle definitions of sine and cosine to prove that

$$\frac{a^2\cos^2 B - b^2\cos^2 A}{a^2 - b^2} = 1$$

83. The following figure shows a semicircle with radius $AO = 1$.

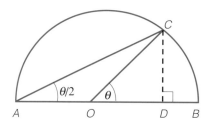

(a) Use the figure to derive the formula

$$\tan \frac{\theta}{2} = \frac{\sin \theta}{1 + \cos \theta}$$

Hint: Show that $CD = \sin \theta$ and $OD = \cos \theta$. Then look at right triangle ADC to find $\tan(\theta/2)$.
(b) Use the formula developed in part (a) to show that

(i) $\tan 15° = \dfrac{1}{2 + \sqrt{3}} = 2 - \sqrt{3}$;

(ii) $\tan 22.5° = \sqrt{2} - 1$.

84. This exercise outlines geometric derivations of formulas for $\sin 2\theta$ and $\cos 2\theta$. Refer to the figure (on the following page), which shows a semicircle with diameter $\overline{SV}$ and radius $UV = 1$. In each case, supply the reasons for the given statement.

(a) $TU = \cos 2\theta$ (b) $TV = 1 + \cos 2\theta$
(c) $WT = \sin 2\theta$
(d) $WV = 2 \cos \theta$ *Hint:*   Use right triangle SWV.
(e) $SW = 2 \sin \theta$
(f) $\sin \theta = \dfrac{\sin 2\theta}{2 \cos \theta}$ *Hint:*   Use $\triangle WTV$ and the results in parts (c) and (d).
(g) $\sin 2\theta = 2 \sin \theta \cos \theta$
(h) $\cos \theta = \dfrac{1 + \cos 2\theta}{2 \cos \theta}$ *Hint:*   Use $\triangle WTV$.
(i) $\cos 2\theta = 2 \cos^2 \theta - 1$

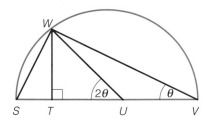

CHAPTER SIX TEST

1. Specify a value for each of the following expressions. (Exact values are required, not calculator approximations.)
   (a) $\tan 30°$ (b) $\sec 45°$ (c) $\sin^2 25° + \cos^2 25°$ (d) $\sin 53° - \cos 37°$

2. Suppose that angle θ is in standard position and the terminal side of the angle is in the third quadrant. Which of the following are positive?
   (a) $\sec \theta$ (b) $\csc \theta$ (c) $\cot \theta$

3. Evaluate each of the following.
   (a) $\sin(-270°)$ (b) $\cos 180°$ (c) $\tan 720°$

4. Factor: $2 \cot^2 \theta + 11 \cot \theta + 12$.

5. Find the area of a triangle in which the lengths of two sides are 8 cm and 9 cm and the angle included between those two sides is $150°$.

6. Evaluate each of the following. (Exact values are required, not calculator approximations.)
   (a) $\sin(-225°)$ (b) $\tan 330°$ (c) $\sec 120°$

7. If $\sin \theta = -\sqrt{5}/5$ and $180° < \theta < 270°$, find $\cos \theta$ and $\tan \theta$.

8. If $\sin \theta = -u/3$ and $270° < \theta < 360°$, express $\cos \theta$ and $\cot \theta$ as functions of u.

9. Simplify: $\dfrac{\dfrac{\cos \theta + 1}{\cos \theta} + 1}{\dfrac{\cos \theta - 1}{\cos \theta} - 1}$.

10. A regular nine-sided polygon is inscribed in a circle of radius 3 m. Find the area of the polygon. Give two forms of the answer: one in terms of a trigonometric function and the other a calculator approximation rounded off to three decimal places.

11. Express each term using sines and cosines and simplify as much as possible:

    $$\frac{\cot^2 \theta}{\csc^2 \theta} + \frac{\tan^2 \theta}{\sec^2 \theta}$$

12. In the figure, $\angle DAB = 25°$, $\angle CAB = 55°$, and $AB = 50$ cm. Find CD. Leave your answer in terms of the trigonometric functions (rather than using a calculator).

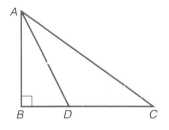

13. Without using a calculator, determine in each case which of the two trigonometric values is the larger. State your reasons. *Hint:* Sketch each pair of angles in standard position.
    (a) sin 5° or sin 85°   (b) sin 5° or cos 5°   (c) tan 175° or tan 185°

14. In △ABC (shown in the figure), ∠A = 20°, AB = 3.25 cm, and AC = 2.75 cm. Complete steps (a) and (b) to determine the length CB.
    (a) Find AD, DB, and CD. Leave your answers in terms of the trigonometric functions.
    (b) Find CB by applying the Pythagorean theorem in △CDB. For your final answer, use a calculator and round off to two decimal places.

CHAPTER SEVEN

TRIGONOMETRIC FUNCTIONS OF REAL NUMBERS

Trigonometric functions were originally introduced and tabulated in calculations in astronomy and navigation and were used extensively in surveying. What is more important [now], they occur whenever we want to describe a periodic process, like the vibration of a string, the tides, planetary motion, alternating currents, or the emission of light by atoms.

From *The Mathematical Sciences,* edited by the Committee on Support of Research in the Mathematical Sciences with the collaboration of George Boehm (Cambridge, Mass.: The M.I.T. Press, 1969)

INTRODUCTION

As the opening quotation indicates, the trigonometric functions play an important role in many of the modern applications of mathematics. In these applications, the inputs for these functions are real numbers rather than angles, as in the previous chapter. In order to make the transition from angle inputs to real-number inputs, we introduce in Section 7.1 radian measure for angles. Then, in Section 7.2, we use radian measure to restate the definitions of the trigonometric functions in such a way that the domains are indeed sets of real numbers. Among other advantages, this allows us to analyze the trigonometric functions using the graphing techniques that we developed in Chapter 3.

We discuss the graphs of the trigonometric functions in Sections 7.3 through 7.5. In Section 7.6, we develop four fundamental trigonometric identities known as the *addition formulas for sine and cosine.* Then, in the section after this, we consider a number of identities that follow rather directly from these addition formulas. In Section 7.8, we use some of these identities (along with our earlier work) to solve trigonometric equations. In the course of solving these equations, we'll find it useful to introduce the *inverse trigonometric functions.* In this first exposure to these functions, our approach is computational, and we use a calculator to evaluate the functions. We approach the inverse trigonometric functions from a more conceptual point of view in the last section of this chapter, Section 7.9.

7.1 RADIAN MEASURE

Degree measure, traditionally used to measure the angles of a geometric figure, has one serious drawback. It is artificial. There is no intrinsic connection between a degree and the geometry of a circle. Why 360 degrees for one revolution? Why not 400? or 100?

Calculus, One and Several Variables, 5th ed., by S. L. Salas, Einar Hille, and John T. Anderson (New York: John Wiley and Sons, Inc., 1986)

For the portion of trigonometry dealing with angles and geometric figures, the units of degrees are quite suitable for measuring angles. However, for the more analytical portions of trigonometry and for calculus, *radian* measure is used. The radian measure of an angle is defined as follows.

FIGURE 1

The radian measure θ is defined by the equation $\theta = s/r$.

DEFINITION   The Radian Measure of an Angle

Place the vertex of the angle at the center of a circle of radius r. Let s denote the length of the arc intercepted by the angle, as indicated in Figure 1. The **radian measure** θ of the angle is the ratio of the arc length s to the radius r. In symbols,

$$\theta = \frac{s}{r} \qquad (1)$$

In this definition, it is assumed that s and r have the same linear units.

EXAMPLE I   Determine the radian measure for each angle in Figure 2.

FIGURE 2

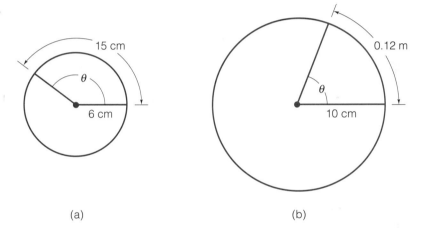

(a)   (b)

Solution   In Figure 2(a) we have $r = 6$ cm, $s = 15$ cm, and therefore

$$\theta = \frac{s}{r} = \frac{15 \text{ cm}}{6 \text{ cm}} = \frac{5}{2}$$

So, for the angle in Figure 2(a), we obtain $\theta = 5/2$ radians. In the calculations just completed, notice that although both s and r have the dimensions of length, the resulting radian measure s/r is simply a real number with no dimensions. (The dimensions "cancel out.")

Before carrying out the calculations for the radian measure in Figure 2(b), we have to arrange matters so that the same unit of length is used for the radius and the arc length. Converting the arc length s to centimeters, we have $s = 12$ cm and, consequently,

$$\theta = \frac{s}{r} = \frac{12 \text{ cm}}{10 \text{ cm}} = \frac{6}{5}$$

Thus, in Figure 2(b), we have $\theta = 6/5$ radians. Again, note that the radian measure s/r is dimensionless. (Check for yourself that the same answer is obtained if we use meters rather than centimeters for the common unit of measurement.)

At first it may appear to you that the radian measure depends on the radius of the particular circle that we use. But as you will see, this is not the case. To

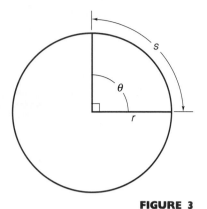

FIGURE 3

gain some experience in working with the definition of radian measure, let's calculate the radian measure of the right angle in Figure 3. We begin with the formula $\theta = s/r$. Now, since θ is a right angle, the arc length s is one-quarter of the entire circumference. Thus,

$$s = \frac{1}{4}(2\pi r) = \frac{\pi r}{2} \tag{2}$$

Using equation (2) to substitute for s in equation (1), we get

$$\theta = \frac{\pi r/2}{r} = \frac{\pi r}{2} \times \frac{1}{r} = \frac{\pi}{2}$$

In other words,

$$90° = \frac{\pi}{2} \text{ radians} \tag{3}$$

Notice that the radius r does not appear in our answer.

For practical reasons, we would like to be able to convert rapidly between degree and radian measure. Multiplying both sides of equation (3) by 2 yields

$$180° = \pi \text{ radians} \tag{4}$$

Equation (4) is useful and should be memorized. For instance, dividing both sides of equation (4) by 6 yields

$$\frac{180°}{6} = \frac{\pi}{6} \text{ radians}$$

or

$$30° = \frac{\pi}{6} \text{ radians}$$

Similarly, dividing both sides of equation (4) by 4 and 3, respectively, yields

$$45° = \frac{\pi}{4} \text{ radians} \qquad \text{and} \qquad 60° = \frac{\pi}{3} \text{ radians}$$

And, multiplying both sides of equation (4) by 2 gives us

$$360° = 2\pi \text{ radians}$$

Figure 4 summarizes some of these results.

FIGURE 4

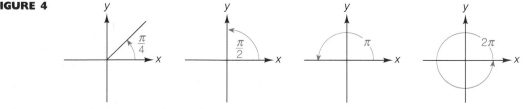

| | | **EXAMPLE 2**

(a) Express $1°$ in radian measure.
(b) Express 1 radian in terms of degrees.

Solution

(a) Dividing both sides of equation (4) by 180 yields

$$1° = \frac{\pi}{180} \text{ radian } (\approx 0.017 \text{ radian})$$

(b) Dividing both sides of equation (4) by π yields

$$\frac{180°}{\pi} = 1 \text{ radian}$$

In other words, 1 radian is approximately $180°/3.14$, or $57.3°$. ▪▪▪

From the results in Example 2, we have the following rules for converting between radians and degrees.

> To convert from degrees to radians, multiply by $\pi/180°$. To convert from radians to degrees, multiply by $180°/\pi$.

| | | **EXAMPLE 3**   Convert $150°$ to radians.

Solution

$$150°\left(\frac{\pi}{180°}\right) = \frac{5\pi}{6} \qquad \text{reducing the fraction}$$

Thus,

$$150° = \frac{5\pi}{6} \text{ radians}$$

▪▪▪

| | | **EXAMPLE 4**   Convert $11\pi/6$ radians to degrees.

Solution

$$\frac{11\pi}{6}\left(\frac{180°}{\pi}\right) = \frac{(11)(180°)}{6} = 11(30°) = 330°$$

So, $11\pi/6$ radians $= 330°$. ▪▪▪

We saw in Example 2(b) that 1 radian is approximately $57°$. It is also important to be able to visualize an angle of 1 radian without thinking in terms of degree measure. This is done as follows. In the equation $\theta = s/r$, we let $\theta = 1$. This yields $1 = s/r$ and, consequently, $r = s$. In other words, in a circle, 1 radian is the central angle that intercepts an arc equal in length to the radius of the circle. In the box that follows we summarize this result. Figure 5 displays angles of 1, 2, and 3 radians.

PROPERTY SUMMARY   RADIAN MEASURE

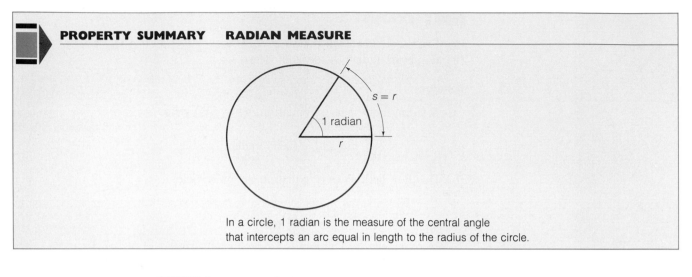

In a circle, 1 radian is the measure of the central angle that intercepts an arc equal in length to the radius of the circle.

FIGURE 5

(a) 1 radian   (b) 2 radians   (c) 3 radians

From now on, when we specify the measure of an angle, we will assume that the units are radians unless the degree symbol is explicitly used. (This convention is also used in calculus.) For instance, the equation $\theta = 2$ means that θ is 2 radians. Similarly, the expression sin $\pi/6$ refers to the sine of $\pi/6$ radians.

EXAMPLE 5 Evaluate the following quantities: (a) $\sin\dfrac{\pi}{6}$;
(b) $\cos 2\pi$.

Solution

(a) $\dfrac{\pi}{6}$ radians = 30°, therefore

$$\sin\frac{\pi}{6} = \sin 30° = \frac{1}{2} \qquad \text{because } \sin 30° = \frac{1}{2}$$

(b) 2π radians = 360°, therefore

$$\cos 2\pi = \cos 360° = 1 \qquad \text{because } \cos 360° = 1$$

EXAMPLE 6 Without using a calculator, determine whether the following values are positive or negative: (a) cos 3; (b) sin 1; (c) tan 6.

Solution

(a) Since π radians is 180°, we estimate that 3 radians is slightly less than 180° (because 3 is slightly less than π). Thus, in standard position, the terminal side of an angle of 3 radians lies in the second quadrant. Therefore, cos 3 is negative.

(b) Since 1 radian is approximately 60° [as we saw in Example 2(b) and in Figure 5], sin 1 is positive.

(c) We know that 6 radians is slightly less than 2π radians, which is one revolution, or 360°. Thus, in standard position, the terminal side of an angle of 6 radians lies in the fourth quadrant. Consequently, tan 6 is negative. ▌▌▌

As we mentioned in Chapter 6, all scientific calculators have keys with which you can indicate whether you are working in degrees or radians. In preparation for the next example, refer to the owner's manual for your calculator to make sure you know how it operates in this respect. (Here's a quick way to find out if you've properly set the calculator in the radian mode. Compute $\sin(\pi/6)$ and see if the calculator reads 0.5, as indeed it should.)

▌■▌ **EXAMPLE 7** Use Figure 6 to approximate to within successive tenths the following trigonometric values. Then use a calculator to check your answers. Round off the calculator values to two decimal places.

(a) cos 3 and sin 3   **(b)** cos(−3) and sin(−3)

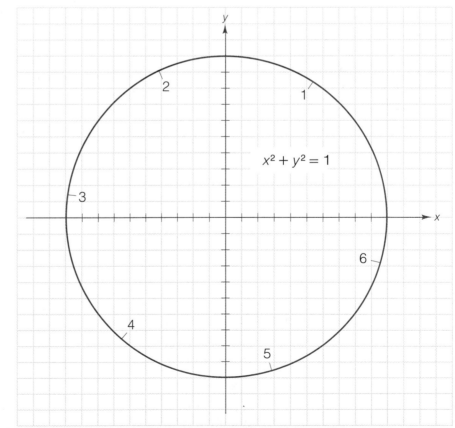

Solution

(a) Figure 6 shows that (the terminal side of) an angle of 3 radians in standard position meets the unit circle at a point in the second quadrant. Letting (x, y) denote the coordinates of that point, we have (using the grid in Figure 6)

$$-1.0 < x < -0.9 \qquad \text{and} \qquad 0.1 < y < 0.2$$

But, by definition, $\cos 3 = x$ and $\sin 3 = y$, so, consequently,

$$-1.0 < \cos 3 < -0.9 \qquad \text{and} \qquad 0.1 < \sin 3 < 0.2$$

The corresponding calculator values here are

$$\cos 3 \approx -0.99 \qquad \text{and} \qquad \sin 3 \approx 0.14$$

Note that the estimates we obtained from Figure 6 are consistent with these calculator values. (When you verify these calculator values for yourself, be sure that the calculator is set in the radian mode.)

(b) In Figure 6, we want to know where the terminal side of an angle of -3 radians intersects the unit circle. There is no marker for this point in Figure 6. But we can locate the point, nevertheless, by using symmetry. Let (x, y) denote the point corresponding to 3 radians, determined in part (a). By reflecting this point in the x-axis, we obtain the point corresponding to -3 radians. This means that the required point will have the same x-coordinate as the point in part (a), while the y-coordinate will be just the negative of that in part (a). So, given that

$$-1.0 < \cos 3 < -0.9 \qquad \text{and} \qquad 0.1 < \sin 3 < 0.2$$

we conclude that

$$-1.0 < \cos(-3) < -0.9 \qquad \text{and} \qquad -0.2 < \sin(-3) < -0.1$$

For the corresponding calculator values here, we obtain

$$\cos(-3) \approx -0.99 \qquad \text{and} \qquad \sin(-3) \approx -0.14$$

Again, note that the estimates from Figure 6 are consistent with these calculator values. When you verify these calculator values for yourself, remember that the calculator should be in the radian mode. Also, you may need to consult your calculator owner's manual about using negative numbers as inputs for these functions. You may find that you cannot simply use the calculator's arithmetic subtraction sign for the input -3. Instead, you may first need to enter "3" and then press a "change sign" key. For instance, on some types of Casio and Texas Instruments calculators (set in the radian mode), the sequence of keystrokes for evaluating $\cos(-3)$ is

3   (+/−)   (COS)

while on a Hewlett-Packard calculator, the sequence is

3   (CHS)   (COS)   ∎

Let us now return to the defining equation for radian measure, $\theta = s/r$, and multiply both sides by r to obtain $s = r\theta$. This gives us a useful formula for arc length on a circle in terms of the radius r and the central angle θ subtended by the arc.

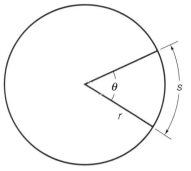

FIGURE 7

Formula for Arc Length

(Refer to Figure 7.) In a circle of radius r, the arc length s determined by a central angle of radian measure θ is given by

$$s = r\theta$$

In this formula, it is assumed that s and r have the same linear units.

■■■ **EXAMPLE 8** Compute the indicated arc lengths in Figures 8(a) and 8(b). Express the answers both in terms of π and as decimal approximations rounded off to two decimal places.

FIGURE 8

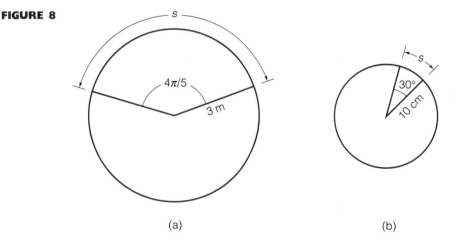

(a)   (b)

Solution

(a) We use the formula $s = r\theta$ with $r = 3$ m and $\theta = 4\pi/5$. This yields

$$s = (3\text{ m})\left(\tfrac{4\pi}{5}\right) = \tfrac{12\pi}{5}\text{ m}$$

Using a calculator, we find that this is approximately 7.54 m.

(b) To apply the formula $s = r\theta$, the angle measure θ must be expressed in radians (because the formula is just a restatement of the definition of radian measure). We've seen previously that $30° = \pi/6$ radians. So we have

$$s = r\theta = (10\text{ cm})\left(\tfrac{\pi}{6}\right) = \tfrac{5\pi}{3}\text{ cm}$$

Before picking up the calculator, note that we can obtain a quick approximation here: since $\pi \approx 3$, we have $s \approx 5(3/3) = 5$ cm. With a calculator now, we obtain $s = \tfrac{5\pi}{3}$ cm ≈ 5.24 cm. ■■■

One of the advantages of using radian measure is that many formulas then take on particularly simple forms. The formulas $s = r\theta$ is one example of this. As another example, let us derive a formula for the area of a *sector* of a circle. (In Figure 9, the shaded region is the sector.) From geometry, we know that the area A of the sector is directly proportional to the measure θ of its central angle. In symbols,

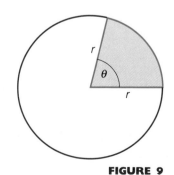

FIGURE 9

$$A = k\theta \tag{5}$$

To determine the constant k, we use the fact that when $\theta = 2\pi$ radians, we have a complete circle. Thus, when $\theta = 2\pi$, the area must be πr^2. Substituting the values $\theta = 2\pi$ and $A = \pi r^2$ in equation (5) yields

$$\pi r^2 = k(2\pi)$$

and therefore

$$\tfrac{1}{2}r^2 = k$$

Using this value for k in equation (5) gives us the following result.

Formula for the Area of a Sector

In a circle of radius r, the area of a sector with central angle of radian measure θ is given by

$$A = \frac{1}{2}r^2\theta$$

FIGURE 10

▮▬▮ **EXAMPLE 9**   Compute the area of the sector in Figure 10.

Solution   We first convert 120° to radians:

$$120°\left(\frac{\pi}{180°}\right) = \frac{2\pi}{3} \text{ radians}$$

We then have

$$A = \tfrac{1}{2}r^2\theta$$
$$= \tfrac{1}{2}(5 \text{ cm})^2\left(\tfrac{2\pi}{3}\right) = \tfrac{25\pi}{3} \text{ cm}^2 \quad (\approx 26.18 \text{ cm}^2) \qquad ▮▮▮$$

The next example indicates how radian measure is used in the study of rotating objects. As a prerequisite for this example, we first define the terms *angular speed* and *linear speed*.

DEFINITION   Angular Speed and Linear Speed

Suppose that a wheel rotates about its axis at a constant rate.

1. Refer to Figure 11(a). If a radial line turns through an angle θ in time t, then the **angular speed** of the wheel [denoted by the Greek letter ω (*omega*)] is defined to be

$$\text{angular speed} = \omega = \frac{\theta}{t}$$

2. Refer to Figure 11(b). If a point P on the rotating wheel travels a distance d in time t, then the **linear speed** of P, denoted by v, is defined to be

$$\text{linear speed} = v = \frac{d}{t}$$

FIGURE 11

(a)   (b)

▮▯▮ **EXAMPLE 10** A circular gear in a motor rotates at the rate of 100 rpm (revolutions per minute).

(a) What is the angular speed of the gear in radians per minute?
(b) Find the linear speed of a point on the gear 4 cm from the center.

Solution

(a) Each revolution of the gear is 2π radians. So in 100 revolutions, there are

$$\theta = 100(2\pi) = 200\pi \text{ radians}$$

Consequently, we have

$$\omega = \frac{\theta}{t} = \frac{200\pi \text{ radians}}{1 \text{ min}} = 200\pi \text{ radians/min}$$

(b) We can use the formula $s = r\theta$ to find the distance traveled by the point in 1 minute. Using $r = 4$ cm and $\theta = 200\pi$, we obtain

$$s = r\theta = 4(200\pi) = 800\pi \text{ cm}$$

The linear speed therefore is

$$v = \frac{d}{t} = \frac{800\pi \text{ cm}}{1 \text{ min}}$$
$$= 800\pi \text{ cm/min } (\approx 2513 \text{ cm/min}) \qquad ▮▮▮$$

EXERCISE SET 7.1

A

In Exercises 1–4, convert to radian measure. Express your answers both in terms of π and as decimal approximations rounded off to two decimal places.

1. (a) 45°   (b) 90°   (c) 135°
2. (a) 30°   (b) 150°   (c) 300°
3. (a) 0°   (b) 360°   (c) 450°
4. (a) 36°   (b) 35°   (c) 720°

In Exercises 5–8, convert the radian measures to degrees.

5. (a) $\pi/12$ (b) $\pi/6$ (c) $\pi/4$
6. (a) 3π (b) $3\pi/2$ (c) 2π
7. (a) $\pi/3$ (b) $5\pi/3$ (c) 4π
8. (a) $5\pi/6$ (b) $11\pi/6$ (c) 0

In Exercises 9 and 10, convert the radian measures to degrees. Round off the answers to two decimal places.

9. (a) 2   (b) 3   (c) π^2

10. (a) 1.32   (b) 0.96   (c) $1/\pi$

11. Suppose that the radian measure of an angle is $\frac{3}{2}$. Without using a calculator or tables, determine if this angle is larger or smaller than a right angle.
    Hint: What is the radian measure of a right angle?

12. Two angles in a triangle have radian measure $\pi/5$ and $\pi/6$. What is the radian measure of the third angle?

For Exercises 13 and 14, refer to the following figure, which shows all of the angles from 0° to 360° that are multiples of 30° or 45°.

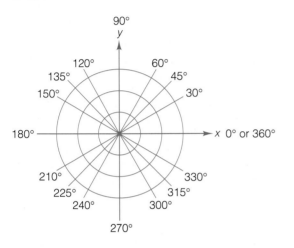

13. In the figure, relabel the angles in quadrants I and II using radian measure.

14. In the figure, relabel the angles in quadrants III and IV using radian measure.

In Exercises 15–18, use the definition $\theta = s/r$ to determine the radian measure of the indicated angle.

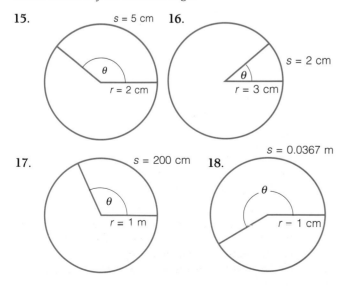

In Exercises 19–28, use Figure 6 to approximate the given trigonometric values to within successive tenths. Then use a calculator to compute the values to the nearest hundredth.

19. cos 1 and sin 1
20. cos 2 and sin 2
21. $\cos(-1)$ and $\sin(-1)$
22. $\cos(-2)$ and $\sin(-2)$
23. cos 4 and sin 4
24. cos 5 and sin 5
25. $\cos(-4)$ and $\sin(-4)$
26. $\cos(-5)$ and $\sin(-5)$
27. $\sin(1 + 2\pi)$
28. $\sin(2 + 2\pi)$

In Exercises 29–32, find the arc length s in each case.

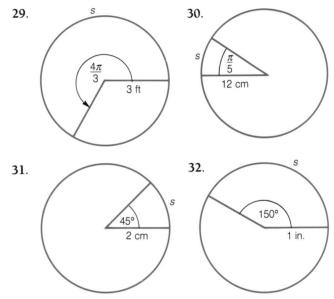

In Exercises 33 and 34, find the area of the sector determined by the given radius r and central angle θ. Express the answer both in terms of π and as a decimal approximation rounded off to two decimal places.

33. (a) $r = 6$ cm; $\theta = 2\pi/3$ (b) $r = 5$ m; $\theta = 80°$
    (c) $r = 24$ m; $\theta = \pi/20$ (d) $r = 1.8$ cm; $\theta = 144°$

34. (a) $r = 4$ cm; $\theta = \pi/10$ (b) $r = 16$ m; $\theta = 5°$
    (c) $r = 21$ ft; $\theta = 11\pi/6$ (d) $r = 4.2$ in.; $\theta = 170°$

35. In a circle of radius 1 cm, the area of a certain sector is $\pi/5$ cm². Find the radian measure of the central angle. Express the answer in terms of π rather than as a decimal approximation.

36. In a circle of radius 3 m, the area of a certain sector is 20 m². Find the degree measure of the central angle. Round off the answer to two decimal places.

In Exercises 37–42, you are given the rate of rotation of a wheel and its radius. In each case, determine the following:
(a) *the angular speed, in units of radians/sec;*
(b) *the linear speed, in units of cm/sec, of a point on the circumference of the wheel;*

(c) *the linear speed, in cm/sec, of a point halfway between the center of the wheel and the circumference.*

37. 6 revolutions/sec; $r = 12$ cm

38. 15 revolutions/sec; $r = 20$ cm

39. 1080°/sec; $r = 25$ cm **40.** 2160°/sec, $r = 60$ cm

41. 500 rpm; $r = 45$ cm **42.** 1250 rpm; $r = 10$ cm

B

43. Suppose that you have two sticks and a piece of wire, each of length 1 ft, fastened at the ends to form an equilateral triangle. See Figure (a). If side $\overline{BC}$ is bent out to form an arc of a circle with center A, then the angle at A will decrease from 60° to something less. See Figure (b). What is the measure of this new angle at A?

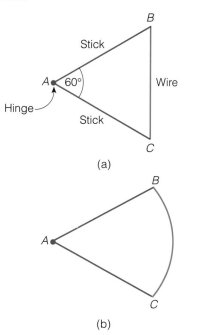

(a)

(b)

44. (a) When a clock reads 4:00, what is the radian measure of the (smaller) angle between the hour hand and minute hand?

(b) When a clock reads 5:30, what is the radian measure of the (smaller) angle between the hour hand and minute hand?

In Exercises 45 and 46, suppose that a belt drives two wheels of radii r and R, as indicated in the figure.

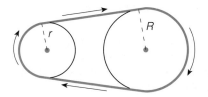

45. If $r = 6$ cm, $R = 10$ cm, and the angular speed of the larger wheel is 100 rpm, determine each of the following.

(a) The angular speed of the larger wheel in radians per minute.

(b) The linear speed of a point on the circumference of the larger wheel.

(c) The angular speed of the smaller wheel in radians per minute. *Hint:* Because of the belt, the linear speed of a point on the circumference of the larger wheel is equal to the linear speed of a point on the circumference of the smaller wheel.

(d) The angular speed of the smaller wheel in rpm.

46. Follow Exercise 45, assuming $r = 5$ cm, $R = 15$ cm, and the angular speed of the larger wheel is 1800 rpm.

The latitude of a point P on the surface of the Earth is specified by means of the angle θ in the figure. For instance, the latitude of Paris, France, is 48°52′ N. The letter N is used here to indicate that the location is north of, rather than south of, the equator. (Recall that the notation 52′ indicates $\frac{52}{60}$ of one degree.) In Exercises 47–52, use the arc length formula (and your calculator) to determine the distance $\overset{\frown}{PE}$ from the given location P to the equator. Assume that the Earth is a sphere with radius $OP = OE = 3960$ miles. Round off each answer to the nearest 10 miles.

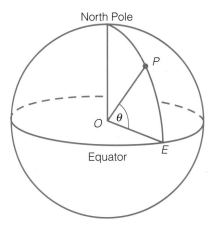

47. Point Barrow, Alaska: 71°23′ N

48. Singapore: 1°17′ N

49. Honolulu: 21°19′ N

50. Lagos: 6°27′ N

51. Washington, D.C.: 38°54′ N

52. Fairbanks, Alaska: 64°51′ N

53. The following figure (on the next page) shows a circular sector with radius r cm and central angle 2θ (radian measure). The perimeter of the sector is 12 cm.

(a) Express r as a function of θ.

(b) Express the area A of the sector as a function of θ.

$$P = \frac{\pi\theta}{4} - \frac{\theta^2}{4}$$

Is this a quadratic function?

(b) For what value of θ is P a maximum?

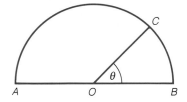

54. The figure (to the right) shows a semicircle of radius 1 unit and two adjacent sectors, AOC and COB.
(a) Show that the product P of the areas of the two sectors is given by

7.2 TRIGONOMETRIC FUNCTIONS OF REAL NUMBERS

The definitions of the trigonometric functions in Section 6.4 are based on the unit circle. Let us look at radian measure in this context. By definition, the radian measure θ is given by $\theta = s/r$. So when $r = 1$, we obtain

$$\theta = \frac{s}{1} = s$$

Thus, in the unit circle, the arc length *is* the radian measure of the angle. (See Figure 1.)

In view of this observation, we can now define the trigonometric functions in such a way that the domains are sets of real numbers, rather than angles. There are two reasons for doing this. First, for analytical work, the domains of the trigonometric functions are sets of real numbers. Second, if the domains are sets of real numbers, then the techniques for graphing that we developed in Chapter 3 can be used to analyze these functions.

In the definitions that follow, you may think of θ as either the measure of an arc length or the radian measure of an angle. But in both cases—and this is the point—θ denotes a real number. The conventions regarding the measurement of arc length on the unit circle are the same as those used previously for angles: We measure from the point $(1, 0)$ and we assume that the positive direction is counterclockwise.

FIGURE 1

In the unit circle, the length of the intercepted arc is the radian measure of the central angle.

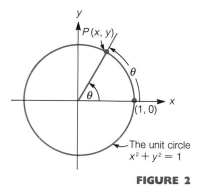

FIGURE 2

Trigonometric Functions of Real Numbers

(Refer to Figure 2.) Let $P(x, y)$ denote the point on the unit circle such that its arc length from $(1, 0)$ is the real number θ. Or, equivalently, let $P(x, y)$ denote the point where the terminal side of the angle with radian measure θ intersects the unit circle. Then the six trigonometric functions are defined as follows.

$$\sin \theta = y \qquad \csc \theta = \frac{1}{y} \quad (y \neq 0)$$

$$\cos \theta = x \qquad \sec \theta = \frac{1}{x} \quad (x \neq 0)$$

$$\tan \theta = \frac{y}{x} \quad (x \neq 0) \qquad \cot \theta = \frac{x}{y} \quad (y \neq 0)$$

Let us immediately note that there is nothing new here as far as evaluating the trigonometric functions. For instance, $\sin \pi/6$ is still equal to $\frac{1}{2}$. [If this seems unfamiliar, review Example 5(a) on page 348]. Also, whether we are working in radians or degrees, our three-step procedure for evaluating the trigonometric functions is applicable. For instance, in Example 4(b) on page 325, we computed $\cos 135°$ ($\cos 3\pi/4$). Here is how that same procedure looks when radian measure is used. Carefully compare each of the corresponding steps.

STEP 1 The reference angle for $3\pi/4$ is $\pi/4$* $\left(\text{since } \pi - \frac{3\pi}{4} = \frac{\pi}{4}\right)$.

STEP 2 $\cos \pi/4 = \sqrt{2}/2$

STEP 3 $\cos \theta$ is the x-coordinate. In the second quadrant, x-coordinates are negative. Thus, $\cos 3\pi/4$ is negative, since the terminal side of $\theta = 3\pi/4$ lies in the second quadrant. (Equivalently, the arc of length $3\pi/4$ terminates in the second quadrant.) Thus we have

$$\cos \frac{3\pi}{4} = -\frac{\sqrt{2}}{2}$$

For the remainder of this section we discuss some of the basic identities for the trigonometric functions of real numbers. First of all, there are five identities that are immediate consequences of the definitions.

$$\csc \theta = \frac{1}{\sin \theta} \qquad \tan \theta = \frac{\sin \theta}{\cos \theta}$$

$$\sec \theta = \frac{1}{\cos \theta} \qquad \cot \theta = \frac{\cos \theta}{\sin \theta}$$

$$\cot \theta = \frac{1}{\tan \theta}$$

The next identities that we consider are the three *Pythagorean identities*.

The Pythagorean Identities

$$\sin^2 \theta + \cos^2 \theta = 1$$
$$\tan^2 \theta + 1 = \sec^2 \theta$$
$$\cot^2 \theta + 1 = \csc^2 \theta$$

To establish the identity $\sin^2 \theta + \cos^2 \theta = 1$, we think of the real number θ, for the moment, as the radian measure of an angle in standard position, as indicated in Figure 2 on page 356. Now we proceed exactly as in the previous chapter, when we proved this identity in the context of angles. Since the point (x, y) in Figure 2 lies on the unit circle, we have

$$x^2 + y^2 = 1$$

Now, replacing x by $\cos \theta$ and y by $\sin \theta$ gives us

$$\cos^2 \theta + \sin^2 \theta = 1$$

---

*Some texts refer to this as the reference *number,* instead of reference angle, to emphasize the fact that $3\pi/4$ is a real number.

which is what we wished to show. As we remarked in Chapter 6, you should also be familiar with two other ways of writing this identity:

$$\cos^2 \theta = 1 - \sin^2 \theta \qquad \text{and} \qquad \sin^2 \theta = 1 - \cos^2 \theta$$

To prove the second of the Pythagorean identities, we begin with $\sin^2 \theta + \cos^2 \theta = 1$ and divide both sides by the quantity $\cos^2 \theta$ to obtain

$$\frac{\sin^2 \theta}{\cos^2 \theta} + \frac{\cos^2 \theta}{\cos^2 \theta} = \frac{1}{\cos^2 \theta} \qquad \text{(assuming } \cos \theta \neq 0)$$

and, consequently,

$$\tan^2 \theta + 1 = \sec^2 \theta \qquad \text{as required}$$

Since the proof of the third Pythagorean identity is similar, we omit it here.

▌�en▌ EXAMPLE 1 If $\sin \theta = \frac{2}{3}$ and $\frac{\pi}{2} < \theta < \pi$, compute $\cos \theta$ and $\tan \theta$.

Solution

$$\cos^2 \theta = 1 - \sin^2 \theta \qquad \text{using the first Pythagorean identity}$$
$$= 1 - \left(\frac{2}{3}\right)^2 \qquad \text{substituting}$$
$$= 1 - \frac{4}{9} = \frac{5}{9}$$

Consequently,

$$\cos \theta = \frac{\sqrt{5}}{3} \qquad \text{or} \qquad \cos \theta = \frac{-\sqrt{5}}{3}$$

Now, since $\frac{\pi}{2} < \theta < \pi$, it follows that $\cos \theta$ is negative. (Why?) Thus,

$$\cos \theta = \frac{-\sqrt{5}}{3} \qquad \text{as required}$$

To compute $\tan \theta$, we use the identity $\tan \theta = \sin \theta / \cos \theta$ to obtain

$$\tan \theta = \frac{2/3}{-\sqrt{5}/3} = \frac{2}{3} \times \frac{3}{-\sqrt{5}}$$
$$= -\frac{2}{\sqrt{5}}$$

If required, we can rationalize the denominator (by multiplying by $\sqrt{5}/\sqrt{5}$) to obtain

$$\tan \theta = -\frac{2\sqrt{5}}{5}$$

▌▌▌

▌▌▌ EXAMPLE 2 If $\sec \theta = -\frac{5}{3}$ and $\pi < \theta < \frac{3\pi}{2}$, compute $\cos \theta$ and $\tan \theta$.

Solution Since $\cos \theta$ is the reciprocal of $\sec \theta$, we have $\cos \theta = -\frac{3}{5}$. We compute $\tan \theta$ using the second Pythagorean identity as follows.

$$\tan^2 \theta = \sec^2 \theta - 1$$

$$= \left(-\frac{5}{3}\right)^2 - 1 = \frac{25}{9} - \frac{9}{9} = \frac{16}{9}$$

Therefore,

$$\tan \theta = \frac{4}{3} \qquad \text{or} \qquad \tan \theta = -\frac{4}{3}$$

Since θ is between π and $3\pi/2$, the quantity $\tan \theta$ is positive. Thus,

$$\tan \theta = \frac{4}{3}$$

The next example shows how certain experessions containing radicals can be simplified through an appropriate trigonometric substitution. (This is often useful in calculus.)

EXAMPLE 3 In the expression $u/\sqrt{u^2 - 1}$, make the substitution $u = \sec \theta$ and show that the resulting expression is equal to $\csc \theta$. $\left(\text{Assume that } 0 < \theta < \frac{\pi}{2}.\right)$

Solution Replacing u by $\sec \theta$ in the given expression yields

$$\frac{u}{\sqrt{u^2 - 1}} = \frac{\sec \theta}{\sqrt{\sec^2 \theta - 1}}$$

$$= \frac{\sec \theta}{\sqrt{\tan^2 \theta}} \qquad \text{using the second Pythagorean identity}$$

$$= \frac{\sec \theta}{\tan \theta} = \frac{1/\cos \theta}{\sin \theta/\cos \theta} = \frac{1}{\cos \theta} \times \frac{\cos \theta}{\sin \theta}$$

$$= \frac{1}{\sin \theta} = \csc \theta$$

QUESTION Where did we use the condition $0 < \theta < \frac{\pi}{2}$?

The next three identities indicate the effects of replacing θ by $-\theta$ in the expressions $\sin \theta$, $\cos \theta$, and $\tan \theta$.

The Opposite-Angle Identities

$$\cos(-\theta) = \cos \theta$$
$$\sin(-\theta) = -\sin \theta$$
$$\tan(-\theta) = -\tan \theta$$

To see why the first two of these identities are true, consider Figure 3. (Although Figure 3 shows an arc of length θ terminating in the first quadrant, the

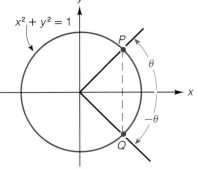

FIGURE 3

same kind of argument we use will work when θ terminates in any quadrant.) By definition, the coordinates of P and Q are as follows:

P: $(\cos\theta, \sin\theta)$

Q: $(\cos(-\theta), \sin(-\theta))$

However, as you can see by looking at Figure 3, the x-coordinates of P and Q are the same, while the y-coordinates are negatives of each other. Thus,

$$\cos(-\theta) = \cos\theta$$

and

$$\sin(-\theta) = -\sin\theta \qquad \text{as we wished to show}$$

Now we can establish the third identity, involving $\tan(-\theta)$, as follows:

$$\tan(-\theta) = \frac{\sin(-\theta)}{\cos(-\theta)} = \frac{-\sin\theta}{\cos\theta}$$

$$= -\frac{\sin\theta}{\cos\theta} = -\tan\theta$$

▮▮▮ EXAMPLE 4

(a) If $\sin t = -0.76$, find $\sin(-t)$.

(b) If $\cos s = -0.29$, find $\cos(-s)$.

(c) If $\sin u = 0.54$, find $\sin^2(-u) + \cos^2(-u)$.

Solution

(a) $\sin(-t) = -\sin t$
$\qquad\quad = -(-0.76) = 0.76$

(b) $\cos(-s) = \cos s = -0.29$

(c) The opposite-angle identities are unnecessary here, and so too is the given value of $\sin u$. For, according to the first Pythagorean identity, we have

$$\sin^2(-u) + \cos^2(-u) = 1 \qquad\qquad ▮▮▮$$

The last identities that we are going to discuss in this section are simply consequences of the fact that the circumference C of the unit circle is 2π. (Substituting $r = 1$ in the formula $C = 2\pi r$ gives us $C = 2\pi$.) Thus, if we begin at any point P on the unit circle and travel a distance of 2π units along the perimeter, we return to the same point P. In other words, arc lengths of θ and $\theta + 2\pi$ [measured from $(1, 0)$, as usual] yield the same terminal point on the unit circle. Since the trigonometric functions are defined in terms of the coordinates of that point P, we obtain the following identities.

Periodicity of Sine and Cosine
$\sin(\theta + 2\pi) = \sin\theta$
$\cos(\theta + 2\pi) = \cos\theta$

These two identities are true for all real numbers θ. As you will see in the next section, they provide important information about the graphs of the sine and cosine functions; the graphs of both functions repeat themselves at intervals of

2π. Similar identities hold for the other trigonometric functions in their respective domains:

$$\tan(\theta + 2\pi) = \tan \theta \qquad \csc(\theta + 2\pi) = \csc \theta$$
$$\cot(\theta + 2\pi) = \cot \theta \qquad \sec(\theta + 2\pi) = \sec \theta$$

EXAMPLE 5 Evaluate $\sin \frac{5\pi}{2}$.

Solution First, simply as a matter of arithmetic, we observe that $\frac{5\pi}{2} = \frac{\pi}{2} + 2\pi$. Thus, in view of our earlier remarks, we have

$$\sin \frac{5\pi}{2} = \sin \frac{\pi}{2} = 1$$

The preceding set of identities can be generalized as follows. If we start at a point P on the unit circle and make two complete counterclockwise revolutions, the arc length we travel is $2\pi + 2\pi = 4\pi$. Similarly, for three complete revolutions, the arc length traversed is $3(2\pi) = 6\pi$. And, in general, if k is an integer, the arc length for k complete revolutions is $2k\pi$. (When k is positive, the revolutions are counterclockwise; when k is negative, the revolutions are clockwise.) Consequently, we have the following identities.

> For any real number θ and any integer k, the following identities hold:
> $$\sin(\theta + 2k\pi) = \sin \theta \qquad \text{and} \qquad \cos(\theta + 2k\pi) = \cos \theta$$

EXAMPLE 6 Evaluate $\cos(-17\pi)$.

Solution $\cos(-17\pi) = \cos(\pi - 18\pi) = \cos \pi = -1$

EXERCISE SET 7.2

A

1. Complete the following table.

θ	$\sin \theta$	$\cos \theta$	$\tan \theta$	$\csc \theta$	$\sec \theta$	$\cot \theta$
0						
$\pi/6$						
$\pi/4$						
$\pi/3$						
$\pi/2$						
$2\pi/3$						
$3\pi/4$						
$5\pi/6$						
π						

2. Complete the following table.

θ	$\sin \theta$	$\cos \theta$	$\tan \theta$	$\csc \theta$	$\sec \theta$	$\cot \theta$
π						
$7\pi/6$						
$5\pi/4$						
$4\pi/3$						
$3\pi/2$						
$5\pi/3$						
$7\pi/4$						
$11\pi/6$						

In Exercises 3–10, check that both sides of the identity are indeed equal for the given values of the variable t. For part (c) of each problem, use your calculator.

3. $\sin^2 t + \cos^2 t = 1$
   (a) $t = \pi/3$
   (b) $t = 5\pi/4$
   (c) $t = -53$

4. $\tan^2 t + 1 = \sec^2 t$
   (a) $t = 3\pi/4$
   (b) $t = -2\pi/3$
   (c) $t = \sqrt{5}$

5. $\cot^2 t + 1 = \csc^2 t$
   (a) $t = -\pi/6$
   (b) $t = 7\pi/4$
   (c) $t = 0.12$

6. $\cos(-t) = \cos t$
   (a) $t = \pi/6$
   (b) $t = -5\pi/3$
   (c) $t = -4$

7. $\sin(-t) = -\sin t$
   (a) $t = 3\pi/2$
   (b) $t = -5\pi/6$
   (c) $t = 13.24$

8. $\tan(-t) = -\tan t$
   (a) $t = -4\pi/3$
   (b) $t = \pi/4$
   (c) $t = 1000$

9. $\sin(t + 2\pi) = \sin t$
   (a) $t = 5\pi/3$
   (b) $t = -3\pi/2$
   (c) $t = \sqrt{19}$

10. $\cos(t + 2\pi) = \cos t$
    (a) $t = -5\pi/3$
    (b) $t = \pi$
    (c) $t = -\sqrt{3}$

In Exercises 11 and 12, show that the equation is not an identity by evaluating both sides using the given value of t and noting that the results are unequal.

11. $\cos 2t = 2 \cos t$; $t = \pi/6$

12. $\sin 2t = 2 \sin t$; $t = \pi/2$

13. If $\sin t = -\frac{3}{5}$ and $\pi < t < \frac{3\pi}{2}$, compute $\cos t$ and $\tan t$.

14. If $\cos t = \frac{5}{13}$ and $\frac{3\pi}{2} < t < 2\pi$, compute $\sin t$ and $\cot t$.

15. If $\sin t = \sqrt{3}/4$ and $\frac{\pi}{2} < t < \pi$, compute $\tan t$.

16. If $\sec s = -\sqrt{13}/2$ and $\sin s > 0$, compute $\tan s$.

17. If $\tan \alpha = \frac{12}{5}$ and $\cos \alpha > 0$, compute $\sec \alpha$, $\cos \alpha$, and $\sin \alpha$.

18. If $\cot \theta = -1/\sqrt{3}$ and $\cos \theta < 0$, compute $\csc \theta$ and $\sin \theta$.

19. In the expression $\sqrt{9 - x^2}$, make the substitution $x = 3 \sin \theta$ $\left(0 < \theta < \frac{\pi}{2}\right)$, and show that the result is $3 \cos \theta$.

20. Make the substitution $u = 2 \cos \theta$ in the expression $1/\sqrt{4 - u^2}$, and simplify the result. (Assume that $0 < \theta < \pi$.)

21. In the expression $1/(u^2 - 25)^{3/2}$, make the substitution $u = 5 \sec \theta$ $\left(0 < \theta < \frac{\pi}{2}\right)$, and show that the result is $(\cot^3 \theta)/125$.

22. In the expression $1/(x^2 + 5)^2$, replace x by $\sqrt{5} \tan \theta$, and show that the result is $(\cos^4 \theta)/25$.

23. In the expression $1/\sqrt{u^2 + 7}$, let $u = \sqrt{7} \tan \theta$, where $0 < \theta < \frac{\pi}{2}$, and simplify the result.

24. In the expression $\sqrt{x^2 - a^2}/x$ $(a > 0)$, let $x = a \sec \theta$ $\left(0 < \theta < \frac{\pi}{2}\right)$, and simplify the result.

25. (a) If $\sin t = \frac{2}{3}$, find $\sin(-t)$.
    (b) If $\sin \phi = -\frac{1}{4}$, find $\sin(-\phi)$.
    (c) If $\cos \alpha = \frac{1}{5}$, find $\cos(-\alpha)$.
    (d) If $\cos s = -\frac{1}{5}$, find $\cos(-s)$.

26. (a) If $\sin t = 0.35$, find $\sin(-t)$.
    (b) If $\sin \phi = -0.47$, find $\sin(-\phi)$.
    (c) If $\cos \alpha = 0.21$, find $\cos(-\alpha)$.
    (d) If $\cos s = -0.56$, find $\cos(-s)$.

27. If $\cos t = -\frac{1}{3}$ $\left(\frac{\pi}{2} < t < \pi\right)$, compute the following:
    (a) $\sin(-t) + \cos(-t)$;   (b) $\sin^2(-t) + \cos^2(-t)$.

28. If $\sin(-s) = \frac{3}{5}$ $\left(\pi < s < \frac{3\pi}{2}\right)$, compute the following:
    (a) $\sin s$;   (b) $\cos(-s)$;   (c) $\cos s$;
    (d) $\tan s + \tan(-s)$.

In Exercises 29 and 30, use one of the identities $\cos(\theta + 2\pi k) = \cos \theta$ or $\sin(\theta + 2\pi k) = \sin \theta$ to evaluate each expression.

29. (a) $\cos\left(\frac{\pi}{4} + 2\pi\right)$ (b) $\sin\left(\frac{\pi}{3} + 2\pi\right)$
    (c) $\sin\left(\frac{\pi}{2} - 6\pi\right)$

30. (a) $\sin \frac{17\pi}{4}$ (b) $\sin\left(\frac{-17\pi}{4}\right)$
    (c) $\cos 11\pi$ (d) $\cos \frac{53\pi}{4}$
    (e) $\tan\left(\frac{-7\pi}{4}\right)$ (f) $\cos \frac{7\pi}{4}$
    (g) $\sec\left(\frac{11\pi}{6} + 2\pi\right)$ (h) $\csc\left(2\pi - \frac{\pi}{3}\right)$

In Exercises 31–34, use the Pythagorean identities to simplify the given expressions.

31. $\dfrac{\sin^2 t + \cos^2 t}{\tan^2 t + 1}$

32. $\dfrac{\sec^2 t - 1}{\tan^2 t}$

33. $\dfrac{\sec^2 \theta - \tan^2 \theta}{1 + \cot^2 \theta}$

34. $\dfrac{\csc^4 \theta - \cot^4 \theta}{\csc^2 \theta + \cot^2 \theta}$

In Exercises 35–42, prove that the equations are identities.

35. $\csc t = \sin t + \cot t \cos t$

36. $\sin^2 t - \cos^2 t = \dfrac{1 - \cot^2 t}{1 + \cot^2 t}$

37. $\dfrac{1}{1 + \sec s} + \dfrac{1}{1 - \sec s} = -2 \cot^2 s$

38. $\dfrac{1 + \tan s}{1 - \tan s} = \dfrac{\sec^2 s + 2 \tan s}{2 - \sec^2 s}$

39. $\cot \theta + \tan \theta + 1 = \dfrac{\cot \theta}{1 - \tan \theta} + \dfrac{\tan \theta}{1 - \cot \theta}$

40. $\dfrac{\sec s + \cot s \csc s}{\cos s} = \csc^2 s \sec^2 s$

41. $(\tan \theta)(1 - \cot^2 \theta) + (\cot \theta)(1 - \tan^2 \theta) = 0$

42. $(\cos \alpha \cos \beta - \sin \alpha \sin \beta) \times (\cos \alpha \cos \beta + \sin \alpha \sin \beta)$
    $= \cos^2 \alpha - \sin^2 \beta$

43. If $\sec t = \frac{13}{5}$ and $\frac{3\pi}{2} < t < 2\pi$, evaluate

$$\frac{2 \sin t - 3 \cos t}{4 \sin t - 9 \cos t}$$

44. If $\sec t = (b^2 + 1)/2b$ and $\pi < t < \frac{3\pi}{2}$, find $\tan t$ and $\sin t$.

B

45. Use the accompanying figure to explain why the following four identities are valid. (The identities can be used to provide an algebraic foundation for the reference number technique that we've used to evaluate the trigonometric functions.)

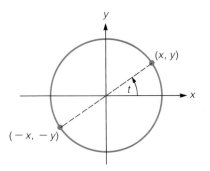

(i) $\sin(t + \pi) = -\sin t$
(ii) $\sin(t - \pi) = -\sin t$
(iii) $\cos(t + \pi) = -\cos t$
(iv) $\cos(t - \pi) = -\cos t$

46. In the equation $x^4 + 6x^2y^2 + y^4 = 32$, make the substitutions $x = X \cos \frac{\pi}{4} - Y \sin \frac{\pi}{4}$ and $y = X \sin \frac{\pi}{4} + Y \cos \frac{\pi}{4}$ and show that the result simplifies to $X^4 + Y^4 = 16$.

47. Consider the equation

$$2 \sin^2 t - \sin t = 2 \sin t \cos t - \cos t$$

(a) Evaluate each side of the equation when $t = \pi/6$.
(b) Evaluate each side of the equation when $t = \pi/4$.
(c) Is the given equation an identity?

48. Suppose that

$$f(t) = (\sin t \cos t)(2 \sin t - 1)(2 \cos t - 1)(\tan t - 1)$$

(a) Compute each of the following: $f(0)$, $f(\pi/6)$, $f(\pi/4)$, $f(\pi/3)$, and $f(\pi/2)$.
(b) Is the equation $f(t) = 0$ an identity?

In Section 7.1, we pointed out that one of the advantages in using radian measure is that many formulas then take on particularly simple forms. Another reason for using radian measure is that the trigonometric functions can be closely approximated by very simple polynomial functions. To see examples of this, complete the tables in Exercises 49 and 50. Note: The approximating polynomials in Exercises 49 and 50 are known as Taylor polynomials, after the English mathematician Brook Taylor (1685–1731). The theory of Taylor polynomials is developed in calculus.

49.

θ	$1 - \dfrac{\theta^2}{2}$	$\cos \theta$
0.1		0.995004. . .
0.2		0.980066. . .
0.3		0.955336. . .

50.

θ	$\theta - \dfrac{\theta^3}{6}$	$\sin \theta$
0.1		0.099833. . .
0.2		0.198669. . .
0.3		0.295520. . .

7.3 GRAPHS OF THE SINE AND COSINE FUNCTIONS

Our focus in this section is on the sine and cosine functions. As preparation for the discussion, we want to understand what is meant by the term *periodic function*. By way of example, both of the functions in Figure 1 (at the bottom of the next page) are **periodic**. That is, their graphs display patterns that repeat themselves at regular intervals.

In Figure 1(a), the graph of the function f repeats itself every six units. We say that the *period* of f is 6. Similarly, the period of g in Figure 1(b) is 2π. In both cases, notice that the period represents the minimum number of units that we must travel along the horizontal axis before the graph begins to repeat itself. Because of this, we can state the definition of a periodic function as follows.

DEFINITION   A Periodic Function and Its Period

A function f is said to be **periodic** if there is a number $p > 0$ such that

$$f(x + p) = f(x)$$

for all x in the domain of f. The smallest such number p is called the **period** of f.

We also want to define the term *amplitude* as it applies to periodic functions. For a function such as g in Figure 1(b), in which the graph is centered about the horizontal axis, the amplitude is simply the maximum height of the graph above the horizontal axis. Thus, the amplitude of g is 4. More generally, we define the amplitude of any periodic function as follows.

DEFINITION   Amplitude

Let f be a periodic function and let m and M denote, respectively, the minimum and maximum values of the function. Then, by definition, the **amplitude** of f is the number

$$\frac{M - m}{2}$$

For the function g in Figure 1(b), this definition tells us that the amplitude is $\dfrac{4 - (-4)}{2} = 4$, which agrees with our previous value. Check for yourself now that the amplitude of the function in Figure 1(a) is 1.

Periodic functions are used throughout the sciences to analyze or describe a variety of phenomena ranging from the vibrations of an electron to the variations in the size of an animal population as it interacts with its environment. Figures 2 through 5 (on pages 365 and 366) display some examples of periodic functions.

FIGURE 1

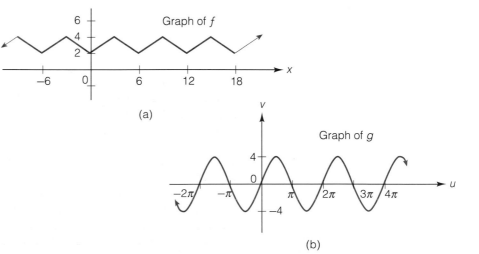

(a)

(b)

FIGURE 2

The surface temperature of the star Delta Cephei is a periodic function of time. The period is 5.367 days. The amplitude is $(6800 - 5400)/2 = 700K$.

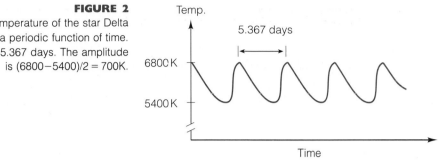

FIGURE 3

Electrical activity of the heart and blood pressure as periodic functions of time. The figure shows a typical ECG (electrocardiogram) and the corresponding graph of arterial blood pressure [adapted from *Physics for the Health Sciences* by C. R. Nave and B. C. Nave (Philadelphia: W. B. Saunders Company, 1975)]

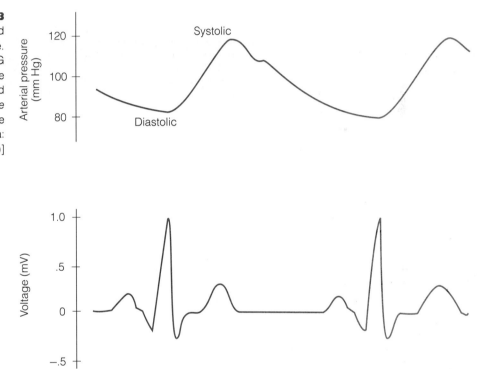

Let us now graph the sine function $f(\theta) = \sin\theta$. We are assuming that θ is in radians, so that the domain of the sine function is the set of all real numbers. Even before making any calculations, we can gain a strong intuitive insight into how the graph must look by carrying out the following experiment. After drawing the unit circle, $x^2 + y^2 = 1$, place your fingertip at the point $(1, 0)$ and then move your finger counterclockwise around the circle. As you do this, keep track of what happens to the y-coordinate of your fingertip. (The y-coordinate is $\sin\theta$.) Figure 6 (on page 367) indicates the general results.

Figure 6 tells us a great deal about the sine function: where the function is increasing and decreasing, where the graph crosses the x-axis, and where the high and low points of the graph occur. Furthermore, since at $\theta = 2\pi$ we've re-

FIGURE 4
The sound wave generated by a note
played on the bamboo flute is a
periodic function. The sound wave is
recorded on an oscilloscope.
(Photograph by Professor Vern Ostdiek)

FIGURE 5
Sound waves

(a) The sound wave generated by a tuning fork vibrating at 440 vibrations per second
(On the piano, this note corresponds to the first A above middle C.)

(b) The sound wave generated by a tuning fork vibrating at 880 vibrations per second
[This note is an A one octave above that in part (a).]

(c) The sound wave generated by simultaneously striking the tuning forks in
parts (a) and (b).

FIGURE 6

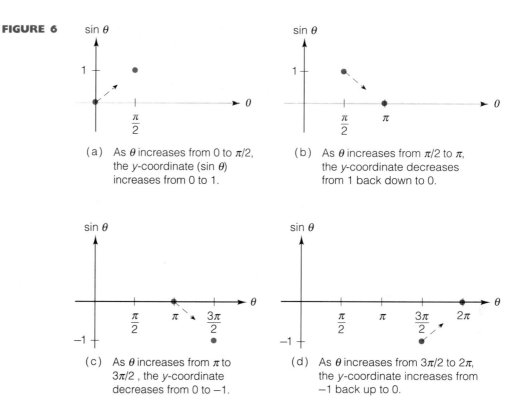

(a) As θ increases from 0 to $\pi/2$, the y-coordinate (sin θ) increases from 0 to 1.

(b) As θ increases from $\pi/2$ to π, the y-coordinate decreases from 1 back down to 0.

(c) As θ increases from π to $3\pi/2$, the y-coordinate decreases from 0 to −1.

(d) As θ increases from $3\pi/2$ to 2π, the y-coordinate increases from −1 back up to 0.

turned to our starting point $(1, 0)$, additional counterclockwise trips around the unit circle will just result in repetitions of the pattern established in Figure 6. In other words, the period of the function $y = \sin \theta$ is 2π.

As informative as Figure 6 is, however, there is not enough information there to tell us the precise shape of the required graph. For instance, all three of the curves in Figure 7 fit the specifications described in Figure 6.

Now let's set up a table so that we can accurately sketch the graph of $y = \sin \theta$. Since the sine function is periodic (as we've just observed), with period 2π, our table need only contain values of θ between 0 and 2π, as shown in Table 1 (on the next page). This will establish the basic pattern for the graph.

In plotting the points obtained from Table 1, we use the approximation $\sqrt{3}/2 \approx 0.87$. Rather than approximating π, however, we mark off units on the horizontal axis in terms of π. The resulting graph is shown in Figure 8. By con-

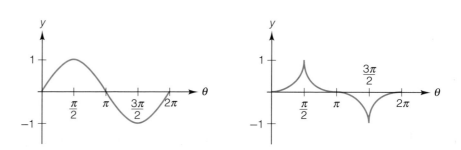

FIGURE 7

TABLE I

θ	0	$\dfrac{\pi}{6}$	$\dfrac{\pi}{3}$	$\dfrac{\pi}{2}$	$\dfrac{2\pi}{3}$	$\dfrac{5\pi}{6}$	π	$\dfrac{7\pi}{6}$	$\dfrac{4\pi}{3}$	$\dfrac{3\pi}{2}$	$\dfrac{5\pi}{3}$	$\dfrac{11\pi}{6}$	2π
$\sin\theta$	0	$\dfrac{1}{2}$	$\dfrac{\sqrt{3}}{2}$	1	$\dfrac{\sqrt{3}}{2}$	$\dfrac{1}{2}$	0	$-\dfrac{1}{2}$	$-\dfrac{\sqrt{3}}{2}$	-1	$-\dfrac{\sqrt{3}}{2}$	$-\dfrac{1}{2}$	0

FIGURE 8

FIGURE 9

tinuing this same pattern to the left and right, we obtain the complete graph of $f(\theta) = \sin\theta$, as indicated in Figure 9.

Before going on to analyze the sine function or to study the graphs of the other trigonometric functions, we are going to make a slight change in the notation we have been using. To conform with common usage, we will use x instead of θ on the horizontal axis, and we will use y for the vertical axis. The sine function is then written simply as $y = \sin x$, where the real number x denotes the radian measure of an angle or, equivalently, the length of the corresponding arc on the unit circle. For reference, we redraw Figures 8 and 9 using this familiar x-y notation, as shown in Figure 10.

FIGURE 10

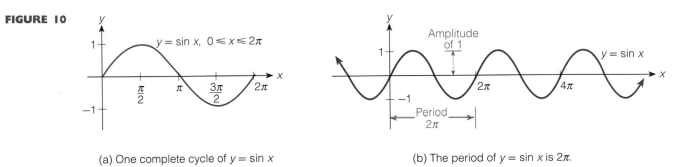

(a) One complete cycle of $y = \sin x$

(b) The period of $y = \sin x$ is 2π.

You should memorize the graph of the sine curve in Figure 10(b) so that you can sketch it without first setting up a table. Of course, once you know the shape and the location of the basic cycle shown in Figure 10(a), you automatically know the graph of the full sine curve in Figure 10(b).

We can use the graphs in Figure 10 to help us list some of the key properties of the sine function.

PROPERTY SUMMARY   **THE SINE FUNCTION:** $y = \sin x$

1. The domain of the sine function is the set of all real numbers. The range of the sine function is the closed interval $[-1, 1]$, and we have

$$-1 \le \sin x \le 1, \qquad \text{for all } x$$

2. The sine function is periodic, with period 2π. The amplitude is 1.

3. The graph of $y = \sin x$ consists of repetitions of the **basic sine wave** shown in Figure 11. The basic sine wave crosses the x-axis at the beginning, middle, and end of the cycle. The curve reaches its highest point one-quarter of the way through the cycle and its lowest point three-quarters of the way through the cycle.

FIGURE 11
The basic sine wave

$y = \sin x, \ \ 0 \le x \le 2\pi$

EXAMPLE 1 Figure 12 shows a portion of the graph of $y = \sin x$. What are the coordinates of the turning points A, B, and C? Give the x-coordinates both in terms of π and as decimal approximations rounded off to three places.

FIGURE 12

Solution   From Figure 11 (which you should also memorize), we know that the coordinates of the point A are $(\pi/2, 1)$. Therefore, since the period of the sine function is 2π, the coordinates of B and C are as follows.

Coordinates of B: $\left(\frac{\pi}{2} + 2\pi, 1\right)$ or $\left(\frac{5\pi}{2}, 1\right)$

Coordinates of C: $\left(\frac{\pi}{2} - 2\pi, 1\right)$ or $\left(-\frac{3\pi}{2}, 1\right)$

As you should check now for yourself, the calculator approximations here are

$$A(1.571, 1) \qquad B(7.854, 1) \qquad C(-4.712, 1)$$ ■■■

There is an interesting fact about the graph of $y = \sin x$ that is useful in numerical work. If you look at the graph in the immediate vicinity of any of its x-intercepts, the graph closely resembles a straight line segment. For example, in Figure 13 we show the graph of $y = \sin x$ and the line $y = x$. Notice that the two graphs appear to be virtually identical in the vicinity of the origin. [Actually, the only point the two graphs have in common is $(0, 0)$. But near the origin, the distance between the two graphs is less than the thickness of the lines we draw—so the two graphs appear to be indistinguishable from one another.]

FIGURE 13
In the immediate vicinity of the origin, the graph of $y = \sin x$ closely resembles the line $y = x$.

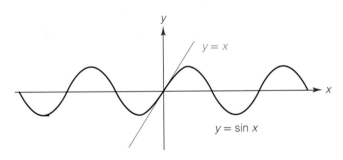

We summarize our observations as follows. (Recall that the symbol $\approx$ means "is approximately equal to.")

$$\sin x \approx x \qquad \text{when } |x| \text{ is close to } 0$$

As examples of this approximation, we have

$$\sin(0.07) \approx 0.07 \qquad \text{and} \qquad \sin(-0.0032) \approx -0.0032$$

You should check for yourself now just how close these approximations are to the values obtained using your calculator. (Remember to set the calculator in the radian mode.)

We could obtain the graph of the cosine function by setting up a table, just as we did with the sine function. A more interesting and informative way to proceed, however, is to use the identity

$$\cos x = \sin\left(x + \frac{\pi}{2}\right) \tag{1}$$

(Exercise 39 at the end of this section outlines a geometric proof of this identity. Also, after studying Section 7.6, you'll see a simple way to prove this identity algebraically.) From our work on graphing functions in Chapter 3, we can interpret equation (1) geometrically. Equation (1) tells us that the graph of $y = \cos x$ is obtained by translating the sine curve $\pi/2$ units to the left. The result is shown in Figure 14(a). Figure 14(b) displays one complete cycle of the cosine curve, from $x = 0$ to $x = 2\pi$.

FIGURE 14

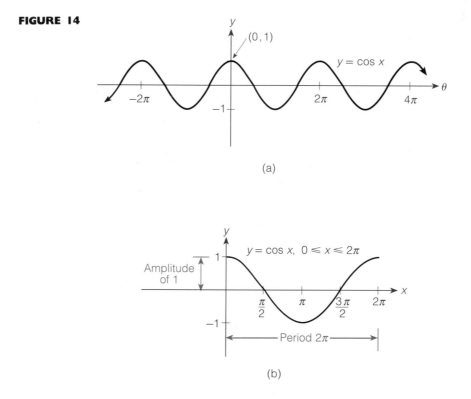

(a)

(b)

As with the sine function, you should memorize the graph and the basic features of the cosine function, which are summarized in the box that follows.

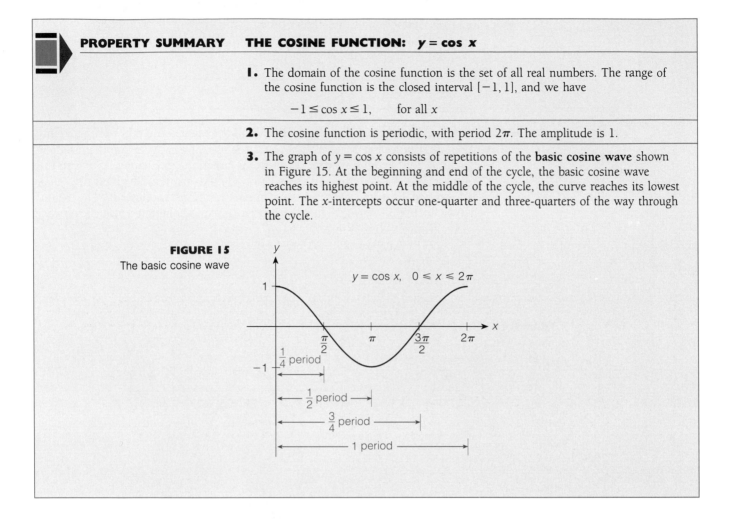

PROPERTY SUMMARY   **THE COSINE FUNCTION:** $y = \cos x$

1. The domain of the cosine function is the set of all real numbers. The range of the cosine function is the closed interval $[-1, 1]$, and we have

$$-1 \leq \cos x \leq 1, \qquad \text{for all } x$$

2. The cosine function is periodic, with period 2π. The amplitude is 1.

3. The graph of $y = \cos x$ consists of repetitions of the **basic cosine wave** shown in Figure 15. At the beginning and end of the cycle, the basic cosine wave reaches its highest point. At the middle of the cycle, the curve reaches its lowest point. The x-intercepts occur one-quarter and three-quarters of the way through the cycle.

FIGURE 15
The basic cosine wave

$y = \cos x, \quad 0 \leq x \leq 2\pi$

EXAMPLE 2   Figure 16 shows a portion of the graph of $y = \cos x$. What are the x-coordinates of the points B and C? Give the values both in terms of π and as decimal approximations rounded off to three places.

FIGURE 16

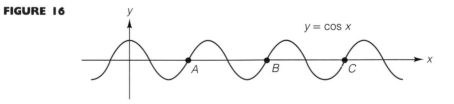

$y = \cos x$

Solution   Comparing Figures 15 and 16, we know that the x-intercept at A is $3\pi/2$. Therefore, since the period of the cosine function is 2π, we have

x-coordinate of B: $\frac{3\pi}{2} + 2\pi = \frac{3\pi}{2} + \frac{4\pi}{2} = \frac{7\pi}{2}$

x-coordinate of C: $\frac{3\pi}{2} + 4\pi = \frac{3\pi}{2} + \frac{8\pi}{2} = \frac{11\pi}{2}$

Using a calculator now, we obtain

x-coordinate of B: $\frac{7\pi}{2} \approx 10.996$

x-coordinate of C: $\frac{11\pi}{2} \approx 17.279$

EXERCISE SET 7.3

A

In Exercises 1–8, specify the period and amplitude for each function.

1.

2.

3.

4.

5.

6.

7.

8.

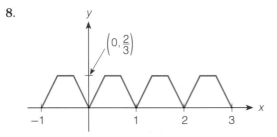

In Exercises 9–18, refer to the graph of $y = \sin x$ in the following figure. Specify the coordinates of the indicated points. Give the x-coodinates both in terms of π and as calculator approximations rounded off to three decimal places.

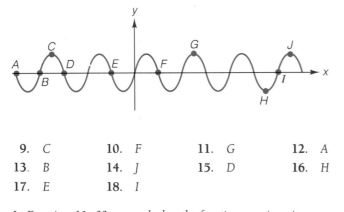

9.	C	10.	F	11.	G	12.	A
13.	B	14.	J	15.	D	16.	H
17.	E	18.	I				

In Exercises 19–22, state whether the function $y = \sin x$ is increasing or decreasing on the given interval. (The terms increasing and decreasing are explained on page 112.)

19. $\frac{3\pi}{2} < x < 2\pi$ 20. $-\frac{\pi}{2} < x < \frac{\pi}{2}$

21. $\frac{5\pi}{2} < x < \frac{7\pi}{2}$ 22. $-\frac{5\pi}{2} < x < -2\pi$

In Exercises 23 and 24, use a calculator to complete the tables. Round off the answers to five decimal places. The results in the third column will tell you something about the accuracy of the approximation $\sin x \approx x$.

23.

x	$\sin x$	$x - \sin x$
0.453		
0.253		
0.0253		
0.00253		

24.

x	$\sin x$	$x - \sin x$
0.526		
0.126		
0.0126		
0.00126		

In Exercises 25–34, refer to the graph of $y = \cos x$ in the following figure. Specify the coordinates of the indicated points. Give the x-coordinates both in terms of π and as calculator approximations rounded off to three decimal places.

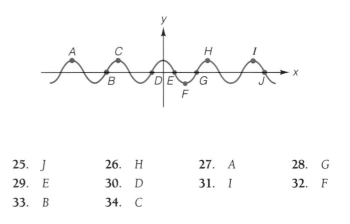

25.	J	26.	H	27.	A	28.	G
29.	E	30.	D	31.	I	32.	F
33.	B	34.	C				

In Exercises 35–38, state whether the function $y = \cos x$ is increasing or decreasing on the given interval.

35. $0 < x < \pi$ 36. $6\pi < x < 7\pi$

37. $-\frac{\pi}{2} < x < 0$ 38. $-\frac{5\pi}{2} < x < -2\pi$

B

39. In the text we used the identity $\cos \theta = \sin(\theta + \pi/2)$ in obtaining the graph of $y = \cos x$ from that of $y = \sin x$. In this exercise you'll derive this identity. Refer to the following figure. (Although the figure shows the angle of radian measure θ in the first quadrant, the proof can be easily carried over for the other quadrants as well.)

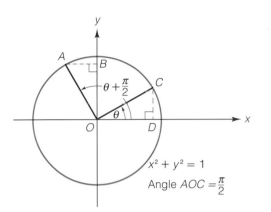

$x^2 + y^2 = 1$

Angle $AOC = \dfrac{\pi}{2}$

(a) What are the coordinates of C?

(b) Show that $\triangle AOB$ and $\triangle COD$ are congruent.

(c) Use the results in parts (a) and (b) to show that the coordinates of A are $(-\sin \theta, \cos \theta)$.

(d) Since the radian measure of $\angle DOA$ is $\theta + \frac{\pi}{2}$, the coordinates of A (by definition) are $\left(\cos\left(\theta + \frac{\pi}{2}\right), \sin\left(\theta + \frac{\pi}{2}\right)\right)$. Now explain why

$$\cos \theta = \sin\left(\theta + \tfrac{\pi}{2}\right)$$

and

$$-\sin \theta = \cos\left(\theta + \tfrac{\pi}{2}\right)$$

7.4 GRAPHS OF $y = A \sin(Bx - C)$ AND $y = A \cos(Bx - C)$

The graphs of $y = \sin x$ and $y = \cos x$ are the building blocks we need for graphing functions of the form

$$y = A \sin(Bx - C) \qquad \text{and} \qquad y = A \cos(Bx - C)$$

As a first example, consider $y = 2 \sin x$. To obtain the graph of $y = 2 \sin x$ from that of $y = \sin x$, we multiply each y-coordinate on the graph of $y = \sin x$ by 2. As indicated in Figure 1, this changes the amplitude from 1 to 2, but it does not affect the period, which remains 2π. As a second example, Figure 2 shows the graphs of $y = \cos x$ and $y = \frac{1}{2} \cos x$. Note that the amplitude of $y = \frac{1}{2} \cos x$ is $\frac{1}{2}$ and the period is, again, 2π. More generally, graphs of functions of the form $y = A \sin x$ and $y = A \cos x$ always have an amplitude of $|A|$ and a period of 2π.

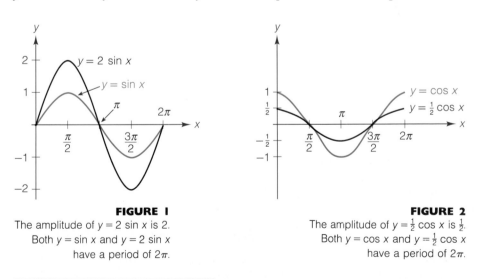

FIGURE I

The amplitude of $y = 2 \sin x$ is 2.
Both $y = \sin x$ and $y = 2 \sin x$
have a period of 2π.

FIGURE 2

The amplitude of $y = \frac{1}{2} \cos x$ is $\frac{1}{2}$.
Both $y = \cos x$ and $y = \frac{1}{2} \cos x$
have a period of 2π.

EXAMPLE I Graph the function $y = -2 \sin x$ over one period. On which interval(s) is the function decreasing?

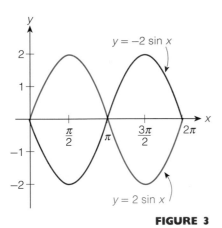

FIGURE 3

Solution In Section 3.3 we saw that the graph of $y = -f(x)$ is obtained from that of $y = f(x)$ by reflection about the x-axis. Thus, we need only take the graph of $y = 2 \sin x$ from Figure 1 and reflect it about the x-axis. See Figure 3. Note that both functions have an amplitude of 2 and a period of 2π. From the graph in Figure 3, we can see that the function $y = -2 \sin x$ is decreasing on the intervals $(0, \pi/2)$ and $(3\pi/2, 2\pi)$. ▪▪▪

We have just seen that functions of the form $y = A \sin x$ and $y = A \cos x$ have an amplitude of $|A|$ and a period of 2π. The next two examples show how to analyze functions of the form $y = A \sin Bx$ and $y = A \cos Bx$ $(B > 0)$. As you'll see, these functions have an amplitude of $|A|$ and a period of $2\pi/B$.

▪▪▪ **EXAMPLE 2** Graph the function $y = \cos 3x$ over one period.

Solution We know that the cosine curve $y = \cos x$ begins its basic pattern when $x = 0$ and completes that pattern when $x = 2\pi$. Thus, $y = \cos 3x$ will begin its basic pattern when $3x = 0$, and it will complete that pattern when $3x = 2\pi$. From the equation $3x = 0$ we conclude that $x = 0$, and from the equation $3x = 2\pi$ we conclude that $x = 2\pi/3$. Thus, the graph of $y = \cos 3x$ begins its basic pattern at $x = 0$ and completes the pattern at $x = 2\pi/3$. This tells us that the period is $2\pi/3$. Next, in preparation for drawing the graph, we divide the period into quarters, as shown in Figure 4(a). In Figure 4(b) we've plotted the points with x-coordinates shown in Figure 4(a). (We've also plotted the point on the curve corresponding to $x = 0$, where the basic pattern is to begin.) From Figure 4(b), we can see that the amplitude is going to be 1. Now, with the points in Figure 4(b) as a guide, we can sketch one cycle of $y = \cos 3x$, as shown in Figure 4(c). ▪▪▪

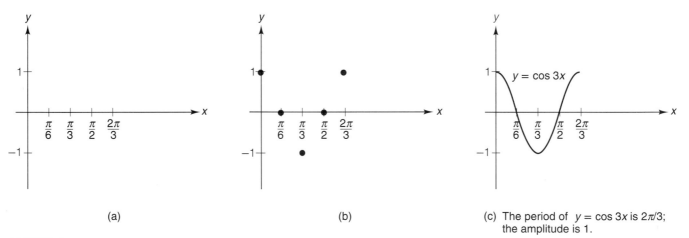

(a)   (b)   (c) The period of $y = \cos 3x$ is $2\pi/3$; the amplitude is 1.

FIGURE 4

▪▪▪ **EXAMPLE 3** Graph each function over one period:

(a) $y = \frac{1}{2} \cos 3x$;   (b) $y = -\frac{1}{2} \cos 3x$.

Solution

(a) In Example 2 we graphed $y = \cos 3x$. To obtain the graph of $y = \frac{1}{2} \cos 3x$ from that of $y = \cos 3x$, we multiply each y-coordinate on the graph of

$y = \cos 3x$ by $\frac{1}{2}$. As indicated in Figure 5(a), this changes the amplitude from 1 to $\frac{1}{2}$, but it does not affect the period, which remains $2\pi/3$.

(b) The graph of $y = -\frac{1}{2}\cos 3x$ is obtained by reflecting the graph of $y = \frac{1}{2}\cos 3x$ about the x-axis, as indicated in Figure 5(b). Both functions have a period of $2\pi/3$ and an amplitude of $\frac{1}{2}$. ▪▪▪

FIGURE 5

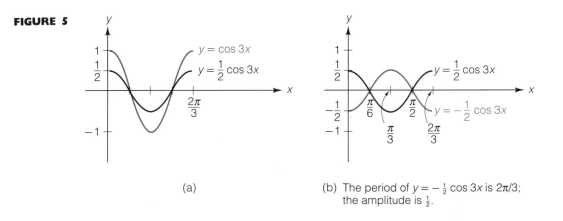

(a)

(b) The period of $y = -\frac{1}{2}\cos 3x$ is $2\pi/3$; the amplitude is $\frac{1}{2}$.

Before looking at more examples, let's take a moment to summarize where we are. Our work in Examples 2 and 3(a) shows how to graph $y = \frac{1}{2}\cos 3x$. The same technique we used in those examples can be applied to any function of the form $y = A\cos Bx$ or $y = A\sin Bx$. As indicated in the box that follows, for both functions, the amplitude is $|A|$ and the period is $2\pi/B$. (Exercise 39 asks you to use the method of Example 2 to show that the period is indeed $2\pi/B$.)

PROPERTY SUMMARY   THE GRAPHS OF $y = A \sin Bx$ AND $y = A \cos Bx$ ($B > 0$)

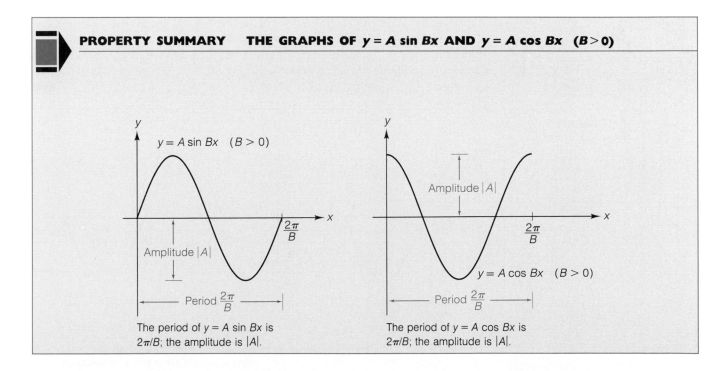

The period of $y = A \sin Bx$ is $2\pi/B$; the amplitude is $|A|$.

The period of $y = A \cos Bx$ is $2\pi/B$; the amplitude is $|A|$.

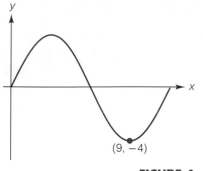

FIGURE 6

▮▮▮▮ **EXAMPLE 4** In Figure 6, a function of the form $y = A \sin Bx$ $(B > 0)$ is graphed for one period. Determine the values of A and B.

Solution From the figure, we see that the amplitude is 4. Also from the figure, we know that three-fourths of the period is 9, so

$$\frac{3}{4}\left(\frac{2\pi}{B}\right) = 9$$

$$\frac{\pi}{2B} = 3$$

$$B = \frac{\pi}{6} \qquad \text{(Check the algebra in the last two lines.)}$$

In summary, we have $A = 4$ and $B = \pi/6$; the equation of the curve is $y = 4 \sin(\pi x/6)$. ▮▮▮

▮▮▮▮ **EXAMPLE 5** Graph the function $y = 4 \sin\left(2x - \frac{2\pi}{3}\right)$ over one period.

Solution The technique here is to factor the quantity within parentheses so that the coefficient of x is 1. We'll then be able to graph the function using a simple translation, as in Chapter 3. We have

$$y = 4 \sin\left(2x - \frac{2\pi}{3}\right)$$
$$= 4 \sin\left[2\left(x - \frac{\pi}{3}\right)\right] \tag{1}$$

Now, note that equation (1) is obtained from $y = 4 \sin 2x$ by replacing x with $x - \frac{\pi}{3}$. Thus, the graph of equation (1) is obtained by translating the graph of $y = 4 \sin 2x$ a distance of $\pi/3$ units to the right. Figure 7(a) shows the graph of $y = 4 \sin 2x$ over one period. By translating this graph $\pi/3$ units to the right, we obtain the required graph, as shown in Figure 7(b). Note that the translated graph has the same amplitude and period as the original graph. Also, as a matter of arithmetic, you should check for yourself that each of the x-coordinates shown in Figure 7(b) is obtained simply by adding $\pi/3$ to a corresponding x-coordinate in Figure 7(a). For example, in Figure 7(a), the cycle ends at $x = \pi$; in Figure 7(b) the cycle ends at $\pi + \frac{\pi}{3} = \frac{4\pi}{3}$. ▮▮▮

FIGURE 7

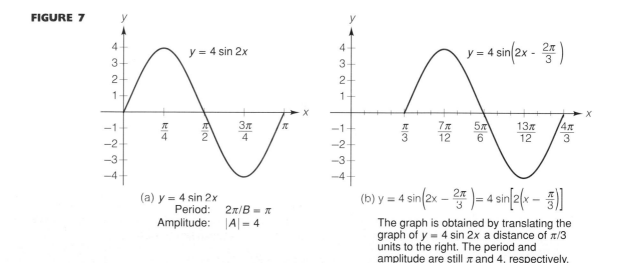

(a) $y = 4 \sin 2x$
  Period: $2\pi/B = \pi$
  Amplitude: $|A| = 4$

(b) $y = 4 \sin\left(2x - \frac{2\pi}{3}\right) = 4 \sin\left[2\left(x - \frac{\pi}{3}\right)\right]$

The graph is obtained by translating the graph of $y = 4 \sin 2x$ a distance of $\pi/3$ units to the right. The period and amplitude are still π and 4, respectively.

In the example just completed, we used translation to graph the function $y = 4\sin\left(2x - \frac{2\pi}{3}\right)$. In particular, we translated the graph of $y = 4\sin 2x$ so that the starting point of the basic cycle was shifted from $x = 0$ to $x = \pi/3$. The number $\pi/3$ in this case is called the *phase shift* of the function. In the box that follows, we define phase shift and we generalize the results of the graphing technique used in Example 5.

PROPERTY SUMMARY $y = A\sin(Bx - C)$ **AND** $y = A\cos(Bx - C)$ $(B > 0, C \neq 0)$

The graphs of $y = A\sin(Bx - C)$ and $y = A\cos(Bx - C)$ are obtained by horizontally translating the graphs of $y = A\sin Bx$ and $y = A\cos Bx$, respectively, so that the starting point of the basic cycle is shifted from $x = 0$ to $x = C/B$. The number C/B is called the **phase shift** for each of the functions $y = A\sin(Bx - C)$ and $y = A\cos(Bx - C)$. The amplitude and the period for these functions are $|A|$ and $2\pi/B$, respectively.

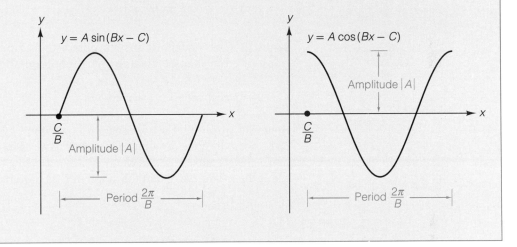

EXAMPLE 6   Specify the amplitude, period, and phase shift for each function.

(a) $f(x) = 3\cos(4x - 5)$;   **(b)** $g(x) = -2\cos\left(\pi x + \frac{2\pi}{3}\right)$

Solution

(a) By comparing the given equation with $y = A\cos(Bx - C)$, we see that $A = 3$, $B = 4$, and $C = 5$. Consequently, we have

$$\text{amplitude} = |A| = 3$$

$$\text{period} = \frac{2\pi}{B} = \frac{2\pi}{4} = \frac{\pi}{2}$$

$$\text{phase shift} = \frac{C}{B} = \frac{5}{4}$$

For purposes of review, let's also calculate the phase shift without explicitly relying on the expression C/B. In the equation $f(x) = 3\cos(4x - 5)$, we can factor a 4 out of the parentheses to obtain

$$f(x) = 3\cos\left[4\left(x - \tfrac{5}{4}\right)\right]$$

This last equation tells us that we can obtain the graph of f by translating the graph of $y = 3 \cos 4x$. In particular, the translation shifts the starting point of the basic cycle from $x = 0$ to $x = \frac{5}{4}$. The number $\frac{5}{4}$ is the phase shift, as obtained previously.

(b) We have $A = -2$, $B = \pi$, and $C = -2\pi/3$, and therefore

$$\text{amplitude} = |A| = 2$$

$$\text{period} = \frac{2\pi}{B} = \frac{2\pi}{\pi} = 2$$

$$\text{phase shift} = \frac{C}{B} = \frac{-2\pi/3}{\pi} = -\frac{2}{3}$$ ▮▮▮

▮▬▮ **EXAMPLE 7**   Graph the following function over one period.

$$g(x) = -2 \cos\left(\pi x + \frac{2\pi}{3}\right)$$

Solution   Our strategy is first to obtain the graph of $y = 2 \cos\left(\pi x + \frac{2\pi}{3}\right)$. The graph of g can then be obtained by a reflection about the x-axis. We have

$$y = 2 \cos\left(\pi x + \frac{2\pi}{3}\right) = 2 \cos\left[\pi\left(x + \frac{2}{3}\right)\right]$$

Now, the graph of this last equation is obtained by translating the graph of $y = 2 \cos \pi x$ a distance of $\frac{2}{3}$ unit to the left. Figures 8(a) and 8(b) show the graphs of $y = 2 \cos \pi x$ and $y = 2 \cos\left[\pi\left(x + \frac{2}{3}\right)\right]$. By reflecting the graph of this last equation about the x-axis, we obtain the graph of g, as required. See Figure 8(c).   ▮▮▮

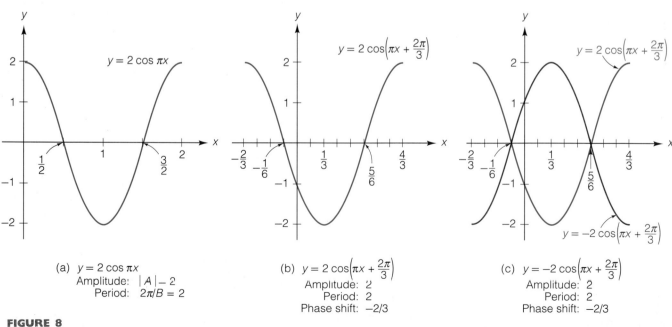

(a) $y = 2 \cos \pi x$
Amplitude: $|A| - 2$
Period: $2\pi/B = 2$

(b) $y = 2 \cos\left(\pi x + \frac{2\pi}{3}\right)$
Amplitude: 2
Period: 2
Phase shift: −2/3

(c) $y = -2 \cos\left(\pi x + \frac{2\pi}{3}\right)$
Amplitude: 2
Period: 2
Phase shift: −2/3

FIGURE 8

EXERCISE SET 7.4

A

In Exercises 1–12, graph the functions for one period. In each case, specify the amplitude, period, x-intercepts, and interval(s) on which the function is increasing.

1. (a) $y = 2\sin x$
   (b) $y = -\sin 2x$

2. (a) $y = 3\sin x$
   (b) $y = \sin 3x$

3. (a) $y = \cos 2x$
   (b) $y = 2\cos 2x$

4. (a) $y = \cos(x/2)$
   (b) $y = -\frac{1}{2}\cos(x/2)$

5. (a) $y = 3\sin(\pi x/2)$
   (b) $y = -3\sin(\pi x/2)$

6. (a) $y = 2\sin \pi x$
   (b) $y = -2\sin \pi x$

7. (a) $y = \cos 2\pi x$
   (b) $y = -4\cos 2\pi x$

8. (a) $y = -2\cos(x/4)$
   (b) $y = -2\cos(\pi x/4)$

9. $y = 1 + \sin 2x$

10. $y = \sin(x/2) - 2$

11. $y = 1 - \cos(\pi x/3)$

12. $y = -2 - 2\cos 3\pi x$

In Exercises 13–28, determine the amplitude, period, and phase shift for the given function. Graph the function over one period. Indicate the x-intercepts and the coordinates of the highest and lowest points on the graph.

13. $f(x) = \sin\left(x - \dfrac{\pi}{6}\right)$

14. $g(x) = \cos\left(x + \dfrac{\pi}{3}\right)$

15. $F(x) = -\cos\left(x + \dfrac{\pi}{4}\right)$

16. $G(x) = -\sin(x + 2)$

17. $y = \sin\left(2x - \dfrac{\pi}{2}\right)$

18. $y = \sin\left(3x + \dfrac{\pi}{2}\right)$

19. $y = \cos(2x - \pi)$

20. $y = \cos\left(x - \dfrac{\pi}{2}\right)$

21. $y = 3\sin\left(\dfrac{1}{2}x + \dfrac{\pi}{6}\right)$

22. $y = -2\sin(\pi x + \pi)$

23. $y = 4\cos\left(3x - \dfrac{\pi}{4}\right)$

24. $y = \cos(x + 1)$

25. $y = \dfrac{1}{2}\sin\left(\dfrac{\pi x}{2} - \pi^2\right)$

26. $y = \cos\left(2x - \dfrac{\pi}{3}\right) + 1$

27. $y = 1 - \cos\left(2x - \dfrac{\pi}{3}\right)$

28. $y = 3\cos\left(\dfrac{2x}{3} + \dfrac{\pi}{6}\right)$

In Exercises 29–34, determine whether the equation for the graph has the form $y = A\sin Bx$ or $y = A\cos Bx$ and then find the value of A and B.

29.

$(3\pi, -2)$

30.

$\left(\dfrac{\pi}{3}, 1.5\right)$

31.

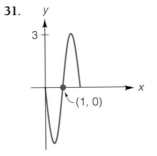

$(1, 0)$

32.

$(5, 1)$

33.

$(5\pi, 4)$

34.

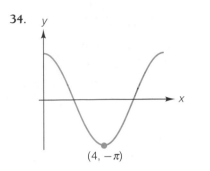

$(4, -\pi)$

B

35. In Section 7.7 you'll see the identity $\sin^2 x = \frac{1}{2} - \frac{1}{2} \cos 2x$. Use this identity to graph the function $y = \sin^2 x$ for one period.

36. In Section 7.7 you'll see the identity $\cos^2 x = \frac{1}{2} + \frac{1}{2} \cos 2x$. Use this identity to graph the function $y = \cos^2 x$ for one period.

37. In Section 7.7 we derive the identity $\sin 2x = 2 \sin x \cos x$. Use this to graph $y = \sin x \cos x$ for one period.

38. In Section 7.7 we derive the identity $\cos 2x = \cos^2 x - \sin^2 x$. Use this to graph $y = \cos^2 x - \sin^2 x$ for one period.

39. In Example 2 we showed that the period of $y = \cos 3x$ is $2\pi/3$. Use the same method to show that the period of $y = A \cos Bx$ is $2\pi/B$.

OPTIONAL TI-81 GRAPHING CALCULATOR EXERCISES

EXERCISES FOR SECTIONS 7.3 AND 7.4

As preparation for this exercise set, press the MODE *key and check that the calculator is set for radians rather than degrees.*

1. In this exercise, we use the TI-81 calculator to obtain several different views of $y = \sin x$ and $y = \cos x$.

(a) Using the Y= key, enter the function $Y_1 = \sin x$. Now press the ZOOM key followed by the 6 key. This yields the graph of $y = \sin x$ in the standard viewing window with $-10 \le x \le 10$ and $-10 \le y \le 10$.

(b) For a better view of $y = \sin x$ (and to see how the RANGE key operates), press the RANGE key and enter the following settings. Then press the GRAPH key.

RANGE

$x_{min} = -8$

$x_{max} = 8$

$x_{scl} = 4$

$y_{min} = -8$

$y_{max} = 8$

$y_{scl} = 4$

$x_{res} = 1$

You should obtain a picture similar to the following (except, of course, we've added some labels here).

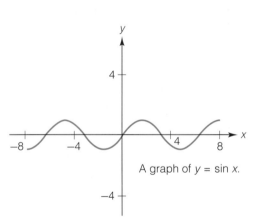

A graph of $y = \sin x$.

(c) As the graph shows, there is an x-intercept between 0 and 4. Using your knowledge of the sine function (and not using a calculator!), what is the exact x-coordinate for this intercept? Similarly, what is the exact coordinate for the x-intercept between 4 and 8?

(d) Now, using the Y= key, deactivate (but don't delete) $Y_1 = \sin x$ and enter $Y_2 = \cos x$. Graph $Y_2 = \cos x$. [Retain the RANGE key settings from part (b).] As you can see from the graph, there is an x-intercept between 0 and 4. What is the exact coordinate of that intercept?

(e) The TI-81 has a setting that allows us to mark off the x-axis in units of $\pi/2$. Press the ZOOM key followed by the 7 key. As you can see in the resulting graph of $Y_2 = \cos x$, the x-axis is indeed marked off in units of $\pi/2$. You can confirm this numerically by pressing the RANGE key and noting that the value for the x-scale is

$$x_{scl} = 1.570796327$$

This is the value of $\pi/2$ rounded off to nine decimal places.

(f) Press the Y= key and reactivate $Y_1 = \sin x$ (so that now both Y_1 and Y_2 are activated). Press the GRAPH key to see $\sin x$ and $\cos x$ on the same set of axes. Notice that the graphs are just translates of one another. By what distance would we have to shift the graph of $Y_1 = \sin x$ to the left in order for it to coincide with the graph of $Y_2 = \cos x$?

2. Press the Y= key and enter the following three functions:

$$Y_1 = \sin x \qquad Y_2 = 2 \sin x \qquad Y_3 = 3 \sin x$$

Now press the ZOOM key followed by the 7 key to obtain the three graphs. What are the amplitudes of the functions? What are the periods?

3. (a) Press the Y= key and enter the following three functions:

$$Y_1 = \cos x \qquad Y_2 = \cos 2x \qquad Y_3 = \cos 3x$$

Now press the ZOOM key followed by the 7 key to obtain the three graphs. Go on to part (b) for a better view of the graphs.

(b) In order to obtain a less cluttered view of the graphs, press the RANGE key and change the value of x_{min} to 0. Now press the GRAPH key to obtain the three graphs on the interval $0 \le x \le 2\pi$. What are the amplitudes of the functions? What are the periods?

Instructions for Problems 4–9.

(a) *First (without using the calculator), determine the amplitude, the period, and (where appropriate) the phase shift for the given function.*

(b) *Next, use the calculator to draw the graph of the function for two complete cycles. [In choosing an appropriate viewing rectangle, you will need to use the information obtained in part (a).]*

(c) *Specify the coordinates of the highest and lowest points on your graph.*

(d) *Use the TRACE key (and then the ARROW keys) on the calculator to check your answers in part (c).*

4. $y = -2.5 \cos(3x + 4)$

5. $y = -2.5 \cos(3\pi x + 4)$

6. $y = \sin(0.5x - 0.75)$

7. $y = \sin(0.5x + 0.75)$

8. $y = 9 \cos(11x - 8)$

9. $y = 0.02 \cos(100\pi x + 4\pi)$

10. Let $F(x) = \sin x$, $G(x) = x^2$, and $H(x) = x^3$. Which, if any, of the following four composite functions have graphs that do not go below the x-axis? First, try to answer without using the calculator, then use the calculator to check yourself. (You will learn more this way than if you were to draw the graphs immediately.)

$$y = G(F(x)) \qquad y = F(G(x))$$
$$y = F(H(x)) \qquad y = H(F(x))$$

11. (a) Enter the following two functions using the Y= key, but do not activate either of them yet.

$$Y_1 = \sin x \qquad \text{and} \qquad Y_2 = \sin(\sin x)$$

Press the ZOOM key and then the 7 key. Now press the RANGE key and set $y_{min} = -1$ and $y_{max} = 1$.

(b) Activate only the function $Y_1 = \sin x$ and draw the graph. Next, deactivate $Y_1 = \sin x$, activate $Y_2 = \sin(\sin x)$, and draw the graph. Are the two graphs identical?

(c) Activate both Y_1 and Y_2 and draw the graphs. As you can see, the two functions do have the same period, and their amplitudes are nearly equal.

(d) Estimate the amplitude of $\sin(\sin x)$ as follows. With only Y_2 activated, regraph $Y_2 = \sin(\sin x)$. Now press the TRACE key and use the ARROW keys to move the cursor to the right until you reach what appears to be the highest point on the graph. What are the coordinates of this point (as indicated at the bottom of the calculator screen)? Is it the x-coordinate or the y-coordinate that gives the amplitude?

(e) Without a calculator, explain why the amplitude of $Y_2 = \sin(\sin x)$ is the number $\sin 1$.

(f) Evaluate $\sin 1$ and use the result to check your approximation in part (d).

7.5   GRAPHS OF THE TANGENT AND THE RECIPROCAL FUNCTIONS

A third . . . function, the tangent of θ, or tan θ, is of secondary importance, in that it is not associated with wave phenomena. Nevertheless, it enters into the body of analysis so prominently that we cannot ignore it.

Samuel E. Urner and William B. Orange in *Elements of Mathematical Analysis* (Boston: Ginn and Company, 1950)

We have seen in the previous sections that the sine and cosine functions are periodic. The remaining four trigonometric functions are also periodic, but their graphs differ significantly from those of sine and cosine. The graphs of $y = \tan x$, $\cot x$, $\csc x$, and $\sec x$ all possess asymptotes.

We'll obtain the graph of the tangent function by a combination of both point-plotting and symmetry considerations. Table 1 displays a list of values for $y = \tan x$ using x-values in the interval $[0, \pi/2)$.

TABLE I

As x increases from 0 toward $\pi/2$, the values of tan x increase slowly at first, then more and more rapidly.

x	0	$\dfrac{\pi}{6}$	$\dfrac{\pi}{4}$	$\dfrac{\pi}{3}$	$\dfrac{5\pi}{12}\,(=75°)$	$\dfrac{17\pi}{36}\,(=85°)$	$\dfrac{89\pi}{180}\,(=89°)$	$\dfrac{\pi}{2}$
$\tan x$	0	$\dfrac{\sqrt{3}}{3} \approx 0.58$	1	$\sqrt{3} \approx 1.73$	3.73	11.43	57.29	undefined

As indicated in Table 1, $\tan x$ is undefined when $x = \pi/2$. This follows from the identity

$$\tan x = \frac{\sin x}{\cos x} \tag{1}$$

When $x = \pi/2$, the denominator in this identity is zero. Indeed, when x is equal to any odd integral multiple of $\pi/2$ (e.g., $\pm 3\pi/2$, $\pm 5\pi/2$), the denominator in equation (1) will be zero and, consequently, $\tan x$ will be undefined.

Because $\tan x$ is undefined when $x = \pi/2$, we want to see how the graph behaves as x gets closer and closer to $\pi/2$. This is why the x-values $5\pi/12$, $17\pi/36$, and $89\pi/180$ are used in Table 1. In Figure 1, we've used the data in Table 1 to draw the graph of $y = \tan x$ for $0 \le x < \pi/2$. As the figure indicates, the vertical line $x = \pi/2$ is an asymptote for the graph.

The graph of $y = \tan x$ can now be completed without further need for tables or a calculator. First, the identity $\tan(-x) = -\tan x$ (from Section 7.2) tells us that the graph of $y = \tan x$ is symmetric about the origin. So, after reflecting the graph in Figure 1 about the origin, we can draw the graph of $y = \tan x$ on the interval $(-\pi/2, \pi/2)$, as shown in Figure 2.

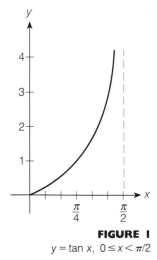

FIGURE I
$y = \tan x$, $0 \le x < \pi/2$

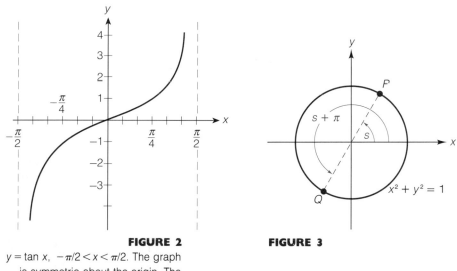

FIGURE 2   **FIGURE 3**

$y = \tan x$, $-\pi/2 < x < \pi/2$. The graph
is symmetric about the origin. The
lines $x = \pm\pi/2$ are asymptotes.

Now, to complete the graph of $y = \tan x$, we'll use the identity

$$\tan(s + \pi) = \tan s \qquad\qquad\qquad (2)$$

Looking at Figure 3, we can see why this identity is valid. By definition, the coordinates of P and Q are

$$P(\cos s, \sin s) \qquad\text{and}\qquad Q\big(\cos(s + \pi), \sin(s + \pi)\big)$$

On the other hand, the points P and Q are symmetric about the origin, and therefore the coordinates of Q are just the negatives of the coordinates of P. That is,

$$\cos(s + \pi) = -\cos s \qquad\text{and}\qquad \sin(s + \pi) = -\sin s$$

Consequently, we have

$$\tan(s + \pi) = \frac{\sin(s + \pi)}{\cos(s + \pi)} = \frac{-\sin s}{-\cos s}$$

$$= \frac{\sin s}{\cos s} = \tan s \qquad\text{as required}$$

[Although Figure 3 shows the angle with radian measure s terminating in the first quadrant, our proof is valid for the other quadrants as well. (Draw a figure for yourself and verify this.)]

Identity (2) tells us that the graph of $y = \tan x$ must repeat itself at intervals of length π. Taking this fact into account, along with Figure 2, we conclude that the period of $y = \tan x$ is exactly π. Our final graph of $y = \tan x$ is shown in the summary box that follows.

PROPERTY SUMMARY   THE TANGENT FUNCTION: $y = \tan x$

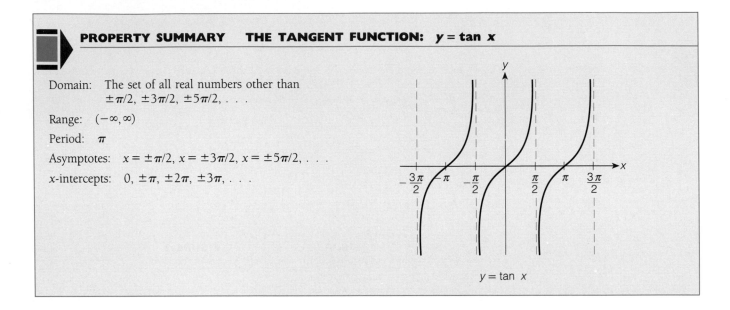

Domain: The set of all real numbers other than $\pm\pi/2, \pm3\pi/2, \pm5\pi/2, \ldots$

Range: $(-\infty, \infty)$

Period: π

Asymptotes: $x = \pm\pi/2, x = \pm3\pi/2, x = \pm5\pi/2, \ldots$

x-intercepts: $0, \pm\pi, \pm2\pi, \pm3\pi, \ldots$

$y = \tan x$

EXAMPLE 1 Graph the following functions for one period. In each case, specify the period, the asymptotes, and the intercepts.

(a) $y = \tan\left(x - \frac{\pi}{4}\right)$ **(b)** $y = -\tan\left(x - \frac{\pi}{4}\right)$

Solution We begin with the graph of one period of $y = \tan x$, as shown in Figure 4(a). By translating this graph $\pi/4$ units to the right, we obtain the graph of $y = \tan\left(x - \frac{\pi}{4}\right)$, in Figure 4(b). With this translation, note that the left asymptote shifts from $x = -\frac{\pi}{2}$ to $x = -\frac{\pi}{2} + \frac{\pi}{4} = -\frac{\pi}{4}$; the right asymptote shifts from $x = \frac{\pi}{2}$ to $x = \frac{\pi}{2} + \frac{\pi}{4} = \frac{3\pi}{4}$; and the x-intercept shifts from 0 to $\frac{\pi}{4}$.

FIGURE 4

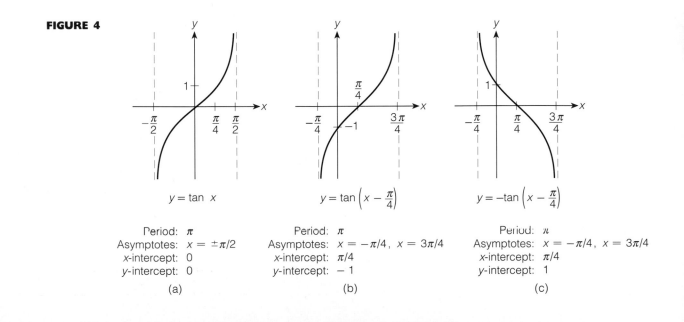

$y = \tan x$

Period: π
Asymptotes: $x = \pm\pi/2$
x-intercept: 0
y-intercept: 0

(a)

$y = \tan\left(x - \frac{\pi}{4}\right)$

Period: π
Asymptotes: $x = -\pi/4, x = 3\pi/4$
x-intercept: $\pi/4$
y-intercept: -1

(b)

$y = -\tan\left(x - \frac{\pi}{4}\right)$

Period: π
Asymptotes: $x = -\pi/4, x = 3\pi/4$
x-intercept: $\pi/4$
y-intercept: 1

(c)

For the y-intercept of $y = \tan\left(x - \frac{\pi}{4}\right)$, replace x with 0 in the equation. This yields

$$y = \tan\left(0 - \frac{\pi}{4}\right) = \tan\left(-\frac{\pi}{4}\right) = -\tan\frac{\pi}{4} = -1$$

Finally, for the graph of $y = -\tan\left(x - \frac{\pi}{4}\right)$, we need only reflect the graph in Figure 4(b) about the x-axis; see Figure 4(c).

EXAMPLE 2 Graph the function $y = \tan(x/2)$ for one period.

Solution First we refer back to Figure 2, which shows the basic pattern for one period of $y = \tan x$. In this basic pattern, the asymptotes occur when the radian measure of the angle equals $-\pi/2$ or $\pi/2$. Consequently, for $y = \tan(x/2)$, the asymptotes occur when $x/2 = -\pi/2$ and when $x/2 = \pi/2$. From the equation $x/2 = -\pi/2$ we conclude that $x = -\pi$, and from the equation $x/2 = \pi/2$ we conclude that $x = \pi$. Thus, the asymptotes for $y = \tan(x/2)$ are $x = -\pi$ and $x = \pi$. The distance between these asymptotes, namely, 2π, is the period of $y = \tan(x/2)$. This is twice the period of $y = \tan x$. So basically, we want to draw a curve with the same general shape as $y = \tan x$ but twice as wide. See Figure 5. Note that the graph in Figure 5 passes through the origin, since when $x = 0$, we have

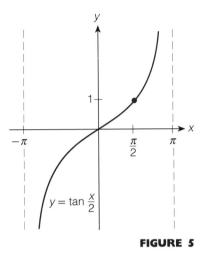

$y = \tan\dfrac{x}{2}$

FIGURE 5

$$y = \tan(0/2) = \tan 0 = 0$$

The graph of the cotangent function can be obtained from that of the tangent function by means of the identity

$$\cot x = -\tan\left(x - \frac{\pi}{2}\right) \tag{3}$$

(Exercise 37 shows how to derive this identity.) According to identity (3), the graph of $y = \cot x$ can be obtained by first translating the graph of $y = \tan x$ to the right $\pi/2$ units and then reflecting the translated graph about the x-axis. When this is done, we obtain the graph shown in the box that follows.

PROPERTY SUMMARY   THE COTANGENT FUNCTION: $y = \cot x$

Domain: The set of all real numbers other than $0, \pm\pi, \pm2\pi, \ldots$

Range: $(-\infty, \infty)$

Period: π (See Figure 6.)

Asymptotes: $x = 0, x = \pm\pi, x = \pm2\pi, \ldots$

x-intercepts: $\pm\dfrac{\pi}{2}, \pm\dfrac{3\pi}{2}, \pm\dfrac{5\pi}{2}, \ldots$

FIGURE 6
$y = \cot x$

EXAMPLE 3   Graph each of the following functions for one period.

(a) $y = \cot \pi x$ **(b)** $y = \frac{1}{2} \cot \pi x$ **(c)** $y = -\frac{1}{2} \cot \pi x$

Solution

(a) Looking at the graph of $y = \cot x$ in Figure 6, we see that one complete pattern or cycle of the graph is bounded by the asymptotes $x = 0$ and $x = \pi$. Now, for the function we are given, x has been replaced by πx. Thus, the corresponding asymptotes occur when $\pi x = 0$ and when $\pi x = \pi$; in other words, when $x = 0$ and when $x = 1$. See Figure 7(a).

(b) The graph of $y = \frac{1}{2} \cot \pi x$ will have the same general shape as that of $y = \cot \pi x$, but each y-coordinate on $y = \frac{1}{2} \cot \pi x$ will be half of the corresponding coordinate on $y = \cot \pi x$. See Figure 7(b).

(c) The graph of $y = -\frac{1}{2} \cot \pi x$ is obtained by reflecting the graph of $y = \frac{1}{2} \cot \pi x$ about the x-axis. See Figure 7(c).

FIGURE 7

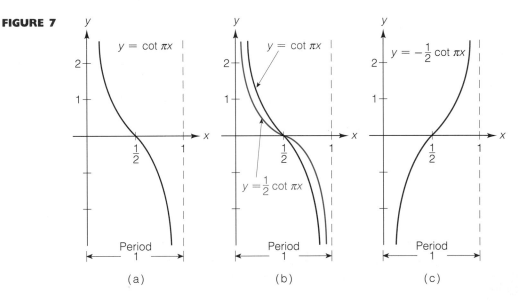

(a)   (b)   (c)

We conclude this section with a discussion of the graphs of $y = \csc x$ and $y = \sec x$. We will obtain these graphs in a series of easy steps, relying on the ideas of symmetry and translation. First consider the function $y = \csc x$. In Figure 8(a), we've set up a table and used it to sketch the graph of $y = \csc x$ on the interval $[\pi/2, \pi)$. Note that $\csc x$ is undefined when $x = \pi$. (Why?) As Figure 8(a) indicates, the vertical line $x = \pi$ is an asymptote for the graph. The graph in Figure 8(a) can be extended to the interval $(0, \pi)$ by means of the identity

$$\csc\left(\tfrac{\pi}{2} + s\right) = \csc\left(\tfrac{\pi}{2} - s\right)$$

(Exercise 38 shows you how to verify this identity.) This identity tells us that, starting at $x = \pi/2$, whether we travel a distance s to the right or a distance s to the left, the value of $y = \csc x$ is the same. In other words, the graph of $y = \csc x$ is symmetric about the line $x = \pi/2$. In view of this symmetry, we can sketch the graph of $y = \csc x$ on the interval $(0, \pi)$, as shown in Figure 8(b).

FIGURE 8

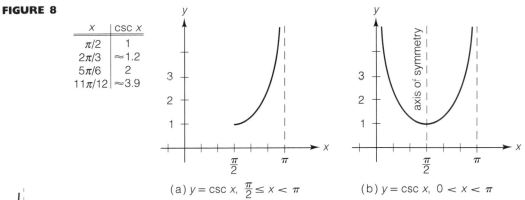

x	csc x
π/2	1
2π/3	≈1.2
5π/6	2
11π/12	≈3.9

(a) $y = \csc x, \ \frac{\pi}{2} \le x < \pi$

(b) $y = \csc x, \ 0 < x < \pi$

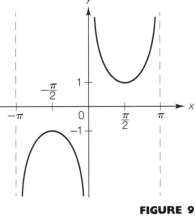

FIGURE 9

$y = \csc x, \quad -\pi < x < \pi$

The graph is symmetric about the origin.

The next step in obtaining the graph of $y = \csc x$ is to use the fact that the graph is symmetric about the origin. To verify this, we need to check that $\csc(-x) = -\csc x$. We have

$$\csc(-x) = \frac{1}{\sin(-x)} = \frac{1}{-\sin x} = -\csc x \qquad \text{as required}$$

Now, taking into account this symmetry about the origin and the portion of the graph that we've already obtained in Figure 8(b), we can sketch the graph of $y = \csc x$ over the interval $(-\pi, \pi)$, as shown in Figure 9.

To complete the graph of $y = \csc x$, we observe that the values of csc x must repeat themselves at intervals of 2π. This is because $\csc x = 1/(\sin x)$, and the sine function has a period of 2π. In view of this, we can draw the graph of $y = \csc x$ as shown in the box that follows.

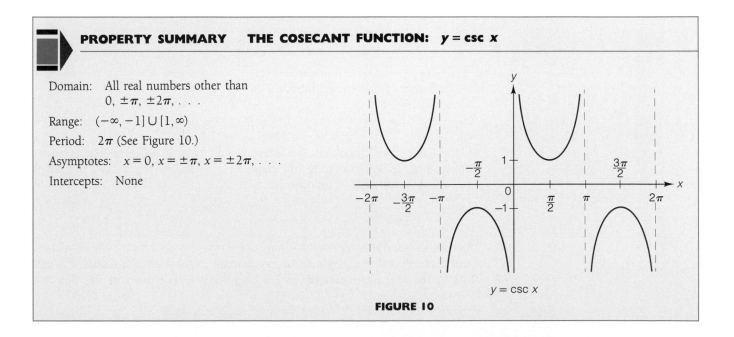

PROPERTY SUMMARY   THE COSECANT FUNCTION: $y = \csc x$

Domain: All real numbers other than
$0, \pm\pi, \pm2\pi, \ldots$

Range: $(-\infty, -1] \cup [1, \infty)$

Period: 2π (See Figure 10.)

Asymptotes: $x = 0, x = \pm\pi, x = \pm2\pi, \ldots$

Intercepts: None

$y = \csc x$

FIGURE 10

▮▮▮ **EXAMPLE 4** Graph the function $y = \csc(x/3)$ for one period.

Solution Since $\csc(x/3) = 1/\sin(x/3)$, it will be helpful first to graph one period of $y = \sin(x/3)$. This is done in Figure 11(a) (using the techniques of the previous section). Note that the period of $y = \sin(x/3)$ is 6π. This is also the period of $y = \csc(x/3)$, because $\csc(x/3)$ and $\sin(x/3)$ are just reciprocals. The asymptotes for $y = \csc(x/3)$, occur when $\sin(x/3) = 0$. From Figure 11(a), we see that $\sin(x/3) = 0$ when $x = 0$, when $x = 3\pi$, and when $x = 6\pi$. These asymptotes are sketched in Figure 11(b). The colored points in Figure 11(b) indicate where the value of $\sin(x/3)$ is 1 or -1. The graph of $y = \csc(x/3)$ must pass through these points. (Why?) Finally, using the points and the asymptotes in Figure 11(b), we can sketch the graph of $y = \csc(x/3)$, as shown in Figure 11(c). ▮▮▮

FIGURE 11

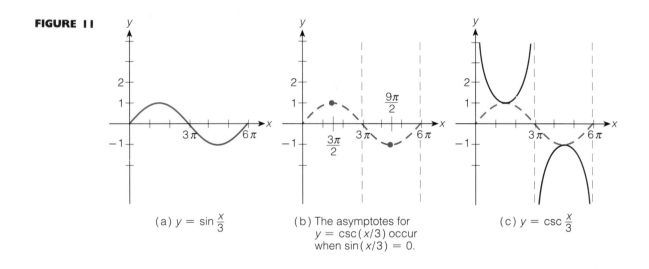

(a) $y = \sin \dfrac{x}{3}$

(b) The asymptotes for $y = \csc(x/3)$ occur when $\sin(x/3) = 0$.

(c) $y = \csc \dfrac{x}{3}$

In Section 7.3, we used the identity

$$\cos x = \sin\left(x + \tfrac{\pi}{2}\right) \tag{4}$$

to graph the cosine function. This identity tells us that the graph of $y = \cos x$ can be obtained by translating the graph of $y = \sin x$ to the left by $\pi/2$ units. Now, from identity (4), it follows that

$$\sec x = \csc\left(x + \tfrac{\pi}{2}\right)$$

This last identity tells us that the graph of $y = \sec x$ can also be obtained by a translation. In this case, by translating the graph of $y = \csc x$ a distance of $\pi/2$ units to the left, we obtain the graph of $y = \sec x$, as shown in the box that follows.

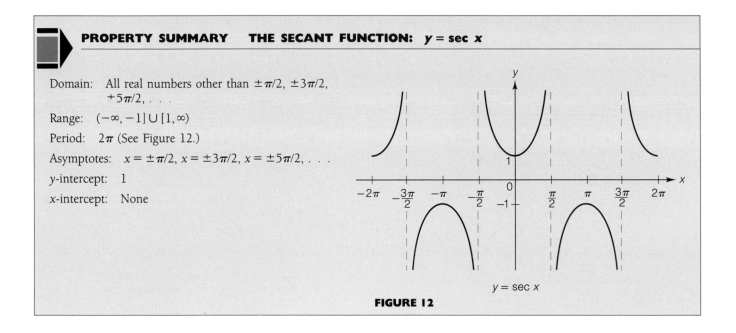

PROPERTY SUMMARY   THE SECANT FUNCTION: $y = \sec x$

Domain: All real numbers other than $\pm\pi/2$, $\pm 3\pi/2$, $+5\pi/2$, . . .

Range: $(-\infty, -1] \cup [1, \infty)$

Period: 2π (See Figure 12.)

Asymptotes: $x = \pm\pi/2$, $x = \pm 3\pi/2$, $x = \pm 5\pi/2$, . . .

y-intercept: 1

x-intercept: None

$y = \sec x$

FIGURE 12

EXERCISE SET 7.5

A

In Exercises 1–24, graph each function for one period, and show (or specify) the intercepts and asymptotes.

1. (a) $y = \tan\left(x + \frac{\pi}{4}\right)$
    (b) $y = -\tan\left(x + \frac{\pi}{4}\right)$

2. (a) $y = \tan\left(x - \frac{\pi}{3}\right)$
    (b) $y = -\tan\left(x - \frac{\pi}{3}\right)$

3. (a) $y = \tan(x/3)$
    (b) $y = -\tan(x/3)$

4. (a) $y = 2 \tan \pi x$
    (b) $y = -2 \tan \pi x$

5. $y = \frac{1}{2} \tan(\pi x/2)$

6. $y = -\frac{1}{2} \tan 2\pi x$

7. $y = \cot(\pi x/2)$

8. $y = \cot 2\pi x$

9. $y = -\cot\left(x - \frac{\pi}{4}\right)$

10. $y = \cot\left(x + \frac{\pi}{6}\right)$

11. $y = \frac{1}{2} \cot 2x$

12. $y = -\frac{1}{2} \cot(x/2)$

13. $y = \csc\left(x - \frac{\pi}{4}\right)$

14. $y = \csc\left(x - \frac{\pi}{6}\right)$

15. $y = -\csc(x/2)$

16. $y = 2 \csc x$

17. $y = \frac{1}{3} \csc \pi x$

18. $y = -\frac{1}{2} \csc 2\pi x$

19. $y = -\sec x$

20. $y = -2 \sec x$

21. $y = \sec(x - \pi)$

22. $y = \sec(x + 1)$

23. $y = 3 \sec(\pi x/2)$

24. $y = -2 \sec(\pi x/3)$

In Exercises 25–28, graph each function for two periods. Specify the intercepts and the asymptotes.

25. (a) $y = 2 \sin\left(3\pi x - \frac{\pi}{6}\right)$
    (b) $y = 2 \csc\left(3\pi x - \frac{\pi}{6}\right)$

26. (a) $y = -\frac{1}{2} \sin\left(\pi x + \frac{\pi}{3}\right)$
    (b) $y = -\frac{1}{2} \csc\left(\pi x + \frac{\pi}{3}\right)$

27. (a) $y = -3 \cos\left(2\pi x - \frac{\pi}{4}\right)$
    (b) $y = -3 \sec\left(2\pi x - \frac{\pi}{4}\right)$

28. (a) $y = \cos\left(3x + \frac{\pi}{3}\right)$
    (b) $y = \sec\left(3x + \frac{\pi}{3}\right)$

B

For Exercises 29–32, six functions are defined as follows:

$$f(x) = \sin x \qquad g(x) = \csc x \qquad h(x) = \pi x - \frac{\pi}{6}$$
$$F(x) = \cos x \qquad G(x) = \sec x \qquad H(x) = \pi x + \frac{\pi}{4}$$

In each case, graph the indicated function for one period.

29. (a) $f \circ h$
    (b) $g \circ h$

30. (a) $F \circ H$
    (b) $G \circ H$

31. (a) $f \circ H$
    (b) $g \circ H$

32. (a) $F \circ h$
    (b) $G \circ h$

For Exercises 33–36, four functions are defined as follows:

$$f(x) = \csc x \qquad T(x) = \tan x$$
$$g(x) = \sec x \qquad A(x) = |x|$$

In each case, graph the indicated function over the interval $[-2\pi, 2\pi]$.

33. $A \circ T$ 34. $A \circ g$

35. $A \circ f$ 36. $f \circ A$

37. In the text, we used the identity

$$\cot s = -\tan\left(s - \frac{\pi}{2}\right)$$

to obtain the graph of $y = \cot x$ from that of $y = \tan x$. The following steps show one way to derive this identity. (Although the accompanying figure shows the angle with radian measure s terminating in the first quadrant, the proof is valid no matter where the angle terminates.) In the figure, $PO \perp QO$.

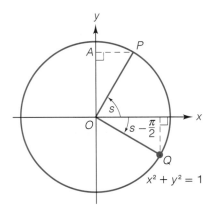

(a) Why are the coordinates of P and Q as follows?

$P(\cos s, \sin s)$ and $Q\left(\cos\left(s - \frac{\pi}{2}\right), \sin\left(s - \frac{\pi}{2}\right)\right)$

(b) Using congruent triangles [and without reference to part (a)], explain why the y-coordinate of Q is the negative of the x-coordinate of P, and the x-coordinate of Q equals the y-coordinate of P.

(c) Use the results in parts (a) and (b) to conclude that

$$\sin\left(s - \frac{\pi}{2}\right) = -\cos s \quad \text{and} \quad \cos\left(s - \frac{\pi}{2}\right) = \sin s$$

(d) Use the result in part (c) to show that

$$\cot s = -\tan\left(s - \frac{\pi}{2}\right)$$

38. In this exercise we verify the identity $\csc\left(\frac{\pi}{2} + s\right) = \csc\left(\frac{\pi}{2} - s\right)$. Refer to the following figure, in which $\angle AOB = \angle BOC = s$ radians. [Although the figure shows s in the interval $0 < s < \pi/2$, a similar proof will work for other intervals as well. (A proof that does not depend upon a picture can be given using the formula for $\sin(s + t)$, which is developed in Section 7.6.)]

(a) Show that the triangles AOE and COD are congruent and, consequently, $AE = CD$.

(b) Explain why the y-coordinates of the points C and A are $\sin\left(\frac{\pi}{2} - s\right)$ and $\sin\left(\frac{\pi}{2} + s\right)$, respectively.

(c) Use parts (a) and (b) to conclude that $\sin\left(\frac{\pi}{2} + s\right) = \sin\left(\frac{\pi}{2} - s\right)$. It follows from this that $\csc\left(\frac{\pi}{2} + s\right) = \csc\left(\frac{\pi}{2} - s\right)$, as required.

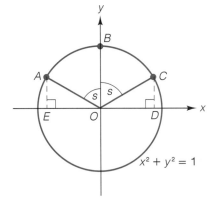

▶ OPTIONAL TI-81 GRAPHING CALCULATOR EXERCISES
EXERCISES FOR SECTION 7.5

1. Using the $Y=$ key, enter the function $Y_1 = \tan x$. Use the following settings. Press the mode key and check that your calculator is in the radian mode. Next, press the ZOOM key followed by the 7 key to mark off the axes in units of $\pi/2$. Check that the graph you obtain is consistent with the graph shown in the text on page 386.

2. In graphing the tangent function in the text, we used the identity $\tan(x + \pi) = \tan x$. Check this identity as follows. Retain the settings from Exercise 1 (including

$Y_1 = \tan x$). Using the $Y=$ key, enter the function $Y_2 = \tan(x + \pi)$. Now, graph Y_1 and Y_2 and observe that the graphs indeed appear to be identical.

3. In the text, we obtained the graph of the cotangent function from that of the tangent function by means of the identity $\cot x = -\tan[x - (\pi/2)]$. Verify this identity graphically as follows. Using the $Y-$ key, enter the functions $Y_1 = \cot x \ (= 1 \div \tan x)$ and $Y_2 = -\tan[x - (\pi/2)]$. Graph Y_1 and Y_2 and observe that the graphs indeed appear to be identical.

4. In this exercise we compare the graphs of $Y_1 = \tan x$ and $Y_2 = x^3$ on the open interval $(-\pi/2, \pi/2)$. Use the Y= key to enter these two functions. Next press ZOOM 7, and then (after waiting for the graphs to conclude) press the RANGE key and set $x_{min} = -1.5708$ and $x_{max} = 1.5708$. Now press the GRAPH key. Use the graphs to answer these questions:
    (a) Which of the two graphs appears to be horizontal as it passes through the origin?
    (b) For x-values in the interval $0 < x < \pi/2$, which quantity is larger, $\tan x$ or x^3?
    (c) For x-values in the interval $-\pi/2 < x < 0$, which quantity is larger, $\tan x$ or x^3?

In Exercises 5–10, graph each function. Adjust the range settings, where necessary, so that the graph is shown for at least two periods.

5. (a) $y = \tan(x/4)$
    (b) $y = \tan(4x)$

6. (a) $y = \cot(2x)$
    (b) $y = \cot(x/2)$

7. (a) $y = 0.5 \tan \pi x$
    (b) $y = 0.5 \tan[\pi x + (\pi/3)]$
    (c) $y = 0.5 \tan(\pi x + 1)$

8. (a) $y = 0.25 \cot x$
    (b) $y = 0.25 \cot[x + (\pi/4)]$
    (c) $y = -0.25 \cot[x + (\pi/4)]$

9. (a) $y = 0.4 \tan(x/2)$
    (b) $y = 0.4 \tan(x/3)$
    (c) $y = 0.4 \tan(x/5)$

10. (a) $y = 0.2 \cot(3x)$
    (b) $y = -0.2 \cot(4x)$
    (c) $y = -0.2 \cot(12x)$

11. Graph the functions $Y_1 = \sin x$ and $Y_2 = \csc x$ using the ZOOM 7 setting and a y-scale value of 1 in the RANGE menu. (To enter csc x, use $1 \div \sin x$.) Observe that $|\sin x| \le 1$ and $|\csc x| \ge 1$. At which points on the graph do we have $\sin x = \csc x$? At which points do we have $\sin x = -\csc x$?

12. Graph the functions $Y_1 = \cos x$ and $Y_2 = \sec x$ using the ZOOM 7 setting and a y-scale value of 1 in the RANGE menu. (To enter sec x, use $1 \div \cos x$.) Observe that $|\cos x| \le 1$ and $|\sec x| \ge 1$. At which points on the graph do we have $\cos x = \sec x$? At which points do we have $\cos x = -\sec x$?

In Exercises 13–16, graph each pair of functions on the same set of axes. Adjust the RANGE and/or ZOOM settings, where necessary, so that the graphs are displayed for at least two periods.

13. $Y_1 = 0.6 \sin(x/2)$
    $Y_2 = 0.6 \csc(x/2)$

14. $Y_1 = -1.2 \sin(\pi x/3)$
    $Y_2 = -1.2 \csc(\pi x/3)$

15. $Y_1 = 3 \cos[2x - (\pi/6)]$
    $Y_2 = 3 \sec[2x - (\pi/6)]$

16. $Y_1 = -1.5 \cos[\pi x - (\pi/6)]$
    $Y_2 = -1.5 \sec[\pi x - (\pi/6)]$

17. Using the ZOOM 7 setting, graph the two functions $Y_1 = \tan^2 x$ and $Y_2 = \sec^2 x - 1$. What do you observe? What does this demonstrate?

18. Using the ZOOM 7 setting, graph the two functions $Y_1 = \cot^2 x$ and $Y_2 = \csc^2 x - 1$. What do you observe? What does this demonstrate?

19. (a) Graph the two functions $Y_1 = \tan x$ and $Y_2 = x$ using the ZOOM 7 setting. Observe that tan x and x are very close to one another when x is close to 0. In fact, the approximation $\tan x \approx x$ is often used in science applications when it is known that x is close to 0.
    (b) To obtain a better view of $Y_1 = \tan x$ and $Y_2 = x$ near the origin, press the RANGE key and set $x_{min} = -1.57$ and $x_{max} = 1.57$. Then press the GRAPH key. Again, note that when x is close to 0, the values of tan x are indeed close to x. To see numerical evidence of this, complete the following table.

x	0.000123	0.01	0.05	0.1	0.2	0.3	0.4	0.5
tan x								

    (c) If you study *Taylor polynomials* in calculus, you'll see that an even better approximation to tan x is

    $$\tan x \approx x + \tfrac{1}{3}x^3 \qquad \text{when x is close to 0}$$

    Using the Y= key, enter the function $Y_3 = x + \tfrac{1}{3}x^3$. Now, after deactivating $Y_2 = x$, press the GRAPH key to obtain the graphs of $Y_1 = \tan x$ and $Y_3 = x + \tfrac{1}{3}x^3$.
    (d) To see numerical evidence of how well tan x is approximated by $x + \tfrac{1}{3}x^3$, add a third row to the table you worked out in part (b): in this third row, show the values for $x + \tfrac{1}{3}x^3$. When you've completed the table, note that the new values are much closer to tan x than were the values of x.

20. Using the Y= key, enter the functions $Y_1 = \tan x$ and

    $$Y_2 = \frac{\tan\left(x + \frac{\pi}{6}\right) - \tan\frac{\pi}{6}}{1 + \tan\left(x + \frac{\pi}{6}\right)\tan\frac{\pi}{6}}$$

    Graph these two functions using the ZOOM 7 setting. What do you observe? *Remark:* After studying Section 7.7, you should be able to prove the identity $Y_1 = Y_2$.

21. Finding a positive root of the equation $\tan x = x$: As preparation for this exercise, press the Y= key and clear (delete) all functions. Next press ZOOM 6. Finally, press the RANGE key and set $x_{min} = 0$ and $x_{max} = 6$.

(a) Using the $Y=$ key, enter the functions $Y_1 = \tan x$ and $Y_2 = x$. Press the GRAPH key. From the graph, estimate the smallest positive root of the equation $\tan x = x$. (That is, estimate the x-coordinate of the point where $Y_1 = \tan x$ and $Y_2 = x$ intersect.)
Answer: Something between 4 and 5, perhaps 4.5.

(b) We can obtain successively closer approximations to the root using the ZOOM key and the ARROW keys as follows. First press the ZOOM key and then the 2 key. If you now look carefully at the x-axis (at $x = 3$, actually), you will see a blinking cursor. Use the ARROW keys to move the cursor to the point where $Y_1 = \tan x$ and $Y_2 = x$ intersect. Now press the ENTER key to obtain a magnified view of that intersection point. Next, use the ARROW keys to move the cursor back onto that intersection point. Record the x-value shown at the bottom of the calculator screen. This is our next approximation [after the one in part (a)] to the required root. By repeating this process, you can obtain better and better approximations to the required root. Continue the process until the first three decimal places of the intersection point remain the same when you progress to the next step.

22. (a) *Without using the graphing calculator:* In what general ways would you expect the graphs of $y = \sec^2 x$ and $y = |\sec x|$ to resemble one another?

(b) Use the calculator (with the ZOOM 7 setting) to draw the graphs of these two functions. Was your response in part (a) correct? In what way do the graphs differ?

$\underline{7.6}$ THE ADDITION FORMULAS FOR SINE AND COSINE

It has long been recognized that the addition formulas are the heart of trigonometry. Indeed, Professor Rademacher and others have shown that the entire body of trigonometry can be derived from the assumption that there exist functions S and C such that

1. $S(x - y) = S(x)C(y) - C(x)S(y)$

2. $C(x - y) = C(x)C(y) + S(x)S(y)$

3. $\displaystyle\lim_{x \to 0^+} \frac{S(x)}{x} = 1$

Professor Frederick H. Young in his article, "The Addition Formulas" from *The Mathematics Teacher,* vol. L (1957), pp. 45–48

For any real numbers r, s, and t, it is always true that $r(s + t) = rs + rt$. This is the so called **distributive law** for real numbers. If f is a function, however, it is not true in general that $f(s + t) = f(s) + f(t)$. By way of example, consider the cosine function. It is not true in general that $\cos(s + t) = \cos s + \cos t$. For instance, with $s = \pi/6$ and $t = \pi/3$, we have

$$\cos\left(\frac{\pi}{6} + \frac{\pi}{3}\right) \overset{?}{=} \cos\frac{\pi}{6} + \cos\frac{\pi}{3}$$

$$\cos\frac{\pi}{2} \overset{?}{=} \cos\frac{\pi}{6} + \cos\frac{\pi}{3}$$

$$0 \overset{?}{=} \frac{\sqrt{3}}{2} + \frac{1}{2} \qquad \text{No!}$$

In this section, we will see just what $\cos(s + t)$ does equal. The correct formula for $\cos(s + t)$ is one of a group of four important trigonometric identities called the **addition formulas.**

The Addition Formulas

$$\sin(s + t) = \sin s \cos t + \cos s \sin t$$
$$\sin(s - t) = \sin s \cos t - \cos s \sin t$$
$$\cos(s + t) = \cos s \cos t - \sin s \sin t$$
$$\cos(s - t) = \cos s \cos t + \sin s \sin t$$

Our strategy for deriving these formulas will be as follows. First we'll prove the fourth formula—this takes some effort. The other three formulas are then relatively easy to derive from the fourth one. As background for our discussion, you will need to recall that the distance d between two points (x_1, y_1) and (x_2, y_2) is given by

$$d = \sqrt{(x_2 - x_1)^2 + (y_2 - y_1)^2}$$

To prove the fourth formula in the box, we use Figure 1. The idea behind the proof is as follows.* We begin in Figure 1(a) with the unit circle and the angles s, t, and $s - t$. Then we rotate $\triangle OPQ$ about the origin until the point P coincides with the point $(1, 0)$, as indicated in Figure 1(b). Although this rotation changes the coordinates for the points P and Q, it certainly has no effect upon the length of the line segment $\overline{PQ}$. Thus, whether we calculate PQ using the coordinates in Figure 1(a) or those in Figure 1(b), the results must be the same. As you'll see, by equating the two expressions for PQ, we will obtain the required formula.

FIGURE 1

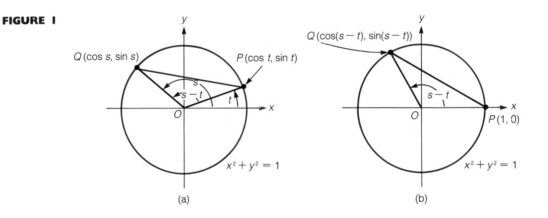

(a)   (b)

Applying the distance formula to Figure 1(a), we have

$$PQ = \sqrt{(\cos s - \cos t)^2 + (\sin s - \sin t)^2}$$
$$= \sqrt{\cos^2 s - 2\cos s \cos t + \cos^2 t + \sin^2 s - 2\sin s \sin t + \sin^2 t}$$
$$= \sqrt{2 - 2\cos s \cos t - 2\sin s \sin t} \quad \text{(Why?)} \tag{1}$$

*This idea for the proof can be traced back to the great French mathematician Augustin-Louis Cauchy (1789–1857).

Next, applying the distance formula to Figure 1(b), we have

$$PQ = \sqrt{[\cos(s-t)-1]^2 + [\sin(s-t)-0]^2}$$
$$= \sqrt{\cos^2(s-t) - 2\cos(s-t) + 1 + \sin^2(s-t)}$$

The right-hand side of this last equation can be simplified using the fact that $\cos^2(s-t) + \sin^2(s-t) = 1$. We then have

$$PQ = \sqrt{2 - 2\cos(s-t)} \qquad\qquad (2)$$

From equations (1) and (2), we conclude that

$$2 - 2\cos(s-t) = 2 - 2\cos s \cos t - 2 \sin s \sin t$$
$$-2\cos(s-t) = -2\cos s \cos t - 2 \sin s \sin t$$
$$\cos(s-t) = \cos s \cos t + \sin s \sin t$$

This completes the proof of the fourth addition formula. Before deriving the other three addition formulas, let's look at some applications of this result.

▌�no▐ EXAMPLE I   Simplify the expression $\cos 2\theta \cos \theta + \sin 2\theta \sin \theta$.

Solution   According to the identity that we just proved, we have

$$\cos 2\theta \cos \theta + \sin 2\theta \sin \theta = \cos(2\theta - \theta) = \cos \theta$$

Thus, the given expression is equal to $\cos \theta$.   ▮▮▮

▌no▐ EXAMPLE 2   Simplify $\cos(\theta - \pi)$.

Solution   We use the formula for $\cos(s-t)$ with s and t replaced by θ and π, respectively. This yields

$$\cos(s - t) = \cos s \cos t + \sin s \sin t$$
$$\updownarrow \quad \updownarrow$$
$$\cos(\theta - \pi) = \cos \theta \cos \pi + \sin \theta \sin \pi$$
$$= (\cos \theta)(-1) + (\sin \theta)(0) = -\cos \theta$$

Thus, the required simplification is $\cos(\theta - \pi) = -\cos \theta$.   ▮▮▮

In Example 2, we found that $\cos(\theta - \pi) = -\cos \theta$. This type of identity is often referred to as a **reduction formula**. The next example develops two basic reduction formulas that we will need to use later in this section.

▌no▐ EXAMPLE 3   Prove the following identities.   **(a)** $\cos\left(\frac{\pi}{2} - \alpha\right) = \sin \alpha$; **(b)** $\sin\left(\frac{\pi}{2} - \beta\right) = \cos \beta$.

Solution

(a) $\cos\left(\frac{\pi}{2} - \alpha\right) = \cos \frac{\pi}{2} \cos \alpha + \sin \frac{\pi}{2} \sin \alpha$
$$= (0)\cos \alpha + (1)\sin \alpha = \sin \alpha$$

This proves the identity. Incidentally, if we use degree measure instead of radian measure, this identity states that

$$\cos(90° - \alpha) = \sin \alpha$$

as we saw earlier, in Chapter 6.

(b) Since the identity $\cos\left(\frac{\pi}{2} - \alpha\right) = \sin \alpha$ holds for all values of α, we can simply replace α by the quantity $\frac{\pi}{2} - \beta$ to obtain

$$\cos\left[\frac{\pi}{2} - \left(\frac{\pi}{2} - \beta\right)\right] = \sin\left(\frac{\pi}{2} - \beta\right)$$
$$\cos\left(\frac{\pi}{2} - \frac{\pi}{2} + \beta\right) = \sin\left(\frac{\pi}{2} - \beta\right)$$
$$\cos \beta = \sin\left(\frac{\pi}{2} - \beta\right)$$

This proves the identity. ∎

The two identities in Example 3 are worth memorizing. For simplicity, we replace both α and β by the single letter θ to get

$$\cos\left(\frac{\pi}{2} - \theta\right) = \sin \theta$$
$$\sin\left(\frac{\pi}{2} - \theta\right) = \cos \theta$$

In Examples 2 and 3, radian measure was used. However, if you look back at Figure 1 and the derivation of the formula for $\cos(s - t)$, you will see that the derivation makes no reference, implicit or explicit, to a specific system of angle measurement. (For instance, the derivation does not involve the formula $s = r\theta$, which does require radian measure.) Thus, the formula for $\cos(s - t)$ is also valid when angles are measured in degrees. In the next example, we apply the formula in just such a case.

EXAMPLE 4 Use the formula for $\cos(s - t)$ to determine the exact value of $\cos 15°$.

Solution First observe that $15° = 45° - 30°$. Then we have

$\cos 15° = \cos(45° - 30°)$

$\qquad = \cos 45° \cos 30° + \sin 45° \sin 30° \qquad$ using the formula for $\cos(s - t)$ with $s = 45°$ and $t = 30°$

$$\qquad = \left(\frac{\sqrt{2}}{2}\right)\left(\frac{\sqrt{3}}{2}\right) + \left(\frac{\sqrt{2}}{2}\right)\left(\frac{1}{2}\right)$$

$$\qquad = \frac{\sqrt{6}}{4} + \frac{\sqrt{2}}{4} = \frac{\sqrt{6} + \sqrt{2}}{4}$$

Thus, the exact value of $\cos 15°$ is $\frac{1}{4}(\sqrt{6} + \sqrt{2})$. ∎

Now let's return to our derivations of the addition formulas. Using the fourth addition formula, we can easily derive the third formula as follows. In the formula

$$\cos(s - t) = \cos s \cos t + \sin s \sin t$$

we replace t by the quantity $-t$. This is permissible because the formula holds for all real numbers. We obtain

$$\cos[s - (-t)] = \cos s \cos(-t) + \sin s \sin(-t)$$

On the right-hand side of this equation, we can use the identities developed for $\cos(-t)$ and $\sin(-t)$ in Section 7.2. Doing this yields

$$\cos(s + t) = (\cos s)(\cos t) + (\sin s)(-\sin t)$$

which is equivalent to

$$\cos(s + t) = \cos s \cos t - \sin s \sin t$$

This is the third addition formula, as we wished to prove.

Next we derive the formula for $\sin(s + t)$. We have

$$\sin(s + t) = \cos\left[\tfrac{\pi}{2} - (s + t)\right] \qquad \text{replacing } \theta \text{ by } s + t \text{ in the identity}$$
$$\sin \theta = \cos\left[\tfrac{\pi}{2} - \theta\right]$$
$$= \cos\left[\left(\tfrac{\pi}{2} - s\right) - t\right]$$
$$= \cos\left(\tfrac{\pi}{2} - s\right) \cos t + \sin\left(\tfrac{\pi}{2} - s\right) \sin t \qquad \text{(Why?)}$$
$$= \sin s \cos t + \cos s \sin t \qquad \text{(Why?)}$$

This proves the first addition formula.

Finally, we can use the first addition formula to prove the second one as follows:

$$\sin(s - t) = \sin[s + (-t)]$$
$$= \sin s \cos(-t) + \cos s \sin(-t)$$
$$= (\sin s)(\cos t) + (\cos s)(-\sin t)$$
$$= \sin s \cos t - \cos s \sin t$$

This completes the proofs of the four addition formulas.

▮ ▆ ▮ **EXAMPLE 5** If $\sin s = \tfrac{3}{5}$, $0 < s < \tfrac{\pi}{2}$, and $\sin t = -\sqrt{3}/4$, $\pi < t < \tfrac{3\pi}{2}$, compute $\sin(s - t)$.

Solution $\sin(s - t) = \underbrace{\sin s} \cos t - \cos s \underbrace{\sin t}$ (3)

$\qquad\qquad\qquad\qquad$ given as $\tfrac{3}{5}$ $\qquad$ given as $-\sqrt{3}/4$

In view of equation (3), we need to find only $\cos t$ and $\cos s$. These can be determined using the Pythagorean identity $\cos^2 \theta = 1 - \sin^2 \theta$. We have

$$\cos^2 t = 1 - \sin^2 t$$
$$= 1 - \left(\frac{-\sqrt{3}}{4}\right)^2 = 1 - \frac{3}{16} = \frac{13}{16}$$

Therefore,

$$\cos t = \frac{\sqrt{13}}{4} \qquad \text{or} \qquad \cos t = \frac{-\sqrt{13}}{4}$$

We choose the negative value here for cosine, since it is given that $\pi < t < 3\pi/2$:

$$\cos t = \frac{-\sqrt{13}}{4}$$

Similarly, to find $\cos s$, we have

$$\cos^2 s = 1 - \sin^2 s = 1 - \left(\frac{3}{5}\right)^2 = \frac{16}{25}$$

Therefore,

$$\cos s = \frac{4}{5} \qquad \cos s \text{ is positive, since } 0 < s < \pi/2$$

Finally, we substitute the values we've obtained for $\cos t$ and $\cos s$, along with the given data, back into equation (3). This yields

$$\sin(s - t) = \left(\frac{3}{5}\right)\left(\frac{-\sqrt{13}}{4}\right) - \left(\frac{4}{5}\right)\left(\frac{-\sqrt{3}}{4}\right)$$

$$= \frac{-3\sqrt{13}}{20} + \frac{4\sqrt{3}}{20} = \frac{-3\sqrt{13} + 4\sqrt{3}}{20} \qquad\qquad \blacksquare\blacksquare\blacksquare$$

EXERCISE SET 7.6

A

In Exercises 1–10, use the addition formulas to simplify the expression.

1. $\sin \theta \cos 2\theta + \cos \theta \sin 2\theta$
2. $\sin \frac{\pi}{6} \cos \frac{\pi}{3} + \cos \frac{\pi}{6} \sin \frac{\pi}{3}$
3. $\sin 3\theta \cos \theta - \cos 3\theta \sin \theta$
4. $\sin 110° \cos 20° - \cos 110° \sin 20°$
5. $\cos 2u \cos 3u - \sin 2u \sin 3u$
6. $\cos 2u \cos 3u + \sin 2u \sin 3u$
7. $\cos \frac{2\pi}{9} \cos \frac{\pi}{18} + \sin \frac{2\pi}{9} \sin \frac{\pi}{18}$
8. $\cos \frac{3\pi}{10} \cos \frac{\pi}{5} - \sin \frac{3\pi}{10} \sin \frac{\pi}{5}$
9. $\sin(A + B) \cos A - \cos(A + B) \sin A$
10. $\cos(s - t) \cos t - \sin(s - t) \sin t$

In Exercises 11–14, simplify each expression (as in Example 2).

11. $\sin\left(\theta - \frac{3\pi}{2}\right)$
12. $\cos\left(\frac{3\pi}{2} + \theta\right)$
13. $\cos(\theta + \pi)$
14. $\sin(\theta - \pi)$
15. Expand $\sin(t + 2\pi)$ using the appropriate addition formula, and check to see that your answer agrees with the formula in the box on page 360.
16. Follow the directions in Exercise 15, but use $\cos(t + 2\pi)$.
17. Use the formula of $\cos(s + t)$ to compute the exact value of $\cos 75°$.
18. Use the formula for $\sin(s - t)$ to compute the exact value of $\sin(\pi/12)$.
19. Use the formula for $\sin(s + t)$ to find $\sin(7\pi/12)$.
20. Determine the exact value of (a) $\sin 105°$; (b) $\cos 105°$.

In Exercises 21–24, use the addition formulas to simplify each expression.

21. $\sin\left(\frac{\pi}{4} + s\right) - \sin\left(\frac{\pi}{4} - s\right)$
22. $\sin\left(t + \frac{\pi}{6}\right) - \sin\left(t - \frac{\pi}{6}\right)$
23. $\cos\left(\frac{\pi}{3} - \theta\right) - \cos\left(\frac{\pi}{3} + \theta\right)$
24. $\cos\left(\theta - \frac{\pi}{4}\right) + \cos\left(\theta + \frac{\pi}{4}\right)$

In Exercises 25–28, compute the indicated quantity using the following data.

$$\sin \alpha = \tfrac{12}{13}, \quad \pi/2 < \alpha < \pi$$
$$\cos \beta = -\tfrac{3}{5}, \quad \pi < \beta < 3\pi/2$$
$$\cos \theta = \tfrac{7}{25}, \quad -2\pi < \theta < -3\pi/2$$

25. (a) $\sin(\alpha + \beta)$
    (b) $\cos(\alpha + \beta)$
26. (a) $\sin(\alpha - \beta)$
    (b) $\cos(\alpha - \beta)$
27. (a) $\sin(\theta - \beta)$
    (b) $\sin(\theta + \beta)$
28. (a) $\cos(\alpha + \theta)$
    (b) $\cos(\alpha - \theta)$
29. Suppose that $\sin \theta = \frac{1}{5}$ and $0 < \theta < \pi/2$.
    (a) Compute $\cos \theta$.
    (b) Compute $\sin 2\theta$. Hint: $\sin 2\theta = \sin(\theta + \theta)$
30. Suppose that $\cos \theta = \frac{12}{13}$ and $3\pi/2 < \theta < 2\pi$.
    (a) Compute $\sin \theta$.
    (b) Compute $\cos 2\theta$. Hint: $\cos 2\theta = \cos(\theta + \theta)$.
31. Given $\tan \theta = -\frac{2}{3}$, $\pi/2 < \theta < \pi$, and $\csc \beta = 2$, $0 < \beta < \pi/2$, find $\sin(\theta + \beta)$ and $\cos(\beta - \theta)$.
32. Given $\sec s = \frac{5}{4}$, $\sin s < 0$, and $\cot t = -1$, $\pi/2 < t < \pi$, find $\sin(s - t)$ and $\cos(s + t)$.

In Exercises 33–35, prove that each equation is an identity.

33. $\sin\left(t + \frac{\pi}{4}\right) = (\sin t + \cos t)/\sqrt{2}$
34. $\cos\left(t + \frac{\pi}{4}\right) = (\cos t - \sin t)/\sqrt{2}$
35. $\sin\left(t + \frac{\pi}{4}\right) + \cos\left(t + \frac{\pi}{4}\right) = \sqrt{2} \cos t$

36. In the following figure, $AB = CD = 1$ and $BC = 2$. Find $\alpha + \beta$ and express your answer using radian measure. *Hint:* Find $\cos(\alpha + \beta)$.

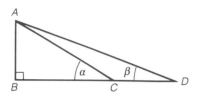

37. (a) Prove that the following equation is an identity:
$$\cos x = \sin\left(x + \tfrac{\pi}{2}\right)$$

   (b) What does the identity in part (a) tell us about the graphs of sine and cosine? (We used this identity in Section 7.3 to obtain the graph of the cosine function.)

38. Prove that the following equation is an identity:
$$\sin x = -\cos\left(x + \tfrac{\pi}{2}\right)$$

B

In Exercises 39–46, prove that each equation is an identity.

39. $\dfrac{\sin(s + t)}{\cos s \cos t} = \tan s + \tan t$

40. $\dfrac{\cos(s - t)}{\cos s \sin t} = \cot t + \tan s$

41. $\cos(A - B) - \cos(A + B) = 2 \sin A \sin B$

42. $\sin(A - B) + \sin(A + B) = 2 \sin A \cos B$

43. $\cos(A + B) \cos(A - B) = \cos^2 A - \sin^2 B$

44. $\sin(A + B) \sin(A - B) = \cos^2 B - \cos^2 A$

45. $\cos(\alpha + \beta) \cos \beta + \sin(\alpha + \beta) \sin \beta = \cos \alpha$

46. $\cos\left(\theta + \tfrac{\pi}{4}\right) + \sin\left(\theta - \tfrac{\pi}{4}\right) = 0$

47. Let $f(t) = \cos^2 t + \cos^2\left(t + \tfrac{2\pi}{3}\right) + \cos^2\left(t - \tfrac{2\pi}{3}\right)$.
   (a) Complete the table. (Use a calculator.)

t	1	2	3	4
$f(t)$				

   (b) On the basis of your results in part (a), make a conjecture about the function f. Prove that your conjecture is correct.

48. Let A, B, and C be the angles of a triangle, so that $A + B + C = \pi$.
   (a) Show that $\sin(A + B) = \sin C$.
   (b) Show that $\cos(A + B) = -\cos C$.
   (c) Show that $\tan(A + B) = -\tan C$.

49. Suppose that A, B, and C are the angles of a triangle, so that $A + B + C = \pi$. Show that
$$\cos^2 A + \cos^2 B + \cos^2 C + 2 \cos A \cos B \cos C = 1$$

50. Prove that
$$\frac{\sin(\alpha - \beta)}{\cos \alpha \cos \beta} + \frac{\sin(\beta - \gamma)}{\cos \beta \cos \gamma} + \frac{\sin(\gamma - \alpha)}{\cos \gamma \cos \alpha} = 0$$

51. Suppose that $a^2 + b^2 = 1$ and $c^2 + d^2 = 1$. Prove that $|ac + bd| \le 1$. *Hint:* Let $a = \cos \theta$, $b = \sin \theta$, $c = \cos \phi$, and $d = \sin \phi$.

In Exercises 52 and 53, simplify the expression.

52. $\cos\left(\tfrac{\pi}{6} + t\right) \cos\left(\tfrac{\pi}{6} - t\right) - \sin\left(\tfrac{\pi}{6} + t\right) \sin\left(\tfrac{\pi}{6} - t\right)$
   Hint: If your solution relies on four separate addition formulas, then you are doing this the hard way.

53. $\sin\left(\tfrac{\pi}{3} - t\right) \cos\left(\tfrac{\pi}{3} + t\right) + \cos\left(\tfrac{\pi}{3} - t\right) \sin\left(\tfrac{\pi}{3} + t\right)$

Exercises 54 and 55 outline simple geometric derivations of the formulas for $\sin(\alpha + \beta)$ and $\cos(\alpha + \beta)$ in the case where α and β are acute angles, with $\alpha + \beta < 90°$. The exercises rely on the accompanying figures, which are constructed as follows. Begin, in Figure A, with $\alpha = \angle GAD$, $\beta = \angle HAG$, and $AH = 1$. Then, from H, draw perpendiculars to $\overline{AD}$ and to $\overline{AG}$, as shown in Figure B. Finally, draw $\overline{FE} \perp \overline{BH}$ and $\overline{FC} \perp \overline{AD}$.

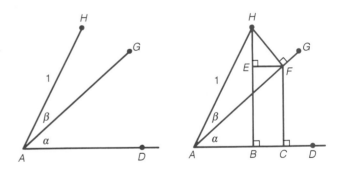

54. *Formula for* $\sin(\alpha + \beta)$ Supply the reasons or steps behind each statement.
   (a) $BH = \sin(\alpha + \beta)$ (b) $FH = \sin \beta$
   (c) $\angle BHF = \alpha$
   (d) $EH = \cos \alpha \sin \beta$
      Hint: Use $\triangle EFH$ and the result in part (b).
   (e) $AF = \cos \beta$ (f) $CF = \sin \alpha \cos \beta$
   (g) $\sin(\alpha + \beta) = \sin \alpha \cos \beta + \cos \alpha \sin \beta$
      Hint: $\sin(\alpha + \beta) = BH = EH + CF$

55. *Formula for* $\cos(\alpha + \beta)$ Supply the reason or steps behind each statement.
   (a) $\cos(\alpha + \beta) = AB$
   (b) $AC = \cos \alpha \cos \beta$
      Hint: Use $\triangle ACF$ and the result in Exercise 54(e).
   (c) $EF = \sin \alpha \sin \beta$

(d) $\cos(\alpha + \beta) = \cos \alpha \cos \beta - \sin \alpha \sin \beta$
   Hint: $AB = AC - BC$

C

56. Prove the following addition formulas for the tangent function.

(a) $\tan(\alpha + \beta) = \dfrac{\tan \alpha + \tan \beta}{1 - \tan \alpha \tan \beta}$

(b) $\tan(\alpha - \beta) = \dfrac{\tan \alpha - \tan \beta}{1 + \tan \alpha \tan \beta}$

57. (a) The angles of a triangle are $A = 20°$, $B = 50°$, and $C = 110°$. Use your calculator to compute the two quantities $\tan A + \tan B + \tan C$ and $\tan A \tan B \tan C$. What do you observe?

(b) The angles of a triangle are $\alpha = \pi/10$, $\beta = 3\pi/10$, and $\gamma = 3\pi/5$. Use your calculator to compute $\tan \alpha + \tan \beta + \tan \gamma$ and $\tan \alpha \tan \beta \tan \gamma$.

(c) If triangle ABC is not a right triangle, prove that $\tan A + \tan B + \tan C = \tan A \tan B \tan C$

58. If triangle ABC is not a right triangle, and $\cos A = \cos B \cos C$, show that $\tan B \tan C = 2$.

7.7 FURTHER IDENTITIES

Ptolemy (c. 150) knew substantially the sine of half an angle . . . and it is probable that Hipparchus (c. 140 B.C.) and certain that Varahamihira (c. 505) knew the relation that we express as $\sin \frac{\phi}{2} = \sqrt{(1 - \cos \phi)/2}$.

David Eugene Smith in *History of Mathematics* (New York: Ginn and Company, 1925)

There are a number of basic identities that follow from the addition formulas for sine and cosine. In the following boxes, we summarize some of the most useful of these.

Addition Formulas for Tangent

a. $\tan(s + t) = \dfrac{\tan s + \tan t}{1 - \tan s \tan t}$

b. $\tan(s - t) = \dfrac{\tan s - \tan t}{1 + \tan s \tan t}$

Double-Angle Formulas

a. $\sin 2\theta = 2 \sin \theta \cos \theta$

b. $\cos 2\theta = \cos^2 \theta - \sin^2 \theta$

c. $\tan 2\theta = \dfrac{2 \tan \theta}{1 - \tan^2 \theta}$

Half-Angle Formulas

a. $\sin \dfrac{s}{2} = \pm \sqrt{\dfrac{1 - \cos s}{2}}$

b. $\cos \dfrac{s}{2} = \pm \sqrt{\dfrac{1 + \cos s}{2}}$

c. $\tan \dfrac{s}{2} = \dfrac{\sin s}{1 + \cos s}$

To prove the formula for $\tan(s + t)$, we begin with

$$\tan(s + t) = \frac{\sin(s + t)}{\cos(s + t)}$$

$$= \frac{\sin s \cos t + \cos s \sin t}{\cos s \cos t - \sin s \sin t} \tag{1}$$

Now we divide both numerator and denominator on the right-hand side of equation (1) by the quantity $\cos s \cos t$. This yields

$$\tan(s + t) = \frac{\dfrac{\sin s \cos t}{\cos s \cos t} + \dfrac{\cos s \sin t}{\cos s \cos t}}{\dfrac{\cos s \cos t}{\cos s \cos t} - \dfrac{\sin s \sin t}{\cos s \cos t}} = \frac{\tan s + \tan t}{1 - \tan s \tan t}$$

This proves the formula for $\tan(s + t)$. The formula for $\tan(s - t)$ can be deduced from this with the aid of the identity $\tan(-t) = -\tan t$, which was derived in Section 7.2. We have

$$\tan(s - t) = \tan[s + (-t)]$$

$$= \frac{\tan s + \tan(-t)}{1 - \tan s \tan(-t)} = \frac{\tan s + (-\tan t)}{1 - (\tan s)(-\tan t)}$$

$$= \frac{\tan s - \tan t}{1 + \tan s \tan t} \qquad \text{as required}$$

■■■ **EXAMPLE 1** Compute $\tan \frac{7\pi}{12}$, using the fact that $\frac{7\pi}{12} = \frac{\pi}{4} + \frac{\pi}{3}$.

Solution

$$\tan \frac{7\pi}{12} = \tan\left(\frac{\pi}{4} + \frac{\pi}{3}\right)$$

$$= \frac{\tan \frac{\pi}{4} + \tan \frac{\pi}{3}}{1 - \tan \frac{\pi}{4} \tan \frac{\pi}{3}} \qquad \text{using the formula for } \tan(s + t) \text{ with } s = \frac{\pi}{4} \text{ and } t = \frac{\pi}{3}$$

$$= \frac{1 + \sqrt{3}}{1 - (1)(\sqrt{3})}$$

Thus, $\tan \dfrac{7\pi}{12} = \dfrac{1 + \sqrt{3}}{1 - \sqrt{3}}$. We can write this answer in a more compact form by rationalizing the denominator. As you can check, the result is

$$\tan \tfrac{7\pi}{12} = -2 - \sqrt{3} \qquad\qquad\qquad ▪▪▪$$

We turn now to the identities for $\sin 2\theta$, $\cos 2\theta$, and $\tan 2\theta$. All of these are derived in the same way: we replace 2θ by $(\theta + \theta)$ and use the appropriate addition formula. For instance, for $\sin 2\theta$ we have

$$\sin 2\theta = \sin(\theta + \theta) = \sin \theta \cos \theta + \cos \theta \sin \theta$$

$$= 2 \sin \theta \cos \theta$$

This establishes the formula for $\sin 2\theta$. (Exercise 37 asks you to carry out the corresponding derivations for $\cos 2\theta$ and $\tan 2\theta$.)

■■■ **EXAMPLE 2** If $\sin \theta = \frac{4}{5}$ and $\pi/2 < \theta < \pi$, find $\cos \theta$, $\sin 2\theta$, and $\cos 2\theta$.

Solution We have

$$\cos^2 \theta = 1 - \sin^2 \theta = 1 - \left(\frac{4}{5}\right)^2 = \frac{9}{25}$$

Consequently, $\cos \theta = 3/5$ or $\cos \theta = -3/5$. We want the negative value for the cosine here, since $\pi/2 < \theta < \pi$. Thus,

$$\cos \theta = -\frac{3}{5}$$

Now that we know the values of $\cos \theta$ and $\sin \theta$, the double-angle formulas can be used to determine $\sin 2\theta$ and $\cos 2\theta$. We have

$$\sin 2\theta = 2 \sin \theta \cos \theta = 2\left(\frac{4}{5}\right)\left(-\frac{3}{5}\right) = -\frac{24}{25}$$

and

$$\cos 2\theta = \cos^2 \theta - \sin^2 \theta$$
$$= \left(-\frac{3}{5}\right)^2 - \left(\frac{4}{5}\right)^2 = \frac{9}{25} - \frac{16}{25} = -\frac{7}{25}$$

The required values are therefore $\cos \theta = -\frac{3}{5}$, $\sin 2\theta = -\frac{24}{25}$, and $\cos 2\theta = -\frac{7}{25}$.

EXAMPLE 3 If $x = 4 \sin \theta$, $0 < \theta < \pi/2$, express $\sin 2\theta$ in terms of x.

Solution The given equation is equivalent to $\sin \theta = x/4$, so we have

$$\sin 2\theta = 2 \sin \theta \cos \theta = 2\left(\frac{x}{4}\right)\cos \theta$$

$$= \frac{x}{2}\sqrt{1 - \sin^2 \theta} \qquad \text{(Why is the positive root appropriate?)}$$

$$= \frac{x}{2}\sqrt{1 - \frac{x^2}{16}} = \frac{x}{2}\sqrt{\frac{16 - x^2}{16}} = \frac{x\sqrt{16 - x^2}}{8}$$

EXAMPLE 4 Prove the following identity:

$$\cos 3\theta = 4 \cos^3 \theta - 3 \cos \theta$$

Solution $\cos 3\theta = \cos(2\theta + \theta)$

$$= \underbrace{\cos 2\theta}_{(\cos^2 \theta - \sin^2 \theta)} \cos \theta - \underbrace{\sin 2\theta}_{(2 \sin \theta \cos \theta)} \sin \theta$$

$$= (\cos^2 \theta - \sin^2 \theta)\cos \theta - (2 \sin \theta \cos \theta)\sin \theta$$
$$= \cos^3 \theta - \sin^2 \theta \cos \theta - 2 \sin^2 \theta \cos \theta$$

Collecting like terms now gives us

$$\cos 3\theta = \cos^3 \theta - 3 \sin^2 \theta \cos \theta$$

Finally, we replace $\sin^2 \theta$ by the quantity $1 - \cos^2 \theta$. This yields

$$\cos 3\theta = \cos^3 \theta - 3(1 - \cos^2 \theta) \cos \theta$$
$$= \cos^3 \theta - 3 \cos \theta + 3 \cos^3 \theta$$
$$= 4 \cos^3 \theta - 3 \cos \theta \quad \text{as required} \quad \blacksquare\blacksquare\blacksquare$$

In the box that follows, we list several alternate ways of writing the formula for $\cos 2\theta$. Formulas (c) and (d) are quite useful in calculus.

Equivalent Forms of the Formula $\cos 2\theta = \cos^2 \theta - \sin^2 \theta$

a. $\cos 2\theta = 2 \cos^2 \theta - 1$ b. $\cos 2\theta = 1 - 2 \sin^2 \theta$

c. $\cos^2 \theta = \dfrac{1 + \cos 2\theta}{2}$ d. $\sin^2 \theta = \dfrac{1 - \cos 2\theta}{2}$

One way to prove identity (a) is as follows:

$$\cos 2\theta = \cos^2 \theta - \sin^2 \theta$$
$$= \cos^2 \theta - (1 - \cos^2 \theta)$$
$$= \cos^2 \theta - 1 + \cos^2 \theta$$
$$= 2 \cos^2 \theta - 1 \quad \text{as required}$$

If, in this last equation, we add 1 to both sides and then divide by 2, the result is identity (c). (Verify this.) The proofs for (b) and (d) are similar; see Exercise 38.

$\blacksquare\ \blacksquare$ **EXAMPLE 5**   Express $\cos^4 t$ in a form that does not involve powers of the trigonometric functions.

Solution $\cos^4 t = (\cos^2 t)^2 = \left(\dfrac{1 + \cos 2t}{2}\right)^2$ using the formula for $\cos^2 \theta$

$$= \dfrac{1 + 2 \cos 2t + \cos^2 2t}{4}$$

$$= \dfrac{1 + 2 \cos 2t + \frac{1}{2}(1 + \cos 4t)}{4} \quad \begin{array}{l}\text{using the formula for } \cos^2 \theta \\ \text{with } \theta = 2t\end{array}$$

An easy way to simplify this last expression is to multiply both the numerator and the denominator by 2. As you should check for yourself, the final result is

$$\cos^4 t = \dfrac{3 + 4 \cos 2t + \cos 4t}{8} \qquad\qquad \blacksquare\blacksquare\blacksquare$$

The last three formulas we are going to prove in this section are the **half-angle formulas:**

$$\cos \frac{s}{2} = \pm \sqrt{\frac{1 + \cos s}{2}}$$

$$\sin \frac{s}{2} = \pm \sqrt{\frac{1 - \cos s}{2}}$$

$$\tan \frac{s}{2} = \frac{\sin s}{1 + \cos s}$$

To derive the formula for cos(s/2), we begin with one of the alternate forms of the cosine double-angle formula:

$$\cos^2 \theta = \frac{1 + \cos 2\theta}{2} \qquad \text{or} \qquad \cos \theta = \pm \sqrt{\frac{1 + \cos 2\theta}{2}}$$

Since this identity holds for all values of θ, we may replace θ by $s/2$ to obtain

$$\cos \frac{s}{2} = \pm \sqrt{\frac{1 + \cos 2(s/2)}{2}} = \pm \sqrt{\frac{1 + \cos s}{2}}$$

This is the required formula for cos(s/2). To derive the formula for sin(s/2), we follow exactly the same procedure, except that we begin with the identity $\sin^2 \theta = \frac{1}{2}(1 - \cos 2\theta)$. [Exercise 38(c) asks you to complete the proof.] In both formulas, the sign before the radical is determined by the quadrant in which the angle or arc $s/2$ terminates.

■ ■ EXAMPLE 6 Evaluate cos 105° using a half-angle formula.

Solution
$$\cos 105° = \cos \frac{210°}{2} = \pm \sqrt{\frac{1 + \cos 210°}{2}} \qquad \text{using the formula for } \cos(s/2) \text{ with } s = 210°$$

$$= \pm \sqrt{\frac{1 + (-\sqrt{3}/2)}{2}}$$

$$= \pm \sqrt{\frac{1 - (\sqrt{3}/2)}{2} \cdot \frac{2}{2}} = \pm \sqrt{\frac{2 - \sqrt{3}}{4}}$$

$$= \frac{\pm \sqrt{2 - \sqrt{3}}}{2}$$

We want to choose the negative value here, since the terminal side of 105° lies in the second quadrant. Thus, we finally obtain

$$\cos 105° = \frac{-\sqrt{2 - \sqrt{3}}}{2}$$

■■■

Our last task is to establish the formula for tan(s/2). To do this, we first prove the equivalent identity, $\tan \theta = \dfrac{\sin 2\theta}{1 + \cos 2\theta}$.

PROOF THAT $\tan \theta = \dfrac{\sin 2\theta}{1 + \cos 2\theta}$

$$\frac{\sin 2\theta}{1 + \cos 2\theta} = \frac{2 \sin \theta \cos \theta}{2 \cos^2 \theta} \qquad \text{using the identity } \cos 2\theta = 2 \cos^2 \theta - 1 \text{ in the denominator}$$

$$= \frac{\sin \theta}{\cos \theta} = \tan \theta$$

If we now replace θ by $s/2$ in the identity $\tan \theta = \dfrac{\sin 2\theta}{1 + \cos 2\theta}$, the result is

$$\tan \frac{s}{2} = \frac{\sin s}{1 + \cos s}$$

This is the half-angle formula for the tangent.

We conclude this section with a summary of the principal trigonometric identities developed in this chapter. For completeness, the list also includes two sets of trigonometric identities that we did not discuss in the text but that are occasionally useful. These are the so called **product-to-sum formulas** and **sum-to-product formulas**. Proofs and applications of these formulas are included in the exercises.

PROPERTY SUMMARY   PRINCIPAL TRIGONOMETRIC IDENTITIES

I. Consequences of the Definitions

(a) $\csc\theta = \dfrac{1}{\sin\theta}$ (b) $\sec\theta = \dfrac{1}{\cos\theta}$ (c) $\cot\theta = \dfrac{1}{\tan\theta}$

(d) $\tan\theta = \dfrac{\sin\theta}{\cos\theta}$ (e) $\cot\theta = \dfrac{\cos\theta}{\sin\theta}$

II. Pythagorean Identities

(a) $\sin^2\theta + \cos^2\theta = 1$ (b) $\tan^2\theta + 1 = \sec^2\theta$ (c) $\cot^2\theta + 1 = \csc^2\theta$

III. Opposite-Angle Formulas

(a) $\sin(-\theta) = -\sin\theta$ (b) $\cos(-\theta) = \cos\theta$ (c) $\tan(-\theta) = -\tan\theta$

IV. Reduction Formulas

(a) $\sin(\theta + 2\pi k) = \sin\theta$ (b) $\cos(\theta + 2\pi k) = \cos\theta$

(c) $\sin\left(\frac{\pi}{2} - \theta\right) = \cos\theta$ (d) $\cos\left(\frac{\pi}{2} - \theta\right) = \sin\theta$

V. Addition Formulas

(a) $\sin(s + t) = \sin s\cos t + \cos s\sin t$ (b) $\sin(s - t) = \sin s\cos t - \cos s\sin t$

(c) $\cos(s + t) = \cos s\cos t - \sin s\sin t$ (d) $\cos(s - t) = \cos s\cos t + \sin s\sin t$

(e) $\tan(s + t) = \dfrac{\tan s + \tan t}{1 - \tan s\tan t}$ (f) $\tan(s - t) = \dfrac{\tan s - \tan t}{1 + \tan s\tan t}$

VI. Double-Angle Formulas

(a) $\sin 2\theta = 2\sin\theta\cos\theta$ (b) $\cos 2\theta = \cos^2\theta - \sin^2\theta$ (c) $\tan 2\theta = \dfrac{2\tan\theta}{1 - \tan^2\theta}$

VII. Half-Angle Formulas

(a) $\sin\dfrac{\theta}{2} = \pm\sqrt{\dfrac{1 - \cos\theta}{2}}$ (b) $\cos\dfrac{\theta}{2} = \pm\sqrt{\dfrac{1 + \cos\theta}{2}}$ (c) $\tan\dfrac{\theta}{2} = \dfrac{\sin\theta}{1 + \cos\theta}$

VIII. Product-to-Sum Formulas

(a) $\sin A\sin B = \frac{1}{2}[\cos(A - B) - \cos(A + B)]$ (b) $\sin A\cos B = \frac{1}{2}[\sin(A + B) + \sin(A - B)]$

(c) $\cos A\cos B = \frac{1}{2}[\cos(A + B) + \cos(A - B)]$

IX. Sum-to-Product Formulas

(a) $\sin\alpha + \sin\beta = 2\sin\dfrac{\alpha + \beta}{2}\cos\dfrac{\alpha - \beta}{2}$ (b) $\sin\alpha - \sin\beta = 2\cos\dfrac{\alpha + \beta}{2}\sin\dfrac{\alpha - \beta}{2}$

(c) $\cos\alpha + \cos\beta = 2\cos\dfrac{\alpha + \beta}{2}\cos\dfrac{\alpha - \beta}{2}$ (d) $\cos\alpha - \cos\beta = -2\sin\dfrac{\alpha + \beta}{2}\sin\dfrac{\alpha - \beta}{2}$

EXERCISE SET 7.7

A

In Exercises 1–12, use the given information to evaluate each expression.

1. $\tan t = \frac{3}{4}$, $\tan s = \frac{7}{24}$
   (a) $\tan(t + s)$ (b) $\tan(t - s)$

2. $\tan \theta = -\frac{3}{4}$
   (a) $\tan\left(\theta + \frac{1}{4}\pi\right)$ (b) $\tan\left(\theta - \frac{1}{4}\pi\right)$

3. $\tan x = \sqrt{2}$
   (a) $\tan\left(x + \frac{3}{4}\pi\right)$ (b) $\tan\left(x - \frac{5}{4}\pi\right)$

4. $\sin x = -\frac{5}{13}$, $3\pi/2 < x < 2\pi$
   (a) $\tan\left(x + \frac{1}{3}\pi\right)$ (b) $\tan\left(x - \frac{1}{3}\pi\right)$

5. $\cos \varphi = \frac{7}{25}$, $0° < \varphi < 90°$
   (a) $\sin 2\varphi$ (b) $\cos 2\varphi$ (c) $\tan 2\varphi$

6. $\cos \varphi = \frac{3}{5}$, $0° < \varphi < 90°$
   (a) $\sin 2\varphi$ (b) $\cos 2\varphi$ (c) $\tan 2\varphi$

7. $\tan u = -4$, $3\pi/2 < u < 2\pi$
   (a) $\sin 2u$ (b) $\cos 2u$ (c) $\tan 2u$

8. $\cot s = 2$, $\pi < s < 3\pi/2$
   (a) $\sin 2s$ (b) $\cos 2s$ (c) $\tan 2s$

9. $\sin \alpha = \sqrt{3}/2$, $0° < \alpha < 90°$
   (a) $\sin(\alpha/2)$ (b) $\cos(\alpha/2)$ (c) $\tan(\alpha/2)$

10. $\cos \beta = -\frac{1}{8}$, $180° < \beta < 270°$
    (a) $\sin(\beta/2)$ (b) $\cos(\beta/2)$ (c) $\tan(\beta/2)$

11. $\cos \theta = -\frac{7}{9}$, $\pi/2 < \theta < \pi$
    (a) $\sin(\theta/2)$ (b) $\cos(\theta/2)$ (c) $\tan(\theta/2)$

12. $\cos \theta = \frac{12}{13}$, $3\pi/2 < \theta < 2\pi$
    (a) $\sin(\theta/2)$ (b) $\cos(\theta/2)$ (c) $\tan(\theta/2)$

In Exercises 13–16, use the given information to compute each of the following: (a) $\sin 2\theta$; (b) $\cos 2\theta$; (c) $\sin(\theta/2)$; (d) $\cos(\theta/2)$.

13. $\sin \theta = \frac{3}{4}$ and $\pi/2 < \theta < \pi$

14. $\cos \theta = \frac{2}{5}$ and $3\pi/2 < \theta < 2\pi$

15. $\cos \theta = -\frac{1}{3}$ and $180° < \theta < 270°$

16. $\sin \theta = -\frac{1}{10}$ and $270° < \theta < 360°$

In Exercises 17–20, use an appropriate half-angle formula to evaluate each quantity.

17. (a) $\sin(\pi/12)$
    (b) $\cos(\pi/12)$
    (c) $\tan(\pi/12)$

18. (a) $\sin(\pi/8)$
    (b) $\cos(\pi/8)$
    (c) $\tan(\pi/8)$

19. (a) $\sin 105°$
    (b) $\cos 105°$
    (c) $\tan 105°$

20. (a) $\sin 165°$
    (b) $\cos 165°$
    (c) $\tan 165°$

In Exercises 21–28, refer to the two triangles and compute the quantities indicated.

21. (a) $\sin 2\theta$ (b) $\cos 2\theta$ (c) $\tan 2\theta$
22. (a) $\sin 2t$ (b) $\cos 2t$ (c) $\tan 2t$
23. (a) $\sin 2\beta$ (b) $\cos 2\beta$ (c) $\tan 2\beta$
24. (a) $\sin 2s$ (b) $\cos 2s$ (c) $\tan 2s$
25. (a) $\sin(\theta/2)$ (b) $\cos(\theta/2)$ (c) $\tan(\theta/2)$
26. (a) $\sin(s/2)$ (b) $\cos(s/2)$ (c) $\tan(s/2)$
27. (a) $\sin(\beta/2)$ (b) $\cos(\beta/2)$ (c) $\tan(\beta/2)$
28. (a) $\sin(t/2)$ (b) $\cos(t/2)$ (c) $\tan(t/2)$

In Exercises 29–32, use the given information to express $\sin 2\theta$ and $\cos 2\theta$ in terms of x.

29. $x = 5 \sin \theta$, $0 < \theta < \pi/2$
30. $x = \sqrt{2} \cos \theta$, $0 < \theta < \pi/2$
31. $x - 1 = 2 \sin \theta$, $0 < \theta < \pi/2$
32. $x + 1 = 3 \sin \theta$, $\pi/2 < \theta < \pi$

In Exercises 33–36, express each quantity in a form that does not involve powers of the trigonometric functions (as in Example 5).

33. $\sin^4 \theta$ 34. $\cos^6 \theta$

35. $\sin^4(\theta/2)$ 36. $\sin^6(\theta/4)$

37. Prove each of the following double-angle formulas. *Hint:* As in the text, replace 2θ with $\theta + \theta$, and use an appropriate addition formula.
    (a) $\cos 2\theta = \cos^2 \theta - \sin^2 \theta$
    (b) $\tan 2\theta = \dfrac{2 \tan \theta}{1 - \tan^2 \theta}$

38. (a) Beginning with the identity $\cos 2\theta = \cos^2 \theta - \sin^2 \theta$, prove that $\cos 2\theta = 1 - 2 \sin^2 \theta$.
    (b) Using the result in part (a), prove that $\sin^2 \theta = \frac{1}{2}(1 - \cos 2\theta)$.
    (c) Derive the formula for $\sin(s/2)$ as follows. Using the identity in part (b), replace θ with $s/2$, and then take square roots.

B

In Exercises 39–54, prove that the given equations are identities.

39. $\cos 2s = \dfrac{1 - \tan^2 s}{1 + \tan^2 s}$

40. $1 + \cos 2t = \cot t \sin 2t$

41. $\cos \theta = 2 \cos^2(\theta/2) - 1$

42. $\dfrac{\sin 2\theta}{\sin \theta} - \dfrac{\cos 2\theta}{\cos \theta} = \sec \theta$

43. $\sin^4 \theta = \dfrac{3 - 4 \cos 2\theta + \cos 4\theta}{8}$

44. $\sin 3\theta = 3 \sin \theta - 4 \sin^3 \theta$

45. $\sin 2\theta = \dfrac{2 \tan \theta}{1 + \tan^2 \theta}$

46. $2 \csc 2\theta = \dfrac{\csc^2 \theta}{\cot \theta}$

47. $\sin 2\theta = 2 \sin^3 \theta \cos \theta + 2 \sin \theta \cos^3 \theta$

48. $\cot \theta = \dfrac{1 + \cos 2\theta}{\sin 2\theta}$

49. $\dfrac{1 + \tan(\theta/2)}{1 - \tan(\theta/2)} = \tan \theta + \sec \theta$

50. $\tan \theta + \cot \theta = 2 \csc 2\theta$

51. $2 \sin^2(45° - \theta) = 1 - \sin 2\theta$

52. $(\sin \theta - \cos \theta)^2 = 1 - \sin 2\theta$

53. $1 + \tan \theta \tan 2\theta = \tan 2\theta \cot \theta - 1$

54. $\tan\left(\frac{\pi}{4} + \theta\right) - \tan\left(\frac{\pi}{4} - \theta\right) = 2 \tan 2\theta$

55. If $\cos(\alpha + \beta) = 0$, show that $\sin(\alpha + 2\beta) = \sin \alpha$.

56. If $\tan \alpha = \frac{1}{11}$ and $\tan \beta = \frac{5}{6}$, find $\alpha + \beta$, given that $0 < \alpha < \pi/2$ and $0 < \beta < \pi/2$. *Hint:* Compute $\tan(\alpha + \beta)$.

57. If $\sin \theta = \dfrac{a^2 - b^2}{a^2 + b^2}$, show that $\tan \dfrac{\theta}{2} = \dfrac{a - b}{a + b}$. Assume that $0 < \theta < \pi/2$ and a and b are positive.

58. Let $z = \tan \theta$. Show that

$$\cos 2\theta = \frac{1 - z^2}{1 + z^2} \quad \text{and} \quad \sin 2\theta = \frac{2z}{1 + z^2}$$

59. The following figure shows a semicircle with radius $AO = 1$.

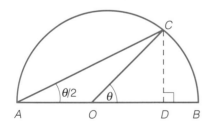

(a) Use the figure to derive the formula

$$\tan \frac{\theta}{2} = \frac{\sin \theta}{1 + \cos \theta} \quad \left(0 < \theta < \frac{\pi}{2}\right)$$

Hint: Show that $CD = \sin \theta$ and $OD = \cos \theta$. Then look at right triangle ADC to find $\tan(\theta/2)$.

(b) Use the formula developed in part (a) to show that

(i) $\tan 15° = \dfrac{1}{2 + \sqrt{3}} = 2 - \sqrt{3}$;

(ii) $\tan(\pi/8) = \sqrt{2} - 1$.

60. In this exercise, we'll use the accompanying figure to prove the following identities:

$$\cos 2\theta = 2 \cos^2 \theta - 1$$
$$\sin 2\theta = 2 \sin \theta \cos \theta$$
$$\cos 3\theta = 4 \cos^3 \theta - 3 \cos \theta$$
$$\sin 3\theta = 3 \sin \theta - 4 \sin^3 \theta$$

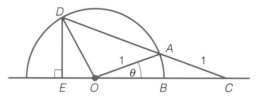

[The figure and technique in this exercise are adapted from the article by Wayne Dancer, "Geometric Proofs of Multiple Angle Formulas," in *American Mathematical Monthly,* vol. 44 (1937) pp. 366–367.] The figure is constructed as follows. We start with $\angle AOB = \theta$ in standard position in the unit circle, as shown. The point C is chosen on the extended diameter such that $CA = 1$. Then $\overline{CA}$ is extended to meet the circle at D, and radius $\overline{DO}$ is drawn. Finally, from D, a perpendicular is drawn to the diameter, as shown.

Supply the reason or reasons that justify each of the following statements.

(a) $\angle ACO = \theta$

(b) $\angle DAO = 2\theta$ *Hint:* $\angle DAO$ is an exterior angle to $\triangle AOC$.

(c) $\angle ODA = 2\theta$

(d) $\angle DOE = 3\theta$

(e) From O, draw a perpendicular to $\overline{CD}$, meeting $\overline{CD}$ at F. From A, draw a perpendicular to $\overline{OC}$, meeting $\overline{OC}$ at G. Then $GC = OG = \cos \theta$ and $FA = DF = \cos 2\theta$.

(f) $\cos \theta = \dfrac{1 + \cos 2\theta}{2 \cos \theta}$ *Hint:*   Use $\triangle CFO$.

(g) $\cos 2\theta = 2 \cos^2 \theta - 1$ *Hint:* In the equation in part (f), solve for $\cos 2\theta$.

(h) $OF = \sin 2\theta$

(i) $\sin 2\theta = 2 \sin \theta \cos \theta$ *Hint:* Find $\sin \theta$ in $\triangle CFO$, and then solve the resulting equation for $\sin 2\theta$.

(j) $DC = 1 + 2 \cos 2\theta$ and $EO = \cos 3\theta$

(k) $\cos \theta = \dfrac{2 \cos \theta + \cos 3\theta}{1 + 2 \cos 2\theta}$ *Hint:* Compute $\cos \theta$ in $\triangle CDE$ and then use part (j).

(l) $\cos 3\theta = 4 \cos^3 \theta - 3 \cos \theta$ *Hint:*   Use the results in parts (k) and (g).

(m) $DE = \sin 3\theta$

(n) $\sin 3\theta = 3 \sin \theta - 4 \sin^3 \theta$ *Hint:*   Compute $\sin \theta$ in $\triangle CDE$.

61. Prove the following identities involving products of cosines. *Suggestion:* In each case, begin with the right-hand side and use the double-angle formula for the sine.

(a) $\cos \theta \cos 2\theta = \dfrac{\sin 4\theta}{4 \sin \theta}$

(b) $\cos \theta \cos 2\theta \cos 4\theta = \dfrac{\sin 8\theta}{8 \sin \theta}$

(c) $\cos \theta \cos 2\theta \cos 4\theta \cos 8\theta = \dfrac{\sin 16\theta}{16 \sin \theta}$

62. Prove the following three formulas, known as the product-to-sum formulas.

(a) $\sin A \sin B = \frac{1}{2}[\cos(A - B) - \cos(A + B)]$

(b) $\sin A \cos B = \frac{1}{2}[\sin(A + B) + \sin(A - B)]$

(c) $\cos A \cos B = \frac{1}{2}[\cos(A + B) + \cos(A - B)]$

63. Express each product as a sum or difference of sines or cosines.

(a) $4 \sin 6\theta \cos 2\theta$ (b) $2 \sin A \sin 3A$

(c) $\cos 4\theta \cos 2\theta$

64. (a) Show that $\sin(\pi/4) \cos(\pi/12) = \frac{1}{4}(\sqrt{3} + 1)$.

(b) Evaluate $2(\sin 82.5°)(\cos 37.5°)$ without using a calculator.

65. Prove the following identities.

(a) $\sin(A + B) \sin(A - B) = \sin^2 A - \sin^2 B$

(b) $\cos(A + B) \cos(A - B) = \cos^2 A - \sin^2 B$

66. In the formula (from Exercise 62)

$$\sin A \cos B = \frac{1}{2}[\sin(A + B) + \sin(A - B)]$$

make the substitutions $A + B = \alpha$ and $A - B = \beta$ and deduce that

$$\sin \alpha + \sin \beta = 2 \sin \frac{\alpha + \beta}{2} \cos \frac{\alpha - \beta}{2}$$

This is one of the sum-to-product formulas mentioned in the text.

67. Prove the following sum-to-product identities. *Hint:* See the previous exercise.

(a) $\sin \alpha - \sin \beta = 2 \cos \dfrac{\alpha + \beta}{2} \sin \dfrac{\alpha - \beta}{2}$

(b) $\cos \alpha + \cos \beta = 2 \cos \dfrac{\alpha + \beta}{2} \cos \dfrac{\alpha - \beta}{2}$

(c) $\cos \alpha - \cos \beta = -2 \sin \dfrac{\alpha + \beta}{2} \sin \dfrac{\alpha - \beta}{2}$

68. (a) Show that $\cos \frac{2\pi}{9} + \cos \frac{\pi}{9} = \sqrt{3} \cos \frac{\pi}{18}$.

(b) Show that $\sin 105° + \sin 15° = \sqrt{6}/2$.

69. Prove that $\dfrac{\sin \theta + \sin 3\theta}{\cos \theta + \cos 3\theta} = \tan 2\theta$.   *Hint:*   Use the sum-to-product formulas.

70. Prove that $\dfrac{\sin 7\theta - \sin 5\theta}{\cos 7\theta + \cos 5\theta} = \tan \theta$.

71. Prove that $\tan(x + y) = \dfrac{\sin 2x + \sin 2y}{\cos 2x + \cos 2y}$.

72. (a) Use your calculator to evaluate the expression $\cos 72° \cos 144°$. Then follow steps (b) through (d) to *prove* that $\cos 72° \cos 144° = -\frac{1}{4}$.

(b) Multiply the expression $\cos 72° \cos 144°$ by the quantity $(\sin 72°)/(\sin 72°)$, which equals 1. Show that the result can be written

$$\frac{\frac{1}{2} \sin 144° \cos 144°}{\sin 72°}$$

(c) Explain why the expression obtained in part (b) is equal to

$$\frac{\frac{1}{4} \sin 288°}{\sin 72°}$$

(d) Use the reference-angle concept to explain why the expression in part (c) is equal to $-\frac{1}{4}$, as required.

73. (a) Use your calculator to evaluate the expression $\cos 72° + \cos 144°$. Then follow steps (b) through (d) to *prove* that $\cos 72° + \cos 144° = -\frac{1}{2}$.

(b) Use the observation

$$\cos 72° + \cos 144° = \cos(108° - 36°) + \cos(108° + 36°)$$

and the addition formulas for cosine to show that

$$\cos 72° + \cos 144° = 2 \cos 108° \cos 36°$$

(c) Use the reference-angle concept to explain why $\cos 108° \cos 36° = \cos 72° \cos 144°$.

(d) Use parts (b) and (c) and the identity in Exercise 72(a) to conclude that $\cos 72° + \cos 144° = -\frac{1}{2}$, as required.

74. In the figure, the points A_1, A_2, A_3, A_4, and A_5 are the vertices of a regular pentagon. Follow steps (a) through (c) to show that

$$(A_1A_2)(A_1A_3)(A_1A_4)(A_1A_5) = 5$$

[This is a particular case of a general result due to Roger Cotes (1682–1716): Suppose that a regular n-gon is inscribed in the unit circle. Let the vertices of the n-gon be A_1, A_2, A_3, . . . , A_n. Then the product $(A_1A_2)(A_1A_3)$. . .(A_1A_n) is equal to n, the number of sides of the polygon.

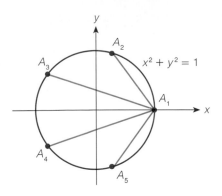

(a) What are the coordinates of the points A_2, A_3, A_4, and A_5? (Give your answers in terms of sines and cosines.)
(b) Show that $(A_1A_2)(A_1A_3)(A_1A_4)(A_1A_5) =$ $(2 - 2\cos 72°)(2 - 2\cos 144°)$.
(c) Show that the expression on the right-hand side of the equation in part (b) is equal to 5.   *Hint:*   Use the equations given in Exercises 72(a) and 73(a).

For Exercises 75 and 76, refer to the following figures. Figure A shows an equilateral triangle inscribed in the unit circle. Figure B shows a regular pentagon inscribed in the unit circle. In both figures, the coordinates of the point P are $(x, 0)$.

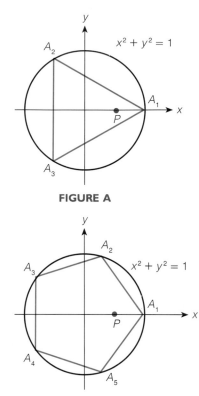

FIGURE A

FIGURE B

75.   For Figure A, show that $(PA_1)(PA_2)(PA_3) = 1 - x^3$.

76.   For Figure B, show that
$(PA_1)(PA_2)(PA_3)(PA_4)(PA_5) = 1 - x^5$.

Hints: First find the coordinates of A_2 and A_3 in terms of sines and cosines. Next, show that
$(PA_1)(PA_2)(PA_3)(PA_4)(PA_5) =$
$(1 - x)[(1 - 2x\cos 72° + x^2)(1 - 2x\cos 144° + x^2)]$.
Then expand the expression within the brackets and simplify using the equations in Exercises 72(a) and 73(a).

Remark: The results in Exercises 75 and 76 are particular cases of the following theorem of Roger Cotes: Suppose that a regular *n*-gon $A_1A_2A_3 . . . A_n$ is inscribed in the unit circle. Suppose that A_1 is the point $(1, 0)$ and that P is a point with coordinates $(x, 0)$, where $0 \leq x \leq 1$. Then the product $(PA_1)(PA_2)(PA_3) . . . (PA_n)$ is equal to $1 - x^n$.

7.8   **TRIGONOMETRIC EQUATIONS**

In this section, we consider some techniques for solving equations involving the trigonometric functions. As usual, by a **solution** of an equation, we mean a value of the variable for which the equation becomes a true statement.

■▮▮▮ **EXAMPLE 1**   Consider the trigonometric equation $\sin x + \cos x = 1$. Is $x = \pi/4$ a solution? Is $x = \pi/2$ a solution?

Solution To see if the value $x = \pi/4$ satisfies the given equation, we write

$$\sin \frac{\pi}{4} + \cos \frac{\pi}{4} \overset{?}{=} 1$$

$$\frac{\sqrt{2}}{2} + \frac{\sqrt{2}}{2} \overset{?}{=} 1$$

$$\sqrt{2} \overset{?}{=} 1 \qquad \text{No!}$$

Thus, $x = \pi/4$ is not a solution. In a similar fashion, we can check to see if $x = \pi/2$ is a solution:

$$\sin \frac{\pi}{2} + \cos \frac{\pi}{2} \overset{?}{=} 1$$

$$1 + 0 \overset{?}{=} 1 \qquad \text{Yes!}$$

Thus, $x = \pi/2$ is a solution. ▮▮▮

The example that we've just concluded serves to remind us of the difference between a *conditional equation* and an *identity*. An identity is true for all values of the variable in its domain. For example, the equation $\sin^2 t + \cos^2 t = 1$ is an identity: it is true for every real number t. In contrast to this, a conditional equation is true only for some (or perhaps even none) of the values of the variable. The equation in Example 1 is a conditional equation; we saw that it is false when $x = \pi/4$ and true when $x = \pi/2$. The equation $\sin t = 2$ is an example of a conditional equation that has no solution. (Why?) The equations that we are going to solve in this section are conditional equations that involve the trigonometric functions. In general, there is no single technique that can be used to solve every trigonometric equation. In the examples that follow, we illustrate some of the more common approaches to solving trigonometric equations.

▮▮▮▮ **EXAMPLE 2** Solve the equation $\sin \theta = \frac{1}{2}$.

Solution First of all, we know that $\sin(\pi/6) = \frac{1}{2}$. Thus, one solution is certainly $\theta = \pi/6$. To find another solution, we note that $\sin \theta$ is positive in the second quadrant, as well as in the first. In the second quadrant, the angle with a reference angle of $\pi/6$ is $5\pi/6$ (150°). Thus, $\sin(5\pi/6) = \frac{1}{2}$, and we conclude that $\theta = 5\pi/6$ is also a solution to the given equation. (See Figure 1.)

Since $\sin \theta$ is negative in the third and fourth quadrants, we needn't look there for solutions. Nevertheless, there are other solutions, because the sine function is periodic, with period 2π. Therefore, since $\theta = \pi/6$ is a solution, we know that $\theta = (\pi/6) + 2\pi$ is another solution. So is $\theta = (\pi/6) + 4\pi$ a solution, and so on. Similarly, adding multiples of 2π to $\theta = 5\pi/6$ produces additional solutions. In general, the solutions of the equation $\sin \theta = \frac{1}{2}$ are given by

$$\frac{\pi}{6} + 2\pi k \qquad \text{and} \qquad \frac{5\pi}{6} + 2\pi k \qquad \text{where } k \text{ is any integer}$$

Figure 2 shows some of these solutions; they are the x-coordinates of the points where the sine curve intersects the line $y = \frac{1}{2}$. If required, we can express the solutions using degree measure. In that case, the solutions are given by

$$30° + 360k° \qquad \text{and} \qquad 150° + 360k° \qquad \text{where } k \text{ is any integer} \qquad ▮▮▮$$

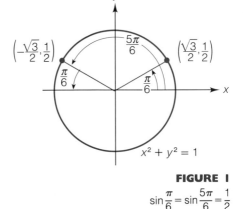

FIGURE I

$$\sin \frac{\pi}{6} = \sin \frac{5\pi}{6} = \frac{1}{2}$$

FIGURE 2
There are infinitely many real numbers x
for which $\sin x = \frac{1}{2}$.

Conceptually, the next example is similar to the one we just completed, but now we'll need to use a calculator. In particular, we will be using the calculator to find a real number x such that $\cos x = 0.351$. After the example is completed, we'll indicate some of the theoretical points that are involved.

EXAMPLE 3 Find all solutions of the equation

$$\cos x = 0.351$$

in the open interval $(0, 2\pi)$. Round off the answers to three decimal places.

Solution As indicated in Figure 3, there are two numbers x_1 and x_2 in the interval $(0, 2\pi)$ that satisfy the equation $\cos x = 0.351$. We can determine x_1 using a calculator set in the radian mode. With some types of Casio calculators, for example, the keystrokes are

0.351 (INV) (COS)

FIGURE 3

Or, for example, with a Texas Instruments TI-81 calculator, we have

(2nd) (COS) 0.351 (ENTER)

In any case, as you should check for yourself, the result (to three decimal places) is

$$x_1 = 1.212$$

For the value of x_2, we can reason as follows. From the graph in Figure 3, we see that x_2 lies in the interval $3\pi/2 < x < 2\pi$. Thinking in terms of angles, x_2 is the radian measure of an angle with terminal side in quadrant IV. Furthermore, x_1 is the measure of the reference angle for x_2. (How do we know, without relying on a calculator, that $x_1 < \pi/2$?) Therefore we have

$$x_2 = 2\pi - x_1 = 5.071 \qquad \text{using a calculator and rounding off}$$

In summary now, the two solutions in the interval $(0, 2\pi)$ are 1.212 and 5.071.

■■■

In the example just concluded, we used a calculator to find a number having a cosine of 0.351. The function that outputs such a number is called the **inverse cosine function**. There are two standard abbreviations for the name of this function, and a third abbreviation that is used on certain brands of calculators. The two standard abbreviations are $\cos^{-1}$ and arccos (read *arc cos*). And, as you saw in Example 3, the notation **INV COS** is used on some calculators. For now, we'll use the notation $\cos^{-1}$.

In the next section, we will analyze the inverse cosine function in some detail. But for our present purposes, you need only know the following definition:

$\cos^{-1}(x)$ denotes the number in the interval $[0, \pi]$ whose cosine is x.

As examples of this notation, we have

$$\cos^{-1}\left(\tfrac{1}{2}\right) = \tfrac{\pi}{3} \qquad \text{because } \cos\tfrac{\pi}{3} = \tfrac{1}{2} \text{ and } 0 < \tfrac{\pi}{3} < \pi$$

$$\cos^{-1}\left(\tfrac{1}{2}\right) \neq \tfrac{5\pi}{3} \qquad \text{because although } \cos\tfrac{5\pi}{3} = \tfrac{1}{2}, \text{ the number } \tfrac{5\pi}{3}$$
$$\text{is not in the required interval } [0, \pi]$$

$$\cos^{-1}\left(-\tfrac{1}{2}\right) = \tfrac{2\pi}{3} \qquad \text{because } \cos\tfrac{2\pi}{3} = -\tfrac{1}{2} \text{ and } 0 < \tfrac{2\pi}{3} < \pi$$

$$\cos^{-1}(0.351) \approx 1.212 \qquad \text{as we saw in Example 3}$$

In Example 3, Figure 3 tells us that the number x_1 is in the interval $[0, \pi]$ but x_2 is not. That is why the sequence of keystrokes

$$0.351 \quad \boxed{\text{COS}^{-1}}$$

gave us the value for x_1 rather than x_2. (We then used the reference angle or reference number concept to obtain x_2.)

For some of the exercises in this section, you will need to use the *inverse sine function* or the *inverse tangent function,* rather than the inverse cosine function that we have just been discussing. For reference, we define the three functions in the box that follows. (All three functions are discussed at greater length in the next section.)

DEFINITION Inverse Trigonometric Functions		
NAME OF FUNCTION	ABBREVIATION	DEFINITION
Inverse cosine	$\cos^{-1}$	$\cos^{-1}(x)$ is the (unique) number in the interval $[0, \pi]$ whose cosine is x.
Inverse sine	$\sin^{-1}$	$\sin^{-1}(x)$ is the (unique) number in the interval $[-\pi/2, \pi/2]$ whose sine is x.
Inverse tangent	$\tan^{-1}$	$\tan^{-1}(x)$ is the (unique) number in the interval $(-\pi/2, \pi/2)$ whose tangent is x.

The next example shows how factoring can be used to solve a trigonometric equation.

EXAMPLE 4   Solve the equation $\cos^2 x + \cos x - 2 = 0$.

Solution By factoring the expression on the left-hand side of the equation, we obtain

$$(\cos x + 2)(\cos x - 1) = 0$$

Therefore,

$$\cos x = -2 \qquad \text{or} \qquad \cos x = 1$$

We discard the result $\cos x = -2$, since the value of $\cos x$ is never less than -1. From the equation $\cos x = 1$, we conclude that $x = 0$ is one solution. After $x = 0$, the next time we have $\cos x = 1$ is when $x = 2\pi$. (You can see this by looking at the graph of $y = \cos x$ or by considering the unit circle definition of the cosine.) In general, then, the solutions of the equation $\cos^2 x + \cos x - 2 = 0$ are given by $x = 2\pi k$, where k is an integer. ▮▮▮

In some equations, more than one trigonometric function is present. A common approach here is to express the various functions in terms of a single one. The next example demonstrates this technique.

▮▬▮ **EXAMPLE 5** Find all solutions of the equation $3 \tan^2 x - \sec^2 x - 5 = 0$.

Solution We use the Pythagorean identity $\tan^2 x + 1 = \sec^2 x$ to substitute for $\sec^2 x$ in the given equation. This gives us

$$3 \tan^2 x - (\tan^2 x + 1) - 5 = 0$$
$$2 \tan^2 x - 6 = 0$$
$$2 \tan^2 x = 6$$
$$\tan^2 x = 3$$
$$\tan x = \pm\sqrt{3}$$

Since the period of the tangent function is π, we need only find values of x between 0 and π that satisfy $\tan x = \pm\sqrt{3}$. The other solutions will then be obtained by adding multiples of π to these solutions. Now, in the first quadrant, the tangent is positive and we know that $\tan x = \sqrt{3}$ when $x = \pi/3$. In the second quadrant, $\tan x$ is negative and we know that $\tan 2\pi/3 = -\sqrt{3}$, since the reference angle for $2\pi/3$ is $\pi/3$. Thus, the solutions between 0 and π are $x = \pi/3$ and $x = 2\pi/3$. It follows that all of the solutions to the equation $3 \tan^2 x - \sec^2 x - 5 = 0$ are given by

$$x = \frac{\pi}{3} + \pi k \qquad \text{and} \qquad x = \frac{2\pi}{3} + \pi k \qquad \text{where } k \text{ is an integer.} \quad ▮▮▮$$

The technique used in Example 5, of expressing the various functions in terms of a single function, is most useful when it does not involve introducing a radical expression. For instance, consider the equation $\sin s + \cos s = 1$. Although we could begin by replacing $\cos s$ by the expression $\pm\sqrt{1 - \sin^2 s}$, it turns out to be easier in this situation to begin by squaring both sides of the given equation. This is done in the next example.

▮▬▮ **EXAMPLE 6** Find all solutions of the equation $\sin s + \cos s = 1$ satisfying $0° \leq s < 360°$.

Solution   Squaring both sides of the equation yields

$$(\sin s + \cos s)^2 = 1^2$$

$$\underset{\underbrace{\qquad\qquad}_{\text{These add to 1}}}{\sin^2 s + 2 \sin s \cos s + \cos^2 s} = 1$$

Consequently, we have

$$2 \sin s \cos s = 0$$

From this last equation, we conclude that $\sin s = 0$ or $\cos s = 0$. When $\sin s = 0$, we know that $s = 0°$ or $s = 180°$. And when $\cos s = 0$, we know that $s = 90°$ or $s = 270°$. Now we must go back and check which (if any) of these values is a solution to the *original* equation. This must be done whenever we square both sides in the process of solving an equation.

$s = 0°$: $\sin 0° + \cos 0° \overset{?}{=} 1$

                      $0 + 1 \overset{?}{=} 1$ True

$s = 90°$: $\sin 90° + \cos 90° \overset{?}{=} 1$

                      $1 + 0 \overset{?}{=} 1$ True

$s = 180°$: $\sin 180° + \cos 180° \overset{?}{=} 1$

                      $0 + (-1) \overset{?}{=} 1$ False

$s = 270°$: $\sin 270° + \cos 270° \overset{?}{=} 1$

                      $-1 + 0 \overset{?}{=} 1$ False

We conclude that the only solutions of the equation $\sin s + \cos s = 1$ on the interval $0° \le s < 360°$ are $s = 0°$ and $s = 90°$.   ∎

In the example that follows, we consider an equation that involves a multiple of the unknown angle.

EXAMPLE 7   Solve the equation $\sin 3x = 1$ on the interval $0 \le x \le 2\pi$.

Solution   We know that $\sin(\pi/2) = 1$. Thus, one solution can be found by writing $3x = \pi/2$, from which we conclude that $x = \pi/6$. We can look for other solutions in the required interval by writing, more generally,

$$3x = \tfrac{\pi}{2} + 2\pi k$$

from which it follows that

$$x = \tfrac{\pi}{6} + \tfrac{2\pi k}{3}$$

Thus, when $k = 1$, we obtain

$$x = \tfrac{\pi}{6} + \tfrac{2\pi}{3} = \tfrac{\pi}{6} + \tfrac{4\pi}{6} = \tfrac{5\pi}{6}$$

When $k = 2$, we obtain

$$x = \tfrac{\pi}{6} + \tfrac{2\pi(2)}{3} = \tfrac{\pi}{6} + \tfrac{8\pi}{6} = \tfrac{3\pi}{2}$$

When $k = 3$, we obtain

$$x = \tfrac{\pi}{6} + \tfrac{2\pi(3)}{3} = \tfrac{\pi}{6} + 2\pi \qquad \text{which is greater than } 2\pi$$

We conclude that the solutions of $\sin 3x = 1$ on the interval $0 \leq x \leq 2\pi$ are $\pi/6$, $5\pi/6$, and $3\pi/2$. ▮▮▮

In Example 7, notice that we did not need to make use of a formula for $\sin 3x$, even though the expression $\sin 3x$ did appear in the given equation. In the next example, however, we do make use of the identity $\sin 2x = 2 \sin x \cos x$.

▮▮▮ **EXAMPLE 8** Solve the equation $\sin x \cos x = 1$.

Solution We could begin by squaring both sides. (Exercise 55 will ask you to use this approach.) However, with the double-angle formula for sine in mind, we can proceed instead as follows. We multiply both sides of the given equation by 2. This yields

$$2 \sin x \cos x = 2$$

and, consequently,

$$\sin 2x = 2 \qquad \text{using the double-angle formula}$$

This last equation has no solution, since the value of the sine function never exceeds 1. Thus, the equation $\sin x \cos x = 1$ has no solution. ▮▮▮

For the last example in this section, we look at another case in which tables or a calculator are required.

▮▮▮ **EXAMPLE 9** Find all angles θ between $0°$ and $360°$ satisfying the equation $\sin \theta = 2 \cos \theta$.

Solution We first want to rewrite the given equation using a single function. The easiest way to do this is to divide both sides by $\cos \theta$. (Nothing is lost here in assuming that $\cos \theta \neq 0$. If $\cos \theta$ were 0, then θ would be $90°$ or $270°$; but neither of those angles is a solution of the given equation.) Dividing through by $\cos \theta$, we obtain

$$\frac{\sin \theta}{\cos \theta} = \frac{2 \cos \theta}{\cos \theta} \qquad \text{or} \qquad \tan \theta = 2$$

From experience, we know that none of the angles with which we are familiar (the multiples of $30°$ and $45°$) has a value of 2 for their tangent. Thus, a calculator or table is required at this point. Using a Casio calculator (set in the degree mode), we find an angle whose tangent is 2 by using the following keystrokes:

$$2 \;\; \boxed{\text{TAN}^{-1}}$$

By so doing, we find that

$$\theta \approx 63.4°$$

(Even if we were to retain all the decimals displayed by the calculator, we would still have only an approximate solution.) To find another value of θ, we note that the tangent is also positive in the third quadrant. Thus, a second (approximate) solution is given by

$$\theta = 180° + 63.4° = 243.4°$$

We conclude that the solutions between $0°$ and $360°$ to the equation $\sin \theta = 2 \cos \theta$ are approximately $63.4°$ and $243.4°$. Note that the calculator provided only one value for θ; we still needed to work out the second value ourselves using the reference-angle concept. ∎

EXERCISE SET 7.8

A

1. Is $\theta = \pi/2$ a solution of the following equation?

   $$2 \cos^2 \theta - 3 \cos \theta = 0$$

2. Is $x = 15°$ a solution of $\left(\sqrt{3}/3\right)\cos 2x + \sin 2x = 1$?
3. Is $x = 3\pi/4$ a solution of $\tan^2 x - 3 \tan x + 2 = 0$?
4. Is $t = 2\pi/3$ a solution of $2 \sin t + 2 \cos t = \sqrt{3} - 1$?

In Exercises 5–22, determine all solutions of the given equations. Express your answers using radian measure.

5. $\sin \theta = \sqrt{3}/2$
6. $\sin \theta = \sqrt{2}/2$
7. $\sin \theta = -\frac{1}{2}$
8. $\sin \theta + \left(\sqrt{2}/2\right) = 0$
9. $\cos \theta = -1$
10. $\cos \theta = \frac{1}{2}$
11. $\tan \theta = \sqrt{3}$
12. $\tan \theta + \left(\sqrt{3}/3\right) = 0$
13. $\tan x = 0$
14. $2 \sin^2 x - 3 \sin x + 1 = 0$
15. $2 \cos^2 \theta + \cos \theta = 0$
16. $\sin^2 x - \sin x - 6 = 0$
17. $\cos^2 t \sin t - \sin t = 0$
18. $\cos \theta + 2 \sec \theta = -3$
19. $2 \cos^2 x - \sin x - 1 = 0$
20. $2 \cot^2 x + \csc^2 x - 2 = 0$
21. $\sqrt{3} \sin t - \sqrt{1 + \sin^2 t} = 0$
22. $\sec \alpha + \tan \alpha = \sqrt{3}$

In Exercises 23–30, determine all of the solutions in the interval $0° \leq \theta < 360°$.

23. $\cos 3\theta = 1$
24. $\tan 2\theta = -1$
25. $\sin 3\theta = -\sqrt{2}/2$
26. $\sin(\theta/2) = \frac{1}{2}$
27. $\sin \theta = \cos(\theta/2)$ *Hint:* $\sin \theta = 2 \sin(\theta/2) \cos(\theta/2)$
28. $2 \sin^2 \theta - \cos 2\theta = 0$
29. $\sin 2\theta = \sqrt{3} \cos 2\theta$ *Hint:* Divide by $\cos 2\theta$.
30. $\sin 2\theta = -2 \cos \theta$

In Exercises 31–36, use a calculator where necessary to find all solutions in the interval $0° \leq \theta < 360°$.

31. $\sin \theta = \frac{1}{4}$
32. $3 \cos \theta = 5$
33. $2 \tan \theta = -4$
34. $3 \sin^2 \theta - 2 \sin \theta - 1 = 0$
35. $\cos^2 x - \cos x - 1 = 0$
36. $2 \sin x = 3$

In Exercises 37–46, use a calculator to find all solutions in the interval $(0, 2\pi)$. Round off the answers to two decimal places.

37. $\cos x = 0.184$
38. $\cos t = -0.567$
39. $\sin x = 1/\sqrt{5}$
40. $\sin t = -0.301$
41. $\tan x = 6$
42. $\tan t = -5.25$
43. $\sin t = 5 \cos t$
44. $\sin x \cos x = 0.035$
45. $\sec t = 2.24$ *Hint:* The equation is equivalent to $\cos t = 1/(2.24)$.
46. $\cot x = -3.27$

B

47. Find all solutions of the equation $\tan 3x - \tan x = 0$ in the interval $0 \leq x < 2\pi$.   *Hint:* Write $\tan 3x = \tan(2x + x)$ and use the addition formula for tangent.
48. Find all solutions of the equation $2 \sin x = 1 - \cos x$ in the interval $0° \leq x < 360°$. Use a calculator and round off the answer(s) to one decimal place.
49. Find all solutions of the equation $\cos(x/2) = 1 + \cos x$ in the interval $0 \leq x < 2\pi$.
50. Find all solutions of the equation

    $$\sin 3x \cos x + \cos 3x \sin x = \sqrt{3}/2$$

    in the interval $0 < x < 2\pi$.
51. Find all real numbers θ for which $\sec 4\theta + 2 \sin 4\theta = 0$.
52. Consider the equation $\sin^2 x - \cos^2 x = \frac{7}{25}$.
    (a) Solve the equation for $\cos x$.

(b) Find all of the solutions of the equation satisfying $0 < x < \pi$.

(c) Solve the original equation by means of a double-angle formula and use the result to check your answers in part (b).

53. Find a solution of the equation $4 \sin \theta - 3 \cos \theta = 2$ in the interval $0° < \theta < 90°$. *Hint:* Add $3 \cos \theta$ to both sides, and then square.

54. Find all solutions of the equation

$$\sin^3 \theta \cos \theta - \sin \theta \cos^3 \theta = -\tfrac{1}{4}$$

in the interval $0 < \theta < \pi$. *Hint:* Factor the left-hand side, and then use the double-angle formulas.

55. Consider the equation $\sin x \cos x = 1$.

(a) Square both sides, and then replace $\cos^2 x$ by $1 - \sin^2 x$. Show that the resulting equation can be written $\sin^4 x - \sin^2 x + 1 = 0$.

(b) Show that the equation $\sin^4 x - \sin^2 x + 1 = 0$ has no solutions. Conclude from this that the original equation has no real-number solutions.

56. The accompanying figure shows a portion of the graph of the periodic function

$$f(x) = \sin(1 + \sin x)$$

Find the x-coordinates of the turning points P, Q, and R. Round off the answers to three decimal places. *Hint:* For Q, the x-coordinate is halfway between the x-coordinates of P and R.

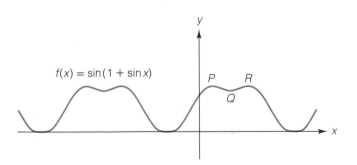

57. (a) Find the smallest solution of $\cos x = 0.412$ in the interval $(1000, \infty)$. (Round off the answer to three decimal places.)

(b) Find the smallest solution of $\cos x = -0.412$ in the interval $(1000, \infty)$.

C

58. In this exercise, you will see how certain cubic equations can be solved by using the following identity (which we proved in Example 4 in Section 7.7):

$$4 \cos^3 \theta - 3 \cos \theta = \cos 3\theta \qquad (1)$$

For example, suppose that we wish to solve the equation

$$8x^3 - 6x - 1 = 0 \qquad (2)$$

To transform this equation into a form where the stated identity is useful, we make the substitution $x = a \cos \theta$, where a is a constant to be determined. With this substitution, equation (2) can be written

$$8a^3 \cos^3 \theta - 6a \cos \theta = 1 \qquad (3)$$

In equation (3), the coefficient of $\cos^3 \theta$ is $8a^3$. Since we want this coefficient to be 4 [as it is in equation (1)], we divide both sides of equation (3) by $2a^3$ to obtain

$$4 \cos^3 \theta - \frac{3}{a^2} \cos \theta = \frac{1}{2a^3} \qquad (4)$$

Next, a comparison of equations (4) and (1) leads us to require that $3/a^2 = 3$. Thus, $a = \pm 1$. For convenience, we choose $a = 1$; equation (4) then becomes

$$4 \cos^3 \theta - 3 \cos \theta = \tfrac{1}{2} \qquad (5)$$

Comparing equation (5) with the identity in (1) leads us to the equation

$$\cos 3\theta = \tfrac{1}{2}$$

As you can check, the solutions here are of the form

$$\theta = 20° + 120k° \qquad \text{and} \qquad \theta = 100° + 120k°$$

Thus,

$$x = \cos(20° + 120k°) \qquad \text{and} \qquad x = \cos(100° + 120k°)$$

Now, however, as you can again check, only three of the angles yield distinct values for $\cos \theta$, namely, $\theta = 20°$, $\theta = 140°$, and $\theta = 260°$. Thus, the solutions of the equation $8x^3 - 6x - 1 = 0$ are given by $x = \cos 20°$, $x = \cos 140°$, and $x = \cos 260°$.

Use the method just described to solve the following equations.

(a) $x^3 - 3x + 1 = 0$
    Answers: $2 \cos 40°$, $-2 \cos 20°$, $2 \cos 80°$

(b) $x^3 - 36x - 72 = 0$

(c) $x^3 - 6x + 4 = 0$ *Answers:* 2, $-1 \pm \sqrt{3}$

(d) $x^3 - 7x - 7 = 0$ (Round off your answers to three decimal places.)

OPTIONAL TI-81 GRAPHING CALCULATOR EXERCISES
EXERCISES FOR SECTION 7.8

1. In Example 3 in Section 7.8, we solved the equation $\cos x = 0.351$ on the interval $(0, 2\pi)$. This exercise shows how to obtain the solutions graphically.
   (a) With the calculator in the radian mode, enter the functions $Y_1 = \cos x$ and $Y_2 = 0.351$. Press ZOOM 7 to obtain the graphs with the x-axis marked off in units of $\pi/2$. Now, to focus on the interval from 0 to 2π (as in Example 3), press the RANGE key and set $x_{min} = 0$. Then press the GRAPH key.
   (b) As you can see in the calculator display, there are two points where the line $Y_2 = 0.351$ meets the curve $Y_1 = \cos x$. The x-coordinates of these points are solutions of the equation $\cos x = 0.351$. In parts (c) through (e), we explain how to obtain these coordinates. But first, just to obtain a better overview of the situation, press the RANGE key and set $y_{min} = -1$ and $y_{max} = 1$. Now press the GRAPH key.
   (c) The graph in part (b) shows that there are two solutions to the equation $\cos x = 0.351$ in the interval $(0, 2\pi)$. Evidently, one solution is a little less than $\pi/2$ and the other is a little more than $3\pi/2$. We can obtain these solutions using the TRACE key and the ZOOM key. (We'll show how to look for the smaller of the two solutions; the other solution is obtained in a similar fashion.) Press the TRACE key. A blinking cursor will appear on the screen. By repeatedly pressing the left ARROW key, move the cursor to the left (along the curve $Y_1 = \cos x$) until the cursor is positioned at the point where the curve $Y_1 = \cos x$ intersects the line $Y_2 = 0.351$. (If you overshoot the mark, use the right ARROW key to back up.) As you can see from the read-out at the bottom of the screen, the x-coordinate of this point is approximately 1.19. This is a first approximation to the solution.
   (d) To obtain a closer approximation, press ZOOM 2, and ENTER. This provides a magnified view of the region around the intersection point. Now press the TRACE key, and then use the arrow keys to move the cursor onto the apparent intersection point. From the read-out at the bottom of the screen, we have $x \approx 1.215$. This is our second approximation to the solution.
   (e) By repeating the process in part (d), you can obtain closer and closer approximations to the

   required solution. Repeat the process until the first three decimal places of the approximation remain the same as you progress to the next step. You should obtain $x \approx 1.212$, in agreement with the result in Example 3.

2. In Example 9 in Section 7.8, we solved the equation $\sin \theta = 2 \cos \theta$ on the interval $0° \le \theta \le 360°$. This exercise shows how to obtain the solutions graphically.
   (a) With the calculator set in the degree mode, enter the functions $Y_1 = \sin x$ and $Y_2 = 2 \cos x$. Next, press the RANGE key and set $x_{min} = 0$, $x_{max} = 360$, $x_{scl} = 90$, $y_{min} = -2$, and $y_{max} = 2$. Now press the GRAPH key. Note that there are two intersection points, one in which x is less than $90°$ and one in which x is between $180°$ and $270°$.
   (b) Use the TRACE and ZOOM technique (explained in Exercise 1) to show that the two solutions are approximately $63.4°$ and $243.4°$. (These are the two values that were obtained in Example 9.)

In Exercises 3–10, solve the equations on the interval $[0, 2\pi]$ using the TRACE and ZOOM technique described in Exercise 1. In each case, carry out the process far enough so that you are certain of the first two decimal places in each answer.

3. $\cos x = 0.623$
4. $\sin x = -0.438$
5. $\cos^2 x = 2 \sin x$
6. $\cos^3 x = 2 \sin x$
7. (a) $\cos 2x + 1 = \cos(2x + 1)$
   (b) $\cos 2x + 0.9 = \cos(2x + 1)$
8. $2 \sin x - 3 \cos x = \tan(x/4)$
9. $\sqrt{x} = \tan x$
10. (a) $\sin(\cos x) = \sin x$
    (b) $\cos(\sin x) = \sin x$

In Exercises 11–16, solve the equations on the interval $0° \le \theta \le 360°$ using the TRACE and ZOOM technique described in Exercises 1 and 2. In each case, carry out the process far enough so that you are certain of the first decimal place in each answer.

11. $\sin \theta = \cos(45° - \theta)$
12. $\sin(\theta/2) = \cos(\theta/3)$
13. $\sin^3 \theta + \cos^3 \theta = \frac{1}{2}$
14. $\sin^3 \theta - \cos^3 \theta = \frac{1}{2}$
15. $\tan \theta = \tan \sqrt{\theta}$
16. (a) $1 - \tan^2 \theta = 2 \sin(\theta/5)$
    (b) $1 - \tan^2 \theta = 2 \sin(\theta/6)$

7.9   **THE INVERSE TRIGONOMETRIC FUNCTIONS**

The notation $\cos^{-1} \theta$ *must not be understood to signify* $\dfrac{1}{\cos\,\theta}$.

John Herschel in *Philosophical Transactions of London,* 1813

In the previous section, we introduced the three inverse trigonometric functions $y = \sin^{-1} x$, $y = \cos^{-1} x$, and $y = \tan^{-1} x$. Because we were using these functions as tools to solve equations, our emphasis there was more computational than theoretical in nature. Now we want to take a second look at these functions, this time from a more conceptual point of view. As background for this work, you should be familiar with the material on inverse functions in Section 3.5.

As indicated in Figure 1(a), the sine function is not one-to-one. Therefore, there is no inverse function. However, let us now consider the **restricted sine function:**

$$y = \sin x, \quad -\pi/2 \le x \le \pi/2$$

As indicated in Figure 1(b), the restricted sine function *is* one-to-one, and therefore the inverse function does exist in this case. We refer to the inverse of the restricted sine function as the **inverse sine function.**

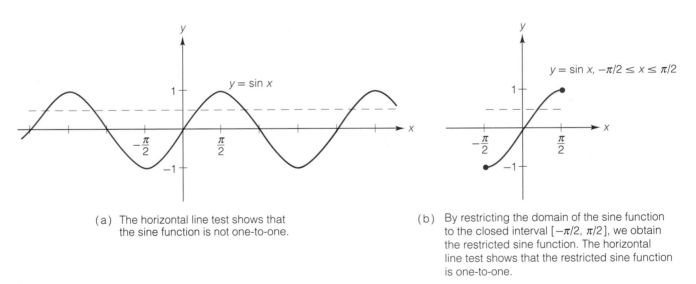

(a) The horizontal line test shows that the sine function is not one-to-one.

(b) By restricting the domain of the sine function to the closed interval $[-\pi/2, \pi/2]$, we obtain the restricted sine function. The horizontal line test shows that the restricted sine function is one-to-one.

FIGURE 1

Two notations are commonly used to denote the inverse sine function:

$$y = \sin^{-1} x \quad \text{and} \quad y = \arcsin x$$

Initially, at least, we will use the notation $y = \sin^{-1} x$. The graph of $y = \sin^{-1} x$ is easily obtained using the fact that the graph of a function and its inverse are reflections of one another about the line $y = x$. Figure 2(a) shows the graph of the restricted sine function and its inverse, $y = \sin^{-1} x$. Figure 2(b) shows the graph of $y = \sin^{-1} x$ alone. From Figure 2(b), we see that the domain of $y = \sin^{-1} x$ is the interval $[-1, 1]$, while the range is $[-\pi/2, \pi/2]$.

FIGURE 2

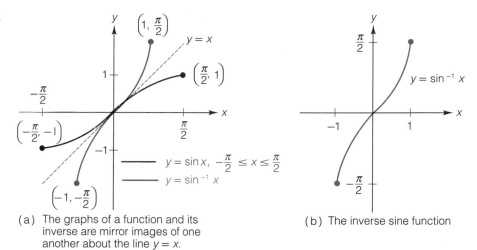

(a) The graphs of a function and its inverse are mirror images of one another about the line $y = x$.

(b) The inverse sine function

Values of $\sin^{-1} x$ are computed according to the following rule.

$\sin^{-1} x$ is that number in the interval $[-\pi/2, \pi/2]$ whose sine is x.

To see why this is so, we begin with the restricted sine function:

$$y = \sin x, \quad -\pi/2 \le x \le \pi/2$$

As explained in Section 3.5, the inverse function is obtained by interchanging x and y (the inputs and the outputs). So, for the inverse function, we have

$$x = \sin y, \quad -\pi/2 \le y \le \pi/2 \tag{1}$$

Equation (1) tells us that y is the number in the interval $[-\pi/2, \pi/2]$ whose sine is x. This is what we wished to show.

EXAMPLE 1 Evaluate: **(a)** $\sin^{-1}\left(\frac{1}{2}\right)$; **(b)** $\sin^{-1}\left(-\frac{1}{2}\right)$.

Solution

(a) $\sin^{-1}\left(\frac{1}{2}\right)$ is that number in the interval $[-\pi/2, \pi/2]$ whose sine is $\frac{1}{2}$. Since $\sin(\pi/6) = \frac{1}{2}$, we conclude that $\sin^{-1}\left(\frac{1}{2}\right) = \pi/6$.

(b) Since $\sin(-\pi/6) = -\frac{1}{2}$ and $-\pi/6$ belongs to the interval $[-\pi/2, \pi/2]$, we conclude that $\sin^{-1}\left(-\frac{1}{2}\right) = -\pi/6$. ∎

We can use the result in Example 1(a) to see why the notation *arcsin* is used for the inverse sine function. With the arcsin notation, the result in Example 1(a) can be written

$$\arcsin \frac{1}{2} = \frac{\pi}{6}$$

Now, as Figure 3 indicates, the arc length with a sine of $\frac{1}{2}$ is $\pi/6$. This is the idea behind the arcsin notation.

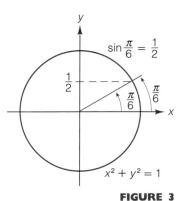

FIGURE 3

EXAMPLE 2 Evaluate arcsin $\frac{3}{4}$.

Solution   The quantity arcsin $\frac{3}{4}$ is that number (or arc length, or angle in radians) in the interval $[-\pi/2, \pi/2]$ whose sine is $\frac{3}{4}$. Since we are not familiar with an angle with a sine of $\frac{3}{4}$, we use a calculator:

0.75   (INV)   (SIN)

The result is

$$\arcsin \frac{3}{4} \approx 0.85$$

∎

▮▮▮ EXAMPLE 3   Show that the following two expressions are not equal:

$$\sin^{-1} 0 \qquad \text{and} \qquad \frac{1}{\sin 0}$$

Solution   The quantity $\sin^{-1} 0$ is that number in the interval $[-\pi/2, \pi/2]$ whose sine is 0. Since $\sin 0 = 0$, we conclude that

$$\sin^{-1} 0 = 0$$

On the other hand, since $\sin 0 = 0$, the expression $1/(\sin 0)$ is not even defined. Thus, the two given expressions certainly are not equal. ∎

If f and f^{-1} are any pair of inverse functions, then by definition,

$$f[f^{-1}(x)] = x \qquad \text{for every } x \text{ in the domain of } f^{-1}$$

and

$$f^{-1}[f(x)] = x \qquad \text{for every } x \text{ in the domain of } f$$

Applying these facts to the restricted sine function, $y = \sin x$, and its inverse, $y = \sin^{-1} x$, we obtain the following two identities.

$\sin(\sin^{-1} x) = x \qquad$ for every x in the inverval $[-1, 1]$

$\sin^{-1}(\sin x) = x \qquad$ for every x in the interval $[-\pi/2, \pi/2]$

The following example indicates that the domain restrictions accompanying these two identities cannot be ignored.

▮▮▮ EXAMPLE 4   Compute each of the following quantities that is defined.

(a) $\sin^{-1}\left(\sin \frac{\pi}{4}\right)$ **(b)** $\sin^{-1}(\sin \pi)$

(c) $\sin(\sin^{-1} 2)$ **(d)** $\sin\left[\sin^{-1}(-1/\sqrt{5})\right]$

Solution

(a) Since $\frac{\pi}{4}$ lies in the domain of the restricted sine function, the identity $\sin^{-1}(\sin x) = x$ is applicable here. Thus,

$$\sin^{-1}\left(\sin \tfrac{\pi}{4}\right) = \tfrac{\pi}{4}$$

Check: $\sin \frac{\pi}{4} = \sqrt{2}/2$. Therefore $\sin^{-1}\left(\sin \frac{\pi}{4}\right) = \sin^{-1}\left(\sqrt{2}/2\right) = \frac{\pi}{4}$.

(b) The number π is not in the domain of the restricted sine function, so the identity $\sin^{-1}(\sin x) = x$ does not apply in this case. However, since $\sin \pi = 0$, we have

$$\sin^{-1}(\sin \pi) = \sin^{-1} 0 = 0$$

Thus, $\sin^{-1}(\sin \pi)$ is equal to 0, not π.

(c) The number 2 is not in the domain of the inverse sine function. Thus, the expression $\sin(\sin^{-1} 2)$ is undefined.

(d) The identity $\sin(\sin^{-1} x) = x$ is applicable here. (Why?) Thus,

$$\sin\left[\sin^{-1}(-1/\sqrt{5})\right] = -1/\sqrt{5}$$

▮▮▮

▮▮▮ EXAMPLE 5 If $\sin \theta = x/3$ and $0 < \theta < \pi/2$, express the quantity $\theta - \sin 2\theta$ as a function of x.

Solution The given conditions tell us that $\theta = \sin^{-1}(x/3)$. That expresses θ in terms of x. To express $\sin 2\theta$ in terms of x, we have

$$\sin 2\theta = 2 \sin \theta \cos \theta = 2 \sin \theta \sqrt{1 - \sin^2 \theta}$$

$$= 2\left(\frac{x}{3}\right)\sqrt{1 - \frac{x^2}{9}} = \frac{2x}{3}\sqrt{\frac{9 - x^2}{9}}$$

$$= \frac{2x\sqrt{9 - x^2}}{9}$$

Now, combining the results, we have

$$\theta - \sin 2\theta = \sin^{-1}\left(\frac{x}{3}\right) - \frac{2x\sqrt{9 - x^2}}{9}$$

▮▮▮

We turn now to the inverse cosine function. We'll use the same general procedure that we followed for the inverse sine to define the inverse cosine. We begin by defining the **restricted cosine function:**

$$y = \cos x, \quad 0 \leq x \leq \pi$$

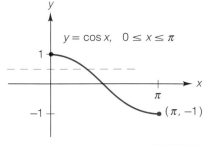

$$y = \cos x, \quad 0 \leq x \leq \pi$$

$(\pi, -1)$

FIGURE 4
The restricted cosine function is one-to-one.

As indicated by the horizontal line test in Figure 4, this function is one-to-one. Since the restricted cosine function is one-to-one, the inverse function exists. We denote the **inverse cosine function** by

$$y = \cos^{-1} x \qquad \text{or} \qquad y = \arccos x$$

The graph of $y = \cos^{-1} x$ is obtained by reflecting the graph of the restricted cosine function about the line $y = x$. Figure 5(a) displays the graph of the restricted cosine function along with $y = \cos^{-1} x$. Figure 5(b) shows the graph of $y = \cos^{-1} x$ alone. From Figure 5(b), we see that the domain of $y = \cos^{-1} x$ is the interval $[-1, 1]$, while the range is $[0, \pi]$.

FIGURE 5

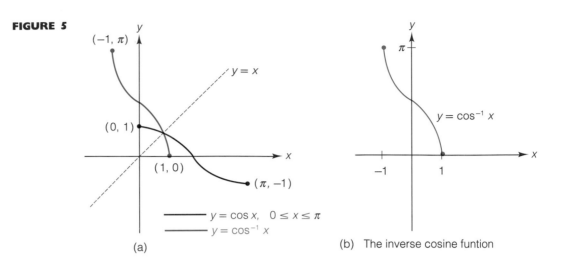

(a)

(b)   The inverse cosine funtion

As we mentioned previously, the defining equations for inverse functions are

$$f(f^{-1}(x)) = x \qquad \text{for every } x \text{ in the domain of } f^{-1}$$

and

$$f^{-1}(f(x)) = x \qquad \text{for every } x \text{ in the domain of } f$$

In the particular case of the restricted cosine function and the inverse cosine function, these two identities read as follows.

$$\cos(\cos^{-1} x) = x \qquad \text{for every } x \text{ in the interval } [-1, 1]$$
$$\cos^{-1}(\cos x) = x \qquad \text{for every } x \text{ in the interval } [0, \pi]$$

As stated in the previous section, the values of $\cos^{-1} x$ are computed according to the following rule.

$\cos^{-1} x$ is that number in the interval $[0, \pi]$ whose cosine is x.

To see why this is so, we begin with the restricted cosine function:

$$y = \cos x, \quad 0 \le x \le \pi$$

Then for the inverse function, we interchange x and y (the inputs and the outputs) to obtain

$$x = \cos y, \quad 0 \le y \le \pi$$

This last equation tells us that y is the number in the interval $[0, \pi]$ whose cosine is x. This is what we wished to show.

EXAMPLE 6   Compute each of the following: **(a)** $\cos^{-1}(0)$; **(b)** $\arccos\left(\cos \frac{2}{5}\right)$.

Solution

(a) $\cos^{-1}(0)$ is that number in the interval $[0, \pi]$ whose cosine is 0. Since $\cos(\pi/2) = 0$, we have

$$\cos^{-1}(0) = \pi/2$$

(b) Since the number $\frac{2}{5}$ is in the domain of the restricted cosine function, the identity $\arccos(\cos x) = x$ is applicable. Thus,

$$\arccos\left(\cos \tfrac{2}{5}\right) = \tfrac{2}{5}$$ ▪▪▪

▮▬▮ **EXAMPLE 7** Show that $\sin(\cos^{-1}x) = \sqrt{1 - x^2}$ for $-1 \le x \le 1$.

Solution We use the identity

$$\sin y = \sqrt{1 - \cos^2 y}, \qquad \text{which is valid for } 0 \le y \le \pi$$

Substituting $\cos^{-1}x$ for y in this identity, we obtain

$$\sin(\cos^{-1}x) = \sqrt{1 - [\cos(\cos^{-1}x)]^2}$$

or

$$\sin(\cos^{-1}x) = \sqrt{1 - x^2} \qquad \text{as required}$$

Before leaving this example, we point out an alternate method of solution that is useful when the restriction on x is $0 < x < 1$. In this case, we let $\theta = \cos^{-1}x$. Then θ is the radian measure of the acute angle whose cosine is x and we can sketch θ as shown in Figure 6. The sides of the triangle in Figure 6 have been labeled in such a way that $\cos \theta = x$. Then, by the Pythagorean theorem, we find that the third side of the triangle is $\sqrt{1 - x^2}$. We therefore have

$$\sin \theta = \frac{\text{opposite}}{\text{hypotenuse}} = \frac{\sqrt{1 - x^2}}{1} = \sqrt{1 - x^2}$$ ▪▪▪

FIGURE 6

Just as there is a basic identity connecting $\sin x$ and $\cos x$, namely, $\sin^2 x + \cos^2 x = 1$, there is also an identity connecting $\sin^{-1}x$ and $\cos^{-1}x$:

$$\sin^{-1}x + \cos^{-1}x = \frac{\pi}{2} \qquad \text{for every } x \text{ in the closed interval } [-1, 1]$$

We can use Figure 7 to see why this identity is valid when $0 < x < 1$. (Exercise 45 shows you how to establish this identity for all values of x from -1 to 1.) In Figure 7, assume that θ and β are the radian measures of the indicated acute angles. Then we have

$$\sin \theta = \frac{x}{1} = x \qquad \text{and} \qquad \cos \beta = \frac{x}{1} = x$$

and therefore

$$\theta = \sin^{-1}x \qquad \text{and} \qquad \beta = \cos^{-1}x$$

But, from Figure 7, we know that $\theta + \beta = \pi/2$. Consequently, we have $\sin^{-1}x + \cos^{-1}x = \pi/2$, as required.

FIGURE 7

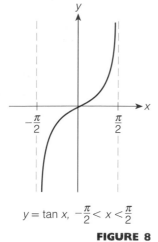

$$y = \tan x, \ -\frac{\pi}{2} < x < \frac{\pi}{2}$$

FIGURE 8
The restricted tangent function

Now let us turn to the definition of the inverse tangent function. We begin by defining the **restricted tangent function**:

$$y = \tan x, \quad -\pi/2 < x < \pi/2$$

Figure 8 shows the graph of the restricted tangent function. As you can check by applying the horizontal line test, the restricted tangent function is one-to-one. This tells us that the inverse function exists. The two common notations for this **inverse tangent function** are

$$y = \tan^{-1}x \qquad \text{or} \qquad y = \arctan x$$

The graph of $y = \tan^{-1}x$ is obtained in the same way we obtained the graphs of the inverse sine and the inverse cosine. That is, we use the fact that the graph of a function and its inverse are mirror images of one another about the line $y = x$. In Figure 9(a), we've reflected the restricted tangent function about the line $y = x$ to obtain the graph of $y = \tan^{-1}x$. In Figure 9(b) we show the graph of $y = \tan^{-1}x$ alone.

FIGURE 9

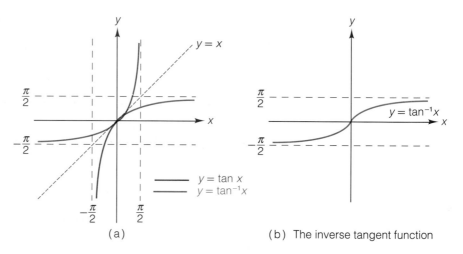

(a)

(b) The inverse tangent function

From the graph in Figure 9(b), we can see that the domain of the inverse tangent function is the set of all real numbers, and the range is the open interval $(-\pi/2, \pi/2)$. Furthermore, if we follow exactly the same line of reasoning used previously for the inverse sine and the inverse cosine, we obtain the following properties of the inverse tangent function.

$\tan^{-1}x$ is that number in the interval $(-\pi/2, \pi/2)$ whose tangent is x.

$\tan(\tan^{-1}x) = x$ for every real number x

$\tan^{-1}(\tan x) = x$ for every x in the open interval $(-\pi/2, \pi/2)$

▌▐▌▌ **EXAMPLE 8**   Evaluate: **(a)** $\tan^{-1}(-1)$;   **(b)** $\tan(\tan^{-1}\sqrt{5})$.

Solution

(a) The quantity $\tan^{-1}(-1)$ is that number in the interval $(-\pi/2, \pi/2)$ whose tangent is -1. Since $\tan(-\pi/4) = -1$, we have $\tan^{-1}(-1) = -\pi/4$.

(b) The identity $\tan(\tan^{-1}x) = x$ holds for all real numbers x. We therefore have $\tan(\tan^{-1}\sqrt{5}) = \sqrt{5}$.

EXAMPLE 9 If $\tan\theta = x/3$ and $0 < \theta < \pi/2$, express the quantity $\theta - \tan 2\theta$ as a function of x.

Solution The given equation tells us that $\theta = \tan^{-1}(x/3)$. That expresses θ in terms of x. To express $\tan 2\theta$ in terms of x, we have

$$\tan 2\theta = \frac{2\tan\theta}{1 - \tan^2\theta} = \frac{2(x/3)}{1 - (x/3)^2}$$

$$= \frac{2x/3}{1 - (x^2/9)} = \frac{6x}{9 - x^2} \qquad \text{multiplying numerator and denominator by 9}$$

Now, combining the results, we have $\theta - \tan 2\theta = \tan^{-1}\dfrac{x}{3} - \dfrac{6x}{9 - x^2}$

EXAMPLE 10 Simplify the quantity $\sec(\tan^{-1}x)$, where $x > 0$.

Solution We let $\theta = \tan^{-1}x$ or, equivalently, $\tan\theta = x$. Now, as shown in Figure 10, we sketch a right triangle with an angle θ (in radians) whose tangent is x. The Pythagorean theorem tells us that the length of the hypotenuse in this triangle is $\sqrt{1 + x^2}$. Consequently, we have

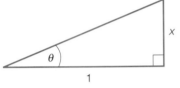

FIGURE 10
$\theta = \tan^{-1}x\,(x > 0)$ or $\tan\theta = x$

$$\sec(\tan^{-1}x) = \sec\theta = \frac{\text{hypotenuse}}{\text{adjacent}} = \frac{\sqrt{1 + x^2}}{1} = \sqrt{1 + x^2}$$

When suitable restrictions are placed on the domains of the secant, cosecant, and cotangent, corresponding inverse function can be defined. However, with the exception of the inverse secant, these are rarely, if ever, encountered in calculus. Thus, we omit a discussion of these functions here. (For the inverse secant function, see Exercise 55.)

EXERCISE SET 7.9

In Exercises 1–20, evaluate each of the quantities that is defined, but do not use a calculator or tables. If a quantity is undefined, say so.

1. $\sin^{-1}(\sqrt{3}/2)$
2. $\cos^{-1}(-1)$
3. $\tan^{-1}\sqrt{3}$
4. $\arccos(-\sqrt{2}/2)$
5. $\arctan(-1/\sqrt{3})$
6. $\arcsin(-1)$
7. $\tan^{-1}1$
8. $\sin^{-1}0$
9. $\cos^{-1}2\pi$
10. $\arctan 0$
11. $\sin[\sin^{-1}(\frac{1}{4})]$
12. $\cos[\cos^{-1}(\frac{4}{3})]$
13. $\cos[\cos^{-1}(\frac{3}{4})]$
14. $\tan(\arctan 3\pi)$
15. $\arctan[\tan(-\pi/7)]$
16. $\sin(\arcsin 2)$
17. $\arcsin[\sin(\pi/2)]$
18. $\arccos[\cos(\pi/8)]$
19. $\arccos(\cos 2\pi)$
20. $\sin^{-1}[\sin(3\pi/2)]$

In Exercises 21–30, evaluate the given quantities without using a calculator or tables.

21. $\tan[\sin^{-1}(\frac{4}{5})]$
22. $\cos(\arcsin\frac{2}{7})$
23. $\sin(\tan^{-1}1)$
24. $\sin[\tan^{-1}(-1)]$
25. $\tan(\arccos\frac{5}{13})$
26. $\cos[\sin^{-1}(\frac{2}{3})]$
27. $\cos(\arctan\sqrt{3})$
28. $\sin[\cos^{-1}(\frac{1}{3})]$
29. $\sin[\arccos(-\frac{1}{3})]$
30. $\tan(\arcsin\frac{20}{29})$
31. Use a calculator to evaluate each of the following quantities. Express your answers to two decimal places; don't round off.
    (a) $\sin^{-1}(\frac{3}{4})$
    (b) $\cos^{-1}(\frac{2}{3})$
    (c) $\tan^{-1}\pi$
    (d) $\tan^{-1}(\tan^{-1}\pi)$

In Exercises 32–34, evaluate the given expressions without using a calculator or tables.

32. $\csc[\sin^{-1}(\frac{1}{2}) - \cos^{-1}(\frac{1}{2})]$
33. $\sec[\cos^{-1}(\sqrt{2}/2) + \sin^{-1}(-1)]$
34. $\cot[\cos^{-1}(-1/2) + \cos^{-1}(0) + \tan^{-1}(1/\sqrt{3})]$

35. Show that $\cos(\sin^{-1} x) = \sqrt{1 - x^2}$ for $-1 \le x \le 1$.
    Suggestion: Use the method of Example 7 in the text.

36. If $\sin \theta = 2x$ and $0 < \theta < \pi/2$, express $\theta + \cos 2\theta$ as a function of x.

37. If $\sin \theta = 3x/2$ and $0 < \theta < \pi/2$, express $\frac{1}{4}\theta - \sin 2\theta$ as a function of x.

38. If $\cos \theta = x - 1$ and $0 < \theta < \pi/2$, express $2\theta - \cos 2\theta$ as a function of x.

39. If $\tan \theta = \frac{1}{2}(x - 1)$ and $0 < \theta < \pi/2$, express $\theta - \cos \theta$ as a function of x.

40. If $\tan \theta = \frac{1}{3}(x + 1)$ and $0 < \theta < \pi/2$, express $2\theta + \tan 2\theta$ as a function of x.

B

41. Evaluate $\sin(2 \tan^{-1} 4)$.   *Hint:* $\sin 2\theta = 2 \sin \theta \cos \theta$.

42. Evaluate $\cos\left[2 \sin^{-1}\left(\frac{5}{13}\right)\right]$.

43. Evaluate $\sin\left(\arccos \frac{3}{5} - \arctan \frac{7}{13}\right)$.   *Suggestion:* Use the formula for $\sin(x - y)$.

44. Show that $\sin\left[\sin^{-1}\left(\frac{1}{3}\right) + \sin^{-1}\left(\frac{1}{4}\right)\right] = \dfrac{\sqrt{15} + 2\sqrt{2}}{12}$.

45. In the text we showed that $\sin^{-1} x + \cos^{-1} x = \pi/2$ for x in the open interval $(0, 1)$. Follow steps (a) and (b) to show that this identity actually holds for *every* x in the closed interval $[-1, 1]$.
    (a) Let $\alpha = \sin^{-1} x$ and $\beta = \cos^{-1} x$. Explain why $-\pi/2 \le \alpha + \beta \le 3\pi/2$.   *Hint:* What are the ranges of the inverse sine and of the inverse cosine?
    (b) Use the addition formula for sine to show that $\sin(\alpha + \beta) = 1$. Conclude [with the help of part (a)] that $\alpha + \beta = \pi/2$, as required.

In Exercises 46–48, solve the given equations.

46. $\cos^{-1} t = \sin^{-1} t$ *Hint:* Compute the cosine of both sides.

47. $\sin^{-1}(3t - 2) = \sin^{-1} t - \cos^{-1} t$

48. $\sin^{-1}(\sin^{-1} x) = \sin^{-1}(\cos^{-1} x)$

49. Show that $\arctan x + \arctan y = \arctan \dfrac{x + y}{1 - xy}$ when x and y are positive and $xy < 1$.

50. Use the identity in Exercise 49 to show that the following equations are correct.
    (a) $\arctan \frac{1}{2} + \arctan \frac{1}{3} = \pi/4$
    (b) $\arctan \frac{1}{4} + \arctan \frac{3}{5} = \pi/4$
    (c) $2 \arctan \frac{1}{3} + \arctan \frac{1}{7} = \pi/4$
        Hint: $2 \arctan \frac{1}{3} = \arctan \frac{1}{3} + \arctan \frac{1}{3}$
    (d) $\arctan \frac{1}{2} + \arctan \frac{1}{5} + \arctan \frac{1}{8} = \pi/4$

In Exercises 51–53, solve the equations.

51. $2 \tan^{-1} x = \tan^{-1}\left(\dfrac{1}{4x}\right)$ *Hint:* Compute the tangent of both sides.

52. $\tan^{-1} x = \cos^{-1} x$

53. $2 \tan^{-1} \sqrt{t - t^2} = \tan^{-1} t + \tan^{-1}(1 - t)$

54. (a) As you can see in the accompanying figure, the graphs of $y = \cos^{-1} x$ and $y = \tan^{-1} x$ intersect at a point in the first quadrant. By solving the equation $\cos^{-1} x = \tan^{-1} x$, show that the x-coordinate of this intersection point is given by
$$x = \sqrt{\dfrac{\sqrt{5} - 1}{2}}$$

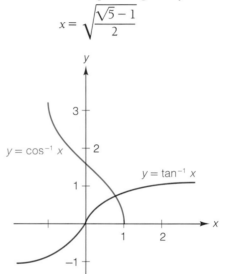

(b) Use the result in part (a) along with your calculator to specify the coordinates of the intersection point. Express both coordinates as decimals, rounded off to three places.

55. (a) The domain of the **restricted secant function** is $[0, \pi/2) \cup [\pi, 3\pi/2)$. Sketch the restricted secant function and note that it is one-to-one.
    (b) The **inverse secant function** can be defined as follows:

    $\sec^{-1} x$ is the unique number in the set $[0, \pi/2) \cup [\pi, 3\pi/2)$ whose secant is x.

    Using the fact that this function is the inverse of the restricted secant function, sketch the graph of $y = \sec^{-1} x$.
    (c) Evaluate $\sec^{-1}(2/\sqrt{3})$ and $\sec^{-1}(-2/\sqrt{3})$.
    (d) Evaluate $\sec^{-1}(\sqrt{2})$ and $\sec^{-1}(-\sqrt{2})$.
    (e) Evaluate $\sec(\sec^{-1} 2)$ and $\sec^{-1}(\sec 0)$.

56. Evaluate each of the following quantities. (The inverse secant function is defined in the previous exercise.)

(a) $\cos\left[\sec^{-1}\left(\frac{4}{3}\right)\right]$ (b) $\sin(\sec^{-1}4)$

(c) $\tan[\sec^{-1}(-3)]$ (d) $\cot\left[\sec^{-1}\left(\frac{13}{12}\right)\right]$

In Exercises 57 and 58, we make use of some simple geometric figures to evaluate sums involving the inverse tangent function. The idea here is taken from the note by Professor Edward M. Harris entitled Behold! Sums of Arctan *(The College Mathematics Journal, vol. 18, no. 2, p. 141).*

57. In this exercise, we use the following figure to show that $\tan^{-1}\left(\frac{2}{3}\right) + \tan^{-1}\left(\frac{1}{5}\right) = \pi/4$.

(a) Show that $\angle ABC = \pi/2$. *Hint:* Compute the slopes of $\overline{AB}$ and $\overline{BC}$. (Assume that the grid lines are marked off at one-unit intervals.)

(b) Show that $AB = BC$ and conclude that $\angle BAC = \pi/4$.

(c) Now explain why $\alpha = \tan^{-1}\left(\frac{2}{3}\right)$, $\beta = \tan^{-1}\left(\frac{1}{5}\right)$, and $\tan^{-1}\left(\frac{2}{3}\right) + \tan^{-1}\left(\frac{1}{5}\right) = \pi/4$.

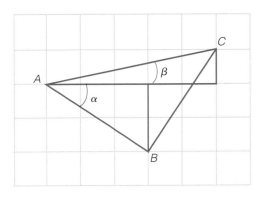

58. In this exercise, we use the accompanying figure to show that $\tan^{-1}1 + \tan^{-1}2 + \tan^{-1}3 = \pi$.

(a) By computing slopes, show that $\overline{DE} \perp \overline{BC}$ and $\overline{AB} \perp \overline{BD}$. (Assume that the grid lines are at one-unit intervals.)

(b) Determine the following lengths: DE, CE, BE, AB, and BD.

(c) Show that $\tan\alpha = 1$, $\tan\beta = 2$, and $\tan\gamma = 3$. Then explain why $\tan^{-1}1 + \tan^{-1}2 + \tan^{-1}3 = \pi$.

OPTIONAL TI-81 GRAPHING CALCULATOR EXERCISES
EXERCISES FOR SECTION 7.9

1. To prepare the graphing screen for this exercise, press ZOOM 7. Then press the RANGE key and set $x_{\min} = -1.57$ and $x_{\max} = 1.57$. (The value of $\pi/2$ is approximately 1.57.)

(a) Press the Y= key and define $Y_1 = x$ and $Y_2 = \sin^{-1}(\sin x)$. First graph only Y_2. Next, graph only Y_1. Finally graph both functions on the same screen. What do you observe?

(b) What do the results in part (a) demonstrate?

2. Make the following adjustments to the RANGE settings from Exercise 1: $x_{\min} = -1$ and $x_{\max} = 1$. Next, press the Y= key. Retain the function $Y_1 = x$ (from Exercise 1), but redefine Y_2 as $Y_2 = \sin(\sin^{-1}x)$. Now graph the functions Y_1 and Y_2. What do you observe? What does this demonstrate?

3. To prepare the graphing screen for this exercise, press ZOOM 7. Then press the RANGE key and set $x_{\min} = -1$ and $x_{\max} = 1$. (These are the same RANGE settings used in Exercise 2.)

(a) Press the Y= key and define $Y_1 = \sin^{-1}x + \cos^{-1}x$. Then, after deactivating any other functions in the menu, press the GRAPH key. What type of function does the resulting graph represent?

(b) What is the *exact* value for the y-intercept of the graph? What identity does the graph in part (a) demonstrate?

4. (a) In Example 7 in Section 7.9, we proved the identity $\sin(\cos^{-1}x) = \sqrt{1-x^2}$, for $-1 \le x \le 1$. Demonstrate this identity graphically by graphing

the two functions $Y_1 = \sin(\cos^{-1}x)$ and $Y_2 = \sqrt{1-x^2}$.

(b) Demonstrate the identity $\cos(\sin^{-1}x) = \sqrt{1-x^2}$, for $-1 \le x \le 1$, by graphing the functions $Y_2 = \sqrt{1-x^2}$ and $Y_3 = \cos(\sin^{-1}x)$ and noting that the graphs appear to be identical.

5. (a) Press the Y= key and clear or delete all functions. Enter the functions $Y_1 = \tan(\tan^{-1}x)$ and $Y_2 = x$. Press ZOOM 7 to obtain the graphs. What do you observe? What does this demonstrate?

(b) Clear or delete the function Y_1 from part (a), and enter $Y_1 = \tan^{-1}(\tan x)$. Next, press the RANGE key and change the settings for $x_{\min}$ and $x_{\max}$ to $x_{\min} = -1.57$ and $x_{\max} = 1.57$. Now press the GRAPH key to obtain the graphs of $Y_1 = \tan^{-1}(\tan x)$ and $Y_2 = x$. What do you observe, and what does this demonstrate?

(c) In part (b), why was it necessary to change the range settings?

6. In Example 10 in Section 7.9 we found that $\sec(\tan^{-1}x) = \sqrt{1+x^2}$, for $x > 0$. Actually, this identity is valid for all real numbers. Demonstrate this graphically as follows. Using the Y= key, enter the

functions $Y_1 = \sec(\tan^{-1}x) = 1/\cos(\tan^{-1}x)$ and $Y_2 = \sqrt{1+x^2}$. Press ZOOM 6 to obtain the graphs, and note that the graphs indeed appear identical.

In Exercises 7–15, solve the equations on the interval $[0, 1]$ using the TRACE and ZOOM technique described in Optional Exercise 1 on page 419. In each case, carry out the process far enough so that you are certain of the first two decimal places in each answer.

7. $\cos^{-1}x = \tan^{-1}x$ 8. $x = \arccos x$

9. $\cos^{-1}x = x^2$

10. (a) $1.3\left(x - \frac{1}{2}\right)^2 = \cos^{-1}x$
    (b) $1.4\left(x - \frac{1}{2}\right)^2 = \cos^{-1}x$

11. $\tan^{-1}x = \sin 3x$

12. (a) $\arccos x = 2 \sin 3x$ 13. (a) $\sin(2.3x) = \arctan x$
    (b) $\arccos x = 2 \sin 4x$ (b) $\sin(2.2x) = \arctan x$

14. (a) $1/(\tan^{-1}x + \sin^{-1}x) = \sin 2x$
    (b) $1/(\tan^{-1}x + \sin^{-1}x) = \sin 3x$

15. (a) $1/(\sin^{-1}x + \cos^{-1}x) = 4x^3$
    (b) $1/(\sin^{-1}x + \cos^{-1}x) = 5x^3$

▦ CHAPTER SEVEN SUMMARY OF PRINCIPAL FORMULAS AND TERMS

FORMULAS OR TERMS	PAGE REFERENCE	COMMENT
1. $\theta = \dfrac{s}{r}$	345	This equation defines the *radian measure* θ of an angle. We assume here that the vertex of the angle is placed at the center of a circle of radius r and that s is the length of the intercepted arc. See Figure 1 in Section 7.1.
2. $s = r\theta$	351	This formula expresses the arc length s on a circle in terms of the radius r and the radian measure θ of the central angle subtended by the arc.
3. $A = \frac{1}{2}r^2\theta$	352	This formula expresses the area of a sector of a circle in terms of the radius r and the radian measure θ of the central angle.
4. $\omega = \dfrac{\theta}{t}$ $\quad v = \dfrac{d}{t}$	352	Suppose that a wheel rotates about its axis at a constant rate. If a radial line on the wheel turns through an angle of measure θ in time t, then the *angular speed* ω of the wheel is defined by $\omega = \theta/t$. If a point on the rotating wheel travels a distance d in time t, then the linear speed v of the point is defined by $v = d/t$.
5. $\cos t$ $\sin t$ $\tan t$ $\sec t$ $\csc t$ $\cot t$	356	Let $P(x, y)$ denote the point on the unit circle such that the arc length from $(1, 0)$ is t. Or equivalently, let $P(x, y)$ denote the point where the terminal side of the angle with radian measure t intersects the unit circle. Then the six trigonometric functions of the real number t are defined as follows.

FORMULAS OR TERMS	PAGE REFERENCE	COMMENT
$\cos t = x \qquad \sec t = \dfrac{1}{x} \quad (x \neq 0)$ $\sin t = y \qquad \csc t = \dfrac{1}{y} \quad (y \neq 0)$ $\tan t = \dfrac{y}{x} \quad (x \neq 0) \qquad \cot t = \dfrac{x}{y} \quad (y \neq 0)$		
6. $\sec t = \dfrac{1}{\cos t} \qquad \csc t = \dfrac{1}{\sin t}$ $\cot t = \dfrac{1}{\tan t} \qquad \tan t = \dfrac{\sin t}{\cos t}$ $\cot t = \dfrac{\cos t}{\sin t}$	357	These five identities are direct consequences of the definitions of the trigonometric functions.
7. $\sin^2 t + \cos^2 t = 1$ $\tan^2 t + 1 = \sec^2 t$ $\cot^2 t + 1 = \csc^2 t$	357	These are the three *Pythagorean* identities.
8. $\cos(-t) = \cos t$ $\sin(-t) = -\sin t$ $\tan(-t) = -\tan t$	359	These three identities are sometimes referred to as the *opposite angle identities*. In each case, a trigonometric function of $-t$ is expressed in terms of that same function of t.
9. Periodic function	363, 364	A function f is periodic if there is a positive number p such that the equation $f(x + p) = f(x)$ holds for all x in the domain of f. The smallest such number p is called the *period* of f. Important examples: The period of $y = \sin x$ is 2π; the period of $y = \cos x$ is 2π; the period of $y = \tan x$ is π.
10. Amplitude	364	Let m and M denote the smallest and the largest values, respectively, of the periodic function f. Then the *amplitude* of f is defined to be the number $\frac{1}{2}(M - m)$. Important examples: The amplitude for both $y = \sin x$ and $y = \cos x$ is 1; amplitude is not defined for the function $y = \tan x$.
11. Period $= \dfrac{2\pi}{B}$	377, 379	This formula gives the period for the functions $y = A \sin(Bx - C)$ and $y = A \cos(Bx - C)$.
12. Phase shift $= \dfrac{C}{B}$	379	The phase shift serves as a guide in graphing functions of the form $y = A \sin(Bx - C)$ or $y = A \cos(Bx - C)$. For instance, to graph one complete cycle of $y = A \sin(Bx - C)$, first sketch one complete cycle of $y = A \sin Bx$, beginning at $x = 0$. Then draw a curve with exactly the same shape, but beginning at $x = C/B$ rather than $x = 0$. This will represent one cycle of $y = A \sin(Bx - C)$.
13. $\sin(s + t) = \sin s \cos t + \cos s \sin t$ $\sin(s - t) = \sin s \cos t - \cos s \sin t$ $\cos(s + t) = \cos s \cos t - \sin s \sin t$ $\cos(s - t) = \cos s \cos t + \sin s \sin t$	395	These identities are referred to as the *addition formulas* for sine and cosine.

FORMULAS OR TERMS	PAGE REFERENCE	COMMENT
14. $\cos(\frac{\pi}{2} - \theta) = \sin\theta$ $\sin(\frac{\pi}{2} - \theta) = \cos\theta$	397	These two identities, called *reduction formulas,* hold for all real numbers θ. In terms of angles, the identities state that the sine of an angle is equal to the cosine of its complement.
15. $\tan(s + t) = \dfrac{\tan s + \tan t}{1 - \tan s \tan t}$ $\tan(s - t) = \dfrac{\tan s - \tan t}{1 + \tan s \tan t}$	401	These two identities are referred to as the *addition formulas* for tangent.
16. $\sin 2\theta = 2\sin\theta\cos\theta$ $\cos 2\theta = \cos^2\theta - \sin^2\theta$ $\tan 2\theta = \dfrac{2\tan\theta}{1 - \tan^2\theta}$	401	These are the *double-angle formulas.* There are four other forms of the double-angle formula for cosine. They appear in the box on page 404.
17. $\sin\dfrac{s}{2} = \pm\sqrt{\frac{1}{2}(1 - \cos s)}$ $\cos\dfrac{s}{2} = \pm\sqrt{\frac{1}{2}(1 + \cos s)}$ $\tan\dfrac{s}{2} = \dfrac{\sin s}{1 + \cos s}$	401	These three identities are referred to as the *half-angle formulas.* In the case of the half-angle formulas for sine and cosine, the sign before the radical is determined by the quadrant in which the angle or arc terminates.
18. Restricted sine function	420	The domain of the function $y = \sin x$ is the set of all real numbers. By allowing inputs only from the closed interval $[-\pi/2, \pi/2]$, we obtain the restricted sine function. (See Figure 1 on page 420.) The motivation for restricting the domain is that the restricted sine function is one-to-one, and therefore the inverse function exists.
19. $\sin^{-1}x$	413, 420	$\sin^{-1}x$ is that number in the interval $[-\pi/2, \pi/2]$ whose sine is x. An alternate but entirely equivalent form of this expression is arcsin x.
20. Restricted cosine function	423	The domain of the function $y = \cos x$ is the set of all real numbers. By allowing inputs only from the closed interval $[0, \pi]$, we obtain the restricted cosine function. (See Figure 4 on page 423.) The motivation for restricting the domain is that the restricted cosine function is one-to-one, and therefore the inverse function exists.
21. $\cos^{-1}x$	413, 423	$\cos^{-1}x$ is that number in the interval $[0, \pi]$ whose cosine is x. An alternate but entirely equivalent form of this expression is arccos x. The graph of the inverse cosine function is on page 424.
22. Restricted tangent function	426	The tangent function, $y = \tan x$, is not one-to-one. However, by allowing inputs only from the open interval $(-\pi/2, \pi/2)$, we obtain the restricted tangent function, which is one-to-one. (See Figure 8 on page 426.) Since the restricted tangent function is one-to-one, the inverse function exists.
23. $\tan^{-1}x$	413, 426	$\tan^{-1}x$ is that number in the interval $(-\pi/2, \pi/2)$ whose tangent is x. An alternate but entirely equivalent form of this expression is arctan x. The graph of the inverse tangent function is on page 426.

WRITING MATHEMATICS

1. Say whether the statement is *true* or *false*. Write out your reason or reasons in complete sentences. If you draw a diagram to accompany your writing, be sure that you clearly label any parts of the diagram to which you refer.

   (a) If t is a real number less than π, then $\sin t$ is positive.

   (b) If t is a real number such that $\pi < t < 3\pi/2$, then the product of $\sin t$ and $\cos t$ is positive.

   (c) In the immediate vicinity of the origin, the graph of $y = \sin x$ actually coincides with the line $y = x$.

   (d) $\dfrac{\sin \pi}{\pi} = \sin$

   (e) Between any two consecutive turning points, the graph of $y = \sin x$ is decreasing.

   (f) If t is a positive real number, then $\sin^2 t$ is a positive real number.

   (g) The equation $\tan^2 t + 1 = \sec^2 t$ is true for every real number t.

   (h) There is no real number x satisfying the equation $\cos(\pi/4 + x) = 2$.

   (i) There is no real number x satisfying the equation $\cos(\pi/4 + x) = \cos x$.

   (j) For every number x in the closed interval $[-1, 1]$, we have $\sin^{-1} x = 1/\sin x$.

   (k) There is no real number x for which $\sin^{-1} x = 1/\sin x$. *Hint:* Draw a careful sketch of the graphs of the inverse sine function and the cosecant function on the interval $0 < x \le 1$.

2. Given a function f and an input x_0, the numbers $f(x_0)$, $f(f(x_0))$, $f(f(f(x_0)))$, ...are called the **iterates** of x_0. For ease of notation we let

   $$x_1 = f(x_0)$$
   $$x_2 = f(f(x_0))$$
   $$x_3 = f(f(f(x_0)))$$

   and so on. For this exercise, we take f to be the function defined by $f(x) = x + \sin x$. This function has an interesting property: the iterates can be used to obtain better and better approximations to the number π. The following table, for instance, shows the first five iterates of $x_0 = 0.8$. As you can see, the fifth iterate is indeed an excellent approximation to π. (In fact, the fifth iterate, x_5, agrees with π through the first 15 decimal places.)

x_1	1.5173560908 . . .
x_2	2.5159285012 . . .
x_3	3.1015642481 . . .
x_4	3.1415819650 . . .
x_5	3.1415926535 . . .

   Iterates of $x_0 = 0.8$ for the function $f(x) = x + \sin x$

   (a) Use your calculator to set up three tables of iterates, similar to the table above, using $x_0 = 0.9$, $x_0 = 1.5$, and $x_0 = 2$. Describe the results.

   (b) Use the figure to explain why the successive iterates approach the number π. *Hint:* What are the coordinates of the point P? What are the y-coordinates of the points A, B, and C?

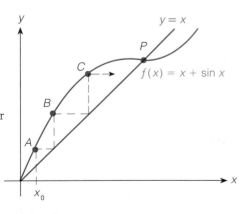

CHAPTER SEVEN REVIEW EXERCISES

1. Evaluate the following: (a) $\sin\frac{5\pi}{3}$; (b) $\cot\frac{11\pi}{6}$.

2. Graph the function $y = 3\cos 3\pi x$ over the interval $-\frac{1}{3} \le x \le \frac{1}{3}$. Specify the x-intercepts and the coordinates of the highest point on the graph.

3. Simplify the following expression:
$$\cos t - \cos(-t) + \sin t - \sin(-t)$$

4. In the expression $1/\sqrt{4 - t^2}$, make the substitution $t = 2\cos x$ and simplify the result. Assume that $0 < x < \pi$.

5. Graph the equation $y = \sec(2\pi x - 3)$ for one complete cycle.

6. If $\sec t = -\frac{5}{3}$ and $\pi < t < 3\pi/2$, compute $\cot t$.

7. Graph the function $y = -\sin(2x - \pi)$ for one complete period. Specify the amplitude, the period, and the phase shift.

8. Graph $y = -\frac{1}{2}\tan(\pi x/3)$ over one period.

In Exercises 9–20, evaluate each expression without using a calculator or tables.

9. $\cos \pi$
10. $\sin(-3\pi/2)$
11. $\csc(2\pi/3)$
12. $\tan(\pi/3)$
13. $\tan(11\pi/6)$
14. $\cos 0$
15. $\sin(\pi/6)$
16. $\sec(3\pi/4)$
17. $\cot(5\pi/4)$
18. $\tan(-7\pi/4)$
19. $\csc(-5\pi/6)$
20. $\sin^2(\pi/7) + \cos^2(\pi/7)$

Exercises 21–30 are calculator exercises. (Set your calculator to the radian mode.) In Exercises 21–26, where numerical answers are required, round off your results to three decimal places.

21. Evaluate $\sin 1$.
22. Evaluate $\cos 2$.
23. Evaluate $\sin(3\pi/2)$.
24. Evaluate $\sin(0.78)$.
25. Evaluate $\sin(\sin 0.0123)$.
26. Evaluate $\sin[\sin(\sin 0.0123)]$.
27. Verify that $\sin^2 1776 + \cos^2 1776 = 1$.
28. Verify that $\sin 14 = 2\sin 7\cos 7$.
29. Verify that $\cos(0.5) = \cos^2(0.25) - \sin^2(0.25)$.
30. Verify that $\cos(0.3) = \left[\frac{1}{2}(1 + \cos 0.6)\right]^{1/2}$.
31. In the expression $\sqrt{25 - x^2}$, make the substitution $x = 5\sin\theta$ $(0 < \theta < \pi/2)$ and simplify the result.
32. In the expression $(49 + x^2)^{1/2}$, make the substitution $x = 7\tan\theta$ $(0 < \theta < \pi/2)$ and simplify the result.
33. In the expression $(x^2 - 100)^{1/2}$, make the substitution $x = 10\sec\theta$ $(0 < \theta < \pi/2)$ and simplify the result.

34. In the expression $(x^2 - 4)^{-3/2}$, make the substitution $x = 2\sec\theta$ $(0 < \theta < \pi/2)$ and simplify the result.
35. In the expression $(x^2 + 5)^{-1/2}$, make the substitution $x = \sqrt{5}\tan\theta$ $(0 < \theta < \pi/2)$ and simplify the result.
36. If $\sin\theta = -\frac{5}{13}$ and $\pi < \theta < 3\pi/2$, compute $\cos\theta$.
37. If $\cos\theta = \frac{8}{17}$, $\sin\theta$ is negative, and $0 < \theta < 2\pi$, compute $\tan\theta$.
38. If $\sec\theta = -\frac{25}{7}$ and $\pi < \theta < 3\pi/2$, compute $\cot\theta$.
39. In the following figure, P is the center of the circle, which has radius $\sqrt{2}$ units. If the radian measure of angle BPA is θ, express the area of the shaded region in terms of θ. Simplify your answer as much as possible.

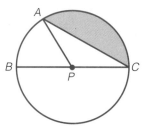

40. Express the area of the shaded region in the following figure in terms of r and θ. (Assume θ is in radians.)

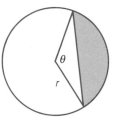

41. In the following figure, $ABCD$ is a square, each side of which is 1 cm. The two arcs are portions of circles with radii of 1 cm and with centers A and C. Find the area of the shaded region. *Hint:* Draw $\overline{BD}$ and use your result from Exercise 40.

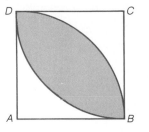

In Exercises 42–46, a function of the form $y = A \sin Bx$ or $y = A \cos Bx$ is graphed for one period. Determine the equation in each case. (Assume $B > 0$.)

42.

$(6, -3)$

43.

4

$(\pi, 0)$

44.

$(3\pi, -6)$

45.

$\left(\frac{\pi}{2}, -2\right)$

46.

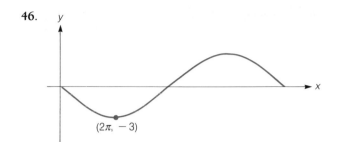

$(2\pi, -3)$

In Exercises 47–52, sketch the graph of each function for one complete cycle. In each case, specify the x-intercepts and the coordinates of the highest and lowest points on the graph.

47. $y = -3 \cos 4x$ **48.** $y = 2 \sin(3\pi x/4)$

49. $y = 2 \sin\left(\frac{\pi x}{2} - \frac{\pi}{4}\right)$ **50.** $y = -\sin(x - 1)$

51. $y = 3 \cos\left(\frac{\pi x}{3} - \frac{\pi}{3}\right)$ **52.** $y = -2 \cos(x + \pi)$

In Exercises 53–56, sketch the graph of each function for one period.

53. (a) $y = \tan(\pi x/4)$ **54.** (a) $y = 2 \cot 2x$
      (b) $y = \cot(\pi x/4)$ (b) $y = 2 \tan 2x$

55. (a) $y = 3 \sec(x/4)$ **56.** (a) $y = \sec \pi x$
      (b) $y = 3 \csc(x/4)$ (b) $y = \csc \pi x$

In Exercises 57–66 simplify the expressions.

57. $\sin[x + (3\pi/2)]$

58. $\sin(\pi - x)$ **59.** $\cos(\pi - x)$

60. $\sin 10° \cos 80° + \cos 10° \sin 80°$

61. $\cos 175° \cos 25° + \sin 175° \sin 25°$

62. $\cos \frac{2\pi}{5} \cos \frac{\pi}{10} - \sin \frac{2\pi}{5} \sin \frac{\pi}{10}$

63. $\cos\left(x - \frac{2\pi}{3}\right) - \cos\left(x + \frac{2\pi}{3}\right)$

64. $\sin\left(x + \frac{\pi}{6}\right) - \sin\left(x - \frac{\pi}{6}\right)$

65. $\tan(x + 45°) \tan(x - 45°)$

66. $\tan x \tan y + \tan y \tan z + \tan z \tan x$, where $y = x + 60°$ and $z = x - 60°$

67. (a) Show that $\tan \frac{\pi}{12} = 2 - \sqrt{3}$.
       Hint: $\frac{\pi}{12} = \frac{\pi}{4} - \frac{\pi}{6}$
       (b) Show that $\cot \frac{\pi}{12} = 2 + \sqrt{3}$.

68. If x, y, and z are acute angles with sines $\frac{3}{5}$, $\frac{5}{13}$, and $\frac{8}{17}$, respectively, compute $\sin(x + y + z)$.

In Exercises 69–100, prove that the equations are identities.

69. $\cot(x + y) = \dfrac{\cot x \cot y - 1}{\cot x + \cot y}$

70. $\cos 2x = \dfrac{1 - \tan^2 x}{1 + \tan^2 x}$

71. $\sin 2x = \dfrac{2 \tan x}{1 + \tan^2 x}$

72. $\sin^2 x - \sin^2 y = \sin(x + y) \sin(x - y)$

73. $\tan^2 x - \tan^2 y = \dfrac{\sin(x + y) \sin(x - y)}{\cos^2 x \cos^2 y}$

74. $2 \csc 2x = \sec x \csc x$

75. $(\sin x)[\tan(x/2) + \cot(x/2)] = 2$

76. $\tan[x + (\pi/4)] = (1 + \tan x)/(1 - \tan x)$

77. $\tan[(\pi/4) + x] - \tan[(\pi/4) - x] = 2 \tan 2x$

78. $\dfrac{\cot x - 1}{\cot x + 1} = \dfrac{1 - \sin 2x}{\cos 2x}$

79. $2 \sin\left(\dfrac{\pi}{4} - \dfrac{x}{2}\right) \cos\left(\dfrac{\pi}{4} - \dfrac{x}{2}\right) = \cos x$

80. $\dfrac{\tan(x + y) - \tan y}{1 + \tan(x + y) \tan y} = \tan x$

81. $\dfrac{1 - \tan(\frac{1}{4}\pi - t)}{1 + \tan(\frac{1}{4}\pi - t)} = \tan t$

82. $\tan(\frac{1}{4}\pi + \frac{1}{2}t) = \tan t + \sec t$

83. $\dfrac{\tan(\alpha - \beta) + \tan \beta}{1 - \tan(\alpha - \beta) \tan \beta} = \tan \alpha$

84. $(1 + \tan \theta)\left[1 + \tan(\frac{1}{4}\pi - \theta)\right] = 2$

85. $\tan 3\theta = \dfrac{3t - t^3}{1 - 3t^2}$, where $t = \tan \theta$

86. $\sin(t + \frac{2}{3}\pi) \cos(t - \frac{2}{3}\pi) + \cos(t + \frac{2}{3}\pi) \sin(t - \frac{2}{3}\pi)$
     $= \sin 2t$

87. $\tan 2x + \sec 2x = \dfrac{\cos x + \sin x}{\cos x - \sin x}$

88. $\cos^4 x - \sin^4 x = \cos 2x$

89. $2 \sin x + \sin 2x = 2 \sin^3 x/(1 - \cos x)$

90. $1 + \tan x \tan(x/2) = \sec x$

91. $\tan \dfrac{x}{2} = \dfrac{1 - \cos x + \sin x}{1 + \cos x + \sin x}$

92. $\dfrac{\sin 3x}{\sin x} - \dfrac{\cos 3x}{\cos x} = 2$

93. $\sin(x + y) \cos y - \cos(x + y) \sin y = \sin x$

94. $\dfrac{\sin x + \sin 2x}{\cos x - \cos 2x} = \cot \dfrac{x}{2}$

95. $\dfrac{1 - \tan^2(x/2)}{1 + \tan^2(x/2)} = \cos x$

96. $4 \sin(x/4) \cos(x/4) \cos(x/2) = \sin x$

97. $\sin 4x = 4 \sin x \cos x - 8 \sin^3 x \cos x$

98. $\cos 4x = 8 \cos^4 x - 8 \cos^2 x + 1$

99. $\sin 5x = 16 \sin^5 x - 20 \sin^3 x + 5 \sin x$

100. $\cos 5x = 16 \cos^5 x - 20 \cos^3 x + 5 \cos x$

In Exercises 101–112, find all solutions of each equation in the interval $[0, 2\pi)$. In cases where a calculator is necessary, round off the answers to two decimal places.

101. $\tan x = 4.26$

102. $\tan x = -4.26$

103. $\csc x = 2.24$

104. $\sin(\sin x) = \pi/6$

105. $\tan^2 x - 3 = 0$

106. $\cot^2 x - \cot x = 0$

107. $\sin x - \cos 2x + 1 = 0$

108. $\sin x + \sin 2x = 0$

109. $3 \csc x - 4 \sin x = 0$

110. $2 \sin^2 x + \sin x - 1 = 0$

111. $2 \sin^4 x - 3 \sin^2 x + 1 = 0$

112. $4 \sin^2 2x + \cos 2x - 2 \cos^2 x - 2 = 0$

In Exercises 113–122, evaluate each expression (without using a calculator or tables).

113. $\cos^{-1}(-\sqrt{2}/2)$

114. $\arctan(\sqrt{3}/3)$

115. $\sin^{-1} 0$

116. $\arcsin \frac{1}{2}$

117. $\sin[\arccos(-\frac{1}{2})]$

118. $\sin[\tan^{-1}(-1)]$

119. $\cot[\cos^{-1}(\frac{1}{2})]$

120. $\sec[\cos^{-1}(\sqrt{2}/3)]$

121. $\sin(\frac{3\pi}{2} + \arccos \frac{3}{5})$

122. $\tan[\frac{\pi}{4} + \sin^{-1}(\frac{5}{13})]$

123. In this exercise we investigate the relationship between the variables x and θ in the accompanying figure. (Assume that θ is in radians.)
(a) Refer to the figure. Show that
$$\theta = \tan^{-1}(10/x) - \tan^{-1}(4/x).$$

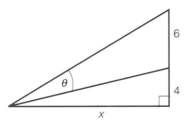

(b) Use your calculator to complete the following table. (Round off the results to two decimal places.) Which x-value in the table yields the largest value for θ?

x	0.1	1	2	3	10	100
θ						

(c) As indicated in the following graph, the value of x that makes θ as large as possible is a number between 5 and 10, closer to 5 than to 10. Using calculus, it can be shown that this value of x is, in fact, $2\sqrt{10}$. Use your calculator to evaluate $2\sqrt{10}$; check that the result is consistent with the given graph. What is the corresponding value of θ in this case? Also, give the coordinates of the highest point on the accompanying graph.

Round off both coordinates to two decimal places.

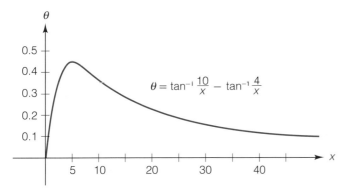

$$\theta = \tan^{-1}\frac{10}{x} - \tan^{-1}\frac{4}{x}$$

(d) What is the degree measure for the angle θ obtained in part (c)? Round off the answer to one decimal place.

In Exercises 124–128, show that each equation is an identity.

124. $\tan(\tan^{-1} x + \tan^{-1} y) = (x + y)/(1 - xy)$

125. $\tan^{-1}\left(x/\sqrt{1 - x^2}\right) = \sin^{-1} x$

126. $\sin(2 \arctan x) = 2x/(1 + x^2)$

127. $\cos(2 \cos^{-1} x) = 2x^2 - 1$

128. $\sin\left[\frac{1}{2} \sin^{-1}(x^2)\right] = \sqrt{\frac{1}{2} - \frac{1}{2}\sqrt{1 - x^4}}$

129. (a) Use a calculator to compute the quantity $\cos 20° \cos 40° \cos 60° \cos 80°$. Give your answer to as many decimal places as is shown on your calculator.
   (b) Now use a product-to-sum formula to *prove* that the display on your calculator is the exact value of the given expression, not an approximation.

130. In this exercise, we will use the following figure to derive the half-angle formula for sine:

$$\sin\tfrac{\theta}{2} = \pm\sqrt{\tfrac{1}{2}(1 - \cos \theta)}$$

Our derivation will make use of the formula for the distance between two points and the identity

$\sin^2 t + \cos^2 t = 1$. However, we will not rely on an addition formula for sine, as we did in Section 7.7.

$$x^2 + y^2 = 1$$

(a) Explain why the coordinates of P and Q are $P\left(\cos\frac{\theta}{2}, \sin\frac{\theta}{2}\right)$, and $Q(\cos \theta, \sin \theta)$.

(b) Use the distance formula to show that

$$(PQ)^2 = 2 - 2 \cos \tfrac{\theta}{2} \cos \theta - 2 \sin \tfrac{\theta}{2} \sin \theta$$

and

$$(PR)^2 = 2 - \cos \tfrac{\theta}{2}$$

(c) Explain why $\triangle POR$ is congruent to $\triangle QOP$.

(d) From part (c), it follows that $(PQ)^2 = (PR)^2$. By equating the expressions for $(PQ)^2$ and $(PR)^2$ [obtained in part (b)], show that

$$\sin \tfrac{\theta}{2} \sin \theta = \left(\cos \tfrac{\theta}{2}\right)(1 - \cos \theta)$$

(e) Square both sides of the equation obtained in part (d); then replace $\cos^2 \frac{\theta}{2}$ by $1 - \sin^2 \frac{\theta}{2}$ and show that the resulting equation can be written

$$\left(\sin^2 \tfrac{\theta}{2}\right)(2 - 2 \cos \theta) = (1 - \cos \theta)^2$$

(f) Solve the equation in part (e) for the quantity $\sin \frac{\theta}{2}$. You should obtain

$$\sin \tfrac{\theta}{2} = \pm\sqrt{(1 - \cos \theta)/2}, \qquad \text{as required}$$

CHAPTER SEVEN TEST

1. Evaluate each expression without using a calculator or tables.
   (a) $\cos(4\pi/3)$ (b) $\csc(-5\pi/6)$ (c) $\sin^2(3\pi/4) + \cos^2(3\pi/4)$

2. In the expression $1/\sqrt{16 - t^2}$, make the substitution $t = 4 \sin u$ and simplify the result. Assume that $0 < u < \pi/2$.

3. Graph the function $y = 0.5 \sec(4\pi x - 1)$ for one complete cycle.

4. Graph the function $y = -\sin\left(3x - \frac{\pi}{4}\right)$ for one complete cycle. Specify the amplitude, period, and phase shift.

5. Graph the function $y = 3 \tan(\pi x/4)$ on the interval $0 \le x \le 4$.

6. (a) Convert $175°$ to radian measure.
    (b) Convert 5 radians to degree measure.

7. Two points B and C are on a circle of radius 5 cm. The center of the circle is A, and angle BAC is $75°$.
    (a) Find the length of the (shorter) arc of the circle from B to C.
    (b) Find the area of the (smaller) sector determined by angle BAC.

8. A wheel rotates about its axis with an angular speed of 25 revolutions/sec.
    (a) Find the angular speed of the wheel in radians/sec.
    (b) Find the linear speed of a point on the wheel that is 5 cm from the center.

9. Prove that the following equation is an identity:

$$\frac{\cot \theta}{1 + \tan(-\theta)} + \frac{\tan \theta}{1 + \cot(-\theta)} = \cot \theta + \tan \theta + 1$$

10. Use an appropriate addition formula to simplify the expression $\sin(\theta + 3\pi/2)$.

11. Compute $\cos 2t$ given that $\sin t = -2\sqrt{5}/5$ and $3\pi/2 < t < 2\pi$.

12. Use a calculator to find all solutions of the equation $\sin x = 3 \cos x$ in the interval $(0, 2\pi)$.

13. Find all solutions of the equation $2 \sin^2 x + 7 \sin x + 3 = 0$ on the interval $0 \le x \le 2\pi$.

14. If $\cos \alpha = 2/\sqrt{5}$ $(3\pi/2 < \alpha < 2\pi)$ and $\sin \beta = \frac{4}{5}$ $(\pi/2 < \beta < \pi)$, compute $\sin(\beta - \alpha)$.

15. Find all solutions of the equation $\sin(x + 30°) = \sqrt{3} \sin x$ on the interval $0° < x < 90°$.

16. If $\csc \theta = -3$ and $\pi < \theta < 3\pi/2$, compute $\sin(\theta/2)$.

17. Compute each of the following quantities.
    (a) $\sin^{-1}[\sin(\pi/10)]$ (b) $\sin^{-1}(\sin 2\pi)$

18. Compute $\cos(\arcsin \frac{3}{4})$.

19. Prove that the following equation is an identity:

$$\tan\left(\frac{\pi}{4} + \frac{\theta}{2}\right) = \frac{1 + \cos \theta + \sin \theta}{1 + \cos \theta - \sin \theta}$$

20. Simplify the following expression: $\sec(\arctan \sqrt{x^2 - 1})$ (assume that $x > 1$).

CHAPTER EIGHT

ADDITIONAL TOPICS IN TRIGONOMETRY

The subject of trigonometry is an excellent example of a branch of mathematics . . . which was motivated by both practical and intellectual interests—surveying, map-making, and navigation on the one hand, and curiosity about the size of the universe on the other. With it the Alexandrian mathematicians triangulated the universe and rendered precise their knowledge about the Earth and the heavens.

Morris Klein in *Mathematics in Western Culture* (New York: Oxford University Press, 1953)

INTRODUCTION

In this chapter we develop six topics that require a background in basic trigonometry. The first two of these topics, discussed in Section 8.1, are the law of sines and the law of cosines. These laws relate the angles and the lengths of the sides in any triangle. The next two sections, 8.2 and 8.3, introduce the important topic of vectors, first from a geometric standpoint, then from an algebraic standpoint. In Sections 8.4 through 8.6 we expand upon some of the ideas in Chapters 2 and 3 on graphs and equations. The topics presented here are parametric equations and polar coordinates.

8.1 THE LAW OF SINES AND THE LAW OF COSINES

The ratio of the sides of a triangle to each other is the same as the ratio of the sines of the opposite angles.

Bartholomaus Pitiscus (1561–1613) in his text, *Trigonometriae sive de dimensione triangulorum libri quinque.* This text was first published in Frankfort, Germany, in 1595, and according to several historians of mathematics, it was the first satisfactory textbook on trigonometry.

In this section we discuss two formulas relating the sides and the angles in any triangle: the *law of sines* and the *law of cosines.* These formulas can be used to determine an unknown side or angle using given information about the triangle. As you will see, which formula to apply in a particular case depends upon what data are initially given. To help you keep track of this, we'll use the following notations from elementary geometry.

NOTATION	EXPLANATION
SAA	One side and two angles are given.
SSA	Two sides and an angle opposite one of those sides are given.
SAS	Two sides and the included angle are given.
SSS	Three sides of the triangle are given.

Also, we will often follow the convention of denoting the angles of a triangle by A, B, and C and the lengths of the corresponding opposite sides by a, b, and c (see Figure 1). With this notation, we are ready to state the law of sines.

Law of Sines

In any triangle, the sines of the angles are proportional to the lengths of the opposite sides:

$$\frac{\sin A}{a} = \frac{\sin B}{b} = \frac{\sin C}{c}$$

FIGURE 1

The proof of the law of sines is easy. We use the following result from Section 6.4: The area of any triangle is equal to half the product of two sides times the sine of the included angle. Thus, with reference to Figure 1, we have

$$\tfrac{1}{2}bc \sin A = \tfrac{1}{2}ac \sin B = \tfrac{1}{2}ab \sin C$$

since each of these three expressions equals the area of triangle ABC. Now we just multiply through by the quantity $2/abc$ to obtain

$$\frac{\sin A}{a} = \frac{\sin B}{b} = \frac{\sin C}{c}$$

which completes the proof.

■■■ EXAMPLE I (SAA) Find the length x in Figure 2.

Solution We can determine x directly by applying the law of sines. We have

$$\frac{\sin 30°}{20} = \frac{\sin 135°}{x}$$

Length of side opposite ⎱——↑   ↑——⎰ Length of side opposite
the 30° angle ⎰   ⎱ the 135° angle

$$x \sin 30° = 20 \sin 135°$$
$$x\left(\frac{1}{2}\right) = 20\left(\frac{\sqrt{2}}{2}\right)$$
$$x = 20\sqrt{2}\ \text{cm}$$

FIGURE 2

■■■ EXAMPLE 2 (SAA) Find the length y in Figure 3 (on the next page).

Solution First we need to determine the angle θ in Figure 3. Since the sum of the angles in any triangle is 180°, we have

$$\theta = 180° - (75° + 62°) = 43°$$

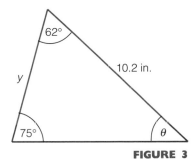

FIGURE 3

Now, using the law of sines, we obtain

$$\frac{\sin 75°}{10.2} = \frac{\sin 43°}{y}$$

$$y \sin 75° = (10.2)\sin 43°$$

$$y = \frac{(10.2)\sin 43°}{\sin 75°}$$

Without the use of a calculator or tables, this is the final form for the answer. On the other hand (as you should check for yourself), a calculator yields

$$y \approx 7.2 \text{ in.}$$

▬▬▬

▮ ▬▬ ▮ **EXAMPLE 3** (SSA) In triangle *ABC*, we are given $\angle C = 45°$, $b = 4\sqrt{2}$ ft, and $c = 8$ ft. Determine the remaining side and angles.

Solution First let's draw a preliminary sketch conveying the given data; see Figure 4. (The sketch must be considered tentative. At the outset, we don't know whether the other angles are acute or even whether the given data are compatible.) To find angle *B*, we have (according to the law of sines)

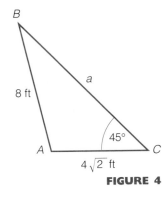

FIGURE 4

$$\frac{\sin B}{4\sqrt{2}} = \frac{\sin 45°}{8}$$

and therefore

$$8 \sin B = 4\sqrt{2} \sin 45° = 4\sqrt{2}\left(\frac{\sqrt{2}}{2}\right) = 4$$

$$\sin B = \frac{1}{2}$$

From our previous work, we know that one possibility for *B* is 30°, since $\sin 30° = \frac{1}{2}$. However, there is another possibility. Since the reference angle for 150° is 30°, we know that sin 150° is also equal to $\frac{1}{2}$. Which angle do we want? For the problem at hand, this is easy to answer. Since angle *C* is given as 45°, angle *B* cannot equal 150°, for the sum of 45° and 150° exceeds 180°. We conclude that

$$\angle B = 30°$$

Next, since $\angle B = 30°$ and $\angle C = 45°$, we have

$$\angle A = 180° - (30° + 45°) = 105°$$

Finally, we use the law of sines to find *a*. From the equation $\dfrac{\sin A}{a} = \dfrac{\sin C}{c}$, we have $a \sin C = c \sin A$, and therefore

$$a = \frac{c \sin A}{\sin C} = \frac{8 \sin 105°}{\sin 45°} = \frac{8 \sin 105°}{1/\sqrt{2}}$$

$$= 8\sqrt{2} \sin 105° \text{ ft} \approx 10.9 \text{ ft}$$

▬▬▬

In the preceding example, two possibilities arose for the angle *B*: both 30° and 150°. However, it turned out that the value 150° was incompatible with the given information in the problem. In Exercise 13 at the end of this section, you will see a case in which both of two possibilities are compatible with the given data. This results in two distinct solutions to the problem. In contrast to this, Exercise 11(a) shows a case in which there is no triangle fulfilling the given conditions. For these reasons, the case SSA is sometimes referred to as the **ambiguous case.**

Now we turn to the law of cosines.

Law of Cosines

In any triangle, the square of the length of any side equals the sum of the squares of the lengths of the other two sides minus twice the product of the lengths of those other two sides times the cosine of their included angle.

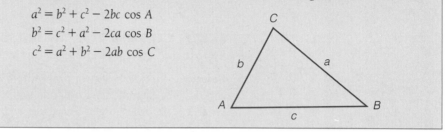

$$a^2 = b^2 + c^2 - 2bc \cos A$$
$$b^2 = c^2 + a^2 - 2ca \cos B$$
$$c^2 = a^2 + b^2 - 2ab \cos C$$

Before looking at a proof of this law, we make two preliminary comments. First, it is important to understand that the three equations in the box all follow the same pattern. For example, look at the first equation:

$$\underset{\text{sides that include this angle}}{\overset{\text{side and opposite angle}}{a^2 = b^2 + c^2 - 2bc \cos A}}$$

Now check for yourself that the other two equations also follow this pattern. It is the pattern that is important here; after all, not every triangle is labeled *ABC*.

The second observation is that the law of cosines is a generalization of the Pythagorean theorem. In fact, look what happens to the equation

$$a^2 = b^2 + c^2 - 2bc \cos A$$

when angle *A* is a right angle:

$$a^2 = b^2 + c^2 - 2bc \underbrace{\cos 90°}_{0}$$

$$a^2 = b^2 + c^2 \qquad \text{which is the Pythagorean theorem}$$

Now let us prove the law of cosines:

$$a^2 = b^2 + c^2 - 2bc \cos A$$

(The other two equations can be proved in the same way. Indeed, just relabeling the figure would suffice.) The proof that we give uses coordinate geometry in a very nice way to complement the trigonometry. We begin by placing angle *A* in standard position, as indicated in Figure 5. (So, in the figure, angle *A* is then identified with angle *CAB*.) Then if *u* and *v* denote the lengths indicated in Figure 5, the coordinates of *C* are (u, v) and we have

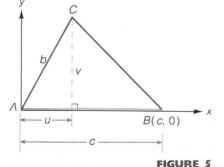

FIGURE 5

$$\cos A = \frac{\text{adjacent}}{\text{hypotenuse}} = \frac{u}{b} \qquad \text{and therefore} \qquad u = b \cos A$$

Similarly, we have

$$\sin A = \frac{\text{opposite}}{\text{hypotenuse}} = \frac{v}{b} \qquad \text{and therefore} \qquad v = b \sin A$$

Thus, the coordinates of C are

$$(b \cos A, b \sin A)$$

(Exercise 44 at the end of this section asks you to check that these represent the coordinates of C even when angle A is not acute.) Now we use the distance formula,

$$d = \sqrt{(x_2 - x_1)^2 + (y_2 - y_1)^2}$$

to compute the required distance a between the points $C(b \cos A, b \sin A)$ and $B(c, 0)$. We have

$$\begin{aligned} a &= \sqrt{(b \cos A - c)^2 + (b \sin A - 0)^2} \\ &= \sqrt{b^2 \cos^2 A - 2bc \cos A + c^2 + b^2 \sin^2 A} \\ &= \sqrt{b^2 \underbrace{(\cos^2 A + \sin^2 A)}_{1} - 2bc \cos A + c^2} \\ &= \sqrt{b^2 + c^2 - 2bc \cos A} \end{aligned}$$

Squaring both sides of this last equation gives us the law of cosines, as we set out to prove.

EXAMPLE 4 (SAS) Compute the length x in Figure 6.

Solution The law of cosines is directly applicable. We have

$$\begin{aligned} x^2 &= 7^2 + 8^2 - 2(7)(8)\cos 120° \\ &= 49 + 64 - 112\left(-\tfrac{1}{2}\right) = 169 \\ x &= \sqrt{169} = 13 \text{ cm} \end{aligned}$$

7 cm

120°

x

8 cm

FIGURE 6

If the equation $a^2 = b^2 + c^2 - 2bc \cos A$ is solved for $\cos A$, the result is

$$\cos A = \frac{b^2 + c^2 - a^2}{2bc}$$

This expresses the cosine of an angle in a triangle in terms of the lengths of the sides. In a similar fashion, we obtain the corresponding formulas

$$\cos B = \frac{c^2 + a^2 - b^2}{2ca} \qquad \text{and} \qquad \cos C = \frac{a^2 + b^2 - c^2}{2ab}$$

These alternative forms for the law of cosines are used in the next example.

EXAMPLE 5 (SSS) In triangle ABC, the sides are $a = 3$ units, $b = 5$ units, and $c = 7$ units. Find the angles.

Solution Figure 7 summarizes the given data. We have

B

$c = 7$

$a = 3$

C

$b = 5$

A

FIGURE 7

$$\cos A = \frac{b^2 + c^2 - a^2}{2bc} = \frac{5^2 + 7^2 - 3^2}{2(5)(7)} = \frac{65}{70} = \frac{13}{14}$$

Now that we know $\cos A = \frac{13}{14}$, we can use a calculator (set in the degree mode) to find $\angle A$:

KEYSTROKES	OUTPUT
13 $\div$ 14 $=$ (INV) (COS)	21.7867893

Thus, $\angle A \approx 21.8°$. In a similar manner, we have

$$\cos B = \frac{c^2 + a^2 - b^2}{2ca} = \frac{7^2 + 3^2 - 5^2}{2(7)(3)} = \frac{33}{42} = \frac{11}{14}$$

So, $\cos B = \frac{11}{14}$ and a calculator then yields

$$\angle B \approx 38.2132107° \approx 38.2°$$

At this point, we can find $\angle C$ in either of two ways. Each has its advantage. The first way is to begin by computing $\cos C$ in the same way that we found $\cos A$ and $\cos B$. As you can check, the result is $\cos C = -\frac{1}{2}$. A calculator is not needed in this case. We know from previous work that $\angle C$ must be 120°. The second method that can be used relies on the fact that the sum of the angles in a triangle is 180°. Thus, we have

$$\angle C = 180° - \angle A - \angle B$$
$$\approx 180° - 21.7867893° - 38.2132107°$$
$$\approx 120.0°$$

This way is quicker than the first method. The disadvantage, however, is that we know that $\angle C$ is only approximately 120°. The first method (using the cosine law) is longer, but it tells us that $\angle C$ is exactly 120°. In summary, then, the three required angles are

$$\angle A \approx 21.8° \qquad \angle B \approx 38.2° \qquad \angle C = 120° \qquad \blacksquare\blacksquare\blacksquare$$

We conclude this section with an example indicating how the law of sines and the law of cosines are used in navigation. In this example, you'll see the term *bearing* used in specifying the location of one point relative to another. To explain this term, we refer to Figure 8.

FIGURE 8

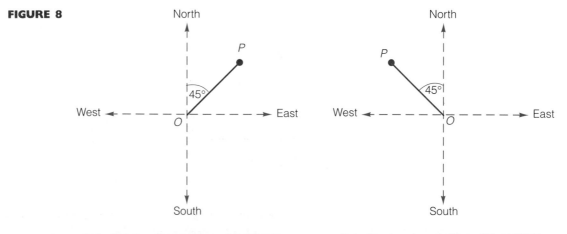

(a) The bearing of P from O is N45°E. (b) The bearing of P from O is N45°W.

In Figure 8(a), the bearing of *P* from *O* is N45°E (read "north, 45° east"). This bearing tells us the *acute* angle between line segment $\overline{OP}$ and the north–south line through *O*. In Figure 8(b), the bearing of *P* from *O* is N45°W (read "north, 45° west"). Again, note that the bearing gives us the acute angle between line segment $\overline{OP}$ and the north–south line through *O*. Figure 9 provides additional examples of this terminology.

FIGURE 9

The bearing is specified by means of the acute angle measured from the north–south line.

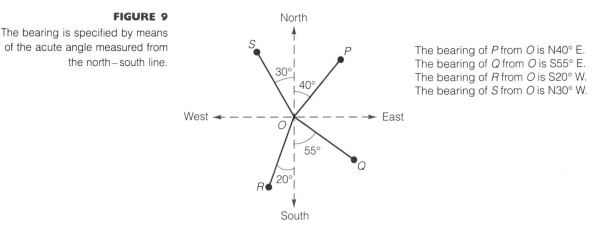

The bearing of *P* from *O* is N40° E.
The bearing of *Q* from *O* is S55° E.
The bearing of *R* from *O* is S20° W.
The bearing of *S* from *O* is N30° W.

EXAMPLE 6 A small fire is sighted from ranger stations *A* and *B*. The bearing of the fire from *A* is N35°E, and the bearing of the fire from *B* is N49°W. Station *A* is 1.3 miles due west of station *B*.

(a) How far is the fire from each ranger station?
(b) At fire station *C*, which is 1.5 miles from *A*, there is a helicopter that can be used to drop water on the fire. If the bearing of *C* from *A* is S42°E, find the distance from *C* to the fire, and find the bearing of the fire from *C*.

Solution

(a) In Figure 10 we have sketched the situation involving ranger stations *A* and *B* and the fire (denoted by *F*). We compute the angles of $\triangle ABF$ as follows.

FIGURE 10

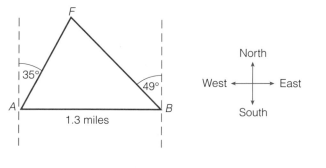

$$\angle FAB = 90° - 35° = 55° \qquad \angle FBA = 90° - 49° = 41°$$
$$\angle F = 180° - (55° + 41°) = 84°$$

We can now use the law of sines in $\triangle ABF$ to compute AF and BF.

$$\frac{\sin(\angle F)}{AB} = \frac{\sin(\angle FBA)}{AF} \qquad\qquad \frac{\sin(\angle F)}{AB} = \frac{\sin(\angle FAB)}{BF}$$

and therefore $\qquad\qquad\qquad\qquad\qquad$ and therefore

$$\frac{\sin 84°}{1.3} = \frac{\sin 41°}{AF} \qquad\qquad\qquad \frac{\sin 84°}{1.3} = \frac{\sin 55°}{BF}$$

or $\qquad\qquad\qquad\qquad\qquad\qquad\qquad$ or

$$AF = \frac{(1.3)\sin 41°}{\sin 84°} \qquad\qquad\qquad BF = \frac{(1.3)\sin 55°}{\sin 84°}$$

We use a calculator to evaluate these expressions for AF and BF. As you should check for yourself, the results (rounded to the nearest tenth of a mile) are

$$AF \approx 0.9 \text{ miles} \qquad \text{and} \qquad BF \approx 1.1 \text{ miles}$$

(b) We draw a sketch of the situation, as shown in Figure 11. In Figure 11, we can compute CF, the distance from the helicopter to the fire, using the law of cosines in $\triangle CAF$. First, note that $\angle CAF = 48° + 55° = 103°$. So we have

$$CF = \sqrt{AC^2 + AF^2 - 2 \cdot AC \cdot AF \cdot \cos 103°}$$
$$= \sqrt{1.5^2 + \left(\frac{1.3 \sin 41°}{\sin 84°}\right)^2 - 2(1.5)\left(\frac{1.3 \sin 41°}{\sin 84°}\right)\cos 103°}$$
$$\approx 1.9 \text{ miles} \qquad \text{using a calculator and rounding to one decimal place}$$

In order to find the bearing of the fire at F from the fire station at C, we need to determine the angle α in Figure 11. First we find the angle β. Using the law of sines in $\triangle CAF$, we have

$$\frac{\sin \beta}{AF} = \frac{\sin 103°}{CF}$$

and therefore

$$\sin \beta = \frac{AF \cdot \sin 103°}{CF}$$

Before using the inverse sine function and our calculator to compute β, we note from Figure 11 that β is an acute angle (because β is an angle in $\triangle CAF$, and in that triangle, $\angle CAF$ is greater than 90°). So in this particular application of the law of sines, there is no ambiguity. We have then

$$\beta = \sin^{-1}\left(\frac{AF \cdot \sin 103°}{CF}\right)$$

Now, on the right-hand side of this last equation, we substitute the expressions we obtained previously for AF and CF; then, using a calculator set in the degree mode, we obtain

$$\beta \approx 26° \qquad \text{rounding off to the nearest degree}$$

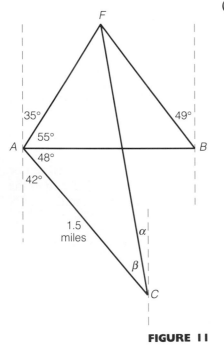

FIGURE 11

(You should verify this calculator value for yourself.) Now that we know β, we can determine the bearing of the fire from station C. From Figure 11, we have

$$\alpha + \beta = 42° \qquad \text{(Why?)}$$

and therefore

$$\alpha = 42° - \beta = 42° - 26° = 16° \qquad \text{(to the nearest one degree)}$$

In summary now, fire station C is approximately 1.9 miles from the fire, and the bearing of the fire from station C is N16°W.   ▰▰

EXERCISE SET 8.1

A

In Exercises 1–8, assume that the vertices and the lengths of the sides of a triangle are labeled as in Figure 1 in the text. For Exercises 1–4, leave your answers in terms of radicals or the trigonometric functions; that is, don't use a calculator. In Exercises 5–8, use a calculator and round off your final answers to one decimal place.

1. If $\angle A = 60°$, $\angle B = 45°$, and $BC = 12$ cm, find AC.
2. If $\angle A = 30°$, $\angle B = 135°$, and $BC = 4$ cm, find AC.
3. If $\angle B = 100°$, $\angle C = 30°$, and $AB = 10$ cm, find BC.
4. If $\angle A = \angle B = 35°$, and $AB = 16$ cm, find AC and BC.
5. If $\angle A = 36°$, $\angle B = 50°$, and $b = 12.61$ cm, find a and c.
6. If $\angle B = 81°$, $\angle C = 55°$, and $b = 6.24$ cm, find c and a.
7. If $a = 29.45$ cm, $b = 30.12$ cm, and $\angle B = 66°$, find the remaining side and angles of the triangle.
8. If $a = 52.15$ cm, $c = 42.90$ cm, and $\angle A = 125°$, find the remaining side and angles of the triangle.

In Exercises 9 and 10, use degree measure for your answers. In parts (c) and (d), use a calculator and round off the results to one decimal place.

9. (a) In $\triangle ABC$, $\sin B = \sqrt{2}/2$. What are the possible values for $\angle B$?
   (b) In $\triangle DEF$, $\cos E = \sqrt{2}/2$. What are the possible values for $\angle E$?
   (c) In $\triangle GHI$, $\sin H = \frac{1}{4}$. What are the possible values for $\angle H$?
   (d) In $\triangle JKL$, $\cos K = -\frac{2}{3}$. What are the possible values for $\angle K$?
10. (a) In $\triangle ABC$, $\sin B = \sqrt{3}/2$. What are the possible values for $\angle B$?
    (b) In $\triangle DEF$, $\cos E = -\sqrt{3}/2$. What are the possible values for $\angle E$?

(c) In $\triangle GHI$, $\sin H = \frac{2}{9}$. What are the possible values for $\angle H$?
(d) In $\triangle JKL$, $\cos K = \frac{2}{3}$. What are the possible values for $\angle K$?

11. (a) Show that there is no triangle satisfying the conditions $a = 2.0$ ft, $b = 6.0$ ft, and $\angle A = 23.1°$. *Hint:* Try computing $\sin B$ using the law of sines.
    (b) If $a = 2.0$ ft, $b = 3.0$ ft, $\angle A = 23.1°$, and $\angle B$ is obtuse, show that $c = 1.1$ ft.

12. (a) Show that there is no triangle with $a = 2$, $b = 3$, and $\angle A = 42°$.
    (b) Is there any triangle in which $a = 2$, $b = 3$, and $\angle A = 41°$?

13. Let $b = 1$, $a = \sqrt{2}$, and $\angle B = 30°$.
    (a) Use the law of sines to show that $\sin A = \sqrt{2}/2$. Conclude that $\angle A = 45°$ or $\angle A = 135°$.
    (b) Assuming that $\angle A = 45°$, determine the remaining parts of $\triangle ABC$.
    (c) Assuming that $\angle A = 135°$, determine the remaining parts of $\triangle ABC$.
    (d) Find the area of the two triangles.

14. Let $a = 30$, $b = 36$, and $\angle A = 20°$.
    (a) Show that $\angle B = 24.23°$ or $\angle B = 155.77°$ (rounding off to two decimal places).
    (b) Determine the remaining parts for each of the two possible triangles. Round off your final results to two decimal places. [However, in your calculations, do *not* work with values that have been rounded off. (Why?)]
    (c) Find the areas of the two triangles.

15. Find the lengths a, b, c, and d in the following figure. Leave your answers in terms of trigonometric functions (rather than using a calculator).

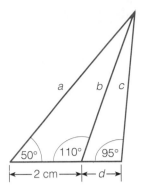

17. Two points P and Q are on opposite sides of a river (see the sketch). From P to another point R on the same side is 300 ft. Angles PRQ and RPQ are found to be 20° and 120°, respectively. Compute the distance from P to Q, across the river. (Round off your answer to the nearest foot.)

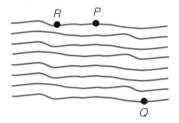

16. In the following figure, $\overline{PQR}$ is a straight line segment. Find the distance $\overline{PR}$. Round off your final answer to two decimal places.

18. Determine the angle θ in the accompanying figure. Round off your answer to two decimal places.

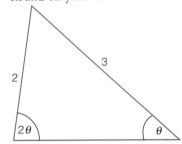

In Exercises 19–22, use the law of cosines to determine the length x in each figure. For Exercises 19 and 20, leave your answers in terms of radicals. In Exercises 21 and 22, use a calculator and round off the answers to one decimal place.

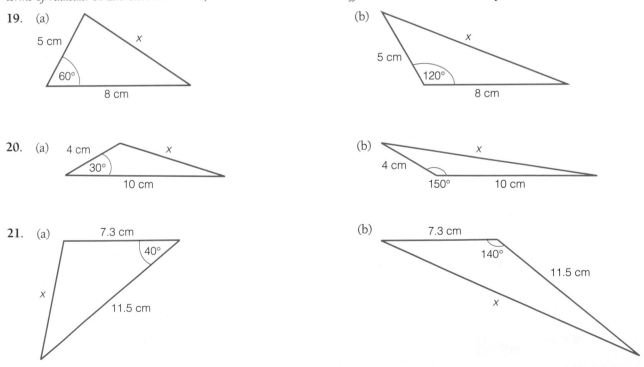

19. (a)

(b)

20. (a)

(b)

21. (a)

(b)

22. (a)

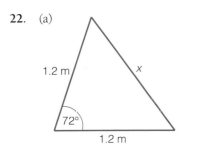

In Exercises 23 and 24, refer to the following figure.

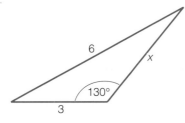

23. In applying the law of cosines to the figure, a student incorrectly writes $x^2 = 3^2 + 6^2 - 2(3)(6)\cos 130°$. Why is this incorrect? What is the correct equation?

24. In applying the law of cosines to the figure, a student writes $6^2 = 3^2 + x^2 + 6x \cos 50°$. Why is this correct?

In Exercises 25 and 26, use the given information to find the cosine of each angle in $\triangle ABC$.

25. $a = 6$ cm, $b = 7$ cm, $c = 10$ cm

26. $a = 17$ cm, $b = 8$ cm, $c = 15$ cm (For this particular triangle, you can check your answers, because there is an alternate method of solution that does not require the law of cosines.)

In Exercises 27–30, compute each angle of the given triangle. Where necessary, use a calculator and round off to one decimal place.

27. $a = 7$, $b = 8$, $c = 13$ **28.** $a = 33$, $b = 7$, $c = 37$

29. $a = b = 2/\sqrt{3}$, $c = 2$ **30.** $a = 36$, $b = 77$, $c = 85$

In Exercises 31–34, round off each answer to one decimal place.

31. A regular pentagon is inscribed in a circle of unit radius. Find the perimeter of the pentagon. *Hint:* First find the length of a side using the law of cosines.

32. Find the perimeter of a regular nine-sided polygon inscribed in a circle of radius 4 cm. (See the hint for Exercise 31.)

33. In triangle ABC, $\angle A = 40°$, $b = 6.1$ cm, and $c = 3.2$ cm.
(a) Find a using the law of cosines.
(b) Find angle C using the law of sines.
(c) Find angle B.

(b)

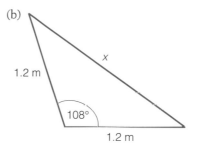

34. In parallelogram $ABCD$, you are given $AB = 6$ in., $AD = 4$ in., and $\angle A = 40°$. Find the length of each diagonal.

35. Town B is 26 miles from town A at a bearing of S15°W. Town C is 54 miles from town A at a bearing of S7°E. Compute the distance from town B to town C. Round off your final answer to the nearest mile.

36. Town C is 5 miles due east of town D. Town E is 12 miles from town C at a bearing (from C) of N52°E.
(a) How far apart are towns D and E? (Round off to the nearest one-half mile.)
(b) Find the bearing of town E from town D. (Round off the angle to the nearest degree.)

37. An airplane crashes in a lake and is spotted by observers at lighthouses A and B along the coast. Lighthouse B is 1.50 miles due east of lighthouse A. The bearing of the airplane from lighthouse A is S20°E; the bearing of the plane from lighthouse B is S42°W. Find the distance from each lighthouse to the crash site. (Round off your final answer to two decimal places.)

38. (Continuation of Exercise 37) A rescue boat is in the lake, three-fourths of a mile from lighthouse B, and at a bearing of S35°E from lighthouse B.
(a) Find the distance from the rescue boat to the airplane. Express your answer using miles and feet, with the portion in feet rounded off to the nearest ten feet.
(b) Find the bearing of the plane from the rescue boat. (Your answer should have the form of Sθ°W. Round off θ to two decimal places.)

39. (Refer to the figure.) When the Sun is viewed from the Earth, it subtends an angle of $\theta = 32'$ $\left(= \frac{32}{60} \text{ degree}\right)$. Assuming that the distance d from the Earth to the Sun is 92,690,000 miles, compute the diameter D of the Sun. Round off the answer to the nearest ten thousand miles.

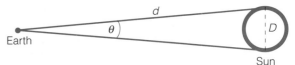

40. Compute the lengths CD and CE in the accompanying figure. Round off the final answers to two decimal places.

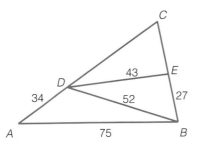

B

41. (a) Let m and n be positive numbers, with $m > n$. Furthermore, suppose that in triangle ABC the lengths a, b, and c are given by

$$a = 2mn + n^2 \qquad b = m^2 - n^2$$
$$c = m^2 + n^2 + mn$$

Show that $\cos C = -\frac{1}{2}$ and conclude that $\angle C = 120°$.

(b) Give an example of a triangle in which the lengths of the sides are whole numbers and one of the angles is 120°. (Specify the three sides; you needn't find the other angles.)

42. If the lengths of two adjacent sides of a parallelogram are a and b, and if the acute angle formed by these two sides is θ, show that the product of the lengths of the two diagonals is given by the expression

$$\sqrt{(a^2 + b^2)^2 - 4a^2b^2 \cos^2 \theta}$$

43. Two trains leave the railroad station at noon. The first train travels along a straight track at 90 mph. The second train travels at 75 mph along another straight track that makes an angle of 130° with the first track. At what time are the trains 400 miles apart? Round off your answer to the nearest minute.

44. In this exercise, you will complete a detail mentioned in the text in the proof of the law of cosines. Let the positive numbers u and v denote the lengths indicated in the following figure, so that the coordinates of C are $(-u, v)$. Show that $u = -b \cos A$ and $v = b \sin A$. Conclude from this that the coordinates of C are

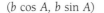

$(b \cos A, b \sin A)$

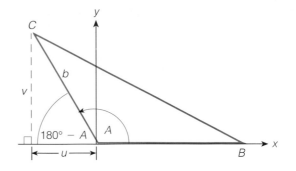

Hint: Use the right-triangle definitions for cosine and sine along with the addition formulas for $\cos(180° - \theta)$ and $\sin(180° - \theta)$.

45. The following figure shows a quadrilateral with sides a, b, c, and d, inscribed in a circle.

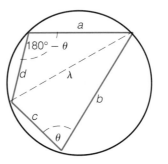

If λ denotes the length of the diagonal indicated in the figure, prove that

$$\lambda^2 = \frac{(ab + cd)(ac + bd)}{bc + ad}$$

This result is known as **Brahmagupta's theorem.** It is named after its discoverer, a seventh-century Hindu mathematician. *Hint:* Assume as given the theorem from geometry stating that when a quadrilateral is inscribed in a circle, the opposite angles are supplementary. Apply the law of cosines in both of the triangles in the figure to obtain expressions for λ^2. Then eliminate $\cos \theta$ from one equation.

46. Prove the following identity for $\triangle ABC$:

$$\frac{\cos A}{a} + \frac{\cos B}{b} + \frac{\cos C}{c} = \frac{a^2 + b^2 + c^2}{2abc}$$

Suggestion: Use the law of cosines to substitute for a^2, for b^2, and for c^2 in the numerator of the expression on the right-hand side.

47. In $\triangle ABC$, suppose that $a^4 + b^4 + c^4 = 2(a^2 + b^2)c^2$. Find $\angle C$. (There are two answers.) *Hint:* Solve the given equation for c^2.

48. In this section we have seen that the cosines of the angles in a triangle can be expressed in terms of the lengths of the sides. For instance, for cos A in $\triangle ABC$, we obtained $\cos A = (b^2 + c^2 - a^2)/2bc$. This exercise shows how to derive corresponding expressions for the sines of the angles. For ease of notation in this exercise, let us agree to use the letter T to denote the following quantity:

$$T = 2(a^2b^2 + b^2c^2 + c^2a^2) - (a^4 + b^4 + c^4)$$

Then the sines of the angles in $\triangle ABC$ are given by

$$\sin A = \frac{\sqrt{T}}{2bc} \qquad \sin B = \frac{\sqrt{T}}{2ac} \qquad \sin C = \frac{\sqrt{T}}{2ab}$$

In the steps that follow, we'll derive the first of these three formulas, the derivations for the other two being entirely similar.

(a) In $\triangle ABC$, why is the positive root always appropriate in the formula $\sin A = \sqrt{1 - \cos^2 A}$?

(b) In the formula in part (a), replace $\cos A$ by $(b^2 + c^2 - a^2)/2bc$ and show that the result can be written

$$\sin A = \frac{\sqrt{4b^2c^2 - (b^2 + c^2 - a^2)^2}}{2bc}$$

(c) On the right-hand side of the equation in part (b), carry out the indicated multiplication. After combining like terms, you should obtain $\sin A = \sqrt{T}/2bc$, as required.

49. In the two easy steps that follow, we derive the law of sines by using the formulas obtained in Exercise 48. (Since the formulas in Exercise 48 were obtained using the law of cosines, we are, in essence, showing how to derive the law of sines from the law of cosines.)

(a) Use the formulas in Exercise 48 to check that each of the three fractions $(\sin A)/a$, $(\sin B)/b$, and $(\sin C)/c$ is equal to $\sqrt{T}/2abc$.

(b) Conclude from part (a) that $(\sin A)/a = (\sin B)/b = (\sin C)/c$.

50. In this exercise you are going to use the law of cosines and the law of sines to determine the area of the shaded equilateral triangle in Figure A. Begin by labeling points, as shown in Figure B.

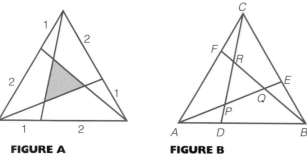

FIGURE A   **FIGURE B**

(a) Apply the law of cosines in $\triangle ABE$ to show that $AE = \sqrt{7}$.

(b) Apply the law of sines in $\triangle ABE$ to show that $\sin(\angle AEB) = (3\sqrt{3})/(2\sqrt{7})$.

(c) Apply the law of sines in $\triangle CFB$ to show that $\sin(\angle FBC) = (\sqrt{3})/(2\sqrt{7})$. *Hint:* By symmetry, $BF = AE$.

(d) Apply the law of sines in $\triangle QEB$ to show that $QE = 1/\sqrt{7}$.

(e) Apply the law of sines in $\triangle QEB$ to show that $QB = 3/\sqrt{7}$.

(f) Show that $PQ = 3\sqrt{7}/7$. *Hint:* $PQ = AE - (QE + AP)$, and by symmetry, $AP = QB$.

(g) Use the result in part (f) to find the area of equilateral triangle PQR.

51. This exercise is adapted from a problem proposed by Professor Norman Schaumberger in the May 1990 issue of *The College Mathematics Journal*. In the accompanying figure, radius $OA = 1$ and $\angle AFC = 60°$. Follow steps (a) through (f) to prove that

$$AD \cdot AB - AE \cdot AC = BC$$

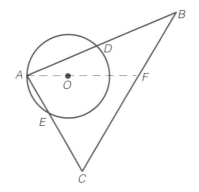

(a) Let $\angle BAF = \alpha$. Show that $AD = 2\cos \alpha$. *Hint:* In isosceles triangle AOD, drop a perpendicular from O to side $\overline{AD}$.

(b) Let $\angle CAF = \beta$. Show that $AE = 2\cos \beta$.

(c) Explain why $\angle B = 60° - \alpha$ and $\angle C = 120° - \beta$.

(d) Using the law of sines, show that

$$AB = \frac{BC \sin(120° - \beta)}{\sin(\alpha + \beta)}$$

and

$$AC = \frac{BC \sin(60° - \alpha)}{\sin(\alpha + \beta)}$$

(e) Using the results in parts (a), (b), and (d), verify that

$$AD \cdot AB - AE \cdot AC =$$

$$BC\left[\frac{2\cos \alpha \sin(120° - \beta) - 2\cos \beta \sin(60° - \alpha)}{\sin(\alpha + \beta)}\right]$$

(f) To complete the proof, you need to show that the quantity in brackets in part (e) is equal to 1. In other words, you want to show that

$$2 \cos \alpha \sin(120° - \beta) - 2 \cos \beta \sin(60° - \alpha)$$
$$= \sin(\alpha + \beta)$$

Use the addition formulas for sine to prove that this last equation is indeed an identity.

C

52. *Heron's formula* Approximately 2000 years ago, Heron of Alexandria derived a formula for the area of a triangle in terms of the lengths of the sides. A more modern derivation of Heron's formula is indicated in the steps that follow.

(a) Use the expression for sin A in Exercise 48(b) to show that

$$\sin^2 A = \frac{(a - b + c)(a + b - c)(b + c - a)(b + c + a)}{4b^2c^2}$$

Hint: Use difference-of-squares factoring repeatedly.

(b) Let s denote one-half of the perimeter of $\triangle ABC$. That is, let $s = \frac{1}{2}(a + b + c)$. Using this notation (which is due to Euler), verify that
(i) $a + b + c = 2s$
(ii) $-a + b + c = 2(s - a)$
(iii) $a - b + c = 2(s - b)$
(iv) $a + b - c = 2(s - c)$
Then, using this notation and the result in part (a), show that

$$\sin A = \frac{2\sqrt{s(s - a)(s - b)(s - c)}}{bc}$$

Note: Since sin A is positive (why?), the positive root is appropriate here.

(c) Use the result in part (b) and the formula area $\triangle ABC = \frac{1}{2}bc \sin A$ to conclude that

$$\text{area} \triangle ABC = \sqrt{s(s - a)(s - b)(s - c)}$$

This is Heron's formula. For historical background and a purely geometric proof, see *An Introduction to the History of Mathematics,* 6th ed., by Howard Eves (Philadelphia: Saunders College Publishing, 1990), pp. 178 and 194.

8.2 VECTORS IN THE PLANE, A GEOMETRIC APPROACH

The idea of a parallelogram of velocities may be found in various ancient Greek authors, and the concept of a parallelogram of forces was not uncommon in the sixteenth and seventeenth centuries.

Michael J. Crowe in *A History of Vector Analysis* (Notre Dame, Ind.: University of Notre Dame Press, 1967)

Vector notation is compact. If we can express a law of physics in vector form we usually find it easier to understand and to manipulate mathematically.

David Halliday and Robert Resnick in *Fundamentals of Physics,* 3rd ed. (New York: John Wiley and Sons, 1988)

Certain quantities, such as temperature, length, and mass, can be specified by means of a single number (assuming that a system of units has been agreed on). We call these quantities **scalars.** On the other hand, quantities such as force and velocity are characterized by both a *magnitude* (a positive number) and a *direction.* We call these quantities **vectors.**

Geometrically, a vector is a directed line segment or arrow. The vector in Figure 1, for instance, represents a wind velocity of 5 mph from the west. The length of this vector represents the magnitude of the wind velocity, while the direction of the vector indicates the direction of the wind velocity. As another example, the vector in Figure 2 represents a force acting on an object: the mag-

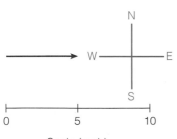

FIGURE 1
A vector representing a wind velocity of 5 mph from the west.

FIGURE 2

A vector representing a force of 3 lb acting at an angle of 135° with the horizontal.

FIGURE 3

nitude of the force is 3 pounds, and the force acts at an angle of 135° with the horizontal.

In a moment, we are going to discuss the important concept of vector addition, but first let us agree on some matters of notation. Suppose that we have a vector drawn from a point P to a point Q, as shown in Figure 3. The point P in Figure 3 is called the **initial point** of the vector, and Q is the **terminal point**. We can denote this vector by

$$\overrightarrow{PQ}$$

The length or magnitude of the vector $\overrightarrow{PQ}$ is denoted by $|\overrightarrow{PQ}|$. On the printed page, vectors are often indicated by boldface letters, such as $\mathbf{a}$, $\mathbf{A}$, and $\mathbf{v}$.

A word about notation. As you know, the notation (a, b) can denote either a point in the x-y plane or an open interval on the number line. In each instance, however, there is usually no danger of confusion; the context makes it clear which meaning is intended. Now we have a similar situation with the notation for the length of a vector. As has perhaps already occurred to you, the same vertical bars that we are using to denote the length of a vector are also used, in another context, to denote the absolute value of a real number. Again, it will be clear from the context which meaning is intended. Some books avoid this situation by using double bars to indicate the length of a vector: $\|\mathbf{v}\|$. In this text, we use the notation $|\mathbf{v}|$ simply because that is the one found in most calculus books.

If two vectors $\mathbf{a}$ and $\mathbf{b}$ have the same length and the same direction, we say that they are *equal* and we write $\mathbf{a} = \mathbf{b}$ (see Figure 4). Notice that our definition for vector equality involves magnitude and direction, but not location. Thus, when it is convenient to do so, we are free to move a given vector to another location, provided we do not alter the magnitude or direction.

FIGURE 4

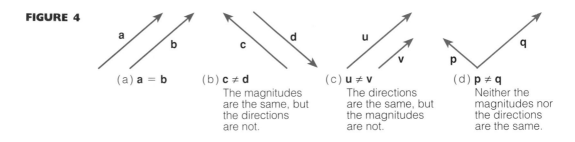

(a) $\mathbf{a} = \mathbf{b}$

(b) $\mathbf{c} \neq \mathbf{d}$
The magnitudes are the same, but the directions are not.

(c) $\mathbf{u} \neq \mathbf{v}$
The directions are the same, but the magnitudes are not.

(d) $\mathbf{p} \neq \mathbf{q}$
Neither the magnitudes nor the directions are the same.

FIGURE 5

As motivation for the definition of vector addition, let us suppose that an object moves from a point P to a point Q. Then we can represent this *displacement* by the vector $\overrightarrow{PQ}$. (Indeed, the word *vector* is derived from the Latin *vectus*, meaning "carried.") Now suppose that after moving from P to Q, the object moves from Q to R. Then, as you can see in Figure 5, the net effect is a displacement from P to R. We say in this case that the vector $\overrightarrow{PR}$ is the **sum** or **resultant** of the vectors $\overrightarrow{PQ}$ and $\overrightarrow{QR}$, and we write

$$\overrightarrow{PQ} + \overrightarrow{QR} = \overrightarrow{PR}$$

These ideas are formalized in the definition that follows.

DEFINITION Vector Addition

Let **u** and **v** be two vectors. Position **v** (without changing its magnitude or direction) so that its initial point coincides with the terminal point of **u**, as in Figure 6(a). Then, as indicated in Figure 6(b), the vector **u** + **v** is the directed line segment from the initial point of **u** to the terminal point of **v**. The vector **u** + **v** is called the **sum** or **resultant** of **u** and **v**.

FIGURE 6

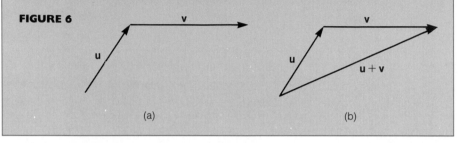

(a) (b)

EXAMPLE 1 Referring to Figure 7, (a) determine the initial and terminal points of **u** + **v**; (b) compute $|\mathbf{u} + \mathbf{v}|$.

FIGURE 7

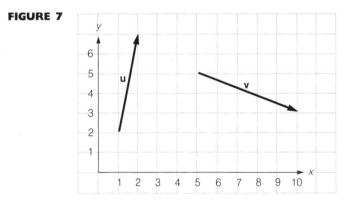

Solution

(a) According to the definition, we first need to move **v** (without changing its length or direction) so that its initial point coincides with the terminal point of **u**. From Figure 7, we see that this can be accomplished by moving each point of **v** three units in the negative x-direction and two units in the positive y-direction. Figure 8(a) shows the new location of **v**, and Figure 8(b) indicates the sum **u** + **v**. From Figure 8(b), we see that the initial and terminal points of **u** + **v** are $(1, 2)$ and $(7, 5)$, respectively.

(b) We can use the distance formula to determine $|\mathbf{u} + \mathbf{v}|$. Using the points $(1, 2)$ and $(7, 5)$ that were obtained in part (a), we have

$$|\mathbf{u} + \mathbf{v}| = \sqrt{(7 - 1)^2 + (5 - 2)^2} = \sqrt{45} = \sqrt{9 \cdot 5} = 3\sqrt{5}$$

FIGURE 8

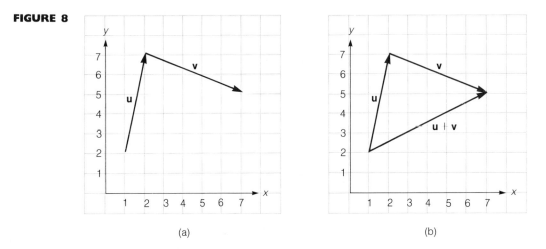

(a)   (b)

One important consequence of our definition for vector addition is that this operation is commutative. That is, for any two vectors **u** and **v**, we have

$$\mathbf{u} + \mathbf{v} = \mathbf{v} + \mathbf{u}$$

Figure 9 indicates why this is so.

In view of Figure 9, vector addition can also be carried out by using the **parallelogram law.** To determine **u** + **v**, position **u** and **v** so that their initial points coincide. Then, as indicated in Figure 10, the vector **u** + **v** is the directed diagonal of the parallelogram determined by **u** and **v**.

FIGURE 9

The upper triangle shows the sum **u** + **v**, while the lower triangle shows the sum **v** + **u**. Since in both cases the resultant is $\overrightarrow{PQ}$, it follows that **u** + **v** = **v** + **u**.

FIGURE 10

The parallelogram law for vector addition.

It is a fact—and it has been verified experimentally—that if two forces **F** and **G** act on an object, the net effect is the same as if just the resultant **F** + **G** acted on the object. In Example 2, we use the parallelogram law to compute the resultant of two forces. Note that the units of force used in this example are **newtons** (N), where $1\,\text{N} \approx 0.2248\,\text{lb}$.

EXAMPLE 2   Two forces **F** and **G** act on an object. As indicated in Figure 11, the force **G** acts horizontally to the right with a magnitude of 12 N, while **F** acts vertically upward with a magnitude of 16 N. Determine the magnitude and direction of the resultant force.

FIGURE 11

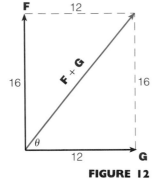

FIGURE 12

Solution We complete the parallelogram, as shown in Figure 12. Now we need to calculate the length of **F** + **G** and the angle θ. Applying the Pythagorean theorem in Figure 12, we have

$$|\mathbf{F} + \mathbf{G}| = \sqrt{12^2 + 16^2} = \sqrt{144 + 256} = \sqrt{400} = 20$$

Also from Figure 12, we have

$$\tan \theta = \frac{16}{12} = \frac{4}{3}$$

Consequently,

$$\theta = \tan^{-1}\frac{4}{3} \approx 53.1° \qquad \text{using a calculator set in the degree mode}$$

Now we can summarize our results. The magnitude of **F** + **G** is 20 N, and the angle θ between **F** + **G** and the horizontal is (approximately) 53.1°. ▮▮▮

In Example 2, we determined the resultant for two perpendicular forces. The next example shows how to compute the resultant when the forces are not perpendicular. Our calculations will make use of both the law of sines and the law of cosines.

FIGURE 13

▮▮▮▮ **EXAMPLE 3** Determine the resultant of the two forces in Figure 13. (Round off the answers to one decimal place.)

Solution As in the previous example, we complete the parallelogram. In Figure 14, the angle in the lower right-hand corner of the parallelogram is 140°. This is because the sum of two adjacent angles in any parallelogram is always 180°. Letting d denote the length of the diagonal in Figure 14, we can use the law of cosines to write

$$d^2 = 15^2 + 5^2 - 2(15)(5)\cos 140°$$
$$d = \sqrt{250 - 150 \cos 140°} = \sqrt{250 + 150 \cos 40°} \qquad \text{(Why?)}$$
$$= \sqrt{25(10 + 6 \cos 40°)} = 5\sqrt{10 + 6 \cos 40°}$$
$$\approx 19.1 \qquad \text{using a calculator}$$

FIGURE 14

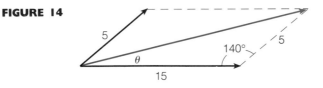

So the magnitude of the resultant is 19.1 N (to one decimal place). To specify the direction of the resultant, we need to determine the angle θ in Figure 14. Using the law of sines, we have

$$\frac{\sin \theta}{5} = \frac{\sin 140°}{d}$$

and, consequently,

$$\sin \theta = \frac{5 \sin 40°}{d} \qquad \text{(Why?)}$$

$$= \frac{5 \sin 40°}{5\sqrt{10 + 6 \cos 40°}} = \frac{\sin 40°}{\sqrt{10 + 6 \cos 40°}}$$

Using a calculator now, we obtain

$$\theta = \sin^{-1}\left(\frac{\sin 40°}{\sqrt{10 + 6 \cos 40°}}\right) \approx 9.7°$$

In summary, the magnitude of the resultant force is about 19.1 N, and the angle θ between the resultant and the 15-N force is approximately 9.7°. ▮▮▮

As background for the next example, we introduce the notion of *components* of a vector. (You will see this concept again in the next section, but in a more algebraic context.) Suppose that the initial point of a vector **v** is located at the origin of a rectangular coordinate system, as shown in Figure 15. Now suppose we draw perpendiculars from the terminal point of **v** to the axes, as indicated by the blue dashed lines in Figure 15. Then the coordinates v_x and v_y are called the **components** of the vector **v** in the *x*- and *y*-directions, respectively. For an example involving components, refer back to Figure 12. The horizontal component of the vector **F** + **G** is 12 N, and the vertical component is 16 N.

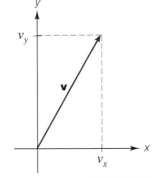

FIGURE 15

▮▮ **EXAMPLE 4** Determine the horizontal and vertical components of the velocity vector **v** in Figure 16.

Solution   From Figure 16, we can write

$$\cos 30° = \frac{\text{adjacent}}{\text{hypotenuse}} = \frac{v_x}{70}$$

and, consequently,

$$v_x = (\cos 30°)(70) = \frac{\sqrt{3}}{2}(70)$$

$$= 35\sqrt{3} \approx 61 \text{ cm/sec} \qquad \text{(to two significant digits)}$$

Similarly, we have

$$\sin 30° = \frac{v_y}{70}$$

and therefore,

$$v_y = (\sin 30°)(70) = 35 \text{ cm/sec}$$

In summary, now, the *x*-component of the velocity is about 61 cm/sec, and the *y*-component is 35 cm/sec. ▮▮▮

FIGURE 16

Our last example in this section will indicate how vectors are used in navigation. First, however, let us introduce some terminology. Suppose that an airplane has a **heading** of due east. This means that the airplane is pointed due east, and if there were no wind effects, the plane would indeed travel due east

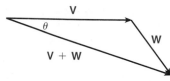

FIGURE 17

with respect to the ground. The **air speed** is the speed of the airplane relative to the air, whereas the **ground speed** is the plane's speed relative to the ground. Again, if there were no wind effects, then the air speed and the ground speed would be equal. Now suppose that the heading and air speed of an airplane are represented by the velocity vector **V** in Figure 17. (The direction of **V** is the heading; the magnitude of **V** is the air speed.) Also suppose that the wind velocity is represented by the vector **W** in Figure 17. Then the vector sum **V + W** represents the actual velocity of the plane with respect to the ground. The direction of **V + W** is called the **course**; it is the direction in which the airplane is moving with respect to the ground. The magnitude of **V + W** is the ground speed (which was defined previously). The angle θ in Figure 17 is called, naturally enough, the **drift angle**.

In navigation, directions are given in terms of the angle measured clockwise from true north. For example, the direction of the velocity vector **v** in Figure 18 is 120°.

FIGURE 18   North

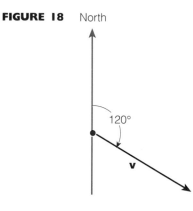

EXAMPLE 5   [Refer to Figure 19(a).] The heading and air speed of an airplane are 60° and 250 mph, respectively. If the wind is 40 mph from 150°, find the ground speed, the drift angle, and the course.

FIGURE 19

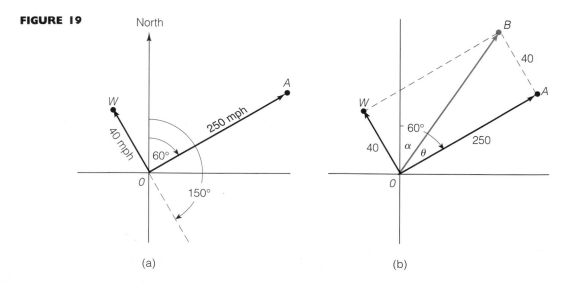

(a)

(b)

Solution In Figure 19(a), the vector $\overrightarrow{OA}$ represents the air speed of 250 mph and the heading of 60°. The vector $\overrightarrow{OW}$ represents a wind of 40 mph from 150°. Also, angle $WOA = 90°$. (Why?) In Figure 19(b), we have completed the parallelogram to obtain the vector sum $\overrightarrow{OA} + \overrightarrow{OW} = \overrightarrow{OB}$. The length of $\overrightarrow{OB}$ represents the ground speed, and θ is the drift angle. Because triangle BOA is a right triangle, we have

$$\tan \theta = \frac{40}{250} = \frac{4}{25} \qquad \Big| \qquad |\overrightarrow{OB}| = \sqrt{250^2 + 40^2}$$
$$\theta = 9.1° \qquad \qquad \approx 253.2$$

From these calculations, we conclude that the ground speed is approximately 253.2 mph and the drift angle is about 9.1°. We still need to compute the course (that is, the direction of vector $\overrightarrow{OB}$). From Figure 19(b), we have

$$\alpha = 60° - \theta \approx 60° - 9.1° = 50.9°$$

Thus, the course is 50.9° (to one decimal place). (We are using the convention, mentioned previously, that directions are given in terms of the angle measured clockwise from true north.) ∎

EXERCISE SET 8.2

A

In Exercises 1–26, assume that the coordinates of the points P, Q, R, S, and O are as follows:

$$P(-1, 3) \qquad Q(4, 6) \qquad R(4, 3)$$
$$S(5, 9) \qquad O(0, 0)$$

For each exercise, draw the indicated vector (using graph paper) and compute its magnitude. In Exercises 7–20, compute the sums using the definition given on page 454. In Exercises 21–26, use the parallelogram law to compute the sums.

1. $\overrightarrow{PQ}$	2. $\overrightarrow{QP}$
3. $\overrightarrow{SQ}$	4. $\overrightarrow{QS}$
5. $\overrightarrow{OP}$	6. $\overrightarrow{PO}$
7. $\overrightarrow{PQ} + \overrightarrow{QS}$	8. $\overrightarrow{SQ} + \overrightarrow{QP}$
9. $\overrightarrow{OP} + \overrightarrow{PQ}$	10. $\overrightarrow{OS} + \overrightarrow{SQ}$
11. $(\overrightarrow{OS} + \overrightarrow{SQ}) + \overrightarrow{QP}$	12. $(\overrightarrow{OS} + \overrightarrow{SP}) + \overrightarrow{PR}$
13. $\overrightarrow{OP} + \overrightarrow{QS}$	14. $\overrightarrow{QS} + \overrightarrow{PO}$
15. $\overrightarrow{SR} + \overrightarrow{PO}$	16. $\overrightarrow{OS} + \overrightarrow{QO}$
17. $\overrightarrow{OP} + \overrightarrow{RQ}$	18. $\overrightarrow{OP} + \overrightarrow{QR}$
19. $\overrightarrow{SQ} + \overrightarrow{RO}$	20. $\overrightarrow{SQ} + \overrightarrow{OR}$
21. $\overrightarrow{OP} + \overrightarrow{OR}$	22. $\overrightarrow{OP} + \overrightarrow{OQ}$
23. $\overrightarrow{RP} + \overrightarrow{RS}$	24. $\overrightarrow{QP} + \overrightarrow{QR}$
25. $\overrightarrow{SO} + \overrightarrow{SQ}$	26. $\overrightarrow{SQ} + \overrightarrow{SR}$

In Exercises 27–32, the vectors **F** and **G** denote two forces that act on an object: **G** acts horizontally to the right, and **F** acts vertically upward. In each case, use the information that is given to compute $|\mathbf{F} + \mathbf{G}|$ and θ, where θ is the angle between **G** and the resultant.

27. $|\mathbf{F}| = 4\,\text{N}, \quad |\mathbf{G}| = 5\,\text{N}$
28. $|\mathbf{F}| = 15\,\text{N}, \quad |\mathbf{G}| = 6\,\text{N}$
29. $|\mathbf{F}| = |\mathbf{G}| = 9\,\text{N}$
30. $|\mathbf{F}| = 28\,\text{N}, \quad |\mathbf{G}| = 1\,\text{N}$
31. $|\mathbf{F}| = 3.22\,\text{N}, \quad |\mathbf{G}| = 7.21\,\text{N}$
32. $|\mathbf{F}| = 4.06\,\text{N}, \quad |\mathbf{G}| = 26.83\,\text{N}$

In Exercises 33–38, the vectors **F** and **G** represent two forces acting on an object, as indicated in the following figure. In each case, use the given information to compute (to two decimal places) the magnitude and direction of the resultant. (Give the direction of the resultant by specifying the angle θ between **F** and the resultant.)

33. $|\mathbf{F}| = 5\,\text{N}$, $|\mathbf{G}| = 4\,\text{N}$, $\alpha = 80°$
34. $|\mathbf{F}| = 8\,\text{N}$, $|\mathbf{G}| = 10\,\text{N}$, $\alpha = 60°$
35. $|\mathbf{F}| = 16\,\text{N}$, $|\mathbf{G}| = 25\,\text{N}$, $\alpha = 35°$
36. $|\mathbf{F}| = 4.24\,\text{N}$, $|\mathbf{G}| = 9.01\,\text{N}$, $\alpha = 45°$
37. $|\mathbf{F}| = 50\,\text{N}$, $|\mathbf{G}| = 25\,\text{N}$, $\alpha = 130°$
38. $|\mathbf{F}| = 1.26\,\text{N}$, $|\mathbf{G}| = 2.31\,\text{N}$, $\alpha = 160°$

In Exercises 39–46, the initial point for each vector is the origin, and θ denotes the angle (measured counterclockwise) from the x-axis to the vector. In each case, compute the horizontal and vertical components of the given vector. (Round off your answers to two decimal places.)

39. The magnitude of $\mathbf{V}$ is 16 cm/sec, and $\theta = 30°$.
40. The magnitude of $\mathbf{V}$ is 40 cm/sec, and $\theta = 60°$.
41. The magnitude of $\mathbf{F}$ is 14 N, and $\theta = 75°$.
42. The magnitude of $\mathbf{F}$ is 23.12 N, and $\theta = 52°$.
43. The magnitude of $\mathbf{V}$ is 1 cm/sec, and $\theta = 135°$.
44. The magnitude of $\mathbf{V}$ is 12 cm/sec, and $\theta = 120°$.
45. The magnitude of $\mathbf{F}$ is 1.25 N, and $\theta = 145°$.
46. The magnitude of $\mathbf{F}$ is 6.34 N, and $\theta = 175°$.

In Exercises 47–50, use the given flight data to compute the ground speed, the drift angle, and the course. (Round off your answers to two decimal places.)

47. The heading and air speed are 30° and 300 mph, respectively; the wind is 25 mph from 120°.
48. The heading and air speed are 45° and 275 mph, respectively; the wind is 50 mph from 135°.
49. The heading and air speed are 100° and 290 mph, respectively; the wind is 45 mph from 190°.
50. The heading and air speed are 90° and 220 mph, respectively; the wind is 80 mph from
(a) 180°; (b) 90°.

B
51. A block weighing 12 lb rests on an inclined plane, as indicated in Figure A. Determine the components of the weight perpendicular to and parallel to the plane. Round off your answers to two decimal places.
Hint: In Figure B, the component of the weight perpendicular to the plane is $|\overrightarrow{OR}|$; the component parallel to the plane is $|\overrightarrow{OP}|$. Why does angle QOR equal 35°?

FIGURE A

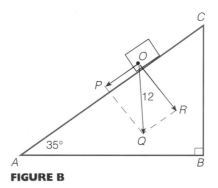

FIGURE B

In Exercises 52 and 53, you are given the weight of a block on an inclined plane, along with the angle θ that the inclined plane makes with the horizontal. In each case, determine the components of the weight perpendicular to and parallel to the plane. (Round off your answers to two decimal places where necessary.)

52. 15 lb; $\theta = 30°$ 53. 12 lb; $\theta = 10°$

54. A block rests on an inclined plane that makes an angle of 20° with the horizontal. The component of the weight parallel to the plane is 34.2 lb.
   (a) Determine the weight of the block. (Round off your answer to one decimal place.)
   (b) Determine the component of the weight perpendicular to the plane. (Round off your answer to one decimal place.)

55. In Section 8.3, we will see that vector addition is associative. That is, for any three vectors $\mathbf{A}$, $\mathbf{B}$, and $\mathbf{C}$, we have $(\mathbf{A} + \mathbf{B}) + \mathbf{C} = \mathbf{A} + (\mathbf{B} + \mathbf{C})$. In this exercise, you are going to check that this property holds in a particular case. Let $\mathbf{A}$, $\mathbf{B}$, and $\mathbf{C}$ be the vectors with initial and terminal points as follows:

Vector	Initial Point	Terminal Point
A	$(-1, 2)$	$(2, 4)$
B	$(1, 2)$	$(3, 0)$
C	$(6, 2)$	$(4, -3)$

(a) Use the definition of vector addition on page 454 to determine the initial and terminal points of (**A** + **B**) + **C**. *Suggestion:* Use graph paper.

(b) Use the definition of vector addition to determine the initial and terminal points of **A** + (**B** + **C**). [Your answers should agree with those in part (a).]

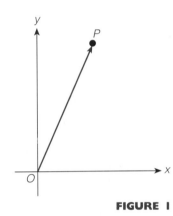

FIGURE 1

8.3 VECTORS IN THE PLANE, AN ALGEBRAIC APPROACH

A great many of the mathematical ideas that apply to physics and engineering are collected in the concept of vector spaces. This branch of mathematics has applications in such practical problems as calculating the vibrations of bridges and airplane wings. Logical extensions to spaces of infinitely many dimensions are widely used in modern theoretical physics as well as in many branches of mathematics itself.

From *The Mathematical Sciences,* edited by the Committee on Support of Research in the Mathematical Sciences with the collaboration of George Boehm (Cambridge, Mass.: The MIT Press, 1969)

The geometric concept of a vector in the plane can be recast in an algebraic setting. This is useful both for computational purposes and (as our opening quotation implies) for more advanced work.

Consider an *x-y* coordinate system and a vector $\overrightarrow{OP}$ with initial point the origin, as shown in Figure 1. We call $\overrightarrow{OP}$ the **position vector** (or **radius vector**) of the point *P*. Most of our work in this section will involve such position vectors. There is no loss of generality in focusing on these types of vectors, for, as indicated in Figure 2, each vector **v** in the plane is equal to a unique position vector $\overrightarrow{OP}$.

FIGURE 2

FIGURE 3
The vector $\overrightarrow{OP}$ is denoted by $\langle a, b \rangle$. The coordinates *a* and *b* are the components of the vector.

If the coordinates of the point *P* are (a, b), we call *a* and *b* the **components** of the vector $\overrightarrow{OP}$, and we use the notation

$$\langle a, b \rangle$$

to denote this vector (see Figure 3). The number *a* is the **horizontal component** or ***x*-component** of the vector; *b* is the **vertical component** or ***y*-component**.

In the previous section, we said that two vectors are equal provided they have the same length and the same direction. For vectors $\langle a, b \rangle$ and $\langle c, d \rangle$, this implies that

$$\langle a, b \rangle = \langle c, d \rangle \qquad \text{if and only if} \qquad a = c \quad \text{and} \quad b = d$$

(Notice the similarity here to equality of complex numbers. Two complex numbers are equal provided their corresponding real and imaginary parts are equal; two vectors are equal provided their corresponding components are equal.)

It's easy to calculate the length of a vector $\mathbf{v}$ when its components are given. Suppose that $\mathbf{v} = \langle v_1, v_2 \rangle$. Applying the Pythagorean theorem in Figure 4, we have

$$|\mathbf{v}|^2 = v_1^2 + v_2^2$$

and, consequently,

$$|\mathbf{v}| = \sqrt{v_1^2 + v_2^2}$$

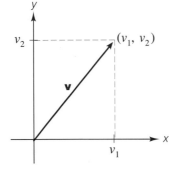

FIGURE 4

Although Figure 4 shows the point (v_1, v_2) in the first quadrant, you can check for yourself that the same formula results when (v_1, v_2) is located in any of the other three quadrants. We therefore have the following general formula.

The Length of a Vector

If $\mathbf{v} = \langle v_1, v_2 \rangle$, then

$$|\mathbf{v}| = \sqrt{v_1^2 + v_2^2}$$

■■■ **EXAMPLE 1** Compute the length of a vector $\mathbf{v} = \langle 2, -4 \rangle$.

Solution $|\mathbf{v}| = \sqrt{v_1^2 + v_2^2} = \sqrt{2^2 + (-4)^2} = \sqrt{20} = 2\sqrt{5}$ ■■■

Even if the initial point of a vector $\mathbf{v}$ is not the origin, we can still find the *components* of $\mathbf{v}$ by determining the components of the equivalent position vector. The formula that we are going to derive for this is as follows.

If the coordinates of the points P and Q are $P(x_1, y_1)$ and $Q(x_2, y_2)$, then

$$\overrightarrow{PQ} = \langle x_2 - x_1, y_2 - y_1 \rangle$$

To derive this formula, we first construct the right triangle shown in Figure 5(a).

FIGURE 5

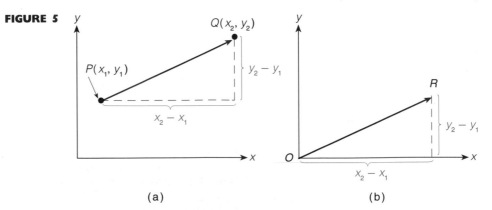

(a)   (b)

Now we let R denote the point $(x_2 - x_1, y_2 - y_1)$ and draw the position vector $\overrightarrow{OR}$ shown in Figure 5(b). Since the right triangles in Figures 5(a) and 5(b) are congruent and have corresponding legs that are parallel, we have

$$\overrightarrow{PQ} = \overrightarrow{OR} = \langle x_2 - x_1, y_2 - y_1 \rangle$$

as required. We can summarize this result as follows. For any vector $\mathbf{v}$, we have

x-component of $\mathbf{v}$:   (x-coordinate of terminal point) $-$ (x-coordinate of initial point)

y-component of $\mathbf{v}$:   (y-coordinate of terminal point) $-$ (y-coordinate of initial point)

▌▌ ▌ EXAMPLE 2 Let P and Q be the points $P(3, 1)$ and $Q(7, 3)$. Find the components of $\overrightarrow{PQ}$.

Solution x-component of $\overrightarrow{PQ} = x$-coordinate of $Q - x$-coordinate of P

$$= 7 - 3 = 4$$

y-component of $\overrightarrow{PQ} = y$-coordinate of $Q - y$-coordinate of P

$$= 3 - 1 = 2$$

Consequently,

$$\overrightarrow{PQ} = \langle 4, 2 \rangle$$ ▌▌▌

Vector addition is particularly simple to carry out when the vectors are in component form. Indeed, we can use the parallelogram law to verify the following result.

> **THEOREM   Vector Addition**
>
> If $\mathbf{u} = \langle u_1, u_2 \rangle$ and $\mathbf{v} = \langle v_1, v_2 \rangle$, then $\mathbf{u} + \mathbf{v} = \langle u_1 + v_1, u_2 + v_2 \rangle$.

This theorem tells us that vector addition can be carried out *componentwise*; in other words, to add two vectors, just add the corresponding components. For example,

$$\langle 1, 2 \rangle + \langle 3, 7 \rangle = \langle 4, 9 \rangle$$

To see why this theorem is valid, consider Figure 6, where we have completed the parallelogram. (*Caution:* In reading the derivation that follows, don't confuse notation such as OA with $\overrightarrow{OA}$; recall that OA denotes the length of the line segment $\overline{OA}$.) Since the x-component of $\mathbf{u}$ is u_1, we have

$$OB = u_1$$

Also,

$$BC = v_1 \qquad \text{(Why?)}$$

Therefore,

$$OC = OB + BC = u_1 + v_1$$

But OC is the x-component of the vector $\overrightarrow{OP}$ and, by the parallelogram law, $\overrightarrow{OP} = \mathbf{u} + \mathbf{v}$. In other words, the x-component of $\mathbf{u} + \mathbf{v}$ is $u_1 + v_1$, as we wished to show. The fact that the y-component of $\mathbf{u} + \mathbf{v}$ is $u_2 + v_2$ is proved in a similar fashion.

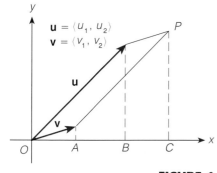

FIGURE 6

The vector $\langle 0, 0 \rangle$ is called the **zero vector**, and it is denoted by **0**. Notice that for any vector $\mathbf{v} = \langle v_1, v_2 \rangle$, we have

$$\mathbf{v} + \mathbf{0} = \mathbf{v} \qquad \text{because} \qquad \langle v_1, v_2 \rangle + \langle 0, 0 \rangle = \langle v_1, v_2 \rangle$$

and

$$\mathbf{0} + \mathbf{v} = \mathbf{v} \qquad \text{because} \qquad \langle 0, 0 \rangle + \langle v_1, v_2 \rangle = \langle v_1, v_2 \rangle$$

So for the operation of vector addition, the zero vector plays the same role as does the real number zero in addition of real numbers. There are, in fact, several other ways in which vector addition resembles ordinary addition of real numbers. We'll return to this point again near the end of this section.

In the box that follows, we define an operation called **scalar multiplication**, in which a vector is "multiplied" by a real number (a **scalar**) to obtain another vector.

DEFINITION   Scalar Multiplication

	EXAMPLES
For each real number k and each vector $\mathbf{v} = \langle x, y \rangle$, we define a vector $k\mathbf{v}$ by the equation $$k\mathbf{v} = k\langle x, y \rangle = \langle kx, ky \rangle$$	If $\mathbf{v} = \langle 2, 1 \rangle$, then $$2\mathbf{v} = \langle 4, 2 \rangle \quad 3\mathbf{v} = \langle 6, 3 \rangle$$ $$0\mathbf{v} = \langle 0, 0 \rangle = \mathbf{0}$$ $$-1\mathbf{v} = \langle -2, -1 \rangle$$

In geometric terms, the length of $k\mathbf{v}$ is $|k|$ times the length of $\mathbf{v}$. The vectors $\mathbf{v}$ and $k\mathbf{v}$ have the same direction if $k > 0$ and opposite directions if $k < 0$. For example, let $\mathbf{v} = \langle 2, 1 \rangle$. In Figure 7(a) we show the vectors $\mathbf{v}$ and $3\mathbf{v}$, while in Figure 7(b) we show $\mathbf{v}$ and $-2\mathbf{v}$.

FIGURE 7

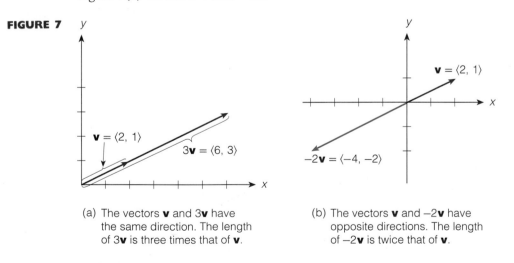

(a) The vectors **v** and 3**v** have the same direction. The length of 3**v** is three times that of **v**.

(b) The vectors **v** and −2**v** have opposite directions. The length of −2**v** is twice that of **v**.

▌�▌ **EXAMPLE 3**   Let $\mathbf{v} = \langle 3, 4 \rangle$ and $\mathbf{w} = \langle -1, 2 \rangle$. Compute each of the following.

(a) $\mathbf{v} + \mathbf{w}$ **(b)** $-2\mathbf{v} + 3\mathbf{w}$ **(c)** $|-2\mathbf{v} + 3\mathbf{w}|$

Solution

(a) $\mathbf{v} + \mathbf{w} = \langle 3, 4 \rangle + \langle -1, 2 \rangle = \langle 3 - 1, \, 4 + 2 \rangle = \langle 2, 6 \rangle$

(b) $-2\mathbf{v} + 3\mathbf{w} = -2\langle 3, 4 \rangle + 3\langle -1, 2 \rangle = \langle -6, -8 \rangle + \langle -3, 6 \rangle$
$$= \langle -9, -2 \rangle$$

(c) $|-2\mathbf{v} + 3\mathbf{w}| = |\langle -9, -2 \rangle| = \sqrt{(-9)^2 + (-2)^2} = \sqrt{85}$ ▮▮▮

For each vector $\mathbf{v}$, we define a vector $-\mathbf{v}$, called the **negative** of $\mathbf{v}$, by the equation

$$-\mathbf{v} = -1\mathbf{v}$$

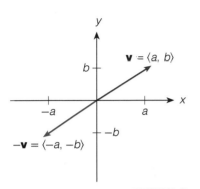

FIGURE 8

Thus, if $\mathbf{v} = \langle a, b \rangle$, then $-\mathbf{v} = \langle -a, -b \rangle$. As indicated in Figure 8, the vectors $\mathbf{v}$ and $-\mathbf{v}$ have the same length but opposite directions.

We can use the ideas in the preceding paragraph to define **vector subtraction**. Given two vectors $\mathbf{u}$ and $\mathbf{v}$, we define a vector $\mathbf{u} - \mathbf{v}$ by the equation

$$\mathbf{u} - \mathbf{v} = \mathbf{u} + (-\mathbf{v}) \tag{1}$$

First let's see what equation (1) is saying in terms of components, then we will indicate a simple geometric interpretation of vector subtraction.

If $\mathbf{u} = \langle u_1, u_2 \rangle$ and $\mathbf{v} = \langle v_1, v_2 \rangle$, then equation (1) tells us that

$$\mathbf{u} - \mathbf{v} = \langle u_1, u_2 \rangle + \langle -v_1, -v_2 \rangle$$
$$= \langle u_1 + (-v_1), u_2 + (-v_2) \rangle$$

That is,

$$\mathbf{u} - \mathbf{v} = \langle u_1 - v_1, \; u_2 - v_2 \rangle \tag{2}$$

In other words, to subtract two vectors, just subtract the corresponding components.

Vector subtraction can be interpreted geometrically. According to the formula in the second box on page 462, the right-hand side of equation (2) represents a vector drawn from the terminal point of $\mathbf{v}$ to the terminal point of $\mathbf{u}$. Figure 9 summarizes this fact, and Figure 10 provides a geometric comparison of vector addition and vector subtraction.

FIGURE 9

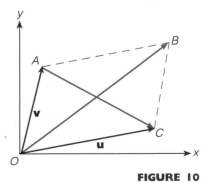

FIGURE 10

$\mathbf{u} + \mathbf{v}$ and $\mathbf{u} - \mathbf{v}$ are the directed diagonals of parallelogram $OABC$: $\mathbf{u} + \mathbf{v} = \overrightarrow{OB}$ and $\mathbf{u} - \mathbf{v} = \overrightarrow{AC}$.

FIGURE 11

EXAMPLE 4   Let $\mathbf{u} = \langle 5, 3 \rangle$ and $\mathbf{v} = \langle -1, 2 \rangle$. Compute $3\mathbf{u} - \mathbf{v}$.

Solution $3\mathbf{u} - \mathbf{v} = 3\langle 5, 3 \rangle - \langle -1, 2 \rangle$

$$= \langle 15, 9 \rangle - \langle -1, 2 \rangle = \langle 16, 7 \rangle$$

The next three examples deal with unit vectors. By definition, any vector with a length of 1 is called a **unit vector**. Two particularly useful unit vectors are

$$\mathbf{i} = \langle 1, 0 \rangle \qquad \text{and} \qquad \mathbf{j} = \langle 0, 1 \rangle$$

These are shown in Figure 11.

Any vector $\mathbf{v} = \langle x, y \rangle$ can be uniquely expressed in terms of the unit vectors $\mathbf{i}$ and $\mathbf{j}$ as follows:

$$\langle x, y \rangle = x\mathbf{i} + y\mathbf{j} \tag{3}$$

To verify equation (3), we have

$$x\mathbf{i} + y\mathbf{j} = x\langle 1, 0 \rangle + y\langle 0, 1 \rangle$$
$$= \langle x, 0 \rangle + \langle 0, y \rangle = \langle x, y \rangle$$

EXAMPLE 5

(a) Express the vector $\langle 3, -7 \rangle$ in terms of the unit vectors $\mathbf{i}$ and $\mathbf{j}$.

(b) Express the vector $\mathbf{v} = -4\mathbf{i} + 5\mathbf{j}$ in component form.

Solution

(a) Using equation (3), we can write

$$\langle 3, -7 \rangle = 3\mathbf{i} + (-7)\mathbf{j} = 3\mathbf{i} - 7\mathbf{j}$$

(b) $\mathbf{v} = -4\mathbf{i} + 5\mathbf{j} = -4\langle 1, 0 \rangle + 5\langle 0, 1 \rangle$

$$= \langle -4, 0 \rangle + \langle 0, 5 \rangle = \langle -4, 5 \rangle$$

Thus, the component form of $\mathbf{v}$ is $\langle -4, 5 \rangle$.

EXAMPLE 6   Find a unit vector $\mathbf{u}$ that has the same direction as the vector $\mathbf{v} = \langle 3, 4 \rangle$.

Solution   First, let us determine the length of $\mathbf{v}$:

$$|\mathbf{v}| = \sqrt{3^2 + 4^2} = \sqrt{25} = 5$$

So we want a vector whose length is one-fifth that of $\mathbf{v}$ and whose direction is the same as that of $\mathbf{v}$. Such a vector is

$$\mathbf{u} = \tfrac{1}{5}\mathbf{v} = \tfrac{1}{5}\langle 3, 4 \rangle = \langle \tfrac{3}{5}, \tfrac{4}{5} \rangle$$

(You should check for yourself now that the length of this vector is 1.)

EXAMPLE 7   The angle from the positive x-axis to the unit vector $\mathbf{u}$ is $\pi/3$, as indicated in Figure 12. Determine the components of $\mathbf{u}$.

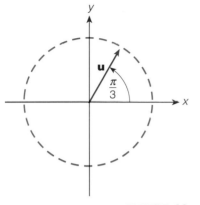

FIGURE 12

Solution Let P denote the terminal point of **u**. Since P lies on the unit circle, the coordinates of P are by definition $\big(\cos(\pi/3),\ \sin(\pi/3)\big)$. We therefore have

$$\mathbf{u} = \left\langle \frac{1}{2}, \frac{\sqrt{3}}{2} \right\rangle$$

It was mentioned earlier that there are several ways in which vector addition resembles ordinary addition of real numbers. In the previous section, for example, we saw that vector addition is commutative. That is, for any two vectors **u** and **v**,

$$\mathbf{u} + \mathbf{v} = \mathbf{v} + \mathbf{u}$$

By using components, we can easily verify that this property holds. (In the previous section, we used a geometric argument to establish this property.) We begin by letting $\mathbf{u} = \langle u_1, u_2 \rangle$ and $\mathbf{v} = \langle v_1, v_2 \rangle$. Then we have

$$\begin{aligned}
\mathbf{u} + \mathbf{v} &= \langle u_1, u_2 \rangle + \langle v_1, v_2 \rangle \\
&= \langle u_1 + v_1, u_2 + v_2 \rangle \\
&= \langle v_1 + u_1, v_2 + u_2 \rangle \qquad \text{Addition of real numbers is commutative.} \\
&= \langle v_1, v_2 \rangle + \langle u_1, u_2 \rangle \\
&= \mathbf{v} + \mathbf{u}
\end{aligned}$$

which is what we wanted to show.

There are a number of other properties of vector addition and scalar multiplication that can be proved in a similar fashion. In the following box, we list a particular collection of these properties, known as the **vector space properties**. (Exercises 55–58 ask that you verify these properties by using components, just as we did for the commutative property.)

Properties of Vector Addition and Scalar Multiplication

For all vectors **u**, **v**, and **w**, and for all scalars (real numbers) a and b, the following properties hold.

1. $\mathbf{u} + (\mathbf{v} + \mathbf{w}) = (\mathbf{u} + \mathbf{v}) + \mathbf{w}$ **5.** $a(\mathbf{u} + \mathbf{v}) = a\mathbf{u} + a\mathbf{v}$

2. $\mathbf{0} + \mathbf{v} = \mathbf{v} + \mathbf{0} = \mathbf{v}$ **6.** $(a + b)\mathbf{v} = a\mathbf{v} + b\mathbf{v}$

3. $\mathbf{v} + (-\mathbf{v}) = \mathbf{0}$ **7.** $(ab)\mathbf{v} = a(b\mathbf{v})$

4. $\mathbf{u} + \mathbf{v} = \mathbf{v} + \mathbf{u}$ **8.** $1\mathbf{v} = \mathbf{v}$

EXERCISE SET 8.3

A

In Exercises 1–6, sketch each vector in an x-y coordinate system, and compute the length of the vector.

1. $\langle 4, 3 \rangle$ **2.** $\langle 5, 12 \rangle$ **3.** $\langle -4, 2 \rangle$

4. $\langle -6, -6 \rangle$ **5.** $\langle \frac{3}{4}, -\frac{1}{2} \rangle$ **6.** $\langle -3, 0 \rangle$

In Exercises 7–12, the coordinates of two points P and Q are given. In each case, determine the components of the vector $\overrightarrow{PQ}$. Write your answers in the form $\langle a, b \rangle$.

7. $P(2, 3)$ and $Q(3, 7)$ **8.** $P(5, 1)$ and $Q(4, 9)$

9. $P(-2, -3)$ and $Q(-3, -2)$

10. $P(0, -4)$ and $Q(0, -8)$ 11. $P(-5, 1)$ and $Q(3, -4)$

12. $P(1, 0)$ and $Q(0, 1)$

In Exercises 13–32, assume that the vectors **a**, **b**, **c**, *and* **d** *are defined as follows:*

$$\mathbf{a} = \langle 2, 3 \rangle \qquad \mathbf{b} = \langle 5, 4 \rangle \qquad \mathbf{c} = \langle 6, -1 \rangle \qquad \mathbf{d} = \langle -2, 0 \rangle$$

Compute each of the indicated quantities.

13. $\mathbf{a} + \mathbf{b}$
14. $\mathbf{c} + \mathbf{d}$
15. $2\mathbf{a} + 4\mathbf{b}$
16. $-2\mathbf{c} + 2\mathbf{d}$
17. $|\mathbf{b} + \mathbf{c}|$
18. $|5\mathbf{b} + 5\mathbf{c}|$
19. $|\mathbf{a} + \mathbf{c}| - |\mathbf{a}| - |\mathbf{c}|$
20. $1/|\mathbf{d}|$
21. $\mathbf{a} + (\mathbf{b} + \mathbf{c})$
22. $(\mathbf{a} + \mathbf{b}) + \mathbf{c}$
23. $3\mathbf{a} + 4\mathbf{a}$
24. $|4\mathbf{b} + 5\mathbf{b}|$
25. $\mathbf{a} - \mathbf{b}$
26. $\mathbf{b} - \mathbf{c}$
27. $3\mathbf{b} - 4\mathbf{d}$
28. $\dfrac{1}{|3\mathbf{b} - 4\mathbf{d}|}(3\mathbf{b} - 4\mathbf{a})$
29. $\mathbf{a} - (\mathbf{b} + \mathbf{c})$
30. $(\mathbf{a} - \mathbf{b}) - \mathbf{c}$
31. $|\mathbf{c} + \mathbf{d}|^2 - |\mathbf{c} - \mathbf{d}|^2$
32. $|\mathbf{a} + \mathbf{b}|^2 + |\mathbf{a} - \mathbf{b}|^2 - 2|\mathbf{a}|^2 - 2|\mathbf{b}|^2$

In Exercises 33–38, express each vector in terms of the unit vectors **i** *and* **j**.

33. $\langle 3, 8 \rangle$
34. $\langle 4, -2 \rangle$
35. $\langle -8, -6 \rangle$
36. $\langle -9, 0 \rangle$
37. $3\langle 5, 3 \rangle + 2\langle 2, 7 \rangle$
38. $|\langle 12, 5 \rangle|\langle 3, 4 \rangle + |\langle 3, 4 \rangle|\langle 12, 5 \rangle$

In Exercises 39–42, express each vector in the form $\langle a, b \rangle$.

39. $\mathbf{i} + \mathbf{j}$
40. $\mathbf{i} - 2\mathbf{j}$
41. $5\mathbf{i} - 4\mathbf{j}$
42. $\dfrac{1}{|\mathbf{i} + \mathbf{j}|}(\mathbf{i} + \mathbf{j})$

In Exercises 43–48, find a unit vector having the same direction as the given vector.

43. $\langle 4, 8 \rangle$ 44. $\langle -3, 3 \rangle$ 45. $\langle 6, -3 \rangle$
46. $\langle -12, 5 \rangle$ 47. $8\mathbf{i} - 9\mathbf{j}$ 48. $\langle 7, 3 \rangle - \mathbf{i} + \mathbf{j}$

In Exercises 49–54, you are given an angle θ *measured counterclockwise from the positive x-axis to a unit vector* $\mathbf{u} = \langle u_1, u_2 \rangle$. *In each case, determine the components* u_1 *and* u_2.

49. $\theta = \pi/6$ 50. $\theta = \pi/4$ 51. $\theta = 2\pi/3$
52. $\theta = 3\pi/4$ 53. $\theta = 5\pi/6$ 54. $\theta = 3\pi/2$

B

In Exercises 55–58, let $\mathbf{u} = \langle u_1, u_2 \rangle$, $\mathbf{v} = \langle v_1, v_2 \rangle$, *and* $\mathbf{w} = \langle w_1, w_2 \rangle$. *Refer to the box on page 467.*

55. Verify properties 1 and 2.
56. Verify properties 3 and 4.

57. Verify properties 5 and 6.
58. Verify properties 7 and 8.

In Exercises 59–75, we study the dot product *of two vectors. Given two vectors* $\mathbf{A} = \langle x_1, y_1 \rangle$ *and* $\mathbf{B} = \langle x_2, y_2 \rangle$, *we define the* **dot product** $\mathbf{A} \cdot \mathbf{B}$ *as follows.*

$$\mathbf{A} \cdot \mathbf{B} = x_1 x_2 + y_1 y_2$$

For example, if $\mathbf{A} = \langle 3, 4 \rangle$ *and* $\mathbf{B} = \langle -2, 5 \rangle$, *then* $\mathbf{A} \cdot \mathbf{B} = (3)(-2) + (4)(5) = 14$. *Notice that the dot product of two vectors is a real number. For this reason, the dot product is also known as the* **scalar product**. *For Exercises 59–61, the vectors* **u**, **v**, *and* **w** *are defined as follows:*

$$\mathbf{u} = \langle -4, 5 \rangle \qquad \mathbf{v} = \langle 3, 4 \rangle \qquad \mathbf{w} = \langle 2, -5 \rangle$$

59. (a) Compute $\mathbf{u} \cdot \mathbf{v}$ and $\mathbf{v} \cdot \mathbf{u}$.
    (b) Compute $\mathbf{v} \cdot \mathbf{w}$ and $\mathbf{w} \cdot \mathbf{v}$.
    (c) Show that for any two vectors **A** and **B**, we have $\mathbf{A} \cdot \mathbf{B} = \mathbf{B} \cdot \mathbf{A}$. That is, show that the dot product is commutative. *Hint:* Let $\mathbf{A} = \langle x_1, y_1 \rangle$ and let $\mathbf{B} = \langle x_2, y_2 \rangle$.

60. (a) Compute $\mathbf{v} + \mathbf{w}$.
    (b) Compute $\mathbf{u} \cdot (\mathbf{v} + \mathbf{w})$.
    (c) Compute $\mathbf{u} \cdot \mathbf{v} + \mathbf{u} \cdot \mathbf{w}$.
    (d) Show that for any three vectors **A**, **B**, and **C**, we have $\mathbf{A} \cdot (\mathbf{B} + \mathbf{C}) = \mathbf{A} \cdot \mathbf{B} + \mathbf{A} \cdot \mathbf{C}$.

61. (a) Compute $\mathbf{v} \cdot \mathbf{v}$ and $|\mathbf{v}|^2$.
    (b) Compute $\mathbf{w} \cdot \mathbf{w}$ and $|\mathbf{w}|^2$.

62. Show that for any vector **A**, we always have $|\mathbf{A}|^2 = \mathbf{A} \cdot \mathbf{A}$. That is, the square of the length of a vector is equal to the dot product of the vector with itself. *Hint:* Let $\mathbf{A} = \langle x, y \rangle$.

Let θ *(where* $0 \le \theta \le \pi$*) denote the angle between the two nonzero vectors* **A** *and* **B**. *Then it can be shown that the cosine of* θ *is given by the formula*

$$\cos \theta = \frac{\mathbf{A} \cdot \mathbf{B}}{|\mathbf{A}| \, |\mathbf{B}|}$$

(See Exercise 75 for the derivation of this result.) In Exercises 63–68, use this formula to find the cosine of the angle between the given pair of vectors. Also, in each case use a calculator to compute the angle. Express the angle using degrees and using radians. Round off the values to two decimal places.

63. $\mathbf{A} = \langle 4, 1 \rangle$ and $\mathbf{B} = \langle 2, 6 \rangle$
64. $\mathbf{A} = \langle 3, -1 \rangle$ and $\mathbf{B} = \langle -2, 5 \rangle$
65. $\mathbf{A} = \langle 5, 6 \rangle$ and $\mathbf{B} = \langle -3, -7 \rangle$
66. $\mathbf{A} = \langle 3, 0 \rangle$ and $\mathbf{B} = \langle 1, 4 \rangle$
67. (a) $\mathbf{A} = \langle -8, 2 \rangle$ and $\mathbf{B} = \langle 1, -3 \rangle$
    (b) $\mathbf{A} = \langle -8, 2 \rangle$ and $\mathbf{B} = \langle -1, 3 \rangle$
68. (a) $\mathbf{A} = \langle 7, 12 \rangle$ and $\mathbf{B} = \langle 1, 2 \rangle$
    (b) $\mathbf{A} = \langle 7, 12 \rangle$ and $\mathbf{B} = \langle -1, -2 \rangle$

69. (a) Compute the cosine of the angle between the vectors $\langle 2, 5 \rangle$ and $\langle -5, 2 \rangle$.
    (b) What can you conclude from your answer in part (a)?
    (c) Draw a sketch to check your conclusion in part (b).

70. Follow Exercise 69, but use the vectors $\langle 6, -8 \rangle$ and $\langle -4, -3 \rangle$.

71. Suppose that **A** and **B** are nonzero vectors and $\mathbf{A} \cdot \mathbf{B} = 0$. Explain why **A** and **B** are perpendicular.

72. Find a value for t such that the vectors $\langle 15, -3 \rangle$ and $\langle -4, t \rangle$ are perpendicular.

73. Find a unit vector that is perpendicular to the vector $\langle -12, 5 \rangle$. (There are two answers.)

C

74. Suppose that **A** and **B** are nonzero vectors such that $|\mathbf{A} + \mathbf{B}|^2 = |\mathbf{A}|^2 + |\mathbf{B}|^2$. Show that **A** and **B** are perpendicular. *Hint:* Let $\mathbf{A} = \langle x_1, y_1 \rangle$ and let $\mathbf{B} = \langle x_2, y_2 \rangle$. After making some calculations, you should be able to apply the result in Exercise 71.

75. Refer to the accompanying figure. In this exercise we are going to derive the following formula for the cosine of the angle θ between two nonzero vectors **A** and **B**:

$$\cos \theta = \frac{\mathbf{A} \cdot \mathbf{B}}{|\mathbf{A}| \, |\mathbf{B}|}$$

(a) Let $\mathbf{A} = \langle x_1, y_1 \rangle$ and $\mathbf{B} = \langle x_2, y_2 \rangle$. Compute the length of the vector **C** in the figure using the fact that $\mathbf{C} = \langle x_2 - x_1, y_2 - y_1 \rangle$.

(b) Using your result in part (a), show that

$$|\mathbf{C}|^2 = |\mathbf{A}|^2 + |\mathbf{B}|^2 - 2\mathbf{A} \cdot \mathbf{B}$$

(c) According to the law of cosines, we have

$$|\mathbf{C}|^2 = |\mathbf{A}|^2 + |\mathbf{B}|^2 - 2|\mathbf{A}| \, |\mathbf{B}| \cos \theta$$

Set this expression for $|\mathbf{C}|^2$ equal to the expression obtained in part (b), and then solve for $\cos \theta$ to obtain the required formula.

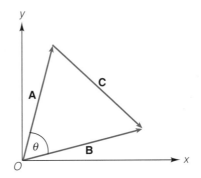

8.4 PARAMETRIC EQUATIONS

We will introduce the idea of parametric equations through a simple example. Suppose that we have a point $P(x, y)$ that moves in the x-y plane, and that the x- and y-coordinates of P at time t (in seconds) are given by the following pair of equations:

$$x = 2t \qquad \text{and} \qquad y = \tfrac{1}{2}t^2 \qquad (t \geq 0)$$

This pair of *parametric equations* tells us the location of the point P at any time t. For instance, when $t = 1$, we have

$$x = 2t = 2(1) = 2 \qquad \text{and} \qquad y = \tfrac{1}{2}t^2 = \tfrac{1}{2}(1)^2 = \tfrac{1}{2}$$

In other words, after one second, the coordinates of P are $\left(2, \tfrac{1}{2}\right)$. Let's see where P is after two seconds. We have

$$x = 2t = 2(2) = 4 \qquad \text{and} \qquad y = \tfrac{1}{2}t^2 = \tfrac{1}{2}(2)^2 = 2$$

So, after two seconds, the location of the point P is $(4, 2)$. In Table 1, we have computed the values of x and y (and hence the location of P) for integral values of t running from 0 to 5. Figure 1(a) shows the points that are determined. In Figure 1(b) we have joined the points with a smooth curve. [If we were not certain of the pattern emerging in Figure 1(a), we could have computed additional points using fractional values for t.]

TABLE I

t	x	y
0	0	0
1	2	$\tfrac{1}{2}$
2	4	2
3	6	$\tfrac{9}{2}$
4	8	8
5	10	$\tfrac{25}{2}$

FIGURE 1

$x = 2t, \ y = \frac{1}{2}t^2 \quad (t \geq 0)$

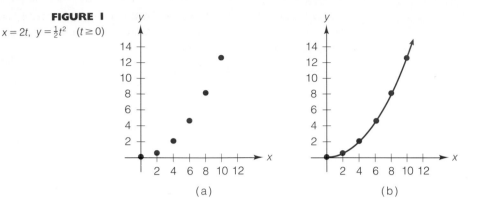

(a)   (b)

The curve in Figure 1(b) appears to be a portion of a parabola. We can confirm this as follows. From the equation $x = 2t$, we obtain $t = x/2$. Now we use this result to substitute for t in the equation $y = \frac{1}{2}t^2$. This yields

$$y = \frac{1}{2}\left(\frac{x}{2}\right)^2 = \frac{1}{8}x^2$$

We conclude from this that the curve in Figure 1(b) is a portion of the parabola $y = \frac{1}{8}t^2$. The equations $x = 2t$ and $y = \frac{1}{2}t^2$ (with the restriction $t \geq 0$) are the **parametric equations** for the curve. The variable t is called the **parameter.** In our initial example, the parameter t represented time. In other cases the parameter might have an interpretation as a slope of a certain line, or as an angle. And in still other cases, there may be no immediate physical interpretation for the parameter.

Before looking at additional examples, we point out two advantages in using parametric equations to describe curves. First, by restricting the values of the parameter (as we did in our initial example), we can focus on specific portions of a curve. Second, parametric equations let us think of a curve as a path traced out by a moving point; as the parameter t increases, a definite direction of motion is established.

In physics, parametric equations are often used to describe the motion of an object. Suppose, for example, that a ball is thrown from a height of 6 feet, with an initial speed of 88 ft/sec and at an angle of 35° with the horizontal, as shown in Figure 2. Then (neglecting air resistance and spin), it can be shown that the parametric equations for the path of the ball are

$$x = (88 \cos 35°)t \tag{1}$$
$$y = 6 + (88 \sin 35°)t - 16t^2 \tag{2}$$

In these equations, x and y are measured in feet, and t is in seconds, with $t = 0$ corresponding to the instant the ball is thrown.

FIGURE 2

The path of a ball thrown from a height of 6 ft, with an initial speed of 88 ft/s, at an angle of 35° with the horizontal.

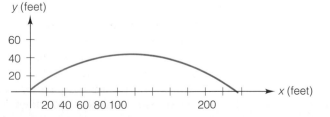

With equations (1) and (2) we can calculate the location of the ball at any time t. For instance, to determine the location when $t = 1$ second, we substitute $t = 1$ in the equations to obtain

$$x = (88 \cos 35°)(1) \approx 72.1 \text{ ft}$$

and

$$y = 6 + (88 \sin 35°)(1) - 16(1)^2 \approx 40.5 \text{ ft}$$

So, after one second, the ball has traveled a horizontal distance of approximately 72.1 ft, and the height of the ball is approximately 40.5 ft. In Figure 3 we display this information along with the results for similar calculations corresponding to $t = 2$ and $t = 3$ seconds. [You should use equations (1) and (2) and your calculator to check the results in Figure 3 for yourself.]

FIGURE 3
The location of the ball after 1, 2, and 3 seconds.

From Figure 3, we can see that the total horizontal distance traveled by the ball in flight is something between 220 ft and 240 ft. Using the parametric equations for the path of the ball, we can determine this distance exactly. We can also find out how long the ball is in the air. When the ball does hit the ground, its y-coordinate is zero. Replacing y by zero in equation (2), we have

$$0 = 6 + (88 \sin 35°)t - 16t^2$$

This is a quadratic equation in the variable t. To solve for t, we use the quadratic formula:

$$t = \frac{-b \pm \sqrt{b^2 - 4ac}}{2a}$$

with the following values for a, b, and c:

$$a = -16 \qquad b = 88 \sin 35° \qquad c = 6$$

As Exercise 19 asks you to check, the results are $t \approx 3.27$ and $t \approx -0.11$. We discard the negative root here because in Figure 3, the values of t are nonnegative. Now, using $t = 3.27$, we can compute x:

$$x = (88 \cos 35°)t \approx (88 \cos 35°)(3.27) \approx 235.7 \text{ ft}$$

In summary, the ball is in the air for about 3.3 seconds, and the total horizontal distance is approximately 236 feet.

Just as in Figure 1, the curve in Figures 2 and 3 is a parabola. Exercise 20 asks you to verify this by solving equation (1) for t in terms of x and then substituting the result in equation (2). This process of obtaining an explicit equation relating x and y from the parametric equations is referred to as *eliminating the parameter*. The next example shows a technique that is often useful in this

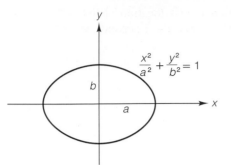

FIGURE 4
The ellipse $(x^2/a^2) + (y^2/b^2) = 1$ is symmetric about both coordinate axes. The intercepts of the ellipse are $x = \pm a$ and $y = \pm b$.

context. As background for this example, we point out that the graph of an equation of the form $\dfrac{x^2}{a^2} + \dfrac{y^2}{b^2} = 1$ is an *ellipse,* as indicated in Figure 4. (We will study this curve in detail in Chapter 10.)

EXAMPLE I The parametric equations of an ellipse are

$$x = 6 \cos t \quad \text{and} \quad y = 3 \sin t \quad (0 \leq t \leq 2\pi)$$

(a) Eliminate the parameter t to obtain an x-y equation for the curve.
(b) Graph the ellipse and indicate the points corresponding to $t = 0$, $\pi/2$, π, $3\pi/2$, and 2π. As t increases, what is the direction of travel along the curve?

Solution

(a) So that we can apply the identity $\cos^2 t + \sin^2 t = 1$, we divide the first equation by 6 and the second by 3. This gives us

$$\frac{x}{6} = \cos t \quad \text{and} \quad \frac{y}{3} = \sin t$$

Squaring and then adding these two equations yields

$$\left(\frac{x}{6}\right)^2 + \left(\frac{y}{3}\right)^2 = \cos^2 t + \sin^2 t = 1$$

or

$$\frac{x^2}{6^2} + \frac{y^2}{3^2} = 1$$

This is the x-y equation for the ellipse.

(b) Figure 5 shows the graph of the ellipse $(x^2/6^2) + (y^2/3^2) = 1$. When $t = 0$, the parametric equations yield

$$x = 6 \cos 0 = 6 \quad \text{and} \quad y = 3 \sin 0 = 0$$

Thus, with $t = 0$, we obtain the point $(6, 0)$ on the ellipse, as indicated in Figure 5. The points corresponding to $t = \pi/2$, π, $3\pi/2$, and 2π are obtained similarly. Note that as t increases, the direction of travel around the ellipse is counterclockwise, and that when $t = 2\pi$, we are back to the starting point $(6, 0)$. ▪▪▪

FIGURE 5
Parametric equations for this ellipse are $x = 6 \cos t$ and $y = 3 \sin t$. The x-y equation is $(x^2/6^2) + (y^2/3^2) = 1$.

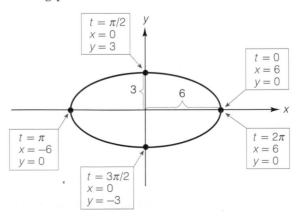

The parametric equations $x = 6 \cos t$ and $y = 3 \sin t$ that we graphed in Figure 5 are by no means the only parametric equations yielding that ellipse. Indeed, there are numerous parametric equations that give rise to the same ellipse. One such pair is

$$x = 6 \sin t \quad \text{and} \quad y = 3 \cos t \quad (0 \le t \le 2\pi)$$

As you can check for yourself, these equations lead to the same x-y equation as before, namely, $(x^2/6^2) + (y^2/3^2) = 1$. The difference now is that in tracing out the curve from the parametric equations, we start (when $t = 0$) with the point $(0, 3)$ on the ellipse and we travel clockwise rather than counterclockwise. Similarly, the equations

$$x = 6 \sin 2t \quad \text{and} \quad y = 3 \cos 2t$$

also produce the ellipse. In this case, as you can check for yourself, as t runs from 0 to 2π, we make two complete trips around the ellipse. If we think of t as time, then the parametric equations $x = 6 \sin 2t$ and $y = 3 \cos 2t$ describe a point traveling around the ellipse $(x^2/6^2) + (y^2/3^2) = 1$ twice as fast as would be the case with the equations $x = 6 \sin t$ and $y = 3 \cos t$.

In Example 1 we saw that the parametric equations $x = 6 \cos t$ and $y = 3 \sin t$ represent an ellipse. The same method we used in Example 1 can be used to establish the following general results.

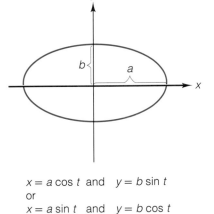

$x = a \cos t$ and $y = b \sin t$
or
$x = a \sin t$ and $y = b \cos t$

FIGURE 6

Parametric Equations for the Ellipse

Let a and b be positive constants. Then the parametric equations

$$x = a \cos t \quad \text{and} \quad y = b \sin t \tag{3}$$

represent the ellipse shown in Figure 6. As t increases, we move around the ellipse in a counterclockwise direction. The parametric equations

$$x = a \sin t \quad \text{and} \quad y = b \cos t$$

also represent the ellipse in Figure 6. In this case, the direction of motion is clockwise.

There is a particular case involving the parametric equations $x = a \cos t$ and $y = b \sin t$ that deserves mention. When a and b are equal, we have $x = a \cos t$ and $y = a \sin t$, from which we deduce that

$$x^2 + y^2 = a^2 \cos^2 t + a^2 \sin^2 t$$
$$= a^2(\cos^2 t + \sin^2 t) = a^2$$

Thus, with $a = b$, the parametric equations describe a circle of radius a. And specializing further still, if $a = b = 1$, we obtain $x = \cos t$ and $y = \sin t$ as parametric equations for the unit circle. This agrees with our unit circle definitions for sine and cosine back in Section 6.4.

▮▮▮ **EXAMPLE 2** The position of a point $P(x, y)$ at time t is given by the parametric equations

$$x = 1 + 2t \quad \text{and} \quad y = 2 + 4t \quad (t \ge 0)$$

Find the x-y equation for the path traced out by the point P.

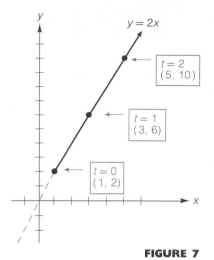

FIGURE 7

The parametric equations $x = 1 + 2t$ and $y = 2 + 4t$, with the restriction $t \geq 0$, describe the portion of the line $y = 2x$ to the right of and including the point $(1, 2)$.

Solution   From the parametric equation for x, we obtain $t = (x - 1)/2$. Using this to substitute for t in the second parametric equation, we have

$$y = 2 + 4\left(\frac{x - 1}{2}\right) = 2 + 2(x - 1) = 2x$$

This tells us that the point P moves along the line $y = 2x$. However, the entire line is not traced out, but only a portion of it. This is because of the original restriction $t \geq 0$. To see which portion of the line is described by the equations, we can successively let $t = 0$, 1, and 2 in the parametric equations. As you can check, the points obtained are $(1, 2)$, $(3, 6)$, and $(5, 10)$. So, as t increases, we move to the right along the line $y = 2x$, starting from the point $(1, 2)$. See Figure 7.   ∎∎∎

One of the recurrent techniques that we have used for analyzing parametric equations in this section has been that of eliminating the parameter. It is important to note, however, that it is not always a simple matter to eliminate the parameter. And indeed, in some cases it may not even be possible. When this occurs we have several techniques available. We can set up a table with t, x, and y and plot points; we can use the techniques of calculus; or we can use a graphing calculator or computer to obtain the graph. Even when we use a graphing calculator, however, the techniques of calculus are helpful in understanding why the graph looks as it does.

EXERCISE SET 8.4

A

In Exercises 1–6, you are given the parametric equations of a curve and a value for the parameter t. Find the coordinates of the point on the curve corresponding to the given value of t.

1. $x = 2 - 4t$, $y = 3 - 5t$; $t = 0$
2. $x = 3 - t^2$, $y = 4 + t^3$; $t = -1$
3. $x = 5 \cos t$, $y = 2 \sin t$; $t = \pi/6$
4. $x = 4 \cos 2t$, $y = 6 \sin 2t$; $t = \pi/3$
5. $x = 3 \sin^3 t$, $y = 3 \cos^3 t$; $t = \pi/4$
6. $x = \sin t - \sin 2t$, $y = \cos t + \cos 2t$; $t = 2\pi/3$

In Exercises 7–16, graph the parametric equations after eliminating the parameter t. Specify the direction on the curve corresponding to increasing values of t.

7. $x = t + 1$, $y = t^2$
8. $x = 2t - 1$, $y = t^2 - 1$
9. $x = t^2 - 1$, $y = t + 1$
10. $x = t - 4$, $y = |t|$
11. $x = 5 \cos t$, $y = 2 \sin t$
12. $x = 2 \sin t$, $y = 3 \cos t$
13. $x = 4 \cos 2t$, $y = 6 \sin 2t$
14. $x = 2 \cos(t/2)$, $y = \sin(t/2)$

15. (a) $x = 2 \cos t$, $y = 2 \sin t$
     (b) $x = 4 \cos t$, $y = 2 \sin t$
16. (a) $x = 3 \sin t$, $y = 3 \cos t$
     (b) $x = 5 \sin t$, $y = 3 \cos t$
17. The accompanying figure shows the parametric equations and the path for a ball thrown from a height of 5 ft, with an initial speed of 100 ft/sec and at an angle of 70° with the horizontal.

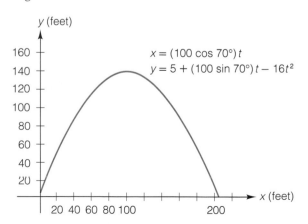

(a) Compute the *x*- and *y*-coordinates of the ball when *t* = 1, 2, and 3 seconds. (Round off the answers to one decimal place.)

(b) How long is the ball in flight? (Round off the answer to two decimal places.) What is the total horizontal distance traveled by the ball before it lands? (Round off to the nearest foot.) Check that your answer is consistent with the figure.

18. The figure at the bottom of this page shows the parametric equations and the path for a ball thrown from a height of 5 ft, with an initial speed of 100 ft/s and at an angle of 45° with the horizontal. (So except for the initial angle, the data is the same as in Exercise 17.)

(a) Compute the *x*- and *y*-coordinates of the ball when *t* = 1, 2, and 3 seconds. (Round off the answers to one decimal place.)

(b) How long is the ball in flight? (Round off the answer to two decimal places.) What is the total horizontal distance traveled by the ball before it lands? (Round off to the nearest foot.)

19. In the text we said that the solutions of the quadratic equation $0 = 6 + (88 \sin 35°)t - 16t^2$ are $t \approx 3.27$ and $t \approx -0.11$. Use the quadratic formula and your calculator to verify these results.

20. Refer to parametric equations (1) and (2) in the text. By eliminating the parameter *t*, show that the equations

describe a parabola. (That is, show that the resulting *x*-*y* equation has the form $y = ax^2 + bx + c$.)

B

21. The curve in the accompanying figure is called an *astroid*. (It is also known as a *hypocycloid of four cusps*.)

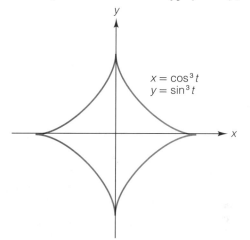

$$x = \cos^3 t$$
$$y = \sin^3 t$$

A pair of parametric equations for the curve is $x = \cos^3 t$, $y = \sin^3 t$. By eliminating the parameter *t*, find the *x*-*y* equation for the curve.   *Hint:*   In each equation, raise both sides to the two-thirds power.

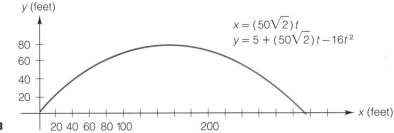

$$x = (50\sqrt{2})t$$
$$y = 5 + (50\sqrt{2})t - 16t^2$$

FIGURE FOR EXERCISE 18

8.5 INTRODUCTION TO POLAR COORDINATES

[Jakob Bernoulli (1654–1705)] *was one of the first to use polar coordinates in a general manner, and not simply for spiral shaped curves.*

Florian Cajori in *A History of Mathematics,* 4th ed. (New York: Chelsea Publishing Company, 1985)

FIGURE I

This polar coordinate graph was discovered by Henri Berger, a student in one of the author's mathematics classes at U.C.L.A. in Spring of 1988.

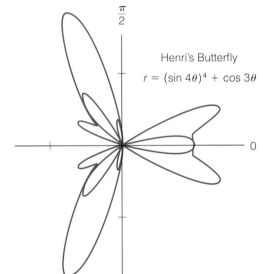

Henri's Butterfly

$r = (\sin 4\theta)^4 + \cos 3\theta$

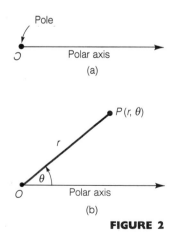

(a)

(b)

FIGURE 2

Up until now, we have always specified the location of a point in the plane by means of a rectangular coordinate system. In this section, we introduce another coordinate system that can be used to locate points in the plane. This is the system of **polar coordinates.** We begin by drawing a half-line or ray emanating from a fixed point O. The fixed point O is called the **pole** or **origin,** and the half-line is called the **polar axis.** As a matter of convention, the polar axis is usually depicted as being horizontal and extending to the right, as indicated in Figure 2(a). Now let P be any point in the plane. As indicated in Figure 2(b), we initially let r denote the distance from O to P and we let θ denote the angle measured from the polar axis to $\overline{OP}$. (Just as in our earlier work with angles in standard position, we take the measure of θ to be positive if the rotation is counterclockwise and negative if it is clockwise.) Then the ordered pair (r, θ) serves to locate the point P with respect to the pole and the polar axis. We refer to r and θ as polar coordinates of P, and we write $P(r, \theta)$ to indicate that P is the point with polar coordinates (r, θ).

Plotting points in polar coordinates is facilitated by the use of polar coordinate graph paper, such as that shown in Figure 3(a). Figure 3(b) shows the points with polar coordinates $A(2, \pi/6)$, $B(3, 2\pi/3)$, and $C(4, 0)$.

FIGURE 3

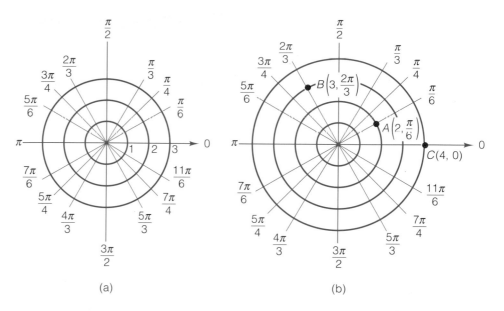

(a)   (b)

There is a minor complication that arises in using polar rather than rectangular coordinates. Consider, for example, the point $C(4,0)$ in Figure 3(b). This point could just as well have been labeled with the coordinates $(4, 2\pi)$, or $(4, 2k\pi)$ for any integral value of k. Similarly, the coordinates (r, θ) and $(r, \theta + 2k\pi)$ represent the same point for all integral values of k. This is in marked contrast to the situation with rectangular coordinates, where the coordinate representation of each point is unique. Also, what polar coordinates should we assign to the origin? For if $r = 0$ in Figure 2(b), we cannot really define an angle θ. The convention agreed on to cover this case is that the coordinates $(0, \theta)$ denote the origin for all values of θ. Finally, we point out that in working with polar coordinates, it is sometimes useful to let r take on negative values. For example, consider the point $P(2, 5\pi/4)$ in Figure 4. The coordinates $(2, 5\pi/4)$ indicate that to reach P from the origin, we go 2 units in the direction $5\pi/4$. Alternately, we can describe this as -2 units in the $\pi/4$ direction. (Refer again to Figure 4.) For reasons such as this, we will adhere to the convention that the polar coordinates (r, θ) and $(-r, \theta + \pi)$ represent the same point. (Since r can now, in fact, be positive, negative, or zero, r is referred to as a *directed distance*.)

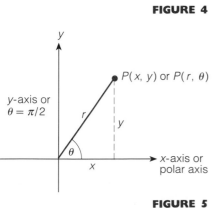

FIGURE 4

It is often useful to consider simultaneously both rectangular and polar coordinates. To do this, we draw the two coordinate systems so that the origins coincide and the positive x-axis coincides with the polar axis (see Figure 5). Suppose now that a point P, other than the origin, has rectangular coordinates (x, y) and polar coordinates (r, θ), as indicated in Figure 5. We wish to find equations relating the two sets of coordinates. From Figure 5, we see that

$$x^2 + y^2 = r^2 \qquad \sin\theta = \frac{y}{r} \qquad \cos\theta = \frac{x}{r} \qquad \tan\theta = \frac{y}{x}$$

Although Figure 5 displays the point P in the first quadrant, it can be shown that these same equations hold when P is in any quadrant. For reference we summarize these equations as follows.

FIGURE 5

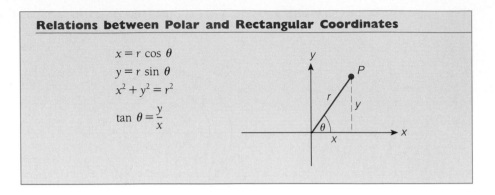

EXAMPLE 1 The polar coordinates of a point are $(5, \pi/6)$. Find the rectangular coordinates.

Solution We are given that $r = 5$ and $\theta = \pi/6$. Thus,

$$x = r \cos \theta = 5 \cos \frac{\pi}{6} = 5\left(\frac{\sqrt{3}}{2}\right)$$

and

$$y = r \sin \theta = 5 \sin \frac{\pi}{6} = 5\left(\frac{1}{2}\right)$$

The rectangular coordinates are therefore $\left(\frac{5}{2}\sqrt{3}, \frac{5}{2}\right)$.

The definition of the graph of an equation in polar coordinates is similar to the corresponding definition for rectangular coordinates. The **graph** of an equation in polar coordinates is the set of all points (r, θ) with coordinates that satisfy the given equation. It is often the case that the equation of a curve is simpler in one coordinate system than in another. The next two examples show instances of this.

EXAMPLE 2 Convert each polar equation to rectangular form.

(a) $r = \cos \theta + 2 \sin \theta$ **(b)** $r^2 = \sin 2\theta$

Solution

(a) In view of the transformation equations $x = r \cos \theta$ and $y = r \sin \theta$, we multiply both sides of the given equation by r to obtain

$$r^2 = r \cos \theta + 2r \sin \theta$$

and therefore

$$x^2 + y^2 = x + 2y \qquad \text{or} \qquad x^2 - x + y^2 - 2y = 0$$

This is the rectangular form of the given equation.
    Question for review: What is the graph of this last equation?
(b) Using the double-angle formula for $\sin 2\theta$, we have

$$r^2 = 2 \sin \theta \cos \theta$$

Now, in order to obtain the expressions $r \sin \theta$ and $r \cos \theta$ on the right-hand side of the equation, we multiply both sides by r^2. This yields

$$r^4 = 2(r \sin \theta)(r \cos \theta)$$

and, consequently,

$$(x^2 + y^2)^2 = 2yx$$

or

$$x^4 + 2x^2y^2 + y^4 - 2xy = 0$$

This is the rectangular form of the given equation. Notice how much simpler the equation is in its polar coordinate form.   ■■■

EXAMPLE 3 Convert the rectangular equation $x^2 + y^2 + ax = a\sqrt{x^2 + y^2}$ to polar form, expressing r as a function of θ. Assume that a is a constant.

Solution   Using the relations $x^2 + y^2 = r^2$ and $x = r \cos \theta$, we obtain

$$r^2 + ar \cos \theta = ar$$

Notice that this equation is satisfied by $r = 0$. In other words, the graph of this equation will pass through the origin. This is consistent with the fact that the original equation is satisfied when x and y are both zero. Now assume for the moment that $r \neq 0$. Then we can divide both sides of the last equation by r to obtain

$$r + a \cos \theta = a \qquad \text{or} \qquad r = a - a \cos \theta$$

This expresses r as a function of θ, as required. Note that when $\theta = 0$, we obtain

$$r = a - a(1) = 0$$

That is, nothing has been lost in dividing through by r; the graph will still pass through the origin.   ■■■

When we work in an x-y coordinate system, one of the basic tools is the formula for the distance between two points: $d = \sqrt{(x_2 - x_1)^2 + (y_2 - y_1)^2}$. There is a corresponding distance formula for use with polar coordinates. In Figure 6, we let d denote the distance between the points with polar coordinates (r_1, θ_1) and (r_2, θ_2). Then, applying the law of cosines, we have $d^2 = r_1^2 + r_2^2 - 2r_1r_2 \cos(\theta_2 - \theta_1)$. This is the distance formula in polar coordinates. For ease of reference, we repeat this result in the box that follows.

FIGURE 6

PROPERTY SUMMARY   DISTANCE FORMULA FOR POLAR COORDINATES

Let d denote the distance between two points with polar coordinates (r_1, θ_1) and (r_2, θ_2). Then

$$d^2 = r_1^2 + r_2^2 - 2r_1 r_2 \cos(\theta_2 - \theta_1)$$

EXAMPLE 4 Compute the distance between the points with polar coordinates $(2, 5\pi/6)$ and $(4, \pi/6)$.

Solution We use the distance formula, taking the points (r_1, θ_1) and (r_2, θ_2) to be $(2, 5\pi/6)$ and $(4, \pi/6)$, respectively. This yields

$$d^2 = r_1^2 + r_2^2 - 2r_1 r_2 \cos(\theta_2 - \theta_1)$$

$$= 2^2 + 4^2 - 2(2)(4)\cos\left(\frac{\pi}{6} - \frac{5\pi}{6}\right)$$

$$= 20 - 16 \cos\left(\frac{-2\pi}{3}\right) = 20 - 16\left(-\frac{1}{2}\right) = 28$$

$$d = \sqrt{28} = 2\sqrt{7}$$

You should check for yourself that the same result is obtained if we choose (r_1, θ_1) and (r_2, θ_2) to be $(4, \pi/6)$ and $(2, 5\pi/6)$, respectively. This is because, in general, $\cos(\theta_2 - \theta_1) = \cos(\theta_1 - \theta_2)$. (Why?)

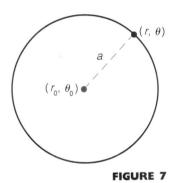

FIGURE 7

The distance formula can be used to find an equation for a circle. Suppose that the radius of the circle is a and that the polar coordinates of the center are (r_0, θ_0). As indicated in Figure 7, we let (r, θ) denote an arbitrary point on the circle. Then, since the distance between the points (r, θ) and (r_0, θ_0) is a, the radius of the circle, we have

$$r^2 + r_0^2 - 2rr_0 \cos(\theta - \theta_0) = a^2 \tag{1}$$

This is the required equation. There is a special case of this equation that is worth noting. If the center of the circle is located at the origin, then $r_0 = 0$ and equation (1) becomes simply $r^2 = a^2$. Therefore

$$r = a \qquad \text{or} \qquad r = -a$$

The graph of each of these equations is a circle of radius a, with center at the origin.

EXAMPLE 5 Determine a polar equation for the circle satisfying the given conditions.

(a) The radius is 2 and the polar coordinates of the center are $(4, \pi/5)$.
(b) The radius is 5 and the center is the origin.

Solution

(a) Using equation (1), we have

$$r^2 + 4^2 - 2r(4)\cos\left(\theta - \frac{\pi}{5}\right) = 2^2$$

FIGURE 8

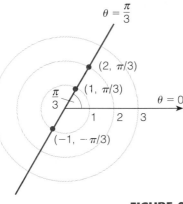

FIGURE 9

and therefore

$$r^2 - 8r \cos\left(\theta - \frac{\pi}{5}\right) = -12$$

(b) A polar equation for a circle of radius a and with center at the origin is $r = a$. So, if the radius is 5, the required equation is $r = 5$. (The equation $r = -5$ would yield the same graph.) ▬▬

We conclude this section by discussing equations of lines in polar coordinates. There are two cases to consider: lines that pass through the origin and lines that do not pass through the origin. First, consider a line through the origin, and suppose that the line makes an angle θ_0 with the positive x-axis; see Figure 8. The polar equation of this line is simply

$$\theta = \theta_0$$

Notice that this equation poses no restrictions on r. In other words, for every real number r, the point with polar coordinates (r, θ_0) lies on this line.

▬▬▬ **EXAMPLE 6** Graph the line with polar equation $\theta = \pi/3$, and locate the points on this line for which $r = 1$, 2, and -1.

Solution See Figure 9. ▬▬

Now let us determine the polar equation for a line $\mathscr{L}$ that does not pass through the origin. As indicated in Figure 10, we assume that the perpendicular distance from the origin to $\mathscr{L}$ is d, and that the polar coordinates of the point N (in Figure 10) are (d, α). The point $P(r, \theta)$ in Figure 10 denotes an arbitrary point on line $\mathscr{L}$. We want to find an equation relating r and θ.

FIGURE 10

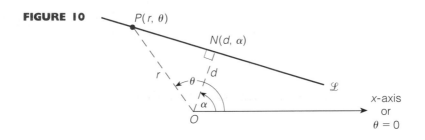

In right triangle ONP, note that $\angle PON = \theta - \alpha$, and therefore

$$\cos(\theta - \alpha) = \frac{d}{r} \qquad \text{or} \qquad r\cos(\theta - \alpha) = d$$

This is the polar equation of the line $\mathscr{L}$. For reference, we summarize our results about lines in the box that follows.

> ### PROPERTY SUMMARY   POLAR EQUATIONS FOR LINES
>
> **1.** LINE THROUGH THE ORIGIN
> Suppose that a line passes through the origin and makes an angle θ_0 with the positive x-axis (as indicated in Figure 8 on page 481). Then an equation for the line is $\theta = \theta_0$.
>
> ---
>
> **2.** LINE NOT PASSING THROUGH THE ORIGIN
> Suppose that the perpendicular distance from the origin to the line is d and that the point (d, α) is the foot of the perpendicular from the origin to the line (as indicated in Figure 10). Then a polar equation for the line is $r \cos(\theta - \alpha) = d$.

■■■ **EXAMPLE 7**   Figure 11 shows a line tangent to the unit circle at the point P. The x-y coordinates of P are $\left(\sqrt{3}/2, \tfrac{1}{2}\right)$.

(a) What are the polar coordinates of the point P?
(b) Find the polar equation for the tangent line.
(c) Use the polar equation to find the x- and y-intercepts of the tangent line.

FIGURE 11

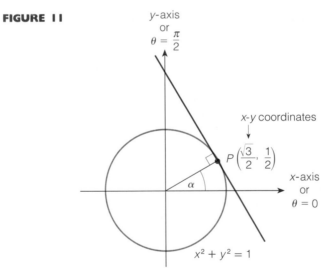

Solution

(a) Using the unit-circle definition for cosine, we have $\cos \alpha = \sqrt{3}/2$ and, consequently, $\alpha = \pi/6$. The polar coordinates of P are therefore $(1, \pi/6)$.

(b) In the general equation $r \cos(\theta - \alpha) = d$, we use the values $\alpha = \pi/6$ and $d = 1$ to obtain

$$r \cos\left(\theta - \frac{\pi}{6}\right) = 1 \qquad (2)$$

This is the polar equation for the tangent line.

(c) The x-axis corresponds to $\theta = 0$. Substituting $\theta = 0$ in equation (2) gives us

$$r \cos\left(0 - \frac{\pi}{6}\right) = 1, \text{ and therefore}$$

$$r \cos \frac{\pi}{6} = 1 \qquad \text{(Why?)}$$

$$r = \frac{1}{\sqrt{3}/2} = \frac{2}{\sqrt{3}} = \frac{2\sqrt{3}}{3}$$

This tells us that the tangent line meets the x-axis at the point with polar coordinates $(2\sqrt{3}/3, 0)$. So, the x-intercept of the line is $2\sqrt{3}/3$. For the y-intercept, we set $\theta = \pi/2$ in equation (2). As you should check for yourself, this yields $r = 2$. Thus, the y-intercept of the tangent line is 2. ■■■

EXERCISE SET 8.5

A

In Exercises 1–3, convert the given polar coordinates to rectangular coordinates.

1. (a) $(3, 2\pi/3)$ (b) $(4, 11\pi/6)$ (c) $(4, -\pi/6)$
2. (a) $(5, \pi/4)$ (b) $(-5, \pi/4)$ (c) $(-5, -\pi/4)$
3. (a) $(1, \pi/2)$ (b) $(1, 5\pi/2)$ (c) $(-1, \pi/8)$

In Exercises 4–6, convert the given rectangular coordinates to polar coordinates. Express your answers in such a way that r is nonnegative and $0 \leq \theta < 2\pi$.

4. $(3, \sqrt{3})$ 5. $(-1, -1)$ 6. $(0, -2)$

In Exercises 7–16, convert to rectangular form.

7. $r = 2 \cos \theta$
8. $2 \sin \theta - 3 \cos \theta = r$
9. $r = \tan \theta$
10. $r = 4$
11. $r = 3 \cos 2\theta$
12. $r = 4 \sin 2\theta$
13. $r^2 = 8/(2 - \sin^2 \theta)$
14. $r^2 = 1/(3 + \cos^2 \theta)$
15. $r \cos\left(\theta - \frac{\pi}{6}\right) = 2$
16. $r \sin\left(\theta + \frac{\pi}{4}\right) = 6$

In Exercises 17–24, convert to polar form.

17. $3x - 4y = 2$
18. $x^2 + y^2 = 25$
19. $y^2 = x^3$
20. $y = x^2$
21. $2xy = 1$
22. $x^2 + 4x + y^2 + 4y = 0$
23. $9x^2 + y^2 = 9$
24. $x^2(x^2 + y^2) = y^2$

In Exercises 25–30, you are given a polar equation and its graph. Use the equation to determine polar coordinates of the points labeled with capital letters. (For the r-values, where necessary, use a calculator and round off to two decimal places.)

25. $r = \dfrac{4}{1 + \sin \theta}$

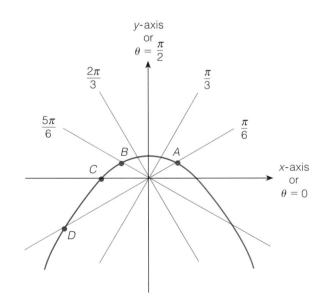

26. $r = \dfrac{2}{\cos\left(\theta - \frac{\pi}{3}\right)}$

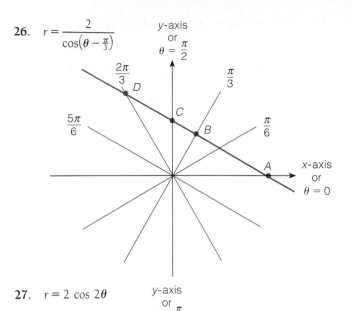

27. $r = 2\cos 2\theta$

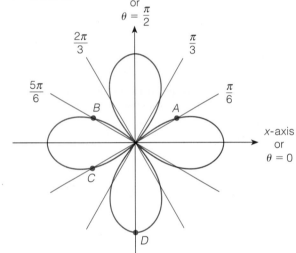

28. $r = -1 - \sin\theta$

29. $r = e^{\theta/6}$

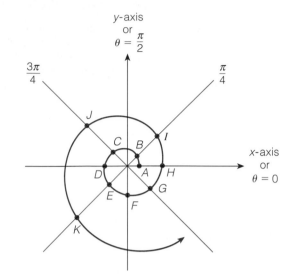

30. $r^2 = \cos 2\theta$

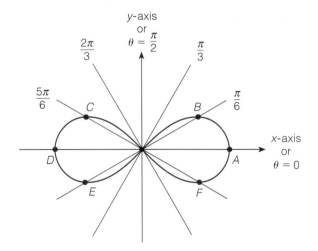

In Exercises 31–34, compute the distance between the given points. (The coordinates are polar coordinates.)

31. $(2, 2\pi/3)$ and $(4, \pi/6)$ 32. $(4, \pi)$ and $(3, 7\pi/4)$

33. $(4, 4\pi/3)$ and $(1, 0)$ 34. $(3, 5\pi/6)$ and $(5, 5\pi/3)$

In Exercises 35–38, determine a polar equation for the circle satisfying the given conditions.

35. The radius is 2 and the polar coordinates of the center are: (a) $(4, 0)$;   (b) $(4, 2\pi/3)$;   (c) $(0, 0)$.

36. The radius is $\sqrt{6}$ and the polar coordinates of the center are: (a) $(2, \pi)$;   (b) $(2, 3\pi/4)$;   (c) $(0, 0)$.

37. The radius is 1 and the polar coordinates of the center are: (a) $(1, 3\pi/2)$;   (b) $(1, \pi/4)$.

38. The radius is 6 and the polar coordinates of the center are: (a) $(-3, 5\pi/4)$; (b) $(-2, \pi/4)$.

In Exercises 39–42, the polar equation of a line is given. In each case, (a) specify the perpendicular distance from the origin to the line; (b) determine the polar coordinates of the points on the line corresponding to $\theta = 0$ and $\theta = \pi/2$; (c) specify the polar coordinates of the foot of the perpendicular from the origin to the line; (d) use the results in parts (a), (b), and (c) to sketch the line.

39. $r \cos\left(\theta - \frac{\pi}{6}\right) = 2$

40. $r \cos\left(\theta - \frac{\pi}{4}\right) = 1$

41. $r \cos\left(\theta + \frac{2\pi}{3}\right) = 4$

42. $r \cos\left(\theta - \frac{\pi}{2}\right) = \sqrt{2}$

43. A line is tangent to the circle $x^2 + y^2 = 4$ at a point P with x-y coordinates $\left(-\sqrt{3}, 1\right)$.
  (a) What are the polar coordinates of P?
  (b) Find a polar equation for the tangent line.
  (c) Use the polar equation to determine the x- and y-intercepts of the line.

44. Follow Exercise 43 using the circle $x^2 + y^2 = 36$ and the point P with x-y coordinates $\left(-3, -3\sqrt{3}\right)$.

Notice that the rectangular form of the polar equation $r \cos\theta = a$ is $x = a$. Thus, the graph of the polar equation $r \cos\theta = a$ is a vertical line with an x-intercept of a. Similarly, the graph of the polar equation $r \sin\theta = b$ is a horizontal line with a y-intercept of b. In Exercises 45 and 46, use these observations to graph the polar equations.

45. (a) $r \cos\theta = 3$
    (b) $r \sin\theta = 3$

46. (a) $r \cos\theta = -2$
    (b) $r \sin\theta = 4$

B

47. The accompanying figure shows the graph of a line $\mathcal{L}$.

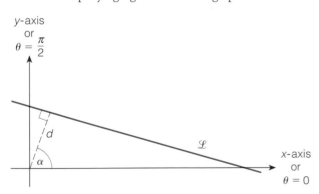

  (a) According to the text, what is the polar equation for $\mathcal{L}$?
  (b) By converting the equation in part (a) to rectangular form, show that the x-y equation for $\mathcal{L}$ is $x \cos\alpha + y \sin\alpha = d$. (This equation is called the **normal form** for a line.) *Hint:* First use one of the addition formulas for cosine.

48. By converting the polar equation
$$r = a \cos\theta + b \sin\theta$$
to rectangular form, show that the graph is a circle, and find the center and the radius.

49. Show that the rectangular form of the equation $r = a \sin 3\theta$ is $(x^2 + y^2)^2 = ay(3x^2 - y^2)$.

50. Show that the rectangular form of the equation
$$r = ab/(1 - a \cos\theta) \quad (a < 1)$$
is $(1 - a^2)x^2 + y^2 - 2a^2bx - a^2b^2 = 0$.

51. Show that the polar form of the equation $(x^2/a^2) - (y^2/b^2) = 1$ is $r^2 = a^2b^2/(b^2 \cos^2\theta - a^2 \sin^2\theta)$.

52. Show that a polar equation for the line passing through the two points $A(r_1, \theta_1)$ and $B(r_2, \theta_2)$ is
$$rr_1 \sin(\theta - \theta_1) - rr_2 \sin(\theta - \theta_2) + r_1r_2 \sin(\theta_1 - \theta_2) = 0$$
Hint: Use the following figure and the observation that
$$\text{area } \triangle OAB = (\text{area } \triangle OBP) + (\text{area } \triangle OPA)$$

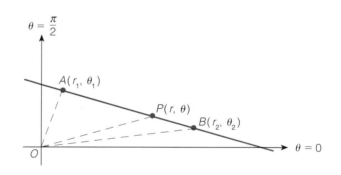

53. The accompanying figure shows an equilateral triangle inscribed in the unit circle.

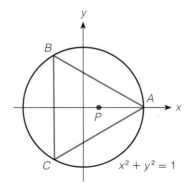

  (a) Specify polar coordinates for the points A, B, and C.
  (b) In the figure, P denotes an arbitrary point between

the origin and A on the x-axis. Let the polar coordinates of P be $(r, 0)$. Use the distance formula for polar coordinates to show that

$$(PA)(PB)(PC) = 1 - r^3$$

[This result is due to the English mathematician Roger Cotes (1682–1716).]

54. The accompanying figure shows an equilateral triangle inscribed in the unit circle. The point Q lies on the bisector of $\angle AOC$, and the distance from O to Q is r. Use the distance formula for polar coordinates to show that

$$(QA)(QB)(QC) = 1 + r^3$$

(This result also is due to Roger Cotes.)

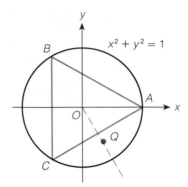

$\dfrac{8.6}{}$ **CURVES IN POLAR COORDINATES**

FIGURE I

This polar coordinate graph was discovered by Oscar Ramirez, a precalculus student at U.C.L.A. in the Fall of 1991.

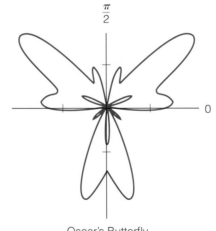

Oscar's Butterfly

$r = \cos^2 5\theta + \sin 3\theta + 0.3$

In principle, the graph of any polar equation $r = f(\theta)$ can be obtained by setting up a table and plotting a sufficient number of points. Indeed, this is just the way a graphing calculator or a computer operates. Figure 1, for example, shows a computer-generated graph of the polar equation $r = \cos^2 5\theta + \sin 3\theta + 0.3$. In this section, we'll concentrate on a few basic types of polar equations and their graphs. With the exception of Example 1, the polar equations that we consider involve trigonometric functions. In order to *understand* why the graphs look as they do, we will often use symmetry and the basic properties of the sine and cosine functions.

■■■ **EXAMPLE 1**   Graph the polar equation $r = \theta/\pi$ for $\theta \geq 0$.

Solution   The equation shows that as θ increases, so does r. Geometrically, this means that as we plot points, moving counterclockwise with increasing θ, the points will be farther and farther from the origin. In Table 1, we have computed values for r corresponding to some convenient values of θ. In Figure 2, we've plotted the points in the table and connected them with a smooth curve. The curve is known as the **spiral of Archimedes**.

TABLE I

θ	0	$\dfrac{\pi}{4}$	$\dfrac{\pi}{2}$	$\dfrac{3\pi}{4}$	π	$\dfrac{5\pi}{4}$	$\dfrac{3\pi}{2}$	$\dfrac{7\pi}{4}$	2π	$\dfrac{5\pi}{2}$	3π	4π
r	0	$\dfrac{1}{4}$	$\dfrac{1}{2}$	$\dfrac{3}{4}$	1	$\dfrac{5}{4}$	$\dfrac{3}{2}$	$\dfrac{7}{4}$	2	$\dfrac{5}{2}$	3	4

FIGURE 2
$r = \theta/\pi \quad (\theta \geq 0)$

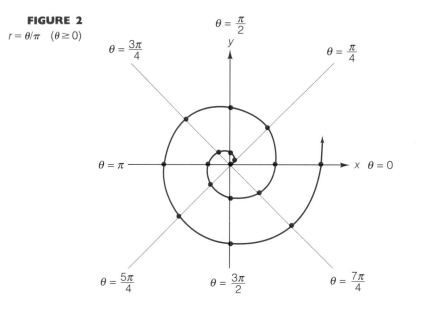

We saw in Chapters 2 and 3 that symmetry considerations can often be used to lessen the amount of work involved in graphing equations. This is also true for polar equations. In the box that follows, we list four tests for symmetry in polar coordinates.

Symmetry Tests in Polar Coordinates

If the following substitution yields an equivalent equation:	then the graph is symmetric about:
1. replacing θ with $-\theta$,	the x-axis;
2. replacing r and θ with $-r$ and $-\theta$, respectively,	the y-axis;
3. replacing θ with $\pi - \theta$,	the y-axis;
4. replacing r with $-r$,	the origin.

CAUTION   As opposed to the case with the x-y symmetry tests (in Chapter 2), if a polar equation fails a symmetry test, then it is still possible for the graph to possess the indicated symmetry. This is a consequence of the fact that the polar coordinates of a point are not unique. (For an example, see Example 3.)

The validity of the first test follows from the fact that the points (r, θ) and $(r, -\theta)$ are reflections of each other about the x-axis [see Figure 3(a)]. The validity of test 2 follows from the fact that the points (r, θ) and $(-r, -\theta)$ are reflections of each other about the y-axis, as indicated in Figure 3(b). The other tests can be justified in a similar manner.

FIGURE 3

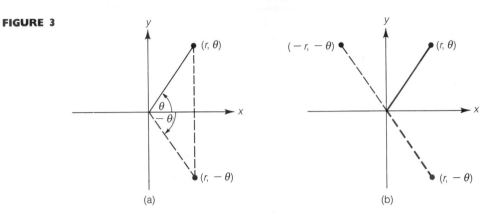

(a)   (b)

▌■■▌ **EXAMPLE 2** Graph the equation $r = 1 - \cos \theta$.

Solution We know (from Section 7.2) that the cosine function satisfies the identity $\cos(-\theta) = \cos \theta$. So, according to symmetry test 1, the graph of $r = 1 - \cos \theta$ will be symmetric about the x-axis. In Figure 4(a), we've set up a table (using a calculator as necessary) with some convenient values of θ running from 0 to π. Plotting the points in the table leads to the curve shown in Figure 4(a). Reflecting this curve in the x-axis, we obtain the graph in Figure 4(b). The fact that the period of the cosine function is 2π implies that we need not consider values of θ beyond 2π. (Why?) Thus, the curve in Figure 4(b) is the required graph of $r = 1 - \cos \theta$. The curve is a **cardioid.**   ▌▌▌

FIGURE 4

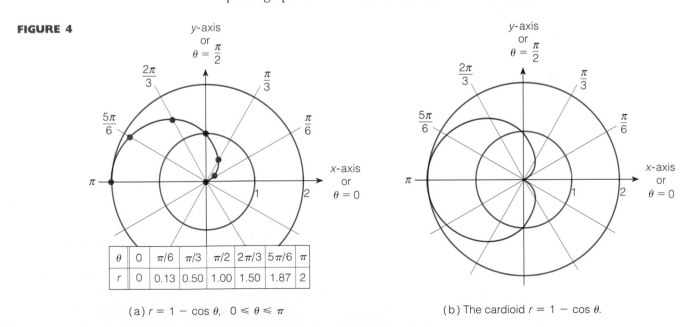

θ	0	$\pi/6$	$\pi/3$	$\pi/2$	$2\pi/3$	$5\pi/6$	π
r	0	0.13	0.50	1.00	1.50	1.87	2

(a) $r = 1 - \cos \theta$, $0 \leqslant \theta \leqslant \pi$

(b) The cardioid $r = 1 - \cos \theta$.

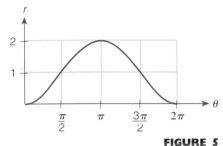

FIGURE 5

The graph of $r = 1 - \cos \theta$ in a rectangular θ-r coordinate system.

In graphing polar equations, just as in graphing Cartesian x-y equations, there is always the question of whether we have plotted a sufficient number of points to reveal the essential features of the graph. As a check on our work in Figure 4(a), we can graph the equation $r = 1 - \cos \theta$ in a *rectangular θ-r coordinate* system, as shown in Figure 5. (This is easy to do using reflection and translation.) From the graph in Figure 5, we see that as θ increases from 0 to π, the values of r increase from 0 to 2. Now, recall that in polar coordinates, r represents the directed distance from the origin. So, on the polar graph, as we go from $\theta = 0$ to π, the points we obtain should be farther and farther from the origin, starting with $r = 0$ and ending with $r = 2$. The polar graph in Figure 4(a) is consistent with these observations.

EXAMPLE 3 Graph the polar curve $r = 1 + 2 \sin \theta$.

Solution Using symmetry test 3, we replace θ by $\pi - \theta$ in the given equation:

$$r = 1 + 2 \sin(\pi - \theta) = 1 + 2(\sin \pi \cos \theta - \cos \pi \sin \theta)$$
$$= 1 + 2 \sin \theta \qquad \text{because } \sin \pi = 0 \quad \text{and} \quad \cos \pi = -1$$

This shows that the graph is symmetric about the y-axis. (Note that symmetry test 2 fails for this equation, yet the graph is symmetric about the y-axis.)

Next, to see what to expect in polar coordinates, we first sketch the graph of the given equation in a rectangular θ-r coordinate system; see Figure 6. In Figure 6, notice that the maximum value of $|r|$ is 3. This implies that the required polar graph will be contained within a circle of radius 3 about the origin. (We'll refer to this sketch several more times in the course of the solution.)

FIGURE 6

The graph of $r = 1 + 2 \sin \theta$ in a rectangular θ-r coordinate system.

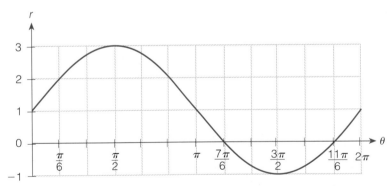

In Figure 7(a), we have set up a table and graphed the given polar equation for θ running from 0 to $\pi/2$. Note that as θ increases from 0 to $\pi/2$, the r-values increase from 1 to 3 (in agreement with Figure 6). By reflecting the curve in Figure 7(a) in the y-axis, we obtain the graph of $r = 1 + 2 \sin \theta$ for $0 \leq \theta \leq \pi$; see Figure 7(b).

FIGURE 7

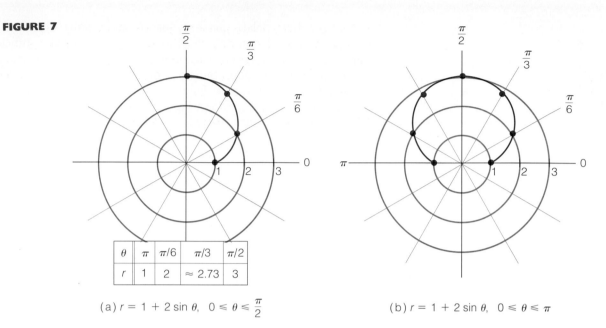

θ	π	$\pi/6$	$\pi/3$	$\pi/2$
r	1	2	≈ 2.73	3

(a) $r = 1 + 2 \sin \theta,\ \ 0 \leqslant \theta \leqslant \dfrac{\pi}{2}$ (b) $r = 1 + 2 \sin \theta,\ \ 0 \leqslant \theta \leqslant \pi$

Next, let us consider the values of r as θ increases from π to $3\pi/2$. Figure 6 (on the previous page) tells us what to expect on this interval:

As θ increases from π to $7\pi/6$, the r-values decrease from 1 to 0; As θ increases from $7\pi/6$ to $3\pi/2$, the r-values decrease from 0 to -1. However (in view of the convention regarding negative r-values), this will mean that the points on the polar graph are, in fact, moving farther and farther from the origin as θ increases from $7\pi/6$ to $3\pi/2$.

FIGURE 8

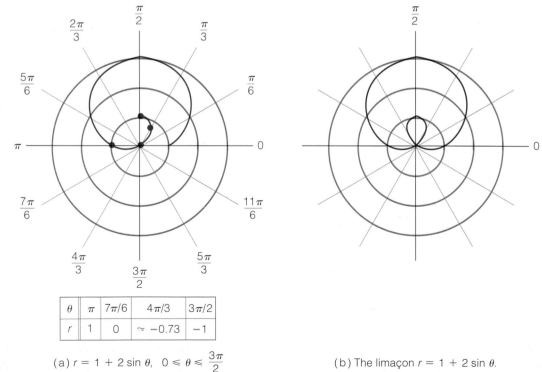

θ	π	$7\pi/6$	$4\pi/3$	$3\pi/2$
r	1	0	≈ -0.73	-1

(a) $r = 1 + 2 \sin \theta,\ \ 0 \leqslant \theta \leqslant \dfrac{3\pi}{2}$ (b) The limaçon $r = 1 + 2 \sin \theta$.

In Figure 8(a), we've set up a table for $\pi \leq \theta \leq 3\pi/2$ and added the corresponding points from Figure 7(b). Finally, again taking into account the y-axis symmetry, we sketch the final graph for $0 \leq \theta \leq 2\pi$, as indicated in Figure 8(b). The curve in Figure 8(b) is a **limaçon**. (Not all limaçons have inner loops. See for example Exercises 23 and 24.) ▮▮▮

▮▮▮ **EXAMPLE 4** Graph the polar equation $r^2 = 4 \cos 2\theta$.

Solution As a guide for graphing the given polar equation, we first sketch the curve $y = 4 \cos 2\theta$ in a rectangular θ-y coordinate system. As indicated in Figure 9, the period of $4 \cos 2\theta$ is π. So, in graphing the given polar equation, we will not need to consider values of θ beyond π. Figure 9 also shows that $4 \cos 2\theta$ is negative on the interval $\frac{\pi}{4} < \theta < \frac{3\pi}{4}$. The value of r^2, however, cannot be negative. Consequently, we do not need to consider values of θ in the open interval from $\frac{\pi}{4}$ to $\frac{3\pi}{4}$. Finally, Figure 9 shows that the absolute value of $4 \cos 2\theta$ is always less than or equal to 4. Consequently, for the polar graph, we have

$$r^2 \leq 4 \qquad \text{and therefore} \qquad |r| \leq 2$$

This tells us that the polar graph will be contained within a circle of radius 2 about the origin.

For the polar graph, we begin by computing a table of values with θ running from 0 to $\frac{\pi}{4}$; see Table 2. Plotting the points in Table 2 and joining them with a smooth curve leads to the graph shown in Figure 10(a). Then, rather than setting up another table with θ running from $\frac{3\pi}{4}$ to π, we rely on symmetry to complete graph. According to the first two symmetry tests, the curve is symmetric about both the x-axis and the y-axis. Thus, we obtain the graph shown in Figure 10(b). The curve is called a **lemniscate**. ▮▮▮

FIGURE 9
$y = 4 \cos 2\theta$

TABLE 2

θ	$r = \pm 2\sqrt{\cos 2\theta}$
0	± 2
$\frac{\pi}{12}$	± 1.86
$\frac{\pi}{8}$	$+1.68$
$\frac{\pi}{6}$	± 1.41
$\frac{\pi}{4}$	0

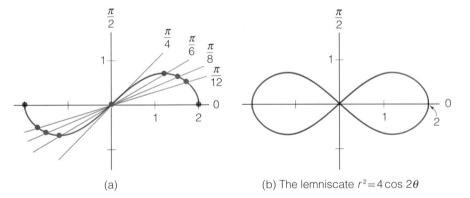

FIGURE 10 (a) (b) The lemniscate $r^2 = 4 \cos 2\theta$

EXERCISE SET 8.6

A

In Exercises 1–26, graph the polar equations.

1. $r = \theta/(2\pi)$, for $\theta \geq 0$
2. $r = \theta/\pi$, for $-4\pi \leq \theta \leq 0$
3. $r = \ln \theta$, for $1 \leq \theta \leq 3\pi$
4. $r = e^{\theta/(2\pi)}$, for $0 \leq \theta \leq 2\pi$
5. $r = e^{-\theta/(2\pi)}$, for $0 \leq \theta \leq 2\pi$
6. $r = \sqrt{\theta}$, $0 \leq \theta \leq 3\pi$
7. $r = 1 + \cos \theta$
8. $r = 1 - \sin \theta$
9. $r = 2 - 2 \sin \theta$
10. $r = 3 + 3 \cos \theta$

11. $r = 1 - 2 \sin \theta$

12. $r = 1 + 2 \cos \theta$

13. $r = 2 + 4 \cos \theta$

14. $r = 2 - 4 \sin \theta$

15. $r^2 = 4 \sin 2\theta$

16. $r^2 = 9 \cos 2\theta$

17. $r^2 = \cos 4\theta$

18. $r^2 = 3 \sin 4\theta$

19. $r = \cos 2\theta$ *(four-leafed rose)*

20. $r = 2 \sin 2\theta$ *(four-leafed rose)*

21. $r = \sin 3\theta$ *(three-leafed rose)*

22. $r = 2 \cos 5\theta$ *(five-leafed rose)*

23. $r = 4 + 2 \sin \theta$ *(limaçon with no inner loop)*

24. $r = 1.5 - \cos \theta$ *(limaçon with no inner loop)*

25. $r = 8 \tan \theta$ *(kappa curve)*

26. $r = \csc \theta + 2$ *(conchoid of Nicomedes)*

In Exercises 27 and 28, match one of the graphs with each equation.   Hint: Compute the values of r when θ = 0, π/2, π, and 3π/2.

27. (a) $r = 3 + 3 \sin \theta$ (b) $r = 3 - 3 \sin \theta$
     (c) $r = 3 - 3 \cos \theta$ (d) $r = 3 + 3 \cos \theta$

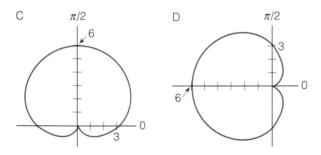

28. (a) $r = 3 + 2 \cos \theta$ (b) $r = 2 + 3 \cos \theta$
     (c) $r = 2 - 3 \cos \theta$ (d) $r = 3 - 2 \cos \theta$

B

29. The following figure shows a graph of the logarithmic spiral $\ln r = a\theta$ for $0 \le \theta \le 2\pi$. (In this equation, a denotes a positive constant.)

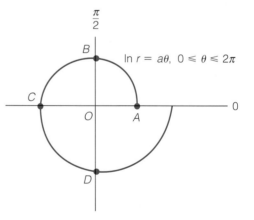

(a) Find the polar coordinates of the points A, B, C, and D, and then show that
$$\frac{OD}{OC} = \frac{OC}{OB} = \frac{OB}{OA} = e^{a\pi/2}$$

(b) Show that $\angle ABC$ and $\angle BCD$ are right angles. *Hint:* Use rectangular coordinates to compute slopes.

30. (a) Graph the polar curves $r = 2 \cos \theta - 1$ and $r = 2 \cos \theta + 1$. What do you observe?

(b) Part (a) shows that algebraically nonequivalent polar equations may have identical graphs. (This is another consequence of the fact that the polar coordinates of a point are not unique.) Show that both equations in part (a) can be written $(x^2 - 2x + y^2)^2 - x^2 - y^2 = 0$.

31. Follow Exercise 30 using the two polar equations $r = \cos(\theta/2)$ and $r = \sin(\theta/2)$.

32. Let k denote a positive constant, and let F_1 and F_2 denote the points with rectangular coordinates $(-k, 0)$ and $(k, 0)$, respectively. A curve known as the *lemniscate of Bernoulli* is defined as the set of points $P(x, y)$ such that $(F_1P) \times (F_2P) = k^2$.
  (a) Show that the rectangular equation of the curve is $(x^2 + y^2)^2 = 2k^2(x^2 - y^2)$.
  (b) Show that the polar equation is $r^2 = 2k^2 \cos 2\theta$.
  (c) Graph the equation $r^2 = 2k^2 \cos 2\theta$.

OPTIONAL TI-81 GRAPHING CALCULATOR EXERCISES
EXERCISES FOR SECTIONS 8.4 THROUGH 8.6

1. In Figure 1 of Section 8.4 we graphed the parametric equations $x = 2t$, $y = \frac{1}{2}t^2$ for $t \geq 0$. This exercise shows how to use the calculator to obtain the graph. As preparation, press ZOOM 6 for the standard viewing rectangle.
  (a) Press the MODE key and choose the parametric mode (rather than the function mode); then press ENTER. Next, press the Y= key and enter $X_1 = 2t$, $Y_1 = \frac{1}{2}t^2$. (The parameter t is typed using the same $X|T$ key that you have always used for graphing ordinary x-y equations.) Finally, press GRAPH.
  (b) To see which values of the parameter t were used in generating the graph in part (a), press the RANGE key. As you can then see, t runs from a minimum of 0 to a maximum of about 6.28 (which is approximately 2π). Change the setting for t_{min} to read $t_{min} = -6.28$. Now press GRAPH. How is the graph changed?
  (c) Press the RANGE key and set $t_{min} = 2$ and $t_{max} = 3$. Now press GRAPH.
  (d) To determine the coordinates of the left-hand endpoint of your graph in part (c), press the TRACE key and then use the left ARROW key to move the cursor to the left-hand endpoint of the graph. The display at the bottom of the screen tells us that the endpoint has the coordinates $(4, 2)$ and that the corresponding value of t is 2.

2. In this exercise we use the calculator to obtain the graph shown in Figure 2 in Section 8.4. The parametric equations are $x = (88 \cos 35°)t$ and $y = 6 + (88 \sin 35°)t - 16t^2$. Before entering these two equations in the Y= menu, press the MODE key and select degree measure rather than radian measure. Now press the Y= key and enter the two given parametric equations. Press ZOOM 6 and then make the following RANGE adjustments: $t_{min} = 0$, $t_{max} = 4$, $t_{step} = 0.5$,

$x_{min} = 0$, $x_{max} = 240$, $x_{scl} = 20$, $y_{min} = 0$, $y_{max} = 60$, $y_{scl} = 20$. Finally, press the GRAPH key. Check that your result is consistent with the graph shown in Figure 2 in Section 8.4. (If you want to improve the appearance of your graph, press the RANGE key and set $t_{step} = 0.01$. Then press GRAPH.)

In Exercises 3–10, graph the parametric equations for the given values of the parameter t. As preparation, press the MODE key and select the radian mode. Press the Y= key and clear all of the functions. Press ZOOM 6 so that you are beginning with the standard viewing rectangle. (In each case, you are on your own as far as making suitable adjustments in the RANGE menu. However, as a general rule, try to choose settings that will utilize as much of the viewing screen as possible. For instance, in graphing the circle given by x = cos t, y = sin t, it would be natural to choose the settings $x_{min} = -1$, $x_{max} = 1$, $y_{min} = -1$, and $y_{max} = 1$. Also, after choosing these settings, the ZOOM 5 command would be useful to ensure that you are seeing the correct proportions in the graph.)

3. $x = 4 \cos t$, $y = 3 \sin t$, $0 \leq t \leq 2\pi$ (an ellipse)

4. $x = 4 \cos t$, $y = -3 \sin t$, $0 \leq t \leq 2\pi$ (the same as the ellipse in Exercise 3, but traced out in the opposite direction)

5. $x = 4 \cos t$, $y = 3 \sin t$, $0 \leq t \leq \pi/2$ (one-quarter of an ellipse)

6. $x = 4 \cos^3 t$, $y = 4 \sin^3 t$, $0 \leq t \leq 2\pi$ [This curve is known as the *astroid* or *hypocycloid of four cusps*. It was first studied by the Danish astronomer Olof Roemer (1644–1710) and by the Swiss mathematician James Bernoulli (1654–1705).]

7. $x = 2 \cos t + \cos 2t$, $y = 2 \sin t - \sin 2t$, $0 \leq t \leq 2\pi$ [This curve is the *deltoid*. It was first studied by the Swiss mathematician Leonhard Euler (1707–1783).]

8. $x = \dfrac{\cos t}{1 + \sin^2 t}$, $y = \dfrac{\sin t \cos t}{1 + \sin^2 t}$, $0 \le t \le 2\pi$ (the *lemniscate of Bernoulli*)

9. $x = 2 \tan t$, $y = 2 \cos^2 t$, $0 \le t \le 2\pi$ [This is the *witch of Agnesi*, named after the Italian mathematician and scientist Maria Gaetana Agnesi (1718–1799). For background on how the curve got its name, see the text by Howard Eves, *An Introduction to the History of Mathematics*, 6th ed., p. 442 (Philadelphia: Saunders College Publishing, 1990).]

10. $x = \dfrac{t^2 - 1}{3t^2 + 1}$, $y = \dfrac{t(t^2 - 1)}{3t^2 + 1}$, $-\infty < t < \infty$ (the *folium of Descartes*)   *Hint:*   Use $t_{min} = -10$, $t_{max} = 10$, $t_{step} = 0.05$.

In Exercises 11–22 you will graph polar equations. A polar equation of the form $r = f(\theta)$ can always be expressed in parametric form by means of the equations

$x = r \cos \theta$ (1)

$y = r \sin \theta$ (2)

In equations (1) and (2), we replace r by $f(\theta)$. This yields

$x = f(\theta) \cos \theta$ (3)

$y = f(\theta) \sin \theta$ (4)

Equations (3) and (4) form a set of parametric equations; the graph of these equations will be the polar curve $r = f(\theta)$.

As preparation for Exercises 11–22, press the MODE key and check that the calculator is set in the radian mode and in the parametric mode. Press the Y= key and clear all functions that are present. Finally press ZOOM 6 so that you are starting with the standard viewing rectangle.

11. In this exercise, we graph the polar equation $r = 2 \cos \theta$. Using equations (3) and (4), we have

$$x = f(\theta) \cos \theta = 2 \cos \theta \cos \theta = 2 \cos^2 \theta$$

and

$$y = f(\theta) \sin \theta = 2 \cos \theta \sin \theta$$

(a) Press the Y= key and enter the parametric equations $X_1 = 2 \cos^2 t$ and $Y_1 = 2 \cos t \sin t$. Now press the GRAPH key. Go on to part (b) for a better view of the graph.

(b) Use the zoom box (ZOOM 1) to obtain a closer look at the graph. (The precise coordinates of the corners of the box are unimportant; what matters is that the entire curve be contained within the box.) After you have obtained a larger picture of the graph, press ZOOM 5 so that the true proportions of the curve are shown. You should obtain a circle.

(c) Convert the equation $r = 2 \cos \theta$ to rectangular form and show that the result can be written

$(x - 1)^2 + y^2 = 1$. Check that this result is consistent with the graph in part (b).

In Exercises 12–18, graph the polar equations.

12. (a) $r = \cos 2\theta$ (*four-leafed rose*)
     (b) $r = \sin 2\theta$

13. (a) $r = \cos 4\theta$ (*eight-leafed rose*)
     (b) $r = \sin 4\theta$

14. (a) $r = \cos 3\theta$ (*three-leafed rose*)
     (b) $r = \sin 3\theta$

15. (a) $r = \cos 5\theta$ (*five-leafed rose*)
     (b) $r = \sin 5\theta$

16. $r = 4(1 + \cos \theta)$ (*cardioid*)

17. $r = \theta$ (*spiral of Archimedes*)   *Suggestion:*   Experiment with different values for t_{max} in the RANGE menu.

18. $r^2 = 4 \cos 2\theta$ (*lemniscate of Bernoulli*)   *Hints:*   The equation is equivalent to the two equations $r = \pm 2\sqrt{\cos 2\theta}$. You'll find, however, that you need to consider only one of these. Also, you can improve the appearance of your final graph if you set $t_{step} = 0.01$. (Don't make this adjustment until the end; the graph will be drawn very slowly.)

19. The *limaçon of Pascal* is defined by the equation $r = a \cos \theta + b$. [The curve is named after Étienne Pascal (1588–1640), the father of Blaise Pascal. According to mathematics historian Howard Eves, however, the curve is misnamed; it appears earlier in the writings of Albrecht Dürer (1471–1528).] When $a = b$, the curve is a cardioid; when $a = 0$ the curve is a circle. Graph the following limaçons.
     (a) $r = 2 \cos \theta + 1$ (b) $r = 2 \cos \theta - 1$
     (c) $r = \cos \theta + 2$ (d) $r = \cos \theta - 2$
     (e) $r = 2 \cos \theta + 2$ (f) $r = \cos \theta$

20. Graph the *cissoid of Diocles*, $r = 10 \sin \theta \tan \theta$. [The Greek mathematician Diocles (ca. 180 B.C.) used the geometric properties of this curve to solve the problem of duplicating the cube. For an explanation of the problem and historical background, see the text by Howard Eves, *An Introduction to the History of Mathematics*, 6th ed., pp. 109–127 (Philadelphia: Saunders College Publishing, 1990).]

21. Graph the *logarithmic spiral* (or *spiral of Descartes*), $r = e^{\theta/10}$, using the standard viewing rectangle (modified with ZOOM 5) and the following settings:
     (a) $t_{max} = 10$;   (b) $t_{max} = 20$;   (c) $t_{max} = 25$.

22. The curve $r = \cos^4(\theta/4)$ has a surprising property that is very difficult to observe without the aid of a graphing calculator or a computer. After entering the parametric equations for this curve, press ZOOM 6. Next, make the following adjustments in the RANGE menu: $t_{max} = 13$, $x_{min} = -1$, $x_{max} = 1$, $y_{min} = -1$,

$y_{max} = 1$. Then press ZOOM 5. As you can see, the graph appears to have a simple inner loop to the left of the origin. This is not the case, however. Use the ZOOM 2 command to zoom in on the origin several times. What do you find?

In Exercises 23 and 24, graph the polar curves.

23. (a) $r = 2|\sin \theta|^{\sin \theta}$
   (b) $r = 2|\sin \theta|^{\sin(\theta/2)}$

24. $r = (\sin \theta)/\theta, \ 0 < \theta \le 4\pi$

▮▮▮ CHAPTER EIGHT   SUMMARY OF PRINCIPAL FORMULAS AND NOTATION

FORMULAS OR NOTATIONS	PAGE REFERENCE	COMMENT
1. $\dfrac{\sin A}{a} = \dfrac{\sin B}{b} = \dfrac{\sin C}{c}$	440	This is the *law of sines*. In words, it states that in any triangle, the sines of the angles are proportional to the lengths of the opposite sides.
2. $a^2 = b^2 + c^2 - 2bc \cos A$	442	This is the *law of cosines*. We can use it to calculate the third side of a triangle when we are given two sides and the included angle. The formula can also be used to calculate the angles of a triangle when we know the three sides, as in Example 5 on page 443.
3. Vector	452 461	Geometrically, a *vector* in the plane is a directed line segment. Vectors can be used to represent quantities that have both magnitude and direction, such as force and velocity. For examples, see Figures 1 and 2 in Section 8.2. Algebraically, a vector in the plane is an ordered pair of real numbers, denoted by $\langle x, y \rangle$. (See Figure 3 in Section 8.3.) The numbers x and y are called the *components* of the vector $\langle x, y \rangle$.
4. Vector equality	453 462	Geometrically, two vectors are said to be equal provided they have the same length and the same direction. In terms of components, this means that two vectors are equal if and only if their corresponding components are equal.
5. Vector addition	454 455 463	Geometrically, two vectors can be added by using the parallelogram law, as in Figure 10 in Section 8.2. Algebraically, this is equivalent to the following componentwise formula for vector addition: $$\langle a, b \rangle + \langle c, d \rangle = \langle a + c, b + d \rangle$$
6. $\|\mathbf{v}\| = \sqrt{v_1^2 + v_2^2}$	453 462	The expression on the left-hand side of the equation denotes the *length* or *magnitude* of the vector $\mathbf{v}$. The expression on the right-hand side of the equation tells us how to compute the length of $\mathbf{v}$ in terms of its respective x- and y-components, v_1 and v_2.
7. $\overrightarrow{PQ} = \langle x_2 - x_1, y_2 - y_1 \rangle$	462	This formula gives the components of the vector $\overrightarrow{PQ}$, where P and Q are the points $P(x_1, y_1)$ and $Q(x_2, y_2)$. See Figure 5 in Section 8.3 for a derivation of this formula.
8. The zero vector	464	The *zero vector*, denoted by $\mathbf{0}$, is the vector $\langle 0, 0 \rangle$. The zero vector plays the same role in vector addition as does the real number zero in addition of real numbers. For any vector $\mathbf{v}$, we have $$\mathbf{0} + \mathbf{v} = \mathbf{v} + \mathbf{0} = \mathbf{v}$$

FORMULAS OR NOTATIONS	PAGE REFERENCE	COMMENT
9. Scalar multiplication	464	For each real number k and each vector $\mathbf{v} = \langle x, y \rangle$, the vector $k\mathbf{v}$ is defined by the equation $$k\mathbf{v} = \langle kx, ky \rangle$$ Geometrically, the length of $k\mathbf{v}$ is $\lvert k \rvert$ times the length of $\mathbf{v}$. The operation that forms the vector $k\mathbf{v}$ from the scalar k and the vector $\mathbf{v}$ is called *scalar multiplication*.
10. $\mathbf{i} = \langle 1, 0 \rangle$ $\mathbf{j} = \langle 0, 1 \rangle$	466	These two equations define the *unit vectors* $\mathbf{i}$ and $\mathbf{j}$; see Figure 11 in Section 8.3. Any vector $\langle x, y \rangle$ can be expressed in terms of the unit vectors $\mathbf{i}$ and $\mathbf{j}$ as follows: $$\langle x, y \rangle = x\mathbf{i} + y\mathbf{j}$$
11. Parametric equations	469 470	Suppose the x- and y-coordinates of the points on a curve are expressed as functions of another variable t: $$x = f(t) \quad \text{and} \quad y = g(t)$$ These equations are parametric equations of the curve, and t is called a *parameter*. An example: A pair of parametric equations for the ellipse $(x^2/a^2) + (y^2/b^2) = 1$ is $x = a \cos t$ and $y = b \sin t$.
12. $x = r \cos \theta$ $y = r \sin \theta$ $x^2 + y^2 = r^2$ $\tan \theta = \dfrac{y}{x}$	478	These formulas relate the rectangular coordinates of a point (x, y) to its polar coordinates (r, θ). See the boxed figure on page 478.
13. $d^2 = r_1^2 + r_2^2 - 2r_1 r_2 \cos(\theta_2 - \theta_1)$	480	This is the formula for the (square of the) distance d between the points with polar coordinates (r_1, θ_1) and (r_2, θ_2).
14. $r^2 + r_0^2 - 2rr_0 \cos(\theta - \theta_0) = a^2$	480	This is a polar equation for the circle with center (r_0, θ_0) and radius a.
15. $r \cos(\theta - \alpha) = d$	482	This is a polar equation for a line not passing through the origin. In the equation, d and α are constants; d is the perpendicular distance from the origin to the line, and the (polar) point (d, α) is the foot of the perpendicular from the origin to the line.

WRITING MATHEMATICS

A regular feature in *Mathematics Magazine* is the "Proof Without Words" section. The idea is to present a proof of a well-known theorem in such a way that the entire proof is "obvious" from a well-chosen picture and at most several equations. Of course, what is obvious to a mathematician may not be so obvious to her or his students (as you well know).

The following clever proof without words for the law of cosines was developed by Professor Sidney H. Kung of Jacksonville University. (It was published in *Mathematics Magazine*, vol. 63, no. 5, December 1990.) Study the proof (or discuss it with friends) until you see how it works. Then, in a paragraph or two, write out the details of the proof as if you were explaining it to a classmate. (To facilitate your explanation, you will probably need to label some of the points in the given figure.)

Proof without words for
the law of cosines, $\theta < 90°$
(by Professor Sidney H. Kung).

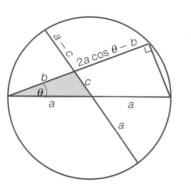

$(2a \cos \theta - b)\, b = (a - c)(a + c)$
$c^2 = a^2 + b^2 - 2ab \cos \theta$

CHAPTER EIGHT   REVIEW EXERCISES

In Exercises 1–8, the lengths of the three sides of a triangle are denoted by a, b, and c; the angles opposite these sides are A, B, and C, respectively. In each exercise, use the given data to find the remaining sides and angles. Use a calculator and round off the final answers to one decimal place.

1. $\angle A = 40°$, $\angle B = 85°$, $c = 16$ cm

2. $\angle C = 84°$, $\angle B = 16°$, $a = 9$ cm

3. (a) $a = 8$ cm, $b = 9$ cm, $\angle A = 52°$, $\angle B < 90°$
    (b) Use the data in part (a), but assume $\angle B > 90°$.

4. (a) $a = 6.25$ cm, $b = 9.44$ cm, $\angle A = 12°$, $\angle B < 90°$
    (b) Use the data in part (a), but assume $\angle B > 90°$.

5. $a = 18$ cm, $b = 14$ cm, $\angle C = 24°$

6. $a = 32.16$ cm, $b = 50.12$ cm, $\angle C = 156°$

7. $a = 4$ cm, $b = 7$ cm, $c = 9$ cm

8. $a = 12.61$ cm, $b = 19.01$ cm, $c = 14.14$ cm

In Exercises 9–18, refer to the following figure. In each case, determine the indicated quantity. Use a calculator, and round off your result to two decimal places.

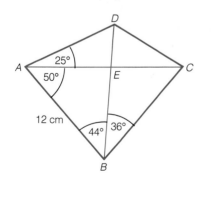

9. BE

10. BC

11. area of $\triangle BCE$

12. BD

13. area of $\triangle ABD$

14. DE

15. CD

16. AD

17. AC

18. area of $\triangle DEC$

In Exercises 19 and 20, refer to the following figure, in which $\overline{BD}$ bisects $\angle ABC$.

19. If $AB = 24$ cm, $BC = 40$ cm, and $AC = 56$ cm, find the length of the angle bisector $\overline{BD}$.

20. If $AB = 105$ cm, $BC = 120$ cm, and $AC = 195$ cm, find the length of the angle bisector $\overline{BD}$.

In Exercises 21 and 22, refer to the following figure, in which $AD = DC$.

21. If $AB = 12$ cm, $BC = 26$ cm, and $AD = DC = 17$ cm, find BD.

22. If $AB = 16$ cm, $BC = 22$ cm, and $AD = DC = 17$ cm, find BD.

23. In $\triangle ABC$, $a = 4$, $b = 5$, and $c = 6$.
    (a) Find $\cos A$ and $\cos C$.
    (b) Using the results in part (a), show that angle C is twice angle A.

24. In $\triangle ABC$, suppose that angle C is twice angle A. Show that $ab = c^2 - a^2$.

In Exercises 25–30, refer to the framework shown in the accompanying figure. (Assume that the figure is drawn to scale.)

In the figure, $\overline{AC}$ is parallel to $\overline{ED}$ and $BE = BD$. In each case, compute the indicated quantity. Round off the final answers to two decimal places.

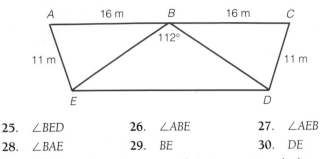

25. $\angle BED$ **26.** $\angle ABE$ **27.** $\angle AEB$

28. $\angle BAE$ **29.** BE **30.** DE

31. The given figure shows a regular pentagon inscribed in a circle. The radius of the circle is 10 cm. Determine the following lengths, rounding off the answers to two decimal places:
(a) one side of the pentagon;   (b) AC.

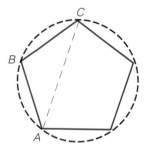

32. A pilot on a training flight is supposed to leave the airport and fly for 100 miles in the direction N40°E. Then he is supposed to turn around and fly directly back to the airport. By error, however, he makes the 100-mile return trip in the direction S25°W. How far is he now from the airport, and in what direction must he fly? (Round off the answers to one decimal place.)

33. The following figures show the same circle of diameter D inscribed first in an equilateral triangle and then in a regular hexagon. Let a denote the length of a side of the triangle and let b denote the length of a side of the hexagon. Show that $ab = D^2$.

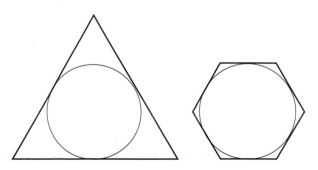

34. Suppose that triangle ABC is a right triangle with the right angle at C. Use the law of cosines to prove the following statements.
(a) The square of the distance from C to the midpoint of the hypotenuse is equal to one-fourth the square of the hypotenuse.
(b) The sum of the squares of the distances from C to the two points that trisect the hypotenuse is equal to five-ninths the square of the hypotenuse.
(c) Let P, Q, and R be points on the hypotenuse such that $AP = PQ = QR = RB$. Derive a result similar to your results in parts (a) and (b) for the sum of the squares of the distances from C to P, Q, and R.

35. The perimeter and the area of the triangle in the figure are 20 cm and $10\sqrt{3}$ cm², respectively. Find a and b. (Assume $a < b$.)

36. In $\triangle ABC$, suppose that $\angle B - \angle A = 90°$.
(a) Show that $c^2 = (b^2 - a^2)^2/(a^2 + b^2)$. *Hint:* Let P be the point on $\overline{AC}$ such that $\angle CBP = 90°$ and $\angle PBA = \angle A$. Show that $CP = (a^2 + b^2)/2b$.
(b) Show that $\dfrac{2}{c^2} = \dfrac{1}{(b+a)^2} + \dfrac{1}{(b-a)^2}$.

37. Two forces **F** and **G** act on an object, **G** horizontally to the right with a magnitude of 15 N, and **F** vertically upward with a magnitude of 20 N. Determine the magnitude and direction of the resultant force. (Use a calculator to determine the angle between the horizontal and the resultant; round off the result to one decimal place.)

38. Determine the resultant of the two forces in the accompanying figure. (Use a calculator, and round off the values you obtain for the magnitude and direction to one decimal place.)

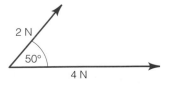

39. Determine the horizontal and vertical components of the velocity vector **v** in the following figure. (Use a calculator, and round off your answers to one decimal place.)

40. The heading and air speed of an airplane are 50° and 220 mph, respectively. If the wind is 60 mph from 140°, find the ground speed, the drift angle, and the course. (Use a calculator, and round off your answers to one decimal place.)

41. A block rests on an inclined plane that makes an angle of 24° with the horizontal. The component of the weight parallel to the plane is 14.8 lb. Determine the weight of the block and the component of the weight perpendicular to the plane. (Use a calculator, and round off your answers to one decimal place.)

42. Find the length of the vector $\langle 20, 99 \rangle$.

43. For which values of b will the vectors $\langle 2, 6 \rangle$ and $\langle -5, b \rangle$ have the same length?

44. Given the points $A(2, 6)$ and $B(-7, 4)$, find the components of the following vectors. Write your answers in the form $\langle x, y \rangle$.

  (a) $\overrightarrow{AB}$ (b) $\overrightarrow{BA}$ (c) $3\overrightarrow{AB}$ (d) $\dfrac{1}{|\overrightarrow{AB}|}\overrightarrow{AB}$

*In Exercises 45–56, compute each of the indicated quantities, given that the vectors **a**, **b**, **c**, and **d** are defined as follows:*

$$\mathbf{a} = \langle 3, 5 \rangle \quad \mathbf{b} = \langle 7, 4 \rangle \quad \mathbf{c} = \langle 2, -1 \rangle \quad \mathbf{d} = \langle 0, 3 \rangle$$

45. $\mathbf{a} + \mathbf{b}$ **46.** $\mathbf{b} - \mathbf{d}$

47. $3\mathbf{c} + 2\mathbf{a}$ **48.** $|\mathbf{a}|$

49. $|\mathbf{b} + \mathbf{d}|^2 - |\mathbf{b} - \mathbf{d}|^2$ **50.** $\mathbf{a} + (\mathbf{b} + \mathbf{c})$

51. $(\mathbf{a} + \mathbf{b}) + \mathbf{c}$ **52.** $\mathbf{a} - (\mathbf{b} - \mathbf{c})$

53. $(\mathbf{a} - \mathbf{b}) - \mathbf{c}$

54. $|\mathbf{a} + \mathbf{b}|^2 + |\mathbf{a} - \mathbf{b}|^2 - 2|\mathbf{a}|^2 - 2|\mathbf{b}|^2$

55. $4\mathbf{c} + 2\mathbf{a} - 3\mathbf{b}$ **56.** $|4\mathbf{c} + 2\mathbf{a} - 3\mathbf{b}|$

57. Express the vector $\langle 7, -6 \rangle$ in terms of the unit vectors **i** and **j**.

58. Express the vector $4\mathbf{i} - 6\mathbf{j}$ in the form $\langle a, b \rangle$.

59. Find a unit vector having the same direction as $\langle 6, 4 \rangle$.

60. Find two unit vectors that are perpendicular to $\langle \cos \theta, \sin \theta \rangle$. Simplify the components as much as possible.

In Exercises 61 and 62, compute the distance between the points with the given polar coordinates. Use a calculator and round off the final answers to two decimal places.

61. $\left(3, \frac{\pi}{12}\right)$ and $\left(2, \frac{17\pi}{18}\right)$ **62.** $(1, 1)$ and $(3, 2)$

In Exercises 63 and 64, determine a polar equation for the circle satisfying the given conditions.

63. The radius is 3 and the polar coordinates of the center are $\left(5, \frac{\pi}{6}\right)$.

64. The radius is 1 and the polar coordinates of the center are $\left(-3, \frac{\pi}{4}\right)$.

In Exercises 65 and 66, the polar equation of a line is given. In each case, (a) specify the perpendicular distance from the origin to the line; (b) determine the polar coordinates of the points on the line corresponding to $\theta = 0$ and $\theta = \pi/2$; (c) specify the polar coordinates of the foot of the perpendicular from the origin to the line; and (d) sketch the line.

65. $r \cos\left(\theta - \frac{\pi}{3}\right) = 3$ **66.** $r \cos(\theta - 1) = \sqrt{5}$

In Exercises 67–74, graph the polar equations.

67. (a) $r = 2 - 2\cos\theta$ **68.** (a) $r^2 = 4\cos 2\theta$
    (b) $r = 2 - 2\sin\theta$ (b) $r^2 = 4\sin 2\theta$

69. (a) $r = 2\cos\theta - 1$ **70.** (a) $r = 2\cos\theta$
    (b) $r = 2\sin\theta - 1$ (b) $r = 2\cos\theta + 1$

71. (a) $r = 4\sin 2\theta$ **72.** (a) $r = 4\sin 3\theta$
    (b) $r = 4\cos 2\theta$ (b) $r = 4\cos 3\theta$

73. (a) $r = 1 + 2\sin(\theta/2)$ **74.** (a) $r = (1.5)^\theta$ $(\theta \ge 0)$
    (b) $r = 1 - 2\cos(\theta/2)$ (b) $r = (1.5)^\theta$ $(\theta \le 0)$

In Exercises 75–80, graph the curve (or line) determined by the parametric equations. Indicate the direction of travel along the curve as t increases.

75. $x = 3 - 5t,\ y = 1 + t$ **76.** $x = t - 1,\ y = 3t$

77. $x = 3\sin t,\ y = 6\cos t$

78. $x = 2\cos t,\ y = 2\sin t$

79. $x = 4\sec t,\ y = 3\tan t$

80. $x = 1 + 3\cos t,\ y = 2 + 4\sin t$

81. In the accompanying figure, P and Q are two points on the polar curve $r = (\sin\theta)/\theta$ such that $\angle AOP = \angle POQ$. Follow steps (a) through (f) to prove that $\overline{PQ} \perp \overline{OQ}$.

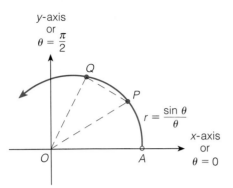

(a) Let $\alpha = \angle AOP = \angle POQ$. What are the polar coordinates of P and Q in terms of α?

(b) What are the rectangular coordinates of P and Q in terms of α?

(c) Using the coordinates determined in part (b), show that the slope of $\overline{OQ}$ is tan 2α.

(d) Using the coordinates determined in part (b), show that the slope of $\overline{PQ}$ is

$$\frac{\sin^2 2\alpha - 2 \sin^2 \alpha}{\sin 2\alpha \cos 2\alpha - 2 \sin \alpha \cos \alpha}$$

(e) Show that the expression in part (d) can be simplified to $-\cot 2\alpha$.   *Hint:*   In the numerator, replace $\sin^2 2\alpha$ by $1 - \cos^2 2\alpha$.

(f) Use the results in parts (c) and (e) to explain why $\overline{PQ} \perp \overline{OQ}$.

CHAPTER EIGHT TEST

1. In triangle ABC, let $A = 120°$, $b = 5$ cm, and $c = 3$ cm. Find a.

2. The sides of a triangle are 2 cm, 3 cm, and 4 cm. Determine the cosine of the angle opposite the longest side. On the basis of your answer, explain whether or not the angle opposite the longest side is an acute angle.

3. Two of the angles in a triangle are 30° and 45°. If the side opposite the 45° angle is $20\sqrt{2}$ cm, find the side opposite the 30° angle.

4. Each side of the square $STUV$ is 8 cm long. The point P lies on diagonal $\overline{SU}$ such that $SP = 2$ cm. Find the distance from P to V.

5. In $\triangle ABC$, $b = 5.8$ cm, $c = 3.2$ cm, and $A = 27°$. Find the remaining sides and angles of the triangle.

6. Two forces **F** and **G** act on an object. The force **G** acts horizontally with a magnitude of 2 N, and **F** acts vertically upward with a magnitude of 4 N.
    (a) Find the magnitude of the resultant.
    (b) Find tan θ, where θ is the angle between **G** and the resultant.

7. Two forces act on an object, as shown in the following figure.

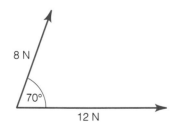

(a) Find the magnitude of the resultant. (Leave your answer in terms of radicals and the trigonometric functions.)

(b) Find sin θ, where θ is the angle between the 12-N force and the resultant.

8. The heading and air speed of an airplane are 40° and 300 mph, respectively. If the wind is 50 mph from 130°, find the ground speed and the tangent of the drift angle. (Leave your answer in terms of radicals and the trigonometric functions.)

9. Let $\mathbf{A} = \langle 2, 4 \rangle$, $\mathbf{B} = \langle 3, -1 \rangle$, and $\mathbf{C} = \langle 4, -4 \rangle$.
    (a) Find $2\mathbf{A} + 3\mathbf{B}$.
    (b) Find $|2\mathbf{A} + 3\mathbf{B}|$.
    (c) Express $\mathbf{C} - \mathbf{B}$ in terms of $\mathbf{i}$ and $\mathbf{j}$.

10. Let P and Q be the points $(4, 5)$ and $(-7, 2)$, respectively. Find a unit vector having the same direction as $\overrightarrow{PQ}$.

11. Convert the polar equation $r^2 = \cos 2\theta$ to rectangular form.

12. Graph the equation $r = 2(1 - \cos \theta)$.

13. Given the parametric equations $x = 4 \sin t$ and $y = 2 \cos t$, eliminate the parameter t and sketch the graph.

14. Compute the distance between the two points with polar coordinates $\left(4, \frac{10\pi}{21}\right)$ and $\left(1, \frac{\pi}{7}\right)$.

15. Find a polar equation for the circle with center (in polar coordinates) $\left(5, \frac{\pi}{2}\right)$ and with radius 2. Does the (polar) point $\left(2, \frac{\pi}{6}\right)$ lie on this circle?

16. Refer to the following figure.

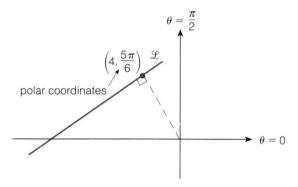

    (a) Determine a polar equation for the line $\mathscr{L}$ in the figure.
    (b) Use an addition formula to show that the polar equation of the line can be written $-r\sqrt{3} \cos \theta + r \sin \theta - 8 = 0$.

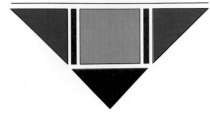

CHAPTER NINE

SYSTEMS OF EQUATIONS

Many problems in a variety of disciplines give rise to . . . [linear] systems. For example, in physics, in order to find the currents in an electrical circuit containing known voltages and resistances, a system of linear equations must be solved. In chemistry, the balancing of chemical equations requires the solution of a system of linear equations. And in economics, the Leontief input–output model reduces problems concerning the production and consumption of goods to systems of linear equations.

Leslie Hogben in *Elementary Linear Algebra* (St. Paul: West Publishing Company, 1987)

INTRODUCTION

In this chapter we consider systems of equations. Roughly speaking, a system of equations is just a collection of equations with a common set of unknowns. In solving such systems, we try to find values for the unknowns that simultaneously satisfy each of the equations in the system. In Section 9.1 we review two techniques that are often presented in intermediate algebra for solving systems involving two linear equations in two unknowns. An important technique for solving larger systems of equations is developed in Section 9.2; this technique is known as *Gaussian elimination*. After that, in Section 9.3, we introduce matrices as a tool for reducing the amount of bookkeeping involved in Gaussian elimination. Two additional techniques for solving systems of linear equations are explained in Sections 9.4 and 9.5. The technique in Section 9.4 uses the idea of an inverse matrix; the technique in Section 9.5 involves determinants and Cramer's rule. Section 9.6 contains a brief discussion of nonlinear systems of equations. In the last section of this chapter, Section 9.7, we consider systems of inequalities.

9.1 SYSTEMS OF TWO LINEAR EQUATIONS IN TWO UNKNOWNS

Both in theory and in applications, it's often necessary to solve two equations in two unknowns. You may have been introduced to the idea of simultaneous equations in a previous course in algebra; however, to put matters on a firm foundation, we begin here with the basic definitions. By a **linear equation in two variables** we mean an equation of the form

$$ax + by = c$$

where the constants a and b are not both zero. The two variables needn't always be denoted by the letters x and y, of course; it is the *form* of the equation that matters. Table 1 displays some examples.

An ordered pair of numbers (x_0, y_0) is said to be a **solution of the linear equation** $ax + by = c$ provided we obtain a true statement when we replace x

TABLE I

Equation in Two Variables	Is It Linear?	
	Yes	No
$3x - 8y = 12$	✓	
$-s + 4t = 0$	✓	
$2x - 3y^2 = 1$		✓
$y = 4 - 2x$	✓	
$\dfrac{4}{u} + \dfrac{5}{v} = 3$		✓

and y in the equation by x_0 and y_0, respectively. For example, the ordered pair $(3, 2)$ is a solution of the equation $x - y = 1$, since $3 - 2 = 1$. On the other hand, $(2, 3)$ is not a solution of $x - y = 1$, since $2 - 3 \neq 1$.

Now consider a **system** of two linear equations in two unknowns:

$$\begin{cases} ax + by = c \\ dx + ey = f \end{cases}$$

If we can find an ordered pair that is a solution to both equations, then we say that ordered pair is a **solution of the system**. Sometimes, to emphasize the fact that a solution must satisfy both equations, we refer to the system as a pair of **simultaneous equations**. A system that has at least one solution is said to be **consistent**. If there are no solutions, the system is **inconsistent**.

EXAMPLE I Consider the system

$$\begin{cases} x + y = 2 \\ 2x - 3y = 9 \end{cases}$$

(a) Is $(1, 1)$ a solution of the system?
(b) Is $(3, -1)$ a solution of the system?

Solution

(a) Although $(1, 1)$ is a solution of the first equation, it is not a solution of the system because it does not satisfy the second equation. (Check this for yourself.)

(b) $(3, -1)$ satisfies the first equation:

$$3 + (-1) \overset{?}{=} 2 \qquad \text{True}$$

$(3, -1)$ satisfies the second equation:

$$2(3) - 3(-1) \overset{?}{=} 9$$
$$6 + 3 \overset{?}{=} 9 \qquad \text{True}$$

Since $(3, -1)$ satisfies both equations, it is a solution of the system. ▪▪▪

We can gain an important perspective on systems of linear equations by looking at Example 1 in graphical terms. Table 2 shows how each of the state-

TABLE 2

Algebraic Idea	Corresponding Geometric Idea
1. The ordered pair $(1, 1)$ is a solution of the equation $x + y = 2$.	1. The point $(1, 1)$ lies on the line $x + y = 2$. See Figure 1.
2. The ordered pair $(1, 1)$ is not a solution of the equation $2x - 3y = 9$.	2. The point $(1, 1)$ does not lie on the line $2x - 3y = 9$. See Figure 1.
3. The ordered pair $(3, -1)$ is a solution of the system $$\begin{cases} x + y = 2 \\ 2x - 3y = 9 \end{cases}$$	3. The point $(3, -1)$ lies on both of the lines $x + y = 2$ and $2x - 3y = 9$. See Figure 1.

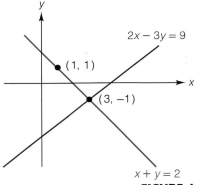

FIGURE 1

ments in that example can be rephrased using geometric ideas with which we are already familiar.

In Example 1, we verified that $(3, -1)$ is a solution of the system

$$\begin{cases} x + y = 2 \\ 2x - 3y = 9 \end{cases}$$

Are there any other solutions of this particular system? No; Figure 1 shows us that there are no other solutions, since $(3, -1)$ is clearly the only point common to both lines. In a moment we'll look at two important methods for solving systems of linear equations in two unknowns. But even before we consider these methods, we can say something about the solutions of linear systems.

PROPERTY SUMMARY   POSSIBILITIES FOR SOLUTIONS OF LINEAR SYSTEMS

Given a system of two linear equations in two unknowns, exactly one of the following cases must occur.

CASE 1   The graphs of the two linear equations intersect in exactly one point. Thus, there is exactly one solution to the system. See Figure 2.

CASE 2   The graphs of the two linear equations are parallel lines. Therefore, the lines do not intersect, and the system has no solution. See Figure 3.

CASE 3   The two equations actually represent the same line. Thus, there are infinitely many points of intersection and correspondingly infinitely many solutions. See Figure 4.

FIGURE 2
A consistent system with exactly one solution.

FIGURE 3
An inconsistent system has no solution.

FIGURE 4
A consistent system with infinitely many solutions.

We are going to review two methods from basic algebra for solving systems of two linear equations in two unknowns. These methods are the **substitution method** and the **addition–subtraction method**. We'll begin by demonstrating the substitution method. Consider the system

$$\begin{cases} 3x + 2y = 17 & \text{(1)} \\ 4x - 5y = -8 & \text{(2)} \end{cases}$$

We first choose one of the two equations and then use it to express one of the variables in terms of the other. In the case at hand, neither equation appears particularly simpler than the other, so let's just start with the first equation and solve for x in terms of y. We have

$$3x = 17 - 2y$$
$$x = \tfrac{1}{3}(17 - 2y) \qquad\qquad \text{(3)}$$

Now we use equation (3) to substitute for x in the equation that we have not used yet, namely, equation (2). This yields

$$4\left[\tfrac{1}{3}(17 - 2y)\right] - 5y = -8$$
$$4(17 - 2y) - 15y = -24 \qquad \text{multiplying by 3}$$
$$-23y = -92$$
$$y = 4$$

The value $y = 4$ that we have just obtained can be used in equation (3) to find x. Replacing y with 4 in equation (3) yields

$$x = \tfrac{1}{3}[17 - 2(4)] = \tfrac{1}{3}(9) = 3$$

We have now found that $x = 3$ and $y = 4$. As you can easily check, this pair of values indeed satisfies both of the original equations. We write our solution as the ordered pair $(3, 4)$. Figure 5 summarizes the situation. It shows that the system is consistent and that $(3, 4)$ is the only solution.

Generally speaking, it is not necessary to graph the equations in a given system in order to decide whether the system is consistent. Rather, this information will emerge as you attempt to follow an algebraic method of solution. Examples 2 and 3 will illustrate this.

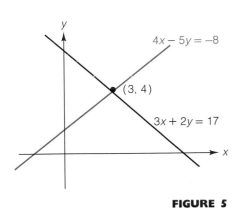

FIGURE 5

■■■ **EXAMPLE 2** Solve the system

$$\begin{cases} \tfrac{3}{2}x - 3y = -9 \\ x - 2y = 4 \end{cases}$$

Solution We use the substitution method. Since it is easy to solve the second equation for x, we begin there:

$$x - 2y = 4$$
$$x = 4 + 2y$$

Now we substitute this result in the first equation of our system to obtain

$$\tfrac{3}{2}(4 + 2y) - 3y = -9$$
$$6 + 3y - 3y = -9$$
$$6 = -9 \qquad \text{False}$$

Since the substitution process leads us to this obviously false statement, we conclude that the given system has no solution; that is, the system is inconsistent.

QUESTION   What can you say about the graphs of the two given equations?

■■■

■■■ **EXAMPLE 3** Solve the system

$$\begin{cases} 3x + 4y = 12 \\ 2y = 6 - \tfrac{3}{2}x \end{cases}$$

Solution We use the method of substitution. Since it is easy to solve the second equation for y, we begin there:

$$2y = 6 - \tfrac{3}{2}x \qquad \text{and therefore} \qquad y = 3 - \tfrac{3}{4}x$$

Now we use this result to substitute for y in the first equation of the original system. The result is

$$3x + 4\left(3 - \tfrac{3}{4}x\right) = 12$$
$$3x + 12 - 3x = 12$$
$$3x - 3x = 12 - 12$$
$$0 = 0 \qquad \text{Always true}$$

This last identity imposes no restrictions on x. Graphically speaking, this says that our two lines intersect for every value of x. In other words, the two lines coincide. We could have foreseen this initially had we solved both equations for y. As you can verify, the result in both cases is

$$y = -\tfrac{3}{4}x + 3$$

Every point on this line yields a solution to our system of equations. In summary then, our system is consistent and the solutions to the system have the form $\left(x, -\tfrac{3}{4}x + 3\right)$, where x can be any real number. For instance, when $x = 0$, we obtain the solution $(0, 3)$. When $x = 1$, we obtain the solution $\left(1, \tfrac{9}{4}\right)$. The idea here is that for *each* value of x, we obtain a solution; thus, there are infinitely many solutions. ▮▮▮

Now let us turn to the addition–subtraction method of solving systems of equations. By way of example, consider the system

$$\begin{cases} 2x + 3y = 5 \\ 4x - 3y = -1 \end{cases}$$

Notice that if we add these two equations, the result will be an equation involving only the unknown x. Carrying this out gives us

$$6x = 4$$
$$x = \tfrac{4}{6} = \tfrac{2}{3}$$

There are now several ways in which the corresponding value of y may be obtained. As you can easily check, substituting the value $x = \tfrac{2}{3}$ in either of the original equations leads to the result $y = \tfrac{11}{9}$.

Another way to find y is by multiplying both sides of the first equation by -2. (You'll see why in a moment.) We display the work this way:

$$2x + 3y = 5 \quad \xrightarrow{\text{Multiply by } -2} \quad -4x - 6y = -10$$

$$4x - 3y = -1 \quad \xrightarrow{\text{No change}} \quad 4x - 3y = -1$$

Adding the last two equations then gives us

$$-9y = -11$$
$$y = \tfrac{11}{9}$$

The required solution is therefore $\left(\tfrac{2}{3}, \tfrac{11}{9}\right)$.

In the previous example, we were able to find x directly by adding the two equations. As the next example shows, it may be necessary first to multiply both sides of each equation by an appropriate constant.

▌▌▌ **EXAMPLE 4** Solve the system

$$\begin{cases} 5x - 3y = 4 \\ 2x + 4y = 1 \end{cases}$$

Solution To eliminate x, we could multiply the second equation by $\frac{5}{2}$ and then subtract the resulting equation from the first equation. However, to avoid working with fractions, we proceed as follows.

$$5x - 3y = 4 \quad \xrightarrow{\text{Multiply by 2}} \quad 10x - 6y = 8 \qquad (4)$$

$$2x + 4y = 1 \quad \xrightarrow{\text{Multiply by 5}} \quad 10x + 20y = 5 \qquad (5)$$

Subtracting equation (5) from equation (4) then yields

$$-26y = 3$$
$$y = -\tfrac{3}{26}$$

To find x, we return to the original system and work in a similar manner:

$$5x - 3y = 4 \quad \xrightarrow{\text{Multiply by 4}} \quad 20x - 12y = 16$$

$$2x + 4y = 1 \quad \xrightarrow{\text{Multiply by 3}} \quad 6x + 12y = 3$$

Upon adding the last two equations, we obtain

$$26x = 19$$
$$x = \tfrac{19}{26}$$

The solution of the given system of equations is therefore $\left(\tfrac{19}{26}, -\tfrac{3}{26}\right)$. ▌▌▌

We conclude this section with some problems that can be solved by using simultaneous equations.

▌▌▌ **EXAMPLE 5** Determine constants b and c so that the parabola $y = x^2 + bx + c$ passes through the points $(-3, 1)$ and $(1, -2)$.

Solution Since the point $(-3, 1)$ lies on the curve $y = x^2 + bx + c$, the coordinates must satisfy the equation. Thus, we have

$$1 = (-3)^2 + b(-3) + c$$
$$-8 = -3b + c \qquad (6)$$

This gives us one equation in two unknowns. We need another equation involving b and c. Since the point $(1, -2)$ also lies on the graph of $y = x^2 + bx + c$, we must have

$$-2 = 1^2 + b(1) + c$$
$$-3 = b + c \qquad (7)$$

Rewriting equations (6) and (7), we have the system

$$\begin{cases} -3b + c = -8 & (8) \\ b + c = -3 & (9) \end{cases}$$

Subtracting equation (9) from (8) then yields

$$-4b = -5 \qquad \text{and therefore} \qquad b = \tfrac{5}{4}$$

One way to obtain the corresponding value of c is to replace b by $\tfrac{5}{4}$ in equation (9). This yields

$$\tfrac{5}{4} + c = -3$$
$$c = -3 - \tfrac{5}{4} = -\tfrac{17}{4}$$

The required values of b and c are therefore

$$b = \tfrac{5}{4}, \qquad c = -\tfrac{17}{4}$$

(Exercise 58 at the end of this section asks you to check that the parabola $y = x^2 + \tfrac{5}{4}x - \tfrac{17}{4}$ indeed passes through the given points.)

▌� ▌ **EXAMPLE 6**   In Figure 6, find the area of the triangular region bounded by the lines $y = 3x$, $y = -\tfrac{1}{2}x + 7$, and the x-axis.

FIGURE 6

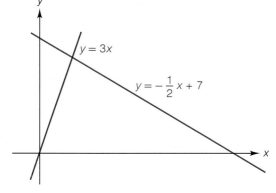

Solution   To compute the area of the triangle, we will need to know the base and the height. To determine the base, we first compute the x-intercept of the line $y = -\tfrac{1}{2}x + 7$:

$$0 = -\tfrac{1}{2}x + 7$$
$$\tfrac{1}{2}x = 7$$
$$x = 14$$

The height of the triangle is the y-coordinate of the intersection point of the lines $y = 3x$ and $y = -\tfrac{1}{2}x + 7$ (see Figure 6). The substitution method can be used to solve this pair of simultaneous equations for y. From the equation $y = 3x$, we obtain $x = \tfrac{1}{3}y$. Then substituting this in the equation $y = -\tfrac{1}{2}x + 7$ yields

$$y = -\tfrac{1}{2}\left(\tfrac{1}{3}y\right) + 7$$
$$6y = -y + 42$$
$$7y = 42$$
$$y = 6$$

Now that we know that the base of the triangle is 14 and the height is 6, we can compute the required area:

$A = \frac{1}{2}bh$

$= \frac{1}{2}(14)(6) = 42$ square units   ▮▮▮

In the next example, we solve a mixture problem using a system of two equations in two unknowns.

▮▮▮ **EXAMPLE 7** Suppose that a chemistry student can obtain two acid solutions from the stockroom. The first solution is 20% acid and the second solution is 45% acid. (The percentages are by volume.) How many milliliters of each solution should the student mix together to obtain 100 ml of a 30% acid solution?

Solution We begin by assigning letters to denote the required quantities.

Let x denote the number of milliliters of the 20% solution to be used.

Let y denote the number of milliliters of the 45% solution to be used.

Then the data can be summarized as in Table 3. Since the final mixture must total 100 ml, we have the equation

$x + y = 100$ \hfill (10)

This gives us one equation in two unknowns. However, we need a second equation. Looking at the data in the right-hand column of Table 3, we can write

$$\underbrace{0.20x}_{} \quad + \quad \underbrace{0.45y}_{} \quad = \quad \underbrace{(0.30)(x+y)}_{}$$

Amount of acid in x ml of the 20% solution   Amount of acid in y ml of the 45% solution   Amount of acid in the final mixture

TABLE 3

Type of Solution	Number of ml	Percent of Acid	Total Acid (ml)
First solution (20% acid)	x	20	$(0.20)x$
Second solution (45% acid)	y	45	$(0.45)y$
Mixture	$x + y$	30	$(0.30)(x+y)$

Thus,

$$0.20x + 0.45y = 0.30(x + y)$$
$$20x + 45y = 30(x + y)$$
$$4x + 9y = 6(x + y) = 6x + 6y$$
$$-2x + 3y = 0 \hspace{4cm} (11)$$

Equations (10) and (11) can be solved by either the substitution method or the addition method. As Exercise 59 at the end of this section asks you to show, the results are

$x = 60$ ml   and $y = 40$ ml   ▮▮▮

EXERCISE SET 9.1

A

1. Which of the following are linear equations in two variables?
   (a) $3x + 3y = 10$ (b) $2x + 4xy + 3y = 1$
   (c) $u - v = 1$ (d) $x = 2y + 6$

2. Which of the following are linear equations in two variables?
   (a) $y = x$ (b) $y = x^2$
   (c) $\dfrac{4}{x} - \dfrac{3}{y} = -1$ (d) $2w + 8z = -4w + 3$

3. Is $(5, 1)$ a solution of the following system?
$$\begin{cases} 2x - 8y = 2 \\ 3x + 7y = 22 \end{cases}$$

4. Is $(14, -2)$ a solution of the following system?
$$\begin{cases} x + y = 12 \\ x - y = 4 \end{cases}$$

5. Is $(0, -4)$ a solution of the following system?
$$\begin{cases} \frac{1}{6}x + \frac{1}{2}y = -2 \\ \frac{2}{3}x + \frac{3}{4}y = 2 \end{cases}$$

6. Is $(12, -8)$ a solution of the system in Exercise 5?

7. Is $(3, -2)$ a solution of the following system?
$$\begin{cases} \frac{2}{7}x + \frac{1}{5}y = \frac{44}{35} \\ \frac{1}{3}x - \frac{5}{4}y = \frac{7}{2} \end{cases}$$

In Exercises 8–18, use the substitution method to find all solutions of each system.

8. $\begin{cases} 4x - y = 7 \\ -2x + 3y = 9 \end{cases}$ 9. $\begin{cases} 3x - 2y = -19 \\ x + 4y = -4 \end{cases}$

10. $\begin{cases} 6x - 2y = -3 \\ 5x + 3y = 4 \end{cases}$ 11. $\begin{cases} 4x + 2y = 3 \\ 10x + 4y = 1 \end{cases}$

12. $\begin{cases} \frac{3}{2}x - 5y = 1 \\ x + \frac{3}{4}y = -1 \end{cases}$ 13. $\begin{cases} 13x - 8y = -3 \\ -7x + 2y = 0 \end{cases}$

14. $\begin{cases} 4x + 6y = 3 \\ -6x - 9y = -\frac{9}{2} \end{cases}$ 15. $\begin{cases} -\frac{2}{5}x + \frac{1}{4}y = 3 \\ \frac{1}{4}x - \frac{2}{5}y = -3 \end{cases}$

16. $\begin{cases} 0.02x - 0.03y = 1.06 \\ 0.75x + 0.50y = -0.01 \end{cases}$

17. $\begin{cases} \sqrt{2}x - \sqrt{3}y = \sqrt{3} \\ \sqrt{3}x - \sqrt{8}y = \sqrt{2} \end{cases}$ 18. $\begin{cases} 7x - 3y = -12 \\ \frac{14}{3}x - 2y = 2 \end{cases}$

In Exercises 19–28, use the addition–subtraction method to find all solutions of each system of equations.

19. $\begin{cases} 5x + 6y = 4 \\ 2x - 3y = -3 \end{cases}$ 20. $\begin{cases} -8x + y = -2 \\ 4x - 3y = 1 \end{cases}$

21. $\begin{cases} 4x + 13y = -5 \\ 2x - 54y = -1 \end{cases}$ 22. $\begin{cases} 16x - 3y = 100 \\ 16x + 10y = 10 \end{cases}$

23. $\begin{cases} \frac{1}{4}x - \frac{1}{3}y = 4 \\ \frac{2}{7}x - \frac{1}{7}y = \frac{1}{10} \end{cases}$

Suggestion: First clear both equations of fractions.

24. $\begin{cases} 2.1x - 3.5y = 1.2 \\ 1.4x + 2.6y = 1.1 \end{cases}$ 25. $\begin{cases} 8x + 16y = 5 \\ 2x + 5y = \frac{5}{4} \end{cases}$

26. $\begin{cases} 8x + 16y = 5 \\ 2x + 4y = 1 \end{cases}$ 27. $\begin{cases} 125x - 40y = 45 \\ \frac{1}{10}x + \frac{1}{10}y = \frac{3}{10} \end{cases}$

28. $\begin{cases} \sqrt{6}x - \sqrt{3}y = 3\sqrt{2} - \sqrt{3} \\ \sqrt{2}x - \sqrt{5}y = \sqrt{6} + \sqrt{5} \end{cases}$

29. Find b and c, given that the parabola $y = x^2 + bx + c$ passes through $(0, 4)$ and $(2, 14)$.

30. Determine the constants a and b, given that the parabola $y = ax^2 + bx + 1$ passes through $(-1, 11)$ and $(3, 1)$.

31. Determine the constants A and B, given that the line $Ax + By = 2$ passes through the points $(-4, 5)$ and $(7, -9)$.

32. Find the area of the triangular region in the first quadrant bounded by the lines $y = 5x$ and $y = -3x + 6$ and the x-axis.

33. Find the area of the triangular region in the first quadrant bounded by the x-axis and the lines $y = 2x - 5$ and $y = -\frac{1}{2}x + 3$.

34. Find the area of the triangular region in the first quadrant bounded by the y-axis and the lines $y = 2x + 2$ and $y = -\frac{3}{2}x + 9$.

35. A student in a chemistry laboratory has access to two acid solutions. The first solution is 10% acid and the second is 35% acid. (The percentages are by volume.) How many cubic centimeters of each should she mix together to obtain 200 cm³ of a 25% acid solution?

36. One salt solution is 15% salt and another is 20% salt. How many cubic centimeters of each solution must be mixed to obtain 50 cm³ of a 16% salt solution?

37. A shopkeeper has two types of coffee beans on hand. One type sells for \$5.20/lb, the other for \$5.80/lb. How many pounds of each type must be mixed to produce 16 lb of a blend that sells for \$5.50/lb?

38. A certain alloy contains 10% tin and 30% copper. (The percentages are by weight.) How many pounds of tin and how many pounds of copper must be melted with 1000 lb of the given alloy to yield a new alloy containing 20% tin and 35% copper?

B

39. Find x and y in terms of a and b:

$$\begin{cases} \dfrac{x}{a}+\dfrac{y}{b}=1 \\[2mm] \dfrac{x}{b}+\dfrac{y}{a}=1 \end{cases}$$

Does your solution impose any conditions on a and b?

40. Solve the following system for x and y in terms of a and b, where $a \neq b$:

$$\begin{cases} ax + by = 1/a \\ b^2 x + a^2 y = 1 \end{cases}$$

41. Solve the following system for x and y in terms of a and b, where $a \neq b$:

$$\begin{cases} ax + a^2 y = 1 \\ bx + b^2 y = 1 \end{cases}$$

Does your solution impose any additional conditions on a and b?

42. Solve the following system for s and t:

$$\begin{cases} \dfrac{3}{s}-\dfrac{4}{t}=2 \\[2mm] \dfrac{5}{s}+\dfrac{1}{t}=-3 \end{cases}$$

Hint: Make the substitutions $1/s = x$ and $1/t = y$ in order to obtain a system of two linear equations.

43. Solve the following system for s and t:

$$\begin{cases} \dfrac{1}{2s}-\dfrac{1}{2t}=-10 \\[2mm] \dfrac{2}{s}+\dfrac{3}{t}=5 \end{cases}$$

(Use the hint in Exercise 42.)

In Exercises 44–53, find all solutions of the given systems. For Exercises 47–52, use a calculator and round off the final answers to two decimal places.

44. $\begin{cases} 0.5x - 0.8y = 0.3 \\ 0.4x - 0.1y = 0.9 \end{cases}$

45. $\begin{cases} \dfrac{2w-1}{3}+\dfrac{z+2}{4}=4 \\[2mm] \dfrac{w+3}{2}-\dfrac{w-z}{3}=3 \end{cases}$

46. $\dfrac{x-y}{2}=\dfrac{x+y}{3}=1$

47. $\begin{cases} 1.03x - 2.54y = 5.47 \\ 3.85x + 4.29y = -1.84 \end{cases}$

48. $\begin{cases} 2.39x + 8.16y = -2.83 \\ 1.01x + 2.98y = 4.41 \end{cases}$

49. $\begin{cases} 2 \ln x - 5 \ln y = 11 \\ \ln x + \ln y = -5 \end{cases}$
Hint: Let $u = \ln x$ and $v = \ln y$.

50. $\begin{cases} 3 \ln x + \ln y = 3 \\ 4 \ln x - 6 \ln y = -7 \end{cases}$

51. $\begin{cases} e^x - 3e^y = 2 \\ 3e^x + e^y = 16 \end{cases}$
Hint: Let $u = e^x$ and $v = e^y$.

52. $\begin{cases} e^x + 2e^y = 4 \\ \frac{1}{2}e^x - e^y = 0 \end{cases}$

53. $\begin{cases} 4\sqrt{x^2 - 3x} - 3\sqrt{y^2 + 6y} = -4 \\ \frac{1}{2}\left(\sqrt{x^2 - 3x} + \sqrt{y^2 + 6y}\right) - 3 \end{cases}$
Hint: Let $u = \sqrt{x^2 - 3x}$ and $v = \sqrt{y^2 + 6y}$.

54. The sum of two numbers is 64. Twice the larger number plus five times the smaller number is 20. Find the two numbers. (Let x denote the larger number and let y denote the smaller number.)

55. In a two-digit number, the sum of the digits is 14. Twice the tens digit exceeds the units digit by one. Find the number.

56. The sum of the digits in a two-digit number is 14. Furthermore, the number itself is 2 greater than 11 times the tens digit. Find the number.

57. The perimeter of a rectangle is 34 in. If the length is 2 in. more than twice the width, find the length and the width.

58. Verify that the parabola $y = x^2 + \frac{5}{4}x - \frac{17}{4}$ indeed passes through $(-3, 1)$ and $(1, -2)$, as stated in the text.

59. Consider the following system from Example 7:

$$\begin{cases} x + y = 100 \\ -2x + 3y = 0 \end{cases}$$

(a) Solve this system using the method of substitution.
   Answer: $(60, 40)$

(b) Solve the system using the addition–subtraction method.

60. Consider the following system:

$$\begin{cases} 2x - 5y = -2 \\ 3x + 4y = 5 \end{cases}$$

(a) Solve the system using the substitution method.

(b) Solve the system using the addition–subtraction method. *Answer:* $\left(\frac{17}{23}, \frac{16}{23}\right)$

(c) Verify that your solution indeed satisfies both equations of the system.

61. Solve the following system for x and y:

$$\begin{cases} by = x + ab \\ cy = x + ac \end{cases}$$

(Assume a, b, and c are constants and that $b \neq c$.)

62. Solve for x and y in terms of a, b, c, d, e, and f:

$$\begin{cases} ax + by = c \\ dx + ey = f \end{cases}$$

(Assume that $ae - bd \neq 0$.)

63. Solve the following system for x and y in terms of a and b, where $a \neq b$:

$$\begin{cases} \dfrac{a}{bx} + \dfrac{b}{ay} = a + b \\ \dfrac{b}{x} + \dfrac{a}{y} = a^2 + b^2 \end{cases}$$

C

64. Solve the following system for x and y in terms of a and b, where $ab \neq -1$:

$$\begin{cases} \dfrac{x + y - 1}{x - y + 1} = a \\ \dfrac{y - x + 1}{x - y + 1} = ab \end{cases}$$

Answer: $\left(\dfrac{a+1}{ab+1}, \dfrac{a(b+1)}{ab+1} \right)$

65. Given that the lines $7x + 5y = 4$, $x + ky = 3$, and $5x + y + k = 0$ are concurrent (pass through a common point), what are the possible values for k?

66. Solve the following system for x and y in terms of a and b:

$$\begin{cases} (a + b)x + (a^2 + b^2)y = a^3 + b^3 \\ (a - b)x + (a^2 - b^2)y = a^3 - b^3 \end{cases}$$

(Assume that a and b are nonzero and that $a \neq \pm b$.)

67. The vertices of triangle ABC are $A(2a, 0)$, $B(2b, 0)$, and $C(0, 2)$.
  (a) Show that the equations of the sides are $x + by = 2b$, $x + ay = 2a$, and $y = 0$.
  (b) Show that the equations of the medians are $x + (2a - b)y = 2a$, $x + (2b - a)y = 2b$, and $2x + (a + b)y = 2(a + b)$.
  (c) Show that all three medians intersect at the point $\left(\frac{2}{3}(a + b), \frac{2}{3}\right)$.
  (d) Show that the equations of the altitudes are $bx - y = 2ab$, $ax - y = 2ab$, and $x = 0$.
  (e) Show that all three altitudes intersect at the point $(0, -2ab)$.
  (f) Show that the equations of the perpendicular bisectors of the sides are $x = a + b$, $bx - y = b^2 - 1$, and $ax - y = a^2 - 1$.
  (g) Show that all three perpendicular bisectors intersect at the point $(a + b, ab + 1)$.
  (h) Show that the three points found in parts (c), (e), and (g) all lie on a line. This is the **Euler line** of the triangle.

9.2   GAUSSIAN ELIMINATION

A method of solution is perfect if we can foresee from the start, and even prove, that following that method we shall obtain our aim.

Gottfried Wilhelm von Leibniz (1646–1716)

… the first electronic computer, the ABC, named after its designers Atanasoff and Berry, was built specifically to solve systems of 29 equations in 29 unknowns, a formidable task without the aid of a computer.

Angela B. Shiflet in *Discrete Mathematics for Computer Science* (St. Paul: West Publishing Company, 1987)

In the previous section, we solved systems of linear equations in two unknowns. In this section, we introduce the technique known as Gaussian elimination for solving systems of linear equations in which there are more than two unknowns.*

*The technique is named after Carl Friedrich Gauss (1777–1855). Early in the nineteenth century, Gauss used this technique (and introduced the method of least squares for minimizing errors) in analyzing the orbit of the asteroid Pallas. However, the essentials of Gaussian elimination were in existence long before Gauss's time. Indeed, a version of the method appears in the Chinese text *Chui-Chang Suan-Shu* ("Nine Chapters on the Mathematical Art"), written approximately two thousand years ago.

As a first example, consider the following system of three linear equations in the three unknowns x, y, and z:

$$\begin{cases} 3x + 2y - z = -3 \\ 5y - 2z = 2 \\ 5z = 20 \end{cases}$$

This system is easy to solve using the process of *back-substitution*. Dividing both sides of the third equation by 5 yields $z = 4$. Then substituting $z = 4$ back into the second equation gives us

$$5y - 2(4) = 2$$
$$5y = 10$$
$$y = 2$$

Finally, substituting the values $z = 4$ and $y = 2$ into the first equation yields

$$3x + 2(2) - 4 = -3$$
$$3x = -3$$
$$x = -1$$

We have now found that $x = -1$, $y = 2$, and $z = 4$. If you go back and check, you will find that these values indeed satisfy all three equations in the given system. Furthermore, the algebra we've just carried out shows that these are the only possible values for x, y, and z satisfying all three equations. We summarize by saying that the **ordered triple** $(-1, 2, 4)$ is the solution of the given system.

The system that we just considered was easy to solve (using back-substitution) because of the special form in which it was written. This form is called **upper-triangular form**. Although the following definition of upper-triangular form refers to systems with three unknowns, the same type of definition can be given for systems with any number of unknowns. Table 1 (on the next page) displays examples of systems in upper-triangular form.

Upper-Triangular Form (Three Variables)

A system of linear equations in x, y, and z is said to be in **upper-triangular** form provided x appears in no equation after the first and y appears in no equation after the second. (It is possible that y may not even appear in the second equation.)

When we were solving linear systems of equations with two unknowns in the previous section, we observed that there were three possibilities: a unique solution, infinitely many solutions, and no solution. As the next three examples indicate, the situation is similar when dealing with larger systems.

■■■ **EXAMPLE 1**   Find all solutions of the system

$$\begin{cases} x + y + 2z = 2 \\ 3y - 4z = -5 \\ 6z = 3 \end{cases}$$

TABLE I

Examples of Systems in Upper-Triangular Form

2 Unknowns: x, y	3 Unknowns: x, y, z	4 Unknowns: x, y, z, t
$\begin{cases} 3x + 5y = 7 \\ 8y = 5 \end{cases}$	$\begin{cases} 4x - 3y + 2z = -5 \\ 7y + z = 9 \\ -4z = 3 \end{cases}$	$\begin{cases} x - y + z - 4t = 1 \\ 3y - 2z + t = -1 \\ 3z - 5t = 4 \\ 6t = 7 \end{cases}$
	$\begin{cases} 15x - 2y + z = 1 \\ 3z = -8 \end{cases}$	$\begin{cases} 2x + y + 2z - t = -3 \\ 4z + 3t = 1 \\ 5t = 6 \end{cases}$
		$\begin{cases} 8x + 3y - z + t = 2 \\ 2y + z - 4t = 1 \end{cases}$

Solution The system is in upper-triangular form, so we can use back-substitution. Dividing the third equation by 6 yields $z = \frac{1}{2}$. Substituting this value for z back into the second equation then yields

$$3y - 4\left(\tfrac{1}{2}\right) = -5$$
$$3y = -3$$
$$y = -1$$

Now, substituting the values $z = \frac{1}{2}$ and $y = -1$ back into the first equation, we obtain

$$x + (-1) + 2\left(\tfrac{1}{2}\right) = 2$$
$$x = 2$$

As you can easily check, the values $x = 2$, $y = -1$, and $z = \frac{1}{2}$ indeed satisfy all three equations. Furthermore, the algebra we've just carried out shows that these are the only possible values for x, y, and z satisfying all three equations. We summarize by saying that the unique solution to our system is the ordered triple $\left(2, -1, \tfrac{1}{2}\right)$. ∎

EXAMPLE 2 Find all solutions of the system

$$\begin{cases} -2x + y + 3z = 6 \\ 2z = 10 \end{cases}$$

Solution Again, the system is in upper-triangular form, and we use back-substitution. Solving the second equation for z yields $z = 5$. Then replacing z by 5 in the first equation gives us

$$-2x + y + 3(5) = 6$$
$$-2x + y = -9$$
$$y = 2x - 9$$

At this point, we've made use of both equations in the given system. There is no third equation to provide additional restrictions on x, y, or z. We know from the previous section that the equation $y = 2x - 9$ has infinitely many solutions, all of the form

$(x, 2x - 9)$ where x is a real number

It follows, then, that there are infinitely many solutions to the given system and they may be written

$(x, 2x - 9, 5)$ where x is a real number

For instance, choosing in succession $x = 0$, $x = 1$, and $x = 2$ yields the solutions $(0, -9, 5)$, $(1, -7, 5)$, and $(2, -5, 5)$. (We remark in passing that any linear system in upper-triangular form in which the number of unknowns exceeds the number of equations will always have infinitely many solutions.)   ▪▪▪

▮▮▮ **EXAMPLE 3**   Find all solutions of the system

$$\begin{cases} 4x - 7y + 3z = 1 \\ 3x + y - 2z = 4 \\ 4x - 7y + 3z = 6 \end{cases}$$

(Note that the system is not in upper-triangular form.)

Solution   Look at the left-hand sides of the first and third equations: they are identical. Thus, if there were values for x, y, and z that satisfied both equations, it would follow that $1 = 6$, which is clearly impossible. We conclude that the given system has no solutions.   ▪▪▪

As Examples 1 and 2 have demonstrated, systems in upper-triangular form can be readily solved. In view of this, it would be useful to have a technique for converting a given system into an equivalent system in upper-triangular form. (The expression **equivalent system** means a system with exactly the same set of solutions as the original system.) **Gaussian elimination** is one such technique. We will demonstrate this technique in Examples 4, 5, and 6. In using Gaussian elimination, we will rely on what are called the three **elementary operations**, listed in the box that follows. These are operations that can be performed on an equation in a system without altering the set of solutions of the system.

The Elementary Operations

1. Multiply both sides of an equation by a nonzero constant.

2. Interchange the order in which two equations of a system are listed.

3. To one equation add a multiple of another equation in the system.

▮▮▮ **EXAMPLE 4**   Find all solutions of the system

$$\begin{cases} x + 2y + z = 3 \\ 2x + y + z = 16 \\ x + y + 2z = 9 \end{cases}$$

Solution First we want to eliminate x from the second and third equations. To eliminate x from the second equation, we add to it -2 times the first equation. The result is the equivalent system

$$\begin{cases} x + 2y + z = 3 \\ \quad -3y - z = 10 \\ x + y + 2z = 9 \end{cases}$$

To eliminate x from the third equation, we add to it -1 times the first equation. The result is the equivalent system

$$\begin{cases} x + 2y + z = 3 \\ \quad -3y - z = 10 \\ \quad -y + z = 6 \end{cases}$$

Now to bring the system into upper-triangular form, we need to eliminate y from the third equation. We could do this by adding $-\frac{1}{3}$ times the second equation to the third equation. However, to avoid working with fractions as long as possible, we proceed instead to interchange the second and third equations to obtain the equivalent system

$$\begin{cases} x + 2y + z = 3 \\ \quad -y + z = 6 \\ \quad -3y - z = 10 \end{cases}$$

Now we add -3 times the second equation to the last equation to obtain the equivalent system

$$\begin{cases} x + 2y + z = 3 \\ \quad -y + z = 6 \\ \quad -4z = -8 \end{cases}$$

The system is now in upper-triangular form, and back-substitution yields, in turn, $z = 2$, $y = -4$, and $x = 9$. (Check this for yourself.) The required solution is therefore $(9, -4, 2)$. ∎

In Table 2 we list some convenient abbreviations that are used in describing the elementary operations. You should plan on using these abbreviations yourself: they'll make it simpler for you (and your instructor) to check your work. In Table 2, the notation E_i stands for the ith equation in a system. For instance, for the initial system in Example 4, the symbol E_1 denotes the first equation: $x + 2y + z = 3$.

TABLE 2

Abbreviations for the Elementary Operations

Abbreviation	Explanation
1. cE_i	Multiply both sides of the ith equation by c.
2. $E_i \leftrightarrow E_j$	Interchange the ith and jth equations.
3. $cE_i + E_j$	To the jth equation, add c times the ith equation.

■■■ **EXAMPLE 5** Solve the system

$$\begin{cases} 4x - 3y + 2z = 40 \\ 5x + 9y - 7z = 47 \\ 9x + 8y - 3z = 97 \end{cases}$$

Solution

$$\begin{cases} 4x - 3y + 2z = 40 \\ 5x + 9y - 7z = 47 \\ 9x + 8y - 3z = 97 \end{cases} \xrightarrow{(-1)E_2 + E_1} \begin{cases} -x - 12y + 9z = -7 \\ 5x + 9y - 7z = 47 \\ 9x + 8y - 3z = 97 \end{cases}$$ adding -1 times the second equation to the first equation

$$\xrightarrow[9E_1 + E_3]{5E_1 + E_2} \begin{cases} -x - 12y + 9z = -7 \\ -51y + 38z = 12 \\ -100y + 78z = 34 \end{cases}$$

$$\xrightarrow{\frac{1}{2}E_3} \begin{cases} -x - 12y + 9z = -7 \\ -51y + 38z = 12 \\ -50y + 39z = 17 \end{cases}$$

$$\xrightarrow{(-1)E_3 + E_2} \begin{cases} -x - 12y + 9z = -7 \\ -y - z = -5 \\ -50y + 39z = 17 \end{cases}$$ to allow working with smaller but integral coefficients

$$\xrightarrow{-50E_2 + E_3} \begin{cases} -x - 12y + 9z = -7 \\ -y - z = -5 \\ 89z = 267 \end{cases}$$

The system is now in upper-triangular form. Solving the third equation, we obtain $z = 3$. Substituting this value back into the second equation yields $y = 2$. (Check this for yourself.) Finally, substituting $z = 3$ and $y = 2$ back into the first equation yields $x = 10$. (Again, check this for yourself.) The solution to the system is therefore $(10, 2, 3)$. ■■■

■■■ **EXAMPLE 6** Solve the system

$$\begin{cases} x + 2y + 4z = 0 \\ x + 3y + 9z = 0 \end{cases}$$

Solution This system is similar to the one in Example 2 in that there are fewer equations than there are unknowns. By subtracting the first equation from the second, we readily obtain an equivalent system in upper-triangular form:

$$\begin{cases} x + 2y + 4z = 0 \\ y + 5z = 0 \end{cases}$$

Although the system is now in upper-triangular form, notice that the second equation does not determine y or z uniquely; that is, there are infinitely many number pairs (y, z) satisfying the second equation. We can solve the second equation for y in terms of z; the result is $y = -5z$. Now we replace y with $-5z$ in the first equation to obtain

$$x + 2(-5z) + 4z = 0 \qquad \text{or} \qquad x = 6z$$

At this point, we've used both of the equations in the system to express x and y in terms of z. Furthermore, there is no third equation in the system to provide additional restrictions on x, y, or z. We therefore conclude that the given system has infinitely many solutions. These solutions have the form

$$(6z, -5z, z) \qquad \text{where } z \text{ is any real number} \qquad \blacksquare$$

■■■ **EXAMPLE 7** Solve the system

$$\begin{cases} x - 4y + z = 3 \\ 3x + 5y - 2z = -1 \\ 7x + 6y - 3z = 2 \end{cases}$$

Solution $\begin{cases} x - 4y + z = 3 \\ 3x + 5y - 2z = -1 \\ 7x + 6y - 3z = 2 \end{cases} \xrightarrow[\;-7E_1 + E_3\;]{-3E_1 + E_2} \begin{cases} x - 4y + z = 3 \\ 17y - 5z = -10 \\ 34y - 10z = -19 \end{cases}$

$\xrightarrow{-2E_2 + E_3} \begin{cases} x - 4y + z = 3 \\ 17y - 5z = -10 \\ 0 = 1 \end{cases}$

From the third equation in this last system, we conclude that this system, and consequently the original system, has no solution. (Reason: If there *were* values for x, y, and z satisfying the original system, then it would follow that $0 = 1$, which is impossible.) ■■■

In the next two examples, we introduce the subject of **partial fractions**. The basic goal here is to write a given fractional expression as a sum or difference of two or more *simpler* fractions. For instance, in Example 8, we will be given the fraction $1/(x - 1)(x + 1)$ and we will be asked to find constants A and B so that

$$\frac{1}{(x - 1)(x + 1)} = \frac{A}{x - 1} + \frac{B}{x + 1}$$

When A and B are determined, we refer to the right-hand side of the preceding equation as the **partial fraction decomposition** of the given fraction. Detailed techniques for working with partial fractions are developed as the need arises in calculus courses. The two examples and the exercises that follow will provide you with ample background for that later work.

■■■ **EXAMPLE 8** Determine constants A and B so that the following equation is an identity (i.e., the equation holds for all values of x for which the denominators are not zero):

$$\frac{1}{(x - 1)(x + 1)} = \frac{A}{x - 1} + \frac{B}{x + 1} \tag{1}$$

Solution First, to clear equation (1) of fractions, we multiply both sides of the equation by the quantity $(x - 1)(x + 1)$. This yields

$$1 = (x + 1)A + (x - 1)B = Ax + A + Bx - B$$
$$1 = (A + B)x + (A - B) \tag{2}$$

Since this last equation is to be an identity, we now reason as follows:

$$\begin{bmatrix} \text{coefficient of } x \text{ on the left-} \\ \text{hand side of equation (2)} \end{bmatrix} = \begin{bmatrix} \text{coefficient of } x \text{ on the right-} \\ \text{hand side of equation (2)} \end{bmatrix}$$

Therefore,

$$0 = A + B \tag{3}$$

Similarly, we have

$$\begin{bmatrix} \text{constant term on the left-} \\ \text{hand side of equation (2)} \end{bmatrix} = \begin{bmatrix} \text{constant term on the right-} \\ \text{hand side of equation (2)} \end{bmatrix}$$

Therefore,

$$1 = A - B \tag{4}$$

Rewriting equations (3) and (4) together gives us a system of two linear equations in the unknowns A and B:

$$\begin{cases} A + B = 0 \\ A - B = 1 \end{cases}$$

Adding these two equations yields

$$2A = 1 \qquad \text{and therefore} \qquad A = \tfrac{1}{2}$$

Subtracting the two equations gives us

$$B - (-B) = 0 - 1 \qquad \text{and therefore} \qquad B = -\tfrac{1}{2}$$

We have now found that $A = \tfrac{1}{2}$ and $B = -\tfrac{1}{2}$, as required. The partial fraction decomposition of $1/(x - 1)(x + 1)$ is therefore

$$\frac{1}{(x - 1)(x + 1)} = \frac{\tfrac{1}{2}}{x - 1} + \frac{-\tfrac{1}{2}}{x + 1}$$

$$= \frac{1}{2(x - 1)} - \frac{1}{2(x + 1)}$$

You should check for yourself that the result of combining the two fractions on the right-hand side of this last equation is indeed $1/(x - 1)(x + 1)$. ∎

In Example 8, we were able to find the partial fraction decomposition by solving a system of two linear equations. There is a shortcut that is sometimes useful in such problems. As before, we multiply both sides of equation (1) by $(x - 1)(x + 1)$ to obtain

$$1 = (x + 1)A + (x - 1)B \tag{5}$$

Now, equation (5) is to hold for all values of x; in particular, it must hold when x is -1. Replacing x with -1 in the equation then yields

$$1 = -2B$$
$$B = -\tfrac{1}{2} \qquad \text{as obtained previously}$$

The value of A can be obtained in a similar way. We replace x by 1 in equation (5). As you can check, this readily yields $A = \tfrac{1}{2}$, which again agrees with the result we obtained previously.

■■■ **EXAMPLE 9** Determine the constants A, B, and C so that the following equation is an identity:

$$\frac{2x + 1}{(x + 2)(x - 3)^2} = \frac{A}{x + 2} + \frac{B}{(x - 3)} + \frac{C}{(x - 3)^2}$$

Solution First, to clear the given equation of fractions, we multiply both sides by the quantity $(x + 2)(x - 3)^2$. After dividing out the common factors, we have

$$2x + 1 = A(x - 3)^2 + B(x + 2)(x - 3) + C(x + 2)$$
$$= A(x^2 - 6x + 9) + B(x^2 - x - 6) + C(x + 2)$$
$$= Ax^2 - 6Ax + 9A + Bx^2 - Bx - 6B + Cx + 2C$$
$$2x + 1 = (A + B)x^2 + (-6A - B + C)x + 9A - 6B + 2C$$

Equating the coefficients of x^2 on both sides of this identity yields

$$A + B = 0$$

Similarly, equating the x-coefficients on both sides yields

$$-6A - B + C = 2$$

Finally, equating the constant terms on both sides gives us

$$9A - 6B + 2C = 1$$

We now have a system of three equations in the three unknowns A, B, and C:

$$\begin{cases} A + B = 0 \\ -6A - B + C = 2 \\ 9A - 6B + 2C = 1 \end{cases}$$

Exercise 12 at the end of this section asks you to verify that the required values are $A = -\tfrac{3}{25}$, $B = \tfrac{3}{25}$, and $C = \tfrac{7}{5}$. The partial fraction decomposition is therefore

$$\frac{2x + 1}{(x + 2)(x - 3)^2} = \frac{-3}{25(x + 2)} + \frac{3}{25(x - 3)} + \frac{7}{5(x - 3)^2} \qquad ■■■$$

Here is an alternative method for working Example 9; it relies on the shortcut mentioned earlier. Again, we first multiply the given equation by $(x + 2)(x - 3)^2$ to obtain

$$2x + 1 = A(x - 3)^2 + B(x + 2)(x - 3) + C(x + 2) \qquad (6)$$

Now we just replace x by 3 in this equation. This yields

$$7 = 5C \qquad \text{or} \qquad C = \tfrac{7}{5} \qquad \text{as obtained previously}$$

The value of A can be obtained in a similar manner. We replace x by -2 in equation (6). As you can check, this readily yields $A = -\frac{3}{25}$, which again agrees with the value obtained previously. We've now computed A and C using the short method. But as you'll see if you try, the value of B cannot be obtained in the same way. One way to find B here is by substituting the values found for A and C in equation (6) to obtain

$$2x + 1 = -\tfrac{3}{25}(x - 3)^2 + B(x + 2)(x - 3) + \tfrac{7}{5}(x + 2)$$

This equation is to hold for all values of x; in particular, it must hold for the convenient value $x = 0$. Substituting $x = 0$ in this last equation yields

$$1 = -\tfrac{3}{25}(-3)^2 + B(2)(-3) + \tfrac{7}{5}(2)$$

$25 = -27 - 150B + 70$	multiplying by 25
$150B = 18$	collecting like terms
$B = \frac{18}{150} = \frac{3}{25}$	dividing by 150 and simplifying

We have now found that $B = \frac{3}{25}$, in agreement with the value obtained previously.

EXERCISE SET 9.2

A

In Exercises 1–10, the systems of linear equations are in upper-triangular form. Find all solutions of each system.

1. $\begin{cases} 2x + y + z = -9 \\ 3y - 2z = -4 \\ 8z = -8 \end{cases}$

2. $\begin{cases} -3x + 7y + 2z = -19 \\ y + z = 1 \\ -2z = -2 \end{cases}$

3. $\begin{cases} 8x + 5y + 3z = 1 \\ 3y + 4z = 2 \\ 5z = 3 \end{cases}$

4. $\begin{cases} 2x + 7z = -4 \\ 5y - 3z = 6 \\ 6z = 18 \end{cases}$

5. $\begin{cases} -4x + 5y = 0 \\ 3y + 2z = 1 \\ 3z = -1 \end{cases}$

6. $\begin{cases} 3x - 2y + z = 4 \\ 3z = 9 \end{cases}$

7. $\begin{cases} -x + 8y + 3z = 0 \\ 2z = 0 \end{cases}$

8. $\begin{cases} -x + y + z + w = 9 \\ 2y - z - w = 9 \\ 3z + 2w = 1 \\ 11w = 22 \end{cases}$

9. $\begin{cases} 2x + 3y + z + w = -6 \\ y + 3z - 4w = 23 \\ 6z - 5w = 31 \\ -2w = 10 \end{cases}$

10. $\begin{cases} 7x - y - z + w = 3 \\ 2y - 3z - 4w = -2 \\ 3w = 6 \end{cases}$

In Exercises 11–30, find all solutions of each system.

11. $\begin{cases} x + y + z = 12 \\ 2x - y - z = -1 \\ 3x + 2y + z = 22 \end{cases}$

12. $\begin{cases} A + B = 0 \\ -6A - B + C = 2 \\ 9A - 6B + 2C = 1 \end{cases}$
Answer: $\left(-\tfrac{3}{25}, \tfrac{3}{25}, \tfrac{7}{5}\right)$

13. $\begin{cases} 2x - 3y + 2z = 4 \\ 4x + 2y + 3z = 7 \\ 5x + 4y + 2z = 7 \end{cases}$

14. $\begin{cases} x + 2z = 5 \\ y - 30z = -16 \\ x - 2y + 4z = 8 \end{cases}$

15. $\begin{cases} 3x + 3y - 2z = 13 \\ 6x + 2y - 5z = 13 \\ 7x + 5y - 3z = 26 \end{cases}$

16. $\begin{cases} 2x + 5y - 3z = 4 \\ 4x - 3y + 2z = 9 \\ 5x + 6y - 2z = 18 \end{cases}$

17. $\begin{cases} x + y + z = 1 \\ -2x + y + z = -2 \\ 3x + 6y + 6z = 5 \end{cases}$

18. $\begin{cases} 7x + 5y - 7z = -10 \\ 2x + y + z = 7 \\ x + y - 3z = -8 \end{cases}$

19. $\begin{cases} 2x - y + z = -1 \\ x + 3y - 2z = 2 \\ -5x + 6y - 5z = 5 \end{cases}$

20. $\begin{cases} -2x + 2y - z = 0 \\ 3x - 4y + z = 1 \\ 5x - 8y + z = 4 \end{cases}$

21. $\begin{cases} 2x - y + z = 4 \\ x + 3y + 2z = -1 \\ 7x + 5z = 11 \end{cases}$

22. $\begin{cases} 3x + y - z = 10 \\ 8x - y - 6z = -3 \\ 5x - 2y - 5z = 1 \end{cases}$

23. $\begin{cases} x + y + z + w = 4 \\ x - 2y - z - w = 3 \\ 2x - y + z - w = 2 \\ x - y + 2z - 2w = -7 \end{cases}$

24. $\begin{cases} x + y - 3z + 2w = 0 \\ -2x - 2y + 6z + w = -5 \\ -x + 3y + 3z + 3w = -5 \\ 2x + y - 3z - w = 4 \end{cases}$

25. $\begin{cases} 2x + 3y + 2z = 5 \\ x + 4y - 3z = 1 \end{cases}$

26. $\begin{cases} 4x - y - 3z = 2 \\ 6x + 5y - z = 0 \end{cases}$

27. $\begin{cases} x - 2y - 2z + 2w = -10 \\ 3x + 4y - z - 3w = 11 \\ -4x - 3y - 3z + 8w = -21 \end{cases}$

28. $\begin{cases} 2x + y + z + w = 1 \\ x + 3y - 3z - 3w = 0 \\ -3x - 4y + 2z + 2w = -1 \end{cases}$

29. $\begin{cases} 4x - 2y + 3z = -2 \\ 6y - 4z = 6 \end{cases}$

30. $\begin{cases} 6x - 2y - 5z + w = 3 \\ 5x - y + z + 5w = 4 \\ -4y - 7z - 9w = -5 \end{cases}$

In Exercises 31–40, determine the constants (denoted by capital letters) so that each equation is an identity.

31. (a) $\dfrac{1}{(x-2)(x+2)} = \dfrac{A}{x-2} + \dfrac{B}{x+2}$

   (b) $\dfrac{x}{(x-2)(x+2)} = \dfrac{A}{x-2} + \dfrac{B}{x+2}$

32. (a) $\dfrac{2}{(x+3)(x+2)} = \dfrac{A}{x+3} + \dfrac{B}{x+2}$

   (b) $\dfrac{x-9}{(x+3)(x+2)} = \dfrac{A}{x+3} + \dfrac{B}{x+2}$

33. $\dfrac{3x}{(x+4)(x-5)} = \dfrac{A}{x+4} + \dfrac{B}{x-5}$

34. $\dfrac{4x+26}{(x-1)(x+5)} = \dfrac{A}{x-1} + \dfrac{B}{x+5}$

35. (a) $\dfrac{3}{(x-4)(x+4)} = \dfrac{A}{x-4} + \dfrac{B}{x+4}$

   (b) $\dfrac{3}{(x-4)(x+4)^2} = \dfrac{A}{x-4} + \dfrac{B}{x+4} + \dfrac{C}{(x+4)^2}$

36. (a) $\dfrac{1}{(x+1)(x+6)} = \dfrac{A}{x+1} + \dfrac{B}{x+6}$

   (b) $\dfrac{1}{(x+1)(x+6)^2} = \dfrac{A}{x+1} + \dfrac{B}{x+6} + \dfrac{C}{(x+6)^2}$

37. $\dfrac{1}{(x+1)(x^2-x+1)} = \dfrac{A}{x+1} + \dfrac{Bx+C}{x^2-x+1}$

38. $\dfrac{1}{(x+1)^2(x^2-x+1)} = \dfrac{A}{x+1} + \dfrac{B}{(x+1)^2} + \dfrac{Cx+D}{x^2-x+1}$

39. $\dfrac{4}{x(1-x)} = \dfrac{A}{x} + \dfrac{B}{1-x}$

40. $\dfrac{4}{x^2(1-x)} = \dfrac{A}{x} + \dfrac{B}{x^2} + \dfrac{C}{1-x}$

B

41. Determine the constants A, B, C, and D so that the following equation is an identity:

$$\dfrac{1}{(x^2+x+1)(x^2-x+1)} = \dfrac{Ax+B}{x^2+x+1} + \dfrac{Cx+D}{x^2-x+1}$$

42. Suppose that the height of an object as a function of time is given by $f(t) = at^2 + bt + c$, where t is time in seconds, $f(t)$ is the height in feet at time t, and a, b, and c are certain constants. If, after 1, 2, and 3 seconds, the corresponding heights are 184 ft, 136 ft, and 56 ft, respectively, find the time at which the object is at ground level (height = 0 ft).

43. Find the equation of the circle passing through the points $(-1, -5)$, $(1, 2)$, and $(-2, 0)$. Write your answer in the form $Ax^2 + Ay^2 + Bx + Cy + D = 0$.

44. Find the equation of a circle that passes through the origin and the points $(8, -4)$ and $(7, -1)$. Specify the center and radius.

45. Solve the following system for x, y, and z:

$$\begin{cases} e^x + e^y - 2e^z = 2a \\ e^x + 2e^y - 4e^z = 3a \\ \frac{1}{2}e^x - 3e^y + e^z = -5a \end{cases}$$

(Assume that $a > 0$.)   *Hint:*   Let $A = e^x$, $B = e^y$, and $C = e^z$. Solve the resulting linear system for A, B, and C.

46. Solve the following system for α, β, and γ:

$$\begin{cases} \ln \alpha - \ln \beta - \ln \gamma = 2 \\ 3 \ln \alpha + 5 \ln \beta - 2 \ln \gamma = 1 \\ 2 \ln \alpha - 4 \ln \beta + \ln \gamma = 2 \end{cases}$$

47. The following figure displays three circles that are mutually tangent. The line segments joining the centers have lengths a, b, and c, as shown. Let r_1, r_2, and r_3 denote the radii of the circles, as indicated in the figure. Show that

$$r_1 = \dfrac{a+c-b}{2} \qquad r_2 = \dfrac{a+b-c}{2}$$

$$r_3 = \dfrac{b+c-a}{2}$$

48. Consider the system

$$\begin{cases} \lambda x + y + z = a \\ x + \lambda y + z = b \\ x + y + \lambda z = c \end{cases}$$

(a) Assuming that the value of the constant λ is neither 1 nor -2, find all solutions of the system.
(b) If $\lambda = -2$, how must a, b, and c be related for the system to have solutions? Find these solutions.
(c) If $\lambda = 1$, how must a, b, and c be related for the system to have solutions? Find these solutions.

In Exercises 49 and 50, the lowercase letters a, b, p, and q denote given constants. In each case, determine the values of A and B so that the equation is an identity.

49. (a) $\dfrac{1}{(x-a)(x-b)} = \dfrac{A}{x-a} + \dfrac{B}{x-b}$ $(a \neq b)$

(b) $\dfrac{px+q}{(x-a)(x-b)} = \dfrac{A}{x-a} + \dfrac{B}{x-b}$ $(a \neq b)$

50. (a) $\dfrac{1}{(x-a)(x+a)} = \dfrac{A}{x-a} + \dfrac{B}{x+a}$ $(a \neq 0)$

(b) $\dfrac{px+q}{(x-a)(x+a)} = \dfrac{A}{x-a} + \dfrac{B}{x+a}$ $(a \neq 0)$

C

51. Express A, B, and C in terms of a, b, c, p, and q so that the following is an identity. (Assume a, b, and c are all unequal.)

$$\frac{x^2 + px + q}{(x-a)(x-b)(x-c)} = \frac{A}{x-a} + \frac{B}{x-b} + \frac{C}{x-c}$$

52. Express A, B, and C in terms of a, b, and c so that the following equation is an identity. (Assume a, b, and c are all unequal).

$$\frac{1}{(1-ax)(1-bx)(1-cx)} = \frac{A}{1-ax} + \frac{B}{1-bx} + \frac{C}{1-cx}$$

The following exercise appears in Algebra for Colleges and Schools *by H. S. Hall and S. R. Knight, revised by F. L. Sevenoak (New York: The Macmillan Company, 1906).*

53. A, B, and C are three towns forming a triangle. A man has to walk from one to the next, ride thence to the next, and drive thence to his starting point. He can walk, ride, and drive a mile in a, b, and c minutes, respectively. If he starts from B he takes $a + c - b$ hours, if he starts from C he takes $b + a - c$ hours, and if he starts from A he takes $c + b - a$ hours. Find the length of the circuit. [Assume that the circuit from A to B to C is counterclockwise.]

9.3 MATRICES

Arthur Cayley (1821–1895) and James Joseph Sylvester (1814–1897), two English mathematicians, invented the matrix ... in the 1850s.... The operations of addition and multiplication of matrices were later defined and the algebra of matrices was then developed. In 1925, Werner Heisenberg, a German physicist, used matrices in developing his theory of quantum mechanics, extending the role of matrices from algebra to the area of applied mathematics.

John K. Luedeman and Stanley M. Lukawecki in *Elementary Linear Algebra* (St. Paul: West Publishing Co., 1986)

Recall that in Gaussian Elimination, row operations are used to change the coefficient matrix to an upper triangular matrix. The solution is then found by back substitution, starting from the last equation in the reduced system.

Steven C. Althoen and Renate McLaughlin in "Gauss–Jordan Reduction: A Brief History," *American Mathematical Monthly*, vol. **94** (1987): pp. 130–142

As you saw in the previous section, there can be a good deal of bookkeeping involved in using Gaussian elimination to solve systems of equations. We can lighten the load somewhat by using matrices, where a **matrix** is simply a rectangular array of numbers, enclosed in parentheses or brackets. Here are three examples:

$$\begin{pmatrix} 2 & 3 \\ -5 & 4 \end{pmatrix} \qquad \begin{pmatrix} -6 & 0 & 1 & \frac{1}{4} \\ \frac{2}{3} & 1 & 5 & 8 \end{pmatrix} \qquad \begin{pmatrix} \pi & 0 & 0 \\ 0 & 1 & 9 \\ -1 & -2 & 3 \\ -4 & 8 & 6 \end{pmatrix}$$

The particular numbers constituting a matrix are called its **entries** or **elements**. For instance, the entries in the matrix $\begin{pmatrix} 2 & 3 \\ -5 & 4 \end{pmatrix}$ are the four numbers 2, 3, −5, and 4. In this section, the entries will always be real numbers. However, it is also possible to consider matrices in which some or all of the entries are nonreal complex numbers.

It is convenient to agree on a standard system for labeling the rows and columns of a matrix. The rows are numbered from top to bottom and the columns from left to right, as indicated in the following example:

$$\begin{array}{cc} \text{column 1} & \text{column 2} \\ \searrow & \swarrow \end{array}$$

$$\begin{array}{l} \text{row 1} \longrightarrow \\ \text{row 2} \longrightarrow \\ \text{row 3} \longrightarrow \end{array} \begin{pmatrix} 5 & -3 \\ 0 & 2 \\ -1 & 16 \end{pmatrix}$$

We express the **size**, or **dimension**, of a matrix by specifying the number of rows and the number of columns, in that order. For instance, we would say that the matrix

$$\begin{pmatrix} 5 & -3 \\ 0 & 2 \\ -1 & 16 \end{pmatrix}$$

is a 3 × 2 (read "3 by 2") matrix, not 2 × 3. The following example will help fix in your mind the terminology that we have introduced.

EXAMPLE 1 Consider the matrix

$$\begin{pmatrix} 1 & 3 & 5 \\ 7 & 9 & 11 \end{pmatrix}$$

(a) List the entries.
(b) What is the size of the matrix?
(c) Which element is in the second row, third column?

Solution

(a) The entries are 1, 3, 5, 7, 9, and 11.
(b) Since there are two rows and three columns, this is a 2 × 3 matrix.
(c) To locate the element in the second row, third column, draw lines through the second row and the third column and see where they intersect:

$$\begin{pmatrix} 1 & 3 & 5 \\ 7 & 9 & 11 \end{pmatrix}$$

Thus, the entry in row 2, column 3 is 11.

There is a natural way to use matrices to describe and solve systems of linear equations. Consider, for example, the following system of linear equations in **standard form** (with the x, y, and z terms lined up on the left-hand side and the constant terms on the right-hand side of each equation):

$$\begin{cases} x + 2y - 3z = 4 \\ 3x \qquad + z = 5 \\ -x - 3y + 4z = 0 \end{cases}$$

The **coefficient matrix** of this system is the matrix

$$\begin{pmatrix} 1 & 2 & -3 \\ 3 & 0 & 1 \\ -1 & -3 & 4 \end{pmatrix}$$

As the name implies, the coefficient matrix of the system is the matrix whose entries are the coefficients of x, y, and z, written in the same relative positions as they appear in the system. Notice the zero appearing in the second row, second column of the matrix. It is there because the coefficient of y in the second equation is in fact zero. The **augmented matrix** of the system of equations considered here is

$$\begin{pmatrix} 1 & 2 & -3 & 4 \\ 3 & 0 & 1 & 5 \\ -1 & -3 & 4 & 0 \end{pmatrix}$$

As you can see, the augmented matrix is formed by *augmenting* the coefficient matrix with the column of constant terms taken from the right-hand side of the given system of equations. To help relate the augmented matrix to the original system of equations, we will write the augmented matrix this way:

$$\left(\begin{array}{ccc|c} 1 & 2 & -3 & 4 \\ 3 & 0 & 1 & 5 \\ -1 & -3 & 4 & 0 \end{array} \right)$$

■ ■ ■ EXAMPLE 2 Write the coefficient matrix and the augmented matrix for the system

$$\begin{cases} 8x - 2y + z = 1 \\ 3x - 4z + y = 2 \\ 12y - 3z - 6 = 0 \end{cases}$$

Solution First we write the system in standard form, with the x, y, and z, terms lined up and the constant terms on the right. This yields

$$\begin{cases} 8x - 2y + z = 1 \\ 3x + y - 4z = 2 \\ 12y - 3z = 6 \end{cases}$$

The coefficient matrix is then

$$\begin{pmatrix} 8 & -2 & 1 \\ 3 & 1 & -4 \\ 0 & 12 & -3 \end{pmatrix}$$

while the augmented matrix is

$$\begin{pmatrix} 8 & -2 & 1 & \vdots & 1 \\ 3 & 1 & -4 & \vdots & 2 \\ 0 & 12 & -3 & \vdots & 6 \end{pmatrix}$$

∎∎∎

In the previous section we used the three elementary operations in solving systems of linear equations. In Table 1 we express these operations in the language of matrices. The matrix operations are called the **elementary row operations.**

Table 2 displays an example of each elementary row operation. In the table, notice that the notation for describing these operations is essentially the same as that introduced in the previous section. For example, in the previous section, $10E_1$ indicated that the first equation in a system was multiplied by 10. Here, $10R_1$ indicates that each entry in the first *row* of a matrix is multiplied by 10.

TABLE 1

Elementary Operations for a System of Linear Equations	Corresponding Elementary Row Operation for a Matrix
1. Multiply both sides of an equation by a nonzero constant.	1′. Multiply each entry in a given row by a nonzero constant.
2. Interchange two equations.	2′. Interchange two rows.
3. To one equation, add a multiple of another equation.	3′. To one row, add a multiple of another row.

TABLE 2

Examples of the Elementary Row Operations		Comments
$\begin{pmatrix} 1 & 2 & 3 \\ 4 & 5 & 6 \\ 7 & 8 & 9 \end{pmatrix}$ $\xrightarrow{\ 10R_1\ }$ $\begin{pmatrix} 10 & 20 & 30 \\ 4 & 5 & 6 \\ 7 & 8 & 9 \end{pmatrix}$		Multiply each entry in row 1 by 10.
$\begin{pmatrix} 1 & 2 & 3 \\ 4 & 5 & 6 \\ 7 & 8 & 9 \end{pmatrix}$ $\xrightarrow{\ R_2 \leftrightarrow R_3\ }$ $\begin{pmatrix} 1 & 2 & 3 \\ 7 & 8 & 9 \\ 4 & 5 & 6 \end{pmatrix}$		Interchange rows 2 and 3.
$\begin{pmatrix} 1 & 2 & 3 \\ 4 & 5 & 6 \\ 7 & 8 & 9 \end{pmatrix}$ $\xrightarrow{\ -4R_1 + R_2\ }$ $\begin{pmatrix} 1 & 2 & 3 \\ 0 & -3 & -6 \\ 7 & 8 & 9 \end{pmatrix}$		To each entry in row 2, add −4 times the corresponding entry in row 1.

We are now ready to use matrices to solve systems of equations. In the example that follows, we'll use the same system of equations used in Example 5 of the previous section. *Suggestion:* After reading the next example, carefully compare each step with the corresponding one taken in Example 5 of the previous section.

∎▬∎ **EXAMPLE 3** Solve the system

$$\begin{cases} 4x - 3y + 2z = 40 \\ 5x + 9y - 7z = 47 \\ 9x + 8y - 3z = 97 \end{cases}$$

Solution

$$\begin{pmatrix} 4 & -3 & 2 & \vdots & 40 \\ 5 & 9 & -7 & \vdots & 47 \\ 9 & 8 & -3 & \vdots & 97 \end{pmatrix} \xrightarrow{(-1)R_2 + R_1} \begin{pmatrix} -1 & -12 & 9 & \vdots & -7 \\ 5 & 9 & -7 & \vdots & 47 \\ 9 & 8 & -3 & \vdots & 97 \end{pmatrix}$$

$$\xrightarrow[9R_1 + R_3]{5R_1 + R_2} \begin{pmatrix} -1 & -12 & 9 & \vdots & -7 \\ 0 & -51 & 38 & \vdots & 12 \\ 0 & -100 & 78 & \vdots & 34 \end{pmatrix}$$

$$\xrightarrow{\frac{1}{2}R_3} \begin{pmatrix} -1 & -12 & 9 & \vdots & -7 \\ 0 & -51 & 38 & \vdots & 12 \\ 0 & -50 & 39 & \vdots & 17 \end{pmatrix}$$

$$\xrightarrow{(-1)R_3 + R_2} \begin{pmatrix} -1 & -12 & 9 & \vdots & -7 \\ 0 & -1 & -1 & \vdots & -5 \\ 0 & -50 & 39 & \vdots & 17 \end{pmatrix}$$

$$\xrightarrow{-50R_2 + R_3} \begin{pmatrix} -1 & -12 & 9 & \vdots & -7 \\ 0 & -1 & -1 & \vdots & -5 \\ 0 & 0 & 89 & \vdots & 267 \end{pmatrix}$$

This last augmented matrix represents a system of equations in upper-triangular form:

$$\begin{cases} -x - 12y + 9z = -7 \\ \quad\quad -y - z = -5 \\ \quad\quad\quad\quad 89z = 267 \end{cases}$$

As you should now check for yourself, this yields the values $z = 3$, then $y = 2$, then $x = 10$. The solution of the original system is therefore $(10, 2, 3)$. ▮▮▮

For the remainder of this section (and in part of the next) we are going to study matrices without reference to systems of equations. As motivation for this, we point out that matrices are essential tools in many fields of study. For example, a knowledge of matrices and their properties is needed for work in computer graphics. To begin, we need to say what it means for two matrices to be equal.

DEFINITION   Equality of Matrices

EXAMPLES

Two matrices are equal provided they have the same size (same number of rows, same number of columns) and the corresponding entries are equal.

$$\begin{pmatrix} 2 & 3 \\ 4 & 5 \end{pmatrix} = \begin{pmatrix} 2 & 3 \\ 4 & 5 \end{pmatrix}$$

$$\begin{pmatrix} 2 & 3 & 0 \\ 4 & 5 & 0 \end{pmatrix} \neq \begin{pmatrix} 2 & 3 \\ 4 & 5 \end{pmatrix}$$

$$\begin{pmatrix} 2 & 3 \\ 4 & 5 \end{pmatrix} \neq \begin{pmatrix} 2 & 3 \\ 5 & 4 \end{pmatrix}$$

Now we can define matrix addition and subtraction. These operations are defined only between matrices of the same size.

DEFINITION Matrix Addition	
	EXAMPLES
To add (or subtract) two matrices of the same size, add (or subtract) the corresponding entries.	$\begin{pmatrix} 2 & 3 \\ -1 & 4 \end{pmatrix} + \begin{pmatrix} 6 & 1 \\ 0 & -4 \end{pmatrix} = \begin{pmatrix} 8 & 4 \\ -1 & 0 \end{pmatrix}$
	$\begin{pmatrix} 2 & 3 \\ -1 & 4 \\ 9 & 10 \end{pmatrix} - \begin{pmatrix} 6 & 1 \\ 0 & -4 \\ 3 & 2 \end{pmatrix} = \begin{pmatrix} -4 & 2 \\ -1 & 8 \\ 6 & 8 \end{pmatrix}$

Many properties of the real numbers also apply to matrices. For instance, matrix addition is *commutative:*

$$A + B = B + A \qquad \text{where } A \text{ and } B \text{ are matrices of the same size}$$

Matrix addition is also *associative:*

$$A + (B + C) = (A + B) + C \qquad \text{where } A, B, \text{ and } C \text{ are matrices of the same size}$$

In the next example, we verify these properties in two specific instances.

EXAMPLE 4 Let $A = \begin{pmatrix} 1 & 2 \\ 3 & 4 \end{pmatrix}$, $B = \begin{pmatrix} 0 & -5 \\ 8 & -1 \end{pmatrix}$, and $C = \begin{pmatrix} 6 & 7 \\ 8 & 9 \end{pmatrix}$.

(a) Show that $A + C = C + A$.
(b) Show that $A + (B + C) = (A + B) + C$.

Solution

(a) $A + C = \begin{pmatrix} 1 & 2 \\ 3 & 4 \end{pmatrix} + \begin{pmatrix} 6 & 7 \\ 8 & 9 \end{pmatrix} = \begin{pmatrix} 7 & 9 \\ 11 & 13 \end{pmatrix}$

$C + A = \begin{pmatrix} 6 & 7 \\ 8 & 9 \end{pmatrix} + \begin{pmatrix} 1 & 2 \\ 3 & 4 \end{pmatrix} = \begin{pmatrix} 7 & 9 \\ 11 & 13 \end{pmatrix}$

This shows that $A + C = C + A$, since both $A + C$ and $C + A$ represent the matrix $\begin{pmatrix} 7 & 9 \\ 11 & 13 \end{pmatrix}$.

(b) First we compute $A + (B + C)$:

$$A + (B + C) = \begin{pmatrix} 1 & 2 \\ 3 & 4 \end{pmatrix} + \left[\begin{pmatrix} 0 & -5 \\ 8 & -1 \end{pmatrix} + \begin{pmatrix} 6 & 7 \\ 8 & 9 \end{pmatrix} \right]$$

$$= \begin{pmatrix} 1 & 2 \\ 3 & 4 \end{pmatrix} + \begin{pmatrix} 6 & 2 \\ 16 & 8 \end{pmatrix}$$

$$= \begin{pmatrix} 7 & 4 \\ 19 & 12 \end{pmatrix}$$

Next we compute $(A + B) + C$:

$$(A + B) + C = \left[\begin{pmatrix} 1 & 2 \\ 3 & 4 \end{pmatrix} + \begin{pmatrix} 0 & -5 \\ 8 & -1 \end{pmatrix} \right] + \begin{pmatrix} 6 & 7 \\ 8 & 9 \end{pmatrix}$$

$$= \begin{pmatrix} 1 & -3 \\ 11 & 3 \end{pmatrix} + \begin{pmatrix} 6 & 7 \\ 8 & 9 \end{pmatrix}$$

$$= \begin{pmatrix} 7 & 4 \\ 19 & 12 \end{pmatrix}$$

We conclude from these calculations that $A + (B + C) = (A + B) + C$, since both sides of that equation represent the matrix $\begin{pmatrix} 7 & 4 \\ 19 & 12 \end{pmatrix}$.   ▮▮▮

We will now define an operation on matrices called **scalar multiplication**. First of all, the word *scalar* here just means *real number*. So we are talking about multiplying a matrix by a real number. (In more advanced work, nonreal complex scalars are also considered.)

DEFINITION   Scalar Multiplication

EXAMPLES

To multiply a matrix by a scalar, multiply each entry in the matrix by that scalar.

$$2\begin{pmatrix} 5 & 9 & 0 \\ -1 & 2 & 3 \end{pmatrix} = \begin{pmatrix} 10 & 18 & 0 \\ -2 & 4 & 6 \end{pmatrix}$$

$$1\begin{pmatrix} 1 & 2 \\ 3 & 4 \end{pmatrix} = \begin{pmatrix} 1 & 2 \\ 3 & 4 \end{pmatrix}$$

There are two simple but useful properties of scalar multiplication that are worth noting at this point. We'll omit the proofs of these two properties; however, Example 5 does ask us to verify them for a particular case.

PROPERTY SUMMARY   **PROPERTIES OF SCALAR MULTIPLICATION**

1. $c(kM) = (ck)M$ for all scalars c and k and any matrix M

2. $c(M + \mathcal{N}) = cM + c\mathcal{N}$, where c is any scalar and M and $\mathcal{N}$ are any matrices of the same size

---

▮▬▮ **EXAMPLE 5**   Let $c = 2$, $k = 3$, $M = \begin{pmatrix} 1 & 2 \\ 3 & 4 \end{pmatrix}$, and $\mathcal{N} = \begin{pmatrix} 5 & 6 \\ 7 & 8 \end{pmatrix}$.

(a) Show that $c(kM) = (ck)M$.
(b) Show that $c(M + \mathcal{N}) = cM + c\mathcal{N}$.

Solution

(a) $c(kM) = 2\left[3\begin{pmatrix} 1 & 2 \\ 3 & 4 \end{pmatrix} \right] = 2\begin{pmatrix} 3 & 6 \\ 9 & 12 \end{pmatrix} = \begin{pmatrix} 6 & 12 \\ 18 & 24 \end{pmatrix}$

$(ck)M = (2 \cdot 3)\begin{pmatrix} 1 & 2 \\ 3 & 4 \end{pmatrix} = 6\begin{pmatrix} 1 & 2 \\ 3 & 4 \end{pmatrix} = \begin{pmatrix} 6 & 12 \\ 18 & 24 \end{pmatrix}$

Thus, $c(kM) = (ck)M$, since in both cases the result is $\begin{pmatrix} 6 & 12 \\ 18 & 24 \end{pmatrix}$.

(b) $c(M + N) = 2\left[\begin{pmatrix} 1 & 2 \\ 3 & 4 \end{pmatrix} + \begin{pmatrix} 5 & 6 \\ 7 & 8 \end{pmatrix}\right]$

$$= 2\begin{pmatrix} 6 & 8 \\ 10 & 12 \end{pmatrix} = \begin{pmatrix} 12 & 16 \\ 20 & 24 \end{pmatrix}$$

$$cM + cN = 2\begin{pmatrix} 1 & 2 \\ 3 & 4 \end{pmatrix} + 2\begin{pmatrix} 5 & 6 \\ 7 & 8 \end{pmatrix}$$

$$= \begin{pmatrix} 2 & 4 \\ 6 & 8 \end{pmatrix} + \begin{pmatrix} 10 & 12 \\ 14 & 16 \end{pmatrix} = \begin{pmatrix} 12 & 16 \\ 20 & 24 \end{pmatrix}$$

Thus, $c(M + N) = cM + cN$, since both sides equal $\begin{pmatrix} 12 & 16 \\ 20 & 24 \end{pmatrix}$. ∎

A matrix with zeros for all of its entries plays the same role in matrix addition as does the number zero in ordinary addition of real numbers. For instance, in the case of 2×2 matrices, we have

$$\begin{pmatrix} a & b \\ c & d \end{pmatrix} + \begin{pmatrix} 0 & 0 \\ 0 & 0 \end{pmatrix} = \begin{pmatrix} a & b \\ c & d \end{pmatrix}$$

and

$$\begin{pmatrix} 0 & 0 \\ 0 & 0 \end{pmatrix} + \begin{pmatrix} a & b \\ c & d \end{pmatrix} = \begin{pmatrix} a & b \\ c & d \end{pmatrix}$$

for all real numbers a, b, c, and d. The matrix $\begin{pmatrix} 0 & 0 \\ 0 & 0 \end{pmatrix}$ is called the **additive identity** for 2×2 matrices. Similarly, any matrix with all zero entries is the additive identity for matrices of that size. It is sometimes convenient to denote an additive identity matrix by a boldface zero: **0**. With this notation, we can write

$$A + \mathbf{0} = \mathbf{0} + A = A \qquad \text{for any matrix } A$$

For this matrix equation, it is understood that the size of the matrix **0** is the same as the size of A. With this notation, we also have

$$A - A = \mathbf{0} \qquad \text{for any matrix } A$$

Our last topic in this section is matrix multiplication. We will begin with the simplest case and then work up to the more general situation. By convention, a matrix with only one row is called a **row vector**. Examples of row vectors are

$$(2 \quad 13), \qquad (-1 \quad 4 \quad 3), \qquad \text{and} \qquad (0 \quad 0 \quad 0 \quad 1)$$

Similarly, a matrix with only one column is called a **column vector**. Examples are

$$\begin{pmatrix} 2 \\ 13 \end{pmatrix}, \qquad \begin{pmatrix} -1 \\ 4 \\ 3 \end{pmatrix}, \qquad \text{and} \qquad \begin{pmatrix} 0 \\ 0 \\ 0 \\ 1 \end{pmatrix}$$

The following definition tells us how to multiply a row vector and a column vector when they have the same number of entries.

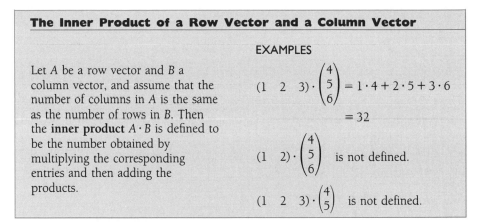

An important observation here is that the end result of taking the inner product is always just a number. The definition of matrix product that we now give depends on this observation.

The Product of Two Matrices

Let A and B be two matrices, and assume that the number of columns in A is the same as the number of rows in B. Then the product matrix AB is computed according to the following rule.

The entry in the ith row and the jth column of AB is the inner product of the ith row of A with the jth column of B.

The matrix AB will have as many rows as A and as many columns as B.

As an example of matrix multiplication, we will compute the product AB, where $A = \begin{pmatrix} 1 & 2 \\ 3 & 4 \end{pmatrix}$ and $B = \begin{pmatrix} 5 & 6 & 0 \\ 7 & 8 & 1 \end{pmatrix}$. In other words, we will compute $\begin{pmatrix} 1 & 2 \\ 3 & 4 \end{pmatrix}\begin{pmatrix} 5 & 6 & 0 \\ 7 & 8 & 1 \end{pmatrix}$. Before we attempt to carry out the calculations of any matrix multiplication, however, we should check on two points:

1. Is the product defined? That is, does the number of columns in A equal the number of rows in B? In this case, yes; the common number is 2.
2. What is the size of the product? According to the definition, the product AB will have as many rows as A and as many columns as B. Thus, the size of AB will be 2×3.

Schematically, then, the situation looks like this:

$$\begin{pmatrix} 1 & 2 \\ 3 & 4 \end{pmatrix}\begin{pmatrix} 5 & 6 & 0 \\ 7 & 8 & 1 \end{pmatrix} = \begin{pmatrix} ? & ? & ? \\ ? & ? & ? \end{pmatrix}$$

We have six positions to fill. The computations are presented in Table 3.

TABLE 3

Position	How to Compute	Computation
row 1, column 1	inner product of row 1 and column 1 $$\begin{pmatrix} 1 & 2 \\ 3 & 4 \end{pmatrix}\begin{pmatrix} 5 & 6 & 0 \\ 7 & 8 & 1 \end{pmatrix}$$	$1 \cdot 5 + 2 \cdot 7 = 19$
row 1, column 2	inner product of row 1 and column 2 $$\begin{pmatrix} 1 & 2 \\ 3 & 4 \end{pmatrix}\begin{pmatrix} 5 & 6 & 0 \\ 7 & 8 & 1 \end{pmatrix}$$	$1 \cdot 6 + 2 \cdot 8 = 22$
row 1, column 3	inner product of row 1 and column 3 $$\begin{pmatrix} 1 & 2 \\ 3 & 4 \end{pmatrix}\begin{pmatrix} 5 & 6 & 0 \\ 7 & 8 & 1 \end{pmatrix}$$	$1 \cdot 0 + 2 \cdot 1 = 2$
row 2, column 1	inner product of row 2 and column 1 $$\begin{pmatrix} 1 & 2 \\ 3 & 4 \end{pmatrix}\begin{pmatrix} 5 & 6 & 0 \\ 7 & 8 & 1 \end{pmatrix}$$	$3 \cdot 5 + 4 \cdot 7 = 43$
row 2, column 2	inner product of row 2 and column 2 $$\begin{pmatrix} 1 & 2 \\ 3 & 4 \end{pmatrix}\begin{pmatrix} 5 & 6 & 0 \\ 7 & 8 & 1 \end{pmatrix}$$	$3 \cdot 6 + 4 \cdot 8 = 50$
row 2, column 3	inner product of row 2 and column 3 $$\begin{pmatrix} 1 & 2 \\ 3 & 4 \end{pmatrix}\begin{pmatrix} 5 & 6 & 0 \\ 7 & 8 & 1 \end{pmatrix}$$	$3 \cdot 0 + 4 \cdot 1 = 4$

Reading from the table, we see that our result is

$$AB = \begin{pmatrix} 1 & 2 \\ 3 & 4 \end{pmatrix}\begin{pmatrix} 5 & 6 & 0 \\ 7 & 8 & 1 \end{pmatrix} = \begin{pmatrix} 19 & 22 & 2 \\ 43 & 50 & 4 \end{pmatrix}$$

EXAMPLE 6 Let $A = \begin{pmatrix} 1 & 2 \\ 3 & 4 \end{pmatrix}$ and $B = \begin{pmatrix} 5 & 6 \\ 7 & 8 \end{pmatrix}$. By computing AB and then BA, show that $AB \neq BA$. This shows that, in general, matrix multiplication is not commutative.

Solution $AB = \begin{pmatrix} 1 & 2 \\ 3 & 4 \end{pmatrix}\begin{pmatrix} 5 & 6 \\ 7 & 8 \end{pmatrix} = \begin{pmatrix} 1 \cdot 5 + 2 \cdot 7 & 1 \cdot 6 + 2 \cdot 8 \\ 3 \cdot 5 + 4 \cdot 7 & 3 \cdot 6 + 4 \cdot 8 \end{pmatrix} = \begin{pmatrix} 19 & 22 \\ 43 & 50 \end{pmatrix}$

$BA = \begin{pmatrix} 5 & 6 \\ 7 & 8 \end{pmatrix}\begin{pmatrix} 1 & 2 \\ 3 & 4 \end{pmatrix} = \begin{pmatrix} 5 \cdot 1 + 6 \cdot 3 & 5 \cdot 2 + 6 \cdot 4 \\ 7 \cdot 1 + 8 \cdot 3 & 7 \cdot 2 + 8 \cdot 4 \end{pmatrix} = \begin{pmatrix} 23 & 34 \\ 31 & 46 \end{pmatrix}$

Comparing the two matrices AB and BA, we conclude that $AB \neq BA$. ▮▮▮

EXERCISE SET 9.3

A

In Exercises 1–4, specify the size of each matrix.

1. (a) $\begin{pmatrix} -4 & 0 & 5 \\ 2 & 8 & -1 \end{pmatrix}$ (b) $\begin{pmatrix} 7 & 1 \\ 4 & -3 \\ 0 & 0 \end{pmatrix}$

2. (a) $\begin{pmatrix} 1 & 0 \\ 0 & -1 \end{pmatrix}$ (b) $\begin{pmatrix} 1 \\ 6 \\ 8 \\ 1 \end{pmatrix}$

3. $\begin{pmatrix} 1 & a & b & c \\ a & 1 & 0 & a \\ b & 0 & 1 & b \\ c & a & b & 1 \\ 0 & 0 & 0 & 1 \end{pmatrix}$

4. $(-3 \quad 1 \quad 6 \quad 0)$

$A = \begin{pmatrix} 2 & 3 \\ -1 & 4 \end{pmatrix} \qquad B = \begin{pmatrix} 1 & -1 \\ 3 & 0 \end{pmatrix} \qquad C = \begin{pmatrix} 1 & 0 \\ 0 & 1 \end{pmatrix}$

$D = \begin{pmatrix} -1 & 2 & 3 \\ 4 & 0 & 5 \end{pmatrix} \qquad E = \begin{pmatrix} 2 & 1 \\ 8 & -1 \\ 6 & 5 \end{pmatrix}$

$F = \begin{pmatrix} 5 & -1 \\ -4 & 0 \\ 2 & 3 \end{pmatrix} \qquad G = \begin{pmatrix} 0 & 0 \\ 0 & 0 \\ 0 & 0 \end{pmatrix}$

In Exercises 5–8, write the coefficient matrix and the augmented matrix for each system.

5. $\begin{cases} 2x + 3y + 4z = 10 \\ 5x + 6y + 7z = 9 \\ 8x + 9y + 10z = 8 \end{cases}$

6. $\begin{cases} 5x - y + z = 0 \\ 4y + 2z = 1 \\ 3x + y + z = -1 \end{cases}$

7. $\begin{cases} x + z + w = -1 \\ x + y + 2w = 0 \\ y + z + w = 1 \\ 2x - y - z = 2 \end{cases}$

8. $\begin{cases} 8x - 8y = 5 \\ x - y + z = 1 \end{cases}$

In Exercises 9–22, use matrices to solve each system of equations.

9. $\begin{cases} x - y + 2z = 7 \\ 3x + 2y - z = -10 \\ -x + 3y + z = -2 \end{cases}$

10. $\begin{cases} 2x - 3y + 4z = 14 \\ 3x - 2y + 2z = 12 \\ 4x + 5y - 5z = 16 \end{cases}$

11. $\begin{cases} x + z = -2 \\ -3x + 2y = 17 \\ x - y - z = -9 \end{cases}$

12. $\begin{cases} 5x + y + 10z = 23 \\ 4x + 2y - 10z = 76 \\ 3x - 4y = 18 \end{cases}$

13. $\begin{cases} x + y + z = -4 \\ 2x - 3y + z = -1 \\ 4x + 2y - 3z = 33 \end{cases}$

14. $\begin{cases} 2x + 3y - 4z = 7 \\ x - y + z = -\frac{3}{2} \\ 6x - 5y - 2z = -7 \end{cases}$

15. $\begin{cases} 3x - 2y + 6z = 0 \\ x + 3y + 20z = 15 \\ 10x - 11y - 10z = -9 \end{cases}$

16. $\begin{cases} 3A - 3B + C = 4 \\ 6A + 9B - 3C = -7 \\ A - 2B - 2C = -3 \end{cases}$

17. $\begin{cases} 4x - 3y + 3z = 2 \\ 5x + y - 4z = 1 \\ 9x - 2y - z = 3 \end{cases}$

18. $\begin{cases} 6x + y - z = -1 \\ -3x + 2y + 2z = 2 \\ 5y + 3z = 1 \end{cases}$

19. $\begin{cases} x - y + z + w = 6 \\ x + y - z + w = 4 \\ x + y + z - w = -2 \\ -x + y + z + w = 0 \end{cases}$

20. $\begin{cases} x + 2y - z - 2w = 5 \\ 2x + y + 2z + w = -7 \\ -2x - y - 3z - 2w = 10 \\ z + w = -3 \end{cases}$

21. $\begin{cases} 15A + 14B + 26C = 1 \\ 18A + 17B + 32C = -1 \\ 21A + 20B + 38C = 0 \end{cases}$

22. $\begin{cases} A + B + C + D + E = -1 \\ 3A - 2B - 2C + 3D + 2E = 13 \\ 3C + 4D - 4E = 7 \\ 5A - 4B + E = 30 \\ C - 2E = 3 \end{cases}$

In Exercises 23–50, the matrices A, B, C, D, E, F, and G are defined as follows:

In each exercise, carry out the indicated matrix operations if they are defined. If an operation is not defined, say so.

23. $A + B$ 　　24. $A - B$ 　　25. $2A + 2B$
26. $2(A + B)$ 　　27. AB 　　28. BA
29. AC 　　30. CA 　　31. $3D + E$
32. $E + F$ 　　33. $2F - 3G$ 　　34. DE
35. ED 　　36. DF 　　37. FD
38. $A + D$ 　　39. $G + A$ 　　40. DG
41. GD 　　42. $(A + B) + C$ 　　43. $A + (B + C)$
44. CD 　　45. DC 　　46. $5E - 3F$
47. $A^2(=AA)$ 　　48. A^2A 　　49. AA^2
50. C^2

51. Let
$$A = \begin{pmatrix} -1 & 3 & 4 \\ 3 & 2 & -3 \\ 9 & 1 & 6 \end{pmatrix} \qquad B = \begin{pmatrix} 7 & 0 & 1 \\ 0 & 0 & 3 \\ -1 & 2 & 4 \end{pmatrix}$$
$$C = \begin{pmatrix} 4 & 6 & 1 \\ 2 & 1 & 3 \\ -1 & -1 & 2 \end{pmatrix}$$
(a) Compute $A(B + C)$. 　　(b) Compute $AB + AC$.
(c) Compute $(AB)C$ 　　(d) Compute $A(BC)$.

B

52. Let $A = \begin{pmatrix} 1 & 2 \\ 3 & 4 \end{pmatrix}$ and $B = \begin{pmatrix} 5 & 6 \\ 7 & 8 \end{pmatrix}$. Let A^2 and B^2 denote the matrix products AA and BB, respectively. Compute each of the following.
(a) $(A + B)(A + B)$ 　　(b) $A^2 + 2AB + B^2$
(c) $A^2 + AB + BA + B^2$

53. Let $A = \begin{pmatrix} 3 & 5 \\ 7 & 9 \end{pmatrix}$ and $B = \begin{pmatrix} 2 & 4 \\ 6 & 8 \end{pmatrix}$. Compute each of the following.
(a) $A^2 - B^2$ 　　(b) $(A - B)(A + B)$
(c) $(A + B)(A - B)$ 　　(d) $A^2 + AB - BA - B^2$

54. Let
$$A = \begin{pmatrix} 1 & 0 \\ 0 & 1 \end{pmatrix} \qquad B = \begin{pmatrix} 1 & 0 \\ 0 & -1 \end{pmatrix}$$
$$C = \begin{pmatrix} -1 & 0 \\ 0 & 1 \end{pmatrix} \qquad D = \begin{pmatrix} -1 & 0 \\ 0 & -1 \end{pmatrix}$$

Complete the following multiplication table.

	A	B	C	D
A				
B			D	
C				
D				

Hint: In the second row, third column, D is the proper entry because (as you can check) $BC = D$.

55. In this exercise, let us agree to write the coordinates (x, y) of a point in the plane as the 2×1 matrix $\begin{pmatrix} x \\ y \end{pmatrix}$.

(a) Let $A = \begin{pmatrix} 1 & 0 \\ 0 & -1 \end{pmatrix}$ and $Z = \begin{pmatrix} x \\ y \end{pmatrix}$. Compute the matrix AZ. After computing AZ, observe that it represents the point obtained by reflecting $\begin{pmatrix} x \\ y \end{pmatrix}$ about the x-axis.

(b) Let $B = \begin{pmatrix} -1 & 0 \\ 0 & 1 \end{pmatrix}$ and $Z = \begin{pmatrix} x \\ y \end{pmatrix}$. Compute the matrix BZ. After computing BZ, observe that it represents the point obtained by reflecting $\begin{pmatrix} x \\ y \end{pmatrix}$ about the y-axis.

(c) Let A, B, and Z represent the matrices defined in parts (a) and (b). Compute the matrix $(AB)Z$, and then interpret it in terms of reflection about the axes.

56. In this exercise, we continue to explore some of the connections between matrices and geometry. As in Exercise 55, we will use 2×1 matrices to specify the coordinates of points in the plane. Let P, S, and T be the matrices defined as follows:

$$P = \begin{pmatrix} \cos x \\ \sin x \end{pmatrix} \qquad S = \begin{pmatrix} \cos \theta & -\sin \theta \\ \sin \theta & \cos \theta \end{pmatrix}$$

$$T = \begin{pmatrix} \cos \beta & -\sin \beta \\ \sin \beta & \cos \beta \end{pmatrix}$$

Notice that the point P lies on the unit circle.

(a) Compute the matrix SP. After computing SP,

observe that it represents the point on the unit circle obtained by rotating P (about the origin) through an angle θ.

(b) Show that

$$ST = TS = \begin{pmatrix} \cos(\theta + \beta) & -\sin(\theta + \beta) \\ \sin(\theta + \beta) & \cos(\theta + \beta) \end{pmatrix}$$

(c) Compute $(ST)P$. What is the angle through which P is rotated?

57. A function f is defined as follows. The domain of f is the set of all 2×2 matrices (with real entries), and if $A = \begin{pmatrix} a & b \\ c & d \end{pmatrix}$, then $f(A) = ad - bc$.

(a) Let $A = \begin{pmatrix} 1 & 2 \\ 3 & 4 \end{pmatrix}$ and $B = \begin{pmatrix} 3 & -1 \\ 5 & 8 \end{pmatrix}$. Compute $f(A)$, $f(B)$, and $f(AB)$. Is it true, in this case, that $f(A) \cdot f(B) = f(AB)$?

(b) Let $A = \begin{pmatrix} a & b \\ c & d \end{pmatrix}$ and $B = \begin{pmatrix} e & f \\ g & h \end{pmatrix}$. Show that $f(A) \cdot f(B) = f(AB)$.

58. The **trace** of a 2×2 matrix $\begin{pmatrix} a & b \\ c & d \end{pmatrix}$ is defined by

$$\mathrm{tr}\begin{pmatrix} a & b \\ c & d \end{pmatrix} = a + d$$

(a) If $A = \begin{pmatrix} 1 & 2 \\ 3 & 4 \end{pmatrix}$ and $B = \begin{pmatrix} 5 & 6 \\ 7 & 8 \end{pmatrix}$, verify that $\mathrm{tr}(A + B) = \mathrm{tr}\, A + \mathrm{tr}\, B$.

(b) If $A = \begin{pmatrix} a & b \\ c & d \end{pmatrix}$ and $B = \begin{pmatrix} e & f \\ g & h \end{pmatrix}$, show that $\mathrm{tr}(A + B) = \mathrm{tr}\, A + \mathrm{tr}\, B$.

59. Let $A = \begin{pmatrix} a & b \\ c & d \end{pmatrix}$. The **transpose** of A is the matrix denoted by A^{T} and defined by $A^{\mathrm{T}} = \begin{pmatrix} a & c \\ b & d \end{pmatrix}$ In other words, A^{T} is obtained by switching the columns and rows of A. Show that the following equations hold for all 2×2 matrices A and B.

(a) $(A + B)^{\mathrm{T}} = A^{\mathrm{T}} + B^{\mathrm{T}}$ (b) $(A^{\mathrm{T}})^{\mathrm{T}} = A$

(c) $(AB)^{\mathrm{T}} = B^{\mathrm{T}}A^{\mathrm{T}}$

60. Find an example of two 2×2 matrices A and B for which $AB = \mathbf{0}$ but neither A nor B is $\mathbf{0}$.

9.4 THE INVERSE OF A SQUARE MATRIX

We have noted Cayley's work in analytic geometry, especially in connection with the use of determinants; but [Arthur] Cayley [1821–1895] also was one of the first men to study matrices, another instance of the British concern for form and structure in algebra.

Carl B. Boyer in *A History of Mathematics,* 2d ed., revised by Uta C. Merzback (New York: John Wiley and Sons, 1991)

A matrix in which there are the same number of rows as columns is called a **square matrix**. So, two examples of square matrices are

$$A = \begin{pmatrix} 1 & 2 \\ 3 & 4 \end{pmatrix} \quad \text{and} \quad B = \begin{pmatrix} -5 & 6 & 7 \\ \frac{1}{2} & 0 & 1 \\ 8 & 4 & -3 \end{pmatrix}$$

The matrix A is said to be a square matrix of **order two** (or, more simply, a 2×2 matrix); B is a square matrix of **order three** (that is, a 3×3 matrix). We will first present the concepts and techniques of this section in terms of square matrices of order two. After that, we'll show how the ideas carry over to larger square matrices.

We begin by defining a special matrix, I_2:

$$I_2 = \begin{pmatrix} 1 & 0 \\ 0 & 1 \end{pmatrix}$$

The matrix I_2 has the following property: For every 2×2 matrix $A = \begin{pmatrix} a & b \\ c & d \end{pmatrix}$, we have (as you can easily verify)

$$\begin{pmatrix} a & b \\ c & d \end{pmatrix}\begin{pmatrix} 1 & 0 \\ 0 & 1 \end{pmatrix} = \begin{pmatrix} a & b \\ c & d \end{pmatrix} \quad \text{and} \quad \begin{pmatrix} 1 & 0 \\ 0 & 1 \end{pmatrix}\begin{pmatrix} a & b \\ c & d \end{pmatrix} = \begin{pmatrix} a & b \\ c & d \end{pmatrix}$$

In other words,

$$AI_2 = A \quad \text{and} \quad I_2A = A$$

This shows that the matrix I_2 plays the same role in the multiplication of 2×2 matrices as does the number 1 in the multiplication of real numbers. For this reason, I_2 is called the **multiplicative identity** for square matrices of order two.

If we have two real numbers a and b such that $ab = 1$, then we say that a and b are (multiplicative) inverses. In the box that follows, we apply this same terminology to matrices.

DEFINITION   The Inverse of a 2 × 2 Matrix

If A and B are 2×2 matrices such that

$$AB = I_2 \quad \text{and} \quad BA = I_2$$

then A and B are said to be **inverses** of one another.

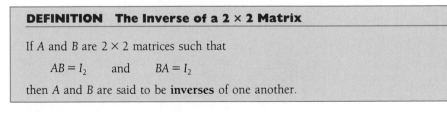

EXAMPLE I   Find the inverse of $A = \begin{pmatrix} 1 & 2 \\ 3 & 4 \end{pmatrix}$.

Solution   We need to find numbers a, b, c, and d such that the following two matrix equations are valid.

$$\begin{pmatrix} 1 & 2 \\ 3 & 4 \end{pmatrix}\begin{pmatrix} a & b \\ c & d \end{pmatrix} = \begin{pmatrix} 1 & 0 \\ 0 & 1 \end{pmatrix} \tag{1}$$

$$\begin{pmatrix} a & b \\ c & d \end{pmatrix}\begin{pmatrix} 1 & 2 \\ 3 & 4 \end{pmatrix} = \begin{pmatrix} 1 & 0 \\ 0 & 1 \end{pmatrix} \tag{2}$$

From equation (1), we have

$$\begin{pmatrix} a + 2c & b + 2d \\ 3a + 4c & 3b + 4d \end{pmatrix} = \begin{pmatrix} 1 & 0 \\ 0 & 1 \end{pmatrix}$$

For this last equation to be valid, the corresponding entries in the two matrices must be equal. (Why?) Consequently, we obtain four equations, two involving a and c, and two involving b and d:

$$\begin{cases} a + 2c = 1 \\ 3a + 4c = 0 \end{cases} \qquad \begin{cases} b + 2d = 0 \\ 3b + 4d = 1 \end{cases}$$

As Exercise 32(a) asks you to show, the solution to the first system is $a = -2$ and $c = \frac{3}{2}$; and for the second system, $b = 1$ and $d = -\frac{1}{2}$. Furthermore (as you can check), these same values are obtained for a, b, c, and d if we begin with matrix equation (2) rather than (1). Thus, the required inverse matrix is

$$\begin{pmatrix} -2 & 1 \\ \frac{3}{2} & -\frac{1}{2} \end{pmatrix}$$

Exercise 32(b) asks you to carry out the matrix multiplication to confirm that we indeed have

$$\begin{pmatrix} 1 & 2 \\ 3 & 4 \end{pmatrix}\begin{pmatrix} -2 & 1 \\ \frac{3}{2} & -\frac{1}{2} \end{pmatrix} = \begin{pmatrix} -2 & 1 \\ \frac{3}{2} & -\frac{1}{2} \end{pmatrix}\begin{pmatrix} 1 & 2 \\ 3 & 4 \end{pmatrix} = \begin{pmatrix} 1 & 0 \\ 0 & 1 \end{pmatrix} \qquad \blacksquare\blacksquare\blacksquare$$

In Example 1 we found the inverse of a square matrix by solving systems of equations. However, as you've seen in previous sections, there are systems of equations that do not have solutions. In the present context, this implies that there are matrices that do not have inverses. For example (as Exercise 33 asks you to show), the matrix $\begin{pmatrix} 2 & 5 \\ 6 & 15 \end{pmatrix}$ does not have an inverse. A matrix that does not have an inverse is called a **singular matrix**. If a matrix does possess an inverse, we say that the matrix is **nonsingular**. So, according to the result in Example 1, the matrix $\begin{pmatrix} 1 & 2 \\ 3 & 4 \end{pmatrix}$ is nonsingular.

It can be proven that if a square matrix has an inverse, then that inverse is unique. (In other words, you can never find two different inverses for the same matrix.) The inverse of a nonsingular matrix A is denoted by A^{-1} (read A *inverse*).

There are several methods that can be used to compute the inverse of a matrix (or to show that there is no inverse). One method involves working with systems of equations, as in Example 1. A more efficient method, however, is the following. (We'll demonstrate the method with an example, but we won't give a formal proof of its validity.)

Suppose that we want to compute the inverse (if it exists) for the matrix

$$A = \begin{pmatrix} 6 & 2 \\ 13 & 4 \end{pmatrix}$$

We begin by writing a larger matrix, formed by merging A and I_2 as follows.

A convenient abbreviation for this matrix is $(A \mid I_2)$. Now we carry out elementary row operations on this matrix until one of the following two situations occurs:

1. The matrix takes on the form

$$\begin{pmatrix} 1 & 0 & \vdots & a & b \\ 0 & 1 & \vdots & c & d \end{pmatrix}$$

$I_2 \longrightarrow$

In this case, $A^{-1} = \begin{pmatrix} a & b \\ c & d \end{pmatrix}$.

2. One of the rows to the left of the dashed line consists entirely of zeros. In this case, A^{-1} does not exist.

The calculations for our example then run as follows.

$$\begin{pmatrix} 6 & 2 & \vdots & 1 & 0 \\ 13 & 4 & \vdots & 0 & 1 \end{pmatrix} \xrightarrow{\;-2R_1 + R_2\;} \begin{pmatrix} 6 & 2 & \vdots & 1 & 0 \\ 1 & 0 & \vdots & -2 & 1 \end{pmatrix}$$

$$\xrightarrow{\;R_1 \leftrightarrow R_2\;} \begin{pmatrix} 1 & 0 & \vdots & -2 & 1 \\ 6 & 2 & \vdots & 1 & 0 \end{pmatrix}$$

$$\xrightarrow{\;-6R_1 + R_2\;} \begin{pmatrix} 1 & 0 & \vdots & -2 & 1 \\ 0 & 2 & \vdots & 13 & -6 \end{pmatrix}$$

$$\xrightarrow{\;\frac{1}{2}R_2\;} \begin{pmatrix} 1 & 0 & \vdots & -2 & 1 \\ 0 & 1 & \vdots & \frac{13}{2} & -3 \end{pmatrix}$$

The inverse matrix can now be read off:

$$A^{-1} = \begin{pmatrix} -2 & 1 \\ \frac{13}{2} & -3 \end{pmatrix}$$

(You should verify for yourself that we indeed have $AA^{-1} = I_2$ and $A^{-1}A = I_2$.)

Inverse matrices can be used in solving certain systems of equations in which the number of unknowns is the same as the number of equations. Before explaining how this works, we describe how a system of equations can be written in matrix form. Consider the system

$$\begin{cases} x + 2y = 8 \\ 3x + 4y = 6 \end{cases} \tag{3}$$

As defined in the previous section, the coefficient matrix A for this system is

$$A = \begin{pmatrix} 1 & 2 \\ 3 & 4 \end{pmatrix}$$

Now we define two matrices X and B:

$$X = \begin{pmatrix} x \\ y \end{pmatrix} \qquad B = \begin{pmatrix} 8 \\ 6 \end{pmatrix}$$

Then system (3) can be written as a single matrix equation:

$$AX = B \qquad\qquad (4)$$

To see why this is so, we expand equation (4) to obtain

$$\begin{pmatrix} 1 & 2 \\ 3 & 4 \end{pmatrix}\begin{pmatrix} x \\ y \end{pmatrix} = \begin{pmatrix} 8 \\ 6 \end{pmatrix} \qquad \text{using the definitions of } A, X, \text{ and } B$$

$$\begin{pmatrix} x + 2y \\ 3x + 4y \end{pmatrix} = \begin{pmatrix} 8 \\ 6 \end{pmatrix} \qquad \text{carrying out the matrix multiplication}$$

By equating the corresponding entries of the matrices in this last equation, we obtain $x + 2y = 8$ and $3x + 4y = 6$, as given initially in system (3).

EXAMPLE 2 Use an inverse matrix to solve the system

$$\begin{cases} x + 2y = 8 \\ 3x + 4y = 6 \end{cases}$$

Solution As explained just prior to this example, the matrix form for this system is

$$AX = B \qquad\qquad (5)$$

where

$$A = \begin{pmatrix} 1 & 2 \\ 3 & 4 \end{pmatrix} \qquad X = \begin{pmatrix} x \\ y \end{pmatrix} \qquad B = \begin{pmatrix} 8 \\ 6 \end{pmatrix}$$

From Example 1, we know that

$$A^{-1} = \begin{pmatrix} -2 & 1 \\ \frac{3}{2} & -\frac{1}{2} \end{pmatrix}$$

Now we multiply both sides of equation (5) by A^{-1} to obtain

$$A^{-1}(AX) = A^{-1}B$$
$$(A^{-1}A)X = A^{-1}B \qquad \text{Matrix multiplication is associative.}$$
$$I_2 X = A^{-1}B$$
$$X = A^{-1}B \qquad \text{(Why?)}$$

Substituting the actual matrices into this last equation, we have

$$\begin{pmatrix} x \\ y \end{pmatrix} = \begin{pmatrix} -2 & 1 \\ \frac{3}{2} & -\frac{1}{2} \end{pmatrix}\begin{pmatrix} 8 \\ 6 \end{pmatrix} = \begin{pmatrix} -10 \\ 9 \end{pmatrix}$$

Therefore $x = -10$ and $y = 9$, as required.

All of the ideas we have discussed for square matrices of order two can be carried over rather directly to larger square matrices. It is easy to check that the following matrices, I_3 and I_4, are the multiplicative identities for 3×3 and 4×4 matrices, respectively.

$$I_3 = \begin{pmatrix} 1 & 0 & 0 \\ 0 & 1 & 0 \\ 0 & 0 & 1 \end{pmatrix} \qquad I_4 = \begin{pmatrix} 1 & 0 & 0 & 0 \\ 0 & 1 & 0 & 0 \\ 0 & 0 & 1 & 0 \\ 0 & 0 & 0 & 1 \end{pmatrix}$$

These identity matrices are described by saying that they have ones down the **main diagonal** and zeros everywhere else. (Larger identity matrices can be defined following this same pattern.) Sometimes, when it is clear from the context or as a matter of convenience, we'll omit the subscript and denote the appropriately sized identity matrix simply by I. (This is done in the box that follows.)

PROPERTY SUMMARY   THE INVERSE OF A SQUARE MATRIX

1. Suppose that A and B are square matrices of the same size, and let I denote the identity matrix of the same size. Then A and B are said to be **inverses** of one another provided

$$AB = I \qquad \text{and} \qquad BA = I$$

2. It can be shown that every square matrix has at most one inverse.

3. A square matrix that has an inverse is said to be **nonsingular** (or **invertible**). A square matrix that does not have an inverse is called **singular**.

4. If the matrix A is nonsingular, then the inverse of A is denoted by A^{-1}. In this case, we have

$$AA^{-1} = I \qquad \text{and} \qquad A^{-1}A = I$$

EXAMPLE 3 Let $A = \begin{pmatrix} 5 & 0 & 2 \\ 2 & 2 & 1 \\ -3 & 1 & -1 \end{pmatrix}$. Use the elementary row operations to compute A^{-1}, if it exists.

Solution Following the method that we described for the 2×2 case, we first write down the matrix $(A \mid I_3)$ formed by merging A and I_3:

$$\begin{pmatrix} 5 & 0 & 2 & \vdots & 1 & 0 & 0 \\ 2 & 2 & 1 & \vdots & 0 & 1 & 0 \\ -3 & 1 & -1 & \vdots & 0 & 0 & 1 \end{pmatrix}$$

Now we carry out the elementary row operations, trying to obtain I_3 to the left of the dashed line. We have

$$\left(\begin{array}{ccc|ccc} 5 & 0 & 2 & 1 & 0 & 0 \\ 2 & 2 & 1 & 0 & 1 & 0 \\ -3 & 1 & -1 & 0 & 0 & 1 \end{array}\right) \xrightarrow{-2R_2 + R_1} \left(\begin{array}{ccc|ccc} 1 & -4 & 0 & 1 & -2 & 0 \\ 2 & 2 & 1 & 0 & 1 & 0 \\ -3 & 1 & -1 & 0 & 0 & 1 \end{array}\right)$$

$$\xrightarrow[\substack{3R_1 + R_3}]{-2R_1 + R_2} \left(\begin{array}{ccc|ccc} 1 & -4 & 0 & 1 & -2 & 0 \\ 0 & 10 & 1 & -2 & 5 & 0 \\ 0 & -11 & -1 & 3 & -6 & 1 \end{array}\right) \xrightarrow{1R_3 + R_2} \left(\begin{array}{ccc|ccc} 1 & -4 & 0 & 1 & -2 & 0 \\ 0 & -1 & 0 & 1 & -1 & 1 \\ 0 & -11 & -1 & 3 & -6 & 1 \end{array}\right)$$

$$\xrightarrow{(-1)R_2} \left(\begin{array}{ccc|ccc} 1 & -4 & 0 & 1 & -2 & 0 \\ 0 & 1 & 0 & -1 & 1 & -1 \\ 0 & -11 & -1 & 3 & -6 & 1 \end{array}\right) \xrightarrow[\substack{11R_2 + R_3}]{4R_2 + R_1} \left(\begin{array}{ccc|ccc} 1 & 0 & 0 & -3 & 2 & -4 \\ 0 & 1 & 0 & -1 & 1 & -1 \\ 0 & 0 & -1 & -8 & 5 & -10 \end{array}\right)$$

$$\xrightarrow{(-1)R_3} \left(\begin{array}{ccc|ccc} 1 & 0 & 0 & -3 & 2 & -4 \\ 0 & 1 & 0 & -1 & 1 & -1 \\ 0 & 0 & 1 & 8 & -5 & 10 \end{array}\right)$$

The required inverse is therefore

$$A^{-1} = \left(\begin{array}{ccc} -3 & 2 & -4 \\ -1 & 1 & -1 \\ 8 & -5 & 10 \end{array}\right)$$

EXERCISE SET 9.4

A

In Exercises 1–4, the matrices A, B, C, and D are defined as follows.

$$A = \left(\begin{array}{cc} 4 & -1 \\ -5 & 2 \end{array}\right) \qquad B = \left(\begin{array}{cc} \frac{1}{2} & 5 \\ 3 & 1 \end{array}\right)$$

$$C = \left(\begin{array}{ccc} 3 & 0 & -2 \\ 0 & 5 & 6 \\ 1 & 4 & -7 \end{array}\right) \qquad D = \left(\begin{array}{ccc} 1 & 2 & 3 \\ 4 & 5 & 6 \\ 7 & 8 & 9 \end{array}\right)$$

1. Compute AI_2 and I_2A to verify that $AI_2 = I_2A = A$.
2. Compute BI_2 and I_2B to verify that $BI_2 = I_2B = B$.
3. Compute CI_3 and I_3C to verify that $CI_3 = I_3C = C$.
4. Compute DI_3 and I_3D to verify that $DI_3 = I_3D = D$.

In Exercises 5–12, compute A^{-1}, if it exists, using the method of Example 1.

5. $A = \left(\begin{array}{cc} 7 & 9 \\ 4 & 5 \end{array}\right)$ 6. $A = \left(\begin{array}{cc} 3 & -8 \\ 2 & -5 \end{array}\right)$

7. $A = \left(\begin{array}{cc} -3 & 1 \\ 5 & 6 \end{array}\right)$ 8. $A = \left(\begin{array}{cc} -4 & 0 \\ 9 & 3 \end{array}\right)$

9. $A = \left(\begin{array}{cc} -2 & 3 \\ -4 & 6 \end{array}\right)$ 10. $A = \left(\begin{array}{cc} \frac{5}{3} & -2 \\ -\frac{2}{3} & 1 \end{array}\right)$

11. $A = \left(\begin{array}{cc} \frac{1}{3} & \frac{1}{3} \\ -\frac{1}{9} & \frac{2}{9} \end{array}\right)$ 12. $A = \left(\begin{array}{cc} -3 & 7 \\ 12 & -28 \end{array}\right)$

In Exercises 13–26, compute the inverse matrix, if it exists, using elementary row operations (as shown in Example 3).

13. $\left(\begin{array}{cc} 2 & 1 \\ 3 & 2 \end{array}\right)$ 14. $\left(\begin{array}{cc} -6 & 5 \\ 18 & -15 \end{array}\right)$

15. $\left(\begin{array}{cc} 0 & -11 \\ 1 & 6 \end{array}\right)$ 16. $\left(\begin{array}{cc} -2 & 13 \\ -4 & 25 \end{array}\right)$

17. $\left(\begin{array}{cc} \frac{2}{3} & -\frac{1}{4} \\ -8 & 3 \end{array}\right)$ 18. $\left(\begin{array}{cc} -\frac{2}{5} & \frac{1}{3} \\ -6 & 5 \end{array}\right)$

19. $\left(\begin{array}{ccc} -5 & 4 & -3 \\ 10 & -7 & 6 \\ 8 & -6 & 5 \end{array}\right)$ 20. $\left(\begin{array}{ccc} 1 & 0 & -2 \\ -3 & -1 & 6 \\ 2 & 1 & -5 \end{array}\right)$

21. $\left(\begin{array}{ccc} 1 & 2 & -1 \\ 0 & 3 & 0 \\ -4 & 0 & 5 \end{array}\right)$ 22. $\left(\begin{array}{ccc} 1 & -4 & -8 \\ 1 & 2 & 5 \\ 1 & 1 & 3 \end{array}\right)$

23. $\left(\begin{array}{ccc} -7 & 5 & 3 \\ 3 & -2 & -2 \\ 3 & -2 & -1 \end{array}\right)$ 24. $\left(\begin{array}{ccc} 2 & -1 & -1 \\ 1 & 0 & -1 \\ -2 & 1 & 2 \end{array}\right)$

25. $\left(\begin{array}{ccc} 1 & 2 & 3 \\ 4 & 5 & 6 \\ 7 & 8 & 9 \end{array}\right)$ 26. $\left(\begin{array}{ccc} 2 & 1 & 3 \\ 4 & 5 & -7 \\ 2 & 1 & 3 \end{array}\right)$

27. If $A = \begin{pmatrix} 3 & 8 \\ 4 & 11 \end{pmatrix}$, then $A^{-1} = \begin{pmatrix} 11 & -8 \\ -4 & 3 \end{pmatrix}$. Use this fact and the method of Example 2 to solve the following systems.

 (a) $\begin{cases} 3x + 8y = 5 \\ 4x + 11y = 7 \end{cases}$ (b) $\begin{cases} 3x + 8y = -12 \\ 4x + 11y = 0 \end{cases}$

28. If $A = \begin{pmatrix} 3 & -7 \\ 4 & -9 \end{pmatrix}$, then $A^{-1} = \begin{pmatrix} -9 & 7 \\ -4 & 3 \end{pmatrix}$. Use this fact and the method of Example 2 to solve the following systems.

 (a) $\begin{cases} 3x - 7y = 30 \\ 4x - 9y = 39 \end{cases}$ (b) $\begin{cases} 3x - 7y = -45 \\ 4x - 9y = -71 \end{cases}$

29. The inverse of the matrix $A = \begin{pmatrix} 3 & 2 & 6 \\ 1 & 1 & 2 \\ 2 & 2 & 5 \end{pmatrix}$ is

 $A^{-1} = \begin{pmatrix} 1 & 2 & -2 \\ -1 & 3 & 0 \\ 0 & -2 & 1 \end{pmatrix}$. Use this fact to solve the following systems.

 (a) $\begin{cases} 3x + 2y + 6z = 28 \\ x + y + 2z = 9 \\ 2x + 2y + 5z = 22 \end{cases}$

 (b) $\begin{cases} 3x + 2y + 6z = -7 \\ x + y + 2z = -2 \\ 2x + 2y + 5z = -6 \end{cases}$

30. The inverse of the matrix $A = \begin{pmatrix} 1 & -1 & 1 \\ 2 & -3 & 2 \\ -4 & 6 & 1 \end{pmatrix}$ is

 $A^{-1} = \begin{pmatrix} 3 & -\frac{7}{5} & -\frac{1}{5} \\ 2 & -1 & 0 \\ 0 & \frac{2}{5} & \frac{1}{5} \end{pmatrix}$. Use this fact to solve the following systems.

 (a) $\begin{cases} x - y + z = 5 \\ 2x - 3y + 2z = -15 \\ -4x + 6y + z = 25 \end{cases}$

 (b) $\begin{cases} x - y + z = 1 \\ 2x - 3y + 2z = -2 \\ -4x + 6y + z = 0 \end{cases}$

B

31. Let $A = \begin{pmatrix} 1 & -6 & 3 \\ 2 & -7 & 3 \\ 4 & -12 & 5 \end{pmatrix}$.

 (a) Compute the matrix product AA. What do you observe?
 (b) Use the result in part (a) to solve the following system.

 $\begin{cases} x - 6y + 3z = \frac{19}{2} \\ 2x - 7y + 3z = 11 \\ 4x - 12y + 5z = 19 \end{cases}$

32. (a) Solve the following two systems and then check to see that your results agree with those given in Example 1.

 $\begin{cases} a + 2c = 1 \\ 3a + 4c = 0 \end{cases}$ $\begin{cases} b + 2d = 0 \\ 3b + 4d = 1 \end{cases}$

 (b) At the end of Example 1, it is asserted that

 $$\begin{pmatrix} 1 & 2 \\ 3 & 4 \end{pmatrix}\begin{pmatrix} -2 & 1 \\ \frac{3}{2} & -\frac{1}{2} \end{pmatrix} = \begin{pmatrix} -2 & 1 \\ \frac{3}{2} & -\frac{1}{2} \end{pmatrix}\begin{pmatrix} 1 & 2 \\ 3 & 4 \end{pmatrix}$$
 $$= \begin{pmatrix} 1 & 0 \\ 0 & 1 \end{pmatrix}$$

 Carry out the indicated matrix multiplications to verify that these equations are valid.

33. Use the technique in Example 1 to show that the matrix $\begin{pmatrix} 2 & 5 \\ 6 & 15 \end{pmatrix}$ does not have an inverse.

34. Use the elementary row operations (as in Example 3) to find the inverse of the following matrix.

 $\begin{pmatrix} 1 & 1 & 1 & 1 \\ 1 & 2 & 3 & 4 \\ 1 & 3 & 6 & 10 \\ 1 & 4 & 10 & 20 \end{pmatrix}$

35. Let $A = \begin{pmatrix} 2 & 3 \\ 4 & 5 \end{pmatrix}$ and $B = \begin{pmatrix} 7 & 8 \\ 6 & 7 \end{pmatrix}$.
 (a) Compute A^{-1}, B^{-1}, and $B^{-1}A^{-1}$.
 (b) Compute $(AB)^{-1}$. What do you observe?

36. Let $A = \begin{pmatrix} a & b \\ c & d \end{pmatrix}$. Compute A^{-1}. (Assume that $ad - bc \neq 0$.)

9.5 DETERMINANTS AND CRAMER'S RULE

The idea of the determinant ... dates back essentially to Leibniz (1693), the Swiss mathematician Gabriel Cramer (1750), and Lagrange (1773); the name is due to Cauchy (1812). Y. Mikami has pointed out that the Japanese mathematician Seki Kōwa had the idea of a determinant sometime before 1683.

Dirk J. Struik in *A Concise History of Mathematics,* 4th ed. (New York: Dover Publications, 1987)

As you saw in the previous section, square matrices and their inverses can be used to solve certain systems of equations. In this section, we are going to associate a number with each square matrix; this number is called the **determinant** of the matrix. As you'll see, this too has an application in solving systems of equations.

The determinant of a matrix A can be denoted by det A or $|A|$. Determinants are also denoted simply by replacing the parentheses of matrix notation with vertical lines. Thus, three examples of determinants are

$$\begin{vmatrix} 1 & 2 \\ 3 & 4 \end{vmatrix} \qquad \begin{vmatrix} 1 & 2 & 3 \\ 4 & 5 & 6 \\ 7 & 8 & 9 \end{vmatrix} \qquad \begin{vmatrix} 3 & 7 & 8 & 9 \\ 5 & 6 & 4 & 3 \\ -9 & 9 & 0 & 1 \\ 1 & 3 & -2 & 1 \end{vmatrix}$$

A determinant with n rows and n columns is said to be an **nth-order determinant.** Therefore, the determinants we've just written are, respectively, second-, third-, and fourth-order determinants. As with matrices, we speak of the numbers in a determinant as its **entries.** We also number the rows and the columns of a determinant as we do with matrices. However, unlike matrices, each determinant has a definite value. The value of a second-order determinant is defined as follows.

DEFINITION   2 × 2 Determinant

$$\begin{vmatrix} a & b \\ c & d \end{vmatrix} = ad - bc$$

Table 1 illustrates how this definition is used to evaluate, or *expand*, a second-order determinant.

TABLE 1

Determinant	Value of Determinant
$\begin{vmatrix} 3 & 7 \\ 5 & 10 \end{vmatrix}$	$\begin{vmatrix} 3 & 7 \\ 5 & 10 \end{vmatrix} = 3(10) - 7(5) = 30 - 35 = -5$
$\begin{vmatrix} 3 & -7 \\ 5 & 10 \end{vmatrix}$	$\begin{vmatrix} 3 & -7 \\ 5 & 10 \end{vmatrix} = 3(10) - (-7)(5) = 30 + 35 = 65$
$\begin{vmatrix} a & a^3 \\ 1 & a^2 \end{vmatrix}$	$\begin{vmatrix} a & a^3 \\ 1 & a^2 \end{vmatrix} = a(a^2) - a^3(1) = a^3 - a^3 = 0$

In general, the value of an nth-order determinant $(n > 2)$ is defined in terms of certain determinants of order $n - 1$. For instance, the value of a third-order determinant is defined in terms of second-order determinants. We'll use the following example to introduce the necessary terminology here:

$$\begin{vmatrix} 8 & 3 & 5 \\ 2 & 4 & 6 \\ 9 & 1 & 7 \end{vmatrix}$$

Pick a given entry—say, 8—and imagine crossing out all of the entries occupying the same row or the same column as 8.

$$\begin{vmatrix} 8 & 3 & 5 \\ 2 & 4 & 6 \\ 9 & 1 & 7 \end{vmatrix}$$

Then we are left with the second-order determinant $\begin{vmatrix} 4 & 6 \\ 1 & 7 \end{vmatrix}$. This second-order determinant is called the **minor** of the entry 8. Similarly, to find the minor of the entry 6 in the original determinant, imagine crossing out all of the entries that occupy the same row or the same column as 6.

$$\begin{vmatrix} 8 & 3 & 5 \\ 2 & 4 & 6 \\ 9 & 1 & 7 \end{vmatrix}$$

We are left with the second-order determinant $\begin{vmatrix} 8 & 3 \\ 9 & 1 \end{vmatrix}$, which by definition is the minor of the entry 6. In the same manner, *the minor of any element is the determinant obtained by crossing out the entries occupying the same row or column as the given element.*

$$\begin{vmatrix} + & - & + \\ - & + & - \\ + & - & + \end{vmatrix}$$

FIGURE 1

Closely related to the minor of an entry in a determinant is the cofactor of that entry. The **cofactor** of an entry is defined as the minor multiplied by $+1$ or -1, according to the scheme displayed in Figure 1.

After looking at an example, we'll give a more formal rule for computing cofactors, one that will not rely on a figure and that will also apply to larger determinants.

▌▌ EXAMPLE 1 Consider the determinant

$$\begin{vmatrix} 1 & 2 & 3 \\ 4 & 5 & 6 \\ 7 & 8 & 9 \end{vmatrix}$$

Compute the minor and the cofactor of the entry 4.

Solution By definition, we have

$$\text{minor of } 4 = \begin{vmatrix} 2 & 3 \\ 8 & 9 \end{vmatrix} = 18 - 24 = -6$$

Thus, the minor of the entry 4 is -6. To compute the cofactor of 4, we first notice that 4 is located in the second row and first column of the given determinant. Upon checking the corresponding position in Figure 1, we see a negative sign. Therefore, we have by definition

$$\text{cofactor of } 4 = (-1)(\text{minor of } 4)$$
$$= (-1)(-6) = 6$$

The cofactor of 4 is therefore 6.   ▪▪▪

The following rule tells us how cofactors can be computed without relying on Figure 1. For reference, we also restate the definition of a minor. (You should verify for yourself that this rule yields results that are consistent with Figure 1.)

DEFINITION   Minors and Cofactors

The **minor** of an entry b in a determinant is the determinant formed by suppressing the entries in the row and in the column in which b appears.

Suppose that the entry b is in the ith row and the jth column. Then the **cofactor** of b is given by the expression

$$(-1)^{i+j}(\text{minor of } b)$$

We are now prepared to state the definition that tells us how to evaluate a third-order determinant. Actually, as you'll see later, the definition is quite general and may be applied to determinants of any size.

DEFINITION   The Value of a Determinant

Multiply each entry in the first row of the determinant by its cofactor and then add the results. The value of the determinant is defined to be this sum.

To see how this definition is used, let us evaluate the determinant

$$\begin{vmatrix} 1 & 2 & 3 \\ 4 & 5 & 6 \\ 7 & 8 & 9 \end{vmatrix}$$

The definition tells us to multiply each entry in the first row by its cofactor and then add the results. Carrying out this procedure, we have

$$\begin{vmatrix} 1 & 2 & 3 \\ 4 & 5 & 6 \\ 7 & 8 & 9 \end{vmatrix} = 1\begin{vmatrix} 5 & 6 \\ 8 & 9 \end{vmatrix} - 2\begin{vmatrix} 4 & 6 \\ 7 & 9 \end{vmatrix} + 3\begin{vmatrix} 4 & 5 \\ 7 & 8 \end{vmatrix}$$
$$= 1(45 - 48) - 2(36 - 42) + 3(32 - 35)$$
$$= 1(-3) - 2(-6) + 3(-3)$$
$$= 0$$

So the value of this particular determinant is zero. The procedure we've used here is referred to as **expanding the determinant along its first row**. The following theorem (stated here without proof) tells us that the value of a determinant can be obtained by expanding along any row or along any column; the results are the same in all cases.

Theorem

Select any row or any column in a determinant and multiply each element in that row or column by its cofactor; add the results. The number obtained will be the value of the determinant. (In other words, the number obtained will be the same as that obtained by expanding the determinant along its first row.)

According to this theorem, we could have evaluated the determinant

$$\begin{vmatrix} 1 & 2 & 3 \\ 4 & 5 & 6 \\ 7 & 8 & 9 \end{vmatrix}$$

by expanding it along any row or any column. Let's expand it along the second column and check to see that the result agrees with the value obtained earlier. Expanding along the second column, we have

$$\begin{vmatrix} 1 & 2 & 3 \\ 4 & 5 & 6 \\ 7 & 8 & 9 \end{vmatrix} = -2\begin{vmatrix} 4 & 6 \\ 7 & 9 \end{vmatrix} + 5\begin{vmatrix} 1 & 3 \\ 7 & 9 \end{vmatrix} - 8\begin{vmatrix} 1 & 3 \\ 4 & 6 \end{vmatrix}$$

$$= -2(36 - 42) + 5(9 - 21) - 8(6 - 12)$$

$$= -2(-6) + 5(-12) - 8(-6)$$

$$= 12 - 60 + 48 = 0 \qquad \text{as obtained previously}$$

We would have obtained the same result had we chosen to begin with any other row or column. (Exercise 13 at the end of this section asks you to verify this.)

There are three basic rules that make it easier to evaluate determinants. These are summarized in the box that follows. (Suggestions for proving these can be found in the exercises.)

PROPERTY SUMMARY   RULES FOR MANIPULATING DETERMINANTS

EXAMPLES

1. If each entry in a given row is multiplied by the constant k, then the value of the determinant is multiplied by k. This is true for columns, also.

$$10\begin{vmatrix} 1 & 3 & 4 \\ 1 & 2 & 3 \\ 4 & 5 & 6 \end{vmatrix} = \begin{vmatrix} 10 & 30 & 40 \\ 1 & 2 & 3 \\ 4 & 5 & 6 \end{vmatrix}$$

$$k\begin{vmatrix} a & b & c \\ d & e & f \\ g & h & i \end{vmatrix} = \begin{vmatrix} a & kb & c \\ d & ke & f \\ g & kh & i \end{vmatrix}$$

2. If a multiple of one row is added to another row, the value of the determinant is not changed. This applies to columns, also.

$$\begin{vmatrix} a & b & c \\ d & e & f \\ g & h & i \end{vmatrix} = \begin{vmatrix} a & b & c \\ d+ka & e+kb & f+kc \\ g & h & i \end{vmatrix}$$

3. If two rows are interchanged, then the value of the determinant is multiplied by -1. This applies to columns, also.

$$\begin{vmatrix} 1 & 2 & 3 \\ 4 & 5 & 6 \\ a & b & c \end{vmatrix} = -\begin{vmatrix} 4 & 5 & 6 \\ 1 & 2 & 3 \\ a & b & c \end{vmatrix}$$

EXAMPLE 2   Evaluate the determinant $\begin{vmatrix} 15 & 14 & 26 \\ 18 & 17 & 32 \\ 21 & 20 & 42 \end{vmatrix}$.

Solution

$$\begin{vmatrix} 15 & 14 & 26 \\ 18 & 17 & 32 \\ 21 & 20 & 42 \end{vmatrix} = (3 \times 2) \begin{vmatrix} 5 & 14 & 13 \\ 6 & 17 & 16 \\ 7 & 20 & 21 \end{vmatrix}$$

using Rule 1 to factor 3 from the first column and 2 from the third column

$$= 6 \begin{vmatrix} 5 & 1 & 13 \\ 6 & 1 & 16 \\ 7 & -1 & 21 \end{vmatrix}$$

using Rule 2 to subtract the third column from the second column

$$= 6 \begin{vmatrix} 12 & 0 & 34 \\ 13 & 0 & 37 \\ 7 & -1 & 21 \end{vmatrix}$$

using Rule 2 to add the third row to the first and second rows

$$= 6 \left[-\left(-1 \begin{vmatrix} 12 & 34 \\ 13 & 37 \end{vmatrix} \right) \right]$$

expanding the determinant along the second column

$$= 6 \begin{vmatrix} 12 & 34 \\ 1 & 3 \end{vmatrix}$$

using Rule 2 to subtract the first row from the second row

$$= 6(36 - 34) = 12$$

The value of the given determinant is therefore 12. Notice the general strategy. We use Rules 1 and 2 until one column (or one row) contains two zeros. Then it is a simple matter to expand the determinant along that column (or row).

■■■

■■■ **EXAMPLE 3** Show that

$$\begin{vmatrix} 1 & 1 & 1 \\ a & b & c \\ a^2 & b^2 & c^2 \end{vmatrix} = (b - a)(c - a)(c - b)$$

Solution

$$\begin{vmatrix} 1 & 1 & 1 \\ a & b & c \\ a^2 & b^2 & c^2 \end{vmatrix} = \begin{vmatrix} 1 & 0 & 0 \\ a & b - a & c - a \\ a^2 & b^2 - a^2 & c^2 - a^2 \end{vmatrix}$$

using Rule 2 to subtract the first column from the second and the third columns

$$= \begin{vmatrix} 1 & 0 & 0 \\ a & b - a & c - a \\ a^2 & (b - a)(b + a) & (c - a)(c + a) \end{vmatrix}$$

$$= (b - a)(c - a) \begin{vmatrix} 1 & 0 & 0 \\ a & 1 & 1 \\ a^2 & b + a & c + a \end{vmatrix}$$

using Rule 1 to factor $(b - a)$ from the second column and $(c - a)$ from the third column

$$= (b - a)(c - a)[(c + a) - (b + a)]$$

expanding the determinant along the first row

$$= (b - a)(c - a)(c - b)$$

■■■

The definition that we gave for third-order determinants can also be applied to define fourth-order (or larger) determinants. Consider, for example, the fourth-order determinant A given by

$$A = \begin{vmatrix} 25 & 40 & 5 & 10 \\ 9 & 0 & 3 & 6 \\ -2 & 3 & 11 & -17 \\ -3 & 4 & 7 & 2 \end{vmatrix}$$

By definition, then, we could evaluate this determinant by selecting the first row, multiplying each entry by its cofactor, and then adding the results. This yields

$$A = 25 \begin{vmatrix} 0 & 3 & 6 \\ 3 & 11 & -17 \\ 4 & 7 & 2 \end{vmatrix} - 40 \begin{vmatrix} 9 & 3 & 6 \\ -2 & 11 & -17 \\ -3 & 7 & 2 \end{vmatrix}$$

$$+ 5 \begin{vmatrix} 9 & 0 & 6 \\ -2 & 3 & -17 \\ -3 & 4 & 2 \end{vmatrix} - 10 \begin{vmatrix} 9 & 0 & 3 \\ -2 & 3 & 11 \\ -3 & 4 & 7 \end{vmatrix}$$

The problem is now reduced to evaluating four third-order determinants. If we had instead expanded down the second column, the ensuing work would be somewhat less because of the zero in that column. Nevertheless, there would still be three third-order determinants to evaluate.

Because of the amount of computation involved, we in fact rarely evaluate fourth-order determinants directly from the definition. Instead, we use the three rules (given just before Example 2) to simplify matters first. In the example that follows, we show how this is done. (We will accept the fact that the three rules are applicable in evaluating determinants of any size.)

EXAMPLE 4 Evaluate the determinant A given by

$$A = \begin{vmatrix} 25 & 40 & 5 & 10 \\ 9 & 0 & 3 & 6 \\ -2 & 3 & 11 & -17 \\ -3 & 4 & 7 & 2 \end{vmatrix}$$

Solution

$$A = (5 \times 3) \begin{vmatrix} 5 & 8 & 1 & 2 \\ 3 & 0 & 1 & 2 \\ -2 & 3 & 11 & -17 \\ -3 & 4 & 7 & 2 \end{vmatrix} = 15 \begin{vmatrix} 2 & 8 & 0 & 0 \\ 3 & 0 & 1 & 2 \\ -35 & 3 & 0 & -39 \\ -24 & 4 & 0 & -12 \end{vmatrix}$$

Factor 5 from the first row and 3 from the second row.

Subtract the second row from the first. Subtract 11 times the second row from the third. Subtract 7 times the second row from the fourth.

$$= (15 \times 2 \times 4) \begin{vmatrix} 1 & 4 & 0 & 0 \\ 3 & 0 & 1 & 2 \\ -35 & 3 & 0 & -39 \\ -6 & 1 & 0 & -3 \end{vmatrix} = 120 \cdot (-1) \begin{vmatrix} 1 & 4 & 0 \\ -35 & 3 & -39 \\ -6 & 1 & -3 \end{vmatrix}$$

Factor 2 from the first row and 4 from the fourth row.

Expand the determinant along the third column.

$$= -120 \cdot (-3) \begin{vmatrix} 1 & 4 & 0 \\ -35 & 3 & 13 \\ -6 & 1 & 1 \end{vmatrix} = 360 \begin{vmatrix} 1 & 0 & 0 \\ -35 & 143 & 13 \\ -6 & 25 & 1 \end{vmatrix}$$

Factor (-3) from the third column.

Subtract 4 times the first column from the second.

$$= 360 \begin{vmatrix} 143 & 13 \\ 25 & 1 \end{vmatrix} = 360 \begin{vmatrix} -7 & 7 \\ 25 & 1 \end{vmatrix}$$

Expand the determinant
along the first row.
————————— Subtract 6 times the second
row from the first row.

$$= 360(-7 - 175) = -65520$$

————————— Do the arithmetic!

The value of the given fourth-order determinant is therefore -65520.   ▮▮▮

Determinants can be used to solve certain systems of linear equations in which there are as many unknowns as there are equations. In the box that follows, we state **Cramer's rule** for solving a system of three linear equations in three unknowns.* A more general but entirely similar version of Cramer's rule holds for n equations in n unknowns.

Cramer's Rule

Consider the system

$$\begin{cases} a_1 x + b_1 y + c_1 z = d_1 \\ a_2 x + b_2 y + c_2 z = d_2 \\ a_3 x + b_3 y + c_3 z = d_3 \end{cases}$$

Let the four determinants D, D_x, D_y, and D_z be defined as follows:

$$D = \begin{vmatrix} a_1 & b_1 & c_1 \\ a_2 & b_2 & c_2 \\ a_3 & b_3 & c_3 \end{vmatrix} \qquad D_x = \begin{vmatrix} d_1 & b_1 & c_1 \\ d_2 & b_2 & c_2 \\ d_3 & b_3 & c_3 \end{vmatrix}$$

$$D_y = \begin{vmatrix} a_1 & d_1 & c_1 \\ a_2 & d_2 & c_2 \\ a_3 & d_3 & c_3 \end{vmatrix} \qquad D_z = \begin{vmatrix} a_1 & b_1 & d_1 \\ a_2 & b_2 & d_2 \\ a_3 & b_3 & d_3 \end{vmatrix}$$

Then if $D \neq 0$, the unique values of x, y, and z satisfying the system are given by

$$x = \frac{D_x}{D} \qquad y = \frac{D_y}{D} \qquad z = \frac{D_z}{D}$$

[If $D = 0$, the solutions (if any) can be found using Gaussian elimination.]

Notice that the determinant D in Cramer's rule is just the determinant of the coefficient matrix of the given system. If you replace the first column of D with the column of numbers on the right side of the given system, you obtain D_x. The determinants D_y and D_z are obtained similarly.

Before actually proving Cramer's rule, let's take a look at how the rule is applied.

---

*The rule is named after one of its discoverers, the Swiss mathematician Gabriel Cramer (1704–1752).

▌■■▌ **EXAMPLE 5** Use Cramer's rule to find all solutions of the following system of equations:

$$\begin{cases} 2x + 2y - 3z = -20 \\ x - 4y + z = 6 \\ 4x - y + 2z = -1 \end{cases}$$

Solution First, we list the determinants D, D_x, D_y, and D_z:

$$D = \begin{vmatrix} 2 & 2 & -3 \\ 1 & -4 & 1 \\ 4 & -1 & 2 \end{vmatrix} \qquad D_x = \begin{vmatrix} -20 & 2 & -3 \\ 6 & -4 & 1 \\ -1 & -1 & 2 \end{vmatrix}$$

$$D_y = \begin{vmatrix} 2 & -20 & -3 \\ 1 & 6 & 1 \\ 4 & -1 & 2 \end{vmatrix} \qquad D_z = \begin{vmatrix} 2 & 2 & -20 \\ 1 & -4 & 6 \\ 4 & -1 & -1 \end{vmatrix}$$

The calculations for evaluating D begin as follows:

$$D = \begin{vmatrix} 2 & 2 & -3 \\ 1 & -4 & 1 \\ 4 & -1 & 2 \end{vmatrix} = \begin{vmatrix} 0 & 10 & -5 \\ 1 & -4 & 1 \\ 0 & 15 & -2 \end{vmatrix}$$

Subtract twice the second row from the first.
Subtract 4 times the second row from the third.

Since we now have two zeros in the first column, it is an easy matter to expand D along that column to obtain

$$D = -1 \begin{vmatrix} 10 & -5 \\ 15 & -2 \end{vmatrix}$$
$$= -1[-20 - (-75)] = -1(55) = -55$$

The value of D is therefore -55. (Since this value is nonzero, Cramer's rule does apply.) As Exercise 31 at the end of this section asks you to verify, the values of the other three determinants are

$$D_x = 144 \qquad D_y = 61 \qquad D_z = -230$$

By Cramer's rule, then, the unique values of x, y, and z that satisfy the system are

$$x = \frac{D_x}{D} = \frac{144}{-55} = -\frac{144}{55} \qquad y = \frac{D_y}{D} = \frac{61}{-55} = -\frac{61}{55} \qquad z = \frac{D_z}{D} = \frac{-230}{-55} = \frac{46}{11}$$

■■■

One way we can prove Cramer's rule is to use Gaussian elimination to solve the system

$$\begin{cases} a_1 x + b_1 y + c_1 z = d_1 \\ a_2 x + b_2 y + c_2 z = d_2 \\ a_3 x + b_3 y + c_3 z = d_3 \end{cases} \qquad (1)$$

A much shorter and simpler proof, however, has been found by D. E. Whitford and M. S. Klamkin.* This is the proof we give here; it makes effective use of the rules employed in this section for manipulating determinants.

Consider the system of equations (1) and assume that $D \neq 0$. We will show that if x, y, and z satisfy the system, then in fact $x = D_x/D$, with similar equations giving y and z. (Exercise 64 at the end of this section then shows how to check that these values indeed satisfy the given system.) We have

$$D_x = \begin{vmatrix} d_1 & b_1 & c_1 \\ d_2 & b_2 & c_2 \\ d_3 & b_3 & c_3 \end{vmatrix} \qquad \text{by definition}$$

$$= \begin{vmatrix} (a_1x + b_1y + c_1z) & b_1 & c_1 \\ (a_2x + b_2y + c_2z) & b_2 & c_2 \\ (a_3x + b_3y + c_3z) & b_3 & c_3 \end{vmatrix} \qquad \begin{array}{l} \text{using the equations in (1)} \\ \text{to substitute for } d_1, d_2, \text{ and } d_3 \end{array}$$

$$= \begin{vmatrix} a_1x & b_1 & c_1 \\ a_2x & b_2 & c_2 \\ a_3x & b_3 & c_3 \end{vmatrix} \qquad \begin{array}{l} \text{subtracting } y \text{ times the second} \\ \text{column as well as } z \text{ times the third} \\ \text{column from the first column} \end{array}$$

$$= x \begin{vmatrix} a_1 & b_1 & c_1 \\ a_2 & b_2 & c_2 \\ a_3 & b_3 & c_3 \end{vmatrix} \qquad \text{factoring } x \text{ out of the first column}$$

$$= xD \qquad \text{by definition}$$

We now have $D_x = xD$, which is equivalent to $x = D_x/D$, as required. The formulas for y and z are obtained similarly.

*The proof was published in the *American Mathematical Monthly*, vol. **60** (1953), pp. 186–187.

EXERCISE SET 9.5

A

In Exercises 1–6, evaluate the determinants.

1. (a) $\begin{vmatrix} 2 & -17 \\ 1 & 6 \end{vmatrix}$ (b) $\begin{vmatrix} 1 & 6 \\ 2 & -17 \end{vmatrix}$

2. (a) $\begin{vmatrix} 1 & 0 \\ 0 & 1 \end{vmatrix}$ (b) $\begin{vmatrix} 0 & 1 \\ 0 & 1 \end{vmatrix}$

3. (a) $\begin{vmatrix} 5 & 7 \\ 500 & 700 \end{vmatrix}$ (b) $\begin{vmatrix} 5 & 500 \\ 7 & 700 \end{vmatrix}$

4. (a) $\begin{vmatrix} -8 & -3 \\ 4 & -5 \end{vmatrix}$ (b) $\begin{vmatrix} -3 & -8 \\ -5 & 4 \end{vmatrix}$

5. $\begin{vmatrix} \cos\theta & -\sin\theta \\ \sin\theta & \cos\theta \end{vmatrix}$ 6. $\begin{vmatrix} \cos 50° & \sin 10° \\ \sin 50° & \cos 10° \end{vmatrix}$

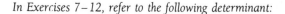 *In Exercises 7–12, refer to the following determinant:*

$$\begin{vmatrix} -6 & 3 & 8 \\ 5 & -4 & 1 \\ 10 & 9 & -10 \end{vmatrix}$$

7. Evaluate the minor of the entry 3.

8. Evaluate the cofactor of the entry 3.

9. Evaluate the minor of -10.

10. Evaluate the cofactor of -10.

11. (a) Multiply each entry in the first row by its minor and find the sum of the results.
    (b) Multiply each entry in the first row by its cofactor and find the sum of the results.
    (c) Does the answer in part (a) or part (b) give you the value of the determinant?

12. (a) Multiply each entry in the first column by its cofactor and find the sum of the results.

(b) Follow the same instructions as in part (a), but use the second column.

(c) Follow the same instructions as in part (a), but use the third column.

13. Let $A = \begin{vmatrix} 1 & 2 & 3 \\ 4 & 5 & 6 \\ 7 & 8 & 9 \end{vmatrix}$. Evaluate A by expanding it along the indicated row or column: (a) second row; (b) third row; (c) first column; (d) third column.

In Exercises 14–21, evaluate the determinants.

14. $\begin{vmatrix} 1 & 2 & -1 \\ 2 & -1 & 1 \\ 4 & 0 & 2 \end{vmatrix}$

15. $\begin{vmatrix} 5 & 10 & 15 \\ 1 & 2 & 3 \\ -9 & 11 & 7 \end{vmatrix}$

16. $\begin{vmatrix} 8 & 4 & 2 \\ 3 & 9 & 3 \\ -2 & 8 & 6 \end{vmatrix}$

17. $\begin{vmatrix} 1 & 2 & -3 \\ 4 & 5 & -9 \\ 0 & 0 & 1 \end{vmatrix}$

18. $\begin{vmatrix} 3 & 0 & 0 \\ 0 & 19 & 0 \\ 0 & 0 & 10 \end{vmatrix}$

19. $\begin{vmatrix} -6 & -8 & 18 \\ 25 & 12 & 15 \\ -9 & 4 & 13 \end{vmatrix}$

20. $\begin{vmatrix} 23 & 0 & 47 \\ -37 & 0 & 18 \\ 14 & 0 & 25 \end{vmatrix}$

21. $\begin{vmatrix} 16 & 0 & -64 \\ -8 & 15 & -12 \\ 30 & -20 & 10 \end{vmatrix}$

22. Use the method illustrated in Example 3 in the text to show that

$$\begin{vmatrix} 1 & 1 & 1 \\ a & b & c \\ a^3 & b^3 & c^3 \end{vmatrix} = (b-a)(c-a)(c^2 - b^2 + ac - ab)$$

$$= (b-a)(c-a)(c-b)(a+b+c)$$

23. Use the method shown in Example 3 to express the determinant $\begin{vmatrix} 1 & x & x^2 \\ 1 & y & y^2 \\ 1 & z & z^2 \end{vmatrix}$ as a product.

24. Show that

$$\begin{vmatrix} 1 & 1 & 1 \\ a^2 & b^2 & c^2 \\ a^3 & b^3 & c^3 \end{vmatrix} = (b-a)(c-a)(bc^2 - b^2c + ac^2 - ab^2)$$

$$= (b-a)(c-a)(c-b)(bc+ac+ab)$$

25. Simplify the determinant $\begin{vmatrix} 1 & 1 & 1 \\ 1 & 1+x & 1 \\ 1 & 1 & 1+y \end{vmatrix}$

26. Use the method shown in Example 3 to express the following determinant as a product of three factors:

$$\begin{vmatrix} 1 & 1 & 1 \\ a & b & c \\ bc & ca & ab \end{vmatrix}$$

In Exercises 27–30, evaluate the determinants.

27. $\begin{vmatrix} 1 & -1 & 0 & 2 \\ 0 & 1 & -1 & 0 \\ 2 & 1 & 0 & -1 \\ -2 & 2 & 1 & 1 \end{vmatrix}$

28. $\begin{vmatrix} 3 & -2 & 3 & 4 \\ 1 & 4 & -3 & 2 \\ 6 & 3 & -6 & -3 \\ -1 & 0 & 1 & 5 \end{vmatrix}$

29. $\begin{vmatrix} 2 & 0 & 0 & 0 \\ 0 & 3 & 0 & 0 \\ 0 & 0 & 4 & 0 \\ 0 & 0 & 0 & 5 \end{vmatrix}$

30. $\begin{vmatrix} 7 & -8 & 1 & 2 \\ 21 & 4 & 3 & -1 \\ -35 & 8 & 3 & -2 \\ 14 & 16 & 0 & 1 \end{vmatrix}$

31. Verify the following statements (from Example 5).

(a) $\begin{vmatrix} -20 & 2 & -3 \\ 6 & -4 & 1 \\ -1 & -1 & 2 \end{vmatrix} = 144$

(b) $\begin{vmatrix} 2 & -20 & -3 \\ 1 & 6 & 1 \\ 4 & -1 & 2 \end{vmatrix} = 61$

(c) $\begin{vmatrix} 2 & 2 & -20 \\ 1 & -4 & 6 \\ 4 & -1 & -1 \end{vmatrix} = -230$

32. Consider the following system:

$$\begin{cases} x + y + 3z = 0 \\ x + 2y + 5z = 0 \\ x - 4y - 8z = 0 \end{cases}$$

(a) Without doing any calculations, find one obvious solution of this system.

(b) Calculate the determinant D.

(c) List all solutions of this system.

In Exercises 33–42, use Cramer's rule to solve those systems for which $D \neq 0$. In cases where $D = 0$, use Gaussian elimination or matrix methods.

33. $\begin{cases} 3x + 4y - z = 5 \\ x - 3y + 2z = 2 \\ 5x - 6z = -7 \end{cases}$

34. $\begin{cases} 3A - B - 4C = 3 \\ A + 2B - 3C = 9 \\ 2A - B + 2C = -8 \end{cases}$

35. $\begin{cases} 3x + 2y - z = -6 \\ 2x - 3y - 4z = -11 \\ x + y + z = 5 \end{cases}$

36. $\begin{cases} 5x - 3y - z = 16 \\ 2x + y - 3z = 5 \\ 3x - 2y + 2z = 5 \end{cases}$

37. $\begin{cases} 2x + 5y + 2z = 0 \\ 3x - y - 4z = 0 \\ x + 2y - 3z = 0 \end{cases}$

38. $\begin{cases} 4u + 3v - 2w = 14 \\ u + 2v - 3w = 6 \\ 2u - v + 4w = 2 \end{cases}$

39. $\begin{cases} 12x - 11z = 13 \\ 6x + 6y - 4z = 26 \\ 6x + 2y - 5z = 13 \end{cases}$

40. $\begin{cases} 3x + 4y + 2z = 1 \\ 4x + 6y + 2z = 7 \\ 2x + 3y + z = 11 \end{cases}$

41. $\begin{cases} x + y + z + w = -7 \\ x - y + z - w = -11 \\ 2x - 2y - 3z - 3w = 26 \\ 3x + 2y + z - w = -9 \end{cases}$

42. $\begin{cases} 2A - B - 3C + 2D = -2 \\ A - 2B + C - 3D = 4 \\ 3A - 4B + 2C - 4D = 12 \\ 2A + 3B - C - 2D = -4 \end{cases}$

43. Find all values of x for which

$$\begin{vmatrix} x - 4 & 0 & 0 \\ 0 & x + 4 & 0 \\ 0 & 0 & x + 1 \end{vmatrix} = 0$$

44. Find all values of x for which

$$\begin{vmatrix} 1 & x & x^2 \\ 1 & 1 & 1 \\ 4 & 5 & 0 \end{vmatrix} = 0$$

45. By expanding the determinant $\begin{vmatrix} a & b & c \\ a & b & c \\ d & e & f \end{vmatrix}$ down the first column, show that its value is zero.

46. By expanding the determinant $\begin{vmatrix} ka & kb & kc \\ d & e & f \\ g & h & i \end{vmatrix}$ along its first row, show that it is equal to $k \begin{vmatrix} a & b & c \\ d & e & f \\ g & h & i \end{vmatrix}$.

B

47. Show that

$$\begin{vmatrix} a_1 + A_1 & b_1 & c_1 \\ a_2 + A_2 & b_2 & c_2 \\ a_3 + A_3 & b_3 & c_3 \end{vmatrix} = \begin{vmatrix} a_1 & b_1 & c_1 \\ a_2 & b_2 & c_2 \\ a_3 & b_3 & c_3 \end{vmatrix} + \begin{vmatrix} A_1 & b_1 & c_1 \\ A_2 & b_2 & c_2 \\ A_3 & b_3 & c_3 \end{vmatrix}$$

48. Show that

$$\begin{vmatrix} a_1 & b_1 & c_1 \\ a_2 & b_2 & c_2 \\ a_3 & b_3 & c_3 \end{vmatrix} = \begin{vmatrix} a_1 + kb_1 & b_1 & c_1 \\ a_2 + kb_2 & b_2 & c_2 \\ a_3 + kb_3 & b_3 & c_3 \end{vmatrix}$$

49. By expanding each determinant along a row or column, show that

$$\begin{vmatrix} a_1 & b_1 & c_1 \\ a_2 & b_2 & c_2 \\ a_3 & b_3 & c_3 \end{vmatrix} = -\begin{vmatrix} a_2 & b_2 & c_2 \\ a_1 & b_1 & c_1 \\ a_3 & b_3 & c_3 \end{vmatrix}$$

50. Solve for x in terms of a, b, and c:

$$\begin{vmatrix} a & a & x \\ c & c & c \\ b & x & b \end{vmatrix} = 0 \quad (c \neq 0)$$

51. Show that

$$\begin{vmatrix} 1 + a & 1 & 1 & 1 \\ 1 & 1 + b & 1 & 1 \\ 1 & 1 & 1 + c & 1 \\ 1 & 1 & 1 & 1 + d \end{vmatrix}$$

$$= abcd\left(\frac{1}{a} + \frac{1}{b} + \frac{1}{c} + \frac{1}{d} + 1\right)$$

52. Show that

$$\begin{vmatrix} 1 & a & a^2 \\ a^2 & 1 & a \\ a & a^2 & 1 \end{vmatrix} = (a^3 - 1)^2$$

53. Consider the determinant $D = \begin{vmatrix} a_1 & b_1 & c_1 \\ a_2 & b_2 & c_2 \\ a_3 & b_3 & c_3 \end{vmatrix}$. Let A_1 denote the cofactor of a_1, let B_1 denote the cofactor of b_1, and so on. Prove that

$$\begin{vmatrix} B_2 & C_2 \\ B_3 & C_3 \end{vmatrix} = a_1 D$$

54. Show that

$$\begin{vmatrix} 1 & bc & b + c \\ 1 & ca & c + a \\ 1 & ab & a + b \end{vmatrix} = (b - c)(c - a)(a - b)$$

55. Find the value of each determinant.

(a) $\begin{vmatrix} 1 & 0 & 0 \\ x & 1 & 0 \\ x & y & 1 \end{vmatrix}$ (b) $\begin{vmatrix} 1 & 0 & 0 & 0 \\ x & 1 & 0 & 0 \\ x & y & 1 & 0 \\ x & y & z & 1 \end{vmatrix}$

56. Evaluate the determinant.

$$\begin{vmatrix} a & b & c \\ a & a + b & a + b + c \\ a & 2a + b & 3a + 2b + c \end{vmatrix}$$

57. Show that

$$\begin{vmatrix} 1 & a & a & a \\ 1 & b & a & a \\ 1 & a & b & a \\ 1 & a & a & b \end{vmatrix} = (b - a)^3$$

58. Show that

$$\begin{vmatrix} a & 1 & 1 & 1 \\ 1 & a & 1 & 1 \\ 1 & 1 & a & 1 \\ 1 & 1 & 1 & a \end{vmatrix} = (a - 1)^3(a + 3)$$

59. Solve the following system for x, y, and z:

$$\begin{cases} ax + by + cz = k \\ a^2x + b^2y + c^2z = k^2 \\ a^3x + b^3y + c^3z = k^3 \end{cases}$$

(Assume that the values of a, b, and c are all distinct and all nonzero.)

60. Show that the equation

$$\begin{vmatrix} x & y & 1 \\ x_1 & y_1 & 1 \\ 1 & m & 0 \end{vmatrix} = 0$$

represents a line with slope m, passing through (x_1, y_1).

For Exercise 61, use the fact that the equation of a line passing through (x_1, y_1) and (x_2, y_2) can be written

$$\begin{vmatrix} x & y & 1 \\ x_1 & y_1 & 1 \\ x_2 & y_2 & 1 \end{vmatrix} = 0$$

For Exercise 62, use the fact that the equation of a circle passing through (x_1, y_1), (x_2, y_2), and (x_3, y_3) can be written

$$\begin{vmatrix} x^2 + y^2 & x & y & 1 \\ x_1^2 + y_1^2 & x_1 & y_1 & 1 \\ x_2^2 + y_2^2 & x_2 & y_2 & 1 \\ x_3^2 + y_3^2 & x_3 & y_3 & 1 \end{vmatrix} = 0$$

61. Find the equation of the line passing through $(-3, -1)$ and $(2, 9)$. Write the answer in the form $y = mx + b$.

62. Find the equation of the circle passing through $(7, 0)$, $(5, -6)$, and $(-1, -4)$. Write the answer in the form $(x - h)^2 + (y - k)^2 = r^2$.

C

63. Show that the area of the triangle in Figure A is $\frac{1}{2}\begin{vmatrix} a & b \\ c & d \end{vmatrix}$. *Hint:* Figure B indicates how the required area can be found using a rectangle and three right triangles.

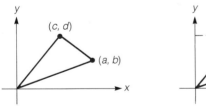

FIGURE A

FIGURE B

64. This exercise completes the derivation of Cramer's rule given in the text. Using the same notation, and assuming $D \neq 0$, we need to show that the values $x = D_x/D$, $y = D_y/D$, and $z = D_z/D$ satisfy the equations in (1) on page 549. We will show that these values satisfy the first equation in (1), the verification for the other equations being entirely similar.

(a) Check that substituting the values $x = D_x/D$, $y = D_y/D$, and $z = D_z/D$ in the first equation of (1) yields an equation equivalent to

$$a_1 D_x + b_1 D_y + c_1 D_z - d_1 D = 0$$

(b) Show that the equation in part (a) can be written

$$a_1 \begin{vmatrix} b_1 & c_1 & d_1 \\ b_2 & c_2 & d_2 \\ b_3 & c_3 & d_3 \end{vmatrix} - b_1 \begin{vmatrix} a_1 & c_1 & d_1 \\ a_2 & c_2 & d_2 \\ a_3 & c_3 & d_3 \end{vmatrix}$$
$$+ c_1 \begin{vmatrix} a_1 & b_1 & d_1 \\ a_2 & b_2 & d_2 \\ a_3 & b_3 & d_3 \end{vmatrix} - d_1 \begin{vmatrix} a_1 & b_1 & c_1 \\ a_2 & b_2 & c_2 \\ a_3 & b_3 & c_3 \end{vmatrix} = 0$$

(c) Show that the equation in part (b) can be written

$$\begin{vmatrix} a_1 & b_1 & c_1 & d_1 \\ a_1 & b_1 & c_1 & d_1 \\ a_2 & b_2 & c_2 & d_2 \\ a_3 & b_3 & c_3 & d_3 \end{vmatrix} = 0$$

(d) Now explain why the equation in (c) indeed holds.

65. Let D denote the determinant of the matrix $\begin{pmatrix} a & b \\ c & d \end{pmatrix}$.

(a) Show that the inverse of this matrix is $\frac{1}{D}\begin{pmatrix} d & -b \\ -c & a \end{pmatrix}$. Assume that $ad - bc \neq 0$.

(b) Use the result in part (a) to find the inverse of the matrix $\begin{pmatrix} -6 & 7 \\ 1 & 9 \end{pmatrix}$.

66. Assuming that $p + q + r = 0$, solve the following equation for x:

$$\begin{vmatrix} p - x & r & q \\ r & q - x & p \\ q & p & r - x \end{vmatrix} = 0$$

Hint: After adding certain rows or columns, the quantity $p + q + r$ will appear.

OPTIONAL TI-81 GRAPHING CALCULATOR EXERCISES
EXERCISES FOR SECTIONS 9.3, 9.4, AND 9.5

Exercise 1 shows how to use the calculator to carry out some basic computations with matrices and determinants. As examples, we will use the following two matrices:

$$A = \begin{pmatrix} 5 & 0 & 2 \\ 2 & 2 & 1 \\ -3 & 1 & -1 \end{pmatrix} \qquad B = \begin{pmatrix} -2 & 1 & -9 \\ 0 & 7 & 3 \\ -1 & 6 & 4 \end{pmatrix}$$

To enter matrix A, press the MATRX key and then the right (or the left) ARROW key. The menu for editing matrices now appears on the screen. Since we wish to edit the matrix A, press the number 1. Next, type 3 ENTER and then type 3 ENTER again. As you can see on the screen, the effect is to change the default setting from six rows and six columns to three rows and three columns.

The blinking cursor is now located on the default value zero for the matrix element a_{11}. Since we want a_{11} to be 5, type 5 ENTER. The blinking cursor is now located on the default value zero for the matrix element a_{12}. Since we want a_{12} to be 0, type 0 ENTER. Continue this process for the remaining entries in A. After you have entered the last element, $a_{33} = -1$, press the MATRX key to return to the matrix screen; then press the right ARROW key to go to the edit menu.

Type 2 so that you can edit the B matrix. Now follow the same procedure that you used in defining A. After you have entered the last element of B, namely, $b_{33} = 4$, press the 2nd key and then the CLEAR key; this takes you back to the home screen.

Before doing any computations, we want to check for typing errors in entering the elements of A and B. To see the matrix A displayed, press the 2nd key, then the [A] key (which is the same as the 1 key), and then the ENTER key. As you can see, the matrix A is now displayed on the screen. If you detect typing errors, press the MATRX key, then the right ARROW key, then 1. You can now retype any entries that need to be fixed. When you are finished, or if there were no errors to correct, exit to the home screen by pressing the 2nd key and then the CLEAR key. Repeat the same procedure for checking the entries in B.

1. (a) Use the following sequence of keystrokes to compute the matrix product AB.

    (2nd) (\[A]) (×) (2nd) (\[B]) (ENTER)

    (b) Use the following sequence of keystrokes to compute the matrix product BA.

    (2nd) (\[B]) (×) (2nd) (\[A]) (ENTER)

    (c) Compute A^{-1} using the following sequence of keystrokes.

    (2nd) (\[A]) (x^{-1}) (ENTER)

Check the answer that the calculator gives you by referring to Example 3 in Section 3 of this chapter.

(d) Compute B^{-1} using the following sequence of keystrokes.

(2nd) (\[B]) (x^{-1}) (ENTER)

The resulting display will show the first column of B^{-1}. Press the right ARROW key to see the other two columns. As you can see, the values that are displayed are ten-place decimal approximations. To see the entries of B^{-1} rounded off to three decimal places, use the following sequence of keystrokes. (In the sequence, the command following the ALPHA key is a comma; this is the same as the decimal point key.)

(MATH) (▶) (1) (2nd) (\[B]) (x^{-1})
(ALPHA) (,) (3) ()) (ENTER)

(e) Use the following keystrokes to compute the determinant of the matrix A.

(MATRX) (5) (2nd) (\[A]) (ENTER)

For Exercises 2–6, enter the following matrices and then use the calculator to carry out the indicated operations. In cases where the results involve decimals, round off each entry to three decimal places.

$$A = \begin{pmatrix} 3 & 2 & 6 \\ 1 & 1 & 2 \\ 2 & 2 & 5 \end{pmatrix} \qquad B = \begin{pmatrix} 5 & 4 & 1 \\ -1 & 2 & 9 \\ 0 & 8 & -6 \end{pmatrix} \qquad C = \begin{pmatrix} -21 \\ 17 \\ 8 \end{pmatrix}$$

2. (a) AB (b) AC (c) A^{-1}
    (d) B^{-1} (e) $(BA)^{-1}$ (f) $A^{-1}B^{-1}$
    Hint: In part (f), first recalculate A^{-1}. [The calculator will automatically store the result in the (temporary) ANS file.] Then use the following keystrokes.

    (2nd) (ANS) (×) (2nd) (\[B]) (x^{-1}) (ENTER)

3. (a) $\det(A)$ (b) $\det(B)$ (c) $\det(AB)$
    (d) $[\det(A)] \times [\det(B)]$

4. (a) $\det(A^{-1})$ (b) $1/\det(A)$

5. (a) $\det(B^{-1})$ (b) $1/\det(B)$

6. (a) A^2 (This indicates the matrix product AA.)
    (b) A^3 *Hint:* Assuming that your result from part (a) is still displayed, use the keystrokes

    (×) (2nd) (\[A]) (ENTER)

    (c) A^4

9.6 NONLINEAR SYSTEMS OF EQUATIONS

In the previous sections of this chapter, we looked at several techniques for solving systems of linear equations. In particular, we saw that if solutions exist, they can always be found by Gaussian elimination. In the present section, we consider **nonlinear systems** of equations, that is, systems in which at least one of the equations is not linear. There is no single technique that serves to solve all nonlinear systems. However, simple substitution will often suffice. The work in this section focuses on examples showing some of the more common approaches. In all the examples and in the exercises, we will be concerned exclusively with solutions (x, y) in which both x and y are real numbers. In Example 1, the system consists of one linear equation and one quadratic equation. Such a system can always be solved by substitution.

EXAMPLE 1 Find all solutions (x, y) of the following system, where x and y are real numbers:

$$\begin{cases} 2x + y = 1 \\ y = 4 - x^2 \end{cases}$$

Solution We use the second equation to substitute for y in the first equation. This will yield an equation with only one unknown:

$$2x + (4 - x^2) = 1$$
$$-x^2 + 2x + 3 = 0$$
$$x^2 - 2x - 3 = 0$$
$$(x - 3)(x + 1) = 0$$

$$x - 3 = 0 \quad | \quad x + 1 = 0$$
$$x = 3 \quad | \quad x = -1$$

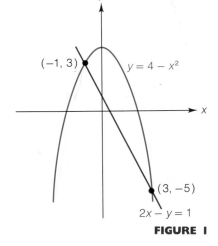

y

$(-1, 3)$ $y = 4 - x^2$

x

$(3, -5)$

$2x - y = 1$

FIGURE 1

The values $x = 3$ and $x = -1$ can now be substituted back into either of the original equations. Substituting $x = 3$ in the equation $y = 4 - x^2$ yields $y = -5$. Similarly, substituting $x = -1$ in the equation $y = 4 - x^2$ gives us $y = 3$. We have now determined the two ordered pairs $(3, -5)$ and $(-1, 3)$. As you can easily check, both of these are solutions of the given system. Figure 1 displays the graphical interpretation of this result. The line $2x + y = 1$ intersects the parabola $y = 4 - x^2$ at the points $(-1, 3)$ and $(3, -5)$.

EXAMPLE 2 Where do the graphs of the parabola $y = x^2$ and the circle $x^2 + y^2 = 1$ intersect? See Figure 2.

Solution The system that we wish to solve is

$$\begin{cases} y = x^2 & (1) \\ x^2 + y^2 = 1 & (2) \end{cases}$$

In view of equation (1), we can replace the x^2-term of equation (2) by y. Doing this yields

$$y + y^2 = 1$$

y

$y = x^2$

x

$x^2 + y^2 = 1$

FIGURE 2

or

$$y^2 + y - 1 = 0$$

This last equation can be solved by using the quadratic formula with $a = 1$, $b = 1$, and $c = -1$. As you can check, the results are

$$y = \frac{-1 + \sqrt{5}}{2} \quad \text{and} \quad y = \frac{-1 - \sqrt{5}}{2}$$

However, from Figure 2, it is clear that the y-coordinate at each intersection point is positive. Therefore, we discard the negative number $y = -\frac{1}{2}(1 + \sqrt{5})$ from further consideration in this context. Substituting the positive number $y = \frac{1}{2}(-1 + \sqrt{5})$ back in the equation $y = x^2$ then gives us

$$x^2 = \tfrac{1}{2}(-1 + \sqrt{5}) \quad \text{and therefore} \quad x = \pm \sqrt{\tfrac{1}{2}(-1 + \sqrt{5})}$$

By choosing the positive square root, we obtain the x-coordinate for the intersection point in the first quadrant. That point is therefore

$$\left(\sqrt{\tfrac{1}{2}(-1 + \sqrt{5})}, \tfrac{1}{2}(-1 + \sqrt{5}) \right) \approx (0.79, 0.62) \qquad \text{using a calculator}$$

Similarly, the negative square root yields the x-coordinate for the intersection point in the second quadrant. That point is

$$\left(-\sqrt{\tfrac{1}{2}(-1 + \sqrt{5})}, \tfrac{1}{2}(-1 + \sqrt{5}) \right) \approx (-0.79, 0.62) \qquad \text{using a calculator}$$

We have now found the two intersection points, as required. ▮▮▮

▮▬▮ **EXAMPLE 3** Find all solutions (x, y) of the following system, where x and y are real numbers:

$$\begin{cases} xy = 1 & (3) \\ y = 3x + 1 & (4) \end{cases}$$

Solution Since these equations are easy to graph, we do so, because that will tell us something about the required solutions. As Figure 3 indicates, there are two intersection points, one in the first quadrant, the other in the third quadrant. One way to begin now would be to solve equation (3) for one unknown in terms of the other. However, to avoid introducing fractions at the outset, let us use equation (4) to substitute for y in equation (3). This yields

$$x(3x + 1) = 1$$

or

$$3x^2 + x - 1 = 0$$

This last equation can be solved by using the quadratic formula. As you should verify, the solutions are

$$x = \frac{-1 + \sqrt{13}}{6} \quad \text{and} \quad x = \frac{-1 - \sqrt{13}}{6}$$

The corresponding y-values can now be obtained by substituting for x in either of the given equations. We will substitute in equation (4). (Exercise 23 at the end of this section asks you to substitute in equation (3) as well and then to show that the very different-looking answers obtained in that way are in fact

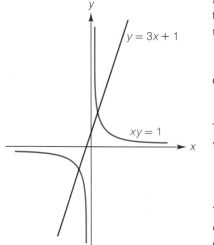

FIGURE 3

equal to those found here.) Substituting $x = \frac{1}{6}(-1 + \sqrt{13})$ in the equation $y = 3x + 1$ gives us

$$y = 3\left(\frac{-1 + \sqrt{13}}{6}\right) + 1$$

$$= \frac{-1 + \sqrt{13}}{2} + \frac{2}{2} = \frac{1 + \sqrt{13}}{2}$$

Thus, one of the intersection points is

$$\left(\tfrac{1}{6}(-1 + \sqrt{13}), \tfrac{1}{2}(1 + \sqrt{13})\right)$$

Notice that this must be the first-quadrant point of intersection, since both coordinates are positive. The intersection point in the third quadrant is obtained in exactly the same manner. As you can check, substituting the value $x = -\frac{1}{6}(1 + \sqrt{13})$ in the equation $y = 3x + 1$ yields $y = \frac{1}{2}(1 - \sqrt{13})$. Thus, the other intersection point is

$$\left(-\tfrac{1}{6}(1 + \sqrt{13}), \tfrac{1}{2}(1 - \sqrt{13})\right)$$

As Figure 3 indicates, there are no other solutions.   ▊▊▊

▊▊▊▊ **EXAMPLE 4** Find all real numbers x and y satisfying the system of equations

$$\begin{cases} y = \sqrt{x} \\ (x + 2)^2 + y^2 = 1 \end{cases}$$

Solution We use the first equation to substitute for y in the second equation. This yields

$$(x + 2)^2 + \left(\sqrt{x}\right)^2 = 1$$

$$x^2 + 4x + 4 + x = 1$$

$$x^2 + 5x + 3 = 0$$

Using the quadratic formula to solve this last equation, we have

$$x = \frac{-5 \pm \sqrt{5^2 - 4(1)(3)}}{2(1)} = \frac{-5 \pm \sqrt{13}}{2}$$

The two values of x are thus

$$-\tfrac{1}{2}(5 - \sqrt{13}) \qquad \text{and} \qquad -\tfrac{1}{2}(5 + \sqrt{13})$$

However, notice that both of these quantities are negative. (The second is obviously negative; without using a calculator, can you explain why the first is negative?) Thus, neither of these quantities is an appropriate x-input in the equation $y = \sqrt{x}$, since we are looking for y-values that are real numbers. We conclude from this that there are no pairs of real numbers x and y satisfying the given system. In geometric terms, this means that the two graphs do not intersect. See Figure 4.   ▊▊▊

FIGURE 4

In the next example, we look at a system that can be reduced to a linear system through appropriate substitutions.

▌▌▌ EXAMPLE 5   Solve the system

$$\begin{cases} \dfrac{2}{x^2} - \dfrac{3}{y^2} = -6 \\[2mm] \dfrac{3}{x^2} + \dfrac{4}{y^2} = 59 \end{cases}$$

Solution   Let $u = 1/x^2$ and $v = 1/y^2$, so that the system becomes

$$\begin{cases} 2u - 3v = -6 \\ 3u + 4v = 59 \end{cases}$$

This is now a linear system. As Exercise 46 asks you to show, the solution is $u = 9$, $v = 8$. In view of the definitions of u and v, then, we have

$$\frac{1}{x^2} = 9 \qquad\qquad \frac{1}{y^2} = 8$$
$$x^2 = 1/9 \qquad\qquad y^2 = 1/8$$
$$x = \pm 1/3 \qquad\quad y = \pm 1/\sqrt{8} = \pm 1/(2\sqrt{2})$$
$$= \pm\sqrt{2}/4$$

This gives us four possible solutions for the original system:

$$\left(\frac{1}{3}, \frac{\sqrt{2}}{4}\right) \quad \left(\frac{1}{3}, -\frac{\sqrt{2}}{4}\right) \quad \left(-\frac{1}{3}, \frac{\sqrt{2}}{4}\right) \quad \left(-\frac{1}{3}, -\frac{\sqrt{2}}{4}\right)$$

As you can check, all four of these pairs satisfy the given system.   **▌▌▌**

▌▌▌ EXAMPLE 6   Determine all solutions (x, y) of the following system, where x and y are real numbers:

$$\begin{cases} y = 3^x & \text{(5)} \\ y = 3^{2x} - 2 & \text{(6)} \end{cases}$$

Solution   We'll use the substitution method. First we rewrite equation (6) as

$$y = (3^x)^2 - 2 \qquad\qquad\qquad\qquad (7)$$

Now, in view of equation (5), we can replace 3^x with y in equation (7) to obtain

$$y = y^2 - 2$$
$$0 = y^2 - y - 2 = (y + 1)(y - 2)$$

From this last equation, we see that $y = -1$ or $y = 2$. With $y = -1$, equation (5) becomes $-1 = 3^x$, contrary to the fact that 3^x is positive for all real numbers x. Thus, we discard the case where $y = -1$. On the other hand, if $y = 2$, equation (5) becomes

$$2 = 3^x$$

We can solve this exponential equation by taking the logarithm of both sides. Using base e logarithms, we have

$$\ln 2 = \ln 3^x = x \ln 3$$

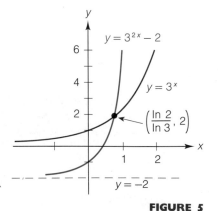

FIGURE 5

and, consequently,

$$x = \frac{\ln 2}{\ln 3} \qquad \text{Caution:} \quad \frac{\ln 2}{\ln 3} \neq \ln 2 - \ln 3$$

We've now found that $x = (\ln 2)/(\ln 3)$ and $y = 2$. Figure 5 displays a graphical interpretation of this result. (Using a calculator for the x-coordinate, we find that $x \approx 0.6$, which is consistent with Figure 5.) ▮▮▮

EXERCISE SET 9.6

A

In Exercises 1–22, find all solutions (x, y) of the given systems, where x and y are real numbers.

1. $\begin{cases} y = 3x \\ y = x^2 \end{cases}$

2. $\begin{cases} y = x + 3 \\ y = 9 - x^2 \end{cases}$ $(x+3)(x-3)$

3. $\begin{cases} x^2 + y^2 = 25 \\ 24y = x^2 \end{cases}$

4. $\begin{cases} 3x + 4y = 12 \\ x^2 - y + 1 = 0 \end{cases}$

5. $\begin{cases} xy = 1 \\ y = -x^2 \end{cases}$

6. $\begin{cases} x + 2y = 0 \\ xy = -2 \end{cases}$

7. $\begin{cases} 2x^2 + y^2 = 17 \\ x^2 + 2y^2 = 22 \end{cases}$

8. $\begin{cases} x - 2y = 1 \\ y^2 - x^2 = 3 \end{cases}$

9. $\begin{cases} y = 1 - x^2 \\ y = x^2 - 1 \end{cases}$

10. $\begin{cases} xy = 4 \\ y = 4x \end{cases}$

11. $\begin{cases} xy = 4 \\ y = 4x + 1 \end{cases}$

12. $\begin{cases} \dfrac{2}{x^2} + \dfrac{5}{y^2} = 3 \\ \dfrac{3}{x^2} - \dfrac{2}{y^2} = 1 \end{cases}$

13. $\begin{cases} \dfrac{1}{x^2} - \dfrac{3}{y^2} = 14 \\ \dfrac{2}{x^2} + \dfrac{1}{y^2} = 35 \end{cases}$

14. $\begin{cases} y = -\sqrt{x} \\ (x - 3)^2 + y^2 = 4 \end{cases}$

15. $\begin{cases} y = -\sqrt{x - 1} \\ (x - 3)^2 + y^2 = 4 \end{cases}$

16. $\begin{cases} y = -\sqrt{x - 6} \\ (x - 3)^2 + y^2 = 4 \end{cases}$

17. $\begin{cases} y = 2^x \\ y = 2^{2x} - 12 \end{cases}$

18. $\begin{cases} y = e^{4x} \\ y = e^{2x} + 6 \end{cases}$

19. $\begin{cases} 2(\log_{10} x)^2 - (\log_{10} y)^2 = -1 \\ 4(\log_{10} x)^2 - 3(\log_{10} y)^2 = -11 \end{cases}$

20. $\begin{cases} y = \log_2(x + 1) \\ y = 5 - \log_2(x - 3) \end{cases}$

21. $\begin{cases} 2^x \cdot 3^y = 4 \\ x + y = 5 \end{cases}$

22. $\begin{cases} a^{2x} + a^{2y} = 10 \\ a^{x+y} = 4 \end{cases}$ $(a > 0)$

Hint: Use the substitutions $a^x = t$ and $a^y = u$.

23. Let $x = \frac{1}{6}(-1 + \sqrt{13})$. Using this x-value, show that the equations $y = 3x + 1$ and $y = 1/x$ yield the same y-value.

24. A sketch shows that the line $y = 100x$ intersects the parabola $y = x^2$ at the origin. Are there any other intersection points? If so, find them. If not, explain why not.

25. Solve the following system for x and y:

$$\begin{cases} ax + by = 2 \\ abxy = 1 \end{cases}$$

(Assume that neither a nor b is zero.)

B

26. Let a, b, and c be constants (with $a \neq 0$), and consider the system

$$\begin{cases} y = ax^2 + bx + c \\ y = k \end{cases}$$

For which value of k (in terms of a, b, and c) will the system have exactly one solution? What is that solution? What is the relationship between the solution you've found and the graph of $y = ax^2 + bx + c$?

27. Find all solutions of the system

$$\begin{cases} x^3 + y^3 = 3473 \\ x + y = 23 \end{cases}$$

28. Solve the following system for x, y, and z:

$$\begin{cases} yz = p^2 \\ zx = q^2 \\ xy = r^2 \end{cases}$$

(Assume that p, q, and r are positive constants.)

29. If the diagonal of a rectangle has length d and the perimeter of the rectangle is $2p$, express the lengths of the sides in terms of d and p.

30. Solve the following system for x and y:

$$\begin{cases} xy + pq = 2px \\ x^2y^2 + p^2q^2 = 2q^2y^2 \end{cases} \quad (pq \neq 0)$$

Hint: Square the first equation, then subtract the second. This results in an equation that can be written $(2px + qy)(px - qy) = 0$.

31. Solve the following system for u and v:

$$\begin{cases} u^2 - v^2 = 9 \\ \sqrt{u + v} + \sqrt{u - v} = 4 \end{cases}$$

Hint: Square the second equation.

32. Solve the given system for x and y in terms of p and q:

$$\begin{cases} q^{\ln x} = p^{\ln y} \\ (px)^{\ln a} = (qy)^{\ln b} \end{cases}$$

(Assume all constants and variables are positive.)

33. The accompanying figure shows the graph of a function $\mathcal{N} = \mathcal{N}_0 e^{kt}$. In each case, determine the constants $\mathcal{N}_0$ and k so that the graph passes through the two given points.

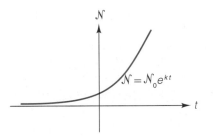

(a) $(2, 3)$ and $(8, 24)$
(b) $\left(\frac{1}{2}, 1\right)$ and $(4, 10)$

34. [As background for this exercise, first work Exercise 33(a).] Measurements (begun in 1958, by Charles D. Keeling of the Scripps Institution of Oceanography) show that the concentration of carbon dioxide in the atmosphere has increased exponentially over the years 1958–1988. In 1958, the average carbon dioxide concentration was $\mathcal{N} = 315$ ppm (parts per million), and in 1988 it was $\mathcal{N} = 351$ ppm.
(a) Use this data to determine the constants k and $\mathcal{N}_0$ in the growth law $\mathcal{N} = \mathcal{N}_0 e^{kt}$. *Hint:* The curve $\mathcal{N} = \mathcal{N}_0 e^{kt}$ passes through the two points $(1958, 315)$ and $(1988, 351)$. Use this to obtain a system of two equations in two unknowns.
(b) Stephen H. Schneider in his article "The Changing Climate" [*Scientific American,* vol. **261** (1989), pp. 70–79] estimates that, under certain circumstances, the concentration of carbon dioxide in the atmosphere could reach 600 ppm by the year 2080. Use your results in part (a) to make a projection for the year 2080. Is your projection higher or lower than Dr. Schneider's?

35. Solve the following system for x, y, and z in terms of p, q, and r:

$$\begin{cases} x(x + y + z) = p^2 \\ y(x + y + z) = q^2 \\ z(x + y + z) = r^2 \end{cases}$$

(Assume that p, q, and r are nonzero.)   *Hint:* Denote $x + y + z$ by w; then add the three equations.

36. (a) Find the points where the line $y = -2x - 2$ intersects the parabola $y = \frac{1}{2}x^2$.
(b) On the same set of axes, sketch the line $y = -2x - 2$ and the parabola $y = \frac{1}{2}x^2$. Be certain that your sketch is consistent with the results obtained in part (a).

37. The area of a right triangle is $180\ \text{cm}^2$, and the hypotenuse is 41 cm. Find the lengths of the two legs.

38. The sum of two numbers is 8, while their product is -128. What are the two numbers?

39. The perimeter of a rectangle is 46 cm, and the area is $60\ \text{cm}^2$. Find the length and the width.

40. Find all right triangles for which the perimeter is 24 units and the area is 24 square units.

41. Solve the following system for x and y using the substitution method:

$$\begin{cases} x^2 + y^2 = 5 & (1) \\ xy = 2 & (2) \end{cases}$$

42. The substitution method in Exercise 41 leads to a quadratic equation. Here is an alternative approach to solving that system; this approach leads to linear equations. Multiply equation (2) by 2 and add the resulting equation to equation (1). Now take square roots to conclude that $x + y = \pm 3$. Next multiply equation (2) by 2 and subtract the resulting equation from equation (1). Take square roots to conclude that $x - y = \pm 1$. You now have the following four linear systems, each of which can be solved (with almost no work) by the addition–subtraction method:

$$\begin{cases} x + y = 3 \\ x - y = 1 \end{cases} \qquad \begin{cases} x + y = 3 \\ x - y = -1 \end{cases}$$

$$\begin{cases} x + y = -3 \\ x - y = 1 \end{cases} \qquad \begin{cases} x + y = -3 \\ x - y = -1 \end{cases}$$

Solve these systems and compare your results with those obtained in Exercise 41.

43. Solve the following system using the method explained in Exercise 42:

$$\begin{cases} x^2 + y^2 = 7 \\ xy = 3 \end{cases}$$

44. Solve the following system using the substitution method:

$$\begin{cases} 3xy - 4x^2 = 2 \\ -5x^2 + 3y^2 = 7 \end{cases}$$

(Begin by solving the first equation for y.)

45. Here is an alternative approach for solving the system in Exercise 44. Let $y = mx$, where m is a constant to be determined. Replace y with mx in both equations of the system to obtain the following pair of equations:

$$x^2(3m - 4) = 2 \qquad\qquad (1)$$
$$x^2(-5 + 3m^2) = 7 \qquad\qquad (2)$$

Now divide equation (2) by equation (1). After clearing fractions and simplifying, you can write the resulting equation as $2m^2 - 7m + 6 = 0$. Solve this last equation by factoring. The values of m can then be used in equation (1) to determine values for x. In each case, the corresponding y-values are determined by the equation $y = mx$.

46. Solve the following system. (You should obtain $u = 9$ and $v = 8$, as stated in Example 5.)

$$\begin{cases} 2u - 3v = -6 \\ 3u + 4v = 59 \end{cases}$$

47. Solve the following system for x and y:

$$\begin{cases} xy + pq = 2px \\ x^2y^2 + p^2q^2 = 2q^2y^2 \end{cases}$$

Hint: Square the first equation, then subtract the second. This results in an equation that can be factored.

C

48. Solve the following system for x and y:

$$\begin{cases} \dfrac{1}{x^2} + \dfrac{1}{xy} = \dfrac{1}{a^2} \\ \dfrac{1}{y^2} + \dfrac{1}{xy} = \dfrac{1}{b^2} \end{cases}$$

(Assume that a and b are positive.)

49. Solve the following system for x and y:

$$\begin{cases} x^4 = y^6 \\ \ln \dfrac{x}{y} = \dfrac{\ln x}{\ln y} \end{cases}$$

50. Find two acute angles such that the sum of their sines is 1 and the sum of their cosines is $\frac{3}{2}$. Express the answers both in degrees and in radians. (Round to two decimal places.)

OPTIONAL TI-81 GRAPHING CALCULATOR EXERCISES

EXERCISES FOR SECTION 9.6

1. The graphing calculator can be used to determine (or at least approximate) the solution of many types of systems of equations in two variables. Consider, for instance, the following system, discussed in Example 1 in the first section of this chapter.

$$\begin{cases} x + y = 2 \\ 2x - 3y = 9 \end{cases}$$

(a) Solve each equation for y. Then press the Y= key and enter the results as Y_1 and Y_2. If you've done the algebra correctly, you should have $Y_1 = 2 - x$ and $Y_2 = \frac{2}{3}x - 3$. Press ZOOM 6 to see the graphs in the standard viewing rectangle.

(b) As you can see, the lines intersect at a point in the fourth quadrant. Press the TRACE key and then use the ARROW keys to move the cursor to what appears to be the intersection point. The readout at the bottom of the screen tells you that the point is close to $(3, -1)$. [From the text, we know that $(3, -1)$ is the precise intersection point.]

(c) To obtain a closer approximation to the intersection point, press ZOOM 2. The blinking cursor will still be on the approximate intersection point. Now, press ENTER to zoom in on that point. Next press the TRACE key and then use the ARROW keys to move the cursor to the intersection point. Note the readout at the bottom of the screen. The values will be closer than they were in part (b) to $x = 3$ and $y = -1$. (Repeating the process will yield a still closer approximation.)

In Exercises 2–10, use the procedure described in Exercise 1 to determine or approximate the solutions of each system. Repeat the procedure until you are sure of the second decimal place in each coordinate of the intersection point.

2. $\begin{cases} 3x - 8y = 63 \\ 2x + y = 4 \end{cases}$

3. $\begin{cases} x - y - 1 = 0 \\ y = 2\sqrt{x + 3} \end{cases}$

4. $\begin{cases} y = e^x \\ y = x^2 \end{cases}$

5. $\begin{cases} y = e^{x/2} \\ y = x^2 \end{cases}$

6. $\begin{cases} y = \ln x \\ y = e^{-x} \end{cases}$

7. $\begin{cases} x^2 - 2x + y = 0 \\ y = 3x^3 - x^2 - 10x \end{cases}$

9. $\begin{cases} y = x^3 \\ y = \frac{1}{2}(e^x + e^{-x}) \end{cases}$ *Hint:* There are two solutions.

8. $\begin{cases} y = \sqrt{x} - 1 \\ y = \ln x \end{cases}$ *Hint:* There are two solutions. You will need a viewing rectangle that is larger than the standard viewing rectangle to see that there is a second solution.

10. $\begin{cases} y = x^3 \\ y = \frac{1}{2}(e^x - e^{-x}) \end{cases}$ *Hint:* There are four solutions. You can cut the amount of work in half by using the fact that both graphs are symmetric about the origin.

9.7 SYSTEMS OF INEQUALITIES

... the variety of problems in business, science, and industry that can be solved by using the linear programming model is impressive, to say the least. Problems in transportation, scheduling, capital budgeting, production planning, inventory control, advertising, and many more areas can be solved by the linear programming approach.

Stanley J. Farlow and Gary M. Haggard in *Finite Mathematics and Its Applications* (New York: Random House, 1988)

In the first section of this chapter, we solved systems of equations in two unknowns. Now let us consider systems of inequalities in two unknowns. The techniques we develop are used in calculus when discussing functions of two variables and in business and economics in the study of linear programming.

Let a, b, and c denote real numbers, and assume that a and b are not both zero. Then all of the following are called **linear inequalities**:

$$ax + by + c < 0 \qquad ax + by + c > 0$$
$$ax + by + c \le 0 \qquad ax + by + c \ge 0$$

An ordered pair of numbers (x_0, y_0) is said to be a **solution** of a given inequality (linear or not) provided we obtain a true statement upon substituting x_0 and y_0 for x and y, respectively. For instance, $(-1, 1)$ is a solution of the inequality $2x + 3y - 6 < 0$, since substitution yields

$$2(-1) + 3(1) - 6 < 0$$
$$-5 < 0 \qquad \text{True}$$

Like a linear equation in two unknowns, a linear inequality has infinitely many solutions. For this reason, we often represent the solutions graphically. When we do this, we say that we are **graphing the inequality**.

For example, let us graph the inequality $y < 2x$. First we observe that the coordinates of the points on the line $y = 2x$ do not, by definition, satisfy this inequality. So it remains to consider points above the line and points below the line. We will in fact show that the required graph consists of all points that lie below the line. Take any point $P(x_0, y_0)$ not on the line $y = 2x$. Let $Q(x_0, 2x_0)$ be the point on $y = 2x$ with the same first coordinate as P. Then, as indicated in Figure 1, the point P lies below the line if and only if the y-coordinate of P is less than the y-coordinate of Q. In other words, (x_0, y_0) lies below the line if and only if $y_0 < 2x_0$. This last statement is equivalent to saying that (x_0, y_0) satisfies the inequality $y < 2x$. This shows that the graph of $y < 2x$ consists of all points below the line $y = 2x$; see Figure 2. The broken line in Figure 2 indicates that

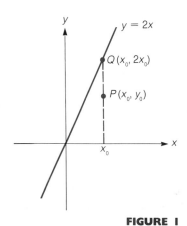

FIGURE 1

the points on $y = 2x$ are not part of the required graph. If the original inequality had been $y \leq 2x$, then we would use a solid line rather than a broken one, indicating that the line is included in the graph. And if the original inequality had been $y > 2x$, the graph would be the region above the line.

Just as the graph of $y < 2x$ is the region below the line $y = 2x$, it is true in general that the graph of $y < f(x)$ is the region below the graph of the function f. For example, Figures 3 and 4 display the graphs of $y < x^2$ and $y \geq x^2$.

Example 1 summarizes the technique developed so far for graphing an inequality. After this example, we will point out a useful alternative method.

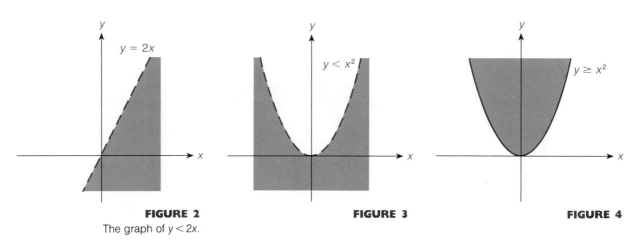

FIGURE 2

The graph of $y < 2x$.

FIGURE 3

FIGURE 4

EXAMPLE I Graph the inequality $4x - 3y \leq 12$.

Solution The graph will include the line $4x - 3y = 12$ and either the region above or the region below the line. To decide which region, solve the inequality for y:

$$4x - 3y \leq 12$$
$$-3y \leq -4x + 12$$
$$y \geq \frac{4}{3}x - 4 \qquad \text{multiplying or dividing by a negative number reverses an inequality}$$

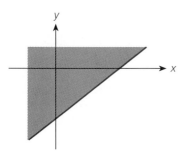

FIGURE 5

The graph of $4x - 3y \leq 12$.

This last inequality tells us that we want the region above the line, as well as the line itself. Figure 5 displays the required graph. ▪▪▪

There is another method that we can use in Example 1 to determine which side of the line we want. This method involves a **test point**. We pick any convenient point that is not on the line $4x - 3y = 12$. Then we test to see if this point satisfies the given inequality. For example, let us pick the point $(0, 0)$. Substituting these coordinates in the inequality $4x - 3y \leq 12$ yields

$$4(0) - 3(0) \overset{?}{\leq} 12$$
$$0 \overset{?}{\leq} 12 \qquad \text{True}$$

We conclude from this that the required side of the line is that which includes the point $(0, 0)$. In agreement with Figure 5, we see this is the region above the line.

Next, we discuss systems of inequalities in two unknowns. As with systems of equations, a **solution** of a system of inequalities is an ordered pair (x_0, y_0) that satisfies all of the inequalities in the system. As a first example, let us graph the points that satisfy the following nonlinear system:

$$\begin{cases} y - x^2 \geq 0 \\ x^2 + y^2 < 1 \end{cases}$$

By writing the first inequality as $y \geq x^2$, we see that it describes the set of points on or above the parabola $y = x^2$. For the second inequality, we must decide whether it describes the points inside or outside the circle $x^2 + y^2 = 1$. One way to do this is to choose $(0, 0)$ as a test point. Substituting the values $x = 0$ and $y = 0$ in the inequality $x^2 + y^2 < 1$ yields the true statement $0^2 + 0^2 < 1$. Since $(0, 0)$ lies within the circle and satisfies the inequality, we conclude that the inequality $x^2 + y^2 < 1$ describes the set of all points within the circle. Now we put our information together. We wish to graph the points that lie on or above the parabola $y = x^2$ but within the circle $x^2 + y^2 = 1$. This is the shaded region shown in Figure 6.

FIGURE 6

[figure: left margin — parabola $y = x^2$ and circle $x^2 + y^2 = 1$ with shaded region]

▌▬▌ **EXAMPLE 2**   Graph the system

$$\begin{cases} -x + 3y \leq 12 \\ x + y \leq 8 \\ x \geq 0 \\ y \geq 0 \end{cases}$$

Solution   Solving the first inequality for y, we have

$$3y \leq x + 12$$
$$y \leq \tfrac{1}{3}x + 4$$

Thus, the first inequality is satisfied by the points on or below the line $-x + 3y = 12$. Similarly, by solving the second inequality for y, we see that it describes the set of points on or below the line $x + y = 8$. The third inequality, $x \geq 0$, describes the points on or to the right of the y-axis. Similarly, the fourth inequality, $y \geq 0$, describes the points on or above the x-axis. We summarize these statements in Figure 7(a). The arrows indicate which side of each line we wish to consider.

FIGURE 7

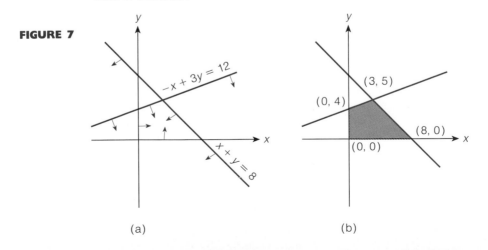

(a)   (b)

Finally, Figure 7(b) shows the graph of the given system. The coordinates of each point in the shaded region satisfy all four of the given inequalities. (Exercise 25 at the end of this section asks you to show that the coordinates listed in the figure are correct.) ∎∎∎

We can use Figure 7(b) to introduce some terminology that is useful in describing sets of points in the plane. The **vertices** of a region are the corners, or points, where the adjacent bounding sides meet. Thus, the vertices of the shaded region in Figure 7(b) are the four points $(0,0)$, $(8,0)$, $(3,5)$, and $(0,4)$. The shaded region in Figure 7(b) is **convex**. This means that given any two points in that region, the straight line segment joining these two points lies wholly within the region. Figure 8(a) displays another example of a convex set, while Figure 8(b) shows a set that is not convex. The shaded region in Figure 8(b) is also an example of a **bounded region**. By this we mean that the region can be wholly contained within some (sufficiently large) circle. Perhaps the simplest example of a region that is not bounded is the entire x-y plane itself. Figure 9 shows another example of an unbounded region.

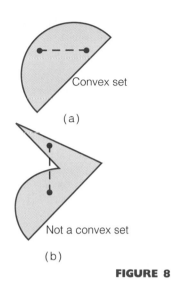

Convex set

(a)

Not a convex set

(b)

FIGURE 8

FIGURE 9

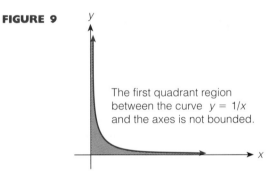

The first quadrant region between the curve $y = 1/x$ and the axes is not bounded.

EXERCISE SET 9.7

A

In Exercises 1 and 2, decide whether or not the ordered pairs are solutions of the given inequality.

1. $4x - 6y + 3 \geq 0$ (a) $(1,2)$ (b) $\left(0, \frac{1}{2}\right)$
2. $5x + 2y < 1$ (a) $(-1,3)$ (b) $(0,0)$

In Exercises 3–16, graph the given inequalities.

3. $2x - 3y > 6$
4. $2x - 3y < 6$
5. $2x - 3y \geq 6$
6. $2x - 3y \leq 6$
7. $x - y < 0$
8. $y \leq \frac{1}{2}x - 1$
9. $x \geq 1$
10. $y < 0$
11. $x > 0$
12. $y \leq \sqrt{x}$
13. $y > x^3 + 1$
14. $y \leq |x - 2|$
15. $x^2 + y^2 \geq 25$
16. $y \geq e^x - 1$

In Exercises 17–22, graph the systems of inequalities.

17. $\begin{cases} y \leq x^2 \\ x^2 + y^2 \leq 1 \end{cases}$

18. $\begin{cases} y < x \\ x^2 + y^2 < 1 \end{cases}$

19. $\begin{cases} y \geq 1 \\ y \leq |x| \end{cases}$

20. $\begin{cases} x \geq 0 \\ y \geq 0 \\ y < \sqrt{x} \\ x \leq 4 \end{cases}$

21. $\begin{cases} x \geq 0 \\ y \geq 0 \\ y \leq 1 - x^2 \end{cases}$

22. $\begin{cases} y < 2x \\ y > \frac{1}{2}x \end{cases}$

In Exercises 23–34, graph the systems of linear inequalities. In each case specify the vertices. Is the region convex? Is the region bounded?

23. $\begin{cases} y \leq x + 5 \\ y \leq -2x + 14 \\ x \geq 0 \\ y \geq 0 \end{cases}$

24. $\begin{cases} y \geq x + 5 \\ y \geq -2x + 14 \\ x \geq 0 \\ y \geq 0 \end{cases}$

25. $\begin{cases} -x + 3y \leq 12 \\ x + y \leq 8 \\ x \geq 0 \\ y \geq 0 \end{cases}$

26. $\begin{cases} y \geq 2x \\ y \geq -x + 6 \end{cases}$

27. $\begin{cases} 0 \leq 2x - y + 3 \\ x + 3y \leq 23 \\ 5x + y \leq 45 \end{cases}$

28. $\begin{cases} 0 \leq 2x - y + 3 \\ x + 3y \leq 23 \\ 5x + y \leq 45 \\ x \geq 0 \\ y \geq 0 \end{cases}$

29. $\begin{cases} 5x + 6y < 30 \\ y > 0 \end{cases}$

30. $\begin{cases} 5x + 6y < 30 \\ x > 0 \end{cases}$

31. $\begin{cases} 5x + 6y < 30 \\ x > 0 \\ y > 0 \end{cases}$

32. $\begin{cases} 2x + 3y \geq 6 \\ 2x + 3y \leq 12 \end{cases}$

33. $\begin{cases} x \geq 0 \\ y \geq 0 \\ 20 - x \geq 0 \\ 30 - y \geq 0 \\ x + y \leq 40 \\ x + y \geq 35 \end{cases}$

34. $\begin{cases} x \geq 0 \\ y \geq 0 \\ 3x - y + 1 \geq 0 \\ 0 \leq x - y + 3 \\ y \leq 5 \\ x \leq \frac{1}{3}(17 - y) \\ x \leq \frac{1}{2}(y + 8) \end{cases}$

B

35. Graph each system of inequalities and specify the vertices.

(a) $\begin{cases} x \geq 0 \\ y \geq e^x \\ y \leq e^{-x} + 1 \end{cases}$

(b) $\begin{cases} 0 \leq x \leq 2\pi \\ y \geq \sin x \\ y \leq \cos x \end{cases}$

A formula such as

$$f(x, y) = \sqrt{2x - y + 1} \qquad (1)$$

defines a function of two variables. The inputs for such a function are ordered pairs (x, y) of real numbers. For example, using the ordered pair $(3, 5)$ as an input, we have

$$f(3, 5) = \sqrt{2(3) - 5 + 1} = \sqrt{2}$$

So the input $(3, 5)$ yields an output of $\sqrt{2}$. We define the domain for this function just as we did in Chapter 3: The domain is the set of all inputs that yield real-number outputs. For instance, the ordered pair $(1, 4)$ is not in the domain of the function we have been discussing because (as you should check for yourself) $f(1, 4) = \sqrt{-1}$, which is not a real number. We can determine the domain of the function in equation (1) by requiring that the quantity under the radical sign be nonnegative. Thus, we require that $2x - y + 1 \geq 0$ and, consequently, $y \leq 2x + 1$. (Check this.) The following figure shows the graph of this inequality; the domain of our function is the set of ordered pairs making up the graph. In Exercises 36–41, follow a similar procedure and sketch the domain of the given function.

36. $f(x, y) = \sqrt{x + y + 2}$

37. $f(x, y) = \sqrt{x^2 + y^2 - 1}$

38. $g(x, y) = \sqrt{25 - x^2 - y^2}$

39. $g(x, y) = \ln(x^2 - y)$

40. $h(x, y) = \ln(xy)$

41. $h(x, y) = \sqrt{x} + \sqrt{y}$

 CHAPTER NINE SUMMARY OF PRINCIPAL TERMS

TERMS	PAGE REFERENCE	COMMENT
1. Linear equation in two variables	502	A linear equation in two variables is an equation of the form $ax + by = c$, where a, b, and c are constants and x and y are variables or unknowns. Similarly, a linear equation in three variables is an equation of the form $ax + by + cz = d$.
2. Solution of a linear equation	502	A solution of the linear equation $ax + by = c$ is an ordered pair of numbers (x_0, y_0) such that $ax_0 + by_0 = c$. Similarly, a solution of the linear equation $ax + by + cz = d$ is an ordered triple of numbers (x_0, y_0, z_0) such that $ax_0 + by_0 + cz_0 = d$.
3. Consistent system; inconsistent system	503	A system of equations is *consistent* if it has at least one solution, otherwise, the system is *inconsistent*. See Figures 2, 3, and 4 in Section 9.1 for a geometric interpretation of these terms.

TERMS	PAGE REFERENCE	COMMENT
4. Upper-triangular form (three variables)	513	A system of linear equations in x, y, and z is said to be in *upper-triangular form* if x appears in no equation after the first, and y appears in no equation after the second. This definition can be extended to include systems with any number of unknowns. See Table 1 in Section 9.2 for examples of systems that are in upper-triangular form. When a system is in upper-triangular form, it is a simple matter to obtain the solutions, see, for instance, Examples 1 and 2 in Section 9.2.
5. Gaussian elimination	512, 515	This is a technique for converting a system of equations to upper-triangular form. See Examples 4 and 5 in Section 9.2.
6. Elementary operations	515	These are operations that can be performed on an equation in a system without altering the solution set. See the box on page 515 for a list of these operations.
7. Matrix	523	A *matrix* is a rectangular array of numbers, enclosed in parentheses or brackets. The numbers constituting the rectangular array are called the *entries* or the *elements* in the matrix. The *size* or *dimension* of a matrix is expressed by specifying the number of rows and the number of columns, in that order. For examples of this terminology, see Example 1 in Section 9.3.
8. Matrix equality	527	Two matrices are said to be *equal* provided they are the same size and their corresponding entries are equal.
9. Matrix addition and subtraction	528	To add two matrices of the same size, add the corresponding entries. Similarly, to subtract two matrices of the same size, subtract the corresponding entries.
10. Matrix multiplication	531	Let A and B be two matrices. The matrix product AB is defined only when the number of columns in A is the same as the number of rows in B. In this case, the matrix AB will have as many rows as A and as many columns as B. The entry in the ith row and jth column of AB is the number formed as follows: Multiply the corresponding entries in the ith row of A and the jth column of B, then add the results.
11. Square matrix	535	A matrix in which there are as many rows as there are columns is called a *square matrix*. An $n \times n$ square matrix is said to be an nth-order square matrix.
12. Multiplicative identity matrix	535, 539	For the set of $n \times n$ matrices, the multiplicative identity matrix I_n is the $n \times n$ matrix with ones down the main diagonal (upper left corner to bottom right corner) and zeros everywhere else. For example, $$I_2 = \begin{pmatrix} 1 & 0 \\ 0 & 1 \end{pmatrix} \quad \text{and} \quad I_3 = \begin{pmatrix} 1 & 0 & 0 \\ 0 & 1 & 0 \\ 0 & 0 & 1 \end{pmatrix}$$ For every $n \times n$ matrix A, we have $AI_n = A$ and $I_n A = A$.
13. Inverse matrix	535, 539	Given an $n \times n$ matrix A, the inverse matrix (if it exists) is denoted by A^{-1}. The defining equations for A^{-1} are $AA^{-1} = I$ and $A^{-1}A = I$. Here, I stands for the multiplicative identity matrix that is the same size as A. If a square matrix has an inverse, it is said to be *nonsingular*; if there is no inverse, then the matrix is *singular*.

TERMS	PAGE REFERENCE	COMMENT
14. Minor	543, 544	The *minor* of an entry b in a determinant is the determinant formed by suppressing the entries in the row and column in which b appears.
15. Cofactor	543, 544	If an entry b appears in the ith row and the jth column of a determinant, then the *cofactor* of b is computed by multiplying the minor of b times the number $(-1)^{i+j}$.
16. Determinant	542, 544, 545	The value of the 2×2 determinant $\begin{vmatrix} a & b \\ c & d \end{vmatrix}$ is defined to be $ad - bc$. For larger determinants, the value can be found as follows. Select any row or column and multiply each entry in that row or column by its cofactor. Then add the results. The resulting sum is the value of the determinant. It can be shown that this value is independent of the particular row or column that is chosen.
17. Cramer's rule	548	This rule yields the solutions of certain systems of linear equations in terms of determinants. For a statement of the rule, see page 548. Example 5 on page 549 shows how the rule is applied. The proof of Cramer's rule begins on page 550.
18. Linear inequality in two variables	562	A linear inequality in two variables is any one of the four types of inequalities that result when the equal sign in the equation $ax + by = c$ is replaced by one of the four symbols $<, \leq, >, \geq$. For any of these linear inequalities, a *solution* is an ordered pair of numbers (x_0, y_0) with the property that a true statement is obtained when x and y (in the inequality) are replaced by x_0 and y_0, respectively.
19. Vertices	565	The *vertices* of a region in the x-y plane are the corners or points where the adjacent bounding sides meet.
20. Convex region	565	A region in the x-y plane is *convex* if, given any two points in the region, the line segment joining those points lies wholly within the region.
21. Bounded region	565	A region in the x-y plane is *bounded* if it can be wholly contained within some (sufficiently large) circle.

WRITING MATHEMATICS

On your own or with a group of classmates, complete the following exercise. Then (strictly on your own) write out a detailed solution. This will involve a combination of English composition and algebra (much like the exposition in this textbook). At each stage, be sure to tell the reader (in complete sentences) where you are headed and why each of the main steps is necessary.

Three integers are said to form a *Pythagorean triple* if the square of the largest is equal to the sum of the squares of the other two. For example, 3, 4, and 5 form a Pythagorean triple because $3^2 + 4^2 = 5^2$. Do you know other Pythagorean triples? This exercise shows how to develop expressions for generating an infinite number of Pythagorean triples.

(a) Refer to the following figure. Suppose that the coordinates of the point P are $(a/c, b/c)$, where a, b, and c are integers. Explain why a, b, and c form a Pythagorean triple.

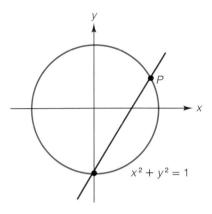

$$x^2 + y^2 = 1$$

(b) Let m denote the slope of the line passing through P and $(0, -1)$. Using the methods of Section 9.6, show that the coordinates of the point P are

$$\left(\frac{2m}{m^2 + 1}, \frac{m^2 - 1}{m^2 + 1} \right)$$

(c) Use the results in parts (a) and (b) to explain why the three numbers $2m$, $m^2 - 1$, and $m^2 + 1$ form a Pythagorean triple. Then complete the following table to obtain examples of Pythagorean triples.

m	$2m$	$m^2 - 1$	$m^2 + 1$
2			
4			
6			
8			

CHAPTER NINE   REVIEW EXERCISES

In Exercises 1–38, solve each system of equations. If there are no solutions in a particular case, say so. In cases where there are literal (rather than numerical) coefficients, specify any restrictions that your solutions impose on those coefficients.

1. $\begin{cases} x + y = -2 \\ x - y = 8 \end{cases}$

2. $\begin{cases} x - y = 1 \\ x + y = 5 \end{cases}$

3. $\begin{cases} 2x + y = 2 \\ x + 2y = 7 \end{cases}$

4. $\begin{cases} 3x + 2y = 6 \\ 5x + 4y = 4 \end{cases}$

5. $\begin{cases} 7x + 2y = 9 \\ 4x + 5y = 63 \end{cases}$

6. $\begin{cases} \dfrac{x}{2} + \dfrac{y}{3} = 9 \\ \dfrac{x}{5} - \dfrac{y}{2} = -4 \end{cases}$

7. $\begin{cases} 2x - \dfrac{y}{2} = -8 \\ \dfrac{x}{3} + \dfrac{y}{8} = -1 \end{cases}$

8. $\begin{cases} 3x - 14y - 1 = 0 \\ -6x + 28y - 3 = 0 \end{cases}$

9. $\begin{cases} 3x + 5y - 1 = 0 \\ 9x - 10y - 8 = 0 \end{cases}$

10. $\begin{cases} 9x + 15y - 1 = 0 \\ 6x + 10y + 1 = 0 \end{cases}$

11. $\begin{cases} \frac{2}{3}x = -\frac{1}{2}y - 12 \\ \dfrac{x}{2} = y + 2 \end{cases}$

12. $\begin{cases} 0.1x + 0.2y = -5 \\ -0.2x - 0.5y = 13 \end{cases}$

13. $\begin{cases} \dfrac{1}{x} + \dfrac{1}{y} = -1 \\ \dfrac{2}{x} + \dfrac{5}{y} = -14 \end{cases}$

14. $\begin{cases} \dfrac{2}{x} + \dfrac{15}{y} = -9 \\ \dfrac{1}{x} + \dfrac{10}{y} = -2 \end{cases}$

15. $\begin{cases} ax + (1 - a)y = 1 \\ (1 - a)x + y = 0 \end{cases}$

16. $\begin{cases} ax - by - 1 = 0 \\ (a - 1)x + by + 2 = 0 \end{cases}$

17. $\begin{cases} 2x - y = 3a^2 - 1 \\ 2y + x = 2 - a^2 \end{cases}$

18. $\begin{cases} 3ax + 2by = 3a^2 - ab + 2b^2 \\ 3bx + 2ay = 2a^2 + 5ab - 3b^2 \end{cases}$

19. $\begin{cases} 5x - y = 4a^2 - 6b^2 \\ 2x + 3y = 5a^2 + b^2 \end{cases}$

20. $\begin{cases} \dfrac{2b}{x} - \dfrac{3}{y} = 7ab \\ \dfrac{4a}{x} + \dfrac{5a}{by} = 3a^2 \end{cases}$

21. $\begin{cases} px - qy = q^2 \\ qx + py = p^2 \end{cases}$

22. $\begin{cases} x - y = \dfrac{a - b}{a + b} \\ x + y = 1 \end{cases}$

23. $\begin{cases} \dfrac{4a}{x} - \dfrac{3b}{y} = a - 7b \\ \dfrac{3a^2}{x} - \dfrac{2b^2}{y} = (3a + b)(a - 2b) \end{cases}$

24. $\begin{cases} 6b^2 x^{-1/2} - 5a^2 y^{-1/2} - a^2 b^2 = 0 \\ 2b^2 x^{-1/2} + a^2 y^{-1/2} - 3a^2 b^2 = 0 \end{cases}$

   Hint: Let $u = x^{-1/2}$ and $v = y^{-1/2}$.

25. $\begin{cases} x + y + z = 9 \\ x - y - z = -5 \\ 2x + y - 2z = -1 \end{cases}$

26. $\begin{cases} x - 4y + 2z = 9 \\ 2x + y + z = 3 \\ 3x - 2y - 3z = -18 \end{cases}$

27. $\begin{cases} 4x - 4y + z = 4 \\ 2x + 3y + 3z = -8 \\ x + y + z = -3 \end{cases}$

28. $\begin{cases} x - 8y + z = 1 \\ 5x + 16y + 3z = 3 \\ 4x - 4y - 4z = -4 \end{cases}$

29. $\begin{cases} -2x + y + z = 1 \\ x - 2y + z = -2 \\ x + y - 2z = 4 \end{cases}$

30. $\begin{cases} -x + y + z = 1 \\ x - y + z = -1 \\ x + y - z = 1 \end{cases}$

31. $\begin{cases} 4x + 2y - 3z = 15 \\ 2x + y + 3z = 3 \end{cases}$

32. $\begin{cases} 3x + 2y + 17z = 1 \\ x + 2y + 3z = 3 \end{cases}$

33. $\begin{cases} x + 2y - 3z = -2 \\ 2x - y + z = 1 \\ 3x - 4y + 5z = 1 \end{cases}$

34. $\begin{cases} 9x + y + z = 0 \\ -3x + y - z = 0 \\ 3x - 5y + 3z = 0 \end{cases}$

35. $\begin{cases} x + y + z = a + b \\ 2x - y + 2z = -a + 5b \\ x - 2y + z = -2a + 4b \end{cases}$

36. $\begin{cases} ax + by - 2az = 4ab + 2b^2 \\ x + y + z = 4a + 2b \\ bx + ay + 4az = 5a^2 + b^2 \end{cases}$

37. $\begin{cases} x + y + z + w = 8 \\ 3x + 3y - z - w = 20 \\ 4x - y - z + 2w = 18 \\ 2x + 5y + 5z - 5w = 8 \end{cases}$

38. $\begin{cases} x - 2y + 3z + w = 1 \\ x + y + z + w = 5 \\ 2x + 3y + 2z - w = 3 \\ 3x + y - z + 2w = 4 \end{cases}$

For Exercises 39–50, determine the constants (denoted by capital letters) so that each equation is an identity.

39. $\dfrac{1}{(x - 10)(x + 10)} = \dfrac{A}{x - 10} + \dfrac{B}{x + 10}$

40. $\dfrac{x}{(x - 5)(x + 5)} = \dfrac{A}{x - 5} + \dfrac{B}{x + 5}$

41. $\dfrac{2x}{(x + 1)^2} = \dfrac{A}{x + 1} + \dfrac{B}{(x + 1)^2}$

42. $\dfrac{x + 2}{(x - 1)^2} = \dfrac{A}{x - 1} + \dfrac{B}{(x - 1)^2}$

43. $\dfrac{5}{x(x - 4)} = \dfrac{A}{x} + \dfrac{B}{x - 4}$

44. $\dfrac{6x}{(x + 2)(x + 3)} = \dfrac{A}{x + 2} + \dfrac{B}{x + 3}$

45. $\dfrac{1}{(x - 1)(x + 3)^2} = \dfrac{A}{x - 1} + \dfrac{B}{x + 3} + \dfrac{C}{(x + 3)^2}$

46. $\dfrac{x^2}{(x + 2)(x - 2)^2} = \dfrac{A}{x + 2} + \dfrac{B}{x - 2} + \dfrac{C}{(x - 2)^2}$

47. $\dfrac{4x^2 + 2x + 15}{(x - 1)(x^2 + x + 5)} = \dfrac{A}{x - 1} + \dfrac{Bx + C}{x^2 + x + 5}$

48. $\dfrac{1}{x(x^2 + 16)} = \dfrac{A}{x} + \dfrac{Bx + C}{x^2 + 16}$

49. $\dfrac{1}{x^3 + 64} = \dfrac{A}{x + 4} + \dfrac{Bx + C}{x^2 - 4x + 16}$

50. $\dfrac{1}{x^3 + 3x^2 + 3x + 1} = \dfrac{A}{x + 1} + \dfrac{B}{(x + 1)^2} + \dfrac{C}{(x + 1)^3}$

In Exercises 51–54, the lowercase letters a and b denote nonzero constants, with $a \neq b$. In each case, determine the values of A and B (and C, if applicable) so that the equation is an identity.

51. $\dfrac{x}{(x - a)^3} = \dfrac{A}{x - a} + \dfrac{B}{(x - a)^2} + \dfrac{C}{(x - a)^3}$

52. $\dfrac{ax(1 - a)}{(x - a)(x + a^2)} = \dfrac{A}{x - a} + \dfrac{B}{x + a^2}$

53. $\dfrac{(a - b)(a + b - x)}{(x - a)(x - b)} = \dfrac{A}{x - a} + \dfrac{B}{x - b}$

54. $\dfrac{4a^2 x + 6a^3}{(x + a)(x + 2a)(x + 3a)} = \dfrac{A}{x + a} + \dfrac{B}{x + 2a} + \dfrac{C}{x + 3a}$

In Exercises 55–62, evaluate each of the determinants.

55. $\begin{vmatrix} 1 & 5 \\ -6 & 4 \end{vmatrix}$

56. $\begin{vmatrix} \frac{1}{6} & 1 \\ 2 & 12 \end{vmatrix}$

57. $\begin{vmatrix} 4 & 0 & 3 \\ -2 & 1 & 5 \\ 0 & 2 & -1 \end{vmatrix}$

58. $\begin{vmatrix} 2 & 6 & 4 \\ 6 & 18 & 24 \\ 15 & 5 & -10 \end{vmatrix}$

59. $\begin{vmatrix} 1 & 5 & 7 \\ 1 & 5 & 7 \\ 17 & 19 & 21 \end{vmatrix}$

60. $\begin{vmatrix} 0 & 2 & 4 & 0 \\ 4 & 0 & 6 & 2 \\ 0 & 0 & 1 & 1 \\ 14 & 7 & 1 & 0 \end{vmatrix}$

61. $\begin{vmatrix} 1 & 0 & 0 & 0 \\ 0 & 2 & 0 & 0 \\ 0 & 0 & 3 & 0 \\ 0 & 0 & 0 & 4 \end{vmatrix}$

62. $\begin{vmatrix} 1 & a & b & c \\ 0 & 2 & d & e \\ 0 & 0 & 3 & f \\ 0 & 0 & 0 & 4 \end{vmatrix}$

63. Show that $\begin{vmatrix} a & b & c \\ b & c & a \\ c & a & b \end{vmatrix} = 3abc - a^3 - b^3 - c^3.$

64. Show that $\begin{vmatrix} 1 & 1 & 1 \\ 1 & 1+x & 1 \\ 1 & 1 & 1+x^2 \end{vmatrix} = x^3.$

65. Show that
$$\begin{vmatrix} a^2 + x & b & c & d \\ -b & 1 & 0 & 0 \\ -c & 0 & 1 & 0 \\ -d & 0 & 0 & 1 \end{vmatrix} = a^2 + b^2 + c^2 + d^2 + x$$

66. In a two-digit number, the sum of the digits is 11. Four times the units digit exceeds the tens digit by 4. Find the number.

67. Determine constants a and b so that the parabola $y = ax^2 + bx - 1$ passes through the points $(-2, 5)$ and $(2, 9)$.

68. Find two numbers that have a sum and a difference of 52 and 10, respectively.

69. The vertices of triangle ABC are $A(-2, 0)$, $B(4, 0)$, and $C(0, 6)$. Let A_1, B_1, and C_1 denote the midpoints of sides $\overline{BC}$, $\overline{AC}$, and $\overline{AB}$, respectively.
   (a) Find the point where the line segments $\overline{AA_1}$ and $\overline{BB_1}$ intersect.   *Answer:* $\left(\frac{2}{3}, 2\right)$
   (b) Follow the instructions given in part (a) using $\overline{BB_1}$ and $\overline{CC_1}$.
   (c) Follow the instructions given in part (a) using $\overline{AA_1}$ and $\overline{CC_1}$.
   (d) Let P denote the point $\left(\frac{2}{3}, 2\right)$ that you found in part (a). Compute each of the following ratios: AP/PA_1, BP/PB_1, CP/PC_1. What do you observe?

70. An *altitude* of a triangle is a line segment drawn from a vertex perpendicular to the opposite side. Suppose that the vertices of triangle ABC are as given in Exercise 69. Find the intersection point for each pair of altitudes. What do you observe about the three answers?

71. This exercise appears in *Plane and Solid Analytic Geometry,* by W. F. Osgood and W. G. Graustein (New York: Macmillan, 1920): Let P be any point (a, a) of the line $x - y = 0$, other than the origin. Through P draw two lines, of arbitrary slopes m_1 and m_2, intersecting in the x-axis in A_1 and A_2, and the y-axis in B_1 and B_2, respectively. Prove that the lines $\overline{A_1 B_2}$ and $\overline{A_2 B_1}$ will, in general, meet on the line $x + y = 0$.

72. Determine constants h, k, and r so that the circle $(x - h)^2 + (y - k)^2 = r^2$ passes through the three points $(0, 0)$, $(0, 1)$, and $(1, 0)$.

73. The vertices of a triangle are the points of intersection of the lines $y = x - 1$, $y = -x - 2$, and $y = 2x + 3$. Find the equation of the circle passing through these three intersection points.

71. The vertices of triangle ABC are $A(0, 0)$, $B(3, 0)$, and $C(0, 4)$.
   (a) Find the center and the radius of the circle that passes through A, B, and C. This circle is called the **circumcircle** for triangle ABC.
   (b) The figure shows the **inscribed circle** for triangle ABC; this is the circle that is tangent to all three sides of the triangle. Find the center and the radius of this circle using the following two facts.
      (i) The center of the inscribed circle is the common intersection point of the three angle bisectors of the triangle.
      (ii) The line that bisects angle B has slope $-\frac{1}{2}$.

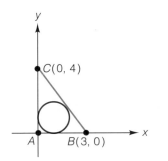

   (c) Verify the following statement for triangle ABC. (The statement actually holds for any triangle; the result is known as *Euler's theorem.*) Let R and r denote the radii of the circles in parts (a) and (b), respectively. Then the distance d between the centers of the circles satisfies the equation
$$d^2 = R^2 - 2rR$$

   (d) Verify the following statement for triangle ABC (the statement actually holds for any triangle): The area of triangle ABC is equal to one-half the product of the perimeter and the radius of the inscribed circle.

   (e) Verify the following statement for triangle ABC (the statement actually holds for any triangle): The sum of the reciprocals of the lengths of the altitudes is equal to the reciprocal of the radius of the inscribed circle.

In Exercises 75–98, the matrices A, B, C, D, E, and F are defined as follows.

$$A = \begin{pmatrix} 3 & -2 \\ 1 & 5 \end{pmatrix} \qquad B = \begin{pmatrix} 2 & 1 \\ 1 & 8 \end{pmatrix} \qquad C = \begin{pmatrix} -1 & 0 \\ 0 & -1 \end{pmatrix}$$

$$D = \begin{pmatrix} -4 & 0 & 6 \\ 1 & 3 & 2 \end{pmatrix} \qquad E = \begin{pmatrix} 3 & -1 \\ 4 & 1 \\ -5 & 9 \end{pmatrix} \qquad F = \begin{pmatrix} 2 & 6 \\ 5 & 3 \\ 5 & 8 \end{pmatrix}$$

In each exercise, carry out the indicated matrix operations if they are defined. If an operation is not defined, say so.

75. $2A + 2B$

76. $2(A + B)$

77. $4B$

78. $B + 4$

79. AB

80. BA

81. $AB - BA$

82. $B + E$

83. $B + C$

84. $A(B + C)$

85. $AB + AC$

86. $(B + C)A$

87. $BA + CA$

88. $D + E$

89. DE

90. $(EE)D$

91. $E(ED)$

92. $E + F$

93. EF

94. $3E - 2F$

95. $(A + B) + C$

96. $A + (B + C)$

97. $(AB)C$

98. $A(BC)$

99. For a square matrix A, the notation A^2 means AA. Similarly, A^3 means AAA. If $A = \begin{pmatrix} 1 & 1 \\ 0 & 1 \end{pmatrix}$, verify that

$$A^2 = \begin{pmatrix} 1 & 2 \\ 0 & 1 \end{pmatrix} \quad \text{and} \quad A^3 = \begin{pmatrix} 1 & 3 \\ 0 & 1 \end{pmatrix}$$

100. Let A be the matrix $\begin{pmatrix} 0 & 0 & 0 \\ a & 0 & 0 \\ b & c & 0 \end{pmatrix}$. Compute A^2 and A^3.

For Exercises 101–104, in each case compute the inverse of the matrix in part (a), and then use that inverse to solve the system of equations in part (b).

101. (a) $\begin{pmatrix} 1 & 5 \\ 2 & 9 \end{pmatrix}$ (b) $\begin{cases} x + 5y = 3 \\ 2x + 9y = -4 \end{cases}$

102. (a) $\begin{pmatrix} 5 & -4 \\ 14 & -11 \end{pmatrix}$ (b) $\begin{cases} 5x - 4y = 2 \\ 14x - 11y = 5 \end{cases}$

103. (a) $\begin{pmatrix} 1 & -2 & 3 \\ 2 & -5 & 10 \\ -1 & 2 & -2 \end{pmatrix}$

(b) $\begin{cases} x - 2y + 3z = -2 \\ 2x - 5y + 10z = -3 \\ -x + 2y - 2z = 6 \end{cases}$

104. (a) $\begin{pmatrix} -1 & 4 & 2 \\ -3 & 10 & 5 \\ 0 & 3 & 1 \end{pmatrix}$

(b) $\begin{cases} -x + 4y + 2z = 8 \\ -3x + 10y + 5z = 0 \\ 3y + z = -12 \end{cases}$

In Exercises 105 and 106, find the inverse of each matrix.

105. $\begin{pmatrix} 5 & 3 & 6 & -7 \\ 3 & -4 & 0 & -9 \\ 0 & 1 & -1 & -1 \\ 2 & 2 & 3 & -2 \end{pmatrix}$

106. $\begin{pmatrix} 1 & -1 & 0 & -3 \\ 5 & -2 & 3 & -11 \\ 2 & 3 & 2 & -3 \\ 4 & 5 & 5 & -5 \end{pmatrix}$

In Exercises 107–114, compute D, D_x, D_y, D_z (and D_w where appropriate) for each system of equations. Use Cramer's rule to solve the systems in which $D \neq 0$. If $D = 0$, solve the system using Gaussian elimination or matrix methods.

107. $\begin{cases} 2x - y + z = 1 \\ 3x + 2y + 2z = 0 \\ x - 5y - 3z = -2 \end{cases}$

108. $\begin{cases} x + 2y - z = -1 \\ 2x - 3y + 3z = 3 \\ 2x + 3y + z = 1 \end{cases}$

109. $\begin{cases} x + 2y + 3z = -1 \\ 4x + 5y + 6z = 2 \\ 7x + 8y + 9z = -3 \end{cases}$

110. $\begin{cases} 3x + 2y - 2z = 0 \\ 2x + 3y - z = 0 \\ 8x + 7y - 5z = 0 \end{cases}$

111. $\begin{cases} 3x + 2y - 2z = 1 \\ 2x + 3y - z = -2 \\ 8x + 7y - 5z = 0 \end{cases}$

112. $\begin{cases} x + y + z + w = 5 \\ x - y - z + w = 3 \\ 2x + 3y + 3z + 2w = 21 \\ 4z - 3w = -7 \end{cases}$

113. $\begin{cases} 2x - y + z + 3w = 15 \\ x + 2y + 2w = 12 \\ 3y + 3z + 4w = 12 \\ -4x + y - 4z = -11 \end{cases}$

114. $\begin{cases} x + y + z = (a + b)^2 \\ \dfrac{bx}{a} + \dfrac{ay}{b} - z = 0 \\ x + y - z = (a - b)^2 \end{cases}$

In Exercises 115–128, find all solutions (x, y) for each system, where x and y are real numbers.

115. $\begin{cases} y = 6x \\ y = x^2 \end{cases}$

116. $\begin{cases} y = 4x \\ y = x^3 \end{cases}$

117. $\begin{cases} y = 9 - x^2 \\ y = x^2 - 9 \end{cases}$

118. $\begin{cases} x^3 - y = 0 \\ xy - 16 = 0 \end{cases}$

119. $\begin{cases} x^2 - y^2 = 9 \\ x^2 + y^2 = 16 \end{cases}$

120. $\begin{cases} 2x + 3y = 6 \\ y = \sqrt{x + 1} \end{cases}$

121. $\begin{cases} x^2 + y^2 = 1 \\ y = \sqrt{x} \end{cases}$

122. $\begin{cases} \dfrac{x}{11} + \dfrac{y}{12} = 2 \\ \dfrac{xy}{132} = 1 \end{cases}$

123. $\begin{cases} x^2 + y^2 = 1 \\ y = 2x^2 \end{cases}$

124. $\begin{cases} x^2 - 3xy + y^2 = -11 \\ 2x^2 + xy - y^2 = 8 \end{cases}$

125. $\begin{cases} x^2 + 2xy + 3y^2 = 68 \\ 3x^2 - xy + y^2 = 18 \end{cases}$

126. $\begin{cases} \dfrac{x^2}{a^2} + \dfrac{y^2}{b^2} = 1 \\ \dfrac{x^2}{b^2} + \dfrac{y^2}{a^2} = 1 \end{cases}$ $\quad (a > b > 0)$

127. $\begin{cases} 2(x-3)^2 - (y+1)^2 = -1 \\ -3(x-3)^2 + 2(y+1)^2 = 6 \end{cases}$
Hint: Let $u = x - 3$ and $v = y + 1$.

128. $\begin{cases} x^4 = y - 1 \\ y - 3x^2 + 1 = 0 \end{cases}$

Exercises 129–134 appear (in German) in an algebra text by Leonhard Euler, first published in 1770. The English versions given here are taken from the translated version, Elements of Algebra, *5th ed., by Leonhard Euler (London: Longman, Orme, and Co., 1840). [This, in turn, has been reprinted by Springer-Verlag (New York, 1984).]*

129. Required two numbers, whose sum may be s, and their proportion as a to b.

Answer: $\dfrac{as}{a+b}$ and $\dfrac{bs}{a+b}$

130. The sum $2a$, and the sum of the squares $2b$, of two numbers being given; to find the numbers.

Answer: $a - \sqrt{b - a^2}$ and $a + \sqrt{b - a^2}$

131. To find three numbers, so that [the sum of] one-half of the first, one-third of the second, and one-quarter of the third, shall be equal to 62; one-third of the first, one-quarter of the second, and one-fifth of the third, equal to 47; and one-quarter of the first, one-fifth of the second, and one-sixth of the third, equal to 38. *Answer:* 24, 60, 120

132. Required two numbers, whose product may be 105, and whose squares [when added] may together make 274.

133. Required two numbers, whose product may be m, and the sum of the squares n $(n \geq 2m)$.

134. Required two numbers such that their sum, their product, and the difference of their squares may all be equal.

In Exercises 135–138, graph each system of inequalities and specify whether the region is convex or bounded.

135. $\begin{cases} x^2 + y^2 \geq 1 \\ y - 4x \leq 0 \\ y - x \geq 0 \\ x \geq 0 \\ y > 0 \end{cases}$

136. $\begin{cases} x^2 + y^2 \leq 1 \\ x \geq 0 \\ y \geq 0 \end{cases}$

137. $\begin{cases} y - \sqrt{x} \leq 0 \\ y \geq 0 \\ x \geq 1 \\ x - 4 \leq 0 \end{cases}$

138. $\begin{cases} y - |x| \leq 0 \\ x + 1 \geq 0 \\ x - 1 \leq 0 \\ y + 1 \geq 0 \end{cases}$

CHAPTER NINE TEST

1. Determine all solutions of the system
$$\begin{cases} 3x + 4y = 12 \\ y = x^2 + 2x + 3 \end{cases}$$

2. Find all solutions of the system
$$\begin{cases} x - 2y = 13 \\ 3x + 5y = -16 \end{cases}$$

3. (a) Find all solutions of the following system using Gaussian elimination:
$$\begin{cases} x + 4y - z = 0 \\ 3x + y + z = -1 \\ 4x - 4y + 5z = -7 \end{cases}$$

   (b) Compute D, D_x, D_y, and D_z for the system in part (a). [Then check your answer in part (a) using Cramer's rule.]

4. Suppose that the matrices A and B are defined as follows:
$$A = \begin{pmatrix} 1 & -3 \\ 2 & -1 \end{pmatrix} \qquad B = \begin{pmatrix} 0 & 4 \\ 1 & 3 \end{pmatrix}$$
   (a) Compute $2A - B$.　　(b) Compute BA.

5. Determine the area of the triangular region in the first quadrant that is bounded by the x-axis and the lines $y = 2x$ and $y = -x + 6$.

6. Find all solutions of the system
$$\begin{cases} \dfrac{1}{2x} + \dfrac{1}{3y} = 10 \\[2mm] -\dfrac{5}{x} - \dfrac{4}{y} = -4 \end{cases}$$

7. Specify the coefficient matrix for the system
$$\begin{cases} x + y - z = -1 \\ 2x - y + 2z = 11 \\ x - 2y + z = 10 \end{cases}$$

   Also specify the augmented matrix for this system.

8. Use matrix methods to find all solutions of the system displayed in the previous problem.

9. Find the equation of a line passing through the point of intersection of the lines $x + y = 11$ and $3x + 2y = 7$ and perpendicular to the line $2x - 4y = 7$.

10. Consider the determinant
$$\begin{vmatrix} 2 & 3 & -1 \\ 0 & 1 & 4 \\ 5 & -2 & 6 \end{vmatrix}$$

   (a) What is the minor of the entry in the third row, second column?
   (b) What is the cofactor of the entry in the third row, second column?

11. Evaluate the determinant
$$\begin{vmatrix} 4 & -5 & 0 \\ -8 & 10 & 7 \\ 16 & 20 & 14 \end{vmatrix}$$

12. Find all solutions of the system
$$\begin{cases} x^2 + y^2 = 15 \\ xy = 5 \end{cases}$$

13. Find the solutions of the system
$$\begin{cases} A + 2B + 3C = 1 \\ 2A - B - C = 2 \end{cases}$$

14. (a) Determine the inverse of the following matrix:
$$\begin{pmatrix} 10 & -2 & 5 \\ 6 & -1 & 4 \\ 1 & 0 & 1 \end{pmatrix}$$

(b) Use the inverse matrix determined in part (a) to solve the following system.

$$\begin{cases} 10u - 2v + 5w = -1 \\ 6u - v + 4w = -2 \\ u + w = 3 \end{cases}$$

15. Graph the inequality $5x - 6y \geq 30$.

16. Determine the constants P and Q, given that the parabola $y = Px^2 + Qx - 5$ passes through the two points $(-2, -1)$ and $(-1, -2)$.

17. Graph the inequality $x^2 - 4x + y^2 + 3 > 0$. Is the solution set bounded? Is it convex?

18. Graph the following system of inequalities and specify the vertices:

$$\begin{cases} x \geq 0 \\ y \geq 0 \\ 2y - x \leq 14 \\ x + 3y \leq 36 \\ 9x + y \leq 99 \end{cases}$$

In Exercises 19 and 20, determine the constants (denoted by capital letters) so that each equation is an identity.

19. $\dfrac{4}{(x - 5)(x + 5)} = \dfrac{A}{x - 5} + \dfrac{B}{x + 5}$

20. $\dfrac{x - 2}{(x + 1)(x - 1)^2} = \dfrac{A}{x + 1} + \dfrac{B}{x - 1} + \dfrac{C}{(x - 1)^2}$

CHAPTER TEN

ANALYTIC GEOMETRY

The Greeks knew the properties of the curves given by cutting a cone with a plane — the ellipse, the parabola and hyperbola. Kepler discovered by analysis of astronomical observations, and Newton proved mathematically ... that the planets move in ellipses. The geometry of Ancient Greece thus became the cornerstone of modern astronomy.

John Lighton Synge (1897–1987)

Navigator, lay in a conic section flight path to the cloud's center.

Captain James T. Kirk in *Star Trek, The Motion Picture*

INTRODUCTION

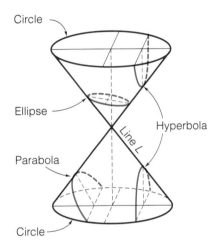

In this chapter we study analytic geometry, with an emphasis on the *conic sections*. The **conic sections** (or **conics**, for short) are the curves formed when a plane intersects the surface of a right cylindrical cone. As indicated in the figure in the margin, these curves are the circle, the ellipse, the hyperbola, and the parabola.

The study of the conics dates back over 2000 years to ancient Greece, where Apollonius of Perga (262–190 B.C.) wrote an eight-volume treatise on the subject. However, until the seventeenth century, the conic sections were studied only as a portion of pure (as opposed to applied) mathematics. Then, in the seventeenth century, it was discovered that the conic sections were crucial in expressing some of the most important laws of nature. This is essentially the observation that is made in the opening quotation by the physicist J. L. Synge.

10.1 THE BASIC EQUATIONS

As background for the work in this chapter, you need to be familiar with the following results from Chapter 2.

1. The distance formula: $d = \sqrt{(x_2 - x_1)^2 + (y_2 - y_1)^2}$

2. The equation for a circle: $(x - h)^2 + (y - k)^2 = r^2$

3. The slope of a line: $m = \dfrac{y_2 - y_1}{x_2 - x_1}$

4. The point–slope formula: $y - y_1 = m(x - x_1)$

5. The slope–intercept formula: $y = mx + b$

6. The condition for two nonvertical lines to be parallel: $m_1 = m_2$

7. The condition for two nonvertical lines to be perpendicular: $m_1 m_2 = -1$

8. The midpoint formula: $(x_0, y_0) = \left(\dfrac{x_1 + x_2}{2}, \dfrac{y_1 + y_2}{2} \right)$

In this section we are going to develop two additional results to supplement those we just listed. The first of these results concerns the slope of a line. We begin by defining the **angle of inclination** (or, simply, the **inclination**) of a line to be the angle θ measured counterclockwise from the positive side or positive direction of the x-axis to the line (see Figure 1). If θ denotes the angle of inclination, we always have $0° \leq \theta < 180°$ if θ is measured in degrees and $0 \leq \theta < \pi$ if θ is in radians. (Notice that when $\theta = 0°$, the line is horizontal.)

As you might suspect, there is a simple relationship between the angle of inclination and the slope of a line. We state this relationship in the box.

PROPERTY SUMMARY   SLOPE AND THE ANGLE OF INCLINATION

The slope m of a line and the angle of inclination θ are related by the equation

$$m = \tan \theta$$

FIGURE 1
The angle of inclination.

FIGURE 2

The formula $m = \tan \theta$ provides a useful connection between elementary coordinate geometry and trigonometry. To derive this formula, we can work with the line $y = mx$ rather than $y = mx + b$ (because these lines are parallel and therefore have equal angles of inclination). As indicated in Figure 2, we let P denote the point where the line $y = mx$ intersects the unit circle. Then the coordinates of P are ($\cos \theta$, $\sin \theta$), and we can compute the slope m using the point P and the origin:

$$m = \frac{y_2 - y_1}{x_2 - x_1} = \frac{\sin \theta - 0}{\cos \theta - 0} = \frac{\sin \theta}{\cos \theta} = \tan \theta \qquad \text{as required}$$

EXAMPLE 1   Determine the acute angle θ between the x-axis and the line $y = 2x - 1$. (See Figure 3.)

Solution   From the equation $y = 2x - 1$, we read directly that $m = 2$. Then, since $\tan \theta = m$, we conclude that

$$\tan \theta = 2 \qquad \text{or} \qquad \theta = \tan^{-1} 2$$

FIGURE 3

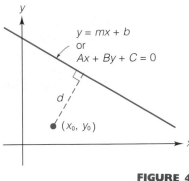

FIGURE 4

If required, we can obtain an approximate numerical value for θ with a calculator. For many types of calculators, the sequence of keystrokes is

2 (INV) (TAN)

When expressed in degrees, the result in this case is

$\theta \approx 63.4°$

In Section 2.1, we reviewed the formula for the distance between two points. Now we consider another kind of distance formula, one that gives the distance d from a point to a line. As indicated in Figure 4, distance in this context means the *shortest* distance, which is the perpendicular distance. In the box that follows, we show two equivalent forms for this formula. Although the second form is more widely known, the first is just as useful and somewhat simpler to derive.

PROPERTY SUMMARY   DISTANCE FROM A POINT TO A LINE

1. The distance d from the point (x_0, y_0) to the line $y = mx + b$ is given by

$$d = \frac{|mx_0 + b - y_0|}{\sqrt{1 + m^2}}$$

2. The distance d from the point (x_0, y_0) to the line $Ax + By + C = 0$ is given by

$$d = \frac{|Ax_0 + By_0 + C|}{\sqrt{A^2 + B^2}}$$

We will derive the first formula in the box with the aid of Figure 5. (Each dashed line in the figure is parallel to the x- or y-axis.)

FIGURE 5

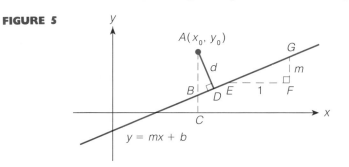

In Figure 5, $\triangle ABD$ is similar to $\triangle EGF$. (Exercise 44 asks you to verify this.) So we have

$$\frac{AB}{EG} = \frac{AD}{EF} \qquad \text{corresponding sides of similar triangles are proportional}$$

and therefore

$$\frac{AB}{\sqrt{1 + m^2}} = \frac{d}{1} \qquad \text{using the Pythagorean theorem and the fact that } EF = 1$$

or

$$d = \frac{AB}{\sqrt{1 + m^2}} \qquad (1)$$

Next, from Figure 5 we have

$$AB = AC - BC$$
$$= y_0 - (mx_0 + b) \qquad \text{(Why?)}$$

Now we can use this last equation to substitute for AB in equation (1). This yields

$$d = \frac{y_0 - (mx_0 + b)}{\sqrt{1 + m^2}}$$

For the general case (in which the point and line may not be situated as in Figure 5), we need to use the absolute value of the quantity in the numerator, to assure that AB and d are nonnegative. We then have

$$d = \frac{|y_0 - (mx_0 + b)|}{\sqrt{1 + m^2}} = \frac{|mx_0 + b - y_0|}{\sqrt{1 + m^2}} \qquad \text{as required}$$

EXAMPLE 2

(a) Find the distance from the point $(-3, 1)$ to the line $y = -2x + 7$; see Figure 6.
(b) Find the equation of the circle that has center $(-3, 1)$ and that is tangent to the line $y = -2x + 7$.

Solution

(a) To find the distance from the point $(-3, 1)$ to the line $y = -2x + 7$, we use the formula $d = |mx_0 + b - y_0|/\sqrt{1 + m^2}$ with $x_0 = -3$, $y_0 = 1$, $m = -2$, and $b = 7$. This yields

$$d = \frac{|(-2)(-3) + 7 - 1|}{\sqrt{1 + (-2)^2}} = \frac{12}{\sqrt{5}} = \frac{12\sqrt{5}}{5} \text{ units}$$

FIGURE 6

(b) The equation of a circle with center (h, k) and radius of length r is $(x - h)^2 + (y - k)^2 = r^2$. We are given here that (h, k) is $(-3, 1)$. Furthermore, the distance determined in part (a) is the length of the radius. (A theorem from geometry tells us that the radius drawn to the point of tangency is perpendicular to the tangent.) Thus, the equation of the required circle is

$$[x - (-3)]^2 + (y - 1)^2 = \left(12\sqrt{5}/5\right)^2$$

or

$$(x + 3)^2 + (y - 1)^2 = \frac{144}{5}$$

EXAMPLE 3   Find the area of triangle ABC in Figure 7.

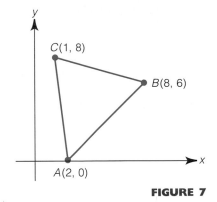

FIGURE 7

Solution The area of any triangle is one-half the product of the base and the height. Let us view $\overline{AB}$ as the base. Then, using the formula for the distance between two points, we have

$$AB = \sqrt{(x_2 - x_1)^2 + (y_2 - y_1)^2}$$
$$= \sqrt{(8 - 2)^2 + (6 - 0)^2}$$
$$= \sqrt{36 + 36} = \sqrt{2 \times 36} = 6\sqrt{2} \text{ units}$$

With $\overline{AB}$ as the base, the height of the triangle is the perpendicular distance from C to $\overline{AB}$. To compute that distance, we first need the equation of the line through A and B. The slope of the line is

$$m = \frac{y_2 - y_1}{x_2 - x_1} = \frac{6 - 0}{8 - 2} = \frac{6}{6} = 1$$

Then, using the point $(2, 0)$ and the slope $m = 1$, we have

$$y - y_1 = m(x - x_1)$$
$$y - 0 = 1(x - 2) \qquad \text{and therefore} \qquad y = x - 2$$

We can now compute the height of the triangle by finding the perpendicular distance from $C(1, 8)$ to $y = x - 2$:

$$\text{height} = \frac{|mx_0 + b - y_0|}{\sqrt{1 + m^2}}$$
$$= \frac{|1(1) + (-2) - 8|}{\sqrt{1 + 1^2}} = \frac{|-9|}{\sqrt{2}} = \frac{9}{\sqrt{2}} \text{ units}$$

Now we're ready to compute the area of the triangle, since we know the base and the height. We have

$$\text{area} = \frac{1}{2}(6\sqrt{2})\frac{9}{\sqrt{2}} = 27 \text{ square units}$$

■■■ ▌ EXAMPLE 4 From the point $(8, 1)$, a line is drawn tangent to the circle $x^2 + y^2 = 20$, as shown in Figure 8. Find the slope of this tangent line.

Solution This is a problem in which the direct approach is not the simplest. The direct approach would be first to determine the coordinates of the point of tangency. Then, using those coordinates along with $(8, 1)$, the required slope could be computed. As it turns out, however, the coordinates of the point of tangency are not very easy to determine. In fact, one of the advantages of the following method is that those coordinates need not be found.

In Figure 8, let m denote the slope of the tangent line. Because the tangent line passes through $(8, 1)$, we can write its equation

$$y - y_1 = m(x - x_1)$$
$$y - 1 = m(x - 8) \qquad \text{or} \qquad y = mx - 8m + 1$$

FIGURE 8

Now, since the distance from the origin to the tangent line is $\sqrt{20}$ units (the radius of the circle), we have

$$\sqrt{20} = \frac{|mx_0 + b - y_0|}{\sqrt{1 + m^2}}$$

$$= \frac{|m(0) - 8m + 1 - 0|}{\sqrt{1 + m^2}} \qquad \text{using } x_0 - 0 = y_0 \text{ and } b = -8m + 1$$

$$= \frac{|-8m + 1|}{\sqrt{1 + m^2}}$$

To solve this equation for m, square both sides to obtain

$$20 = \frac{64m^2 - 16m + 1}{1 + m^2}$$

or

$$20(1 + m^2) = 64m^2 - 16m + 1$$

$$0 = 44m^2 - 16m - 19$$

$$0 = (22m - 19)(2m + 1)$$

From this we see that the two roots of the equation are $\frac{19}{22}$ and $-\frac{1}{2}$. Because the slope of the tangent line specified in Figure 8 is negative, we choose the value $m = -\frac{1}{2}$; this is the required slope. ∎

The last formula that we consider in this section is

$$d = \frac{|Ax_0 + By_0 + C|}{\sqrt{A^2 + B^2}}$$

As we pointed out earlier, this formula gives the distance from the point (x_0, y_0) to the line $Ax + By + C = 0$. To derive the formula, first note that the slope and the y-intercept of the line $Ax + By + C = 0$ are

$$m = -\frac{A}{B} \qquad \text{and} \qquad b = -\frac{C}{B}$$

So we have

$$d = \frac{|mx_0 + b - y_0|}{\sqrt{1 + m^2}} = \frac{|(-A/B)x_0 + (-C/B) - y_0|}{\sqrt{1 + (-A/B)^2}}$$

Now, to complete the derivation, we need to show that when this last expression is simplified, the result is $|Ax_0 + By_0 + C|/\sqrt{A^2 + B^2}$. Exercise 43 asks you to carry out the details.

■□■ **EXAMPLE 5** Use the formula $d = |Ax_0 + By_0 + C|/\sqrt{A^2 + B^2}$ to compute the distance from the point $(-3, 1)$ to the line $y = -2x + 7$. *Note:* In Example 2, we computed this quantity using the distance formula $d = |mx_0 + b - y_0|/\sqrt{1 + m^2}$.

Solution First, we write the given equation $y = -2x + 7$ in the form $Ax + By + C = 0$:

$$2x + y - 7 = 0$$

From this we see that $A = 2$, $B = 1$, and $C = -7$. Thus, we have

$$d = \frac{|Ax_0 + By_0 + C|}{\sqrt{A^2 + B^2}}$$

$$= \frac{|2(-3) + 1(1) + (-7)|}{\sqrt{2^2 + 1^2}} = \frac{12}{\sqrt{5}} = \frac{12\sqrt{5}}{5}$$

The required distance is therefore $12\sqrt{5}/5$ units, as we obtained previously in Example 2(a). ▪▪▪

EXERCISE SET 10.1

A

Exercises 1–12 are review exercises. To solve these problems, you will need to utilize the formulas listed at the beginning of this section.

1. Find the distance between the points $(-5, -6)$ and $(3, -1)$.

2. Find the equation of the line that passes through $(2, -4)$ and is parallel to the line $3x - y = 1$. Write your answer in the form $y = mx + b$.

3. Find the equation of a line that is perpendicular to the line $4x - 5y - 20 = 0$ and has the same y-intercept as the line $x - y + 1 = 0$. Write your answer in the form $Ax + By + C = 0$.

4. Find the equation of the line passing through the points $(6, 3)$ and $(1, 0)$. Write your answer in the form $y = mx + b$.

5. Find the equation of the line that is the perpendicular bisector of the line segment joining the points $(2, 1)$ and $(6, 7)$. Write your answer in the form $Ax + By + C = 0$.

6. Find the area of the circle $(x - 12)^2 + (y + \sqrt{5})^2 = 49$.

7. Find the x- and the y-intercepts of the circle with center $(1, 0)$ and radius 5.

8. Find the equation of the line with a positive slope that is tangent to the circle $(x - 1)^2 + (y - 1)^2 = 4$ at one of its y-intercepts. Write your answer in the form $y = mx + b$.

9. Suppose that the coordinates of A, B, and C are $A(1, 2)$, $B(6, 1)$, and $C(7, 8)$. Find the equation of the line passing through C and through the midpoint of the line segment $\overline{AB}$. Write your answer in the form $ax + by + c = 0$.

10. Find the equation of the line passing through the point $(-4, 0)$ and through the point of intersection of the lines $2x - y + 1 = 0$ and $3x + y - 16 = 0$. Write your answer in the form $y = mx + b$.

11. Find the perimeter of $\triangle ABC$ in the following figure.

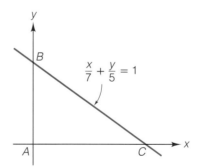

12. Find the sum of the x- and y-intercepts of the line

$$(\csc^2 \alpha)x + (\sec^2 \alpha)y = 1$$

(Assume that α is a constant.)

In Exercises 13–16, determine the angle of inclination of each line. Express the answer in both radians and degrees. In cases where a calculator is necessary, round the answer to two decimal places.

13. $y = \sqrt{3}x + 4$

14. $x + \sqrt{2}y - 2 = 0$

15. (a) $y = 5x + 1$
    (b) $y = -5x + 1$

16. (a) $3x - y - 3 = 0$
    (b) $3x + y - 3 = 0$

In Exercises 17–20, find the distance from the point to the line using (a) the formula $d = |mx_0 + b - y_0|/\sqrt{1 + m^2}$; (b) the formula $d = |Ax_0 + By_0 + C|/\sqrt{A^2 + B^2}$.

17. $(1, 4)$; $y = x - 2$

18. $(-2, -3)$; $y = -4x + 1$

19. $(-3, 5)$; $4x + 5y + 6 = 0$

20. $(0, -3)$; $3x - 2y = 1$

21. (a) Find the equation of the circle that has center $(-2, -3)$ and is tangent to the line $2x + 3y = 6$.
    (b) Find the radius of the circle that has center $(1, 3)$ and is tangent to the line $y = \frac{1}{2}x + 5$.

22. Find the area of the triangle with vertices $(3, 1)$, $(-2, 7)$, and $(6, 2)$.   *Hint:* Use the method shown in Example 3.

23. Find the area of the quadrilateral $ABCD$ with vertices $A(0, 0)$, $B(8, 2)$, $C(4, 7)$, and $D(1, 6)$. *Suggestion:* Draw a diagonal and use the method shown in Example 3 for the two resulting triangles.

24. From the point $(7, -1)$, tangent lines are drawn to the circle $(x - 4)^2 + (y - 3)^2 = 4$. Find the slopes of these lines.

25. From the point $(0, -5)$, tangent lines are drawn to the circle $(x - 3)^2 + y^2 = 4$. Find the slope of each tangent.

26. Find the distance between the two parallel lines $y = 2x - 1$ and $y = 2x + 4$.   *Hint:* Draw a sketch; then find the distance from the origin to each line.

27. Find the distance between the two parallel lines $3x + 4y = 12$ and $3x + 4y = 24$.

28. Find the equation of the line that passes through $(3, 2)$ and whose x- and y-intercepts are equal. (There are two answers.)

29. Find the equation of the line that passes through the point $(2, 6)$ in such a way that the segment of the line cut off between the axes is bisected by the point $(2, 6)$.

30. Find the equation of the line whose angle of inclination is $60°$ and whose distance from the origin is four units. (There are two answers.)

B

31. Find the equation of the angle bisector in the accompanying figure.   *Hint:* Let (x, y) be a point on the angle bisector. Then (x, y) is equidistant from the two given lines.

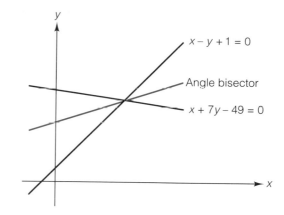

32. Find the center and the radius of the circle that passes through the points $(-2, 7)$, $(0, 1)$, and $(2, -1)$.

33. (a) Find the center and the radius of the circle passing through the points $A(-12, 1)$, $B(2, 1)$ and $C(0, 7)$.
    (b) Let R denote the radius of the circle in part (a). In $\triangle ABC$, let a, b, and c be the lengths of the sides opposite angles A, B, and C, respectively. Show that the area of $\triangle ABC$ is equal to $abc/4R$.

34. (Continuation of Exercise 33.)
    (a) Let H denote the point where the altitudes of $\triangle ABC$ intersect. Find the coordinates of H.
    (b) Let d denote the distance from H to the center of the circle in Exercise 33(a). Show that
    $$d^2 = 9R^2 - (a^2 + b^2 + c^2)$$

35. Suppose the line $x - 7y + 44 = 0$ intersects the circle $x^2 - 4x + y^2 - 6y = 12$ at points P and Q. Find the length of chord $\overline{PQ}$.

36. The point $(1, -2)$ is the midpoint of a chord of the circle $x^2 - 4x + y^2 + 2y = 15$. Find the length of the chord.

37. Show that the product of the distances from the point $(0, c)$ to the lines $ax + y = 0$ and $x + by = 0$ is
    $$\frac{|bc^2|}{\sqrt{a^2 + a^2b^2 + b^2 + 1}}$$

38. Suppose that the point (x_0, y_0) lies on the circle $x^2 + y^2 = a^2$. Show that the equation of the line tangent to the circle at (x_0, y_0) is $x_0x + y_0y = a^2$.

39. The vertices of $\triangle ABC$ are $A(0, 0)$, $B(8, 0)$, and $C(8, 6)$.
    (a) Find the equations of the three lines that bisect the angles in triangle ABC.   *Hint:* Make use of the identity $\tan(\theta/2) = (\sin \theta)/(1 + \cos \theta)$.
    (b) Find the points where each pair of angle bisectors intersect. What do you observe?

40. Show that the equations of the lines with slope m that are tangent to the circle $x^2 + y^2 = a^2$ are

$$y = mx + a\sqrt{1 + m^2} \quad \text{and}$$
$$y = mx - a\sqrt{1 + m^2}$$

41. The point (x, y) is equidistant from the point $\left(0, \frac{1}{4}\right)$ and the line $y = -\frac{1}{4}$. Show that x and y satisfy the equation $y = x^2$.

42. The point (x_0, y_0) is equidistant from the line $x + 2y = 0$ and the point $(3, 1)$. Find (and simplify) an equation relating x_0 and y_0.

43. (a) Find the slope m and the y-intercept b of the line $Ax + By + C = 0$.
   (b) Use the formula $d = |mx_0 + b - y_0|/\sqrt{1 + m^2}$ to show that the distance from the point (x_0, y_0) to the line $Ax + By + C = 0$ is given by

$$d = \frac{|Ax_0 + By_0 + C|}{\sqrt{A^2 + B^2}}$$

44. Refer to Figure 5 on page 578. Show that $\triangle ABD$ is similar to $\triangle EGF$.

C
45. Show that the distance of the point (x_1, y_1) from the line passing through the two points (x_2, y_2) and (x_3, y_3) is given by $d = |D|/\sqrt{(x_2 - x_3)^2 + (y_2 - y_3)^2}$, where

$$D = \begin{vmatrix} x_1 & y_1 & 1 \\ x_2 & y_2 & 1 \\ x_3 & y_3 & 1 \end{vmatrix}$$

46. Let (a_1, b_1), (a_2, b_2), and (a_3, b_3) be three noncollinear points. Show that the equation of the circle passing through these three points is

$$\begin{vmatrix} x^2 + y^2 & x & y & 1 \\ a_1^2 + b_1^2 & a_1 & b_1 & 1 \\ a_2^2 + b_2^2 & a_2 & b_2 & 1 \\ a_3^2 + b_3^2 & a_3 & b_3 & 1 \end{vmatrix} = 0$$

47. Find the equation of the circle that passes through the points $(6, 3)$ and $(-4, -3)$ and that has its center on the line $y = 2x - 7$.

48. Let a be a positive number and suppose that the coordinates of points P and Q are $P(a \cos \theta, a \sin \theta)$ and $Q(a \cos \beta, a \sin \beta)$. Show that the distance from the origin to the line passing through P and Q is

$$a\left| \cos\left(\frac{\theta - \beta}{2} \right) \right|$$

49. Find the equation of a circle that has radius 5 and is tangent to the line $2x + 3y = 26$ at the point $(4, 6)$. Write your answer in standard form. (There are two answers.)

50. Find the equation of the circle passing through $(2, -1)$ and tangent to the line $y = 2x + 1$ at $(1, 3)$. Write your answer in standard form.

51. For the last exercise in this section, you will prove an interesting property of the curve $x^3 + y^3 = 6xy$. This curve, known as the *folium of Descartes*, is shown in Figure A. Figure B and the accompanying caption indicate the property that we will establish.

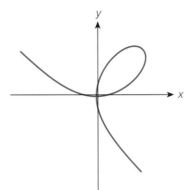

The Folium of Descartes, $x^3 + y^3 = 6xy$

FIGURE A

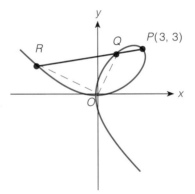

A property of the folium: Suppose that a line through the point $P(3, 3)$ meets the folium again at points Q and R. Then $\angle ROQ$ is a right angle.

FIGURE B

(a) Suppose that the slope of the line segment $\overline{OQ}$ is t. By solving the system of equations

$$\begin{cases} y = tx \\ x^3 + y^3 = 6xy \end{cases}$$

show that the coordinates of the point Q are $x = 6t/(1 + t^3)$ and $y = 6t^2/(1 + t^3)$.
(b) Show that the slope of the line joining the points P and Q is $(t^2 - t - 1)/(t^2 + t - 1)$.

(c) Suppose that the slope of the line segment $\overline{OR}$ is u. By repeating the procedure used in parts (a) and (b), you'll find that the slope of the line joining the points P and R is $(u^2 - u - 1)/(u^2 + u - 1)$. Now, since the points P, Q, and R are collinear (lie on the same line), it must be the case that

$$\frac{t^2 - t - 1}{t^2 + t - 1} = \frac{u^2 - u - 1}{u^2 + u - 1}$$

Working from this equation, show that $tu = -1$. This shows that $\angle ROQ$ is a right angle, as required.

10.2 THE PARABOLA

In Section 4.2, we saw that the graph of a quadratic function $y = ax^2 + bx + c$ is a symmetric U-shaped curve called a parabola. In this section, we give a more general definition of the parabola, a definition that emphasizes the geometric properties of the curve.

> ### The Parabola
>
> A **parabola** is the set of all points in the plane equally distant from a fixed line and a fixed point not on the line. The fixed line is called the **directrix**. The fixed point is called the **focus**.

Let us initially suppose that the focus of the parabola is the point $(0, p)$ and the directrix is the line $y = -p$. We will assume throughout this section that p is positive. To understand the geometric content of the definition of the parabola, the special graph paper displayed in Figure 1(a) is useful. The common center of the concentric circles in Figure 1(a) is the focus $(0, p)$. Thus, all the points on a given circle are at a fixed distance from the focus. The radii of the circles increase in increments of p units. Similarly, the broken horizontal lines in the figure are drawn at intervals that are multiples of p units from the directrix $y = -p$. By considering the points where the circles intersect the horizontal lines, we can find a number of points equally distant from the focus $(0, p)$ and the directrix $y = -p$ [see Figure 1(b)].

FIGURE I

(a)　　　　　　　　　　　　　　　　　　(b)

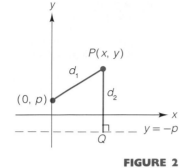

FIGURE 2

The graphical method just described lets us locate many points equally distant from the focus $(0, p)$ and the directrix $y = -p$. Figure 1(b) shows that the points on the parabola are symmetric about a line, in this case the y-axis. Also, by studying the figure, you should be able to convince yourself that in this case there can be no points below the x-axis that satisfy the stated condition. However, to describe the required set of points completely, and to show that our new definition is consistent with the old, Figure 1(b) is inadequate. We need to bring algebraic methods to bear on the problem. Thus, let d_1 denote the distance from the point $P(x, y)$ to the focus $(0, p)$ and let d_2 denote the distance from $P(x, y)$ to the directrix $y = -p$, as shown in Figure 2. The distance d_1 is then

$$d_1 = \sqrt{(x - 0)^2 + (y - p)^2} = \sqrt{x^2 + y^2 - 2py + p^2}$$

The distance d_2 in Figure 2 is just the vertical distance between the points P and Q. Thus,

$$d_2 = (y\text{-coordinate of } P) - (y\text{-coordinate of } Q)$$
$$= y - (-p) = y + p$$

(Absolute value signs are unnecessary here for, as noted earlier, P cannot lie below the x-axis.) The condition that P be equally distant from the focus $(0, p)$ and the directrix $y = -p$ can be expressed by the equation

$$d_1 = d_2$$

Using the expressions we've found for d_1 and d_2, this last equation becomes

$$\sqrt{x^2 + y^2 - 2py + p^2} = y + p$$

We can obtain a simpler but equivalent equation by squaring both sides. (Two nonnegative quantities are equal if and only if their squares are equal.) Thus,

$$x^2 + y^2 - 2py + p^2 = y^2 + 2py + p^2$$

After combining like terms, this equation becomes

$$x^2 = 4py$$

This is the equation of a parabola with focus $(0, p)$ and directrix $y = -p$. In the box that follows, we summarize the properties of the parabola $x^2 = 4py$. As Figure 3(b) indicates, the terminology we've introduced applies equally well to an arbitrary parabola for which the axis of symmetry is not necessarily parallel to one of the coordinate axes and the vertex is not necessarily the origin.

PROPERTY SUMMARY   THE PARABOLA

1. A **parabola** is the set of points equidistant from a fixed line called the **directrix** and a fixed point, not on the line, called the **focus**.

2. The **axis** of a parabola is the line drawn through the focus and perpendicular to the directrix.

3. The **vertex** of a parabola is the point where the parabola intersects its axis. The vertex is located halfway between the focus and the directrix. See Figure 3.

(a)   (b)

FIGURE 3

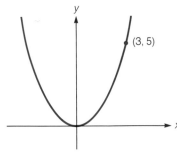

FIGURE 4

■ ■ ■ ■ EXAMPLE 1 Determine the equation of the parabola in Figure 4, given that the curve passes through the point $(3, 5)$. Specify the focus and the directrix.

Solution We have just seen that the general equation for a parabola in this position is $x^2 = 4py$. Since the point $(3, 5)$ lies on the curve, its coordinates must satisfy the equation $x^2 = 4py$. Thus,

$$3^2 = 4p(5)$$
$$9 = 20p$$
$$\frac{9}{20} = p$$

With $p = \frac{9}{20}$, the equation $x^2 = 4py$ becomes

$$x^2 = 4\left(\frac{9}{20}\right)y \quad \text{or} \quad x^2 = \frac{9}{5}y \quad \text{as required}$$

Furthermore, since $p = \frac{9}{20}$, the focus is $\left(0, \frac{9}{20}\right)$ and the directrix is $y = -\frac{9}{20}$.   **■ ■ ■**

We have seen that the equation of a parabola with focus $(0, p)$ and directrix $y = -p$ is $x^2 = 4py$. By following the same method, we can obtain general equations for parabolas with other orientations. The basic results are summarized in Figure 5.

PROPERTY SUMMARY   BASIC EQUATIONS FOR THE PARABOLA

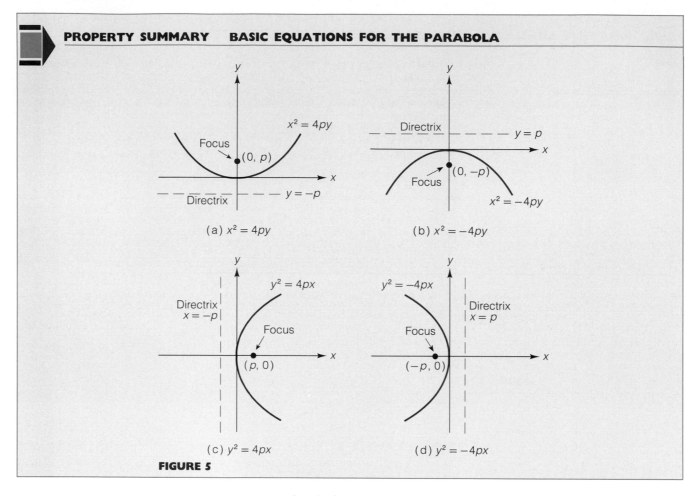

(a) $x^2 = 4py$

(b) $x^2 = -4py$

(c) $y^2 = 4px$

(d) $y^2 = -4px$

FIGURE 5

By a **chord** of a parabola, we mean a straight line segment joining any two points on the curve. If the chord passes through the focus, it is called a **focal chord**. For purposes of graphing, it is useful to know the length of the focal chord perpendicular to the axis of the parabola. This is the length of the horizontal line segment $\overline{AB}$ in Figure 6 and the vertical line segment $\overline{A'B'}$ in Figure 7. We will call this length the **focal width**.

FIGURE 6
Focal width = $AB = 4p$

FIGURE 7
Focal width = $A'B' = 4p$

In Figure 6, the distance from A to F is the same as the distance from A to the line $y = -p$. (Why?) But the distance from A to the line $y = -p$ is $2p$. Therefore, $AF = 2p$ and AB is twice this, or $4p$. We have shown that the focal width of the parabola is $4p$. In other words, given a parabola $x^2 = 4py$, the width at its focus is $4p$, the coefficient of y. In the same way, the length of the focal chord $\overline{A'B'}$ in Figure 7 is also $4p$, the coefficient of x in that case.

■■ ■ EXAMPLE 2 Find the focus and directrix of the parabola $y^2 = -4x$, and sketch the graph.

Solution Comparing the basic equation $y^2 = -4px$ [in Figure 5(d)] with the equation at hand, we see that $4p = 4$ and thus $p = 1$. The focus is therefore $(-1, 0)$, and the directrix is $x = 1$. The basic form of the graph will be as in Figure 5(d). For purposes of graphing, we note that the focal width is 4 (the absolute value of the coefficient of x). This, along with the fact that the vertex is $(0, 0)$, gives us enough information to draw the graph; see Figure 8. **■■■**

$y^2 = -4x$

Focus
$(-1, 0)$

Directrix
$x = 1$

FIGURE 8

In Examples 1 and 2 and in Figure 5, the vertex of each parabola is located at the origin. Now we want to consider parabolas that are translated (shifted) from this standard position. As background for this, we review and generalize the results about translation in Section 3.3. Consider, as an example, the two equations

$$y = (x - 1)^2 \qquad \text{and} \qquad y = x^2 + 1$$

As we saw in Section 3.3, the graphs of these two equations can each be obtained by translating the graph of $y = x^2$. To graph $y = (x - 1)^2$, translate the graph of $y = x^2$ one unit to the right; to graph $y = x^2 + 1$, translate $y = x^2$ one unit up.

To see the underlying pattern here, rewrite the equation $y = x^2 + 1$ as $y - 1 = x^2$. Then the situation is this:

EQUATION	HOW GRAPH OBTAINED
$y = (x - 1)^2$	translation in positive x-direction
$y - 1 = x^2$	translation in positive y-direction

Observation *In the equation $y = x^2$, the effect of replacing x with $x - 1$ is to translate the graph one unit in the positive x-direction. Similarly, replacing y with $y - 1$ translates the graph one unit in the positive y-direction.*

As a second example before we generalize, consider the equations

$$y = (x + 1)^2 \qquad \text{and} \qquad y = x^2 - 1$$

The first equation involves a translation in the negative x-direction; the second involves a translation in the negative y-direction. As before, to see the underlying pattern, we rewrite the second equation as $y + 1 = x^2$. Now we have the following situation:

EQUATION	HOW GRAPH OBTAINED
$y = (x + 1)^2$	translation in negative x-direction
$y + 1 = x^2$	translation in negative y-direction

Observation   *In the equation $y = x^2$, the effect of replacing x with $x + 1$ is to translate the graph one unit in the negative x-direction. Similarly, replacing y with $y + 1$ translates the graph one unit in the negative y-direction.*

Both of the examples that we have just considered are specific instances of the following basic result. (The result is valid whether or not the given equation and graph represent a function.)

PROPERTY SUMMARY   **TRANSLATION AND COORDINATES**

Suppose that we have an equation that determines a graph in the x-y plane, and let h and k denote positive numbers. Then, replacing x with $x - h$ or $x + h$, or replacing y with $y - k$ or $y + k$, has the following effects on the graph of the original equation.

REPLACEMENT	RESULTING TRANSLATION
1. x replaced with $x - h$	h units in the positive x-direction
2. y replaced with $y - k$	k units in the positive y-direction
3. x replaced with $x + h$	h units in the negative x-direction
4. y replaced with $y + k$	k units in the negative y-direction

---

EXAMPLE 3   Graph the parabola $(y + 1)^2 = -4(x - 2)$. Specify the vertex, the focus, the directrix, and the axis of symmetry.

Solution   The given equation is obtained from $y^2 = -4x$ (which we graphed in the previous example) by replacing x and y with $x - 2$ and $y + 1$, respectively. So the required graph is obtained by translating the parabola in Figure 8 two units to the right and one unit down. In particular, this means that the vertex moves from $(0, 0)$ to $(2, -1)$; the focus moves from $(-1, 0)$ to $(1, -1)$; the directrix moves from $x = 1$ to $x = 3$; and the axis of symmetry moves from $y = 0$ (which is the x-axis) to $y = -1$. The required graph is shown in Figure 9.

FIGURE 9

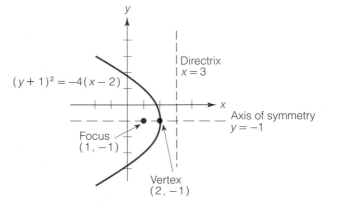

EXAMPLE 4   Graph the parabola $2y^2 - 4y - x + 5 = 0$, and specify each of the following: vertex, focus, directrix, axis of symmetry, and focal width.

Solution Just as we did in Section 4.2, we use the technique of completing the square:

$$2(y^2 - 2y \quad) = x - 5$$
$$2(y^2 - 2y + 1) = x - 5 + 2 \qquad \text{adding 2 to both sides}$$
$$(y - 1)^2 = \tfrac{1}{2}(x - 3)$$

The graph of this last equation is obtained by translating the graph of $y^2 = \tfrac{1}{2}x$ "right 3, up 1." This moves the vertex from $(0, 0)$ to $(3, 1)$. Now, for $y^2 = \tfrac{1}{2}x$, the focus and directrix are determined by setting $4p = \tfrac{1}{2}$. Therefore $p = \tfrac{1}{8}$, and consequently the focus and directrix of $y^2 = \tfrac{1}{2}x$ are $\left(\tfrac{1}{8}, 0\right)$ and $x = -\tfrac{1}{8}$, respectively. Thus, the focus of the translated curve is $\left(3\tfrac{1}{8}, 1\right)$ and the directrix is $x = 2\tfrac{7}{8}$. Figure 10(a) shows the graph of $y^2 = \tfrac{1}{2}x$, and Figure 10(b) shows the translated graph. (You should check for yourself that the information accompanying Figure 10(b) is correct.) ∎

FIGURE 10

(a)

Focus: $\left(\tfrac{1}{8}, 0\right)$
Directrix: $x = -\tfrac{1}{8}$
Focal width: $\tfrac{1}{2}$

(b)

Vertex: $(3, 1)$
Focus: $\left(\tfrac{25}{8}, 1\right)$
Directrix: $x = \tfrac{23}{8}$
Axis of symmetry: $y = 1$
Focal width: $\tfrac{1}{2}$

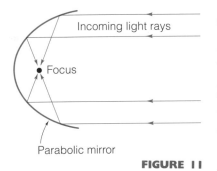

FIGURE 11

There are numerous applications of the parabola in the sciences. Figure 11 shows a cross section of a parabolic mirror in a telescope. As indicated in Figure 11, light rays coming in parallel to the axis of the parabola are reflected through the focus. (The word *focus* comes from a Latin word meaning "fireplace.") Parabolic reflectors are also used in communication and surveillance systems, in radio telescopes, and in automobile headlights. In an automobile headlight, the bulb is located at the focus. To diagram the parabolic mirror of an automobile headlight, just reverse the directions of the arrows in Figure 11. [For a more complete description of the uses of the parabola, see the article, "The Standup Conic Presents: The Parabola and Applications" by Lee Whitt in *The Journal of Undergraduate Mathematics and its Applications*, vol. III, no. 3 (1982)].

EXERCISE SET 10.2

A

In Exercises 1–22, graph the parabolas. In each case, specify the focus, the directrix, and the focal width. For Exercises 13–22, also specify the vertex.

1. $x^2 = 4y$
2. $x^2 = 16y$
3. $y^2 = -8x$
4. $y^2 = 12x$
5. $x^2 = -20y$
6. $x^2 - y = 0$

7. $y^2 + 28x = 0$
8. $4y^2 + x = 0$
9. $x^2 = 6y$
10. $y^2 = -10x$
11. $4x^2 = 7y$
12. $3y^2 = 4x$
13. $y^2 - 6y - 4x + 17 = 0$
14. $y^2 + 2y + 8x + 17 = 0$
15. $x^2 - 8x - y + 18 = 0$
16. $x^2 + 6y + 18 = 0$
17. $y^2 + 2y - x + 1 = 0$
18. $2y^2 - x + 1 = 0$
19. $2x^2 - 12x - y + 18 = 0$
20. $y + \sqrt{2} = (x - 2\sqrt{2})^2$
21. $2x^2 - 16x - y + 33 = 0$
22. $\frac{1}{4}y^2 - y - x + 1 = 0$

B

For Exercises 23 and 24, refer to the following figure.

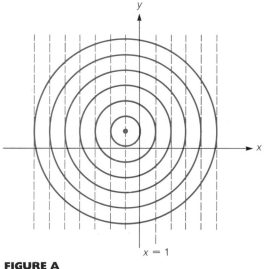

$x = 1$

FIGURE A

23. Make a photocopy of Figure A. Then, in your copy of Figure A, indicate (by drawing dots) eleven points that are equidistant from the point $(-1, 1)$ and the line $x = 1$. What is the equation of the line of symmetry for the set of dots?

24. The eleven dots that you located in Exercise 23 are part of a parabola. Find the equation of that parabola and sketch its graph. Specify the vertex, the focus and directrix, and the focal width.

For Exercises 25–30, find the equation of the parabola satisfying the given conditions. In each case, assume that the vertex is at the origin.

25. The focus is $(0, 3)$.

26. The directrix is $y - 8 = 0$.

27. The directrix is $x + 32 = 0$.

28. The focus lies on the y-axis, and the parabola passes through the point $(7, -10)$.

29. The parabola is symmetric about the x-axis, the x-coordinate of the focus is negative, and the length of the focal chord perpendicular to the x-axis is 9.

30. The focus is the smaller of the two x-intercepts of the circle $x^2 - 8x + y^2 - 6y + 9 = 0$.

31. Let P denote the point $(8, 8)$ on the parabola $x^2 = 8y$, and let $\overline{PQ}$ be a focal chord.
    (a) Find the equation of the line through the point $(8, 8)$ and the focus.
    (b) Find the coordinates of Q.
    (c) Find the length of $\overline{PQ}$.
    (d) Find the equation of the circle with this focal chord as a diameter.
    (e) Show that the circle determined in part (d) intersects the directrix of the parabola in only one point. Conclude from this that the directrix is tangent to the circle. Draw a sketch of the situation.

32. The following figure shows a parabolic cross section of a reflecting mirror. Find the distance from the vertex of the parabola to the focus. *Hint:* Choose a convenient coordinate system.

2 in.   12 in.

33. An arch is in the shape of a parabola with a vertical axis. The arch is 15 ft high at the center and 40 ft wide at the base. At what height above the base is the width 20 ft? *Hint:* Choose a convenient coordinate system in which the vertex of the parabola is at the origin.

34. The following figure depicts the cable of a suspension bridge. The cable is in the form of a parabola with a vertical axis. The horizontal highway on the bridge is 300 ft long. The longest of the vertical supporting wires is 100 ft; the shortest is 40 ft. Find the length of a supporting wire that is 50 ft from the middle.

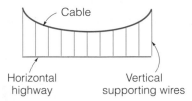

Cable

Horizontal highway

Vertical supporting wires

35. The segments $\overline{AA'}$ and $\overline{BB'}$ are focal chords of the parabola $x^2 = 2y$. The coordinates of A and B are $(4, 8)$ and $(-2, 2)$, respectively.
    (a) Find the equation of the line through A and B'.
    (b) Find the equation of the line through B and A'.
    (c) Show that the two lines you have found intersect at a point on the directrix.

36. Let $\overline{PQ}$ be the horizontal focal chord of the parabola $x^2 = 8y$. Let R denote the point where the directrix of the parabola meets the y-axis. Show that $\overline{PR}$ is perpendicular to $\overline{QR}$.

37. Suppose $\overline{PQ}$ is a focal chord of the parabola $y = x^2$ and that the coordinates of P are $(2, 4)$.
    (a) Find the coordinates of Q.
    (b) Find the coordinates of M, the midpoint of $\overline{PQ}$.
    (c) A perpendicular is drawn from M to the y-axis, meeting the y-axis at S. Also, a line perpendicular to the focal chord is drawn through M, meeting the y-axis at T. Find ST and verify that it is equal to one-half the focal width of the parabola.

38. Let F be the focus of the parabola $x^2 = 8y$, and let P denote the point on the parabola with coordinates $(8, 8)$. Let $\overline{PQ}$ be a focal chord. If V denotes the vertex of the parabola, verify that

    $$PF \cdot FQ = VF \cdot PQ$$

39. In the following figure, triangle OAB is equilateral and $\overline{AB}$ is parallel to the x-axis. Find the length of a side and the area of triangle OAB.

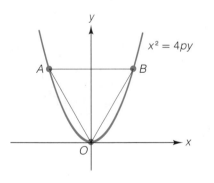

C

40. If $\overline{PQ}$ is a focal chord of the parabola $x^2 = 4py$ and the coordinates of P are (x_0, y_0), show that the coordinates of Q are

    $$\left(\frac{-4p^2}{x_0}, \frac{p^2}{y_0} \right)$$

41. If $\overline{PQ}$ is a focal chord of the parabola $y^2 = 4px$ and the coordinates of P are (x_0, y_0), show that the coordinates of Q are

    $$\left(\frac{p^2}{x_0}, \frac{-4p^2}{y_0} \right)$$

42. Let F and V denote the focus and the vertex, respectively, of the parabola $x^2 = 4py$. If $\overline{PQ}$ is a focal chord of the parabola, show that

    $$PF \cdot FQ = VF \cdot PQ$$

43. Let $\overline{PQ}$ be a focal chord of the parabola $y^2 = 4px$, and let M be the midpoint of $\overline{PQ}$. A perpendicular is drawn from M to the x-axis, meeting the x-axis at S. Also from M, a line segment is drawn that is perpendicular to $\overline{PQ}$ and that meets the x-axis at T. Show that the length of $\overline{ST}$ is one-half the focal width of the parabola.

44. Let $\overline{AB}$ be a chord (not necessarily a focal chord) of the parabola $y^2 = 4px$, and suppose that $\overline{AB}$ subtends a right angle at the vertex. (In other words, $\angle AOB = 90°$, where O is the origin in this case.) Find the x-intercept of the segment $\overline{AB}$. What is surprising about this result? *Hint:* Begin by writing the coordinates of A and B as $A(a^2/4p, a)$ and $B(b^2/4p, b)$.

45. Let $\overline{PQ}$ be a focal chord of the parabola $x^2 = 4py$. Complete the following steps to prove that the circle with $\overline{PQ}$ as a diameter is tangent to the directrix of the parabola. Let the coordinates of P be (x_0, y_0).
    (a) Show that the coordinates of Q are

    $$\left(\frac{-4p^2}{x_0}, \frac{p^2}{y_0} \right)$$

    (b) Show that the midpoint of $\overline{PQ}$ is

    $$\left(\frac{x_0^2 - 4p^2}{2x_0}, \frac{y_0^2 + p^2}{2y_0} \right)$$

    (c) Show that the length of $\overline{PQ}$ is $(y_0 + p)^2/y_0$. *Suggestion:* This can be done using the formula for the distance between two points, but the following is simpler. Let F be the focus. Then $PQ = PF + FQ$. Now, both PF and FQ can be determined by using the definition of the parabola rather than the distance formula.

    (d) Show that the distance from the center of the circle to the directrix equals the radius of the circle. How does this complete the proof?

OPTIONAL TI-81 GRAPHING CALCULATOR EXERCISES

EXERCISES FOR SECTION 10.2

In Exercises 1–4, graph the parabolas. These graphs provide examples of the four basic orientations shown in Figure 5 on page 588.

1. $x^2 = 8y$ *Hint:*   Enter this as $\frac{1}{8}x^2$.

2. $x^2 = -8y$

3. $y^2 = 8x$ *Hint:*   Enter this as two functions, $\sqrt{8x}$ and $-\sqrt{8x}$.

4. $y^2 = -8x$

5. In this exercise we graph the parabola
    $4x + y^2 + 2y - 7 = 0$.
    (a) By completing the square, show that the equation can be written $(y + 1)^2 = -4(x - 2)$.
    (b) Solving the equation in part (a) for y gives us the two functions $y = -1 \pm \sqrt{-4(x - 2)}$. Enter these two functions in the $Y=$ menu and press ZOOM 6. Now check that your graph is consistent with Figure 9 on page 590.

6. (a) Complete the square to determine the vertex of the parabola $y = 2x^2 - 16x + 33$.
    (b) Use the calculator to graph this function. Check that the general appearance of the graph you obtain is consistent with your result in part (a).
    (c) Use the ZOOM and TRACE keys to estimate the coordinates of the vertex. How close are the calculator values to those you found in part (a)?

Exercises 7–10 deal with lines that are tangent to a parabola. To determine the equations for the tangent lines in these problems, use the following result, which was known (in a purely geometric form) to Archimedes.

7. As a first example, consider the parabola $x^2 = 8y$. Note that the point $(4, 2)$ lies on this parabola. (Why?) According to the statement accompanying the figure, the tangent to the parabola at $(4, 2)$ will have a y-intercept of $y_0 = -2$. Use this fact to find the equation of the tangent line. Then use the calculator to graph the parabola and the tangent line.

8. Graph the parabola $x^2 = 2y$ and the tangent to the curve at the point $(-2, 2)$.

9. Graph the parabola $x^2 = y$ and the tangent to the curve at the point $(-3, 9)$.

10. In this exercise, we use the graphing calculator to verify a particular case of the following general result: Suppose $\overline{PQ}$ is a focal chord of a parabola. Then the tangents at P and Q are perpendicular to each other, and they intersect at a point on the directrix.
    (a) The line segment $\overline{PQ}$ is a focal chord of the parabola $x^2 = 8y$ and the coordinates of P are $(8, 8)$. Find the coordinates of Q. Also find the equation of the directrix.
    (b) Find the equations of the tangent lines at P and Q.
    (c) Graph the parabola, the two tangent lines, and the directrix. (Use ZOOM 6, then ZOOM 5.) If your calculations are correct, the tangent lines will meet at a point on the directrix and they will be perpendicular.

In Exercises 11–14, graph the equations and use the TRACE and ARROW keys to estimate the coordinates of the intersection points. Also, in Exercises 11 and 12, use the techniques of Section 9.6 to determine the exact coordinates of the intersection points.

11. $x^2 = 5y;\ y^2 = 5x$

12. $(y + 1)^2 = \frac{1}{2}(x - 1);\ 2x + y = 5$

13. $x^2 = y;\ y = \sin x$ 14. $x^2 = \frac{1}{3}y;\ y = e^x$

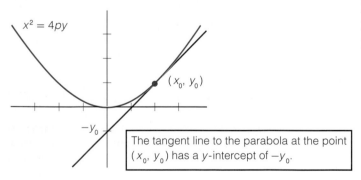

$x^2 = 4py$

(x_0, y_0)

$-y_0$

The tangent line to the parabola at the point (x_0, y_0) has a y-intercept of $-y_0$.

10.3 TANGENTS TO PARABOLAS (OPTIONAL)

Many of the more important properties of the parabola relate to the tangent to the curve. For instance, there is a close connection between the optical property illustrated in Figure 11 on page 591 and the tangents to the parabola. (See Exercise 19 at the end of this section for details.) In general, the techniques of calculus are required to deal with tangents to curves. However, for curves with equations and graphs as simple as the parabola, the methods of algebra are often adequate. The following discussion shows how we can determine the tangent to a parabola without using calculus.

To begin with, we need a definition for the tangent to a parabola. As motivation, recall from geometry that a tangent line to a circle is defined to be a line that intersects the circle in exactly one point (see Figure 1). However, this definition is not quite adequate for the parabola. For instance, the y-axis intersects the parabola $y = x^2$ in exactly one point, but surely it is not tangent to the curve. With this in mind, we adopt the definition in the following box for the tangent to a parabola. Along with this definition, we make the assumption that through each point P on a parabola, there is only one tangent line that can be drawn.

Tangent
Not a tangent

FIGURE 1

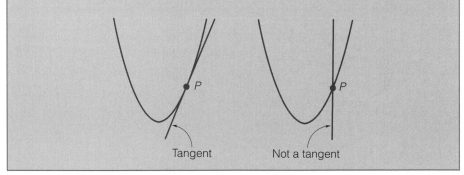

> **DEFINITION Tangent to a Parabola**
>
> Let P be a point on a parabola. Then a line through P is said to be **tangent** to the parabola at P provided that the line intersects the parabola only at P and the line is not parallel to the axis of the parabola.
>
> Tangent   Not a tangent

We'll demonstrate how to find tangents to parabolas by means of an example. The method used here can be used for any parabola. Suppose that we want to find the equation of the tangent to the parabola $x^2 = 4y$ at the point $(2, 1)$ on the curve. (See Figure 2.) Let m denote the slope of the tangent line. Since the line must pass through the point $(2, 1)$, its equation is

$$y - 1 = m(x - 2)$$

This is the tangent line, so it intersects the parabola in only one point, namely, $(2, 1)$. Algebraically, this means that the ordered pair $(2, 1)$ is the only solution of the following system of equations:

$$\begin{cases} y - 1 = m(x - 2) & (1) \\ x^2 = 4y & (2) \end{cases}$$

Tangent line
$m = ?$

$x^2 = 4y$

$(2, 1)$

FIGURE 2

The strategy now is to solve this system of equations in terms of m. Then we'll reconcile the results with the fact that $(2, 1)$ is known to be the unique solution. This will allow us to determine m. From equation (2), we have $y = x^2/4$. Then, substituting for y in equation (1), we obtain

$$\frac{x^2}{4} - 1 = m(x - 2)$$

$$x^2 - 4 = 4m(x - 2) \qquad \text{multiplying by 4}$$

$$(x - 2)(x + 2) = 4m(x - 2) \qquad \text{factoring}$$

$$(x - 2)(x + 2) - 4m(x - 2) = 0 \qquad \begin{array}{l}\text{subtracting } 4m(x - 2) \text{ from}\\ \text{both sides}\end{array}$$

Observe now that the quantity $x - 2$ appears twice on the left-hand side of the last equation, so we can factor it out to obtain

$$(x - 2)[(x + 2) - 4m] = 0$$

$$\begin{array}{c|c} x - 2 = 0 & x + 2 - 4m = 0 \\ x = 2 & x = 4m - 2 \end{array}$$

Now look at these two x-values that we've just determined. The first value, $x = 2$, yields no new information, since we knew from the start that this was the x-coordinate of the intersection point. However, the second value, $x = 4m - 2$, must also equal 2, since the system has but one solution. Thus, we have

$$4m - 2 = 2$$

$$4m = 4$$

$$m = 1$$

Therefore, the slope of the tangent line is 1. Substituting this value of m in equation (1) yields

$$y - 1 = 1(x - 2)$$

or

$$y = x - 1 \qquad \text{as required}$$

EXERCISE SET 10.3

A

In Exercises 1–8, use the method shown in the text to find the equation of the tangent to the parabola at the given point. In each case, include a sketch with your answer.

1. $x^2 = y$; $(2, 4)$
2. $x^2 = -2y$; $(2, -2)$
3. $x^2 = 8y$; $(4, 2)$
4. $x^2 = 12y$; $(6, 3)$
5. $x^2 = -y$; $(-3, -9)$
6. $x^2 = -6y$; $\left(\sqrt{6}, -1\right)$
7. $y^2 = 4x$; $(1, 2)$
8. $y^2 = -8x$; $(-8, -8)$

B

In Exercises 9–11, find the slope of the tangent to the curve at the indicated point. (Use the method shown in the text for parabolas.)

9. $y = \sqrt{x}$; $(4, 2)$ *Hint:* $x - 4 = \left(\sqrt{x} - 2\right)\left(\sqrt{x} + 2\right)$
10. $y = 1/x$; $\left(3, \frac{1}{3}\right)$
11. $y = x^3$; $(2, 8)$
12. Consider the parabola $x^2 = 4y$, and let (x_0, y_0) denote a point on the parabola in the first quadrant.

(a) Find the y-intercept of the line tangent to the parabola at the point on the parabola where $y_0 = 1$.
(b) Repeat part (a) using $y_0 = 2$.
(c) Repeat part (a) using $y_0 = 3$.
(d) On the basis of your results in parts (a)–(c), make a conjecture about the y-intercept of the line that is tangent to the parabola at the point (x_0, y_0) on the curve. Verify your conjecture by computing this y-intercept.

C

13. Let (x_0, y_0) be a point on the parabola $x^2 = 4py$. Using the method explained in the text, show that the equation of the line tangent to the parabola at (x_0, y_0) is

$$y = \frac{x_0}{2p}x - y_0$$

Thus, the y-intercept of the line tangent to $x^2 = 4py$ at (x_0, y_0) is just $-y_0$.

14. Let (x_0, y_0) be a point on the parabola $y^2 = 4px$. Show that the equation of the line tangent to the parabola at (x_0, y_0) is

$$y = \frac{2p}{y_0}x + \frac{y_0}{2}$$

Show that the x-intercept of this line is $-x_0$.

Exercises 15–22 contain results about tangents to parabolas. In working these problems, you'll find it convenient to use the facts developed in Exercises 13 and 14. Also assume, as given, the results about focal chords in Exercises 40 and 41 of Exercise Set 10.2. In some of the problems, reference is made to the normal line. The normal line or normal to a parabola at the point (x_0, y_0) on the parabola is defined as the line through (x_0, y_0) that is perpendicular to the tangent at (x_0, y_0).

15. Let $\overline{PQ}$ be a focal chord of the parabola $x^2 = 4py$.
(a) Show that the tangents to the parabola at P and Q are perpendicular to each other.
(b) Show that the tangents to the parabola at P and Q intersect at a point on the directrix.
(c) Let D be the intersection point of the tangents at P and Q. Show that the line segment from D to the focus is perpendicular to $\overline{PQ}$.

16. Let $P(x_0, y_0)$ be a point [other than $(0, 0)$] on the parabola $y^2 = 4px$. Let A be the point where the normal line to the parabola at P meets the axis of the parabola. Let B be the point where the line drawn from P perpendicular to the axis of the parabola meets the axis. Show that $AB = 2p$.

17. Let $P(x_0, y_0)$ be a point on the parabola $y^2 = 4px$. Let A be the point where the normal line to the parabola at P meets the axis of the parabola. Let F be the focus of

the parabola. If a line is drawn from A perpendicular to $\overline{FP}$, meeting $\overline{FP}$ at Z, show that $ZP = 2p$.

18. Let $\overline{AB}$ be a chord (not necessarily a focal chord) of the parabola $x^2 = 4py$. Let M be the midpoint of the chord and let C be the point where the tangents at A and B intersect.
(a) Show that $\overline{MC}$ is parallel to the axis of the parabola.
(b) If D is the point where the parabola meets $\overline{MC}$, show that $CD = DM$.

19. In this exercise, we prove the *reflection property* of parabolas. The following figure shows a line tangent to the parabola $y^2 = 4px$ at $P(x_0, y_0)$. The line through H and P is parallel to the axis of the parabola. We wish to prove that $\alpha = \beta$. That is, we wish to prove the reflection property: the angle of incidence equals the angle of reflection.

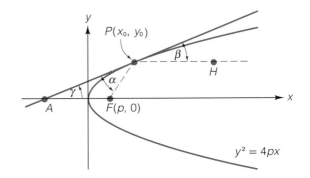

(a) Show that $FA = x_0 + p = FP$. *Hint:* Regarding FP, it is easier to rely on the definition of a parabola than on the distance formula.
(b) Why does $\alpha = \gamma$? Why does $\gamma = \beta$?
(c) Conclude that $\alpha = \beta$, as required.

20. From a point $P(x_0, y_0)$ on the parabola $x^2 = 4py$, a tangent line is drawn meeting the axis of the parabola at A. From the focus F, a line is drawn perpendicular to $\overline{AP}$, meeting $\overline{AP}$ at B. Finally, from P, a line is drawn perpendicular to the directrix, meeting the directrix at C.
(a) Show that B lies on the line that is tangent to the parabola at the vertex.
(b) Show that B is the midpoint of $\overline{FC}$.

21. Verify that the point $A(3, 2)$ lies on the line that is tangent to the parabola $x^2 = 4y$ at $P(4, 4)$. Let F be the focus of the parabola. A perpendicular is drawn from A to $\overline{FP}$, meeting $\overline{FP}$ at B. Also from A, a perpendicular is drawn to the directrix, meeting the directrix at C. Show that $FB = FC$. (This result is known as *Adams'*

theorem; it holds for any parabola and any point on a tangent line.)

22. The segment $\overline{AB}$ is a focal chord of the parabola $y^2 = 4x$ and the coordinates of A are $(4, 4)$. Normals are drawn through A and B that meet the parabola again at A' and B', respectively. Prove that $\overline{A'B'}$ is three times as long as $\overline{AB}$.

10.4   THE ELLIPSE

We have here apparently [in the work of Anthemius of Tralles (a sixth-century Greek architect and mathematician)] the first mention of the construction of an ellipse by means of a string stretched tight round the foci.

Sir Thomas Heath in *A History of Greek Mathematics,* vol. II (Oxford: The Clarendon Press, 1921)

The true orbit of Mars was even less of a circle than the Earth's. It took almost two years for Kepler to realize that its orbital shape is that of an ellipse. An ellipse is the shape of a circle when viewed at an angle.

Phillip Flower in *Understanding the Universe* (St. Paul: West Publishing Company, 1990)

The heavenly motions are nothing but a continuous song for several voices, to be perceived by the intellect, not by the ear.

Johannes Kepler (1571–1630)

In this section, we discuss the symmetric, oval-shaped curve known as the *ellipse*. As Kepler discovered and Newton later proved, this is the curve described by the planets in their motions around the sun.

The Ellipse

An **ellipse** is the set of all points in the plane, the sum of whose distances from two fixed points is constant. Each fixed point is called a **focus** (plural: **foci**).

Subsequently, we will derive an equation describing the ellipse, just as we found an equation for the parabola in Section 10.2. But first let us consider some rather immediate consequences of the definition. In fact, we can learn a great deal about the ellipse even before we derive its equation.

There is a simple mechanical method for constructing an ellipse that arises directly from the definition of the curve. Mark the given foci, say F_1 and F_2, on a drawing board and insert thumbtacks at those points. Now take a piece of string whose length is greater than the distance from F_1 to F_2 and tie the ends of the string to the tacks. Next, pull the string taut with a pencil point and touch the pencil point to the drawing board. Then if you move the pencil while keeping the string taut, the curve traced out will be an ellipse, as indicated in Figure 1. The reason the curve is an ellipse is that for each point on the curve, the sum of the distances from the foci is constant, the constant being the length of the string. By actually carrying out this construction for yourself several times, each time varying the distance between the foci or the length of the string, you can learn a great deal about the ellipse. For instance, when the distance between the foci is small compared to the length of the string, the ellipse begins to resemble

FIGURE 1

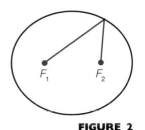

FIGURE 2

When the distance between the foci is small compared to the length of the string, the ellipse resembles a circle.

FIGURE 3

When the distance between the foci is nearly equal to the length of the string, the ellipse is relatively flat.

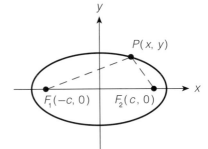

FIGURE 4

$F_1P + F_2P = 2a$

a circle, as in Figure 2. On the other hand, when the distance between the foci is nearly equal to the length of the string, the ellipse becomes relatively flat, as in Figure 3.

We now derive one of the standard forms for the equation of an ellipse. As indicated in Figure 4, we assume that the foci are $F_1(-c, 0)$ and $F_2(c, 0)$ and that the sum of the distances from a point $P(x, y)$ on the ellipse to the foci is $2a$. Since the point P lies on the ellipse, we have

$$F_1P + F_2P = 2a$$

and therefore

$$\sqrt{(x + c)^2 + (y - 0)^2} + \sqrt{(x - c)^2 + (y - 0)^2} = 2a$$

or

$$\sqrt{(x - c)^2 + y^2} = 2a - \sqrt{(x + c)^2 + y^2} \qquad (1)$$

Now, by following a straightforward but lengthy process of squaring and simplifying (as outlined in detail in Exercise 40), we find that equation (1) becomes

$$(a^2 - c^2)x^2 + a^2y^2 = a^2(a^2 - c^2) \qquad (2)$$

To write equation (2) in a more symmetric form, we define the positive number b by the equation

$$b^2 = a^2 - c^2 \qquad (3)$$

Note For this definition to make sense, we need to know that the right-hand side of equation (3) is positive. (See Exercise 41 at the end of this section for details.)

Finally, using equation (3) to substitute for $a^2 - c^2$ in equation (2), we obtain

$$b^2x^2 + a^2y^2 = a^2b^2$$

which can be written

$$\frac{x^2}{a^2} + \frac{y^2}{b^2} = 1 \qquad (a > b) \qquad (4)$$

We have now shown that the coordinates of each point on the ellipse satisfy equation (4). Conversely, it can be shown that if x and y satisfy equation (4), then the point (x, y) indeed lies on the ellipse. We refer to equation (4) as the **standard form** for the equation of an ellipse with foci $(-c, 0)$ and $(c, 0)$. For an ellipse in this form, it will always be the case that a is greater than b; this follows from equation (3).

For purposes of graphing, we want to know the intercepts of the ellipse. To find the x-intercepts, we set $y = 0$ in equation (4) to obtain

$$\frac{x^2}{a^2} = 1$$

$$x^2 = a^2 \qquad \text{or} \qquad x = \pm a$$

The x-intercepts are therefore a and $-a$. In a similar fashion, you can check that the y-intercepts are b and $-b$. Also (according to the symmetry tests in Section 2.4), note that the graph of equation (4) must be symmetric about both coordi-

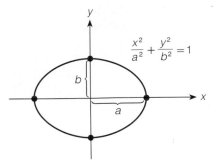

FIGURE 5

The intercepts of the ellipse $(x^2/a^2) + (y^2/b^2) = 1$ are $x = \pm a$ and $y = \pm b$.

nate axes. Figure 5 shows the graph of the ellipse $(x^2/a^2) + (y^2/b^2) = 1$. (Calculator exercises at the end of this section will help convince you that the general shape of the curve in Figure 5 is correct.)

In the box that follows, we define several terms that are useful in describing and analyzing the ellipse.

DEFINITION Terminology for the Ellipse

1. The **focal axis** is the line passing through the foci of the ellipse.

2. The **center** is the point midway between the foci. This is the point C in Figure 6.

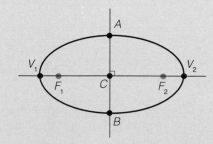

FIGURE 6

3. The **vertices** (singular: **vertex**) are the two points where the focal axis meets the ellipse. These are the points V_1 and V_2 in Figure 6.

4. The **major axis** is the line segment joining the vertices. This is the line segment $\overline{V_1V_2}$ in Figure 6. The **minor axis** is the line segment through the center of the ellipse, perpendicular to the major axis, and with endpoints on the ellipse. This is the segment $\overline{AB}$ in Figure 6.

5. The **eccentricity** is the ratio $\dfrac{c}{a} = \dfrac{\sqrt{a^2 - b^2}}{a}$.

The eccentricity (as defined in the box) provides a numerical measure of how much the ellipse deviates from being a circle. As Figure 7 indicates, the closer the eccentricity is to zero, the more the ellipse resembles a circle. In the other direction, as the eccentricity approaches 1, the ellipse becomes increasingly flat.

FIGURE 7

Eccentricity is a number between 0 and 1 that describes the shape of an ellipse. The narrowest ellipse in this figure has the same proportions as the orbit of Halley's comet. By way of contrast, the eccentricity of Earth's orbit is 0.0017; if an ellipse with this eccentricity were included in Figure 7, it would appear indistinguishable from a circle.

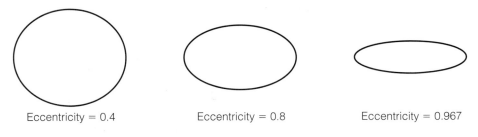

Eccentricity = 0.4   Eccentricity = 0.8   Eccentricity = 0.967

In the next box, we summarize our discussion up to this point. (Note that we use the letter e to denote the eccentricity. This is the conventional choice, even though the same letter is used with a very different meaning in connection with exponential functions and logarithms.)

PROPERTY SUMMARY   **THE ELLIPSE** $\dfrac{x^2}{a^2} + \dfrac{y^2}{b^2} = 1$ **$(a > b)$**

Foci: $(\pm c, 0)$, where $c^2 = a^2 - b^2$

Center: $(0, 0)$

Vertices: $(\pm a, 0)$

Length of major axis: $2a$

Length of minor axis: $2b$

Eccentricity: $e = \dfrac{c}{a} = \dfrac{\sqrt{a^2 - b^2}}{a}$

EXAMPLE 1   Find the lengths of the major and minor axes of the ellipse $9x^2 + 16y^2 = 144$. Specify the coordinates of the foci and the eccentricity. Graph the ellipse.

Solution   To convert the equation $9x^2 + 16y^2 = 144$ to standard form, divide both sides by 144. This yields

$$\frac{9x^2}{144} + \frac{16y^2}{144} = \frac{144}{144}$$

$$\frac{x^2}{16} + \frac{y^2}{9} = 1 \qquad \text{or} \qquad \frac{x^2}{4^2} + \frac{y^2}{3^2} = 1$$

This is the standard form. Comparing this equation with $(x^2/a^2) + (y^2/b^2) = 1$, we see that $a = 4$ and $b = 3$. Thus, the major and minor axes are 8 and 6 units, respectively. Next, to determine the foci, we use the equation $c^2 = a^2 - b^2$. We have

$$c^2 = 4^2 - 3^2 = 7 \qquad \text{and therefore} \qquad c = \sqrt{7}$$

(We choose the positive square root because $c > 0$.) It follows that the coordinates of the foci are $\left(-\sqrt{7}, 0\right)$ and $\left(\sqrt{7}, 0\right)$. We now calculate the eccentricity using the formula $e = c/a$. This yields $e = \sqrt{7}/4$. Figure 8 shows the graph along with the required information.

FIGURE 8

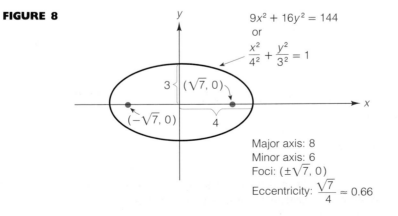

$9x^2 + 16y^2 = 144$
or
$\dfrac{x^2}{4^2} + \dfrac{y^2}{3^2} = 1$

Major axis: 8
Minor axis: 6
Foci: $(\pm\sqrt{7}, 0)$
Eccentricity: $\dfrac{\sqrt{7}}{4} \approx 0.66$

■ ■ EXAMPLE 2 The foci of an ellipse are $(-1, 0)$ and $(1, 0)$, and the eccentricity is $\frac{1}{3}$. Find the equation of the ellipse (in standard form), and specify the lengths of the major and minor axes.

Solution Since the foci are $(\pm 1, 0)$, we have $c = 1$. Using the equation $e = c/a$ with $e = \frac{1}{3}$ and $c = 1$, we obtain

$$\frac{1}{3} = \frac{1}{a} \qquad \text{and therefore} \qquad a = 3$$

Recall now that b^2 is defined by the equation $b^2 = a^2 - c^2$. In view of this, we have

$$b^2 = 3^2 - 1^2 = 8$$
$$b = \sqrt{8} = 2\sqrt{2}$$

The equation of the ellipse in standard form is therefore

$$\frac{x^2}{3^2} + \frac{y^2}{(2\sqrt{2})^2} = 1$$

Furthermore, since $a = 3$ and $b = 2\sqrt{2}$, the lengths of the major and minor axes are 6 and $4\sqrt{2}$ units, respectively.   **■■■**

■ ■ EXAMPLE 3 The eccentricity of an ellipse is $\frac{4}{5}$, and the sum of the distances from a point P on the ellipse to the foci is 10 units. Compute the distance between the foci F_1 and F_2. See Figure 9.

Solution We are required to find the distance between the foci F_1 and F_2. By definition, this is the quantity $2c$. Since the sum of the distances from a point on the ellipse to the foci is 10 units, we have

$$2a = 10 \qquad \text{by definition of } 2a$$
$$a = 5$$

Now we substitute the values $a = 5$ and $e = \frac{4}{5}$ in the formula $e = c/a$:

$$\frac{4}{5} = \frac{c}{5} \qquad \text{and therefore} \qquad c = 4$$

It now follows that the required distance $2c$ is 8 units.   **■■■**

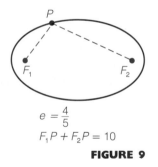

$e = \dfrac{4}{5}$

$F_1 P + F_2 P = 10$

FIGURE 9

Suppose now that we translate the graph of $(x^2/a^2) + (y^2/b^2) = 1$ by h units in the x-direction and k units in the y-direction, as shown in Figure 10. Then the equation of the translated ellipse is

$$\frac{(x - h)^2}{a^2} + \frac{(y - k)^2}{b^2} = 1 \tag{5}$$

Equation (5) is another **standard form** for the equation of an ellipse. As indicated in Figure 10, the center of this ellipse is the point (h, k). In the next example, we use the technique of completing the square to convert an equation for an ellipse to standard form. Once the equation is in standard form, the graph is readily obtained.

FIGURE 10

∎∎ ∎ **EXAMPLE 4** Determine the center, foci, and eccentricity of the ellipse

$$4x^2 + 9y^2 - 8x - 54y + 49 = 0$$

Graph the ellipse.

Solution We will convert the given equation to standard form by using the technique of completing the square. We have

$$4x^2 - 8x + 9y^2 - 54y = -49$$
$$4(x^2 - 2x) + 9(y^2 - 6y) = -49$$
$$4(x^2 - 2x + 1) + 9(y^2 - 6y + 9) = -49 + (4)(1) + 9(9)$$
$$4(x - 1)^2 + 9(y - 3)^2 = 36$$

Dividing this last equation by 36, we obtain

$$\frac{(x - 1)^2}{9} + \frac{(y - 3)^2}{4} = 1$$

or

$$\frac{(x - 1)^2}{3^2} + \frac{(y - 3)^2}{2^2} = 1$$

This last equation represents an ellipse with center at $(1, 3)$ and with $a = 3$ and $b = 2$. We can calculate c using the formula $c^2 = a^2 - b^2$:

$$c^2 = 3^2 - 2^2 = 5 \qquad \text{and therefore} \qquad c = \sqrt{5}$$

Since the center of this ellipse is $(1, 3)$, the foci are therefore $\left(1 + \sqrt{5}, 3\right)$ and $\left(1 - \sqrt{5}, 3\right)$. Finally, the eccentricity is c/a, which is $\sqrt{5}/3$. Figure 11 shows the graph of this ellipse. ∎∎∎

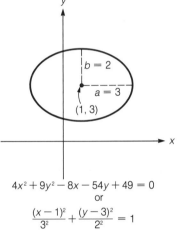

$4x^2 + 9y^2 - 8x - 54y + 49 = 0$
or
$\frac{(x - 1)^2}{3^2} + \frac{(y - 3)^2}{2^2} = 1$

FIGURE 11

In developing the equation $(x^2/a^2) + (y^2/b^2) = 1$, we assumed that the foci of the ellipse were located on the x-axis at the points $(-c, 0)$ and $(c, 0)$. If, instead, the foci are located on the y-axis at the points $(0, c)$ and $(0, -c)$, then the same method we used in the previous case will show the equation of the ellipse to be

$$\frac{x^2}{b^2} + \frac{y^2}{a^2} = 1 \qquad (a > b)$$

We still assume that $2a$ represents the sum of the distances from a point on the ellipse to the foci. In the box that follows, we summarize the situation for the ellipse with foci $(0, c)$ and $(0, -c)$.

PROPERTY SUMMARY   **THE ELLIPSE $\dfrac{x^2}{b^2} + \dfrac{y^2}{a^2} = 1$ $(a > b)$**

Foci: $(0, \pm c)$, where $c^2 = a^2 - b^2$

Center: $(0, 0)$

Vertices: $(0, \pm a)$

Length of major axis: $2a$

Length of minor axis: $2b$

Eccentricity: $e = \dfrac{c}{a} = \dfrac{\sqrt{a^2 - b^2}}{a}$

We now have two standard forms for the ellipse:

Foci on the x-axis at $(\pm c, 0)$: $\dfrac{x^2}{a^2} + \dfrac{y^2}{b^2} = 1$ $(a > b)$

Foci on the y-axis at $(0, \pm c)$: $\dfrac{x^2}{b^2} + \dfrac{y^2}{a^2} = 1$ $(a > b)$

Because a is greater than b in both standard forms, it is always easy to determine by inspection whether the foci lie on the x-axis or the y-axis. Consider, for instance, the equation $(x^2/5^2) + (y^2/6^2) = 1$. In this case, since $6 > 5$, we have $a = 6$. And since 6^2 appears under y^2, we conclude that the foci lie on the y-axis.

■▬■ EXAMPLE 5 The point $(5, 3)$ lies on an ellipse with vertices $\left(0, \pm 2\sqrt{21}\right)$. Find the equation of the ellipse. Write the answer both in standard form and in the form $Ax^2 + By^2 = C$.

Solution Since the vertices are $\left(0, \pm 2\sqrt{21}\right)$, the standard form for the equation in this case is $(x^2/b^2) + (y^2/a^2) = 1$. Furthermore, in view of the coordinates of the vertices, we have $a = 2\sqrt{21}$. Therefore,

$$\frac{x^2}{b^2} + \frac{y^2}{84} = 1$$

Now, since the point $(5, 3)$ lies on the ellipse, its coordinates must satisfy this last equation. We thus have

$$\frac{5^2}{b^2} + \frac{3^2}{84} = 1$$

$$\frac{25}{b^2} + \frac{3}{28} = 1$$

$$\frac{25}{b^2} = \frac{25}{28}$$

From this last equation, we see that $b^2 = 28$ and therefore

$$b = \sqrt{28} = 2\sqrt{7}$$

Now that we've determined a and b, we can write the equation of the ellipse in standard form. It is

$$\frac{x^2}{(2\sqrt{7})^2} + \frac{y^2}{(2\sqrt{21})^2} = 1$$

As you should verify for yourself, this equation can also be written in the equivalent form

$$3x^2 + y^2 = 84$$ ▐▐▐

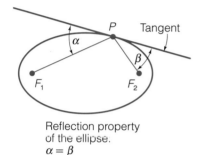

Reflection property
of the ellipse.
$\alpha = \beta$

FIGURE 12

As with the parabola, many of the interesting properties of the ellipse are related to tangent lines. We define a **tangent to an ellipse** as a line that intersects the ellipse in exactly one point. Figure 12 shows a line tangent to an ellipse at an arbitrary point P on the curve. Line segments $\overline{F_1P}$ and $\overline{F_2P}$ are drawn from the foci to the point of tangency. These two segments are called **focal radii**. One of the most basic properties of the ellipse is the **reflection property**. According to this property, the focal radii drawn to the point of tangency make equal angles with the tangent. (The steps for proving this fact are outlined in detail in Exercise 65 at the end of this section. For a similar result concerning the parabola, see Exercise 19 in the previous section.)

If $P(x_1, y_1)$ is a point on the ellipse $(x^2/a^2) + (y^2/b^2) = 1$, then the equation of the line tangent to the ellipse at $P(x_1, y_1)$ is

$$\frac{x_1 x}{a^2} + \frac{y_1 y}{b^2} = 1$$

This equation is easy to remember because it so closely resembles the equation of the ellipse itself. However, do notice that the equation $(x_1x/a^2) + (y_1y/b^2) = 1$ is indeed linear, since x_1, y_1, a, and b all denote constants. This equation for the tangent to an ellipse can be derived using the same technique we employed with the parabola in the previous section. (See Exercise 64 at the end of this section for the details.) The two examples that follow show how this equation is used.

▐▐▐ **EXAMPLE 6** Find the equation of the line that is tangent to the ellipse $x^2 + 3y^2 = 57$ at the point $(3, 4)$ on the ellipse. See Figure 13. Write the answer in the form $y = mx + b$.

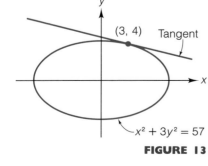

FIGURE 13

Solution We know that the equation of the line tangent to the ellipse at the point (x_1, y_1) is $(x_1x/a^2) + (y_1y/b^2) = 1$. We are given that $x_1 = 3$ and $y_1 = 4$. Thus, we need to determine a^2 and b^2. To do that, we convert the equation $x^2 + 3y^2 = 57$ to standard form by dividing through by 57. This yields

$$\frac{x^2}{57} + \frac{y^2}{19} = 1$$

It follows that $a^2 = 57$ and $b^2 = 19$. The equation of the tangent line then becomes

$$\frac{3x}{57} + \frac{4y}{19} = 1$$

$$\frac{x}{19} + \frac{4y}{19} = 1$$

$$x + 4y = 19$$

Solving for y, we obtain

$$y = -\frac{1}{4}x + \frac{19}{4}$$

■■■

FIGURE 14

■■■ EXAMPLE 7   Show that the slope of the tangent line in Figure 14 is equal to the eccentricity of the ellipse, $e = c/a$.

Solution   We first determine the coordinates of the point P in Figure 14. Evidently, the x-coordinate of P is $-c$. Substituting this value for x in the equation of the ellipse yields

$$\frac{c^2}{a^2} + \frac{y^2}{b^2} = 1$$

To solve this equation for y, let us first clear the equation of fractions by multiplying by $a^2 b^2$. This yields

$$b^2 c^2 + a^2 y^2 = a^2 b^2$$

$$a^2 y^2 = a^2 b^2 - b^2 c^2 = b^2(a^2 - c^2)$$

$$= b^2(b^2) \qquad \text{by definition of } b^2$$

$$= b^4$$

$$y^2 = \frac{b^4}{a^2}$$

$$y = \frac{b^2}{a}$$

The coordinates of P are therefore $(-c, b^2/a)$. Substituting these coordinates for x_1 and y_1 in the equation $(x_1 x/a^2) + (y_1 y/b^2) = 1$, we obtain

$$-\frac{cx}{a^2} + \frac{(b^2/a)y}{b^2} = 1$$

$$-\frac{cx}{a^2} + \frac{y}{a} = 1$$

Therefore,

$$\frac{y}{a} = \frac{c}{a^2}x + 1 \qquad \text{and} \qquad y = \frac{c}{a}x + a$$

The slope of the line is c/a, the coefficient of x. This is what we wanted to show.

■■■

FIGURE 15

A consequence of the reflection property of the ellipse. A light ray or sound emitted from one focus is reflected through the other focus.

We conclude this section by mentioning a few of the applications of the ellipse. Each planet in the solar system moves in an elliptical orbit with the sun at one focus. Some gears in machines are elliptical rather than circular. (In certain brands of racing bikes, this is true of one of the gears in front.) As with the parabola, the reflection property of the ellipse has applications in optics and acoustics. As indicated in Figure 15, a light ray or sound emitted from one focus of an ellipse is always reflected through the other focus. This property is used in the design of "whispering galleries." These are rooms (with elliptical cross sections) in which a person standing at focus F_2 can hear a whisper from focus F_1 while others closer to F_1 might hear nothing. Statuary Hall in the Capitol building in Washington, D.C., is a whispering gallery. This idea is used in a modern medical device known as the *lithotripter,* in which high-energy sound waves are used to break up kidney stones. The patient is positioned with the kidney stone at one focus in an elliptical water bath while sound waves are emitted from the other focus.

EXERCISE SET 10.4

A

In Exercises 1–24, graph the ellipses. In each case, specify the lengths of the major and minor axes, the foci, and the eccentricity. For Exercises 13–24, also specify the center of the ellipse.

1. $4x^2 + 9y^2 = 36$
2. $4x^2 + 25y^2 = 100$
3. $x^2 + 16y^2 = 16$
4. $9x^2 + 25y^2 = 225$
5. $x^2 + 2y^2 = 2$
6. $2x^2 + 3y^2 = 3$
7. $16x^2 + 9y^2 = 144$
8. $25x^2 + y^2 = 25$
9. $15x^2 + 3y^2 = 5$
10. $9x^2 + y^2 = 4$
11. $2x^2 + y^2 = 4$
12. $36x^2 + 25y^2 = 400$
13. $\dfrac{(x-5)^2}{5^2} + \dfrac{(y+1)^2}{3^2} = 1$
14. $\dfrac{(x-1)^2}{2^2} + \dfrac{(y+4)^2}{3^2} = 1$
15. $\dfrac{(x-1)^2}{1^2} + \dfrac{(y-2)^2}{2^2} = 1$
16. $\dfrac{x^2}{4^2} + \dfrac{(y-3)^2}{2^2} = 1$
17. $\dfrac{(x+3)^2}{3^2} + \dfrac{y^2}{1^2} = 1$
18. $\dfrac{(x-2)^2}{2^2} + \dfrac{(y-2)^2}{2^2} = 1$
19. $3x^2 + 4y^2 - 6x + 16y + 7 = 0$
20. $16x^2 + 64x + 9y^2 - 54y + 1 = 0$
21. $5x^2 + 3y^2 - 40x - 36y + 188 = 0$
22. $x^2 + 16y^2 - 160y + 384 = 0$
23. $16x^2 + 25y^2 - 64x - 100y + 564 = 0$
24. $4x^2 + 4y^2 - 32x + 32y + 127 = 0$

In Exercises 25–32, find the equation of the ellipse satisfying the given conditions. Write the answer both in standard form and in the form $Ax^2 + By^2 = C$.

25. Foci $(\pm 3, 0)$; vertices $(\pm 5, 0)$
26. Foci $(0, \pm 1)$; vertices $(0, \pm 4)$
27. Vertices $(\pm 4, 0)$; eccentricity $\frac{1}{4}$
28. Foci $(0, \pm 2)$; endpoints of minor axis $(\pm 5, 0)$
29. Foci $(0, \pm 2)$; endpoints of major axis $(0, \pm 5)$
30. Endpoints of major axis $(\pm 10, 0)$; endpoints of minor axis $(0, \pm 4)$
31. Center at the origin; vertices on the x-axis; length of major axis twice the length of minor axis; $\left(1, \sqrt{2}\right)$ lies on ellipse
32. Eccentricity $\frac{3}{5}$; one endpoint of minor axis $(-8, 0)$; center at the origin
33. Find the equation of the tangent to the ellipse $x^2 + 3y^2 = 76$ at each of the following points. Write your answers in the form $y = mx + b$.
    (a) $(8, 2)$ (b) $(-7, 3)$ (c) $(1, -5)$
34. (a) Find the equation of the line tangent to the ellipse $x^2 + 3y^2 = 84$ at the point $(3, 5)$ on the ellipse. Write your answer in the form $y = mx + b$.
    (b) Repeat part (a), but at the point $(-3, -5)$ on the ellipse.
    (c) Are the lines determined in (a) and (b) parallel?
35. A line is drawn tangent to the ellipse $3x^2 + y^2 = 52$ at the point $(4, 2)$ on the ellipse.
    (a) Find the equation of this tangent line.
    (b) Find the area of the first-quadrant triangle bounded by the axes and this tangent line.
36. Tangent lines are drawn to the ellipse $x^2 + 3y^2 = 12$ at the points $(3, -1)$ and $(-3, -1)$ on the ellipse.

(a) Find the equation of each tangent line. Write your answers in the form $y = mx + b$.

(b) Find the point where the tangent lines intersect.

37. Let F_1 and F_2 be the foci of the ellipse $9x^2 + 25y^2 = 225$. Let P be the point $(1, 6\sqrt{6}/5)$.

(a) Show that P lies on the ellipse.

(b) Find the equation of the line that is tangent to the ellipse at P. Write your answer in the form $Ax + By + C = 0$.

(c) Let d_1 denote the distance from F_1 to the tangent line determined in part (b). Similarly let d_2 denote the distance from F_2 to this tangent line. Compute d_1 and d_2.

(d) Verify that $d_1 d_2 = b^2$, where b is half the length of the minor axis.

38. Solve the equation $(x^2/3^2) + (y^2/2^2) = 1$ for y to obtain $y = \pm\frac{1}{3}\sqrt{36 - 4x^2}$. Then complete the following table and use the results to graph the given equation. (Use the fact that the graph must be symmetric about the y-axis.)

x	0	0.5	1.0	1.5	2.0	2.5	3
y	± 2						0

39. Solve the equation $(x^2/1^2) + (y^2/4^2) = 1$ for y to obtain $y = \pm 4\sqrt{1 - x^2}$. Then complete the following table and use the results to graph the given equation. (Use the fact that the graph must be symmetric about the y-axis.)

x	0	0.1	0.2	0.3	0.4	0.5
y	± 4					

x	0.6	0.7	0.8	0.9	1.0
y					0

40. This exercise outlines the steps needed to complete the derivation of the equation $(x^2/a^2) + (y^2/b^2) = 1$.

(a) Square both sides of equation (1) on page 599. After simplifying, you should obtain
$$a\sqrt{(x + c)^2 + y^2} = a^2 + xc.$$

(b) Square both sides of the equation in part (a). Show that the result can be written
$$a^2 x^2 - x^2 c^2 + a^2 y^2 = a^4 - a^2 c^2.$$

(c) Verify that the equation in part (b) is equivalent to
$$(a^2 - c^2)x^2 + a^2 y^2 = a^2(a^2 - c^2).$$

(d) Using the equation in part (c), replace the quantity $a^2 - c^2$ with b^2. Then show that the resulting equation can be rewritten $(x^2/a^2) + (y^2/b^2) = 1$, as required.

41. In the text we defined the positive number b^2 by the equation $b^2 = a^2 - c^2$. For this definition to make sense, we need to show that the quantity $a^2 - c^2$ is positive. This can be done as follows. First, recall that in any triangle, the sum of the lengths of two sides is always greater than the length of the third side. Now apply this fact to triangle $F_1 P F_2$ in Figure 4 to show that $2a > 2c$. Conclude from this that $a^2 - c^2 > 0$, as required.

42. A line is drawn tangent to the ellipse $(x^2/a^2) + (y^2/b^2) = 1$ at the point (x_1, y_1) on the ellipse. Let P and Q denote the points where the tangent meets the y- and the x-axis, respectively. Show that the midpoint of PQ is $(a^2/2x_1, b^2/2y_1)$.

43. A *normal* to an ellipse is a line drawn perpendicular to the tangent at the point of tangency. Show that the equation of the normal to the ellipse $(x^2/a^2) + (y^2/b^2) = 1$ at the point (x_1, y_1) can be written
$$a^2 y_1 x - b^2 x_1 y = (a^2 - b^2)x_1 y_1$$

44. Let (x_1, y_1) be any point on the ellipse $(x^2/a^2) + (y^2/b^2) = 1$, $a > b$, other than one of the endpoints of the major or minor axis. Show that the normal at (x_1, y_1) does not pass through the origin. *Hint:* Find the y-intercept of the normal.

45. Find the points of intersection of the ellipses
$$\frac{x^2}{a^2} + \frac{y^2}{b^2} = 1 \quad \text{and} \quad \frac{x^2}{b^2} + \frac{y^2}{a^2} = 1 \quad (a > b)$$

Include a sketch with your answer.

46. Let F_1 and F_2 denote the foci of the ellipse $(x^2/a^2) + (y^2/b^2) = 1$, $a > b$. Suppose that P is one of the endpoints of the minor axis and angle $F_1 P F_2$ is a right angle. Compute the eccentricity of the ellipse.

47. Let $P(x_1, y_1)$ be a point on the ellipse $(x^2/a^2) + (y^2/b^2) = 1$. Let $\mathcal{N}$ be the point where the normal through P meets the x-axis, and let F be the focus $(-c, 0)$. Show that $F\mathcal{N}/FP = e$, where e denotes the eccentricity.

48. Let $P(x_1, y_1)$ be a point on the ellipse $b^2 x^2 + a^2 y^2 = a^2 b^2$. Suppose that the tangent to the ellipse at P meets the y-axis at A and the x-axis at B. If $AP = PB$, what are x_1 and y_1 (in terms of a and b)?

49. (a) Verify that the points $A(5, 1)$, $B(4, -2)$, and $C(-1, 3)$ all lie on the ellipse $x^2 + 3y^2 = 28$.

(b) Find a point D on the ellipse such that $\overline{CD}$ is parallel to $\overline{AB}$.

(c) If O denotes the center of the ellipse, show that the

triangles OAC and OBD have equal areas.
Suggestion: In computing the areas, the formula given in Exercise 14(b) of Exercise Set 2.1 is useful.

50. Find the points of intersection of the parabola $y = x^2$ and the ellipse $b^2x^2 + a^2y^2 = a^2b^2$.

51. Recall that the two line segments joining a point on the ellipse to the foci are called *focal radii*. These are the segments $\overline{F_1P}$ and $\overline{F_2P}$ in the following figure.

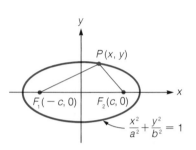

(a) Show that $F_1P = a + ex$. *Hint:* If you try to do this from scratch, it can involve a rather lengthy calculation. Begin instead with the equation $a\sqrt{(x+c)^2 + y^2} = a^2 + xc$ [from Exercise 40(a)], and divide both sides by a.

(b) Show that $F_2P = a - ex$. *Hint:* Make use of the result in part (a), along with the fact that the sum of F_1P and F_2P is, by definition, $2a$.

52. Find the coordinates of a point P in the first quadrant on the ellipse $9x^2 + 25y^2 = 225$ such that angle F_2PF_1 is a right angle.

53. The accompanying figure shows the two tangent lines drawn from the point (h, k) to the ellipse $b^2x^2 + a^2y^2 = a^2b^2$. Follow steps (a) through (d) to show that the equation of the line passing through the two points of tangency is $b^2hx + a^2ky = a^2b^2$.

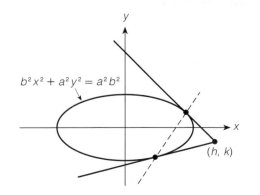

(a) Let (x_1, y_1) be one of the points of tangency. Check that the equation of the tangent line through this point is $b^2x_1x + a^2y_1y = a^2b^2$.

(b) Using the result in part (a), explain why $b^2x_1h + a^2y_1k = a^2b^2$.

(c) In a similar fashion, show that $b^2x_2h + a^2y_2k = a^2b^2$, where (x_2, y_2) is the other point of tangency.

(d) The equation $b^2hx + a^2ky = a^2b^2$ represents a line. Explain why this line must pass through the points (x_1, y_1) and (x_2, y_2).

C

54. The normal to the ellipse $b^2x^2 + a^2y^2 = a^2b^2$ at $P(x_1, y_1)$ meets the x-axis at A and the y-axis at B. Show that

$$PA \cdot PB = F_1P \cdot F_2P$$

where, as usual, F_1 and F_2 are the foci.

55. In the following figure, $P(x, y)$ denotes an arbitrary point on the ellipse $(x^2/a^2) + (y^2/b^2) = 1$, and R is the foot of the perpendicular drawn from P to the line $x = a/e$. Show that $F_2P/PR = e$. The line $x = a/e$ is called a *directrix* of the ellipse. *Hint:* This result is easy to prove if you make use of the formula for F_2P in Exercise 51(b).

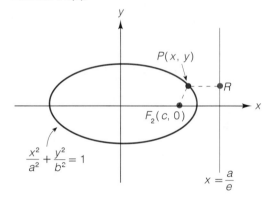

56. Let $P(x, y)$ be a point on the ellipse $(x^2/a^2) + (y^2/b^2) = 1$. Let Q be the foot of the perpendicular drawn from P to the line $x = -a/e$. Let F_1 be the focus located at $(-c, 0)$. Show that $F_1P/PQ = e$. The line $x = -a/e$ is called a *directrix* of the ellipse.

57. Let D denote the distance from the center of the ellipse $(x^2/a^2) + (y^2/b^2) = 1$ to the directrix $x = a/e$, and let d denote the distance from the focus $F_2(c, 0)$ to the directrix. Show that $D/d = a^2/b^2$.

58. At the point $P(x_1, y_1)$ on the ellipse $(x^2/a^2) + (y^2/b^2) = 1$, a tangent line is drawn that meets the directrix $x = a/e$ at Q. If F denotes the focus located at $(c, 0)$, show that angle PFQ is a right angle.

59. (a) Show that the point $(c, b^2/a)$ lies on the ellipse $(x^2/a^2) + (y^2/b^2) = 1$, where c has its usual meaning.

(b) A tangent is drawn to the ellipse $(x^2/a^2) + (y^2/b^2) = 1$

at the point $(c, b^2/a)$. Find the point where this tangent intersects the directrix $x = a/e$.

60. The *auxiliary circle* of an ellipse is the circle centered at the center of the ellipse and with radius half the length of the major axis of the ellipse.
    (a) Find the equation of the auxiliary circle of the ellipse $x^2 + 3y^2 = 12$. Sketch the circle and the ellipse.
    (b) Verify that the point $P(3, 1)$ lies on the ellipse.
    (c) Find the equation of the line tangent to the ellipse at P.
    (d) Let T be the point where the perpendicular drawn from a focus to the tangent meets the tangent. Show that T lies on the auxiliary circle.

61. The points P and Q have the same x-coordinate. The point P lies on the circle $x^2 + y^2 = a^2$, and Q lies on the ellipse $(x^2/a^2) + (y^2/b^2) = 1$. Show that the tangent to the circle at P meets the tangent to the ellipse at Q at a point on the x-axis. (Assume that P and Q are distinct, and that neither point is an endpoint of the major or minor axis of the ellipse.)

62. A line segment that passes through the center of an ellipse is called a *diameter*.
    (a) If (x_1, y_1) is an endpoint of a diameter of the ellipse $(x^2/a^2) + (y^2/b^2) = 1$, show that the other endpoint is $(-x_1, -y_1)$.
    (b) If (x_1, y_1) is an endpoint of a diameter of the ellipse $(x^2/a^2) + (y^2/b^2) = 1$, show that the length of the diameter is $2\sqrt{(a^2 - b^2)x_1^2 + a^2 b^2}/a$.

63. Let $\overline{PQ}$ be a diameter of the ellipse $(x^2/a^2) + (y^2/b^2) = 1$. Show that the tangents to the ellipse at P and Q are parallel.

64. This exercise outlines the steps required to show that the equation of the tangent to the ellipse $(x^2/a^2) + (y^2/b^2) = 1$ at the point (x_1, y_1) on the ellipse is $(x_1 x/a^2) + (y_1 y/b^2) = 1$.
    (a) Show that the equation $(x^2/a^2) + (y^2/b^2) = 1$ is equivalent to

    $$b^2 x^2 + a^2 y^2 = a^2 b^2 \qquad (1)$$

    Conclude that (x_1, y_1) lies on the ellipse if and only if

    $$b^2 x_1^2 + a^2 y_1^2 = a^2 b^2 \qquad (2)$$

    (b) Subtract equation (2) from (1) to show that

    $$b^2(x^2 - x_1^2) + a^2(y^2 - y_1^2) = 0 \qquad (3)$$

    Equation (3) is equivalent to equation (1) provided only that (x_1, y_1) lies on the ellipse. In the following steps, we will find the algebra much simpler if we use equation (3) to represent the ellipse, rather than the equivalent and perhaps more familiar equation (1).

(c) Let the equation of the line tangent to the ellipse at (x_1, y_1) be

$$y - y_1 = m(x - x_1)$$

Explain why the following system of equations must have exactly one solution, namely, (x_1, y_1):

$$\begin{cases} b^2(x^2 - x_1^2) + a^2(y^2 - y_1^2) = 0 & (4) \\ y - y_1 = m(x - x_1) & (5) \end{cases}$$

(d) Solve equation (5) for y and then substitute for y in equation (4) to obtain

$$b^2(x^2 - x_1^2) + a^2 m^2(x - x_1)^2 + 2a^2 m y_1(x - x_1) = 0 \quad (6)$$

(e) Show that equation (6) can be written

$$(x - x_1)[b^2(x + x_1) + a^2 m^2(x - x_1) + 2a^2 m y_1] = 0 \quad (7)$$

(f) Equation (7) is a quadratic equation in x, but as pointed out earlier, $x = x_1$ must be the only solution. (That is, $x = x_1$ is a double root.) Thus, the factor in brackets must equal zero when x is replaced by x_1. Use this observation to show that

$$m = -\frac{b^2 x_1}{a^2 y_1}$$

This represents the slope of the line tangent to the ellipse at (x_1, y_1).

(g) Using this value for m, show that equation (5) becomes

$$b^2 x_1 x + a^2 y_1 y = b^2 x_1^2 + a^2 y_1^2 \qquad (8)$$

(h) Now use equation (2) to show that equation (8) can be written

$$\frac{x_1 x}{a^2} + \frac{y_1 y}{b^2} = 1$$

which is what we set out to show.

65. This exercise outlines a proof of the reflection property of the ellipse. We want to prove that right triangles $F_1 AP$ and $F_2 BP$ in the accompanying figure are similar.

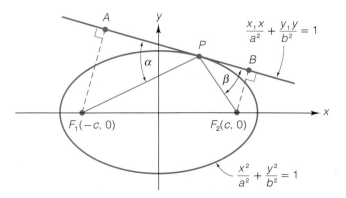

This will imply that $\alpha = \beta$, as required, since corresponding angles in similar triangles are equal. To show that the right triangles F_1AP and F_2BP are similar, it is enough to demonstrate that

$$\frac{F_1A}{F_1P} = \frac{F_2B}{F_2P}$$

(a) The quantity F_1A is the distance from the point $(-c, 0)$ to the line $(x_1x/a^2) + (y_1y/b^2) = 1$. Verify that this distance is given by

$$F_1A = \frac{\left| \dfrac{x_1c}{a^2} + 1 \right|}{\sqrt{A^2 + B^2}}$$

where $A = x_1/a^2$ and $B = y_1/b^2$.

(b) In the preceding expression for F_1A, replace c by ae. Then, by making use of the expression for F_1P obtained in Exercise 51, show that

$$F_1A = \frac{F_1P}{a\sqrt{A^2 + B^2}}$$

(c) Show similarly that

$$F_2B = \frac{F_2P}{a\sqrt{A^2 + B^2}}$$

(d) Now use the expressions for F_1A and F_2B obtained in parts (b) and (c) to verify that

$$\frac{F_1A}{F_1P} = \frac{F_2B}{F_2P}$$

This implies that right triangles F_1AP and F_2BP are similar and, consequently, that $\alpha = \beta$, as required.

▶ OPTIONAL TI-81 GRAPHING CALCULATOR EXERCISES
EXERCISES FOR SECTION 10.4

1. On page 600 in the text was the statement that calculator exercises at the end of that section would help to convince you that the general shape of the ellipse shown in Figure 5 (on page 600) is correct. Now let's make that "graphing calculator exercises" rather than merely "calculator exercises."
   (a) Take the equation $(x^2/3^2) + (y^2/2^2) = 1$, as given in Exercise 38 on page 608, and solve for y to obtain

   $$y = \pm\tfrac{1}{3}\sqrt{36 - 4x^2}$$

   Press the $Y=$ key, enter these two functions, press ZOOM 6, and then ZOOM 5. Which types of symmetry does the ellipse possess?
   (b) Take the equation $(x^2/1^2) + (y^2/4^2) = 1$, as given in Exercise 39 on page 608, and solve for y to obtain

   $$y = \pm 4\sqrt{1 - x^2}$$

   Press the $Y=$ key, enter these two functions, press ZOOM 6, then ZOOM 5. Which types of symmetry does the ellipse possess?

2. (a) Solve the equation $4x^2 + 9y^2 = 36$ for y, and then graph this ellipse (as in Exercise 1).
   (b) If we translate the ellipse in part (a) one unit to the right and three units up, what is the equation of the new ellipse?
   (c) Check that if you solve the equation obtained in

part (b) for y, the result can be written

$$y = \pm\tfrac{1}{3}\sqrt{36 - 4(x - 1)^2} + 3$$

(d) Press the $Y=$ key and enter the two functions given in part (c); then press ZOOM 6 to see the graph in the standard viewing rectangle. What is the center of this ellipse?

3. In this exercise we graph four ellipses. Each ellipse will have the same major axis, but the eccentricities will vary from 0 to 0.8. (Actually, we'll only graph the top half of each ellipse, the portion corresponding to $y \geq 0$, because we can enter at most four functions in the $Y=$ menu.)
   (a) The equation $b^2x^2 + 16y^2 = 16b^2$ represents an ellipse. Show that the length of the horizontal axis of this ellipse is 8.
   (b) Check that the following values for b^2 yield the indicated eccentricities.

b^2	16	13.44	10.24	5.76
eccentricity	0	0.4	0.6	0.8

   (c) Solve the equation $b^2x^2 + 16y^2 = 16b^2$ for y. Show that the result can be written

   $$y = \pm(0.25)\sqrt{b^2(16 - x^2)}$$

Now use the calculator to graph the (top half) of the four ellipses corresponding to the four values of b^2 in the table. Use the following settings: first press ZOOM 6; then press the RANGE key and set $x_{min} = -4$, $x_{max} = 4$, $y_{min} = 0$, and $y_{max} = 4$; finally, press ZOOM 5 so that the same scale is used on both axes. [Without this last step, the circle (that is, the ellipse with eccentricity 0) will not look like a circle.] If you have done things correctly, your graphs should be similar to the following figure.

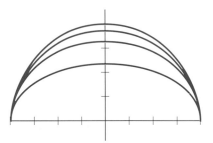

4. The *auxiliary circle* of an ellipse is defined to be the circle with diameter the same as the major axis of the ellipse. Determine the equation of the auxiliary circle for the ellipse $9x^2 + 25y^2 = 225$, then use the calculator to graph the ellipse and the circle.

In Exercises 5 and 6, determine the equation of the tangent to the ellipse at the given point. Then graph the ellipse and the tangent line.

5. $3x^2 + y^2 = 12$; $(1, -3)$

6. $x^2 + 3y^2 = 28$; $(4, 2)$

7. (As background for this exercise, you need to have completed Exercises 4 and 5.) In this exercise you're going to illustrate an interesting result concerning an ellipse and its auxiliary circle.
    (a) Graph the ellipse $x^2 + 3y^2 = 12$.

(b) Find the values of a, b, and c for this ellipse.
(c) Find the equation of the auxiliary circle for this ellipse. Add the graph of this circle to your picture from part (a).
(d) Verify that the point $P(3, 1)$ lies on the ellipse. Then find the equation of the tangent line to the ellipse at P. Next, we want to add this tangent line to the picture in part (c), but we already have four functions defined in the $Y=$ menu. Delete the two functions in the $Y=$ menu that correspond to the bottom halves of the ellipse and the circle. Now you can add the tangent line to the picture.
(e) In part (b) you determined the value of c. Find the equation of the line passing through $(-c, 0)$ and perpendicular to the tangent in part (d). Then add the graph of this line to your picture. If you've done things correctly, you should obtain a figure similar to the following one. The figure provides an example of this general result: The line through the focus and perpendicular to the tangent meets the tangent at a point on the auxiliary circle.

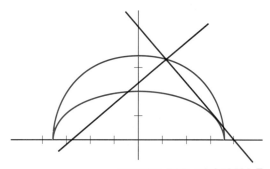

FIGURE FOR PROBLEM 7
The line through the focus and perpendicular to a tangent meets the tangent at a point on the auxiliary circle.

10.5   **THE HYPERBOLA**

... Menaechmus [c. 350 B.C.] is reputed to have discovered the curves that were later known as the ellipse, the parabola, and the hyperbola.

Carl B. Boyer in *A History of Mathematics*, 2nd ed., revised by Uta C. Merzbach (New York: John Wiley and Sons, 1991)

The ellipse is the general shape of any closed orbit.... It is also possible to have orbits that are not closed, whose shapes are represented by open curves. Even if the two bodies are not bound together by their mutual gravitational attraction, their gravitational attraction for each other still affects their relative motion. Then the general shape of their orbit is a hyperbola.

Theodore P. Snow in *The Dynamic Universe: An Introduction to Astronomy,* 4th ed. (St. Paul: West Publishing Company, 1990)

In the previous section, we defined an ellipse as the set of points P such that the sum of the distances from P to two fixed points is constant. By considering the difference instead of the sum, we are led to the definition of the hyperbola.

The Hyperbola

A **hyperbola** is the set of all points in the plane, the difference of whose distances from two fixed points is a positive constant. Each fixed point is called a **focus**.

As with the ellipse, we label the foci F_1 and F_2. Before obtaining an equation for the hyperbola, we can see the general features of the curve by using two sets of concentric circles, with centers F_1 and F_2, to locate points satisfying the definition of a hyperbola. In Figure 1, we've plotted a number of points P such that either $F_1P - F_2P = 3$ or $F_2P - F_1P = 3$. By joining these points, we obtain the graph of the hyperbola shown in Figure 1.

FIGURE 1

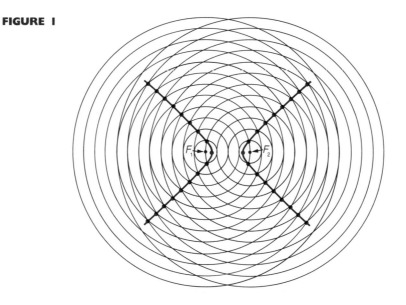

Unlike the parabola or the ellipse, the hyperbola is composed of two distinct parts, or **branches**. As you can check, the left branch in Figure 1 corresponds to the equation $F_2P - F_1P = 3$, while the right branch corresponds to the equation $F_1P - F_2P = 3$. Figure 1 also reveals that the hyperbola possesses two types of symmetry. First, it is symmetric about the line passing through the two foci F_1 and F_2; this line is referred to as the **focal axis** of the hyperbola. Second, the hyperbola is symmetric about the line that is the perpendicular bisector of the segment $\overline{F_1F_2}$.

To derive an equation for the hyperbola, let us initially assume that the foci are located at the points $F_1(-c, 0)$ and $F_2(c, 0)$, as indicated in Figure 2. We will

FIGURE 2

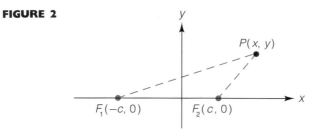

use $2a$ to denote the positive constant referred to in the definition of the hyperbola. By definition, then, $P(x, y)$ lies on the hyperbola if and only if

$$|F_1P - F_2P| = 2a$$

or, equivalently,

$$F_1P - F_2P = \pm 2a$$

If we use the formula for the distance between two points, this last equation becomes

$$\sqrt{(x + c)^2 + (y - 0)^2} - \sqrt{(x - c)^2 + (y - 0)^2} = \pm 2a$$

or

$$\sqrt{(x + c)^2 + y^2} - \sqrt{(x - c)^2 + y^2} = \pm 2a$$

We can simplify this equation by carrying out the same procedure that we used for the ellipse in the previous section. As Exercise 42 asks you to verify, the resulting equation is

$$(c^2 - a^2)x^2 - a^2 y^2 = a^2(c^2 - a^2) \tag{1}$$

Before further simplifying equation (1), we point out that the quantity $c^2 - a^2$ [which appears twice in equation (1)] is positive. To see why this is so, refer back to Figure 2. In triangle $F_1 F_2 P$ (as in any triangle), the length of any side is less than the sum of the lengths of the other two sides. Thus

$$F_1P < F_1F_2 + F_2P \qquad \text{and} \qquad F_2P < F_1F_2 + F_1P$$

and therefore

$$F_1P - F_2P < F_1F_2 \qquad \text{and} \qquad F_2P - F_1P < F_1F_2$$

These last two equations tell us that

$$|F_1P - F_2P| < F_1F_2$$

Therefore, in view of the definitions of $2a$ and $2c$, we have

$$0 < 2a < 2c \qquad \text{or} \qquad 0 < a < c$$

This last inequality tells us that $c^2 - a^2$ is positive, as we wished to show.

Now, since $c^2 - a^2$ is positive, we may define the positive number b by the equation

$$b^2 = c^2 - a^2$$

With this notation, equation (1) becomes

$$b^2x^2 - a^2y^2 = a^2b^2$$

Dividing by a^2b^2, we obtain

$$\frac{x^2}{a^2} - \frac{y^2}{b^2} = 1 \qquad (2)$$

We have now shown that the coordinates of every point on the hyperbola satisfy equation (2). Conversely, it can be shown that if the coordinates of a point satisfy equation (2), then the point satisfies the original definition of the hyperbola. Equation (2) is the **standard form** for the equation of a hyperbola with foci $F_1(-c, 0)$ and $F_2(c, 0)$.

The intercepts of the hyperbola are readily obtained from equation (2). To find the x-intercepts, we set y equal to zero to obtain

$$\frac{x^2}{a^2} = 1$$

$$x^2 = a^2 \qquad \text{or} \qquad x = \pm a$$

Thus, the hyperbola crosses the x-axis at the points $(-a, 0)$ and $(a, 0)$. On the other hand, the curve does not cross the y-axis, for if we set x equal to zero in equation (2), we obtain $-y^2/b^2 = 1$ or

$$y^2 = -b^2 \qquad (b > 0)$$

Since the square of any real number y is nonnegative, this last equation has no solution. Therefore, the graph does not cross the y-axis. Finally, let us note that (according to the symmetry tests in Section 2.4) the graph of equation (2) must be symmetric about both coordinate axes.

Before graphing the hyperbola $(x^2/a^2) - (y^2/b^2) = 1$, we point out the important fact that the two lines $y = (b/a)x$ and $y = -(b/a)x$ are asymptotes for the curve. We can see why as follows. First we solve equation (2) for y:

$$\frac{x^2}{a^2} - \frac{y^2}{b^2} = 1$$

$$b^2x^2 - a^2y^2 = a^2b^2 \qquad \text{multiplying by } a^2b^2$$

$$-a^2y^2 = a^2b^2 - b^2x^2$$

$$y^2 = \frac{b^2x^2 - a^2b^2}{a^2} = \frac{b^2(x^2 - a^2)}{a^2}$$

$$y = \pm\frac{b}{a}\sqrt{x^2 - a^2} \qquad (3)$$

TABLE 1

x	$\sqrt{x^2 - 5^2}$
100	99.875
1000	999.987
10000	9999.999

Now, as x grows arbitrarily large, the value of the quantity $\sqrt{x^2 - a^2}$ becomes closer and closer to x itself. Table 1 provides some empirical evidence for this statement in the case when $a = 5$. (A formal proof of the statement properly belongs to calculus.) In summary, then, we have the approximation $\sqrt{x^2 - a^2} \approx x$ as x grows arbitrarily large. So, in view of equation (3), we have

$$y = \pm \frac{b}{a} \sqrt{x^2 - a^2} \approx \pm \frac{b}{a} x \qquad \text{as } x \text{ grows arbitrarily large}$$

In other words, the two lines $y = \pm(b/a)x$ are asymptotes for the hyperbola.

A simple way to sketch the two asymptotes and then graph the hyperbola is as follows. First draw the rectangle with vertices (a, b), $(-a, b)$, $(-a, -b)$, and $(a, -b)$, as indicated in Figure 3(a). The slopes of the diagonals in this rectangle are b/a and $-b/a$. Thus by extending these diagonals as in Figure 3(b), we obtain the two asymptotes $y = \pm(b/a)x$. Now, since the x-intercepts of the hyperbola are a and $-a$, we can sketch the curve as shown in Figure 3(c).

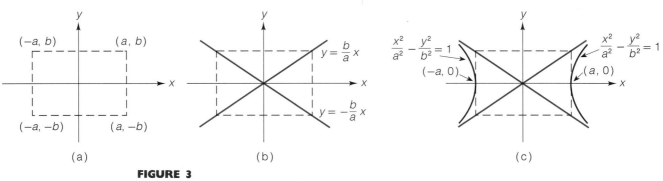

FIGURE 3
Steps in graphing the hyperbola
$(x^2/a^2) - (y^2/b^2) = 1$ and its asymptotes.

For the hyperbola in Figure 3(c), the two points $(\pm a, 0)$, where the curve meets the x-axis, are referred to as **vertices**. The midpoint of the line segment joining the two vertices is called the **center** of the hyperbola. (Equivalently, we can define the center as the point of intersection of the two asymptotes.) For the hyperbola in Figure 3(c), the center coincides with the origin. The line segment joining the vertices of a hyperbola is the **transverse axis** of the hyperbola. For reference, in the box that follows we summarize our work up to this point on the hyperbola. *Note:* Several new terms describing the hyperbola are also given in the box.

PROPERTY SUMMARY   THE HYPERBOLA $\dfrac{x^2}{a^2} - \dfrac{y^2}{b^2} = 1$

1. The **foci** are the points $F_1(-c, 0)$ and $F_2(c, 0)$. The hyperbola is the set of points P such that $|F_1P - F_2P| = 2a$.

2. The **focal axis** is the line passing through the foci.

3. The **vertices** are the points at which the hyperbola intersects its focal axis. In Figure 4, these are the two points $V_1(-a, 0)$ and $V_2(a, 0)$.

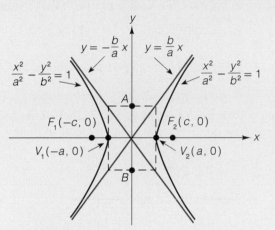

FIGURE 4

4. The **center** is the point on the focal axis midway between the foci. The center of the hyperbola in Figure 4 is the origin.

5. The **transverse axis** is the line segment joining the two vertices. In Figure 4, the length of the transverse axis $\overline{V_1V_2}$ is $2a$.

6. The **conjugate axis** is the line segment perpendicular to the transverse axis, passing through the center and extending a distance b on either side of the center. In Figure 4, this is the segment $\overline{AB}$.

7. The **eccentricity** e is defined by $e = c/a$, where $c^2 = a^2 + b^2$.

In general, if A, B, and C are positive numbers, then the graph of an equation of the form

$$Ax^2 - By^2 = C$$

will be a hyperbola of the type shown in Figure 4. Example 1 shows why this is so.

EXAMPLE 1   Graph the hyperbola $16x^2 - 9y^2 = 144$ after determining the following: vertices, foci, eccentricity, lengths of the transverse and conjugate axes, and asymptotes.

Solution   First convert the given equation to standard form by dividing through by 144. This yields

$$\frac{x^2}{9} - \frac{y^2}{16} = 1 \qquad \text{or} \qquad \frac{x^2}{3^2} - \frac{y^2}{4^2} = 1$$

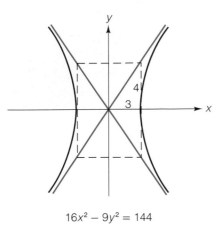

$16x^2 - 9y^2 = 144$
or
$$\frac{x^2}{3^2} - \frac{y^2}{4^2} = 1$$

FIGURE 5

By comparing this with the equation $(x^2/a^2) - (y^2/b^2) = 1$, we see that $a = 3$ and $b = 4$. The value of c can be determined by using the equation $c^2 = a^2 + b^2$. We have

$$c^2 = 3^2 + 4^2 = 25 \quad \text{and therefore} \quad c = 5$$

Now that we know the values of a, b, and c, we can list the required information:

Vertices: $(\pm 3, 0)$ Length of transverse axis ($= 2a$): 6
Foci: $(\pm 5, 0)$ Length of conjugate axis ($= 2b$): 8
Eccentricity: $\dfrac{5}{3}$ Asymptotes: $y = \pm\dfrac{4}{3}x$

The graph of the hyperbola is shown in Figure 5. ▮▮▮

The same method that we used to derive the equation for the hyperbola with foci $(\pm c, 0)$ can be used when the foci are instead located on the y-axis at the points $(0, \pm c)$. The equation of the hyperbola in this case is

$$\frac{y^2}{a^2} - \frac{x^2}{b^2} = 1$$

We summarize the basic properties of this hyperbola in the following box.

PROPERTY SUMMARY   **THE HYPERBOLA $\dfrac{y^2}{a^2} - \dfrac{x^2}{b^2} = 1$**

Foci: $(0, \pm c)$, where $c^2 = a^2 + b^2$
Vertices: $(0, \pm a)$
Asymptotes: $y = \pm(a/b)x$
Length of transverse axis: $2a$
Length of conjugate axis: $2b$
Eccentricity: $e = c/a$

FIGURE 6

▮▮▮ **EXAMPLE 2** Use the technique of completing the square to show that the graph of the following equation is a hyperbola:

$$9y^2 - 54y - 25x^2 + 200x - 544 = 0$$

Graph the hyperbola and specify the center, the vertices, the foci, the length of the transverse axis, and the equations of the asymptotes.

Solution

$$9(y^2 - 6y \quad) - 25(x^2 - 8x \quad) = 544 \qquad \text{factoring}$$

$$9(y^2 - 6y + 9) - 25(x^2 - 8x + 16) = 544 + 81 - 400 \qquad \text{completing the squares}$$

$$9(y - 3)^2 - 25(x - 4)^2 = 225$$

$$\frac{(y-3)^2}{5^2} - \frac{(x-4)^2}{3^2} = 1 \qquad \text{dividing by 225}$$

Now, the graph of this last equation is obtained by translating the graph of the equation

$$\frac{y^2}{5^2} - \frac{x^2}{3^2} = 1 \tag{4}$$

four units to the right and three units up. So first we analyze the graph of equation (4). The general form of the graph is shown in Figure 6. In our case, we have $a = 5$, $b = 3$, and

$$c = \sqrt{a^2 + b^2} = \sqrt{5^2 + 3^2} = \sqrt{34} \ (\approx 5.8)$$

Consequently, the vertices [for equation (4)] are $(0, \pm 5)$; the foci are $(0, \pm\sqrt{34})$; and the asymptotes are $y = \pm\frac{5}{3}x$. Figure 7 shows the graph of this hyperbola. Finally, by translating the graph in Figure 7 to the right four units and up three units, we obtain the graph of the original equation, as shown in Figure 8. You should verify for yourself that the information accompanying Figure 8 is correct. ■■■

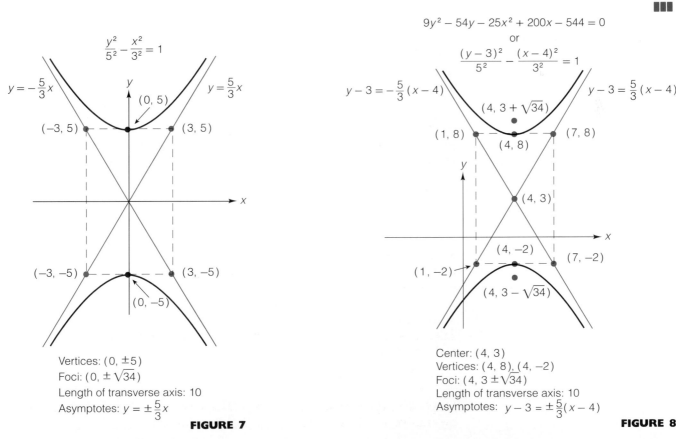

Vertices: $(0, \pm 5)$
Foci: $(0, \pm\sqrt{34})$
Length of transverse axis: 10
Asymptotes: $y = \pm\frac{5}{3}x$

FIGURE 7

Center: $(4, 3)$
Vertices: $(4, 8), (4, -2)$
Foci: $(4, 3 \pm\sqrt{34})$
Length of transverse axis: 10
Asymptotes: $y - 3 = \pm\frac{5}{3}(x - 4)$

FIGURE 8

If you reread the example we just completed, you'll see that it was not necessary to know in advance that the given equation represented a hyperbola. Rather, this fact emerged naturally after we completed the square. Indeed, completing the square is a useful technique for identifying the graph of any equation of the form

$$Ax^2 + Cy^2 + Dx + Ey + F = 0$$

■ ▮ ▌ **EXAMPLE 3** Identify the graph of the equation

$$4x^2 - 32x - y^2 + 2y + 63 = 0$$

Solution As before, we complete the squares:

$$4(x^2 - 8x) - (y^2 - 2y) = -63$$
$$4(x^2 - 8x + 16) - (y^2 - 2y + 1) = -63 + 4(16) - 1$$
$$4(x - 4)^2 - (y - 1)^2 = 0$$

Since the right-hand side of this last equation is 0, dividing both sides by 4 will not bring the equation into one of the standard forms. Indeed, if we factor the left-hand side of the equation as a difference of two squares, we obtain

$$[2(x - 4) - (y - 1)][2(x - 4) + (y - 1)] = 0$$
$$(2x - y - 7)(2x + y - 9) = 0$$

$$\begin{array}{c|c} 2x - y - 7 = 0 & 2x + y - 9 = 0 \\ -y = -2x + 7 & y = -2x + 9 \\ y = 2x - 7 & \end{array}$$

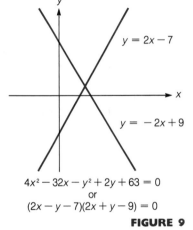

$4x^2 - 32x - y^2 + 2y + 63 = 0$
or
$(2x - y - 7)(2x + y - 9) = 0$

FIGURE 9

Thus the given equation is equivalent to the two linear equations $y = 2x - 7$ and $y = -2x + 9$. These two lines, taken together, constitute the graph. See Figure 9. ■ ▮ ▌

The two lines that we graphed in Figure 9 are actually the asymptotes for the hyperbola $4(x - 4)^2 - (y - 1)^2 = 1$. (Verify this for yourself.) For that reason, the graph in Figure 9 is referred to as a **degenerate hyperbola**. There are other cases similar to this that can arise in graphing equations of the form $Ax^2 + Cy^2 + Dx + Ey + F = 0$. For instance, as you can check for yourself by completing the squares, the graph of the equation

$$x^2 - 2x + 4y^2 - 16y + 17 = 0$$

consists of the single point $(1, 2)$. We refer to the graph in this case as a **degenerate ellipse**. Similarly, as you can check by completing the squares, the equation $x^2 - 2x + 4y^2 - 16y + 18 = 0$ has no graph; there are no points with coordinates that satisfy the equation.

We can obtain the equation of a tangent line to a hyperbola using the same ideas employed for the parabola and the ellipse in the previous sections. The result is this: The equation of the line tangent to the hyperbola $(x^2/a^2) - (y^2/b^2) = 1$ at the point (x_1, y_1) on the curve is

$$\frac{x_1 x}{a^2} - \frac{y_1 y}{b^2} = 1$$

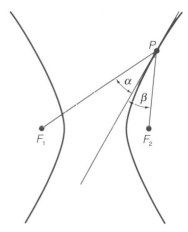

FIGURE 10

Reflection property of the hyperbola: The tangent at P bisects the angle formed by the focal radii drawn to P, so $\alpha = \beta$.

(See Exercise 53 at the end of this section for an outline of the derivation.) As with the parabola and the ellipse, many of the interesting properties of the hyperbola are related to the tangent lines. For instance, the hyperbola has a reflection property that is similar to the reflection properties of the parabola and ellipse. To state this property, we first define a **focal radius** of a hyperbola as a line segment drawn from a focus to a point on the hyperbola. Then the **reflection property** of the hyperbola is that the tangent line bisects the angle formed by the focal radii drawn to the point of tangency, as indicated in Figure 10. (See Exercise 56 for the proof.)

We conclude this section by listing several applications of the hyperbola. Some comets have hyperbolic orbits. Unlike Halley's comet, which has an elliptical orbit, these comets pass through the solar system once and never return. The Cassegrain telescope (invented by the Frenchman Sieur Cassegrain in 1672) uses both a hyperbolic mirror and a parabolic mirror. The Hubble Space Telescope, launched into orbit in April 1990, utilizes a Cassegrain telescope. The hyperbola is also used in some navigation systems. In the LORAN (LOng RAnge Navigation) system, an airplane or a ship at a point P receives radio signals that were transmitted simultaneously from two locations, F_1 and F_2. The time difference between the two signals is converted to a difference in distances: $F_1P - F_2P$. This locates the ship along one branch of a hyperbola. Then, using data from a second set of signals, the ship is located along a second hyperbola. The intersection of the two hyperbolas then determines the location of the airplane or ship.

EXERCISE SET 10.5

A

For Exercises 1–24, graph the hyperbolas. In each case in which the hyperbola is nondegenerate, specify the following: vertices, foci, lengths of transverse and conjugate axes, eccentricity, and equations of the asymptotes. In Exercises 11–24, also specify the centers.

1. $x^2 - 4y^2 = 4$
2. $y^2 - x^2 = 1$
3. $y^2 - 4x^2 = 4$
4. $25x^2 - 9y^2 = 225$
5. $16x^2 - 25y^2 = 400$
6. $9x^2 - y^2 = 36$
7. $2y^2 - 3x^2 = 1$
8. $x^2 - y^2 = 9$
9. $4y^2 - 25x^2 = 100$
10. $x^2 - 3y^2 = 3$
11. $\dfrac{(x-5)^2}{5^2} - \dfrac{(y+1)^2}{3^2} = 1$
12. $\dfrac{(x-5)^2}{3^2} - \dfrac{(y+1)^2}{5^2} = 1$
13. $\dfrac{(y-2)^2}{2^2} - \dfrac{(x-1)^2}{1^2} = 1$
14. $\dfrac{(y-3)^2}{2^2} - \dfrac{x^2}{1^2} = 1$

15. $\dfrac{(x+3)^2}{4^2} - \dfrac{(y-4)^2}{4^2} = 1$
16. $\dfrac{(x+1)^2}{5^2} - \dfrac{(y+2)^2}{3^2} = 1$
17. $x^2 - y^2 + 2y - 5 = 0$
18. $16x^2 - 32x - 9y^2 + 90y - 353 = 0$
19. $x^2 - y^2 - 4x + 2y - 6 = 0$
20. $x^2 - 8x - y^2 + 8y - 25 = 0$
21. $y^2 - 25x^2 + 8y - 9 = 0$
22. $9y^2 - 18y - 4x^2 - 16x - 43 = 0$
23. $x^2 + 7x - y^2 - y + 12 = 0$
24. $9x^2 + 9x - 16y^2 + 4y + 2 = 0$
25. Let $P(x, y)$ be a point in the first quadrant on the hyperbola $(x^2/2^2) - (y^2/1^2) = 1$. Let Q be the point in the first quadrant with the same x-coordinate as P and lying on an asymptote to the hyperbola. Show that $PQ = \frac{1}{2}(x - \sqrt{x^2 - 4})$.
26. The distance PQ in Exercise 25 represents the vertical distance between the hyperbola and the asymptote.

Complete the following table to see numerical evidence that this separation distance approaches zero as x gets larger and larger. (Round off each entry to one significant digit.)

x	10	50	100	500	1000	10000
PQ						

In Exercises 27–36, determine the equation of the hyperbola satisfying the given conditions. Write each answer in the form $Ax^2 - By^2 = C$ or in the form $Ay^2 - Bx^2 = C$.

27. Foci $(\pm 4, 0)$; vertices $(\pm 1, 0)$
28. Foci $(0, \pm 5)$; vertices $(0, \pm 3)$
29. Asymptotes $y = \pm\frac{1}{2}x$; vertices $(\pm 2, 0)$
30. Asymptotes $y = \pm x$; foci $(0, \pm 1)$
31. Asymptotes $y = \pm\sqrt{10}x/5$; foci $(\pm\sqrt{7}, 0)$
32. Length of transverse axis 6; eccentricity $\frac{4}{3}$; center $(0,0)$; focal axis horizontal
33. Vertices $(0, \pm 7)$; passing through the point $(1, 9)$
34. Eccentricity 2; foci $(\pm 1, 0)$
35. Length of transverse axis 6; length of conjugate axis 2; foci on the y-axis; center at the origin
36. Asymptotes $y = \pm 2x$; passing through $(1, \sqrt{3})$
37. Show that the two asymptotes of the hyperbola $x^2 - y^2 = 16$ are perpendicular to each other.
38. (a) Verify that the point $P(6, 4\sqrt{3})$ lies on the hyperbola $16x^2 - 9y^2 = 144$.
    (b) In Example 1, we found that the foci of this hyperbola were $F_1(-5, 0)$ and $F_2(5, 0)$. Compute the lengths F_1P and F_2P, where P is the point $(6, 4\sqrt{3})$.
    (c) Verify that $|F_1P - F_2P| = 2a$.
39. (a) Verify that the point $P(5, 6)$ lies on the hyperbola $5y^2 - 4x^2 = 80$.
    (b) Find the foci.
    (c) Compute the lengths of the line segments $\overline{F_1P}$ and $\overline{F_2P}$, where P is the point $(5, 6)$.
    (d) Verify that $|F_1P - F_2P| = 2a$.

B
40. (a) Let e_1 denote the eccentricity of the hyperbola $(x^2/4^2) - (y^2/3^2) = 1$, and let e_2 denote the eccentricity of the hyperbola $(x^2/3^2) - (y^2/4^2) = 1$. Verify that $e_1^2e_2^2 = e_1^2 + e_2^2$.
    (b) Let e_1 and e_2 denote the eccentricities of the two hyperbolas $(x^2/A^2) - (y^2/B^2) = 1$ and $(y^2/B^2) - (x^2/A^2) = 1$, respectively. Verify that $e_1^2e_2^2 = e_1^2 + e_2^2$.

41. (a) If the hyperbola $(x^2/a^2) - (y^2/b^2) = 1$ has perpendicular asymptotes, show that $a = b$. What is the eccentricity in this case?
    (b) Show that the asymptotes of the hyperbola $(x^2/a^2) - (y^2/a^2) = 1$ are perpendicular. What is the eccentricity of this hyperbola?
42. Derive equation (1) in this section from the equation that precedes it.
43. Let $P(x, y)$ be a point on the right-hand branch of the hyperbola $(x^2/a^2) - (y^2/b^2) = 1$. As usual, let F_2 denote the focus located at $(c, 0)$. The following steps outline a proof of the fact that the length of the line segment $\overline{F_2P}$ in this case is given by $F_2P = xe - a$.
    (a) Explain why
    $$\sqrt{(x+c)^2 + y^2} - \sqrt{(x-c)^2 + y^2} = 2a$$
    (b) In the preceding equation, add the quantity $\sqrt{(x-c)^2 + y^2}$ to both sides and then square both sides. Show that the result can be written
    $$xc - a^2 = a\sqrt{(x-c)^2 + y^2}$$
    or
    $$xc - a^2 = a(F_2P)$$
    (c) Divide both sides of the preceding equation by a to show that $xe - a = F_2P$, as required.
44. Let $P(x, y)$ be a point on the right-hand branch of the hyperbola $(x^2/a^2) - (y^2/b^2) = 1$. As usual, let F_1 denote the focus located at $(-c, 0)$. Show that $F_1P = xe + a$. *Hint:* Use the result in Exercise 43 along with the fact that the right-hand branch is defined by the equation $F_1P - F_2P = 2a$.
45. Let P be a point on the right-hand branch of the hyperbola $x^2 - y^2 = k^2$. If d denotes the distance from P to the center of the hyperbola, show that
    $$d^2 = (F_1P)(F_2P)$$
46. Let $P(x, y)$ be a point on the right-hand branch of the hyperbola $(x^2/a^2) - (y^2/b^2) = 1$. From P, a line segment is drawn perpendicular to the line $x = a/e$, meeting this line at D. If F_2 denotes (as usual) the focus located at $(c, 0)$, show that
    $$\frac{F_2P}{PD} = e$$
    The line $x = a/e$ is called the *directrix* of the hyperbola corresponding to the focus F_2. *Hint:* Use the expression for F_2P developed in Exercise 43.
47. Consider the hyperbola $(x^2/a^2) - (y^2/b^2) = 1$, and let D be the point where the asymptote $y = (b/a)x$ meets the directrix $x = a/e$. If F denotes the focus corresponding to this directrix, show that the line segment $\overline{FD}$ is perpendicular to the asymptote.

48. Consider the hyperbola $(x^2/a^2) - (y^2/b^2) = 1$ and suppose that a perpendicular is drawn from the focus $F_2(c, 0)$ to the asymptote $y = (b/a)x$, meeting the asymptote at A.
    (a) Show that $F_2A = b$.
    (b) Show that the distance from the center of the hyperbola to A is equal to a.

49. By solving the system
    $$\begin{cases} y = \frac{4}{3}x - 1 \\ 16x^2 - 9y^2 = 144 \end{cases}$$
    show that the line and the hyperbola intersect in exactly one point. Draw a sketch of the situation. This demonstrates that a line that intersects a hyperbola in exactly one point does not have to be a tangent line.

In Exercises 50–52, find the equation of the line that is tangent to the hyperbola at the given point. Write your answer in the form $y = mx + b$.

50. $x^2 - 4y^2 = 16$; $\left(5, \frac{3}{2}\right)$
51. $3x^2 - y^2 = 12$; $(4, 6)$
52. $16x^2 - 25y^2 = 400$; $\left(10, 4\sqrt{3}\right)$

C

53. We define a *tangent line to a hyperbola* as a line that is not parallel to an asymptote and that intersects the hyperbola in exactly one point. Show that the equation of the line tangent to the hyperbola $(x^2/a^2) - (y^2/b^2) = 1$ at the point (x_1, y_1) on the curve is
    $$\frac{x_1 x}{a^2} - \frac{y_1 y}{b^2} = 1$$

    Hint: Allow for signs, but follow exactly the same steps as were supplied in Exercise 64 of Exercise Set 10.4, where we found the tangent to the ellipse. You should find that the slope in the present case is $m = (b^2 x_1/a^2 y_1)$. Explain why this slope cannot equal the slope of an asymptote as long as (x_1, y_1) is on the hyperbola.

54. (The results in this exercise will be used in Exercise 56 to prove the reflection property of the hyperbola.) Let $P(x_1, y_1)$ be a point on the right-hand branch of the hyperbola $(x^2/a^2) - (y^2/b^2) = 1$. Suppose that the tangent to the hyperbola at P meets the x-axis at A.
    (a) Show that the coordinates of A are $(a^2/x_1, 0)$.
    (b) Show that $F_1A = a(x_1 e + a)/x_1$.
    (c) Show that $F_2A = a(x_1 e - a)/x_1$.

55. (The result in this exercise will be used in Exercise 56 to prove the reflection property of the hyperbola.) Refer to the following figure. If $F_1P/F_1A = F_2P/F_2A$, show that $\alpha = \beta$.

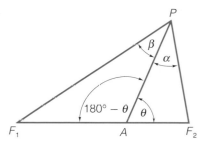

Hint: Use the law of sines in triangle AF_2P to show that $\dfrac{F_2P}{F_2A} = \dfrac{\sin \theta}{\sin \alpha}$. Then carry out the same procedure in triangle AF_1P to show that $\dfrac{F_1P}{F_1A} = \dfrac{\sin \theta}{\sin \beta}$.

56. According to the reflection property of the hyperbola, the tangent to the hyperbola at P bisects the angle formed by the focal radii at P. This exercise outlines a proof of this property. The following figure shows a point $P(x_1, y_1)$ on the right-hand branch of the hyperbola $(x^2/a^2) - (y^2/b^2) = 1$. The tangent at P cuts the x-axis at A.

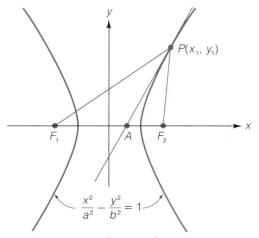

    (a) Show that $F_1P/F_1A = x_1/a$. *Hint:* This is easy if you use the expressions for F_1P and F_1A developed in Exercises 44 and 54, respectively.
    (b) Show that $F_2P/F_2A = x_1/a$. *Hint:* Use the expressions for F_2P and F_2A developed in Exercises 43 and 54, respectively.
    (c) Use the results in parts (a) and (b) along with the result of Exercise 55 to conclude that angle F_1PA equals angle F_2PA, as required.

57. The *normal line* to a hyperbola at a point P on the hyperbola is the line through P that is perpendicular to the tangent at P. If the coordinates of P are (x_1, y_1), show that the equation of the normal line is
    $$a^2 y_1 x + b^2 x_1 y = x_1 y_1(a^2 + b^2)$$

58. Suppose the point $P(x_1, y_1)$ is on the hyperbola $(x^2/a^2) - (y^2/b^2) = 1$. A tangent is drawn at P, meeting the x-axis at A and the y-axis at B. Also, perpendiculars are drawn from P to the x- and y-axes, meeting these axes at C and D, respectively. If O denotes the origin, show that   (a) $OA \cdot OC = a^2$;   (b) $OB \cdot OD = b^2$.

59. Suppose the point $P(x_1, y_1)$ is in the first quadrant on the hyperbola $(x^2/a^2) - (y^2/b^2) = 1$. The normal line through P meets the x- and y-axes at Q and R, respectively. If O denotes the origin, show that the area of triangle OQR is $\dfrac{x_1 y_1 (a^2 + b^2)^2}{2a^2 b^2}$.

60. Let $P(x_1, y_1)$ be a point in the first quadrant on the hyperbola $(x^2/a^2) - (y^2/b^2) = 1$. Let D be the point where the tangent at P meets the line $x = a/e$. Show that angle DF_2P is a right angle.

61. The tangent to the hyperbola $(x^2/a^2) - (y^2/b^2) = 1$ at the point P meets the asymptotes at A and B. Show that P is the midpoint of $\overline{AB}$.

62. At the point P on the hyperbola $(x^2/a^2) - (y^2/b^2) = 1$, a tangent line is drawn that meets the lines $x = a$ and $x = -a$ at S and T, respectively. Show that the circle with $\overline{ST}$ as a diameter passes through the two foci of the hyperbola.

OPTIONAL TI-81 GRAPHING CALCULATOR EXERCISES

EXERCISES FOR SECTION 10.5

1. In this exercise you are going to look at the graph of the hyperbola $16x^2 - 9y^2 = 144$ from two perspectives. As preparation, clear all functions from the $Y=$ menu and press ZOOM 6 for the standard viewing rectangle.
   (a) Solve the equation $16x^2 - 9y^2 = 144$ for y, enter the two functions that you obtain in the $Y=$ menu, and press GRAPH.
   (b) Determine the equations of the two asymptotes for this hyperbola. Then enter the equations for the asymptotes in the $Y=$ menu and press GRAPH.
   (c) Looking at the graph obtained in part (b), you can see that the hyperbola seems to be drawing closer to its asymptotes as $|x|$ gets large. To see more dramatic evidence of this, press ZOOM 3 and ENTER. At this scale, the hyperbola is virtually indistinguishable from its asymptotes. (Use the RANGE key to see what scale is involved.)

2. In this exercise we graph the hyperbola
   $$\frac{(y-3)^2}{5^2} - \frac{(x-4)^2}{3^2} = 1$$
   (a) Show that the equation is equivalent to
   $$y = 3 \pm 5\sqrt{1 + (x-4)^2/9}$$

   Enter these two functions in the $Y=$ menu and press ZOOM 6 to see the hyperbola in the standard viewing rectangle. Then, for a better view, make the following adjustments in the RANGE menu: $x_{min} = -20$, $x_{max} = 20$, $x_{scl} = 2$, $y_{min} = -20$, $y_{max} = 20$, and $y_{scl} = 2$.
   (b) Show that the equations of the asymptotes are
   $$y = \pm \tfrac{5}{3}(x - 4) + 3$$

   Enter these two equations in the $Y=$ menu and press GRAPH. Check to see that your result is consistent with the graph shown in Figure 8 on page 619.

3. In this exercise we graph a hyperbola in which the axes of the curve are not parallel to the coordinate axes. The equation is $x^2 + 4xy - 2y^2 = 6$.
   (a) Use the quadratic formula to solve the equation for y in terms of x. Show that the result can be written
   $$y = x \pm \tfrac{1}{2}\sqrt{6x^2 - 12}$$
   (b) Enter these two functions in the $Y=$ menu. Press ZOOM 6 and then make the following adjustments in the RANGE menu: $x_{min} = -6$, $x_{max} = 6$, $x_{scl} = 0.5$. Finally, press GRAPH.

10.6 THE FOCUS–DIRECTRIX PROPERTY OF CONICS

Here then [in the Conics, written by Apollonius of Perga (ca. 262–ca. 190 B.C.)] we have the properties of the three curves expressed in the precise language of the Pythagorean application of areas, and the curves are named accordingly: parabola (παραβολη) where the rectangle is exactly applied [equal], *hyperbola (υπερβολη) where it exceeds, and ellipse (ελλειψιζ) where it falls short.*

Sir Thomas Heath in *A History of Greek Mathematics,* vol. II (Oxford: The Clarendon Press, 1921)

The most important contribution which Pappus [ca. 300] *made to our knowledge of the conics was his publication of the focus, directrix, eccentricity theorem.*

Julian Lowell Coolidge in *A History of the Conic Sections and Quadric Surfaces* (London: Oxford University Press, 1946)

We begin by recalling the focus–directrix property that we used in defining the parabola. A point P is on a parabola if and only if the distance from P to the focus is equal to the distance from P to the directrix. For the parabola in Figure 1, this means that for each point P on the parabola, we have $FP = PD$ or, equivalently,

$$\frac{FP}{PD} = 1$$

The ellipse and the hyperbola also can be characterized by focus–directrix properties. We'll begin with the ellipse. To help us in subsequent computations, we need to know the lengths of the line segments $\overline{F_1P}$ and $\overline{F_2P}$ in Figure 2.

Parabola

Focus

Directrix

FIGURE 1

FIGURE 2

Line segments drawn from a point on the ellipse to the foci are called focal radii. Here $\overline{F_1P}$ and $\overline{F_2P}$ are the focal radii.

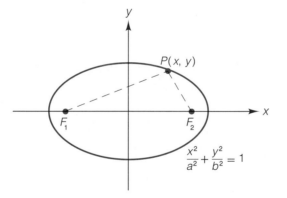

These line segments joining a point on the ellipse to the foci are called **focal radii.** In the box that follows, we give a formula for the length of each focal radius. In the formulas, e ($= c/a$) denotes the eccentricity of the ellipse. (For the derivation of these formulas, see Exercise 15 at the end of this section.)

The Focal Radii of the Ellipse $\dfrac{x^2}{a^2} + \dfrac{y^2}{b^2} = 1$
As indicated in Figure 2, let $P(x, y)$ be a point on the ellipse. Then the lengths of the focal radii are $$F_1P = a + ex \qquad \text{and} \qquad F_2P = a - ex$$

■ ■ **EXAMPLE 1** Figure 3 shows the ellipse $x^2 + 3y^2 = 28$. Compute the lengths of the focal radii drawn to the point $P(-1, 3)$.

FIGURE 3

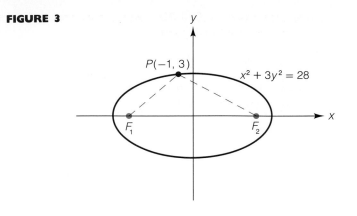

$P(-1, 3)$

$x^2 + 3y^2 = 28$

F_1 F_2

Solution To apply the formulas in the box, we need to know the values of a and e for the given ellipse. (We already know that $x = -1$; that is given.) As you should verify for yourself using the techniques of Section 10.4, we have $a = 2\sqrt{7}$ and $e = \sqrt{6}/3$. Therefore

$$F_1P = a + ex$$

$$= 2\sqrt{7} + \left(\frac{\sqrt{6}}{3}\right)(-1) = \frac{6\sqrt{7} - \sqrt{6}}{3}$$

and

$$F_2P = a - ex$$

$$= 2\sqrt{7} - \left(\frac{\sqrt{6}}{3}\right)(-1) = \frac{6\sqrt{7} + \sqrt{6}}{3}$$

In Figure 4 we show the ellipse $(x^2/a^2) + (y^2/b^2) = 1$ and the horizontal line $x = a/e$. The line $x = a/e$ is a **directrix** of the ellipse. In Figure 4, F_2P is the distance from the focus F_2 to a point P on the ellipse, and PD is the

FIGURE 4

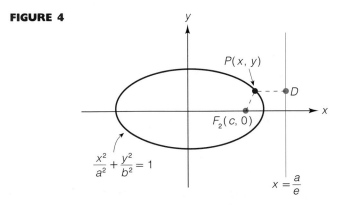

$P(x, y)$

D

$F_2(c, 0)$

$\frac{x^2}{a^2} + \frac{y^2}{b^2} = 1$

$x = \dfrac{a}{e}$

distance from P to the directrix $x = a/e$. We are going to prove the following remarkable *focus–directrix property of the ellipse:* For any point $P(x, y)$ on the ellipse, the ratio of F_2P to PD is equal to e, the eccentricity of the ellipse. To prove this, we proceed as follows:

$$\frac{F_2P}{PD} = \frac{a - ex}{PD}$$ using our formula for F_2P

$$= \frac{a - ex}{\dfrac{a}{e} - x}$$ using Figure 4

$$= \frac{e(a - ex)}{a - ex}$$ multiplying both numerator and denominator by e

$$= e$$

So, $F_2P/PD = e$, as we wished to show.

The directrix $x = a/e$ is associated with the focus $F_2(c, 0)$. What about the focus $F_1(-c, 0)$? From the symmetry of the ellipse, it follows that the line $x = -a/e$ is the directrix associated with this focus. More specifically, referring to Figure 5, not only do we have $F_2P/PD = e$, we also have $F_1P/PE = e$.

FIGURE 5

For every point P on the ellipse, we have $F_1P/PE = e$ and $F_2P/PD = e$.

We have seen that each point P on the ellipse in Figure 5 satisfies the following two equations:

$$\frac{F_1P}{PE} = e \tag{1}$$

$$\frac{F_2P}{PD} = e \tag{2}$$

Now, conversely, suppose that the point $P(x, y)$ in Figure 6 satisfies equations (1) and (2). We are going to show that $P(x, y)$ must, in fact, lie on the ellipse $(x^2/a^2) + (y^2/b^2) = 1$.*

*The short proof given here was communicated to the author by Professor Ray Redheffer. For a proof using equation (2) but not equation (1), see Exercise 16.

FIGURE 6

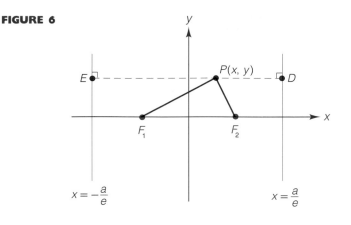

$$x = -\frac{a}{e} \qquad\qquad x = \frac{a}{e}$$

In Figure 6, the distance from E to D is $2a/e$. Therefore

$$PE + PD = 2a/e$$

and, consequently,

$$\frac{F_1P}{e} + \frac{F_2P}{e} = \frac{2a}{e} \qquad \text{using equations (1) and (2) to substitute for } PE \text{ and } PD$$

or

$$F_1P + F_2P = 2a \qquad \text{multiplying by } e$$

Thus, by definition, P lies on the ellipse $(x^2/a^2) + (y^2/b^2) = 1$, as we wished to show.

In the box that follows, we summarize the focus–directrix properties of the ellipse $(x^2/a^2) + (y^2/b^2) = 1$. For reference, we also include the original defining property of the ellipse from Section 10.4.

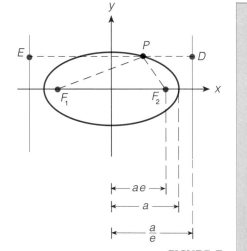

FIGURE 7
The ellipse $(x^2/a^2) + (y^2/b^2) = 1$ with foci $F_1(-c, 0)$ and $F_2(c, 0)$ and directrices $x = \pm a/e$.

Focus and Focus–Directrix Properties of the Ellipse $\dfrac{x^2}{a^2} + \dfrac{y^2}{b^2} = 1$

Refer to Figure 7.

1. A point P is on the ellipse if and only if the sum of the distances from P to the foci $F_1(-c, 0)$ and $F_2(c, 0)$ is $2a$.

2. A point P is on the ellipse if and only if

$$\frac{F_2P}{PD} = e$$

where F_2P is the distance from P to the focus F_2, PD is the distance from P to the directrix $x = a/e$, and $e\,(= c/a)$ is the eccentricity of the ellipse.

3. A point P is on the ellipse if and only if

$$\frac{F_1P}{PE} = e$$

where F_1P is the distance from P to the focus F_1 and PE is the distance from P to the directrix $x = -a/e$.

▮▯▮ **EXAMPLE 2** The foci of an ellipse are $(\pm 3, 0)$ and the directrix corresponding to the focus $(3, 0)$ is $x = 5$. Find the equation of the ellipse. Write the answer in the form $Ax^2 + By^2 = C$.

Solution Using the given data (and referring to Figure 7), we have

$$ae = 3 \qquad \text{and} \qquad a/e = 5$$

From the first equation we obtain $e = 3/a$. Substituting this value of e in the second equation yields

$$\frac{a}{3/a} = 5 \qquad \text{and therefore} \qquad a^2 = 15$$

Now we can calculate b^2 by using the relation $b^2 = a^2 - c^2$ and the given information $c = 3$:

$$b^2 = a^2 - c^2 = 15 - 9 = 6$$

Substituting the values we've obtained for a^2 and b^2 in the equation $(x^2/a^2) + (y^2/b^2) = 1$ yields

$$\frac{x^2}{15} + \frac{y^2}{6} = 1$$

or

$$2x^2 + 5y^2 = 30 \qquad \text{multiplying by 30}$$ ▮▮▮

The focus–directrix property can be developed for the hyperbola in just the same way that we have proceeded for the ellipse. In fact, the algebra is so similar that we shall omit the details here and simply summarize the results. The hyperbola $(x^2/a^2) - (y^2/b^2) = 1$ has two directrices: they are the vertical lines $x = a/e$ in Figure 8(a) and $x = -a/e$ in Figure 8(b). Notice that these equations are identical to those for the directrices of the ellipse. In the box following Figure 8, we summarize the focus and the focus–directrix properties of the hyperbola.

FIGURE 8
The hyperbola $(x^2/a^2) - (y^2/b^2) = 1$ has two directrices. The directrix $x = a/e$ corresponds to the focus $F_2(c, 0)$. The directrix $x = -a/e$ corresponds to the focus $F_1(-c, 0)$.

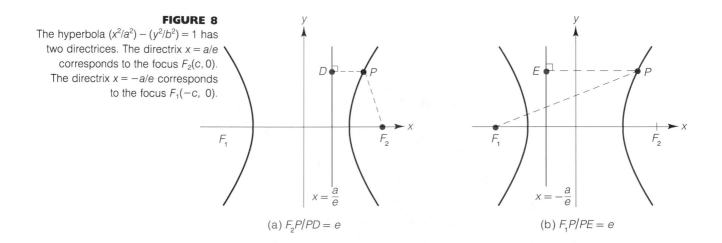

(a) $F_2P/PD = e$ (b) $F_1P/PE = e$

Focus and Focus–Directrix Properties of the Hyperbola
$$\frac{x^2}{a^2} - \frac{y^2}{b^2} = 1$$

Refer to Figure 8.

1. A point P is on the hyperbola if and only if the difference of the distances from P to the foci $F_1(-c, 0)$ and $F_2(c, 0)$ is $2a$.

2. A point P is on the hyperbola if and only if

$$\frac{F_2P}{PD} = e$$

where F_2P is the distance from P to the focus F_2, PD is the distance from P to the directrix $x = a/e$, and $e\,(= c/a)$ is the eccentricity of the hyperbola.

3. A point P is on the hyperbola if and only if

$$\frac{F_1P}{PE} = e$$

where F_1P is the distance from P to the focus F_1 and PE is the distance from P to the directrix $x = -a/e$.

▌▬▌ EXAMPLE 3 Determine the foci, the eccentricity, and the directrices for the hyperbola $9x^2 - 16y^2 = 144$.

Solution   To convert the given equation to standard form, we divide both sides by 144. This yields

$$\frac{x^2}{4^2} - \frac{y^2}{3^2} = 1$$

So, we have $a = 4$, $b = 3$, and

$$c = \sqrt{a^2 + b^2} = \sqrt{4^2 + 3^2} = 5$$

The foci are therefore $(0, \pm 5)$, and the eccentricity is

$$e = \frac{c}{a} = \frac{5}{4}$$

The directrices are

$$x = \pm\frac{a}{e} = \pm\frac{4}{5/4} = \pm\frac{16}{5} \qquad\qquad \blacksquare\blacksquare\blacksquare$$

▌▬▌ EXAMPLE 4 Figure 9 shows the graph of the hyperbola $9x^2 - 16y^2 = 144$. (This is the hyperbola discussed in the previous example.)

(a) Verify that the point $P\left(-5, \frac{9}{4}\right)$ lies on this hyperbola.

(b) Compute F_2D and PD, and verify that

$$\frac{F_2P}{PD} = e$$

FIGURE 9

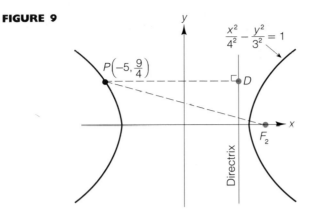

Solution

(a) Substituting the values $x = -5$ and $y = \frac{9}{4}$ in the equation of the hyperbola yields

$$9(-5)^2 - 16\left(\tfrac{9}{4}\right)^2 = 144$$

or

$$225 - 81 = 144$$

Since this last equation is correct, we conclude that the point $P\left(-5, \frac{9}{4}\right)$ indeed lies on the hyperbola.

(b) From the previous example, we know that the coordinates of the focus F_2 are $(5, 0)$. So, using the distance formula, we have

$$F_2P = \sqrt{\left(5 - (-5)\right)^2 + \left(0 - \frac{9}{4}\right)^2}$$

$$= \sqrt{100 + \frac{81}{16}} = \sqrt{\frac{1681}{16}} = \frac{41}{4}$$

Next, we use the fact (from Example 3) that the equation of the directrix in Figure 9 is $x = \frac{16}{5}$. Thus, in Figure 9,

$$PD = |-5| + \frac{16}{5} = \frac{41}{5}$$

Finally, we compute the ratio F_2P/PD:

$$\frac{F_2P}{PD} = \frac{41/4}{41/5} = \frac{5}{4}$$

This is the same number that we obtained for the eccentricity in Example 3. So, we have verified in this case that the ratio of F_2P to PD is equal to the eccentricity.   ∎

The focus–directrix property provides a unified approach to the parabola, the ellipse, and the hyperbola. For both the ellipse and the hyperbola, we've seen that for any point P on the curve, we have

$$\frac{\text{distance from } P \text{ to a focus}}{\text{distance from } P \text{ to the corresponding directrix}} = e \qquad (3)$$

where e is a positive constant, the eccentricity of the curve. For the ellipse we have $e < 1$, and for the hyperbola we have $e > 1$. Equation (3) also holds for the parabola if we agree on the convention that the eccentricity of a parabola is 1. This is because with $e = 1$, equation (3) tells us that the distance from P to the focus equals the distance from P to the directrix, which is the defining condition for the parabola. The following theorem summarizes these remarks.

Theorem

Refer to Figure 10. Let $\mathscr{L}$ be a fixed line, F a fixed point, and e a positive constant. Consider the set of points P satisfying the condition $FP/PD = e$. (The point D is defined in the figure.) Then

(a) If $e = 1$, the set of points is a *parabola* with focus F and directrix $\mathscr{L}$;

(b) If $0 < e < 1$, the set of points is an *ellipse* with focus F, corresponding directrix $\mathscr{L}$, and eccentricity e;

(c) If $e > 1$, the set of points is a *hyperbola* with focus F, corresponding directrix $\mathscr{L}$, and eccentricity e.

FIGURE 10

Note   In our development in this section, for simplicity, we've always considered cases in which the directrix is vertical. However, the preceding theorem is valid for any orientation of the line $\mathscr{L}$.

EXERCISE SET 10.6

A

In Exercises 1–4, you are given an ellipse and a point P on the ellipse. Find F_1P and F_2P, the lengths of the focal radii.

1. $x^2 + 3y^2 = 76$; $P(-8, 2)$
2. $x^2 + 3y^2 = 57$; $P(3, -4)$

3. $(x^2/15^2) + (y^2/5^2) = 1$; $P(9, 4)$
4. $2x^2 + 3y^2 = 14$; $P(-1, -2)$

In Exercises 5–10, determine the foci, the eccentricity, and the directrices for each ellipse and hyperbola.

5. (a) $(x^2/4^2) + (y^2/3^2) = 1$
  (b) $(x^2/4^2) - (y^2/3^2) = 1$

6. (a) $x^2 + 4y^2 = 1$
  (b) $x^2 - 4y^2 = 1$

7. (a) $12x^2 + 13y^2 = 156$
  (b) $12x^2 - 13y^2 = 156$

8. (a) $x^2 + 2y^2 = 2$
  (b) $x^2 - 2y^2 = 2$

9. (a) $25x^2 + 36y^2 = 900$
  (b) $25x^2 - 36y^2 = 900$

10. (a) $4x^2 + 25y^2 = 100$
  (b) $4x^2 - 25y^2 = 100$

In Exercises 11 and 12, use the given information to find the equation of the ellipse. Write the answer in the form $Ax^2 + By^2 = C$.

11. The foci are $(\pm 1, 0)$ and the directrices are $x = \pm 4$.

12. The foci are $\left(\pm\sqrt{3}, 0\right)$ and the eccentricity is $\frac{2}{3}$.

In Exercises 13 and 14, use the given information to find the equation of the hyperbola. Write the answer in the form $Ax^2 - By^2 = C$.

13. The foci are $(\pm 2, 0)$ and the directrices are $x = \pm 1$.

14. The foci are $(\pm 3, 0)$ and the eccentricity is 2.

B

15. In this exercise we show that the focal radii of the ellipse $(x^2/a^2) + (y^2/b^2) = 1$ are $F_1P = a + ex$ and $F_2P = a - ex$. The method used here, which avoids the use of radicals, appears in an eighteenth-century text, *Traité analytique des sections coniques* (Paris: 1707), by the Marquis de l'Hospital [1661–1704]. (For another method, one that does use radicals, see Exercise 51 in Section 10.4.) For convenience, let $d_1 = F_1P$ and $d_2 = F_2P$, as indicated in the accompanying figure.

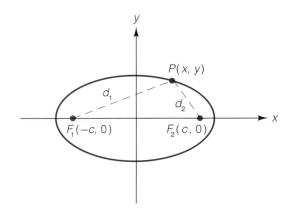

(a) Using the distance formula, verify that
$$d_1^2 = (x + c)^2 + y^2 \quad \text{and} \quad d_2^2 = (x - c)^2 + y^2$$

(b) Use the two equations in part (a) to show that $d_1^2 - d_2^2 = 4cx$.

(c) Explain why $d_1 + d_2 = 2a$.

(d) Factor the left-hand side of the equation in part (b) and substitute for one of the factors using the equation in part (c). Show that the result can be written $d_1 - d_2 = 2cx/a$.

(e) Add the equations in parts (c) and (d). Show that the resulting equation can be written $d_1 = a + ex$, as required.

(f) Use the equation in part (c) and the result in part (e) to show that $d_2 = a - ex$.

16. In this exercise we show that if the point $P(x, y)$ in the accompanying figure satisfies the condition
$$\frac{F_2P}{PD} = e$$
then, in fact, the point $P(x, y)$ lies on the ellipse $(x^2/a^2) + (y^2/b^2) = 1$.

(a) From the given equation we have $(F_2P)^2 = e^2(PD)^2$. Use the distance formula and the figure to deduce from this equation that
$$(x - c)^2 + y^2 = e^2\left(\frac{a}{e} - x\right)^2$$

(b) In the equation in part (a), replace e with c/a. After carrying out the indicated operations and simplifying, show that the equation can be written
$$a^2x^2 - c^2x^2 + a^2y^2 = a^4 - a^2c^2$$

(c) The equation in part (b) is equivalent to $(a^2 - c^2)x^2 + a^2y^2 = a^2(a^2 - c^2)$. Now replace the quantity $a^2 - c^2$ by b^2 and show that the resulting equation can be written $(x^2/a^2) + (y^2/b^2) = 1$; thus, P lies on the ellipse, as we wished to show.

10.7 THE CONICS IN POLAR COORDINATES

The use of polar coordinates permits a unified treatment of the conic sections, and it is the polar coordinate equations for these curves that are used in celestial mechanics.

Professor Bernice Kastner in her text *Space Mathematics,* published in 1985 by NASA (National Aeronautics and Space Administration)

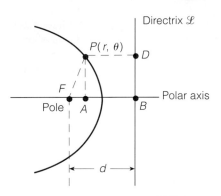

FIGURE 1
A conic with focus *F* and directrix *ℒ*.

As indicated in the opening quotation, the polar coordinate equations of the parabola, ellipse, and hyperbola are useful in applications. We will develop these equations by using the focus–directrix property of the conics that we discussed in Section 10.6. Suppose that we have a conic with focus *F*, directrix *ℒ*, and eccentricity *e*. Then, as shown in Figure 1, we can set up a polar coordinate system in which the focus of the conic is the origin or pole, and the directrix is perpendicular to the polar axis. In Figure 1, we've used *d* to denote the distance from the focus to the directrix.

Using the focus–directrix property for the conic in Figure 1, we have

$$\frac{FP}{PD} = e$$

and therefore

$$\frac{r}{FB - FA} = e$$

In this last equation, we can replace *FB* by *d* and *FA* by $r \cos \theta$ (why?). This yields

$$\frac{r}{d - r \cos \theta} = e$$

Now, as Exercise 22 asks you to verify, when we solve this equation for *r*, we obtain

$$r = \frac{ed}{1 + e \cos \theta}$$

This is the polar form for the equation of the conic in Figure 1, in which the directrix is vertical and to the right of the focus. By the same technique, we can obtain similar equations when the directrix is to the left of the focus and when the directrix is horizontal. We summarize the results in the following box.

Polar Equations of the Conics

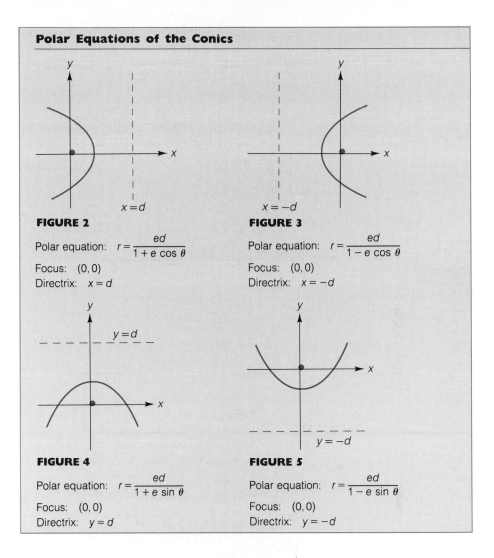

FIGURE 2

Polar equation: $r = \dfrac{ed}{1 + e\cos\theta}$

Focus: $(0, 0)$
Directrix: $x = d$

FIGURE 3

Polar equation: $r = \dfrac{ed}{1 - e\cos\theta}$

Focus: $(0, 0)$
Directrix: $x = -d$

FIGURE 4

Polar equation: $r = \dfrac{ed}{1 + e\sin\theta}$

Focus: $(0, 0)$
Directrix: $y = d$

FIGURE 5

Polar equation: $r = \dfrac{ed}{1 - e\sin\theta}$

Focus: $(0, 0)$
Directrix: $y = -d$

For the example that follows and for the exercises at the end of this section, it will be convenient to have formulas that express b in terms of e and a for the ellipse and the hyperbola. For the ellipse, we have

$$b^2 = a^2 - c^2 = a^2 - (ae)^2 = a^2(1 - e^2)$$

and therefore

$$b = a\sqrt{1 - e^2} \qquad \text{for the ellipse} \tag{1}$$

Similarly, for the hyperbola, we have

$$b^2 = c^2 - a^2 = (ae)^2 - a^2 = a^2(e^2 - 1)$$

and therefore

$$b = a\sqrt{e^2 - 1} \qquad \text{for the hyperbola} \tag{2}$$

EXAMPLE I Sketch the graph of the conic $r = 8/(4 - 3\cos\theta)$.

Solution When we compare the given equation with the four basic types shown in the box, we see that the appropriate standard equation is the one associated with Figure 3:

$$r = \frac{ed}{1 - e \cos \theta} \tag{3}$$

To bring the given equation into this form (in which the first term in the denominator is 1), we divide both numerator and denominator on the right-hand side by 4. This yields

$$r = \frac{2}{1 - \frac{3}{4} \cos \theta} \tag{4}$$

Comparing equations (3) and (4), we see that

$$e = \tfrac{3}{4} \qquad \text{and} \qquad ed = 2$$

Therefore

$$d = \frac{2}{e} = \frac{2}{3/4} = \frac{8}{3}$$

The eccentricity e is $\frac{3}{4}$, which is less than 1, so the conic is an ellipse. From the result $d = \frac{8}{3}$ (and the graph in Figure 3), we conclude that the directrix corresponding to the focus $(0, 0)$ is the vertical line $x = -\frac{8}{3}$. (Actually, as you'll see, we won't need this information about the directrix in drawing the graph.) Also from Figure 3, we know that the major axis of the ellipse lies along the polar or x-axis. Perhaps the simplest way to proceed now is to compute the value of r when $\theta = 0$, $\pi/2$, π, and $3\pi/2$. In Figure 6 we show the results of these computations and the four points that are determined.

FIGURE 6
Four points on the ellipse
$r = 8/(4 - 3 \cos \theta)$.

$r = 8/(4 - 3\cos\theta)$

θ	0	$\pi/2$	π	$3\pi/2$
r	8	2	8/7	2

Since the major axis of this ellipse lies along the polar axis, the length of the major axis is

$$2a = 8 + \frac{8}{7} = \frac{64}{7} \qquad \text{and therefore} \qquad a = \frac{32}{7}$$

For the x-coordinate of the center of the ellipse, we want the number (on the x-axis) that is halfway between $-\frac{8}{7}$ and 8. As you can check (by averaging the two numbers), this x-coordinate is $\frac{24}{7}$.

The last piece of information that we need for drawing an accurate sketch is the value of b for this ellipse. Using equation (1), we have

$$b = a\sqrt{1 - e^2}$$
$$= \frac{32}{7}\sqrt{1 - \left(\frac{3}{4}\right)^2} = \frac{32}{7}\sqrt{\frac{7}{16}} = \frac{8}{7}\sqrt{7}$$

With a calculator, we find that $b \approx 3.02$. We can now draw the graph, as shown in Figure 7 ■■■

FIGURE 7

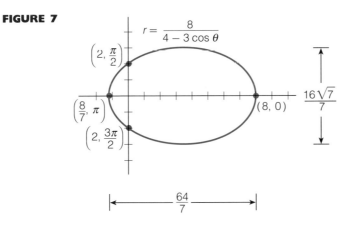

■■■ **EXAMPLE 2** Find the polar equation for the parabola in Figure 8. The focus of the parabola is $(0, 0)$ and the (rectangular) equation of the directrix is $y = -9/4$.

Solution The parabola in Figure 8 is a conic of the type shown in Figure 5, so the required equation must be of the form

$$r = \frac{ed}{1 - e \sin \theta}$$

For a parabola, the eccentricity e is 1. From Figure 8, we see that the distance d from the focus to the directrix is $\frac{9}{4}$. Using these values for e and d, the equation becomes

$$r = \frac{1(9/4)}{1 - (1)\sin \theta}$$

or

$$r = \frac{9}{4 - 4 \sin \theta} \qquad \text{multiplying both numerator and denominator by 4} \qquad ■■■$$

y

Parabola

Focus

x

Directrix

$y = -\dfrac{9}{4}$

FIGURE 8

■■■ **EXAMPLE 3** Graph the conic $r = 9/(4 + 5 \cos \theta)$.

Solution To bring this equation into standard form, we divide both the numerator and the denominator of the fraction by 4. This yields

$$r = \frac{9/4}{1 + (5/4)\cos \theta}$$

Comparing this with the standard form

$$r = \frac{ed}{1 + e \cos \theta}$$

we see that the eccentricity e is $\frac{5}{4}$, which is greater than 1, so the conic is a hyperbola. Figure 2 on page 635 shows us the general form for one branch of this hyperbola. From Figure 2, we see that the transverse axis of the hyperbola must lie along the polar or x-axis. Now, just as we did in Example 1, we compute the values of r corresponding to $\theta = 0$, $\pi/2$, π, and $3\pi/2$. Figure 9(a) shows the four points that are obtained. The two points $(1,0)$ and $(-9,\pi)$ both lie on the transverse axis of the hyperbola, so they must be the vertices. This allows us to draw the rough sketch in Figure 9(b). ▮▮▮

FIGURE 9

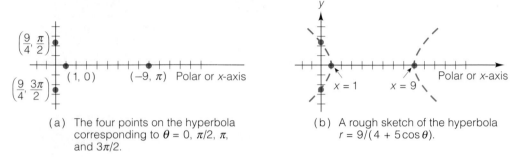

(a) The four points on the hyperbola corresponding to $\theta = 0$, $\pi/2$, π, and $3\pi/2$.

(b) A rough sketch of the hyperbola $r = 9/(4 + 5\cos\theta)$.

For a more accurate drawing of the hyperbola, we need to determine the values of a and b. Using Figure 9(b), we have

$$2a = 9 - 1 = 8 \qquad \text{and therefore} \qquad a = 4$$

The center of the hyperbola lies on the x-axis, halfway between the points with rectangular coordinates $(1,0)$ and $(9,0)$. Thus the center is $(5,0)$ and consequently $c = 5$. (We are using the fact that the origin is a focus.) Now that we know the values of a and c, we can calculate b:

$$b^2 = c^2 - a^2 = 5^2 - 4^2 = 9 \qquad \text{and therefore} \qquad b = 3$$

(Another way to calculate b is to use the formula on page 635: $b = a\sqrt{e^2 - 1}$). In summary, then, we have the following information to use in drawing the hyperbola in the x-y coordinate system: the rectangular coordinates of the vertices are $(1,0)$ and $(9,0)$; the center is $(5,0)$; and the values of a and b are 4 and 3, respectively. This allows us to draw the graph as shown in Figure 10.

FIGURE 10
The hyperbola $r = 9/(4 + 5\cos\theta)$: the left-hand focus is located at the origin, and the coordinates shown are x-y coordinates.

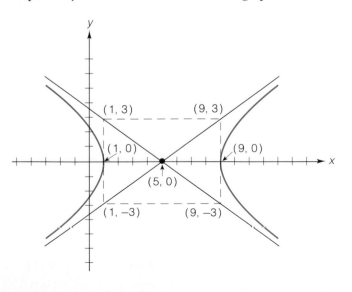

EXERCISE SET 10.7

A

In Exercises 1 and 2, graph each ellipse. Specify the eccentricity, the center, and the endpoints of the major and the minor axes.

1.　(a) $r = \dfrac{6}{3 + 2 \cos \theta}$

　　(b) $r = \dfrac{6}{3 - 2 \cos \theta}$

2.　(a) $r = \dfrac{12}{5 + 3 \sin \theta}$

　　(b) $r = \dfrac{12}{5 - 3 \sin \theta}$

In Exercises 3 and 4, graph each parabola. Specify the (rectangular) coordinates of the vertex and the equation of the directrix.

3.　(a) $r = \dfrac{5}{2 + 2 \cos \theta}$

　　(b) $r = \dfrac{5}{2 - 2 \cos \theta}$

4.　(a) $r = \dfrac{2}{1 + \sin \theta}$

　　(b) $r = \dfrac{2}{1 - \sin \theta}$

In Exercises 5 and 6, graph each hyperbola. Specify the eccentricity, the center, and values of a, b, and c.

5.　(a) $r = \dfrac{3}{2 + 4 \cos \theta}$

　　(b) $r = \dfrac{3}{2 - 4 \cos \theta}$

6.　(a) $r = \dfrac{3}{3 + 4 \sin \theta}$

　　(b) $r = \dfrac{3}{3 - 4 \sin \theta}$

In Exercises 7–18, graph each conic section. If the conic is a parabola, specify (using rectangular coordinates) the vertex and the directrix. If the conic is an ellipse, specify the center, the eccentricity, and the lengths of the major and minor axes. If the conic is a hyperbola, specify the center, the eccentricity, and the lengths of the transverse and conjugate axes.

7.　$r = \dfrac{24}{2 - 3 \cos \theta}$

8.　$r = \dfrac{16}{10 + 5 \sin \theta}$

9.　$r = \dfrac{8}{5 + 3 \sin \theta}$

10.　$r = \dfrac{9}{1 + 2 \cos \theta}$

11.　$r = \dfrac{12}{5 - 5 \sin \theta}$

12.　$r = \dfrac{5}{3 - 2 \sin \theta}$

13.　$r = \dfrac{12}{7 + 5 \cos \theta}$

14.　$r = \dfrac{5}{3 + 3 \cos \theta}$

15.　$r = \dfrac{4}{5 + 5 \sin \theta}$

16.　$r = \dfrac{14}{7 - 8 \cos \theta}$

17.　$r = \dfrac{9}{1 - 2 \cos \theta}$

18.　$r = \dfrac{2}{\sqrt{2} - \sqrt{2} \sin \theta}$

B

For Exercises 19–21, refer to the following figure. The line segment $\overline{PQ}$ is called a focal chord of the conic.

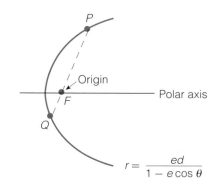

$$r = \frac{ed}{1 - e \cos \theta}$$

19.　Show that $\dfrac{1}{FP} + \dfrac{1}{FQ} = \dfrac{2}{ed}$. What is remarkable about this result? *Hint:* Denote the polar coordinates of P by (r, θ), where $r = ed/(1 - e \cos \theta)$. Now, what are the polar coordinates of Q?

20.　If the coordinates of P are (r, θ), show that

$$PQ = \frac{2ed}{1 - e^2 \cos^2 \theta}$$

　　Hint: $PQ = FP + FQ$.

21.　In the given figure (preceding Exercise 19), suppose that we draw a focal chord $\overline{AB}$ that is perpendicular to $\overline{PQ}$. Show that the sum $\dfrac{1}{PQ} + \dfrac{1}{AB}$ is a constant.

22.　Solve the equation $r/(d - r \cos \theta) = e$ for r. [As stated in the text, you should obtain $r = ed/(1 + e \cos \theta)$.]

10.8 ROTATION OF AXES

In Section 10.5 we saw that the equation

$$Ax^2 + Cy^2 + Dx + Ey + F = 0$$

can represent, in general, one of three curves: a parabola, an ellipse, or a hyperbola. (We use the phrase "in general" here to allow for the so-called degenerate cases.) In the present section, we will find that the second-degree equation

$$Ax^2 + Bxy + Cy^2 + Dx + Ey + F = 0 \qquad (1)$$

also represents, in general, one of these three curves. The difference now is that due to the xy-term in equation (1), the axes of the curves will no longer be parallel or perpendicular to the x- and y-axes. To study the curves defined by equation (1), it is useful first to introduce the technique known as **rotation of axes**.

Suppose that the x- and y-axes are rotated through a positive angle θ to yield a new x'-y' coordinate system, as shown in Figure 1. This procedure is referred to as a rotation of axes. We wish to obtain formulas relating the old and new coordinates. Let P be a given point with coordinates (x, y) in the original coordinate system and (x', y') in the new coordinate system. In Figure 2, let r denote the distance OP and α the angle measured from the x-axis to $\overline{OP}$. From Figure 2, we have

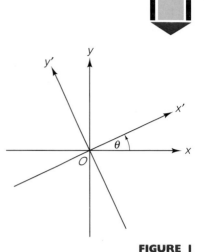

FIGURE 1

$$\cos \alpha = \frac{\text{adjacent}}{\text{hypotenuse}} = \frac{x}{r} \qquad \text{and} \qquad \sin \alpha = \frac{\text{opposite}}{\text{hypotenuse}} = \frac{y}{r}$$

Thus,

$$x = r \cos \alpha \qquad \text{and} \qquad y = r \sin \alpha \qquad (2)$$

Again from Figure 2, we have

$$\cos(\alpha - \theta) = \frac{x'}{r} \qquad \text{and} \qquad \sin(\alpha - \theta) = \frac{y'}{r}$$

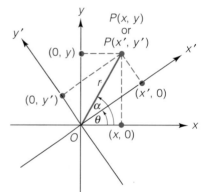

FIGURE 2 Thus,

$$x' = r \cos(\alpha - \theta) \qquad \text{and} \qquad y' = r \sin(\alpha - \theta)$$

or

$$x' = r \cos \alpha \cos \theta + r \sin \alpha \sin \theta$$

and

$$y' = r \sin \alpha \cos \theta - r \cos \alpha \sin \theta$$

With the aid of equations (2), this last pair of equations can be rewritten

$$x' = x \cos \theta + y \sin \theta \qquad \text{and} \qquad y' = -x \sin \theta + y \cos \theta$$

These two equations tell us how to express the new coordinates (x', y') in terms of the original coordinates (x, y) and the angle of rotation θ. On the other hand, it is also useful to express x and y in terms of x', y', and θ. This can be accomplished by treating the two equations we've just derived as a system of two equations in the unknowns x and y. As Exercise 41 at the end of this section asks you to verify, the results of solving this system for x and y are $x = x' \cos \theta - y' \sin \theta$ and $y = x' \sin \theta + y' \cos \theta$.

Formulas for the Rotation of Axes

$$\begin{cases} x = x' \cos \theta - y' \sin \theta \\ y = x' \sin \theta + y' \cos \theta \end{cases} \qquad \begin{cases} x' = x \cos \theta + y \sin \theta \\ y' = -x \sin \theta + y \cos \theta \end{cases}$$

■■ EXAMPLE 1 Suppose that the angle of rotation from the x-axis to the x'-axis is $45°$. If the coordinates of a point P are $(2, 0)$ with respect to the x'-y' coordinate system, what are the coordinates of P with respect to the x-y system?

Solution Substitute the values $x' = 2$, $y' = 0$, and $\theta = 45°$ in the formulas

$$\begin{cases} x = x' \cos \theta - y' \sin \theta \\ y = x' \sin \theta + y' \cos \theta \end{cases}$$

This yields

$$\begin{aligned} x &= 2 \cos 45° - 0 \sin 45° & y &= 2 \sin 45° + 0 \cos 45° \\ &= 2(\sqrt{2}/2) = \sqrt{2} & &= 2(\sqrt{2}/2) = \sqrt{2} \end{aligned}$$

Thus, the coordinates of P in the x-y system are $\left(\sqrt{2}, \sqrt{2}\right)$. ■■■

■■ EXAMPLE 2 Suppose that the angle of rotation from the x-axis to the x'-axis is $45°$. Write the equation $xy = 1$ in terms of the x'-y' coordinate system and then sketch the graph of this equation.

Solution With $\theta = 45°$, the rotation formulas for x and y become

$$\begin{cases} x = x' \cos 45° - y' \sin 45° \\ y = x' \sin 45° + y' \cos 45° \end{cases}$$

Thus we have

$$x = x'\left(\frac{\sqrt{2}}{2}\right) - y'\left(\frac{\sqrt{2}}{2}\right) = \frac{\sqrt{2}}{2}(x' - y')$$

and

$$y = x'\left(\frac{\sqrt{2}}{2}\right) + y'\left(\frac{\sqrt{2}}{2}\right) = \frac{\sqrt{2}}{2}(x' + y')$$

If we now substitute these expressions for x and y in the equation $xy = 1$, we obtain

$$\left[\frac{\sqrt{2}}{2}(x' - y')\right]\left[\frac{\sqrt{2}}{2}(x' + y')\right] = 1$$

$$\frac{1}{2}[(x')^2 - (y')^2] = 1$$

$$\frac{(x')^2}{\left(\sqrt{2}\right)^2} - \frac{(y')^2}{\left(\sqrt{2}\right)^2} = 1$$

This last equation represents a hyperbola in the x'-y' coordinate system. With respect to this x'-y' system, the hyperbola can be analyzed using the techniques developed in Section 10.5. The results (as you should verify for yourself) are as follows:

As noted, the preceding specifications are in terms of the x'-y' coordinate system. However, since the original equation, $xy = 1$, is given in terms of the x-y system, we would like to express these specifications in terms of the same x-y coordinate system. This can be done by the method shown in Example 1. The results are as follows:

$$xy = 1$$
or
$$\frac{(x')^2}{(\sqrt{2})^2} - \frac{(y')^2}{(\sqrt{2})^2} = 1$$

Focal axis: $y = x$
Center: origin
Vertices: $(1, 1)$ and $(-1, -1)$
Foci: $(\sqrt{2}, \sqrt{2})$ and $(-\sqrt{2}, -\sqrt{2})$
Asymptotes: x- and y-axes

> These specifications are in terms of the x-y coordinate system.

Figure 3 displays the graph of this hyperbola.

FIGURE 3
$xy = 1$

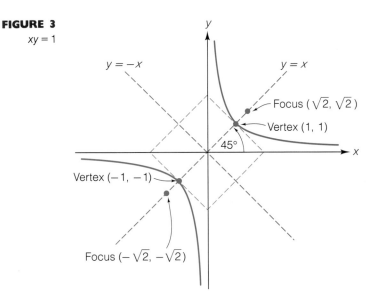

In Example 2, we saw that a rotation of 45° reduced the given equation to one of the standard forms with which we are already familiar. Now let us consider the situation in greater generality. We begin with the second-degree equation

$$Ax^2 + Bxy + Cy^2 + Dx + Ey + F = 0 \qquad (B \neq 0) \tag{3}$$

If we rotate the axes through an angle θ, equation (3) will, after some simplification, take on the form

$$A'(x')^2 + B'x'y' + C'(y')^2 + D'x' + E'y' + F' = 0 \tag{4}$$

for certain constants A', B', C', D', E', and F'. We wish to determine an angle of rotation θ for which $B' = 0$. The reason we want to do this is that if $B' = 0$, we will be able to analyze equation (4) using the techniques of the previous sections. We begin with the rotation formulas:

$$x = x' \cos \theta - y' \sin \theta$$
$$y = x' \sin \theta + y' \cos \theta$$

Substituting these expressions for x and y in equation (3) yields

$$A(x' \cos \theta - y' \sin \theta)^2 + B(x' \cos \theta - y' \sin \theta)(x' \sin \theta + y' \cos \theta)$$
$$+ C(x' \sin \theta + y' \cos \theta)^2 + D(x' \cos \theta - y' \sin \theta)$$
$$+ E(x' \sin \theta + y' \cos \theta) + F = 0$$

We can simplify this equation by performing the indicated operations and then collecting like terms. As Exercise 44 asks you to verify, the resulting equation is

$$A'(x')^2 + B'x'y' + C'(y')^2 + D'x' + E'y' + F' = 0 \qquad (5)$$

where

$$A' = A \cos^2 \theta + B \sin \theta \cos \theta + C \sin^2 \theta$$
$$B' = 2(C - A)\sin \theta \cos \theta + B(\cos^2 \theta - \sin^2 \theta)$$
$$C' = A \sin^2 \theta - B \sin \theta \cos \theta + C \cos^2 \theta$$
$$D' = D \cos \theta + E \sin \theta$$
$$E' = E \cos \theta - D \sin \theta$$
$$F' = F$$

Thus, B' will be zero provided that

$$2(C - A)\sin \theta \cos \theta + B(\cos^2 \theta - \sin^2 \theta) = 0$$

By using the double-angle formulas, we can write this last equation as

$$(C - A)\sin 2\theta + B \cos 2\theta = 0$$

or

$$B \cos 2\theta = (A - C)\sin 2\theta$$

Now, dividing by $B \sin 2\theta$, we obtain

$$\frac{\cos 2\theta}{\sin 2\theta} = \frac{A - C}{B}$$

or

$$\cot 2\theta = \frac{A - C}{B}$$

We have now shown that if θ satisfies the condition $\cot 2\theta = (A - C)/B$, then equation (5) will contain no $x'y'$ term. Although we will not prove it here, it can be shown that there is always a value of θ in the range $0° < \theta < 90°$ for which $\cot 2\theta = (A - C)/B$. In subsequent examples, we will always choose θ in this range.

■■■ **EXAMPLE 3** Graph the equation $2x^2 + \sqrt{3}xy + y^2 = 2$.

Solution We first rotate the axes through an angle θ so that the new equations will contain no $x'y'$ term. To choose an appropriate value of θ, we require that

$$\cot 2\theta = \frac{A - C}{B} = \frac{2 - 1}{\sqrt{3}} = \frac{1}{\sqrt{3}}$$

Thus, $\cot 2\theta = 1/\sqrt{3}$, from which we conclude that $2\theta = 60°$, or $\theta = 30°$. With this value of θ, the rotation formulas become

$$x = x'\left(\frac{\sqrt{3}}{2}\right) - y'\left(\frac{1}{2}\right) \quad \text{and} \quad y = x'\left(\frac{1}{2}\right) + y'\left(\frac{\sqrt{3}}{2}\right)$$

Now we use these formulas to substitute for x and y in the given equation. This yields

$$2\left(\frac{x'\sqrt{3}}{2} - \frac{y'}{2}\right)^2 + \sqrt{3}\left(\frac{x'\sqrt{3}}{2} - \frac{y'}{2}\right)\left(\frac{x'}{2} + \frac{y'\sqrt{3}}{2}\right) + \left(\frac{x'}{2} + \frac{y'\sqrt{3}}{2}\right)^2 = 2$$

After simplification, this last equation becomes

$$(y')^2 + 5(x')^2 = 4$$

or

$$\frac{(y')^2}{2^2} + \frac{(x')^2}{(2/\sqrt{5})^2} = 1$$

We recognize this as the equation of an ellipse in the x'-y' coordinate system. The focal axis is the y'-axis. The values of a and b are 2 and $2/\sqrt{5}$, respectively. The ellipse can now be sketched as in Figure 4. ■■■

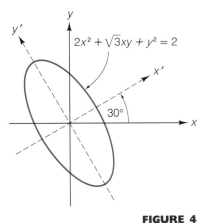

FIGURE 4

In Example 3, we were able to determine the angle θ directly, since we recognized the quantity $1/\sqrt{3}$ as the value of $\cot 60°$. However, this is the exception rather than the rule. In most problems, the value of $(A - C)/B$ is not so easily identified as the cotangent of one of the more familiar angles. The next examples demonstrate a technique that can be used in such cases. The technique relies on the following three trigonometric identities:

1. $\sec^2 \beta = 1 + \tan^2 \beta$

2. $\sin \theta = \sqrt{\dfrac{1 - \cos 2\theta}{2}}$

The positive square roots are appropriate, since $0° < \theta < 90°$.

3. $\cos \theta = \sqrt{\dfrac{1 + \cos 2\theta}{2}}$

■■■ **EXAMPLE 4** Graph the equation $x^2 + 4xy - 2y^2 = 6$.

Solution We have $A = 1$, $B = 4$, and $C = -2$. Therefore,

$$\cot 2\theta = \frac{A - C}{B} = \frac{1 - (-2)}{4} = \frac{3}{4}$$

Since $\cot 2\theta = \frac{3}{4}$, it follows that $\tan 2\theta = \frac{4}{3}$. Therefore

$$\sec^2 2\theta = 1 + \left(\frac{4}{3}\right)^2 \qquad \text{(Why?)}$$

$$= \frac{9}{9} + \frac{16}{9} = \frac{25}{9}$$

Thus,

$$\sec 2\theta = \pm\frac{5}{3}$$

At this point, we need to decide whether the positive or negative sign is appropriate. Since we are assuming that $0° < \theta < 90°$, the angle 2θ must lie in either the first or the second quadrant. To decide which, we note that the value determined for $\cot 2\theta$ was positive. That rules out the possibility that 2θ might lie in the second quadrant. We conclude that in this case, $0° < 2\theta < 90°$. Therefore, the sign of $\sec 2\theta$ must be positive, and we have

$$\sec 2\theta = \frac{5}{3} \qquad \text{or} \qquad \cos 2\theta = \frac{3}{5}$$

The values of $\sin\theta$ and $\cos\theta$ can now be obtained as follows:

$$\sin\theta = \sqrt{\frac{1 - \cos 2\theta}{2}} \qquad \qquad \cos\theta = \sqrt{\frac{1 + \cos 2\theta}{2}}$$

$$= \sqrt{\frac{1 - \frac{3}{5}}{2}} \qquad \qquad \qquad = \sqrt{\frac{1 + \frac{3}{5}}{2}}$$

$$= \frac{1}{\sqrt{5}} \quad \text{after simplifying} \qquad = \frac{2}{\sqrt{5}} \quad \text{after simplifying}$$

With these values for $\sin\theta$ and $\cos\theta$, the rotation formulas become

$$x = x'\left(\frac{2}{\sqrt{5}}\right) - y'\left(\frac{1}{\sqrt{5}}\right) = \frac{1}{\sqrt{5}}(2x' - y')$$

and

$$y = x'\left(\frac{1}{\sqrt{5}}\right) + y'\left(\frac{2}{\sqrt{5}}\right) = \frac{1}{\sqrt{5}}(x' + 2y')$$

We now substitute these expressions for x and y in the original equation, $x^2 + 4xy - 2y^2 = 6$. This yields

$$\left(\frac{2x' - y'}{\sqrt{5}}\right)^2 + 4\left(\frac{2x' - y'}{\sqrt{5}}\right)\left(\frac{x' + 2y'}{\sqrt{5}}\right) - 2\left(\frac{x' + 2y'}{\sqrt{5}}\right)^2 = 6$$

As Exercise 42 asks you to verify, this equation can be simplified to

$$2(x')^2 - 3(y')^2 = 6$$

or

$$\frac{(x')^2}{(\sqrt{3})^2} - \frac{(y')^2}{(\sqrt{2})^2} = 1$$

This last equation represents a hyperbola with center at the origin of the x'-y' coordinate system and with a focal axis that coincides with the x'-axis. We can sketch the hyperbola using the methods of Section 10.5, but it is first necessary to know the angle θ between the x- and x'-axes. We have

$$\sin \theta = 1/\sqrt{5}$$
$$\theta = \sin^{-1}(1/\sqrt{5})$$
$$\theta \approx 27° \qquad \text{using a calculator}$$

Figure 5 shows the required graph.

FIGURE 5
$x^2 + 4xy - 2y^2 = 6$

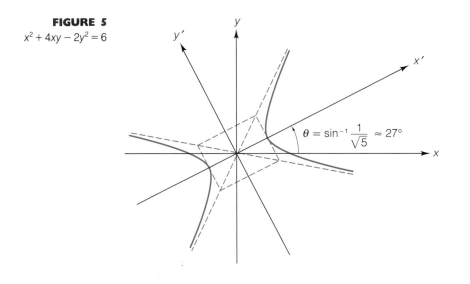

In Examples 3 and 4, we graphed equations of the form

$$Ax^2 + Bxy + Cy^2 + F = 0$$

The technique used in those examples is equally effective in graphing equations of the form $Ax^2 + Bxy + Cy^2 + Dx + Ey + F = 0$, in which the x- and y-terms are present. This is demonstrated in Example 5. Since the general technique employed in Example 5 is the same as in the previous examples, we will merely outline the procedure and the results in the solution, leaving the detailed calculations to Exercise 43 at the end of this section.

▌███▌ **EXAMPLE 5** Graph the equation $16x^2 - 24xy + 9y^2 + 110x - 20y + 100 = 0$.

Outline of Solution

$$\cot 2\theta = \frac{A - C}{B} = -\frac{7}{24}$$

Now, proceeding as in the last example, we find that $\cos 2\theta = -\frac{7}{25}$, $\cos \theta = \frac{3}{5}$, and $\sin \theta = \frac{4}{5}$. Thus, the rotation formulas become

$$x = x'\left(\frac{3}{5}\right) - y'\left(\frac{4}{5}\right) = \frac{1}{5}(3x' - 4y')$$

and

$$y = x'\left(\frac{4}{5}\right) + y'\left(\frac{3}{5}\right) = \frac{1}{5}(4x' + 3y')$$

Next we substitute for x and y in the given equation. After straightforward but lengthy computations, we obtain

$$(y')^2 + 2x' - 4y' + 4 = 0$$

We graphed equations of this form in Section 10.2 by completing the square. Using that technique here, we have

$$(y')^2 - 4y' = -2x' - 4$$
$$(y')^2 - 4y' + 4 = -2x' - 4 + 4$$
$$(y' - 2)^2 = -2x'$$

This is the equation of a parabola. With respect to the x'-y' system, the vertex is $(0, 2)$ and the axis of the parabola is the line $y' = 2$. In terms of the x-y system, the vertex is $\left(-\frac{8}{5}, \frac{6}{5}\right)$ and the axis of the parabola is the line $-4x + 3y = 10$. Finally, the angle of rotation is $\theta = \sin^{-1}\left(\frac{4}{5}\right)$. Using a calculator, we find this to be approximately $53°$. The required graph is shown in Figure 6.　▮▮▮

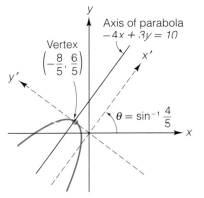

FIGURE 6

$16x^2 - 24xy + 9y^2 + 110x$
$- 20y + 100 = 0.$

EXERCISE SET 10.8

A

In Exercises 1–3, an angle of rotation is specified, followed by the coordinates of a point in the x'-y' system. Find the coordinates of each point with respect to the x-y system.

1. $\theta = 30°$; $(x', y') = (\sqrt{3}, 2)$
2. $\theta = 60°$; $(x', y') = (-1, 1)$
3. $\theta = 45°$; $(x', y') = (\sqrt{2}, -\sqrt{2})$

In Exercises 4–6, an angle of rotation is specified, followed by the coordinates of a point in the x-y system. Find the coordinates of each point with respect to the x'-y' system.

4. $\theta = 45°$; $(x, y) = (0, -2)$
5. $\theta = \sin^{-1}\left(\frac{5}{13}\right)$; $(x, y) = (-3, 1)$
6. $\theta = 15°$; $(x, y) = (1, 0)$

In Exercises 7–14, find $\sin\theta$ and $\cos\theta$, where θ is the (acute) angle of rotation that eliminates the $x'y'$ term. Note: *You are not asked to graph the equation.*

7. $25x^2 - 24xy + 18y^2 + 1 = 0$
8. $x^2 + 24xy + 8y^2 - 8 = 0$
9. $x^2 - 24xy + 8y^2 - 8 = 0$
10. $220x^2 + 119xy + 100y^2 = 0$
11. $x^2 - 2\sqrt{3}xy - y^2 = 3$

12. $5x^2 + 12xy - 4 = 0$
13. $161xy - 240y^2 - 1 = 0$
14. $4x^2 - 5xy + 4y^2 + 2 = 0$
15. Suppose that the angle of rotation is $45°$. Write the equation $2xy = 9$ in terms of the x'-y' coordinate system and then graph the equation.
16. Suppose that the angle of rotation is $45°$. Write the equation $5x^2 - 6xy + 5y^2 + 16 = 0$ in terms of the x'-y' system.

In Exercises 17–40, graph the equations.

17. $7x^2 + 8xy + y^2 - 1 = 0$
18. $2x^2 - \sqrt{3}xy + y^2 - 20 = 0$
19. $x^2 + 4xy + 4y^2 = 1$
20. $x^2 + 4xy + 4y^2 = 0$
21. $9x^2 - 24xy + 16y^2 - 400x - 300y = 0$
22. $8x^2 + 12xy + 13y^2 = 34$
23. $4xy + 3y^2 + 4x + 6y = 1$
24. $x^2 - 2xy + y^2 + x - y = 0$
25. $3x^2 - 2xy + 3y^2 - 6\sqrt{2}x + 2\sqrt{2}y + 4 = 0$
26. $x^2 + 3xy + y^2 = 1$
27. $(x - y)^2 = 8(y - 6)$

28. $4x^2 - 4xy + y^2 - 4x + 2y + 1 = 0$
29. $3x^2 + 4xy + 6y^2 = 7$
30. $x^2 + 2\sqrt{3}xy + 3y^2 + 12\sqrt{3}x - 12y - 24 = 0$
31. $17x^2 - 12xy + 8y^2 - 80 = 0$
32. $7x^2 - 2\sqrt{3}xy + 5y^2 = 32$
33. $3xy - 4y^2 + 18 = 0$
34. $x^2 + y^2 = 2xy + 4x + 4y - 8$
35. $(x + y)^2 + 4\sqrt{2}(x - y) = 0$
36. $41x^2 - 24xy + 9y^2 = 45$
37. $3x^2 - \sqrt{15}xy + 2y^2 = 3$
38. $3x^2 + 10xy + 3y^2 - 2\sqrt{2}x + 2\sqrt{2}y - 10 = 0$
39. $3x^2 - 2xy + 3y^2 + 2 = 0$
40. $(x + y)(x + y + 1) = 2$

B

41. Solve for x and y:
$$\begin{cases} (\cos\theta)x + (\sin\theta)y = x' \\ (-\sin\theta)x + (\cos\theta)y = y' \end{cases}$$

      Answer: $\quad x = x' \cos\theta - y' \sin\theta$
                      $y = x' \sin\theta + y' \cos\theta$

42. Simplify the equation:
$$\left(\frac{2x' - y'}{\sqrt{5}}\right)^2 + 4\left(\frac{2x' - y'}{\sqrt{5}}\right)\left(\frac{x' + 2y'}{\sqrt{5}}\right) - 2\left(\frac{x' + 2y'}{\sqrt{5}}\right)^2 = 6$$

      Answer: $\quad 2(x')^2 - 3(y')^2 = 6$

43. Refer to Example 5 in the text.
   (a) Show that $\cos 2\theta = -\frac{7}{25}$.

(b) Show that $\cos\theta = \frac{3}{5}$ and $\sin\theta = \frac{4}{5}$.
(c) Make the substitutions $x = \frac{1}{5}(3x' - 4y')$ and
   $y = \frac{1}{5}(4x' + 3y')$ in the given equation
   $16x^2 - 24xy + 9y^2 + 110x - 20y + 100 = 0$ and
   show that the resulting equation simplifies to
   $(y')^2 + 2x' - 4y' + 4 = 0$.

44. Make the substitutions $x = x' \cos\theta - y' \sin\theta$ and
   $y = x' \sin\theta + y' \cos\theta$ in the equation
   $Ax^2 + Bxy + Cy^2 + Dx + Ey + F = 0$ and show
   that the result is
   $A'(x')^2 + B'x'y' + C'(y')^2 + D'x' + E'y' + F' = 0$, where

     $A' = A \cos^2\theta + B \sin\theta \cos\theta + C \sin^2\theta$

     $B' = 2(C - A)\sin\theta \cos\theta + B(\cos^2\theta - \sin^2\theta)$

     $C' = A \sin^2\theta - B \sin\theta \cos\theta + C \cos^2\theta$

     $D' = D \cos\theta + E \sin\theta$

     $E' = E \cos\theta - D \sin\theta$

     $F' = F$

45. (Refer to Exercise 44.) Show that $A + C = A' + C'$.

46. Complete the following steps to derive the equation
   $(B')^2 - 4A'C' = B^2 - 4AC$.
   (a) Show that $A' - C' = (A - C)\cos 2\theta + B \sin 2\theta$.
   (b) Show that $B' = B \cos 2\theta - (A - C)\sin 2\theta$.
   (c) Square the equations in parts (a) and (b), then add
      the two resulting equations to show that
      $(A' - C')^2 + (B')^2 = (A - C)^2 + B^2$.
   (d) Square the equation given in Exercise 45, then
      subtract the result from the equation in part (c).
      The result can be written $(B')^2 - 4A'C' = B^2 - 4AC$, as required.

![logo] **CHAPTER TEN**   **SUMMARY OF PRINCIPAL TERMS AND FORMULAS**

TERMS AND FORMULAS	PAGE REFERENCE	COMMENT
1. Conic sections	576	These are the curves that are formed when a plane intersects the surface of a right circular cone. As indicated in the figure at the beginning of the chapter, these curves are the circle, the ellipse, the hyperbola, and the parabola.
2. Angle of inclination	577	The angle of inclination of a line is the angle between the x-axis and the line, measured counterclockwise from the positive side or positive direction of the x-axis to the line.
3. $m = \tan\theta$	577	The slope of a line is equal to the tangent of the angle of inclination.
4. $d = \dfrac{\lvert mx_0 + b - y_0 \rvert}{\sqrt{1 + m^2}}$	578	This is the formula for the (perpendicular) distance d from the point (x_0, y_0) to the line $y = mx + b$.

TERMS AND FORMULAS	PAGE REFERENCE	COMMENT
5. $d = \dfrac{\lvert Ax_0 + By_0 + C \rvert}{\sqrt{A^2 + B^2}}$	578, 581	This is the formula for the (perpendicular) distance d from the point (x_0, y_0) to the line $Ax + By + C = 0$.
6. Parabola	585	A parabola is the set of all points in the plane equally distant from a fixed line and a fixed point not on the line. The fixed line is called the *directrix*, and the fixed point is called the *focus*.
7. Axis (of a parabola)	587	This is the line drawn through the focus of the parabola, perpendicular to the directrix.
8. Vertex (of a parabola)	587	This is the point at which the parabola intersects its axis.
9. Focal chord (of a parabola)	588	A focal chord of a parabola is a line segment passing through the focus, with endpoints on the parabola.
10. Focal width (of a parabola)	588	The focal width of a parabola is the length of the focal chord that is perpendicular to the axis of the parabola. For a given value of p, the two parabolas $x^2 = 4py$ and $y^2 = 4px$ have the same focal width; it is $4p$.
11. Tangent line (to a parabola)	595	A line that is not parallel to the axis of a parabola is tangent to the parabola provided that it intersects the parabola in exactly one point.
12. Ellipse	598	An ellipse is the set of all points in the plane, the sum of whose distances from two fixed points is constant. Each fixed point is called a *focus* of the ellipse.
13. Eccentricity (of an ellipse)	600	The eccentricity is a number that measures how much the ellipse deviates from being a circle. See Figure 7 in Section 10.4. The eccentricity e is defined by the formula $e = c/a$, where c and a are defined by the following conventions. The distance between the foci is denoted by $2c$. The sum of the distances from a point on the ellipse to the two foci is denoted by $2a$.
14. Focal axis (of an ellipse)	600	This is the line passing through the two foci.
15. Center (of an ellipse)	600	This is the midpoint of the line segment joining the foci.
16. Vertices (of an ellipse)	600	The two points at which an ellipse meets its focal axis are called the *vertices* of the ellipse.
17. Major axis (of an ellipse)	600	This is the line segment joining the two vertices of the ellipse.
18. Minor axis (of an ellipse)	600	This is the line segment through the center of the ellipse, perpendicular to the major axis, and with endpoints on the ellipse.
19. $\dfrac{x^2}{a^2} + \dfrac{y^2}{b^2} = 1 \quad (a > b)$	601	This is the standard form for the equation of an ellipse with foci $(\pm c, 0)$.
$\dfrac{x^2}{b^2} + \dfrac{y^2}{a^2} = 1 \quad (a > b)$	604	This is the standard form for the equation of an ellipse with foci $(0, \pm c)$.
20. Tangent line to an ellipse	605	A tangent to an ellipse is a line that intersects the ellipse in exactly one point. See Figure 12 in Section 10.4.
21. $\dfrac{x_1 x}{a^2} + \dfrac{y_1 y}{b^2} = 1$	605	This is the equation of the line that is tangent to the ellipse $(x^2/a^2) + (y^2/b^2) = 1$ at the point (x_1, y_1) on the ellipse.

TERMS AND FORMULAS	PAGE REFERENCE	COMMENT
22. Hyperbola	613	A hyperbola is the set of all points in the plane, the difference of whose distances from two fixed points is a positive constant. The two fixed points are the *foci,* and the line passing through the foci is the *focal axis.*
23. $\dfrac{x^2}{a^2} - \dfrac{y^2}{b^2} = 1$	615	This is the standard form for the equation of a hyperbola with foci $(\pm c, 0)$.
$\dfrac{y^2}{a^2} - \dfrac{x^2}{b^2} = 1$	618	This is the standard form for the equation of a hyperbola with foci $(0, \pm c)$.
24. Asymptote	615	A line is said to be an *asymptote* for a curve if the distance between the line and the curve approaches zero as we move out farther and farther along the line. The asymptotes for the hyperbola $(x^2/a^2) - (y^2/b^2) = 1$ are the two lines $y = \pm(b/a)x$. For the hyperbola $(y^2/a^2) - (x^2/b^2) = 1$, the asymptotes are $y = \pm(a/b)x$.
25. Focal axis (of a hyperbola)	617	This is the line passing through the foci.
26. Vertices (of a hyperbola)	617	The two points at which the hyperbola intersects its focal axis are called *vertices.*
27. Center (of a hyperbola)	617	This is the point on the focal axis midway between the foci.
28. Transverse axis	617	This is the line segment joining the two vertices of a hyperbola.
29. Conjugate axis	617	This is the line segment perpendicular to the transverse axis of the hyperbola, passing through the center and extending a distance $b\left(= \sqrt{c^2 - a^2}\right)$ on either side of the center.
30. Eccentricity (of a hyperbola)	617	For both of the standard forms for the hyperbola, the eccentricity e is defined by $e = c/a$.
31. $\dfrac{x_1 x}{a^2} - \dfrac{y_1 y}{b^2} = 1$	620	This is the equation of the tangent to the hyperbola $(x^2/a^2) - (y^2/b^2) = 1$ at the point (x_1, y_1) on the curve. (See Exercise 53 in Exercise Set 10.5 for the definition of a tangent to a hyperbola.)
32. Focus–directrix property of conics	632	Refer to Figure 10 on page 632. Let $\mathscr{L}$ be a fixed line, F a fixed point, and e a positive constant. Consider the set of points P satisfying the condition $FP/PD = e$, where D is defined by Figure 10. Then: **(a)** If $e = 1$ the set of points is a *parabola* with focus F and directrix $\mathscr{L}$; **(b)** If $0 < e < 1$, the set of points is an *ellipse* with focus F, corresponding directrix $\mathscr{L}$, and eccentricity e; **(c)** If $e > 1$, the set of points is a *hyperbola* with focus F, corresponding directrix $\mathscr{L}$, and eccentricity e.
33. $r = \dfrac{ed}{1 \pm e \cos \theta}$ $r = \dfrac{ed}{1 \pm e \sin \theta}$	635	These are polar equations of conics with the focus at the pole or origin, as indicated in Figures 2–5 on page 635. The eccentricity is e, and d is the distance from the focus to the directrix.
34. $\begin{cases} x = x' \cos \theta - y' \sin \theta \\ y = x' \sin \theta + y' \cos \theta \end{cases}$ $\begin{cases} x' = x \cos \theta + y \sin \theta \\ y' = -x \sin \theta + y \cos \theta \end{cases}$	641	These are the formulas that relate the coordinates of a point in the x-y system to the coordinates in the rotated x'-y' system. See Figures 1 and 2 in Section 10.8.

TERMS AND FORMULAS	PAGE REFERENCE	COMMENT
35. $\cos 2\theta = \dfrac{A - C}{B}$	643	This formula determines an angle of rotation θ. When the equation $$Ax^2 + Bxy + Cy^2 + Dx + Ey + F = 0 \qquad (B \neq 0)$$ is written in the x'-y' system, the resulting equation does not contain an $x'y'$ term. The graph can then be analyzed by means of the technique of completing the square.

WRITING MATHEMATICS

Write out your answers to the following questions in complete sentences. If you draw a diagram to accompany your writing, or if you use equations, be sure that you clearly label any elements to which you refer.

1. Refer to Figure 1 on page 598. If the thumbtacks are 3 in. apart, what length of string should be used to produce an ellipse with eccentricity $\frac{2}{3}$?

2. Refer to Figure 1 on page 598 to explain each of the following.
   (a) When the eccentricity of an ellipse is close to 1, the ellipse is very flat.
   (b) When the eccentricity of an ellipse is close to 0, the ellipse resembles a circle.

3. Investigate the geometric significance of the eccentricity of a hyperbola by completing the three steps that follow. Then, write a report telling what you have done, what patterns you have observed, and what relationship you have found between the eccentricity and the shape of the hyperbola.
   (a) Use the definition of the eccentricity e to show that $e = \sqrt{1 + (b/a)^2}$. (This shows that the eccentricity is always greater than 1.)
   (b) Compute the eccentricity for each of the following hyperbolas. (Use a calculator.)
      (i) $(0.0201)x^2 - y^2 = 0.0201$
      (ii) $3x^2 - y^2 = 3$
      (iii) $8x^2 - y^2 = 8$
      (iv) $15x^2 - y^2 = 15$
      (v) $99x^2 - y^2 = 99$
   (c) On the same set of axes, sketch the first-quadrant portion of each of the hyperbolas in part (b).

CHAPTER TEN   REVIEW EXERCISES

In Exercises 1–15, refer to the following figure and show that the given statements are correct.

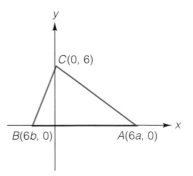

1. The equations of the lines forming the sides of △ABC are $x + by = 6b$, $x + ay = 6a$, and $y = 0$.

2. The equations of the lines forming the medians of △ABC are $2x + (a + b)y = 6(a + b)$, $x − (a − 2b)y = 6b$, and $x − (b − 2a)y = 6a$. (A *median* is a line segment drawn from a vertex of a triangle to the midpoint of the opposite side.)

3. Each pair of medians of △ABC intersect at the point $G(2a + 2b, 2)$. (The point G is called the *centroid* of △ABC.)

4. The equations of the lines forming the altitudes of triangle ABC are $y = ax − 6ab$, $y = bx − 6ab$, and $x = 0$. (An *altitude* is a line segment drawn from a vertex to the opposite side, perpendicular to that side.)

5. Each pair of altitudes intersects at the point $H(0, −6ab)$. (The point H is the *orthocenter* of △ABC.)

6. The equations of the perpendicular bisectors of the sides of △ABC are $x = 3a + 3b$, $bx − y = 3b^2 − 3$, and $ax − y = 3a^2 − 3$.

7. Each pair of perpendicular bisectors intersect at the point $O(3a + 3b, 3ab + 3)$. (The point O is called the *circumcenter* of △ABC.)

8. The distance from the circumcenter O to each vertex is $3\sqrt{(a^2 + 1)(b^2 + 1)}$. (This distance, denoted by R, is the *circumradius* of △ABC. Note that the circle with center O and radius R passes through the points A, B, and C.)

9. In △ABC, let p, q, and r denote the lengths of $\overline{BC}$, $\overline{AC}$, and $\overline{AB}$, respectively. Then the area of △ABC is $pqr/4R$.

10. $AH^2 + BC^2 = 4(OA)^2$

11. $OH^2 = 9R^2$ $(p^2 + q^2 + r^2)$ *Hint:* See Exercise 9.

12. $GH^2 = 4R^2 − \frac{4}{9}(p^2 + q^2 + r^2)$

13. $HA^2 + HB^2 + HC^2 = 12R^2 − (p^2 + q^2 + r^2)$

14. The points H, G, and O are collinear. (The line through these three points is the *Euler line* of triangle ABC.)

15. $GH = 2(GO)$

In Exercises 16–18, find the angle of inclination for each line. Use a calculator to express your answers in degrees. (Round off to one decimal place.)

16. $y = 4x − 3$ 17. $2x + 3y = 6$ 18. $y = 2x$

19. Find the distance from the point $(−1, −3)$ to the line $5x + 6y = 30$.

20. Find the distance from the point $(2, 1)$ to the line $y = \frac{1}{2}x + 4$.

21. The vertices of an equilateral triangle are $(\pm 6, 0)$ and $\left(0, 6\sqrt{3}\right)$. Verify that the sum of the three distances from the point $(1, 2)$ to the sides of the triangle is equal to the height of the triangle. (It can be shown that, for any point inside an equilateral triangle, the sum of the distances to the sides is equal to the height of the triangle. This is *Viviani's theorem*.)

22. A tangent line is drawn from the point $(−12, −1)$ to the circle $x^2 + y^2 = 20$. Find the slope of this line, given that its y-intercept is positive.

In Exercises 23–26, find the equation of the parabola satisfying the given conditions. In each case, assume the vertex is $(0, 0)$.

23. (a) The focus is $(4, 0)$.   (b) The focus is $(0, 4)$.

24. The focus lies on the x-axis, and the curve passes through the point $(3, 1)$.

25. The parabola is symmetric about the y-axis, the y-coordinate of the focus is positive, and the length of the focal chord perpendicular to the y-axis is 12.

26. The focus of the parabola is the center of the circle $x^2 − 8x + y^2 + 15 = 0$.

In Exercises 27–29, find the equation of the ellipse satisfying the given conditions. Write your answer in the form $Ax^2 + By^2 = C$.

27. Foci $(\pm 2, 0)$; endpoints of the major axis $(\pm 8, 0)$

28. Foci $(0, \pm 1)$; endpoints of the minor axis $(\pm 4, 0)$

29. Eccentricity $\frac{4}{5}$; one end of the minor axis at $(−6, 0)$; center at the origin

30. For any point P on a certain ellipse, the sum of the distances from $(1, 2)$ and $(−1, −2)$ is 12. Find the equation of the ellipse.   *Hint:* Use the distance formula and the definition of an ellipse.

In Exercises 31–34, find the equation of the hyperbola satisfying the given conditions. Write each answer in the form $Ax^2 − By^2 = C$ or in the form $Ay^2 − Bx^2 = C$.

31. Foci $(\pm 6, 0)$; vertices $(\pm 2, 0)$

32. Asymptotes $y = \pm 2x$; foci $(0, \pm 3)$

33. Eccentricity 4; foci $(\pm 3, 0)$

34. Length of transverse axis 3; eccentricity $\frac{5}{4}$; center $(0, 0)$; focal axis horizontal

In Exercises 35–40, graph the parabolas, and in each case specify the vertex, the focus, the directrix, and the focal width.

35. $x^2 = 10y$

36. $x^2 = 5y$

37. $x^2 = -12(y - 3)$

38. $x^2 = -8(y + 1)$

39. $(y - 1)^2 = -4(x - 1)$

40. $(y + 3)^2 = 2(x - 1)$

In Exercises 41–46, graph the ellipses, and in each case specify the center, the foci, the lengths of the major and minor axes, and the eccentricity.

41. $x^2 + 2y^2 = 4$

42. $4x^2 + 9y^2 = 144$

43. $49x^2 + 9y^2 = 441$

44. $9x^2 + y^2 = 9$

45. $\dfrac{(x - 1)^2}{5^2} + \dfrac{(y + 2)^2}{3^2} = 1$

46. $\dfrac{(x + 3)^2}{3^2} + \dfrac{y^2}{3^2} = 1$

In Exercises 47–52, graph the hyperbolas. In each case specify the center, the vertices, the foci, the equations of the asymptotes, and the eccentricity.

47. $x^2 - 2y^2 = 4$

48. $4x^2 - 9y^2 = 144$

49. $49y^2 - 9x^2 = 441$

50. $9y^2 - x^2 = 9$

51. $\dfrac{(x - 1)^2}{5^2} - \dfrac{(y + 2)^2}{3^2} = 1$

52. $\dfrac{(y + 3)^2}{3^2} - \dfrac{x^2}{3^2} = 1$

In Exercises 53–66, use the technique of completing the square to graph the given equation. If the graph is a parabola, specify the vertex, axis, focus, and directrix. If the graph is an ellipse, specify the center, foci, and lengths of the major and minor axes. If the graph is a hyperbola, specify the center, vertices, foci, and equations of the asymptotes. Finally, if the equation has no graph, say so.

53. $3x^2 + 4y^2 - 6x + 16y + 7 = 0$

54. $y^2 - 16x - 8y + 80 = 0$

55. $y^2 + 4x + 2y - 15 = 0$

56. $16x^2 + 64x + 9y^2 - 54y + 1 = 0$

57. $16x^2 - 32x - 9y^2 + 90y - 353 = 0$

58. $x^2 + 6x - 12y + 33 = 0$

59. $5x^2 + 3y^2 - 40x - 36y + 188 = 0$

60. $x^2 - y^2 - 4x + 2y - 6 = 0$

61. $9x^2 - 90x - 16y^2 + 32y + 209 = 0$

62. $x^2 + 2y - 12 = 0$

63. $y^2 - 25x^2 + 8y - 9 = 0$

64. $x^2 + 16y^2 - 160y + 384 = 0$

65. $16x^2 + 25y^2 - 64x - 100y + 564 = 0$

66. $16x^2 - 25y^2 - 64x + 100y - 36 = 0$

67. Let F_1 and F_2 denote the foci of the hyperbola $5x^2 - 4y^2 = 80$.
  (a) Verify that the point P with coordinates $(6, 5)$ lies on the hyperbola.
  (b) Compute the quantity $(F_1P - F_2P)^2$.

68. Show that the coordinates of the vertex of the parabola $Ax^2 + Dx + Ey + F = 0$ are given by
$$x = -\frac{D}{2A} \quad \text{and} \quad y = \frac{D^2 - 4AF}{4AE}$$

69. If the equation $Ax^2 + Cy^2 + Dx + Ey + F = 0$ represents an ellipse or a hyperbola, show that the center is the point $(-D/2A, -E/2C)$.

70. The figure shows the parabola $x^2 = 4py$ and a circle with center at the origin and diameter $3p$. If V and F denote the vertex and focus of the parabola, respectively, show that the common chord of the circle and parabola bisects the line segment $\overline{VF}$.

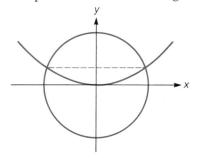

71. The following figure shows an ellipse and a parabola. As indicated in the figure, the curves are symmetric about the x-axis and they both have an x-intercept of 5. Find the equation of the ellipse and the parabola, given that the point $(3, 0)$ is a focus for both curves.

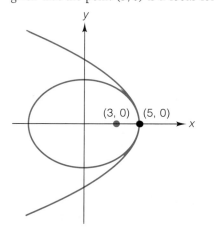

In Exercises 72–75, find sin θ and cos θ, where θ is the (acute) angle of the rotation that eliminates the x'y' term. Then graph the equation.

72. $4x^2 + 2xy + 4y^2 - 15 = 0$

73. $4x^2 + 4xy + y^2 + 20x - 10y = 0$

74. $2x^2 + 4xy - y^2 - 12 = 0$

75. $13x^2 + 10xy + 13y^2 + 16x - 16y - 272 = 0$

76. Show that the equation of a line tangent to the circle $(x - h)^2 + (y - k)^2 = r^2$ at the point (a, b) on the circle is

$$(a - h)(x - h) + (b - k)(y - k) = r^2$$

In Exercises 77–82 graph the conics.

77. $r = \dfrac{3}{2 - 4\cos\theta}$

78. $r = \dfrac{5}{4 + 4\cos\theta}$

79. $r = \dfrac{4}{4 - \cos\theta}$

80. $r = \dfrac{9}{1 + 2\sin\theta}$

81. $r = \dfrac{20}{3 - 5\sin\theta}$

82. $r = \dfrac{10}{1 + \sin\theta}$

83. In the following figure, A, and B are fixed points, and the point $P(x, y)$ moves in such a way that $\tan\theta \tan\beta = 2$. Show that the point P lies on the ellipse $2x^2 + y^2 = 2a^2$.

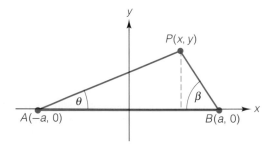

84. The area A of an ellipse $(x^2/a^2) + (y^2/b^2) = 1$ is given by the formula $A = \pi ab$. Use this formula to compute the area of the ellipse $5x^2 + 6y^2 = 60$.

85. As you know, there is a simple expression for the circumference of a circle of radius a, namely, $2\pi a$. However, there is no similar type of elementary expression for the circumference of an ellipse. (The circumference of an ellipse can be computed to as many decimal places as required using the methods of calculus.) Nevertheless, there are some interesting elementary formulas that allow us to approximate the circumference of an ellipse quite closely. Three such formulas follow, along with the names of their discoverers and the dates of discovery. Each formula

yields an approximate value for the circumference of the ellipse $(x^2/a^2) + (y^2/b^2) = 1$.

$$C_1 = \pi\left[a + b + \tfrac{1}{2}(\sqrt{a} - \sqrt{b})^2\right]$$
Giuseppe Peano, 1887

$$C_2 = \pi\left[3(a + b) - \sqrt{(a + 3b)(3a + b)}\right]$$
Srinivasa Ramanujan, 1914

$$C_3 = \frac{\pi}{2}\left[a + b + \sqrt{2(a^2 + b^2)}\right]$$
R. A. Johnson, 1930

Use these formulas to complete the following table of approximations for the circumference of the ellipse $(x^2/5^2) + (y^2/3^2) = 1$. Round off the values of C_1, C_2, and C_3 to six decimal places. To complete the right-hand column of the table, you need two facts. First, the actual circumference of the ellipse, rounded off to six decimal places, is 25.526999. Second, percentage error in an approximation is given by

$$\frac{|\text{true value} - \text{approximation}|}{\text{true value}} \times 100$$

Round off the percentage errors to two significant digits. Which of the three approximations is the best?

	Approximation Obtained	Percentage Error
C_1		
C_2		
C_3		

86. In the following figure, V is the vertex of the parabola $y = ax^2 + bx + c$. If r_1 and r_2 are the roots of the equation $ax^2 + bx + c = 0$, show that

$$VO^2 - VA^2 = r_1 r_2$$

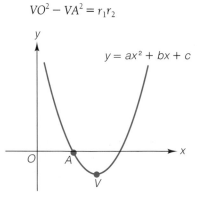

87. The equations of the sides of a triangle are $2px - 2y = qr$, $2qx - 2y = pr$, and $2rx - 2y = pq$.
(a) Find the vertices of the triangle.

(b) Show that the area of the triangle is

$$\frac{|(p - q)(q - r)(r - p)|}{8}$$

88. Suppose that the asymptote $y = (b/a)x$ of the hyperbola $(x^2/a^2) - (y^2/b^2) = 1$ makes an acute angle θ with the positive x-axis. Show that the eccentricity of the hyperbola is sec θ.

CHAPTER TEN TEST

1. Find the focus and the directrix of the parabola $y^2 = -12x$, and sketch the graph.

2. Graph the hyperbola $x^2 - 4y^2 = 4$. Specify the foci and the asymptotes.

3. (a) Determine an angle of rotation θ so that there is no $x'y'$ term present when the equation

   $$x^2 + 2\sqrt{3}xy + 3y^2 - 12\sqrt{3}x + 12y = 0$$

   is transformed to the x'-y' coordinate system.
   (b) Graph the equation

   $$x^2 + 2\sqrt{3}xy + 3y^2 - 12\sqrt{3}x + 12y = 0$$

4. Determine the angle of inclination for the line $y = (1/\sqrt{3})x - 4$.

5. The foci of an ellipse are $(0, \pm 2)$ and the eccentricity is $\frac{1}{2}$. Determine the equation of the ellipse. Write your answer in standard form.

6. Tangents are drawn from the point $(-4, 0)$ to the circle $x^2 + y^2 = 1$. Find the slopes of the tangents.

7. The x-intercept of a line is 2, and its angle of inclination is $60°$. Find the equation of the line. Write your answer in the form $Ax + By + C = 0$.

8. Determine the equation of the hyperbola with foci $(\pm 2, 0)$ and with asymptotes $y = \pm(1/\sqrt{3})x$. Write your answer in standard form.

9. Let F_1 and F_2 denote the foci of the hyperbola $5x^2 - 4y^2 = 80$.
   (a) Verify that the point $P(6, 5)$ lies on the hyperbola.
   (b) Compute the quantity $(F_1P - F_2P)^2$.

10. Graph the ellipse $4x^2 + 25y^2 = 100$. Specify the foci and the lengths of the major and minor axes.

11. Compute the distance from the point $(-1, 0)$ to the line $2x - y - 1 = 0$.

12. Graph the equation $16x^2 + y^2 - 64x + 2y + 65 = 0$.

13. Graph the equation $\dfrac{(x + 4)^2}{3^2} - \dfrac{(y - 4)^2}{1^2} = 1$.

14. Graph the equation $r = 9/(5 - 4 \cos \theta)$. Which type of conic is this?

15. Graph the parabola $(x - 1)^2 = 8(y - 2)$. Specify the focal width and the vertex.

16. Consider the ellipse $(x^2/6^2) + (y^2/5^2) = 1$.
    (a) What are the equations of the directrices?
    (b) If P is a point on the ellipse in the first quadrant such that the x-coordinate of P is 3, compute F_1P and F_2P, the lengths of the focal radii.

17. Determine the equation of the line that is tangent to the ellipse $x^2 + 3y^2 = 52$ at the point $(-2, 4)$. Write your answer in the form $y = mx + b$.

18. Find the equation of the line that is tangent to the parabola $x^2 = 2y$ at the point $(4, 8)$. Write your answer in the form $y = mx + b$.

ROOTS OF POLYNOMIAL EQUATIONS

It is necessary that I make some general statements concerning the nature of equations,....

René Descartes (1596–1650) in *La Géométrie* (1637)

INTRODUCTION

In this chapter we continue the work begun in Section 1.4 on solving polynomial equations. The first two sections of the chapter are preparatory. In Section 11.1 we review the complex number system, and in Section 11.2 we discuss the division process for polynomials. Section 11.3 presents two theorems about polynomials: the remainder theorem and the factor theorem. The remainder theorem and the factor theorem are then applied repeatedly in the next section (Section 11.4) to develop some fundamental results regarding polynomial equations and their solutions. You can view much of the material in this section as a kind of generalization of what you already know about quadratic equations. The last three sections of this chapter present additional results that are useful in actually solving polynomial equations or in determining the nature of their solutions.

11.1 THE COMPLEX NUMBER SYSTEM

In the following I shall denote the expression $\sqrt{-1}$ by the letter i so that ii = −1.

Leonhard Euler in a paper presented to the Saint Petersburg Academy in 1777.

In elementary algebra, complex numbers appear as expressions a + bi, where a and b are ordinary real numbers and $i^2 = −1$.... Complex numbers are manipulated by the usual rules of algebra, with the convention that i^2 is to be replaced by −1 whenever it occurs.

Ralph Boas in *Invitation to Complex Analysis* (New York: Random House, 1987)

The preceding quotation from Professor Boas summarizes the basic approach we will follow in this section. Near the end of the section, after you've become accustomed to working with complex numbers, we'll present a formal list of some of the basic definitions and properties that can be used to develop the subject more rigorously, as is required in more advanced courses.

When we solve equations in this chapter, you'll see instances in which the real-number system proves to be inadequate. In particular, since the square of a real number is never negative, there is no real number x such that $x^2 = −1$. To overcome this inconvenience, mathematicians define the symbol i by the equation

$$i^2 = -1$$

For reasons that are more historical than mathematical, i is referred to as the **imaginary unit**. This name is unfortunate in a sense, because to an engineer or a mathematician, i is neither less "real" nor less tangible than any real number. Having said this, however, we do have to admit that i does not belong to the real-number system.

Algebraically, we operate with the symbol i as if it were any letter in a polynomial expression. However, when we see i^2, we must remember to replace it by -1. Here are four sample calculations involving i.

1. $3i + 2i = 5i$
2. $-2i^2 + 6i = -2(-1) + 6i = 2 + 6i$
3. $(-i)^2 = i^2 = -1$
4. $0i = 0$

An expression of the form $a + bi$, where a and b are real numbers, is called a **complex number**.* Thus, four examples of complex numbers are:

$2 + 3i$

$4 + (-5)i$ (usually written $4 - 5i$)

$1 - \sqrt{2}\, i$ (also written $1 - i\sqrt{2}$)

$\dfrac{1}{2} + \dfrac{3}{2} i$ $\left(\text{also written } \dfrac{1 + 3i}{2}\right)$

Given a complex number $a + bi$, we say that a is the **real part** of $a + bi$, and b is the **imaginary part** of $a + bi$. For example, the real part of $3 - 4i$ is 3, and the imaginary part is -4. Observe that both the real part and the imaginary part of a complex number are themselves real numbers.

We define the notion of *equality* for complex numbers in terms of their real and imaginary parts. Two complex numbers are said to be **equal** if their corresponding real and imaginary parts are equal. We can write this definition symbolically as follows:

$$a + bi = c + di \quad \text{if and only if} \quad a = c \text{ and } b = d$$

▋▋ EXAMPLE 1 Determine the real numbers c and d such that $10 + 4i = 2c + di$.

Solution Equating the real parts of the two complex numbers gives us $2c = 10$, and therefore $c = 5$. Similarly, equating the imaginary parts yields $d = 4$. These are the required values for c and d. ▋▋▋

As the next example indicates, addition, subtraction, and multiplication of complex numbers are carried out using the usual rules of algebra, with the un-

---

*The phrase *complex number* is attributed to Carl Friedrich Gauss (1777–1855). The phrase *imaginary number* originated with René Descartes (1596–1650).

derstanding (as mentioned before) that i^2 is always to be replaced with -1. (We'll discuss division of complex numbers subsequently.)

■ ■ EXAMPLE 2 Let $z = 2 + 5i$ and $w = 3 - 4i$. Compute each of the following.

(a) $w + z$ **(b)** $3z$ **(c)** $w - 3z$ **(d)** zw **(e)** wz

Solution

(a) $w + z = (3 - 4i) + (2 + 5i) = (3 + 2) + (-4i + 5i) = 5 + i$
(b) $3z = 3(2 + 5i) = 6 + 15i$
(c) $w - 3z = (3 - 4i) - (6 + 15i) = (3 - 6) + (-4i - 15i) = -3 - 19i$
(d) $zw = (2 + 5i)(3 - 4i) = 6 - 8i + 15i - 20i^2$
$$= 6 + 7i - 20(-1) = 26 + 7i$$
(e) $wz = (3 - 4i)(2 + 5i) = 6 + 15i - 8i - 20i^2$
$$= 6 + 7i - 20(-1) = 26 + 7i$$ ■■■

If you look over the result of part (a) in Example 2, you can see that the sum is obtained simply by adding the corresponding real and imaginary parts of the given numbers. Likewise in part (c), the difference is obtained by subtracting the corresponding real and imaginary parts. In part (d), however, notice that the product is *not* obtained in a similar fashion. (Exercise 54 provides some perspective on this.) Finally, notice that the results in parts (d) and (e) are identical. In fact, it can be shown that for any two complex numbers z and w, we always have $zw = wz$. In other words, just as with real numbers, multiplication of complex numbers is commutative. Furthermore, along the same lines, it can be shown that all of the properties of real numbers listed in Appendix A.3 continue to hold for complex numbers. We'll return to this point at the end of this section and in the exercises.

As background for the discussion of division of complex numbers, we introduce the notion of a *complex conjugate*, or simply a *conjugate*.*

DEFINITION Complex Conjugate

	EXAMPLES
Let $z = a + bi$. The **complex conjugate** of z, denoted by $\bar{z}$, is defined by $\bar{z} = a - bi$	If $z = 3 + 4i$, then $\bar{z} = 3 - 4i$. If $w = 9 - 2i$, then $\bar{w} = 9 + 2i$.

■ ■ EXAMPLE 3

(a) If $z = 6 - 3i$, compute $z\bar{z}$. **(b)** If $w = a + bi$, compute $w\bar{w}$.

Solution

(a) $z\bar{z} = (6 - 3i)(6 + 3i)$ **(b)** $w\bar{w} = (a + bi)(a - bi)$
    $= 36 + 18i - 18i - 9i^2$ $= a^2 - abi + abi - b^2i^2$
    $= 36 + 9 = 45$ $= a^2 + b^2$

---

*The term "conjugates" (*conjuguées*) was introduced by the nineteenth-century French mathematician A. L. Cauchy in his text *Cours d'Analyse Algébrique* (Paris: 1821).

Note The result in part (b) shows that the product of a complex number and its conjugate is always a real number. ◆ ▮▮▮

Quotients of complex numbers are easy to compute using conjugates. Suppose, for example, that we wish to compute the quotient

$$\frac{5-2i}{3+4i}$$

To do this, take the conjugate of the denominator, namely, $3-4i$, and then multiply the given fraction by $\dfrac{3-4i}{3-4i}$, which equals 1 (assuming that the usual rules of algebra are in force). This yields

$$\frac{5-2i}{3+4i} = \frac{5-2i}{3+4i} \cdot \frac{3-4i}{3-4i}$$

$$= \frac{15-26i+8i^2}{9-16i^2}$$

$$= \frac{7-26i}{25} \qquad \text{since } 8i^2 = -8 \text{ and } -16i^2 = 16$$

$$= \frac{7}{25} - \frac{26}{25}i \qquad \text{as required}$$

In the box that follows, we summarize our procedure for computing quotients. The condition $w \neq 0$ means that w is any complex number other than $0 + 0i$.

Procedure for Computing Quotients

Let z and w be two complex numbers, $w \neq 0$. Then z/w is computed as follows:

$$\frac{z}{w} = \frac{z}{w} \cdot \frac{\overline{w}}{\overline{w}}$$

▮▮▮ **EXAMPLE 4** Let $z = 3 + 4i$ and $w = 1 - 2i$. Compute each of the following quotients.

(a) $\dfrac{1}{z}$ **(b)** $\dfrac{z}{w}$

Solution

(a) $\dfrac{1}{z} = \dfrac{1}{z} \cdot \dfrac{\overline{z}}{\overline{z}} = \dfrac{1}{3+4i} \cdot \dfrac{3-4i}{3-4i}$

$= \dfrac{3-4i}{9-16i^2} = \dfrac{3-4i}{25}$

$= \dfrac{3}{25} - \dfrac{4}{25}i$

(b) $\dfrac{z}{w} = \dfrac{z}{w} \cdot \dfrac{\overline{w}}{\overline{w}} = \dfrac{3+4i}{1-2i} \cdot \dfrac{1+2i}{1+2i}$

$= \dfrac{3+10i+8i^2}{1-4i^2} = \dfrac{-5+10i}{5}$

$= -1 + 2i$ ∎

We began this section by defining i with the equation $i^2 = -1$. This can be rewritten

$$i = \sqrt{-1}$$

provided that we agree to certain conventions regarding principal square roots and negative numbers. In dealing with the principal square root of a negative real number, say, $\sqrt{-5}$, we shall write

$$\sqrt{-5} = \sqrt{(-1)(5)} = \sqrt{-1}\sqrt{5} = i\sqrt{5}$$

In other words, we are allowing the use of the rule $\sqrt{ab} = \sqrt{a}\sqrt{b}$ when a is -1 and b is a positive real number. However, the rule $\sqrt{ab} = \sqrt{a}\sqrt{b}$ *cannot* be used when both a and b are negative. If that were allowed, we could write

$$1 = (-1)(-1)$$

and then

$$\sqrt{1} = \sqrt{(-1)(-1)} = \sqrt{-1}\sqrt{-1} = (i)(i)$$

Consequently,

$$1 = \sqrt{1} = i^2 = -1 \qquad \text{or} \qquad 1 = -1$$

which is a contradiction. Again, the point here is that the rule $\sqrt{ab} = \sqrt{a}\sqrt{b}$ cannot be applied when both a and b are negative.

EXAMPLE 5 Simplify: **(a)** i^4; **(b)** i^{101}; **(c)** $\sqrt{-12} + \sqrt{-27}$; **(d)** $\sqrt{-9}\sqrt{-4}$.

Solution

(a) We make use of the defining equation for i, which is $i^2 = -1$. Thus, we have

$$i^4 = (i^2)^2 = (-1)^2 = 1$$

(The result, $i^4 = 1$, is worth remembering.)

(b) $i^{101} = i^{100}i = (i^4)^{25}i = 1^{25}i = i$

(c) $\sqrt{-12} + \sqrt{-27} = \sqrt{12}\sqrt{-1} + \sqrt{27}\sqrt{-1}$
$= \sqrt{4}\sqrt{3}\,i + \sqrt{9}\sqrt{3}\,i$
$= 2\sqrt{3}\,i + 3\sqrt{3}\,i = 5\sqrt{3}\,i$

(d) $\sqrt{-9}\sqrt{-4} = (3i)(2i) = 6i^2 = -6$

Note: $\sqrt{-9}\sqrt{-4} \neq \sqrt{36}$; why? ∎

Near the beginning of this section, we mentioned that we'd eventually present a formal list of some of the basic definitions and properties that can be used to develop the subject more rigorously. These are given in the two boxes that follow. As you'll see in some of the exercises, these definitions and properties are indeed consistent with our work in this section and with the properties of real numbers listed in Appendix A.3.

DEFINITION **Addition, Subtraction, Multiplication, and Division for Complex Numbers**

Let $z = a + bi$ and $w = c + di$. Then $z + w$, $z - w$, and zw are defined as follows.

1. $z + w = (a + bi) + (c + di) = (a + c) + (b + d)i$

2. $z - w = (a + bi) - (c + di) = (a - c) + (b - d)i$

3. $zw = (a + bi)(c + di) = (ac - bd) + (ad + bc)i$

Furthermore, if $w \neq 0$, then z/w is defined as follows.

4. $\dfrac{z}{w} = \dfrac{a + bi}{c + di} \cdot \dfrac{c - di}{c - di} = \left(\dfrac{ac + bd}{c^2 + d^2}\right) + \left(\dfrac{bc - ad}{c^2 + d^2}\right)i$

PROPERTY SUMMARY **PROPERTIES OF COMPLEX CONJUGATES**

If $z = a + bi$, the complex conjugate, $\bar{z} = a - bi$, has the following properties.

1. $\bar{\bar{z}} = z$

2. $z = \bar{z}$ if and only if z is a real number

3. $\overline{z + w} = \bar{z} + \bar{w}$; $\overline{z - w} = \bar{z} - \bar{w}$

4. $\overline{zw} = \bar{z}\,\bar{w}$; $\dfrac{\bar{z}}{\bar{w}} = \overline{\left(\dfrac{z}{w}\right)}$

5. $(\bar{z})^n = \overline{(z^n)}$ for each natural number n

---

EXERCISE SET 11.1

A

1. Complete the table.

i^2	i^3	i^4	i^5	i^6	i^7	i^8
-1		1	i	-1		1

2. Simplify the following expression, and write the answer in the form $a + bi$.

$$1 + 3i - 5i^2 + 4 - 2i - i^3$$

For Exercises 3 and 4, specify the real and imaginary parts of each complex number.

3. (a) $4 + 5i$ (b) $4 - 5i$
   (c) $\frac{1}{2} - i$ (d) $16i$

4. (a) $-2 + \sqrt{7}i$ (b) $1 + 5^{1/3}i$
   (c) $-3i$ (d) 0

5. Determine the real numbers c and d such that

$$8 - 3i = 2c + di$$

6. Determine the real numbers a and b such that
$$27 - 64i = a^3 - b^3i$$

7. Simplify each of the following.
   (a) $(5 - 6i) + (9 + 2i)$ (b) $(5 - 6i) - (9 + 2i)$

8. If $z = 1 + 4i$, compute $z - 10i$.

9. Compute each of the following.
   (a) $(3 - 4i)(5 + i)$ (b) $(5 + i)(3 - 4i)$
   (c) $\dfrac{3 - 4i}{5 + i}$ (d) $\dfrac{5 + i}{3 - 4i}$

10. Compute each of the following.
   (a) $(2 + 7i)(2 - 7i)$ (b) $\dfrac{-1 + 3i}{2 + 7i}$
   (c) $\dfrac{1}{2 + 7i}$ (d) $\dfrac{1}{2 + 7i} \cdot (-1 + 3i)$

In Exercises 11–36, evaluate each expression using the values $z = 2 + 3i$, $w = 9 - 4i$, and $w_1 = -7 - i$.

11. (a) $z + w$ (b) $\bar{z} + w$ (c) $z + \bar{z}$
12. (a) $\bar{z} + \bar{w}$ (b) $\overline{(z + w)}$ (c) $w - \bar{w}$
13. $(z + w) + w_1$ 14. $z + (w + w_1)$
15. zw 16. wz
17. $z\bar{z}$ 18. $w\bar{w}$
19. $z(ww_1)$ 20. $(zw)w_1$
21. $z(w + w_1)$ 22. $zw + zw_1$
23. $z^2 - w^2$ 24. $(z - w)(z + w)$
25. $(zw)^2$ 26. z^2w^2
27. z^3 28. z^4
29. z/w 30. w/z
31. $\bar{z}/\bar{w}$ 32. $\overline{(z/w)}$
33. $z/\bar{z}$ 34. $\bar{z}/z$
35. $(w - \bar{w})/(2i)$ 36. $(w + \bar{w})/2$

For Exercises 37–40, compute each quotient.

37. $\dfrac{i}{5 + i}$ 38. $\dfrac{1 - i\sqrt{3}}{1 + i\sqrt{3}}$

39. $\dfrac{1}{i}$ 40. $\dfrac{i + i^2}{i^3 + i^4}$

In Exercises 41–48, simplify each expression.

41. $\sqrt{-49} + \sqrt{-9} + \sqrt{-4}$
42. $\sqrt{-25} + i$
43. $\sqrt{-20} - 3\sqrt{-45} + \sqrt{-80}$
44. $\sqrt{-4}\sqrt{-4}$
45. $1 + \sqrt{-36}\sqrt{-36}$ 46. $i - \sqrt{-100}$
47. $3\sqrt{-128} - 4\sqrt{-18}$ 48. $64 + \sqrt{-64}\sqrt{-64}$

49. Let $z = a + bi$ and $w = c + di$. Compute each of the following quantities, and then check that your results agree with the definitions in the first box on page 661.
   (a) $z + w$ (b) $z - w$ (c) zw (d) $\dfrac{z}{w}$

B

50. Show that $\left(\dfrac{-1 + i\sqrt{3}}{2}\right)^2 + \left(\dfrac{-1 - i\sqrt{3}}{2}\right)^2 = -1$.

51. Let $z = \dfrac{-1 + i\sqrt{3}}{2}$ and $w = \dfrac{-1 - i\sqrt{3}}{2}$. Verify the following statements.
   (a) $z^3 = 1$ and $w^3 = 1$ (b) $zw = 1$
   (c) $z = w^2$ and $w = z^2$
   (d) $(1 - z + z^2)(1 + z - z^2) = 4$

52. Let $z = a + bi$ and $w = c + di$.
   (a) Show that $\bar{\bar{z}} = z$.
   (b) Show that $\overline{(z + w)} = \bar{z} + \bar{w}$.

53. Show that the complex number $0 (= 0 + 0i)$ has the following properties.
   (a) $0 + z = z$ and $z + 0 = z$, for all complex numbers z.
      Hint: Let $z = a + bi$.
   (b) $0 \cdot z = 0$ and $z \cdot 0 = 0$, for all complex numbers z.

54. This exercise indicates one of the reasons why multiplication of complex numbers is not carried out simply by multiplying the corresponding real and imaginary parts of the numbers. (Recall that addition and subtraction *are* carried out in this manner.) Suppose for the moment that we were to define multiplication in this seemingly less complicated way:
$$(a + bi)(c + di) = ac + (bd)i \qquad (*)$$
   (a) Compute $(2 + 3i)(5 + 4i)$, assuming that multiplication is defined by $(*)$.
   (b) Still assuming that multiplication is defined by $(*)$, find two complex numbers z and w such that $z \neq 0$, $w \neq 0$, but $zw = 0$ (where 0 denotes the complex number $0 + 0i$).
   Now notice that the result in part (b) is contrary to our expectation or desire that the product of two nonzero numbers should be nonzero, as is the case for real numbers. On the other hand, it can be shown that when multiplication is carried out as described in the text, then the product of two complex numbers is nonzero if and only if both factors are nonzero.

55. (a) Show that addition of complex numbers is commutative. That is, show that $z + w = w + z$ for all complex numbers z and w.   *Hint:* Let $z = a + bi$ and $w = c + di$.
   (b) Show that multiplication of complex numbers is commutative. That is, show that $zw = wz$ for all complex numbers z and w.

56. Let $z = a + bi$.
(a) Show that $(\bar{z})^2 = \overline{(z^2)}$
(b) Show that $(\bar{z})^3 = \overline{(z^3)}$.

C

57. Let a and b be real numbers. Find the real and imaginary parts of the quantity

$$\frac{a + bi}{a - bi} + \frac{a - bi}{a + bi}$$

58. Find the real and imaginary parts of the quantity
$$\left(\frac{a + bi}{a - bi}\right)^2 - \left(\frac{a - bi}{a + bi}\right)^2.$$

59. Find the real part of $\dfrac{(a + bi)^2}{a - bi} - \dfrac{(a - bi)^2}{a + bi}$.

60. (a) Let $\alpha = \left(\dfrac{\sqrt{a^2 + b^2} + a}{2}\right)^{1/2}$ and let

$\beta = \left(\dfrac{\sqrt{a^2 + b^2} - a}{2}\right)^{1/2}$. Show that the square of the complex number $\alpha + \beta i$ is $a + bi$.

(b) Use the result in part (a) to find a complex number z such that $z^2 = i$.

(c) Use the result in part (a) to find a complex number z such that $z^2 = -7 + 24i$.

One of the basic properties of real numbers is that if a product is equal to zero, then at least one of the factors is zero. Exercises 61 and 62 show that this property also holds for complex numbers. For Exercises 61 and 62, assume that $zw = 0$, where $z = a + bi$ and $w = c + di$.

61. If $a \neq 0$, prove that $w = 0$. (That is, prove that $c = d = 0$.)

62. If $b \neq 0$, prove that $w = 0$. (That is, prove that $c = d = 0$.)

11.2 DIVISION OF POLYNOMIALS

Although the process of long division for polynomials is often taught in elementary algebra courses, it usually does not receive sufficient emphasis there. As with ordinary long division for numbers, the process is best learned by first watching someone do examples and then practicing on your own. The terms *quotient, remainder, divisor,* and *dividend* will be used here in the same way they are used in ordinary division of numbers. For instance, when 7 is divided by 2, the quotient is 3 and the remainder is 1. We write this

$$\frac{7}{2} = 3 + \frac{1}{2}$$

or, equivalently,

$$7 = 2 \times 3 + 1$$

dividend, divisor, quotient, remainder

The process of long division for polynomials follows the same four-step cycle used in ordinary long division of numbers: divide, multiply, subtract, bring down. As a first example, we divide $2x^2 - 7x + 8$ by $x - 2$. Notice that in setting up the division, we write both the dividend and the divisor in decreasing powers of x.

$$
\begin{array}{r}
2x - 3 \\
x - 2 \overline{\smash{\big)}\ 2x^2 - 7x + 8} \\
\underline{2x^2 - 4x} \\
-3x + 8 \\
\underline{-3x + 6} \\
2
\end{array}
$$

1. Divide the first term of the dividend by the first term of the divisor: $\dfrac{2x^2}{x} = 2x$. The result becomes the first term of the quotient, as shown.

2. Multiply the divisor $x - 2$ by the term $2x$ obtained in the previous step. This yields the quantity $2x^2 - 4x$, which is written below the dividend, as shown.

3. From the quantity $2x^2 - 7x$ in the dividend, subtract the quantity $2x^2 - 4x$. This yields $-3x$.

4. Bring down the $+8$ in the dividend, as shown. The resulting quantity, $-3x + 8$, is now treated as the dividend and the entire process is repeated.

We've now found that when $2x^2 - 7x + 8$ is divided by $x - 2$, the quotient is $2x - 3$ and the remainder is 2. This is summarized by writing either

$$\frac{2x^2 - 7x + 8}{x - 2} = 2x - 3 + \frac{2}{x - 2} \tag{1}$$

or, after multiplying through by $x - 2$,

$$\underbrace{2x^2 - 7x + 8}_{\text{dividend}} = \underbrace{(x - 2)}_{\text{divisor}} \underbrace{(2x - 3)}_{\text{quotient}} + \underbrace{2}_{\text{remainder}} \tag{2}$$

There are two observations to be made here. First, notice that equation (2) is valid for all real numbers x, whereas equation (1) carries the implicit restriction that x may not equal 2. For this reason, we often prefer to write our results in the form of equation (2). Second, notice that the degree of the remainder is less than the degree of the divisor. This is very similar to the situation with ordinary division of positive integers, where the remainder is always less than the divisor.

As another example of the long division process, we divide $3x^4 - 2x^3 + 2$ by $x^2 - 1$. Notice in what follows that we've inserted the terms in the divisor and dividend that have coefficients of zero. These terms act as place holders.

$$
\begin{array}{r}
3x^2 - 2x + 3 \\
x^2 + 0x - 1 \overline{\smash{\big)}\ 3x^4 - 2x^3 + 0x^2 + 0x + 2} \\
\underline{3x^4 + 0x^3 - 3x^2} \\
-2x^3 + 3x^2 + 0x \\
\underline{-2x^3 + 0x^2 + 2x} \\
3x^2 - 2x + 2 \\
\underline{3x^2 + 0x - 3} \\
-2x + 5
\end{array}
$$

We can write this result as

$$\frac{3x^4 - 2x^3 + 2}{x^2 - 1} = 3x^2 - 2x + 3 + \frac{-2x + 5}{x^2 - 1}$$

or, multiplying through by $x^2 - 1$,

$$\underbrace{3x^4 - 2x^3 + 2}_{\text{dividend}} = \underbrace{(x^2 - 1)}_{\text{divisor}} \underbrace{(3x^2 - 2x + 3)}_{\text{quotient}} + \underbrace{(-2x + 5)}_{\text{remainder}}$$

Notice that this last equation holds for all values of x, whereas the previous equation carries the restrictions that x may be neither 1 nor -1. Also, as in our previous example, observe that the degree of the remainder is less than that of the divisor.

There is a theorem, commonly referred to as the **division algorithm,** that summarizes rather nicely the key results of the long division process. We state the theorem here without proof.

The Division Algorithm

Let $p(x)$ and $d(x)$ be polynomials, and assume that $d(x)$ is not the zero polynomial. Then there are unique polynomials $q(x)$ and $R(x)$ such that

$$p(x) = d(x) \cdot q(x) + R(x)$$

where either $R(x)$ is the zero polynomial or the degree of $R(x)$ is less than the degree of $d(x)$.

The polynomials $p(x)$, $d(x)$, $q(x)$, and $R(x)$ are referred to, respectively, as the dividend, divisor, quotient, and remainder. When $R(x) = 0$, we have $p(x) = d(x) \cdot q(x)$ and we say that $d(x)$ and $q(x)$ are **factors** of $p(x)$. Also, since $d(x)$ is not the zero polynomial, notice that the equation $p(x) = d(x) \cdot q(x) + R(x)$ implies that the degree of $q(x)$ is less than or equal to the degree of $p(x)$. (Why?)

■ ■ ■ **EXAMPLE 1**   Let $p(x) = x^3 + 2x^2 - 4$ and $d(x) = x - 3$. Use the long division process to find the polynomials $q(x)$ and $R(x)$ such that

$$p(x) = d(x) \cdot q(x) + R(x)$$

where either $R(x) = 0$ or the degree of $R(x)$ is less than the degree of $d(x)$.

Solution   After inserting the term $0x$ in the dividend $p(x)$, we use long division to divide $p(x)$ by $d(x)$:

$$
\begin{array}{r}
x^2 + 5x + 15 \\
x - 3 \overline{\smash{\big)}\, x^3 + 2x^2 + 0x - 4} \\
\underline{x^3 - 3x^2} \\
5x^2 + 0x \\
\underline{5x^2 - 15x} \\
15x - 4 \\
\underline{15x - 45} \\
41
\end{array}
$$

We now have

$$\underbrace{x^3 + 2x^2 - 4}_{p(x)} = \underbrace{(x - 3)}_{d(x)}\underbrace{(x^2 + 5x + 15)}_{q(x)} + \underbrace{41}_{R(x)}$$

Thus, $q(x) = x^2 + 5x + 15$ and $R(x) = 41$. Notice that the degree of $R(x)$ is less than the degree of $d(x)$.   ■■■

The long division procedure for polynomials can be streamlined when the divisor is of the form $x - r$. This shortened version, known as **synthetic division,** will be useful in subsequent sections when we are solving polynomial equations.

We can explain the idea behind synthetic division by using the long division carried out in Example 1. The basic idea is that in the long division process, it is the *coefficients* of the various polynomials that carry all the necessary information. In our example, for instance, the quotient and remainder can be abbreviated by writing down a sequence of four numbers:

$$1 \quad 5 \quad 15 \quad 41$$

By studying the long division process, you will find that these numbers are obtained through the following four steps:

STEP 1 Write down the first coefficient of the dividend. This will be the first coefficient of the quotient.

Result $\boxed{1}$

STEP 2 Multiply the 1 obtained in the previous step by the -3 in the divisor. Then subtract the result from the second coefficient of the dividend:

$$-3 \times 1 = -3 \quad\quad 2 - (-3) = 5 \quad\quad\quad\quad\quad Result \quad \boxed{5}$$

STEP 3 Multiply the 5 obtained in the previous step by the -3 in the divisor. Then subtract the result from the third coefficient of the dividend:

$$-3 \times 5 = -15 \quad\quad 0 - (-15) = 15 \quad\quad\quad\quad Result \quad \boxed{15}$$

STEP 4 Multiply the 15 obtained in the previous step by the -3 in the divisor. Then subtract the result from the fourth coefficient of the dividend:

$$-3 \times 15 = -45 \quad\quad -4 - (-45) = 41 \quad\quad\quad Result \quad \boxed{41}$$

A convenient format for setting up this process involves writing the constant term of the divisor and the coefficients of the dividend as follows:

$$\underline{-3|} \quad 1 \quad\quad 2 \quad\quad 0 \quad -4$$

Now, using this format, let us again go through the four steps we have just described:

STEP 1 Bring down the 1.

$$
\begin{array}{r|rrrr}
-3 & 1 & 2 & 0 & -4 \\
\hline
& 1 & & &
\end{array}
$$

STEP 2 $-3 \times 1 = -3$
$2 - (-3) = 5$

$$
\begin{array}{r|rrrr}
-3 & 1 & 2 & 0 & -4 \\
& & -3 & & \\
\hline
& 1 & 5 & &
\end{array}
$$

STEP 3 $-3 \times 5 = -15$
$0 - (-15) = 15$

$$
\begin{array}{r|rrrr}
-3 & 1 & 2 & 0 & -4 \\
& & -3 & -15 & \\
\hline
& 1 & 5 & 15 &
\end{array}
$$

STEP 4 $-3 \times 15 = -45$
$-4 - (-45) = 41$

$$
\begin{array}{r|rrrr}
-3 & 1 & 2 & 0 & -4 \\
& & -3 & -15 & -45 \\
\hline
& 1 & 5 & 15 & 41
\end{array}
$$

Although we have now obtained the required sequence of numbers, 1 5 15 41, there is one further simplification that can be made. In Steps 2 through 4, we can add instead of subtract if we use 3 instead of -3 in the initial format. (You will see the motivation for this in the next section when we discuss the remainder theorem.) With this change, let us now summarize the technique of synthetic division using the example with which we've been working. The method is applicable for any polynomial division where the divisor has the form $x - a$.

To Divide $x^3 + 2x^2 - 4$ by $x - 3$ Using Synthetic Division

		COMMENTS
Format	$\underline{3\rfloor\ 1\quad 2\quad\ \ 0\quad -4}$	Since the divisor is $x - 3$, the format begins with 3. The coefficients from the dividend are written in the order corresponding to decreasing powers of x. A zero coefficient is inserted as a place holder.
Procedure	$\begin{array}{r} 3\rfloor\ 1\quad 2\quad\ \ 0\quad -4 \\ \underline{\quad 3\quad 15\quad 45} \\ 1\quad 5\quad 15\quad 41 \end{array}$	Step 1: Bring down the 1. Step 2: $3 \times 1 = 3$; $2 + 3 = 5$ Step 3: $3 \times 5 = 15$; $0 + 15 = 15$ Step 4: $3 \times 15 = 45$; $-4 + 45 = 41$
Answer	Quotient: $x^2 + 5x + 15$ Remainder: 41	The degree of the first term in the quotient is one less than the degree of the first term of the dividend.

◼▦◼ EXAMPLE 2 Use synthetic division to divide $x^3 - 6x + 4$ by $x - 2$.

Solution $\begin{array}{r} 2\rfloor\ 1\quad\ \ 0\quad -6\quad\ \ 4 \\ \underline{\quad\ 2\quad\ \ 4\quad -4} \\ 1\quad\ \ 2\quad -2\quad\ \ 0 \end{array}$

Looking at the third row of numbers in the synthetic division we've carried out, we see that the quotient is $x^2 + 2x - 2$ and the remainder is 0. In other words, both $x - 2$ and $x^2 + 2x - 2$ are factors of $x^3 - 6x + 4$, and we have

$$x^3 - 6x + 4 = (x - 2)(x^2 + 2x - 2)$$

◼◼◼

◼▦◼ EXAMPLE 3 Use synthetic division to divide $x^5 - a^5$ by $x - a$.

Solution $\begin{array}{r} a\rfloor\ 1\quad\ \ 0\quad\ \ 0\quad\ \ 0\quad\ \ 0\quad -a^5 \\ \underline{\quad\ a\quad a^2\quad a^3\quad a^4\quad a^5} \\ 1\quad\ \ a\quad a^2\quad a^3\quad a^4\quad\ \ 0 \end{array}$

As before, we read off the quotient and remainder from the third row of numbers. The quotient is

$$x^4 + ax^3 + a^2x^2 + a^3x + a^4$$

and the remainder is zero. So we have

$$\underbrace{x^5 - a^5}_{\text{dividend}} = \underbrace{(x - a)}_{\text{divisor}}\underbrace{(x^4 + ax^3 + a^2x^2 + a^3x + a^4)}_{\text{quotient}} + \underbrace{0}_{\text{remainder}}$$

or

$$x^5 - a^5 = (x - a)(x^4 + ax^3 + a^2x^2 + a^3x + a^4)$$ ▮▮▮

The last equation in Example 3 tells us how to factor $x^5 - a^5$. One factor is $x - a$. Notice the pattern in the second factor $x^4 + ax^3 + a^2x^2 + a^3x + a^4$.

first term: a^0x^{5-1}

second term: a^1x^{5-2}

third term: a^2x^{5-3}

fourth term: a^3x^{5-4}

fifth term: a^4x^{5-5}

In the same way that we've found a factorization for $x^5 - a^5$, we can find a factorization for $x^n - a^n$ for any positive integer $n \geq 2$. In each case, the first factor is $x - a$, while the second factor follows the same pattern just described for $x^5 - a^5$. We state the general result in the box that follows. The result can be proved by using *mathematical induction* (discussed in Chapter 12) or by using the *remainder theorem* (discussed in the next section).

Factorization of $x^n - a^n$

$$x^n - a^n = (x - a)(x^{n-1} + ax^{n-2} + a^2x^{n-3} + \cdots + a^{n-1})$$

In our development of synthetic division, we assumed that the form of the divisor was $x - r$. The next example shows what to do when the form of the divisor is $x + r$.

▮▮▮ **EXAMPLE 4**   Use synthetic division to divide $x^4 - 2x^3 + 5x^2 - 4x + 3$ by $x + 1$.

Solution   We first need to write the divisor $x + 1$ in the form $x - r$. We have

$$x + 1 = x - (-1)$$

In other words, r is -1, and this is the value we use to set up the synthetic division format. The format then is

$$\underline{-1|} \quad 1 \quad -2 \quad 5 \quad -4 \quad 3$$

Now we carry out the synthetic division procedure:

$$
\begin{array}{r|rrrrr}
-1 & 1 & -2 & 5 & -4 & 3 \\
   &   & -1 & 3 & -8 & 12 \\
\hline
   & 1 & -3 & 8 & -12 & 15
\end{array}
$$

The quotient is therefore $x^3 - 3x^2 + 8x - 12$, and the remainder is 15. We can summarize this result by writing

$$x^4 - 2x^3 + 5x^2 - 4x + 3 = (x + 1)(x^3 - 3x^2 + 8x - 12) + 15$$

(Notice that the degree of the remainder is less than the degree of the divisor, in agreement with the division algorithm.)   ▮▮▮

EXERCISE SET 11.2

A

In Exercises 1–20, use long division to find the quotient and remainder. Also, write each answer in the form $p(x) = d(x) \cdot q(x) + R(x)$, as in equation (2) in the text.

1. $\dfrac{x^2 - 8x + 4}{x - 3}$

2. $\dfrac{x^3 - 4x^2 + x - 2}{x - 5}$

3. $\dfrac{x^2 - 6x - 2}{x + 5}$

4. $\dfrac{3x^2 + 4x - 1}{x - 1}$

5. $\dfrac{6x^3 - 2x + 3}{2x + 1}$

6. $\dfrac{x^4 - 4x^3 + 6x^2 - 4x + 1}{x - 1}$

7. $\dfrac{x^5 + 2}{x + 3}$

8. $\dfrac{4x^3 - x^2 + 8x - 1}{x^2 - x + 1}$

9. $\dfrac{x^6 - 64}{x - 2}$

10. $\dfrac{x^6 + 64}{x - 2}$

11. $\dfrac{5x^4 - 3x^2 + 2}{x^2 - 3x + 5}$

12. $\dfrac{8x^6 - 36x^4 + 54x^2 - 27}{2x^2 - 3}$

13. $\dfrac{3y^3 - 4y^2 - 3}{y^2 + 5y + 2}$

14. $\dfrac{4y^4 - y^3 + 2y - 1}{2y^2 - 3y - 4}$

15. $\dfrac{t^4 - 4t^3 + 4t^2 - 16}{t^2 - 2t + 4}$

16. $\dfrac{2t^5 - 6t^4 - t^2 + 2t + 3}{t^3 - 2}$

17. $\dfrac{z^5 - 1}{z - 1}$

18. $\dfrac{1 + z + z^2 + z^3}{1 + z + z^2}$

19. $\dfrac{ax^2 + bx + c}{x - r}$

20. $\dfrac{ax^3 + bx^2 + cx + d}{x - r}$

In Exercises 21–40, use synthetic division to find the quotients and remainders. Also, in each case, write the result of the division in the form $p(x) = d(x) \cdot q(x) + R(x)$, as in equation (2) in the text.

21. $\dfrac{x^2 - 6x - 2}{x - 5}$

22. $\dfrac{3x^2 + 4x - 1}{x - 1}$

23. $\dfrac{4x^2 - x - 5}{x + 1}$

24. $\dfrac{x^2 - 1}{x + 2}$

25. $\dfrac{6x^3 - 5x^2 + 2x + 1}{x - 4}$

26. $\dfrac{x^4 - 4x^3 + 6x^2 - 4x + 1}{x - 1}$

27. $\dfrac{x^3 - 1}{x - 2}$

28. $\dfrac{x^3 - 8}{x - 2}$

29. $\dfrac{x^5 - 1}{x + 2}$

30. $\dfrac{x^3 - 8x^2 - 1}{x + 3}$

31. $\dfrac{x^4 - 6x^3 + 2}{x + 4}$

32. $\dfrac{3x^3 - 2x^2 + x + 1}{x - \frac{1}{2}}$

33. $\dfrac{x^3 - 4x^2 - 3x + 6}{x - 10}$

34. $\dfrac{1 + 3x + 3x^2 + x^3}{x + 1}$

35. $\dfrac{x^3 - x^2}{x + 5}$

36. $\dfrac{5x^4 - 4x^3 + 3x^2 - 2x + 1}{x + \frac{1}{2}}$

37. $\dfrac{14 - 27x - 27x^2 + 54x^3}{x - \frac{2}{3}}$

38. $\dfrac{14 - 27x - 27x^2 + 54x^3}{x + \frac{2}{3}}$

39. $\dfrac{x^4 + 3x^2 + 12}{x - 3}$

40. (a) $\dfrac{x^4 - 16}{x - 2}$ (b) $\dfrac{x^4 + 16}{x + 2}$

In Exercises 41–44, each expression has the form $x^n - a^n$. Write each expression as a product of two factors (as in the box on page 668).

41. $x^5 - 32$

42. $y^6 - 1$

43. $z^4 - 81$

44. $x^7 - y^7$

B

In Exercises 45–48, use synthetic division to determine the quotient $q(x)$ and the remainder $R(x)$ in each case.

45. $\dfrac{6x^2 - 8x + 1}{3x - 4}$ *Hint:*   Divide both numerator and denominator by 3. (Why?)

46. $\dfrac{4x^3 + 6x^2 - 6x - 5}{2x - 3}$

47. $\dfrac{6x^3 + 1}{2x + 1}$

48. $\dfrac{5x^3 - 3x^2 + 1}{3x + 1}$

49. When $x^3 + kx + 1$ is divided by $x + 1$, the remainder is -4. Find k.

50. (a) Show that when $x^3 + kx + 6$ is divided by $x + 3$, the remainder is $-21 - 3k$.
    (b) Determine a value of k such that $x + 3$ will be a factor of $x^3 + kx + 6$.

51. When $x^2 + 2px - 3q^2$ is divided by $x - p$, the remainder is zero. Show that $p^2 = q^2$.

52. Given that $x - 3$ is a factor of $x^3 - 2x^2 - 4x + 3$, solve the equation $x^3 - 2x^2 - 4x + 3 = 0$.

The process of synthetic division applies equally well when some or all of the coefficients are nonreal complex numbers. In Exercises 53–56, use synthetic division to determine the quotient $q(x)$ and the remainder $R(x)$ in each case.

53. $\dfrac{x^2 - 4x + 1}{x - i}$

54. $\dfrac{x^3 - 2x^2 - 4}{x - 3i}$

55. $\dfrac{x^2 - 2x + 2}{x - (1 + i)}$

56. $\dfrac{x^3 - x^2 + 4x - 4}{x + 2i}$

57. Given that the identity $f(x) = d(x) \cdot q(x) + R(x)$ holds for the following polynomials,

$$f(x) = 2x^5 + 5x^4 - 8x^3 + 7x^2 - 9 \qquad d(x) = x^2 - 3$$
$$q(x) = 2x^3 + 5x^2 - 2x + 22 \qquad R(x) = -6x + 57$$

evaluate $f(\sqrt{3})$. *Hint (of sorts):* There's an easy way and a tedious way.

58. Given that the identity $f(t) = d(t) \cdot q(t) + R(t)$ holds for the following polynomials,

$$f(t) = t^5 - 3t^4 + 2t^3 - 5t^2 + 6t - 7 \qquad d(t) = t - 4$$
$$q(t) = t^4 + t^3 + 6t^2 + 19t + 82 \qquad R(t) = 321$$

evaluate $f(4)$.

59. Find the remainder when $t^5 - 5a^4t + 4a^5$ is divided by $t - a$.

60. When $f(x)$ is divided by $(x - a)(x - b)$, the remainder is $Ax + B$. Apply the division algorithm to show that

$$A = \frac{f(a) - f(b)}{a - b} \qquad \text{and} \qquad B = \frac{bf(a) - af(b)}{b - a}$$

11.3 THE REMAINDER THEOREM AND THE FACTOR THEOREM

Descartes recommended [in his La Géométrie of 1637] that all terms [of an equation] should be taken to one side and equated with zero. Though he was not the first to suggest this, he was the earliest writer to realize the advantage to be gained. He pointed out that a polynomial $f(x)$ was divisible by $(x - a)$ if and only if a was a root of $f(x)$.

David M. Burton in *The History of Mathematics, An Introduction,* 2nd ed. (Dubuque, Iowa: Wm. C. Brown Publishers, 1991)

The techniques for solving polynomial equations of degree 2 were discussed in Section 1.4. Now we want to extend those ideas. Our focus in this section and in the remainder of the chapter is on solving polynomial equations of any degree, that is, equations of the form

$$f(x) = a_n x^n + a_{n-1} x^{n-1} + \cdots + a_1 x + a_0 = 0 \tag{1}$$

Here, as in Chapter 1, a **root** or **solution** of equation (1) is a number r that when substituted for x leads to a true statement. Thus, r is a root of equation (1) provided that $f(r) = 0$. We also refer to the number r in this case as a **zero** of the function f.

▌▌▌ EXAMPLE 1

(a) Is -3 a zero of the function f defined by $f(x) = x^4 + x^2 - 6$?
(b) Is $\sqrt{2}$ a root of the equation $x^4 + x^2 - 6 = 0$?

Solution

(a) By definition, -3 will be a zero of f if $f(-3) = 0$. We have

$$f(-3) = (-3)^4 + (-3)^2 - 6 = 81 + 9 - 6 = 84 \neq 0$$

Thus, -3 is not a zero of the function f.

(b) To check if $\sqrt{2}$ is a root of the given equation, we have

$$(\sqrt{2})^4 + (\sqrt{2})^2 - 6 \overset{?}{=} 0$$
$$4 + 2 - 6 \overset{?}{=} 0$$
$$0 \overset{?}{=} 0 \qquad \text{True}$$

Thus, $\sqrt{2}$ is a root of the equation $x^4 + x^2 - 6 = 0$. ∎

There are cases in which a root of an equation is what we call a **repeated root.** Consider, for instance, the equation $x(x-1)(x-1) = 0$. We have

$$x(x-1)(x-1) = 0$$

$$x = 0 \quad | \quad x - 1 = 0 \quad | \quad x - 1 = 0$$
$$\qquad\qquad x = 1 \qquad\qquad x = 1$$

The roots of the equation are therefore 0, 1, and 1. The repeated root here is $x = 1$. We say in this case that 1 is a **double root** or, equivalently, that 1 is a **root of multiplicity 2.** More generally, if a root is repeated k times, we call it a **root of multiplicity k.**

■■■ **EXAMPLE 2** State the multiplicity of each root of the equation

$$(x-4)^2(x-5)^3 = 0$$

Solution We have $(x-4)(x-4)(x-5)(x-5)(x-5) = 0$. By setting each factor equal to zero, we obtain the roots 4, 4, 5, 5, and 5. From this we see that the root 4 has multiplicity 2, while the root 5 has multiplicity 3. Notice that it is not really necessary to write out all the factors as we did here; the exponents of the factors in the original equation give us the required multiplicities. ∎

There are two simple but important theorems that will form the basis for much of our subsequent work with polynomials: the *remainder theorem* and the *factor theorem.* We begin with a statement of the remainder theorem.

The Remainder Theorem

When a polynomial $f(x)$ is divided by $x - r$, the remainder is $f(r)$.

Before turning to a proof of the remainder theorem, let's see what the theorem is saying in two particular cases. First, suppose that we divide the polynomial $f(x) = 2x^2 - 3x + 4$ by $x - 1$. Then according to the remainder theorem, the remainder in this case should be the number $f(1)$. Let's check:

$$
\begin{array}{r|rrr}
1 & 2 & -3 & 4 \\
  &   & 2 & -1 \\
\hline
  & 2 & -1 & ③
\end{array}
\qquad
\begin{aligned}
f(x) &= 2x^2 - 3x + 4 \\
f(1) &= 2(1)^2 - 3(1) + 4 \\
f(1) &= ③
\end{aligned}
$$

As the calculations show, the remainder is indeed equal to $f(1)$. As a second example, let us divide the polynomial $g(x) = ax^2 + bx + c$ by $x - r$. According to the remainder theorem, the remainder should be $g(r)$. Again, let us check:

$$
\begin{array}{r|rrr}
r & a & b & c \\
  &   & ar & ar^2 + br \\
\hline
  & a & ar+b & \boxed{ar^2 + br + c}
\end{array}
\qquad
\begin{aligned}
g(x) &= ax^2 + bx + c \\
\\
g(r) &= \boxed{ar^2 + br + c}
\end{aligned}
$$

The calculations show that the remainder is equal to $g(r)$, as we wished to check.

In the example just concluded, we verified that the remainder theorem holds for any quadratic polynomial $g(x) = ax^2 + bx + c$. A general proof of the remainder theorem can easily be given along these same lines. The only drawback is that it becomes slightly cumbersome to carry out the synthetic division process when the dividend is

$$a_n x^n + a_{n-1} x^{n-1} + \cdots + a_1 x + a_0$$

For this reason, mathematicians often prefer to base the proof of the remainder theorem on the division algorithm given in the previous section. This is the path we will follow here.

To prove the remainder theorem, we must show that when the polynomial $f(x)$ is divided by $x - r$, the remainder is $f(r)$. Now according to the division algorithm, we can write

$$f(x) = (x - r) \cdot q(x) + R(x) \qquad (2)$$

In this identity, either $R(x)$ is the zero polynomial or the degree of $R(x)$ is less than that of $x - r$. But since the degree of $x - r$ is 1, we must have in this case that the degree of $R(x)$ is zero. Thus, in *either* case, the remainder $R(x)$ is a constant. Denoting this constant by c, we can rewrite equation (2) as

$$f(x) = (x - r) \cdot q(x) + c$$

Now if we set $x = r$ in this identity, we obtain

$$f(r) = (r - r) \cdot q(r) + c = c = R(x)$$

We have now shown that $f(r) = c$. But by definition, c is the remainder $R(x)$. Thus, $f(r) = R(x)$. This proves the remainder theorem.

▮▮▮ EXAMPLE 3   Let $f(x) = 2x^3 - 5x^2 + x - 6$.

(a) Use the remainder theorem to evaluate $f(3)$.
(b) Is $x - 3$ a factor of $f(x) = 2x^3 - 5x^2 + x - 6$?

Solution

(a) According to the remainder theorem, $f(3)$ is the remainder when $f(x)$ is divided by $x - 3$. Using synthetic division, we have

$$
\begin{array}{r|rrrr}
3 & 2 & -5 & 1 & -6 \\
  &   & 6 & 3 & 12 \\
\hline
  & 2 & 1 & 4 & 6
\end{array}
$$

The remainder is 6, and therefore $f(3) = 6$.

(b) By definition, $x - 3$ is a factor of $f(x)$ if we obtain a zero remainder when $f(x)$ is divided by $x - 3$. But from our work in part (a), we know that the remainder is 6, not 0. So, $x - 3$ is not a factor of $f(x)$.   ▮▮▮

▮▮▮ EXAMPLE 4   In the previous section it was stated, but not proved, that $x - a$ is a factor of $f(x) = x^n - a^n$. Use the remainder theorem to prove this fact.

Solution   We need to show that when $f(x) = x^n - a^n$ is divided by $x - a$, the remainder is 0. By the remainder theorem, the remainder is equal to $f(a)$, which is easy to find:

$$f(a) = a^n - a^n = 0$$

Thus, the remainder is zero, and $x - a$ is a factor of $x^n - a^n$, as required.   ▮▮▮

From our experience with quadratic equations, we know that there is a close connection between factoring a quadratic polynomial $f(x)$ and solving the polynomial equation $f(x) = 0$. The factor theorem states this relationship between roots and factors in a precise form. Furthermore, the factor theorem tells us that this relationship holds for polynomials of all degrees, not just quadratics.

The Factor Theorem
Let $f(x)$ be a polynomial. If $f(r) = 0$, then $x - r$ is a factor of $f(x)$. Conversely, if $x - r$ is a factor of $f(x)$, then $f(r) = 0$.

In terms of roots, we can summarize the factor theorem by saying that r is a root of the equation $f(x) = 0$ if and only if $x - r$ is a factor of $f(x)$. To prove the factor theorem, let us begin by assuming that $f(r) = 0$. We want to show that $x - r$ is a factor of $f(x)$. Now, according to the remainder theorem, if $f(x)$ is divided by $x - r$, the remainder is $f(r)$. So we can write

$$f(x) = (x - r) \cdot q(x) + f(r) \qquad \text{for some polynomial } q(x)$$

But since $f(r)$ is zero, this equation becomes

$$f(x) = (x - r) \cdot q(x)$$

This last equation tells us that $x - r$ is a factor of $f(x)$, as we wished to prove.

Now, conversely, let us assume that $x - r$ is a factor of $f(x)$. We want to show that $f(r) = 0$. Since $x - r$ is a factor of $f(x)$, we can write

$$f(x) = (x - r) \cdot q(x) \qquad \text{for some polynomial } q(x)$$

If we now let $x = r$ in this last equation, we obtain

$$f(r) = (r - r) \cdot q(r) = 0$$

as we wished to show.

The example that follows indicates how the factor theorem can be used to solve equations.

▮▮▮▮ **EXAMPLE 5**   Solve the equation $x^3 - 2x + 1 = 0$, given that one root is $x = 1$.

Solution   Since $x = 1$ is a root, the factor theorem tells us that $x - 1$ is a factor of $x^3 - 2x + 1$. In other words,

$$x^3 - 2x + 1 = (x - 1) \cdot q(x) \qquad \text{for some polynomial } q(x)$$

To determine this other factor $q(x)$, we divide $x^3 - 2x + 1$ by $x - 1$:

$$
\begin{array}{r|rrrr}
1 & 1 & 0 & -2 & 1 \\
  &   & 1 & 1 & -1 \\
\hline
  & 1 & 1 & -1 & 0 \\
\end{array}
$$

Thus, $q(x) = x^2 + x - 1$ and we have the factorization

$$x^3 - 2x + 1 = (x - 1)(x^2 + x - 1)$$

Using this identity, the original equation becomes

$$(x - 1)(x^2 + x - 1) = 0$$

Now the problem is reduced to solving the linear equation $x - 1 = 0$ and the quadratic equation $x^2 + x - 1 = 0$. The linear equation yields $x = 1$, but we already knew that 1 was a root. So if we are to find any additional roots, they must come from the equation $x^2 + x - 1 = 0$. Using the quadratic formula, we obtain

$$x = \frac{-1 \pm \sqrt{(1)^2 - 4(1)(-1)}}{2(1)} = \frac{-1 \pm \sqrt{5}}{2}$$

We now have three roots of the cubic equation $x^3 - 2x + 1 = 0$. They are 1, $\frac{1}{2}(-1 + \sqrt{5})$, and $\frac{1}{2}(-1 - \sqrt{5})$. As you will see in the next section, a cubic equation can have at most three roots. So in this case, we have determined all the roots; that is, we have solved the equation. ■■■

Before going on to other examples, let's take a moment to summarize the technique we used in Example 5. We want to solve a polynomial equation $f(x) = 0$, given that one root is $x = r$. Since r is a root, the factor theorem tells us that $x - r$ is a factor of $f(x)$. Then, with the aid of synthetic division, we obtain a factorization

$$f(x) = (x - r) \cdot q(x) \qquad \text{for some polynomial } q(x)$$

This gives rise to the two equations, $x - r = 0$ and $q(x) = 0$. Since the first of these only reasserts that $x = r$ is a root, we try to solve the second equation, $q(x) = 0$. We refer to the equation $q(x) = 0$ as the **reduced equation.** Example 5 showed you the idea behind this terminology; the degree of $q(x)$ is one less than that of $f(x)$. If, as in Example 5, the reduced equation happens to be a quadratic equation, then we can always determine the remaining roots by factoring or by the quadratic formula. In subsequent sections, we will look at techniques that are helpful in cases where $q(x)$ is not quadratic.

■ ▨ ▨ **EXAMPLE 6** Solve the equation $x^4 + 2x^3 - 7x^2 - 20x - 12 = 0$, given that $x = 3$ and $x = -2$ are roots.

Solution Since $x = 3$ is a root, the factor theorem tells us that $x - 3$ is a factor of the polynomial $x^4 + 2x^3 - 7x^2 - 20x - 12$. That is,

$$x^4 + 2x^3 - 7x^2 - 20x - 12 = (x - 3) \cdot q(x) \qquad \text{for some polynomial } q(x)$$

As in Example 5, we can find $q(x)$ by synthetic division. We have

$$
\begin{array}{r|rrrrr}
3 & 1 & 2 & -7 & -20 & -12 \\
  &   & 3 & 15 & 24 & 12 \\
\hline
  & 1 & 5 & 8 & 4 & 0
\end{array}
$$

Thus, $q(x) = x^3 + 5x^2 + 8x + 4$, and our original equation is equivalent to

$$(x - 3)(x^3 + 5x^2 + 8x + 4) = 0$$

Now, $x = -2$ is also a root of this equation. But $x = -2$ surely is not a root of the equation $x - 3 = 0$. Therefore, $x = -2$ must be a root of the reduced equation $x^3 + 5x^2 + 8x + 4 = 0$. Again the factor theorem is applicable. Since $x = -2$ is a root of the reduced equation, $x + 2$ must be a factor of $x^3 + 5x^2 + 8x + 4$. That is,

$$x^3 + 5x^2 + 8x + 4 = (x + 2) \cdot p(x) \qquad \text{for some polynomial } p(x)$$

We can use synthetic division to determine $p(x)$:

$$
\begin{array}{r|rrrr}
-2 & 1 & 5 & 8 & 4 \\
   &   & -2 & -6 & -4 \\
\hline
   & 1 & 3 & 2 & 0
\end{array}
$$

Thus, $p(x) = x^2 + 3x + 2$, and our reduced equation can be written

$$(x + 2)(x^2 + 3x + 2) = 0$$

This gives rise to a second reduced equation.

$$x^2 + 3x + 2 = 0$$

In this case, the roots are readily obtained by factoring. We have

$$x^2 + 3x + 2 = 0$$
$$(x + 2)(x + 1) = 0$$

$$
\begin{array}{c|c}
x + 2 = 0 & x + 1 = 0 \\
x = -2 & x = -1
\end{array}
$$

Now, of the two roots we've just found, the -2 happens to be one of the roots that we were initially given in the statement of the problem. The -1, on the other hand, is a distinct additional root. In summary, then, we have three distinct roots: 3, -2, and -1. The root -2 has multiplicity 2. As you will see in the next section, a fourth-degree equation can have at most four (not necessarily distinct) roots. So in the case at hand, we have found all the roots of the given equation.   ■■■

■■■ **EXAMPLE 7**   In each case, find a polynomial equation $f(x) = 0$ satisfying the given conditions. If there is no such equation, say so.

(a) The numbers -1, 4, and 5 are roots.
(b) A factor of $f(x)$ is $x - 3$, and -4 is a root of multiplicity 2.
(c) The degree of f is 4, the number -5 is a root of multiplicity 3, and 6 is a root of multiplicity 2.

Solution

(a) If $f(x)$ is any polynomial containing the factors $(x + 1)$, $(x - 4)$, and $(x - 5)$, then the equation $f(x) = 0$ will certainly be satisfied when $x = -1$, $x = 4$, or $x = 5$. The simplest polynomial equation in this case is therefore

$$(x + 1)(x - 4)(x - 5) = 0$$

This is a polynomial equation with the required roots. If required, we can carry out the multiplication on the left-hand side of the equation. As you can check, this yields $x^3 - 8x^2 + 11x + 20 = 0$.

(b) According to the factor theorem, since -4 is a root, $x + 4$ must be a factor of $f(x)$. In fact, since -4 is a root of multiplicity 2, the quantity $(x + 4)^2$ must be a factor of $f(x)$. The following equation therefore satisfies the given conditions:

$$(x - 3)(x + 4)^2 = 0$$

(c) We are given two roots, one with multiplicity 3, the other with multiplicity 2. Thus the degree of $f(x)$ must be at least $3 + 2 = 5$. (Why?) But this then contradicts the given condition that the degree of f should be 4. Consequently, there is no polynomial equation that satisfies the given conditions. ▪▪▪

EXERCISE SET 11.3

A

In Exercises 1–6, determine whether the given value for the variable is a root of the equation.

1. $12x - 8 = 112$; $x = 10$

2. $12x^2 - x - 20 = 0$; $x = \frac{5}{4}$

3. $x^2 - 2x - 4 = 0$; $x = 1 - \sqrt{5}$

4. $1 - x + x^2 - x^3 = 0$; $x = -1$

5. $2x^2 - 3x + 1 = 0$; $x = \frac{1}{2}$

6. $(x - 1)(x - 2)(x - 3) = 0$; $x = 4$

In Exercises 7–14, determine whether the given value is a zero of the function.

7. $f(x) = 3x - 2$; $x = \frac{2}{3}$

8. $g(x) = 1 + x^2$; $x = -1$

9. $h(x) = 5x^3 - x^2 + 2x + 8$; $x = -1$

10. $F(x) = -2x^5 + 3x^4 + 8x^3$; $x = 0$

11. $f(t) = 1 + 2t + t^3 - t^5$; $t = 2$

12. $f(t) = 1 + 2t + t^3 - t^5$; $t = \sqrt{2}$

13. $f(x) = 2x^3 - 3x + 1$: (a) $x = \frac{1}{2}(\sqrt{3} - 1)$; (b) $x = \frac{1}{2}(\sqrt{3} + 1)$

14. $g(x) = x^4 + 8x^3 + 9x^2 - 8x - 10$: (a) $x = 1$; (b) $x = \sqrt{6} - 4$; (c) $x = \sqrt{6} + 4$

15. List the distinct roots of each of the following equations. In the case of a repeated root, give its multiplicity.
    (a) $(x - 1)(x - 2)^3(x - 3) = 0$
    (b) $(x - 1)(x - 1)(x - 1) = 0$
    (c) $(x - 5)^6(x + 1)^4 = 0$ (d) $x^5(x - 1) = 0$

16. In this exercise, we verify that the remainder theorem is valid for the cubic polynomial
    $$g(x) = ax^3 + bx^2 + cx + d$$

(a) Compute $g(r)$.
(b) Using synthetic division, divide $g(x)$ by $x - r$. Check that the remainder you obtain is the same as the answer in part (a).

In Exercises 17–24, use the remainder theorem (as in Example 3) to evaluate $f(x)$ for the given value of x.

17. $f(x) = 4x^3 - 6x^2 + x - 5$; $x = -3$

18. $f(x) = 2x^3 - x - 4$; $x = 4$

19. $f(x) = 6x^4 + 5x^3 - 8x^2 - 10x - 3$; $x = \frac{1}{2}$

20. $f(x) = x^5 - x^4 - x^3 - x^2 - x - 1$; $x = -2$

21. $f(x) = x^2 + 3x - 4$; $x = -\sqrt{2}$

22. $f(x) = -3x^3 + 8x^2 - 12$; $x = 5$

23. $f(x) = \frac{1}{2}x^3 - 5x^2 - 13x - 10$; $x = 12$

24. $f(x) = x^7 - 7x^6 + 5x^4 + 1$; $x = -3$

In Exercises 25–38, you are given a polynomial equation and one or more roots. Solve each equation using the method shown in Examples 5 and 6. To help you decide if you have found all the roots in each case, you may rely on the following theorem, discussed in the next section: A polynomial equation of degree n has at most n (not necessarily distinct) roots.

25. $x^3 - 4x^2 - 9x + 36 = 0$; -3 is a root.

26. $x^3 + 7x^2 + 11x + 5 = 0$; -1 is a root.

27. $x^3 + x^2 - 7x + 5 = 0$; 1 is a root.

28. $x^3 + 8x^2 - 3x - 24 = 0$; -8 is a root.

29. $3x^3 - 5x^2 - 16x + 12 = 0$; -2 is a root.

30. $2x^3 - 5x^2 - 46x + 24 = 0$; 6 is a root.

31. $2x^3 + x^2 - 5x - 3 = 0$; $-\frac{3}{2}$ is a root.

32. $6x^4 - 19x^3 - 25x^2 + 18x + 8 = 0$; 4 and $-\frac{1}{3}$ are roots.

33. $x^4 - 15x^3 + 75x^2 - 125x = 0$; 5 is a root.

34. $2x^3 + 5x^2 - 8x - 20 = 0$; 2 is a root.

35. $x^4 + 2x^3 - 23x^2 - 24x + 144 = 0$; -4 and 3 are roots.

36. $6x^5 + 5x^4 - 29x^3 - 25x^2 - 5x = 0$; $\sqrt{5}$ and $-\frac{1}{3}$ are roots.

37. $x^3 + 7x^2 - 19x - 9 = 0$; -9 is a root.

38. $4x^5 - 15x^4 + 8x^3 + 19x^2 - 12x - 4 = 0$, 1, 2, and -1 are roots.

In Exercises 39 and 40, use the given information to solve each equation.

39. $f(x) = 4x^4 - 12x^3 + 5x^2 + 6x + 1 = 0$ and $2x^2 - 3x - 1$ is a factor of $f(x)$.

40. $f(x) = 3x^5 - 3x^4 - x^3 - 24x^2 + 24x + 8 = 0$ and $3x^2 - 3x - 1$ is a factor of $f(x)$.

For Exercises 41 and 42, refer to the following computer-generated tables for the functions f and g.

$f(t) = t^3 - 4t + 3$		$g(t) = t^5 + 2t^4 + t^3 + 2t^2 - t - 2$	
t	$f(t)$	t	$g(t)$
0.500	1.125000	-3.00	-89.0
0.750	0.421875	-2.50	-22.15625
1.000	0.000000	-2.25	-7.4228515625
1.250	-0.046875	-2.00	0.0
1.500	0.375000	-1.75	2.8603515625

41. (a) What is the remainder when $f(t)$ is divided by $t - \frac{1}{2}$?

(b) What is the remainder when $f(t)$ is divided by $t - 1.25$?

(c) Specify a linear factor of $f(t)$.

(d) Solve the equation $f(t) = 0$.

42. (a) What is the remainder when $g(t)$ is divided by $t + 3$?

(b) What is the remainder when $g(t)$ is divided by $t + \frac{5}{2}$?

(c) Specify a linear factor of $g(t)$.

In Exercises 43–50, find a polynomial equation $f(x) = 0$ satisfying the given conditions. If no such equation is possible, say so.

43. Has degree 3; the coefficient of x^3 is 1; three of the roots are 3, -4, and 5.

44. Has degree 3; the coefficients are integers; $\frac{1}{2}$, $\frac{2}{5}$, and $-\frac{3}{4}$ are roots.

45. Has degree 3; the number -1 is a root of multiplicity two; $x + 6$ is a factor of $f(x)$.

46. Has degree 3; the number 4 is a root of multiplicity two; -1 is a root of multiplicity two.

47. Has degree 4; the number $\frac{1}{2}$ is a root of multiplicity three; $x^2 - 3x - 4$ is a factor of $f(x)$.

48. Has degree 3; both -2 and -3 are roots.

49. Has degree 4; both $x^2 - 3x + 1$ and $x + 6$ are factors of $f(x)$.

50. Has degree 4; the coefficients are integers; $\frac{1}{2}$ is a root of multiplicity two; $2x^2 - 4x - 1$ is a factor of $f(x)$.

In Exercises 51–54, use the remainder theorem (as in Example 3) to evaluate $f(x)$ for the given value of x. Use a calculator, and round off your answers to two decimal places.

51. $f(x) = x^3 - 3x^2 + 12x + 9$; $x = 1.16$

52. $f(x) = x^3 - 2x - 5$: (a) $x = 2.09$; (b) $x = 2.094$; (c) $x = 2.0945$

53. $f(x) = x^3 - 5x - 2$: (a) $x = 2.41$; (b) $x = 2.42$

54. $f(x) = x^4 - 2x^3 - 5x^2 + 10x - 3$: (a) $x = -2.3$; (b) $x = -2.302$; (c) $x = -2.30277$

55. A computer was used to generate the following information about the function $f(x) = x^3 + x^2 - 18x + 10$.

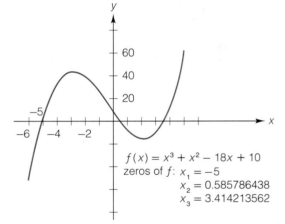

$f(x) = x^3 + x^2 - 18x + 10$
zeros of f: $x_1 = -5$
$x_2 = 0.585786438$
$x_3 = 3.414213562$

(a) Determine the exact values for the zeros x_2 and x_3.

(b) Use your calculator to evaluate the answers in part (a), and check that they are consistent with the computer values given with the figure.

56. A computer was used to generate the following information about the function $g(x) = -2x^3 + 7x^2 - 2x - 6$.

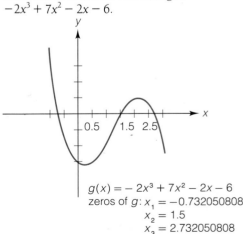

$g(x) = -2x^3 + 7x^2 - 2x - 6$
zeros of g: $x_1 = -0.732050808$
$x_2 = 1.5$
$x_3 = 2.732050808$

(a) Determine the exact values for the zeros x_1 and x_3.
(b) Use your calculator to evaluate the answers in part (a), and check that they are consistent with the computer values given with the figure.

B

In Exercises 57–60, determine whether the given value is a zero of the function.

57. $f(x) = \frac{1}{2}x^2 + bx + c; \quad x = -b - \sqrt{b^2 - 2c}$

58. $Q(x) = ax^2 + bx + c; \quad x = \left(-b + \sqrt{b^2 - 4ac}\right)/2a$
Hint: Look before you leap!

59. $F(x) = 2x^4 + 4x + 1; \quad x = \frac{1}{2}\left(-\sqrt{2} + \sqrt{2\sqrt{2} - 2}\right)$

60. $f(x) = x^3 - 3x^2 + 3x - 3$:   (a) $x = \sqrt[3]{2} - 1$;
(b) $x = \sqrt[3]{2} + 1$

61. Determine values for a and b such that $x - 1$ is a factor of both $x^3 + x^2 + ax + b$ and $x^3 - x^2 - ax + b$.

62. Determine a quadratic equation with the given roots.
(a) $a/b, \ -b/a$ (b) $-a + 2\sqrt{2b}, \ -a - 2\sqrt{2b}$

63. One root of the equation $x^2 + bx + 1 = 0$ is twice the other; find b. (There are two answers.)

64. Determine a value for a such that one root of the equation $ax^2 + x - 1 = 0$ is five times the other.

C

65. Solve the equation $x^3 - 12x + 16 = 0$, given that one of the roots has multiplicity 2.

11.4 THE FUNDAMENTAL THEOREM OF ALGEBRA

One of the most intriguing problems was the question of the number of roots of an equation, which brought in negative and imaginary quantities, and led to the conclusion, in the work of Girard and Descartes, that an equation of degree n can have no more than n roots. The more precise statement, that an equation of degree n always has one root, and hence always has n roots (allowing for multiple roots), became known as the fundamental theorem of algebra. After several attempts by D'Alembert, Euler, and others, the proof was finally given by Gauss in 1799.

From *A Source Book in Mathematics, 1200–1800*, D. J. Struik, ed. (Princeton: Princeton University Press, 1986)

Every equation of algebra has as many solutions as the exponent of the highest term indicates.

Albert Girard, 1629, as quoted by David M. Burton in *The History of Mathematics, An Introduction,* 2nd ed., (Dubuque, Iowa: Wm. C. Brown Publishers, 1991)

. . . we arrive from all the investigations explained above at the rigorous proof of the theorem that every integral rational algebraic function of one variable can be decomposed into real factors of the first and second degree.

Carl Friedrich Gauss in his doctoral dissertation (1799)

Does every polynomial equation (of degree at least 1) have a root? Or, to put the question another way, is it possible to write a polynomial equation that has no solution? Certainly, if we consider only real roots (i.e., roots that are real numbers), then it is easy to specify an equation with no real root:

$$x^2 = -1$$

This equation has no real root because the square of a real number is never negative. On the other hand, if we expand our base of operations from the real-number system to the complex number system, then $x^2 = -1$ does indeed have a root. In fact, both i and $-i$ are roots in this case.

As it turns out, the situation just described for the equation $x^2 = -1$ holds quite generally. That is, within the complex number system, every polynomial equation of degree at least 1 has at least one root. This is the substance of a remarkable theorem that was first proved by the great mathematician Carl Friedrich Gauss in 1799. (Gauss was only 22 years old at the time.) Although there are many fundamental theorems in algebra, Gauss's result has come to be known as *the* **fundamental theorem of algebra.**

The Fundamental Theorem of Algebra

Every polynomial equation of the form

$$a_n x^n + a_{n-1} x^{n-1} + \cdots + a_1 x + a_0 = 0 \qquad (n \geq 1,\ a_n \neq 0)$$

has at least one root among the complex numbers. (This root may be a real number.)

Although most of this section deals with polynomials with real coefficients, Gauss's theorem still applies in cases in which some or all of the coefficients are nonreal complex numbers. The proof of Gauss's theorem is usually given in the post-calculus course called complex variables.

The fundamental theorem of algebra asserts that every polynomial equation of degree at least 1 has a root. There are two initial observations to be made here. First, notice that the theorem says nothing about actually finding the root. Second, notice that the theorem deals only with polynomial equations. Indeed, it is easy to specify a nonpolynomial equation that does not have a root; such an equation is $1/x = 0$. The expression on the left-hand side of this equation can never be zero because the numerator is 1.

■■■ **EXAMPLE 1** Which of the following equations has at least one root?

(a) $x^3 - 17x^2 + 6x - 1 = 0$ **(b)** $\sqrt{2}x^{47} - \pi x^{25} + \sqrt{3} = 0$
(c) $x^2 - 2ix + (3 + i) = 0$

Solution All three equations are polynomial equations. So, according to the fundamental theorem of algebra, each equation has at least one root. ■■■

In Chapter 1 we used factoring as a tool for solving quadratic equations. The next theorem, a consequence of the fundamental theorem of algebra, tells us that (in principle, at least) any polynomial of degree n can be factored into a product of n linear factors. In proving the **linear factors theorem,** we will need to use the factor theorem. If you reread the proof of that theorem in the previous section, you will see that it makes no difference whether the number r appearing in the factor $x - r$ is a real number or a nonreal complex number. Thus, the factor theorem is valid in either case.

The Linear Factors Theorem

Let $f(x) = a_n x^n + a_{n-1} x^{n-1} + \cdots + a_1 x + a_0$ $(n \geq 1,\ a_n \neq 0)$. Then $f(x)$ can be expressed as a product of n linear factors:

$$f(x) = a_n(x - r_1)(x - r_2) \cdots (x - r_n)$$

(The complex numbers r_k appearing in these factors are not necessarily all distinct, and some or all of the r_k may be real numbers.)

PROOF OF THE LINEAR FACTORS THEOREM   According to the fundamental theorem of algebra, the equation $f(x) = 0$ has a root. Let us call this root r_1. By the factor theorem, $x - r_1$ is a factor of $f(x)$, and we can write

$$f(x) = (x - r_1) \cdot Q_1(x)$$

for some polynomial $Q_1(x)$ that has degree $n - 1$ and leading coefficient a_n. If the degree of $Q_1(x)$ happens to be zero, we are done. On the other hand, if the degree of $Q_1(x)$ is at least 1, another application of the fundamental theorem of algebra followed by the factor theorem gives us

$$Q_1(x) = (x - r_2) \cdot Q_2(x)$$

where the degree of $Q_2(x)$ is $n - 2$ and the leading coefficient of $Q_2(x)$ is a_n. We now have

$$f(x) = (x - r_1)(x - r_2) \cdot Q_2(x)$$

We continue this process until the quotient is $Q_n(x) = a_n$. As a result, we obtain

$$f(x) = (x - r_1)(x - r_2) \cdots (x - r_n)a_n$$
$$= a_n(x - r_1)(x - r_2) \cdots (x - r_n)$$

as we wished to show.

The linear factors theorem tells us that any polynomial can be expressed as a product of linear factors. The theorem gives us no information, however, as to how those factors can actually be obtained. The next example demonstrates a case in which the factors are readily obtainable; this is always the case with quadratic polynomials.

▌■■▌ **EXAMPLE 2**   Express each of the following second-degree polynomials in the form $a_n(x - r_1)(x - r_2)$:   **(a)** $3x^2 - 5x - 2$;   **(b)** $x^2 - 4x + 5$.

Solution

(a) A factorization for $3x^2 - 5x - 2$ can be found by simple trial and error. We have

$$3x^2 - 5x - 2 = (3x + 1)(x - 2)$$

We now write the factor $3x + 1$ as $3\left(x + \frac{1}{3}\right)$. This, in turn, can be written $3\left[x - \left(-\frac{1}{3}\right)\right]$. The final factorization is then

$$3x^2 - 5x - 2 = 3\left[x - \left(-\frac{1}{3}\right)\right](x - 2)$$

(b) From the factor theorem, or from our more elementary work with quadratic equations, we know that if r_1 and r_2 are the roots of the equation $x^2 - 4x + 5 = 0$, then $x - r_1$ and $x - r_2$ are the factors of $x^2 - 4x + 5$. That is, $x^2 - 4x + 5 = (x - r_1)(x - r_2)$. The values for r_1 and r_2 in this case are readily obtained by using the quadratic formula. As you can check, the results are

$$r_1 = 2 + i \qquad \text{and} \qquad r_2 = 2 - i$$

The required factorization is therefore

$$x^2 - 4x + 5 = [x - (2 + i)][x - (2 - i)]$$

■■■

Using the linear factors theorem, we can show that every polynomial equation of degree n ($n \geq 1$) has exactly n roots. To help you follow the reasoning, we make two preliminary comments. First, we agree that a root of multiplicity k will be counted as k roots. For example, although the third-degree equation $(x - 1)(x - 4)^2 = 0$ has only two distinct roots, namely, 1 and 4, it has three roots if we agree to count the repeated root, 4, two times. The second preliminary comment concerns the *zero-product property*, which states that $pq = 0$ if and only if $p = 0$ or $q = 0$. When we stated this in Section 1.4, we were working within the real-number system. However, the property is also valid within the complex-number system. Now let's state and prove our theorem.

Theorem

Every polynomial equation of degree $n \geq 1$ has exactly n roots, where a root of multiplicity k is counted k times.

PROOF OF THEOREM   Using the linear factors theorem, we can write the nth-degree polynomial equation $f(x) = a_n x^n + a_{n-1} x^{n-1} + \cdots + a_0 = 0$ as

$$f(x) = a_n(x - r_1)(x - r_2) \cdots (x - r_n) = 0 \qquad (a_n \neq 0) \tag{1}$$

By the factor theorem, each of the numbers $r_1, r_2, \ldots, r_n$ is a root. Some of these numbers may in fact be equal; in other words, we may have repeated roots in this list. In any case, if we agree to count a root of multiplicity k as k roots, then we obtain exactly n roots from the list $r_1, r_2, \ldots, r_n$. Furthermore, the equation $f(x) = 0$ can have no other roots, as we now show. Suppose that r is any number distinct from all the numbers $r_1, r_2, \ldots, r_n$. Replacing x with r in equation (1) yields

$$f(r) = a_n(r - r_1)(r - r_2) \cdots (r - r_n)$$

But the expression on the right-hand side of this last equation cannot be zero, because none of the factors is zero. Thus, $f(r)$ is not zero, and so r is not a root. This completes the proof of the theorem.

▌▌   EXAMPLE 3   Find a polynomial $f(x)$ with leading coefficient 1 such that the equation $f(x) = 0$ has only those roots specified in Table 1. What is the degree of this polynomial?

Solution   The expressions $(x - 3)^2$, $(x + 2)$, and $(x - 0)^2$ all must appear as factors of $f(x)$. Furthermore, no other linear factor can appear. The form of $f(x)$ is therefore

$$f(x) = a_n(x - 3)^2(x + 2)(x - 0)^2$$

Since the leading coefficient a_n is to be 1, we can rewrite this last equation as

$$f(x) = x^2(x - 3)^2(x + 2)$$

This is the required polynomial. The degree here is 5. This can be seen either by multiplying out the factors or by simply adding the multiplicities of the roots in Table 1.   ▟▟▟

TABLE 1

Root	Multiplicity
3	2
−2	1
0	2

■■■ **EXAMPLE 4** Find a quadratic function f that has zeros of 3 and 5 and a graph that passes through the point $(2, -9)$.

Solution The general form of a quadratic function with 3 and 5 as zeros is $f(x) = a_n(x - 3)(x - 5)$. Since the graph passes through $(2, -9)$, we have

$$-9 = a_n(2 - 3)(2 - 5) = a_n(3)$$
$$-3 = a_n$$

The required function is therefore

$$f(x) = -3(x - 3)(x - 5)$$

If we wish, we can carry out the multiplication and rewrite this as

$$f(x) = -3x^2 + 24x - 45$$

■■■

The next example shows how we can use the factored form of a polynomial $f(x)$ to determine the relationships between the coefficients of the polynomial and the roots of the equation $f(x) = 0$.

■■■ **EXAMPLE 5** Let r_1 and r_2 be the roots of the equation $x^2 + bx + c = 0$. Show that

$$r_1 r_2 = c \qquad \text{and} \qquad r_1 + r_2 = -b$$

Solution Since r_1 and r_2 are the roots of the equation $x^2 + bx + c = 0$, we have the identity

$$x^2 + bx + c = (x - r_1)(x - r_2)$$

After multiplying out the right-hand side, we can rewrite this identity as

$$x^2 + bx + c = x^2 - (r_1 + r_2)x + r_1 r_2$$

By equating coefficients, we readily obtain $r_1 r_2 = c$ and $r_1 + r_2 = -b$, as required.

■■■

The technique used in Example 5 can be used to obtain similar relationships between the roots and the coefficients of polynomial equations of any given degree. In Table 2, for instance, we show the relationships obtained in Example 5, along with the corresponding relationships that can be derived for a cubic equation. (Exercise 31 at the end of this section asks you to verify the results for the cubic equation.)

TABLE 2

Equation	Roots	Relationships between Roots and Coefficients
$x^2 + bx + c = 0$	r_1, r_2	$r_1 + r_2 = -b$ $r_1 r_2 = c$
$x^3 + bx^2 + cx + d = 0$	r_1, r_2, r_3	$r_1 + r_2 + r_3 = -b$ $r_1 r_2 + r_2 r_3 + r_3 r_1 = c$ $r_1 r_2 r_3 = -d$

We conclude this section with some remarks concerning the solving of polynomial equations by formulas. You know that the roots of the quadratic equation $ax^2 + bx + c = 0$ are given by the formula

$$x = \frac{-b \pm \sqrt{b^2 - 4ac}}{2a}$$

The question is, are there similar formulas for the solutions of higher-degree equations? By "similar" we mean a formula involving the coefficients and radicals. To answer this question, we look at a bit of history. As early as 1700 B.C., Babylonian mathematicians were able to solve quadratic equations. This is clear from the study of the clay tablets with cuneiform numerals that archeologists have found. The ancient Greeks also were able to solve quadratic equations. Like the Babylonians, the Greeks worked without the aid of algebra as we know it. The mathematicians of ancient Greece used geometric constructions to solve equations. Of course, since all quantities were interpreted geometrically, negative roots were never considered.

The general quadratic formula was known to the Moslem mathematicians sometime before A.D. 1000. For the next 500 years, mathematicians searched for, but did not discover, a formula to solve the general cubic equation. Indeed, in 1494, Luca Pacioli stated in his text *Summa di Arithmetica* that the general cubic equation could not be solved by the algebraic techniques then available. All of this was to change, however, within the next several decades.

Around 1515, the Italian mathematician Scipione del Ferro solved the cubic equation $x^3 + px + q = 0$ using algebraic techniques. This essentially constituted a solution of the seemingly more general equation $x^3 + bx^2 + cx + d = 0$. The reason for this is that if we make the substitution $x = y - b/3$ in the latter equation, the result is a cubic equation with no y^2-term. By 1540, the Italian mathematician Ludovico Ferrari had solved the general fourth-degree equation. Actually, at that time in Renaissance Italy, there was considerable controversy regarding exactly who discovered the various formulas first. Details of the dispute can be found in any text on the history of mathematics. But for our purposes, the point is simply that by the middle of the sixteenth century, all polynomial equations of degree 4 or less could be solved by the formulas that had been discovered. The common feature of these formulas was that they involved the coefficients, the four basic operations of arithmetic, and various radicals. For example, a formula for the solution of the equation $x^3 + px + q = 0$ is as follows:

$$x = \sqrt[3]{\frac{-q}{2} + \sqrt{\frac{q^2}{4} + \frac{p^3}{27}}} + \sqrt[3]{\frac{-q}{2} - \sqrt{\frac{q^2}{4} + \frac{p^3}{27}}}$$

To get some idea of the practical difficulties inherent in computing with this formula, try using it to show that $x = -2$ is a root of the cubic equation $x^3 + 4x + 16 = 0$.

For more than 200 years after the cubic and quartic (fourth-degree) equations had been solved, mathematicians continued to search for a formula that would yield the solutions of the general fifth-degree equation. The first breakthrough, if it can be called that, occurred in 1770, when the French mathematician Joseph Louis Lagrange found a technique that served to unify and summarize all of the previous methods used for the equations of degrees 2, 3, and 4.

However, Lagrange then showed that his technique could not work in the case of the general fifth-degree equation. While we will not describe the details of Lagrange's work here, it is worth pointing out that he relied on the types of relationships between roots and coefficients that we looked at in Table 2 of this section.

Finally, in 1828, the Norwegian mathematician Niels Henrik Abel proved that for the general polynomial equation of degree 5 or higher, there could be no formula yielding the solutions in terms of the coefficients and radicals. This is not to say that such equations do not possess solutions. In fact they must, as we saw earlier in this section. It is just that we cannot in every case express the solutions in terms of the coefficients and radicals. For example, it can be shown that the equation $x^5 - 6x + 3 = 0$ has a real root between 0 and 1, but this number cannot be expressed in terms of the coefficients and radicals. (We can, however, compute the root to as many decimal places as we wish, as you will see in the next section.) In 1830, the French mathematician Evariste Galois completed matters by giving conditions for determining exactly which polynomial equations can be solved in terms of coefficients and radicals.

EXERCISE SET 11.4

A

According to the fundamental theorem of algebra, which of the equations in Exercises 1 and 2 have at least one root?

1. (a) $x^5 - 14x^4 + 8x + 53 = 0$
   (b) $(4.17)x^3 + (2.06)x^2 + (0.01)x + 1.23 = 0$
   (c) $ix^2 + (2 + 3i)x - 17 = 0$
   (d) $x^{2.1} + 3x^{0.3} + 1 = 0$
2. (a) $\sqrt{3}x^{17} + \sqrt{2}x^{13} + \sqrt{5} = 0$
   (b) $17x^{\sqrt{3}} + 13x^{\sqrt{2}} + \sqrt{5} = 0$
   (c) $1/(x^2 + 1) = 0$ (d) $2^{3x} - 2^x - 1 = 0$

In Exercises 3–10, express each polynomial in the form $a_n(x - r_1)(x - r_2) \cdots (x - r_n)$.

3. $x^2 - 2x - 3$
4. $x^3 - 2x^2 - 3x$
5. $4x^2 + 23x - 6$
6. $6x^2 + x - 12$
7. $x^2 - 5$
8. $x^2 + 5$
9. $x^2 - 10x + 26$
10. $x^3 + 2x^2 - 3x - 6$

In Exercises 11–16, find a polynomial $f(x)$ with leading coefficient 1 such that the equation $f(x) = 0$ has the given roots and no others. If the degree of $f(x)$ is 7 or more, express $f(x)$ in factored form; otherwise, express $f(x)$ in the form $a_n x^n + a_{n-1} x^{n-1} + \cdots + a_1 x + a_0$.

11.
Root	1	−3
Multiplicity	2	1

12.
Root	0	4
Multiplicity	2	1

13.
Root	2	−2	2i	−2i
Multiplicity	1	1	1	1

14.
Root	2 + i	2 − i
Multiplicity	1	1

15.
Root	$\sqrt{3}$	$-\sqrt{3}$	4i	−4i
Multiplicity	2	2	1	1

16.
Root	5	1	1 − i	1 + i
Multiplicity	2	3	1	1

In Exercises 17–20, express the polynomial $f(x)$ in the form $a_n x^n + a_{n-1} x^{n-1} + \cdots + a_1 x + a_0$.

17. A quadratic function that has zeros −4 and 9 and a graph that passes through the point $(3, 5)$.

18. A quadratic function that has a maximum value of 2 and that has −2 and 4 as zeros.

19. A third-degree polynomial function that has zeros -5, 2, and 3 and a graph that passes through the point $(0, 1)$.

20. A fourth-degree polynomial function that has zeros $\sqrt{2}$, $-\sqrt{2}$, 1, and -1 and a graph that passes through $(2, -20)$.

In Exercises 21–28, find a quadratic equation with the given roots and no others. Write your answers in the form $Ax^2 + Bx + C = 0$. Suggestion: Make use of Table 2.

21. $r_1 = -i, \ r_2 = -\sqrt{3}$

22. $r_1 = 1 + i\sqrt{3}, \ r_2 = 1 - i\sqrt{3}$

23. $r_1 = 9, \ r_2 = -6$ 24. $r_1 = 5, \ r_2 = \frac{3}{4}$

25. $r_1 = 1 + \sqrt{5}, \ r_2 = 1 - \sqrt{5}$

26. $r_1 = 6 - 5i, \ r_2 = 6 + 5i$

27. $r_1 = a + \sqrt{b}, \ r_2 = a - \sqrt{b} \ (b > 0)$

28. $r_1 = a + bi, \ r_2 = a - bi$

B

29. Express the polynomial $x^4 + 64$ as a product of four linear factors. *Hint:* First add and subtract the quantity $16x^2$ so that you can use the difference-of-squares factoring formula.

30. Suppose that p and q are positive integers with $p > q$. Find a quadratic equation with integer coefficients and with roots $\sqrt{p}/(\sqrt{p} \pm \sqrt{p-q})$.

31. Let r_1, r_2, and r_3 be the roots of the equation $x^3 + bx^2 + cx + d = 0$. Use the method shown in Example 5 to verify the following relationships:

$$r_1 + r_2 + r_3 = -b$$
$$r_1r_2 + r_2r_3 + r_3r_1 = c$$
$$r_1r_2r_3 = -d$$

32. Let r_1, r_2, r_3, and r_4 be the roots of the equation $x^4 + bx^3 + cx^2 + dx + e = 0$. Use the method shown in Example 5 to prove the following facts:

$$r_1 + r_2 + r_3 + r_4 = -b$$
$$r_1r_2 + r_2r_3 + r_3r_4 + r_4r_1 + r_2r_4 + r_3r_1 = c$$
$$r_1r_2r_3 + r_2r_3r_4 + r_3r_4r_1 + r_4r_1r_2 = -d$$
$$r_1r_2r_3r_4 = e$$

33. Solve the equation $x^3 - 4x^2 - 9x + 36 = 0$, given that the sum of two of the roots is 0. *Suggestion:* Use Table 2.

34. Solve the equation $x^3 - 75x + 250 = 0$, given that two of the roots are equal. *Suggestion:* Use Table 2.

35. Let α and β be the roots of the polynomial equation $x^2 + bx + c = 0$.

(a) Show that $\alpha^2 + \beta^2 = b^2 - 2c$. *Hint:* Use Table 2 along with the identity $\alpha^2 + \beta^2 = (\alpha + \beta)^2 - 2\alpha\beta$.

(b) Show that $\dfrac{1}{\alpha^2} + \dfrac{1}{\beta^2} = \dfrac{b^2 - 2c}{c^2}$.

(c) Show that $\alpha^3 + \beta^3 = -b(b^2 - 3c)$.

C

36. (a) Let r_1, r_2, r_3, and r_4 be four real roots of the equation $x^4 + ax^2 + bx + c = 0$. Show that $r_1 + r_2 + r_3 + r_4 = 0$. *Hint:* Use the first formula in Exercise 32.

(b) Suppose a circle intersects the parabola $y = x^2$ in the points $(x_1, y_1), \ldots, (x_4, y_4)$. Show that $x_1 + x_2 + x_3 + x_4 = 0$. *Hint:* Use the result in part (a).

37. (a) Let r_1, r_2, and r_3 be three distinct numbers that are roots of the equation $f(x) = Ax^2 + Bx + C = 0$. Show that $f(x) = 0$ for all values of x. *Hint:* You need to show that $A = B = C = 0$. First show that both A and B are zero as follows. If either A or B were nonzero, then the equation $f(x) = 0$ would be a polynomial equation of degree at most 2 with three distinct roots. Why is that impossible?

(b) Use the result in part (a) to prove the following identity.

$$\frac{a^2 - x^2}{(a - b)(a - c)} + \frac{b^2 - x^2}{(b - c)(b - a)}$$
$$+ \frac{c^2 - x^2}{(c - a)(c - b)} - 1 = 0$$

Hint: Let $f(x)$ denote the quadratic expression on the left-hand side of the equation. Compute $f(a)$, $f(b)$, and $f(c)$.

38. Prove that the following equation is an identity:

$$\frac{(x - a)(x - b)c^2}{(c - a)(c - b)} + \frac{(x - b)(x - c)a^2}{(a - b)(a - c)}$$
$$+ \frac{(x - c)(x - a)b^2}{(b - c)(b - a)} - x^2 = 0$$

Hint: Use the result in Exercise 37(a).

39. For this exercise, assume as given the following two identities (which were derived in Exercises 72 and 73 in Section 7.7).

$$\cos 72° + \cos 144° = -\frac{1}{2} \quad \text{and}$$

$$(\cos 72°)(\cos 144°) = -\frac{1}{4}$$

(a) Find a quadratic equation, with integer coefficients, whose roots are the numbers $\cos 72°$ and $\cos 144°$. *Hint:* Use the result in Example 5 (or Table 2).

(b) Use the quadratic formula to solve the equation in part (a). Conclude that

$$\cos 72° = \frac{1}{4}(-1 + \sqrt{5}) \text{ and } \cos 144° = \frac{1}{4}(-1 - \sqrt{5}).$$

(c) Use a calculator to check the results in part (b).

40. (a) In the trigonometric identity $\cos 3\theta = 4\cos^3\theta - 3\cos\theta$, make the substitution $\theta = 20°$ and conclude that the number $\cos 20°$ is a root of the equation $8x^3 - 6x - 1 = 0$.

(b) Follow the method in part (a) to show that each of the numbers $\cos 100°$ and $\cos 140°$ is a root of $8x^3 - 6x - 1 = 0$.

(c) Use Table 2 to explain why $(\cos 20°)(\cos 100°)(\cos 140°) = \frac{1}{8}$.

(d) Use the result in part (c) and the reference angle concept to show that $(\cos 20°)(\cos 40°)(\cos 80°) = \frac{1}{8}$.

(e) Use a calulator to check the result in part (d).

11.5 RATIONAL AND IRRATIONAL ROOTS

Jacques Peletier (1517–1582), a French man of letters, poet, and mathematician, had observed as early as 1558, that the root of an equation is a divisor of the last term.

Florian Cajori in *A History of Mathematics,* 4th ed. (New York: Chelsea Publishing Co., 1985)

Of the two leading commentators on Descartes in the 17th century, the first (1649) was his warm personal friend, Florimond de Beaune, an officeholder at Blois. He also wrote on algebra, being one of the first to treat scientifically of the superior and inferior limits of the roots of a numerical equation.

David Eugene Smith in *History of Mathematics,* vol. I (New York: Ginn and Co., 1923)

As we saw in the previous section, not every polynomial equation has a real root. Furthermore, even if a polynomial equation does possess a real root, that root isn't necessarily a rational number. (The equation $x^2 = 2$ provides a simple example.) If a polynomial equation with integer coefficients does have a rational root, however, we can find that root by applying the **rational roots theorem**, which we now state.

The Rational Roots Theorem

Consider the polynomial equation

$$a_n x^n + a_{n-1} x^{n-1} + \cdots + a_1 x + a_0 = 0 \qquad (n \geq 1, \, a_n \neq 0)$$

and suppose that all the coefficients are integers. Let p/q be a rational number, where p and q have no common factors other than ± 1. If p/q is a root of the equation, then p is a factor of a_0 and q is a factor of a_n.

A proof of the rational roots theorem is outlined in Exercise 39 at the end of this section. For the moment, though, let's just see why the theorem is plausible. Suppose that the two rational numbers a/b and c/d are the roots of a certain quadratic equation. Then, from our experience with quadratics (or by the linear factors theorem), we know that the equation can be written in the form

$$k\left(x - \frac{a}{b}\right)\left(x - \frac{c}{d}\right) = 0 \tag{1}$$

where k is a constant. Now, as Exercise 28 will ask you to check, if we carry out the multiplication and clear of fractions, equation (1) becomes

$$(kbd)x^2 - (kad + kbc)x + kac = 0 \tag{2}$$

Observe that a and c (the numerators of the two roots) are factors of the constant term kac in equation (2), just as the rational roots theorem asserts. Furthermore, b and d (the denominators of the roots) are factors of the coefficient of the x^2 term in equation (2), again as the theorem asserts.

The following example shows how the rational roots theorem can be used to solve a polynomial equation.

EXAMPLE 1 Find the rational roots (if any) of the equation $2x^3 - x^2 - 9x - 4 = 0$. Then solve the equation.

Solution First we list the factors of a_0, the factors of a_n, and the possibilities for rational roots:

factors of $a_0 = -4$: $\pm 1, \pm 2, \pm 4$

factors of $a_3 = 2$: $\pm 1, \pm 2$

possible rational roots: $\pm\dfrac{1}{1}, \pm\dfrac{1}{2}, \pm\dfrac{2}{1}, \pm\dfrac{4}{1}$

Now we can use synthetic division to test whether or not any of these possibilities is a root. (A zero remainder will tell us that we have a root.) As you can check, the first three possibilities $\left(1, -1, \text{and } \frac{1}{2}\right)$ are not roots. However, using $-\frac{1}{2}$, we have

$$
\begin{array}{r|rrrr}
-1/2 & 2 & -1 & -9 & -4 \\
     &   & -1 & 1 & 4 \\
\hline
     & 2 & -2 & -8 & 0
\end{array}
$$

Thus, $x = -\frac{1}{2}$ is a root. We could now continue to check the remaining possibilities in this same manner. At this point, however, it is simpler to consider the reduced equation $2x^2 - 2x - 8 = 0$, or $x^2 - x - 4 = 0$. Since this is a quadratic equation, it can be solved directly. We have

$$x = \frac{-(-1) \pm \sqrt{(-1)^2 - 4(1)(-4)}}{2(1)} = \frac{1 \pm \sqrt{17}}{2}$$

We have now determined three distinct roots. Since the degree of the original equation is 3, there can be no other roots. We conclude that $x = -\frac{1}{2}$ is the only rational root. The three roots of the equation are $-\frac{1}{2}$ and $\left(1 \pm \sqrt{17}\right)/2$. ∎

As Example 1 indicates, the number of possibilities for rational roots can be relatively large, even for rather simple equations. The next theorem we develop allows us to reduce the number of possibilities. We say that a real number B is an **upper bound** for the roots of an equation if every real root is less than or equal to B. Similarly, a real number b is a **lower bound** if every real root is greater than or equal to b. The following theorem tells us how synthetic division can be used in determining upper and lower bounds for roots.

The Upper and Lower Bound Theorem for Real Roots

Consider the polynomial equation

$$f(x) = a_n x^n + a_{n-1} x^{n-1} + \cdots + a_1 x + a_0 = 0$$

where all of the coefficients are real numbers and a_n is positive.

1. If we use synthetic division to divide $f(x)$ by $x - B$, where $B > 0$, and we obtain a third row containing no negative numbers, then B is an upper bound for the real roots of $f(x) = 0$.

2. If we use synthetic division to divide $f(x)$ by $x - b$, where $b < 0$, and we obtain a third row in which the numbers are alternately positive and negative, then b is a lower bound for the real roots of $f(x) = 0$. (In determining whether the signs alternate in the third row, zeros may be suitably denoted by $+0$ or -0.)

CAUTION   A number may fail the lower bound test but still be a lower bound. For an example, see Exercise 52.

We will prove the first part of this theorem. A proof of the second part can be developed along similar lines. To prove the first part of the theorem, we use the division algorithm to write

$$f(x) = (x - B) \cdot Q(x) + R \tag{3}$$

The remainder R here is a constant that may be zero. To show that B is an upper bound, we must show that any number greater than B is not a root. Toward this end, let p be a number that is greater than B. Note that p must be positive, since B is positive. Then with $x = p$, equation (3) becomes

$$f(p) = (p - B) \cdot Q(p) + R \tag{4}$$

We are now going to show that the right-hand side of equation (4) is positive. This will tell us that p is not a root. First, look at the factor $(p - B)$. This is positive, since p is greater than B. Next consider $Q(p)$. By hypothesis, the coefficients of $Q(x)$ are all nonnegative. Furthermore, the leading coefficient of $Q(x)$ is a number a_n that is positive. Since p is also positive, it follows that $Q(p)$ must be positive. Finally, the number R is nonnegative because, in the synthetic division of $f(x)$ by $x - B$, all the numbers in the third row are nonnegative. It now follows that the right-hand side of equation (4) is positive. Consequently, $f(p)$ is not zero and p is not a root of the equation $f(x) = 0$. This is what we wished to show.

▮ ▮ EXAMPLE 2   Determine the rational roots, or show that none exist, for the equation

$$\frac{1}{4}x^4 - \frac{3}{4}x^3 + \frac{17}{4}x^2 + 4x + 5 = 0$$

Solution   We will use the rational roots theorem along with the upper and lower bound theorem. First of all, if we are to apply the rational roots theorem, then our equation must have integer coefficients. In view of this, we multiply both sides of the given equation by 4 to obtain

$$x^4 - 3x^3 + 17x^2 + 16x + 20 = 0$$

As in the previous example, we list the factors of a_0, the factors of a_n, and the possibilities for rational roots:

factors of $a_0 = 20$: $\pm 1, \pm 2, \pm 4, \pm 5, \pm 10, \pm 20$

factors of $a_4 = 1$: ± 1

possible rational roots: $\pm 1, \pm 2, \pm 4, \pm 5, \pm 10, \pm 20$

Our strategy here will be first to check for positive roots, beginning with 1 and working upward. The checks for $x = 1$, $x = 2$, and $x = 4$ are as follows:

$$
\underline{1\,|}\;
\begin{array}{rrrrr}
1 & -3 & 17 & 16 & 20 \\
  & 1 & -2 & 15 & 31 \\
\hline
1 & -2 & 15 & 31 & 51
\end{array}
\qquad
\underline{2\,|}\;
\begin{array}{rrrrr}
1 & -3 & 17 & 16 & 20 \\
  & 2 & -2 & 30 & 92 \\
\hline
1 & -1 & 15 & 46 & 112
\end{array}
\qquad
\underline{4\,|}\;
\begin{array}{rrrrr}
1 & -3 & 17 & 16 & 20 \\
  & 4 & 4 & 84 & 400 \\
\hline
1 & 1 & 21 & 100 & 420
\end{array}
$$

As you can see, none of the remainders here is zero. However, notice that in the division corresponding to $x = 4$, all of the numbers appearing in the third row are nonnegative. It therefore follows that 4 is an upper bound for the roots of the given equation. In view of this, we needn't bother to check the remaining values $x = 5$, $x = 10$, and $x = 20$, since none of those can be roots. At this point, we can conclude that the given equation has no positive rational roots.

Next we check for negative rational roots, beginning with -1 and working downward (if necessary). Checking $x = -1$, we have

$$
\underline{-1\,|}\;
\begin{array}{rrrrr}
1 & -3 & 17 & 16 & 20 \\
  & -1 & 4 & -21 & 5 \\
\hline
1 & -4 & 21 & -5 & 25
\end{array}
$$

Two conclusions can be drawn from this synthetic division. First, $x = -1$ is not a root of the equation. Second, -1 is a lower bound for the roots because the signs in the third row of the synthetic division alternate. This means that we needn't bother to check whether any of the numbers -2, -4, -5, -15, or -20 are roots; none of them can be roots, since they are all less than -1.

Let's summarize our results. We have shown that the given equation has no positive rational roots and no negative rational roots. Furthermore, by inspection we see that zero is not a root of the equation. Thus, the given equation possesses no rational roots. ▮▮▮

We conclude this section by demonstrating a method for approximating irrational roots. The method depends on the **location theorem.**

> **The Location Theorem**
>
> Let $f(x)$ be a polynomial, all of whose coefficients are real numbers. If a and b are real numbers such that $f(a)$ and $f(b)$ have opposite signs, then the equation $f(x) = 0$ has at least one real root between a and b.

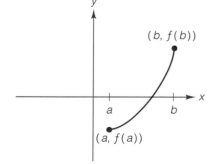

FIGURE I

Figure 1 indicates why this theorem is plausible. If the point $(a, f(a))$ lies below the x-axis and $(b, f(b))$ lies above the x-axis, then it certainly seems that the graph of f must cross the x-axis at some point x_0 between a and b. At this intercept, we have $f(x_0) = 0$; that is, x_0 is a root of the equation $f(x) = 0$. (The location theorem is a special case of the so-called intermediate value theorem; the proof is usually discussed in calculus courses.)

Our technique for approximating (or "locating") irrational roots uses the **method of successive approximations.** We will demonstrate this method in Example 3.

▌▬▌ EXAMPLE 3 The following equation has exactly one positive root. Locate this root between successive hundredths.

$$f(x) = x^3 + 2x - 4 = 0$$

Solution First we need to find two numbers a and b such that $f(a)$ and $f(b)$ have opposite signs. By inspection (or by trial and error), we find that $f(1) = -1$ and $f(2) = 8$. Thus, according to the location theorem, the equation $f(x) = 0$ has a real root in the interval $(1, 2)$.

Now that we have located the root between successive integers, we can locate it between successive tenths. We compute $f(1.0)$, $f(1.1)$, $f(1.2)$, and so on, up through $f(2.0)$ if necessary, until we find a sign change. As you can verify, using synthetic division and a calculator, the results are as follows:

$$f(1.0) = -1 \qquad \left.\begin{array}{l} f(1.1) = -0.469 \\ f(1.2) = 0.128 \end{array}\right\} \longleftarrow \text{sign change}$$

This shows that the root lies between 1.1 and 1.2.

Having located the root between successive tenths, we follow a similar process to locate the root between successive hundredths. Using synthetic division and a calculator, we obtain these results:

$$\begin{array}{lll}
f(1.10) = -0.469 & f(1.13) \approx -0.297 & f(1.16) \approx -0.119 \\
f(1.11) \approx -0.412 & f(1.14) \approx -0.238 & \left.\begin{array}{l}f(1.17) \approx -0.058 \\ f(1.18) \approx 0.003\end{array}\right\} \longleftarrow \text{sign change} \\
f(1.12) \approx -0.355 & f(1.15) \approx -0.179 &
\end{array}$$

This shows that the root lies between 1.17 and 1.18, so we've located the root between successive hundredths, as required. **▬▬▬**

The procedure described in Example 3 could be continued to yield closer and closer approximations for the required root. We note in passing that such calculations are easily handled with the aid of a programmable calculator.

EXERCISE SET 11.5

A

In Exercises 1–6, list the possibilities for rational roots.

1. $4x^3 - 9x^2 - 15x + 3 = 0$

2. $x^4 - x^3 + 10x^2 - 24 = 0$

3. $8x^5 - x^2 + 9 = 0$

4. $18x^4 - 10x^3 + x^2 - 4 = 0$

5. $\frac{2}{3}x^3 - x^2 - 5x + 2 = 0$

6. $\frac{1}{2}x^4 - 5x^3 + \frac{4}{3}x^2 + 8x - \frac{1}{3} = 0$

In Exercises 7–12, show that each equation has no rational roots.

7. $x^3 - 3x + 1 = 0$

8. $x^3 + 8x^2 - 1 = 0$

9. $x^3 + x^2 - x + 1 = 0$

10. $12x^4 - x^2 - 6 = 0$

11. $x^4 + 4x^3 + 4x^2 - 16 = 0$

12. $4x^5 - x^4 - x^3 - x^2 + x - 8 = 0$

For Exercises 13–25, find the rational roots of each equation and then solve the equation. (Use the rational roots theorem and the upper and lower bound theorem, as in Example 2.)

13. $x^3 + 3x^2 - x - 3 = 0$

14. $2x^3 - 5x^2 - 3x + 9 = 0$

15. $4x^3 + x^2 - 20x - 5 = 0$

16. $3x^3 - 16x^2 + 17x - 4 = 0$

17. $9x^3 + 18x^2 + 11x + 2 = 0$

18. $4x^3 - 10x^2 - 25x + 4 = 0$

19. $x^4 + x^3 - 25x^2 - x + 24 = 0$

20. $10x^4 + 107x^3 + 301x^2 + 171x + 23 = 0$

21. $x^4 - 4x^3 + 6x^2 - 4x + 1 = 0$

22. $24x^3 - 46x^2 + 29x - 6 = 0$

23. $x^3 - \frac{5}{2}x^2 - 23x + 12 = 0$

24. $x^3 - \frac{17}{3}x^2 - \frac{10}{3}x + 8 = 0$

25. $2x^4 - \frac{9}{10}x^3 - \frac{29}{10}x^2 + \frac{27}{20}x - \frac{3}{20} = 0$

In Exercises 26 and 27, determine integral upper and lower bounds for the real roots of the equations. (Follow the method used within the solution of Example 2.)

26. (a) $x^3 + 2x^2 - 5x + 20 = 0$
    (b) $x^5 - 3x^2 + 100 = 0$

27. (a) $5x^4 - 10x - 12 = 0$
    (b) $3x^4 - 4x^3 + 5x^2 - 2x - 4 = 0$
    (c) $2x^4 - 7x^3 - 5x^2 + 28x - 12 = 0$

28. Referring to equation (1) in this section, multiply out the left-hand side and then clear the equation of fractions. Check that your result agrees with equation (2).

In Exercises 29–34, each equation has exactly one positive root. In each case, locate the root between successive hundredths. Use a calculator.

29. $x^3 + x - 1 = 0$ 30. $x^3 - 2x - 5 = 0$

31. $x^5 - 200 = 0$ 32. $x^3 - 3x^2 + 3x - 26 = 0$

33. $x^3 - 8x^2 + 21x - 22 = 0$

34. $2x^4 - x^3 - 12x^2 - 16x - 8 = 0$

In Exercises 35–38, each equation has exactly one negative root. In each case, use a calculator to locate the root between successive hundredths.

35. $x^3 + x^2 - 2x + 1 = 0$ 36. $x^5 + 100 = 0$

37. $x^3 + 2x^2 + 2x + 101 = 0$

38. $x^4 + 4x^3 - 6x^2 - 8x - 3 = 0$

B

39. This exercise outlines a proof of the rational roots theorem. At one point in the proof, we will need to rely on the following fact, which is proved in courses on number theory.

    FACT FROM NUMBER THEORY Suppose that A, B, and C are integers and that A is a factor of the number BC.

Then if A has no factor in common with C (other than ± 1), A must be a factor of B.

(a) Let $A = 2$, $B = 8$, and $C = 5$. Verify that the fact from number theory is correct here.

(b) Let $A = 20$, $B = 8$, and $C = 5$. Note that A is a factor of BC, but A is not a factor of B. Why doesn't this contradict the fact from number theory?

(c) Now we're ready to prove the rational roots theorem. We begin with a polynomial equation with integer coefficients:

$$a_n x^n + a_{n-1} x^{n-1} + \cdots + a_1 x + a_0 = 0$$
$$(n \geq 1, a_n \neq 0)$$

We assume that the rational number p/q is a root of the equation and that p and q have no common factors other than 1. Why is the following equation now true?

$$a_n \left(\frac{p}{q}\right)^n + a_{n-1} \left(\frac{p}{q}\right)^{n-1} + \cdots + a_1 \left(\frac{p}{q}\right) + a_0 = 0$$

(d) Show that the last equation in part (c) can be written

$$p(a_n p^{n-1} + a_{n-1} q p^{n-2} + \cdots + a_1 q^{n-1}) = -a_0 q^n$$

Since p is a factor of the left-hand side of this last equation, p must also be a factor of the right-hand side. That is, p must be a factor of $a_0 q^n$. But since p and q have no common factors, neither do p and q^n. Our fact from number theory now tells us that p must be a factor of a_0, as we wished to show. (The proof that q is a factor of a_n is carried out in a similar manner.)

40. The location theorem asserts that the polynomial equation $f(x) = 0$ has a root in the open interval (a, b) whenever $f(a)$ and $f(b)$ have unlike signs. If $f(a)$ and $f(b)$ have the same sign, can the equation $f(x) = 0$ have a root between a and b? *Hint:* Look at the graph of $f(x) = x^2 - 2x + 1$ with $a = 0$ and $b = 2$.

Use a calculator for Exercises 41–43. Round off your answers to two decimal places.

41. On the same set of axes, sketch the graphs of $y = x^3$ and $y = 1 - 3x$. Find the x-coordinate of the point where the graphs intersect.

42. On the same set of axes, sketch the graphs of $y = x^3$ and $y = x + 1$. Find the x-coordinate of the point where the graphs intersect.

43. Find the x-coordinate of the point where the curves $y = x^2 - 1$ and $y = \sqrt{x}$ meet.

44. In a note that appeared in *The Two-Year College Mathematics Journal* (vol. 12, 1981, pp. 334–336), Professors Warren Page and Leo Chosid explain how

the process of testing for rational roots can be shortened. In essence, their result is as follows. Suppose that we have a polynomial with integer coefficients and we are testing for a possible root p/q. Then, if a noninteger is generated at any point in the synthetic division process, p/q cannot be a root of the polynomial. For example, suppose we want to know if $\frac{4}{3}$ is a root of $6x^4 - 10x^3 + 2x^2 - 9x + 8 = 0$. The first few steps of the synthetic division are as follows.

$$\frac{4}{3} \big] \quad \begin{array}{rrrrr} 6 & -10 & 2 & -9 & 8 \\ & 8 & -\frac{8}{3} & & \\ \hline 6 & -2 & & & \end{array}$$

Since the noninteger $-\frac{8}{3}$ has been generated in the synthetic division process, the process can stop; $\frac{4}{3}$ is not a root of the polynomial. Use this idea to shorten your work in testing to see if any of $\frac{3}{4}$, $\frac{1}{8}$, and $-\frac{3}{2}$ are roots of the equation $8x^5 - 5x^4 + 3x^2 - 2x - 6 = 0$.

45. In a note that appeared in *The College Mathematics Journal* (vol. 20, 1989, pp. 139–141), Professor Don Redmond proved the following interesting result.

> Consider the polynomial equation $f(x) = a_n x^n + a_{n-1} x^{n-1} + \cdots + a_1 x + a_0 = 0$, and suppose that the degree of $f(x)$ is at least two and that all of the coefficients are integers. If the three numbers a_0, a_n, and $f(1)$ are all odd, then the given equation has no rational roots.

Use this result to show that the following equations have no rational roots.
(a) $9x^5 - 8x^4 + 3x^2 - 2x + 27 = 0$
(b) $5x^5 + 5x^4 - 11x^2 - 3x - 25 = 0$

C

In Exercises 46–50, you need to know that a prime number is a positive integer greater than 1 with no factors other than itself and 1. Thus, the first seven prime numbers are 2, 3, 5, 7, 11, 13, and 17.

46. Find all prime numbers p for which the equation $x^2 + x - p = 0$ has a rational root.

47. Find all prime numbers p for which the equation $x^3 + x^2 + x - p = 0$ has at least one rational root. For each value of p that you find, find the corresponding *real* roots of the equation.

48. Consider the equation $x^3 + px - q = 0$, where p and q are prime numbers. Observe that there are only four possible rational roots here: 1, -1, q, and $-q$.
(a) Show that if $x = 1$ is a root, then we must have $q = 3$ and $p = 2$. What are the remaining roots in this case?
(b) Show that none of the numbers -1, q, and $-q$ can be a root of the equation.   *Hint:*   For each case, assume the contrary and deduce a contradiction.

49. Consider the equation $x^2 + x - pq = 0$, where p and q are prime numbers. If this equation has rational roots, show that these roots must be -3 and 2.   *Suggestion:* The possible rational roots are ± 1, $\pm p$, $\pm q$, and $\pm pq$. In each case, assume that the given number is a root and see where that leads.

50. If p and q are prime numbers, show that the equation $x^3 + px - pq = 0$ has no rational roots.

51. Find all integral values of b for which the equation $x^3 - b^2 x^2 + 3bx - 4 = 0$ has a rational root.

52. Let $f(x) = x^3 + 3x^2 - x - 3$.
(a) Factor $f(x)$ by using the basic factoring techniques in Section 1.3.
(b) Sketch the graph of $f(x) = x^3 + 3x^2 - x - 3$. Note that -3 is a lower bound for the roots.
(c) Show that the number -3 fails the lower bound test. This shows that a number may fail the lower bound test and yet be a lower bound. (We say that the lower bound test provides a *sufficient* but not a *necessary* condition for a lower bound.)

11.6 CONJUGATE ROOTS AND DESCARTES'S RULE OF SIGNS

> *We now proceed to investigate a remarkable theorem, implicit in the work of [Thomas] Harriot [1560–1621] but first used explicitly by Descartes (1637), which limits the number of positive or negative roots of an equation....*
>
> *Remarkable as the Harriot–Descartes's Rule of Signs is, it still leaves uncertainty as to the exact number of real roots in an equation: it only gives an upper limit to them. The problem of finding an exact test ... was finally solved in 1829 by [Jacques Charles François] Sturm.*
>
> H. W. Turnbull in *Theory of Equations* (Edinburgh: Oliver and Boyd, 1939)

An equation can have as many true [positive] roots as it contains changes of sign, from plus to minus or from minus to plus; and as many false [negative] roots as the number of times two plus signs or two minus signs are found in succession.

René Descartes (1637)

As you know from earlier work involving quadratic equations with real coefficients, when nonreal complex roots occur, they occur in conjugate pairs. For instance, as you can check by means of the quadratic formula, the roots of the equation $x^2 - 2x + 5 = 0$ are $1 + 2i$ and $1 - 2i$. The **conjugate roots theorem** tells us that the situation is the same for all polynomial equations with real coefficients.

The Conjugate Roots Theorem

Let $f(x)$ be a polynomial, with all real coefficients. Suppose that $a + bi$ is a root of the equation $f(x) = 0$, where a and b are real and $b \neq 0$. Then $a - bi$ is also a root of the equation.

To prove the conjugate roots theorem, we use four of the properties of complex conjugates listed in Section 11.1:

Property 1: $\bar{z}_1 \bar{z}_2 = \overline{z_1 z_2}$

Property 2: $(\bar{z})^m = \overline{z^m}$

Property 3: $\bar{r} = r$ for every real number r

Property 4: $\bar{z}_1 + \bar{z}_2 = \overline{z_1 + z_2}$

Using these properties, we can prove the theorem as follows. We begin with a polynomial with real coefficients:

$$f(x) = a_n x^n + a_{n-1} x^{n-1} + \cdots + a_1 x + a_0$$

We must show that if $z = a + bi$ is a root of $f(x) = 0$, then $\bar{z} = a - bi$ is also a root. We have

$$f(\bar{z}) = a_n \bar{z}^n + a_{n-1} \bar{z}^{n-1} + \cdots + a_1 \bar{z} + a_0$$

$$= \bar{a}_n \bar{z}^n + \overline{a_{n-1}} \bar{z}^{n-1} + \cdots + \bar{a}_1 \bar{z} + \bar{a}_0 \qquad \text{Properties 3 and 2}$$

$$= \overline{a_n z^n} + \overline{a_{n-1} z^{n-1}} + \cdots + \overline{a_1 z} + \bar{a}_0 \qquad \text{Property 1}$$

$$= \overline{a_n z^n + a_{n-1} z^{n-1} + \cdots + a_1 z + a_0} \qquad \text{Property 4}$$

$$= \overline{f(z)} = \bar{0} \qquad f(z) = 0, \text{ since } z \text{ is a root}$$

$$= 0 \qquad \text{Property 3}$$

We have now shown that $f(\bar{z}) = 0$, given that $f(z) = 0$. Thus, $\bar{z}$ is a root, as we wished to show.

Although the conjugate roots theorem concerns nonreal complex roots, it can nevertheless be used to obtain information about real roots, as the next two examples demonstrate.

EXAMPLE 1 Solve the equation $f(x) = 2x^4 - 3x^3 + 12x^2 + 22x - 60 = 0$, given that one of the roots is $1 + 3i$.

Solution Since all the coefficients of $f(x)$ are real numbers, we know that the conjugate of $1 + 3i$ must also be a root. Thus, $1 + 3i$ and $1 - 3i$ are roots, from which it follows that $[x - (1 + 3i)]$ and $[x - (1 - 3i)]$ are factors of $f(x)$. As you can check, the product of these two factors is $x^2 - 2x + 10$. Thus, we must have

$$f(x) = (x^2 - 2x + 10) \cdot Q(x)$$

for some polynomial $Q(x)$. We compute $Q(x)$, using long division:

$$
\begin{array}{r}
2x^2 + \ x \ - \ 6 \\
x^2 - 2x + 10 \overline{\smash{\big)}\ 2x^4 - 3x^3 + 12x^2 + 22x - 60} \\
\underline{2x^4 - 4x^3 + 20x^2} \\
x^3 - \ 8x^2 + 22x \\
\underline{x^3 - \ 2x^2 + 10x} \\
-6x^2 + 12x - 60 \\
\underline{-6x^2 + 12x - 60} \\
0
\end{array}
$$

Thus, $Q(x) = 2x^2 + x - 6$, and the original equation becomes

$$f(x) = (x^2 - 2x + 10)(2x^2 + x - 6) = 0$$

We can now find any additional roots by solving the equation $2x^2 + x - 6 = 0$. We have

$$2x^2 + x - 6 = 0$$
$$(2x - 3)(x + 2) = 0$$

$$
\begin{array}{c|c}
2x - 3 = 0 & x + 2 = 0 \\
x = 3/2 & x = -2
\end{array}
$$

We now have four distinct roots of the original equation: $1 + 3i$, $1 - 3i$, $\frac{3}{2}$, and -2. Since the degree of the equation is 4, there can be no other roots. ▮▮▮

▮▮ ▮ **EXAMPLE 2** Explain why the equation $x^3 - 2x^2 + x - 1 = 0$ has at least one irrational root.

Solution Allowing for multiple roots, the given equation has three roots. Therefore, since nonreal complex roots occur in pairs, the equation can have either two nonreal complex roots or no nonreal complex roots. In either case, then, the remaining root or roots must be real. Furthermore, the equation has no rational roots; this is easily checked using the rational roots theorem. It follows now that the equation must have at least one irrational root. (The reasoning here relies on the fact that each real number is either rational or irrational, but not both.) ▮▮▮

There is a theorem, similar to the conjugate roots theorem, that tells us about irrational roots of the form $a + b\sqrt{c}$. As background for this theorem, let's look at two preliminary examples. First, we consider the equation $x^2 - 2x - 5 = 0$. As you can check, the roots in this case are $1 + \sqrt{6}$ and $1 - \sqrt{6}$. However, it is not true in general that irrational roots such as these always occur in pairs. Consider as a second example the quadratic equation

$$(x + 2)(x - \sqrt{3}) = 0$$

or
$$x^2 + (2 - \sqrt{3})x - 2\sqrt{3} = 0$$

Here, one of the roots is $\sqrt{3}$, yet $-\sqrt{3}$ is not a root. This type of behavior can occur in polynomial equations where not all of the coefficients are rational. On the other hand, when the coefficients are all rational, we do have the following theorem. (See Exercise 41 at the end of this section for a proof.)

Theorem

Let $f(x)$ be a polynomial in which all the coefficients are rational. Suppose that $a + b\sqrt{c}$ is a root of the equation $f(x) = 0$, where a, b, and c are rational and $\sqrt{c}$ is irrational. Then $a - b\sqrt{c}$ is also a root of the equation.

■■■ EXAMPLE 3 Find a quadratic equation with rational coefficients and a leading coefficient of 1 such that one of the roots is $r_1 = 4 + 5\sqrt{3}$.

Solution If one root is $r_1 = 4 + 5\sqrt{3}$, then the other must be $r_2 = 4 - 5\sqrt{3}$. We denote the required equation by $x^2 + bx + c = 0$. Thus, according to Table 2 in Section 11.4, we have

$$b = -(r_1 + r_2) = -\left[(4 + 5\sqrt{3}) + (4 - 5\sqrt{3})\right] = -8$$

and

$$c = r_1 r_2 = (4 + 5\sqrt{3})(4 - 5\sqrt{3}) = 16 - 75 = -59$$

The required equation is therefore $x^2 - 8x - 59 = 0$. This answer can also be obtained without using the table. Since the roots are $4 \pm 5\sqrt{3}$, we can write the required equation as $\left[x - (4 + 5\sqrt{3})\right]\left[x - (4 - 5\sqrt{3})\right] = 0$. As you can now check by multiplying out the two factors, this equation is equivalent to $x^2 - 8x - 59 = 0$, as obtained previously.   **■■■**

We conclude this section with a discussion of **Descartes's rule of signs**. This rule, published by Descartes in 1637, provides us with information about the types of roots an equation can have, even before we attempt to solve the equation. In order to state Descartes's rule of signs, we first explain what is meant by a variation in sign in a polynomial with real coefficients. Suppose that $f(x)$ is a polynomial with real coefficients, written in descending (or ascending) powers of x. For example, let $f(x) = 2x^3 - 4x^2 - 3x + 1$. Then we say that there is a **variation in sign** if two successive coefficients have opposite signs. In the case of $f(x) = 2x^3 - 4x^2 - 3x + 1$, there are two variations in sign, the first occurring as we go from 2 to -4 and the second occurring as we go from -3 to 1. In looking for variations in sign, we ignore terms with zero coefficients. Table 1 shows a few more examples of how we count variations in sign.

We now state Descartes's rule of signs and look at some examples. Although the proof of this theorem is not difficult, it is rather lengthy and we shall omit it here.

TABLE 1

Polynomial	Number of Variations in Sign
$x^2 + 4x$	0
$-3x^5 + x^2 + 1$	1
$x^3 + 3x^2 - x + 6$	2

> **Descartes's Rule of Signs**
>
> Let $f(x)$ be a polynomial with all real coefficients, and consider the equation $f(x) = 0$. Then:
>
> **a.** The number of positive roots either is equal to the number of variations in sign of $f(x)$ or is less than that by an even integer.
>
> **b.** The number of negative roots either is equal to the number of variations in sign of $f(-x)$ or is less than that by an even integer.

■■■ EXAMPLE 4 Use Descartes's rule of signs to obtain information regarding the roots of the equation $x^3 + 8x + 5 = 0$.

Solution Let $f(x) = x^3 + 8x + 5$. Then, since there are no variations in sign for $f(x)$, we see from part (a) of Descartes's rule that the given equation has no positive roots. Next we compute $f(-x)$ to learn about the possibilities for negative roots: we have $f(-x) = -x^3 - 8x + 5$. So $f(-x)$ has one sign change, and consequently [by part (b) of Descartes's rule] the original equation has one negative root. Furthermore, notice that zero is not a root of the equation. Thus, the equation has only one real root, a negative root. Since the equation has a total of three roots, we can conclude that we have one negative root and two nonreal complex roots, where the two nonreal roots will be complex conjugates.   **■■■**

■■■ EXAMPLE 5 Use Descartes's rule to obtain information regarding the roots of the equation $x^4 + 3x^2 - 7x - 5 = 0$.

Solution Let $f(x) = x^4 + 3x^2 - 7x - 5$. Then $f(x)$ has one variation in sign. So according to part (a) of Descartes's rule, the equation has one positive root. That leaves us three roots still to account for, since the degree of the equation is 4. We have $f(-x) = x^4 + 3x^2 + 7x - 5$. Since $f(-x)$ has one sign change, we know from part (b) of Descartes's rule that the equation has one negative root. Noting now that zero is not a root, we conclude that the two remaining roots must be nonreal complex roots. In summary, then, the equation has one positive root, one negative root, and two nonreal complex (conjugate) roots.   **■■■**

■■■ EXAMPLE 6 Use Descartes's rule to obtain information regarding the roots of the equation $f(x) = x^3 - x^2 + 3x + 2 = 0$.

Solution Since $f(x)$ has two variations in sign, the given equation has either two positive roots or no positive roots. To see how many negative roots are possible, we compute

$$f(-x) = (-x)^3 - (-x)^2 + 3(-x) + 2$$
$$= -x^3 - x^2 - 3x + 2$$

Since $f(-x)$ has one variation in sign, we conclude from part (b) of Descartes's rule that the equation has exactly one negative root. In summary, then, there are two possibilities:

either: one negative root and two positive roots;

or: one negative root and two nonreal complex roots.   **■■■**

By using Descartes's rule in Examples 4 and 5, we were able to determine the exact numbers of positive roots and negative roots for the given equations. As Example 6 indicates, however, there are cases in which a direct application of Descartes's rule provides several distinct possibilities for the types of roots, rather than a single definitive result. (Exercises 55 and 56 in the Review Exercises for this chapter illustrate a technique that is sometimes useful in gaining additional information from Descartes's rule. In particular, Exercise 56 will show you that the equation in Example 6 has no positive roots.)

EXERCISE SET 11.6

A

In Exercises 1–16, an equation is given, followed by one or more roots of the equation. In each case, determine the remaining roots.

1. $x^2 - 14x + 53 = 0$; $x = 7 - 2i$
2. $x^2 - x - \frac{1535}{4} = 0$; $x = \frac{1}{2} + 8\sqrt{6}$
3. $x^3 - 13x^2 + 59x - 87 = 0$; $x = 5 + 2i$
4. $x^4 - 10x^3 + 30x^2 - 10x - 51 = 0$; $x = 4 + i$
5. $x^4 + 10x^3 + 38x^2 + 66x + 45 = 0$; $x = -2 + i$
6. $2x^3 + 11x^2 + 30x - 18 = 0$; $x = -3 - 3i$
7. $4x^3 - 47x^2 + 232x + 61 = 0$; $x = 6 - 5i$
8. $9x^4 + 18x^3 + 20x^2 - 32x - 64 = 0$; $x = -1 + \sqrt{3}\,i$
9. $4x^4 - 32x^3 + 81x^2 - 72x + 162 = 0$; $x = 4 + \sqrt{2}\,i$
10. $2x^4 - 17x^3 + 137x^2 - 57x - 65 = 0$; $x = 4 - 7i$
11. $x^4 - 22x^3 + 140x^2 - 128x - 416 = 0$; $x = 10 + 2i$
12. $4x^4 - 8x^3 + 24x^2 - 20x + 25 = 0$; $x = \frac{1}{2}(1 + 3i)$
13. $15x^3 - 16x^2 + 9x - 2 = 0$; $x = \frac{1}{3}(1 + \sqrt{2}i)$
14. $x^5 - 5x^4 + 30x^3 + 18x^2 + 92x - 136 = 0$; $x = -1 + i\sqrt{3}$, $x = 3 - 5i$
15. $x^7 - 3x^6 - 4x^5 + 30x^4 + 27x^3 - 13x^2 - 64x + 26 = 0$; $x = 3 - 2i$, $x = -1 + i$, $x = 1$
16. $x^6 - 2x^5 - 2x^4 + 2x^3 + 2x + 1 = 0$; $x = 1 + \sqrt{2}$

In Exercises 17–20, find a quadratic equation with rational coefficients and a root that is the given number. Write your answer so that the coefficient of x^2 is 1. Use either of the methods shown in Example 3.

17. $r_1 = 1 + \sqrt{6}$
18. $r_1 = 2 - \sqrt{3}$
19. $r_1 = \frac{1}{3}(2 + \sqrt{10})$
20. $r_1 = \frac{1}{2} + \frac{1}{4}\sqrt{5}$

In Exercises 21–36, use Descartes's rule of signs to obtain information regarding the roots of the equations.

21. $x^3 + 5 = 0$
22. $x^4 + x^2 + 1 = 0$
23. $2x^5 + 3x + 4 = 0$
24. $x^3 + 8x - 3 = 0$
25. $5x^4 + 2x - 7 = 0$
26. $x^3 - 4x^2 + x - 1 = 0$
27. $x^3 - 4x^2 - x - 1 = 0$
28. $x^8 + 4x^6 + 3x^4 + 2x^2 + 5 = 0$
29. $3x^8 + x^6 - 2x^2 - 4 = 0$
30. $12x^4 - 5x^3 - 7x^2 - 4 = 0$
31. $x^9 - 2 = 0$
32. $x^9 + 2 = 0$
33. $x^8 - 2 = 0$
34. $x^8 + 2 = 0$
35. $x^6 + x^2 - x - 1 = 0$
36. $x^7 + x^2 - x - 1 = 0$

B

37. Consider the equation $x^4 + cx^2 + dx - e = 0$, where c, d, and e are positive. Show that the equation has one positive root, one negative root, and two nonreal complex roots.

38. Consider the equation $x^n - 1 = 0$.
    (a) Show that the equation has $n - 2$ nonreal complex roots when n is even.
    (b) How many nonreal complex roots are there when n is odd?

39. Find the polynomial $f(x)$ of lowest degree with rational coefficients and with leading coefficient 1, such that $\sqrt{3} + 2i$ is a root of the equation $f(x) = 0$.

40. Find a quadratic polynomial $f(x)$ with integer coefficients, such that $(2 + \sqrt{3})/(2 - \sqrt{3})$ is a root of the equation $f(x) = 0$. *Hint:* First rationalize the given root.

C

41. Let $f(x)$ be a polynomial with rational coefficients. Suppose that $a + b\sqrt{c}$ is a root of $f(x) = 0$, where a, b, and c are rational and $\sqrt{c}$ is irrational. Complete the following steps to prove that $a - b\sqrt{c}$ is also a root of the equation $f(x) = 0$.
    (a) If $b = 0$, we're done. Why?
    (b) (From now on we'll assume that $b \neq 0$.) Let

$d(x) = [x - (a + b\sqrt{c})][x - (a - b\sqrt{c})]$. Explain why $d(a + b\sqrt{c}) = 0$.

(c) Verify that $d(x) = (x - a)^2 - b^2c$. Thus, $d(x)$ is a quadratic polynomial with rational coefficients.

(d) Now suppose that we use the long division process to divide the polynomial $f(x)$ by the quadratic polynomial $d(x)$. We'll obtain a quotient $Q(x)$ and a remainder. Since the degree of $d(x)$ is 2, our remainder will be of degree 1 or less. In other words, the general form of this remainder will be $Cx + D$. Furthermore, C and D will have to be rational, because all of the coefficients in $f(x)$ and

in $d(x)$ are rational. In summary, we have the identity

$$f(x) = d(x) \cdot Q(x) + (Cx + D)$$

Now make the substitution $x = a + b\sqrt{c}$ in this identity, and conclude that $C = D = 0$.

(e) Using the result in part (d), we have

$$f(x) = [x - (a + b\sqrt{c})][x - (a - b\sqrt{c})] \cdot Q(x)$$

Let $x = a - b\sqrt{c}$ in this last identity and conclude that $a - b\sqrt{c}$ is a root of $f(x) = 0$, as required.

42. Find a polynomial $f(x)$ with integer coefficients, such that one root of $f(x) = 0$ is $x = \sqrt{2} + \sqrt[3]{2}$.

▶ OPTIONAL TI-81 GRAPHING CALCULATOR EXERCISES
EXERCISES FOR SECTIONS 11.3–11.6

In Exercises 1–9, you are given a polynomial equation $f(x) = 0$. In parentheses following each equation is the example number and the section in which that equation was considered. For each equation $f(x) = 0$, (a) graph the corresponding function $y = f(x)$, and (b) use the ZOOM and TRACE keys to determine (or at least approximate) the x-intercepts of the graph. These x-intercepts are the roots of the equation $f(x) = 0$. In each case, check that your values are consistent with the results in the text.

1. $x^4 + x^2 - 6 = 0$ (Example 1 in 11.3)
2. $2x^3 - 5x^2 + x - 6 = 0$ (Example 3 in 11.3)
3. $x^3 - 2x + 1 = 0$ (Example 5 in 11.3)
4. $x^4 + 2x^3 - 7x^2 - 20x - 12 = 0$ (Example 6 in 11.3)
5. $2x^3 - x^2 - 9x - 4 = 0$ (Example 1 in 11.5)
6. $\frac{1}{4}x^4 - \frac{3}{4}x^3 + \frac{17}{4}x^2 + 4x + 5 = 0$ (Example 2 in 11.5)
7. $x^3 + 2x - 4 = 0$ (Example 3 in 11.5)
8. $2x^4 - 3x^3 + 12x^2 + 22x - 60 = 0$ (Example 1 in 11.6)
9. $x^3 - 2x^2 + x - 1 = 0$ (Example 2 in 11.6)

10. According to Descartes's rule, the equation $x^3 + 8x + 5 = 0$ has one negative root and two nonreal complex roots. (See Example 4 in Section 11.6.) Verify this result by graphing the function $f(x) = x^3 + 8x + 5$.

11. According to Descartes's rule, the equation $x^4 + 3x^2 - 7x - 5 = 0$ has one negative root, one positive root, and two nonreal complex roots. (See Example 5 in Section 11.6.) Verify this result by graphing the function $f(x) = x^4 + 3x^2 - 7x - 5$.

12. Consider the equation $x^3 - x^2 + 3x + 2 = 0$. In Example 6 of Section 11.6 we found that there were two possible cases for the roots:

    Case 1:   one negative root and two positive roots

    Case 2:   one negative root and two nonreal complex roots

    (a) By graphing the function $f(x) = x^3 - x^2 + 3x + 2$, determine which of the two cases, in fact, describes the roots.
    (b) Use the ZOOM and the TRACE keys to approximate the real root(s).

11.7   DEMOIVRE'S THEOREM

Go to Mr. DeMoivre; he knows these things better than I do.

Isaac Newton [according to Boyer's *A History of Mathematics* (New York: John Wiley and Sons, Inc., 1968)]

In this section, we explore one of the many important connections between trigonometry and the complex number system. We begin with an observation that

FIGURE 1

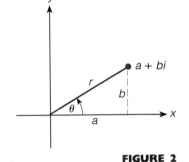

FIGURE 2

was first made in the year 1797 by the Norwegian surveyor and mathematician Caspar Wessel. He realized, essentially, that the complex numbers can be visualized as points in the x-y plane by identifying the complex number $a + bi$ with the point (a, b). In this context, we often refer to the x-y plane as the **complex plane**, and we refer to the complex numbers as points in this plane.

■■ ▍ **EXAMPLE 1**   Plot the point $2 + 3i$ in the complex plane.

Solution   The complex number $2 + 3i$ is identified with the point $(2, 3)$. See Figure 1.   ■■■

As indicated in Figure 2, the distance from the origin to the point $a + bi$ is denoted by r. We call the distance r the **modulus** of the complex number $a + bi$. The angle θ in Figure 2 (measured counterclockwise from the positive x-axis) is referred to as the **argument** of the complex number $a + bi$. (Using the terminology of Section 8.5, r and θ are the **polar coordinates** of the point $a + bi$.)

From Figure 2, we have the following three equations relating the quantities a, b, r, and θ. (Although Figure 2 shows $a + bi$ in the first quadrant, the equations remain valid for the other quadrants as well.)

$$r = \sqrt{a^2 + b^2} \tag{1}$$

$$a = r \cos \theta \tag{2}$$

$$b = r \sin \theta \tag{3}$$

If we have a complex number $z = a + bi$, we can use equations (2) and (3) to write

$$z = (r \cos \theta) + (r \sin \theta)i = r(\cos \theta + i \sin \theta)$$

That is,

$$z = r(\cos \theta + i \sin \theta) \tag{4}$$

The expression that appears on the right-hand side of equation (4) is called the **trigonometric** (or **polar**) **form** of the complex number z. In contrast, the expression $a + bi$ is referred to as the **rectangular form** of the complex number z.

■■ ▍ **EXAMPLE 2**   Express the complex number $z = 3\left(\cos \frac{\pi}{3} + i \sin \frac{\pi}{3}\right)$ in rectangular form.

Solution

$$z = 3\left(\cos \tfrac{\pi}{3} + i \sin \tfrac{\pi}{3}\right)$$
$$= 3\left(\tfrac{1}{2} + \tfrac{1}{2}\sqrt{3}i\right) = \tfrac{3}{2} + \tfrac{3}{2}\sqrt{3}i$$

The rectangular form is therefore $\tfrac{3}{2} + \tfrac{3}{2}\sqrt{3}i$.   ■■■

■■ ▍ **EXAMPLE 3**   Find the trigonometric form for the complex number $-\sqrt{2} + i\sqrt{2}$.

Solution   We are asked to write the given number in the form $r(\cos \theta + i \sin \theta)$, so we need to find r and θ. Using equation (1) and the values $a = -\sqrt{2}$, $b = \sqrt{2}$, we have

$$r = \sqrt{(-\sqrt{2})^2 + (\sqrt{2})^2} = \sqrt{2 + 2} = \sqrt{4} = 2$$

Now that we know r, we can use equations (2) and (3) to determine θ. From equation (2), we obtain

$$\cos \theta = \frac{a}{r} = \frac{-\sqrt{2}}{2} \tag{5}$$

Similarly, equation (3) gives us

$$\sin \theta = \frac{b}{r} = \frac{\sqrt{2}}{2} \tag{6}$$

One angle satisfying both of equations (5) and (6) is $\theta = 3\pi/4$. (There are other angles, and we'll return to this point in a moment.) In summary, then, we have $r = 2$ and $\theta = 3\pi/4$, so the required trigonometric form is

$$2\left(\cos \tfrac{3\pi}{4} + i \sin \tfrac{3\pi}{4}\right)$$ ▮▮▮

In the example we just completed, we noted that $\theta = 3\pi/4$ was only one angle satisfying the conditions $\cos \theta = -\sqrt{2}/2$ and $\sin \theta = \sqrt{2}/2$. Another such angle is $\frac{3\pi}{4} + 2\pi$. Indeed, any angle of the form $\frac{3\pi}{4} + 2\pi k$, where k is an integer, would do just as well. The upshot of this is that θ, the argument of a complex number, is not uniquely determined. In Example 3, we followed a common convention in converting to trigonometric form: we picked θ in the interval $0 \leq \theta < 2\pi$. Furthermore, although it won't cause us any difficulties in this section, you might also note that the argument θ is undefined for the complex number $0 + 0i$. (Why?)

We are now ready to derive a formula that will make it easy to multiply two complex numbers in trigonometric form. Suppose that the two complex numbers are

$$r(\cos \alpha + i \sin \alpha) \qquad \text{and} \qquad R(\cos \beta + i \sin \beta)$$

Then their product is

$$rR[(\cos \alpha + i \sin \alpha)(\cos \beta + i \sin \beta)]$$
$$= rR[(\cos \alpha \cos \beta - \sin \alpha \sin \beta) + i(\sin \alpha \cos \beta + \cos \alpha \sin \beta)]$$
$$= rR[\cos(\alpha + \beta) + i \sin(\alpha + \beta)] \qquad \text{using the addition formulas from Section 7.6}$$

Notice that the modulus of the product is rR, which is the product of the two original moduli. Also, the argument is $\alpha + \beta$, which is the *sum* of the two original arguments. So, to multiply two complex numbers, just multiply the moduli and add the arguments. There is a similar rule for obtaining the quotient of two complex numbers: Divide their moduli and subtract the arguments. These two rules are stated more precisely in the box that follows. (For a proof of the division rule, see Exercise 75.)

Let $z = r(\cos \alpha + i \sin \alpha)$ and $w = R(\cos \beta + i \sin \beta)$. Then

$$zw = rR[\cos(\alpha + \beta) + i \sin(\alpha + \beta)]$$

Also, if $R \neq 0$, then

$$\frac{z}{w} = \frac{r}{R}[\cos(\alpha - \beta) + i \sin(\alpha - \beta)]$$

EXAMPLE 4 Let $z = 8\left(\cos \frac{5\pi}{3} + i \sin \frac{5\pi}{3}\right)$ and $w = 4\left(\cos \frac{2\pi}{3} + i \sin \frac{2\pi}{3}\right)$. Compute **(a)** zw; **(b)** z/w.
Express each answer in both trigonometric and rectangular form.

Solution

(a) $zw = (8)(4)\left[\cos\left(\frac{5\pi}{3} + \frac{2\pi}{3}\right) + i \sin\left(\frac{5\pi}{3} + \frac{2\pi}{3}\right)\right]$

$\qquad = 32\left(\cos \frac{7\pi}{3} + i \sin \frac{7\pi}{3}\right) = 32\left(\cos \frac{\pi}{3} + i \sin \frac{\pi}{3}\right)$ $\qquad$ trigonometric form

$\qquad = 32\left(\frac{1}{2} + \frac{1}{2}\sqrt{3}i\right) = 16 + 16\sqrt{3}i$ $\qquad$ rectangular form

(b) $\dfrac{z}{w} = \dfrac{8}{4}\left[\cos\left(\frac{5\pi}{3} - \frac{2\pi}{3}\right) + i \sin\left(\frac{5\pi}{3} - \frac{2\pi}{3}\right)\right]$

$\qquad = 2(\cos \pi + i \sin \pi)$ $\qquad$ trigonometric form

$\qquad = 2(-1 + i \cdot 0) = -2$ $\qquad$ rectangular form

EXAMPLE 5 Compute z^2, where $z = r(\cos \theta + i \sin \theta)$.

Solution $z^2 = [r(\cos \theta + i \sin \theta)][r(\cos \theta + i \sin \theta)]$

$\qquad = r^2[\cos(\theta + \theta) + i \sin(\theta + \theta)]$

$\qquad = r^2(\cos 2\theta + i \sin 2\theta)$

The result in Example 5 is a particular case of an important theorem attributed to Abraham DeMoivre (1667–1754). In the box that follows, we state *DeMoivre's theorem*. (The theorem can be proved using mathematical induction, which is discussed in Chapter 12.)

DeMoivre's Theorem

Let n be a natural number. Then

$$[r(\cos \theta + i \sin \theta)]^n = r^n(\cos n\theta + i \sin n\theta)$$

EXAMPLE 6 Use DeMoivre's theorem to compute $\left(-\sqrt{2} + i\sqrt{2}\right)^5$.
Express your answer in rectangular form.

Solution In Example 3, we saw that the trigonometric form of $-\sqrt{2} + i\sqrt{2}$ is given by

$$-\sqrt{2} + i\sqrt{2} = 2\left(\cos \frac{3\pi}{4} + i \sin \frac{3\pi}{4}\right)$$

Therefore,

$$(-\sqrt{2} + i\sqrt{2})^5 = 2^5(\cos \tfrac{15\pi}{4} + i \sin \tfrac{15\pi}{4})$$
$$= 32[\tfrac{1}{2}\sqrt{2} + i(-\tfrac{1}{2}\sqrt{2})]$$
$$= 16\sqrt{2} - 16\sqrt{2}i$$

The next two examples show how DeMoivre's theorem is used in computing roots. If n is a natural number and $z^n = w$, then we say that z is an **nth root** of w. The work in the examples also relies on the following observation about equality between nonzero complex numbers in trigonometric form: If $r(\cos \theta + i \sin \theta) = R(\cos A + i \sin A)$, then $r = R$ and $\theta = A + 2\pi k$, where k is an integer.

I   EXAMPLE 7 Find the cube roots of $8i$.

Solution First we express $8i$ in trigonometric form. As can be seen from Figure 3,

$$8i = 8(\cos \tfrac{\pi}{2} + i \sin \tfrac{\pi}{2})$$

Now we let $z = r(\cos \theta + i \sin \theta)$ denote a cube root of $8i$. Then the equation $z^3 = 8i$ becomes

$$r^3(\cos 3\theta + i \sin 3\theta) = 8(\cos \tfrac{\pi}{2} + i \sin \tfrac{\pi}{2}) \tag{7}$$

From equation (7), we conclude that $r^3 = 8$ and, consequently, that $r = 2$. Also from equation (7), we have

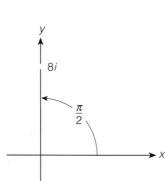

FIGURE 3

$$3\theta = \tfrac{\pi}{2} + 2\pi k$$

or

$$\theta = \tfrac{\pi}{6} + \tfrac{2\pi k}{3} \qquad \text{dividing by 3} \tag{8}$$

If we let $k = 0$, equation (8) yields $\theta = \pi/6$. Thus, one of the cube roots of $8i$ is

$$2(\cos \tfrac{\pi}{6} + i \sin \tfrac{\pi}{6}) = 2(\tfrac{1}{2}\sqrt{3} + i \cdot \tfrac{1}{2}) = \sqrt{3} + i$$

Next we let $k = 1$ in equation (8). As you can check, this yields $\theta = 5\pi/6$. So another cube root of $8i$ is

$$2(\cos \tfrac{5\pi}{6} + i \sin \tfrac{5\pi}{6}) = 2(-\tfrac{1}{2}\sqrt{3} + i \cdot \tfrac{1}{2}) = -\sqrt{3} + i$$

Similarly, using $k = 2$ in equation (8) yields $\theta = 3\pi/2$. (Verify this.) Consequently, a third cube root is

$$2(\cos \tfrac{3\pi}{2} + i \sin \tfrac{3\pi}{2}) = 2[0 + i(-1)] = -2i$$

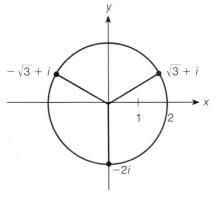

FIGURE 4

We have now found three distinct cube roots of $8i$. If we were to continue the process, using $k = 3$, for example, we would find that no additional roots are obtained in this manner. We therefore conclude that there are exactly three cube roots of $8i$. We have plotted these cube roots in Figure 4. Notice that the points lie equally spaced on a circle of radius 2 about the origin.

When we used DeMoivre's theorem in Example 7, we found that the number $8i$ had three distinct cube roots. Along these same lines, it is true in general that any nonzero number $a + bi$ possesses exactly n distinct nth roots. This is true even if $b = 0$. In the next example, for instance, we compute the five fifth roots of 2.

■▨■ EXAMPLE 8 Compute the five fifth roots of 2.

Solution We will follow the procedure we used in Example 7. In trigonometric form, the number 2 becomes $2(\cos 0 + i \sin 0)$. Now we let $z = r(\cos \theta + i \sin \theta)$ denote a fifth root of 2. Then the equation $z^5 = 2$ becomes

$$r^5(\cos 5\theta + i \sin 5\theta) = 2(\cos 0 + i \sin 0) \qquad (9)$$

From equation (9), we see that $r^5 = 2$ and, consequently, $r = 2^{1/5}$. Also from equation (9), we have

$$5\theta = 0 + 2\pi k \qquad \text{or} \qquad \theta = \frac{2\pi k}{5}$$

Using the values $k = 0, 1, 2, 3,$ and 4 in succession, we obtain the following results:

k	0	1	2	3	4
θ	0	$2\pi/5$	$4\pi/5$	$6\pi/5$	$8\pi/5$

The five fifth roots of 2 are therefore

$$z_1 = 2^{1/5}(\cos 0 + i \sin 0) = 2^{1/5} \qquad z_4 = 2^{1/5}\left(\cos \tfrac{6\pi}{5} + i \sin \tfrac{6\pi}{5}\right)$$
$$z_2 = 2^{1/5}\left(\cos \tfrac{2\pi}{5} + i \sin \tfrac{2\pi}{5}\right) \qquad z_5 = 2^{1/5}\left(\cos \tfrac{8\pi}{5} + i \sin \tfrac{8\pi}{5}\right)$$
$$z_3 = 2^{1/5}\left(\cos \tfrac{4\pi}{5} + i \sin \tfrac{4\pi}{5}\right)$$

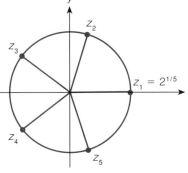

FIGURE 5

In Figure 5, we have plotted these five fifth roots. Notice that the points are equally spaced on a circle of radius $2^{1/5}$ about the origin. ■■■

EXERCISE SET 11.7

A

In Exercises 1–8, plot each point in the complex plane.

1. $4 + 2i$
2. $4 - 2i$
3. $-5 + i$
4. $-3 - 5i$
5. $1 - 4i$
6. i
7. $-i$
8. $1 (= 1 + 0i)$

In Exercises 9–18, convert each complex number to rectangular form.

9. $2\left(\cos \tfrac{1}{4}\pi + i \sin \tfrac{1}{4}\pi\right)$
10. $6\left(\cos \tfrac{5}{3}\pi + i \sin \tfrac{5}{3}\pi\right)$
11. $4\left(\cos \tfrac{5}{6}\pi + i \sin \tfrac{5}{6}\pi\right)$
12. $3\left(\cos \tfrac{3}{2}\pi + i \sin \tfrac{3}{2}\pi\right)$
13. $\sqrt{2}(\cos 225° + i \sin 225°)$
14. $\tfrac{1}{2}(\cos 240° + i \sin 240°)$
15. $\sqrt{3}\left(\cos \tfrac{1}{2}\pi + i \sin \tfrac{1}{2}\pi\right)$
16. $5(\cos \pi + i \sin \pi)$
17. $4(\cos 75° + i \sin 75°)$ *Hint:* Use the addition formulas from Section 7.6 to evaluate $\cos 75°$ and $\sin 75°$.

18. $2\left(\cos \tfrac{1}{8}\pi + i \sin \tfrac{1}{8}\pi\right)$ *Hint:* Use the half-angle formulas from Section 7.7 to evaluate $\cos(\pi/8)$ and $\sin(\pi/8)$.

In Exercises 19–28, convert from rectangular to trigonometric form. (In each case, choose an argument θ such that $0 \le \theta < 2\pi$.)

19. $\tfrac{1}{2}\sqrt{3} + \tfrac{1}{2}i$
20. $\sqrt{2} + \sqrt{2}i$
21. $-1 + \sqrt{3}i$
22. -4
23. $-2\sqrt{3} - 2i$
24. $-3\sqrt{2} - 3\sqrt{2}i$
25. $-6i$
26. $-4 - 4\sqrt{3}i$
27. $\tfrac{1}{4}\sqrt{3} - \tfrac{1}{4}i$
28. 16

In Exercises 29–54, carry out the indicated operations. Express your results in rectangular form for those cases in which the trigonometric functions are readily evaluated without tables or a calculator.

29. $2(\cos 22° + i \sin 22°) \times 3(\cos 38° + i \sin 38°)$

30. $4(\cos 5° + i \sin 5°) \times 6(\cos 130° + i \sin 130°)$

31. $\sqrt{2}\left(\cos \frac{1}{3}\pi + i \sin \frac{1}{3}\pi\right) \times \sqrt{2}\left(\cos \frac{4}{3}\pi + i \sin \frac{4}{3}\pi\right)$

32. $\left(\cos \frac{1}{5}\pi + i \sin \frac{1}{5}\pi\right) \times \left(\cos \frac{1}{20}\pi + i \sin \frac{1}{20}\pi\right)$

33. $3\left(\cos \frac{1}{7}\pi + i \sin \frac{1}{7}\pi\right) \times \sqrt{2}\left(\cos \frac{1}{7}\pi + i \sin \frac{1}{7}\pi\right)$

34. $\sqrt{3}(\cos 3° + i \sin 3°) \times \sqrt{3}(\cos 38° + i \sin 38°)$

35. $6(\cos 50° + i \sin 50°) \div 2(\cos 5° + i \sin 5°)$

36. $\sqrt{3}(\cos 140° + i \sin 140°) \div 3(\cos 5° + i \sin 5°)$

37. $2^{4/3}\left(\cos \frac{5}{12}\pi + i \sin \frac{5}{12}\pi\right) \div 2^{1/3}\left(\cos \frac{1}{4}\pi + i \sin \frac{1}{4}\pi\right)$

38. $\sqrt[3]{6}\left(\cos \frac{16}{9}\pi + i \sin \frac{16}{9}\pi\right) \div \sqrt{2}\left(\cos \frac{1}{9}\pi + i \sin \frac{1}{9}\pi\right)$

39. $\left(\cos \frac{2}{5}\pi + i \sin \frac{2}{5}\pi\right) \div \left(\cos \frac{2}{5}\pi + i \sin \frac{2}{5}\pi\right)$

40. $\left(\cos \frac{2}{5}\pi + i \sin \frac{2}{5}\pi\right) \div \left(\cos \frac{1}{10}\pi + i \sin \frac{1}{10}\pi\right)$

41. $\left[3\left(\cos \frac{1}{3}\pi + i \sin \frac{1}{3}\pi\right)\right]^5$

42. $\left[\sqrt{2}\left(\cos \frac{5}{6}\pi + i \sin \frac{5}{6}\pi\right)\right]^4$

43. $\left[\frac{1}{2}\left(\cos \frac{1}{24}\pi + i \sin \frac{1}{24}\pi\right)\right]^6$

44. $\left[\sqrt{3}(\cos 70° + i \sin 70°)\right]^3$

45. $[2^{1/5}(\cos 63° + i \sin 63°)]^{10}$

46. $\left[2\left(\cos \frac{1}{5}\pi + i \sin \frac{1}{5}\pi\right)\right]^3$

47. $2(\cos 200° + i \sin 200°) \times \sqrt{2}(\cos 20° + i \sin 20°)$
    $\times \frac{1}{2}(\cos 5° + i \sin 5°)$

48. $\left[\dfrac{\cos(\pi/8) + i \sin(\pi/8)}{\cos(-\pi/8) + i \sin(-\pi/8)}\right]^5$

49. $\left[\frac{1}{2}\left(1 - \sqrt{3}i\right)\right]^5$ *Hint:* Convert to trigonometric form.

50. $(1 - i)^3$ 51. $(-2 - 2i)^5$

52. $\left(-\frac{1}{2} + \frac{1}{2}\sqrt{3}i\right)^6$ 53. $\left(-2\sqrt{3} - 2i\right)^4$

54. $(1 + i)^{16}$

In Exercises 55–61, use DeMoivre's theorem to find the indicated roots. Express the results in rectangular form.

55. Cube roots of $-27i$ 56. Cube roots of 2

57. Eighth roots of 1   58. Square roots of i

59. Cube roots of 64   60. Square roots of $-\frac{1}{2} - \frac{1}{2}\sqrt{3}i$

61. Sixth roots of 729

Use a calculator to complete Exercises 62–65.

62. Compute $(9 + 9i)^6$.   63. Compute $(7 - 7i)^8$.

64. Compute the cube roots of $1 + 2i$. Express your answers in rectangular form, with the real and imaginary parts rounded off to two decimal places.

65. Compute the fifth roots of i. Express your answers in rectangular form, with the real and imaginary parts rounded off to two decimal places.

B

In Exercises 66–68, find the indicated roots. Express the results in rectangular form.

66. Find the fourth roots of i.   *Hint:* Use the half-angle formulas from Section 7.7.

67. Find the fourth roots of $8 - 8\sqrt{3}i$.   *Hint:* Use the addition formulas or the half-angle formulas.

68. Find the square roots of $7 + 24i$.   *Hint:* You'll need to use the half-angle formulas from Section 7.7.

69. (a) Compute the three cube roots of 1.
    (b) Let z_1, z_2, and z_3 denote the three cube roots of 1. Verify that $z_1 + z_2 + z_3 = 0$ and also that $z_1 z_2 + z_2 z_3 + z_3 z_1 = 0$.

70. (a) Compute the four fourth roots of 1.
    (b) Verify that the sum of these four fourth roots is 0.

71. Evaluate $\left(-\frac{1}{2} + \frac{1}{2}\sqrt{3}i\right)^5 + \left(-\frac{1}{2} - \frac{1}{2}\sqrt{3}i\right)^5$.   *Hint:* Use DeMoivre's theorem.

72. Show that $\left(-\frac{1}{2} + \frac{1}{2}\sqrt{3}i\right)^6 + \left(-\frac{1}{2} - \frac{1}{2}\sqrt{3}i\right)^6 = 2$.

73. Compute $(\cos \theta + i \sin \theta)(\cos \theta - i \sin \theta)$.

74. In the identity $(\cos \theta + i \sin \theta)^2 = \cos 2\theta + i \sin 2\theta$, carry out the multiplication on the left-hand side of the equation. Then equate the corresponding real parts and imaginary parts from each side of the equation that results. What do you obtain?

75. Show that
$$\frac{r(\cos \alpha + i \sin \alpha)}{R(\cos \beta + i \sin \beta)} = \frac{r}{R}[\cos(\alpha - \beta) + i \sin(\alpha - \beta)]$$
Suggestion: Begin with the quantity on the left side and multiply it by $(\cos \beta - i \sin \beta)/(\cos \beta - i \sin \beta)$.

76. Show that
$$1 + \cos \theta + i \sin \theta = 2 \cos\left(\tfrac{\theta}{2}\right)\left[\cos\left(\tfrac{\theta}{2}\right) + i \sin\left(\tfrac{\theta}{2}\right)\right]$$

77. If $z = r(\cos \theta + i \sin \theta)$, show that
$$\frac{1}{z} = \frac{1}{r}(\cos \theta - i \sin \theta)$$
Hint: $1/z = [1(\cos 0 + i \sin 0)]/[r(\cos \theta + i \sin \theta)]$.

78. Show that $\dfrac{1 + \sin \theta + i \cos \theta}{1 + \sin \theta - i \cos \theta} = \sin \theta + i \cos \theta$.
Assume that $\theta \neq \frac{3}{2}\pi + 2\pi k$, where k is an integer.
Hint: Work with the left-hand side; first "rationalize" the denominator by multiplying by the quantity
$$\frac{(1 + \sin \theta) + i \cos \theta}{(1 + \sin \theta) + i \cos \theta}$$

CHAPTER ELEVEN SUMMARY OF PRINCIPAL TERMS AND NOTATION **705**

CHAPTER ELEVEN   SUMMARY OF PRINCIPAL TERMS AND NOTATION

TERMS OR NOTATIONS	PAGE REFERENCE	COMMENT
1. $i^2 = -1$	657	This is the defining equation for the symbol i. The quantity i is not a real number.
2. Complex numbers	657	An expression of the form $a + bi$, where a and b are real numbers, is a complex number. With the exception of division, complex numbers are combined using the usual rules of algebra, along with the convention $i^2 = -1$. Division is carried out using complex conjugates (defined below), as indicated in the box on page 659.
3. If $z = c + di$, then $\bar{z} = c - di$.	658	$\bar{z}$ is called the *conjugate* of z.
4. Division algorithm	665	This theorem summarizes the results of the long division process for polynomials. Suppose that $p(x)$ and $d(x)$ are polynomials and $d(x)$ is not the zero polynomial. Then according to the *division algorithm*, there are unique polynomials $q(x)$ and $R(x)$ such that $$p(x) = d(x) \cdot q(x) + R(x)$$ where either $R(x)$ is the zero polynomial or the degree of $R(x)$ is less than the degree of $d(x)$.
5. Root, solution, zero	670	A *root*, or *solution*, of a polynomial equation $f(x) = 0$ is a number r such that $f(r) = 0$. The root r is also called a *zero* of the function f.
6. The remainder theorem	671	The *remainder theorem* asserts that when a polynomial $f(x)$ is divided by $x - r$, the remainder is $f(r)$. Example 3 in Section 11.3 shows how this theorem can be used with synthetic division to evaluate a polynomial.
7. The factor theorem	673	The *factor theorem* makes two statements about a polynomial $f(x)$. First, if $f(r) = 0$, then $x - r$ is a factor of $f(x)$. And second, if $x - r$ is a factor of $f(x)$, then $f(r) = 0$.
8. The fundamental theorem of algebra	679	Let $f(x)$ be a polynomial of degree 1 or greater. The *fundamental theorem of algebra* asserts that the equation $f(x) = 0$ has at least one root among the complex numbers. (See Example 1 in Section 11.4.)
9. The linear factors theorem	679	Let $f(x) = a_n x^n + a_{n-1} x^{n-1} + \cdots + a_0$, where $n \geq 1$ and $a_n \neq 0$. Then this theorem asserts that $f(x)$ can be expressed as a product of n linear factors: $$f(x) = a_n(x - r_1)(x - r_2) \cdots (x - r_n)$$ (The complex numbers $r_1, r_2, \ldots, r_n$ are not necessarily all distinct, and some or all of them may be real numbers.) On page 681, the linear factors theorem is used to prove that every polynomial equation of degree $n \geq 1$ has exactly n roots, where a root of multiplicity k is counted k times.
10. The rational roots theorem	686	Given a polynomial equation, this theorem tells us which rational numbers are candidates for roots of the equation. For a statement of the theorem, see page 686. A proof of the

TERMS OR NOTATIONS	PAGE REFERENCE	COMMENT
		theorem is outlined in Exercise 39, Exercise Set 11.5. For an example of how the theorem is applied, see Example 1 in Section 11.5.
11. Upper bound (for roots); lower bound (for roots)	687	A real number B is an upper bound for the roots of an equation if every real root is less than or equal to B. Similarly, a real number b is a lower bound for the roots if every real root is greater than or equal to b.
12. The upper and lower bound theorem for real roots	688	This theorem tells how synthetic division can be used in determining upper and lower bounds for roots of equations. For the statement and proof of the theorem, see page 688 in Section 11.5. For a demonstration of how the theorem is applied, see Example 2 in Section 11.5.
13. The location theorem	689	Let $f(x)$ be a polynomial with real coefficients. If a and b are real numbers such that $f(a)$ and $f(b)$ have opposite signs, then the equation $f(x) = 0$ has at least one root between a and b. To see how this theorem is applied, see Example 3 in Section 11.5.
14. The conjugate roots theorem	693	Let $f(x)$ be a polynomial with real coefficients, and suppose that $a + bi$ is a root of the equation $f(x) = 0$, where a and b are real numbers and $b \neq 0$. Then this theorem asserts that $a - bi$ is also a root of the equation. (In other words, for polynomial equations in which all of the coefficients are real numbers, when complex nonreal roots occur, they occur in conjugate pairs.) For illustrations of how this theorem is applied, see Examples 1 and 2 in Section 11.6.
15. Variation in sign	695	Suppose that $f(x)$ is a polynomial with real coefficients, written in descending or ascending powers of x. Then a variation in sign occurs whenever two successive coefficients have opposite signs. For examples, see Table 1 in Section 11.6.
16. Descartes's rule of signs	696	Let $f(x)$ be a polynomial, all of whose coefficients are real numbers, and consider the equation $f(x) = 0$. Then, according to Descartes's rule: **a.** The number of positive roots either is equal to the number of variations in sign of $f(x)$ or is less than that by an even integer. **b.** The number of negative roots either is equal to the number of variations in sign of $f(-x)$ or is less than that by an even integer. Examples 4–6 in Section 11.6 show how Descartes's rule is applied.
17. The complex plane	699	Refer to Figure 2 on page 699. A complex number $a + bi$ can be identified with the point (a, b) in the x-y plane. In this context, the x-y plane is referred to as the *complex plane*. The distance r in the figure is the *modulus* of the complex number $a + bi$; the angle θ is an *argument* of $a + bi$.

TERMS OR NOTATIONS	PAGE REFERENCE	COMMENT
18. $z = r(\cos\theta + i\sin\theta)$	699	The expression on the right-hand side is the *trigonometric form* of the complex number $z = a + bi$. (The expression $a + bi$ is the *rectangular form* of the complex number z.) The trigonometric form and the rectangular form are related by the following equations: $$r = \sqrt{a^2 + b^2} \qquad a = r\cos\theta \qquad b = r\sin\theta$$
19. DeMoivre's theorem	701	Let n be a natural number and let $z = r(\cos\theta + i\sin\theta)$. Then DeMoivre's theorem states that $$z^n = r^n(\cos n\theta + i\sin n\theta)$$ For applications of this theorem, see Examples 6, 7, and 8 in Section 11.7.

WRITING MATHEMATICS

Discuss each of the following statements with a classmate and decide whether it is true or false. Then (on your own), write out the reason (or reasons) for each decision.

1. Every equation has a root.
2. Every polynomial equation of degree 4 has four distinct roots.
3. No cubic equation can have a root of multiplicity 4.
4. The degree of the polynomial $x(x-1)(x-2)(x-3)$ is 3.
5. The degree of the polynomial $6(x+1)^2(x-5)^4$ is 6.
6. Every polynomial of degree n, where $n \geq 1$, can be written in the form $(x - r_1)(x - r_2)\cdots(x - r_n)$.
7. The sum of the roots of the polynomial equation $x^2 - px + q = 0$ is p.
8. The product of the roots of the polynomial equation $2x^3 - x^2 + 3x - 1 = 0$ is 1.
9. Although a polynomial equation of degree n may have n distinct roots, the fundamental theorem of algebra tells us how to find only one of the roots.
10. Every polynomial equation of degree $n \geq 1$ has at least one real root.
11. If all the coefficients in a polynomial equation are real, then at least one of the roots must be real.
12. Every cubic equation whose roots are $\sqrt{5}$, $\sqrt{6}$, and $\sqrt{7}$ can be written in the form
$$a_n(x - \sqrt{5})(x - \sqrt{6})(x - \sqrt{7}) = 0$$
13. Every polynomial equation of degree at least 1 has at least one real root.
14. According to the rational roots theorem, there are four possibilities for the rational roots of the equation $x^5 + 6x^2 - 2 = 0$.
15. According to the rational roots theorem, there are only two possibilities for the rational roots of the equation $\frac{1}{3}x^3 - 5x + 1 = 0$.
16. The sum of the roots of the equation $x^2 - 12x + 16 = 0$ is -12.
17. According to the location theorem, if $f(x)$ is a polynomial and $f(a)$ and $f(b)$ have the same sign, then the equation $f(x) = 0$ has at least one root between a and b.

18. According to Descartes's rule, the equation $x^7 - 4x + 3 = 0$ has two positive roots.

CHAPTER ELEVEN   REVIEW EXERCISES

In Exercises 1 and 2, you are given polynomials p(x) and d(x). In each case, use synthetic division to determine polynomials q(x) and R(x) such that

$$p(x) = d(x) \cdot q(x) + R(x)$$

where either R(x) = 0 or the degree of R(x) is less than the degree of d(x).

1. $p(x) = x^4 + 3x^3 - x^2 - 5x + 1;\ d(x) = x + 2$

2. $p(x) = 4x^4 + 2x + 1;\ d(x) = x - 2$

In Exercises 3–8, use synthetic division to find the quotients and the remainders.

3. $\dfrac{x^4 - 2x^2 + 8}{x - 3}$

4. $\dfrac{x^3 - 1}{x - 2}$

5. $\dfrac{2x^3 - 5x^2 - 6x - 3}{x + 4}$

6. $\dfrac{x^3 + x - 3\sqrt{2}}{x - \sqrt{2}}$

7. $\dfrac{5x^2 - 19x - 4}{x + 0.2}$

8. $\dfrac{x^3 - 3a^2x^2 - 4a^4x + 9a^6}{x - a^2}$

In Exercises 9–16, use synthetic division and the remainder theorem to find the indicated values of the functions. Use a calculator for Exercises 15 and 16.

9. $f(x) = x^5 - 10x + 4;\ f(10)$

10. $f(x) = x^4 + 2x^3 - x;\ f(-2)$

11. $f(x) = x^3 - 10x^2 + x - 1;\ f\left(\frac{1}{10}\right)$

12. $f(x) = x^4 - 2a^2x^2 + 3a^3x - a^4;\ f(-a)$

13. $f(x) = x^3 + 3x^2 + 3x + 1;\ f(a - 1)$

14. $f(x) = x^3 - 1;\ f(1.1)$

15. $f(x) = x^4 + 4x^3 - 6x^2 - 8x - 2;$
   (a) $f(-0.3)$ (Round off the result to two decimal places.)
   (b) $f(-0.39)$ (Round off the result to three decimal places.)
   (c) $f(-0.394)$ (Round off the result to five decimal places.)

16. $f(-4.907378)$, where f is the function in Exercise 15. (Round off the result to three decimal places.)

17. Find a value for a such that 3 is a root of the equation $x^3 - 4x^2 - ax - 6 = 0$.

18. For which values of b will -1 be a root of the equation $x^3 + 2b^2x^2 + x - 48 = 0$?

19. For which values of a will $x - 1$ be a factor of the polynomial $a^2x^3 + 3ax^2 + 2$?

20. Use synthetic division to verify that $\sqrt{2} - 1$ is a root of the equation $x^6 + 14x^3 - 1 = 0$.

21. Let $f(x) = ax^3 + bx^2 + cx + d$ and suppose that r is a root of the equation $f(x) = 0$.
   (a) Show that $r - h$ is a root of $f(x + h) = 0$.
   (b) Show that $-r$ is a root of the equation $f(-x) = 0$.
   (c) Show that kr is a root of the equation $f(x/k) = 0$.

22. Suppose that r is a root of the equation $a_2x^2 + a_1x + a_0 = 0$. Show that mr is a root of the quadratic equation $a_2x^2 + ma_1x + m^2a_0 = 0$.

In Exercises 23–28, list the possibilities for the rational roots of the equations.

23. $x^5 - 12x^3 + x - 18 = 0$

24. $x^5 - 12x^3 + x - 17 = 0$

25. $2x^4 - 125x^3 + 3x^2 - 8 = 0$

26. $\frac{2}{5}x^3 - 8x^2 - \frac{1}{2}x + \frac{3}{2} = 0$

27. $x^3 + x - p = 0$, where p is a prime number

28. $x^3 + x - pq = 0$, where both p and q are prime numbers

In Exercises 29–36, each equation has at least one rational root. Solve the equations. Suggestion: Use the upper and lower bound theorem to eliminate some of the possibilities for rational roots.

29. $2x^3 + x^2 - 7x - 6 = 0$

30. $x^3 + 6x^2 - 8x - 7 = 0$

31. $2x^3 - x^2 - 14x + 10 = 0$

32. $2x^3 + 12x^2 + 13x + 15 = 0$

33. $\frac{3}{2}x^3 + \frac{1}{2}x^2 + \frac{1}{2}x - 1 = 0$

34. $x^4 - 2x^3 - 13x^2 + 38x - 24 = 0$

35. $x^5 + x^4 - 14x^3 - 14x^2 + 49x - 49 = 0$

36. $8x^5 + 12x^4 + 14x^3 + 13x^2 + 6x + 1 = 0$

37. Solve the equation $x^3 - 9x^2 + 24x - 20 = 0$, using the fact that one of the roots has multiplicity 2.

38. One root of the equation $x^2 + kx + 2k = 0$ $(k \neq 0)$ is twice the other. Find k and find the roots of the equation.

39. State each of the following theorems.
(a) The division algorithm
(b) The remainder theorem
(c) The factor theorem
(d) The fundamental theorem of algebra

40. Find a quadratic equation with roots $a - \sqrt{a^2 - 1}$ and $a + \sqrt{a^2 - 1}$ $(a > 1)$.

In Exercises 41–44, write each polynomial in the form

$$a_n(x - r_1)(x - r_2) \cdots (x - r_n)$$

41. $6x^2 + 7x - 20$ **42.** $x^2 + x - 1$

43. $x^4 - 4x^3 + 5x - 20$ **44.** $x^4 - 4x^2 - 5$

Each of Exercises 45–48 gives an equation, followed by one or more roots. Solve the equation.

45. $x^3 - 7x^2 + 25x - 39 = 0$, $x = 2 - 3i$

46. $x^3 + 6x^2 - 24x + 160 = 0$; $x = 2 + 2i\sqrt{3}$

47. $x^4 - 2x^3 - 4x^2 + 14x - 21 = 0$; $x = 1 + i\sqrt{2}$

48. $x^5 + x^4 - x^3 + x^2 + x - 1 = 0$; $x = \frac{1}{2}(1 + i\sqrt{3})$,
$x = \frac{1}{2}(-1 - \sqrt{5})$

In Exercises 49–54, use Descartes's rule of signs to obtain information regarding the roots of the equations.

49. $x^3 + 8x - 7 = 0$ **50.** $3x^4 + x^2 + 4x - 2 = 0$

51. $x^3 + 3x + 1 = 0$ **52.** $2x^6 + 3x^2 + 6 = 0$

53. $x^4 - 10 = 0$ **54.** $x^4 + 5x^2 - x + 2 = 0$

55. Consider the equation $x^3 + x^2 + x + 1 = 0$.
(a) Use Descartes's rule to show that the equation has either one or three negative roots.
(b) Now show that the equation cannot have three negative roots. Hint: Multiply both sides of the equation by $x - 1$. Then simplify the left-hand side and reapply Descartes's rule to the new equation.
(c) Actually, the original equation can be solved using only the basic algebraic techniques discussed in Chapter 1. Solve the equation in this manner.

56. Use Descartes's rule to show that the equation $x^3 - x^2 + 3x + 2 = 0$ has no positive roots.
Hint: Multiply both sides of the equation by $x + 1$ and apply Descartes's rule to the resulting equation.

57. Let P be the point in the first quadrant where the curve $y = x^3$ intersects the circle $x^2 + y^2 = 1$. Using a calculator, locate the x-coordinate of P within successive hundredths.

58. Let P be the point in the first quadrant where the parabola $y = 4 - x^2$ intersects the curve $y = x^3$. Using a calculator, locate the x-coordinate of P within successive hundredths.

59. Consider the equation $x^3 - 36x - 84 = 0$.
(a) Use Descartes's rule to check that this equation has exactly one positive root.

(b) Use the upper and lower bound theorem to show that 7 is an upper bound for the positive root.
(c) Using a calculator, locate the positive root within successive hundredths,

60. Consider the equation $x^3 - 3x + 1 = 0$.
(a) Use Descartes's rule to check that this equation has exactly one negative root.
(b) Use the upper and lower bound theorem to show that -2 is a lower bound for the negative root.
(c) Using a calculator, locate the negative root within successive hundredths.

In Exercises 61–64, find polynomial equations that have integer coefficients and the given values as roots.

61. $4 - \sqrt{5}$

62. $a + b$ and $a - b$ (Here a and b are integers.)

63. $6 - 2i$ and $\sqrt{5}$

64. $\dfrac{5 + \sqrt{6}}{5 - \sqrt{6}}$ Hint: First rationalize the expression.

65. Find a fourth-degree polynomial equation with integer coefficients, such that $x = 1 + \sqrt{2} + \sqrt{3}$ is a root.
Hint: Begin by writing the given relationship as $x - 1 = \sqrt{2} + \sqrt{3}$; then square both sides.

66. Find a cubic equation with integer coefficients, such that $x = 1 + \sqrt[3]{2}$ is a root. Is $1 - \sqrt[3]{2}$ also a root of the equation?

In Exercises 67–70, first determine the zeros of each function; then sketch the graph.

67. $y = x^3 - 2x^2 - 3x$ **68.** $y = x^4 + 3x^3 + 3x^2 + x$

69. $y = x^4 - 4x^2$ **70.** $y = x^3 + 6x^2 + 5x - 12$

For Exercises 71–78, carry out the indicated operations and express your answer in the standard form $a + bi$.

71. $(3 - 2i)(3 + 2i) + (1 + 3i)^2$

72. $2i(1 + i)^2$

73. $(1 + i\sqrt{2})(1 - i\sqrt{2}) + (\sqrt{2} + i)(\sqrt{2} - i)$

74. $\dfrac{2 + 3i}{1 + i}$ **75.** $\dfrac{3 - i\sqrt{3}}{3 + i\sqrt{3}}$

76. $\dfrac{1 + i}{1 - i} + \dfrac{1 - i}{1 + i}$

77. $-\sqrt{-2}\sqrt{-9} + \sqrt{-8} - \sqrt{-72}$

78. $\dfrac{\sqrt{-4} - \sqrt{-3}\sqrt{-3}}{\sqrt{-100}}$

79. The **real part** of a complex number z is denoted by $\text{Re}(z)$. For instance, $\text{Re}(2 + 5i) = 2$. Show that for any complex number z, we have $\text{Re}(z) = \frac{1}{2}(z + \bar{z})$.
Hint: Let $z = a + bi$.

80. The **imaginary part** of a complex number z is denoted by $\text{Im}(z)$. For instance, $\text{Im}(2 + 5i) = 5$. Show that for any complex number z, we have $\text{Im}(z) = \dfrac{1}{2i}(z - \bar{z})$.

81. The **absolute value** of the complex number $a + bi$ is defined by $|a + bi| = \sqrt{a^2 + b^2}$.
    (a) Compute $|6 + 2i|$ and $|6 - 2i|$.
    (b) As you know, the absolute value of the real number -3 is 3. Now write -3 in the form $-3 + 0i$ and compute its absolute value using the new definition. (The point here is to observe that the two results agree.)
    (c) Let $z = a + bi$. Show that $z\bar{z} = |z|^2$.

In Exercises 82–86, verify that the formulas are correct by carrying out the operations indicated on the left-hand side of the equations. (This list of formulas appears in A Treatise on Algebra, by George Peacock, published in 1845.)

82. $\dfrac{1}{a + bi} + \dfrac{1}{a - bi} = \dfrac{2a}{a^2 + b^2}$

83. $\dfrac{1}{a - bi} - \dfrac{1}{a + bi} = \dfrac{2bi}{a^2 + b^2}$

84. $\dfrac{a + bi}{a - bi} + \dfrac{a - bi}{a + bi} = \dfrac{2(a^2 - b^2)}{a^2 + b^2}$

85. $\dfrac{a + bi}{a - bi} - \dfrac{a - bi}{a + bi} = \dfrac{4abi}{a^2 + b^2}$

86. $\dfrac{a + bi}{c + di} + \dfrac{a - bi}{c - di} = \dfrac{2(ac + bd)}{c^2 + d^2}$

In Exercises 87–90, convert each complex number to rectangular form.

87. $3\left(\cos \tfrac{1}{3}\pi + i \sin \tfrac{1}{3}\pi\right)$

88. $\cos \tfrac{1}{6}\pi + i \sin \tfrac{1}{6}\pi$

89. $2^{1/4}\left(\cos \tfrac{7}{4}\pi + i \sin \tfrac{7}{4}\pi\right)$

90. $5\left[\cos\left(-\tfrac{1}{4}\pi\right) + i \sin\left(-\tfrac{1}{4}\pi\right)\right]$

In Exercises 91–94, express the complex numbers in trigonometric form.

91. $\tfrac{1}{2}(1 + \sqrt{3}i)$

92. $3i$

93. $-3\sqrt{2} - 3\sqrt{2}i$

94. $2\sqrt{3} - 2i$

In Exercises 95–102, carry out the indicated operations. Express your results in rectangular form for those cases in which the trigonometric functions are readily evaluated without tables or a calculator.

95. $5\left(\cos \tfrac{1}{7}\pi + i \sin \tfrac{1}{7}\pi\right) \times 2\left(\cos \tfrac{3}{28}\pi + i \sin \tfrac{3}{28}\pi\right)$

96. $4\left(\cos \tfrac{1}{12}\pi + i \sin \tfrac{1}{12}\pi\right) \times 3\left(\cos \tfrac{1}{12}\pi + i \sin \tfrac{1}{12}\pi\right)$

97. $8\left(\cos \tfrac{1}{12}\pi + i \sin \tfrac{1}{12}\pi\right) \div 4\left(\cos \tfrac{1}{3}\pi + i \sin \tfrac{1}{3}\pi\right)$

98. $4(\cos 32° + i \sin 32°) \div 2^{1/2}(\cos 2° + i \sin 2°)$

99. $\left[3^{1/4}\left(\cos \tfrac{1}{36}\pi + i \sin \tfrac{1}{36}\pi\right)\right]^{12}$

100. $\left[2\left(\cos \tfrac{2}{15}\pi + i \sin \tfrac{2}{15}\pi\right)\right]^5$

101. $\left(\sqrt{3} + i\right)^{10}$

102. $\left(\sqrt{2} - \sqrt{2}i\right)^{15}$

In Exercises 103–106, use DeMoivre's theorem to find the indicated roots. Express your results in rectangular form.

103. Sixth roots of 1

104. Cube roots of $-64i$

105. Square roots of $\sqrt{2} - \sqrt{2}i$

106. Fourth roots of $1 + \sqrt{3}i$

107. Find the five fifth roots of $1 + i$. Use a calculator to express the roots in rectangular form. (Round off each decimal to two places in the final answer.)

108. If $z = r(\cos \theta + i \sin \theta)$, show that
$$r[\cos(\theta + \pi) + i \sin(\theta + \pi)] = -z$$

CHAPTER ELEVEN TEST

1. Let $f(x) = 6x^4 - 5x^3 + 7x^2 - 2x - 2$. Make use of the remainder theorem and synthetic division to compute $f\left(\tfrac{1}{2}\right)$.

2. Solve the equation $x^3 + x^2 - 11x - 15 = 0$, given that one of the roots is -3.

3. List the possibilities for the rational roots of the equation $2x^5 - 4x^3 + x - 6 = 0$.

4. Find a quadratic function with zeros 1 and -8 and with a y-intercept of -24.

5. Use synthetic division to divide $4x^3 + x^2 - 8x + 3$ by $x + 1$.

6. (a) State the factor theorem.
   (b) State the fundamental theorem of algebra.

7. (a) The equation $x^3 - 2x^2 - 1 = 0$ has just one positive root. Use the upper and lower bound theorem to determine the smallest integer that is an upper bound for that root.
   (b) Locate the root between successive tenths. (Use a calculator.)

8. Solve the equation $x^5 - 6x^4 + 11x^3 + 16x^2 - 50x + 52 = 0$, given that two of the roots are $1 + i$ and $3 - 2i$.

9. Let $p(x) = x^4 + 2x^3 - x + 6$ and $d(x) = x^2 + 1$. Find polynomials $q(x)$ and $R(x)$ such that $p(x) = d(x) \cdot q(x) + R(x)$.

10. Express the polynomial $2x^2 - 6x + 5$ in the factored form $a_n(x - r_1)(x - r_2)$.

11. Consider the equation $x^4 - x^3 + 24 = 0$.
    (a) List the possibilities for rational roots.
    (b) Use the upper and lower bound theorem to show that 2 is an upper bound for the roots.
    (c) In view of parts (a) and (b), what possibilities now remain for positive rational roots?
    (d) Which (if any) of the possibilities in part (c) are actually roots?

12. (a) Find the rational roots of the cubic equation $2x^3 - x^2 - x - 3 = 0$.
    (b) Find all solutions of the equation in part (a).

13. Use Descartes's rule of signs to obtain information regarding the roots of the following equation: $3x^4 + x^2 - 5x - 1 = 0$.

14. Find a cubic polynomial $f(x)$ with integer coefficients, such that $1 - 3i$ and -2 are roots of the equation $f(x) = 0$.

15. Find a polynomial $f(x)$ with leading coefficient 1, such that the equation $f(x) = 0$ has the following roots and no other:

Root	Multiplicity
2	1
$3i$	3
$1 + \sqrt{2}$	2

    Write your answer in the form $a_n(x - r_1)(x - r_2) \cdots (x - r_n)$.

16. Simplify the following expression. Write the answer in the form $a + bi$.

$$(4 + 2i)(4 - 2i) + \frac{3 + i}{1 + 2i}$$

17. Find the rectangular form for the complex number $z = 2\left(\cos \frac{2}{3}\pi + i \sin \frac{2}{3}\pi\right)$.

18. Find the trigonometric form of the complex number $\sqrt{2} - \sqrt{2}i$.

19. Let $z = 3\left(\cos \frac{2\pi}{9} + i \sin \frac{2\pi}{9}\right)$ and $w = 5\left(\cos \frac{\pi}{9} + i \sin \frac{\pi}{9}\right)$. Compute the product zw and express your answer in rectangular form.

20. Compute the cube roots of $64i$.

CHAPTER TWELVE

ADDITIONAL TOPICS IN ALGEBRA

... in the ordinary treatise on the elements of algebra, these [additional] topics are either completely omitted or treated carelessly. For this reason, I am certain that the material I have gathered in this book is quite sufficient to remedy that defect.

Leonhard Euler (1707–1783) in his classic text, *Introductio in analysis infinitorum* (1748)

INTRODUCTION

A strong background in algebra is an important prerequisite for courses in calculus and in probability and statistics. In this final chapter we develop several additional topics that help provide a foundation in those areas of study. We begin in Section 12.1 with the principle of mathematical induction. This gives us a framework for proving statements about the natural numbers. In Section 12.2 we discuss the binomial theorem, which is used to analyze and expand expressions of the form $(a + b)^n$. As you'll see, the proof of the binomial theorem uses mathematical induction. Section 12.3 introduces the related (but distinct) concepts of sequences and series. Then in the next two sections (Sections 12.4 and 12.5), we study arithmetic and geometric sequences and series. The last topic introduced in Section 12.5 concerns finding sums of infinite geometric series. In a sense, this is an appropriate topic for our last chapter, for it is closely related to the idea of a *limit*, which is the starting point for calculus.

12.1 MATHEMATICAL INDUCTION

Mathematical induction is not a method of discovery but a technique of proving rigorously what has already been discovered.

David M. Burton in *The History of Mathematics, An Introduction* (Boston: Allyn and Bacon, 1985)

In Pascal's treatise [Traité du triangle arithmétique, written in 1653] ... appears one of the earliest acceptable statements of the method of mathematical induction.

Howard Eves in *An Introduction to the History of Mathematics*, 5th ed. (Philadelphia: Saunders College Publishing, 1983)

TABLE I

n	$1 + 3 + 5 + \cdots + (2n - 1)$	
1	1	$= 1$
2	$1 + 3$	$= 4$
3	$1 + 3 + 5$	$= 9$
4	$1 + 3 + 5 + 7$	$= 16$
5	$1 + 3 + 5 + 7 + 9$	$= 25$

Is mathematics an experimental science? The answer to this question is both yes and no, as the following example illustrates. Consider the problem of determining a formula for the sum of the first n odd natural numbers:

$$1 + 3 + 5 + \cdots + (2n - 1)$$

We begin by doing some calculations in the hope that this may shed some light on the problem. Table 1 shows the results of calculating the sum of the first n

odd natural numbers for values of n ranging from 1 to 5. Upon inspecting the table, we observe that each sum in the right-hand column is the square of the corresponding entry in the left-hand column. For instance, for $n = 5$, we see that

$$\underbrace{1 + 3 + 5 + 7 + 9}_{\text{five terms}} = 5^2$$

Now let's try the next case, where $n = 6$, and see if the pattern persists. That is, we want to know if it is true that

$$\underbrace{1 + 3 + 5 + 7 + 9 + 11}_{\text{six terms}} = 6^2$$

As you can easily check, this last equation is true. Thus, based on the experimental (or empirical) evidence, we are led to the following conjecture:

CONJECTURE   The sum of the first n odd natural numbers is n^2. That is, $1 + 3 + 5 + \cdots + (2n - 1) = n^2$, for each natural number n.

At this point, the "law" we've discovered is indeed really only a conjecture. After all, we've checked it only for values of n ranging from 1 to 6. It is conceivable at this point (although we may feel it is unlikely) that the conjecture is false for certain values of n. For the conjecture to be useful, we must be able to prove that it holds without exception for *all* natural numbers n. In fact, we will subsequently prove that this conjecture is valid. But before explaining the method of proof to be used, let's look at one more example.

Again, let n denote a natural number. Then consider the following question: Which quantity is the larger, 2^n or $(n + 1)^2$? As before, we begin by doing some calculations. This is the experimental stage of our work. According to Table 2, the quantity $(n + 1)^2$ is larger than 2^n for each value of n up through $n = 5$. Thus, we make the following conjecture:

CONJECTURE $(n + 1)^2 > 2^n$ for all natural numbers n.

Again, we note that this is only a conjecture at this point. Indeed, if we try the case where $n = 6$, we find that the pattern does not persist. That is, when $n = 6$, we find that 2^n is 64, while $(n + 1)^2$ is only 49. So in this example, the conjecture is not true in general; we have found a value of n for which it fails.

The preceding examples show that experimentation does have a place in mathematics, but we must be careful with the results. When experimentation leads to a conjecture, proof is required before the conjecture can be viewed as a valid law. For the remainder of this section, we shall discuss one such method of proof—*mathematical induction*.

In order to state the principle of mathematical induction, we first introduce some notation. Suppose that for each natural number n we have a statement P_n to be proved. Consider, for instance, the first conjecture we arrived at:

$$1 + 3 + 5 + \cdots + (2n - 1) = n^2$$

TABLE 2

n	2^n	$(n + 1)^2$
1	2	4
2	4	9
3	8	16
4	16	25
5	32	36

If we denote this statement by P_n, then we have, for example, that

P_1 is the statement that $1 = 1^2$

P_2 is the statement that $1 + 3 = 2^2$

P_3 is the statement that $1 + 3 + 5 = 3^2$

With this notation, we can now state the **principle of mathematical induction**.

Principle of Mathematical Induction

Suppose that for each natural number n, we have a statement P_n for which the following two conditions hold:

1. P_1 is true.

2. For each natural number k, if P_k is true, then P_{k+1} is true.

Then all the statements are true; that is, P_n is true for all natural numbers n.

The idea behind mathematical induction is a simple one. Think of each statement P_n as the rung of a ladder to be climbed. Then we make the analogy shown in Table 3.

TABLE 3

Mathematical Induction		Ladder Analogy	
Hypotheses	1. P_1 is true. 2. If P_k is true, then P_{k+1} is true, for any k.	Hypotheses	1′. You can reach the first rung. 2′. If you are on the kth rung, you can reach the $(k+1)$st rung, for any k.
Conclusion	3. P_n is true for all n.	Conclusion	3′. You can climb the entire ladder.

According to the principle of mathematical induction, we can prove that a statement or formula P_n is true for all n if we carry out the following two steps:

STEP 1 Show that P_1 is true.

STEP 2 Assume that P_k is true, and on the basis of this assumption, show that P_{k+1} is true.

In Step 2, the assumption that P_k is true is referred to as the **induction hypothesis**. (In computer science, Step 1 is sometimes referred to as the **initialization step**.) Let us now turn to some examples of proof by mathematical induction.

EXAMPLE 1 Use mathematical induction to prove that

$$1 + 3 + 5 + \cdots + (2n - 1) = n^2$$

for all natural numbers n.

Solution Let P_n denote the statement that $1 + 3 + 5 + \cdots + (2n - 1) = n^2$. Then we want to show that P_n is true for all natural numbers n.

STEP 1 We must check that P_1 is true. But P_1 is just the statement that $1 = 1^2$, which is true.

STEP 2 Assuming that P_k is true, we must show that P_{k+1} is true. Thus, we assume that

$$1 + 3 + 5 + \cdots + (2k - 1) = k^2 \qquad (1)$$

That is the induction hypothesis. We must now show that

$$1 + 3 + 5 + \cdots + (2k - 1) + [2(k + 1) - 1] = (k + 1)^2 \qquad (2)$$

In order to derive equation (2) from equation (1), we add the quantity $[2(k + 1) - 1]$ to both sides of equation (1). (The motivation for this stems from the observation that the left-hand sides of equations (1) and (2) differ only by the quantity $[2(k + 1) - 1]$.) We obtain

$$1 + 3 + 5 + \cdots + (2k - 1) + [2(k + 1) - 1] = k^2 + [2(k + 1) - 1]$$
$$= k^2 + 2k + 1$$
$$= (k + 1)^2$$

This last equation is what we wanted to show. And having now carried out Steps 1 and 2, we conclude by the principle of mathematical induction that P_n is true for all natural numbers n. ■■■

■■■ **EXAMPLE 2** Use mathematical induction to prove that

$$2^3 + 4^3 + 6^3 + \cdots + (2n)^3 = 2n^2(n + 1)^2$$

for all natural numbers n.

Solution Let P_n denote the statement that

$$2^3 + 4^3 + 6^3 + \cdots + (2n)^3 = 2n^2(n + 1)^2$$

Then we want to show that P_n is true for all natural numbers n.

STEP 1 We must check that P_1 is true. P_1 is the statement that

$$2^3 = 2(1^2)(1 + 1)^2 \qquad \text{or} \qquad 8 = 8$$

Thus, P_1 is true.

STEP 2 Assuming that P_k is true, we must show that P_{k+1} is true. Thus, we assume that

$$2^3 + 4^3 + 6^3 + \cdots + (2k)^3 = 2k^2(k + 1)^2 \qquad (3)$$

We must now show that

$$2^3 + 4^3 + 6^3 + \cdots + (2k)^3 + [2(k + 1)]^3 = 2(k + 1)^2(k + 2)^2 \qquad (4)$$

Adding $[2(k + 1)]^3$ to both sides of equation (3) yields

$$2^3 + 4^3 + 6^3 + \cdots + (2k)^3 + [2(k + 1)]^3 = 2k^2(k + 1)^2 + [2(k + 1)]^3$$
$$= 2k^2(k + 1)^2 + 8(k + 1)^3$$
$$= 2(k + 1)^2[k^2 + 4(k + 1)]$$
$$= 2(k + 1)^2(k^2 + 4k + 4)$$
$$= 2(k + 1)^2(k + 2)^2$$

We have now derived equation (4) from equation (3), as we wished to do. Having carried out Steps 1 and 2, we conclude by the principle of mathematical induction that P_n is true for all natural numbers n. ■■■

TABLE 4

n	$2^{2n} - 1$
1	$3 \, (= 3 \cdot 1)$
2	$15 \, (= 3 \cdot 5)$
3	$63 \, (= 3 \cdot 21)$
4	$255 \, (= 3 \cdot 85)$

■▒■ EXAMPLE 3 As indicated in Table 4, the number 3 is a factor of $2^{2n} - 1$ when $n = 1, 2, 3$, and 4. Use mathematical induction to show that 3 is a factor of $2^{2n} - 1$ for all natural numbers n.

Solution Let P_n denote the statement that 3 is a factor of $2^{2n} - 1$. We want to show that P_n is true for all natural numbers n.

STEP 1 We must check that P_1 is true. But P_1 in this case is just the statement that 3 is a factor of $2^{2(1)} - 1$; that is, 3 is a factor of 3, which is surely true.

STEP 2 Assuming that P_k is true, we must show that P_{k+1} is true. Thus, we assume that

$$3 \text{ is a factor of } 2^{2k} - 1 \tag{5}$$

and we must show that

$$3 \text{ is a factor of } 2^{2(k+1)} - 1$$

The strategy here will be to rewrite the expression $2^{2(k+1)} - 1$ in such a way that the induction hypothesis, statement (5), can be applied. We have

$$\begin{aligned} 2^{2(k+1)} - 1 &= 2^{2k+2} - 1 \\ &= 2^2 \cdot 2^{2k} - 1 \\ &= 4 \cdot 2^{2k} - 4 + 3 \\ &= 4(2^{2k} - 1) + 3 \end{aligned} \tag{6}$$

Now, look at the right-hand side of equation (6). By the induction hypothesis, 3 is a factor of $2^{2k} - 1$. Thus, 3 is a factor of $4(2^{2k} - 1)$, from which it certainly follows that 3 is a factor of $4(2^{2k} - 1) + 3$. In summary, then, 3 is a factor of the right-hand side of equation (6). Consequently, 3 must be a factor of the left-hand side of equation (6), which is what we wished to show.

Having now completed Steps 1 and 2, we conclude by the principle of mathematical induction that P_n is true for all natural numbers n. In other words, 3 is a factor of $2^{2n} - 1$ for all natural numbers n. ■■■

There are instances in which a given statement P_n is false for certain initial values of n, but true thereafter. An example of this is provided by the statement

$$2^n > (n + 1)^2$$

As you can easily check, this statement is false for $n = 1, 2, 3, 4$, and 5. But, as Example 4 shows, the statement is true for $n \geq 6$. In Example 4, we adapt the principle of mathematical induction by beginning in Step 1 with a consideration of P_6 rather than P_1.

■▒■ EXAMPLE 4 Use mathematical induction to prove that

$$2^n > (n + 1)^2 \qquad \text{for all natural numbers } n \geq 6$$

Solution

STEP 1 We must first check that P_6 is true. But P_6 is simply the assertion that

$$2^6 > (6 + 1)^2 \qquad \text{or} \qquad 64 > 49$$

Thus, P_6 is true.

STEP 2 Assuming that P_k is true, where $k \geq 6$, we must show that P_{k+1} is true. Thus, we assume that

$$2^k > (k + 1)^2 \qquad \text{where } k \geq 6 \tag{7}$$

We must show that

$$2^{k+1} > (k + 2)^2$$

Multiplying both sides of inequality (7) by 2 gives us

$$2(2^k) > 2(k + 1)^2 = 2k^2 + 4k + 2$$

This can be rewritten

$$2^{k+1} > k^2 + 4k + (k^2 + 2)$$

However, since $k \geq 6$, it is certainly true that

$$k^2 + 2 > 4$$

We therefore have

$$2^{k+1} > k^2 + 4k + 4 \qquad \text{or} \qquad 2^{k+1} > (k + 2)^2$$

as we wished to show.

Having now completed Steps 1 and 2, we conclude that P_n is true for all natural numbers $n \geq 6$. ■■■

EXERCISE SET 12.1

A

In Exercises 1–18, use the principle of mathematical induction to show that the statements are true for all natural numbers.

1. $1 + 2 + 3 + \cdots + n = \frac{1}{2}n(n + 1)$

2. $2 + 4 + 6 + \cdots + 2n = n(n + 1)$

3. $1 + 4 + 7 + \cdots + (3n - 2) = \frac{1}{2}n(3n - 1)$

4. $5 + 9 + 13 + \cdots + (4n + 1) = n(2n + 3)$

5. $1^2 + 2^2 + 3^2 + \cdots + n^2 = \frac{1}{6}n(n + 1)(2n + 1)$

6. $2^2 + 4^2 + 6^2 + \cdots + (2n)^2 = \frac{2}{3}n(n + 1)(2n + 1)$

7. $1^2 + 3^2 + 5^2 + \cdots + (2n - 1)^2 = \frac{1}{3}n(2n - 1)(2n + 1)$

8. $2 + 2^2 + 2^3 + \cdots + 2^n = 2^{n+1} - 2$

9. $3 + 3^2 + 3^3 + \cdots + 3^n = \frac{1}{2}(3^{n+1} - 3)$

10. $e^x + e^{2x} + e^{3x} + \cdots + e^{nx} = \dfrac{e^{(n+1)x} - e^x}{e^x - 1} \quad (x \neq 0)$

11. $1^3 + 2^3 + 3^3 + \cdots + n^3 = \left[\frac{1}{2}n(n + 1)\right]^2$

12. $2^3 + 4^3 + 6^3 + \cdots + (2n)^3 = 2n^2(n + 1)^2$

13. $1^3 + 3^3 + 5^3 + \cdots + (2n - 1)^3 = n^2(2n^2 - 1)$

14. $1 \cdot 2 + 3 \cdot 4 + 5 \cdot 6 + \cdots + (2n - 1)(2n)$
$$= \frac{1}{3}n(n + 1)(4n - 1)$$

15. $1 \cdot 3 + 3 \cdot 5 + 5 \cdot 7 + \cdots + (2n - 1)(2n + 1)$
$$= \frac{1}{3}n(4n^2 + 6n - 1)$$

16. $\dfrac{1}{1 \times 3} + \dfrac{1}{2 \times 4} + \dfrac{1}{3 \times 5} + \cdots + \dfrac{1}{n(n + 2)}$
$$= \dfrac{n(3n + 5)}{4(n + 1)(n + 2)}$$

17. $1 + \dfrac{3}{2} + \dfrac{5}{2^2} + \dfrac{7}{2^3} + \cdots + \dfrac{2n - 1}{2^{n-1}} = 6 - \dfrac{2n + 3}{2^{n-1}}$

18. $1 + 2 \cdot 2 + 3 \cdot 2^2 + 4 \cdot 2^3 + \cdots + n \cdot 2^{n-1}$
$$= (n - 1)2^n + 1$$

19. Show that $n \leq 2^{n-1}$ for all natural numbers n.

20. Show that 3 is a factor of $n^3 + 2n$ for all natural numbers n.

21. Show that $n^2 + 4 < (n + 1)^2$ for all natural numbers $n \geq 2$.

22. Show that $n^3 > (n + 1)^2$ for all natural numbers $n \geq 3$.

In Exercises 23–26, prove that the statement is true for all natural numbers in the specified range. Use a calculator to carry out Step 1.

23. $(1.5)^n > 2n$; $n \geq 7$

24. $(1.25)^n > n$; $n \geq 11$

25. $(1.1)^n > n$; $n \geq 39$

26. $(1.1)^n > 5n$; $n \geq 60$

B

27. Let $f(n) = \dfrac{1}{1 \cdot 2} + \dfrac{1}{2 \cdot 3} + \dfrac{1}{3 \cdot 4} + \cdots + \dfrac{1}{n(n+1)}$.

(a) Complete the following table.

n	1	2	3	4	5
$f(n)$					

(b) On the basis of the results in the table, what would you guess to be the value of $f(6)$? Compute $f(6)$ to see if this is correct.

(c) Make a conjecture about the value of $f(n)$, and prove it using mathematical induction.

28. Let $f(n) = \dfrac{1}{1 \times 3} + \dfrac{1}{3 \times 5} + \dfrac{1}{5 \times 7} + \cdots$
$$+ \dfrac{1}{(2n-1)(2n+1)}.$$

(a) Complete the following table.

n	1	2	3	4
$f(n)$				

(b) On the basis of the results in the table, what would you guess to be the value of $f(5)$? Compute $f(5)$ to see if your guess is correct.

(c) Make a conjecture about the value of $f(n)$, and prove it using mathematical induction.

29. Suppose that a function f satisfies the following conditions:
$$f(1) = 1$$
$$f(n) = f(n-1) + 2\sqrt{f(n-1)} + 1 \qquad \text{for } n \geq 2$$

(a) Complete the table.

n	1	2	3	4	5
$f(n)$					

(b) On the basis of the results in the table, what would you guess to be the value of $f(6)$? Compute $f(6)$ to see if your guess is correct.

(c) Make a conjecture about the value of $f(n)$ when n is a natural number, and prove the conjecture using mathematical induction.

30. This exercise demonstrates the necessity of carrying out both Step 1 and Step 2 before considering an induction proof valid.

(a) Let P_n denote the statement that $n^2 + 1$ is even. Check that P_1 is true. Then give an example showing that P_n is not true for all n.

(b) Let Q_n denote the statement that $n^2 + n$ is odd. Show that Step 2 of an induction proof can be completed in this case, but not Step 1.

31. A *prime number* is a natural number that has no factors other than itself and 1. For technical reasons, 1 is not considered a prime. Thus, the list of the first seven primes looks like this: 2, 3, 5, 7, 11, 13, 17. Let P_n be the statement that $n^2 + n + 11$ is prime. Check that P_n is true for all values of n less than 10. Check that P_{10} is false.

32. Prove that if $x \neq 1$,
$$1 + 2x + 3x^2 + \cdots + nx^{n-1} = \frac{1-x^n}{(1-x)^2} - \frac{nx^n}{1-x}$$
for all natural numbers n.

33. If $r \neq 1$, show that $1 + r + r^2 + \cdots + r^{n-1} = \dfrac{(r^n - 1)}{(r - 1)}$ for all natural numbers n.

34. Use mathematical induction to show that
$$x^n - 1 = (x - 1)(1 + x + x^2 + \cdots + x^{n-1})$$
for all natural numbers n.

35. Prove that 5 is a factor of $n^5 - n$ for all natural numbers $n \geq 2$.

36. Prove that 4 is a factor of $5^n + 3$ for all natural numbers n.

37. Prove that 5 is a factor of $2^{2n+1} + 3^{2n+1}$ for all nonnegative integers n.

38. Prove that 8 is a factor of $3^{2n} - 1$ for all natural numbers n.

39. Prove that 3 is a factor of $2^{n+1} + (-1)^n$ for all nonnegative integers n.

40. Prove that 6 is a factor of $n^3 + 3n^2 + 2n$ for all natural numbers n.

41. Use mathematical induction to show that $x - y$ is a factor of $x^n - y^n$ for all natural numbers n. *Suggestion for Step 2:* Verify and then use the fact that $x^{k+1} - y^{k+1} = x^k(x - y) + (x^k - y^k)y$.

In Exercises 42 and 43, use mathematical induction to prove that the formulas hold for all natural numbers n.

42. $\log_{10}(a_1 a_2 \cdots a_n) = \log_{10} a_1 + \log_{10} a_2 + \cdots + \log_{10} a_n$

43. $(1 + p)^n \geq 1 + np$, where $p > -1$

44. $(\cos \theta + i \sin \theta)^n = \cos n\theta + i \sin n\theta$ (This is *Demoivre's theorem.*)

C

In Exercises 45–47, use mathematical induction to prove that the statements hold for all natural numbers n.

45. $1^5 + 2^5 + 3^5 + \cdots + n^5 = \dfrac{n^2(n + 1)^2(2n^2 + 2n - 1)}{12}$

46. $(\cos x)(\cos 2x)(\cos 4x) \cdots (\cos 2^{n-1}x) = \dfrac{\sin 2^n x}{2^n \sin x}$

47. $\dfrac{0^3 + 1^3 + 2^3 + \cdots + n^3}{n^3 + n^3 + n^3 + \cdots + n^3} = \dfrac{1}{4} + \dfrac{1}{4n}$

48. (a) Without using induction, prove the following trigonometric identity:

$\sin^2[(n + 1)\theta] - \sin^2 n\theta = \sin[(2n + 1)\theta]\sin \theta$

Hint: On the left-hand side, use difference-of-squares factoring, followed by the sum-to-product formulas.

(b) Use mathematical induction to prove that the following equation holds for all natural numbers n:

$$\sin \theta + \sin 3\theta + \cdots + \sin[(2n - 1)\theta] = \frac{\sin^2 n\theta}{\sin \theta}$$

Hint: In the induction step, use the identity in part (a).

(c) Use the identity in part (b) and your calculator to evaluate the sum

$$\sin 1 + \sin 3 + \sin 5 + \cdots + \sin 25$$

Round off your answer to four decimal places.

12.2 THE BINOMIAL THEOREM

A mathematician, like a painter or a poet, is a maker of patterns.

G. H. Hardy (1877–1947)

[Blaise Pascal] *made numerous discoveries relating to this array and set them forth in his* Traité du triangle arithmétique, *published posthumously in 1665, and among these was essentially our present Binomial Theorem for positive integral exponents.*

David Eugene Smith in *History of Mathematics,* vol. II (New York: Ginn and Co., 1925)

If you look back at Section 1.3, you will see that two of the special products listed there are

$$(a + b)^2 = a^2 + 2ab + b^2$$

and

$$(a + b)^3 = a^3 + 3a^2b + 3ab^2 + b^3$$

Our present goal is to develop a general formula, known as the *binomial theorem,* for expanding any product of the form $(a + b)^n$, when n is a natural number.

We begin by looking for patterns in the expansion of $(a + b)^n$. To do this, let's list the expansions of $(a + b)^n$ for $n = 1, 2, 3, 4$, and 5. (Exercises 1 and 2 ask you to verify these results simply by repeated multiplication.)

$$(a + b)^1 = a + b$$
$$(a + b)^2 = a^2 + 2ab + b^2$$
$$(a + b)^3 = a^3 + 3a^2b + 3ab^2 + b^3$$
$$(a + b)^4 = a^4 + 4a^3b + 6a^2b^2 + 4ab^3 + b^4$$
$$(a + b)^5 = a^5 + 5a^4b + 10a^3b^2 + 10a^2b^3 + 5ab^4 + b^5$$

After surveying these results, we note the following patterns.

PROPERTY SUMMARY	PATTERNS OBSERVED IN $(a + b)^n$ for $n = 1, 2, 3, 4, 5$
GENERAL STATEMENT	EXAMPLE
There are $n + 1$ terms.	There are $4 (= 3 + 1)$ terms in the expansion of $(a + b)^3$.
The expansion begins with a^n and ends with b^n.	$(a + b)^3$ begins with a^3 and ends with b^3.
The sum of the exponents in each term is n.	The sum of the exponents in each term of $(a + b)^3$ is 3.
The exponents of a decrease by 1 from term to term.	$(a + b)^3 = a^{③} + 3a^{②}b + 3a^{①}b^2 + a^{⓪}b^3$
The exponents of b increase by 1 from term to term.	$(a + b)^3 = a^3b^{⓪} + 3a^2b^{①} + 3ab^{②} + b^{③}$
When n is even, the coefficients are symmetric about the middle term.	The sequence of coefficients for $(a + b)^4$ is 1, 4, 6, 4, 1.
When n is odd, the coefficients are symmetric about the two middle terms.	The sequence of coefficients for $(a + b)^5$ is 1, 5, 10, 10, 5, 1.

FIGURE 1
"Pascal's" triangle,
by Chu-Shi-Kie,
A.D. 1303

The patterns we have just observed for $(a + b)^n$ persist for all natural numbers n. (This will follow from the binomial theorem, which is proved at the end of this section.) Thus, for example, the form of $(a + b)^6$ must be as follows:

$$(a + b)^6 = a^6 + \underline{?}a^5b + \underline{?}a^4b^2 + \underline{?}a^3b^3 + \underline{?}a^2b^4 + \underline{?}ab^5 + b^6$$

The problem now is to find the proper coefficient for each term. To do this, we need to discover additional patterns in the expansion of $(a + b)^n$.

We have already written out the expansions of $(a + b)^n$ for values of n ranging from 1 to 5. Let us now write only the coefficients in those expansions. The resulting triangular array of numbers is known as **Pascal's triangle.*** For reasons of symmetry, we begin with $(a + b)^0$ rather than $(a + b)^1$.

$$(a + b)^0 \dots\dots\dots\dots\dots\dots\dots 1$$
$$(a + b)^1 \dots\dots\dots\dots\dots\dots 1 \quad 1$$
$$(a + b)^2 \dots\dots\dots\dots\dots 1 \quad 2 \quad 1$$
$$(a + b)^3 \dots\dots\dots\dots 1 \quad 3 \quad 3 \quad 1$$
$$(a + b)^4 \dots\dots\dots 1 \quad 4 \quad 6 \quad 4 \quad 1$$
$$(a + b)^5 \dots\dots 1 \quad 5 \quad 10 \quad 10 \quad 5 \quad 1$$

The key observation regarding Pascal's triangle is this: *Each entry in the array (other than the 1's along the sides) is the sum of the two numbers diagonally above it.* For instance, the 6 that appears in the fifth row is the sum of the two 3's diagonally above it. Using this observation, we can form as many additional rows as we please. The coefficients for $(a + b)^n$ will then appear in the $(n + 1)$st

*The array is named after Blaise Pascal, a seventeenth-century French mathematician and philosopher. However, as Figure 1 indicates, the Pascal triangle was known to Chinese mathematicians centuries earlier.

row of the array.[†] For instance, to obtain the row corresponding to $(a + b)^6$, we have

sixth row, $(a + b)^5$:   1 ⁔ 5 ⁔ 10 ⁔ 10 ⁔ 5 ⁔ 1
seventh row, $(a + b)^6$:   1   6   15   20   15   6   1

Thus, the sequence of coefficients for $(a + b)^6$ is 1, 6, 15, 20, 15, 6, 1. This answers the question raised earlier about the expansions of $(a + b)^6$. We have

$$(a + b)^6 = a^6 + 6a^5b + 15a^4b^2 + 20a^3b^3 + 15a^2b^4 + 6ab^5 + b^6$$

For analytical work or for larger values of the exponent n, it is inefficient to rely on Pascal's triangle. For this reason, we point out another pattern in the expansions of $(a + b)^n$.

In the expansion of $(a + b)^n$, the coefficient of any term after the first can be generated as follows. In the *previous* term, multiply the coefficient by the exponent of a and then divide by the number of that previous term.

To see how this observation is used, let's compute the second, third, and fourth coefficients in the expansion of $(a + b)^6$. To compute the coefficient of the second term, we go back to the first term, which is a^6. We have

coefficient of first term
exponent of a in first term

$$\text{coefficient of second term} = \frac{1 \cdot 6}{1} = 6$$

number of first term

Thus, the second term is $6a^5b$ and, consequently, we have

coefficient of second term
exponent of a in second term

$$\text{coefficient of third term} = \frac{6 \cdot 5}{2} = 15$$

number of second term

Continuing now with this method, you should check for yourself that the coefficient of the fourth term in the expansion of $(a + b)^6$ is 20.

Note We now know that the first four coefficients are 1, 6, 15, and 20. By symmetry, it follows that the complete sequence of coefficients for this expansion is 1, 6, 15, 20, 15, 6, 1. No additional calculation for the coefficients is necessary.

▮▮▮ **EXAMPLE 1**   Expand $(2x - y^2)^7$.

[†]That these numbers actually are the appropriate coefficients follows from the binomial theorem, which is proved at the end of this section.

Solution First we write the expansion of $(a + b)^7$ using the method explained just prior to this example, or using Pascal's triangle. As you should check for yourself, the expansion is

$$(a + b)^7 = a^7 + 7a^6b + 21a^5b^2 + 35a^4b^3 + 35a^3b^4 + 21a^2b^5 + 7ab^6 + b^7$$

Now we make the substitutions $a = 2x$ and $b = -y^2$. This yields

$$\begin{aligned}[2x + (-y^2)]^7 &= (2x)^7 + 7(2x)^6(-y^2) + 21(2x)^5(-y^2)^2 \\ &\quad + 35(2x)^4(-y^2)^3 + 35(2x)^3(-y^2)^4 \\ &\quad + 21(2x)^2(-y^2)^5 + 7(2x)(-y^2)^6 + (-y^2)^7 \\ &= 128x^7 - 448x^6y^2 + 672x^5y^4 - 560x^4y^6 + 280x^3y^8 \\ &\quad - 84x^2y^{10} + 14xy^{12} - y^{14}\end{aligned}$$

This is the required expansion. Notice how the signs alternate in the final answer; this is characteristic of all expansions of the form $(a - b)^n$. ∎

In preparation for the binomial theorem, we introduce two notations that are used not only in connection with the binomial theorem, but in many other areas of mathematics as well. The first of these notations is $n!$ (read "n factorial").

DEFINITION   The Factorial Symbol

	EXAMPLES
$n! = 1 \cdot 2 \cdot 3 \cdots n$	$3! = 1 \cdot 2 \cdot 3 = 6$
where n is a natural number	$\dfrac{6!}{4!} = \dfrac{6 \cdot 5 \cdot 4 \cdot 3 \cdot 2 \cdot 1}{4 \cdot 3 \cdot 2 \cdot 1}$
$0! = 1$	$= 6 \times 5 = 30$

EXAMPLE 2   Simplify the expression $\dfrac{(n + 1)!}{(n - 1)!}$.

Solution $\dfrac{(n + 1)!}{(n - 1)!} = \dfrac{(n + 1) \cdot n \cdot (n - 1)!}{(n - 1)!} = (n + 1) \cdot n = n^2 + n$ ∎

The second notation that we introduce in preparation for the binomial theorem is $\binom{n}{k}$. This notation is read "n choose k" (because it can be shown that $\binom{n}{k}$ is equal to the number of ways of choosing a subset of k elements from a set with n elements).

DEFINITION   The Binomial Coefficient $\binom{n}{k}$

EXAMPLE

Let n and k be nonnegative integers with $k \leq n$. Then the binomial coefficient $\binom{n}{k}$ is defined by

$$\binom{n}{k} = \frac{n!}{k!(n-k)!}$$

$$\binom{5}{2} = \frac{5!}{2!(5-2)!} = \frac{5!}{2!3!}$$

$$= \frac{5 \cdot 4 \cdot 3 \cdot 2 \cdot 1}{(2 \cdot 1)(3 \cdot 2 \cdot 1)}$$

$$= \frac{5 \cdot 4}{2 \cdot 1} = 10$$

The binomial coefficients are so named because they are indeed the coefficients in the expansion of $(a + b)^n$. More precisely, the relationship is this:

The coefficients in the expansion of $(a + b)^n$ are the $n + 1$ numbers

$$\binom{n}{0}, \binom{n}{1}, \binom{n}{2}, \ldots, \binom{n}{n}$$

Subsequently, we will see why this statement is true. For now, however, let's look at an example. Consider the binomial coefficients $\binom{3}{0}, \binom{3}{1}, \binom{3}{2},$ and $\binom{3}{3}$. According to our statement, these four quantities should be the coefficients in the expansion of $(a + b)^3$. Let's check:

$$\binom{3}{0} = \frac{3!}{0!(3-0)!} = \frac{3!}{1(3!)} = 1$$

$$\binom{3}{1} = \frac{3!}{1!(3-1)!} = \frac{3 \cdot 2 \cdot 1}{1(2 \cdot 1)} = 3$$

$$\binom{3}{2} = \frac{3!}{2!(3-2)!} = \frac{3 \cdot 2 \cdot 1}{(2 \cdot 1)1} = 3$$

$$\binom{3}{3} = \frac{3!}{3!(3-3)!} = \frac{3!}{3!0!} = 1$$

The values of $\binom{3}{0}, \binom{3}{1}, \binom{3}{2},$ and $\binom{3}{3}$ are thus 1, 3, 3, and 1, respectively. But these last four numbers are indeed the coefficients in the expansion of $(a + b)^3$, as we wished to check.

We are now in a position to state the binomial theorem, after which we will look at several applications. Finally, at the end of this section, we will use mathematical induction to prove the theorem. In the statement of the theorem that follows, we are assuming that the exponent n is a natural number.

Binomial Theorem

$$(a + b)^n = \binom{n}{0}a^n + \binom{n}{1}a^{n-1}b + \binom{n}{2}a^{n-2}b^2 + \cdots + \binom{n}{n-1}ab^{n-1} + \binom{n}{n}b^n$$

One of the uses of the binomial theorem is in identifying specific terms in an expansion without computing the entire expansion. This is particularly helpful when the exponent n is relatively large. Looking back at the statement of the binomial theorem, there are three observations we can make. First, the coefficient of the rth term is $\binom{n}{r-1}$. For instance, the coefficient of the third term is $\binom{n}{3-1} = \binom{n}{2}$. The second observation is that the exponent for a in the rth term is $n - (r - 1)$. For instance, the exponent for a in the third term is $n - (3 - 1) = n - 2$. Finally, we observe that the exponent for b in the rth term is $r - 1$, the same quantity that appears in the lower position of the corresponding binomial coefficient. For instance, the exponent for b in the third term is $r - 1 = 3 - 1 = 2$. We summarize these three observations with the following statement.

The rth term in the expansion of $(a + b)^n$ is
$$\binom{n}{r-1} a^{n-r+1} b^{r-1}$$

EXAMPLE 3   Find the fifteenth term in the expansion of $\left(x^2 - \dfrac{1}{x}\right)^{18}$.

Solution   Using the values $r = 15$, $n = 18$, $a = x^2$, and $b = -1/x$, we have

$$\binom{n}{r-1} a^{n-r+1} b^{r-1} = \binom{18}{15-1} (x^2)^{18-15+1} \left(\frac{-1}{x}\right)^{15-1}$$

$$= \binom{18}{14} x^8 \cdot \frac{1}{x^{14}}$$

$$= \frac{18 \times 17 \times 16 \times 15 \times (14!)}{14!(4 \times 3 \times 2 \times 1)} x^{-6}$$

$$= \frac{18 \times 17 \times 16 \times 15}{4 \times 3 \times 2 \times 1} x^{-6}$$

$$= 3060 x^{-6} \qquad \text{as required}$$

EXAMPLE 4   Find the coefficient of the term containing x^4 in the expression of $(x + y^2)^{30}$.

Solution   Again we use the fact that the rth term in the expansion of $(a + b)^n$ is $\binom{n}{r-1} a^{n-r+1} b^{r-1}$. In this case, n is 30 and x plays the role of a. The exponent for x is then $n - r + 1$ or $30 - r + 1$. To see when this exponent is 4, we write

$$30 - r + 1 = 4 \qquad \text{and therefore} \qquad r = 27$$

The required coefficient is thus $\binom{30}{27-1}$. We then have

$$\binom{30}{26} = \frac{30!}{26!(30-26)!} = \frac{30 \times 29 \times 28 \times 27}{4 \times 3 \times 2 \times 1}$$

After carrying out the indicated arithmetic, we find that $\binom{30}{26} = 27,405$. This is the required coefficient. ∎

I ▬ **I** **EXAMPLE 5** Find the coefficient of the term containing a^9 in the expansion of $(a + 2\sqrt{a})^{10}$.

Solution The rth term in this expansion is

$$\binom{10}{r-1} a^{10-r+1} (2\sqrt{a})^{r-1}$$

We can rewrite this as

$$\binom{10}{r-1} a^{10-r+1} (2^{r-1})(a^{1/2})^{r-1} = 2^{r-1}\binom{10}{r-1} a^{10-r+1+(r-1)/2}$$

This shows that the general form of the coefficient we wish to find is $2^{r-1}\binom{10}{r-1}$. We now need to determine r when the exponent of a is 9. Thus, we require that

$$10 - r + 1 + \frac{r-1}{2} = 9$$

$$-r + \frac{r-1}{2} = -2$$

$$-2r + r - 1 = -4 \qquad \text{multiplying by 2}$$

$$r = 3$$

The required coefficient is now obtained by substituting $r = 3$ in the expression $2^{r-1}\binom{10}{r-1}$. Thus, the required coefficient is

$$2^2\binom{10}{2} = \frac{4 \cdot 10!}{2!(10-2)!} = 2 \cdot 10 \cdot 9 = 180$$ ∎

There are three simple identities involving the binomial coefficients that will simplify our proof of the binomial theorem.

IDENTITY 1 $\quad \binom{r}{0} = 1 \qquad$ for all nonnegative integers r

IDENTITY 2 $\quad \binom{r}{r} = 1 \qquad$ for all nonnegative integers r

IDENTITY 3 $\quad \binom{k}{r} + \binom{k}{r-1} = \binom{k+1}{r} \qquad$ for all natural numbers k and r with $r \leq k$

All three of these identities can be proved directly from the definitions of the binomial coefficients, without the need for mathematical induction. The proofs of the first two are straightforward, and we omit them here. The proof of Identity 3 runs as follows:

$$\binom{k}{r} + \binom{k}{r-1} = \frac{k!}{r!(k-r)!} + \frac{k!}{(r-1)!(k-r+1)!}$$

$$= \frac{k!}{r(r-1)!(k-r)!} + \frac{k!}{(r-1)!(k-r+1)(k-r)!}$$

Now, the common denominator on the right-hand side of the last equation is $r(r-1)!(k-r+1)(k-r)!$ Thus, we have

$$\binom{k}{r} + \binom{k}{r-1} = \frac{k!(k-r+1)}{r(r-1)!(k-r+1)(k-r)!} + \frac{k!r}{r(r-1)!(k-r+1)(k-r)!}$$

$$= \frac{k!(k-r+1) + k!r}{r(r-1)!(k-r+1)(k-r)!}$$

$$= \frac{k!(k-r+1+r)}{r(r-1)!(k-r+1)(k-r)!} = \frac{k!(k+1)}{r!(k-r+1)!}$$

$$= \frac{(k+1)!}{r![(k+1)-r]!}$$

$$= \binom{k+1}{r} \qquad \text{as required}$$

Taken together, the three identities show why the $(n+1)$st row of Pascal's triangle consists of the numbers $\binom{n}{0}, \binom{n}{1}, \binom{n}{2}, \ldots, \binom{n}{n}$. Identities 1 and 2 tell us that this row of numbers begins and ends with 1. Identity 3 is then just a statement of the fact that each entry in the row, other than the initial and final 1, is generated by adding the two entries diagonally above it.

We conclude this section by using mathematical induction to prove the binomial theorem. The statement P_n that we wish to prove for all natural numbers n is this:

$$(a+b)^n = \binom{n}{0}a^n + \binom{n}{1}a^{n-1}b + \cdots + \binom{n}{n-1}ab^{n-1} + \binom{n}{n}b^n$$

First, we check that P_1 is true. The statement P_1 asserts that

$$(a+b)^1 = \binom{1}{0}a^1 + \binom{1}{1}a^0b$$

However, in view of Identities 1 and 2, this last equation becomes

$$(a+b)^1 = 1 \cdot a + 1 \cdot b$$

which is surely true. Now let us assume that P_k is true and, on the basis of this assumption, show that P_{k+1} is true. The statement P_k is

$$(a+b)^k = \binom{k}{0}a^k + \binom{k}{1}a^{k-1}b + \cdots + \binom{k}{k-1}ab^{k-1} + \binom{k}{k}b^k$$

Multiplying both sides of this equation by the quantity $(a + b)$ yields

$$(a + b)^{k+1} = (a + b)\left[\binom{k}{0}a^k + \binom{k}{1}a^{k-1}b + \cdots + \binom{k}{k-1}ab^{k-1} + \binom{k}{k}b^k\right]$$

$$= a\left[\binom{k}{0}a^k + \binom{k}{1}a^{k-1}b + \cdots + \binom{k}{k-1}ab^{k-1} + \binom{k}{k}b^k\right]$$

$$+ b\left[\binom{k}{0}a^k + \binom{k}{1}a^{k-1}b + \cdots + \binom{k}{h-1}ab^{k-1} + \binom{k}{k}b^k\right]$$

$$= \binom{k}{0}a^{k+1} + \binom{k}{1}a^k b + \cdots + \binom{k}{k-1}a^2 b^{k-1} + \binom{k}{k}ab^k$$

$$+ \binom{k}{0}a^k b + \binom{k}{1}a^{k-1}b^2 + \cdots + \binom{k}{k-1}ab^k + \binom{k}{k}b^{k+1}$$

$$= \binom{k}{0}a^{k+1} + \left[\binom{k}{1} + \binom{k}{0}\right]a^k b + \cdots + \left[\binom{k}{k} + \binom{k}{k-1}\right]ab^k + \binom{k}{k}b^{k+1}$$

We can now make some substitutions on the right-hand side of this last equation. The initial binomial coefficient $\binom{k}{0}$ can be replaced by $\binom{k+1}{0}$, for both are equal to 1 according to Identity 1. Similarly, the binomial coefficient $\binom{k}{k}$ appearing at the end of the equation can be replaced by $\binom{k+1}{k+1}$, since both are equal to 1 according to Identity 2. Finally, we can use Identity 3 to simplify each of the sums in the brackets. We obtain

$$(a + b)^{k+1} = \binom{k+1}{0}a^{k+1} + \binom{k+1}{1}a^k b + \cdots + \binom{k+1}{k}ab^k + \binom{k+1}{k+1}b^{k+1}$$

But this last equation is just the statement P_{k+1}; that is, we have derived P_{k+1} from P_k, as we wished to do. The induction proof is now complete.

EXERCISE SET 12.2

A

In Exercises 1 and 2, verify each statement directly, without using the techniques developed in this section.

1. (a) $(a + b)^2 = a^2 + 2ab + b^2$
   (b) $(a + b)^3 = a^3 + 3a^2b + 3ab^2 + b^3$
       Hint: $(a + b)^3 = (a + b)(a + b)^2$

2. (a) $(a + b)^4 = a^4 + 4a^3b + 6a^2b^2 + 4ab^3 + b^4$
       Hint: Use the result in Exercise 1(b) and the fact that $(a + b)^4 = (a + b)(a + b)^3$.
   (b) $(a + b)^5 = a^5 + 5a^4b + 10a^3b^2 + 10a^2b^3 + 5ab^4 + b^5$

In Exercises 3–28, carry out the indicated expansions.

3. $(a + b)^9$
4. $(a - b)^9$
5. $(2A + B)^3$
6. $(1 + 2x)^6$
7. $(1 - 2x)^6$
8. $(3x^2 - y)^5$
9. $(\sqrt{x} + \sqrt{y})^4$
10. $(\sqrt{x} - \sqrt{y})^4$
11. $(x^2 + y^2)^5$
12. $(5A - B^2)^3$
13. $[1 - (1/x)]^6$
14. $(3x + y^2)^4$
15. $[(x/2) - (y/3)]^3$
16. $(1 - z^2)^7$
17. $(ab^2 + c)^7$
18. $[x - (1/x)]^8$
19. $(x + \sqrt{2})^8$
20. $(4A - \frac{1}{2})^5$
21. $(\sqrt{2} - 1)^3$
22. $(1 + \sqrt{5})^4$
23. $(\sqrt{2} + \sqrt{3})^5$
24. $(\frac{1}{2} - 2a)^6$
25. $(2\sqrt[3]{2} - \sqrt[3]{4})^3$
26. $(x + y + 1)^4$ *Suggestion:* Rewrite the expression as $[(x + y) + 1]^4$.

27. $(x^2 - 2x - 1)^5$ *Suggestion:*   Rewrite the expression as $[x^2 - (2x + 1)]^5$.

28. $[x^2 - 2x - (1/x)]^6$

In Exercises 29–38, evaluate or simplify each expression.

29. $5!$

30. (a) $3! + 2!$
(b) $(3 + 2)!$

31. $\binom{7}{3}\binom{3}{2}$

32. $\dfrac{20!}{18!}$

33. (a) $\binom{5}{3}$ (b) $\binom{5}{4}$

34. (a) $\binom{7}{7}$ (b) $\binom{7}{0}$

35. $\dfrac{(n + 2)!}{n!}$

36. $\dfrac{n[(n - 2)!]}{(n + 1)!}$

37. $\binom{6}{4} + \binom{6}{3} - \binom{7}{4}$

38. $(3!)! + (3!)^2$

39. Find the fifteenth term in the expansion of $(a + b)^{16}$.

40. Find the third term in the expansion of $(a - b)^{30}$.

41. Find the one hundredth term in the expansion of $(1 + x)^{100}$.

42. Find the twenty-third term in the expansion of $[x - (1/x^2)]^{25}$.

43. Find the coefficient of the term containing a^4 in the expansion of $\left(\sqrt{a} - \sqrt{x}\right)^{10}$.

44. Find the coefficient of the term containing a^4 in the expansion of $(3a - 5x)^{12}$.

45. Find the coefficient of the term containing y^8 in the expansion of $[(x/2) - 4y]^9$.

46. Find the coefficient of the term containing x^6 in the expansion of $[x^2 + (1/x)]^{12}$.

47. Find the coefficient of the term containing x^3 in the expansion of $\left(1 - \sqrt{x}\right)^8$.

48. Find the coefficient of the term containing a^8 in the expansion of $\left[a - (2/\sqrt{a})\right]^{14}$.

49. Find the term that does not contain A in the expansion of $[(1/A) + 3A^2]^{12}$.

50. Find the coefficient of B^{-10} in the expansion of $[(B^2/2) - (3/B^3)]^{10}$.

B

51. Show that the coefficient of x^n in the expansion of $(1 + x)^{2n}$ is $(2n)!/(n!)^2$.

52. Find n so that the coefficients of the eleventh and thirteenth terms in $(1 + x)^n$ are the same.

53. (a) Complete the following table.

k	0	1	2	3	4	5	6	7	8
$\binom{8}{k}$									

(b) Use the results in part (a) to verify that
$$\binom{8}{0} + \binom{8}{1} + \binom{8}{2} + \cdots + \binom{8}{8} = 2^8.$$

(c) By taking $a = b = 1$ in the expansion of $(a + b)^n$, show that $\binom{n}{0} + \binom{n}{1} + \binom{n}{2} + \cdots + \binom{n}{n} = 2^n$.

C

54. Two real numbers A and B are defined by $A = \sqrt[99]{99!}$ and $B = \sqrt[100]{100!}$. Which number is larger, A or B? *Hint:* Compare A^{9900} and B^{9900}.

55. This exercise outlines a proof of the identity
$$\binom{n}{0}^2 + \binom{n}{1}^2 + \binom{n}{2}^2 + \cdots + \binom{n}{n}^2 = \binom{2n}{n}.$$

(a) Verify that
$$(1 + x)^n\left(1 + \frac{1}{x}\right)^n = \frac{(1 + x)^{2n}}{x^n} \tag{1}$$

(This requires only basic algebra, not the binomial theorem.)

(b) Show that the coefficient of the term independent of x on the right side of equation (1) is $\binom{2n}{n}$.

(c) Use the binomial theorem to expand $(1 + x)^n$. Then show that the coefficient of the term independent of x on the left side of equation (1) is
$$\binom{n}{0}^2 + \binom{n}{1}^2 + \binom{n}{2}^2 + \cdots + \binom{n}{n}^2$$

12.3   INTRODUCTION TO SEQUENCES AND SERIES

Students of calculus do not always understand that infinite series are primarily tools for the study of functions.

George F. Simmons in *Calculus with Analytic Geometry* (New York: McGraw-Hill, 1985)

This section and the next two sections in this chapter deal with numerical sequences. We will begin with a somewhat informal definition of this concept. Then, after looking at some examples and terminology, we will present a more formal definition. A **numerical sequence** is an ordered list of numbers. Here are four examples:

example A: $1, \sqrt{2}, 10$

example B: $2, 4, 6, 8, \ldots$

example C: $1, \frac{1}{2}, \frac{1}{4}, \frac{1}{8}, \ldots$

example D: $1, 1, 1, 1, \ldots$

The individual entries in a numerical sequence are called the **terms** of the sequence. In this chapter, the terms in each sequence will always be real numbers, so for convenience, we will drop the adjective *numerical* and refer simply to *sequences*. (It is worth pointing out, however, that in more advanced courses, sequences are studied in which the individual terms are functions.) Any sequence possessing only a finite number of terms is called a **finite sequence**. Thus, the sequence in example A is a finite sequence. On the other hand, examples B, C, and D are examples of what we call **infinite sequences**; each contains infinitely many terms. As example D indicates, it is not necessary that all the terms in a sequence be distinct. In this chapter, all the sequences we discuss will be infinite sequences.

In examples B, C, and D, the three dots are read "and so on." In using this notation, we are assuming that it is clear what the subsequent terms of the sequence are. Toward this end, we often specify a formula for the *n*th term in a sequence. Example B in this case would appear this way:

$$2, 4, 6, 8, \ldots, 2n, \ldots$$

A letter with subscripts is often used to denote the various terms in a sequence. For instance, if we denote the sequence in example B by $a_1, a_2, a_3, \ldots$, then we have $a_1 = 2$, $a_2 = 4$, $a_3 = 6$, and, in general, $a_n = 2n$.

EXAMPLE 1 Consider the sequence $a_1, a_2, a_3, \ldots$, in which the *n*th term a_n is given by

$$a_n = \frac{n}{n+1}$$

Compute the first three terms of the sequence, as well as the one thousandth term.

Solution To obtain the first term, we replace *n* by 1 in the given formula. This yields

$$a_1 = \frac{1}{1+1} = \frac{1}{2}$$

The other terms are similarly obtained. We have

$$a_2 = \frac{2}{2+1} = \frac{2}{3}$$

$$a_3 = \frac{3}{3+1} = \frac{3}{4}$$

$$a_{1000} = \frac{1000}{1000+1} = \frac{1000}{1001}$$

In the example just concluded, we were given an explicit formula for the nth term of the sequence. The next example shows a different way of specifying a sequence, in which each term except the first is defined in terms of the previous term. This is an example of a **recursive definition.** (Recursive definitions are particularly useful in computer programming.)

■■■ EXAMPLE 2   Compute the first three terms of the sequence b_1, b_2, b_3, ..., which is defined recursively by

$$b_1 = 4$$
$$b_n = 2(b_{n-1} - 1) \qquad \text{for } n \geq 2$$

Solution   We are given the first term: $b_1 = 4$. To find b_2, we replace n by 2 in the formula $b_n = 2(b_{n-1} - 1)$ to obtain

$$b_2 = 2(b_1 - 1) = 2(4 - 1) = 6$$

Thus, $b_2 = 6$. Next we use this value of b_2 in the formula $b_n = 2(b_{n-1} - 1)$ to obtain b_3. Replacing n by 3 in this formula yields

$$b_3 = 2(b_2 - 1) = 2(6 - 1) = 10$$

We have now found the first three terms of the sequence: $b_1 = 4$, $b_2 = 6$, and $b_3 = 10$.   ■■■

If you think about the central idea behind the concept of a sequence, you can see that a function is involved: For each input n, we have an output a_n. For this reason, the formal definition of a sequence is phrased as follows.

DEFINITION   Sequence

A **sequence** is a function whose domain is the set of natural numbers.

If we denote the function by f for the moment, then $f(1)$ would denote what we have been calling a_1, the first term of the sequence. Similarly, $f(2) = a_2$, $f(3) = a_3$, and so on.

■■■ EXAMPLE 3   Consider the sequence with nth term $a_n = 1/2^n$. Graph this sequence for $n = 1, 2, 3,$ and 4.

Solution   First, we compute the required values of a_n, as shown in the following table.

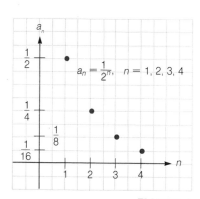

n	1	2	3	4
a_n	$\frac{1}{2}$	$\frac{1}{4}$	$\frac{1}{8}$	$\frac{1}{16}$

FIGURE I
The graph of the sequence
$a_n = 1/2^n$ for $n = 1, 2, 3, 4$

Now, locating the inputs n on the horizontal axis and the outputs a_n on the vertical axis, we obtain the graph shown in Figure 1.   ■■■

We will often be interested in the sum of certain terms of a sequence. Consider, for example, the sequence

$$10, \; 20, \; 30, \; 40, \; \ldots$$

in which the nth term is $10n$. The sum of the first four terms in this sequence is

$$10 + 20 + 30 + 40 = 100$$

More generally, the sum of the first n terms in this sequence is indicated by

$$10 + 20 + 30 + \cdots + 10n$$

This expression is an example of a **finite series**, which simply means a sum of a finite number of terms.

We can indicate the sum of the first n terms of the sequence a_1, a_2, a_3, ... by

$$a_1 + a_2 + a_3 + \cdots + a_n$$

Another way to indicate this sum uses what is called **sigma notation**, which we now introduce. The capital Greek letter sigma is written $\sum$. We define the notation $\sum\limits_{k=1}^{n} a_k$ by the equation

$$\sum_{k=1}^{n} a_k = a_1 + a_2 + a_3 + \cdots + a_n$$

For example, $\sum\limits_{k=1}^{3} a_k$ stands for the sum $a_1 + a_2 + a_3$, the idea in this case being to replace the subscript k successively by 1, 2, and 3 and then add the results.

For a more concrete example, let's evaluate the expression $\sum\limits_{k=1}^{4} k^2$. We have

$$\sum_{k=1}^{4} k^2 = 1^2 + 2^2 + 3^3 + 4^2$$

$$= 1 + 4 + 9 + 16 = 30$$

There is nothing special about the choice of the letter k in the expression $\sum\limits_{k=1}^{4} k^2$. For instance, we could equally well write

$$\sum_{j=1}^{4} j^2 = 1^2 + 2^2 + 3^2 + 4^2 = 30$$

The letter k in the expression $\sum\limits_{k=1}^{4} k^2$ is called the **index of summation**. Similarly, the letter j appearing in $\sum\limits_{j=1}^{4} j^2$ is the index of summation in that case. As we have seen, the choice of the letter used for the index of summation has no effect on the value of the indicated sum. For this reason, the index of summation is referred to as a *dummy variable*. The next two examples provide further practice with the sigma notation.

■ ▬ ■ EXAMPLE 4 Express each of the following sums without sigma notation.

(a) $\displaystyle\sum_{k=1}^{3} (3k-2)^2$ 　　(b) $\displaystyle\sum_{i=1}^{4} ix^{i-1}$ 　　(c) $\displaystyle\sum_{j=1}^{5} (a_{j+1}-a_j)$

Solution

(a) The notation $\displaystyle\sum_{k=1}^{3} (3k-2)^2$ directs us to replace k successively by 1, 2, and 3 in the expression $(3k-2)^2$ and then add the results. We thus obtain

$$\sum_{k=1}^{3} (3k-2)^2 = 1^2 + 4^2 + 7^2 = 66$$

(b) The notation $\displaystyle\sum_{i=1}^{4} ix^{i-1}$ directs us to replace i successively by 1, 2, 3, and 4 in the expression ix^{i-1} and then add the results. We have

$$\sum_{i=1}^{4} ix^{i-1} = 1x^0 + 2x^1 + 3x^2 + 4x^3$$

$$= 1 + 2x + 3x^2 + 4x^3$$

(c) To expand $\displaystyle\sum_{j=1}^{5} (a_{j+1}-a_j)$, we replace j successively by 1, 2, 3, 4, and 5 in the expression $(a_{j+1}-a_j)$ and then add. We obtain

$$\sum_{j=1}^{5} (a_{j+1}-a_j) = (a_2 - a_1) + (a_3 - a_2) + (a_4 - a_3) + (a_5 - a_4) + (a_6 - a_5)$$

$$= a_2 - a_1 + a_3 - a_2 + a_4 - a_3 + a_5 - a_4 + a_6 - a_5$$

Combining like terms, we have

$$\sum_{j=1}^{5} (a_{j+1}-a_j) = a_6 - a_1$$

Sums such as $\displaystyle\sum_{j=1}^{5} (a_{j+1}-a_j)$ are known as **collapsing** or **telescoping sums**.
　　　　　　　　　　　　　　　　　　　　　　　　　　　　　　　■■■

■ ▬ ■ EXAMPLE 5 Use sigma notation to rewrite each sum.

(a) $\dfrac{x}{1!} + \dfrac{x^2}{2!} + \dfrac{x^3}{3!} + \cdots + \dfrac{x^{12}}{12!}$ 　　(b) $\dfrac{x}{2!} + \dfrac{x^2}{3!} + \dfrac{x^3}{4!} + \cdots + \dfrac{x^n}{(n+1)!}$

Solution

(a) Since the exponents on x run from 1 to 12, we choose a dummy variable, say k, running from 1 to 12. Also, we notice that if the numerator of a given term in the sum is x^k, then the corresponding denominator is $k!$. Consequently, the sum can be written

$$\frac{x}{1!} + \frac{x^2}{2!} + \frac{x^3}{3!} + \cdots + \frac{x^{12}}{12!} = \sum_{k=1}^{12} \frac{x^k}{k!}$$

(**b**) Since the exponents on x run from 1 to n, we choose a dummy variable, say k, running from 1 to n. (Note that both of the letters n and x would be inappropriate here as dummy variables.) Also, we notice that if the numerator of a given term in the sum is x^k, then the corresponding denominator is $(k+1)!$. Thus, the given sum can be written

$$\frac{x}{2!} + \frac{x^2}{3!} + \frac{x^3}{4!} + \cdots + \frac{x^n}{(n+1)!} - \sum_{k=1}^{n} \frac{x^k}{(k+1)!}$$

▮▮▮

EXERCISE SET 12.3

A

In Exercises 1–20, compute the first five terms in each sequence. [In Exercises 1–12, where the nth term is given by an explicit formula, for instance, $a_n = n/(n+1)$, assume that the equation holds for all natural numbers n.]

1. $a_n = \dfrac{n}{n+1}$

2. $a_n = n^3$

3. $a_n = (n-1)^2$

4. $b_n = n!$

5. $b_n = [1 + (1/n)]^n$

6. $b_n = \dfrac{n^n}{n!}$

7. $u_n = (-1)^n$

8. $a_n = (-1)^n \sqrt{n^n}$

9. $a_n = (-1)^{n+1}/n!$

10. $S_n = 1/[(n+1)!]$

11. $a_n = 3n$

12. $b_n = 3$

13. $a_1 = 1;\ a_n = (1 + a_{n-1})^2,\ n \ge 2$

14. $a_1 = 2;\ a_n = \sqrt{a_{n-1}^2 + 1},\ n \ge 2$

15. $a_1 = 2;\ a_2 = 2;\ a_n = a_{n-1}a_{n-2},\ n \ge 3$

16. $F_1 = 1;\ F_2 = 1;\ F_n = F_{n-1} + F_{n-2},\ n \ge 3$

17. $a_1 = 1;\ a_{n+1} = na_n,\ n \ge 1$

18. $a_1 = 1;\ a_2 = 2;\ a_n = a_{n-1}/a_{n-2},\ n \ge 3$

19. $a_1 = 0;\ a_n = 2^{a_{n-1}},\ n \ge 2$

20. $a_1 = 0;\ a_2 = 1;\ a_n = \frac{1}{2}(a_{n-1} + a_{n-2}),\ n \ge 3$

In Exercises 21–26, graph the sequences for the indicated values of n.

21. $a_n = (-1)^n;\ n = 1, 2, 3, 4$

22. $a_n = 2n - 1;\ n = 1, 2, 3$

23. $a_n = 1/n;\ n = 1, 2, 3, 4$

24. $a_n = (-1)^n/n;\ n = 1, 2, 3, 4$

25. $a_n = (n-1)/(n+1);\ n = 1, 2, 3$

26. $a_n = [1 + (1/n)]^n;\ n = 1, 2, 3$

In Exercises 27–32, find the sum of the first five terms in the sequence with an nth term as given.

27. $a_n = 2^n$

28. $b_n = 2^{-n}$

29. $a_n = n^2 - n$

30. $b_n = (n-1)!$

31. $a_n = (-1)^n/n!$

32. $a_1 = 1;\ a_n = \dfrac{1}{n-1} - \dfrac{1}{n+1},\ n \ge 2$

33. Find the sum of the first five terms of the sequence that is defined recursively by $a_1 = 1;\ a_2 = 2;$ $a_n = a_{n-1}^2 + a_{n-2}^2,\ n \ge 3$.

34. Find the sum of the first four terms of the sequence defined recursively by $a_1 = 2;\ a_n = n/a_{n-1},\ n \ge 2$.

35. Find the sum of the first four terms of the sequence defined recursively by $a_1 = 2;\ a_n = (a_{n-1})^2,\ n \ge 2$.

36. Find the sum of the first six terms of the sequence defined recursively by $a_1 = 1;\ a_2 = 2;\ a_{n+1} = a_n a_{n-1},$ $n \ge 2$.

In Exercises 37–48, express each of the sums without using sigma notation. Simplify your answers where possible.

37. $\displaystyle\sum_{k=1}^{3} (k-1)$

38. $\displaystyle\sum_{k=1}^{5} k$

39. $\displaystyle\sum_{k=4}^{5} k^2$

40. $\displaystyle\sum_{k=2}^{6} (1 - 2k)$

41. $\displaystyle\sum_{n=1}^{3} x^n$

42. $\displaystyle\sum_{n=1}^{3} (n-1)x^{n-2}$

43. $\displaystyle\sum_{n=1}^{4} \frac{1}{n}$

44. $\displaystyle\sum_{n=0}^{4} 3^n$

45. $\displaystyle\sum_{j=1}^{9} \log_{10} \frac{j}{j+1}$

46. $\displaystyle\sum_{j=2}^{5} \log_{10} j$

47. $\displaystyle\sum_{j=1}^{6} \left(\frac{1}{j} - \frac{1}{j+1}\right)$

48. $\displaystyle\sum_{j=1}^{5} (x^{j+1} - x^j)$

In Exercises 49–58, rewrite the sums using sigma notation.

49. $5 + 5^2 + 5^3 + 5^4$

50. $5 + 5^2 + 5^3 + \cdots + 5^n$

51. $x + x^2 + x^3 + x^4 + x^5 + x^6$

52. $x + 2x^2 + 3x^3 + 4x^4 + 5x^5 + 6x^6$

53. $\frac{1}{1} + \frac{1}{2} + \frac{1}{3} + \cdots + \frac{1}{12}$

54. $\frac{1}{1} + \frac{1}{2} + \frac{1}{3} + \cdots + \frac{1}{n}$

55. $2 - 2^2 + 2^3 - 2^4 + 2^5$

56. $\dbinom{10}{3} + \dbinom{10}{4} + \dbinom{10}{5} + \cdots + \dbinom{10}{10}$

57. $1 - 2 + 3 - 4 + 5$

58. $\frac{1}{2} - \frac{1}{4} + \frac{1}{8} - \frac{1}{16} + \frac{1}{32} - \frac{1}{64} + \frac{1}{128}$

B
59. The *Fibonacci numbers* $F_1, F_2, F_3, \ldots$ are defined as follows:

$$F_1 = 1; \quad F_2 = 1; \quad F_{n+2} = F_n + F_{n+1} \qquad (n \geq 1)$$

(a) Verify that the first ten Fibonacci numbers are as follows.

F_1	F_2	F_3	F_4	F_5	F_6	F_7	F_8	F_9	F_{10}
1	1	2	3	5	8	13	21	34	55

(b) Complete the following table.

n	$F_1 + F_2 + F_3 + \cdots + F_n$	$F_{n+2} - 1$
1		
2		
3		
4		
5		

(c) Use mathematical induction to prove that
$$F_1 + F_2 + F_3 + \cdots + F_n = F_{n+2} - 1$$
for all natural numbers n.

(d) Use mathematical induction to prove that
$$F_n \geq n \qquad \text{for all natural numbers } n \geq 5$$

(e) Use mathematical induction to show that
$$F_1^2 + F_2^2 + F_3^2 + \cdots + F_n^2 = F_n F_{n+1}$$
for all natural numbers n.

(f) Use mathematical induction to show that
$$F_{n+1}^2 = F_n F_{n+2} + (-1)^n$$
for all natural numbers n. *Hint for Step 2:* Add $F_{k+1}F_{k+2}$ to both sides of the equation in the induction hypothesis. Then factor F_{k+1} from the left-hand side and factor F_{k+2} from the first two terms on the right-hand side.

(g) This part of the exercise requires a knowledge of matrix multiplication. Show that
$$\begin{pmatrix} 1 & 1 \\ 1 & 0 \end{pmatrix}^n = \begin{pmatrix} F_{n+1} & F_n \\ F_n & F_{n-1} \end{pmatrix} \qquad \text{for } n \geq 2$$

12.4 ARITHMETIC SEQUENCES AND SERIES

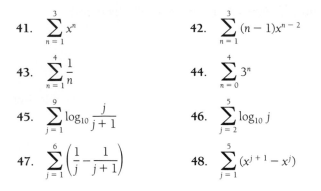

One of the most natural ways to generate a sequence is to begin with a fixed number a and then repeatedly add a fixed constant d. This yields the sequence

$$a, a+d, a+2d, a+3d, \ldots$$

Such a sequence is called an **arithmetic sequence** or an **arithmetic progression**. Notice that the difference between any two consecutive terms is the constant d. We call d the **common difference**. Here are several examples of arithmetic sequences:

example A: 1, 2, 3, ...
example B: 3, 7, 11, 15, ...
example C: 10, 5, 0, −5, ...

In example A, the first term is $a = 1$ and the common difference is $d = 1$. For example B, we have $a = 3$. The value of d in this example is found by subtracting any two consecutive terms; thus, $d = 4$. Finally, in example C, we have $a = 10$ and $d = −5$. Notice when the common difference is negative, the terms of the sequence decrease.

There is a simple formula for the nth term in an arithmetic sequence

$$a, a + d, a + 2d, a + 3d, \ldots$$

In this sequence, notice that

$$a_1 = a + 0d$$
$$a_2 = a + 1d$$
$$a_3 = a + 2d$$
$$a_4 = a + 3d$$

Following this pattern, it appears that the formula for a_n should be

$$a_n = a + (n − 1)d$$

Indeed, this is the correct formula, and Exercise 32 asks you to verify it using mathematical induction.

The nth Term of an Arithmetic Sequence

The nth term of an arithmetic sequence $a, a + d, a + 2d, \ldots$ is given by

$$a_n = a + (n − 1)d$$

EXAMPLE 1   Determine the one hundredth term of the arithmetic sequence

$$7, 10, 13, 16, \ldots$$

Solution   The first term is $a = 7$, and the common difference is $d = 3$. Substituting these values in the formula $a_n = a + (n − 1)d$ yields

$$a_n = 7 + (n − 1)3 = 3n + 4$$

To find the 100th term, we replace n by 100 in this last equation to obtain

$$a_{100} = 3(100) + 4 = 304 \qquad \text{as required}$$

EXAMPLE 2   Determine the arithmetic sequence in which the second term is −2 and the eighth term is 40.

Solution   We are given that the second term is -2. Using this information in the formula $a_n = a + (n-1)d$, we have

$$-2 = a + (2-1)d = a + d$$

This gives us one equation in two unknowns. We are also given that the eighth term is 40. Therefore,

$$40 = a + (8-1)d = a + 7d$$

We now have a system of two equations in two unknowns:

$$\begin{cases} -2 = a + d \\ 40 = a + 7d \end{cases}$$

Subtracting the first equation from the second gives us

$$42 = 6d \qquad \text{or} \qquad 7 = d$$

To find a, we replace d by 7 in the first equation of the system. This yields

$$-2 = a + 7 \qquad \text{or} \qquad -9 = a$$

We have now determined the sequence, since we know that the first term is -9 and the common difference is 7. The first four terms of the sequence are -9, -2, 5, and 12.   ▪▪▪

Next we would like to derive a formula for the sum of the first n terms of an arithmetic sequence. Such a sum is referred to as an **arithmetic series.** If we use S_n to denote the required sum, we have

$$S_n = a + (a + d) + \cdots + [a + (n-2)d] + [a + (n-1)d] \qquad (1)$$

Of course, we must obtain the same sum if we add the terms from right to left rather than left to right. That is, we must have

$$S_n = [a + (n-1)d] + [a + (n-2)d] + \cdots + (a + d) + a \qquad (2)$$

Let us now add equations (1) and (2). Adding the left-hand sides is easy; we obtain $2S_n$. Now we add the corresponding terms on the right-hand sides. For the first terms, we have

$$\underset{\substack{\uparrow \\ \text{first term} \\ \text{in equation (1)}}}{a} \quad + \quad \underset{\substack{\text{first term} \\ \text{in equation (2)}}}{\underbrace{[a + (n-1)d]}} = 2a + (n-1)d$$

Next we add the second terms:

$$\underset{\substack{\uparrow \\ \text{second term} \\ \text{in equation (1)}}}{(a + d)} \quad + \quad \underset{\substack{\text{second term} \\ \text{in equation (2)}}}{\underbrace{[a + (n-2)d]}} = 2a + d + (n-2)d = 2a + (n-1)d$$

Notice that the sum of the second terms is again $2a + (n-1)d$, the same quantity we arrived at with the first terms. As you can check, this pattern continues all the way through to the last terms. For instance,

$$\underset{\substack{\text{last term} \\ \text{in equation (1)}}}{\underbrace{[a + (n-1)d]}} \quad + \quad \underset{\substack{\uparrow \\ \text{last term} \\ \text{in equation (2)}}}{a} \quad = 2a + (n-1)d$$

We conclude from these observations that by adding the right-hand sides of equations (1) and (2), the quantity $2a + (n - 1)d$ is added a total of n times. Therefore,

$$2S_n = n[2a + (n - 1)d]$$

$$S_n = \frac{n}{2}[2a + (n - 1)d]$$

This gives us the desired formula for the sum of the first n terms in an arithmetic sequence. There is an alternate form of this formula, which now follows rather quickly:

$$S_n = \frac{n}{2}[2a + (n - 1)d]$$

$$= \frac{n}{2}\{a + [a + (n - 1)d]\}$$

$$= \frac{n}{2}(a + a_n) = n\left(\frac{a + a_n}{2}\right)$$

This last equation is easy to remember. It says that the sum of an arithmetic series is obtained by averaging the first and last terms and then multiplying this average by n, the number of terms. For reference, we summarize both formulas as follows.

Formulas for the Sum of an Arithmetic Series
$S_n = \frac{n}{2}[2a + (n - 1)d]$ $S_n = n\left(\frac{a + a_n}{2}\right)$

EXAMPLE 3 Find the sum of the first 30 terms of the arithmetic sequence 2, 6, 10, 14, ...

Solution We have $a = 2$, $d = 4$, and $n = 30$. Substituting these values in the formula $S_n = (n/2)[2a + (n - 1)d]$ then yields

$$S_{30} = \frac{30}{2}[2(2) + (30 - 1)4]$$

$$= 15[4 + 29(4)]$$

$$= 1800 \qquad \text{(Check the arithmetic)}$$

EXAMPLE 4 In a certain arithmetic sequence, the first term is 6 and the fortieth term is 71. Find the sum of the first 40 terms and also the common difference d for the sequence.

Solution We have $a = 6$ and $a_{40} = 71$. Using these values in the formula $S_n = n(a + a_n)/2$ yields

$$S_{40} = \frac{40}{2}(6 + 71) = 20(77) = 1540$$

The sum of the first 40 terms is thus 1540. The value of d can now be found by using the formula $a_n = a + (n - 1)d$. We have $a_{40} = a + (40 - 1)d$, and therefore

$$71 = 6 + 39d$$
$$39d = 65$$
$$d = \frac{65}{39} = \frac{5}{3}$$

The required value of d is therefore $\frac{5}{3}$. ∎

EXAMPLE 5 Show that $\displaystyle\sum_{k=1}^{50} (3k - 2)$ represents an arithmetic series, and compute the sum.

Solution There are two different ways to see that $\displaystyle\sum_{k=1}^{50} (3k - 2)$ is an arithmetic series. One way is simply to write out the first few terms and look at the pattern. We then have

$$\sum_{k=1}^{50} (3k - 2) = 1 + 4 + 7 + 10 + \cdots + 148$$

From this it is clear that we are indeed summing the terms in an arithmetic sequence in which $d = 3$ and $a = 1$.

Another, more formal way to show that $\displaystyle\sum_{k=1}^{50} (3k - 2)$ represents an arithmetic series is to prove that the difference between successive terms in the indicated sum is a constant. Now, the form of a typical term in this sum is $(3k - 2)$. Thus, the form of the next term must be $[3(k + 1) - 2]$. The difference between these terms is then

$$[3(k + 1) - 2] - (3k - 2) = 3k + 3 - 2 - 3k + 2 = 3$$

The difference therefore is constant, as we wished to show.

To evaluate $\displaystyle\sum_{k=1}^{50} (3k - 2)$, we can use either of the two formulas for the sum of an arithmetic series. Using the formula $S_n = (n/2)[2a + (n - 1)d]$, we obtain

$$S_{50} = \frac{50}{2}[2(1) + 49(3)] = 25(149) = 3725$$

Thus, the required sum is 3725. You should check for yourself that the same value is obtained using the formula $S_n = n(a + a_n)/2$. ∎

EXERCISE SET 12.4

A

1. Find the common difference d for each of the following arithmetic sequences.
   (a) $1, 3, 5, 7, \ldots$ (b) $10, 6, 2, -2, \ldots$
   (c) $\frac{2}{3}, 1, \frac{4}{3}, \frac{5}{3}, \ldots$
   (d) $1, 1 + \sqrt{2}, 1 + 2\sqrt{2}, 1 + 3\sqrt{2}, \ldots$

2. Which of the following are arithmetic sequences?
   (a) $2, 4, 8, 16, \ldots$ (b) $5, 9, 13, 17, \ldots$
   (c) $3, \frac{11}{5}, \frac{7}{5}, \frac{3}{5}, \ldots$ (d) $-1, -1, -1, -1, \ldots$
   (e) $-1, 1, -1, 1, \ldots$

In Exercises 3–8, find the indicated term in each sequence.

3. $10, 21, 32, 43, \ldots; a_{12}$ **4.** $7, 2, -3, -8, \ldots; a_{20}$

5. $6, 11, 16, 21, \ldots; a_{100}$ **6.** $\frac{2}{5}, \frac{4}{5}, \frac{6}{5}, \frac{8}{5}, , \ldots; a_{30}$

7. $-1, 0, 1, 2, \ldots; a_{1000}$

8. $42, 1, -40, -81, \ldots; a_{15}$

9. The fourth term in an arithmetic sequence is -6, and the tenth term is 5. Find the common difference and the first term.

10. The fifth term in an arithmetic sequence is $\frac{1}{2}$, and the twentieth term is $\frac{7}{8}$. Find the first three terms of the sequence.

11. The sixtieth term in an arithmetic sequence is 105, and the common difference is 5. Find the first term.

12. Find the common difference in an arithmetic sequence in which $a_{10} - a_{20} = 70$.

13. Find the common difference in an arithmetic sequence in which $a_{15} - a_7 = -1$.

14. Find the sum of the first 16 terms in the sequence 2, 11, 20, 29, $\ldots$.

15. Find the sum of the first 1000 terms in the sequence 1, 2, 3, 4, $\ldots$.

16. Find the sum of the first 50 terms in an arithmetic series that has first term -8 and fiftieth term 139.

17. Find the sum: $\dfrac{\pi}{3} + \dfrac{2\pi}{3} + \pi + \dfrac{4\pi}{3} + \cdots + \dfrac{13\pi}{3}$.

18. Find the sum: $\dfrac{1}{e} + \dfrac{3}{e} + \dfrac{5}{e} + \cdots + \dfrac{21}{e}$.

19. Determine the first term of an arithmetic sequence in which the common difference is 5 and the sum of the first 38 terms is 3534.

20. The sum of the first 12 terms in an arithmetic sequence is 156. What is the sum of the first and twelfth terms?

21. In a certain arithmetic sequence, the first term is 4 and the sixteenth term is -100. Find the sum of the first

16 terms and also the common difference for the sequence.

22. The fifth and fiftieth terms of an arithmetic sequence are 3 and 30, respectively. Find the sum of the first 10 terms.

23. The eighth term in an arithmetic sequence is 5, and the sum of the first ten terms is 20. Find the common difference and the first term of the sequence.

In Exercises 24–26, find each sum.

24. $\displaystyle\sum_{i=1}^{10} (2i - 1) = 1 + 3 + 5 + \cdots + 19$

25. $\displaystyle\sum_{k=1}^{20} (4k + 3)$ **26.** $\displaystyle\sum_{n=5}^{100} (2n - 1)$

27. The sum of three consecutive terms in an arithmetic sequence is 30, and their product is 360. Find the three terms. *Suggestion:* Let x denote the *middle* term and d the common difference.

28. The sum of three consecutive terms in an arithmetic sequence is 21, and the sum of their squares is 197. Find the three terms.

29. The sum of three consecutive terms in an arithmetic sequence is 6, and the sum of their cubes is 132. Find the three terms.

30. In a certain arithmetic sequence, $a = -4$ and $d = 6$. If $S_n = 570$, find n.

31. Let $a_1 = \dfrac{1}{1 + \sqrt{2}}$, $a_2 = -1$, and $a_3 = \dfrac{1}{1 - \sqrt{2}}$.
   (a) Show that $a_2 - a_1 = a_3 - a_2$.
   (b) Find the sum of the first six terms in the arithmetic sequence
   $$\frac{1}{1 + \sqrt{2}}, -1, \frac{1}{1 - \sqrt{2}}, \ldots$$

32. Using mathematical induction, prove that the nth term of the sequence $a, a + d, a + 2d, \ldots$ is given by
   $$a_n = a + (n - 1)d$$

B

33. Let b denote a positive constant. Find the sum of the first n terms in the sequence.
   $$\frac{1}{1 + \sqrt{b}}, \frac{1}{1 - b}, \frac{1}{1 - \sqrt{b}}, \ldots$$

34. The sum of the first n terms in a certain arithmetic sequence is given by $S_n = 3n^2 - n$. Show that the rth term is given by $a_r = 6r - 4$.

35. Let $a_1, a_2, a_3, \ldots$ be an arithmetic sequence, and let S_k denote the sum of the first k terms. If $S_n/S_m = n^2/m^2$, show that $\dfrac{a_n}{a_m} = \dfrac{2n-1}{2m-1}$.

36. If the common difference in an arithmetic sequence is twice the first term, show that $\dfrac{S_n}{S_m} = \dfrac{n^2}{m^2}$.

37. The lengths of the sides of a right triangle form three consecutive terms in an arithmetic sequence. Show that the triangle is similar to the $3-4-5$ right triangle.

38. Suppose that $1/a$, $1/b$, and $1/c$ are three consecutive terms in an arithmetic sequence. Show that:
    (a) $\dfrac{a}{c} = \dfrac{a-b}{b-c}$; (b) $b = \dfrac{2ac}{a+c}$.

39. Suppose that a, b, and c are three positive numbers with $a > c > 2b$. If $1/a$, $1/b$, and $1/c$ are consecutive terms in an arithmetic sequence, show that
$$\ln(a+c) + \ln(a - 2b + c) = 2 \ln(a - c)$$

12.5 GEOMETRIC SEQUENCES AND SERIES

A **geometric sequence** or a **geometric progression** is a sequence of the form

$$a, ar, ar^2, ar^3, \ldots \qquad \text{where } a \text{ and } r \text{ are constants}$$

As you can see, each term after the first in a geometric sequence is obtained by multiplying the previous term by r. The number r is called the **common ratio**, since the ratio of any term to the previous one is always r. For instance, the ratio of the fourth term to the third is $ar^3/ar^2 = r$. Here are two examples of geometric sequences:

$$1, \frac{1}{2}, \frac{1}{4}, \frac{1}{8}, \ldots$$
$$10, -100, 1000, -10{,}000, \ldots$$

In the first example, we have $a = 1$ and $r = \frac{1}{2}$; in the second example, we have $a = 10$ and $r = -10$.

■□□■ **EXAMPLE 1** In a certain geometric sequence, the first term is 2, the third term is 3, and the common ratio is negative. Find the second term.

Solution Let x denote the second term, so that the sequence begins

$$2, x, 3, \ldots$$

By definition, the ratios $3/x$ and $x/2$ must be equal. Thus, we have

$$\frac{3}{x} = \frac{x}{2}$$
$$x^2 = 6$$
$$x = \pm\sqrt{6}$$

Now, the second term must be negative, since the first term is positive and the common ratio is negative. Thus, the second term is $x = -\sqrt{6}$. ■■■

The formula for the nth term of a geometric sequence is easily deduced by considering Table 1. The table indicates that the exponent on r is one less than the value of n in each case. On the basis of this observation, it appears that the nth term must be given by $a_n = ar^{n-1}$. Indeed, it can be shown by mathematical

TABLE 1

n	a_n
1	ar^0
2	ar^1
3	ar^2
4	ar^3
⋮	⋮

induction that this formula does hold for all natural numbers n. (Exercise 30 asks you to carry out the proof.) We summarize this result as follows:

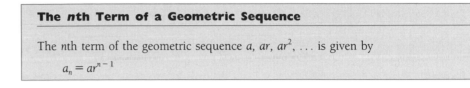

The *n*th Term of a Geometric Sequence

The nth term of the geometric sequence a, ar, ar^2, ... is given by

$$a_n = ar^{n-1}$$

EXAMPLE 2   Find the seventh term in the geometric sequence 2, 6, 18, ...

Solution   We can find the common ratio r by dividing the second term by the first; thus, $r = 3$. Now, using $a = 2$, $r = 3$, and $n = 7$ in the formula $a_n = ar^{n-1}$, we have

$$a_7 = 2(3)^6 = 2(729) = 1458$$

The seventh term of the sequence is therefore 1458.

Suppose that we begin with a geometric sequence a, ar, ar^2, ..., in which $r \neq 1$. If we add the first n terms and denote the sum by S_n, we have

$$S_n = a + ar + ar^2 + \cdots + ar^{n-2} + ar^{n-1} \tag{1}$$

This sum is called a **finite geometric series**. We would like to find a formula for S_n. To do this, we multiply equation (1) by r to obtain

$$rS_n = ar + ar^2 + ar^3 + \cdots + ar^{n-1} + ar^n \tag{2}$$

We now subtract equation (2) from equation (1). This yields (after combining like terms)

$$S_n - rS_n = a - ar^n$$
$$S_n(1 - r) = a(1 - r^n)$$
$$S_n = \frac{a(1 - r^n)}{1 - r} \qquad (r \neq 1)$$

This is the formula for the sum of a finite geometric series. We summarize this result in the box that follows.

Formula for the Sum of a Geometric Series

Let S_n denote the sum $a + ar + ar^2 + \cdots + ar^{n-1}$, and assume that $r \neq 1$. Then

$$S_n = \frac{a(1 - r^n)}{1 - r}$$

EXAMPLE 3   Evaluate the sum $\dfrac{1}{2^1} + \dfrac{1}{2^2} + \dfrac{1}{2^3} + \cdots + \dfrac{1}{2^{10}}$.

Solution This is a finite geometric series with $a = \frac{1}{2}$, $r = \frac{1}{2}$, and $n = 10$. Using these values in the formula for S_n yields

$$S_{10} = \frac{\frac{1}{2}\left[1 - \left(\frac{1}{2}\right)^{10}\right]}{1 - \frac{1}{2}} = \frac{\frac{1}{2}\left[1 - \frac{1}{1024}\right]}{\frac{1}{2}}$$

$$= 1 - \frac{1}{1024} = \frac{1023}{1024}$$

We would now like to attach a meaning to certain expressions of the form

$$a + ar + ar^2 + \cdots$$

Such an expression is called an **infinite geometric series**. The three dots indicate (intuitively at least) that the additions are to be carried out indefinitely, without end. To see how to proceed here, let's look at some examples involving finite geometric series. In particular, we will consider the series

$$\frac{1}{2^1} + \frac{1}{2^2} + \frac{1}{2^3} + \cdots + \frac{1}{2^n}$$

for increasing values of n. The idea is to look for a pattern as n grows ever larger. Let $S_1 = \frac{1}{2}$, $S_2 = \frac{1}{2^1} + \frac{1}{2^2}$, $S_3 = \frac{1}{2^1} + \frac{1}{2^2} + \frac{1}{2^3}$, and, in general,

$$S_n = \frac{1}{2^1} + \frac{1}{2^2} + \cdots + \frac{1}{2^n}$$

Then we can compute S_n for any given value of n by means of the formula for the sum of a finite geometric series. From Table 2, which displays the results of such calculations, it seems clear that as n grows larger and larger, the value of S_n grows ever closer to 1. More precisely (but leaving the details for calculus), it can be shown that the value of S_n can be made as close to 1 as we please, provided only that n is sufficiently large. For this reason, we say that the *sum of the infinite geometric series* $\frac{1}{2^1} + \frac{1}{2^2} + \frac{1}{2^3} + \cdots$ is 1, and we write

$$\frac{1}{2^1} + \frac{1}{2^2} + \frac{1}{2^3} + \cdots = 1$$

TABLE 2

n	$S_n = \frac{1}{2} + \frac{1}{2^2} + \cdots + \frac{1}{2^n}$
1	0.5
2	0.75
5	0.96875
10	0.999023437...
15	0.999969482...
20	0.999999046...
25	0.999999970...

We can arrive at this last result another way. First we compute the sum of the finite geometric series $\frac{1}{2^1} + \frac{1}{2^2} + \cdots + \frac{1}{2^n}$. As you can check, the result is

$$S_n = 1 - \left(\frac{1}{2}\right)^n$$

Now, as n grows larger and larger, the value of $\left(\frac{1}{2}\right)^n$ gets closer and closer to zero. Thus, as n grows ever larger, the value of S_n will more and more resemble $1 - 0$, or 1.

Now let us repeat our reasoning to obtain a formula for the sum of the infinite geometric series

$$a + ar + ar^2 + \cdots \qquad \text{where } |r| < 1$$

First we consider the finite geometric series

$$a + ar + ar^2 + \cdots + ar^{n-1}$$

The sum S_n in this case is

$$S_n = \frac{a(1-r^n)}{1-r}$$

We want to know how S_n behaves as n grows ever larger. This is where the assumption $|r| < 1$ is crucial. Just as $\left(\frac{1}{2}\right)^n$ approaches zero as n grows larger and larger, so will r^n approach zero as n increases. Thus, as n grows ever larger, the sum S_n will more and more resemble

$$\frac{a(1-0)}{1-r} = \frac{a}{1-r}$$

For this reason, we say that the sum of the infinite geometric series is $a/(1-r)$. We will make free use of this result in the subsequent examples. However, a more rigorous development of infinite series properly belongs to calculus.

Formula for the Sum of an Infinite Geometric Series

Suppose that $|r| < 1$. Then the sum S of the infinite geometric series $a + ar + ar^2 + \cdots$ is given by

$$S = \frac{a}{1-r}$$

■▦■ EXAMPLE 4 Find the sum of the infinite geometric series $1 + \frac{2}{3} + \frac{4}{9} + \cdots$.

Solution In this case, we have $a = 1$ and $r = \frac{2}{3}$. Thus,

$$S = \frac{a}{1-r} = \frac{1}{1-\frac{2}{3}} = \frac{1}{\frac{1}{3}} = 3$$

The sum of the series is 3.   ■■■

■▦■ EXAMPLE 5 Find a fraction equivalent to the repeating decimal $0.2\overline{35}$.

Solution Let $S = 0.2\overline{35}$. Then we have

$$S = 0.2353535\ldots$$
$$= \frac{2}{10} + \frac{35}{1000} + \frac{35}{100{,}000} + \frac{35}{10{,}000{,}000} + \cdots$$

Now, the expression following $\frac{2}{10}$ on the right-hand side of this last equation is an infinite geometric series in which $a = \frac{35}{1000}$ and $r = \frac{1}{100}$. Thus,

$$S = \frac{2}{10} + \frac{a}{1 - r}$$

$$= \frac{2}{10} + \frac{\frac{35}{1000}}{1 - \frac{1}{100}} = \frac{233}{990} \qquad \text{(Check the arithmetic!)}$$

The given decimal is therefore equivalent to 233/990.

EXERCISE SET 12.5

A

1. Find the second term in a geometric sequence in which the first term is 9, the third term is 4, and the common ratio is positive.

2. Find the fifth term in a geometric sequence in which the fourth term is 4, the sixth term is 6, and the common ratio is negative.

3. The product of the first three terms in a geometric sequence is 8000. If the first term is 4, find the second and third terms.

In Exercises 4–8, find the indicated term of the geometric sequence.

4. $9, 81, 729, \ldots$; a_7
5. $-1, 1, -1, 1, \ldots$; a_{100}
6. $\frac{1}{2}, \frac{1}{4}, \frac{1}{8}, \ldots$; a_9
7. $\frac{2}{3}, \frac{4}{9}, \frac{8}{27}, \ldots$; a_8
8. $1, -\sqrt{2}, 2, \ldots$; a_6

9. Find the common ratio in a geometric sequence in which the first term is 1 and the seventh term is 4096.

10. Find the first term in a geometric sequence in which the common ratio is $\frac{4}{3}$ and the tenth term is $\frac{16}{9}$.

11. Find the sum of the first ten terms of the sequence 7, 14, 28, ...

12. Find the sum of the first five terms of the sequence $-\frac{1}{2}, \frac{3}{10}, -\frac{9}{50}, \ldots$.

13. Find the sum: $1 + \sqrt{2} + 2 + \cdots + 32$.

14. Find the sum of the first 12 terms in the sequence $-4, -2, -1, \ldots$.

In Exercises 15–17, evaluate each sum.

15. $\displaystyle\sum_{k=1}^{6} \left(\frac{3}{2}\right)^k$
16. $\displaystyle\sum_{k=1}^{6} \left(\frac{2}{3}\right)^{k+1}$
17. $\displaystyle\sum_{k=2}^{6} \left(\frac{1}{10}\right)^k$

In Exercises 18–22, determine the sum of each infinite geometric series.

18. $\dfrac{1}{4} + \dfrac{1}{4^2} + \dfrac{1}{4^3} + \cdots$
19. $\dfrac{2}{3} - \dfrac{4}{9} + \dfrac{8}{27} - \cdots$
20. $\dfrac{9}{10} + \dfrac{9}{100} + \dfrac{9}{1000} + \cdots$
21. $1 + \dfrac{1}{1.01} + \dfrac{1}{(1.01)^2} + \cdots$
22. $-1 - \dfrac{1}{\sqrt{2}} - \dfrac{1}{2} - \cdots$

In Exercises 23–27, express each repeating decimal as a fraction.

23. $0.555\ldots$
24. $0.\overline{47}$
25. $0.1\overline{23}$
26. $0.050505\ldots$
27. $0.4\overline{32}$

B

28. The lengths of the sides in a right triangle form three consecutive terms of a geometric sequence. Find the common ratio of the sequence. (There are two distinct answers.)

29. The product of three consecutive terms in a geometric sequence is -1000, and their sum is 15. Find the common ratio. (There are two answers.) *Suggestion:* Denote the terms by a/r, a, and ar.

30. Use mathematical induction to prove that the nth term of the geometric sequence $a, ar, ar^2, \ldots$ is ar^{n-1}.

31. Show that the sum of the following infinite geometric series is $\frac{3}{2}$:

$$\frac{\sqrt{3}}{\sqrt{3}+1} + \frac{\sqrt{3}}{\sqrt{3}+3} + \cdots$$

32. Let A_1 denote the area of an equilateral triangle, each side of which is 1 unit long. A second equilateral triangle is formed by joining the midpoints of the sides of the first triangle. Let A_2 denote the area of this second triangle. This process is then repeated to form a third triangle with area A_3; and so on. Find the sum of the areas: $A_1 + A_2 + A_3 + \cdots$.

33. Let $a_1, a_2, a_3, \ldots$ be a geometric sequence such that $r \neq 1$. Let $S = a_1 + a_2 + a_3 + \cdots + a_n$, and let $T = \dfrac{1}{a_1} + \dfrac{1}{a_2} + \cdots + \dfrac{1}{a_n}$. Show that $\dfrac{S}{T} = a_1 a_n$.

34. Suppose that a, b, and c are three consecutive terms in a geometric sequence. Show that $\dfrac{1}{a+b}$, $\dfrac{1}{2b}$, and $\dfrac{1}{c+b}$ are three consecutive terms in an arithmetic sequence.

35. A ball is dropped from a height of 6 ft. Assuming that on each bounce the ball rebounds to one-third of its previous height, find the total distance traveled by the ball.

◢ CHAPTER TWELVE SUMMARY OF PRINCIPAL TERMS AND FORMULAS

TERMS OR FORMULAS	PAGE REFERENCE	COMMENT
1. Principle of mathematical induction	714	This principle can be stated as follows. Suppose that for each natural number n we have a statement P_n. Suppose P_1 is true. Also suppose that P_{k+1} is true whenever P_k is true. Then according to the *principle of mathematical induction*, all the statements are true; that is, P_n is true for all natural numbers n. The principle of mathematical induction has the status of an axiom; that is, we accept its validity without proof.
2. Pascal's triangle	720	Pascal's triangle refers to the triangular array of numbers displayed on page 720. Additional rows can be added to the triangle according to the following rule: Each entry in the array (other than the 1's along the sides) is the sum of the two numbers diagonally above it. The numbers in the nth row of Pascal's triangle are the coefficients of the terms in the expansion of $(a+b)^{n-1}$
3. $n!$ (read: n factorial)	722	This denotes the product of the first n natural numbers. For example, $4! = (4)(3)(2)(1) = 24$.
4. $\dbinom{n}{k}$ (read: n choose k)	723	Let n and k be nonnegative integers with $k \leq n$. Then the *binomial coefficient* $\dbinom{n}{k}$ is defined by $$\binom{n}{k} = \frac{n!}{k!(n-k)!}$$
5. Binomial theorem	723	The *binomial theorem* is a formula that allows us to analyze and expand expressions of the form $(a+b)^n$. If we use the sigma notation, then the statement of the binomial theorem that appears on page 723 can be abbreviated to read $$(a+b)^n = \sum_{k=0}^{n} \binom{n}{k} a^{n-k} b^k$$
6. Sequence	729, 730	In the context of the present chapter, a *sequence* is an ordered list of real numbers: $a_1, a_2, a_3, \ldots$. The numbers a_k are called the *terms* of the sequence. A sequence can also be defined as a *function* with domain the set of natural numbers.

TERMS OR FORMULAS	PAGE REFERENCE	COMMENT
7. $\sum_{k=1}^{n} a_k$	731	This expression stands for the sum $a_1 + a_2 + a_3 + \cdots + a_n$. The letter k in the expression is referred to as the *index of summation*.
8. Arithmetic sequence	734	A sequence in which the successive terms differ by a constant is called an *arithmetic sequence*. The general form of an arithmetic sequence is $$a, a + d, a + 2d, a + 3d, \ldots$$ In this sequence, d is referred to as the *common difference*.
9. $a_n = a + (n - 1)d$	735	This is the formula for the nth term of an arithmetic sequence.
10. $S_n = \dfrac{n}{2}[2a + (n - 1)d]$ $S_n = n\left(\dfrac{a + a_n}{2}\right)$	737	These are the formulas for the sum of the first n terms of an arithmetic sequence.
11. Geometric sequence	740	A sequence in which the ratio of successive terms is constant is called a *geometric sequence*. The general form of a geometric sequence is $$a, ar, ar^2, \ldots$$ The number r is referred to as the *common ratio*.
12. $a_n = ar^{n-1}$	741	This is the formula for the nth term of a geometric sequence.
13. $S_n = \dfrac{a(1 - r^n)}{1 - r}$	741	This is the formula for the sum of the first n terms of a geometric sequence.
14. $S = \dfrac{a}{1 - r}$	743	This is the formula for the sum of the infinite geometric series $a + ar + ar^2 + \cdots$, where $|r| < 1$.

WRITING MATHEMATICS

1. A sequence is defined recursively as follows: $a_0 = \pi$; $a_{n+1} = \dfrac{a_n}{a_n + 1}$ for $n \geq 0$.

   Compute a_0, a_1, and a_2. Describe (in a complete sentence or two) what pattern you observe. Make a conjecture about a general formula for a_n. Then use mathematical induction to prove that the formula is valid. (As in the examples in the text, use complete sentences in writing and explaining the proof.)

2. A sequence is defined recursively as follows.

   $$a_1 = 3/2$$
   $$a_{n+1} = a_n^2 - 2a_n + 2 \quad \text{for } n \geq 1$$

   Use mathematical induction to show that $1 < a_n < 2$ for all natural numbers n.

3. Study the procedure used in the text (on page 741) to obtain the formula for the sum of a finite geometric series. Work with a classmate or your instructor to apply the procedure to find the sum of the series $1 + 2 \sin \theta + 3 \sin^2 \theta + 4 \sin^3 \theta + \cdots + n \sin^{n-1} \theta$. Then, on your own, write a paragraph or two explaining in detail what you've done.

4. Study the following purported "proof" by mathematical induction, and explain the fallacy.

   Let P_n denote the following statement: In any set of n numbers, the numbers are all equal.

   STEP 1 P_1 is clearly true.

   STEP 2 Assume that P_k is true. We must show that P_{k+1} is true. That is, given a set of $k + 1$ numbers $a_1, a_2, a_3, \ldots, a_{k+1}$, we must show that $a_1 = a_2 = a_3 = \cdots = a_{k+1}$. By the inductive hypothesis, however, we have $a_1 = a_2 = a_3 = \cdots = a_k$ and also $a_2 = a_3 = a_4 = \cdots = a_{k+1}$. These last two sets of equations imply that $a_1 = a_2 = a_3 = \cdots = a_k = a_{k+1}$, as we wished to show. Having now completed Steps 1 and 2, we conclude that P_n is true for every n.

CHAPTER TWELVE   REVIEW EXERCISES

In Exercises 1–10, use the principle of mathematical induction to show that each statement is true for all natural numbers.

1. $5 + 10 + 15 + \cdots + 5n = \frac{5}{2}n(n + 1)$

2. $10 + 10^2 + 10^3 + \cdots + 10^n = \frac{10}{9}(10^n - 1)$

3. $1 \cdot 2 + 2 \cdot 3 + 3 \cdot 4 + \cdots + n(n + 1) = \frac{1}{3}n(n + 1)(n + 2)$

4. $\dfrac{1}{2} + \dfrac{2}{2^2} + \dfrac{3}{2^3} + \cdots + \dfrac{n}{2^n} = 2 - \dfrac{2 + n}{2^n}$

5. $1 + 3 \cdot 2 + 5 \cdot 2^2 + 7 \cdot 2^3 + \cdots + (2n - 1) \cdot 2^{n-1}$
   $= 3 + (2n - 3) \cdot 2^n$

6. $\dfrac{1}{1 \cdot 4} + \dfrac{1}{4 \cdot 7} + \dfrac{1}{7 \cdot 10} + \cdots$
   $+ \dfrac{1}{(3n - 2)(3n + 1)} = \dfrac{n}{3n + 1}$

7. $1 + 2^2 \cdot 2 + 3^2 \cdot 2^2 + 4^2 \cdot 2^3 + \cdots + n^2 \cdot 2^{n-1}$
   $= (n^2 - 2n + 3)2^n - 3$

8. 9 is a factor of $n^3 + (n + 1)^3 + (n + 2)^3$.

9. 3 is a factor of $7^n - 1$.

10. 8 is a factor of $9^n - 1$.

In Exercises 11–20, expand the given expressions.

11. $(3a + b^2)^4$

12. $(5a - 2b)^3$

13. $(x + \sqrt{x})^4$

14. $(1 - \sqrt{3})^6$

15. $(x^2 - 2y^2)^5$

16. $\left(\dfrac{1}{a} + \dfrac{2}{b}\right)^3$

17. $\left(1 + \dfrac{1}{x}\right)^5$

18. $\left(x^3 + \dfrac{1}{x^2}\right)^6$

19. $(a\sqrt{b} - b\sqrt{a})^4$

20. $(x^{-2} + y^{5/2})^8$

21. Find the fifth term in the expansion of $(3x + y^2)^5$.

22. Find the eighth term in the expansion of $(2x - y)^9$.

23. Find the coefficient of the term containing a^5 in the expansion of $(a - 2b)^7$.

24. Find the coefficient of the term containing b^8 in the expansion of $\left(2a - \dfrac{b}{3}\right)^{10}$.

25. Find the coefficient of the term containing x^3 in the expansion of $(1 + \sqrt{x})^8$.

26. Expand $(1 + \sqrt{x} + x)^6$. *Suggestion:* Rewrite the expression as $[(1 + \sqrt{x}) + x]^6$.

In Exercises 27–32, verify each assertion by computing the indicated binomial coefficients.

27. $\dbinom{2}{0}^2 + \dbinom{2}{1}^2 + \dbinom{2}{2}^2 = \dbinom{4}{2}$

28. $\dbinom{3}{0}^2 + \dbinom{3}{1}^2 + \dbinom{3}{2}^2 + \dbinom{3}{3}^2 = \dbinom{6}{3}$

29. $\dbinom{4}{0}^2 + \dbinom{4}{1}^2 + \dbinom{4}{2}^2 + \dbinom{4}{3}^2 + \dbinom{4}{4}^2 = \dbinom{8}{4}$

30. $\dbinom{2}{0} + \dbinom{2}{1} + \dbinom{2}{2} = 2^2$

31. $\dbinom{3}{0} + \dbinom{3}{1} + \dbinom{3}{2} + \dbinom{3}{3} = 2^3$

32. $\dbinom{4}{0} + \dbinom{4}{1} + \dbinom{4}{2} + \dbinom{4}{3} + \dbinom{4}{4} = 2^4$

In Exercises 33–38, compute the first four terms in each sequence. Also, in Exercises 33–35, graph the sequences for n = 1, 2, 3, and 4.

33. $a_n = \dfrac{2n}{n+1}$ **34.** $a_n = \dfrac{3n-2}{3n+2}$

35. $a_n = (-1)^n\left(1 - \dfrac{1}{n+1}\right)$

36. $a_0 = 4$; $a_n = 2a_{n-1}$, $n \geq 1$

37. $a_0 = -3$; $a_n = 4a_{n-1}$, $n \geq 1$

38. $a_0 = 1$; $a_1 = 2$; $a_n = 3a_{n-1} + 2a_{n-2}$, $n \geq 2$

In Exercises 39 and 40, evaluate each sum.

39. (a) $\displaystyle\sum_{k=1}^{3} (-1)^k(2k+1)$ (b) $\displaystyle\sum_{k=0}^{8} \left(\dfrac{1}{k+1} - \dfrac{1}{k+2}\right)$

40. (a) $\displaystyle\sum_{j=1}^{4} \dfrac{(-1)^j}{j}$ (b) $\displaystyle\sum_{n=1}^{5} \left(\dfrac{1}{n} - \dfrac{1}{n+1}\right)$

In Exercises 41 and 42, rewrite each sum using sigma notation.

41. $5/3 + 5/3^2 + 5/3^3 + 5/3^4 + 5/3^5$

42. $1/2 - 2/2^2 + 3/2^3 - 4/2^4 + 5/2^2 - 6/2^6$

In Exercises 43–46, find the indicated term in each sequence.

43. $5, 9, 13, 17, \ldots$; a_{18}

44. $5, \frac{9}{2}, 4, \frac{7}{2}, \ldots$; a_{20}

45. $10, 5, \frac{5}{2}, \frac{5}{4}, \ldots$; a_{12}

46. $\sqrt{2} + 1, 1, \sqrt{2} - 1, 3 - 2\sqrt{2}, \ldots$; a_{10}

47. Determine the sum of the first 12 terms of an arithmetic sequence in which the first term is 8 and the twelfth term is $\frac{43}{2}$.

48. Find the sum of the first 45 terms in the sequence $10, \frac{29}{3}, \frac{28}{3}, 9, \ldots$.

49. Find the sum of the first 10 terms in the sequence $7, 70, 700, \ldots$.

50. Find the sum of the first 12 terms in the sequence $\frac{1}{3}, -\frac{2}{9}, \frac{4}{27}, -\frac{8}{81}, \ldots$.

51. In a certain geometric sequence, the third term is 4 and the fifth term is 10. Find the sixth term, given that the common ratio is negative.

52. For a certain infinite geometric series, the first term is 2 and the sum is $\frac{18}{11}$. Find the common ratio r.

In Exercises 53–56, find the sum of each infinite geometric series.

53. $\frac{3}{5} + \frac{3}{25} + \frac{3}{125} + \cdots$ **54.** $\frac{7}{10} + \frac{7}{100} + \frac{7}{1000} + \cdots$

55. $\frac{1}{9} - \frac{1}{81} + \frac{1}{729} - \cdots$

56. $1 + \dfrac{1}{1+\sqrt{2}} + \dfrac{1}{(1+\sqrt{2})^2} + \cdots$

57. Find a fraction equivalent to $0.\overline{45}$.

58. Find a fraction equivalent to $0.2\overline{13}$.

In Exercises 59–62, verify each equation using the formula for the sum of an arithmetic series:

$$S_n = \dfrac{n}{2}[2a + (n-1)d]$$

(The formulas given in Exercises 59–62 appear in Elements of Algebra *by Leonhard Euler, first published in 1770.)*

59. $1 + 2 + 3 + \cdots + n = n + n(n-1)/2$

60. $1 + 3 + 5 + \cdots$ (to n terms) $= n + 2n(n-1)/2$

61. $1 + 4 + 7 + \cdots$ (to n terms) $= n + 3n(n-1)/2$

62. $1 + 5 + 9 + \cdots$ (to n terms) $= n + 4n(n-1)/2$

63. In this exercise, you will use the following (remarkably simple) formula for approximating sums of powers of integers:

$$1^k + 2^k + 3^k + \cdots + n^k \approx \dfrac{\left(n + \frac{1}{2}\right)^{k+1}}{k+1} \qquad (1)$$

[This formula appears in the article, "Sums of Powers of Integers" by B. L. Burrows and R. F. Talbot, published in the *American Mathematical Monthly*, vol. **91** (1984): p. 394.]

(a) Use formula (1) to estimate the sum $1^2 + 2^2 + 3^2 + \cdots + 50^2$. Round off your answer to the nearest integer.

(b) Compute the exact value of the sum in part (a) using the formula $\displaystyle\sum_{k=1}^{n} k^2 = \frac{1}{6}n(n+1)(2n+1)$.

(This formula can be proved using mathematical induction.) Then compute the percent error for the approximation obtained in part (a). The percent error is given by

$$\dfrac{|\text{actual value} - \text{approximate value}|}{\text{actual value}} \times 100$$

(c) Use formula (1) to estimate the sum $1^4 + 2^4 + 3^4 + \cdots + 200^4$. Round off your answer to six significant digits.

(d) The following formula for the sum $1^4 + 2^4 + \cdots + n^4$ can be proved using mathematical induction:

$$\sum_{k=1}^{n} k^4 = \dfrac{n(n+1)(2n+1)(3n^2 + 3n - 1)}{30}$$

Use this formula to compute the sum in part (c). Round off your answer to six significant digits. Then use this result to compute the percent error for the approximation in part (c).

64. According to *Stirling's formula* [named after James Stirling (1692–1770)], the quantity $n!$ can be approximated as follows:

$$n! \approx \sqrt{2\pi n}\left(\frac{n}{e}\right)^n$$

In this formula, e is the constant $2.718\ldots$ (discussed in Section 5.2). Use a calculator to complete the following table. Round off your answers to five significant digits. As you will see, the numbers in the right-hand column approach 1 as n increases. This shows that, in a certain sense, the approximation improves as n increases.

n	$n!$	$\sqrt{2\pi n}(n/e)^n$	$\dfrac{n!}{\sqrt{2\pi n}(n/e)^n}$
10			
20			
30			
40			
50			
60			
65			

65. The nonzero numbers a, b, and c are consecutive terms in a geometric sequence, and $a + b + c = 70$. Furthermore, $4a$, $5b$, and $4c$ are consecutive terms in an arithmetic sequence. Find a, b, and c.

66. The nonzero numbers a, b, and c are consecutive terms in a geometric sequence, and a, $2b$, and c are consecutive terms in an arithmetic sequence. Show that the common ratio in the geometric sequence must be either $2 + \sqrt{3}$ or $2 - \sqrt{3}$.

67. If the numbers $\dfrac{1}{b+c}$, $\dfrac{1}{c+a}$, and $\dfrac{1}{a+b}$ are consecutive terms in an arithmetic sequence, show that a^2, b^2, and c^2 are also consecutive terms in an arithmetic sequence.

68. If a, b, and c are consecutive terms in a geometric sequence, prove that $\dfrac{1}{a+b}$, $\dfrac{1}{2b}$, and $\dfrac{1}{c+b}$ are consecutive terms in an arithmetic sequence.

69. (a) Find a value for x such that $3 + x$, $4 + x$, and $5 + x$ are consecutive terms in a geometric sequence.

(b) Given three numbers a, b, and c, find a value for x (in terms of a, b, and c) such that $a + x$, $b + x$, and $c + x$ are consecutive terms in a geometric sequence.

70. If $\ln(A + C) + \ln(A + C - 2B) = 2\ln(A - C)$, show that $1/A$, $1/B$, and $1/C$ are consecutive terms in an arithmetic sequence.

71. Let a_1, a_2, a_3, $\ldots$ be an arithmetic sequence with common difference d, and let $r \, (\neq 1)$ be a real number. In this exercise we develop a formula for the sum of the series

$$a_1 + ra_2 + r^2 a_3 + \cdots + r^{n-1} a_n$$

The method we use here is essentially the same as the method used in the text to derive the formula for the sum of a finite geometric series. So, as background for this exercise, you should review the derivation on page 711 for the sum of a geometric series.

(a) Let S denote the required sum. Show that

$$S - rS = a_1 + rd + r^2 d + \cdots + r^{n-1} d - r^n a_n$$

$$= a_1 + \frac{d(r - r^n)}{1 - r} - r^n a_n$$

(b) Show that

$$S = \frac{a_1 - r^n a_n}{1 - r} + \frac{d(r - r^n)}{(1 - r)^2}$$

72. Use your calculator and the formula in Exercise 71(b) to find the sum of each of the following series.

(a) $1 + 2 \times 2 + 2^2 \times 3 + 2^3 \times 4 + \cdots + 2^{13} \times 14$

(b) $2 + 4 \times 5 + 4^2 \times 8 + 4^3 \times 11 + \cdots + 4^6 \times 20$

(c) $3 - \dfrac{1}{2}\cdot 5 + \dfrac{1}{2^2}\cdot 7 - \dfrac{1}{2^3}\cdot 9 + \cdots +$ (to 10 terms)

73. Let a_1, a_2, a_3, a_4, $\ldots$ be an arithmetic sequence in which all of the terms are positive. Show that the following equations are valid.

(a) $\dfrac{1}{\sqrt{a_1} + \sqrt{a_2}} + \dfrac{1}{\sqrt{a_2} + \sqrt{a_3}} = \dfrac{2}{\sqrt{a_1} + \sqrt{a_3}}$

(b) $\dfrac{1}{\sqrt{a_1} + \sqrt{a_2}} + \dfrac{1}{\sqrt{a_2} + \sqrt{a_3}} + \dfrac{1}{\sqrt{a_3} + \sqrt{a_4}}$

$$= \dfrac{3}{\sqrt{a_1} + \sqrt{a_4}}$$

(c) $\dfrac{1}{\sqrt{a_1} + \sqrt{a_2}} + \dfrac{1}{\sqrt{a_2} + \sqrt{a_3}} +$

$$\cdots + \dfrac{1}{\sqrt{a_{n-1}} + \sqrt{a_n}} = \dfrac{n-1}{\sqrt{a_1} + \sqrt{a_n}}$$

74. For our last exercise in this book, we take the following problem from the classic text by Isaac Todhunter, *Plane Trigonometry for the Use of Colleges and Schools*, 5th ed. (London: Macmillan & Co., 1874):

> If the cotangents of the angles of a triangle be in arithmetical progression, the squares of the sides will also be in arithmetical progression.

Hint: Write the given information as an equation involving sines and cosines. Use the formulas on page 443 to express the cosines in terms of the lengths of the sides. For the sines, use the formulas in Exercise 48 on page 451.

CHAPTER TWELVE TEST

1. Use the principle of mathematical induction to show that the following formula is valid for all natural numbers n: $1^2 + 2^2 + 3^2 + \cdots + n^2 = \dfrac{n(n+1)(2n+1)}{6}$.

2. Express each of the following sums without using sigma notation, and then evaluate each sum: (a) $\displaystyle\sum_{k=0}^{2} (10k - 1)$;   (b) $\displaystyle\sum_{k=1}^{3} (-1)^k k^2$.

3. (a) Write the formula for the sum S_n of a finite geometric series.
   (b) Evaluate the sum $\dfrac{3}{2} + \dfrac{3^2}{2^2} + \dfrac{3^3}{2^3} + \cdots + \dfrac{3^{10}}{2^{10}}$.

4. (a) Determine the coefficient of the term containing a^3 in the expansion of $(a - 2b^3)^{11}$.
   (b) Find the fifth term of the expansion in part (a).

5. Expand the expression $(3x^2 + y^3)^5$.

6. Determine the sum of the first 12 terms of an arithmetic sequence in which the first term is 8 and the twelfth term is $\frac{43}{2}$.

7. Find the sum of the following infinite geometric series: $\frac{7}{10} + \frac{7}{100} + \frac{7}{1000} + \cdots$.

8. A sequence is defined recursively as follows: $a_1 = 1$, $a_2 = 1$, and $a_n = (a_{n-1})^2 + a_{n-2}$ for $n \geq 3$. Determine the fourth and the fifth terms in this sequence.

9. In a certain geometric sequence, the third term is 4 and the fifth term is 10. Find the sixth term, given that the common ratio is negative.

10. What is the twentieth term in the arithmetic sequence $-61, -46, -31, \ldots$?

APPENDIX

A.I   SIGNIFICANT DIGITS AND CALCULATORS

Many of the numbers that we use in scientific work and in daily life are approximations. In some cases the approximations arise because the numbers are obtained through measurements or experiments. Consider, for example, the following statement from an astronomy textbook:

The diameter of the Moon is 3476 km.

We interpret this statement as meaning that the actual diameter D is closer to 3476 km than it is to either 3475 km or 3477 km. In other words,

$$3475.5 \text{ km} \leq D \leq 3476.5 \text{ km}$$

The interval $[3475.5, 3476.5]$ in this example provides information about the accuracy of the measurement. Another way to indicate accuracy in an approximation is by specifying the number of *significant digits* it contains. The measurement 3476 km has four significant digits. In general, the number of significant digits in a given number is found as follows.

Significant Digits		
The number of significant digits in a given number is determined by counting the digits from left to right, beginning with the leftmost nonzero digit.	EXAMPLES	
	Number	Number of Significant Digits
	1.43	3
	0.52	2
	0.05	1
	4837	4
	4837.0	5

Numbers obtained through measurements are not the only source of approximations in scientific work. For example, to five significant digits, we have the following approximation for the irrational number π:

$$\pi \approx 3.1416$$

This statement tells us that π is closer to 3.1416 than it is to either 3.1415 or 3.1417. In other words,

$$3.14155 \le \pi \le 3.14165$$

Table 1 provides some additional examples of the ideas we've introduced.

TABLE 1

Number	Number of Significant Digits	Range of Measurement
37	2	[36.5, 37.5]
37.0	3	[36.95, 37.05]
268.1	4	[268.05, 268.15]
1.036	4	[1.0355, 1.0365]
0.036	2	[0.0355, 0.0365]

There is an ambiguity involving zero that can arise in counting significant digits. Suppose that someone measures the width w of a rectangle and reports the result as 30 cm. How many significant digits are there? If the value 30 cm was obtained by measuring to the nearest 10 cm, then only the digit 3 is significant, and we can conclude only that the width w lies in the range $25 \text{ cm} \le w \le 35 \text{ cm}$. On the other hand, if the 30 cm was obtained by measuring to the nearest 1 cm, then both the digits 3 and 0 are significant and we have $29.5 \text{ cm} \le w \le 30.5 \text{ cm}$.

By using **scientific notation** we can avoid the type of ambiguity discussed in the previous paragraph. A number written in the form

$$b \times 10^n \qquad \text{where } 1 \le b < 10 \text{ and } n \text{ is an integer}$$

is said to be expressed in scientific notation. For the example in the previous paragraph, then, we would write

$$w = 3 \times 10^1 \text{ cm} \qquad \text{if the measurement were to the nearest 10 cm}$$

and

$$w = 3.0 \times 10^1 \text{ cm} \qquad \text{if the measurement were to the nearest 1 cm}$$

As the figures in Table 2 indicate, for a number $b \times 10^n$ in scientific notation, the number of significant digits is just the number of digits in b. (This is one of the advantages in using scientific notation; the number of significant digits, and hence the accuracy of the measurement, is readily apparent.)

TABLE 2

Measurement	Number of Significant Digits	Range of Measurement
Mass of the earth:		
6×10^{27} g	1	$[5.5 \times 10^{27} \text{ g}, 6.5 \times 10^{27} \text{ g}]$
6.0×10^{27} g	2	$[5.95 \times 10^{27} \text{ g}, 6.05 \times 10^{27} \text{ g}]$
5.974×10^{27} g	4	$[5.9735 \times 10^{27} \text{ g}, 5.9745 \times 10^{27} \text{ g}]$
Mass of a proton:		
1.67×10^{-24} g	3	$[1.665 \times 10^{-24} \text{ g}, 1.675 \times 10^{-24} \text{ g}]$

There are a number of exercises in this text in which a calculator either is required or is extremely useful. Although you could work many of these exercises using tables instead of a calculator, this author recommends that you purchase a scientific calculator. *Scientific calculator* is a generic term describing a calculator with (at least) the following features or functions beyond the usual arithmetic functions:

1. Memory
2. Scientific notation
3. Powers and roots
4. Logarithms (base ten and base e)
5. Trigonometric functions
6. Inverse functions

Because of the variety and differences in the calculators that are available, no specific instructions are provided in this section of the appendix for operating a calculator. When you buy a calculator, read the owner's manual carefully and work through some of the examples in it. In general, learn to use the memory capabilities of the calculator so that, as far as possible, you don't need to write down the results of the intermediate steps in a given calculation.

Many of the numerical exercises in the text ask that you round off the answers to a specified number of decimal places. Our rules for rounding off are as follows.

Rules for Rounding Off a Number (with More Than n Decimal Places) to n Decimal Places

1. If the digit in the $(n + 1)$st decimal place is greater than 5, increase the digit in the nth place by 1. If the digit in the $(n + 1)$st place is less than 5, leave the nth digit unchanged.

2. If the digit in the $(n + 1)$st decimal place is 5 and there is at least one nonzero digit to the right of this 5, increase the digit in the nth decimal place by 1.

3. If the digit in the $(n + 1)$st decimal place is 5 and there are no nonzero digits to the right of this 5, then increase the digit in the nth decimal place by 1 only if this results in an even digit.

The examples in Table 3 illustrate the use of these rules.

TABLE 3

Number	Rounded to One Decimal Place	Rounded to Three Decimal Places
4.3742	4.4	4.374
2.0515	2.1	2.052
2.9925	3.0	2.992

These same rules can be adapted for rounding off a result to a specified number of significant digits. As examples of this, we have

2347	rounded off to two significant digits is $2300 = 2.3 \times 10^3$
2347	rounded off to three significant digits is $2350 = 2.35 \times 10^3$
975	rounded off to two significant digits is $980 = 9.8 \times 10^2$
0.985	rounded off to two significant digits is $0.98 = 9.8 \times 10^{-1}$

FIGURE 1

In calculator exercises that ask you to round off your answers, it's important that you postpone rounding until the final calculation is carried out. For example, suppose that you are required to determine the hypotenuse x of the right triangle in Figure 1 to two significant digits. Using the Pythagorean theorem, we have

$$x = \sqrt{(1.36)^2 + (2.46)^2}$$
$$= 2.8 \qquad \text{using a calculator and rounding off the}$$
$$\qquad\qquad \text{final result to two significant digits}$$

On the other hand, if we first round off each of the given lengths to two significant digits, we obtain

$$x = \sqrt{(1.4)^2 + (2.5)^2}$$
$$= 2.9 \qquad \text{(to two significant digits)}$$

TABLE 4

Number	Range of Measurement
1.36	[1.355, 1.365]
2.46	[2.455, 2.465]

This last result is inappropriate and we can see why as follows. As Table 4 shows, the maximum possible values for the sides are 1.365 and 2.465, respectively. Thus, the maximum possible value for the hypotenuse must be

$$\sqrt{(1.365)^2 + (2.465)^2} = 2.817\ldots \quad \text{(calculator display)}$$
$$= 2.8 \qquad \text{(to two significant digits)}$$

This shows that the value 2.9 is indeed inappropriate, as we stated previously.

An error often made by people working with calculators and approximations is to report a final answer with a greater degree of accuracy than the data warrant. Consider, for example, the right triangle in Figure 2. Using the Pythagorean theorem and a calculator with an eight-digit display, we obtain

$$h = 3.6055513 \text{ cm}$$

FIGURE 2

This value for h is inappropriate, since common sense tells us that the answer should be no more accurate than the data used to obtain that answer. In particular, since the given sides of the triangle apparently were measured only to the nearest tenth of a centimeter, we certainly should not expect any improvement in accuracy for the resulting value of the hypotenuse. An appropriate form for the value of h here would be $h = 3.6$ cm. In general, for calculator exercises in this text that do not specify a required number of decimal places or significant digits in the final results, you should use the following guidelines.

Guidelines for Computing with Approximations

1. *For adding and subtracting:* Round off the final result so that it contains only as many decimal places as there are in the data with the fewest decimal places.

2. *For multiplying and dividing:* Round off the final result so that it contains only as many significant digits as there are in the data with the fewest significant digits.

3. *For powers and roots:* In computing a power or a root of a real number b, round off the result so that it contains as many significant digits as there are in b.

A.2 USING THE TI-81

Prepared by M. G. Settle © 1991 MR Creations

In the coming weeks, you will find that graphing functions greatly increases your ability to learn concepts in mathematics. The TI-81 Graphics Calculator can increase your efficiency in producing these graphs. Before learning how to use the TI-81, take a moment to learn its layout and basic features.

The Display Screen   At the top of the TI-81 is an 8-line-by-16-character display screen. (See area ①.) On this screen, computations are entered, and the results are displayed. Additionally, this screen is used to display the graph of the function.

The Graph Keys   Just below the display screen of the TI-81 are five keys. (See area ②.) These keys are used to define, display, manipulate, and interpret the graph of a function.

The 2nd Key   The top leftmost key below the Graph Keys is the **2nd** key. (See area ③.) This key is used to access the functions displayed on the left above some keys (such as "10^x" and "$\sqrt{\ }$").

The ALPHA Key   Just below the **2nd** key is the **ALPHA** key. (See area ④.) This key is used to access the characters displayed on the right above some keys (such as "A" and "?").

Directional Arrow Keys   The four blue keys at the top right (see area ⑤) are used to move the cursor during editing and to trace along the curve of a graph.

The Function Keys   The remaining black keys are used to access various mathematical functions ("SIN", "LOG", "x^2", etc.), to edit entries (the **DEL** and **INS** keys), and to display menus (the **MATH**, **MATRX**, **PRGM**, and **VARS** keys).

The Numeric Keypad   The gray keys are used to type numeric values. (See area ⑥.)

The Operation Keys   On the lower right are the four operation keys (**+**, **−**, **×**, **÷**). (See area ⑦.)

The ENTER Key   At the bottom right is the **ENTER** key. Pressing this key informs the TI-81 to execute the command (instruction) that you have typed. (See area ⑧.)

The opening screen
and the cursor. ⑨

Comment

If you have trouble seeing
the display, press the 2nd
key and then hold down the
down-arrow key (▼) or the
up-arrow key (▲) to change
the contrast.

Order of operations yields − 17.

Use parentheses to override
the order of operations.

Place the cursor here, using
the left-arrow key.

GETTING STARTED

With the TI-81 in hand, find the **ON** key (at the left of the bottom row) and press it to turn on the TI-81. The TI-81 responds by presenting the cursor (a blinking square character). If you have trouble seeing the screen, try tilting the calculator slightly. If you still can't see the cursor, press the **2nd** key and hold the down-arrow key (▼) to set the contrast. (See frame ⑨.)

DOING CALCULATIONS

Because the TI-81 is different from a standard calculator, we will take a moment to observe some common computations on the TI-81.

Example: Compute $3 - 4 \times 5$.

Solution: Type $3 - 4 \times 5$ and then press the **ENTER** key. As a result, the multiplication is performed to yield $3 - 20$, and then the subtraction is performed. The number -17 is displayed to the right of the computation entered. (See frame ⑩.)

As you would expect, the TI-81 follows the standard algebraic order of operations (precedence) rules. To override these rules, parentheses can be entered. The parentheses keys are located just above the gray numeric keypad section.

Example: Compute $(3 - 4) \times 5$.

Solution: Type $(3 - 4) \times 5$ and then press the **ENTER** key. As a result, the computation in the parentheses is performed first (subtraction in this case), and then the multiplication is performed to yield -5. (See frame ⑪.)

EDITING COMPUTATIONS: THE INS KEY

In the above examples, the TI-81 operates like a calculator. However, the TI-81 is different from a standard scientific calculator. The first major difference is the TI-81's ability to edit computations. After a computation is typed, it can be edited (changed) before pressing the **ENTER** key. Consider the following example.

Example: Evaluate $(3x - 2)/(x + 5)$ for x equal to 10.

Solution: Type $(3 \times 10 - 2) \div 10 + 5$ but *do not* press the **ENTER** key yet! As typed, the computations in the parentheses will be performed, divided by 10, and then 5 will be added to the results. This is *not* what we want. We want the computation in the parentheses divided by the *sum* of 10 and 5 (not just 10). To edit (change) the computation entered, the directional-arrow keys are used.

Press the left-arrow key (◀) until the cursor is under the 1 of the second 10, as shown in frame ⑫. Now, press the **INS** key to enter the Insert mode. The cursor changes to a blinking underscore character.

Press the left parenthesis key, and the (character is inserted.

⑬ Press the left parenthesis key, (, to insert the left parenthesis. (See frame ⑬.)

Press the right-arrow key (▶) to move the cursor to the right of the 5.

Press the right parenthesis key,), to insert the right parenthesis. (See frame ⑭.)

Finally, press the **ENTER** key to execute the computation and display the results. (See frame ⑮.)

Place the cursor using the right-arrow key. Press the right parenthesis key, and the) character is inserted.

After the **ENTER** key is pressed, the computation is performed and the results are displayed.

With the **ENTRY** command (the **2nd** key followed by the **ENTER** key), the last entry is displayed and the cursor is placed at the right end of the computation. You can now edit the displayed computation.

EDITING COMPUTATIONS: THE DEL KEY

As just demonstrated, text can be inserted in a computation by use of the **INS** key (the parentheses in the preceding example). Text can also be deleted from a computation by using the **DEL** key. In the following example, you will use the **replay** feature to display the preceding computation for editing. Next, you will use the **DEL** key to delete the inside parentheses. Finally, you will use the **ENTER** key to execute and display the resulting computation.

Example: Compute $(3 \times 10 - 2 \div 10 + 5)$.

Solution: To do this computation, we can simply edit the existing computation, $(3 \times 10 - 2) \div (10 + 5)$, by deleting the inside parentheses.

Press the **2nd** key (the cursor changes to a blinking up-arrow) followed by the **ENTER** key (the ENTRY command) to **replay** the last command entered (the computation above). (See frame ⑯.)

Press the left-arrow key to move the cursor to the) character after the 2.

Press the **DEL** key, and the) character is deleted. (See frame ⑰.)

Now move the cursor to the (character after the / character. Press the **DEL** key to delete the (character.

Finally, press the **ENTER** key to execute and display the results of the new computation. (See frame ⑱.)

Pressing the **DEL** key removes the) character.

EXERCISE SET I

1. Do the following computations using TI-81.
   (a) $5 - 14/8$ (b) $2/3 + 1/5$
   (c) $1.56 - (3 + 13/5)$

2. Evaluate $x^2 - 4x + 5$ for $x = 3$.

3. Type the computation $(5 - 3 \times 6) \div (12 + 3 \times 8)$.
   a. Press the **ENTER** key and record the results.
   b. Use the ENTRY command to redisplay the computation. Edit the computation to $(5 - 3) * 6/(12 + 3 * 8)$. Press the **ENTER** key and record the results.
   c. Use the ENTRY command to redisplay the computation. Edit the computation to $5 - 3 * (6/12 + 3 * 8)$. Press the **ENTER** key and record the results.

COMPUTATIONS: THE (−), ANS, AND CLEAR KEYS

To clear the TI-81 screen, the **CLEAR** key is used. That is, press the **CLEAR** key, and the screen is erased. To redisplay the *results* of the last computation, the **ANS** key [the **2nd** key followed by the (−) key] followed by the **ENTER** key can be pressed. Actually, **ANS** is a storage location in the memory of the TI-81. Pressing this key displays the characters "Ans" on the screen. Pressing the **ENTER** key results in the TI-81 displaying the contents of the **ANS** memory location. As you will see, **ANS** can be used in a computation. Finally, *to enter a negative number, the (−) key must be used.* To see how to use these keys, consider the following example.

Example: For $f(x) = x^2 - 3x + 4$, find $f(-2)$.

Solution: First, $f(-2)$ means evaluate $x^2 - 3x + 4$ for x equal to -2. Before doing this evaluation, you need to learn how to square a negative number. To do this, start by pressing the **CLEAR** key to clear the display.

Now, press the (−) key, followed by the **2** key. (See frame ⑲.)

Press the $\mathbf{x^2}$ key and then press the **ENTER** key. (See frame ⑳.) As you can see, the (−) key produces the opposite of a value. Order of operations resulted in the computation $-1 \times 2 \times 2$, which is not what we want. To square -2, we need to type $(-2)^2$. Now, to find $f(-2)$ for $f(x) = x^2 - 3x + 4$, do the following.

Press the **(** key, the (−) key, the **2** key, the **)** key, and the $\mathbf{x^2}$ key to display $(-2)^2$.

Press the **−** key, the **3** key, the **(** key, the (−) key, the **2** key, and the **)** key to display $(-2)^2 - 3(-2)$.

Press the **+** key and the **4** key to finish the entry. (See frame ㉑.)

Finally, press the **ENTER** key to compute and then display the results. (See frame ㉒.) Thus, $f(-2)$ is 14.

Comment

A shortcut for the **ENTRY** command is the up-arrow key. That is, press the up-arrow key and the last command executed is displayed.

CLEAR   This key erases the screen.

⑲

After pressing the (−) key and the **2** key.

⑳

After pressing the $\mathbf{x^2}$ key and the **ENTER** key

㉑

The (−) key.

The − (subtraction) key.

The (−) key.

㉒

Comment

The TI-81 understands juxtaposition, as in $3(-2)$, to mean multiplication. You could have written $3 * -2$ instead of $3(-2)$.

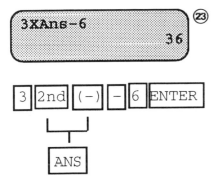

3 2nd (-) – 6 ENTER

ANS

To demonstrate the use of the **ANS** key, consider the following example.

Example: For $f(x) = x^2 - 3x + 4$, find $3f(-2) - 6$.

Solution: Rather than find $f(-2)$ again, we can use the **ANS** key, because the value of $f(-2)$ is currently stored in the **ANS** memory location, which holds (stores) the value of the last calculation, even after the calculator is turned off. Press the **3** key, the **×** key, the **ANS** key, (**2nd** key followed by (**−**) key), the **−** key, the **6** key, and then the **ENTER** key. (See frame ㉓.) So, $3f(-2) - 6$ equals 36.

Comment   If you now pressed the **ANS** key followed by the x^2 key and then pressed the **ENTER** key, the displayed results would be 1296. This is because 36 is currently stored in the **ANS** memory location, and 36 squared is 1296.

EXERCISE SET 2

Match the following:

A. ENTRY	F. ENTER
B. 2nd	G. INS
C. ANS	H. DEL
D. DEL	I. CLEAR
E. (−)	J. ▼

_____ 1. A key pressed to compute and display a computation entered.

_____ 2. A key pressed to indicate the opposite of a number.

_____ 3. A key pressed to erase the display.

_____ 4. A key pressed to move the cursor on the screen.

_____ 5. A key pressed to remove (delete) a character from a computation displayed on the screen.

_____ 6. A key pressed to access the function listed on the top of a key (such as ENTRY or the 10^x function).

_____ 7. A key pressed to provide access to the most recently displayed computed results.

8. If you enter the computation $3 \times 5 - 8$, press the **ENTER** key, and then press the **2nd** key followed by the (**−**) key (the **ANS** key), followed by the x^2 key and the **ENTER** key, what number will be displayed?

9. For $f(x) = 3 - \dfrac{x}{x + 5}$, find $f(-1)$.

10. For $f(x) = -x^2 + 3x - 1$, find $f(5)$.

Press the **RANGE** key.

```
RANGE
Xmin:-10
Xmax:10
Xscl:1
Ymin:-10
Ymax:10
Yscl:1
Xres=1
```

The default-range display.

Press the **GRAPH** key.

The graph display.

Use the **QUIT** key to return to the text display.

$$\boxed{2\text{nd}}\ \boxed{\text{CLEAR}} \rightarrow \boxed{\text{QUIT}}$$

Press the **Y=** key. Type

$$\boxed{3}\ \boxed{\text{X}|\text{T}}\ \boxed{-}\ \boxed{2}$$

```
:Y1 =3X-2
:Y2 =
:Y3 =
:Y4 =
```

Press the **GRAPH** key.

The graph of $f(x) = 3x - 2$.

A FIRST GRAPH

㉔ A major feature of the TI-81 is its ability to display the graph of a function. In the following examples, you will learn to use the **RANGE** key (to specify the *x*-axis interval and the *y*-axis interval), the **Y=** key (to display a prompt for entering the rule for a function), the **GRAPH** key (to switch to a graph display), and the QUIT command (to return to the text display).

The Right Range   Before displaying a first graph, press the **RANGE** key and check that the values displayed are as shown in frame ㉔.

Comment   If the values displayed are different than those shown in ㉔, press the **ZOOM** key followed by the **6** key and then the **CLEAR** key to return to the default values shown in frame ㉔. Later, you will learn to set
㉕ any range values you desire. Press the **RANGE** key to bring the display back to the range display shown in frame ㉔.

The Graph Screen   Before inserting the rule for a function to be graphed, we will view the graph screen. Press the **GRAPH** key and you will see a set of axes with the default range settings displayed. (See frame ㉕.) The leftmost mark on the *x*-axis represents -10, and the rightmost mark on the *x*-axis represents 10. Likewise, the topmost mark on the *y*-axis represents 10, and the bottommost mark on the *y*-axis represents -10. The scale for both the *x*-axis and the *y*-axis is 1. That is, each mark represents 1 unit. Use the QUIT command (the **2nd** key followed by the **CLEAR** key) to bring the display back to the text display.

Graphing a Function   We will now graph the linear function $f(x) = 3x - 2$. The graph of this function is a straight line crossing the *y*-axis at $(0, -2)$. Let's work through the following example.

Example: Graph $f(x) = 3x - 2$.

Solution: With the TI-81 in text display, press the **Y=** key. The screen clears, and the prompt ":Y1 =" is displayed.

Press the **3** key.

㉖ Press the **X|T** key to place the X on the display.

Press the **−** key.

Press the **2** key. (See frame ㉖.)

Press the **GRAPH** key to have the graph drawn and displayed on the graph screen. The TI-81 display changes to the graph screen, and the ordered pairs defined by the function rule $f(x) = 3x - 2$ are plotted on the axis. (See
㉗ frame ㉗.)

Use the QUIT command (the **2nd** key followed by the **CLEAR** key) to switch back to the home screen (the text display).

Clearing the function definition(s) clears the graphics screen.

CLEAR   clears the home screen.

Graph Y₁=X^3-X+1   (28)

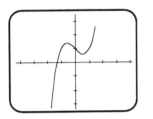

The graph of
$f(x) = x^3 - x + 1$.

ZOOM
1:Box
2:Zoom In
3:Zoom Out
4:Set Factors
5:Square
6:Standard
7↓Trig

The ZOOM menu.

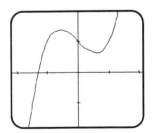

The graph of
$f(x) = x^3 - x + 1$ after zooming in (ZOOM 2).

Checking the Displays   At this point, you can press the **RANGE** key and display the current ranges for the x-axis and y-axis. Pressing the **GRAPH** key again toggles the display back to the graph screen. Also, the **Y=** key can be pressed to display the function definition (rule). The QUIT command can be used to switch back to the home screen. Finally, you can clear the home screen by pressing the **CLEAR** key. To clear the graphics screen, press the **Y=** key, and then with the cursor on the function rule, press the **CLEAR** key.

As another example of graphing a function, consider the following.

Example: Graph $f(x) = x^3 - x + 1$.

Solution: First, press the **Y=** key followed by the **CLEAR** key to clear the graphics screen.

Press the **X|T** key to place X on the screen.

Press the ˆ key followed by the **3** key to get x^3. [The screen displays "Y1 = X^3", which means $f(x) = x^3$.]

(29)   Press the − key to get "Y1 = X^3 −".

Press the **X|T** key to get "Y1 = X^3 − X".

Press the **+** key and then the **1** key to get "Y1 = X^3−X+1". (See frame (28).)

Press the **GRAPH** key to display the graph of the function $f(x) = x^3 - x + 1$. (See frame (29).)

ZOOM IN, ZOOM OUT

A major feature TI-81 graphics feature is its ability to enlarge (*zoom in on*) or reduce (*zoom out on*) a graph. To zoom in on a graph, the **ZOOM** key is pressed, followed by the **2** key. To zoom out on a graph, the **ZOOM** key is pressed, followed by the **3** key.

Example: Graph $f(x) = x^3 - x + 1$ with $-2.25 < x < 2.25$.

Solution: For the graph in this example, the function domain is $-10 < x < 10$, the default-range values. If we zoom in on the graph of this function, the new x-values will be approximately $-2.25 < x < 2.25$ (one-fourth of -10 and one-fourth of 10).

With the graph of $f(x) = x^3 - x + 1$ on the screen, press the **ZOOM** key.

A menu of options is presented. (See frame (30).)

(31)   Press the **2** key to select the Zoom In option. You are returned to the graph screen with a blinking dot in the center of the screen. The blowup will be centered on this blinking dot, which can be moved using the directional-arrow keys.

Press the **ENTER** key. The new graph appears as shown in frame (31).

```
RANGE
Xmin:-2.36315789
Xmax:2.3736842107
Xscl:1
Ymin:-2.14285714
Ymax:2.87142858
Yscl:1
Xres=1
```

The new range display after
zooming in on the graph of
$f(x) = x^3 - x + 1$.

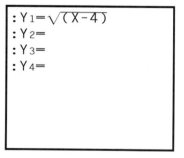

```
:Y1=√(X-4)
:Y2=
:Y3=
:Y4=
```

The graph of
$f(x) = \sqrt{(x-4)}$.

(32) Press the **RANGE** key to check that the x-values for the function $f(x) = x^3 - x + 1$ are approximately between -2.25 and 2.25. (See frame (32).) As you can see, one-fourth of the x- and y-ranges were computed based on the location of the blinking dot [which the TI-81 didn't consider to be exactly at $(0,0)$]. To set x between -2.25 and 2.25 exactly, do the following.

With the cursor on the $-$ in the XMin line, type -2.25 and then press the **ENTER** key.

With the cursor on the 2 in the XMax line, type 2.25 and then press the **ENTER** key.

Press the **GRAPH** key to display the graph of $f(x) = x^3 - x + 1$ with this set of values for x.

Use the **QUIT** command to return to the text screen.

Comment   If you now zoom out (press the **ZOOM** key followed by the **3** key), the x-values will be approximately between -10 and 10. If you zoom out again, the x-values will be approximately between -40 and 40. Remember, ZOOM 6 (the Standard option on the ZOOM menu) returns you to the default-range settings.

EXERCISE SET 3

Use the TI-81 to display the following graphs, and then transfer the information to graph paper.
1. Graph $f(x) = x^3 + x^2 - 2$ for $-5 \le x \le 5$ and $-5 \le y \le 5$.
2. Graph $f(x) = x^2 - 2x + 1$ for $-2.25 \le x \le 2.25$ and $-1 \le y \le 5$.
3. Graph $f(x) = -x^2 - 2x + 2$ for $-10 \le x \le 10$ and $-10 \le y \le 10$.

THE ZOOM BOX

(33) When you use the Zoom In and Zoom Out options on the ZOOM menu, the TI-81 automatically adjusts the range of values displayed. These values are computed based on the location of the blinking dot on the screen and the four factor (one-fourth for Zoom In and four times for Zoom Out). This factor can be changed by selecting the Set Factors option from the ZOOM menu. Of course, you can set exactly the ranges you want by entering them on the RANGE display. One of the best features of the TI-81 is the Box option on the ZOOM menu. To see for yourself how this feature operates, work through the following (34) example.

Example: Graph $f(x) = \sqrt{(x-4)}$.

Solution: Press the **Y=** key, and then use the **CLEAR** key to clear the function definiton for Y1 (and thus clear the graphics screen). Use the Standard option on the ZOOM menu to reset the range values to their default settings.

Enter the function $f(x) = \sqrt{(x-4)}$ by pressing the **Y=** key, the √ key, the (key, the **X|T** key, the $-$ key, the **4** key, and the) key. (See frame (33).) Press the **GRAPH** key. (See frame (34).)

X=3 Y=.15873016

(35) With the graph of $f(x) = \sqrt{(x-4)}$ on the screen, press the **ZOOM** key and select the box option. (Press the **ENTER** key or the **1** key.) The message X=.10526316 Y=.15873016 appears at the bottom of the graph screen. A blinking dot appears at the origin. Use the directional-arrow keys to move the blinking dot as close as possible to the point $(3, 3)$. Press the **ENTER** key. Notice that the dot changes shape. (See frame **(35)**.)

Now, press the right-arrow key until the X= message displays X = 10.

Press the down-arrow key until the Y= message reads Y = -1.111111. Press the **ENTER** key. Notice the box drawn on the screen. This is known as a *zoom box*.

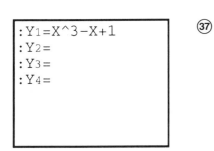

Press the **ENTER** key, and the graph is enlarged to fill the zoom box. (See frame **(36)**)

EXERCISE SET 4

Use the TI-81 to graph the following functions over the given intervals. Start with the default ranges, then use the zoom box feature.

1. Graph $f(x) = x^2 - 2x - 1$ for $-2 \le x \le 4$ and $-3 \le y \le 8$.
2. Graph $f(x) = \sqrt{x} - 4$ for $-1 \le x \le 8$ and $-4 \le y \le 1$.
3. Graph $f(x) = \sqrt{(4-x)}$ for $-5 \le x \le 5$ and $-1 \le y \le 4$.
4. Graph $f(x) = (3x - 1)/(x + 2)$ for $-5 \le x \le 5$ and $-5 \le y \le 8$.

THE TRACE FEATURE

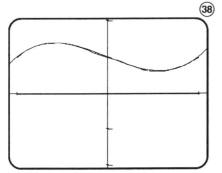

:Y1=X^3−X+1
:Y2=
:Y3=
:Y4=

(37) Often you will want to find the coordinates of a point on a graph drawn by the TI-81. To do so, you can use the *trace* feature as in the next example.

Example: Find the relative maximum for the function $f(x) = x^3 - x + 1$ over the interval $-1 \le x \le 1$.

Solution: Although you could graph the function and visually sight the point where the graph is at its highest point (where the function value or y-coordinate is greater than that of any other point in the specified interval), using the trace feature provides more accuracy.

(38) Press the **ZOOM** key and select the Standard option to set the range values to their default settings.

Press the **Y=** key, and then press the **CLEAR** key to erase the function definition for Y1 (and thus the graphics screen).

Enter the function $f(x) = x^3 - x + 1$. (See frame **(37)**.)

Press the **GRAPH** key to display the graph of $f(x) = x^3 - x + 1$.

Use the zoom box to display the portion of the graph from $-1 \le x \le 1$ and $-0.5 \le y \le 2$. (See frame **(38)**.)

X=-.5728532 Y=1.3848652

Solve: $\begin{cases} 3x - 5y = 10 \\ 2x + y = 3 \end{cases}$

Solution: $(\square, \square)$

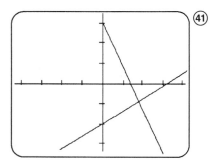

The graphs of
$y = \frac{3}{5}x - 2$ and $y = -2x + 3$.

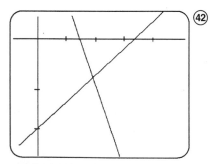

A blowup of the graphs of
$y = \frac{3}{5}x - 2$ and $y = -2x + 3$.

(39) Press the **TRACE** key. A blinking dot will appear on the curve.

Press the left-arrow key until the X= message displays X = −0.5728532. (See frame **(39)**.) Thus, the function $f(x) = x^3 - x + 1$ over the interval $-1 \le x \le 1$ reaches its relative maximum at −0.5728532, and the maximum value is 1.3848652 (approximately).

Comment   If you now use the Zoom In option, the enlargement will center on the current location of the blinking dot displayed by the trace feature. Pressing the **TRACE** key again will improve the accuracy of the approximation of a point's coordinates.

Solving a System of Equations   As another example of the use of the trace feature of the TI-81, we now consider solving a *system* of linear equations, such as the one shown in frame **(40)**. This example will also demonstrate how to display two graphs on the TI-81 graphics screen.

Example: Solve the system of equations

$$3x - 5y = 10 \qquad \text{and} \qquad 2x + y = 3$$

Solution: To solve the system graphically on the TI-81, the equations need to be rewritten in function form (or slope and y-intercept form). The equation $3x - 5y = 10$ becomes $y = \frac{3}{5}x - 2$, and the equation $2x + y = 3$ becomes $y = -2x + 3$.

Press the **Y=** key and use the **CLEAR** key to clear the Y1 definition.

Use the **RANGE** key to set the ranges to the values Xmin = −4, Xmax = 4, Ymin = −3.2, Ymax = 3.2.

Press the **Y=** key and type (3 ÷ 5)X − 2 for Y1; then press the **ENTER** key.

Type −2X + 3 for Y2.

Press the **GRAPH** key to display the graphs of the two linear functions. (See frame **(41)**.)

With the graph of the two linear functions on the screen, press the **TRACE** key. The message "X = .04210526   Y = −1.974737" will appear at the bottom of the screen.

Press the right-arrow key until the blinking dot appears at the intersection of the two lines. (The screen should display "X = 1.8947368   Y = −.8631579".)

Zoom in on the graph by pressing the **ZOOM** key followed by the **2** key and then the **ENTER** key. (See frame **(42)**.)

Press the **TRACE** key. The message "X = 1.9052632   Y = −.8568421" will appear on the screen.

Press the right-arrow key once to place the blinking dot at the intersection of the two lines (X = 1.9263158). Thus, using graphics, the solution to the system is (1.9263158, −0.8442105).

Comment   The actual solution to the system of equations is $(\frac{25}{13}, -\frac{11}{13})$ or (1.9231, −0.8461), accurate to four decimal places. Continuing this process of Trace and Zoom will yield an x-coordinate of 1.9230 and a y-coordinate of −0.8462.

EXERCISE SET 5

1. For $f(x) = x^3 - x^2 + 1$, use the trace and zoom features to find the minimal value in the interval $0 \le x \le 1$.
2. Find the point of intersection of the graph of $f(x) = x^2 - 3x - 1$ and that of $f(x) = 0.3x - 2$ with an x-coordinate that is between 0 and 1.
3. Solve: $3x - y = 4$ and $2x + 3y = 3$.
4. Solve: $y = x^2 - x$ and $y = -2x + 1$.

SUMMARY

When the TI-81 is turned on (using the **ON** key at the bottom left of the calculator), the screen is blank, with the cursor at the top right. If no entry is made for six minutes, the calculator automatically turns off. Power can be restored by pressing the **ON** key. To turn the TI-81 off, use the OFF command. (Press the **2nd** key followed by the **ON** key.)

Computation   Order-of-operations rules are followed in all computations. To override these rules, parentheses can be used.

To enter a negative value, the $(-)$ key must be used. To enter the subtraction operation, the $-$ key must be used.

```
ERROR 06 SYNTAX
1:Goto Error
2:Quit
```

If an error message (menu) appears on the screen, respond by pressing the **ENTER** key to select option 1 (try to edit the offending command), or press the **2** key to QUIT (stop trying to execute the command).

An entry can be edited (before pressing the **ENTER** key) by using the directional-arrow keys, the **DEL** key, and the **INS** key.

The ENTRY command (the **2nd** key followed by the **ENTER** key or the up-arrow key) is used to *replay* the last command entered.

The **CLEAR** key is used to clear the text screen.

The **ENTER** key is used to carry out a command (usually to compute and then to display the results of a computation).

Graphs   The graph screen is cleared by clearing the function definitions.

The **RANGE** key is used to display the current range settings (the portion of the x-axis, the portion of the y-axis, and the scales for the axes). These range values can be altered to the values desired.

The Standard option of the ZOOM menu is used to reset the range values to their default settings $(-10, 10, 1, -10, 10, 1)$.

The **Y=** key is used to enter the function to be graphed. The prompt :Y1 = is displayed. At this point, you enter the function rule that will determine the ordered pairs to be graphed. Pressing the **GRAPH** key draws the graph using the current range settings.

To type the independent variable *x* when entering a function rule, press the **X|T** key.

The **2nd** key followed by the **CLEAR** key (the QUIT command) is used to switch the view back to the text screen.

With a graph displayed on the screen, the **ZOOM** key can be used to zoom in (reduce the graph viewing area) or zoom out (enlarge the viewing area).

With a graph displayed on the screen, the **TRACE** key can be used to place a blinking dot (pixel) on the graphed curve. The *x*-coordinate and *y*-coordinate of the dot are displayed at the bottom of the display. Pressing the left/right directional-arrow keys moves the dot along the graph, changing the coordinates displayed.

YOU MIGHT WANT TO KNOW

Q1 Is there a way to have computed values reported to a fixed number of decimal places (for example, accurate to hundredths)?

A1 Yes. Press the **MODE** key. The menu shown in frame ④③ is displayed. Press the down-arrow key once to make the word Float start flashing. Press the right-arrow key to place the cursor on 2 (for 2 decimal places). Press the **ENTER** key. The 2 starts flashing. (See frame ④④.) Use the QUIT command (the **2nd** key followed by the **CLEAR** key) to return to the text screen.

Q2 If I am to graph the function $f(t) = t^2 - 3t + 2$, do I use the **ALPHA** key and the **4** key to type $T^2 - 3T + 2$ in response to the :Y1 = prompt?

A2 No. You must use the **X|T** key to type $X^2 - 3X + 2$.

Q3 What is the fifth root of 27, and how do I find it?

A3 You want a number such that five times itself yields 27. To use the TI-81 to estimate that number, type 27^(1 ÷ 5), and then press the **ENTER** key. The number 1.933182045 is displayed. Remember from algebra, $X^{p/q}$ means $\sqrt[q]{X^p}$. Thus, $27^{1/5}$ is $\sqrt[5]{27}$.

If you were looking for the third root of 27, you could press the **MATH** key to display the MATH menu. (See frame ④⑤.) Press the **4** key to select option 4 (the $\sqrt[3]{}$). You are returned to the text screen with the $\sqrt[3]{}$ symbol on the screen. Type 27, and then press the **ENTER** key. The number 3 is displayed, because (3)(3)(3) is 27.

Q4 How can I use the TI-81 to graph a line through the points $(-3, 5)$ and $(4, -1)$?

A4 By using the DRAW menu: first, use the **ZOOM** key and select the Standard option to set the ranges to their default values $(-10, 10, 1, -10, 10, 1)$. Also, you might want to press the **Y=** key and then use the **CLEAR** key to erase any existing function definitions and clear the graphics screen.

Make sure you are on the home screen (use QUIT). See the comment in the margin.

Margin frames

④③
```
Norm Sci Eng
Float 0123456789
Rad Deg
Function Param
Connected Dot
Sequence Simul
Grid Off Grid On
Rect Polar
```

④④
```
Norm Sci Eng
Float 0123456789
Rad Deg
Function Param
Connected Dot
Sequence Simul
Grid Off Grid On
Rect Polar
```

④⑤
```
MATH NUM HYP PRB
1:R▸P(
2:P▸R(
3:³
4:³√
5:!
6:°
7↓r
```

Comment

If you are on the graphic screen when you select the DRAW menu, the points $(-3, 5)$ and $(4, -1)$ are entered as if you were defining a zoom box.

Warning

If you mistakenly press the **2nd** key followed by the **+** key, you will get the RESET screen. Press the **ENTER** key to return to the home screen.

(16)

```
DRAW
1:ClrDraw
2:Line(
3:PT-On(
4:PT-Off(
5:PT-CHg(
6:DrawF
7:Shade(
```

(47)

The output of the command Line (−3, 5, 4, −1).

(48)

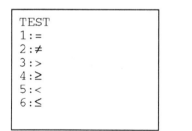

```
TEST
1:=
2:≠
3:>
4:≥
5:<
6:≤
```

(49)

Now press the **2nd** key followed by the **PRGM** key, and the **DRAW** menu is presented. (See frame (46).) Press the **2** key to select the Line option. The prompt Line(is presented on the text screen. Press the (−) key and the **3** key to get Line(−3. Now press the **ALPHA** key followed by the **.** key to get Line(−3,. Press the **5** key, the comma key, the **4** key, the comma key, the (−) key, the **1** key, and then the **)** key to get Line(−3, 5, 4, −1). Press the **ENTER** key. (See frame (47).)

As you can see, the line segment from (−3, 5) to (4, −1) was drawn. You may now use a zoom box to zoom in on the area of interest. However, this may be difficult. When the TI-81 *draws* points on the screen (using the **DRAW** menu options), these points are not handled like function definitions (from the **Y=** key). If you zoom in on a line segment, the graph screen is redrawn, and any points from the **DRAW** options are lost. But if you use the **ENTRY** command (the **2nd** key followed by the **ENTER** key, or the up-arrow key) to *replay* the last command, the line segment will be redrawn on the graph screen with the new range values.

To clear any points placed on the graph screen as a result of the **DRAW** options, you can select Option 1 (ClrDraw) from the **DRAW** menu.

Q5 Is there a way to have the TI-81 graph a piecewise-defined function such as the following?

$$f(x) = \begin{cases} x - 1 & \text{for } x \leq 1 \\ x^2 - 2x - 2 & \text{for } x > 1 \end{cases}$$

A5 Yes. First, set the range values to default using the Standard option from the **ZOOM** menu. Also, press the **MODE** key and select the Dot mode. (Move the flashing rectangle cursor down to line 5, then over to Dot, press the **ENTER** key, and then use the **QUIT** command.) Press the **Y=** key and type the following for the definition of function Y1:

$$:Y1 = (X - 1)(X \leq 1) + (X^2 - 2X - 2)(X > 1)$$

To get the inequalities ≤ and >, use the **TEST** command (the **2nd** key followed by the **MATH** key) and select the one you want from the **TEST** menu. (See frame (48).) With the graph on the screen (see frame (49)), you can use a zoom box to look at the portion of interest of the graph. When you are finished with the graph, change the **MODE** setting back to Connected.

Q6 I tried to graph the function $f(x) = \sin x$, but all I got was a straight (horizontal) line below the negative x-axis. When I zoomed in on the graph, I got a horizontal line just above the positive x-axis. What's going on?

A6 You have graphed the sine function from −10 degrees to 10 degrees. To correct this, press the **MODE** key. Make sure the Rad option is selected in line 3. (If not, use the directional-arrow keys to place the flashing rectangle on Rad and then press the **ENTER** key.) Press the **RANGE** key and set the values: −6.28, 6.28, 1, −2, 2, 1. Press the **GRAPH** key.

Note: If you wish to graph the function $f(x) = \sin x$ using degree measure for x, select the Deg option on line 3 of the **MODE** menu, but set the x-range values from −360 to 360.

A-19

A.3 SOME PROOFS FOR SECTION 1.1

In this section, we list the most basic properties of the real-number system. Then we use those properties to prove the familiar rules of algebra for working with signed numbers.

The set of real numbers is **closed** with respect to the operations of addition and multiplication. This simply means that when we add or multiply two real numbers, the result (i.e., the **sum** or **product**, respectively) is again a real number. Some of the other basic properties and definitions for the real-number system are listed in the following box. Note that the lowercase letters a, b, and c denote arbitrary real numbers.

▶ PROPERTY SUMMARY	SOME FUNDAMENTAL PROPERTIES OF THE REAL NUMBERS
Commutative properties	$a + b = b + a \qquad ab = ba$
Associative properties	$a + (b + c) = (a + b) + c$ $a(bc) = (ab)c$
Identity properties	**1.** There is a unique real number 0 (called **zero** or the **additive identity**) such that $\qquad a + 0 = a \qquad$ and $\qquad 0 + a = a$ **2.** There is a unique real number 1 (called **one** or the **multiplicative identity**) such that $\qquad a \cdot 1 = a \qquad$ and $\qquad 1 \cdot a = a$
Inverse properties	**3.** For each real number a, there is a real number $-a$ (called the **additive inverse** of a or the **negative** of a) such that $\qquad a + (-a) = 0 \qquad$ and $\qquad (-a) + a = 0$ **4.** For each real number $a \neq 0$, there is a real number denoted by $\dfrac{1}{a}$ (or $1/a$ or a^{-1}) and called the **multiplicative inverse** or **reciprocal** of a, such that $\qquad a \cdot \dfrac{1}{a} = 1 \qquad$ and $\qquad \dfrac{1}{a} \cdot a = 1$
Distributive properties	$a(b + c) = ab + ac \qquad (b + c)a = ba + ca$

After reading this list of properties for the first time, many students ask the natural question, "Why do we even bother to list such obvious properties?" One reason is that all the other laws of arithmetic and algebra (including the "not-so-obvious" ones) can be derived from our rather short list. For example, the rule $0 \cdot a = 0$ can be proved using the distributive property, as can the rule that the product of two negative numbers is a positive number. We will now state and prove these properties as well as several others.

Theorem

Let a and b be real numbers. Then

(a) $a \cdot 0 = 0$ **(d)** $a(-b) = -ab$

(b) $-a = (-1)a$ **(e)** $(-a)(-b) = ab$

(c) $-(-a) = a$

PROOF OF PART (a)

$$a \cdot 0 = a \cdot (0 + 0) \qquad \text{additive identity property}$$
$$= a \cdot 0 + a \cdot 0 \qquad \text{distributive property}$$

Now, since $a \cdot 0$ is a real number, it has an additive inverse, $-(a \cdot 0)$. Adding this to both sides of the last equation, we obtain

$$a \cdot 0 + [-(a \cdot 0)] = (a \cdot 0 + a \cdot 0) + [-(a \cdot 0)]$$
$$a \cdot 0 + [-(a \cdot 0)] = a \cdot 0 + \{a \cdot 0 + [-(a \cdot 0)]\} \qquad \text{associative property of addition}$$
$$0 = a \cdot 0 + 0 \qquad \text{additive inverse property}$$
$$0 = a \cdot 0 \qquad \text{additive identity property}$$

Thus $a \cdot 0 = 0$, as we wished to show.

PROOF OF PART (b)

$$0 = 0 \cdot a \qquad \text{using part (a) and the commutative property of multiplication}$$
$$= [1 + (-1)]a \qquad \text{additive inverse property}$$
$$= 1 \cdot a + (-1)a \qquad \text{distributive property}$$
$$= a + (-1)a \qquad \text{multiplicative identity property}$$

Now, by adding $-a$ to both sides of this last equation, we obtain

$$-a + 0 = -a + [a + (-1)a]$$
$$-a = (-a + a) + (-1)a \qquad \text{additive identity property and associative property of addition}$$
$$= 0 + (-1)a \qquad \text{additive inverse property}$$
$$= (-1)a \qquad \text{additive identity property}$$

This last equation asserts that $-a = (-1)a$, as we wished to show.

PROOF OF PART (c)

$$-(-a) + (-a) = 0 \qquad \text{additive inverse property}$$

By adding a to both sides of this last equation, we obtain

$$[-(-a) + (-a)] + a = 0 + a$$

$$-(-a) + (-a + a) = a \qquad \text{associative property of addition and additive identity property}$$

$$-(-a) + 0 = a \qquad \text{additive inverse property}$$

$$-(-a) = a \qquad \text{additive identity property}$$

This last equation states that $-(-a) = a$, as we wished to show.

PROOF OF PART (d)

$$a(-b) = a[(-1)b] \qquad \text{using part (b)}$$

$$= [a(-1)]b \qquad \text{associative property of multiplication}$$

$$= [(-1)a]b \qquad \text{commutative property of multiplication}$$

$$= (-1)(ab) \qquad \text{associative property of multiplication}$$

$$= -(ab) \qquad \text{using part (b)}$$

Thus $a(-b) = -ab$, as we wished to show.

PROOF OF PART (e)

$$(-a)(-b) = -[(-a)b] \qquad \text{using part (d)}$$

$$= -[b(-a)] \qquad \text{commutative property of multiplication}$$

$$= -[-(ba)] \qquad \text{using part (d)}$$

$$= ba \qquad \text{using part (c)}$$

$$= ab \qquad \text{commutative property of multiplication}$$

We've now shown that $(-a)(-b) = ab$, as required.

A.4 $\sqrt{2}$ IS IRRATIONAL

The proof is by reductio ad absurdum, *and* reductio ad absurdum, *which Euclid loved so much, is one of a mathematician's finest weapons.*

G. H. Hardy (1877–1947)

We will use an *indirect proof* to show that the square root of 2 is an irrational number. The strategy is as follows.

1. We suppose that $\sqrt{2}$ is a rational number.
2. Using (1) and the usual rules of logic and algebra, we derive a contradiction.
3. On the basis of the contradiction in (2), we conclude that the supposition in (1) is untenable; that is, we conclude that $\sqrt{2}$ is irrational.

In carrying out the proof, we'll assume that the following three statements are known:

If x is an even natural number, then $x = 2k$, for some natural number k.

Any rational number can be written in the form a/b, where the integers a and

b have no common integral factors other than ± 1. (In other words, any fraction can be reduced to lowest terms.)

If x is a natural number and x^2 is even, then x is even.

Our indirect proof now proceeds as follows. Suppose that $\sqrt{2}$ were a rational number. Then we would be able to write

$$\sqrt{2} = \frac{a}{b} \qquad \text{where } a \text{ and } b \text{ are natural numbers with no common factor other than 1} \tag{1}$$

Since both sides of equation (1) are positive, we can square both sides to obtain the equivalent equation

$$2 = \frac{a^2}{b^2} \qquad \text{or} \qquad 2b^2 = a^2 \tag{2}$$

Since the left-hand side of equation (2) is an even number, the right-hand side must be even. But if a^2 is even then a is even, and so

$$a = 2k \qquad \text{for some natural number } k$$

Using this last equation to substitute for a in equation (2), we have

$$2b^2 = (2k)^2 = 4k^2$$

or

$$b^2 = 2k^2$$

Hence (reasoning as before) b^2 is even, and therefore b is even. But then we have that both b and a are even, contrary to our hypothesis that b and a have no common factor other than 1. We conclude from this that equation (1) cannot hold; that is, there is no rational number a/b such that $\sqrt{2} = a/b$. Thus $\sqrt{2}$ is irrational, as we wished to prove.

A.5　REVIEW OF INTEGER EXPONENTS AND nTH ROOTS

In algebra, you became familiar with the definition and basic properties of the **exponential notation** a^n, where a is a real number and n is a natural number. Since that definition and those properties are so fundamental, we repeat them here in the two boxes that follow.

DEFINITION I　Base and Exponent

Given a real number a and a natural number n, we define a^n by

$$a^n = \underbrace{a \cdot a \cdot a \cdot \cdots \cdot a}_{n \text{ factors}}$$

In the expression a^n, a is the **base** and n is the **exponent**.

PROPERTY	EXAMPLES
1. $a^m a^n = a^{m+n}$	$a^5 a^6 = a^{11}$; $(x+1)(x+1)^2 = (x+1)^3$
2. $(a^m)^n = a^{mn}$	$(2^3)^4 = 2^{12}$; $[(x+1)^2]^3 = (x+1)^6$
3. $\dfrac{a^m}{a^n} = \begin{cases} a^{m-n} & \text{if } m > n \\ \dfrac{1}{a^{n-m}} & \text{if } m < n \\ 1 & \text{if } m = n \end{cases}$	$\dfrac{a^6}{a^2} = a^4$; $\dfrac{a^2}{a^6} = \dfrac{1}{a^4}$; $\dfrac{a^5}{a^5} = 1$
4. $(ab)^m = a^m b^m$; $\left(\dfrac{a}{b}\right)^m = \dfrac{a^m}{b^m}$	$(2x^2)^3 = 2^3 \cdot (x^2)^3 = 8x^6$; $\left(\dfrac{x^2}{y^3}\right)^4 = \dfrac{x^8}{y^{12}}$

Each of these properties is a direct consequence of the definition of a^n. For instance, according to the first property we have $a^2 a^3 = a^5$. To verify that this is indeed correct, we note that

$$a^2 a^3 = (aa)(aaa) = a^5$$

Now we want to extend our definition of a^n to allow for exponents that are integers but not necessarily natural numbers. We begin by defining a^0.

DEFINITION 2 The Zero Exponent

EXAMPLES

For any nonzero real number a,

$$a^0 = 1$$

(0^0 is not defined).

a. $2^0 = 1$

b. $(-\pi)^0 = 1$

c. $\left(\dfrac{3}{1 + a^2 + b^2}\right)^0 = 1$

It's easy to see the motivation for defining a^0 to be 1. Assuming that the exponent zero is to have the same properties as do exponents that are natural numbers, we can write

$$a^0 a^n = a^{0+n} \qquad \text{and therefore} \qquad a^0 a^n = a^n$$

Now we divide both sides of this last equation by a^n to obtain $a^0 = 1$, which agrees with our definition.

Our next definition assigns a meaning to the expression a^{-n} when n is a natural number.

DEFINITION 3   Negative Exponents

For any nonzero real number a and natural number n,

$$a^{-n} = \frac{1}{a^n}$$

EXAMPLES

a. $2^{-1} = \frac{1}{2^1} = \frac{1}{2}$

b. $\left(\frac{1}{10}\right)^{-1} = \frac{1}{\left(\frac{1}{10}\right)^1} = 10$

c. $x^{-2} = \frac{1}{x^2}$

d. $(a^2b)^{-3} = \frac{1}{(a^2b)^3} = \frac{1}{a^6b^3}$

e. $\frac{1}{2^{-3}} = \frac{1}{\frac{1}{2^3}} = 2^3 = 8$

Again, it's easy to see the motivation for this definition. We have

$$a^na^{-n} = a^{n+(-n)} = a^0 \qquad \text{and therefore} \qquad a^na^{-n} = 1$$

Now divide both sides of this last equation by a^n to obtain $a^{-n} = 1/a^n$, in agreement with Definition 3.

It can be shown that the four properties of exponents that we listed earlier continue to hold for all integer exponents. We make use of this fact in the next two examples.

EXAMPLE 1 Simplify the expression $(a^2b^3)^2(a^5b)^{-1}$. Write the answer in such a way that only positive exponents are used.

Solution FIRST METHOD

$$(a^2b^3)^2(a^5b)^{-1} = (a^2b^3)^2 \cdot \frac{1}{a^5b}$$

$$= \frac{a^4b^6}{a^5b} = \frac{b^{6-1}}{a^{5-4}} = \frac{b^5}{a}$$

ALTERNATE METHOD

$$(a^2b^3)^2(a^5b)^{-1} = (a^4b^6)(a^{-5}b^{-1})$$
$$= a^{4-5}b^{6-1} = a^{-1}b^5$$
$$= \frac{b^5}{a}$$

EXAMPLE 2 Simplify the expression $\left(\dfrac{a^{-5}b^2c^0}{a^3b^{-1}}\right)^3$, writing the answer so that negative exponents are not used.

Solution $\left(\dfrac{a^{-5}b^2c^0}{a^3b^{-1}}\right)^3 = \dfrac{a^{-15}b^6}{a^9b^{-3}} = \dfrac{b^{6-(-3)}}{a^{9-(-15)}}$

$$= \frac{b^9}{a^{24}}$$

Now that we've completed our review of integer exponents, we are ready to consider roots. Here is the basic definition.

EXAMPLES

Let n be a natural number. If a and b are real numbers and

$$a^n = b$$

then we say that a is an **nth root** of b. When $n = 2$ and when $n = 3$, we refer to it as a **square root** and a **cube root**, respectively.

Both 3 and -3 are square roots of 9 because $3^2 = 9$ and $(-3)^2 = 9$.

Both 2 and -2 are fourth roots of 16 because $2^4 = 16$ and $(-2)^4 = 16$.

2 is a cube root of 8 because $2^3 = 8$.

-3 is a fifth root of -243 because $(-3)^5 = -243$.

As the examples in the box suggest, square roots, fourth roots, and all *even* roots of positive numbers occur in pairs, one positive and one negative. In these cases, we use the notation $\sqrt[n]{b}$ to denote the positive, or **principal**, nth root of b. As examples of this notation, we can write

$$\sqrt[4]{81} = 3 \qquad \text{(The principal fourth root of 81 is 3.)}$$
$$\sqrt[4]{81} \neq -3 \qquad \text{(The principal fourth root of 81 is not } -3.)$$
$$-\sqrt[4]{81} = -3 \qquad \text{(The negative of the principal fourth root of 81 is } -3.)$$

The symbol $\sqrt{\ }$ is called a **radical sign**, and the number under the radical sign is the **radicand**. The natural number n used in the notation $\sqrt[n]{\ }$ is called the **index** of the radical. For square roots we suppress the index and simply write $\sqrt{\ }$ rather than $\sqrt[2]{\ }$. So, for example, $\sqrt{25} = 5$.

As we saw in the examples, cube roots, fifth roots, and, in fact, all *odd* roots occur singly, not in pairs. In these cases we again use the notation $\sqrt[n]{b}$ for the nth root. The definition and examples in the following box summarize our discussion up to this point.

EXAMPLES

1. Let n be a natural number. If a and b are nonnegative real numbers, then

$$\sqrt[n]{b} = a \quad \text{if and only if} \quad b = a^n$$

The number a is the **principal nth root** of b.

2. If a and b are negative and n is an odd natural number, then

$$\sqrt[n]{b} = a \quad \text{if and only if} \quad b = a^n$$

$\sqrt{25} = 5; \quad \sqrt{25} \neq -5$

$-\sqrt{25} = -5$

$\sqrt[4]{\dfrac{1}{16}} = \dfrac{1}{2}; \quad \sqrt[3]{8} = 2$

$\sqrt[5]{-\dfrac{1}{32}} = -\dfrac{1}{2}$

Five properties of nth roots are frequently used in simplifying certain expressions. The first four are similar to the properties of square roots that are developed in elementary algebra. For reference, we list these properties side by side in the following box. Property 5 is listed here only for the sake of completeness; we'll postpone discussing it until Appendix A.6.

	PROPERTIES OF nTH ROOTS	CORRESPONDING PROPERTIES FOR SQUARE ROOTS
	Suppose that x and y are real numbers and that m and n are natural numbers. Then each of the following properties holds, provided only that the expressions on both sides of the equation are defined (and so represent real numbers).	
1.	$\left(\sqrt[n]{x}\right)^n = x$	$\left(\sqrt{x}\right)^2 = x$
2.	$\sqrt[n]{xy} = \sqrt[n]{x}\,\sqrt[n]{y}$	$\sqrt{xy} = \sqrt{x}\,\sqrt{y}$
3.	$\sqrt[n]{\dfrac{x}{y}} = \dfrac{\sqrt[n]{x}}{\sqrt[n]{y}}$	$\sqrt{\dfrac{x}{y}} = \dfrac{\sqrt{x}}{\sqrt{y}}$
4.	n even: $\sqrt[n]{x^n} = \lvert x \rvert$ n odd: $\sqrt[n]{x^n} = x$	$\sqrt{x^2} = \lvert x \rvert$
5.	$\sqrt[m]{\sqrt[n]{x}} = \sqrt[mn]{x}$	

These properties are used in simplifying expressions involving nth roots. In general, we try to factor the expression under the radical so that one factor is the largest perfect nth power that we can find. Then we apply Property 2 or Property 3. For instance, we can simplify the expression $\sqrt{72}$ as follows:

$$\sqrt{72} = \sqrt{(36)(2)} = \sqrt{36}\,\sqrt{2} = 6\sqrt{2}$$

In this procedure, we begain by factoring 72 as $(36)(2)$. Note that 36 is the largest factor of 72 that is a perfect square. If we had begun instead with a different factorization, say $72 = (9)(8)$, we would still have arrived at the same answer, but it would have taken longer. (Check this for yourself.)

As another example, let's simplify $\sqrt[3]{40}$. First, what (if any) is the largest perfect cube factor of 40? Since the first few perfect cubes are

$$1^3 = 1 \qquad 2^3 = 8 \qquad 3^3 = 27 \qquad 4^3 = 64$$

we see that 8 is a perfect cube factor of 40, and we write

$$\sqrt[3]{40} = \sqrt[3]{(8)(5)} = \sqrt[3]{8}\,\sqrt[3]{5} = 2\sqrt[3]{5}$$

EXAMPLE 3 Simplify: **(a)** $\sqrt{12} + \sqrt{75}$; **(b)** $\sqrt{\dfrac{162}{49}}$.

Solution

(a) $\begin{aligned}\sqrt{12} + \sqrt{75} &= \sqrt{(4)(3)} + \sqrt{(25)(3)}\\ &= \sqrt{4}\,\sqrt{3} + \sqrt{25}\,\sqrt{3}\\ &= 2\sqrt{3} + 5\sqrt{3}\\ &= 7\sqrt{3}\end{aligned}$

(b) $\sqrt{\dfrac{162}{49}} = \dfrac{\sqrt{162}}{\sqrt{49}} = \dfrac{\sqrt{81}\,\sqrt{2}}{7} = \dfrac{9\sqrt{2}}{7}$

EXAMPLE 4 Simplify: $\sqrt[3]{16} + \sqrt[3]{250} - \sqrt[3]{128}$.

Solution $\sqrt[3]{16} + \sqrt[3]{250} - \sqrt[3]{128} = \sqrt[3]{8}\,\sqrt[3]{2} + \sqrt[3]{125}\,\sqrt[3]{2} - \sqrt[3]{64}\,\sqrt[3]{2}$
$$= 2\sqrt[3]{2} + 5\sqrt[3]{2} - 4\sqrt[3]{2}$$
$$= 3\sqrt[3]{2}$$

EXAMPLE 5 Simplify each expression by removing the largest possible perfect square factor from within the radical.

(a) $\sqrt{8x^2}$ (b) $\sqrt{8x^2}$ assuming $x \geq 0$

(c) $\sqrt{18a^7}$ assuming $a \geq 0$ (d) $\sqrt[3]{16y^5}$

Solution

(a) $\sqrt{8x^2} = \sqrt{(4)(2)(x^2)} = \sqrt{4}\,\sqrt{2}\,\sqrt{x^2} = 2\sqrt{2}\,|x|$

(b) $\sqrt{8x^2} = \sqrt{4}\,\sqrt{2}\,\sqrt{x^2} = 2x\sqrt{2}$ $\sqrt{x^2} = x$ because $x \geq 0$

(c) $\sqrt{18a^7} = \sqrt{(9a^6)(2a)} = \sqrt{9a^6}\,\sqrt{2a} = 3a^3\sqrt{2a}$

(d) $\sqrt[3]{16y^5} = \sqrt[3]{8y^3}\,\sqrt[3]{2y^2} = 2y\sqrt[3]{2y^2}$

EXAMPLE 6 Simplify: $\dfrac{1}{\sqrt{2}} - 3\sqrt{50}$.

Solution First, we rationalize the denominator in the fraction $1/\sqrt{2}$:

$$\frac{1}{\sqrt{2}} = \frac{1}{\sqrt{2}} \cdot 1 = \frac{1}{\sqrt{2}} \cdot \frac{\sqrt{2}}{\sqrt{2}} = \frac{\sqrt{2}}{2}$$

Next, we simplify the expression $3\sqrt{50}$:

$$3\sqrt{50} = 3\sqrt{(25)(2)} = 3\sqrt{25}\,\sqrt{2} = (3)(5)\sqrt{2} = 15\sqrt{2}$$

Now, putting things together, we have

$$\frac{1}{\sqrt{2}} - 3\sqrt{50} = \frac{\sqrt{2}}{2} - 15\sqrt{2} = \frac{\sqrt{2}}{2} - \frac{30\sqrt{2}}{2}$$
$$= \frac{\sqrt{2} - 30\sqrt{2}}{2} = \frac{(1-30)\sqrt{2}}{2} = \frac{-29\sqrt{2}}{2}$$

EXAMPLE 7 Rationalize the denominator: $\dfrac{4}{2 + \sqrt{3}}$.

Solution We multiply by 1, writing 1 as $\dfrac{2 - \sqrt{3}}{2 - \sqrt{3}}$:

$$\frac{4}{2 + \sqrt{3}} \cdot 1 = \frac{4}{2 + \sqrt{3}} \cdot \frac{2 - \sqrt{3}}{2 - \sqrt{3}}$$
$$= \frac{4(2 - \sqrt{3})}{4 - (\sqrt{3})^2} = \frac{4(2 - \sqrt{3})}{4 - 3} = \frac{8 - 4\sqrt{3}}{1}$$
$$= 8 - 4\sqrt{3}$$

Note Check for yourself that multiplying the original fraction by $\dfrac{2 + \sqrt{3}}{2 + \sqrt{3}}$ does *not* eliminate radicals in the denominator. ▪▪▪

In the next example, we are asked to rationalize the numerator rather than the denominator. This is useful in calculus.

EXAMPLE 8 Rationalize the *numerator*: $\dfrac{\sqrt{x} - \sqrt{3}}{x - 3}$ $(x \geq 0, x \neq 3)$.

Solution $\dfrac{\sqrt{x} - \sqrt{3}}{x - 3} \cdot 1 = \dfrac{\sqrt{x} - \sqrt{3}}{x - 3} \cdot \dfrac{\sqrt{x} + \sqrt{3}}{\sqrt{x} + \sqrt{3}}$

$$= \dfrac{(\sqrt{x})^2 - (\sqrt{3})^2}{(x - 3)(\sqrt{x} + \sqrt{3})} = \dfrac{x - 3}{(x - 3)(\sqrt{x} + \sqrt{3})}$$

$$= \dfrac{1}{\sqrt{x} + \sqrt{3}}$$ ▪▪▪

The strategy for rationalizing numerators or denominators involving *n*th roots is similar to that used for square roots. To rationalize a numerator or a denominator involving an *n*th root, we multiply that numerator or denominator by a factor that will yield a product that is a perfect *n*th power. The next example displays two instances of this.

EXAMPLE 9

(a) Rationalize the denominator: $\dfrac{6}{\sqrt[3]{7}}$.

(b) Rationalize the denominator: $\dfrac{ab}{\sqrt[4]{a^2 b^3}}$ $(a > 0, b > 0)$.

Solution

(a) $\dfrac{6}{\sqrt[3]{7}} \cdot 1 = \dfrac{6}{\sqrt[3]{7}} \cdot \dfrac{\sqrt[3]{7^2}}{\sqrt[3]{7^2}}$

$$= \dfrac{6\sqrt[3]{7^2}}{\sqrt[3]{7^3}} = \dfrac{6\sqrt[3]{49}}{7}$$

(b) $\dfrac{ab}{\sqrt[4]{a^2 b^3}} \cdot 1 = \dfrac{ab}{\sqrt[4]{a^2 b^3}} \cdot \dfrac{\sqrt[4]{a^2 b}}{\sqrt[4]{a^2 b}}$

$$= \dfrac{ab\sqrt[4]{a^2 b}}{\sqrt[4]{a^4 b^4}} = \dfrac{ab\sqrt[4]{a^2 b}}{ab} = \sqrt[4]{a^2 b}$$ ▪▪▪

EXERCISE SET A.5

A

For Exercises 1–18, simplify the expression. For exercises where negative exponents appear, write the answers without using negative exponents.

1. (a) $x^3 x^{12}$
   (c) $(x + 1)^3 (x + 1)^{12}$
   (b) $(x^3)^{12}$
   (d) $\left[(x + 1)^3\right]^{12}$

2. (a) $y^2 y^6$
   (c) $(y^2 y^6)^3$
   (b) $(y^2)^6$
   (d) $\left[(y + 1)^2\right]^6$

3. (a) $\dfrac{a^{15}}{a^9}$
   (b) $\dfrac{(a + 1)^{15}}{(a + 1)^9}$

4. (a) $\dfrac{(x^2 a)^5}{(x^3 a^2)^4}$
   (b) $\dfrac{(x^2 + 1)^5}{(x^2 + 1)^4}$

5. (a) $(64)^0$ (b) $(64^3)^0$

6. (a) $(2^0 + 3^0)^3$ (b) $(2 + 3)^0$ (c) $\left[\left(\frac{1}{10}\right)^0\right]^3$

7. (a) $10^{-1} + 10^{-2}$ (b) $(10^{-1} + 10^{-2})^{-1}$ (c) $(10^{-1})^{-2}$

8. (a) $4^{-2} + 4^{-1}$ (b) $\left[\left(\frac{1}{4}\right)^{-1} + \left(\frac{1}{4}\right)^{-2}\right]^{-1}$ (c) $\left[\left(\frac{1}{4}\right)^{-1}\right]^{-2}$

9. (a) $(a^2bc^0)^{-3}$ (b) $(a^3b)^3(a^2b^4)^{-1}$
(c) $(a^{-3}b^{-1}c^3)^{-2}$

10. (a) $(x^3y^{-2}z^{-1})^{-4}$ (b) $(x^2y^3)^{-3}(x^2y^4)^{-2}$
(c) $(x^{-1}y^{-1})^{-1}/xy$

11. $(2^{-2} + 2^{-1} + 2^0)^{-2}$ **12.** $(10^{-2} + 10^{-1} + 10^2)^{-1}$

13. $\left(\dfrac{x^3y^{-2}z}{xy^2z^{-3}}\right)^{-3}$ **14.** $\left(\dfrac{x^4y^{-8}z^2}{xy^2z^{-6}}\right)^2$

15. $\left(\dfrac{a^{-2}b^{-3}c^{-4}}{a^2b^3c^4}\right)^2$ **16.** $(-2x)^{-3}(-2x^{-2})^{-1}$

17. $\dfrac{x^2}{y^{-3}} \div \dfrac{x^2}{y^3}$ **18.** $(2x^2)^{-3} - 2(x^2)^{-3}$

In Exercises 19–24, determine whether the statement is true or false.

19. (a) $\sqrt{81} = -9$ (b) $\sqrt{81} = 9$
(c) $-\sqrt{81} = -9$

20. (a) $-\sqrt{121} = -11$ (b) $\sqrt{121} = \pm 11$
(c) $\sqrt{121} = 11$

21. (a) $\sqrt{9 + 16} = \sqrt{9} + \sqrt{16}$
(b) $\sqrt{(9)(16)} = \sqrt{9}\sqrt{16}$

22. (a) $\sqrt{10 + 6} = \sqrt{10} + \sqrt{6}$
(b) $\sqrt{10}\sqrt{6} = \sqrt{(10)(6)}$

23. (a) $\sqrt{(-5)^2} = -5$ (b) $\sqrt{x^2} = x$ $(x < 0)$

24. (a) $\sqrt{(-6)^2} = 6$ (b) $\sqrt{y^2} = |y|$

In Exercises 25–32, evaluate the expression. If the expression is undefined (i.e., does not represent a real number), say so.

25. (a) $\sqrt[3]{-64}$ (b) $\sqrt[4]{-64}$

26. (a) $\sqrt[5]{32}$ (b) $\sqrt[5]{-32}$

27. (a) $\sqrt[3]{8/125}$ (b) $\sqrt[3]{-8/125}$

28. (a) $\sqrt[3]{-1/1000}$ (b) $\sqrt[6]{-1/1000}$

29. (a) $\sqrt{-16}$ (b) $\sqrt[4]{-16}$

30. (a) $-\sqrt[4]{16}$ (b) $-\sqrt[4]{-16}$

31. (a) $\sqrt[4]{256/81}$ (b) $\sqrt[3]{-27/125}$

32. (a) $\sqrt[6]{64}$ (b) $\sqrt[6]{-64}$

In Exercises 33–54, simplify the expression. (Unless otherwise specified, assume that all letters in Exercises 45–54 represent positive numbers.)

33. (a) $\sqrt{18}$ (b) $\sqrt[3]{54}$

34. (a) $\sqrt{150}$ (b) $\sqrt[3]{375}$

35. (a) $\sqrt{98}$ (b) $\sqrt[5]{-64}$

36. (a) $\sqrt{27}$ (b) $\sqrt[3]{-108}$

37. (a) $\sqrt{25/4}$ (b) $\sqrt[4]{16/625}$

38. (a) $\sqrt{225/49}$ (b) $\sqrt[5]{-256/243}$

39. (a) $\sqrt{2} + \sqrt{8}$ (b) $\sqrt[3]{2} + \sqrt[3]{16}$

40. (a) $4\sqrt{3} - 2\sqrt{27}$ (b) $2\sqrt[3]{81} + 3\sqrt[3]{24}$

41. (a) $4\sqrt{50} - 3\sqrt{128}$ (b) $\sqrt[4]{32} + \sqrt[4]{162}$

42. (a) $\sqrt{3} - \sqrt{12} + \sqrt{48}$
(b) $\sqrt[5]{-2} + \sqrt[5]{-64} - \sqrt[5]{486}$

43. $\sqrt{\sqrt{64}}$ **44.** $\sqrt[3]{\sqrt{4096}}$

45. (a) $\sqrt{36x^2}$ (b) $\sqrt{36y^2}$ $(y < 0)$

46. (a) $\sqrt{225x^4y^3}$ (b) $\sqrt[4]{16a^4}$ $(a < 0)$

47. (a) $\sqrt{ab^2}\sqrt{a^2b}$ (b) $\sqrt{ab^3}\sqrt{a^3b}$

48. (a) $\sqrt[3]{125x^6}$ (b) $\sqrt[4]{64y^4}$ $(y < 0)$

49. $\sqrt[4]{16a^4b^5}$ $(a < 0)$ **50.** $\sqrt[3]{8a^4b^6}$

51. $\sqrt[3]{16a^{12}b^2/c^9}$ **52.** $\sqrt[4]{ab^3}\sqrt[4]{a^3b}$

53. $\sqrt[6]{\dfrac{5a^7}{a^{-5}b^6}}$ $(b < 0)$ **54.** $\sqrt[3]{a^2b}\sqrt[3]{ab}\sqrt[3]{b^4}$

In Exercises 55–70, rationalize the denominator and simplify where possible. (Assume that all letters represent positive quantities.)

55. (a) $4/\sqrt{7}$ (b) $3/\sqrt{3}$ (c) $\sqrt{2}/\sqrt{5}$

56. (a) $3/\sqrt{2}$ (b) $6/\sqrt{6}$ (c) $\sqrt{10}/\sqrt{2}$

57. (a) $\dfrac{1}{1 + \sqrt{5}}$ (b) $\dfrac{1}{1 - \sqrt{5}}$ (c) $\dfrac{1 + \sqrt{5}}{1 - \sqrt{5}}$

58. (a) $\dfrac{3}{2 - \sqrt{3}}$ (b) $\dfrac{3}{2 + \sqrt{3}}$ (c) $\dfrac{2 - \sqrt{3}}{2 + \sqrt{3}}$

59. $\dfrac{1}{\sqrt{5}} + 4\sqrt{45}$ **60.** $\dfrac{3}{\sqrt{8}} - \sqrt{450}$

61. $1/\sqrt[3]{25}$ **62.** $4/\sqrt[3]{16}$

63. $1/\sqrt[4]{2ab^5}$ **64.** $1/\sqrt[3]{4a^2b^8}$

65. $3/\sqrt[5]{16a^4b^9}$ **66.** $3/\sqrt[4]{27a^5b^{11}}$

67. $\dfrac{x}{\sqrt{x} - 2}$ **68.** $\dfrac{\sqrt{x}}{5 - \sqrt{x}}$

69. $\dfrac{\sqrt{x} - \sqrt{a}}{\sqrt{x} + \sqrt{a}}$ **70.** $\dfrac{1}{\sqrt{x+h} - \sqrt{x}}$

In Exercises 71–74, rationalize the numerator.

71. $\dfrac{\sqrt{x} - \sqrt{5}}{x - 5}$ **72.** $\dfrac{\sqrt{a} - \sqrt{b}}{a - b}$

73. $\dfrac{\sqrt{2 + h} - \sqrt{2}}{h}$ **74.** $\dfrac{\sqrt{x+h} - \sqrt{x}}{h}$

B

75. (a) Use a calculator to evaluate $\sqrt{8 - 2\sqrt{7}}$ and $\sqrt{7} - 1$. What do you observe?

(b) Prove that $\sqrt{8 - 2\sqrt{7}} = \sqrt{7} - 1$. *Hint:* In view of the definition of a principal square root, you need to check that $(\sqrt{7} - 1)^2 = 8 - 2\sqrt{7}$.

76. Use a calculator to provide empirical evidence that both of the following equations may be correct:

$$\frac{2 - \sqrt{3}}{\sqrt{2} - \sqrt{2 - \sqrt{3}}} + \frac{2 + \sqrt{3}}{\sqrt{2} + \sqrt{2 + \sqrt{3}}} = \sqrt{2}$$

$$\sqrt{\sqrt{97.5 - \frac{1}{11}}} = \pi$$

The point of this exercise is to remind you that, as useful as calculators may be, there is still the need for proofs in mathematics. In fact, it can be shown that the first equation is indeed correct, but the second is not.

77. Simplify: $\dfrac{x^{3a+2b-c}}{(x^{2a})(x^b)} \cdot x^{3c-a-b}$.

C

78. Let a and b be nonnegative integers, and consider the following two quantities:

$$\sqrt{a} + \sqrt{b} \qquad \text{and} \qquad \sqrt{a + b}$$

(a) Carry out a few computations to find out which quantity seems, in general, to be the larger.

(b) Write an inequality to summarize your results in part (a).

(c) Prove your statement in part (b). *Hint:* If x and y are nonnegative quantities, the condition $x \leq y$ is equivalent to $x^2 \leq y^2$.

A.6 REVIEW OF RATIONAL EXPONENTS

... Nicole Oresme, a bishop in Normandy (about 1323–1382), first conceived the notion of fractional powers, afterwards rediscovered by [Simon] Stevin [1548–1620]....

Florian Cajori in *A History of Elementary Mathematics* (New York: Macmillan Company, 1917)

If Descartes ... had discarded the radical sign altogether and had introduced the notation for fractional as well as integral exponents, ... it is conceivable that generations upon generations of pupils would have been saved the necessity of mastering the operations with two ... notations when one alone (the exponential) would have answered all purposes.

Florian Cajori, *A History of Mathematical Notations*, vol. 1 (La Salle, Illinois: The Open Court Publishing Company, 1928)

We can use the concept of an nth root to give a meaning to fractional exponents. First, by way of motivation, suppose that we want to assign a value to $5^{1/3}$. Assuming that the usual properties of exponents continue to apply here, we can write

$$(5^{1/3})^3 = 5^1 = 5$$

That is,

$$(5^{1/3})^3 = 5 \qquad \text{or} \qquad 5^{1/3} = \sqrt[3]{5}$$

By replacing 5 and 3 with b and n, respectively, we can see that we want to define $b^{1/n}$ to mean $\sqrt[n]{b}$. Also, by thinking of $b^{m/n}$ as $(b^{1/n})^m$, we see that the definition for $b^{m/n}$ ought to be $(\sqrt[n]{b})^m$. These definitions are formalized in the box that follows.

EXAMPLES

Let b denote a real number and n a natural number. We define $b^{1/n}$ by

$$b^{1/n} = \sqrt[n]{b}$$

$$4^{1/2} = \sqrt{4} = 2$$

$$(-8)^{1/3} = \sqrt[3]{-8} = -2$$

(If n is even, we require that $b \geq 0$.)

Let m/n be a rational number reduced to lowest terms. Assume that n is positive and that $\sqrt[n]{b}$ exists. Then,

$$b^{m/n} = \left(\sqrt[n]{b}\right)^m$$

$$8^{2/3} = \left(\sqrt[3]{8}\right)^2 = 2^2 = 4$$

or, equivalently,

or, equivalently,

$$b^{m/n} = \sqrt[n]{b^m}$$

$$8^{2/3} = \sqrt[3]{8^2} = \sqrt[3]{64} = 4$$

It can be shown that the four properties of exponents that we listed in Appendix A.5 continue to hold for rational exponents in general. In fact, we'll take this for granted rather than follow the lengthy argument needed for its verification. We will also assume that these properties apply to irrational exponents. So, for instance, we have

$$\left(2^{\sqrt{5}}\right)^{\sqrt{5}} = 2^5 = 32$$

(The definition of irrational exponents is discussed in Section 5.1.) In the next examples, we display the basic techniques for working with rational exponents.

EXAMPLE 1 Simplify each of the following quantities. Express the answers using positive exponents. If an expression does not represent a real number, say so.

(a) $49^{1/2}$ **(b)** $-49^{1/2}$ **(c)** $(-49)^{1/2}$ **(d)** $49^{-1/2}$

Solution

(a) $49^{1/2} = \sqrt{49} = 7$

(b) $-49^{1/2} = -(49^{1/2}) = -\sqrt{49} = -7$

(c) The quantity $(-49)^{1/2}$ does not represent a real number because there is no real number x such that $x^2 = -49$.

(d) $49^{-1/2} = \sqrt{49^{-1}} = \sqrt{\dfrac{1}{49}} = \dfrac{\sqrt{1}}{\sqrt{49}} = \dfrac{1}{7}$

Alternatively, we have

$$49^{-1/2} = (49^{1/2})^{-1} = 7^{-1} = \frac{1}{7}$$

EXAMPLE 2 Simplify each of the following. Write the answers using positive exponents. (Assume that $a > 0$.)

(a) $(5a^{2/3})(4a^{3/4})$ **(b)** $\sqrt[5]{\dfrac{16a^{1/3}}{a^{1/4}}}$ **(c)** $(x^2 + 1)^{1/5}(x^2 + 1)^{4/5}$

Solution

(a) $(5a^{2/3})(4a^{3/4}) = 20a^{(2/3)+(3/4)}$

$$= 20a^{17/12} \qquad \text{because } \frac{2}{3} + \frac{3}{4} = \frac{17}{12}$$

(b) $\sqrt[5]{\dfrac{16a^{1/3}}{a^{1/4}}} = \left(\dfrac{16a^{1/3}}{a^{1/4}}\right)^{1/5}$

$$= (16a^{1/12})^{1/5} \qquad \text{because } \frac{1}{3} - \frac{1}{4} = \frac{1}{12}$$

$$= 16^{1/5}a^{1/60}$$

(c) $(x^2 + 1)^{1/5}(x^2 + 1)^{4/5} = (x^2 + 1)^1 = x^2 + 1$

■■■

■■■ **EXAMPLE 3** Simplify: **(a)** $32^{-2/5}$; **(b)** $(-8)^{4/3}$.

Solution

(a) $32^{-2/5} = \left(\sqrt[5]{32}\right)^{-2} = 2^{-2} = \dfrac{1}{2^2} = \dfrac{1}{4}$

Alternatively, we have

$$32^{-2/5} = (2^5)^{-2/5} = 2^{-2} = \frac{1}{2^2} = \frac{1}{4}$$

(b) $(-8)^{4/3} = \left(\sqrt[3]{-8}\right)^4 = (-2)^4 = 16$

Alternatively, we can write

$$(-8)^{4/3} = [(-2)^3]^{4/3} = (-2)^4 = 16$$

■■■

Rational exponents can be used to simplify certain expressions containing radicals. For example, one of the properties of nth roots, which was listed but not discussed in the previous section, is $\sqrt[m]{\sqrt[n]{x}} = \sqrt[mn]{x}$. Using exponents, it is easy to verify this property. We have

$$\sqrt[m]{\sqrt[n]{x}} = (x^{1/n})^{1/m}$$
$$= x^{1/mn} = \sqrt[mn]{x} \qquad \text{as we wished to show.}$$

■■■ **EXAMPLE 4** Consider the expression $\sqrt{x}\,\sqrt[3]{y^2}$, where x and y are positive.

(a) Rewrite the expression using rational exponents.
(b) Rewrite the expression using only one radical sign.

Solution

(a) $\sqrt{x}\,\sqrt[3]{y^2} = x^{1/2}y^{2/3}$

(b) $\sqrt{x}\,\sqrt[3]{y^2} = x^{1/2}y^{2/3}$

$$= x^{3/6}y^{4/6} \qquad \text{rewriting the fractions using a common denominator}$$
$$= \sqrt[6]{x^3}\,\sqrt[6]{y^4} = \sqrt[6]{x^3y^4}$$

■■■

EXAMPLE 5 Rewrite the following expression using rational exponents. (Assume that x, y, and z are positive.)

Solution

$$\sqrt{\frac{\sqrt[3]{x}\sqrt[4]{y^3}}{\sqrt[5]{z^4}}} = \left(\frac{\sqrt[3]{x}\sqrt[4]{y^3}}{\sqrt[5]{z^4}}\right)^{1/2} = \left(\frac{x^{1/3}y^{3/4}}{z^{4/5}}\right)^{1/2}$$

$$= \frac{x^{1/6}y^{3/8}}{z^{2/5}} = x^{1/6}y^{3/8}z^{-2/5}$$

EXERCISE SET A.6

A

Exercises 1–14 are warmup exercises to help you become familiar with the definition $b^{m/n} = \left(\sqrt[n]{b}\right)^m = \sqrt[n]{b^m}$. In Exercises 1–8, write the expression in the two equivalent forms $\left(\sqrt[n]{b}\right)^m$ and $\sqrt[n]{b^m}$. In Exercises 9–14, write the expression in the form $b^{m/n}$.

1. $a^{3/5}$
2. $x^{3/7}$
3. $5^{2/3}$
4. $10^{4/5}$
5. $(x^2 + 1)^{3/4}$
6. $(a + b)^{7/10}$
7. $2^{xy/3}$
8. $2^{(a+1)/b}$
9. $\sqrt[3]{p^2}$
10. $\sqrt[5]{R^3}$
11. $\sqrt[7]{(1+u)^4}$
12. $\sqrt[6]{(1+u)^5}$
13. $\sqrt[p]{(a^2 + b^2)^3}$
14. $\sqrt[3p]{(a^2 + b^2)^{2p}}$

In Exercises 15–50, evaluate or simplify each expression. Express the answers using positive exponents. If an expression is undefined (i.e., does not represent a real number), say so. (Assume that all letters represent positive numbers.)

15. $16^{1/2}$
16. $100^{1/2}$
17. $\left(\frac{1}{36}\right)^{1/2}$
18. $0.09^{1/2}$
19. $(-16)^{1/2}$
20. $(-1)^{1/2}$
21. $625^{1/4}$
22. $\left(\frac{1}{81}\right)^{1/4}$
23. $8^{1/3}$
24. $0.001^{1/3}$
25. $8^{2/3}$
26. $64^{2/3}$
27. $(-32)^{1/5}$
28. $\left(-\frac{1}{125}\right)^{1/3}$
29. $(-1000)^{1/3}$
30. $(-243)^{1/5}$
31. $49^{-1/2}$
32. $121^{-1/2}$
33. $(-49)^{-1/2}$
34. $(-64)^{-1/3}$
35. $36^{-3/2}$
36. $(-0.001)^{-2/3}$
37. $125^{2/3}$
38. $125^{-2/3}$
39. $(-1)^{3/5}$
40. $27^{4/3} + 27^{-4/3} + 27^0$
41. $32^{4/5} - 32^{-4/5}$
42. $64^{1/2} + 64^{-1/2} - 64^{4/3}$
43. $\left(\frac{9}{16}\right)^{-5/2} - \left(\frac{1000}{27}\right)^{4/3}$
44. $(256^{-3/4})^{4/3}$
45. $(2a^{1/3})(3a^{1/4})$
46. $\sqrt[5]{3a^4}$
47. $\sqrt[4]{\frac{64a^{2/3}}{a^{1/3}}}$

48. $(x^2 + 1)^{2/3} (x^2 + 1)^{4/3}$
49. $\frac{(x^2 + 1)^{3/4}}{(x^2 + 1)^{-1/4}}$
50. $\frac{(2x^2 + 1)^{-6/5}(2x^2 + 1)^{6/5} (x^2 + 1)^{-1/5}}{(x^2 + 1)^{9/5}}$

For Exercises 51–58, follow Example 4 in the text to rewrite each expression in two ways: (a) using rational exponents; (b) using only one radical sign. (Assume x, y, and z are positive.)

51. $\sqrt{3}\sqrt[3]{6}$
52. $\sqrt{5}\sqrt[3]{7}$
53. $\sqrt[3]{6}\sqrt[4]{2}$
54. $\sqrt[3]{2}\sqrt[5]{2}$
55. $\sqrt[3]{x^2}\sqrt[5]{y^4}$
56. $\sqrt{x}\sqrt[3]{y}\sqrt[4]{z}$
57. $\sqrt[4]{x^a}\sqrt[3]{x^b}\sqrt{x^{a/6}}$
58. $\sqrt[3]{27\sqrt{64x}}$

In Exercises 59–66, rewrite each expression using rational exponents rather than radicals. (Assume that x, y, and z are positive.)

59. $\sqrt[3]{(x+1)^2}$
60. $\frac{1}{\sqrt{x}} + \sqrt{x}$
61. $\left(\sqrt[5]{x+y}\right)^2$
62. $\sqrt{\sqrt{x}}$
63. $\sqrt[3]{\sqrt{x}} + \sqrt{\sqrt[3]{x}}$
64. $\sqrt[3]{\sqrt{2}}$
65. $\sqrt{\sqrt[3]{x}\sqrt[4]{y}}$
66. $\sqrt[5]{\frac{\sqrt{x}\sqrt[3]{y}}{\sqrt[4]{z^2}}}$

In Exercises 67–72, refer to the following table. The left-hand column of the table lists six errors to avoid in working with expressions containing fractional exponents. In each case, give a numerical example showing that the expressions on the two sides of the equation are not equal. Use a calculator as necessary.

Errors to Avoid	Numerical example showing that the formula is not, in general, valid
67. $(a + b)^{1/2} \not\Xi a^{1/2} + b^{1/2}$	
68. $(x^2 + y^2)^{1/2} \not\Xi x + y$	
69. $(u + v)^{1/3} \not\Xi u^{1/3} + v^{1/3}$	
70. $(p^3 + q^3)^{1/3} \not\Xi p + q$	
71. $x^{1/m} \not\Xi \dfrac{1}{x^m}$	
72. $x^{-1/2} \not\Xi \dfrac{1}{x^2}$	

B

73. *Without* using a calculator, decide which number is larger in each case.
    (a) $2^{2/3}$ or $2^{3/2}$ (b) $5^{1/2}$ or 5^{-2}
    (c) $2^{1/2}$ or $2^{1/3}$ (d) $\left(\frac{1}{2}\right)^{1/2}$ or $\left(\frac{1}{2}\right)^{1/3}$
    (e) $10^{1/10}$ or $\left(\frac{1}{10}\right)^{10}$

74. *Without* using a calculator, decide which number is closer to zero: $(0.5)^{1/3}$ or $(0.5)^{1/4}$. Then use a calculator to check your answer.

75. Show that $\dfrac{a - b}{a + b} \sqrt{\dfrac{a + b}{a - b}} = \left(\dfrac{a - b}{a + b}\right)^{1/2}$. $(a > b > 0)$.

TABLES

TABLE I

Squares and Square Roots

NO.	SQ.	SQ. RT.	NO.	SQ.	SQ. RT.
1	1	1.000	51	2,601	7.141
2	4	1.414	52	2,704	7.211
3	9	1.732	53	2,809	7.280
4	16	2.000	54	2,916	7.348
5	25	2.236	55	3,025	7.416
6	36	2.449	56	3,136	7.483
7	49	2.646	57	3,249	7.550
8	64	2.828	58	3,364	7.616
9	81	3.000	59	3,481	7.681
10	100	3.162	60	3,600	7.746
11	121	3.317	61	3,721	7.810
12	144	3.464	62	3,844	7.874
13	169	3.606	63	3,969	7.937
14	196	3.742	64	4,096	8.000
15	225	3.873	65	4,225	8.062
16	256	4.000	66	4,356	8.124
17	289	4.123	67	4,489	8.185
18	324	4.243	68	4,624	8.246
19	361	4.359	69	4,761	8.307
20	400	4.472	70	4,900	8.367
21	441	4.583	71	5,041	8.426
22	484	4.690	72	5,184	8.485
23	529	4.796	73	5,329	8.544
24	576	4.899	74	5,476	8.602
25	625	5.000	75	5,625	8.660
26	676	5.099	76	5,776	8.718
27	729	5.196	77	5,929	8.775
28	784	5.292	78	6,084	8.832
29	841	5.385	79	6,241	8.888
30	900	5.477	80	6,400	8.944
31	961	5.568	81	6,561	9.000
32	1,024	5.657	82	6,724	9.055
33	1,089	5.745	83	6,889	9.110
34	1,156	5.831	84	7,056	9.165
35	1,225	5.916	85	7,225	9.220
36	1,296	6.000	86	7,396	9.274
37	1,369	6.083	87	7,569	9.327
38	1,444	6.164	88	7,744	9.381
39	1,521	6.245	89	7,921	9.434
40	1,600	6.325	90	8,100	9.487
41	1,681	6.403	91	8,281	9.539
42	1,764	6.481	92	8,464	9.592
43	1,849	6.557	93	8,649	9.644
44	1,936	6.633	94	8,836	9.695
45	2,025	6.708	95	9,025	9.747
46	2,116	6.782	96	9,216	9.798
47	2,209	6.856	97	9,409	9.849
48	2,304	6.928	98	9,604	9.899
49	2,401	7.000	99	9,801	9.950
50	2,500	7.071	100	10,000	10.000

TABLE 2

Exponential Functions

x	e^x	e^{-x}	x	e^x	e^{-x}
0.00	1.0000	1.0000	1.5	4.4817	0.2231
0.01	1.0101	0.9901	1.6	4.9530	0.2019
0.02	1.0202	0.9802	1.7	5.4739	0.1827
0.03	1.0305	0.9704	1.8	6.0496	0.1653
0.04	1.0408	0.9608	1.9	6.6859	0.1496
0.05	1.0513	0.9512	2.0	7.3891	0.1353
0.06	1.0618	0.9418	2.1	8.1662	0.1225
0.07	1.0725	0.9324	2.2	9.0250	0.1108
0.08	1.0833	0.9231	2.3	9.9742	0.1003
0.09	1.0942	0.9139	2.4	11.023	0.0907
0.10	1.1052	0.9048	2.5	12.182	0.0821
0.11	1.1163	0.8958	2.6	13.464	0.0743
0.12	1.1275	0.8869	2.7	14.880	0.0672
0.13	1.1388	0.8781	2.8	16.445	0.0608
0.14	1.1503	0.8694	2.9	18.174	0.0550
0.15	1.1618	0.8607	3.0	20.086	0.0498
0.16	1.1735	0.8521	3.1	22.198	0.0450
0.17	1.1853	0.8437	3.2	24.533	0.0408
0.18	1.1972	0.8353	3.3	27.113	0.0369
0.19	1.2092	0.8270	3.4	29.964	0.0334
0.20	1.2214	0.8187	3.5	33.115	0.0302
0.21	1.2337	0.8106	3.6	36.598	0.0273
0.22	1.2461	0.8025	3.7	40.447	0.0247
0.23	1.2586	0.7945	3.8	44.701	0.0224
0.24	1.2712	0.7866	3.9	49.402	0.0202
0.25	1.2840	0.7788	4.0	54.598	0.0183
0.30	1.3499	0.7408	4.1	60.340	0.0166
0.35	1.4191	0.7047	4.2	66.686	0.0150
0.40	1.4918	0.6703	4.3	73.700	0.0136
0.45	1.5683	0.6376	4.4	81.451	0.0123
0.50	1.6487	0.6065	4.5	90.017	0.0111
0.55	1.7333	0.5769	4.6	99.484	0.0101
0.60	1.8221	0.5488	4.7	109.95	0.0091
0.65	1.9155	0.5220	4.8	121.51	0.0082
0.70	2.0138	0.4966	4.9	134.29	0.0074
0.75	2.1170	0.4724	5.0	148.41	0.0067
0.80	2.2255	0.4493	5.5	244.69	0.0041
0.85	2.3396	0.4274	6.0	403.43	0.0025
0.90	2.4596	0.4066	6.5	665.14	0.0015
0.95	2.5857	0.3867	7.0	1096.6	0.0009
1.0	2.7183	0.3679	7.5	1808.0	0.0006
1.1	3.0042	0.3329	8.0	2981.0	0.0003
1.2	3.3201	0.3012	8.5	4914.8	0.0002
1.3	3.6693	0.2725	9.0	8103.1	0.0001
1.4	4.0552	0.2466	10.0	22026	0.00005

TABLE 3

Logarithms to the Base Ten

x	0	1	2	3	4	5	6	7	8	9
1.0	.0000	.0043	.0086	.0128	.0170	.0212	.0253	.0294	.0334	.0374
1.1	.0414	.0453	.0492	.0531	.0569	.0607	.0645	.0682	.0719	.0755
1.2	.0792	.0828	.0864	.0899	.0934	.0969	.1004	.1038	.1072	.1106
1.3	.1139	.1173	.1206	.1239	.1271	.1303	.1335	.1367	.1399	.1430
1.4	.1461	.1492	.1523	.1553	.1584	.1614	.1644	.1673	.1703	.1732
1.5	.1761	.1790	.1818	.1847	.1875	.1093	.1931	.1959	.1987	.2014
1.6	.2041	.2068	.2095	.2122	.2148	.2175	.2201	.2227	.2253	.2279
1.7	.2304	.2330	.2355	.2380	.2405	.2430	.2455	.2480	.2504	.2529
1.8	.2553	.2577	.2601	.2625	.2648	.2672	.2695	.2718	.2742	.2765
1.9	.2788	.2810	.2833	.2856	.2878	.2900	.2923	.2945	.2967	.2989
2.0	.3010	.3032	.3054	.3075	.3096	.3118	.3139	.3160	.3181	.3201
2.1	.3222	.3243	.3263	.3284	.3304	.3324	.3345	.3365	.3385	.3404
2.2	.3424	.3444	.3464	.3483	.3502	.3522	.3541	.3560	.3579	.3598
2.3	.3617	.3636	.3655	.3674	.3692	.3711	.3729	.3747	.3766	.3784
2.4	.3802	.3820	.3838	.3856	.3874	.3892	.3909	.3927	.3945	.3962
2.5	.3979	.3997	.4014	.4031	.4048	.4065	.4082	.4099	.4116	.4133
2.6	.4150	.4166	.4183	.4200	.4216	.4232	.4249	.4265	.4281	.4298
2.7	.4314	.4330	.4346	.4362	.4378	.4393	.4409	.4425	.4440	.4456
2.8	.4472	.4487	.4502	.4518	.4533	.4548	.4564	.4579	.4594	.4609
2.9	.4624	.4639	.4654	.4669	.4683	.4698	.4713	.4728	.4742	.4757
3.0	.4771	.4786	.4800	.4814	.4829	.4843	.4857	.4871	.4886	.4900
3.1	.4914	.4928	.4942	.4955	.4969	.4983	.4997	.5011	.5024	.5038
3.2	.5051	.5065	.5079	.5092	.5105	.5119	.5132	.5145	.5159	.5172
3.3	.5185	.5198	.5211	.5224	.5237	.5250	.5263	.5276	.5289	.5302
3.4	.5315	.5328	.5340	.5353	.5366	.5378	.5391	.5403	.5416	.5428
3.5	.5441	.5453	.5465	.5478	.5490	.5502	.5514	.5527	.5539	.5551
3.6	.5563	.5575	.5587	.5599	.5611	.5623	.5635	.5647	.5658	.5670
3.7	.5682	.5694	.5705	.5717	.5729	.5740	.5752	.5763	.5775	.5786
3.8	.5798	.5809	.5821	.5832	.5843	.5855	.5866	.5877	.5888	.5899
3.9	.5911	.5922	.5933	.5944	.5955	.5966	.5977	.5988	.5999	.6010
4.0	.6021	.6031	.6042	.6053	.6064	.6075	.6085	.6096	.6107	.6117
4.1	.6128	.6138	.6149	.6160	.6170	.6180	.6191	.6201	.6212	.6222
4.2	.6232	.6243	.6253	.6263	.6274	.6284	.6294	.6304	.6314	.6325
4.3	.6335	.6345	.6355	.6365	.6375	.6385	.6395	.6405	.6415	.6425
4.4	.6435	.6444	.6454	.6464	.6474	.6484	.6493	.6503	.6513	.6522
4.5	.6532	.6542	.6551	.6561	.6571	.6580	.6590	.6599	.6609	.6618
4.6	.6628	.6637	.6646	.6656	.6665	.6675	.6684	.6693	.6702	.6712
4.7	.6721	.6730	.6739	.6749	.6758	.6767	.6776	.6785	.6794	.6803
4.8	.6812	.6821	.6830	.6839	.6848	.6857	.6866	.6875	.6884	.6893
4.9	.6902	.6911	.6920	.6928	.6937	.6946	.6955	.6964	.6972	.6981
5.0	.6990	.6998	.7007	.7016	.7024	.7033	.7042	.7050	.7059	.7067
5.1	.7076	.7084	.7093	.7101	.7110	.7118	.7126	.7135	.7143	.7152
5.2	.7160	.7168	.7177	.7185	.7193	.7202	.7210	.7218	.7226	.7235
5.3	.7243	.7251	.7259	.7267	.7275	.7284	.7292	.7300	.7308	7316
5 4	7324	7332	.7340	.7348	.7356	.7364	.7372	.7380	.7388	.7396

x	0	1	2	3	4	5	6	7	8	9

TABLE 3

Logarithms to the Base Ten *(continued)*

x	0	1	2	3	4	5	6	7	8	9
5.5	.7404	.7412	.7419	.7427	.7435	.7443	.7451	.7459	.7466	.7474
5.6	.7482	.7490	.7497	.7505	.7513	.7520	.7528	.7536	.7543	.7551
5.7	.7559	.7566	.7574	.7582	.7589	.7597	.7604	.7612	.7619	.7627
5.8	.7634	.7642	.7649	.7657	.7664	.7672	.7679	.7686	.7694	.7701
5.9	.7709	.7716	.7723	.7731	.7738	.7745	.7752	.7760	.7767	.7774
6.0	.7782	.7789	.7796	.7803	.7810	.7818	.7825	.7832	.7839	.7846
6.1	.7853	.7860	.7868	.7875	.7882	.7889	.7896	.7903	.7910	.7917
6.2	.7924	.7931	.7938	.7945	.7952	.7959	.7966	.7973	.7980	.7987
6.3	.7993	.8000	.8007	.8014	.8021	.8028	.8035	.8041	.8048	.8055
6.4	.8062	.8069	.8075	.8082	.8089	.8096	.8102	.8109	.8116	.8122
6.5	.8129	.8136	.8142	.8149	.8156	.8162	.8169	.8176	.8182	.8189
6.6	.8195	.8202	.8209	.8215	.8222	.8228	.8235	.8241	.8248	.8254
6.7	.8261	.8267	.8274	.8280	.8287	.8293	.8299	.8306	.8312	.8319
6.8	.8325	.8331	.8338	.8344	.8351	.8357	.8363	.8370	.8376	.8382
6.9	.8388	.8395	.8401	.8407	.8414	.8420	.8426	.8432	.8439	.8445
7.0	.8451	.8457	.8463	.8470	.8476	.8482	.8488	.8494	.8500	.8506
7.1	.8513	.8519	.8525	.8531	.8537	.8543	.8549	.8555	.8561	.8567
7.2	.8573	.8579	.8585	.8591	.8597	.8603	.8609	.8615	.8621	.8627
7.3	.8633	.8639	.8645	.8651	.8657	.8663	.8669	.8675	.8681	.8686
7.4	.8692	.8698	.8704	.8710	.8716	.8722	.8727	.8733	.8739	.8745
7.5	.8751	.8756	.8762	.8768	.8774	.8779	.8785	.8791	.8797	.8802
7.6	.8808	.8814	.8820	.8825	.8831	.8837	.8842	.8848	.8854	.8859
7.7	.8865	.8871	.8876	.8882	.8887	.8893	.8899	.8904	.8910	.8915
7.8	.8921	.8927	.8932	.8938	.8943	.8949	.8954	.8960	.8965	.8971
7.9	.8976	.8982	.8987	.8993	.8998	.9004	.9009	.9015	.9020	.9025
8.0	.9031	.9036	.9042	.9047	.9053	.9058	.9063	.9069	.9074	.9079
8.1	.9085	.9090	.9096	.9101	.9106	.9112	.9117	.9122	.9128	.9133
8.2	.9138	.9143	.9149	.9154	.9159	.9165	.9170	.9175	.9180	.9186
8.3	.9191	.9196	.9201	.9206	.9212	.9217	.9222	.9227	.9232	.9238
8.4	.9243	.9248	.9253	.9258	.9263	.9269	.9274	.9279	.9284	.9289
8.5	.9294	.9299	.9304	.9309	.9315	.9320	.9325	.9330	.9335	.9340
8.6	.9345	.9350	.9355	.9360	.9365	.9370	.9375	.9380	.9385	.9390
8.7	.9395	.9400	.9405	.9410	.9415	.9420	.9425	.9430	.9435	.9440
8.8	.9445	.9450	.9455	.9460	.9465	.9469	.9474	.9479	.9484	.9489
8.9	.9494	.9499	.9504	.9509	.9513	.9518	.9523	.9528	.9533	.9538
9.0	.9542	.9547	.9552	.9557	.9562	.9566	.9571	.9576	.9581	.9586
9.1	.9590	.9595	.9600	.9605	.9609	.9614	.9619	.9624	.9628	.9633
9.2	.9638	.9643	.9647	.9652	.9657	.9661	.9666	.9671	.9675	.9680
9.3	.9685	.9689	.9694	.9699	.9703	.9708	.9713	.9717	.9722	.9727
9.4	.9731	.9736	.9741	.9745	.9750	.9754	.9759	.9763	.9768	.9773
9.5	.9777	.9782	.9786	.9791	.9795	.9800	.9805	.9809	.9814	.9818
9.6	.9823	.9827	.9832	.9836	.9841	.9845	.9850	.9854	.9859	.9863
9.7	.9868	.9872	.9877	.9881	.9886	.9890	.9894	.9899	.9903	.9908
9.8	.9912	.9917	.9921	.9926	.9930	.9934	.9939	.9943	.9948	.9952
9.9	.9956	.9961	.9965	.9969	.9974	.9978	.9983	.9987	.9991	.9996
x	0	1	2	3	4	5	6	7	8	9

TABLE 4

Natural Logarithms

n	$\ln n$	n	$\ln n$	n	$\ln n$
		4.5	1.5041	9.0	2.1972
0.1	−2.3026	4.6	1.5261	9.1	2.2083
0.2	−1.6094	4.7	1.5476	9.2	2.2192
0.3	−1.2040	4.8	1.5686	9.3	2.2300
0.4	−0.9163	4.9	1.5892	9.4	2.2407
0.5	−0.6931	5.0	1.6094	9.5	2.2513
0.6	−0.5108	5.1	1.6292	9.6	2.2618
0.7	−0.3567	5.2	1.6487	9.7	2.2721
0.8	−0.2231	5.3	1.6677	9.8	2.2824
0.9	−0.1054	5.4	1.6864	9.9	2.2925
1.0	0.0000	5.5	1.7047	10	2.3026
1.1	0.0953	5.6	1.7228	11	2.3979
1.2	0.1823	5.7	1.7405	12	2.4849
1.3	0.2624	5.8	1.7579	13	2.5649
1.4	0.3365	5.9	1.7750	14	2.6391
1.5	0.4055	6.0	1.7918	15	2.7081
1.6	0.4700	6.1	1.8083	16	2.7726
1.7	0.5306	6.2	1.8245	17	2.8332
1.8	0.5878	6.3	1.8405	18	2.8904
1.9	0.6419	6.4	1.8563	19	2.9444
2.0	0.6931	6.5	1.8718	20	2.9957
2.1	0.7419	6.6	1.8871	25	3.2189
2.2	0.7885	6.7	1.9021	30	3.4012
2.3	0.8329	6.8	1.9169	35	3.5553
2.4	0.8755	6.9	1.9315	40	3.6889
2.5	0.9163	7.0	1.9459	45	3.8067
2.6	0.9555	7.1	1.9601	50	3.9120
2.7	0.9933	7.2	1.9741	55	4.0073
2.8	1.0296	7.3	1.9879	60	4.0943
2.9	1.0647	7.4	2.0015	65	4.1744
3.0	1.0986	7.5	2.0149	70	4.2485
3.1	1.1314	7.6	2.0281	75	4.3175
3.2	1.1632	7.7	2.0412	80	4.3820
3.3	1.1939	7.8	2.0541	85	4.4427
3.4	1.2238	7.9	2.0669	90	4.4998
3.5	1.2528	8.0	2.0794	95	4.5539
3.6	1.2809	8.1	2.0919	100	4.6052
3.7	1.3083	8.2	2.1041		
3.8	1.3350	8.3	2.1163		
3.9	1.3610	8.4	2.1282		
4.0	1.3863	8.5	2.1401		
4.1	1.4110	8.6	2.1518		
4.2	1.4351	8.7	2.1633		
4.3	1.4586	8.8	2.1748		
4.4	1.4816	8.9	2.1861		

TABLE 5

Values of Trigonometric Functions

Angle θ									
Degrees	Radians	sin θ	csc θ	tan θ	cot θ	sec θ	cos θ		
0°00′	.0000	.0000	No value	.0000	No value	1.000	1.0000	1.5708	90°00′
10	029	029	343.8	029	343.8	000	000	679	50
20	058	058	171.9	058	171.9	000	000	650	40
30	087	087	114.6	087	114.6	000	1.0000	621	30
40	116	116	85.95	116	85.94	000	.9999	592	20
50	145	145	68.76	145	68.75	000	999	563	10
1°00′	.0175	.0175	57.30	.0175	57.29	1.000	.9998	1.5533	89°00′
10	204	204	49.11	204	49.10	000	998	504	50
20	233	233	42.98	233	42.96	000	997	475	40
30	262	262	38.20	262	38.19	000	997	446	30
40	291	291	34.38	291	34.37	000	996	417	20
50	320	320	31.26	320	31.24	001	995	388	10
2°00′	.0349	.0349	28.65	.0349	28.64	1.001	.9994	1.5359	88°00′
10	378	378	26.45	378	26.43	001	993	330	50
20	407	407	24.56	407	24.54	001	992	301	40
30	436	436	22.93	437	22.90	001	990	272	30
40	465	465	21.49	466	21.47	001	989	243	20
50	495	494	20.23	495	20.21	001	988	213	10
3°00′	.0524	.0523	19.11	.0524	19.08	1.001	.9986	1.5184	87°00′
10	553	552	18.10	553	18.07	002	985	155	50
20	582	581	17.20	582	17.17	002	983	126	40
30	611	610	16.38	612	16.35	002	981	097	30
40	640	640	15.64	641	15.60	002	980	068	20
50	669	669	14.96	670	14.92	002	978	039	10
4°00′	.0698	.0698	14.34	.0699	14.30	1.002	.9976	1.5010	86°00′
10	727	727	13.76	729	13.73	003	974	981	50
20	756	756	13.23	758	13.20	003	971	952	40
30	785	785	12.75	787	12.71	003	969	923	30
40	814	814	12.29	816	12.25	003	967	893	20
50	844	843	11.87	846	11.83	004	964	864	10
5°00′	.0873	.0872	11.47	.0875	11.43	1.004	.9962	1.4835	85°00′
10	902	901	11.10	904	11.06	004	959	806	50
20	931	929	10.76	934	10.71	004	957	777	40
30	960	958	10.43	963	10.39	005	954	748	30
40	.0989	.0987	10.13	.0992	10.08	005	951	719	20
50	.1018	.1016	9.839	.1022	9.788	005	948	690	10
6°00′	.1047	.1045	9.567	.1051	9.514	1.006	.9945	1.4661	84°00′
10	076	074	9.309	080	9.255	006	942	632	50
20	105	103	9.065	110	9.010	006	939	603	40
30	134	132	8.834	139	8.777	006	936	573	30
40	164	161	8.614	169	8.556	007	932	544	20
50	193	190	8.405	198	8.345	007	929	515	10
7°00′	.1222	.1219	8.206	.1228	8.144	1.008	.9925	1.4486	83°00′
	cos θ	sec θ	cot θ	tan θ	csc θ	sin θ	Radians	Degrees	
								Angle θ	

TABLE 5

Values of Trigonometric Functions (continued)

Angle θ									
Degrees	Radians	$\sin\theta$	$\csc\theta$	$\tan\theta$	$\cot\theta$	$\sec\theta$	$\cos\theta$		
7°00′	.1222	.1219	8.206	.1228	8.144	1.008	.9925	1.4486	83°00′
10	251	248	8.016	257	7.953	008	922	457	50
20	280	276	7.834	287	7.770	008	918	428	40
30	309	305	7.661	317	7.596	009	914	399	30
40	338	334	7.496	346	7.429	009	911	370	20
50	367	363	7.337	376	7.269	009	907	341	10
8°00′	.1396	.1392	7.185	.1405	7.115	1.010	.9903	1.4312	82°00′
10	425	421	7.040	435	6.968	010	899	283	50
20	454	449	6.900	465	827	011	894	254	40
30	484	478	765	495	691	011	890	224	30
40	513	507	636	524	561	012	886	195	20
50	542	536	512	554	435	012	881	166	10
9°00′	.1571	.1564	6.392	.1584	6.314	1.012	.9877	1.4137	81°00′
10	600	593	277	614	197	013	872	108	50
20	629	622	166	644	6.084	013	868	079	40
30	658	650	6.059	673	5.976	014	863	050	30
40	687	679	5.955	703	871	014	858	1.4021	20
50	716	708	855	733	769	015	853	1.3992	10
10°00′	.1745	.1736	5.759	.1763	5.671	1.015	.9848	1.3963	80°00′
10	774	765	665	793	576	016	843	934	50
20	804	794	575	823	485	016	838	904	40
30	833	822	487	853	396	017	833	875	30
40	862	851	403	883	309	018	827	846	20
50	891	880	320	914	226	018	822	817	10
11°00′	.1920	.1908	5.241	.1944	5.145	1.019	.9816	1.3788	79°00′
10	949	937	164	.1974	5.066	019	811	759	50
20	.1978	965	089	.2004	4.989	020	805	730	40
30	.2007	.1994	5.016	035	915	020	799	701	30
40	036	.2022	4.945	065	843	021	793	672	20
50	065	051	876	095	773	022	787	643	10
12°00′	.2094	.2079	4.810	.2126	4.705	1.022	.9781	1.3614	78°00′
10	123	108	745	156	638	023	775	584	50
20	153	136	682	186	574	024	769	555	40
30	182	164	620	217	511	024	763	526	30
40	211	193	560	247	449	025	757	497	20
50	240	221	502	278	390	026	750	468	10
13°00′	.2269	.2250	4.445	.2309	4.331	1.026	.9744	1.3439	77°00′
10	298	278	390	339	275	027	737	410	50
20	327	306	336	370	219	028	730	381	40
30	356	334	284	401	165	028	724	352	30
40	385	363	232	432	113	029	717	323	20
50	414	391	182	462	061	030	710	294	10
14°00′	.2443	.2419	4.134	.2493	4.011	1.031	.9703	1.3265	76°00′
		$\cos\theta$	$\sec\theta$	$\cot\theta$	$\tan\theta$	$\csc\theta$	$\sin\theta$	Radians	Degrees

| Angle θ |

TABLE 5

Values of Trigonometric Functions (*continued*)

Angle θ									
Degrees	Radians	sin θ	csc θ	tan θ	cot θ	sec θ	cos θ		
14°00′	.2443	.2419	4.134	.2493	4.011	1.031	.9703	1.3265	76°00′
10	473	447	086	524	3.962	031	696	235	50
20	502	476	4.039	555	914	032	689	206	40
30	531	504	3.994	586	867	033	681	177	30
40	560	532	950	617	821	034	674	148	20
50	589	560	906	648	776	034	667	119	10
15°00′	.2618	.2588	3.864	.2679	3.732	1.035	.9659	1.3090	75°00′
10	647	616	822	711	689	036	652	061	50
20	676	644	782	742	647	037	644	032	40
30	705	672	742	773	606	038	636	1.3003	30
40	734	700	703	805	566	039	628	1.2974	20
50	763	728	665	836	526	039	621	945	10
16°00′	.2793	.2756	3.628	.2867	3.487	1.040	.9613	1.2915	74°00′
10	822	784	592	899	450	041	605	886	50
20	851	812	556	931	412	042	596	857	40
30	880	840	521	962	376	043	588	828	30
40	909	868	487	.2944	340	044	580	799	20
50	938	896	453	.3026	305	045	572	770	10
17°00′	.2967	.2924	3.420	.3057	3.271	1.046	.9563	1.2741	73°00′
10	.2996	952	388	089	237	047	555	712	50
20	.3025	.2979	357	121	204	048	546	683	40
30	054	.3007	326	153	172	048	537	654	30
40	083	035	295	185	140	049	528	625	20
50	113	062	265	217	108	050	520	595	10
18°00′	.3142	.3090	3.236	.3249	3.078	1.051	.9511	1.2566	72°00′
10	171	118	207	281	047	052	502	537	50
20	200	145	179	314	3.018	053	492	508	40
30	229	173	152	346	2.989	054	483	479	30
40	258	201	124	378	960	056	474	450	20
50	287	228	098	411	932	057	465	421	10
19°00′	.3316	.3256	3.072	.3443	2.904	1.058	.9455	1.2392	71°00′
10	345	283	046	476	877	059	446	363	50
20	374	311	3.021	508	850	060	436	334	40
30	403	338	2.996	541	824	061	426	305	30
40	432	365	971	574	798	062	417	275	20
50	462	393	947	607	773	063	407	246	10
20°00′	.3491	.3420	2.924	.3640	2.747	1.064	.9397	1.2217	70°00′
10	520	448	901	673	723	065	387	188	50
20	549	475	878	706	699	066	377	159	40
30	578	502	855	739	675	068	367	130	30
40	607	529	833	772	651	069	356	101	20
50	636	557	812	805	628	070	346	072	10
21°00′	.3665	.3584	2.790	.3839	2.605	1.071	.9336	1.2043	69°00′
	cos θ	sec θ	cot θ	tan θ	csc θ	sin θ		Radians	Degrees

| Angle θ | | | | | | | | | |

TABLE 5

Values of Trigonometric Functions (continued)

Degrees	Radians	sin θ	csc θ	tan θ	cot θ	sec θ	cos θ		
21°00′	.3665	.3584	2.790	.3839	2.605	1.071	.9336	1.2043	69°00′
10	694	611	769	872	583	072	325	1.2014	50
20	723	638	749	906	560	074	315	1.1985	40
30	752	665	729	939	539	075	304	956	30
40	782	692	709	.3973	517	076	293	926	20
50	811	719	689	.4006	496	077	283	897	10
22°00′	.3840	.3746	2.669	.4040	2.475	1.079	.9272	1.1868	68°00′
10	869	773	650	074	455	080	261	839	50
20	898	800	632	108	434	081	250	810	40
30	927	827	613	142	414	082	239	781	30
40	956	854	595	176	394	084	228	752	20
50	985	881	577	210	375	085	216	723	10
23°00′	.4014	.3907	2.559	.4245	2.356	1.086	.9205	1.1694	67°00′
10	043	934	542	279	337	088	194	665	50
20	072	961	525	314	318	089	182	636	40
30	102	.3987	508	348	300	090	171	606	30
40	131	.4014	491	383	282	092	159	577	20
50	160	041	475	417	264	093	147	548	10
24°00′	.4189	.4067	2.459	.4452	2.246	1.095	.9135	1.1519	66°00′
10	218	094	443	487	229	096	124	490	50
20	247	120	427	522	211	097	112	461	40
30	276	147	411	557	194	099	100	432	30
40	305	173	396	592	177	100	088	403	20
50	334	200	381	628	161	102	075	374	10
25°00′	.4363	.4226	2.366	.4663	2.145	1.103	.9063	1.1345	65°00′
10	392	253	352	699	128	105	051	316	50
20	422	279	337	734	112	106	038	286	40
30	451	305	323	770	097	108	026	257	30
40	480	331	309	806	081	109	013	228	20
50	509	358	295	841	066	111	.9001	199	10
26°00′	.4538	.4384	2.281	.4877	2.050	1.113	.8988	1.1170	64°00′
10	567	410	268	913	035	114	975	141	50
20	596	436	254	950	020	116	962	112	40
30	625	462	241	.4986	2.006	117	949	083	30
40	654	488	228	.5022	1.991	119	936	054	20
50	683	514	215	059	977	121	923	1.1025	10
27°00′	.4712	.4540	2.203	.5095	1.963	1.122	.8910	1.0996	63°00′
		cos θ	sec θ	cot θ	tan θ	csc θ	sin θ	Radians	Degrees

Angle θ

TABLE 5

Values of Trigonometric Functions (*continued*)

Angle θ									
Degrees	Radians	sin θ	csc θ	tan θ	cot θ	sec θ	cos θ		
27°00′	.4712	.4540	2.203	.5095	1.963	1.122	.8910	1.0996	63°00′
10	741	566	190	132	949	124	897	966	50
20	771	592	178	169	935	126	884	937	40
30	800	617	166	206	921	127	870	908	30
40	829	643	154	243	907	129	857	879	20
50	858	669	142	280	894	131	843	850	10
28°00′	.4887	.4695	2.130	.5317	1.881	1.133	.8829	1.0821	62°00′
10	916	720	118	354	868	134	816	792	50
20	945	746	107	392	855	136	802	763	40
30	.4974	772	096	430	842	138	788	734	30
40	.5003	797	085	467	829	140	774	705	20
50	032	823	074	505	816	142	760	676	10
29°00′	.5061	.4848	2.063	.5543	1.804	1.143	.8746	1.0647	61°00′
10	091	874	052	581	792	145	732	617	50
20	120	899	041	619	780	147	718	588	40
30	149	924	031	658	767	149	704	559	30
40	178	950	020	696	756	151	689	530	20
50	207	.4975	010	735	744	153	675	501	10
30°00′	.5236	.5000	2.000	.5774	1.732	1.155	.8660	1.0472	60°00
10	265	025	1.990	812	720	157	646	443	50
20	294	050	980	851	709	159	631	414	40
30	323	075	970	890	698	161	616	385	30
40	352	100	961	930	686	163	601	356	20
50	381	125	951	.5969	675	165	587	327	10
31°00′	.5411	.5150	1.942	.6009	1.664	1.167	.8572	1.0297	59°00′
10	440	175	932	048	653	169	557	268	50
20	469	200	923	088	643	171	542	239	40
30	498	225	914	128	632	173	526	210	30
40	527	250	905	168	621	175	511	181	20
50	556	275	896	208	611	177	496	152	10
32°00′	.5585	.5299	1.887	.6249	1.600	1.179	.8480	1.0123	58°00′
10	614	324	878	289	590	181	465	094	50
20	643	348	870	330	580	184	450	065	40
30	672	373	861	371	570	186	434	036	30
40	701	398	853	412	560	188	418	1.0007	20
50	730	422	844	453	550	190	403	.9977	10
33°00′	.5760	.5446	1.836	.6494	1.540	1.192	.8387	.9948	57°00′
		cos θ	sec θ	cot θ	tan θ	csc θ	sin θ	Radians	Degrees
								Angle θ	

TABLE 5

Values of Trigonometric Functions (continued)

	Angle θ								
Degrees	Radians	sin θ	csc θ	tan θ	cot θ	sec θ	cos θ		
33°00′	.5760	.5446	1.836	.6494	1.540	1.192	.8387	.9948	57°00′
10	789	471	828	536	530	195	371	919	50
20	818	495	820	577	520	197	355	890	40
30	847	519	812	619	511	199	339	861	30
40	876	544	804	661	501	202	323	832	20
50	905	568	796	703	492	204	307	803	10
34°00′	.5934	.5592	1.788	.6745	1.483	1.206	.8290	.9774	56°00′
10	963	616	781	787	473	209	274	745	50
20	.5992	640	773	830	464	211	258	716	40
30	.6021	644	766	873	455	213	241	687	30
40	050	688	758	916	446	216	225	657	20
50	080	712	751	.6959	437	218	208	628	10
35°00′	.6109	.5736	1.743	.7002	1.428	1.221	.8192	.9599	55°00′
10	138	760	736	046	419	223	175	570	50
20	167	783	729	089	411	226	158	541	40
30	196	807	722	133	402	228	141	512	30
40	225	831	715	177	393	231	124	483	20
50	254	854	708	221	385	233	107	454	10
36°00′	.6283	.5878	1.701	.7265	1.376	1.236	.8090	.9425	54°00′
10	312	901	695	310	368	239	073	396	50
20	341	925	688	355	360	241	056	367	40
30	370	948	681	400	351	244	039	338	30
40	400	972	675	445	343	247	021	308	20
50	429	.5995	668	490	335	249	.8004	279	10
37°00′	.6458	.6018	1.662	.7536	1.327	1.252	.7986	.9250	53°00′
10	487	041	655	581	319	255	969	221	50
20	516	065	649	627	311	258	951	192	40
30	545	088	643	673	303	260	934	163	30
40	574	111	636	720	295	263	916	134	20
50	603	134	630	766	288	266	898	105	10
38°00′	.6632	.6157	1.624	.7813	1.280	1.269	.7880	.9076	52°00′
10	661	180	618	860	272	272	862	047	50
20	690	202	612	907	265	275	844	.9018	40
30	720	225	606	.7954	257	278	826	.8988	30
40	749	248	601	.8002	250	281	808	959	20
50	778	271	595	050	242	284	790	930	10
39°00′	.6807	.6293	1.589	.8098	1.235	1.287	.7771	.8901	51°00′
		cos θ	sec θ	cot θ	tan θ	csc θ	sin θ	Radians	Degrees
								Angle θ	

TABLE 5

Values of Trigonometric Functions (*continued*)

	Angle θ								
Degrees	Radians	sin θ	csc θ	tan θ	cot θ	sec θ	cos θ		
39°00′	.6807	.6293	1.589	.8098	1.235	1.287	.7771	.8901	51°00′
10	836	316	583	146	228	290	753	872	50
20	865	338	578	195	220	293	735	843	40
30	894	361	572	243	213	296	716	814	30
40	923	383	567	292	206	299	698	785	20
50	952	406	561	342	199	302	679	756	10
40°00′	.6981	.6428	1.556	.8391	1.192	1.305	.7660	.8727	50°00′
10	.7010	450	550	441	185	309	642	698	50
20	039	472	545	491	178	312	623	668	40
30	069	494	540	541	171	315	604	639	30
40	098	517	535	591	164	318	585	610	20
50	127	539	529	642	157	322	566	581	10
41°00′	.7156	.6561	1.524	.8693	1.150	1.325	.7547	.8552	49°00′
10	185	583	519	744	144	328	528	523	50
20	214	604	514	796	137	332	509	494	40
30	243	626	509	847	130	335	490	465	30
40	272	648	504	899	124	339	470	436	20
50	301	670	499	.8952	117	342	451	407	10
42°00′	.7330	.6691	1.494	.9004	1.111	1.346	.7431	.8378	48°00′
10	359	713	490	057	104	349	412	348	50
20	389	734	485	110	098	353	392	319	40
30	418	756	480	163	091	356	373	290	30
40	447	777	476	217	085	360	353	261	20
50	476	799	471	271	079	364	333	232	10
43°00′	.7505	.6820	1.466	.9325	1.072	1.367	.7314	.8203	47°00′
10	534	841	462	380	066	371	294	174	50
20	563	862	457	435	060	375	274	145	40
30	592	884	453	490	054	379	254	116	30
40	621	905	448	545	048	382	234	087	20
50	650	926	444	601	042	386	214	058	10
44°00′	.7679	.6947	1.440	.9657	1.036	1.390	.7193	.8029	46°00′
10	709	967	435	713	030	394	173	.7999	50
20	738	.6988	431	770	024	398	153	970	40
30	767	.7009	427	827	018	402	133	941	30
40	796	030	423	884	012	406	112	912	20
50	825	050	418	.9942	006	410	092	883	10
45°00′	.7854	.7071	1.414	1.000	1.000	1.414	.7071	.7854	45°00′
		cos θ	sec θ	cot θ	tan θ	csc θ	sin θ	Radians	Degrees

| | | | | | | | | Angle θ | |

ANSWERS TO SELECTED EXERCISES

CHAPTER ONE

Exercise Set 1.1

1. a. natural number, integer, rational number **b.** integer, rational number
3. a. rational number **b.** irrational number **5. a.** natural number, integer, rational number **b.** rational number **7. a.** rational number **b.** rational number **9.** irrational number **11.** irrational number

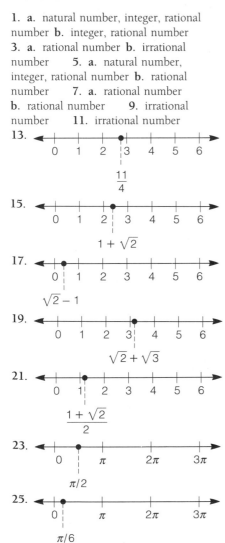

35. false **37.** true **39.** false
41. false **43.** true

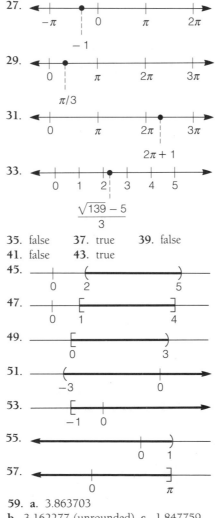

59. a. 3.863703
b. 3.162277 (unrounded) **c.** 1.847759
61. a. one decimal place **b.** two decimal places **c.** six decimal places **d.** nine decimal places **63. a.** $a = \sqrt{2}$ and $b = \sqrt{8}$ (other answers are possible)

b. $a = \sqrt{2}$ and $b = \sqrt{3}$ (other answers are possible) **65. a.** $2^{1/2} = \sqrt{2}$
b. $\left(\sqrt{2}\right)^2 = 2$

Exercise Set 1.2

1. 3 **3.** 6 **5.** 2 **7.** 0
9. 0 **11.** 17 **13.** 1 **15.** 0
17. 25 **19.** -1 **21.** 0 **23.** 3
25. $\sqrt{2} - 2$ **27.** $x - 3$
29. $t^2 + 1$ **31.** $\sqrt{3} + 4$
33. $-2x + 7$ **35.** 1 **37.** $3x + 11$
39. $|x - 4| = 8$ **41.** $|x - 1| = \frac{1}{2}$
43. $|x - 1| \geq \frac{1}{2}$ **45.** $|y - (-4)| < 1$, or $|y + 4| < 1$ **47.** $|y - 0| < 3$, or $|y| < 3$
49. $|x^2 - a^2| < M$

c. The interval in (b) does not include 2.
65. a. $\min(6, 1) = 1$
b. $\min(1, -6) = -6$
c. $\min(-6, -6) = -6$

71. a. Property 1(b) **d.** Since
$a + b \le |a| + |b|$ and $-(a + b) \le |a| + |b|$,
then $|a + b| \le |a| + |b|$.

Exercise Set 1.3

1. a. all real numbers **b.** all nonnegative
real numbers   **3. a.** all real numbers
b. all real numbers
5. a. all nonnegative real numbers
b. all positive real numbers except 1
7. a. degree: 0; (nonzero) coefficient: 4
b. degree: 3; (nonzero) coefficient: 4
c. degree: 6; (nonzero) coefficients: 1, 4,
$-1, 2$ **11. a.** $x^2 - y^2$ **b.** $x^4 - 25$
13. a. $A^2 - 16$ **b.** $a^2 + 2ab + b^2 - 16$
15. a. $x^2 - 16x + 64$ **b.** $4x^4 - 20x^2 + 25$
17. a. $x + 2\sqrt{xy} + y$
b. $2x + y - 2\sqrt{x^2 + xy}$
19. a. $a^3 + 3a^2 + 3a + 1$
b. $27x^6 - 54x^4a^2 + 36x^2a^4 - 8a^6$
21. a. $x^3 + 1$ **b.** $x^6 + 1$
23. a. $(x + 8)(x - 8)$ **b.** $7x^2(x^2 + 2)$
c. $z(11 + z)(11 - z)$ **d.** $(ab + c)(ab - c)$
25. a. $(x + 3)(x - 1)$ **b.** $(x - 3)(x + 1)$
c. irreducible
d. $(-x + 3)(x + 1)$ or $-(x - 3)(x + 1)$
27. a. $(x + 1)(x^2 - x + 1)$
b. $(x + 6)(x^2 - 6x + 36)$
c. $8(5 - x^2)(25 + 5x^2 + x^4)$
d. $(4ax - 5)(16a^2x^2 + 20ax + 25)$
29. a. $(12 + x)(12 - x)$ **b.** irreducible
c. $(9 + y)(15 - y)$
31. a. $h^3(1 + h)(1 - h)$
b. $h^3(10 + h)(10 - h)$
c. $(h + 1)^3(11 + h)(9 - h)$
33. a. $(x - 8)(x - 5)$ **b.** irreducible
35. a. $(x + 9)(x - 4)$ **b.** $(x - 9)(x - 4)$
37. a. $(3x + 2)(x - 8)$ **b.** irreducible
39. a. $(3x - 1)(2x + 5)$
b. $(6x + 5)(x - 1)$ **41. a.** $(t^2 + 1)^2$
b. $(t + 1)^2(t - 1)^2$ **c.** irreducible
43. a. $x(4x^2 - 20x - 25)$ **b.** $x(2x - 5)^2$
45. a. $(a - c)(b + a)$
b. $(u + v - y)(x + u + v)$
47. $(xz + t)(xz + y)$ **49.** $(a^2 - 2b^2c^2)^2$
51. irreducible
53. $(x + 4)(x^2 - 4x + 16)$
55. $x(x^2 + 3xy + 3y^2)$
57. $(x - y)(x^2 + xy + y^2 + 1)$
59. a. $(p^2 + 1)(p + 1)(p - 1)$
b. $(p^4 + 1)(p^2 + 1)(p + 1)(p - 1)$
61. $(x + 1)^3$ **63.** irreducible

65. $\left(\frac{5}{4} + c\right)\left(\frac{5}{4} - c\right)$
67. $\left(z^2 + \frac{9}{4}\right)\left(z + \frac{3}{2}\right)\left(z - \frac{3}{2}\right)$
69. $\left(\frac{5}{mn} - 1\right)\left(\frac{25}{m^2n^2} + \frac{5}{mn} + 1\right)$
71. $\left(\frac{1}{2}x + y\right)^2$
73. $(x - a)(8x - 8a + 1)(8x - 8a - 1)$
75. $(ax + b)(x + 1)$ **77.** $-x(x + 1)^{1/2}$
79. $x(x + 1)^{-3/2}$ **81.** $-\frac{2}{3}x(2x + 3)^{1/2}$
83. $(A + B)^3$
85. $x(a^2 + x^2)^{-3/2}(2a^2 + x^2)$
87. $(y^2 - pq)(y^2 + pq - py - qy)$
89. a.

x	$\dfrac{x^2 - 16}{x - 4}$	x	$\dfrac{x^2 - 16}{x - 4}$
3.9	7.9	4.1	8.1
3.99	7.99	4.01	8.01
3.999	7.999	4.001	8.001
3.9999	7.9999	4.0001	8.0001
3.99999	7.99999	4.00001	8.00001

b. As x approaches 4, the value of the
expression $\dfrac{x^2 - 16}{x - 4}$ approaches 8.
c. $\dfrac{x^2 - 16}{x - 4} = \dfrac{(x + 4)(x - 4)}{x - 4} = x + 4$ if
$x \ne 4$ **91.** $(x^2 + 4x + 8)(x^2 - 4x + 8)$
93. $2(x - t)(x + y + z + t)$
95. $3(b - a)(c - a)(c - b)$

Exercise Set 1.4

1. yes   **3.** no   **5.** no   **7.** $6, -1$
9. $10, -10$ **11.** $\frac{6}{5}$ **13.** $-\frac{1}{5}, \frac{3}{2}$
15. $1, -3$ **17.** $\frac{1}{2}, 6$ **19.** $0, -13,$
12   **21.** $\frac{1}{2}, \pm\frac{1}{2}\sqrt{3}$
23. $1, \pm\sqrt{2}$ **25.** $-1, -9$
27. $(1 \pm \sqrt{21})/2$ **29.** $(-3 \pm \sqrt{41})/4$
31. $-\frac{1}{6}, -\frac{5}{2}$ **33.** $(1 \pm \sqrt{41})/4$
35. $(6 \pm \sqrt{42})/6$ **37.** $-1.47, -1.53$
39. $-67, -89$ **41.** $\frac{2}{3}, 1$
43. $(-2 \pm \sqrt{13})/3$ **45.** $\pm 2\sqrt{6}$
47. $(1 \pm \sqrt{5})/2$ **49.** $(3 \pm \sqrt{15})/3$
51. $\sqrt{5}/2, -2\sqrt{5}/5$ **53.** $x = 1/(2y)$ or
$x = 1/y$ **55.** $x = 0$ or $x = p + q$
57. $x = -\frac{5}{4}a$ or $x = \frac{4}{3}a$
59. $r = (-h \pm \sqrt{h^2 + 40})/2$

61. $t = 0$ or $t = v_0/16$ **63.** $x = 0,$
$x = -2b$, or $x = b$ **65.** two real roots
67. two real roots   **69.** one real root
71. two real roots   **73.** 36
75. $\pm 2\sqrt{5}$ **77. a.** 24
b. no real-number solutions   **79.** -1
81. $\pm 2, \pm 3$ **83.** $-4, -\frac{20}{9}$
85. no real-number solutions
87. no real-number solutions
89. $\frac{5}{2}, -4$ **91.** $\frac{1}{3}, \frac{1}{2}$ **93.** 1
95. $-3 \pm 2\sqrt{2}$ **97.** Any value of x
(except $x = -1$ or $x = -3$) is a solution
to the equation.   **99.** $x = a + b$
101. $x = a^2b^2/(2a + b)^2$
103. $(1 \pm \sqrt{21})/2$ **105.** 1 m
107. a. 266 people **b.** 8 days
109. $\frac{1}{4}L$ in. and $\frac{3}{4}L$ in.
111. b. $k = -\frac{1}{2}$ **c.** $x = (-1 \pm \sqrt{5})/2$
113. $A = 0$ and $B = 0$, $A = 1$ and $B = -2$
119. $x = \dfrac{a + b}{ab}$ or $x = \dfrac{-2}{a + b}$

Exercise Set 1.5

1. $(-\infty, -1)$ **3.** $\left[\frac{1}{3}, \infty\right)$
5. $\left(-\infty, -\frac{9}{2}\right)$ **7.** $[-6, \infty)$
9. $\left(-\infty, \frac{5}{2}\right)$ **11.** $\left[-\frac{17}{23}, \infty\right)$
13. $[4, 6]$ **15.** $\left[-\frac{1}{2}, 1\right]$
17. $(3.98, 3.998)$ **19. a.** $\left[-\frac{1}{2}, \frac{1}{2}\right]$
b. $\left(-\infty, -\frac{1}{2}\right) \cup \left(\frac{1}{2}, \infty\right)$
21. a. $(-\infty, 0) \cup (0, \infty)$ **b.** no solution
23. a. $(-\infty, 3)$ **b.** $(1, 3)$
c. $(-\infty, 1) \cup (3, \infty)$ **25. a.** $[-4, \infty)$
b. $[-4, 6]$ **c.** $(-\infty, -4) \cup (6, \infty)$
27. a. $(a - c, \infty)$ **b.** $(a - c, a + c)$
c. $(-\infty, a - c] \cup [a + c, \infty)$
29. $(-10, 14)$ **31.** $(-11, 1)$
33. a. $(-2h, h)$ **b.** $(h, -2h)$
35. $-175° \le C \le 125°$
37. $[-13°, 887°]$
39. $[9200°, 11700°]$

41. a.

a	b	$\sqrt{ab}$ (G.M.)	$\dfrac{a+b}{2}$ (A.M.)	Which is larger, G.M. or A.M.?
1	2	1.4142	1.5	A.M.
1	3	1.7320	2.0	A.M.
1	4	2.0000	2.5	A.M.
2	3	2.4495	2.5	A.M.
3	4	3.4641	3.5	A.M.
5	10	7.0711	7.5	A.M.
9	10	9.4868	9.5	A.M.
99	100	99.4987	99.5	A.M.
999	1000	999.4999	999.5	A.M.

43. a.

x	y	$\dfrac{x}{y}+\dfrac{y}{x}$	$\dfrac{x}{y}+\dfrac{y}{x}\ge 2$ True or False?
1	1	2.0000	true
2	3	2.1667	true
3	5	2.2667	true
4	7	2.3214	true
5	9	2.3556	true
9	10	2.0111	true
49	50	2.0004	true
99	100	2.0001	true

49. c. $x = 5$; the rectangle is a square with dimensions 5 ft by 5 ft.

Exercise Set 1.6

1. $(-3, 2)$ **3.** $(-\infty, 2) \cup (9, \infty)$
5. $(-\infty, 4] \cup [5, \infty)$
7. $(-\infty, -4] \cup [4, \infty)$ **9.** no solution
11. $(-7, -6) \cup (0, \infty)$
13. $[-15, 0] \cup [15, \infty)$ **15.** $(-\infty, \infty)$
17. $\left(-\infty, -\frac{3}{4}\right) \cup \left(-\frac{2}{3}, 0\right)$

19. $\left(-\infty, \dfrac{-1-\sqrt{5}}{2}\right) \cup \left(\dfrac{-1+\sqrt{5}}{2}, \infty\right)$
21. $\left[4-\sqrt{14}, 4+\sqrt{14}\right]$
23. $[-4, -3] \cup [1, \infty)$
25. $(-\infty, -6) \cup (-5, -4)$
27. $\left(-\infty, -\frac{1}{3}\right) \cup \left(\frac{1}{3}, 2\right) \cup (2, \infty)$
29. $\left(-\infty, -\frac{2}{3}\right] \cup \left\{3, -\frac{1}{2}\right\}$
31. $\left(-\infty, -\sqrt{5}\right] \cup [-2, 2] \cup \left[\sqrt{5}, \infty\right)$
33. $(-3, 3) \cup (4, \infty)$
35. $\left(-\sqrt{14}, -2\sqrt{2}\right) \cup \left(2\sqrt{2}, \sqrt{14}\right)$
37. $\left(-\infty, \dfrac{3-\sqrt{11}}{2}\right) \cup \left(\dfrac{3+\sqrt{11}}{2}, 4\right)$
39. $(-2, -1) \cup (1, \infty)$ **41.** $(-1, 1]$
43. $\left(-\infty, \frac{3}{2}\right) \cup [2, \infty)$
45. $(-\infty, -1) \cup (0, 9)$
47. $\left(-\frac{7}{2}, -\frac{4}{3}\right) \cup (-1, 0) \cup (1, \infty)$
49. $(-\infty, -1)$ **51.** $(-1, 0)$
53. $(-5, \infty)$ **55.** $[-1, 1) \cup (2, 4]$
57. $\left(-1, 2-\sqrt{5}\right) \cup \left(1, 2+\sqrt{5}\right)$
59. $\left(-\frac{3}{2}, 0\right)$ **61.** $(-\infty, -1) \cup (0, \infty)$
63. a. $(-\infty, -1] \cup [5, \infty)$
b. $(-\infty, -1) \cup (5, \infty)$
65. $(-\infty, -2] \cup [2, \infty)$
67. $(-\infty, 0) \cup \left(\frac{1}{2}, \infty\right)$ **69.** $\left(\frac{1}{5}, \frac{4}{5}\right)$
71. $(0, 2)$
73. a.

x	1	2	5	10
$\dfrac{x^2 + 1000}{2x^2 + x}$	333.67	100.40	18.64	5.24

x	100	200	10,000
$\dfrac{x^2 + 1000}{2x^2 + x}$	0.55	0.51	0.50

b. $(2000, \infty)$ **c.** $n = 2001$
75. $\left(-\infty, \dfrac{3a + 5b}{8}\right)$

Chapter One Review Exercises

1. $(a - 4b)(a + 4b)$
3. $(1 - a)(7 + 4a + a^2)$ **5.** $x(ax + b)^2$
7. $(4x + 1)(2x + 1)$
9. $a^4x^4(1 - xa)(1 + xa)(1 + x^2a^2)$
11. $(2 + a)^3$ **13.** $z^3(4xy + 1)(xy - 1)$
15. $(1 - x)(1 + x + x^2)(1 + x)(1 - x + x^2)$
17. $(ax + b - 2abx)(ax + b + 2abx)$
19. $(a - b)(a + b + c + ab)$

21. $81x^2 - y^2$ **23.** $4a^2b^2 - 12ab + 9$
25. $8a^3 + 36a^2 + 54a + 27$
27. $1 - 27a^3$ **29.** $x - y$
31. $x + 3x^{2/3}y^{1/3} + 3x^{1/3}y^{2/3} + y$
33. $x^4 - 6x^3 + 11x^2 - 6x + 1$
35. $|x - 6| = 2$ **37.** $|a - b| = 3$
39. $|x - 0| > 10$ or $|x| > 10$
41. $\sqrt{6} - 2$ **43.** $y^4 + y^2 + 1$
45. a. $-2x + 5$ **b.** 1 **c.** $2x - 5$
47.

49.

51.

53.

55.

57. a.

b.

59.

x	$1 + \frac{1}{2}x^2 + \frac{3}{8}x^4$	$1/\sqrt{1 - x^2}$
0.1	1.00504	1.00504
0.01	1.00005	1.00005

61. $\frac{1}{3}, -\frac{1}{2}$ **63.** $-6, 4$ **65.** $10, -1$
67. $\frac{2}{3}, 3$ **69.** $\pm\sqrt{1 + \sqrt{3}}, 0$
71. -7 **73.** $\frac{1}{4}$ **75.** 10 **77.** 2
79. $(-2, 1)$ **81.** $\left[-\frac{1}{2}, \frac{1}{2}\right]$
83. $\left(-\frac{41}{10}, -\frac{39}{10}\right)$
85. $(-\infty, -2] \cup [3, \infty)$ **87.** $(-8, 5)$
89. $\left(3 - \sqrt{10}, 3 + \sqrt{10}\right)$
91. $(-5, -3) \cup (3, 5)$ **93.** $(-2, \infty)$
95. $(-\infty, -1) \cup [1, 9]$
97. $\left(-\infty, -\frac{1}{2}\right) \cup \left[\frac{1}{6}, \infty\right)$ **99.** $(0, 25]$
101. $\left[-\frac{1}{2}, \frac{1}{2}\right]$
103. $\left(-\infty, 3 - 2\sqrt{2}\right] \cup \left[3 + 2\sqrt{2}, \infty\right)$
105. $(16\,\text{cm}, \infty)$ **107. a.** $\frac{1}{4}$ sec, $\frac{15}{4}$ sec
b. between 1.75 sec and 2.25 sec
109. $16(t_1 + t_2)$ ft/sec

Chapter One Test

1. a. $(6 - x)(12 + x^2)$ **b.** irreducible
2. a. irreducible **b.** $z(3xy - 1)(2xy + 1)$
3. $-3x\left(1 - x^2\right)^{-3/2}(2x^2 + x - 2)$

4. $(3a + 2b)(3x - 2y)$
5. $8a^3 + 36a^2b^2 + 54ab^4 + 27b^6$ **6.** 1
7. a. $\left(\frac{39}{10}, \frac{41}{10}\right)$ **b.** $[-2, \infty)$ **8. a.** $-5, 1$ **b.**
b. $-2 \pm \sqrt{5}$ **9.** 2 **10.** $0, -\frac{3}{2}, 1$
11. $-3, \pm 2\sqrt{2}$ **12. a.** $1, -\frac{1}{3}$
b. $[7, 9]$ **13.** $(-\infty, 3]$ **14.** $\left(\frac{27}{10}, \frac{31}{10}\right)$
15. $[-8, \infty)$

16. $\left(-2, \dfrac{-3 - \sqrt{3}}{3}\right] \cup$
$\left(-1, \dfrac{-3 + \sqrt{3}}{3}\right] \cup (0, \infty)$

CHAPTER TWO

Exercise Set 2.1

1.

3. a.

b. 6 square units **5. a.** 5 **b.** 13
7. a. 10 **b.** 9 **9.** 4 **11. a.** $\left(4, \frac{1}{2}\right)$
b. $(-6, 7)$ **13. a.** yes **b.** yes **c.** no
15. 0; the three points are collinear.
17. center: $(3, 1)$; radius: 5;
y-intercepts: 5 and -3
19. center: $(0, 0)$; radius: $\sqrt[4]{2}$;
y-intercepts: $\pm \sqrt[4]{2}$
21. center: $(-4, 3)$; radius: 1;
no y-intercepts **23.** center: $\left(-3, \frac{1}{3}\right)$;
radius: $\sqrt{2}$; no y-intercepts
25. a. $\frac{4}{3}$ **b.** $-\frac{4}{3}$ **27. a.** $-\sqrt{2}/2$
b. $\sqrt{2}/2$ **29. a.** $\sqrt{5}/3$ **b.** $-\sqrt{5}/3$
31. a. 0.758 **b.** 0.758
33. $(x - 1)^2 + (y - 1)^2 = 34$

35. a. $(6, 5)$ **b.** $\left(\frac{1}{2}, -\frac{3}{2}\right)$ **c.** $(1, -4)$
37. $(11, -8)$
39. a.

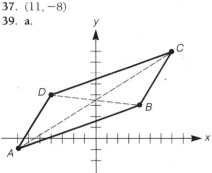

b. $\left(0, \frac{7}{2}\right)$; $\left(0, \frac{7}{2}\right)$ **c.** The diagonals of a
parallelogram bisect each other.
41. a. 128 **b.** 96 **c.** $\frac{128}{96} = \frac{4}{3}$
43. $(x - 3)^2 + (y - 2)^2 = 169$
45. $(x - 3)^2 + (y - 5)^2 = 9$
47. $a = \sqrt{2}$; $b = \sqrt{3}$; $c = 2$; $d = \sqrt{5}$;
$e = \sqrt{6}$; $f = \sqrt{7}$; $g = \sqrt{8} = 2\sqrt{2}$
49. $(x - 3)^2 + (y - 2)^2 = 13$
51. 7 and -3
53. b. $\left(\dfrac{a + b}{2}, \dfrac{c}{2}\right)$; $\left(\dfrac{a + b}{2}, \dfrac{c}{2}\right)$
61. a. Each side of the outermost
quadrilateral is the hypotenuse of the
triangle in Figure A, so each side has a
length of c. Also, since each angle of the
outermost quadrilateral is the sum of the
two acute angles from the triangle, the
measure of each angle is $90°$. Since the
quadrilateral has angles of $90°$ and sides
of equal length, it must be a square.
b. $b - a$ **c.** The area is equal to $a^2 + b^2$.
63. c. $u = 5$, $v = 12$, and $w = 13$

Optional TI-81 Graphing Calculator
Exercises for Section 2.1

1. a. $y = \pm \sqrt{4 - x^2}$

b.

c.

d. The graph is identical to that in part (c).
3.

5.

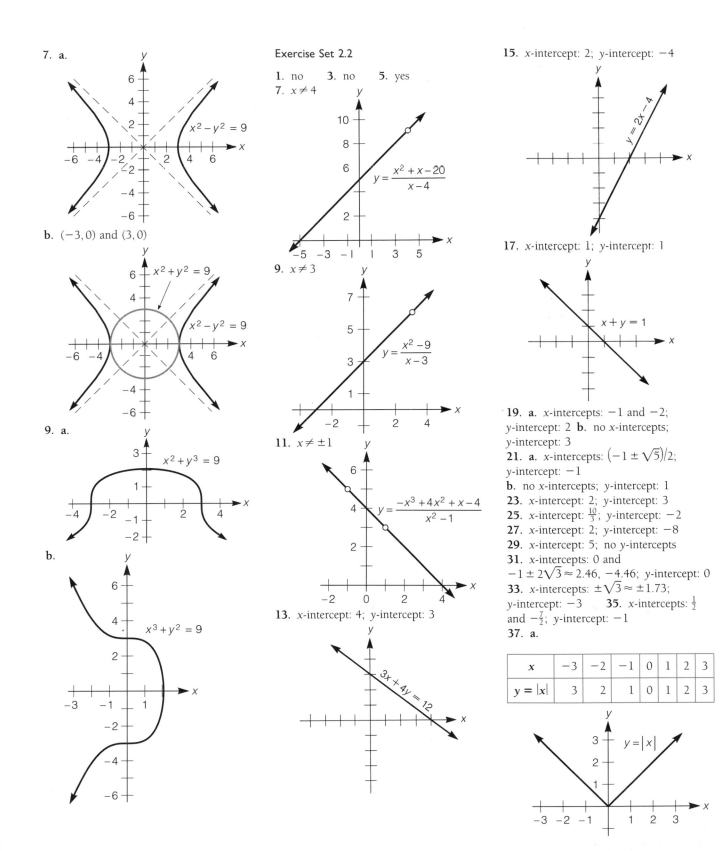

7. a.

$x^2 - y^2 = 9$

b. $(-3, 0)$ and $(3, 0)$

$x^2 + y^2 = 9$

$x^2 - y^2 = 9$

9. a.

$x^2 + y^3 = 9$

b.

$x^3 + y^2 = 9$

Exercise Set 2.2

1. no **3.** no **5.** yes
7. $x \neq 4$

$y = \dfrac{x^2 + x - 20}{x - 4}$

9. $x \neq 3$

$y = \dfrac{x^2 - 9}{x - 3}$

11. $x \neq \pm 1$

$y = \dfrac{-x^3 + 4x^2 + x - 4}{x^2 - 1}$

13. x-intercept: 4; y-intercept: 3

$3x + 4y = 12$

15. x-intercept: 2; y-intercept: -4

$y = 2x - 4$

17. x-intercept: 1; y-intercept: 1

$x + y = 1$

19. a. x-intercepts: -1 and -2; y-intercept: 2 **b.** no x-intercepts; y-intercept: 3
21. a. x-intercepts: $(-1 \pm \sqrt{5})/2$; y-intercept: -1
b. no x-intercepts; y-intercept: 1
23. x-intercept: 2; y-intercept: 3
25. x-intercept: $\frac{10}{3}$; y-intercept: -2
27. x-intercept: 2; y-intercept: -8
29. x-intercept: 5; no y-intercepts
31. x-intercepts: 0 and
$-1 \pm 2\sqrt{3} \approx 2.46, -4.46$; y-intercept: 0
33. x-intercepts: $\pm\sqrt{3} \approx \pm 1.73$;
y-intercept: -3 **35.** x-intercepts: $\frac{1}{2}$
and $-\frac{7}{2}$; y-intercept: -1
37. a.

x	-3	-2	-1	0	1	2	3
$y = |x|$	3	2	1	0	1	2	3

$y = |x|$

b.

x	-3	-2	-1	0	1	2	3
$y = x^2$	9	4	1	0	1	4	9

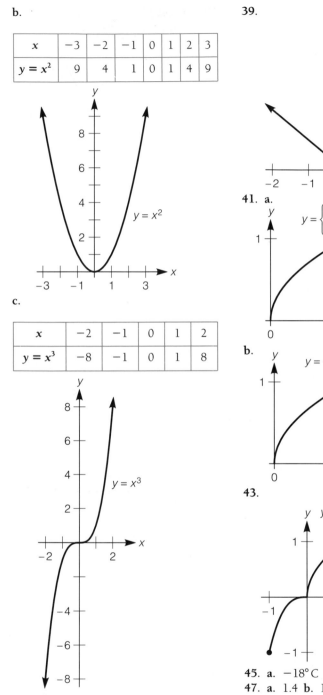

$y = x^2$

c.

x	-2	-1	0	1	2
$y = x^3$	-8	-1	0	1	8

$y = x^3$

39.

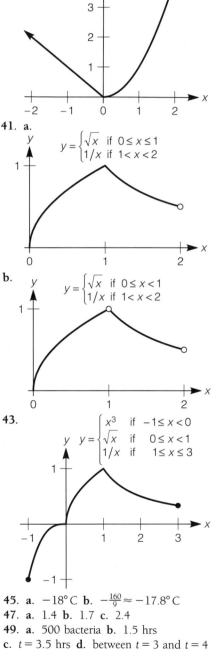

$y = \begin{cases} |x| & \text{if } x \le 0 \\ x^2 & \text{if } x > 0 \end{cases}$

41. a.

$y = \begin{cases} \sqrt{x} & \text{if } 0 \le x \le 1 \\ 1/x & \text{if } 1 < x < 2 \end{cases}$

b.

$y = \begin{cases} \sqrt{x} & \text{if } 0 \le x < 1 \\ 1/x & \text{if } 1 < x < 2 \end{cases}$

43.

$y = \begin{cases} x^3 & \text{if } -1 \le x < 0 \\ \sqrt{x} & \text{if } 0 \le x < 1 \\ 1/x & \text{if } 1 \le x \le 3 \end{cases}$

45. a. $-18°\text{C}$ **b.** $-\frac{160}{9} \approx -17.8°\text{C}$

47. a. 1.4 **b.** 1.7 **c.** 2.4

49. a. 500 bacteria **b.** 1.5 hrs
c. $t = 3.5$ hrs **d.** between $t = 3$ and $t = 4$

51.

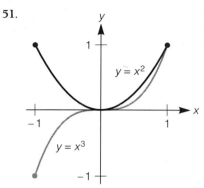

$y = x^2$

$y = x^3$

53. point A: 1; point B: 3

Optional TI-81 Graphing Calculator
Exercises for Section 2.2

1. a.

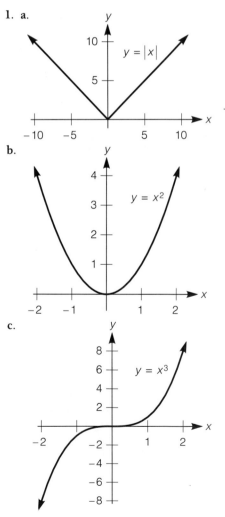

$y = |x|$

b.

$y = x^2$

c.

$y = x^3$

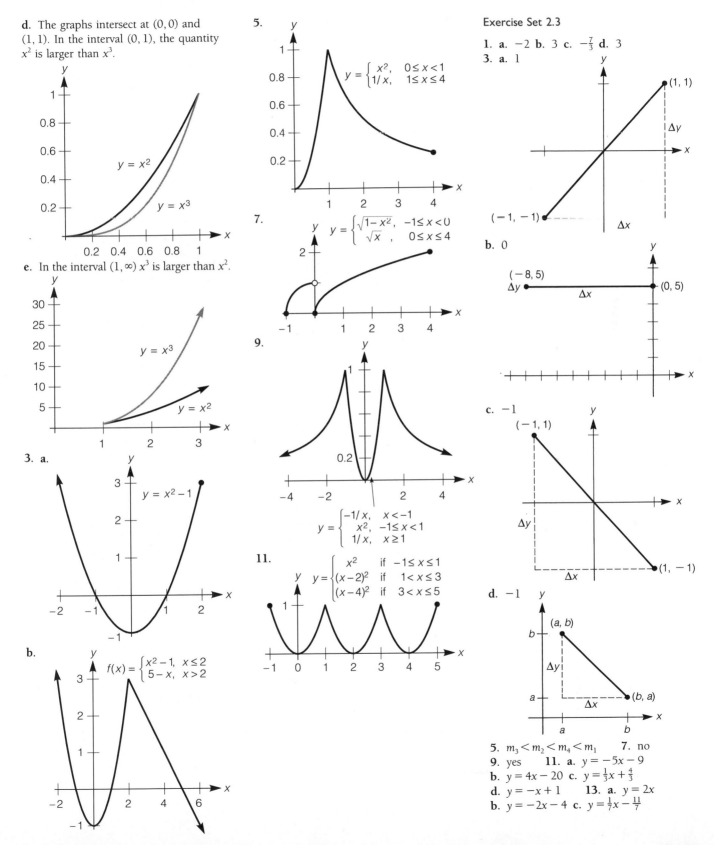

d. The graphs intersect at $(0,0)$ and $(1,1)$. In the interval $(0,1)$, the quantity x^2 is larger than x^3.

$y = x^2$

$y = x^3$

e. In the interval $(1,\infty)$ x^3 is larger than x^2.

$y = x^3$

$y = x^2$

3. a.

$y = x^2 - 1$

b.

$f(x) = \begin{cases} x^2 - 1, & x \le 2 \\ 5 - x, & x > 2 \end{cases}$

5.

$y = \begin{cases} x^2, & 0 \le x < 1 \\ 1/x, & 1 \le x \le 4 \end{cases}$

7.

$y = \begin{cases} \sqrt{1-x^2}, & -1 \le x < 0 \\ \sqrt{x}, & 0 \le x \le 4 \end{cases}$

9.

$y = \begin{cases} -1/x, & x < -1 \\ x^2, & -1 \le x < 1 \\ 1/x, & x \ge 1 \end{cases}$

11.

$y = \begin{cases} x^2 & \text{if } -1 \le x \le 1 \\ (x-2)^2 & \text{if } 1 < x \le 3 \\ (x-4)^2 & \text{if } 3 < x \le 5 \end{cases}$

15. $x = -3$ **17.** $y = 4$ **19.** $x = 0$

21. a. $y = -4x + 7$ **b.** $y = 2x + \frac{3}{2}$
c. $y = -\frac{4}{3}x + 14$

23. a. $y = 4x + 11$

b. $y = \frac{1}{2}x - \frac{5}{4}$

c. $y = -\frac{5}{6}x + 5$

d. $y = \frac{3}{4}x + \frac{3}{2}$

e. $y = 4x - 2$

25. $y = x + 2$

27. $y - 4 = 0$

29. a. x-intercept: 5; y-intercept: 3;
area: $\frac{15}{2}$; perimeter: $8 + \sqrt{34}$
b. x-intercept: 5; y-intercept: -3;
area: $\frac{15}{2}$; perimeter: $8 + \sqrt{34}$
31. a. neither **b.** parallel
c. perpendicular **d.** perpendicular
e. neither **f.** parallel
33. $y = \frac{2}{5}x + \frac{12}{5}$; $2x - 5y + 12 = 0$
35. $y = -\frac{4}{3}x + \frac{16}{3}$; $4x + 3y - 16 = 0$
37. $y = \frac{3}{5}x + 11$; $3x - 5y + 55 = 0$

39. a.

b. $y = -\frac{4}{3}x - \frac{25}{3}$
47.

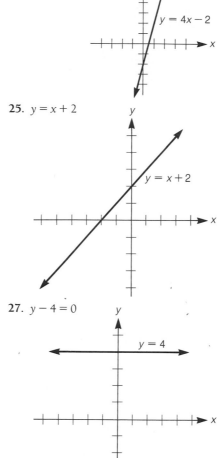

x	2.5	2.9	2.99	2.999	2.9999
y	6.25	8.41	8.9401	8.994001	8.99940001
Δx	0.5	0.1	0.01	0.001	0.0001
Δy	2.75	0.59	0.0599	0.005999	0.00059999
m	5.5	5.9	5.99	5.999	5.9999

The slope of these lines approaches 6, so
we estimate that the slope of the tangent
line to the curve $y = x^2$ at $T(3, 9)$ is 6.
49. $\left(-\frac{1}{2}, -\frac{1}{8}\right)$ **51.** $\left(8, \frac{1}{8}\right)$
53. 44.1 square units
55. a.

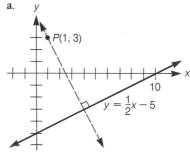

b. $y = -2x + 5$ **c.** $(4, -3)$ **d.** $3\sqrt{5}$
59. a. $x = 410 - \frac{2}{3}P$ **b.** 230 units
c. \$307.50/unit **65. a.** $y = 2x$
b. $y = \sqrt{mn}\,x$

**Optional TI-81 Graphing Calculator
Exercises for Section 2.3**

1. a. $y = x$; $y = 2x$; $y = 3x$; $y = 10x$
As the positive slopes increase, the lines
become steeper.

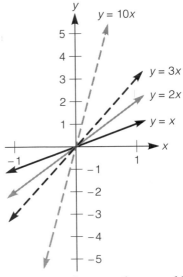

b. $y = -x$; $y = -2x$; $y = -3x$; $y = -10x$
As the negative slopes decrease, the lines
become steeper.

3.

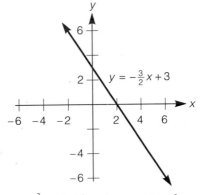

5. a. $y = -\frac{3}{2}x + 3$; The x-intercept is 2
and the y-intercept is 3.

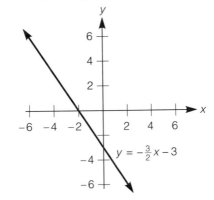

b. $y = -\frac{3}{2}x - 3$; The x-intercept is -2
and the y-intercept is -3.

c. $y = -\frac{5}{6}x + 5$; The x-intercept is 6 and
the y-intercept is 5.

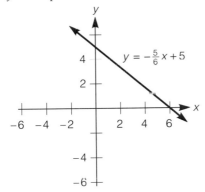

d. $y = -\frac{5}{6}x - 5$; The x-intercept is -6
and the y-intercept is -5.

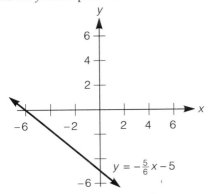

e. $(x/a) + (y/b) = 1$ is a line with an
x-intercept of a and y-intercept of b.
7. a.

b. $M_{AB} = (-1, 0)$; $M_{BC} = (1, 3)$; $M_{AC} = (-2, 3)$; The medians appear to intersect at the point $(-1, 2)$.

c. $y = \frac{3}{5}x + \frac{12}{5}$; $y = -\frac{3}{4}x + \frac{3}{2}$; $y = 6x + 6$ centroid: $\left(-\frac{2}{3}, 2\right)$, which agrees with our estimate from part (b), to the nearest integer

Exercise Set 2.4

1. a.

b.

c.

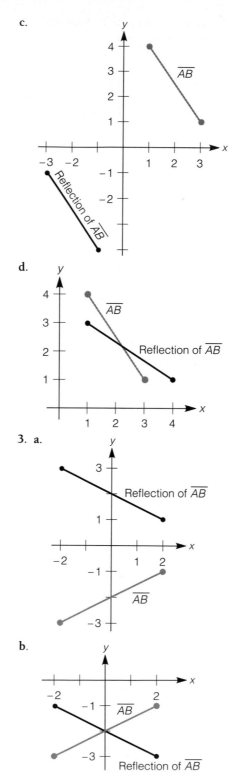

d.

3. a.

b.

c.

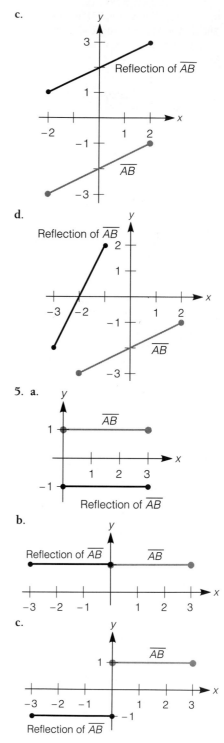

d.

5. a.

b.

c.

d.

7.

	Symmetric about the x-axis	Symmetric about the y-axis	Symmetric about the Origin
$y = x^2$	no	yes	no
$y = x^3$	no	no	yes
$y = \sqrt{x}$	no	no	no

9. x-intercepts: 2 and −2; y-intercept: 4; symmetric about the y-axis

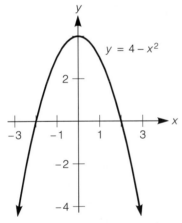

11. no x- or y-intercepts; symmetric about the origin

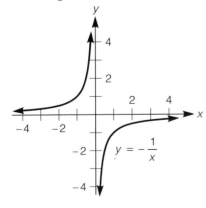

13. x- and y-intercepts: 0; symmetric about the y-axis

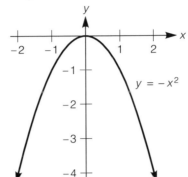

15. no x- or y-intercepts; symmetric about the origin

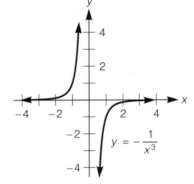

17. x- and y-intercepts: 0; symmetric about the y-axis

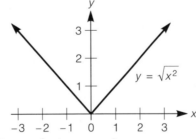

19. x- and y-intercepts: 1; no symmetry

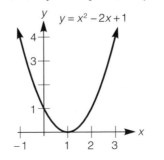

21. x-intercept: 2; no y-intercept; symmetric about the x-axis

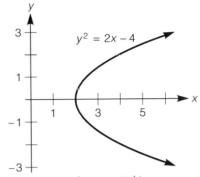

23. x-intercepts: $(-1 \pm \sqrt{33})/4$; y-intercept: −4; no symmetry

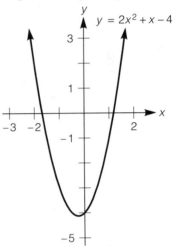

25. a. x- and y-intercepts: 2; no symmetry

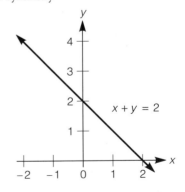

b. x-intercepts: ± 2; y-intercept: 2; symmetric about the y-axis

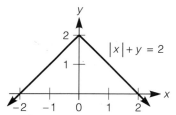

27. a. The x-intercept for each graph is $\frac{8}{3}$, the y-intercept for $y = \frac{3}{4}x - 2$ is -2, and the y-intercept for $y = \left|\frac{3}{4}x - 2\right|$ is 2.
b. $\left[\frac{8}{3}, \infty\right)$ **c.** The graph of $y = \left|\frac{3}{4}x - 2\right|$ can be obtained by reflecting $y = \frac{3}{4}x - 2$ about the x-axis for the interval $\left(-\infty, \frac{8}{3}\right)$. For the interval $\left[\frac{8}{3}, \infty\right)$, no reflection is necessary. **33.** $y = 2x - 6$
35. x-intercepts: 0, $\pm 3\sqrt{3} \approx \pm 5.2$; y-intercept: 0; symmetric with respect to the origin

x	0	0.5	1	1.5	2	2.5	3
y	0	-13.4	-26	-37.1	-46	-51.9	-54

x	3.5	4	4.5	5	5.5	6
y	-51.6	-44	-30.4	-10	-17.9	54

37. $m = \frac{2}{3}$ and $b = 1$
39. b. $\left(\dfrac{2mb - (m^2 - 1)a}{m^2 + 1}, \dfrac{2ma + (m^2 - 1)b}{m^2 + 1}\right)$

Optional TI-81 Graphing Calculator Exercises for Section 2.4

1. no symmetry

3. no symmetry

5. y-axis symmetry

7. origin symmetry

9. no symmetry

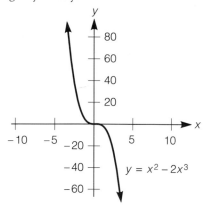

11. a. The graph "appears" to possess origin symmetry.

b.

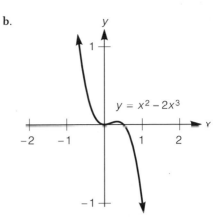

$y = x^2 - 2x^3$

Chapter Two Review Exercises

1. $y = -2x - 6$ 3. $y = \frac{1}{4}x - \frac{5}{2}$
5. $y = 2x + 8$ 7. $y = -2$
9. $y = x + 1$ 11. $9x - 4y + 14 = 0$
13. $2x + y = 0$ 15. $3x - 4y = 0$
17. $x + y - 1 = 0$ and $x - 2y - 4 = 0$
19. 13   21. $(15, 6)$
23. x-axis: no; y-axis: yes; origin: no
25. x-axis: no; y-axis: no; origin: yes
27. x-axis: no; y-axis: no; origin: yes
29. x-axis: no; y-axis: no, origin: yes
31. x-axis: no; y-axis: no; origin: yes
33. x-axis   35. x-axis, y-axis, origin,
line $y = x$ 37. x-axis   39. x-axis
41. $\left(-\frac{27}{13}, \frac{31}{13}\right)$ and $(3, -1)$
43. $(-2, 4)$ and $(1, 1)$

45.

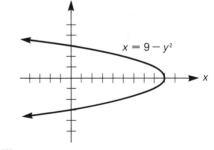

$x = 9 - y^2$

47.

$x^2 + y^2 = 1$

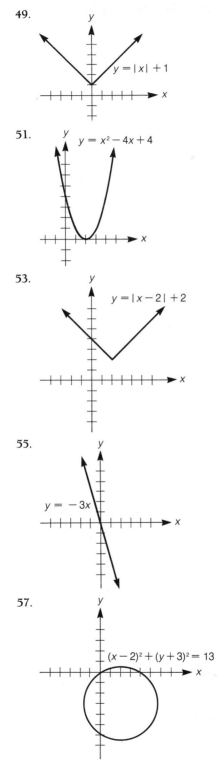

49.

$y = |x| + 1$

51.

$y = x^2 - 4x + 4$

53.

$y = |x - 2| + 2$

55.

$y = -3x$

57.

$(x - 2)^2 + (y + 3)^2 = 13$

59.

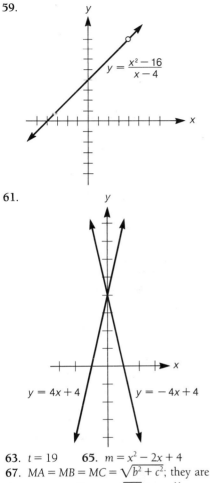

$y = \dfrac{x^2 - 16}{x - 4}$

61.

$y = 4x + 4$ $y = -4x + 4$

63. $t = 19$ 65. $m = x^2 - 2x + 4$
67. $MA = MB = MC = \sqrt{b^2 + c^2}$; they are
the same   69. **a.** $\sqrt{130}$ **b.** $-\frac{11}{3}$
c. $\left(\frac{7}{2}, -\frac{1}{2}\right)$ 71. **a.** 2 **b.** $\sqrt{3}/3$
c. $(0, 0)$ 73. $\frac{49}{6}$ square units
75. **a.** $\left(\frac{5}{3}, \frac{11}{3}\right)$; They all intersect at the
same point. **b.** $\left(\dfrac{2a}{3} + \dfrac{2b}{3}, \dfrac{2c}{3}\right)$; they
intersect at the same point. All medians
of a triangle intersect at a point that is $\frac{2}{3}$
of the distance from each vertex to the
midpoint on the opposite side.
77. $6\sqrt{2}$

81. a.

Reflection of $\overline{AB}$

$\overline{AB}$

b.

Reflection of $\overline{AB}$

$\overline{AB}$

c.

$\overline{AB}$

Reflection of $\overline{AB}$

d.

Reflection of $\overline{AB}$ $\overline{AB}$

83. $13\sqrt{5}/5$ **85.** $17\sqrt{13}/13$

87. a. $(x-1)^2 + (y-1)^2 = 1$
b. $\overline{AT}$: $y = \frac{9}{8}x$; $\overline{BU}$: $y = -\frac{1}{4}x + 1$;
$\overline{CS}$: $y = -3x + 3$ **c.** $\left(\frac{8}{11}, \frac{9}{11}\right)$; $\left(\frac{8}{11}, \frac{9}{11}\right)$; the point of intersection is the same in both cases.

Chapter Two Test

1. $y = 5x - 7$ **2.** $y = \frac{6}{5}x - \frac{17}{5}$
3. $y = -3x$ **4.** $y = \frac{4}{5}x$ **5.** The

point $(3, 9)$ is farther from the origin.
6. a. x-axis: no; y-axis: no; origin: yes
b. x-axis: no; y-axis: yes; origin: no
c. x-axis: yes; y-axis: no; origin: no
7. x-intercept: 5; y-intercept: -3

$3x - 5y = 15$

8. x-intercepts: $\left(5 \pm \sqrt{59}\right)/2$;
y-intercepts: $\left(-3 \pm \sqrt{43}\right)/2$

$\left(x - \frac{5}{2}\right)^2 + \left(y + \frac{3}{2}\right)^2 = 17$

9. no x-intercepts; y-intercept: 3

$y = |x + 1| + 2$

10. $t = 1$ **11.** $(-4, 16)$ and $(3, 9)$
12. yes

13. a.

$y = \dfrac{4x^2 - 1}{2x + 1}$

b.

$y = \begin{cases} \sqrt{1 - x^2} & \text{if } -1 \le x \le 0 \\ |x| & \text{if } 0 < x \le 2 \end{cases}$

c.

$y = \begin{cases} \sqrt{x} & \text{if } 0 \le x \le 1 \\ 1/x & \text{if } 1 < x \le 4 \end{cases}$

14. a.

Reflection of $\overline{AB}$

$\overline{AB}$

b.

c.

d.

15. $\frac{121}{5}$ square units

CHAPTER THREE

Exercise Set 3.1

1. a. $(-\infty, \infty)$ **b.** $\left(-\infty, \frac{1}{5}\right]$
c. $\left(-\infty, \frac{1}{5}\right) \cup \left(\frac{1}{5}, \infty\right)$ **3. a.** $(-\infty, \infty)$
b. $(-\infty, 4] \cup [6, \infty)$ **c.** $[0, 4) \cup (6, \infty)$
5. a. $(-\infty, 4) \cup (4, \infty)$
b. $(-\infty, -2) \cup (-2, 2) \cup (2, \infty)$
c. $(-\infty, -2) \cup (2, \infty)$
7. a. $(-\infty, -1] \cup [5, \infty)$
b. $(-\infty, -1) \cup (5, \infty)$ **c.** $(-\infty, \infty)$
9. a. $(-\infty, 1) \cup (1, \infty)$
b. $(-\infty, 1) \cup (1, \infty)$ **11. a.** $[4, \infty)$
b. $(-\infty, \infty)$ **13.** $f, g, F,$ and H
15. a. range of f: $\{1, 2, 3\}$;
range of g: $\{2, 3\}$; range of F: $\{1\}$;
range of H: $\{1, 2\}$ **b.** range of g: $\{i, j\}$;
range of F: $\{i, j\}$; range of G: $\{k\}$
17. a. $y = (x - 3)^2$ **b.** $y = x^2 - 3$
c. $y = (3x)^2$ **d.** $y = 3x^2$ **19. a.** -1
b. 1 **c.** 5 **d.** $-\frac{5}{4}$ **e.** $z^2 - 3z + 1$

f. $x^2 - x - 1$ **g.** $a^2 - a - 1$
h. $x^2 + 3x + 1$ **i.** 1 **j.** $4 - 3\sqrt{3}$
k. $1 - \sqrt{2}$ **l.** 2 **21. a.** $12x^2$ **b.** $6x^2$
c. $3x^4$ **d.** $9x^4$ **e.** $\frac{3}{4}x^2$ **f.** $\frac{3}{2}x^2$
23. a. 1 **b.** -7 **c.** -3 **d.** $-\frac{7}{18}$
e. $-2x^2 - 4x - 1$ **f.** $1 - 2x^2 - 4xh - 2h^2$
g. $-4xh - 2h^2$ **h.** $-4x - 2h$
25. a. domain: $(-\infty, 2) \cup (2, \infty)$;
range: $(-\infty, 2) \cup (2, \infty)$ **b.** $\frac{1}{2}$ **c.** 0 **d.** 1
e. $\frac{2x^2 - 1}{x^2 - 2}$ **f.** $\frac{2 - x}{1 - 2x}$ **g.** $\frac{2a - 1}{a - 2}$
h. $\frac{2x - 3}{x - 3}$ **27. a.** $d(1) = 80$;
$d\left(\frac{3}{2}\right) - 108$; $d(2) - 128$;
$d(t_0) = -16t_0^2 + 96t_0$ **b.** $t = 0, 6$
c. $(12 \pm \sqrt{143})/4$ **29.** $g(3) = 1$;
$g(x + 4) = |x|$ **31. a.** $2x + h$
b. $4x + 2h - 3$ **c.** $3x^2 + 3xh + h^2$
33. a. $-1/(x - 1)(a - 1)$ **b.** $\dfrac{-1}{2(x - 1)}$
c. $\dfrac{-1}{(x - 1)(x + h - 1)}$ **d.** $\dfrac{-1}{2(2 + h)}$
35. a. $x_0 = \frac{11}{5}$ **b.** $x_0 = \pm 2$ **c.** $x_0 = 0, 1$
d. $x_0 = \frac{3}{2}, -1$ **37. a.** \$125.51
b. \$363.76
39. a.

n	2	3	4	5
$g(n)$	1.4142	1.4422	1.4142	1.3797

n	6	7	8
$g(n)$	1.3480	1.3205	1.2968

b. $n = 15$ **41. a.** $a = 1$ and $b = 2$
(other answers are possible)
43. $a = 2$ (other answers are possible)
45. a. $f(a) = 0$; $f(2a) = \frac{1}{3}$; $f(3a) = \frac{1}{2}$;
no **49.** $a = \frac{1}{2}$ and $b = -\frac{3}{2}$
55. $k = -\frac{1}{3}$ **57. a.** 0 **b.** 1 **c.** 2
d. 6 **e.** -1 **f.** -2 **g.** -6 **h.** $\frac{1}{2}$
59. 0 **61. a.** x **b.** $\frac{22}{7}$ **63.** Note
that $G(4) = 3$ but also $G(4) = 5$, so G is
not a function. Define G to assign the
closest prime number less than or equal
to x (or greater than or equal to x).
65. $P(1) = 17$; $P(2) = 19$; $P(3) = 23$;
$P(4) = 29$; yes ($x = 17$)

Exercise Set 3.2

1. domain: $[-4, 2]$; range: $[-3, 3]$
3. domain: $[-4, -1) \cup (-1, 4]$;

range: $[-2, 3)$ **5.** domain: $[-4, 4]$;
range: $[-2, 2)$ **7.** domain: $[-4, 3]$;
range: $\{2\}$ **9. a.** 1 **b.** -3 **c.** no
d. $x = 2$ **c.** -1 **11. a.** positive
b. $f(-2) = 4$; $f(1) = 1$; $f(2) = 2$;
$f(3) = 0$ **c.** $f(2)$ **d.** -3 **e.** 3
f. domain: $[-2, 4]$; range: $[-2, 4]$
13. a. $g(-2)$ **b.** 5 **c.** $f(2) - g(2)$
d. $x = -2$ or $x = 3$ **e.** range of f
15.

Function	$|x|$	x^2	x^3
Domain	$(-\infty, \infty)$	$(-\infty, \infty)$	$(-\infty, \infty)$
Range	$[0, \infty)$	$[0, \infty)$	$(-\infty, \infty)$
Turning Point	$(0, 0)$	$(0, 0)$	none
Maximum Value	none	none	none
Minimum Value	0	0	none
Interval(s) Where Increasing	$(0, \infty)$	$(0, \infty)$	$(-\infty, \infty)$
Interval(s) Where Decreasing	$(-\infty, 0)$	$(-\infty, 0)$	none

17. a. $[-1, 1]$ **b.** 1 (occurring at $x = 1$)
c. -1 (occurring at $x = 3$) **d.** $[0, 1]$ and
$[3, 4]$ **e.** $[1, 3]$ **19. a.** $[-3, 0]$
b. 0 (occurring at $x = 0$ and $x = 4$)
c. -3 (occurring at $x = 2$) **d.** $[2, 4]$
e. $[0, 2]$ **21. a.** yes **b.** no **c.** no
d. yes
23.

25.

27. a.

b.

29.

31.

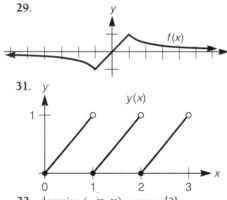

33. domain: $(-\infty, \infty)$; range: $\{3\}$
35. 7: this represents the average rate of change of f on the interval $[3, 4]$.
37. 10 **39.** 0 **41.** 2
43. $-3/ab$ **45.** $-2(a^2 + ab + b^2)$
47. a. $\frac{1}{3}°$ C/min **b.** $\frac{4}{3}°$ C/min **c.** $\frac{1}{2}°$ C/min
49. $b = 5$

51. a.

Interval	$[1, 1.1]$	$[1, 1.01]$	$[1, 1.001]$
$\Delta s/\Delta t$	33.6	32.16	32.016

Interval	$[1, 1.0001]$	$[1, 1.00001]$
$\Delta s/\Delta t$	32.0016	32.00016

b. 32 ft/sec   **53. a.** 4° F/hr
b. $-2°$ F/hr **c.** 0° F/hr **d.** 1° F/hr
55. a.

t	0	0.25	0.5	0.75	1	1.25	1.5
$S(t)$	0	0.25	0.5	0.75	1	1	1

t	1.75	2	2.25	2.5	2.75	3
$S(t)$	1	1	1	1	1	1

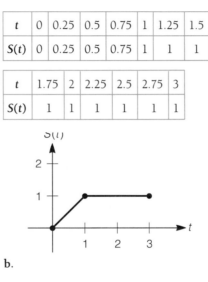

b.

t	3	3.25	3.5	3.75	4
$S(t)$	1	0.75	0.5	0.25	0

t	4.25	4.5	4.75	5
$S(t)$	-0.25	-0.5	-0.75	-1

c.

d.

The graph is a displacement of the graph from part (c), eight units to the right.
57. a.

x	0	0.1	0.5	0.9	1.0
$[x]$	0	0	0	0	1

b.

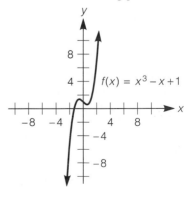

c. The range is the set of all integers.

Optional TI-81 Graphing Calculator
Exercises for Section 3.2

1. a. There are two turning points.

b.

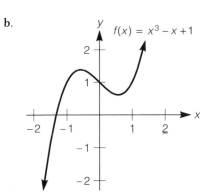
$f(x) = x^3 - x + 1$

c. $(0.568, 0.615)$ and $(-0.568, 1.385)$
d. $x \approx \pm 0.577$; one decimal place
e. increasing: $(-\infty, -1/\sqrt{3})$ and $(1/\sqrt{3}, \infty)$; decreasing: $(-1/\sqrt{3}, 1/\sqrt{3})$
3. a. There are two turning points.

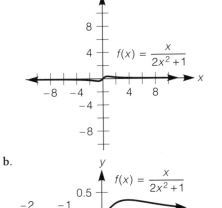
$f(x) = \dfrac{x}{2x^2 + 1}$

b.

$f(x) = \dfrac{x}{2x^2 + 1}$

c. $(0.695, 0.353)$ and $(-0.695, -0.353)$
d. $x \approx \pm 0.707$; one decimal place
e. increasing: $\left(-\sqrt{2}/2, \sqrt{2}/2\right)$; decreasing: $\left(-\infty, -\sqrt{2}/2\right)$ and $\left(\sqrt{2}/2, \infty\right)$
5. a. This is the slope from $(a, f(a))$ to $(b, f(b))$. **b.** 3   **7. a.** 13 **b.** 0
9. a. approximately 1.1750
b. approximately 1.0016
c. approximately 0.9999

Exercise Set 3.3

1. a. C **b.** F **c.** I **d.** A **e.** J **f.** K **g.** D
h. B **i.** E **j.** H **k.** G

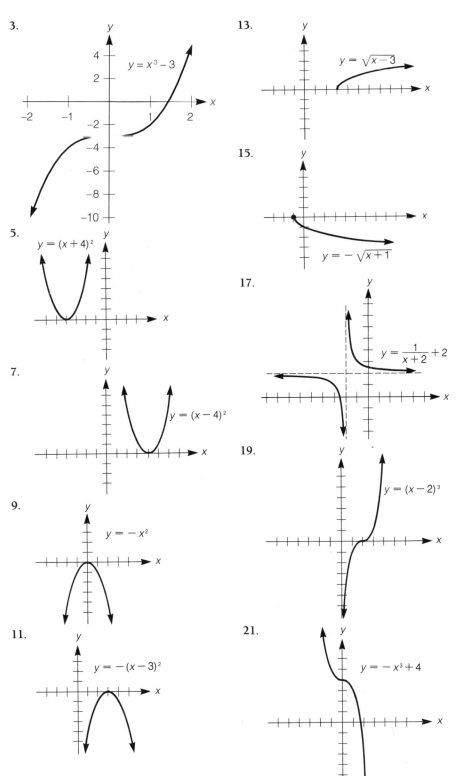

3.
$y = x^3 - 3$

5.
$y = (x + 4)^2$

7.
$y = (x - 4)^2$

9.
$y = -x^2$

11.
$y = -(x - 3)^2$

13.
$y = \sqrt{x - 3}$

15.
$y = -\sqrt{x + 1}$

17.
$y = \dfrac{1}{x + 2} + 2$

19.
$y = (x - 2)^3$

21.
$y = -x^3 + 4$

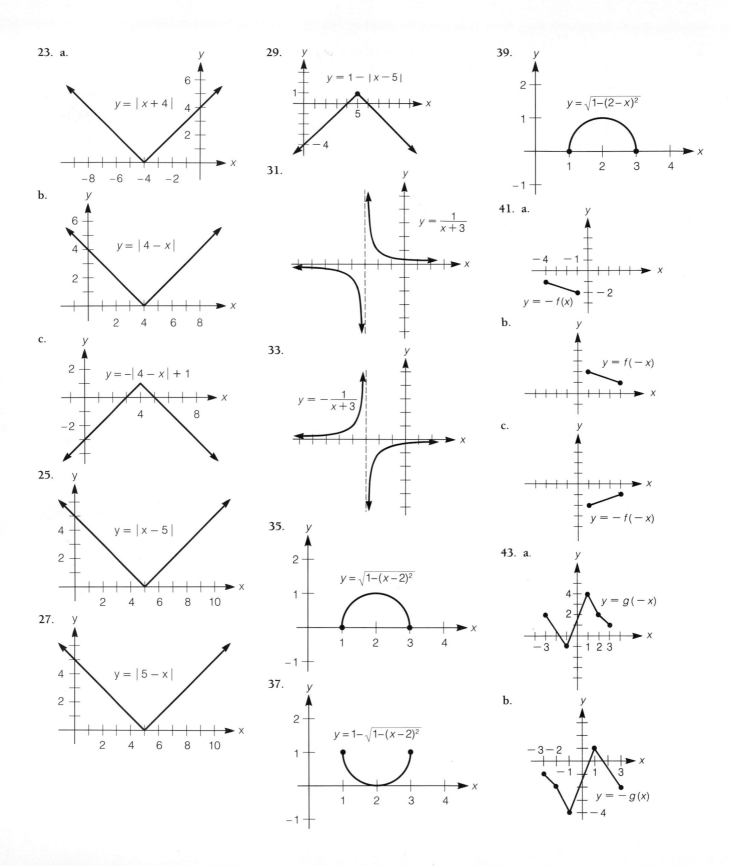

23. a.

$y = |x + 4|$

b.

$y = |4 - x|$

c.

$y = -|4 - x| + 1$

25.

$y = |x - 5|$

27.

$y = |5 - x|$

29.

$y = 1 - |x - 5|$

31.

$y = \dfrac{1}{x + 3}$

33.

$y = -\dfrac{1}{x + 3}$

35.

$y = \sqrt{1 - (x - 2)^2}$

37.

$y = 1 - \sqrt{1 - (x - 2)^2}$

39.

$y = \sqrt{1 - (2 - x)^2}$

41. a.

$y = -f(x)$

b.

$y = f(-x)$

c.

$y = -f(-x)$

43. a.

$y = g(-x)$

b.

$y = -g(x)$

A-68

c.

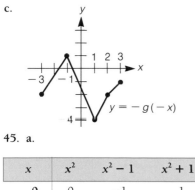

$y = -g(-x)$

45. a.

x	x^2	$x^2 - 1$	$x^2 + 1$
0	0	-1	1
± 1	1	0	2
± 2	4	3	5
± 3	9	8	10

b. The graphs are displacements of $y = x^2$ up/down 1 unit.

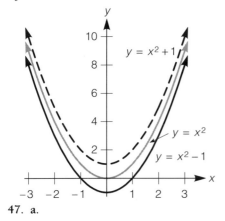

$y = x^2 + 1$

$y = x^2$

$y = x^2 - 1$

47. a.

x	$\sqrt{x}$	$-\sqrt{x}$
0	0.0	0.0
1	1.0	-1.0
2	1.4	-1.4
3	1.7	-1.7
4	2.0	-2.0
5	2.2	-2.2

b. The graph is a reflection of $y = \sqrt{x}$ across the x-axis.

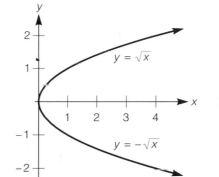

$y = \sqrt{x}$

$y = -\sqrt{x}$

49. a. $(a + 3, b)$ **b.** $(a, b - 3)$
c. $(a + 3, b - 3)$ **d.** $(a, -b)$ **e.** $(-a, b)$
f. $(-a, -b)$ **g.** $(-a + 3, b)$
h. $(-a + 3, -b + 1)$

51.

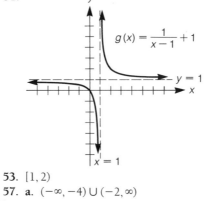

$g(x) = \dfrac{1}{x - 1} + 1$

$y = 1$

$x = 1$

53. $[1, 2)$
57. a. $(-\infty, -4) \cup (-2, \infty)$
b. $(-\infty, 2) \cup (4, \infty)$

Optional TI-81 Graphing Calculator Exercises for Section 3.3

1. a.

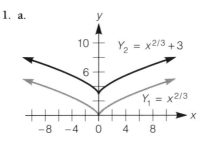

$Y_2 = x^{2/3} + 3$

$Y_1 = x^{2/3}$

b.

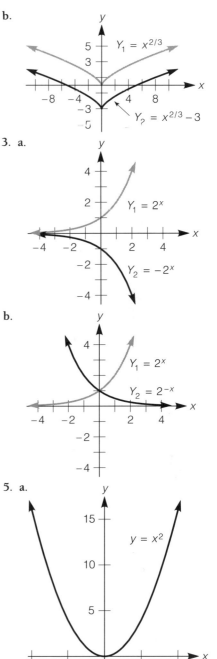

$Y_1 = x^{2/3}$

$Y_2 = x^{2/3} - 3$

3. a.

$Y_1 = 2^x$

$Y_2 = -2^x$

b.

$Y_1 = 2^x$

$Y_2 = 2^{-x}$

5. a.

$y = x^2$

b.

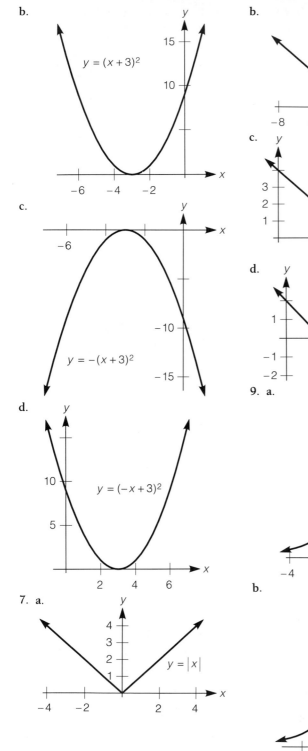

$y = (x + 3)^2$

c.

$y = -(x + 3)^2$

d.

$y = (-x + 3)^2$

7. a.

$y = |x|$

b.

$y = |x + 4|$

c.

$y = |-x + 4|$

d.

$y = |-x + 4| - 2$

9. a.

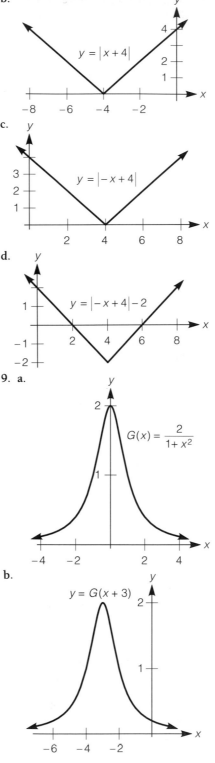

$G(x) = \dfrac{2}{1 + x^2}$

b.

$y = G(x + 3)$

c.

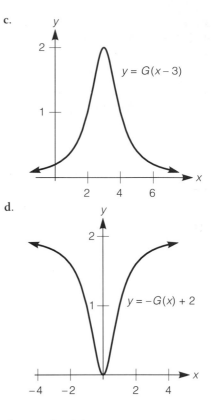

$y = G(x - 3)$

d.

$y = -G(x) + 2$

Exercise Set 3.4

1. a. $x^2 - x - 7$ **b.** $-x^2 + 5x + 5$ **c.** 5

3. a. $x^2 - 2x - 8$ **b.** $-x^2 + 2x + 8$

5. a. $4x - 2$ **b.** $4x - 2$ **c.** -4

7. a. $\dfrac{-x^4 + 22x^2 - 4x - 80}{2x^3 - x^2 - 18x + 9}$ **b.** $-\frac{80}{9}$

9. a. $-x^5 + 9x^3 + 2x^2 - 18$

b. $-x^5 + 9x^3 + 2x^2 - 18$ **c.** -24

11. a. $-6x - 14$ **b.** -74 **c.** $-6x - 7$

d. -67 **13. a.** $(f \circ g)(x) = -x$;

$(f \circ g)(-2) = 2$; $(g \circ f)(x) = 2 - x$;

$(g \circ f)(-2) = 4$

b. $(f \circ g)(x) = 9x^2 - 3x - 6$;

$(f \circ g)(-2) = 36$;

$(g \circ f)(x) = -3x^2 + 9x + 14$;

$(g \circ f)(-2) = -16$

c. $(f \circ g)(x) = (1 - x^4)/3$;

$(f \circ g)(-2) = -5$; $(g \circ f)(x) = 1 - (x^4/81)$;

$(g \circ f)(-2) = \frac{65}{81}$ **d.** $(f \circ g)(x) = 2^{x^2+1}$;

$(f \circ g)(-2) = 32$; $(g \circ f)(x) = 2^{2x} + 1$;

$(g \circ f)(-2) = \frac{17}{16}$ **e.** $(f \circ g)(x) = 3x^5 - 4x^2$;

$(f \circ g)(-2) = -112$; $(g \circ f)(x) = 3x^5 - 4x^2$;

$(g \circ f)(-2) = -112$

f. $(f \circ g)(x) = x$; $(f \circ g)(-2) = -2$;

$(g \circ f)(x) = x$; $(g \circ f)(-2) = -2$

15. a. $(-x + 7)/6x$ **b.** $(-t + 7)/6t$ **c.** $\frac{5}{12}$
d. $(1 - 6x)/7$ **e.** $(1 - 6y)/7$ **f.** $-\frac{11}{7}$
17. a. $M(7) = \frac{13}{5}$; $M[M(7)] = 7$ **b.** x
c. 7 **19. a.** 1 **b.** -3 **c.** -1 **d.** 2
e. 2 **f.** -3
21. a. $(T \circ I)(x) = 4x^3 - 3x^2 + 6x - 1$;
$(I \circ T)(x) = 4x^3 - 3x^2 + 6x - 1$
b. $(G \circ I)(x) = ax^2 + bx + c$;
$(I \circ G)(x) = ax^2 + bx + c$
c. $(f \circ I)(x) = f(x)$; $(I \circ f)(x) = f(x)$
23.

x	0	1	2	3	4	-1
$(f \circ g)(x)$	1	3	2	undef.	2	0

x	0	1	2	3	4
$(g \circ f)(x)$	0	3	4	2	undef.

25. a. $(f \circ g)(x) = 6x - 7$

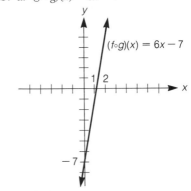

b. $(g \circ f)(x) = 6x - 1$

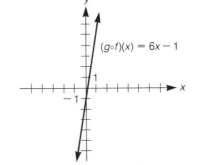

27. a. domain: $[0, \infty)$; range: $[-3, \infty)$

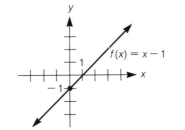

b. domain: $(-\infty, \infty)$; range $(-\infty, \infty)$

c. $(f \circ g)(x) = \sqrt{x} - 4$; domain: $[0, \infty)$;
range: $[-4, \infty)$

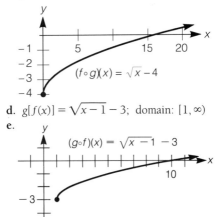

d. $g[f(x)] = \sqrt{x - 1} - 3$; domain: $[1, \infty)$
e.

29. $C(x) = (f \circ g)(x)$
31. a. $F(x) = (f \circ g)(x)$, where $f(x) = \sqrt[3]{x}$
and $g(x) = 3x + 4$ **b.** $G(x) = (f \circ g)(x)$,
where $f(x) = |x|$ and $g(x) = 2x - 3$
c. $H(x) = (f \circ g)(x)$, where $f(x) = x^5$ and
$g(x) = ax + b$ **d.** $T(x) = (f \circ g)(x)$, where
$f(x) = 1/x$ and $g(x) = \sqrt{x}$
33. a. $f(x) = (b \circ c)(x)$ **b.** $g(x) = (a \circ d)(x)$
c. $h(x) = (c \circ d)(x)$ **d.** $K(x) = (c \circ b)(x)$
e. $l(x) = (c \circ a)(x)$ **f.** $m(x) = (a \circ c)(x)$
g. $n(x) = (b \circ d)(x)$ or $n(x) = (d \circ b)(x)$
35. $(C \circ f)(t) = 2\pi/(t^2 + 1)$; $\pi/5$ ft
37. a. $(C \circ f)(t) = 100 + 450t - 25t^2$
b. \$1225 **c.** no **39.** $f(x) = (x + 6)/4$
41. $a = -\frac{1}{2}$ and $b = \frac{1}{2}$
43. a. $2x + 2a - 2$ **b.** $4x + 4a - 4$
45. a. $(x^2/2) + 1$ **b.** $(x + 1)^2/2$

c. $(x^2/4) + 1$ **d.** $(x + 1)^2/4$ **e.** $(x^2 + 1)/2$
47. a. $p(x) = (g \circ f \circ h)(x)$
b. $q(x) = (h \circ g \circ f)(x)$
c. $r(x) = (f \circ g \circ h)(x)$
d. $s(x) = (h \circ f \circ g)(x)$ **49. a.** P: $f(a)$;
Q: $f(f(a))$; R: $f(f(f(a)))$
51. a. 4, 2, 1.41, 1.19, 1.09, 1.04
b. 1.022, 1.011, 1.005, 1.003, 1.001,
1.001; they are approaching 1.
53. a.

∘	i	a	b	c
i	i	a	b	c
a	a	i	c	b
b	b	c	i	a
c	c	b	a	i

b. yes **c.** $a^2 = i$; $b^2 = i$; $c^2 = i$; $c^3 = c$
d. yes **e.** {i} **g.** yes

Optional TI-81 Graphing Calculator
Exercises for Section 3.4

1.

x	0.1	0.2	0.3
$f(g(x))$	194.672	148.176	109.744

3. a.

x	7	8	9
$f(g(x))$	524	1614	3714

b.

x	7	8	9
$g(f(x))$	-20015	-34403	-55453

5. a. 2.25, 2.236111111, and
2.236067978; reached in three iterations
b. 2.333333333, 2.238095238,
2.236068896, and 2.236067978; reached
in four iterations **c.** 12.6, 6.498412698,
3.63391568, 2.504920969, 2.250495964,
2.236114227, and 2.236067978; reached
in seven iterations **d.** $\sqrt{5} \approx 2.236067978$
7. a. 3.733333333, 3.206363379,
3.110265742, 3.107235463, and
3.107232506; reached in five iterations

b. 16.6826667, 11.15770876,
7.518797361, 5.189421557,
3.830946132, 3.235341778,
3.112238272, 3.107240553, and
3.107232506; reached in nine iterations
c. $\sqrt[3]{30} \approx 3.107232506$

Exercise Set 3.5

3. a. 4 **b.** −1 **c.** $\sqrt{2}$ **d.** $t + 1$
e. $f(0) = 1$ and $f^{-1}(1) = 0$
f. $f(-1) = -2$ and $f^{-1}(-2) = -1$
5. a. $f^{-1}(x) = (x + 1)/3$
c.

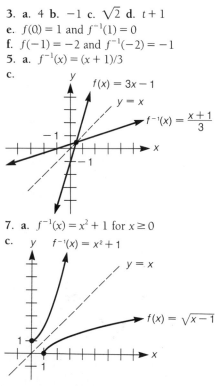

7. a. $f^{-1}(x) = x^2 + 1$ for $x \geq 0$
c.

9. a. domain: $(-\infty, 3) \cup (3, \infty)$;
range: $(-\infty, 1) \cup (1, \infty)$

b. $f^{-1}(x) = \dfrac{3x + 2}{x - 1}$

c. domain of f^{-1}: $(-\infty, 1) \cup (1, \infty)$;
range of f^{-1}: $(-\infty, 3) \cup (3, \infty)$
domain of f = range of f^{-1},
range of f = domain of f^{-1}
11. $f^{-1}(x) = \sqrt[3]{(x - 1)/2}$
13.

15. a. $f^{-1}(x) = \sqrt[3]{x + 1} + 3$
b.

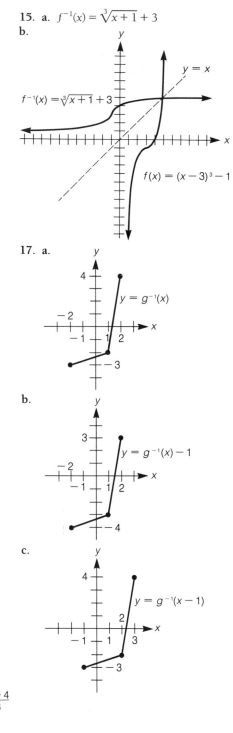

17. a.

b.

c.

d.

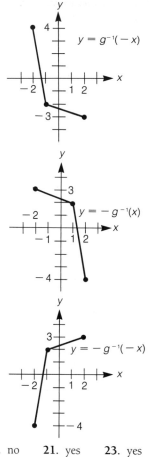

e.

f.

19. no   **21.** yes   **23.** yes
25. no   **27.** no   **29.** yes
33. a. $x = 7$ **b.** $x = -3$ **35.** $t = \frac{9}{4}$
37. a. $f^{-1}(x) = x^2$; domain: $[0, \infty)$
b. (i) graph of f; (ii) graph of f^{-1};
(iii) graph of f; (iv) graph of f^{-1};
(v) graph of f; (vi) graph of f^{-1};
(vii) graph of f^{-1}; (viii) graph of f
39. $A:(a, f(a))$; $B:(f(a), f(a))$;
$C:(f(f(a)), f(a))$; $D:(f(a), f(f(a)))$

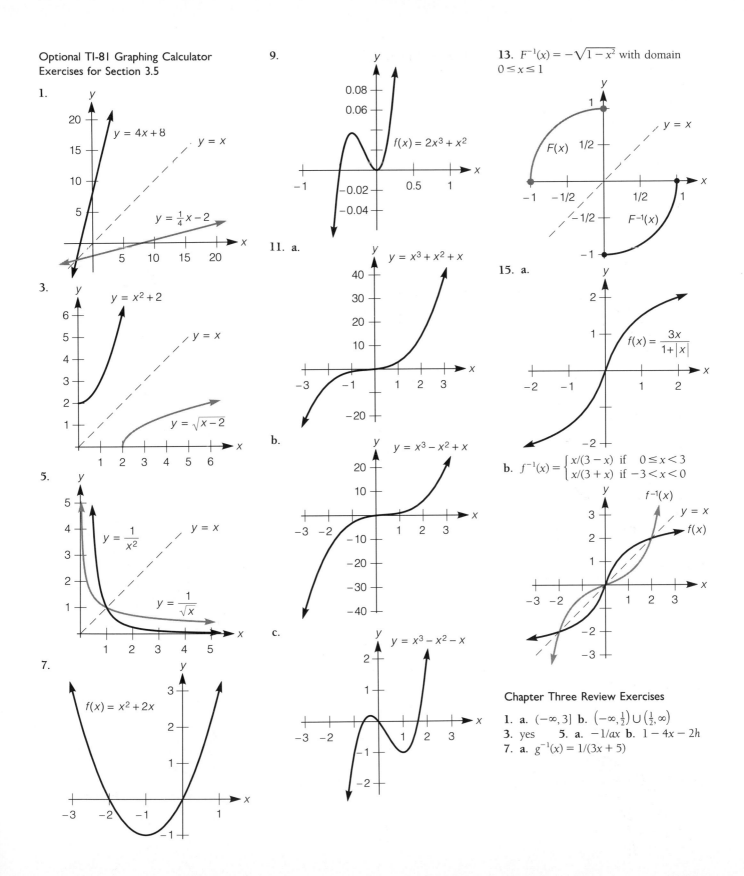

Optional TI-81 Graphing Calculator
Exercises for Section 3.5

1.

$y = 4x + 8$
$y = x$
$y = \frac{1}{4}x - 2$

3.

$y = x^2 + 2$
$y = x$
$y = \sqrt{x - 2}$

5.

$y = \frac{1}{x^2}$
$y = x$
$y = \frac{1}{\sqrt{x}}$

7.

$f(x) = x^2 + 2x$

9.

$f(x) = 2x^3 + x^2$

11. a.

$y = x^3 + x^2 + x$

b.

$y = x^3 - x^2 + x$

c.

$y = x^3 - x^2 - x$

13. $F^{-1}(x) = -\sqrt{1 - x^2}$ with domain $0 \le x \le 1$

$y = x$
$F(x)$
$F^{-1}(x)$

15. a.

$f(x) = \dfrac{3x}{1 + |x|}$

b. $f^{-1}(x) = \begin{cases} x/(3 - x) & \text{if } 0 \le x < 3 \\ x/(3 + x) & \text{if } -3 < x < 0 \end{cases}$

$f^{-1}(x)$
$y = x$
$f(x)$

Chapter Three Review Exercises

1. a. $(-\infty, 3]$ **b.** $\left(-\infty, \frac{1}{2}\right) \cup \left(\frac{1}{2}, \infty\right)$
3. yes **5. a.** $-1/ax$ **b.** $1 - 4x - 2h$
7. a. $g^{-1}(x) = 1/(3x + 5)$

b.

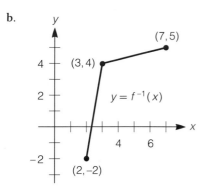

$y = f^{-1}(x)$

(7,5)

(3,4)

(2,−2)

9. a. x-intercepts: -5, 1; y-intercept: -1

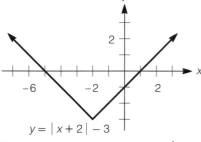

$y = |x + 2| - 3$

b. x-intercept: -1; y-intercept; $-\frac{1}{2}$

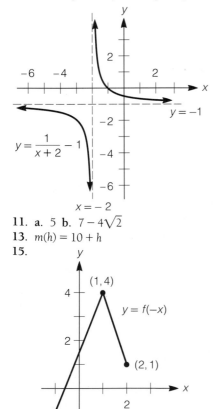

$y = \dfrac{1}{x + 2} - 1$

$y = -1$

$x = -2$

11. a. 5 **b.** $7 - 4\sqrt{2}$
13. $m(h) = 10 + h$
15.

(1,4)

$y = f(-x)$

(2,1)

(−1,−1)

17. a. E **b.** C **c.** L **d.** A **e.** J **f.** G
g. B **h.** M **i.** K **j.** D **k.** I **l.** H **m.** N
n. F   **19.** x-intercepts: $1 \pm \sqrt{2}$;
y-intercept: 1

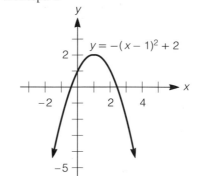

$y = -(x - 1)^2 + 2$

21. no x-intercept; y-intercept: 1

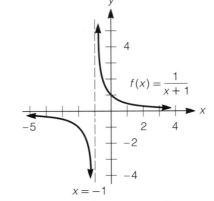

$f(x) = \dfrac{1}{x + 1}$

$x = -1$

23. x-intercept: -3; y-intercept: 3

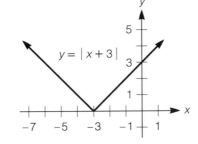

$y = |x + 3|$

25. x-intercepts: ± 1; y-intercept: 1

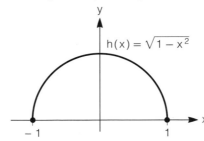

$h(x) = \sqrt{1 - x^2}$

27. x-intercept: 0; y-intercept: 0

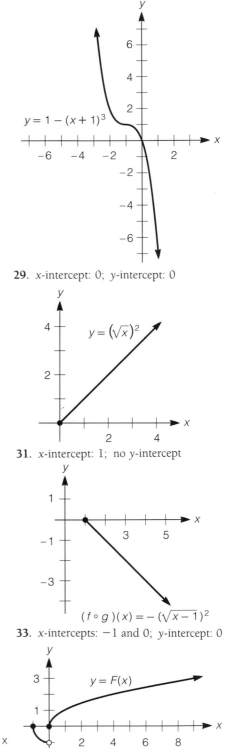

$y = 1 - (x + 1)^3$

29. x-intercept: 0; y-intercept: 0

$y = \left(\sqrt{x}\right)^2$

31. x-intercept: 1; no y-intercept

$(f \circ g)(x) = -\left(\sqrt{x - 1}\right)^2$

33. x-intercepts: -1 and 0; y-intercept: 0

$y = F(x)$

35. no *x*- or *y*-intercepts

37. *x*-intercept: $\frac{1}{2}$; *y*-intercept: -1

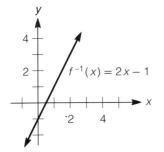

$f^{-1}(x) = 2x - 1$

39. *x*-intercept: 0; *y*-intercept: 0

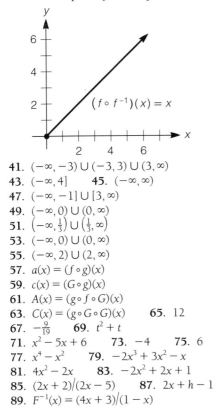

$(f \circ f^{-1})(x) = x$

41. $(-\infty, -3) \cup (-3, 3) \cup (3, \infty)$
43. $(-\infty, 4]$ **45.** $(-\infty, \infty)$
47. $(-\infty, -1] \cup [3, \infty)$
49. $(-\infty, 0) \cup (0, \infty)$
51. $\left(-\infty, \frac{1}{3}\right) \cup \left(\frac{1}{3}, \infty\right)$
53. $(-\infty, 0) \cup (0, \infty)$
55. $(-\infty, 2) \cup (2, \infty)$
57. $a(x) = (f \circ g)(x)$
59. $c(x) = (G \circ g)(x)$
61. $A(x) = (g \circ f \circ G)(x)$
63. $C(x) = (g \circ G \circ G)(x)$ **65.** 12
67. $-\frac{9}{19}$ **69.** $t^2 + t$
71. $x^2 - 5x + 6$ **73.** -4 **75.** 6
77. $x^4 - x^2$ **79.** $-2x^3 + 3x^2 - x$
81. $4x^2 - 2x$ **83.** $-2x^2 + 2x + 1$
85. $(2x + 2)/(2x - 5)$ **87.** $2x + h - 1$
89. $F^{-1}(x) = (4x + 3)/(1 - x)$

91. *x* for $x \neq 4$ **93.** *x*
95. $(1 + x)/2$ **97.** $\frac{22}{7}$ **99.** negative
101. -1 **103.** -1
105. $(0, -2)$ and $(5, 1)$
107. $[-6, 0]$ and $[5, 8]$
109. 0 (occurring at $x = 2$) **111.** no
113. $x = 4$ **115. a.** $x = 10$ **b.** $x = 0$
117. a. 5 **b.** -3 **c.** 4 **d.** $\frac{1}{4}$
119. $(f \circ f)(10)$ **121.** $[0, 4]$
123. 5 **125.** $(1, 3) \cup (6, 10)$
127. $(4, 7)$

Chapter Three Test

1. $(-\infty, -1] \cup [6, \infty)$
2. $\left(-\infty, \frac{2}{3}\right) \cup \left(\frac{2}{3}, \infty\right)$
3. a. $2x^2 - 2x - 2$ **b.** $2x^2 - 5x + 2$
c. 54 **4.** $-2/at$ **5.** $4x + 2h - 5$
6. $g^{-1}(x) = -x/(6x + 4)$ **7.** $-\frac{8}{5}$
8. a. *x*-intercepts: 4 and 2;
y-intercept: -2

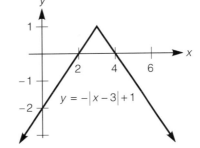

$y = -|x - 3| + 1$

b. *x*-intercept: $-\frac{5}{2}$; *y*-intercept: $-\frac{5}{3}$

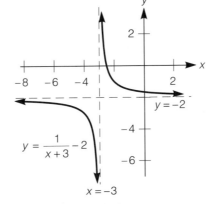

$y = \dfrac{1}{x + 3} - 2$

9. a. $[-3, 1]$ **b.** $(-2, 1)$
c. -3 (when $x = 2$) **d.** 1 (when $x = -2$)
e. $(-2, 2)$ **f.** $-\frac{1}{3}$ **10. a.** $\frac{23}{4}$
b. $12 - 7\sqrt{3}$

11. domain: $(-\infty, -1) \cup (-1, \infty)$

$y = F(x)$

12.

$(-1, 3)$

$(-5, -2)$

13. $t = 1$ **14.** $b = 10$
15. a. $\Delta F/\Delta t = 0.16t_1 - 0.8$
b. $-0.16°$ C/hr

CHAPTER FOUR

Exercise Set 4.1

1. $f(x) = \frac{2}{3}x + \frac{2}{3}$ **3.** $g(x) = \sqrt{2}x$
5. $f(x) = x - \frac{7}{2}$ **7.** $f(x) = \sqrt{3}$
9. $f(x) = \frac{1}{2}x - 2$ **11.** yes
13. $V(t) = -2375t + 20000$
15. a. $V(t) = -12000t + 60000$

b.

End of Year	Yearly Depreciation	Accumulated Depreciation	Value V
0	0	0	60,000
1	12,000	12,000	48,000
2	12,000	24,000	36,000
3	12,000	36,000	24,000
4	12,000	48,000	12,000
5	12,000	60,000	0

17. a. $530 **b.** $538 **c.** $8/fan
19. a. 50 **b.** $50/player
c. The answers are the same.
21. a. $\frac{4}{5}$ ft/sec **b.** 0 cm/sec **c.** 8 mph
23. a. B is traveling faster.
b. A is farther to the right. **c.** $t = 8$ sec
25. a.

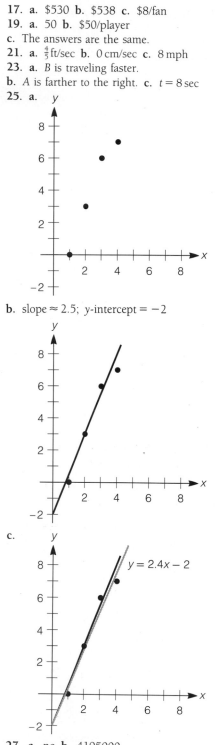

b. slope ≈ 2.5; y-intercept $= -2$

c.

$y = 2.4x - 2$

27. a. no **b.** 4195000
29. a. 3.617̸4 billion dollars **b.** no
31. a. 2.5, 3.25, and 3.625
b. The y-coordinate of point A is

$f(1) = 2.5$. The y-coordinate of point B is
$f(2.5) = 3.25$. The y-coordinate of point
C is $f(3.25) = 3.625$. **c.** 4 **d.** 3.813,
3.906, 3.953, 3.977, 3.988, 3.994, and
3.997 **33. a.** A **b.** A^2 **c.** AC **d.** AC
35. a. $1/AC$ **b.** $1/AC$
39. $f(x) = 3x + 1$
41. a. $\Sigma x = 10$; $\Sigma y = 16$
b. $\Sigma x^2 = 30$; $\Sigma xy = 52$ **c.** $m = 2.4$,
$b = -2$ **43.** $y = 2.4x - 0.4$
45. $y = 0.084x + 37.241$

Optional TI-81 Graphing Calculator
Exercises for Section 4.1

1. The y-coordinate is 4250, which
matches the result from the text.

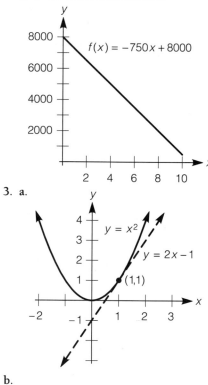

$f(x) = -750x + 8000$

3. a.

$y = x^2$

$y = 2x - 1$

$(1,1)$

b.

x	0.9	0.99	0.999
$2x - 1$	0.8	0.98	0.998
x^2	0.81	0.9801	0.998001

x	1.1	1.01	1.001
$2x - 1$	1.2	1.02	1.002
x^2	1.21	1.0201	1.002001

Exercise Set 4.2

1. vertex: $(-2, 0)$;
axis of symmetry: $x = -2$;
minimum value: 0; x-intercept: -2;
y-intercept: 4

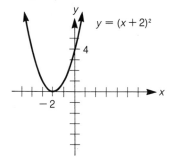

$y = (x + 2)^2$

3. vertex: $(-2, 0)$;
axis of symmetry: $x = -2$;
minimum value: 0; x-intercept: -2;
y-intercept: 8

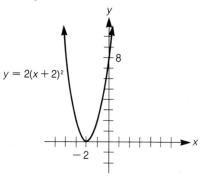

$y = 2(x + 2)^2$

5. vertex: $(-2, 4)$;
axis of symmetry: $x = -2$;
maximum value: 4;
x-intercepts: $-2 \pm \sqrt{2}$; y-intercept: -4

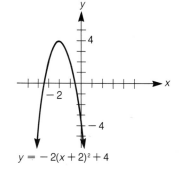

$y = -2(x + 2)^2 + 4$

7. vertex: $(2, -4)$;
axis of symmetry: $x = 2$;
minimum value: -4; x-intercepts: 0, 4;
y-intercept: 0

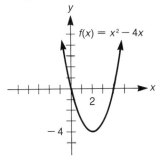

9. vertex: $(0, 1)$; axis of symmetry: $x = 0$;
maximum value: 1; x-intercepts: ± 1;
y-intercept: 1

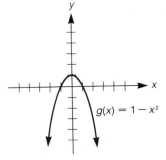

11. vertex: $(1, -4)$;
axis of symmetry: $x = 1$;
minimum value: -4;
x-intercepts: 3 and -1; y-intercept: -3

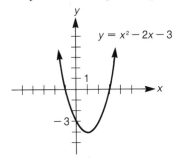

13. vertex: $(3, 11)$;
axis of symmetry: $x = 3$;
maximum value: 11;
x-intercepts: $3 \pm \sqrt{11}$; y-intercept: 2

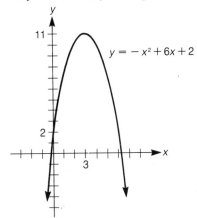

15. vertex: $(0, 0)$; axis of symmetry: $t = 0$;
minimum value: 0; t-intercept: 0;
s-intercept: 0

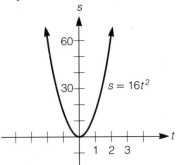

17. vertex: $\left(\frac{1}{6}, \frac{9}{4}\right)$; axis of symmetry: $t = \frac{1}{6}$;
maximum value: $\frac{9}{4}$; t-intercepts: $-\frac{1}{3}, \frac{2}{3}$;
s-intercept: 2

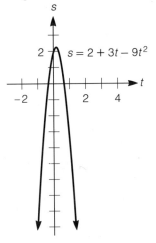

19. The input $x = 1$ will yield a minimum output value.

21. The input $x = \frac{3}{2}$ will yield a maximum output value.

23. The input $x = 0$ will yield a minimum output value.

25. minimum value of -13

27. maximum value of $\frac{25}{8}$

29. maximum value of 1000

31. 5 units   **33.** quadratic

35. neither   **37.** linear

39. a. minimum value at $(3, 8)$

b. mininum value at $(3, 4)$

c. minimum value at $\left(\pm\sqrt{3}, 64\right)$

41. a. 4 **b.** $2\sqrt[3]{2}$ **c.** 16

43. $ax_1 + ax_2 + b$

49. a. The y-coordinate of point A is $f(x_0)$, the first iterate of x_0. The y-coordinate of B is $f(f(x_0))$, the second iterate of x_0. The y-coordinate of C is $f(f(f(x_0)))$, the third iterate of x_0. The y-coordinate of D is $f(f(f(f(x_0))))$, the fourth iterate of x_0. **b.** -0.56, -0.2864, -0.518, and -0.332

51. $y = -\frac{1}{2}(x - 2)^2 + 2$

53. $y = \frac{1}{4}(x - 3)^2 - 1$

55. $(x + 3)^2 + (y + 4)^2 = 25$

Optional TI-81 Graphing Calculator Exercises for Section 4.2

1. a. The graph "narrows" and produces larger y-coordinates.

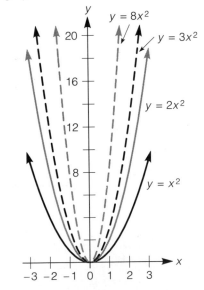

b. The graph should be "flatter" than the others.

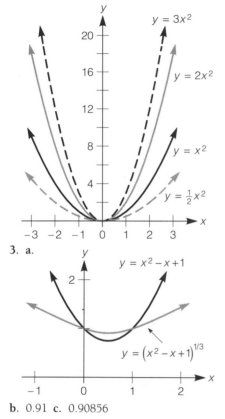

$y = 3x^2$
$y = 2x^2$
$y = x^2$
$y = \frac{1}{2}x^2$

3. a.

$y = x^2 - x + 1$

$y = \left(x^2 - x + 1\right)^{1/3}$

b. 0.91 **c.** 0.90856

Exercise Set 4.3

1. a. $A(x) = 8x - x^2$
b. $P(x) = 2x + 170/x$
3. a. $D(x) = \sqrt{x^4 + 3x^2 + 1}$
b. $m(x) = (x^2 + 1)/x$
5. a. $A(y) = \pi y^2/4$ **b.** $A(y) = \pi^2 y^2/16$
7. a. $P(x) = 16x - x^2$
b. $S(x) = 2x^2 - 32x + 256$
c. $D(x) = (16 - x)^3 - x^3$
d. $A(x) = 8$; notice that the average does not depend on what the two numbers are.
9. $R(x) = -\frac{1}{4}x^2 + 8x$
11. a.

x	1	2	3	4
$P(x)$	17.88	19.49	20.83	21.86

x	5	6	7
$P(x)$	22.49	22.58	21.75

b. 22.58, corresponding to $x = 6$
c. 22.63; this is larger than any of our table values.   **13. a.** $h(s) = \sqrt{3}s$
b. $A(s) = \sqrt{3}s^2$ **c.** $4\sqrt{3}$ cm **d.** $\frac{25}{4}\sqrt{3}$ in.^{2}
15. $V(r) = 2\pi r^3$ **17. a.** $h(r) = 12/r^2$
b. $S(r) = 2\pi r^2 + (24\pi/r)$
19. $V(S) = S\sqrt{\pi S}/6\pi$
21. $A(x) = \frac{1}{2}x\sqrt{400 - x^2}$
23. $d(x) = (x + 4)\sqrt{x^2 + 25}/x$
25. $x = 25$ yields the largest area; $L = 25$

x	5	10	20	24	24.8	24.9
$A(x)$	225	400	600	624	624.96	624.99

x	25	25.1	25.2	45
$A(x)$	625	624.99	624.96	225

27. a. The value $x = 2$ yields the largest area.

x	1	2	3	4
A	7.5	12	10.5	0

The value $x = 2.25$ yields the largest area.

x	1.75	2.00	2.25
A	11.3203	12.0000	12.3047

x	2.50	2.75
A	12.1875	11.6016

The value $x = 2.30$ yields the largest area.

x	2.15	2.20	2.25
A	12.2308	12.2760	12.3047

x	2.30	2.35
A	12.3165	12.3111

b. $x = 2.30$ is the closest x-value; area = 12.3168
29. $A(x) = \frac{17}{144}x^2 - \frac{1}{3}x + \frac{1}{2}$; yes
31. a. $V(r) = \sqrt{3}\pi r^3/3$ **b.** $S(r) = 2\pi r^2$
33. a. $r(h) = 3h/\sqrt{h^2 - 9}$
b. $h(r) = 3r/\sqrt{r^2 - 9}$

35. $A(x) = [4x^2 + \pi(14 - x)^2]/16\pi$
37. $A(r) = r(1 - 4\pi r)/4$
39. $A(x) = \pi x^2/3$
41. $V(x) = 4x^3 - 28x^2 + 48x$
43. a. $A(r) = 32r - 2r^2 - (\pi r^2/2)$
b. downward; yes
45. a. $y(s) = 3s/\sqrt{1 - s^2}$
b. $s(y) = y/\sqrt{y^2 + 9}$
c. $z(s) = 3/\sqrt{1 - s^2}$ **d.** $s(z) = \sqrt{z^2 - 9}/z$
47. a. $m(a) = (a^2 + 1)/a$
49. $A(x) = (8x - x^2)/4$
51. $A(m) = (2m^2 - 8m + 8)/(m^2 - 4m)$
53. $A(m) = (ma - b)^2/(-2m)$

Exercise Set 4.4

1. $\frac{25}{4}$ **3.** $\frac{1}{2}$ **5.** $\frac{25}{4}$ m by $\frac{25}{4}$ m
7. 1250 in.^{2} **9. a.** 18 **b.** $\frac{23}{4}$ **c.** $\frac{47}{8}$
d. $\frac{95}{16}$ **11. a.** $h(1) = 16$ ft; $h\left(\frac{3}{2}\right) = 12$ ft
b. maximum height = 16 ft, attained after 1 sec **c.** $t = \frac{7}{4}$ sec or $t = \frac{1}{4}$ sec
13. point: $\left(\frac{7}{2}, (2 + \sqrt{6})/2\right)$;
distance: $\sqrt{7}/2$ **15. a.** $\frac{1}{2}$ **b.** $\frac{1}{4}$
17. 125 ft by 250 ft **19.** $x = 40$
21. $x = 60$; revenue = \$900; price = \$15
23. a. $\frac{36}{13}$ **b.** $6\sqrt{13}/13$; this is the square root of the answer from (a).
25. a. $\frac{225}{2}$ **27.** area = $\frac{1}{2}$ when
$x = \sqrt{2}/2$ **29.** $\frac{49}{12}$
33. 100 yd by 150 yd
37. a. $p(x) = -2x + 500$
b. maximum revenue = \$31,250;
price = \$250 **39.** $t = \pm\sqrt{2}/2$
41. $t = \sqrt{3}$ or $t = -\sqrt{3}$
43. a. $A(x) = \dfrac{4 + \pi}{16\pi}x^2 - 2x + 16$
b. $x = \dfrac{16\pi}{4 + \pi}$ **c.** $\pi/4$
45. a. $A(x) = \dfrac{4 + \pi}{16\pi}x^2 - \dfrac{L}{8}x + \dfrac{L^2}{16}$
b. $x = \dfrac{\pi L}{4 + \pi}$ **c.** $\pi/4$ **47.** 2
51. 2 **55. a.** 1.143

Optional TI-81 Graphing Calculator Exercises for Section 4.4

1. a. The value was obtained by completing the square, and noting that the vertex was $(4.5, 20.25)$.

b.

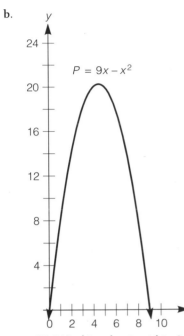

$P = 9x - x^2$

c. $x \approx 20.249$; this value is within 0.001 of the actual value of 20.25.

3. a.

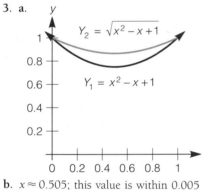

$Y_2 = \sqrt{x^2 - x + 1}$

$Y_1 = x^2 - x + 1$

b. $x \approx 0.505$; this value is within 0.005 of the actual value of 1/2.

5. a.

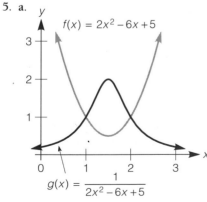

$f(x) = 2x^2 - 6x + 5$

$g(x) = \dfrac{1}{2x^2 - 6x + 5}$

b. 1.998 **c.** 2; our estimate is within 0.002 of this actual value.

Exercise Set 4.5

1. This graph has four turning points, but a polynomial function of degree 3 can have at most two turning points.
3. As $|x|$ gets very large, our function should be similar to $f(x) = a_3x^3$. But $f(x)$ does not have a parabolic shape like the given graph. **5.** As $|x|$ gets very large with x negative, the graph should resemble $2x^5$. But the y-values of $2x^5$ are always negative when x is negative, contrary to the given graph. **7.** This graph has a corner, which cannot occur in the graph of a polynomial function.
9. no x-intercepts; y-intercept: 5

$y = (x - 2)^2 + 1$

11. x-intercept: 1; y-intercept: -1

$y = -(x - 1)^4$

13. x-intercept: $4 + \sqrt[3]{2}$; y-intercept: -66

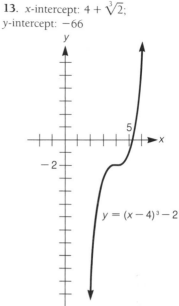

$y = (x - 4)^3 - 2$

15. x-intercept: -5; y-intercept: -1250

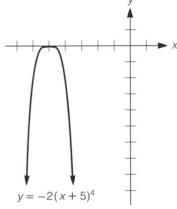

$y = -2(x + 5)^4$

17. x-intercept: -1; y-intercept: $\frac{1}{2}$

$y = \frac{1}{2}(x + 1)^5$

19. x- and y-intercepts: 0

$y = -(x - 1)^3 - 1$

21. a. x-intercepts: 2, 1 and -1;
y-intercept: 2

b.

$y = (x - 2)(x - 1)(x + 1)$

23. a. x-intercepts: 0, 2 and 1;
y-intercept: 0

b.

$y = 2x(x - 2)(x - 1)$

25. a. x-intercepts: 0, 5 and -1;
y-intercept: 0

b.

$y = x^3 - 4x^2 - 5x$

27. a. x-intercepts: -3, -2 and 2;
y-intercept: -12

b.

$y = x^3 + 3x^2 - 4x - 12$

29. a. x-intercepts: 0 and -2;
y-intercept: 0

b.

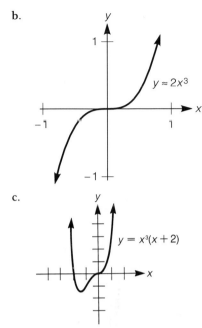

$y \approx 2x^3$

c.

$y = x^3(x + 2)$

31. a. x-intercepts: 1 and 4;
y-intercept: 128

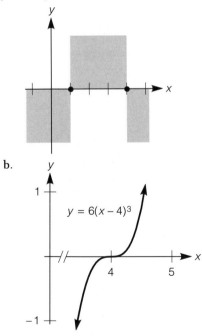

b.

$y = 6(x - 4)^3$

c.

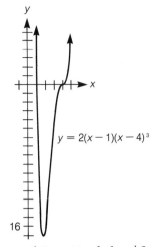

$y = 2(x - 1)(x - 4)^3$

33. a. x-intercepts: -1, 1 and 3;
y-intercept: 3

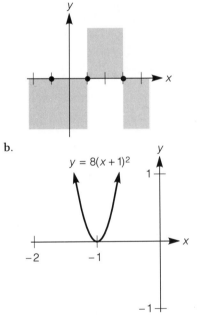

b.

$y = 8(x + 1)^2$

c.

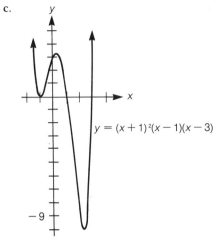

$y = (x + 1)^2(x - 1)(x - 3)$

35. a. x-intercepts: 0, 4 and -2;
y-intercept: 0

b.

$y = 8x^3$

c.

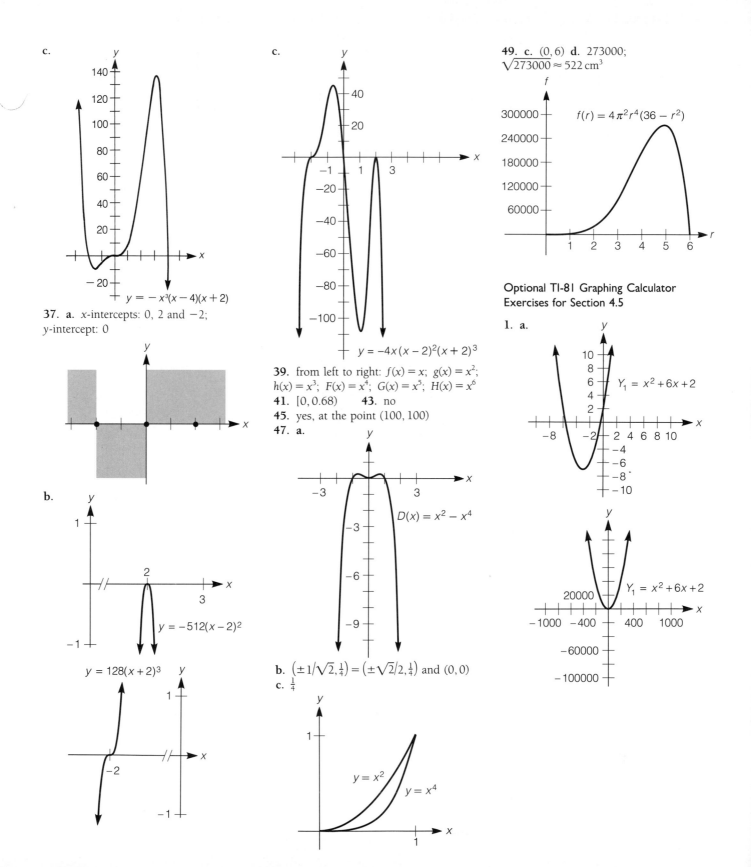

$$y = -x^3(x-4)(x+2)$$

37. a. x-intercepts: 0, 2 and −2;
y-intercept: 0

b.

$$y = 128(x+2)^3$$

$$y = -512(x-2)^2$$

c.

$$y = -4x(x-2)^2(x+2)^3$$

39. from left to right: $f(x) = x$; $g(x) = x^2$;
$h(x) = x^3$; $F(x) = x^4$; $G(x) = x^5$; $H(x) = x^6$

41. $[0, 0.68]$ **43.** no

45. yes, at the point $(100, 100)$

47. a.

$$D(x) = x^2 - x^4$$

b. $\left(\pm 1/\sqrt{2}, \frac{1}{4}\right) = \left(\pm\sqrt{2}/2, \frac{1}{4}\right)$ and $(0,0)$

c. $\frac{1}{4}$

$$y = x^2$$

$$y = x^4$$

49. c. $(0, 6)$ **d.** 273000;
$\sqrt{273000} \approx 522\ \text{cm}^3$

$$f(r) = 4\pi^2 r^4(36 - r^2)$$

Optional TI-81 Graphing Calculator
Exercises for Section 4.5

1. a.

$$Y_1 = x^2 + 6x + 2$$

$$Y_1 = x^2 + 6x + 2$$

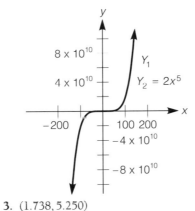

$Y_2 = x^2$

$Y_1 = x^2 + 6x + 2$

20000

−1000 −400   400   1000   x

−60000

−100000

b.

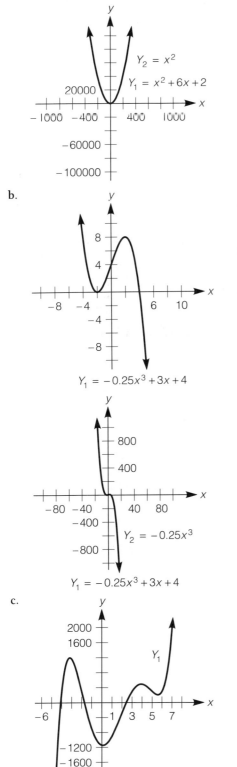

8

4

−8   −4   6   10   x

−4

−8

$Y_1 = -0.25x^3 + 3x + 4$

800

400

−80 −40   40   80   x

−400

$Y_2 = -0.25x^3$

−800

$Y_1 = -0.25x^3 + 3x + 4$

c.

2000
1600

Y_1

−6   −1   3   5   7   x

−1200
−1600
−2000

8×10^{10}

4×10^{10}

Y_1

$Y_2 = 2x^5$

−200   100 200   x

-4×10^{10}

-8×10^{10}

3. $(1.738, 5.250)$

5. 81.872 cubic inches

Exercise Set 4.6

1. domain: $(-\infty, 3) \cup (3, \infty)$;
x-intercept: -5; y-intercept: $-\frac{5}{4}$
3. domain: $(-\infty, -2) \cup (-2, 3) \cup (3, \infty)$;
x-intercepts: 9 and -1; y-intercept: $\frac{3}{2}$
5. domain: $(-\infty, 0) \cup (0, \infty)$;
x-intercepts: -2, 2, and 1;
no y-intercepts
7. no x-intercept; y-intercept: $\frac{1}{4}$;
horizontal asymptote: $y = 0$;
vertical asymptote: $x = -4$

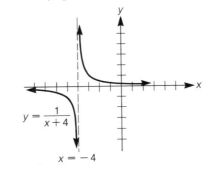

$y = \dfrac{1}{x+4}$

$x = -4$

9. no x-intercept; y-intercept: $\frac{3}{2}$;
horizontal asymptote: $y = 0$;
vertical asymptote: $x = -2$

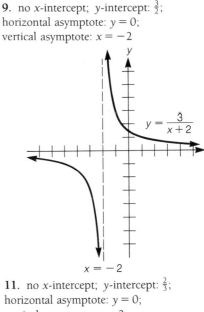

$y = \dfrac{3}{x+2}$

$x = -2$

11. no x-intercept; y-intercept: $\frac{2}{3}$;
horizontal asymptote: $y = 0$;
vertical asymptote: $x = 3$

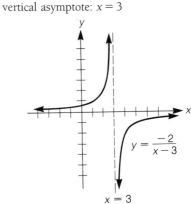

$y = \dfrac{-2}{x-3}$

$x = 3$

13. x-intercept: 3; y-intercept: 3;
horizontal asymptote: $y = 1$;
vertical asymptote: $x = 1$

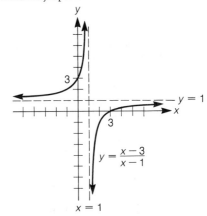

3

$y = 1$

3

$y = \dfrac{x-3}{x-1}$

$x = 1$

15. x-intercept: $\frac{1}{2}$; y-intercept: -2;
horizontal asymptote: $y = 2$;
vertical asymptote: $x = -\frac{1}{2}$

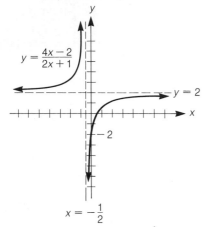

17. no x-intercept; y-intercept: $\frac{1}{4}$;
horizontal asymptote: $y = 0$;
vertical asymptote: $x = 2$

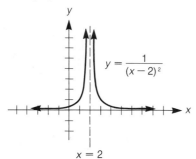

19. no x-intercept; y-intercept: 3;
horizontal asymptote: $y = 0$;
vertical asymptote: $x = -1$

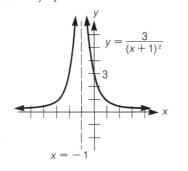

21. no x-intercept; y-intercept: $\frac{1}{8}$;
horizontal asymptote: $y = 0$;
vertical asymptote: $x = -2$

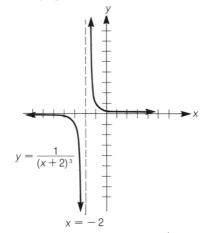

23. no x-intercept; y-intercept: $-\frac{4}{125}$;
horizontal asymptote: $y = 0$;
vertical asymptote: $x = -5$

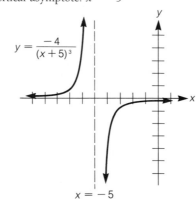

25. x- and y-intercepts: 0;
horizontal asymptote: $y = 0$;
vertical asymptotes: $x = -2$ and $x = 2$

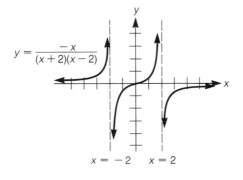

27. x- and y-intercepts: 0;
horizontal asymptote: $y = 0$;
vertical asymptotes: $x = -3$ and $x = 1$

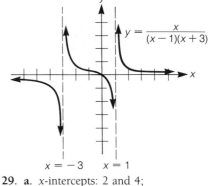

29. a. x-intercepts: 2 and 4;
no y-intercept;
horizontal asymptote: $y = 1$;
vertical asymptotes: $x = 0$ and $x = 1$

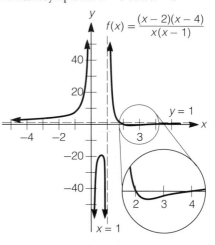

b. *x*-intercepts: 2 and 4; no *y*-intercept;
horizontal asymptote: $y = 1$;
vertical asymptotes: $x = 0$ and $x = 3$

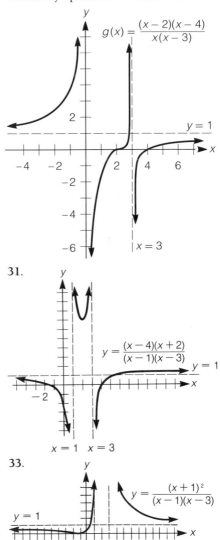

$g(x) = \dfrac{(x-2)(x-4)}{x(x-3)}$

$y = 1$

31.

$y = \dfrac{(x-4)(x+2)}{(x-1)(x-3)}$

$y = 1$

$x = 1 \quad x = 3$

33.

$y = \dfrac{(x+1)^2}{(x-1)(x-3)}$

$y = 1$

$x = 3$
$x = 1$

37. a.

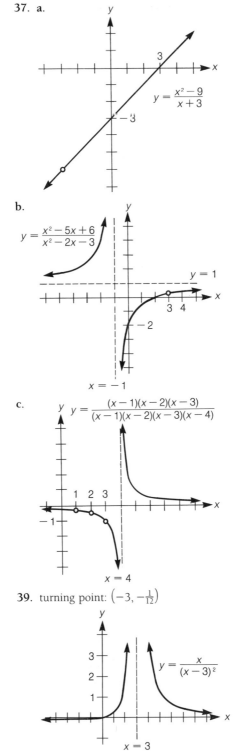

$y = \dfrac{x^2 - 9}{x + 3}$

-3

b.

$y = \dfrac{x^2 - 5x + 6}{x^2 - 2x - 3}$

$y = 1$

$3 \quad 4$

-2

$x = -1$

c. $y = \dfrac{(x-1)(x-2)(x-3)}{(x-1)(x-2)(x-3)(x-4)}$

$1 \ 2 \ 3$

-1

$x = 4$

39. turning point: $\left(-3, -\frac{1}{12}\right)$

3

2

1

$y = \dfrac{x}{(x-3)^2}$

$x = 3$

41. b.

x	$x + 4$	$\dfrac{x^2 + x - 6}{x - 3}$
10	14	14.8571
100	104	104.0619
1000	1004	1004.0060

x	$x + 4$	$\dfrac{x^2 + x - 6}{x - 3}$
-10	-6	-6.4615
-100	-96	-96.0583
-1000	-996	-996.0600

c. vertical asymptote: $x = 3$;
x-intercepts: -3 and 2; *y*-intercept: 2

d.

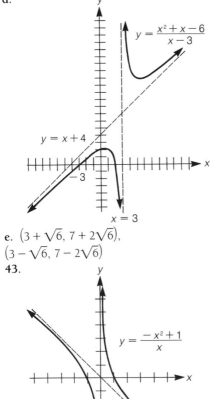

$y = \dfrac{x^2 + x - 6}{x - 3}$

$y = x + 4$

-3

$x = 3$

e. $\left(3 + \sqrt{6}, \ 7 + 2\sqrt{6}\right)$,
$\left(3 - \sqrt{6}, \ 7 - 2\sqrt{6}\right)$

43.

$y = \dfrac{-x^2 + 1}{x}$

$y = -x$

Optional TI-81 Graphing Calculator
Exercises for Section 4.6

1. $m \approx -0.5$

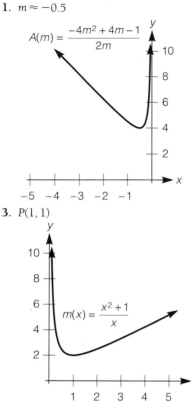

$$A(m) = \frac{-4m^2 + 4m - 1}{2m}$$

3. $P(1, 1)$

$$m(x) = \frac{x^2 + 1}{x}$$

The line is tangent to the parabola.

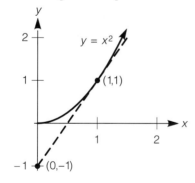

$y = x^2$

(1,1)

(0,−1)

5. a. $P(x) = (2x^2 + 12)/x$
b. $x \approx 2.45$, length $= 2.45$ m

$$P(x) = \frac{2x^2 + 12}{x}$$

7. a. $h = 54/r^2$
b. $S(r) = (2\pi r^3 + 108\pi)/r$ **c.** $r = 3$ in.

$$S(r) = \frac{2\pi r^3 + 108\pi}{r}$$

d. $h = 6$ in.; The value of h is double that of r.

Chapter Four Review Exercises

1. -5 **3.** $20{,}000
5.

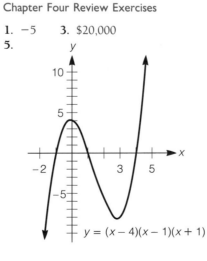

$y = (x - 4)(x - 1)(x + 1)$

7. $V(t) = -180t + 1000$
9. x-intercept: $-\frac{5}{3}$; y-intercept: $\frac{5}{2}$;
horizontal asymptote: $y = 3$;
vertical asymptote: $x = -2$

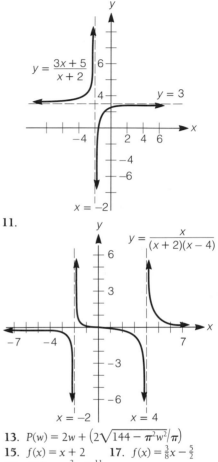

$$y = \frac{3x + 5}{x + 2}$$

$y = 3$

$x = -2$

11.

$$y = \frac{x}{(x + 2)(x - 4)}$$

$x = -2$ $x = 4$

13. $P(w) = 2w + \left(2\sqrt{144 - \pi^2 w^2}/\pi\right)$
15. $f(x) = x + 2$ **17.** $f(x) = \frac{3}{8}x - \frac{5}{2}$
19. $f(x) = -\frac{3}{4}x + \frac{11}{4}$
21. vertex: $(-1, -4)$;
x-intercepts: 1 and -3; y-intercept: -3

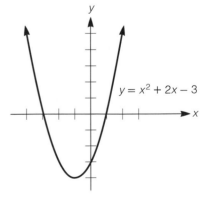

$y = x^2 + 2x - 3$

23. vertex: $\left(\sqrt{3}, 6\right)$;
x-intercepts: $\sqrt{3} + \sqrt{6}$ and $\sqrt{3} - \sqrt{6}$;
y-intercept: 3

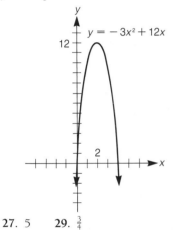

$y = -x^2 + 2\sqrt{3}x + 3$

25. vertex: $(2, 12)$; x-intercepts: 0 and 4;
y-intercept: 0

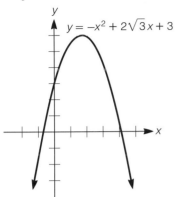

$y = -3x^2 + 12x$

27. 5   **29.** $\frac{3}{4}$

31. a. maximum height: $\frac{v_0^2}{64}$ ft;

$t = \frac{v_0}{32}$ sec **b.** $t = \frac{v_0}{16}$ sec

33. a. $d = \sqrt{x^4 - 3x^2 + 4}$

b. $\left(-\sqrt{6}/2, \frac{3}{2}\right)$ **35.** $b = \frac{17}{3}, -11$

37. 1   **39.** $\frac{225}{4}$ cm^2 **41.** $a = 1$

43. $x = 400$ units; $p = \$80$

45. x-intercepts: -4 and 2;
y-intercept: -8

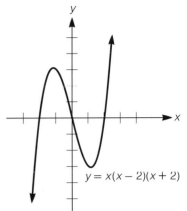

$y = (x + 4)(x - 2)$

47. x-intercept: -5; y-intercept: -125

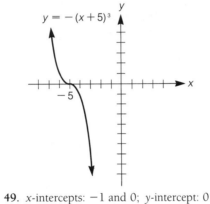

$y = -(x + 5)^3$

49. x-intercepts: -1 and 0; y-intercept: 0

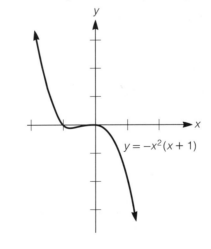

$y = -x^2(x + 1)$

51. x-intercepts: 0, 2 and -2;
y-intercept: 0

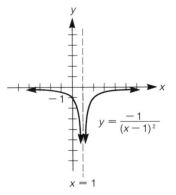

$y = x(x - 2)(x + 2)$

53. x-intercept: $-\frac{1}{3}$; no y-intercept;
horizontal asymptote: $y = 3$;
vertical asymptote: $x = 0$

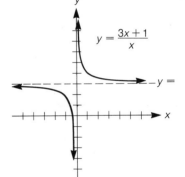

$y = \frac{3x + 1}{x}$

55. no x-intercept; y-intercept: -1;
horizontal asymptote: $y = 0$;
vertical asymptote: $x = 1$

$y = \frac{-1}{(x - 1)^2}$

$x = 1$

57. x-intercept: 2; y-intercept: $\frac{2}{3}$;
horizontal asymptote: $y = 1$;
vertical asymptote: $x = 3$

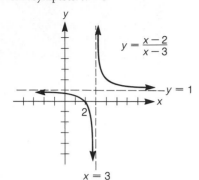

$$y = \frac{x-2}{x-3}$$

59. x-intercept: 1; y-intercept: $\frac{1}{4}$;
horizontal asymptote: $y = 1$;
vertical asymptote: $x = 2$

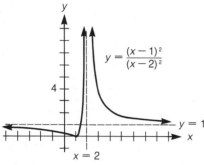

$$y = \frac{(x-1)^2}{(x-2)^2}$$

61. a. 1 unit **b.** $\sqrt{13}$ units
c. $b = \pm\sqrt{2}/2$ **63.** $k = 6$
65. $\left(-\infty, 4 - 2\sqrt{3}\right] \cup \left[4 + 2\sqrt{3}, \infty\right)$
67. $A(m) = m/2$
69. $A(x) = (1 - x)\sqrt{1 - x^2}$
71. a. 349.254 quadrillion Btu
b. too high, by approximately 4.19
quadrillion Btu
73. a.

b. slope ≈ -770;
y-intercept ≈ 1.5 million

c. 2100 **d.** $f^{-1}(x) = (1541090 - x)/771.4$
e. 1996 **75. a.** The graphs are
reflections of each other about the y-axis.

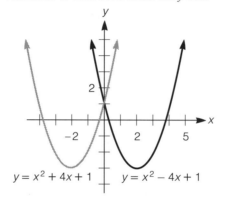

$$y = x^2 + 4x + 1 \qquad y = x^2 - 4x + 1$$

b. $ax^2 - bx + c$ **c.** It will reflect the
graph of $f(x)$ across the y-axis.

Chapter Four Test

1. $-\frac{18}{7}$ **2. a.** $(-\infty, 1)$; 2 **b.** $t = 0$
3. a. x-intercepts: 3 and -4;
y-intercept: -48

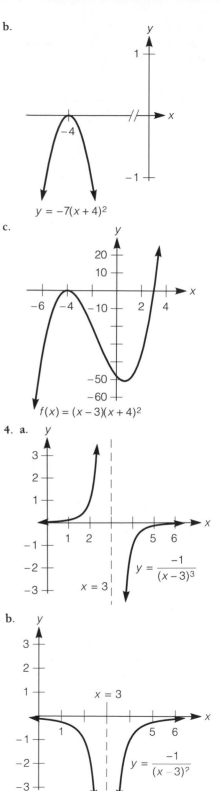

b.

$$y = -7(x + 4)^2$$

c.

$$f(x) = (x - 3)(x + 4)^2$$

4. a.

$$y = \frac{-1}{(x - 3)^3}$$

b.

$$y = \frac{-1}{(x - 3)^2}$$

5. turning point: $\left(\frac{7}{2}, \frac{73}{4}\right)$;
x-intercepts: $\left(7 \pm \sqrt{73}\right)/2$; y-intercept: 6;
axis of symmetry: $x = \frac{7}{2}$

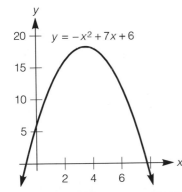

$y = -x^2 + 7x + 6$

6. maximum possible revenue: $9600;
$p = \$40$/unit
7. $V(t) = -1325t + 14000$
8. x-intercept: 3; y-intercept: $-\frac{9}{2}$

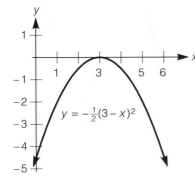

$y = -\frac{1}{2}(3 - x)^2$

9. x-intercept: $\frac{3}{2}$; y-intercept: -3;
vertical asymptote: $x = -1$;
horizontal asymptote: $y = 2$

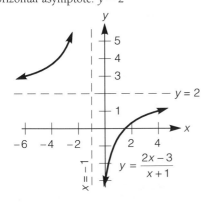

$y = 2$
$x = -1$
$y = \dfrac{2x - 3}{x + 1}$

10. a. $L(x) = \sqrt{10x^2 - 22x + 17}$
b. $x = 1.1$
11. a. vertical asymptotes: $x = 3$ and
$x = -3$; horizontal asymptote: $y = 1$
b.

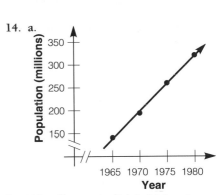

$y = \frac{2}{9}x$
$y = -\frac{2}{5}(x - 2)$

c.

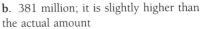

$x = -3$ $x = 3$

d.

y $f(x) = \dfrac{x(x - 2)}{x^2 - 9}$

$y = 1$

$x = -3$ $x = 3$

12. $A(w) = \frac{1}{2}w\sqrt{64 - w^2}$
13. a. As $|x|$ increases in size for positive
values of x, the values of $-x^3$ should be
large negative, contrary to the graph.
b. This graph has four turning points,
and a polynomial function with highest-
degree term $-x^3$ can have at most two
turning points.

14. a.

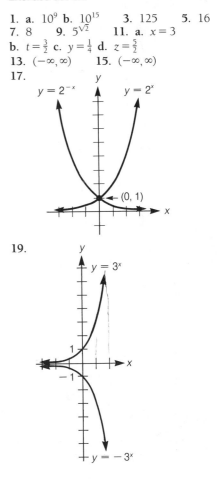

b. 381 million; it is slightly higher than
the actual amount

CHAPTER FIVE

Exercise Set 5.1

1. a. 10^9 **b.** 10^{15} **3.** 125   **5.** 16
7. 8   **9.** $5^{\sqrt{2}}$ **11. a.** $x = 3$
b. $t = \frac{3}{2}$ **c.** $y = \frac{1}{4}$ **d.** $z = \frac{5}{2}$
13. $(-\infty, \infty)$ **15.** $(-\infty, \infty)$
17.

$y = 2^{-x}$ $y = 2^x$

$(0, 1)$

19.

$y = 3^x$

$y = -3^x$

21.

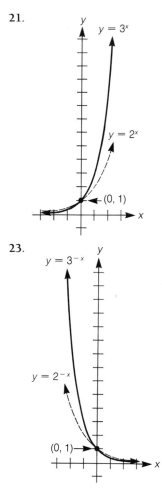

$y = 3^x$
$y = 2^x$
(0, 1)

23.

$y = 3^{-x}$
$y = 2^{-x}$
(0, 1)

25. domain: $(-\infty, \infty)$; range: $(-\infty, 1)$; x-intercept: 0; y-intercept: 0; asymptote: $y = 1$

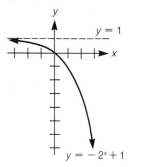

$y = 1$
$y = -2^x + 1$

27. domain: $(-\infty, \infty)$; range: $(1, \infty)$; no x-intercept; y-intercept: 2; asymptote: $y = 1$

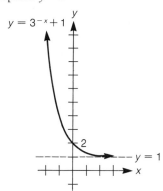

$y = 3^{-x} + 1$
2
$y = 1$

29. domain: $(-\infty, \infty)$; range: $(0, \infty)$; no x-intercept; y-intercept: $\frac{1}{2}$; asymptote: $y = 0$

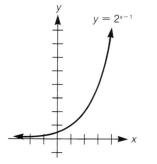

$y = 2^{x-1}$

31. domain: $(-\infty, \infty)$; range: $(1, \infty)$; no x-intercept; y-intercept: 4; asymptote: $y = 1$

$y = 3^{x+1} + 1$
4
$y = 1$

33. $x = -\frac{1}{3}$ **35.** $x = \frac{3}{2}, 1$
39. a.

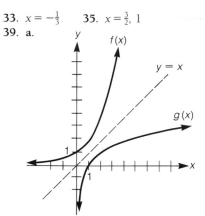

$f(x)$
$y = x$
$g(x)$
1
1

b. domain: $(0, \infty)$; range: $(-\infty, \infty)$; x-intercept: 1; no y-intercept; asymptote: $x = 0$ **41. a.** 1.4 **b.** 1.41
43. a. 1.5 **b.** 1.52   **45. a.** 1.7
b. 1.73   **47. a.** 1.6 **b.** 1.62
49. $x \approx 0.3$ **51.** $x \approx 0.7$
53.

x	$\log_{10} x$
1	0.00
2	0.30
3	0.48
4	0.60
5	0.70
6	0.78
7	0.85
8	0.90
9	0.95
10	1.00

Exercise Set 5.2

1. domain: $(-\infty, \infty)$; range: $(0, \infty)$; no x-intercept; y-intercept: 1; asymptote: $y = 0$

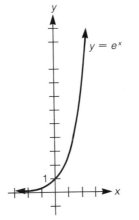

3. domain: $(-\infty, \infty)$; range: $(-\infty, 0)$; no x-intercept; y-intercept: -1; asymptote: $y = 0$

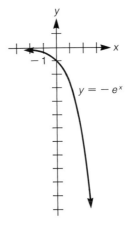

5. domain: $(-\infty, \infty)$; range: $(1, \infty)$; no x-intercept; y-intercept: 2; asymptote: $y = 1$

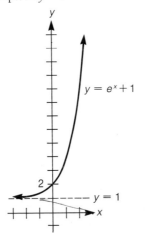

7. domain: $(-\infty, \infty)$; range: $(1, \infty)$; no x-intercept; y-intercept: $e + 1$; asymptote: $y = 1$

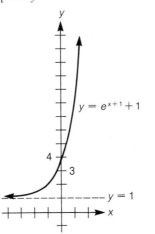

9. domain: $(-\infty, \infty)$; range: $(-\infty, 0)$; no x-intercept; y-intercept: $-1/e^2$; asymptote: $y = 0$

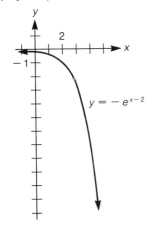

11. domain: $(-\infty, \infty)$; range: $(-\infty, e)$; x-intercept: 1; y-intercept: $e - 1$; asymptote: $y = e$

13.

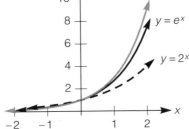

15. false   **17.** false   **19.** true
21. false   **23.** 1.2; 1.22
25. 2.2; 2.23   **27.** 1.6; 1.65
29. -0.2 **31. a.** $\Delta f/\Delta x = 65$;
$\Delta g/\Delta x = 4$; $\Delta h/\Delta x \approx 12.70$; $\Delta f/\Delta x$ is 5
times as large as $\Delta h/\Delta x$
b. $\Delta f/\Delta x = 2465$; $\Delta g/\Delta x = 256$;
$\Delta h/\Delta x \approx 5122.13$; $\Delta h/\Delta x$ is more than
twice as large as $\Delta f/\Delta x$
c. $\Delta f/\Delta x = 12{,}209$; $\Delta g/\Delta x = 16{,}384$;
$\Delta h/\Delta x \approx 2{,}066{,}413$; $\Delta h/\Delta x$ is 126 times
as large as $\Delta g/\Delta x$ **33. a.** e^π is slightly
larger **b.** e^π is larger   **37. a.** $(-\infty, \infty)$
b. 1 **c.** The graph is symmetric about
the y-axis.
d.

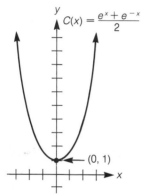

$C(x) = \dfrac{e^x + e^{-x}}{2}$

(0, 1)

39. a.

x	1000	10^4	10^5
$\left(1 + \dfrac{2}{x}\right)^x$	7.374312	7.387579	7.388908

x	10^6	10^7
$\left(1 + \dfrac{2}{x}\right)^x$	7.389041	7.389055

b. 7.389056; the values in the table are
approaching e^2.
41. a.

$f(x) = e^x$

$y = x$

$L(x)$

b. domain: $(0, \infty)$; range: $(-\infty, \infty)$;
x-intercept: 1; asymptote: $x = 0$
c. (i) x-intercept: 1; asymptote: $x = 0$

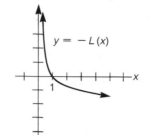

$y = -L(x)$

(ii) x-intercept: -1; asymptote: $x = 0$

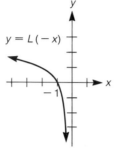

$y = L(-x)$

(iii) x-intercept: 2; asymptote: $x = 1$

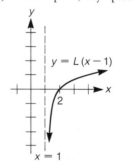

$y = L(x-1)$

$x = 1$

Optional TI-81 Graphing Calculator
Exercises for Sections 5.1 and 5.2

1. a.

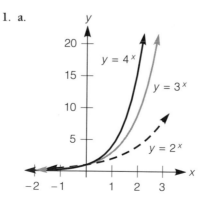

$y = 4^x$
$y = 3^x$
$y = 2^x$

b.

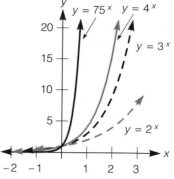

$y = 75^x$ $y = 4^x$
$y = 3^x$
$y = 2^x$

3. a. x-intercept: 2; y-intercept: $\frac{8}{9}$;
asymptote: $y = 1$
b.

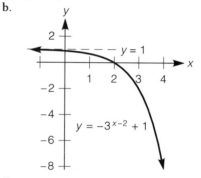

$y = 1$
$y = -3^{x-2} + 1$

5. a. no x-intercept; y-intercept: 5;
asymptote: $y = 4$
b.

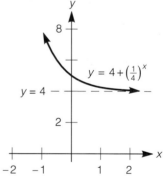

$y = 4 + \left(\frac{1}{4}\right)^x$
$y = 4$

7. a. x-intercept: $1 - \log_{10} 2$;
y-intercept: 0.4; asymptote: $y = 0.5$

A-92

b.

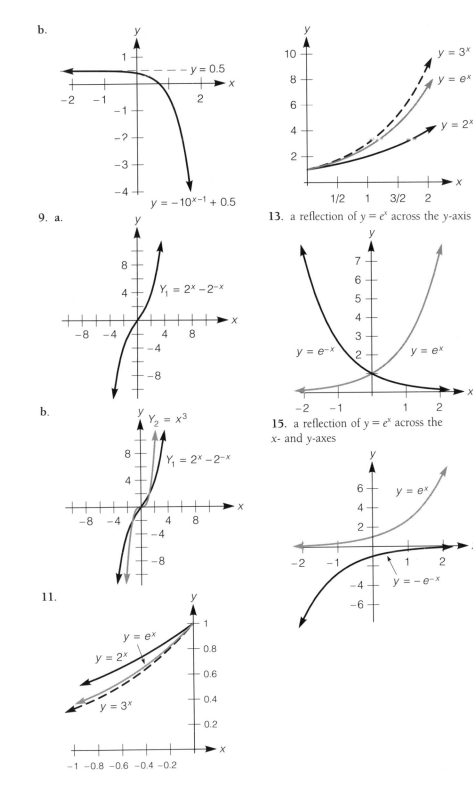

$y = 0.5$

$y = -10^{x-1} + 0.5$

9. a.

$Y_1 = 2^x - 2^{-x}$

b.

$Y_2 = x^3$

$Y_1 = 2^x - 2^{-x}$

11.

$y = e^x$

$y = 2^x$

$y = 3^x$

$y = 3^x$

$y = e^x$

$y = 2^x$

13. a reflection of $y = e^x$ across the y-axis

$y = e^{-x}$

$y = e^x$

15. a reflection of $y = e^x$ across the x- and y-axes

$y = e^x$

$y = -e^{-x}$

17. a. a translation of $y = e^x$ to the right one unit

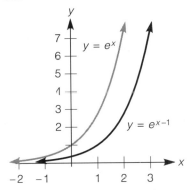

$y = e^x$

$y = e^{x-1}$

b. a translation of $y = e^x$ to the right 1 unit, then a reflection across the y-axis

$y = e^{-x-1}$ y $y = e^x$

19. a.

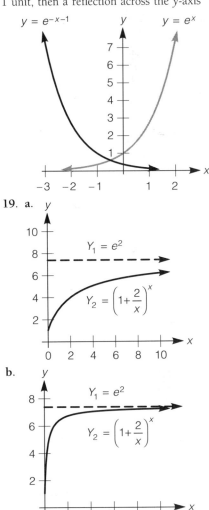

$Y_1 = e^2$

$Y_2 = \left(1 + \dfrac{2}{x}\right)^x$

b.

$Y_1 = e^2$

$Y_2 = \left(1 + \dfrac{2}{x}\right)^x$

21.

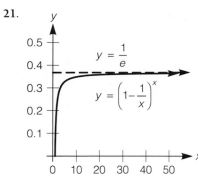

Exercise Set 5.3

1. a. no **b.** yes **c.** yes

3. a. $f^{-1}(x) = (4x + 1)/(2 - 3x)$
b. $(3x + 4)/(2x - 1)$ **c.** $\frac{1}{2}$ **d.** -4

5. $(-1, 3)$ and $(6, -1)$

7. a. $\log_3 9 = 2$ **b.** $\log_{10} 1000 = 3$
c. $\log_7 343 = 3$ **d.** $\log_2 \sqrt{2} = \frac{1}{2}$

9. a. $2^5 = 32$ **b.** $10^0 = 1$ **c.** $e^{1/2} = \sqrt{e}$
d. $3^{-4} = 1/81$ **e.** $t^v = u$

11. The quantity $\log_5 30$ is larger.

13. a. $\frac{3}{2}$ **b.** $-\frac{5}{2}$ **c.** $\frac{3}{2}$ **15. a.** $x = \frac{1}{16}$
b. $x = e^{-2} \approx 0.14$ **17. a.** $(0, \infty)$
b. $\left(-\infty, \frac{3}{4}\right)$ **c.** $(-\infty, 0) \cup (0, \infty)$ **d.** $(0, \infty)$
e. $(-\infty, -5) \cup (5, \infty)$

19. $A(0, 1)$; $B(1, 0)$; $C(4, 2)$; $D(2, 4)$

21. a. domain: $(0, \infty)$; range: $(-\infty, \infty)$;
x-intercept: 1; no y-intercept;
asymptote: $x = 0$

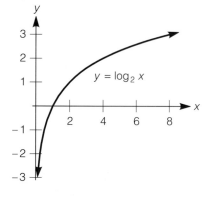

b. domain: $(0, \infty)$; range: $(-\infty, \infty)$;
x-intercept: 1; no y-intercept;
asymptote: $x = 0$

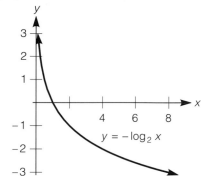

c. domain: $(-\infty, 0)$; range: $(-\infty, \infty)$;
x-intercept: -1; no y-intercept;
asymptote: $x = 0$

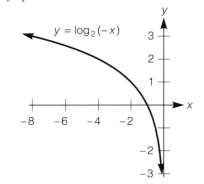

d. domain: $(-\infty, 0)$; range: $(-\infty, \infty)$;
x-intercept: -1; no y-intercept;
asymptote: $x = 0$

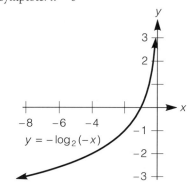

23. domain: $(2, \infty)$; range: $(-\infty, \infty)$;
x-intercept: 5; no y-intercept;
asymptote: $x = 2$

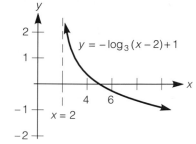

25. domain: $(-e, \infty)$; range: $(-\infty, \infty)$;
x-intercept: $-e + 1$; y-intercept: 1;
asymptote: $x = -e$

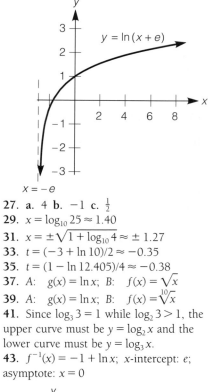

27. a. 4 **b.** -1 **c.** $\frac{1}{2}$
29. $x = \log_{10} 25 \approx 1.40$
31. $x = \pm\sqrt{1 + \log_{10} 4} \approx \pm 1.27$
33. $t = (-3 + \ln 10)/2 \approx -0.35$
35. $t = (1 - \ln 12.405)/4 \approx -0.38$
37. A: $g(x) = \ln x$; B: $f(x) = \sqrt{x}$
39. A: $g(x) = \ln x$; B: $f(x) = \sqrt[10]{x}$
41. Since $\log_3 3 = 1$ while $\log_2 3 > 1$, the upper curve must be $y = \log_2 x$ and the lower curve must be $y = \log_3 x$.
43. $f^{-1}(x) = -1 + \ln x$; x-intercept: e; asymptote: $x = 0$

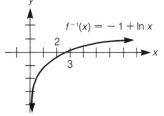

45. The area is less than the rectangle, which has an area of two square units.

47. $x = \ln 6$ **49.** $x \approx 10^{30}$

51. a. 3.5; acid **b.** 0; acid

53. a. $P(10) = 4$; $P(18) = 7$; $P(19) = 8$

b.

x	$P(x)$	$\dfrac{x}{\ln x}$	$\dfrac{P(x)}{x/\ln x}$
10^2	25	22	1.151
10^4	1229	1086	1.132
10^6	78498	72382	1.084
10^8	5761455	5428681	1.061
10^9	50847534	48254942	1.054
10^{10}	455052512	434294482	1.048

c. In the following table, $L(x)$ denotes the quantity $\ln x - 1.08366$.

x	$P(x)$	$\dfrac{x}{L(x)}$	$\dfrac{P(x)}{x/L(x)}$
10^2	25	28	0.8804
10^4	1229	1231	0.9988
10^6	78498	78543	0.9994
10^8	5761455	5768004	0.9989
10^9	50847534	50917519	0.9986
10^{10}	455052512	455743004	0.9985

55. $n = 27$

57.

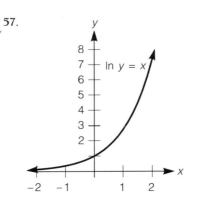

Optional TI-81 Graphing Calculator
Exercises for Section 5.3

1. a. reflections across the line $y = x$

b.

3. a. $(-\infty, \infty)$

b. $(0, \infty)$

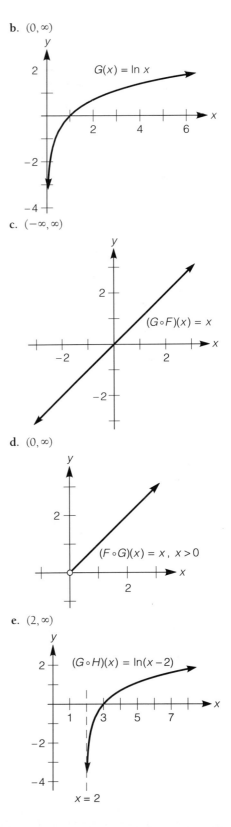

c. $(-\infty, \infty)$

d. $(0, \infty)$

e. $(2, \infty)$

f. $(0, \infty)$

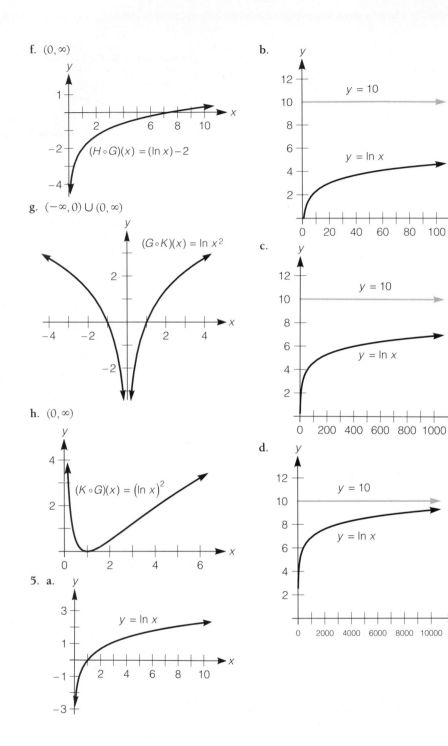

$(H \circ G)(x) = (\ln x) - 2$

g. $(-\infty, 0) \cup (0, \infty)$

$(G \circ K)(x) = \ln x^2$

h. $(0, \infty)$

$(K \circ G)(x) = (\ln x)^2$

5. a.

$y = \ln x$

b.

$y = 10$

$y = \ln x$

c.

$y = 10$

$y = \ln x$

d.

$y = 10$

$y = \ln x$

e. $x = e^{10} \approx 22026$

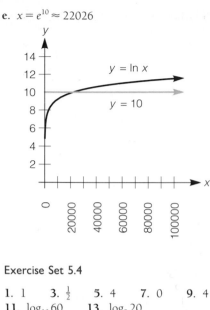

$y = \ln x$

$y = 10$

Exercise Set 5.4

1. 1   **3.** $\frac{1}{2}$ **5.** 4   **7.** 0   **9.** 4
11. $\log_{10} 60$ **13.** $\log_5 20$
15. a. $\ln 6$ **b.** $\ln\left(\frac{3}{16384}\right)$
17. $\log_b\left[\dfrac{4(1+x)^3}{(1-x)^{3/2}}\right]$
19. $\log_{10}\left[\dfrac{27\sqrt{x+1}}{(x^2+1)^6}\right]$
21. a. $2\log_{10} x - \log_{10}(1+x^2)$
b. $2\ln x - \frac{1}{2}\ln(1+x^2)$
23. a. $\frac{1}{2}\log_{10}(3+x) + \frac{1}{2}\log_{10}(3-x)$
b. $\frac{1}{2}\ln(2+x) + \frac{1}{2}\ln(2-x) - \ln(x-1) - \frac{3}{2}\ln(x+1)$ **25. a.** $\frac{1}{2}\log_b x - \frac{1}{2}$
b. $\ln(1+x^2) + \ln(1+x^4) + \ln(1+x^6)$
27. a. $a + 2b + 3c$ **b.** $1 + \frac{1}{2}a$
c. $\frac{1}{2}(1+a+b+c)$ **d.** $1 + a - \frac{1}{2}b - \frac{1}{2}c$
29. a. $1 + t$ **b.** $u - t$ **c.** $\frac{3}{2}t + \frac{1}{2}u - 1$
d. $2 + t + \frac{1}{2}u$
31. a. x-intercept: $(\ln 5)/(\ln 2) \approx 2.32$

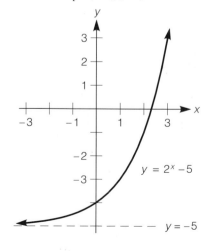

$y = 2^x - 5$

$y = -5$

b. x-intercept: $(2 \ln 5)/(\ln 2) \approx 4.64$

33. $x = (\ln 5 - \ln 2 + 1)/2$
35. $x = (\ln 13)/(\ln 2)$
37. $x = 1/(\ln 10)$ 39. $x = \frac{1}{2}$
41. $x = 3$ 43. $x = \frac{203}{99}$ 45. $x = 3$
47. $x = 7$ 49. a. $x = 10^y/[3(10^y) - 1]$
b. $x = (1 - y)/2$
51. $(\log_{10} 5)/(\log_{10} 2)$
53. $(\log_{10} 3)/(\log_{10} e)$
55. $(\log_{10} 2)/(\log_{10} b)$
57. $(\ln 6)/(\ln 10)$ 59. $1/(\ln 10)$
61. $[\ln(\ln x) - \ln(\ln 10)]/(\ln 10)$
63. a. true b. true c. true d. false
e. true f. false g. true h. false i. true
j. false k. false l. true m. true
65. f. 2345.6 g. 0.123456
67.

x	0.1	0.05
ln(1 + x)	0.095310	0.048790

x	0.005	0.0005
ln(1 + x)	0.004987	0.000499

69. $a = 4$, $b = -(\ln 243)/2.5$
71. a. $x = (\log_{10} 3)/(\log_{10} 2)$
b. $(\log_{10} 3)/(\log_{10} 2) \approx 1.585$;
$(\ln 3)/(\ln 2) \approx 1.585$ 75. x^3 77. $\frac{1}{2}$
79. $x = \beta e^{\alpha/3}$
87. $f^{-1}(x) = (e^{2x} - 1)/2e^x$
89.

x	10^2	10^3	10^6
y	0.4234	0.6589	0.9654

x	10^{20}	10^{50}	10^{99}
y	1.3428	1.5573	1.6918

Exercise Set 5.5

1. $1009.98   3. 8.45%
5. $767.27   7. a. $3869.68
b. $4006.39   9. 13 quarters
11. $3487.50   13. $2610.23
15. 5.83%   17. the 6% investment
19. a. 14 years b. 13.86 years
c. 1.01% of the actual doubling time
21. $26.5 trillion   23. a. 14 years
b. A (dollars)

25. a.

t (years)	1	2	3
A (account #1)	2081	2166	2254
A (account #2)	2082	2167	2255

t (years)	4	10
A (account #1)	2345	2978
A (account #2)	2347	2984

b. account #1: 17.4 years;
account #2: 17.3 years c. $88/year
d. $88/year

Exercise Set 5.6

1. a. $k = (\ln 1.9)/2 \approx 0.3209$
b. 9951 bacteria c. 5.0 hrs
3. $\mathcal{N}_0 = 2000$; $k = (\ln 1.7)/3 \approx 0.1769$
5.

	1992 Population (billions)	% Population in 1992	Growth Rate (% per year)
World	5.420	100	1.7
More Dev.	1.224	22.6	0.5
Less Dev.	4.196	77.4	2.0

	Year 2000 Population (billions)	% of World Population in 2000
World	6.210	100
More Dev.	1.274	20.50
Less Dev.	4.924	79.29

7. a.

Country	1992 Population (millions)	Relative Growth Rate (percent per year)
Iraq	18.2	3.7
United Kingdom	57.8	0.3

Country	Year 2000 Population (millions)	Percent Increase in Population
Iraq	24.5	34.62
United Kingdom	59.2	2.42

b. 2026   9. a. Cyprus: 0.8
million; Gaza: 1.0 million b. year 2025;
3.2 million
11. a. $k = (\ln 1.43)/25 \approx 0.0143$
b. $k = (\ln 1.43)/108 \approx 0.0033$
13. a. $k = (\ln 2.7142)/50 \approx 0.0200$
b. 170,853,155 c. It is smaller; slower.
15. a. 0.684% b. 0.955%
c. 21,731,258 d. Our prediction is
higher than the actual population.
17. a.

Region	1990 Population (millions)	Growth Rate (%)	2025 Population
North America	275.2	0.7	351.6
Soviet Union	291.3	0.7	372.2
Europe	499.5	0.2	535.7
Nigeria	113.3	3.1	335.3

b. 222.0 million c. 193.5 million
d. For Nigeria, our results support this
projection.

19. a.

t (seconds)	0	550	1100	1650	2200
$\mathcal{N}$ (grams)	8	4	2	1	0.5

b.

t (years)	0	4.9×10^9	9.8×10^9
$\mathcal{N}$ (grams)	10	5	2.5

t (years)	14.7×10^9	19.6×10^9
$\mathcal{N}$ (grams)	1.25	0.625

21. $0.55\,g$ **23. a.** $4.29\,g$ **b.** 79 hrs
25. a.

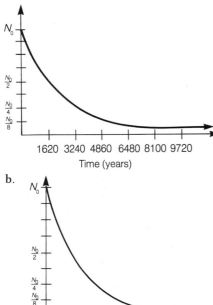

b.

27. a. $1.53\,g$ **b.** 43 yrs
29. a. 73,000 yrs **b.** 10 half-lives
c. 10 half-lives
31. a. $k = (\ln 0.5)/28 \approx -0.0248$
b. 279 yrs **c.** 280 yrs
33. b. $14.4\,\text{sec}$ **c.** $0.05\,g$
35. a. 2020 **b.** 2035 **c.** 2019
37. a. 201 yrs **b.** 132 yrs
39. a. $E = 10^{11.4} \cdot 10^{1.5M}$ **b.** 11,220
41. $k \approx -1.4748 \times 10^{-11}$
43. 4.181 billion years old
45. 15,505 years old **47.** 11,500 yrs

Chapter Five Review Exercises

1. $\log_5 126$ is larger
3. $(4 \ln 1.5)/(\ln 1.25)$ hrs
5. The function f is not one-to-one.

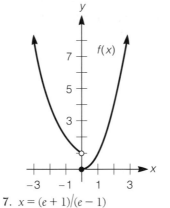

7. $x = (e + 1)/(e - 1)$
9. $y = e^x$: domain is $(-\infty, \infty)$;
range is $(0, \infty)$; $y = \ln x$: domain is $(0, \infty)$;
range is $(-\infty, \infty)$

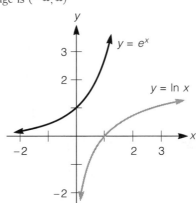

11. $-\frac{3}{2}$ **13.** $k \approx -0.05$
15. $x = 2 - \ln \frac{12}{5}$ **17.** 2010
19. 7 years

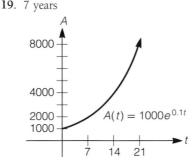

21. horizontal asymptote: $y = 0$;
no vertical asymptote; no x-intercept;
y-intercept: 1

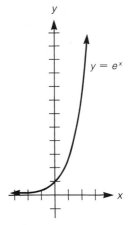

23. no horizontal asymptote;
vertical asymptote: $x = 0$; x-intercept: 1;
no y-intercept

25. horizontal asymptote: $y = 1$;
no vertical asymptote; no x-intercept;
y-intercept: 3

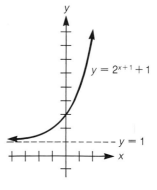

27. horizontal asymptote: $y = 0$; no vertical asymptote; no x-intercept; y-intercept: 1

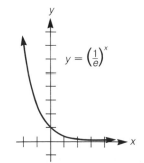

$y = \left(\dfrac{1}{e}\right)^x$

29. horizontal asymptote: $y = 1$; no vertical asymptote; no x-intercept; y-intercept: $e + 1$

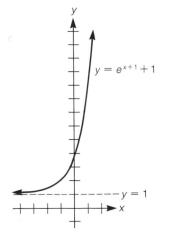

$y = e^{x+1} + 1$

$y = 1$

31. no asymptotes; x- and y-intercepts: 0

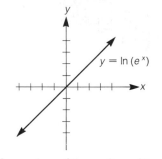

$y = \ln(e^x)$

33. $x = 4$ **35.** $x = 3$ **37.** $x = \frac{2}{3}$
39. $x = \sqrt[3]{3}$ **41.** $x = (1 - 2\ln 3)/10$
43. $x = \frac{200}{99}$ **45.** $x = 2$ **47.** $(0, \infty)$
49. $x = 1$ **51.** $\frac{1}{2}$ **53.** $\frac{1}{5}$
55. -1 **57.** 16 **59.** 4 **61.** 2
63. 2 **65.** $\frac{9}{14}$ **67.** $2a + 3b + (c/2)$
69. $8a + 4b$
71. $\log_{10} 209$ lies between 2 and 3.
73. $\log_6 100$ lies between 2 and 3.

75. $\log_{10} 0.003$ lies between -2 and -3.
77. a. third quadrant

$y = \ln(-x) - 1$

$y = \ln(x + 2)$

$x = -2$

b. $x = -2e/(e + 1) \approx -1.46$
79. $k = \left(\ln \frac{1}{2}\right)/T$
81. 6.25% remaining
83. $[d \ln(c/b)]/\left(\ln \frac{1}{2}\right)$ days **85.** $\log_{10} 2$
87. $\ln 10$ **89.** $\ln x^a y^b$
91. $\frac{1}{2}\ln(x - 3) + \frac{1}{2}(x + 4)$
93. $3 \log_{10} x - \frac{1}{2}\log_{10}(1 + x)$
95. $\frac{1}{3}\log_{10} x - \frac{2}{3}$
97. $3\ln(1 + 2e) - 3\ln(1 - 2e)$
99. $(\ln 2)/\ln[1 + (R/100)]$ yrs
101. 9.92% **103.** $(100 \ln 2)/R$ yrs
105. a. $7\frac{3}{4}$ yrs
b. $(\ln n)/\ln[1 + (R/400)]$ yrs
107. a. $(0, \infty)$ **b.** $[1, \infty)$
109. a. $(-\infty, -3) \cup (-3, 5) \cup (5, \infty)$
b. $(-\infty, -3) \cup (5, \infty)$
111. $(-\infty, -1) \cup (1, \infty)$ **113.** -0.7
115. -2.2 **117.** 4.3 **119.** 2.7
121. 1.6 **123. a.** 1.0986
b. 0.00112% **c.** $\ln \sqrt{3} \approx 0.5493$;
$\ln 9 \approx 2.1972$; $\ln \frac{1}{3} \approx -1.0986$

Chapter Five Test

1. domain: $(-\infty, \infty)$; range: $(-3, \infty)$;
x-intercept: $-(\ln 3)/(\ln 2)$;
y-intercept: -2; asymptote: $y = -3$

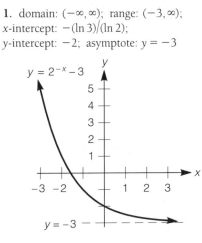

$y = 2^{-x} - 3$

$y = -3$

2. $\left(\ln \frac{5}{3}\right)/(\ln 1.033) \approx 16$ hrs
3. The quantity $\log_2 17$ is larger.
4. $(\ln 15)/(\ln 2)$ **5.** 10^{12} **6.** 4 yrs
7. a. $(0, \infty)$
b.

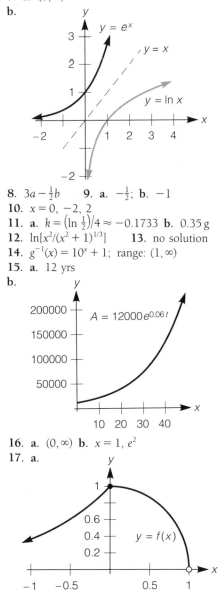

$y = e^x$

$y = x$

$y = \ln x$

8. $3a - \frac{1}{2}b$ **9. a.** $-\frac{1}{2}$; **b.** -1
10. $x = 0, -2, 2$
11. a. $k = \left(\ln \frac{1}{2}\right)/4 \approx -0.1733$ **b.** 0.35 g
12. $\ln[x^2/(x^2 + 1)^{1/3}]$ **13.** no solution
14. $g^{-1}(x) = 10^x + 1$; range: $(1, \infty)$
15. a. 12 yrs
b.

$A = 12000 e^{0.06 t}$

16. a. $(0, \infty)$ **b.** $x = 1, e^2$
17. a.

$y = f(x)$

b. no **18.** $x = 4$

CHAPTER SIX

Exercise Set 6.1

1. a. $\sin \theta = \frac{15}{17}$; $\cos \theta = \frac{8}{17}$; $\tan \theta = \frac{15}{8}$; $\cot \theta = \frac{8}{15}$; $\sec \theta = \frac{17}{8}$; $\csc \theta = \frac{17}{15}$
b. $\sin \beta = \frac{8}{17}$; $\cos \beta = \frac{15}{17}$; $\tan \beta = \frac{8}{15}$; $\cot \beta = \frac{15}{8}$; $\sec \beta = \frac{17}{15}$; $\csc \beta = \frac{17}{8}$
3. a. $\sin \theta = \sqrt{5}/5$; $\cos \theta = 2\sqrt{5}/5$; $\tan \theta = \frac{1}{2}$; $\cot \theta = 2$; $\sec \theta = \sqrt{5}/2$; $\csc \theta = \sqrt{5}$ **b.** $\sin \beta = 2\sqrt{5}/2$; $\cos \beta = \sqrt{5}/5$; $\tan \beta = 2$; $\cot \beta = \frac{1}{2}$; $\sec \beta = \sqrt{5}$; $\csc \beta = \sqrt{5}/2$
5. a. $\cos A = 3\sqrt{13}/13$; $\sin A = 2\sqrt{13}/13$; $\tan A = \frac{2}{3}$
b. $\sec B = \sqrt{13}/2$; $\csc B = \sqrt{13}/3$; $\cot B = \frac{2}{3}$ **7.** $\sin B = \frac{12}{13}$; $\cos B = \frac{5}{13}$; $\tan B = \frac{12}{5}$; $\cot B = \frac{5}{12}$; $\sec B = \frac{13}{5}$; $\csc B = \frac{13}{12}$ **9. a.** $\sin B = \frac{4}{5}$; $\cos A = \frac{4}{5}$ **b.** $\sin A = \frac{3}{5}$; $\cos B = \frac{3}{5}$ **c.** 1
11. a. $\cos A = \frac{24}{25}$; $\sin A = \frac{7}{25}$; $\tan A = \frac{7}{24}$
b. $\cos B = \frac{7}{25}$; $\sin B = \frac{24}{25}$; $\tan B = \frac{24}{7}$
c. 1 **13.** $\sin \theta \approx 0.906$; $\cos \theta \approx 0.423$; $\tan \theta \approx 2.145$
15. $\sin \theta \approx 0.623$; $\cos \theta \approx 0.783$; $\tan \theta \approx 0.795$ **17.** $\sin \theta \approx 0.985$; $\cos \theta \approx 0.173$; $\tan \theta \approx 5.706$
19. $\sec \theta \approx 1.064$; $\csc \theta \approx 2.924$; $\cot \theta \approx 2.747$ **21.** $\sec \theta \approx 1.049$; $\csc \theta \approx 3.326$; $\cot \theta \approx 3.172$
23. $\sec \theta \approx 1.000$; $\csc \theta \approx 57.299$; $\cot \theta \approx 57.290$ **25.** Each side is $\frac{1}{2}$.
27. Each side is $\frac{3}{2}$. **29.** Each side is $\sqrt{3}/2$. **31.** Each side is $\frac{1}{2}$.
33. Each side is $\sqrt{3}/3$.
35. Each side is 2.
37. a. $\cos 30° \approx 0.8660254038$; $\cos 45° \approx 0.7071067812$
b. $\sqrt{3}/2 \approx 0.8660254038$; $\sqrt{2}/2 \approx 0.7071067812$ **39. a.** $\cos \theta$
b. $\sec \beta$ **43.** $\sin 3° \approx 0.0523359562$; $\sin 6° \approx 0.1045284633$; $\sin 9° \approx 0.1564344650$; $\sin 12° \approx 0.2079116908$; $\sin 15° \approx 0.2588190451$; $\sin 18° \approx 0.3090169944$

Exercise Set 6.2

1. a. $11SC$ **b.** $11 \sin \theta \cos \theta$
3. a. $-8C^3S$ **b.** $-8 \cos^3 \theta \sin \theta$
5. a. $1 + 2T + T^2$
b. $1 + 2 \tan \theta + \tan^2 \theta$
7. a. $T^2 + T - 6$ **b.** $\tan^2 \theta + \tan \theta - 6$

9. a. -1 **b.** -1 **11. a.** $(CS + 2)/S$
b. $(\cos A \sin A + 2)/(\sin A)$
13. a. $(T - 1)(T + 9)$
b. $(\tan \beta - 1)(\tan \beta + 9)$
15. a. $(2C + 1)(2C - 1)$
b. $(2 \cos B + 1)(2 \cos B - 1)$
17. a. $3ST^2(3ST + 2)$
b. $(3 \sec B \tan^2 B)(3 \sec B \tan B + 2)$
19. $\cos \theta = \sqrt{7}/4$; $\tan \theta = 3\sqrt{7}/7$; $\cot \theta = \sqrt{7}/3$; $\sec \theta = 4\sqrt{7}/7$; $\csc \theta = \frac{4}{3}$
21. $\sin \beta = \sqrt{22}/5$; $\tan \beta = \sqrt{66}/3$; $\cot \beta = \sqrt{66}/22$; $\sec \beta = 5\sqrt{3}/3$; $\csc \beta = 5\sqrt{22}/22$ **23.** $\cos A = \frac{12}{13}$; $\tan A = \frac{5}{12}$; $\cot A = \frac{12}{5}$; $\sec A = \frac{13}{12}$; $\csc A = \frac{13}{5}$ **25.** $\sin B = \frac{4}{5}$; $\cos B = \frac{3}{5}$; $\cot B = \frac{3}{4}$; $\sec B = \frac{5}{3}$; $\csc B = \frac{5}{4}$
27. $\sin C = \sqrt{5}/3$; $\cos C = \frac{2}{3}$; $\tan C = \sqrt{5}/2$; $\cot C = 2\sqrt{5}/5$; $\csc C = 3\sqrt{5}/5$ **29.** $\sin \alpha = \sqrt{3}/2$; $\cos \alpha = \frac{1}{2}$; $\tan \alpha = \sqrt{3}$; $\sec \alpha = 2$; $\csc \alpha = 2\sqrt{3}/3$ **31.** $\sin \theta \approx 0.887$; $\tan \theta \approx 1.916$; $\cot \theta \approx 0.522$; $\sec \theta \approx 2.162$; $\csc \theta \approx 1.128$
33. $\sin \theta \approx 0.768$; $\cos \theta \approx 0.640$; $\cot \theta \approx 0.833$; $\sec \theta \approx 1.562$; $\csc \theta \approx 1.302$ **35.** $\sin A + \cos A$
37. $\csc \theta$ **39.** $\cos^2 B$
41. $\cos A + 4$ **43.** $2 \csc \theta$ **45.** 0
47. 1 **49.** $\tan \theta$ **51.** -1
53. $\sin^2 \theta$ **55.** $-2 \cos \theta - 1$

Exercise Set 6.3

1. $BC = 30$ cm; $AC = 30\sqrt{3}$ cm
3. $AB = \frac{32}{3}\sqrt{3}$ cm; $BC = \frac{16}{3}\sqrt{3}$ cm
5. $AC \approx 11.5$ cm; $BC \approx 9.6$ cm
7. a. 15.59 ft **b.** 9 ft
9. 34 million miles **11.** 141.1 m
13. a. 27.3 ft **b.** 906.9 ft^2
15. 1.50 in.^{2}
17. $\frac{7}{2} \sin(360°/7) \approx 2.736$ square units
19. $\pi - 2\sqrt{2} \approx 0.313$ square units
21. $\sqrt{3} \approx 1.732$ square units
25. 10,660 ft **27.** 136 m
29. $18(\sqrt{3} - 1)$ cm
31. a. $\angle BOA = 90° - \theta$; $\angle OAB = \theta$; $\angle BAP = 90° - \theta$; $\angle BPA = \theta$
b. $AO = \sin \theta$; $AP = \cos \theta$; $OB = \sin^2 \theta$; $BP = \cos^2 \theta$ **33. a.** $BC = 5/(\sin \theta)$
b. $AB = 4/(\cos \theta)$
c. $AC = 4 \sec \theta + 5 \csc \theta$
35. a. $DE = \sin \theta$ **b.** $OE = \cos \theta$

c. $CF = \tan \theta$ **d.** $OC = \sec \theta$
e. $AB = \cot \theta$ **f.** $OB = \csc \theta$
37. b. 1080 miles
39. b.

n	5	10	50	100
A_n	2.38	2.94	3.1333	3.1395

n	1,000	5,000	10,000
A_n	3.141572	3.1415918	3.1415924

c. As n gets larger, A_n becomes closer to the area of the circle, which is π.

Exercise Set 6.4

1. a. reference angle $= 70°$

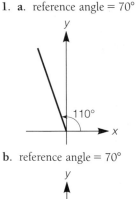

b. reference angle $= 70°$

3. a. reference angle $= 20°$

b. reference angle $= 20°$

5. a. reference angle = 60°

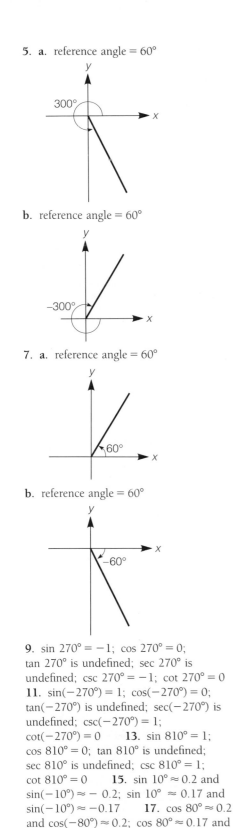

b. reference angle = 60°

7. a. reference angle = 60°

b. reference angle = 60°

9. $\sin 270° = -1$; $\cos 270° = 0$; $\tan 270°$ is undefined; $\sec 270°$ is undefined; $\csc 270° = -1$; $\cot 270° = 0$
11. $\sin(-270°) = 1$; $\cos(-270°) = 0$; $\tan(-270°)$ is undefined; $\sec(-270°)$ is undefined; $\csc(-270°) = 1$; $\cot(-270°) = 0$ **13.** $\sin 810° = 1$; $\cos 810° = 0$; $\tan 810°$ is undefined; $\sec 810°$ is undefined; $\csc 810° = 1$; $\cot 810° = 0$ **15.** $\sin 10° \approx 0.2$ and $\sin(-10°) \approx -0.2$; $\sin 10° \approx 0.17$ and $\sin(-10°) \approx -0.17$ **17.** $\cos 80° \approx 0.2$ and $\cos(-80°) \approx 0.2$; $\cos 80° \approx 0.17$ and

$\cos(-80°) \approx 0.17$ **19.** $\sin 120° \approx 0.9$ and $\sin(-120°) \approx -0.9$; $\sin 120° \approx 0.87$ and $\sin(-120°) \approx -0.87$
21. $\sin 150° = 0.5$ and $\sin(-150°) = -0.5$; $\sin 150° = 0.5$ and $\sin(-150°) = -0.5$
23. $\cos 220° \approx -0.8$ and $\cos(-220°) \approx -0.8$; $\cos 220° \approx -0.77$ and $\cos(-220°) \approx -0.77$
25. $\cos 310° \approx 0.6$ and $\cos(-310°) \approx 0.6$; $\cos 310° \approx 0.64$ and $\cos(-310°) \approx 0.64$
27. $\sin(40° + 360°) \approx 0.6$; $\sin(40° + 360°) \approx 0.64$ **29. a.** $\sqrt{2}/2$
b. $\sqrt{2}/2$ **c.** $-\sqrt{2}/2$ **d.** $\sqrt{2}/2$
31. a. $\frac{1}{2}$ **b.** $\frac{1}{2}$ **c.** $-\sqrt{3}/2$ **d.** $\sqrt{3}/2$
33. a. $-\sqrt{3}/2$ **b.** $-\sqrt{3}/2$ **c.** $-\frac{1}{2}$ **d.** $\frac{1}{2}$
35. a. $\sqrt{3}/2$ **b.** $\sqrt{3}/2$ **c.** $\frac{1}{2}$ **d.** $-\frac{1}{2}$
37. a. -2 **b.** $2\sqrt{3}/3$ **c.** $\sqrt{3}$
d. $-\sqrt{3}/3$
39.

θ	$\sin\theta$	$\cos\theta$	$\tan\theta$
0°	0	1	0
30°	1/2	$\sqrt{3}/2$	$\sqrt{3}/3$
45°	$\sqrt{2}/2$	$\sqrt{2}/2$	1
60°	$\sqrt{3}/2$	1/2	$\sqrt{3}$
90°	1	0	undefined
120°	$\sqrt{3}/2$	$-1/2$	$-\sqrt{3}$
135°	$\sqrt{2}/2$	$-\sqrt{2}/2$	-1
150°	1/2	$-\sqrt{3}/2$	$-\sqrt{3}/3$
180°	0	-1	0

41. $\frac{35}{4}\sqrt{3}$ cm² **43.** 302.7 cm²
45. $108\sqrt{3} \approx 187.06$ cm²
47. a.

Terminal Side of Angle θ Lies In			
$\sin\theta$	positive	positive	negative	negative
$\cos\theta$	positive	negative	negative	positive
$\tan\theta$	positive	negative	positive	negative

49. a. $\sin 70°$ is larger. **b.** $\cos 10°$ is larger. **51. a.** $\cos 100°$ is larger.
b. $\sin 100°$ is larger.
53. a.

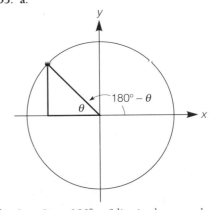

b. θ **c.** Since $180° - \theta$ lies in the second quadrant where all y-coordinates are positive, $\sin(180° - \theta) = \sin\theta$.
55. d. $\sin 75° = (\sqrt{2} + \sqrt{6})/4$
f. $\sin 105° \approx 0.9659$;
$\sin 45° + \sin 60° \approx 1.5731 \neq \sin 105°$
57. a. $\sqrt{3}$ **b.** $-\sqrt{3}$ **c.** $\sqrt{3}/3$
59. a. 0.510 **b.** 0.510 **c.** 1

Exercise Set 6.5

1. a. $\cos\theta = -2\sqrt{6}/5$;
$\tan\theta = -\sqrt{6}/12$; $\cot\theta = -2\sqrt{6}$;
$\sec\theta = -5\sqrt{6}/12$; $\csc\theta = 5$
b. $\cos\theta = -2\sqrt{6}/5$; $\tan\theta = \sqrt{6}/12$;
$\cot\theta = 2\sqrt{6}$; $\sec\theta = -5\sqrt{6}/12$;
$\csc\theta = -5$ **3. a.** $\sin\theta = \frac{12}{13}$;
$\tan\theta = \frac{12}{5}$; $\cot\theta = \frac{5}{12}$; $\sec\theta = \frac{13}{5}$;
$\csc\theta = \frac{13}{12}$ **b.** $\sin\theta = -\frac{12}{13}$; $\tan\theta = \frac{12}{5}$;
$\cot\theta = \frac{5}{12}$; $\sec\theta = -\frac{13}{5}$; $\csc\theta = -\frac{13}{12}$
5. $\sin A = -\frac{1}{3}$; $\cos A = 2\sqrt{2}/3$;
$\tan A = -\sqrt{2}/4$; $\cot A = -2\sqrt{2}$;
$\sec A = 3\sqrt{2}/4$ **7.** $\cos B = -\frac{2}{3}$;
$\sin B = -\sqrt{5}/3$; $\tan B = \sqrt{5}/2$;
$\cot B = 2\sqrt{5}/5$; $\csc B = -3\sqrt{5}/5$
9. $\sin\theta = -\sqrt{9 - t^2}/3$;
$\tan\theta = -\sqrt{9 - t^2}/t$;
$\cot\theta = -t\sqrt{9 - t^2}/(9 - t^2)$; $\sec\theta = 3/t$;
$\csc\theta = -3\sqrt{9 - t^2}/(9 - t^2)$

11. $\cos\theta = -\sqrt{1 - 9u^2}$; $\tan\theta = 3u\sqrt{1 - 9u^2}/(1 - 9u^2)$; $\cot\theta = \sqrt{1 - 9u^2}/(3u)$; $\sec\theta = -\sqrt{1 - 9u^2}/(1 - 9u^2)$; $\csc\theta = -1/(3u)$
13. $\sin\theta = \sqrt{9 - 3u^2}/3$; $\tan\theta = \sqrt{3 - u^2}/u$; $\cot\theta = u\sqrt{3 - u^2}/(3 - u^2)$; $\sec\theta = \sqrt{3}/u$; $\csc\theta = \sqrt{9 - 3u^2}/(3 - u^2)$
37. a. not an identity: let $\alpha = 30°$ **b.** is an identity
45. a. $(\cos\theta - \sin\theta)(1 + \cos\theta\sin\theta)$

Chapter Six Review Exercises

1. $\sqrt{2}/2$ **3.** $-\sqrt{3}$ **5.** -2
7. -1 **9.** $\sqrt{2}/2$ **11.** 1
13. $-2\sqrt{3}/3$ **15.** 2 **17.** 1
19. $a = \frac{1}{2}$, $C = \sqrt{3}/2$ **21.** $b = \frac{35}{2}$
23. $\sin A = \sqrt{55}/8$; $\cot A = 3\sqrt{55}/55$
25. 5 square units **27.** $\frac{1}{5}$
29. $a = 16\sqrt{3}/3$ **31.** $c = 9$; area $= 9\sqrt{19}/4$ **35.** $\cos\theta = \dfrac{p^4 - q^4}{p^4 + q^4}$; $\tan\theta = \dfrac{2p^2q^2}{p^4 - q^4}$ **37.** $\sin A \cos A$
39. $\sin^2 A$ **41.** $\cos A + \sin A$
43. $\sin A \cos A$ **45.** $2\sin A \cos A$
47. $\cos 10° \approx 0.9848$
49. $\sin 10° \approx 0.1736$
51. $\sin\theta = \frac{4}{5}$, $\tan\theta = \frac{4}{3}$
53. $\tan\theta = -\frac{24}{7}$ **55.** $\cot\theta = \frac{5}{12}$
57. $\tan(90° - \theta) = \dfrac{5t\sqrt{1 - 25t^2}}{1 - 25t^2}$
59. $\sin\theta = (2\sqrt{3} - \sqrt{6})/6$
63. 2521 cm² **79. a.** $PN = \sin\theta$
b. $ON = \cos\theta$ **c.** $PT = \tan\theta$
d. $OT = \sec\theta$ **e.** $NA = 1 - \cos\theta$
f. $NT = \sin\theta\tan\theta$
81. $A = (1 - \cos\theta)^2/(\sin\theta\cos\theta)$

Chapter Six Test

1. a. $\sqrt{3}/3$ **b.** $\sqrt{2}$ **c.** 1 **d.** 0
2. a. negative **b.** negative **c.** positive
3. a. 1 **b.** -1 **c.** 0
4. $(2\cot\theta + 3)(\cot\theta + 4)$ **5.** 18 cm²
6. a. $\sqrt{2}/2$ **b.** $-\sqrt{3}/3$ **c.** -2
7. $\cos\theta = -2\sqrt{5}/5$, $\tan\theta = \frac{1}{2}$
8. $\cos\theta = \sqrt{9 - u^2}/3$, $\cot\theta = -\sqrt{9 - u^2}/u$

9. $-2\cos\theta - 1$
10. $\frac{81}{2}\sin 40° \approx 26.033$ m² **11.** 1
12. $CD = 50(\tan 55° - \tan 25°)$
13. a. Since 85° has a much larger y-coordinate than 5°, $\sin 85°$ is larger. **b.** Since the x-coordinate is much larger at 5° than the y-coordinate, $\cos 5°$ is larger. **c.** Since the terminal side for 175° lies in quadrant II, $\tan 175° < 0$. Since the terminal side for 185° lies in quadrant III, $\tan 185° > 0$. Thus $\tan 185°$ is larger.
14. a. $AD = 2.75\cos 20°$; $CD = 2.75\sin 20°$; $DB = 3.25 - 2.75\cos 20°$
b. $CB \approx 1.15$ cm

CHAPTER SEVEN

Exercise Set 7.1

1. a. $\pi/4 \approx 0.79$ radians
b. $\pi/2 \approx 1.57$ radians
c. $3\pi/4 \approx 2.36$ radians
3. a. 0 radians **b.** $2\pi \approx 6.28$ radians
c. $5\pi/2 \approx 7.85$ radians
5. a. 15° **b.** 30° **c.** 45°
7. a. 60° **b.** 300° **c.** 720°
9. a. 114.59° **b.** 171.89° **c.** 565.49°
11. smaller than a right angle
13. $30° = \pi/6$ radians, $45° = \pi/4$ radians, $60° = \pi/3$ radians, $120° = 2\pi/3$ radians, $135° = 3\pi/4$ radians, $150° = 5\pi/6$ radians
15. 2.5 radians **17.** 2 radians
19. $0.5 < \cos 1 < 0.6$; $0.8 < \sin 1 < 0.9$; $\cos 1 \approx 0.54$; $\sin 1 \approx 0.84$
21. $0.5 < \cos(-1) < 0.6$; $-0.9 < \sin(-1) < -0.8$; $\cos(-1) \approx 0.54$; $\sin(-1) \approx -0.84$
23. $-0.7 < \cos 4 < -0.6$; $-0.8 < \sin 4 < -0.7$; $\cos 4 \approx -0.65$; $\sin 4 \approx -0.76$
25. $-0.7 < \cos(-4) < -0.6$; $0.7 < \sin(-4) < 0.8$; $\cos(-4) \approx -0.65$; $\sin(-4) \approx 0.76$
27. $0.8 < \sin(1 + 2\pi) < 0.9$; $\sin(1 + 2\pi) \approx 0.84$ **29.** 4π ft
31. $\pi/2$ cm
33. a. 12π cm² ≈ 37.70 cm²
b. $\frac{50}{9}\pi$ m² ≈ 17.45 m²
c. $\frac{72}{5}\pi$ m² ≈ 45.24 m²
d. 1.296π cm² ≈ 4.07 cm²
35. $2\pi/5$ radians
37. a. 12π radians/sec **b.** 144π cm/sec
c. 72π cm/sec **39. a.** 6π radians/sec

b. 150π cm/sec **c.** 75π cm/sec
41. a. $50\pi/3$ radians/sec
b. 750π cm/sec **c.** 375π cm/sec
43. $180°/\pi \approx 57.30°$
45. a. 200π radians/min
b. 2000π cm/min
c. $1000\pi/3$ radians/min **d.** $\frac{500}{3}$ rpm
47. 4930 miles **49.** 1470 miles
51. 2690 miles **53. a.** $r = 6/(1 + \theta)$
b. $A = 36\theta/(1 + \theta)^2$

Exercise Set 7.2

1.

θ	$\sin\theta$	$\cos\theta$	$\tan\theta$	$\csc\theta$	$\sec\theta$	$\cot\theta$
0	0	1	0	undef.	1	undef.
$\pi/6$	$1/2$	$\sqrt{3}/2$	$\sqrt{3}/3$	2	$2\sqrt{3}/3$	$\sqrt{3}$
$\pi/4$	$\sqrt{2}/2$	$\sqrt{2}/2$	1	$\sqrt{2}$	$\sqrt{2}$	1
$\pi/3$	$\sqrt{3}/2$	$1/2$	$\sqrt{3}$	$2\sqrt{3}/3$	2	$\sqrt{3}/3$
$\pi/2$	1	0	undef.	1	undef.	0
$2\pi/3$	$\sqrt{3}/2$	$-1/2$	$-\sqrt{3}$	$2\sqrt{3}/3$	-2	$-\sqrt{3}/3$
$3\pi/4$	$\sqrt{2}/2$	$-\sqrt{2}/2$	-1	$\sqrt{2}$	$-\sqrt{2}$	-1
$5\pi/6$	$1/2$	$-\sqrt{3}/2$	$-\sqrt{3}/3$	2	$-2\sqrt{3}/3$	$-\sqrt{3}$
π	0	-1	0	undef.	-1	undef.

13. $\cos t = -\frac{4}{5}$; $\tan t = \frac{3}{4}$
15. $\cos t = -\sqrt{13}/4$; $\tan t = -\sqrt{39}/13$
17. $\sec\alpha = \frac{13}{5}$; $\cos\alpha = \frac{5}{13}$; $\sin\alpha = \frac{12}{13}$
23. $(\sqrt{7}\cos\theta)/7$ **25. a.** $-\frac{2}{3}$ **b.** $\frac{1}{4}$
c. $\frac{1}{5}$ **d.** $-\frac{1}{5}$ **27. a.** $-(1 + 2\sqrt{2})/3$
b. 1 **29. a.** $\sqrt{2}/2$ **b.** $\sqrt{3}/2$ **c.** 1
31. $\cos^2 t$ **33.** $\sin^2\theta$ **43.** $\frac{13}{31}$
47. a. 0 **b.** $(2 - \sqrt{2})/2$ **c.** no
49.

θ	$1 - \dfrac{\theta^2}{2}$	$\cos\theta$
0.1	0.995	0.995004...
0.2	0.980	0.980066...
0.3	0.955	0.955336...

Exercise Set 7.3

1. period: 2; amplitude: 1
3. period: 4; amplitude: 6
5. period: 4; amplitude: 2
7. period: 6; amplitude: $\frac{3}{2}$
9. $(-7\pi/2, 1) \approx (-10.996, 1)$
11. $(5\pi/2, 1) \approx (7.854, 1)$
13. $(-4\pi, 0) \approx (-12.566, 0)$
15. $(-3\pi, 0) \approx (-9.425, 0)$
17. $(-\pi, 0) \approx (-3.142, 0)$
19. increasing   21. decreasing
23.

x	$\sin x$	$x - \sin x$
0.453	0.43766	0.01534
0.253	0.25031	0.00269
0.0253	0.02530	0.00000
0.00253	0.00253	0.00000

25. $(9\pi/2, 0) \approx (14.137, 0)$
27. $(-4\pi, 1) \approx (-12.566, 1)$
29. $(\pi/2, 0) \approx (1.571, 0)$
31. $(4\pi, 1) \approx (12.566, 1)$
33. $(-5\pi/2, 0) \approx (-7.854, 0)$
35. decreasing   37. increasing
39. **a.** $C(\cos\theta, \sin\theta)$

Exercise Set 7.4

1. **a.** amplitude: 2; period: 2π;
x-intercepts: 0, π, 2π;
increasing: $(0, \pi/2)$ and $(3\pi/2, 2\pi)$

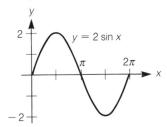

b. amplitude: 1; period: π;
x-intercepts: 0, $\pi/2$, π;
increasing: $(\pi/4, 3\pi/4)$

3. **a.** amplitude: 1; period: π;
x-intercepts: $\pi/4$, $3\pi/4$;
increasing: $(\pi/2, \pi)$

b. amplitude: 2; period: π;
x-intercepts: $\pi/4$, $3\pi/4$;
increasing: $(\pi/2, \pi)$

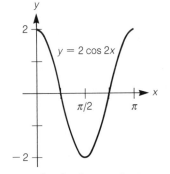

5. **a.** amplitude: 3; period: 4;
x-intercepts: 0, 2, 4;
increasing: $(0, 1)$ and $(3, 4)$

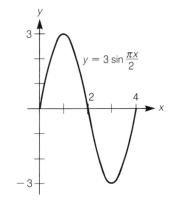

b. amplitude: 3; period: 4;
x-intercepts: 0, 2, 4; increasing: $(1, 3)$

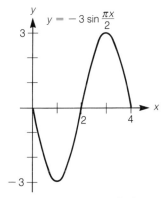

7. **a.** amplitude: 1; period: 1;
x-intercepts: $\frac{1}{4}$, $\frac{3}{4}$; increasing: $\left(\frac{1}{2}, 1\right)$

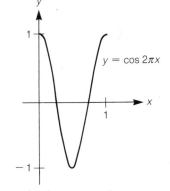

b. amplitude: 4; period: 1;
x-intercepts: $\frac{1}{4}$, $\frac{3}{4}$; increasing: $\left(0\,\frac{1}{2}\right)$

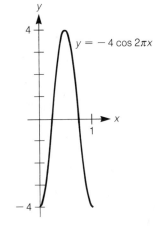

9. amplitude: 1; period: π;
x-intercept: $3\pi/4$;
increasing: $(0, \pi/4)$ and $(3\pi/4, \pi)$

11. amplitude: 1; period: 6;
x-intercepts: 0, 6; increasing: $(0, 3)$

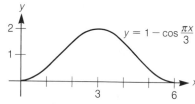

13. amplitude: 1; period: 2π;
phase shift: $\pi/6$;
x-intercepts: $\pi/6, 7\pi/6, 13\pi/6$;
high point: $(2\pi/3, 1)$;
low point: $(5\pi/3, -1)$

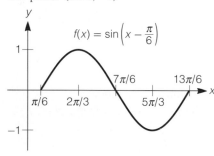

15. amplitude: 1; period: 2π;
phase shift: $-\pi/4$;
x-intercepts: $\pi/4, 5\pi/4$;
high point: $(3\pi/4, 1)$;
low points: $(-\pi/4, -1)$ and $(7\pi/4, -1)$

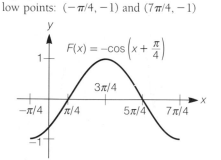

17. amplitude: 1; period: π;
phase shift: $\pi/4$;
x-intercepts: $\pi/4, 3\pi/4, 5\pi/4$;
high point: $(\pi/2, 1)$; low point: $(\pi, -1)$

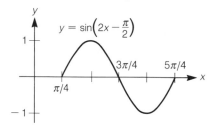

19. amplitude: 1; period: π;
phase shift: $\pi/2$; x-intercepts: $3\pi/4$,
$5\pi/4$; high points: $(\pi/2, 1)$ and $(3\pi/2, 1)$;
low point: $(\pi, -1)$

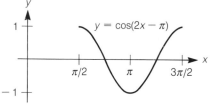

21. amplitude: 3; period: 4π;
phase shift: $-\pi/3$;
x-intercepts: $-\pi/3, 5\pi/3, 11\pi/3$;
high point: $(2\pi/3, 3)$;
low point: $(8\pi/3, -3)$

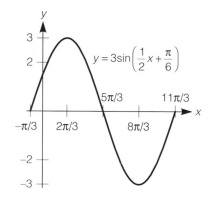

23. amplitude: 4; period: $2\pi/3$;
phase shift: $\pi/12$;
x-intercepts: $\pi/4, 7\pi/12$;
high points: $(\pi/12, 4)$ and $(3\pi/4, 4)$;
low point: $(5\pi/12, -4)$

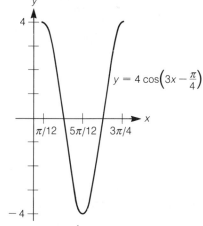

25. amplitude: $\frac{1}{2}$; period: 4;
phase shift: 2π;
x-intercepts: $2\pi, 2\pi + 2, 2\pi + 4$;
high point: $(2\pi + 1, \frac{1}{2})$;
low point: $(2\pi + 3, -\frac{1}{2})$

27. amplitude: 1; period: π;
phase shift: $\pi/6$;
x-intercepts: $\pi/6$ and $7\pi/6$;
high point: $(2\pi/3, 2)$;
low points: $(\pi/6, 0)$ and $(7\pi/6, 0)$

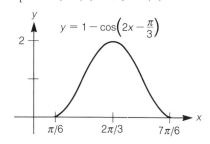

29. $A = 2$, $B = \frac{1}{2}$ **31.** $A = -3$, $B = \pi$

33. $A = -4$, $B = \frac{1}{5}$

35.

c. π, 2π **d.** $\pi/2$

b. amplitude (all three): 1;
period of Y_1: 2π; period of Y_2: π;
period of Y_3: $2\pi/3$

5. a. amplitude: 2.5; period: $\frac{2}{3}$;
phase shift: $-4/(3\pi)$

b.

c. highest points:
$\left(\frac{1}{3} - \frac{4}{3\pi}, 2.5\right) \approx (-0.09, 2.5)$ and
$\left(1 - \frac{4}{3\pi}, 2.5\right) \approx (0.58, 2.5)$
lowest points:
$\left(-\frac{4}{3\pi}, -2.5\right) \approx (-0.42, -2.5)$,
$\left(\frac{2}{3} - \frac{4}{3\pi}, -2.5\right) \approx (0.24, -2.5)$,
$\left(\frac{4}{3} - \frac{4}{3\pi}, -2.5\right) \approx (0.91, -2.5)$

7. a. amplitude: 1; period: 4π;
phase shift: -1.5

b.

Optional TI-81 Graphing Calculator
Exercises for Sections 7.3 and 7.4

1. a.

b.

e.

f. $\pi/2$ units

3. a.

c. highest points:
$(-1.5 + \pi, 1) \approx (1.64, 1)$ and
$(-1.5 + 5\pi, 1) \approx (14.2, 1)$
lowest points:
$(-1.5 + 3\pi, -1) \approx (7.92, -1)$ and
$(-1.5 + 7\pi, -1) \approx (20.49, -1)$
9. a. amplitude: 0.02; period: 0.02;
phase shift: -0.04
b.

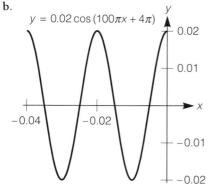

$y = 0.02 \cos(100\pi x + 4\pi)$

c. highest points: $(-0.04, 0.02)$,
$(-0.02, 0.02)$, and $(0, 0.02)$
lowest points: $(-0.03, -0.02)$ and
$(-0.01, -0.02)$
11. b. The two graphs are not identical.

c.

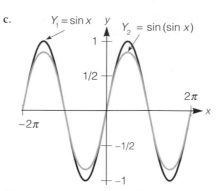

d. $(1.58, 0.84)$; the y-coordinate gives
the amplitude. **e.** Since $\sin x$ is largest
when $x = \pi/2$ (then $\sin x = 1$), and $\sin x$
is increasing for $0 < x < \pi/2$, $\sin(\sin x)$
will have its largest value when $x = \pi/2$.
Its value then is $\sin[\sin(\pi/2)] = \sin 1$.
f. $\sin 1 \approx 0.84$

Exercise Set 7.5

1. a. x-intercept: $-\pi/4$; y-intercept: 1;
asymptotes: $x = -3\pi/4$ and $x = \pi/4$

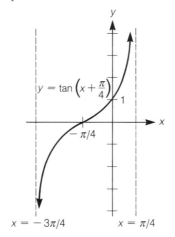

b. x-intercept: $-\pi/4$; y-intercept: -1;
asymptotes: $x = -3\pi/4$ and $x = \pi/4$

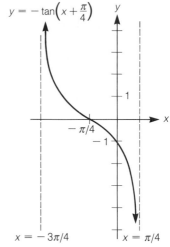

3. a. x- and y-intercepts: 0;
asymptotes: $x = -3\pi/2$ and $x = 3\pi/2$

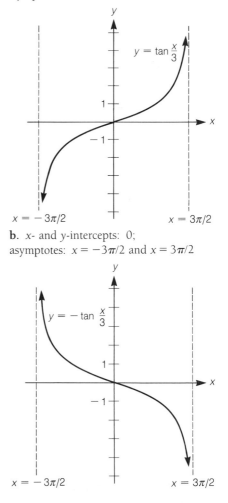

b. x- and y-intercepts: 0;
asymptotes: $x = -3\pi/2$ and $x = 3\pi/2$

5. x- and y-intercepts: 0; asymptotes: $x = -1$ and $x = 1$

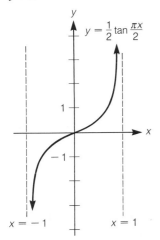

7. x-intercept: 1; no y-intercept; asymptotes: $x = 0$ and $x = 2$

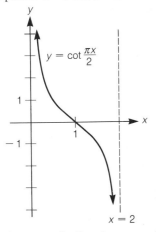

9. x-intercept: $3\pi/4$; y-intercept: 1; asymptotes: $x = \pi/4$ and $x = 5\pi/4$

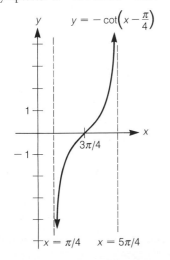

11. x-intercept: $\pi/4$; no y-intercept; asymptotes: $x = 0$ and $x = \pi/2$

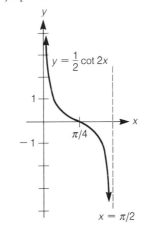

13. no x-intercept; y-intercept: $-\sqrt{2}$; asymptotes: $x = -3\pi/4$, $x = \pi/4$, and $x = 5\pi/4$

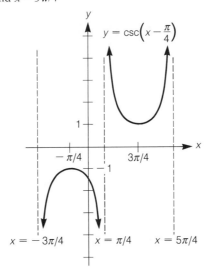

15. no x- or y-intercepts; asymptotes: $x = -2\pi$, $x = 0$, and $x = 2\pi$

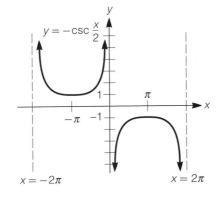

17. no x- or y-intercepts; asymptotes: $x = -1$, $x = 0$, and $x = 1$

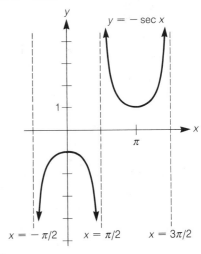

19. no x-intercept; y-intercept: -1; asymptotes: $x = -\pi/2$, $x = \pi/2$, and $x = 3\pi/2$

21. no x-intercept; y-intercept: -1; asymptotes: $x = \pi/2$, $x = 3\pi/2$, and $x = 5\pi/2$

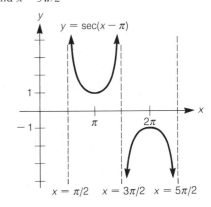

23. no x-intercept; y-intercept: 3;
asymptotes: $x = -1$, $x = 1$, and $x = 3$

$$y = 3 \sec \frac{\pi x}{2}$$

$x = -1 \quad x = 1 \quad x = 3$

25. a. x-intercepts: $-\frac{11}{18}$, $-\frac{5}{18}$, $\frac{1}{18}$, $\frac{7}{18}$,
and $\frac{13}{18}$; y-intercept: -1; no asymptotes

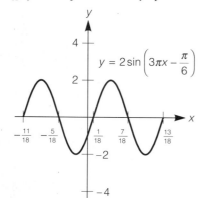

$$y = 2 \sin \left(3\pi x - \frac{\pi}{6} \right)$$

b. no x-intercepts; y-intercept: -4;
asymptotes: $x = -\frac{11}{18}$, $x = -\frac{5}{18}$, $x = \frac{1}{18}$,
$x = \frac{7}{18}$, and $x = \frac{13}{18}$

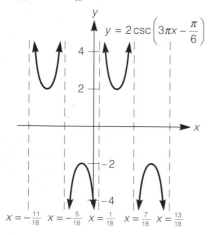

$$y = 2 \csc \left(3\pi x - \frac{\pi}{6} \right)$$

$x = -\frac{11}{18} \quad x = -\frac{5}{18} \quad x = \frac{1}{18} \quad x = \frac{7}{18} \quad x = \frac{13}{18}$

27. a. x-intercepts: $-\frac{5}{8}$, $-\frac{1}{8}$, $\frac{3}{8}$, and $\frac{7}{8}$;
y-intercept: $-3\sqrt{2}/2$; no asymptotes

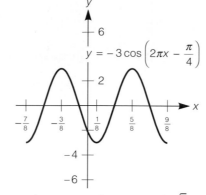

$$y = -3 \cos \left(2\pi x - \frac{\pi}{4} \right)$$

b. no x-intercepts; y-intercept: $-3\sqrt{2}$;
asymptotes: $x = -\frac{5}{8}$, $x = -\frac{1}{8}$, $x = \frac{3}{8}$,
and $x = \frac{7}{8}$

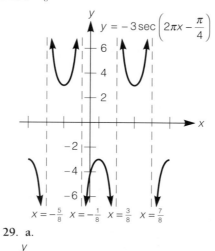

$$y = -3 \sec \left(2\pi x - \frac{\pi}{4} \right)$$

$x = -\frac{5}{8} \quad x = -\frac{1}{8} \quad x = \frac{3}{8} \quad x = \frac{7}{8}$

29. a.

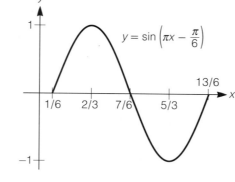

$$y = \sin \left(\pi x - \frac{\pi}{6} \right)$$

b.

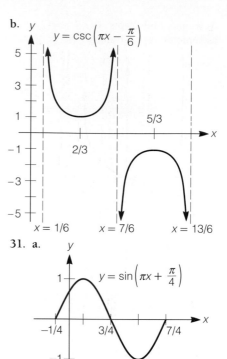

$$y = \csc \left(\pi x - \frac{\pi}{6} \right)$$

$x = 1/6 \quad x = 7/6 \quad x = 13/6$

31. a.

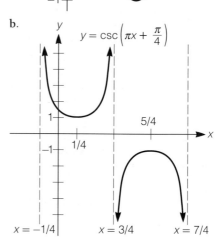

$$y = \sin \left(\pi x + \frac{\pi}{4} \right)$$

b.

$$y = \csc \left(\pi x + \frac{\pi}{4} \right)$$

$x = -1/4 \quad x = 3/4 \quad x = 7/4$

33.

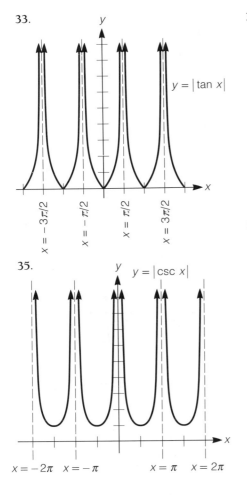

$y = |\tan x|$

$x = -3\pi/2$ $x = -\pi/2$ $x = \pi/2$ $x = 3\pi/2$

35.

$y = |\csc x|$

$x = -2\pi$ $x = -\pi$ $x = \pi$ $x = 2\pi$

Optional TI-81 Graphing Calculator
Exercises for Section 7.5

1.

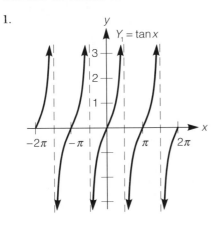

$Y_1 = \tan x$

-2π $-\pi$ π 2π

3.

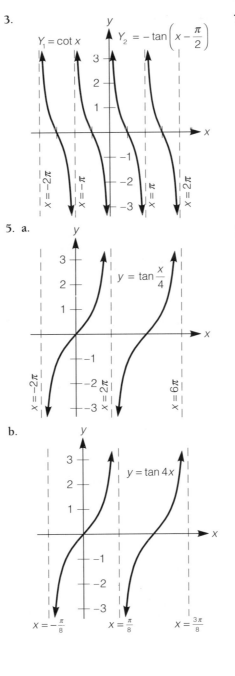

$Y_1 = \cot x$ $Y_2 = -\tan\left(x - \dfrac{\pi}{2}\right)$

$x = -2\pi$ $x = -\pi$ $x = \pi$ $x = 2\pi$

5. a.

$y = \tan\dfrac{x}{4}$

$x = -2\pi$ $x = 2\pi$ $x = 6\pi$

b.

$y = \tan 4x$

$x = -\dfrac{\pi}{8}$ $x = \dfrac{\pi}{8}$ $x = \dfrac{3\pi}{8}$

7. a.

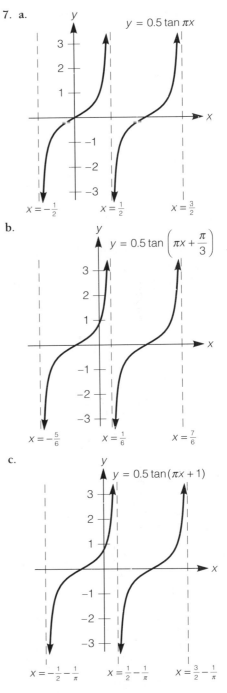

$y = 0.5\tan \pi x$

$x = -\dfrac{1}{2}$ $x = \dfrac{1}{2}$ $x = \dfrac{3}{2}$

b.

$y = 0.5\tan\left(\pi x + \dfrac{\pi}{3}\right)$

$x = -\dfrac{5}{6}$ $x = \dfrac{1}{6}$ $x = \dfrac{7}{6}$

c.

$y = 0.5\tan(\pi x + 1)$

$x = -\dfrac{1}{2} - \dfrac{1}{\pi}$ $x = \dfrac{1}{2} - \dfrac{1}{\pi}$ $x = \dfrac{3}{2} - \dfrac{1}{\pi}$

9. a.

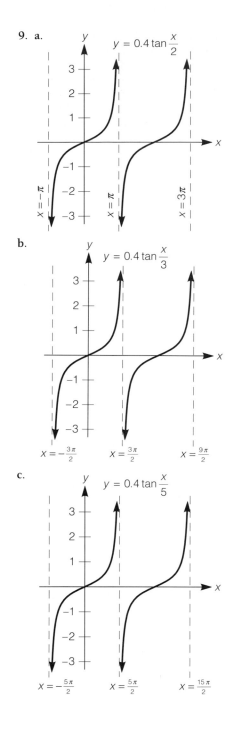

$y = 0.4 \tan \dfrac{x}{2}$

$x = -\pi$ $x = \pi$ $x = 3\pi$

b.

$y = 0.4 \tan \dfrac{x}{3}$

$x = -\dfrac{3\pi}{2}$ $x = \dfrac{3\pi}{2}$ $x = \dfrac{9\pi}{2}$

c.

$y = 0.4 \tan \dfrac{x}{5}$

$x = -\dfrac{5\pi}{2}$ $x = \dfrac{5\pi}{2}$ $x = \dfrac{15\pi}{2}$

11. $\sin x = \csc x$ at $-3\pi/2$, $-\pi/2$, $\pi/2$, and $3\pi/2$; there are no points at which $\sin x = -\csc x$.

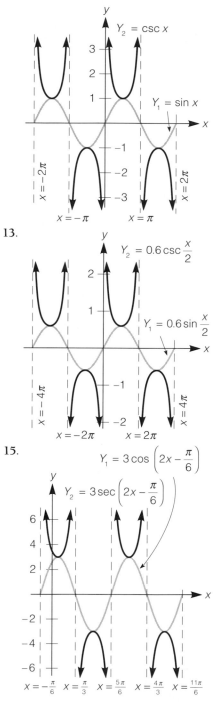

$Y_2 = \csc x$

$Y_1 = \sin x$

$x = -2\pi$ $x = 2\pi$

$x = -\pi$ $x = \pi$

13.

$Y_2 = 0.6 \csc \dfrac{x}{2}$

$Y_1 = 0.6 \sin \dfrac{x}{2}$

$x = -4\pi$ $x = 4\pi$

$x = -2\pi$ $x = 2\pi$

15.

$Y_1 = 3 \cos\left(2x - \dfrac{\pi}{6}\right)$

$Y_2 = 3 \sec\left(2x - \dfrac{\pi}{6}\right)$

$x = -\dfrac{\pi}{6}$ $x = \dfrac{\pi}{3}$ $x = \dfrac{5\pi}{6}$ $x = \dfrac{4\pi}{3}$ $x = \dfrac{11\pi}{6}$

17. The two graphs are identical. This demonstrates that $\tan^2 x = \sec^2 x - 1$ is a trigonometric identity.

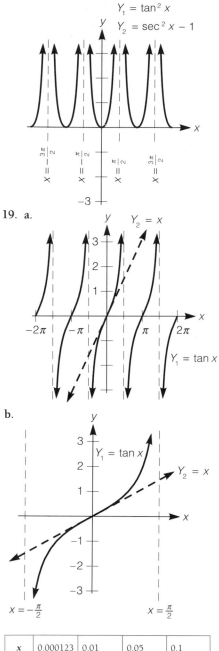

$Y_1 = \tan^2 x$

$Y_2 = \sec^2 x - 1$

$x = -\dfrac{3\pi}{2}$ $x = -\dfrac{\pi}{2}$ $x = \dfrac{\pi}{2}$ $x = \dfrac{3\pi}{2}$

19. a.

$Y_2 = x$

-2π $-\pi$ π 2π

$Y_1 = \tan x$

b.

$Y_1 = \tan x$

$Y_2 = x$

$x = -\dfrac{\pi}{2}$ $x = \dfrac{\pi}{2}$

x	0.000123	0.01	0.05	0.1
an x	0.000123	0.010000	0.050042	0.100335

x	0.2	0.3	0.4	0.5
tan x	0.202710	0.309336	0.422793	0.546302

c.

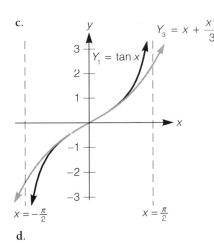

$Y_3 = x + \dfrac{x^3}{3}$

$Y_1 = \tan x$

$x = -\dfrac{\pi}{2}$ $\qquad$ $x = \dfrac{\pi}{2}$

d.

x	0.000123	0.01	0.05	0.1
$\tan x$	0.000123	0.010000	0.050042	0.100335
$x + (x^3/3)$	0.000123	0.010000	0.050042	0.100333

x	0.2	0.3	0.4	0.5
$\tan x$	0.202710	0.309336	0.422793	0.546302
$x + (x^3/3)$	0.202667	0.309	0.421333	0.541667

21. a. The graphs intersect at $x \approx 4.5$.

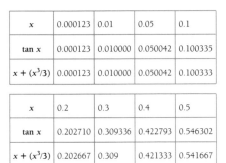

$Y_1 = \tan x$ $\qquad$ $Y_2 = x$

b. The graphs intersect at $x \approx 4.493$.

Exercise Set 7.6

1. $\sin 3\theta$ $\qquad$ **3.** $\sin 2\theta$ $\qquad$ **5.** $\cos 5u$
7. $\sqrt{3}/2$ $\qquad$ **9.** $\sin B$ $\qquad$ **11.** $\cos \theta$
13. $-\cos \theta$ $\qquad$ **15.** $\sin t$
17. $(\sqrt{6} - \sqrt{2})/4$ $\qquad$ **19.** $(\sqrt{6} + \sqrt{2})/4$
21. $\sqrt{2} \sin s$ $\qquad$ **23.** $\sqrt{3} \sin \theta$
25. a. $-\frac{16}{65}$ **b.** $\frac{63}{65}$ $\qquad$ **27. a.** $-\frac{44}{125}$ **b.** $-\frac{4}{5}$
29. a. $2\sqrt{6}/5$ **b.** $4\sqrt{6}/25$

---

31. $\sin(\theta + \beta) = \dfrac{2\sqrt{39} - 3\sqrt{13}}{26}$;

$\cos(\beta - \theta) = \dfrac{2\sqrt{13} - 3\sqrt{39}}{26}$

37. a. The graph of $y = \cos x$ is obtained by translating $y = \sin x$ by $\pi/2$ units to the left.

47. a.

t	1	2	3	4
$f(t)$	1.5	1.5	1.5	1.5

b. $f(t) = 1.5$ $\qquad$ **53.** $\sqrt{3}/2$
57. a. They both equal -1.1918.
b. They both equal -1.3764.

Exercise Set 7.7

1. a. $\frac{4}{3}$ **b.** $\frac{44}{117}$ $\qquad$ **3. a.** $3 - 2\sqrt{2}$
b. $3 - 2\sqrt{2}$ $\qquad$ **5. a.** $\frac{336}{625}$ **b.** $-\frac{527}{625}$
c. $-\frac{336}{527}$ $\qquad$ **7. a.** $-\frac{8}{17}$ **b.** $-\frac{15}{17}$ **c.** $\frac{8}{15}$
9. a. $\frac{1}{2}$ **b.** $\sqrt{3}/2$ **c.** $\sqrt{3}/3$
11. a. $2\sqrt{2}/3$ **b.** $\frac{1}{3}$ **c.** $2\sqrt{2}$
13. a. $-3\sqrt{7}/8$ **b.** $-\frac{1}{8}$
c. $\sqrt{8 + 2\sqrt{7}}/4$ **d.** $\sqrt{8 - 2\sqrt{7}}/4$
15. a. $4\sqrt{2}/9$ **b.** $-\frac{7}{9}$ **c.** $\sqrt{6}/3$
d. $-\sqrt{3}/3$ $\qquad$ **17. a.** $\sqrt{2 - \sqrt{3}}/2$
b. $\sqrt{2 + \sqrt{3}}/2$ **c.** $2 - \sqrt{3}$
19. a. $\sqrt{2 + \sqrt{3}}/2$ **b.** $-\sqrt{2 - \sqrt{3}}/2$
c. $-2 - \sqrt{3}$ $\qquad$ **21. a.** $\frac{24}{25}$ **b.** $\frac{7}{25}$ **c.** $\frac{24}{7}$
23. a. $\frac{24}{25}$ **b.** $-\frac{7}{25}$ **c.** $-\frac{24}{7}$
25. a. $\sqrt{10}/10$ **b.** $3\sqrt{10}/10$ **c.** $\frac{1}{3}$
27. a. $\sqrt{5}/5$ **b.** $2\sqrt{5}/5$ **c.** $\frac{1}{2}$
29. $\sin 2\theta = 2x\sqrt{25 - x^2}/25$;
$\cos 2\theta = (25 - 2x^2)/25$
31. $\sin 2\theta = (x - 1)\sqrt{3 + 2x - x^2}/2$;
$\cos 2\theta = (1 + 2x - x^2)/2$
33. $\frac{1}{8}(3 - 4\cos 2\theta + \cos 4\theta)$
35. $\frac{1}{8}(3 - 4\cos \theta + \cos 2\theta)$
63. a. $2 \sin 8\theta + 2 \sin 4\theta$
b. $\cos 2A - \cos 4A$ **c.** $\frac{1}{2}\cos 6\theta + \frac{1}{2}\cos 2\theta$
73. a. -0.5

Exercise Set 7.8

1. yes $\qquad$ **3.** no $\qquad$ **5.** $(\pi/3) + 2\pi k$ or $(2\pi/3) + 2\pi k$, where k is any integer
7. $(7\pi/6) + 2\pi k$ or $(11\pi/6) + 2\pi k$, where k is any integer $\qquad$ **9.** $\pi + 2\pi k$, where k is any integer $\qquad$ **11.** $(\pi/3) + \pi k$, where k is any integer $\qquad$ **13.** πk, where k is

---

any integer $\qquad$ **15.** $(\pi/2) + \pi k$, $(2\pi/3) + 2\pi k$, or $(4\pi/3) + 2\pi k$, where k is any integer $\qquad$ **17.** πk, where k is any integer $\qquad$ **19.** $(\pi/6) + 2\pi k$, $(5\pi/6) + 2\pi k$, or $(3\pi/2) + 2\pi k$, where k is any integer $\qquad$ **21.** $(\pi/4) + 2\pi k$ or $(3\pi/4) + 2\pi k$, where k is any integer
23. $0°$, $120°$, or $240°$ $\qquad$ **25.** $75°$, $105°$, $195°$, $225°$, $315°$, or $345°$ $\qquad$ **27.** $60°$, $180°$, or $300°$ $\qquad$ **29.** $30°$, $120°$, $210°$, or $300°$ $\qquad$ **31.** $14.5°$ or $165.5°$
33. $116.6°$ or $296.6°$ $\qquad$ **35.** $128.2°$ or $231.8°$ $\qquad$ **37.** 1.39 or 4.90
39. 0.46 or 2.68 $\qquad$ **41.** 1.41 or 4.55
43. 1.37 or 4.51 $\qquad$ **45.** 1.11 or 5.18
47. 0 or π $\qquad$ **49.** $2\pi/3$ or π
51. $(3\pi/16) + (k\pi/4)$, where k is any integer $\qquad$ **53.** $60.45°$
57. a. 1000.173 **b.** 1001.022

Optional TI-81 Graphing Calculator
Exercises for Section 7.8

1. a.

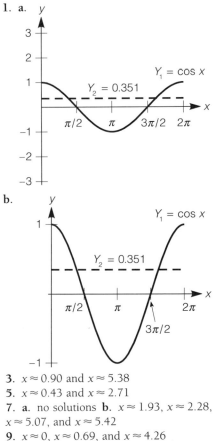

$Y_1 = \cos x$

$Y_2 = 0.351$

b.

$Y_1 = \cos x$

$Y_2 = 0.351$

3. $x \approx 0.90$ and $x \approx 5.38$
5. $x \approx 0.43$ and $x \approx 2.71$
7. a. no solutions **b.** $x \approx 1.93$, $x \approx 2.28$, $x \approx 5.07$, and $x \approx 5.42$
9. $x \approx 0$, $x \approx 0.69$, and $x \approx 4.26$
11. $x \approx 67.5°$ and $x \approx 247.5°$

13. $x \approx 120.8°$ and $x \approx 329.2°$
15. $x = 0°$, $x \approx 1.0°$, and $x \approx 193.9°$

Exercise Set 7.9

1. $\frac{\pi}{3}$ 3. $\frac{\pi}{3}$ 5. $-\frac{\pi}{6}$ 7. $\frac{\pi}{4}$
9. undefined   11. $\frac{1}{4}$ 13. $\frac{3}{4}$
15. $-\frac{\pi}{7}$ 17. $\frac{\pi}{2}$ 19. 0   21. $\frac{4}{3}$
23. $\frac{\sqrt{2}}{2}$ 25. $\frac{12}{5}$ 27. $\frac{1}{2}$
29. $\frac{2\sqrt{2}}{3}$ 31. a. 0.84 b. 0.84
c. 1.26 d. 0.90   33. $\sqrt{2}$
37. $\frac{1}{4}\sin^{-1}(3x/2) - \frac{3}{2}x\sqrt{4 - 9x^2}$
39. $\tan^{-1}\left(\dfrac{x-1}{2}\right) - 2/\sqrt{5 - 2x + x^2}$
41. $\frac{8}{17}$ 43. $31\sqrt{218}/1090$
47. $\frac{1}{2}$, 1   51. $\pm\frac{1}{3}$ 53. $\frac{1}{2}$
55. a.

b.

c. $\sec^{-1}\left(\frac{2}{\sqrt{3}}\right) = \frac{\pi}{6}$; $\sec^{-1}\left(-\frac{2}{\sqrt{3}}\right) = \frac{7\pi}{6}$
d. $\sec^{-1}\left(\sqrt{2}\right) = \frac{\pi}{4}$; $\sec^{-1}\left(-\sqrt{2}\right) = \frac{5\pi}{4}$.
e. $\sec(\sec^{-1} 2) = 2$; $\sec^{-1}(\sec 0) = 0$

Optional TI-81 Graphing Calculator Exercises for Section 7.9

1. a. The graphs are the same.

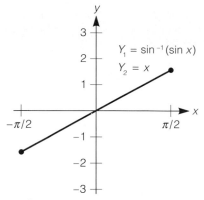

b. $\sin^{-1}(\sin x) = x$, and thus $\sin^{-1} x$ is the inverse function of $\sin x$ on the interval $-\pi/2 \le x \le \pi/2$.
3. a. The graph appears to be that of a constant function.

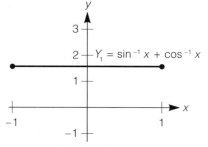

b. $\pi/2$; $\sin^{-1} x + \cos^{-1} x = \pi/2$
5. a. The graphs are identical. This demonstrates that $\tan(\tan^{-1} x) = x$, at least for the values of x that are graphed.

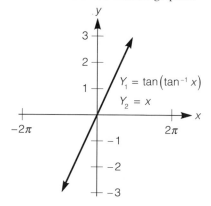

b. The graphs are identical. This demonstrates that $\tan^{-1}(\tan x) = x$ for $-\pi/2 < x < \pi/2$.

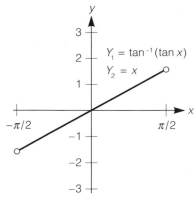

c. It was necessary to change the range settings, since $\tan x$ is one-to-one on the interval $-\pi/2 < x < \pi/2$.   7. $x \approx 0.79$
9. $x \approx 0.80$ 11. $x = 0$, $x \approx 0.80$
13. a. $x = 0$, $x \approx 0.98$ b. $x = 0$
15. a. $x \approx 0.54$ b. $x \approx 0.50$

Chapter Seven Review Exercises

1. a. $-\sqrt{3}/2$ b. $-\sqrt{3}$ 3. $2\sin t$
5.

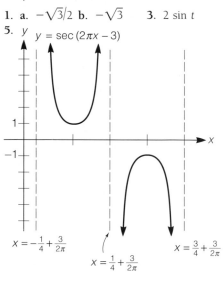

7. amplitude: 1; period: π; phase shift: $\pi/2$

$$y = -\sin(2x - \pi)$$

9. -1 **11.** $2\sqrt{3}/3$ **13.** $-\sqrt{3}/3$
15. $\frac{1}{2}$ **17.** 1 **19.** -2
21. 0.841 **23.** -1 **25.** 0.0123
31. $5\cos\theta$ **33.** $10\tan\theta$
35. $\left(\sqrt{5}/5\right)\cos\theta$ **37.** $-\frac{15}{8}$
39. $A = \pi - \theta - \sin\theta$
41. $(\pi - 2)/2$ cm^2 **43.** $y = 4\sin x$
45. $y = -2\cos 4x$
47. x-intercepts: $\pi/8$ and $3\pi/8$;
high point: $(\pi/4, 3)$;
low points: $(0, -3)$ and $(\pi/2, -3)$

$$y = -3\cos 4x$$

49. x-intercepts: $\frac{1}{2}$, $\frac{5}{2}$, and $\frac{9}{2}$;
high point: $\left(\frac{3}{2}, 2\right)$; low point: $\left(\frac{7}{2}, -2\right)$

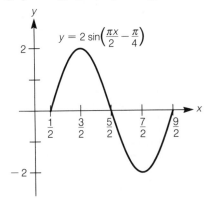
$$y = 2\sin\left(\frac{\pi x}{2} - \frac{\pi}{4}\right)$$

51. x-intercepts: $\frac{5}{2}$ and $\frac{11}{2}$;
high points: $(1, 3)$ and $(7, 3)$;
low point: $(4, -3)$

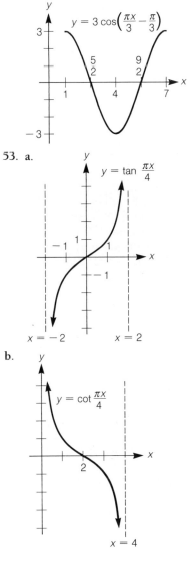
$$y = 3\cos\left(\frac{\pi x}{3} - \frac{\pi}{3}\right)$$

53. a.
$$y = \tan\frac{\pi x}{4}$$

b.
$$y = \cot\frac{\pi x}{4}$$

55. a.

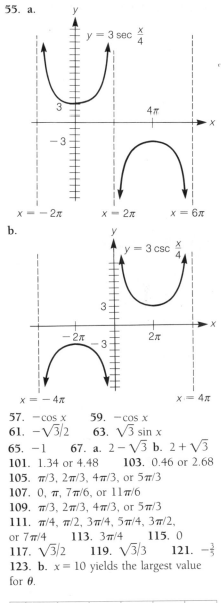
$$y = 3\sec\frac{x}{4}$$

b.
$$y = 3\csc\frac{x}{4}$$

57. $-\cos x$ **59.** $-\cos x$
61. $-\sqrt{3}/2$ **63.** $\sqrt{3}\sin x$
65. -1 **67. a.** $2 - \sqrt{3}$ **b.** $2 + \sqrt{3}$
101. 1.34 or 4.48 **103.** 0.46 or 2.68
105. $\pi/3$, $2\pi/3$, $4\pi/3$, or $5\pi/3$
107. 0, π, $7\pi/6$, or $11\pi/6$
109. $\pi/3$, $2\pi/3$, $4\pi/3$, or $5\pi/3$
111. $\pi/4$, $\pi/2$, $3\pi/4$, $5\pi/4$, $3\pi/2$,
or $7\pi/4$ **113.** $3\pi/4$ **115.** 0
117. $\sqrt{3}/2$ **119.** $\sqrt{3}/3$ **121.** $-\frac{3}{5}$
123. b. $x = 10$ yields the largest value for θ.

x	0.1	1	2	3	10	100
θ	0.01	0.15	0.27	0.35	0.40	0.06

c. $2\sqrt{10} \approx 6.32$; $\theta \approx 0.44$; $(6.32, 0.44)$
129. a. 0.0625

Chapter Seven Test

1. **a.** $-\frac{1}{2}$ **b.** -2 **c.** 1 **2.** $\frac{1}{4}\csc u$

3.

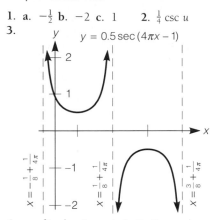

$y = 0.5\sec(4\pi x - 1)$

$x = -\frac{1}{8} + \frac{1}{4\pi}$ $x = \frac{1}{8} + \frac{1}{4\pi}$ $x = \frac{3}{8} + \frac{1}{4\pi}$

4. amplitude: 1; period: $2\pi/3$; phase shift: $\pi/12$

$y = -\sin\left(3x - \frac{\pi}{4}\right)$

5.

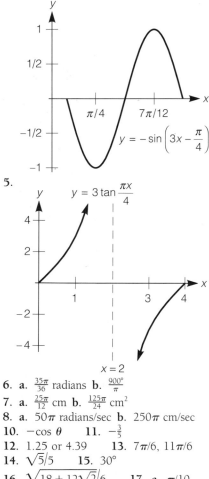

$y = 3\tan\frac{\pi x}{4}$

$x = 2$

6. **a.** $\frac{35\pi}{36}$ radians **b.** $\frac{900°}{\pi}$

7. **a.** $\frac{25\pi}{12}$ cm **b.** $\frac{125\pi}{24}$ cm^2

8. **a.** 50π radians/sec **b.** 250π cm/sec

10. $-\cos\theta$ **11.** $-\frac{3}{5}$

12. 1.25 or 4.39 **13.** $7\pi/6$, $11\pi/6$

14. $\sqrt{5}/5$ **15.** $30°$

16. $\sqrt{18 + 12\sqrt{2}}/6$ **17. a.** $\pi/10$

b. 0 **18.** $\sqrt{7}/4$ **20.** x

CHAPTER EIGHT

Exercise Set 8.1

1. $4\sqrt{6}$ cm **3.** 20 sin 50° cm
5. $a \approx 9.7$ cm; $c \approx 16.4$ cm
7. $A \approx 63.3°$; $C \approx 50.7°$; $c \approx 25.5$ cm
9. **a.** 45° or 135° **b.** 45° **c.** 14.5° or 165.5° **d.** 131.8° **13. b.** $\angle C = 105°$; $c \approx 1.93$ **c.** $\angle C = 15°$; $c \approx 0.52$
d. 0.68 and 0.18

15. $a = \dfrac{2\sin 70°}{\sin 20°}$ cm; $b = \dfrac{2\sin 50°}{\sin 20°}$ cm;

$c = \dfrac{2\sin 50° \sin 70°}{\sin 20° \sin 85°}$ cm;

$d = \dfrac{2\sin 50° \sin 15°}{\sin 20° \sin 85°}$ cm; **17.** 160 ft

19. **a.** 7 cm **b.** $\sqrt{129}$ cm
21. **a.** 7.5 cm **b.** 17.7 cm
23. This is incorrect because x is not the length of the side opposite the 130° angle. The correct equation is $6^2 = x^2 + 3^2 - 2(x)(3)\cos 130°$.
25. $\cos A = \frac{113}{140}$; $\cos B = \frac{29}{40}$; $\cos C = -\frac{5}{28}$
27. $A \approx 27.8°$; $B \approx 32.2°$; $C \approx 120°$
29. $A = 30°$; $B = 30°$; $C = 120°$
31. 5.9 units **33. a.** $a \approx 4.2$ cm
b. $C \approx 29.3°$ **c.** $B \approx 110.7°$
35. approximately 31 miles
37. The distance from the plane to lighthouse A is 1.26 miles and to lighthouse B is 1.60 miles.
39. 860,000 miles **41. b.** $a = 5$, $b = 3$, and $c = 7$ (using $m = 2$ and $n = 1$)
43. 2:40 P.M. **47.** 45° or 135°

Exercise Set 8.2

1. magnitude: $\sqrt{34}$

3. magnitude: $\sqrt{10}$

5. magnitude: $\sqrt{10}$

7. magnitude: $6\sqrt{2}$

9. magnitude: $2\sqrt{13}$

15. magnitude: 9

21. magnitude: $3\sqrt{5}$

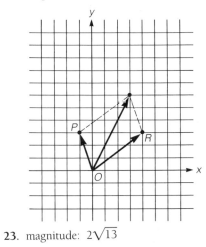

11. magnitude: $\sqrt{10}$

17. magnitude: $\sqrt{37}$

23. magnitude: $2\sqrt{13}$

13. magnitude: 6

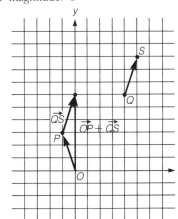

19. magnitude: $\sqrt{61}$

25. magnitude: $6\sqrt{5}$

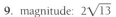

27. $|\mathbf{F} + \mathbf{G}| = \sqrt{41}$ N;
$\theta = \tan^{-1}\left(\frac{4}{5}\right) \approx 38.7°$
29. $|\mathbf{F} + \mathbf{G}| = 9\sqrt{2}$ N; $\theta = \tan^{-1}(1) = 45°$
31. $|\mathbf{F} + \mathbf{G}| = \sqrt{62.3525} \approx 7.90$ N;
$\theta = \tan^{-1}\left(\frac{3.22}{7.21}\right) \approx 24.1°$
33. magnitude ≈ 6.92 N; $\theta \approx 34.67°$
35. magnitude ≈ 39.20 N; $\theta \approx 21.46°$
37. magnitude ≈ 38.96 N; $\theta \approx 29.44°$
39. horizontal ≈ 13.86 cm/sec;
vertical $= 8$ cm/sec
41. horizontal ≈ 3.62 N;
vertical ≈ 13.52 N
43. horizontal ≈ -0.71 cm/sec;
vertical ≈ 0.71 cm/sec
45. horizontal ≈ -1.02 N;
vertical ≈ 0.72 N
47. ground speed ≈ 301.04 mph;
drift angle $\approx 4.76°$; bearing $\approx 25.24°$
49. ground speed ≈ 293.47 mph;
drift angle $\approx 8.82°$; bearing $\approx 91.18°$
51. perpendicular ≈ 9.83 lb;
parallel ≈ 6.88 lb
53. perpendicular ≈ 11.82 lb;
parallel ≈ 2.08 lb   **55. a.** The initial
point of $(\mathbf{A} + \mathbf{B}) + \mathbf{C}$ is $(-1, 2)$ and the
terminal point is $(2, -3)$. **b.** The initial
point of $\mathbf{A} + (\mathbf{B} + \mathbf{C})$ is $(-1, 2)$ and the
terminal point is $(2, -3)$.

Exercise Set 8.3

1. length: 5

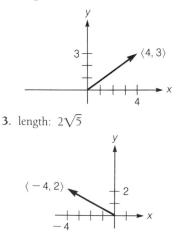

3. length: $2\sqrt{5}$

5. length: $\sqrt{13}/4$

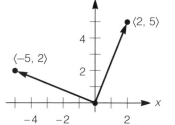

$\left\langle \dfrac{3}{4}, -\dfrac{1}{2} \right\rangle$

7. $\langle 1, 4 \rangle$ **9.** $\langle -1, 1 \rangle$ **11.** $\langle 8, -5 \rangle$
13. $\langle 7, 7 \rangle$ **15.** $\langle 24, 22 \rangle$ **17.** $\sqrt{130}$
19. $2\sqrt{17} - \sqrt{13} - \sqrt{37}$
21. $\langle 13, 6 \rangle$ **23.** $\langle 14, 21 \rangle$
25. $\langle -3, -1 \rangle$ **27.** $\langle 23, 12 \rangle$
29. $\langle -9, 0 \rangle$ **31.** -48 **33.** $3\mathbf{i} + 8\mathbf{j}$
35. $-8\mathbf{i} - 6\mathbf{j}$ **37.** $19\mathbf{i} + 23\mathbf{j}$
39. $\langle 1, 1 \rangle$ **41.** $\langle 5, -4 \rangle$
43. $\langle \sqrt{5}/5, 2\sqrt{5}/5 \rangle$
45. $\langle 2\sqrt{5}/5, -\sqrt{5}/5 \rangle$
47. $(8\sqrt{145}/145)\mathbf{i} - (9\sqrt{145}/145)\mathbf{j}$
49. $u_1 = \sqrt{3}/2$, $u_2 = 1/2$
51. $u_1 = -1/2$, $u_2 = \sqrt{3}/2$
53. $u_1 = -\sqrt{3}/2$, $u_2 = 1/2$
59. a. Both are 8. **b.** Both are -14.
61. a. Both are 25. **b.** Both are 29.
63. $\cos \theta = 7/\sqrt{170}$;
$\theta \approx 57.53°$ or $\theta \approx 1.00$ radian
65. $\cos \theta = -57/\sqrt{3538}$;
$\theta \approx 163.39°$ or $\theta \approx 2.85$ radians
67. a. $\cos \theta = -7/\sqrt{170}$; $\theta \approx 122.47°$
or $\theta \approx 2.14$ radians **b.** $\cos \theta = 7/\sqrt{170}$;
$\theta \approx 57.53°$ or $\theta \approx 1.00$ radian
69. a. 0 **b.** The vectors are
perpendicular. **c.**

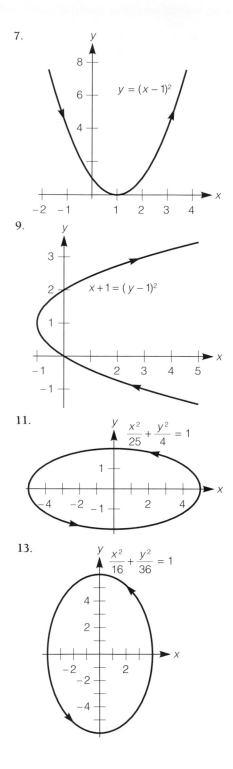

71. Since $\cos \theta = (\mathbf{A} \cdot \mathbf{B})/(|\mathbf{A}||\mathbf{B}|)$, then
$\mathbf{A} \cdot \mathbf{B} = 0$ implies $\cos \theta = 0$. Thus $\theta = 90°$,
and the vectors are perpendicular.
73. $\langle \frac{5}{13}, \frac{12}{13} \rangle$ and $\langle -\frac{5}{13}, -\frac{12}{13} \rangle$

Exercise Set 8.4

1. $(2, 3)$ **3.** $(5\sqrt{3}/2, 1)$
5. $(3\sqrt{2}/4, 3\sqrt{2}/4)$

15. a.

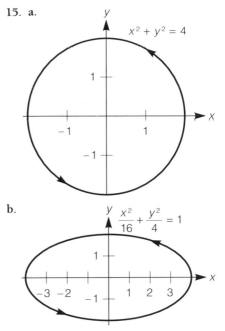

$x^2 + y^2 = 4$

b.

$\dfrac{x^2}{16} + \dfrac{y^2}{4} = 1$

17. a. $t = 1$: $x \approx 34.2$; $y \approx 83.0$;
$t = 2$: $x \approx 68.4$; $y \approx 128.9$;
$t = 3$: $x \approx 102.6$; $y \approx 142.9$
b. approximately 5.93 sec, approximately
203 ft   **21.** $x^{2/3} + y^{2/3} = 1$

Exercise Set 8.5

1. a. $\left(-\frac{3}{2}, 3\sqrt{3}/2\right)$ **b.** $\left(2\sqrt{3}, -2\right)$
c. $\left(2\sqrt{3}, -2\right)$ **3. a.** $(0, 1)$ **b.** $(0, 1)$
c. $\left(-\frac{\sqrt{2+\sqrt{2}}}{2}, -\frac{\sqrt{2-\sqrt{2}}}{2}\right)$
5. $\left(\sqrt{2}, 5\pi/4\right)$ **7.** $(x-1)^2 + y^2 = 1$
9. $x^4 + x^2y^2 - y^2 = 0$
11. $(x^2 + y^2)^3 = 9(x^2 - y^2)^2$
13. $(x^2/4) + (y^2/8) = 1$
15. $y = -\sqrt{3}x + 4$
17. $r = 2/(3\cos\theta - 4\sin\theta)$
19. $r = \tan^2\theta \sec\theta$ **21.** $r^2 = \csc 2\theta$
23. $r^2 = 9/(9\cos^2\theta + \sin^2\theta)$
25. $A\left(\frac{8}{3}, \frac{\pi}{6}\right)$, $B\left(\frac{8}{3}, \frac{5\pi}{6}\right)$, $C(4, \pi)$, $D\left(8, \frac{7\pi}{6}\right)$
27. $A\left(1, \frac{\pi}{6}\right)$, $B\left(1, \frac{5\pi}{6}\right)$, $C\left(1, \frac{7\pi}{6}\right)$, $D\left(-2, \frac{\pi}{2}\right)$
29. $A(1, 0)$, $B\left(1.14, \frac{\pi}{4}\right)$, $C\left(1.48, \frac{3\pi}{4}\right)$,
$D(1.69, \pi)$, $E\left(1.92, \frac{5\pi}{4}\right)$, $F\left(2.19, \frac{3\pi}{2}\right)$,
$G\left(2.50, \frac{7\pi}{4}\right)$, $H(2.85, 2\pi)$, $I\left(3.25, \frac{9\pi}{4}\right)$,
$J\left(4.22, \frac{11\pi}{4}\right)$, $K\left(5.48, \frac{13\pi}{4}\right)$ **31.** $2\sqrt{5}$
33. $\sqrt{21}$ **35. a.** $r^2 - 8r\cos\theta = -12$
b. $r^2 - 8r\cos\left(\theta - \frac{2\pi}{3}\right) = -12$ **c.** $r^2 = 4$
37. a. $r = 2\cos\left(\theta - \frac{3\pi}{2}\right)$
b. $r = 2\cos\left(\theta - \frac{\pi}{4}\right)$ **39. a.** 2

b. $\left(4\sqrt{3}/3, 0\right)$, $\left(4, \frac{\pi}{2}\right)$ **c.** $\left(2, \frac{\pi}{6}\right)$
d. $\theta = \pi/2$

$r\cos\left(\theta - \frac{\pi}{6}\right) = 2$
$\theta = 0$

41. a. 4 **b.** $(-8, 0)$, $\left(-\frac{8}{3}\sqrt{3}, \frac{\pi}{2}\right)$
c. $\left(4, -\frac{2\pi}{3}\right)$
d. $\theta = \pi/2$

$r\cos\left(\theta + \frac{2\pi}{3}\right) = 4$
$\theta = 0$

43. a. $\left(2, \frac{5\pi}{6}\right)$ **b.** $r\cos\left(\theta - \frac{5\pi}{6}\right) = 2$
c. x-intercept: $-4\sqrt{3}/3$; y-intercept: 4
45. a.

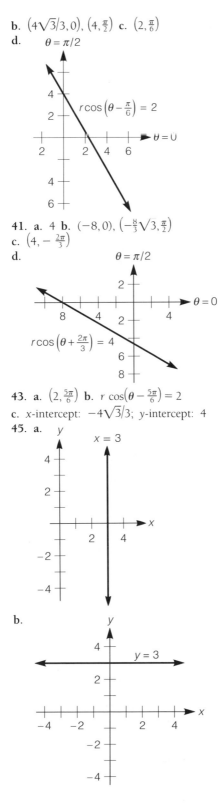

$x = 3$

b.

$y = 3$

47. a. $r\cos(\theta - \alpha) = d$
53. a. $A(1, 0)$, $B\left(1, \frac{2\pi}{3}\right)$, $C\left(1, \frac{4\pi}{3}\right)$

Exercise Set 8.6

1. $\theta = \pi/2$

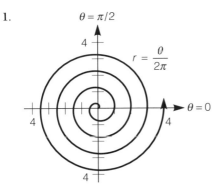

$r = \dfrac{\theta}{2\pi}$

$\theta = 0$

3. $\theta = \pi/2$

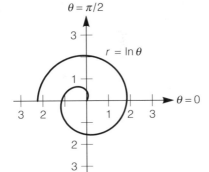

$r = \ln\theta$

$\theta = 0$

5. $\theta = \pi/2$

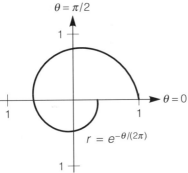

$r = e^{-\theta/(2\pi)}$

$\theta = 0$

7.

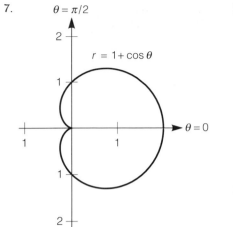

$r = 1 + \cos\theta$

9.

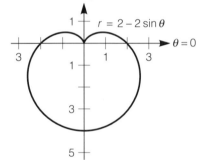

$r = 2 - 2\sin\theta$

11.

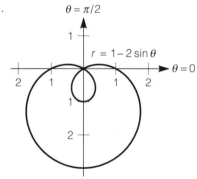

$r = 1 - 2\sin\theta$

13.

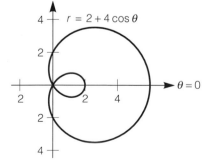

$r = 2 + 4\cos\theta$

15.

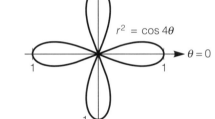

$r^2 - 4\sin 2\theta$

17.

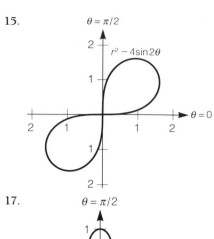

$r^2 = \cos 4\theta$

19.

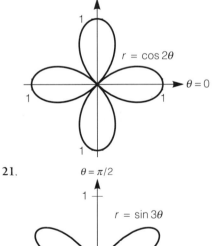

$r = \cos 2\theta$

21.

$r = \sin 3\theta$

23.

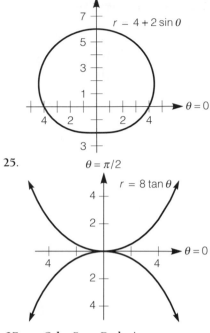

$r - 4 + 2\sin\theta$

25.

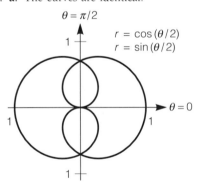

$r = 8\tan\theta$

27. a. C **b.** B **c.** D **d.** A

29. a. $A(1, 0)$, $B\!\left(e^{a\pi/2}, \frac{\pi}{2}\right)$, $C(e^{a\pi}, \pi)$, $D\!\left(e^{3a\pi/2}, \frac{3\pi}{2}\right)$

31. a. The curves are identical.

$r = \cos(\theta/2)$
$r = \sin(\theta/2)$

b. The rectangular form for both curves is $(x^2 + y^2)(2x^2 + 2y^2 - 1)^2 - x^2 = 0$.

Optional TI-81 Graphing Calculator
Exercises for Sections 8.4–8.6

1. a.

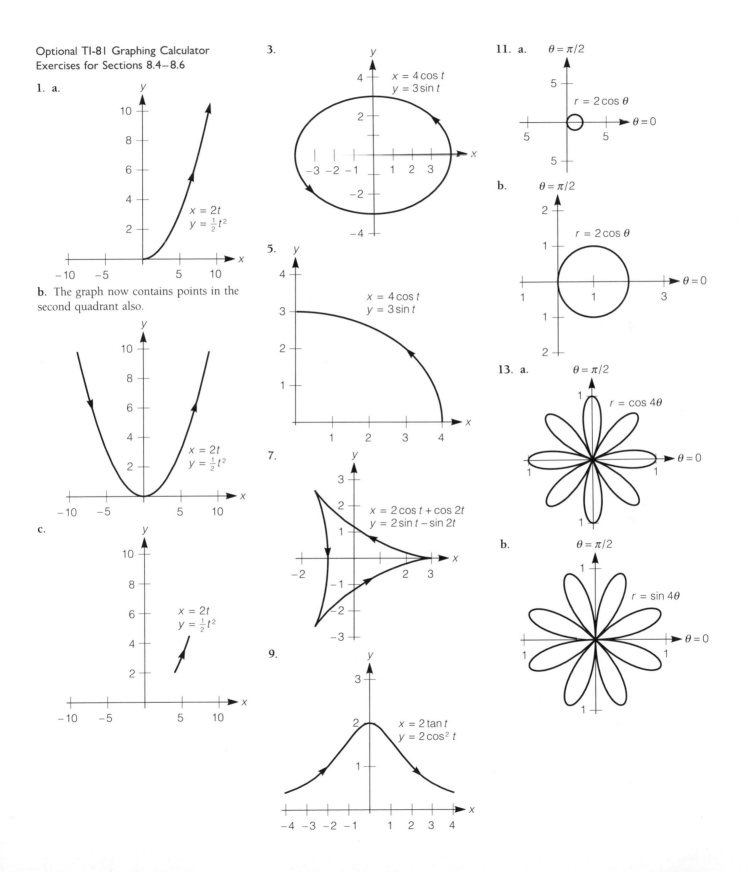

$x = 2t$
$y = \frac{1}{2}t^2$

b. The graph now contains points in the second quadrant also.

$x = 2t$
$y = \frac{1}{2}t^2$

c.

$x = 2t$
$y = \frac{1}{2}t^2$

3.

$x = 4\cos t$
$y = 3\sin t$

5.

$x = 4\cos t$
$y = 3\sin t$

7.

$x = 2\cos t + \cos 2t$
$y = 2\sin t - \sin 2t$

9.

$x = 2\tan t$
$y = 2\cos^2 t$

11. a. $\theta = \pi/2$

$r = 2\cos\theta$
$\theta = 0$

b. $\theta = \pi/2$

$r = 2\cos\theta$
$\theta = 0$

13. a. $\theta = \pi/2$

$r = \cos 4\theta$
$\theta = 0$

b. $\theta = \pi/2$

$r = \sin 4\theta$
$\theta = 0$

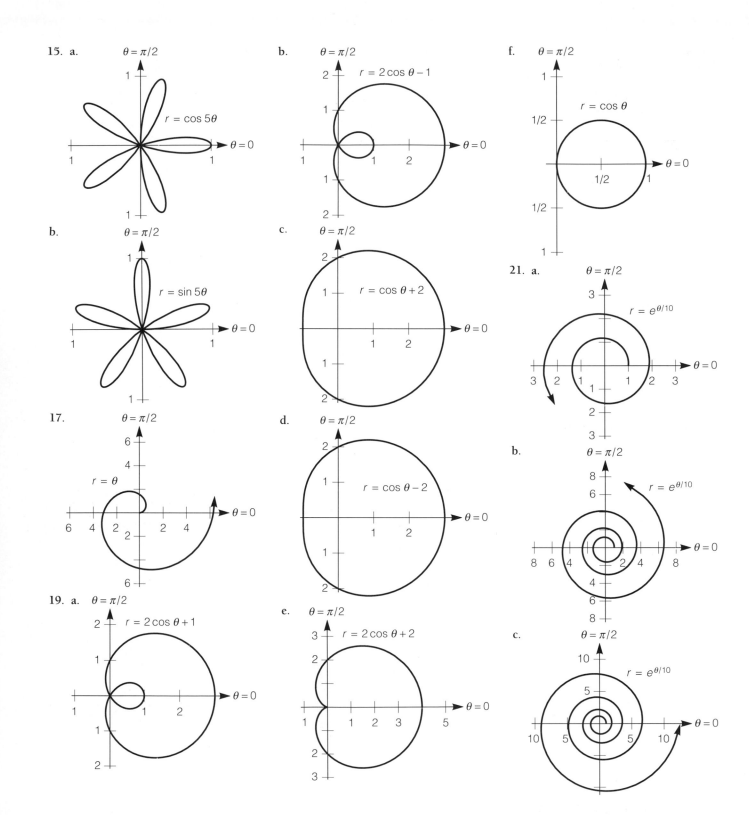

15. a. $\theta = \pi/2$

$r = \cos 5\theta$

b. $\theta = \pi/2$

$r = 2\cos\theta - 1$

f. $\theta = \pi/2$

$r = \cos\theta$

b. $\theta = \pi/2$

$r = \sin 5\theta$

c. $\theta = \pi/2$

$r = \cos\theta + 2$

21. a. $\theta = \pi/2$

$r = e^{\theta/10}$

17. $\theta = \pi/2$

$r = \theta$

d. $\theta = \pi/2$

$r = \cos\theta - 2$

b. $\theta = \pi/2$

$r = e^{\theta/10}$

19. a. $\theta = \pi/2$

$r = 2\cos\theta + 1$

e. $\theta = \pi/2$

$r = 2\cos\theta + 2$

c. $\theta = \pi/2$

$r = e^{\theta/10}$

23. a.

$r = 2|\sin\theta|^{\sin\theta}$

b.

$r = 2|\sin\theta|^{\sin(\theta/2)}$

d.

$r\cos\left(\theta - \frac{\pi}{3}\right) = 3$

67. a.

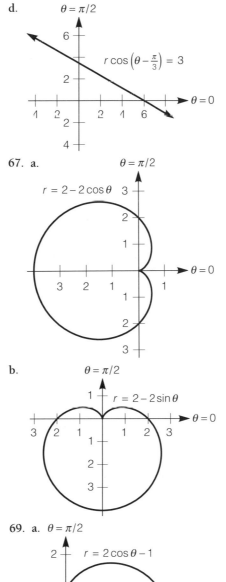

$r = 2 - 2\cos\theta$

b.

$r = 2 - 2\sin\theta$

69. a.

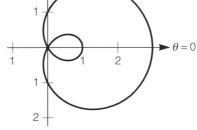

$r = 2\cos\theta - 1$

b.

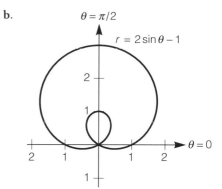

$r = 2\sin\theta - 1$

71. a.

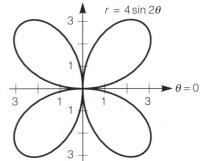

$r = 4\sin 2\theta$

b.

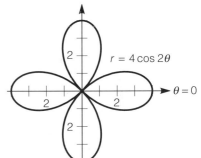

$r = 4\cos 2\theta$

73. a.

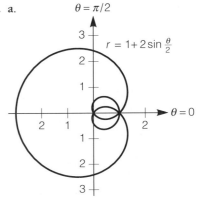

$r = 1 + 2\sin\frac{\theta}{2}$

Chapter Eight Review Exercises

1. $\angle C = 55°$; $a \approx 12.6$ cm; $b \approx 19.5$ cm
3. a. $\angle B \approx 62.4°$; $\angle C \approx 65.6°$;
$c \approx 9.2$ cm **b.** $\angle B \approx 117.6°$; $\angle C \approx 10.4°$;
$c \approx 1.8$ cm **5.** $c \approx 7.7$ cm;
$\angle A \approx 108.5°$; $\angle B \approx 47.5°$
7. $\angle C \approx 106.6°$; $\angle B \approx 48.2°$; $\angle A \approx 25.2°$
9. 9.21 cm **11.** 32.48 cm^2
13. 55.23 cm^2 **15.** 7.89 cm
17. 15.43 cm **19.** 15 cm
21. 11 cm **23. a.** $\cos A = \frac{3}{4}$;
$\cos C = \frac{1}{8}$ **25.** 36° **27.** 121.24°
29. 7.24 m **31. a.** 11.76 cm
b. 19.02 cm **35.** $a = 5$ and $b = 8$
37. magnitude = 25 N; direction $\approx 53.1°$
39. horizontal ≈ 41.0 cm/sec;
vertical ≈ 28.7 cm/sec
41. weight ≈ 36.4 lb;
perpendicular ≈ 33.2 lb **43.** $\pm\sqrt{15}$
45. $\langle 10, 9\rangle$ **47.** $\langle 12, 7\rangle$ **49.** 48
51. $\langle 12, 8\rangle$ **53.** $\langle -6, 2\rangle$
55. $\langle -7, -6\rangle$ **57.** $7\mathbf{i} - 6\mathbf{j}$
59. $\langle 3\sqrt{13}/13, 2\sqrt{13}/13\rangle$
61. $\sqrt{23.8757} \approx 4.89$
63. $r^2 - 10r\cos\left(\theta - \frac{\pi}{6}\right) = -16$
65. a. 3 **b.** $(6, 0)$, $\left(2\sqrt{3}, \frac{\pi}{2}\right)$ **c.** $\left(3, \frac{\pi}{3}\right)$

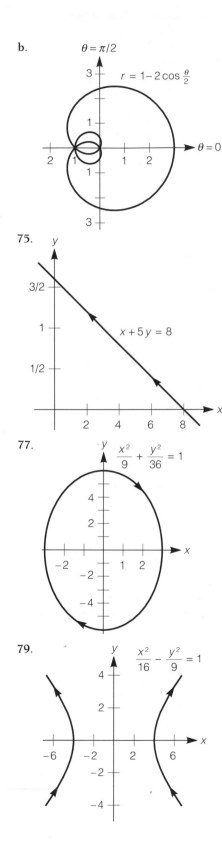

b.

$\theta = \pi/2$

$r = 1 - 2\cos\frac{\theta}{2}$

$\theta = 0$

75.

$x + 5y = 8$

77.

$\dfrac{x^2}{9} + \dfrac{y^2}{36} = 1$

79.

$\dfrac{x^2}{16} - \dfrac{y^2}{9} = 1$

---

81. a. $P\left(\dfrac{\sin\alpha}{\alpha}, \alpha\right)$, $Q\left(\dfrac{\sin 2\alpha}{2\alpha}, 2\alpha\right)$

b. $P\left(\dfrac{\sin\alpha\cos\alpha}{\alpha}, \dfrac{\sin^2\alpha}{\alpha}\right)$,

$Q\left(\dfrac{\sin 2\alpha\cos 2\alpha}{2\alpha}, \dfrac{\sin^2 2\alpha}{2\alpha}\right)$ **f.** Since the product of these slopes is -1, we have $\overline{OQ} \perp \overline{PQ}$.

Chapter Eight Test

1. 7 cm   **2.** $\cos\theta = -\frac{1}{4}$; the angle opposite the 4-cm side must be obtuse (not acute), since its cosine is negative.
3. 20 cm   **4.** $2\sqrt{17 - 4\sqrt{2}}$ cm
5. $a \approx 3.3$ cm; $C \approx 26.2°$; $B \approx 126.8°$
6. a. $2\sqrt{5}$ N **b.** $\tan\theta = 2$
7. a. $d = 4\sqrt{13 - 12\cos 110°}$ N
b. $\sin\theta = \dfrac{2\sin 110°}{\sqrt{13 - 12\cos 110°}}$
8. ground speed $= 50\sqrt{37}$ mph; $\tan\theta = \frac{1}{6}$ **9. a.** $\langle 13, 5 \rangle$ **b.** $\sqrt{194}$
c. $\mathbf{i} - 3\mathbf{j}$
10. $\langle -11\sqrt{130}/130, -3\sqrt{130}/130 \rangle$
11. $(x^2 + y^2)^2 = x^2 - y^2$
12.

$\theta = \pi/2$

$r = 2(1 - \cos\theta)$

$\theta = 0$

13.

y $\dfrac{x^2}{16} + \dfrac{y^2}{4} = 1$

14. $\sqrt{13}$
15. $r^2 - 10r\cos\left(\theta - \frac{\pi}{2}\right) = -21$; no
16. a. $r\cos\left(\theta - \frac{5\pi}{6}\right) = 4$

CHAPTER NINE

Exercise Set 9.1

1. a. yes **b.** no **c.** yes **d.** yes
3. yes   **5.** no   **7.** yes

---

9. $\left(-6, \frac{1}{2}\right)$ **11.** $\left(-\frac{5}{2}, \frac{13}{2}\right)$
13. $\left(\frac{1}{5}, \frac{7}{10}\right)$ **15.** $\left(-\frac{60}{13}, \frac{60}{13}\right)$
17. $\left(\sqrt{6}, 1\right)$ **19.** $\left(-\frac{2}{9}, \frac{23}{27}\right)$
21. $\left(-\frac{283}{242}, -\frac{3}{121}\right)$ **23.** $\left(-\frac{226}{25}, -\frac{939}{50}\right)$
25. $\left(\frac{5}{8}, 0\right)$ **27.** $(1, 2)$
29. $b = 3$; $c = 4$
31. $A = -28$; $B = -22$
33. $\frac{49}{20}$ square units   **35.** 80 cc of 10% acid; 120 cc of 35% acid   **37.** 8 lb of $5.20 coffee; 8 lb of $5.80 coffee

39. $x = \dfrac{ab}{a+b}$; $y = \dfrac{ab}{a+b}$; $a \ne \pm b$

41. $x = \dfrac{a+b}{ab}$; $y = -\dfrac{1}{ab}$; $a \ne 0$; $b \ne 0$

43. $\left(-\frac{1}{11}, \frac{1}{9}\right)$ **45.** $(5, 2)$
47. $(1.32, -1.62)$ **49.** $(0.14, 0.05)$
51. $(1.61, 0)$ **53.** $(4, -8)$, $(4, 2)$, $(-1, -8)$, $(-1, 2)$ **55.** 59
57. length 12 in.; width 5 in.
59. a. $(60, 40)$ **b.** $(60, 40)$ **61.** $(0, a)$
63. $(1/b, 1/a)$ **65.** $k = -\frac{28}{7}, 2$

Exercise Set 9.2

1. $(-3, -2, -1)$ **3.** $\left(-\frac{1}{60}, -\frac{2}{15}, \frac{3}{5}\right)$
5. $\left(\frac{25}{36}, \frac{5}{9}, -\frac{1}{3}\right)$ **7.** $(x, x/8, 0)$, for any real number x **9.** $(-1, 0, 1, -5)$
11. $\left(\frac{11}{3}, \frac{8}{3}, \frac{17}{3}\right)$ **13.** $(1, 0, 1)$
15. $(2, 3, 1)$ **17.** inconsistent (no solution)   **19.** $\left(-\frac{1}{7}(z+1), \frac{5}{7}(z+1), z\right)$ for any real number z
21. $\left(\dfrac{11-5z}{7}, \dfrac{-3z-6}{7}, z\right)$ for any real number z **23.** $(4, 1, -3, 2)$
25. $\left(\dfrac{17-17z}{5}, \dfrac{8z-3}{5}, z\right)$ for any real number z
27.
$\left(\dfrac{12+10w}{11}, \dfrac{146+19w}{55}, \dfrac{159+61w}{55}, w\right)$ for any real number w
29. $\left(-\dfrac{5z}{12}, \dfrac{2z+3}{3}, z\right)$ for any real number z **31. a.** $A = \frac{1}{4}$; $B = -\frac{1}{4}$
b. $A = \frac{1}{2}$; $B = \frac{1}{2}$ **33.** $A = \frac{4}{3}$; $B = \frac{5}{3}$
35. a. $A = \frac{3}{8}$; $B = -\frac{3}{8}$
b. $A = \frac{3}{64}$; $B = -\frac{3}{64}$; $C = -\frac{3}{8}$
37. $A = \frac{1}{3}$; $B = -\frac{1}{3}$; $C = \frac{2}{3}$
39. $A = 4$; $B = 4$
41. $A = \frac{1}{2}$; $B = \frac{1}{2}$; $C = -\frac{1}{2}$; $D = \frac{1}{2}$
43. $17x^2 + 17y^2 - 49x + 65y - 166 = 0$

45. $x = \ln a$, $y = \ln 2a$, $z = \ln(a/2)$
49. a. $A = 1/(a-b)$; $B = 1/(b-a)$
b. $A = \dfrac{ap+q}{a-b}$, $B = \dfrac{bp+q}{b-a}$

51. $A = \dfrac{a^2 + pa + q}{(a-b)(a-c)}$,

$B = \dfrac{b^2 + pb + q}{(b-a)(b-c)}$, $C = \dfrac{c^2 + pc + q}{(c-a)(c-b)}$

53. 60 miles

Exercise Set 9.3

1. a. 2×3 **b.** 3×2 **3.** 5×4

5. coefficient matrix: $\begin{pmatrix} 2 & 3 & 4 \\ 5 & 6 & 7 \\ 8 & 9 & 10 \end{pmatrix}$;

augmented matrix: $\begin{pmatrix} 2 & 3 & 4 & 10 \\ 5 & 6 & 7 & 9 \\ 8 & 9 & 10 & 8 \end{pmatrix}$

7. coefficient matrix:

$\begin{pmatrix} 1 & 0 & 1 & 1 \\ 1 & 1 & 0 & 2 \\ 0 & 1 & 1 & 1 \\ 2 & -1 & -1 & 0 \end{pmatrix}$;

augmented matrix:

$\begin{pmatrix} 1 & 0 & 1 & 1 & -1 \\ 1 & 1 & 0 & 2 & 0 \\ 0 & 1 & 1 & 1 & 1 \\ 2 & -1 & -1 & 0 & 2 \end{pmatrix}$

9. $(-1,-2,3)$ **11.** $(-5,1,3)$
13. $(3,0,-7)$ **15.** $(8,9,-1)$

17. $\left(\dfrac{9z+5}{19}, \dfrac{31z-6}{19}, z \right)$ for any real

number z **19.** $(2,-1,0,3)$

21. no solution **23.** $\begin{pmatrix} 3 & 2 \\ 2 & 4 \end{pmatrix}$

25. $\begin{pmatrix} 6 & 4 \\ 4 & 8 \end{pmatrix}$ **27.** $\begin{pmatrix} 11 & -2 \\ 11 & 1 \end{pmatrix}$

29. $\begin{pmatrix} 2 & 3 \\ -1 & 4 \end{pmatrix}$ **31.** undefined

33. $\begin{pmatrix} 10 & -2 \\ -8 & 0 \\ 4 & 6 \end{pmatrix}$

35. $\begin{pmatrix} 2 & 4 & 11 \\ -12 & 16 & 19 \\ 14 & 12 & 43 \end{pmatrix}$

37. $\begin{pmatrix} -9 & 10 & 10 \\ 4 & -8 & -12 \\ 10 & 4 & 21 \end{pmatrix}$

39. undefined **41.** $\begin{pmatrix} 0 & 0 & 0 \\ 0 & 0 & 0 \\ 0 & 0 & 0 \end{pmatrix}$

43. $\begin{pmatrix} 4 & 2 \\ 2 & 5 \end{pmatrix}$ **45.** undefined

47. $\begin{pmatrix} 1 & 18 \\ -6 & 13 \end{pmatrix}$ **49.** $\begin{pmatrix} -16 & 75 \\ -25 & 34 \end{pmatrix}$

51. a. $\begin{pmatrix} -13 & 1 & 40 \\ 43 & 17 & 0 \\ 89 & 61 & 60 \end{pmatrix}$

b. $\begin{pmatrix} 13 & 1 & 10 \\ 43 & 17 & 0 \\ 89 & 61 & 60 \end{pmatrix}$

c. $\begin{pmatrix} -52 & -82 & 61 \\ 87 & 141 & 0 \\ 216 & 318 & 165 \end{pmatrix}$

d. $\begin{pmatrix} -52 & -82 & 61 \\ 87 & 141 & 0 \\ 216 & 318 & 165 \end{pmatrix}$

53. a. $\begin{pmatrix} 16 & 20 \\ 24 & 28 \end{pmatrix}$ **b.** $\begin{pmatrix} 18 & 26 \\ 18 & 26 \end{pmatrix}$

c. $\begin{pmatrix} 14 & 14 \\ 30 & 30 \end{pmatrix}$ **d.** $\begin{pmatrix} 18 & 26 \\ 18 & 26 \end{pmatrix}$

55. a. $\begin{pmatrix} x \\ -y \end{pmatrix}$ **b.** $\begin{pmatrix} -x \\ y \end{pmatrix}$ **c.** $\begin{pmatrix} -x \\ -y \end{pmatrix}$,

reflection about the origin
57. a. $f(A) = -2$; $f(B) = 29$;
$f(AB) = -58$; yes

Exercise Set 9.4

1. $\begin{pmatrix} 4 & -1 \\ -5 & 2 \end{pmatrix} = A$

3. $\begin{pmatrix} 3 & 0 & -2 \\ 0 & 5 & 6 \\ 1 & 4 & -7 \end{pmatrix} = C$

5. $\begin{pmatrix} -5 & 9 \\ 4 & -7 \end{pmatrix}$ **7.** $\begin{pmatrix} -\frac{6}{23} & \frac{1}{23} \\ \frac{5}{23} & \frac{3}{23} \end{pmatrix}$

9. does not exist **11.** $\begin{pmatrix} 2 & -3 \\ 1 & 3 \end{pmatrix}$

13. $\begin{pmatrix} 2 & -1 \\ -3 & 2 \end{pmatrix}$ **15.** $\begin{pmatrix} \frac{6}{11} & 1 \\ -\frac{1}{11} & 0 \end{pmatrix}$

17. does not exist

19. $\begin{pmatrix} -1 & 2 & -3 \\ 2 & 1 & 0 \\ 4 & -2 & 5 \end{pmatrix}$

21. $\begin{pmatrix} 5 & -\frac{10}{3} & 1 \\ 0 & \frac{1}{3} & 0 \\ 4 & -\frac{8}{3} & 1 \end{pmatrix}$

23. $\begin{pmatrix} 2 & 1 & 4 \\ 3 & 2 & 5 \\ 0 & -1 & 1 \end{pmatrix}$

25. does not exist **27. a.** $x = -1$,
$y = 1$ **b.** $x = -132$, $y = 48$

29. a. $x = 2$, $y = -1$, $z = 4$ **b.** $x = 1$,
$y = 1$, $z = -2$

31. a. $I_3 = \begin{pmatrix} 1 & 0 & 0 \\ 0 & 1 & 0 \\ 0 & 0 & 1 \end{pmatrix}$ so $A^{-1} = A$.

b. $x = \frac{1}{2}$, $y = -1$, $z = 1$

35. a. $A^{-1} = \begin{pmatrix} -\frac{5}{2} & \frac{3}{2} \\ 2 & -1 \end{pmatrix}$;

$B^{-1} = \begin{pmatrix} 7 & -8 \\ -6 & 7 \end{pmatrix}$;

$B^{-1}A^{-1} = \begin{pmatrix} -\frac{67}{2} & \frac{37}{2} \\ 29 & -16 \end{pmatrix}$

b. $\begin{pmatrix} -\frac{67}{2} & \frac{37}{2} \\ 29 & -16 \end{pmatrix}$; $(AB)^{-1} = B^{-1}A^{-1}$

Exercise Set 9.5

1. a. 29 **b.** -29 **3. a.** 0 **b.** 0
5. 1 **7.** -60 **9.** 9
11. a. 314 **b.** 674 **c.** part (b)
13. a. 0 **b.** 0 **c.** 0 **d.** 0 **15.** 0
17. -3 **19.** 6848 **21.** 17120
23. $(y-x)(z-x)(z-y)$ **25.** xy
27. 20 **29.** 120 **33.** $(1,1,2)$
35. $(2,-3,6)$ **37.** $(0,0,0)$
39. $\left(13 - \frac{11}{3}y, y, 13 - 4y\right)$ for any real
number y **41.** $(1,0,-10,2)$
43. $4, -4, -1$ **55. a.** 1 **b.** 1

59. $\left(\dfrac{k(k-b)(k-c)}{a(a-b)(a-c)}, \dfrac{k(k-a)(k-c)}{b(b-a)(b-c)}, \right.$

$\left. \dfrac{k(k-a)(k-b)}{c(c-a)(c-b)} \right)$

61. $y = 2x + 5$ **65. b.** $\begin{pmatrix} -\frac{9}{61} & \frac{7}{61} \\ \frac{1}{61} & \frac{6}{61} \end{pmatrix}$

**Optional TI-81 Graphing Calculator
Exercises for Sections 9.3, 9.4, and 9.5**

1. a. $AB = \begin{pmatrix} -12 & 17 & -37 \\ -5 & 22 & -8 \\ 7 & -2 & 26 \end{pmatrix}$

b. $BA = \begin{pmatrix} 19 & -7 & 6 \\ 5 & 17 & 4 \\ -5 & 16 & 0 \end{pmatrix}$

c. $A^{-1} = \begin{pmatrix} -3 & 2 & -4 \\ -1 & 1 & -1 \\ 8 & -5 & 10 \end{pmatrix}$

d. $B^{-1} = \begin{pmatrix} -0.116 & 0.674 & -0.767 \\ 0.035 & 0.198 & -0.070 \\ -0.081 & -0.128 & 0.163 \end{pmatrix}$

e. 1 **3. a.** 1 **b.** -452 **c.** -452

d. -452 **5. a.** -0.0022123894
b. -0.0022123894

Exercise Set 9.6

1. $(0,0)$, $(3,9)$ **3.** $(2\sqrt{6},1)$,
$(-2\sqrt{6},1)$ **5.** $(-1,1)$ **7.** $(2,3)$;
$(2,-3)$; $(-2,3)$; $(-2,-3)$ **9.** $(1,0)$;
$(-1,0)$

11. $\left(\dfrac{-1+\sqrt{65}}{8},\dfrac{1+\sqrt{65}}{2}\right)$;

$\left(\dfrac{-1-\sqrt{65}}{8},\dfrac{1-\sqrt{65}}{2}\right)$;

13. $\left(\sqrt{17}/17,\pm1\right)$; $\left(-\sqrt{17}/17,\pm1\right)$

15. $(1,0)$; $\left(4,-\sqrt{3}\right)$ **17.** $(2,4)$

19. $(100,1000)$; $\left(100,\frac{1}{1000}\right)$; $\left(\frac{1}{100},1000\right)$;
$\left(\frac{1}{100},\frac{1}{1000}\right)$

21. $\left(\dfrac{2\ln 2-5\ln 3}{\ln 2-\ln 3},\dfrac{3\ln 2}{\ln 2-\ln 3}\right)$

25. $(1/a,1/b)$ **27.** $(9,14)$; $(14,9)$

29. $\dfrac{p-\sqrt{2d^2-p^2}}{2}$ by $\dfrac{p+\sqrt{2d^2-p^2}}{2}$

31. $(5,4)$; $(5,-4)$ **33. a.** $k=\frac{1}{6}\ln 8$,
$\mathcal{N}_0=\frac{3}{2}$ **b.** $k=\frac{2}{7}\ln 10$, $\mathcal{N}_0=10^{-1/7}$

35. $\left(\dfrac{p^2}{A},\dfrac{q^2}{A},\dfrac{r^2}{A}\right)$ and $\left(-\dfrac{p^2}{A},-\dfrac{q^2}{A},-\dfrac{r^2}{A}\right)$,
where $A=\sqrt{p^2+q^2+r^2}$ **37.** 9 cm,
40 cm   **39.** 3 cm by 20 cm

41. $(1,2)$; $(-1,-2)$; $(2,1)$; $(-2,-1)$

43. $\left(\dfrac{1+\sqrt{13}}{2},\dfrac{-1+\sqrt{13}}{2}\right)$;

$\left(\dfrac{-1+\sqrt{13}}{2},\dfrac{1+\sqrt{13}}{2}\right)$;

$\left(\dfrac{1-\sqrt{13}}{2},\dfrac{-1-\sqrt{13}}{2}\right)$;

$\left(\dfrac{-1-\sqrt{13}}{2},\dfrac{1-\sqrt{13}}{2}\right)$

45. $(2,3)$; $(-2,-3)$; $(1,2)$; $(-1,-2)$

47. (q,p); $\left((-1+\sqrt{3})q/2,(1-\sqrt{3})p\right)$;
$\left((-1-\sqrt{3})q/2,(1+\sqrt{3})p\right)$

49. $(e^{9/2},e^3)$

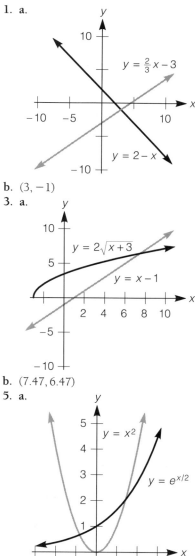

Optional TI-81 Graphing Calculator
Exercises for Section 9.6

1. a.

b. $(3,-1)$

3. a.

b. $(7.47,6.47)$

5. a.

b. $(-0.82,0.67)$, $(1.43,2.04)$

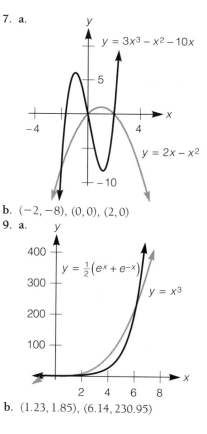

7. a.

b. $(-2,-8)$, $(0,0)$, $(2,0)$

9. a.

b. $(1.23,1.85)$, $(6.14,230.95)$

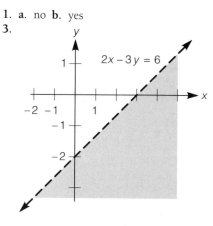

Exercise Set 9.7

1. a. no **b.** yes

3.

5.

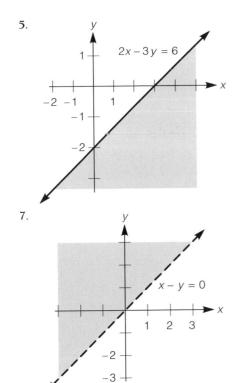

$2x - 3y = 6$

7.

$x - y = 0$

9.

$x = 1$

11.

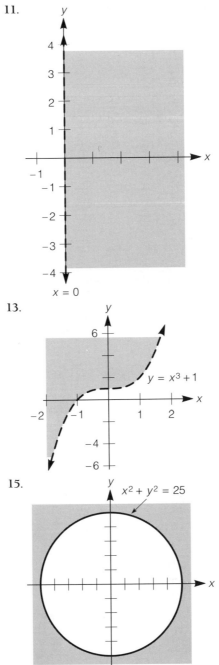

$x = 0$

13.

$y = x^3 + 1$

15.

$x^2 + y^2 = 25$

17.

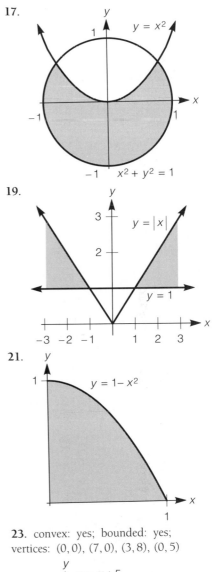

$y = x^2$

$x^2 + y^2 = 1$

19.

$y = |x|$

$y = 1$

21.

$y = 1 - x^2$

23. convex: yes; bounded: yes;
vertices: $(0, 0)$, $(7, 0)$, $(3, 8)$, $(0, 5)$

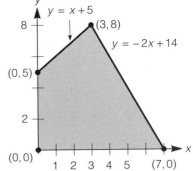

$y = x + 5$

$(3, 8)$

$y = -2x + 14$

$(0, 5)$

$(0, 0)$

$(7, 0)$

25. convex: yes; bounded: yes;
vertices: $(0,0)$, $(0,4)$, $(3,5)$, $(8,0)$

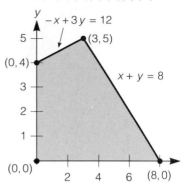

27. convex: yes; bounded: no;
vertices: $(2,7)$, $(8,5)$

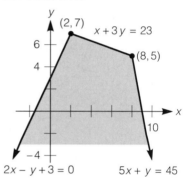

29. convex: yes; bounded: no;
vertex: $(6,0)$

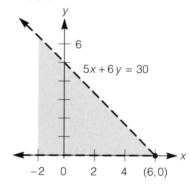

31. convex: yes; bounded: yes;
vertices: $(0,0)$, $(0,5)$, $(6,0)$

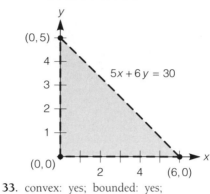

33. convex: yes; bounded: yes;
vertices: $(5,30)$, $(10,30)$, $(20,15)$,
$(20,20)$

35. a. vertices: $(0,1)$, $(0,2)$,
$\left(\ln\dfrac{1+\sqrt{5}}{2}, \dfrac{1+\sqrt{5}}{2}\right)$

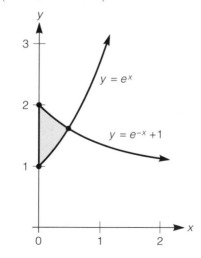

b. vertices: $(0,0)$, $(0,1)$, $\left(\frac{\pi}{4}, \sqrt{2}/2\right)$,
$\left(\frac{5\pi}{4}, -\sqrt{2}/2\right)$, $(2\pi,0)$, $(2\pi,1)$

37.

39.

A-126

41.

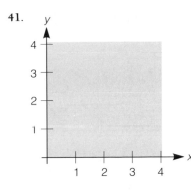

Chapter Nine Review Exercises

1. $(3, -5)$ **3.** $(-1, 4)$

5. $(-3, 15)$ **7.** $\left(-\frac{18}{5}, \frac{8}{5}\right)$

9. $\left(\frac{2}{3}, -\frac{1}{5}\right)$ **11.** $(-12, -8)$

13. $\left(\frac{1}{3}, -\frac{1}{4}\right)$

15. $\left(\frac{-1}{a^2 - 3a + 1}, \frac{1 - a}{a^2 - 3a + 1}\right)$,
$a \neq \left(3 \pm \sqrt{5}\right)/2$ **17.** $(a^2, 1 - a^2)$

19. $(a^2 - b^2, a^2 + b^2)$

21. $\left(\frac{pq(p + q)}{p^2 + q^2}, \frac{p^3 - q^3}{p^2 + q^2}\right)$, where p and q

are not both 0 **23.** $\left(\frac{a}{a - b}, \frac{b}{a + b}\right)$,

$ab \neq 0$, $a \neq \pm b$ **25.** $(2, 3, 4)$

27. $(-1, -2, 0)$ **29.** no solution

31. $(x, 6 - 2x, -1)$ for any real number x

33. no solution **35.** $(2b - z, a - b, z)$

for any real number z

37. $(4, 3, -1, 2)$ **39.** $A = \frac{1}{20}$, $B = -\frac{1}{20}$

41. $A = 2$, $B = -2$ **43.** $A = -\frac{5}{4}$, $B = \frac{5}{4}$

45. $A = \frac{1}{16}$, $B = -\frac{1}{16}$, $C = -\frac{1}{4}$

47. $A = 3$, $B = 1$, $C = 0$ **49.** $A = \frac{1}{48}$,

$B = -\frac{1}{48}$, $C = \frac{1}{6}$ **51.** $A = 0$, $B = 1$,

$C = a$ **53.** $A = b$, $B = -a$ **55.** 34

57. -56 **59.** 0 **61.** 24

67. $a = 2$, $b = 1$ **69. a.** $\left(\frac{2}{3}, 2\right)$

b. $\left(\frac{2}{3}, 2\right)$ **c.** $\left(\frac{2}{3}, 2\right)$ **d.** They all equal 2.

73. $\left(x + \frac{17}{6}\right)^2 + \left(y + \frac{8}{3}\right)^2 = \frac{245}{36}$

75. $\begin{pmatrix} 10 & -2 \\ 4 & 26 \end{pmatrix}$ **77.** $\begin{pmatrix} 8 & 4 \\ 4 & 32 \end{pmatrix}$

79. $\begin{pmatrix} 4 & -13 \\ 7 & 41 \end{pmatrix}$ **81.** $\begin{pmatrix} -3 & -14 \\ -4 & 3 \end{pmatrix}$

83. $\begin{pmatrix} 1 & 1 \\ 1 & 7 \end{pmatrix}$ **85.** $\begin{pmatrix} 1 & -11 \\ 6 & 36 \end{pmatrix}$

87. $\begin{pmatrix} 4 & 3 \\ 10 & 33 \end{pmatrix}$ **89.** $\begin{pmatrix} -42 & 58 \\ 5 & 20 \end{pmatrix}$

91. undefined **93.** undefined

95. $\begin{pmatrix} 4 & -1 \\ 2 & 12 \end{pmatrix}$ **97.** $\begin{pmatrix} -4 & 13 \\ -7 & -41 \end{pmatrix}$

101. a. $\begin{pmatrix} -9 & 5 \\ 2 & -1 \end{pmatrix}$ **b.** $(-47, 10)$

103. a. $\begin{pmatrix} 10 & -2 & 5 \\ 6 & -1 & 4 \\ 1 & 0 & 1 \end{pmatrix}$ **b.** $x = 16$,

$y - 15$, $z - 4$

105. $\begin{pmatrix} -65 & 20 & 9 & 133 \\ 3 & -1 & 0 & -6 \\ 26 & -8 & -4 & -53 \\ -23 & 7 & 3 & 47 \end{pmatrix}$

107. $D = -20$; $D_x = 12$; $D_y = 13$;

$D_z = -31$; $\left(-\frac{3}{5}, -\frac{13}{20}, \frac{31}{20}\right)$

109. $D = 0$; no solution **111.** $D = 0$;

$(-5 - 4y, y, -8 - 5y)$ for any real

number y **113.** $D = 149$; $D_x = 596$;

$D_y = 149$; $D_z = -149$; $D_w = 447$;

$(4, 1, -1, 3)$ **115.** $(0, 0)$, $(6, 36)$

117. $(3, 0)$, $(-3, 0)$

119. $\left(\frac{5\sqrt{2}}{2}, \pm\frac{\sqrt{14}}{2}\right)$,

$\left(-\frac{5\sqrt{2}}{2}, \pm\frac{\sqrt{14}}{2}\right)$

121. $\left(\frac{-1 + \sqrt{5}}{2}, \frac{\sqrt{-2 + 2\sqrt{5}}}{2}\right)$

123. $\left(\pm\frac{\sqrt{-2 + 2\sqrt{17}}}{4}, \frac{-1 + \sqrt{17}}{4}\right)$

125. $\left(\sqrt{2}, 3\sqrt{2}\right)$, $\left(-\sqrt{2}, -3\sqrt{2}\right)$,
$\left(\frac{7}{33}\sqrt{22}, \frac{31}{33}\sqrt{22}\right)$, $\left(-\frac{7}{33}\sqrt{22}, -\frac{31}{33}\sqrt{22}\right)$

127. $(5, 2)$, $(5, -4)$, $(1, 2)$, $(1, -4)$

129. $as/(a + b)$ and $bs/(a + b)$

131. 24, 60, 120

133. $\frac{1}{2}\left(\sqrt{n + 2m} + \sqrt{n - 2m}\right)$ and
$\frac{1}{2}\left(\sqrt{n + 2m} - \sqrt{n - 2m}\right)$,
$\frac{1}{2}\left(\sqrt{n - 2m} - \sqrt{n + 2m}\right)$ and
$-\frac{1}{2}\left(\sqrt{n - 2m} + \sqrt{n + 2m}\right)$

135. convex: no; bounded: no

137. convex: yes; bounded: yes

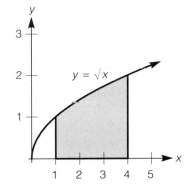

Chapter Nine Test

1. $(0, 3)$, $\left(-\frac{11}{4}, \frac{81}{16}\right)$ **2.** $(3, -5)$

3. a. $(1, -1, -3)$ **b.** $D = -19$;

$D_x = -19$; $D_y = 19$; $D_z = 57$; $(1, -1, -3)$

4. a. $\begin{pmatrix} 2 & -10 \\ 3 & -5 \end{pmatrix}$ **b.** $\begin{pmatrix} 8 & -4 \\ 7 & -6 \end{pmatrix}$

5. 12 square units **6.** $\left(\frac{1}{116}, -\frac{1}{144}\right)$

7. coefficient matrix: $\begin{pmatrix} 1 & 1 & -1 \\ 2 & -1 & 2 \\ 1 & -2 & 1 \end{pmatrix}$;

augmented matrix: $\begin{pmatrix} 1 & 1 & -1 & -1 \\ 2 & -1 & 2 & 11 \\ 1 & -2 & 1 & 10 \end{pmatrix}$

8. $(3, -3, 1)$ **9.** $y = -2x - 4$

10. a. 8 **b.** -8 **11.** -1120

12. $\left(\frac{5 + \sqrt{5}}{2}, \frac{5 - \sqrt{5}}{2}\right)$,

$\left(\frac{5 - \sqrt{5}}{2}, \frac{5 + \sqrt{5}}{2}\right)$,

$\left(\frac{-5 + \sqrt{5}}{2}, \frac{-5 - \sqrt{5}}{2}\right)$, and

$\left(\frac{-5 - \sqrt{5}}{2}, \frac{-5 + \sqrt{5}}{2}\right)$

13. $\left(1 - \frac{1}{5}C, -\frac{7}{5}C, C\right)$, where C is any real

number **14. a.** $\begin{pmatrix} 1 & -2 & 3 \\ 2 & -5 & 10 \\ -1 & 2 & -2 \end{pmatrix}$

b. $(12, 38, -9)$

15.

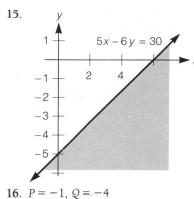

$5x - 6y = 30$

16. $P = -1$, $Q = -4$

17. bounded: no; convex: no

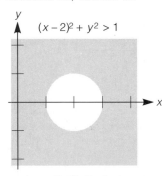

$(x-2)^2 + y^2 > 1$

18. vertices: $(0,0)$, $(0,7)$, $(6,10)$, $\left(\frac{261}{26}, \frac{225}{26}\right)$, $(11,0)$

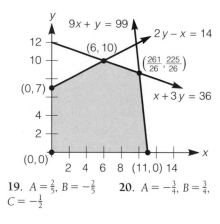

$9x + y = 99$
$(6,10)$
$2y - x = 14$
$\left(\frac{261}{26}, \frac{225}{26}\right)$
$(0,7)$
$x + 3y = 36$
$(0,0)$
$(11,0)$

19. $A = \frac{2}{5}$, $B = -\frac{2}{5}$ **20.** $A = -\frac{3}{4}$, $B = \frac{3}{4}$, $C = -\frac{1}{2}$

CHAPTER TEN

Exercise Set 10.1

1. $\sqrt{89}$ **3.** $5x + 4y - 4 = 0$
5. $2x + 3y - 20 = 0$
7. x-intercepts: 6 and -4; y-intercepts: $\pm 2\sqrt{6}$
9. $13x - 7y - 35 = 0$ **11.** $12 + \sqrt{74}$
13. $\frac{\pi}{3}$ or $60°$ **15. a.** 1.37 or $78.69°$
b. 1.77 or $101.31°$ **17. a.** $5\sqrt{2}/2$

b. $5\sqrt{2}/2$ **19. a.** $19\sqrt{41}/41$
b. $19\sqrt{41}/41$
21. a. $(x+2)^2 + (y+3)^2 = \frac{361}{13}$ **b.** $\sqrt{5}$
23. $\frac{65}{2}$ **25.** $\left(15 \pm 2\sqrt{30}\right)/5$
27. $\frac{12}{5}$ **29.** $y = -3x + 12$
31. $y = \frac{1}{3}x + \frac{9}{2}$
33. a. center: $(-5, 2)$; radius: $5\sqrt{2}$
35. $5\sqrt{2}$ **39. a.** A: $y = \frac{1}{3}x$;
B: $y = -x + 8$; C: $y = 2x - 10$
b. The angle bisectors intersect at the point $(6, 2)$.   **43. a.** slope: $-A/B$;
y-intercept: $-C/B$
47. $\left(x - \frac{26}{11}\right)^2 + \left(y + \frac{25}{11}\right)^2 = \frac{4964}{121}$
49. $\left(x - 4 - \frac{10}{13}\sqrt{13}\right)^2 +$
$$\left(y - 6 - \frac{15}{13}\sqrt{13}\right)^2 = 25;$$
$\left(x - 4 + \frac{10}{13}\sqrt{13}\right)^2 +$
$$\left(y - 6 + \frac{15}{13}\sqrt{13}\right)^2 = 25$$

Exercise Set 10.2

1. focus: $(0, 1)$; directrix: $y = -1$; focal width: 4

$x^2 = 4y$

3. focus: $(-2, 0)$; directrix: $x = 2$; focal width: 8

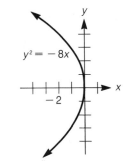

$y^2 = -8x$

5. focus: $(0, -5)$; directrix: $y = 5$; focal width: 20

$x^2 = -20y$

7. focus: $(-7, 0)$; directrix: $x = 7$; focal width: 28

$y^2 = -28x$

9. focus: $\left(0, \frac{3}{2}\right)$; directrix: $y = -\frac{3}{2}$; focal width: 6

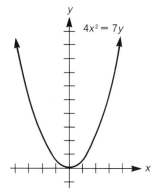

$x^2 = 6y$

11. focus: $\left(0, \frac{7}{16}\right)$; directrix: $y = -\frac{7}{16}$; focal width: $\frac{7}{4}$

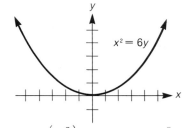

$4x^2 = 7y$

13. vertex: $(2,3)$; focus: $(3,3)$; directrix: $x = 1$; focal width: 4

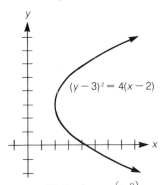

$(y-3)^2 = 4(x-2)$

15. vertex: $(4,2)$; focus: $\left(4, \frac{9}{4}\right)$; directrix: $y = \frac{7}{4}$; focal width: 1

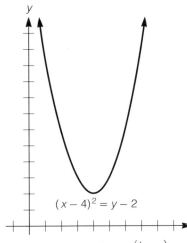

$(x-4)^2 = y-2$

17. vertex: $(0,-1)$; focus: $\left(\frac{1}{4}, -1\right)$; directrix: $x = -\frac{1}{4}$; focal width: 1

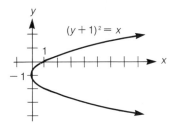

$(y+1)^2 = x$

19. vertex: $(3,0)$; focus: $\left(3, \frac{1}{8}\right)$; directrix: $y = -\frac{1}{8}$; focal width: $\frac{1}{2}$

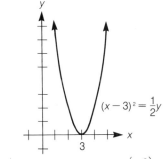

$(x-3)^2 = \frac{1}{2}y$

21. vertex: $(4,1)$; focus: $\left(4, \frac{9}{8}\right)$; directrix: $y = \frac{7}{8}$; focal width: $\frac{1}{2}$

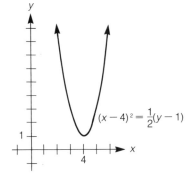

$(x-4)^2 = \frac{1}{2}(y-1)$

23. line of symmetry: $y = 1$

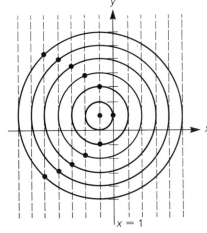

$x = 1$

25. $x^2 = 12y$ **27.** $y^2 = 128x$
29. $y^2 = -9x$ **31. a.** $y = \frac{3}{4}x + 2$
b. $Q\left(-2, \frac{1}{2}\right)$ **c.** $\frac{25}{2}$
d. $(x-3)^2 + \left(y - \frac{17}{4}\right)^2 = \frac{625}{16}$

e.

$P(8,8)$

Q

$y = -2$

33. $11.25\,\text{ft}$ **35. a.** $y = \frac{9}{4}x - 1$
b. $y = -\frac{9}{8}x - \frac{1}{4}$ **37. a.** $Q\left(-\frac{1}{8}, \frac{1}{64}\right)$
b. $\left(\frac{15}{16}, \frac{257}{128}\right)$ **c.** $ST = \frac{1}{2} = \frac{1}{2}(1)$
39. length of side: $8\sqrt{3}p$ units; area: $48\sqrt{3}p^2$ sq. units

Optional TI-81 Graphing Calculator
Exercises for Section 10.2

1.

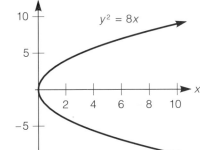

$x^2 = 8y$

3.

$y^2 = 8x$

5. b. The graph is consistent with Figure 9.

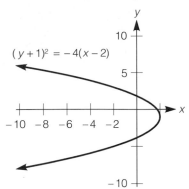

7. tangent line: $y = x - 2$

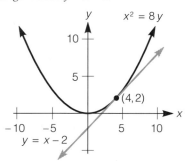

9. tangent line: $y = -6x - 9$

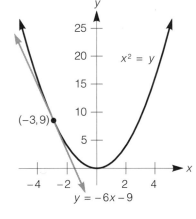

11. intersection points (exact): $(0,0)$ and $(5,5)$

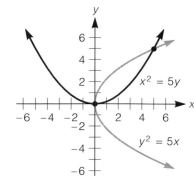

13. intersection point (exact): $(0,0)$; intersection point (approximate): $(0.88, 0.77)$

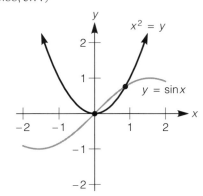

Exercise Set 10.3

1. tangent line: $y = 4x - 4$

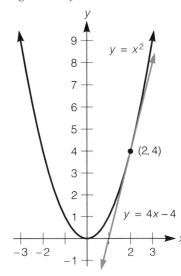

3. tangent line: $y = x - 2$

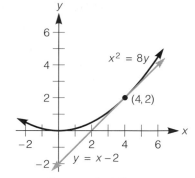

5. tangent line: $y = 6x + 9$

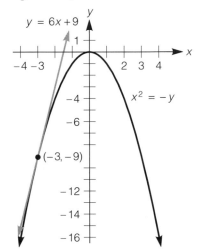

7. tangent line: $y = x + 1$

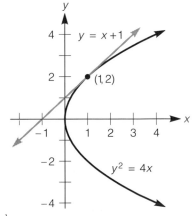

9. $\frac{1}{4}$ **11.** 12   **19. b.** Since $AF = FP$, the triangle AFP is isosceles, and consequently $\alpha = \gamma$. Also, since HP is parallel to the x-axis, we have $\beta = \gamma$.

1. length of major axis: 6; length of minor axis: 4; foci: $(\pm\sqrt{5}, 0)$; eccentricity: $\sqrt{5}/3$

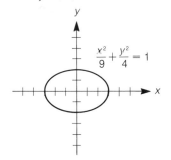

$$\frac{x^2}{9} + \frac{y^2}{4} = 1$$

3. length of major axis: 8; length of minor axis: 2; foci: $(\pm\sqrt{15}, 0)$; eccentricity: $\sqrt{15}/4$

$$\frac{x^2}{16} + \frac{y^2}{1} = 1$$

5. length of major axis: $2\sqrt{2}$; length of minor axis: 2; foci: $(\pm 1, 0)$; eccentricity: $\sqrt{2}/2$

$$\frac{x^2}{2} + \frac{y^2}{1} = 1$$

7. length of major axis: 8; length of minor axis: 6; foci: $(0, \pm\sqrt{7})$; eccentricity: $\sqrt{7}/4$

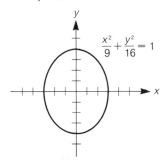

$$\frac{x^2}{9} + \frac{y^2}{16} = 1$$

9. length of major axis: $2\sqrt{15}/3$; length of minor axis: $2\sqrt{3}/3$; foci: $(0, \pm 2\sqrt{3}/3)$; eccentricity: $2\sqrt{5}/5$

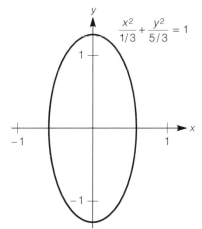

$$\frac{x^2}{1/3} + \frac{y^2}{5/3} = 1$$

11. length of major axis: 4; length of minor axis: $2\sqrt{2}$; foci: $(0, \pm\sqrt{2})$; eccentricity: $\sqrt{2}/2$

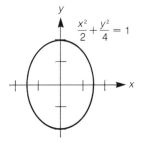

$$\frac{x^2}{2} + \frac{y^2}{4} = 1$$

13. center: $(5, -1)$; length of major axis: 10; length of minor axis: 6; foci: $(9, -1)$ and $(1, -1)$; eccentricity: $\frac{4}{5}$

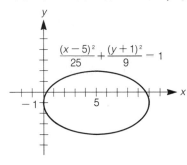

$$\frac{(x-5)^2}{25} + \frac{(y+1)^2}{9} = 1$$

15. center: $(1, 2)$; length of major axis: 4; length of minor axis: 2; foci: $(1, 2 \pm \sqrt{3})$; eccentricity: $\sqrt{3}/2$

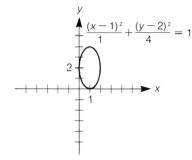

$$\frac{(x-1)^2}{1} + \frac{(y-2)^2}{4} = 1$$

17. center: $(-3, 0)$; length of major axis: 6; length of minor axis: 2; foci: $(-3 \pm 2\sqrt{2}, 0)$; eccentricity: $2\sqrt{2}/3$

$$\frac{(x+3)^2}{9} + \frac{y^2}{1} = 1$$

19. center: $(1, -2)$; length of major axis: 4; length of minor axis: $2\sqrt{3}$; foci: $(2, -2)$ and $(0, -2)$; eccentricity: $\frac{1}{2}$

$$\frac{(x-1)^2}{4} + \frac{(y+2)^2}{3} = 1$$

21. center: $(4, 6)$

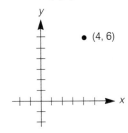

$\bullet\ (4, 6)$

23. no graph   **25.** $\dfrac{x^2}{25}+\dfrac{y^2}{16}=1$, or

$16x^2+25y^2=400$ **27.** $\dfrac{x^2}{16}+\dfrac{y^2}{15}=1$,

or $15x^2+16y^2=240$

29. $\dfrac{x^2}{21}+\dfrac{y^2}{25}=1$, or $25x^2+21y^2=525$

31. $\dfrac{x^2}{9}+\dfrac{y^2}{9/4}=1$, or $x^2+4y^2=9$

33. a. $y=-\frac{4}{3}x+\frac{38}{3}$ **b.** $y=\frac{7}{9}x+\frac{76}{9}$

c. $y=\frac{1}{15}x-\frac{76}{15}$ **35. a.** $y=-6x+26$

b. $\frac{169}{3}$ **37. b.** $3x+10\sqrt{6}\,y-75=0$

c. $d_1=87/\sqrt{609}$; $d_2=63/\sqrt{609}$

39.

x	0	0.1	0.2	0.3	0.4
y	±4	±3.98	±3.92	±3.82	±3.67

x	0.5	0.6	0.7	0.8	0.9	1.0
y	±3.46	±3.2	±2.86	±2.4	±1.74	0

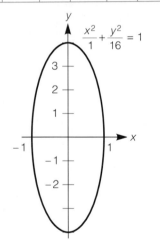

45. $\left(\dfrac{ab}{A},\dfrac{ab}{A}\right)$, $\left(\dfrac{ab}{A},-\dfrac{ab}{A}\right)$, $\left(-\dfrac{ab}{A},\dfrac{ab}{A}\right)$,

$\left(-\dfrac{ab}{A},-\dfrac{ab}{A}\right)$, where $A=\sqrt{a^2+b^2}$

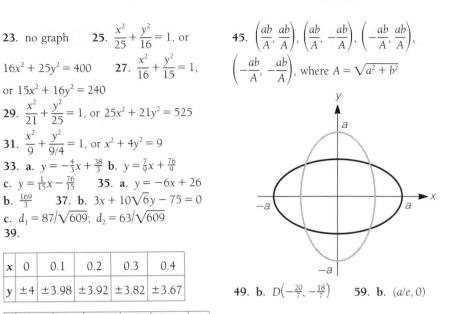

49. b. $D\left(-\frac{20}{7},-\frac{18}{7}\right)$ **59. b.** $(a/e,0)$

Optional TI-81 Graphing Calculator Exercises for Section 10.4

1. a. The graph appears to have symmetry with respect to the x-axis, the y-axis, and the origin.

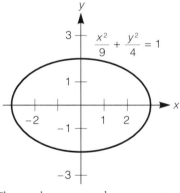

b. The graph appears to have symmetry with respect to the x-axis, the y-axis, and the origin.

3. c.

5. tangent line: $y=x-4$

7. a.

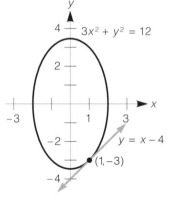

b. $a=2\sqrt{3}$, $b=2$, $c=2\sqrt{2}$
c. auxiliary circle: $x^2+y^2=12$

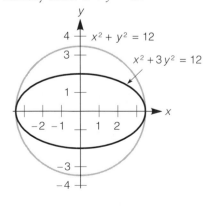

d. tangent line: $x + y = 4$

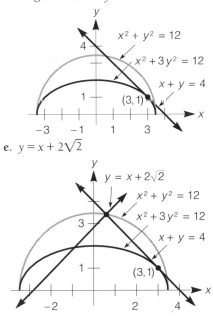

e. $y = x + 2\sqrt{2}$

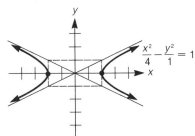

Exercise Set 10.5

1. vertices: $(\pm 2, 0)$; length of transverse axis: 4; length of conjugate axis: 2; asymptotes: $y = \pm \frac{1}{2}x$; foci: $(\pm \sqrt{5}, 0)$; eccentricity: $\sqrt{5}/2$

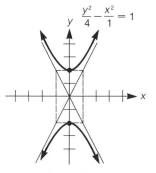

3. vertices: $(0, \pm 2)$; length of transverse axis: 4; length of conjugate axis: 2; asymptotes: $y = \pm 2x$; foci: $(0, \pm \sqrt{5})$; eccentricity: $\sqrt{5}/2$

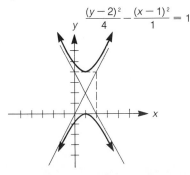

5. vertices: $(\pm 5, 0)$; length of transverse axis: 10; length of conjugate axis: 8; asymptotes: $y = \pm \frac{4}{5}x$; foci: $(\pm \sqrt{41}, 0)$; eccentricity: $\sqrt{41}/5$

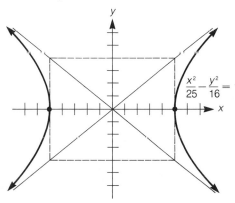

7. vertices: $(0, \pm \sqrt{2}/2)$; length of transverse axis: $\sqrt{2}$; length of conjugate axis: $2\sqrt{3}/3$; asymptotes: $y = \pm \frac{1}{2}\sqrt{6}x$; foci: $(0, \pm \sqrt{30}/6)$; eccentricity: $\sqrt{15}/3$

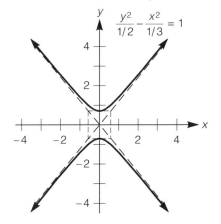

9. vertices: $(0, \pm 5)$; length of transverse axis: 10; length of conjugate axis: 4; asymptotes: $y = \pm \frac{5}{2}x$; foci: $(0, \pm \sqrt{29})$; eccentricity: $\sqrt{29}/5$

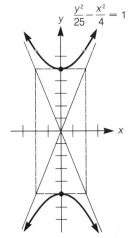

11. center: $(5, -1)$; vertices: $(10, -1)$ and $(0, -1)$; length of transverse axis: 10; length of conjugate axis: 6; asymptotes: $y = \frac{3}{5}x - 4$ and $y = -\frac{3}{5}x + 2$; foci: $(5 \pm \sqrt{34}, -1)$; eccentricity: $\sqrt{34}/5$

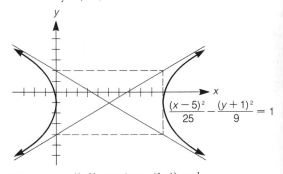

13. center: $(1, 2)$; vertices: $(1, 4)$ and $(1, 0)$; length of transverse axis: 4; length of conjugate axis: 2; asymptotes: $y = 2x$ and $y = -2x + 4$; foci: $(1, 2 \pm \sqrt{5})$; eccentricity: $\sqrt{5}/2$

15. center: $(-3, 4)$; vertices: $(1, 4)$ and $(-7, 4)$; length of transverse axis: 8; length of conjugate axis: 8; asymptotes: $y = x + 7$ and $y = -x + 1$; foci: $\left(-3 \pm 4\sqrt{2}, 4\right)$; eccentricity: $\sqrt{2}$

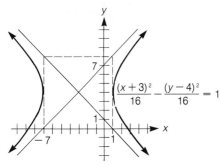

$$\frac{(x+3)^2}{16} - \frac{(y-4)^2}{16} = 1$$

17. center: $(0, 1)$; vertices: $(2, 1)$ and $(-2, 1)$; lengths of both transverse and conjugate axes: 4; asymptotes: $y = x + 1$ and $y = -x + 1$; foci: $\left(\pm 2\sqrt{2}, 1\right)$; eccentricity: $\sqrt{2}$

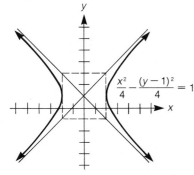

$$\frac{x^2}{4} - \frac{(y-1)^2}{4} = 1$$

19. center: $(2, 1)$; vertices: $(5, 1)$ and $(-1, 1)$; lengths of both transverse and conjugate axes: 6; asymptotes: $y = x - 1$ and $y = -x + 3$; foci: $\left(2 \pm 3\sqrt{2}, 1\right)$; eccentricity: $\sqrt{2}$

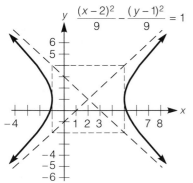

$$\frac{(x-2)^2}{9} - \frac{(y-1)^2}{9} = 1$$

21. center: $(0, -4)$; vertices: $(0, 1)$ and $(0, -9)$; length of transverse axis: 10; length of conjugate axis: 2; asymptotes: $y = 5x - 4$ and $y = -5x - 4$; foci: $\left(0, -4 \pm \sqrt{26}\right)$; eccentricity: $\sqrt{26}/5$

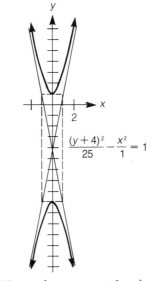

$$\frac{(y+4)^2}{25} - \frac{x^2}{1} = 1$$

23. two lines: $y = x + 3$ and $y = -x - 4$

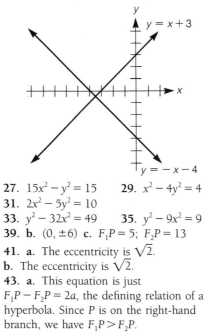

27. $15x^2 - y^2 = 15$ **29.** $x^2 - 4y^2 = 4$
31. $2x^2 - 5y^2 = 10$
33. $y^2 - 32x^2 = 49$ **35.** $y^2 - 9x^2 = 9$
39. b. $(0, \pm 6)$ **c.** $F_1 P = 5$; $F_2 P = 13$
41. a. The eccentricity is $\sqrt{2}$.
b. The eccentricity is $\sqrt{2}$.
43. a. This equation is just $F_1 P - F_2 P = 2a$, the defining relation of a hyperbola. Since P is on the right-hand branch, we have $F_1 P > F_2 P$.

49. $\left(\frac{51}{8}, \frac{15}{2}\right)$ is the intersection point.

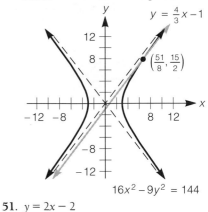

$$16x^2 - 9y^2 = 144$$

51. $y = 2x - 2$

Optional TI-81 Graphing Calculator Exercises for Section 10.5

1. a. $y = \pm \frac{1}{3}\sqrt{16x^2 - 144}$

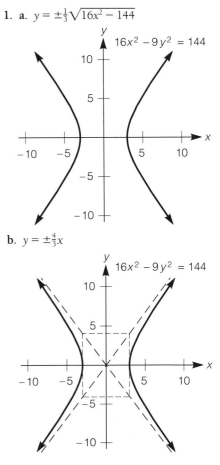

b. $y = \pm \frac{4}{3}x$

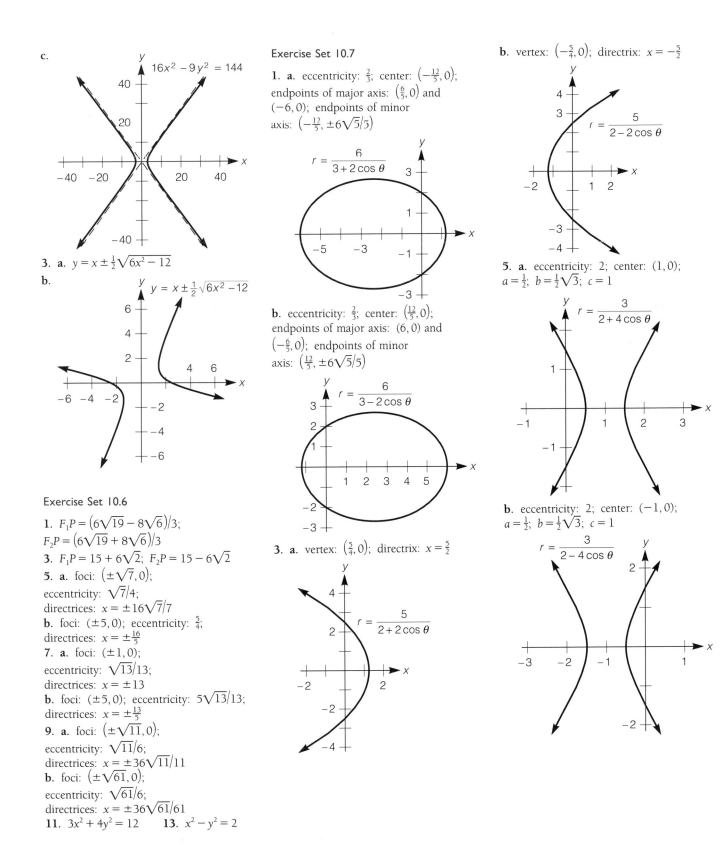

c.

$16x^2 - 9y^2 = 144$

3. a. $y = x \pm \frac{1}{2}\sqrt{6x^2 - 12}$

b.

$y = x \pm \frac{1}{2}\sqrt{6x^2 - 12}$

Exercise Set 10.6

1. $F_1P = (6\sqrt{19} - 8\sqrt{6})/3$;
$F_2P = (6\sqrt{19} + 8\sqrt{6})/3$

3. $F_1P = 15 + 6\sqrt{2}$; $F_2P = 15 - 6\sqrt{2}$

5. a. foci: $(\pm\sqrt{7}, 0)$;
eccentricity: $\sqrt{7}/4$;
directrices: $x = \pm 16\sqrt{7}/7$

b. foci: $(\pm 5, 0)$; eccentricity: $\frac{5}{4}$;
directrices: $x = \pm\frac{16}{5}$

7. a. foci: $(\pm 1, 0)$;
eccentricity: $\sqrt{13}/13$;
directrices: $x = \pm 13$

b. foci: $(\pm 5, 0)$; eccentricity: $5\sqrt{13}/13$;
directrices: $x = \pm\frac{13}{5}$

9. a. foci: $(\pm\sqrt{11}, 0)$;
eccentricity: $\sqrt{11}/6$;
directrices: $x = \pm 36\sqrt{11}/11$

b. foci: $(\pm\sqrt{61}, 0)$;
eccentricity: $\sqrt{61}/6$;
directrices: $x = \pm 36\sqrt{61}/61$

11. $3x^2 + 4y^2 = 12$ **13.** $x^2 - y^2 = 2$

Exercise Set 10.7

1. a. eccentricity: $\frac{2}{3}$; center: $(-\frac{12}{5}, 0)$;
endpoints of major axis: $(\frac{6}{5}, 0)$ and
$(-6, 0)$; endpoints of minor
axis: $(-\frac{12}{5}, \pm 6\sqrt{5}/5)$

$r = \dfrac{6}{3 + 2\cos\theta}$

b. eccentricity: $\frac{2}{3}$; center: $(\frac{12}{5}, 0)$;
endpoints of major axis: $(6, 0)$ and
$(-\frac{6}{5}, 0)$; endpoints of minor
axis: $(\frac{12}{5}, \pm 6\sqrt{5}/5)$

$r = \dfrac{6}{3 - 2\cos\theta}$

3. a. vertex: $(\frac{5}{4}, 0)$; directrix: $x = \frac{5}{2}$

$r = \dfrac{5}{2 + 2\cos\theta}$

b. vertex: $(-\frac{5}{4}, 0)$; directrix: $x = -\frac{5}{2}$

$r = \dfrac{5}{2 - 2\cos\theta}$

5. a. eccentricity: 2; center: $(1, 0)$;
$a = \frac{1}{2}$; $b = \frac{1}{2}\sqrt{3}$; $c = 1$

$r = \dfrac{3}{2 + 4\cos\theta}$

b. eccentricity: 2; center: $(-1, 0)$;
$a = \frac{1}{2}$; $b = \frac{1}{2}\sqrt{3}$; $c = 1$

$r = \dfrac{3}{2 - 4\cos\theta}$

7. center: $\left(-\frac{72}{5}, 0\right)$; eccentricity: $\frac{3}{2}$; length of transverse axis: $\frac{96}{5}$; length of conjugate axis: $\frac{48}{5}\sqrt{5}$

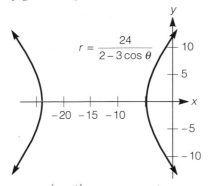

$$r = \frac{24}{2 - 3\cos\theta}$$

9. center: $\left(0, -\frac{3}{2}\right)$; eccentricity: $\frac{3}{5}$; length of major axis: 5; length of minor axis: 4

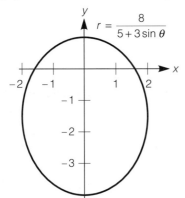

$$r = \frac{8}{5 + 3\sin\theta}$$

11. vertex: $\left(0, -\frac{6}{5}\right)$; directrix: $y = -\frac{12}{5}$

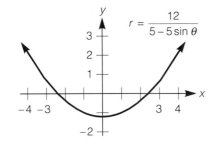

$$r = \frac{12}{5 - 5\sin\theta}$$

13. center: $\left(-\frac{5}{2}, 0\right)$; eccentricity: $\frac{5}{7}$; length of major axis: 7; length of minor axis: $2\sqrt{6}$

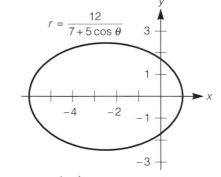

$$r = \frac{12}{7 + 5\cos\theta}$$

15. vertex: $\left(0, \frac{2}{5}\right)$; directrix: $y = \frac{4}{5}$

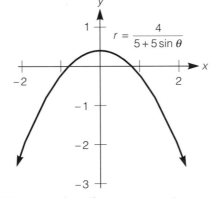

$$r = \frac{4}{5 + 5\sin\theta}$$

17. center: $(-6, 0)$; eccentricity: 2; length of transverse axis: 6; length of conjugate axis: $6\sqrt{3}$

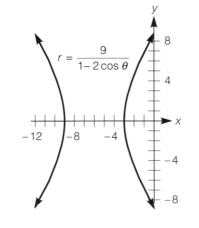

$$r = \frac{9}{1 - 2\cos\theta}$$

Exercise Set 10.8

1. $\left(\frac{1}{2}, 3\sqrt{3}/2\right)$ **3.** $(2, 0)$

5. $\left(-\frac{31}{13}, \frac{27}{13}\right)$ **7.** $\sin\theta = \frac{4}{5}$, $\cos\theta = \frac{3}{5}$

9. $\sin\theta = \frac{3}{5}$, $\cos\theta = \frac{4}{5}$

11. $\sin\theta = \sqrt{3}/2$, $\cos\theta = \frac{1}{2}$

13. $\sin\theta = \frac{7}{34}\sqrt{2}$, $\cos\theta = \frac{23}{34}\sqrt{2}$

15. $(x')^2 - (y')^2 = 9$

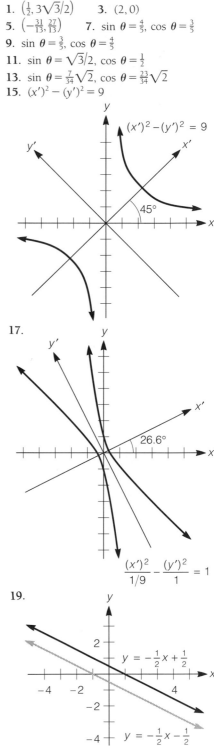

17.

$$\frac{(x')^2}{1/9} - \frac{(y')^2}{1} = 1$$

19.

$y = -\frac{1}{2}x + \frac{1}{2}$

$y = -\frac{1}{2}x - \frac{1}{2}$

21.

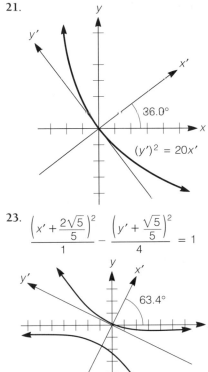

$36.0°$

$(y')^2 = 20x'$

23.
$$\frac{\left(x' + \frac{2\sqrt{5}}{5}\right)^2}{1} - \frac{\left(y' + \frac{\sqrt{5}}{5}\right)^2}{4} = 1$$

$63.4°$

25.

$45°$

$$\frac{(x' - 1)^2}{1} + \frac{(y' + 1)^2}{1/2} = 1$$

27.
$$(y' - \sqrt{2})^2 = 2\sqrt{2}\left(x' - \frac{11\sqrt{2}}{2}\right)$$

$45°$

2

29.

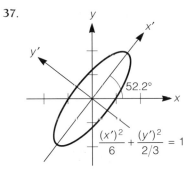

$63.4°$

$$\frac{(x')^2}{1} + \frac{(y')^2}{7/2} = 1$$

31.

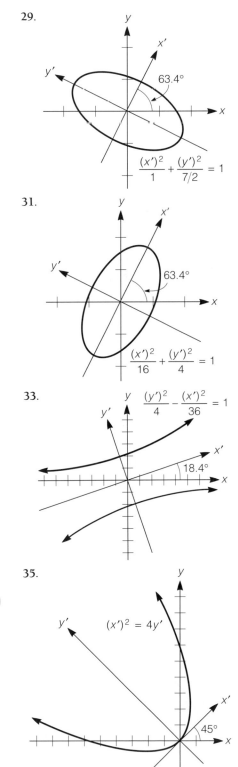

$63.4°$

$$\frac{(x')^2}{16} + \frac{(y')^2}{4} = 1$$

33.

$$\frac{(y')^2}{4} - \frac{(x')^2}{36} = 1$$

$18.4°$

35.

$(x')^2 = 4y'$

$45°$

37.

$52.2°$

$$\frac{(x')^2}{6} + \frac{(y')^2}{2/3} = 1$$

39. no graph
41. $x = x' \cos\theta - y' \sin\theta$;
$y = x' \sin\theta + y' \cos\theta$

Chapter Ten Review Exercises

17. $146.3°$ **19.** $53\sqrt{61}/61$
23. a. $y^2 = 16x$ **b.** $x^2 = 16y$
25. $x^2 = 12y$ **27.** $15x^2 + 16y^2 = 960$
29. $25x^2 + 9y^2 = 900$
31. $8x^2 - y^2 = 32$
33. $240x^2 - 16y^2 = 135$
35. vertex: $(0,0)$; focus: $\left(0, \frac{5}{2}\right)$;
directrix: $y = -\frac{5}{2}$; focal width: 10

$x^2 = 10y$

37. vertex: $(0, 3)$; focus: $(0, 0)$;
directrix: $y = 6$; focal width: 12

$x^2 = -12(y - 3)$

39. vertex: $(1, 1)$; focus: $(0, 1)$;
directrix: $x = 2$; focal width: 4

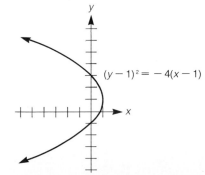

$(y - 1)^2 = -4(x - 1)$

41. center: $(0,0)$; foci: $\left(\pm\sqrt{2},0\right)$; length of major axis: 4; length of minor axis: $2\sqrt{2}$; eccentricity: $\sqrt{2}/2$

$$\frac{x^2}{4}+\frac{y^2}{2}=1$$

43. center: $(0,0)$; foci: $\left(0,\pm2\sqrt{10}\right)$; length of major axis: 14; length of minor axis: 6; eccentricity: $2\sqrt{10}/7$

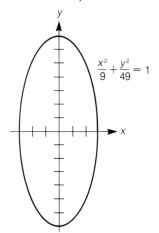

$$\frac{x^2}{9}+\frac{y^2}{49}=1$$

45. center: $(1,-2)$; foci: $(5,-2)$ and $(-3,-2)$; length of major axis: 10; length of minor axis: 6; eccentricity: $\frac{4}{5}$

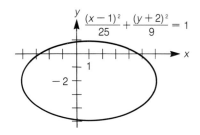

$$\frac{(x-1)^2}{25}+\frac{(y+2)^2}{9}=1$$

47. center: $(0,0)$; vertices: $(\pm2,0)$; foci: $(\pm6,0)$; asymptotes: $y=\pm\frac{1}{2}\sqrt{2}x$; eccentricity: $\sqrt{6}/2$

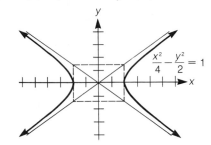

$$\frac{x^2}{4}-\frac{y^2}{2}=1$$

49. center: $(0,0)$; vertices: $(0,\pm3)$; foci: $\left(0,\pm\sqrt{58}\right)$; asymptotes: $y=\pm\frac{3}{7}x$; eccentricity: $\sqrt{58}/3$

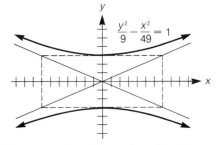

$$\frac{y^2}{9}-\frac{x^2}{49}=1$$

51. center: $(1,-2)$; vertices: $(6,-2)$ and $(-4,-2)$; foci: $\left(1\pm\sqrt{34},-2\right)$; asymptotes: $y=\frac{3}{5}x-\frac{13}{5}$ and $y=-\frac{3}{5}x-\frac{7}{5}$; eccentricity: $\sqrt{34}/5$

$$\frac{(x-1)^2}{25}-\frac{(y+2)^2}{9}=1$$

53. center: $(1,-2)$; foci: $(2,-2)$ and $(0,-2)$; length of major axis: 4; length of minor axis: $2\sqrt{3}$

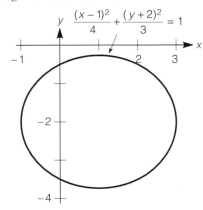

$$\frac{(x-1)^2}{4}+\frac{(y+2)^2}{3}=1$$

55. vertex: $(4,-1)$; axis of symmetry: $y=-1$; focus: $(3,-1)$; directrix: $x=5$

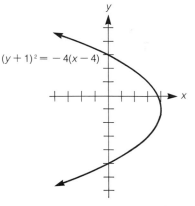

$$(y+1)^2=-4(x-4)$$

57. center: $(1,5)$; vertices: $(4,5)$ and $(-2,5)$; foci: $(6,5)$ and $(-4,5)$; asymptotes: $y=\frac{4}{3}x+\frac{11}{3}$ and $y=-\frac{4}{3}x+\frac{19}{3}$

$$\frac{(x-1)^2}{9}-\frac{(y-5)^2}{16}=1$$

59. point: $(4,6)$

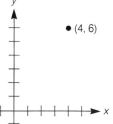

$\bullet\ (4,6)$

61. two lines: $3x - 4y = 11$ and $3x + 4y = 19$

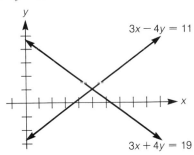

63. center: $(0, -4)$; vertices: $(0, 1)$ and $(0, -9)$; foci: $(0, -4 \pm \sqrt{26})$; asymptotes: $y = 5x - 4$ and $y = -5x - 4$

$$\frac{(y+4)^2}{25} - \frac{x^2}{1} = 1$$

65. no graph   **67. b.** 64

71. parabola: $y^2 = -8(x - 5)$; ellipse: $(x^2/25) + (y^2/16) = 1$

73. $\sin \theta = \sqrt{5}/5$; $\cos \theta = 2\sqrt{5}/5$

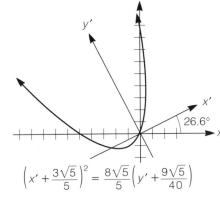

$$\left(x' + \frac{3\sqrt{5}}{5}\right)^2 = \frac{8\sqrt{5}}{5}\left(y' + \frac{9\sqrt{5}}{40}\right)$$

75. $\sin \theta = \sqrt{2}/2$; $\cos \theta = \sqrt{2}/2$

$$\frac{(x')^2}{16} + \frac{\left(y' - \sqrt{2}\right)^2}{36} = 1$$

77.

79.

81.

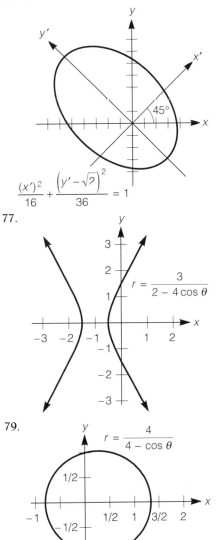

85. C_2 is the best approximation.

	Approximation Obtained	Percentage Error
C_1	25.531776	0.019
C_2	25.526986	0.000049
C_3	25.519489	0.029

87. a. $A\left(-\frac{1}{2}r, -\frac{1}{2}r(p + q)\right)$, $B\left(-\frac{1}{2}q, -\frac{1}{2}q(p + r)\right)$, $C\left(-\frac{1}{2}p, -\frac{1}{2}p(q + r)\right)$

Chapter Ten Test

1. focus: $(-3, 0)$; directrix: $x = 3$

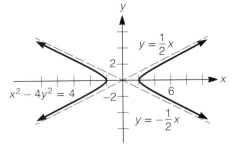

2. foci: $(\pm\sqrt{5}, 0)$; asymptotes: $y = \pm\frac{1}{2}x$

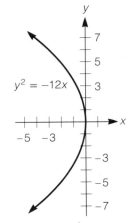

3. a. $\theta = 60°$

b.

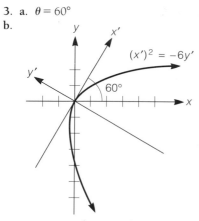

$(x')^2 = -6y'$

4. $30°$ **5.** $(x^2/12) + (y^2/16) = 1$

6. $\pm\sqrt{15}/15$

7. $\sqrt{3}x - y - 2\sqrt{3} = 0$

8. $(x^2/3) - (y^2/1) = 1$ **9. b.** 64

10. foci: $(\pm\sqrt{21}, 0)$; length of major axis: 10; length of minor axis: 4

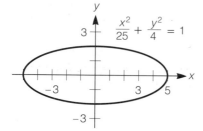

$\dfrac{x^2}{25} + \dfrac{y^2}{4} = 1$

11. $3\sqrt{5}/5$ **12.** point: $(2, -1)$

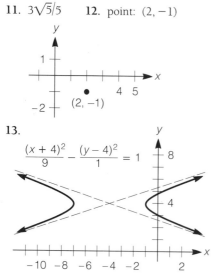

$(2, -1)$

13.

$\dfrac{(x+4)^2}{9} - \dfrac{(y-4)^2}{1} = 1$

14. ellipse

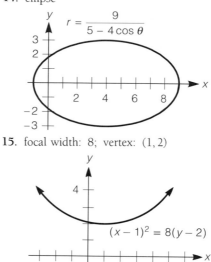

$r = \dfrac{9}{5 - 4\cos\theta}$

15. focal width: 8; vertex: $(1, 2)$

$(x-1)^2 = 8(y-2)$

16. a. $x = \pm\frac{36}{11}\sqrt{11}$

b. $F_1P = (12 + \sqrt{11})/2$,

$F_2P = (12 - \sqrt{11})/2$ **17.** $y = \frac{1}{6}x + \frac{13}{3}$

18. $y = 4x - 8$

CHAPTER ELEVEN

Exercise Set 11.1

1.

i^2	i^3	i^4	i^5	i^6	i^7	i^8
-1	$-i$	1	i	-1	$-i$	1

3. a. real part: 4; imaginary part: 5

b. real part: 4; imaginary part: -5

c. real part: $\frac{1}{2}$; imaginary part: -1

d. real part: 0; imaginary part: 16

5. $c = 4$, $d = -3$ **7. a.** $14 - 4i$

b. $-4 - 8i$ **9. a.** $19 - 17i$

b. $19 - 17i$ **c.** $\frac{11}{26} - \frac{23}{26}i$ **d.** $\frac{11}{25} + \frac{23}{25}i$

11. a. $11 - i$ **b.** $11 - 7i$ **c.** 4

13. $4 - 2i$ **15.** $30 + 19i$ **17.** 13

19. $-191 - 163i$ **21.** $19 - 4i$

23. $-70 + 84i$ **25.** $539 + 1140i$

27. $-46 + 9i$ **29.** $\frac{6}{97} + \frac{35}{97}i$

31. $\frac{6}{97} - \frac{35}{97}i$ **33.** $-\frac{5}{13} + \frac{12}{13}i$

35. -4 **37.** $\frac{1}{26} + \frac{5}{26}i$ **39.** $-i$

41. $12i$ **43.** $-3\sqrt{5}i$ **45.** -35

47. $12\sqrt{2}i$ **49. a.** $(a + c) + (b + d)i$

b. $(a - c) + (b - d)i$

c. $(ac - bd) + (bc + ad)i$

d. $\dfrac{ac + bd}{c^2 + d^2} + \dfrac{bc - ad}{c^2 + d^2}i$

57. real part: $\dfrac{2a^2 - 2b^2}{a^2 + b^2}$;

imaginary part: 0   **59.** 0

Exercise Set 11.2

1. quotient: $x - 5$; remainder: -11;

$x^2 - 8x + 4 = (x - 3)(x - 5) - 11$

3. quotient: $x - 11$; remainder: 53;

$x^2 - 6x - 2 = (x + 5)(x - 11) + 53$

5. quotient: $3x^2 - \frac{3}{2}x - \frac{1}{4}$; remainder: $\frac{13}{4}$;

$6x^3 - 2x + 3 = (2x + 1)\left(3x^2 - \frac{3}{2}x - \frac{1}{4}\right) + \frac{13}{4}$

7. quotient: $x^4 - 3x^3 + 9x^2 - 27x + 81$;

remainder: -241; $x^5 + 2 =$

$(x + 3)(x^4 - 3x^3 + 9x^2 - 27x + 81) - 241$

9. quotient: $x^5 + 2x^4 + 4x^3 + 8x^2 +$

$16x + 32$; remainder: 0; $x^6 - 64 =$

$(x - 2)(x^5 + 2x^4 + 4x^3 + 8x^2 + 16x + 32) + 0$

11. quotient: $5x^2 + 15x + 17$;

remainder: $-24x - 83$; $5x^4 - 3x^2 + 2 =$

$(x^2 - 3x + 5)(5x^2 + 15x + 17) +$

$(-24x - 83)$ **13.** quotient: $3y - 19$;

remainder: $89y + 35$; $3y^3 - 4y^2 - 3 =$

$(y^2 + 5y + 2)(3y - 19) + (89y + 35)$

15. quotient: $t^2 - 2t - 4$; remainder: 0;

$t^4 - 4t^3 + 4t^2 - 16 =$

$(t^2 - 2t + 4)(t^2 - 2t - 4) + 0$

17. quotient: $z^4 + z^3 + z^2 + z + 1$;

remainder: 0; $z^5 - 1 =$

$(z - 1)(z^4 + z^3 + z^2 + z + 1) + 0$

19. quotient: $ax + (b + ar)$;

remainder: $c + r(b + ar) = ar^2 + br + c$;

$ax^2 + bx + c = (x - r)[ax + (b + ar)] +$

$(ar^2 + br + c)$ **21.** quotient: $x - 1$;

remainder: -7;

$x^2 - 6x - 2 = (x - 5)(x - 1) - 7$

23. quotient: $4x - 5$; remainder: 0;

$4x^2 - x - 5 = (x + 1)(4x - 5) + 0$

25. quotient: $6x^2 + 19x + 78$;

remainder: 313; $6x^3 - 5x^2 + 2x + 1 =$

$(x - 4)(6x^2 + 19x + 78) + 313$

27. quotient: $x^2 + 2x + 4$; remainder: 7;

$x^3 - 1 = (x - 2)(x^2 + 2x + 4) + 7$

29. quotient: $x^4 - 2x^3 + 4x^2 - 8x + 16$;

remainder -33; $x^5 - 1 =$

$(x + 2)(x^4 - 2x^3 + 4x^2 - 8x + 16) - 33$

31. quotient: $x^3 - 10x^2 + 40x - 160$;

remainder: 642; $x^4 - 6x^3 + 2 =$

$(x + 4)(x^3 - 10x^2 + 40x - 160) + 642$

33. quotient: $x^2 + 6x + 57$;

remainder: 576; $x^3 - 4x^2 - 3x + 6 =$

$(x - 10)(x^2 + 6x + 57) + 576$

35. quotient: $x^2 - 6x + 30$; remainder: -150; $x^3 - x^2 = (x + 5)(x^2 - 6x + 30) - 150$

37. quotient: $54x^2 + 9x - 21$; remainder: 0; $54x^3 - 27x^2 - 27x + 14 = \left(x - \frac{2}{3}\right)(54x^2 + 9x - 21) + 0$

39. quotient: $x^3 + 3x^2 + 12x + 36$; remainder: 120; $x^4 + 3x^2 + 12 - (x - 3)(x^3 + 3x^2 + 12x + 36) + 120$

41. $(x - 2)(x^4 + 2x^3 + 4x^2 + 8x + 16)$

43. $(z - 3)(z^3 + 3z^2 + 9z + 27)$

45. quotient: $2x$; remainder: 1

47. quotient: $3x^2 - \frac{3}{2}x + \frac{3}{4}$; remainder: $\frac{1}{4}$

49. $k = 4$ **53.** $q(x) = x + (-4 + i)$; $R(x) = -4i$ **55.** $q(x) = x + (-1 + i)$; $R(x) = 0$ **57.** $-6\sqrt{3} + 57$ **59.** 0

Exercise Set 11.3

1. yes **3.** yes **5.** yes **7.** yes **9.** yes **11.** no **13. a.** yes **b.** no **15. a.** 1, 2 (multiplicity 3), 3 **b.** 1 (multiplicity 3) **c.** 5 (multiplicity 6), -1 (multiplicity 4) **d.** 0 (multiplicity 5), 1 **17.** -170 **19.** -9 **21.** $-3\sqrt{2} - 2$ **23.** -22 **25.** ± 3, 4 **27.** 1, $-1 \pm \sqrt{6}$ **29.** $-2, \frac{2}{3}, 3$ **31.** $-\frac{3}{2}, \frac{1}{2}\left(1 \pm \sqrt{5}\right)$ **33.** 0, 5 **35.** -4, 3 **37.** $-9, 1 \pm \sqrt{2}$ **39.** $\frac{1}{4}\left(3 \pm \sqrt{17}\right)$ (multiplicity 2) **41. a.** 1.125 **b.** -0.046875 **c.** $t - 1$ **d.** 1, $\frac{1}{2}\left(-1 \pm \sqrt{13}\right)$ **43.** $x^3 - 4x^2 - 17x + 60 = 0$ **45.** $x^3 + 8x^2 + 13x + 6 = 0$ **47.** No such polynomial equation exists. **49.** $ax^4 + (3a + b)x^3 + (-17a + 3b)x^2 + (6a - 17b)x + 6b = 0$, for any real numbers a and b, where $a \neq 0$ **51.** 20.44 **53. a.** -0.05 **b.** 0.07 **55. a.** $x_2 = 2 - \sqrt{2}$; $x_3 = 2 + \sqrt{2}$ **57.** yes **59.** yes **61.** $a = -1$, $b = -1$ **63.** $b = \pm \frac{3}{2}\sqrt{2}$ **65.** 2 (multiplicity 2), -4

Exercise Set 11.4

1. a, b, c **3.** $[x - (-1)](x - 3)$ **5.** $4\left(x - \frac{1}{4}\right)[x - (-6)]$ **7.** $\left[x - \left(-\sqrt{5}\right)\right]\left(x - \sqrt{5}\right)$ **9.** $[x - (5 + i)][x - (5 - i)]$ **11.** $f(x) = x^3 + x^2 - 5x + 3$ **13.** $f(x) = x^4 - 16$ **15.** $f(x) = x^6 + 10x^4 - 87x^2 + 144$

17. $f(x) = -\frac{5}{42}x^2 + \frac{25}{42}x + \frac{30}{7}$ **19.** $f(x) = \frac{1}{30}x^3 - \frac{19}{30}x + 1$ **21.** $x^2 + \left(i + \sqrt{3}\right)x + i\sqrt{3} = 0$ **23.** $x^2 - 3x - 54 = 0$ **25.** $x^2 - 2x - 4 = 0$ **27.** $x^2 - 2ax + a^2 - b = 0$ **29.** $(x + 2 - 2i)(x + 2 + 2i) \cdot (x - 2 - 2i)(x - 2 + 2i)$ **33.** 4, 3, -3 **39. a.** $4x^2 + 2x - 1 = 0$ **b.** $x = \frac{1}{4}\left(-1 \pm \sqrt{5}\right)$ **c.** $\cos 72° \approx 0.3090$; $\cos 144° \approx -0.8090$

Exercise Set 11.5

1. $\pm 1, \pm \frac{1}{2}, \pm \frac{1}{4}, \pm 3, \pm \frac{3}{2}, \pm \frac{3}{4}$ **3.** $\pm 1, \pm \frac{1}{2}, \pm \frac{1}{4}, \pm \frac{1}{8}, \pm 3, \pm \frac{3}{2}, \pm \frac{3}{4}, \pm \frac{3}{8}, \pm 9, \pm \frac{9}{2}, \pm \frac{9}{4}, \pm \frac{9}{8}$ **5.** $\pm 1, \pm \frac{1}{2}, \pm 2, \pm 3, \pm \frac{3}{2}, \pm 6$ **13.** 1, -1, -3 **15.** $-\frac{1}{4}, \pm \sqrt{5}$ **17.** $-1, -\frac{2}{3}, -\frac{1}{3}$ **19.** 1, $-1, \left(-1 \pm \sqrt{97}\right)/2$ **21.** 1 (multiplicity 4) **23.** $\frac{1}{2}$, 6, -4 **25.** $\frac{1}{4}, \frac{1}{5}, \pm \sqrt{6}/2$ **27. a.** upper bound: 2; lower bound: -1 **b.** upper bound: 2; lower bound: -1 **c.** upper bound: 6; lower bound: -2 **29.** between 0.68 and 0.69 **31.** between 2.88 and 2.89 **33.** between 4.31 and 4.32 **35.** between -2.15 and -2.14 **37.** between -5.27 and -5.26 **41.** $x \approx 0.32$

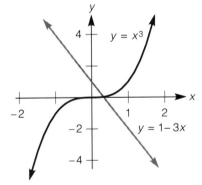

43. $x \approx 1.49$ **45.** $p = 3$, $x = 1$ **51.** $b = 2$

Exercise Set 11.6

1. $7 + 2i$ **3.** $5 - 2i$, 3 **5.** $-2 - i$, -3 (multiplicity 2) **7.** $6 + 5i$, $-\frac{1}{4}$ **9.** $4 - \sqrt{2}i$, $\pm \frac{3}{2}i$ **11.** $10 - 2i$, $1 \pm \sqrt{5}$ **13.** $\left(1 - i\sqrt{2}\right)/3, \frac{2}{5}$ **15.** $3 + 2i$, $-1 - i$, $-1 \pm \sqrt{2}$ **17.** $x^2 - 2x - 5 = 0$ **19.** $x^2 - \frac{4}{3}x - \frac{2}{3} = 0$ **21.** 2 complex roots, 1 negative real root **23.** 4 complex roots, 1 negative real root **25.** 2 complex roots, 1 positive real root, 1 negative real root **27.** either 1 positive real root and 2 negative real roots, or 1 positive real root and 2 complex roots **29.** 1 positive real root, 1 negative real root, 6 complex roots **31.** 1 positive real root, 8 complex roots **33.** 1 positive real root, 1 negative real root, 6 complex roots **35.** 1 positive real root, 1 negative real root, 4 complex roots **39.** $f(x) = x^4 + 2x^2 + 49$

Optional TI-81 Graphing Calculator Exercises for Sections 11.3–11.6

1. a.

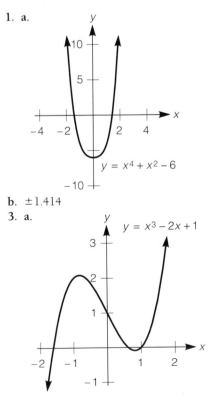

b. ± 1.414

3. a.

b. $-1.618, 0.618, 1$

5. a.

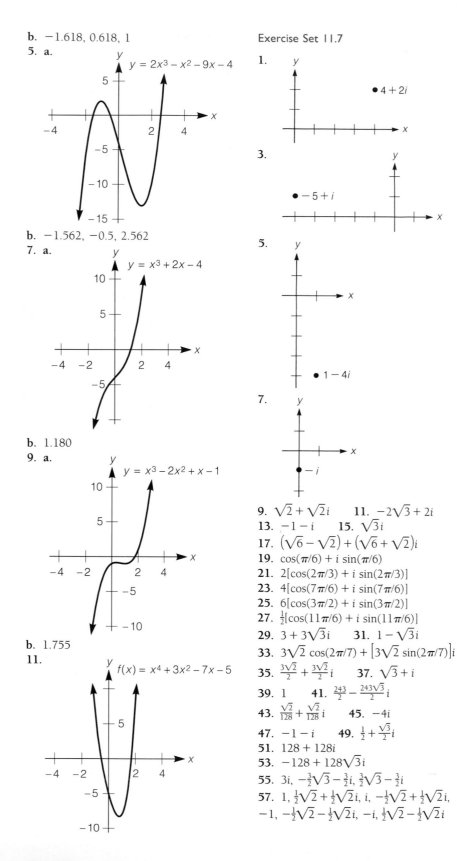

$y = 2x^3 - x^2 - 9x - 4$

b. $-1.562, -0.5, 2.562$

7. a.

$y = x^3 + 2x - 4$

b. 1.180

9. a.

$y = x^3 - 2x^2 + x - 1$

b. 1.755

11.

$f(x) = x^4 + 3x^2 - 7x - 5$

Exercise Set 11.7

1.

$\bullet\, 4 + 2i$

3.

$\bullet\, -5 + i$

5.

$\bullet\, 1 - 4i$

7.

$\bullet\, -i$

9. $\sqrt{2} + \sqrt{2}i$ **11.** $-2\sqrt{3} + 2i$

13. $-1 - i$ **15.** $\sqrt{3}i$

17. $\left(\sqrt{6} - \sqrt{2}\right) + \left(\sqrt{6} + \sqrt{2}\right)i$

19. $\cos(\pi/6) + i\sin(\pi/6)$

21. $2[\cos(2\pi/3) + i\sin(2\pi/3)]$

23. $4[\cos(7\pi/6) + i\sin(7\pi/6)]$

25. $6[\cos(3\pi/2) + i\sin(3\pi/2)]$

27. $\frac{1}{2}[\cos(11\pi/6) + i\sin(11\pi/6)]$

29. $3 + 3\sqrt{3}i$ **31.** $1 - \sqrt{3}i$

33. $3\sqrt{2}\cos(2\pi/7) + \left[3\sqrt{2}\sin(2\pi/7)\right]i$

35. $\frac{3\sqrt{2}}{2} + \frac{3\sqrt{2}}{2}i$ **37.** $\sqrt{3} + i$

39. 1 **41.** $\frac{243}{2} - \frac{243\sqrt{3}}{2}i$

43. $\frac{\sqrt{2}}{128} + \frac{\sqrt{2}}{128}i$ **45.** $-4i$

47. $-1 - i$ **49.** $\frac{1}{2} + \frac{\sqrt{3}}{2}i$

51. $128 + 128i$

53. $-128 + 128\sqrt{3}i$

55. $3i, -\frac{3}{2}\sqrt{3} - \frac{3}{2}i, \frac{3}{2}\sqrt{3} - \frac{3}{2}i$

57. $1, \frac{1}{2}\sqrt{2} + \frac{1}{2}\sqrt{2}i, i, -\frac{1}{2}\sqrt{2} + \frac{1}{2}\sqrt{2}i,$
$-1, -\frac{1}{2}\sqrt{2} - \frac{1}{2}\sqrt{2}i, -i, \frac{1}{2}\sqrt{2} - \frac{1}{2}\sqrt{2}i$

---

59. $4, -2 + 2\sqrt{3}i, -2 - 2\sqrt{3}i$

61. $3, \frac{3}{2} + \frac{3}{2}\sqrt{3}i, -\frac{3}{2} + \frac{3}{2}\sqrt{3}i, -3,$
$-\frac{3}{2} - \frac{3}{2}\sqrt{3}i, \frac{3}{2} - \frac{3}{2}\sqrt{3}i$

63. $92{,}236{,}816$ **65.** $0.95 + 0.31i, i,$
$-0.95 + 0.31i, -0.59 - 0.81i,$
$0.59 - 0.81i$

67. $\frac{1}{2}\left(\sqrt{6} - \sqrt{2}\right) + \frac{1}{2}\left(\sqrt{6} + \sqrt{2}\right)i,$
$-\frac{1}{2}\left(\sqrt{6} + \sqrt{2}\right) + \frac{1}{2}\left(\sqrt{6} - \sqrt{2}\right)i,$
$\frac{1}{2}\left(\sqrt{2} - \sqrt{6}\right) - \frac{1}{2}\left(\sqrt{2} + \sqrt{6}\right)i,$ and
$\frac{1}{2}\left(\sqrt{2} + \sqrt{6}\right) + \frac{1}{2}\left(\sqrt{2} - \sqrt{6}\right)i$

69. a. $1, -\frac{1}{2} + \frac{1}{2}\sqrt{3}i, -\frac{1}{2} - \frac{1}{2}\sqrt{3}i$

71. -1 **73.** 1

Chapter Eleven Review Exercises

1. $q(x) = x^3 + x^2 - 3x + 1$; $R(x) = -1$

3. quotient: $x^3 + 3x^2 + 7x + 21$;
remainder: 71

5. quotient: $2x^2 - 13x + 46$;
remainder: -187

7. quotient: $5x - 20$; remainder: 0

9. $99{,}904$ **11.** $-\frac{999}{1000}$ **13.** a^3

15. a. -0.24 **b.** -0.007 **c.** 0.00003

17. $a = -5$ **19.** $a = -1$ or $a = -2$

23. $\pm 1, \pm 2, \pm 3, \pm 6, \pm 9, \pm 18$

25. $\pm 1, \pm\frac{1}{2}, \pm 2, \pm 4, \pm 8$

27. $\pm p, \pm 1$ **29.** $2, -\frac{3}{2}, -1$

31. $\frac{5}{2}, -1 \pm \sqrt{3}$ **33.** $\frac{2}{3}, \frac{1}{2}(-1 \pm i\sqrt{3})$

35. $-1, -\sqrt{7}$ (multiplicity 2),
$\sqrt{7}$ (multiplicity 2)

37. 2 (multiplicity 2), 5 **39. a.** Let
$p(x)$ and $d(x)$ be polynomials such that
$d(x) \neq 0$. Then there are unique
polynomials $q(x)$ and $R(x)$ such that
$p(x) = d(x) \cdot q(x) + R(x)$, where either
$R(x) = 0$ or the degree of $R(x)$ is less than
the degree of $d(x)$. **b.** When a
polynomial $f(x)$ is divided by $x - r$, the
remainder is $f(r)$. **c.** Let $f(x)$ be a
polynomial. If $f(r) = 0$, then $x - r$ is a
factor of $f(x)$. Conversely, if $x - r$ is a
factor of $f(x)$, then $f(r) = 0$. **d.** Every
polynomial equation of the form

$$a_n x^n + a_{n-1}x^{n-1} + \cdots + a_1 x + a_0 = 0$$
$$(n \geq 1, \, a_n \neq 0)$$

has at least one root among the complex
numbers. (This root may be a real
number.)   **41.** $6\left(x - \frac{4}{3}\right)\left[x - \left(-\frac{5}{2}\right)\right]$

43. $(x - 4)\left[x - \left(-\sqrt[3]{-5}\right)\right] \cdot$
$\left[x - \frac{1}{2}\left(\sqrt[3]{5} + i\sqrt{3\sqrt[3]{25}}\right)\right] \cdot$
$\left[x - \frac{1}{2}\left(\sqrt[3]{5} - i\sqrt{3\sqrt[3]{25}}\right)\right]$

45. $2 \pm 3i$, 3   **47.** $1 \pm i\sqrt{2}$, $\pm\sqrt{7}$
49. 1 positive real root, 2 complex roots
51. 1 negative real root, 2 complex roots
53. 1 positive real root, 1 negative real root, 2 complex roots   **55. c.** -1, $\pm i$
57. between 0.82 and 0.83
59. c. between 6.93 and 6.94
61. $x^2 - 8x + 11 = 0$
63. $x^4 - 12x^3 + 35x^2 + 60x - 200 = 0$
65. $x^4 - 4x^3 - 4x^2 + 16x - 8 = 0$
67. zeros: 0, 3, -1

$y = x^3 - 2x^2 - 3x$

69. zeros: 0, -2, 2

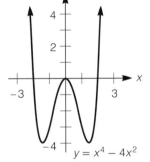

$y = x^4 - 4x^2$

71. $5 + 6i$ **73.** 6
75. $\frac{1}{2}(1 - \sqrt{3}i)$ **77.** $3\sqrt{2} - 4\sqrt{2}i$
81. a. Both are $2\sqrt{10}$. **b.** 3
87. $\frac{3}{2} + \frac{3}{2}\sqrt{3}i$ **89.** $2^{-1/4} - 2^{-1/4}i$
91. $1[\cos(\pi/3) + i\sin(\pi/3)]$
93. $6[\cos(5\pi/4) + i\sin(5\pi/4)]$
95. $5\sqrt{2} + 5\sqrt{2}i$ **97.** $\sqrt{2} - \sqrt{2}i$
99. $\frac{27}{2} + \frac{27}{2}\sqrt{3}i$ **101.** $512 - 512\sqrt{3}i$
103. $1, \frac{1}{2} + \frac{1}{2}\sqrt{3}i, -\frac{1}{2} + \frac{1}{2}\sqrt{3}i, -1,$
$-\frac{1}{2} - \frac{1}{2}\sqrt{3}i$, and $\frac{1}{2} - \frac{1}{2}\sqrt{3}i$
105. $-\dfrac{\sqrt{4+2\sqrt{2}}}{2} + \dfrac{\sqrt{4-2\sqrt{2}}}{2}i$ and
$\dfrac{\sqrt{4+2\sqrt{2}}}{2} - \dfrac{\sqrt{4-2\sqrt{2}}}{2}i$

107. $1.06 + 0.17i$, $0.17 + 1.06i$, $-0.95 + 0.49i$, $-0.76 - 0.76i$, and $0.49 - 0.95i$

Chapter Eleven Test

1. $f\left(\frac{1}{2}\right) = -\frac{3}{2}$ **2.** $-3, 1 + \sqrt{6},$
$1 - \sqrt{6}$ **3.** $\pm 1, \pm\frac{1}{7}, \pm 2, \pm 3,$
$\pm\frac{3}{2}, \pm 6$ **4.** $f(x) = 3x^2 + 21x - 24$
5. quotient: $4x^2 - 3x - 5$; remainder: 8
6. a. Let $f(x)$ be a polynomial. If $f(r) = 0$, then $x - r$ is a factor of $f(x)$. Conversely, if $x - r$ is a factor of $f(x)$, then $f(r) = 0$. **b.** Every polynomial equation of the form

$$a_n x^n + a_{n-1}x^{n-1} + \cdots + a_1 x + a_0 = 0$$
$$(n \geq 1, \ a_n \neq 0)$$

has at least one root among the complex numbers. (This root may be a real number.)   **7. a.** 3 **b.** between 2.2 and 2.3   **8.** $1 \pm i$, $3 \pm 2i$, -2
9. $q(x) = x^2 + 2x - 1$; $R(x) = -3x + 7$
10. $2\left[x - \frac{1}{2}(3 + i)\right]\left[x - \frac{1}{2}(3 - i)\right]$
11. a. $\pm 1, \pm 2, \pm 3, \pm 4, \pm 6, \pm 8,$ $\pm 12, \pm 24$ **c.** 1 **d.** none   **12. a.** $\frac{3}{2}$
b. $\frac{3}{2}, \frac{1}{2}(-1 \pm i\sqrt{3})$ **13.** 1 positive real root, 1 negative real root, and 2 complex roots   **14.** $x^3 + 6x + 20 = 0$
15. $f(x) = (x - 2)(x - 3i)^3\left[x - \left(1 + \sqrt{2}\right)\right]^2$
16. $21 - i$ **17.** $-1 + \sqrt{3}i$
18. $2\left[\cos\left(\frac{7\pi}{4}\right) + i\sin\left(\frac{7\pi}{4}\right)\right]$
19. $\frac{15}{2} + \frac{15}{2}\sqrt{3}i$ **20.** $2\sqrt{3} + 2i$, $-2\sqrt{3} + 2i$, and $-4i$

CHAPTER TWELVE

Exercise Set 12.1

27. a.

n	1	2	3	4	5
$f(n)$	1/2	2/3	3/4	4/5	5/6

b. $\frac{6}{7}$ **c.** $f(n) = n/(n + 1)$
29. a.

n	1	2	3	4	5
$f(n)$	1	4	9	16	25

b. 36 **c.** $f(n) = n^2$

Exercise Set 12.2

1. a. $a^2 + 2ab + b^2$
b. $a^3 + 3a^2 b + 3ab^2 + b^3$
3. $a^9 + 9a^8 b + 36a^7 b^2 + 84a^6 b^3$
$\qquad + 126a^5 b^4 + 126a^4 b^5 + 84a^3 b^6$
$\qquad + 36a^2 b^7 + 9ab^8 + b^9$
5. $8A^3 + 12A^2 B + 6AB^2 + B^3$
7. $1 - 12x + 60x^2 - 160x^3 + 240x^4$
$\qquad\qquad\qquad\qquad - 192x^5 + 64x^6$
9. $x^2 + 4x\sqrt{xy} + 6xy + 4y\sqrt{xy} + y^2$
11. $x^{10} + 5x^8 y^2 + 10x^6 y^4 + 10x^4 y^6$
$\qquad\qquad\qquad\qquad + 5x^2 y^8 + y^{10}$
13. $1 - \dfrac{6}{x} + \dfrac{15}{x^2} - \dfrac{20}{x^3} + \dfrac{15}{x^4} - \dfrac{6}{x^5} + \dfrac{1}{x^6}$
15. $\dfrac{x^3}{8} - \dfrac{x^2 y}{4} + \dfrac{xy^2}{6} - \dfrac{y^3}{27}$
17. $a^7 b^{14} + 7a^6 b^{12}c + 21a^5 b^{10}c^2 + 35a^4 b^8 c^3$
$\qquad + 35a^3 b^6 c^4 + 21a^2 b^4 c^5 + 7ab^2 c^6 + c^7$
19. $x^8 + 8\sqrt{2}x^7 + 56x^6 + 112\sqrt{2}x^5$
$\qquad + 280x^4 + 224\sqrt{2}x^3 + 224x^2$
$\qquad + 64\sqrt{2}x + 16$
21. $5\sqrt{2} - 7$ **23.** $89\sqrt{3} + 109\sqrt{2}$
25. $12 - 24\sqrt[3]{2} + 12\sqrt[3]{4}$
27. $x^{10} - 10x^9 + 35x^8 - 40x^7 - 30x^6$
$\qquad + 68x^5 + 30x^4 - 40x^3 - 35x^2$
$\qquad - 10x - 1$
29. 120   **31.** 105   **33. a.** 10
b. 5   **35.** $n^2 + 3n + 2$ **37.** 0
39. $120a^2 b^{14}$ **41.** $100x^{99}$ **43.** 45
45. 294,912   **47.** 28   **49.** 40,095
53. a.

k	0	1	2	3	4	5	6	7	8
$\binom{8}{k}$	1	8	28	56	70	56	28	8	1

Exercise Set 12.3

1. $\frac{1}{2}, \frac{2}{3}, \frac{3}{4}, \frac{4}{5}, \frac{5}{6}$ **3.** 0, 1, 4, 9, 16
5. $2, \frac{9}{4}, \frac{64}{27}, \frac{625}{256}, \frac{7776}{3125}$ **7.** $-1, 1, -1, 1,$
-1 **9.** $1, -\frac{1}{2}, \frac{1}{6}, -\frac{1}{24}, \frac{1}{120}$ **11.** 3,
6, 9, 12, 15   **13.** 1, 4, 25, 676,
458329   **15.** 2, 2, 4, 8, 32   **17.** 1,
1, 2, 6, 24   **19.** 0, 1, 2, 4, 16
21. a_n

23.

25.

27. 62   **29.** 40   **31.** $-\frac{19}{30}$

33. 903   **35.** 278   **37.** 3

39. 41   **41.** $x + x^2 + x^3$ **43.** $\frac{25}{12}$

45. -1 **47.** $\frac{6}{7}$ **49.** $\sum_{j=1}^{4} 5^j$

51. $\sum_{j=1}^{6} x^j$ **53.** $\sum_{k=1}^{12} \frac{1}{k}$

55. $\sum_{j=1}^{5} (-1)^{j+1} 2^j$ **57.** $\sum_{j=1}^{5} (-1)^{j+1} j$

59. b.

n	$F_1 + F_2 + \ldots + F_n$	$F_{n+2} - 1$
1	1	1
2	2	2
3	4	4
4	7	7
5	12	12

Exercise Set 12.4

1. a. 2 **b.** -4 **c.** $\frac{1}{3}$ **d.** $\sqrt{2}$ **3.** 131

5. 501   **7.** 998   **9.** $d = \frac{11}{6}$; $a = -\frac{23}{2}$

11. -190 **13.** $-\frac{1}{8}$ **15.** 500500

17. $91\pi/3$ **19.** $\frac{1}{2}$

21. $S_{16} = -768$, $d = -\frac{104}{15}$

23. $d = \frac{6}{5}$, $a = -\frac{17}{5}$ **25.** 900

27. 2,10,18 or 18,10,2   **29.** $-1,2,5$
or 5,2,-1 **31. b.** $-6 - 9\sqrt{2}$

33. $\frac{n}{2(1-b)}\left[2 + (n-3)\sqrt{b}\right]$

Exercise Set 12.5

1. 6   **3.** 20, 100   **5.** 1   **7.** $\frac{256}{6561}$

9. ± 4 **11.** 7161   **13.** $63 + 31\sqrt{2}$

15. $\frac{1995}{64}$ **17.** 0.011111   **19.** $\frac{2}{5}$

21. 101   **23.** $\frac{5}{9}$ **25.** $\frac{61}{495}$ **27.** $\frac{16}{37}$

29. $r = -\frac{1}{2}, -2$ **35.** 12 ft

Chapter Twelve Review Exercises

11. $81a^4 + 108a^3b^2 + 54a^2b^4 + 12ab^6 + b^8$

13. $x^4 + 4x^3\sqrt{x} + 6x^3 + 4x^2\sqrt{x} + x^2$

15. $x^{10} - 10x^8y^2 + 40x^6y^4 - 80x^4y^6$
$\qquad\qquad\qquad\qquad + 80x^2y^8 - 32y^{10}$

17. $1 + \frac{5}{x} + \frac{10}{x^2} + \frac{10}{x^3} + \frac{5}{x^4} + \frac{1}{x^5}$

19. $a^4b^2 - 4a^3b^2\sqrt{ab} + 6a^3b^3$
$\qquad\qquad\qquad - 4a^2b^3\sqrt{ab} + a^2b^4$

21. $15xy^8$ **23.** 84   **25.** 28

33. $1, \frac{4}{3}, \frac{3}{2}, \frac{8}{5}$

35. $-\frac{1}{2}, \frac{2}{3}, -\frac{3}{4}, \frac{4}{5}$

37. $-12, -48, -192, -768$

39. a. -5 **b.** $\frac{9}{10}$ **41.** $\sum_{k=1}^{5} \frac{5}{3^k}$

43. 73   **45.** $\frac{5}{1024}$ **47.** 177

49. 7,777,777,777   **51.** $-5\sqrt{10}$

53. $\frac{3}{4}$ **55.** $\frac{1}{10}$ **57.** $\frac{5}{11}$

63. a. 42,929 **b.** 42,925; 0.00932%
c. 6.48040×10^{10} **d.** 6.48027×10^{10};
2×10^{-3}%   **65.** 10, 20, 40 or
40, 20, 10   **69. a.** No value exists.

b. $x = \dfrac{b^2 - ac}{a + c - 2b}$

Chapter Twelve Test

2. a. 27 **b.** -6

3. a. $S_n = \dfrac{a(1 - r^n)}{1 - r}$ **b.** $\frac{174075}{1024}$

4. a. 42240 **b.** $5280a^7b^{12}$

5. $243x^{10} + 405x^8y^3 + 270x^6y^6$
$\qquad\qquad + 90x^4y^9 + 15x^2y^{12} + y^{15}$

6. 177   **7.** $\frac{7}{9}$ **8.** $a_4 = 5$, $a_5 = 27$

9. $-5\sqrt{10}$ **10.** 224

APPENDIX

Exercise Set A.5

1. a. x^{15} **b.** x^{36} **c.** $(x+1)^{15}$ **d.** $(x+1)^{36}$

3. a. a^6 **b.** $(a+1)^6$ **5. a.** 1 **b.** 1

7. a. $\frac{11}{100}$ **b.** $\frac{100}{11}$ **c.** 100   **9. a.** $1/a^6b^3$

b. a^7/b **c.** a^6b^2/c^6 **11.** $\frac{16}{49}$

13. y^{12}/x^6z^{12} **15.** $1/a^8b^{12}c^{16}$ **17.** y^6

19. a. false **b.** true **c.** true

21. a. false **b.** true   **23. a.** false

b. false   **25. a.** -4 **b.** undefined

27. a. $\frac{2}{5}$ **b.** $-\frac{2}{5}$ **29. a.** undefined

b. undefined   **31. a.** $\frac{4}{3}$ **b.** $-\frac{3}{5}$

33. a. $3\sqrt{2}$ **b.** $3\sqrt[3]{2}$ **35. a.** $7\sqrt{2}$

b. $-2\sqrt[5]{2}$ **37. a.** $\frac{5}{2}$ **b.** $\frac{2}{5}$

39. a. $3\sqrt{2}$ **b.** $3\sqrt[3]{2}$ **41. a.** $-4\sqrt{2}$

b. $5\sqrt[4]{2}$ **43.** $2\sqrt{2}$ **45. a.** $6x$

b. $-6y$ **47. a.** $ab\sqrt{ab}$ **b.** a^2b^2

49. $-2ab\sqrt[4]{b}$ **51.** $2a^4\sqrt[3]{2b^2}/c^3$

53. $-a^2\sqrt[6]{5}/b$ **55. a.** $4\sqrt{7}/7$ **b.** $\sqrt{3}$

c. $\sqrt{10}/5$ **57. a.** $(\sqrt{5} - 1)/4$

b. $-(1 + \sqrt{5})/4$ **c.** $-(3 + \sqrt{5})/2$

59. $61\sqrt{5}/5$ **61.** $\sqrt[3]{5}/5$

63. $\sqrt[4]{8a^3b^3}/2ab$ **65.** $3\sqrt[5]{2ab}/2ab^2$

67. $x(\sqrt{x} + 2)/(x - 4)$

69. $(x - 2\sqrt{ax} + a)/(x - a)$

71. $1/(\sqrt{x} + \sqrt{5})$

73. $1/(\sqrt{2+h} + \sqrt{2})$ **75. a.** Both
values are approximately 1.645751.
77. x^{2c}

Exercise Set A.6

1. $\sqrt[5]{a^3}$; $(\sqrt[5]{a})^3$ **3.** $\sqrt[3]{5^2}$; $(\sqrt[3]{5})^2$

5. $\sqrt[4]{(x^2+1)^3}$; $(\sqrt[4]{x^2+1})^3$

7. $\sqrt[3]{2^{xy}}$; $(\sqrt[3]{2})^{xy}$ **9.** $p^{2/3}$

11. $(1 + u)^{4/7}$ **13.** $(a^2 + b^2)^{3/p}$

15. 4   **17.** $\frac{1}{6}$ **19.** undefined

21. 5   **23.** 2   **25.** 4   **27.** -2

29. -10 **31.** $\frac{1}{7}$ **33.** undefined

35. $\frac{1}{216}$ **37.** 25   **39.** -1

41. $\frac{255}{16}$ **43.** $-\frac{28976}{243}$ **45.** $\frac{7}{12}$

47. $2^{3/2}a^{1/12}$ **49.** $x^2 + 1$

51. a. $2^{1/3}3^{5/6}$ **b.** $\sqrt[6]{972}$

53. a. $2^{7/12}3^{1/3}$ **b.** $\sqrt[12]{10368}$

55. a. $x^{2/3}y^{4/5}$ **b.** $\sqrt[15]{x^{10}y^{12}}$

57. a. $x^{(a+b)/3}$ **b.** $\sqrt[3]{x^{a+b}}$

59. $(x+1)^{2/3}$ **61.** $(x+y)^{2/5}$

63. $2x^{1/6}$ **65.** $x^{1/6}y^{1/8}$

67. $a = 9$ and $b = 16$ (other answers possible) **69.** $u = 1$ and $v = 8$ (other answers possible) **71.** $x = 4$ and $m = 2$ (other answers possible)

73. a. $2^{3/2}$ is larger. **b.** $5^{1/2}$ is larger. **c.** $2^{1/2}$ is larger. **d.** $\left(\frac{1}{2}\right)^{1/3}$ is larger. **e.** $10^{1/10}$ is larger.

INDEX

FACTORING TECHNIQUES

Technique	Example or Formula
Common factor	$3x^4 + 6x^3 - 12x^2 = 3x^2(x^2 + 2x - 4)$
	$4(x^2 + 1) - x(x^2 + 1) = (x^2 + 1)(4 - x)$
Difference of squares	$x^2 - a^2 = (x - a)(x + a)$
Trial and error	$x^2 + 2x - 3 = (x + 3)(x - 1)$
Difference of cubes	$x^3 - a^3 = (x - a)(x^2 + ax + a^2)$
Sum of cubes	$x^3 + a^3 = (x + a)(x^2 - ax + a^2)$
Grouping	$x^3 - x^2 + x - 1 = (x^3 - x^2) + (x - 1)$ $= x^2(x - 1) + (x - 1) \cdot 1$ $= (x - 1)(x^2 + 1)$

THE GREEK ALPHABET

A	α	Alpha	N	ν	Nu
B	β	Beta	Ξ	ξ	Xi
Γ	γ	Gamma	O	o	Omicron
Δ	δ	Delta	Π	π	Pi
E	ϵ	Epsilon	P	ρ	Rho
Z	ζ	Zeta	Σ	σ	Sigma
H	η	Eta	T	τ	Tau
Θ	θ	Theta	Υ	υ	Upsilon
I	ι	Iota	Φ	ϕ	Phi
K	κ	Kappa	X	χ	Chi
Λ	λ	Lambda	Ψ	ψ	Psi
M	μ	Mu	Ω	ω	Omega

FORMULAS FROM COORDINATE GEOMETRY

1. The distance formula:
$$d = \sqrt{(x_2 - x_1)^2 + (y_2 - y_1)^2}$$

2. The equation for a circle:
$$(x - h)^2 + (y - k)^2 = r^2$$

3. The definition of slope:
$$m = \frac{y_2 - y_1}{x_2 - x_1}$$

4. The point-slope formula:
$$y - y_1 = m(x - x_1)$$

5. The slope-intercept formula:
$$y = mx + b$$

6. The condition for two nonvertical lines to be parallel:
$$m_1 = m_2$$

7. The condition for two nonvertical lines to be perpendicular:
$$m_1 m_2 = -1$$

8. The midpoint formula:
$$(x_0, y_0) = \left(\frac{x_1 + x_2}{2}, \frac{y_1 + y_2}{2} \right)$$